D0164432

# The Microbial World

FIFTH EDITION

Roger Y. Stanier

John L. Ingraham
*University of California*
*Davis, California*

Mark L. Wheelis
*University of California*
*Davis, California*

Page R. Painter
*University of California*
*Davis, California*

PRENTICE-HALL, Englewood Cliffs, New Jersey 07632

***Library of Congress Cataloging-in-Publication Data***

Main entry under title:

The Microbial world.

Rev. ed. of: The microbial world / Roger Y. Stanier,
Edward A. Adelberg, John L. Ingraham, 4th ed. c1976.
Includes bibliographies and index.
L. Microbiology. I. Stanier, Roger Y. II. Stanier,
Roger Y. Microbial world.
QR41.2.M464 1986 576 85-28189
ISBN 0-13-581042-6

Cover art: Stylized rendition
of a heliozoan protist.

*Editorial/production supervision: Virginia Huebner*
*Interior and cover design: Christine Gerhring-Wolf*
*Cover illustration: Jim Kinstrey*
*Manufacturing buyer: John Hall*



Printed in the United States of America

10 9 8 7 6 5 4 3 2 1

ISBN 0-13-581042-6 01



Prentice-Hall International (UK) Limited, *London*
Prentice-Hall of Australia Pty. Limited, *Sydney*
Prentice-Hall Canada Inc., *Toronto*
Prentice-Hall Hispanoamericana, S.A., *Mexico*
Prentice-Hall of India Private Limited, *New Delhi*
Prentice-Hall of Japan, Inc., *Tokyo*
Prentice-Hall of Southeast Asia Pte. Ltd., *Singapore*
Editora Prentice-Hall do Brasil, Ltda., *Rio de Janeiro*
Whitehall Books Limited, *Wellington, New Zealand*

# Contents

## Chapter 3
## The Nature of the Microbial World 43



## Chapter 4
## Microbial Metabolism: Fuelling Reactions 78

## Chapter 5
## Microbial Metabolism: Biosynthesis, Polymerization, Assembly 102



## Chapter 6
## The Relation Between Structure and Function in Procaryotic Cells 145



## Chapter 7
## Microbial Growth 183

## Chapter 8
## Effect of the Environment on Microbial Growth 196



## Chapter 9
## The Viruses 213



## Chapter 10
## Microbial Genetics: Gene Function and Mutation 235



## Chapter 11
## Microbial Genetics: Genetic Exchange and Recombination

## *Chapter 12*
## *Regulation 286*



## *Chapter 13*
## *The Classification and Phylogeny of Bacteria 311*



## *Chapter 14*
## *The Archaebacteria 33*



## *Chapter 15*
## *The Photosynthetic Eubacteria 344*

## Chapter 21
## Gram-Negative Eubacteria: Spirochetes, Rickettsias and Chlamydias 464



## Chapter 22
## Gram-positive Eubacteria: Unicellular Endospore formers 475



## Chapter 23
## Gram-positive Fermentative Eubacteria 495



## Chapter 24
## Gram-positive Eubacteria: The Actinomycetes 505



## Chapter 25
## The Mollicutes 520



## Chapter 26
## The Protists 525

## Chapter 33
## The Exploitation of Microorganisms by Humans 657

# Preface to the Fifth Edition

In 1957 Roger Y. Stanier, Michael Doudoroff and Edward A. Adelberg, the three members of the Department of Bacteriology at Berkeley with the responsibility for instruction in general microbiology, published the first edition of *The Microbial World* with, as they stated later, "a frankly propagandist purpose—that of accelerating this change (toward the unification of microbiology with the rest of biology) by presenting microbiology in the framework of the facts and concepts of general biology." Some 28 years later we find the unification essentially complete with studies on microorganisms continuing to contribute to the understanding of important biological principles as well as to increase our knowledge of these organisms that are so interesting in their own right. Still, the original purpose of this book has continued to be an invaluable guiding principle during the preparation of the Fifth Edition, the first such revision in which none of the original authors fully participated.

Michael Doudoroff died before the Fourth Edition was written. Edward Adelberg decided before the present revision was begun that the many other demands on his time precluded his involvement in yet another revision (the four previous experiences must have taught him well what a major job it is). Roger Stanier died before the actual writing of the Fifth Edition began, but he was actively engaged in its planning and in encouraging us to carry his plans to completion. He remains the senior author in recognition of his planning role and of the substantial amount of his writing that remains in this edition. Some of Edward Adelberg's writing also remains.

*We affectionately dedicate this Fifth Edition of the Microbial World to its original three authors, Roger Y. Stanier, Michael Doudoroff and Edward A. Adelberg.*

Major changes in the science of Microbiology have occurred since the publication of the Fourth Edition, both in terms of a nearly explosive expansion of factual detail and improved methodology as well as of fundamental changes in our perceptions of the relationships among bacteria. Hence, most of this volume is completely rewritten.

Changes will probably be most apparent in the chapters dealing with the major microbial groups; 12 chapters replace the 8 found in the Fourth Edition. Among the new chapters is one dealing exclusively with the Archaebacteria. Reflecting the fundamental advances made in microbial pathogenecity, this section also has been expanded, now containing 4 chapters rather than 2. Of necessity, other sections, including the one dealing with symbioses have been condensed, we hope without materially detracting from the richness of this interesting topic.

We should like to express our thanks to the many people who helped us in the preparation of this edition. Germaine Stanier was actively involved in all stages of the project; her assistance has been substantial. Wille Brown, Jack Campbell, Martin Dworkin, John Fitzgerald, James Frea, George Hegeman, John Holt, Daniel O'Kane, and Sidney Rittenberg read all or much of the manuscript; each made numerous helpful comments. Steven Krawiec, Joan Macy, Jack Meeks, David Nagle, Jr., Paul Phinney and Carl Woese read one or several specific chapters, and their advice has also been most helpful. Ciaran Condon, Marjorie Ingraham, Pamela Painter and Michael Whitt helped to read proofs. Many colleagues responded generously to our requests for new illustrative material; they are acknowledged individually in figure legends.

# Chapter 1
# The Beginnings of Microbiology

Microbiology is the study of organisms, called *microorganisms* that are too small to be perceived clearly by the unaided human eye. If an object has a diameter of less than 0.1 mm, the eye cannot perceive it at all, and very little detail can be perceived in an object with a diameter of 1 mm. Roughly speaking, therefore, organisms with a diameter of 1 mm or less are microorganisms and fall into the broad domain of microbiology. Microorganisms have a wide taxonomic distribution; they include some metazoan animals, protozoa, many algae and fungi, bacteria, and viruses. The existence of this microbial world was unknown until the invention of microscopes, optical instruments that serve to magnify objects so small that they cannot be clearly seen by the unaided human eye. Microscopes, invented at the beginning of the seventeenth century, opened the biological realm of the very small to systematic scientific exploration.

Early microscopes were of two kinds. The first were *simple microscopes* with a single lens of very short focal length, consequently capable of a high magnification; such instruments did not differ in optical principle from ordinary magnifying glasses able to increase an image severalfold, which had been known since antiquity. The second were *compound microscopes* with a double lens system consisting of an ocular and objective. The compound microscope, with its greater intrinsic power of magnification, eventually displaced completely the simple instrument; all our contemporary microscopes are of the compound type. However, nearly all the great original microscopic discoveries were made with simple microscopes.

## THE DISCOVERY OF THE MICROBIAL WORLD

The discoverer of the microbial world was a Dutch merchant, Anton van Leeuwenhoek (Figure 1.1). His scientific activities were fitted into a life well filled with business affairs and civic duties. In this, he was no exception for his time; many of the great discoveries of this period in all fields of science were made by amateurs who earned their living in other ways, or who were freed from the necessity of earning a living because of their personal wealth.

**FIGURE 1.1**

Anton van Leeuwenhoek (1632–1723). In this portrait, he is holding one of his microscopes. Courtesy of the Rijksmuseum, Amsterdam.

However, Leeuwenhoek differed from his scientific contemporaries in one respect: he had little formal education and never attended a university. This was probably no disadvantage scientifically, since the scientific training then available would have provided little basis for his life's work. More serious handicaps, insofar as the communication of his discoveries went, were his lack of connections in the learned world and his ignorance of any language except Dutch. Nevertheless, through a fortunate chance, his work became widely known in his own lifetime, and its importance was immediately recognized. About the time that Leeuwenhoek began his observations, the Royal Society had been established in England for the communication and publication of scientific work. The Society invited Leeuwenhoek to communicate his observations to its members and a few years later (1680) elected him as a Fellow. For almost 50 years, until his death in 1723, Leeuwenhoek transmitted his discoveries to the Royal Society in the form of a long series of letters written in Dutch. Most of these letters were translated and published in English in the *Proceedings of the Royal Society*, and so became quickly and widely disseminated.

Leeuwenhoek's microscopes (Figure 1.2) bore little resemblance to the instruments with which we are familiar. The almost spherical lens (a) was mounted between two small metal plates. The specimen was placed on the point of a blunt pin (b) attached to the back plate and was brought into focus by manipulating two screws (c) and (d), which varied the position of the pin relative to the lens. During this operation the observer held the instrument with its other face very close to his eye and squinted through the lens. No change of magnification was possible, the magnifying power of each microscope being an intrinsic property of its lens. Despite the simplicity of their construction, Leeuwenhoek's microscopes were able to give clear images at magnifications that ranged, depending on the focal length of the lens, from about 50 to nearly 300 diameters. The highest magnification that he could obtain was consequently somewhat less than one-third of the highest magnification that is obtainable with a modern compound light microscope. Leeuwenhoek constructed hundreds of such instruments, a few of which survive today.

Leeuwenhoek's place in scientific history depends not so much on his skill as a microscope maker, essential though this was, as on the extraordinary range and skill of his microscopic observations. He was endowed with an unusual degree of curiosity and studied almost every conceivable object that could be looked at through a microscope. He made magnificent observations on the microscopic structure of the seeds and embryos of plants and on small invertebrate animals. He discovered the existence of spermatozoa and of red blood cells and was thus the founder of animal histology. By discovering and describing capillary circulation he completed the work on the circu-

**FIGURE 1.2**

A drawing to show the construction of one of Leeuwenhoek's microscopes: (a) lens, (b) mounting pin, (c) and (d) focusing screws. After C. E. Dobell, *Antony van Leeuwenhoek and His Little Animals* (New York: Russell and Russell, Inc., 1932).

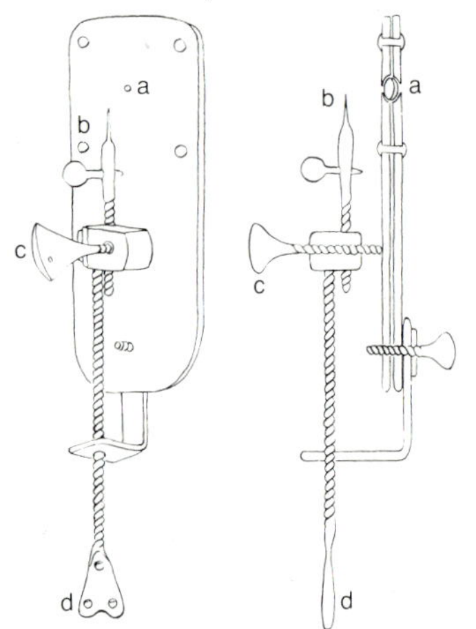

lation of blood begun by Harvey half a century before. Indeed, it would be easy to fill a page with a mere list of his major discoveries about the structure of higher plants and animals. His greatest claim to fame rests, however, on his discovery of the microbial world: the world of "animalcules," or little animals, as he and his contemporaries called them. A new dimension was thus added to biology. All the main kinds of unicellular microorganisms that we know today—protozoa, algae, yeasts, and bacteria—were first described by Leeuwenhoek, often with such accuracy that it is possible to identify individual species from his accounts of them. In addition to the diversity of this microbial world, Leeuwenhoek emphasized its incredible abundance. For example, in one letter describing for the first time the characteristic bacteria of the human mouth, he wrote:

> *I have had several gentlewomen in my house, who were keen on seeing the little eels in vinegar; but some of them were so disgusted at the spectacle, that they vowed they'd never use vinegar again. But what if one should tell such people in the future that there are more animals living in the scum on the teeth in a man's mouth, than there are men in a whole kingdom?*

Although Leeuwenhoek's contemporaries marveled at his scientific discoveries, the microscopic exploration of the microbial world which he had so brilliantly begun was not appreciably extended for over a century after his death. The principal reasons for this long delay seem to have been technical ones. Simple microscopes of high magnification are both difficult and tiring to use, and the manufacture of their very small lenses is an operation that requires great skill. Consequently, most of Leeuwenhoek's contemporaries and immediate successors used compound microscopes. Despite the intrinsic superiority of compound microscopes, the ones available in the seventeenth and eighteenth centuries suffered from serious optical defects,which made them less effective working instruments than Leeuwenhoek's simple microscopes. Thus, Leeuwenhoek's English contemporary, Robert Hooke, a very capable and careful observer, could not repeat with his own compound microscope many of the finer observations reported by Leeuwenhoek.

The major optical improvements that were eventually to lead to compound microscopes of the quality that we use today began about 1820 and extended through the succeeding half century. These improvements were closely followed by resumed exploration of the microbial world and resulted, by the end of the nineteenth century, in a detailed knowledge of its constituent groups. In the meantime, however, the science of microbiology had been developing in other ways, which led to the discovery of the roles that microorganisms play in the transformations of matter and in the causation of disease.

## THE CONTROVERSY OVER SPONTANEOUS GENERATION

After Leeuwenhoek had revealed the vast numbers of microscopic creatures present in nature, scientists began to wonder about their origin. From the beginning there were two schools of thought. Some believed that animalcules were formed spontaneously from nonliving materials, whereas others (Leeuwenhoek included) believed that they were formed from the "seeds" or "germs" of these animalcules, which were always present in the air. The belief in the spontaneous formation of living beings from nonliving matter is known as the doctrine of *spontaneous generation*, or *abiogenesis*, and has had a long existence. In ancient times it was considered self-evident that many plants and animals can be generated spontaneously under special conditions. The doctrine of spontaneous generation was accepted without question until the Renaissance.

As knowledge of living organisms accumulated, it gradually became evident that the spontaneous generation of plants and animals simply does not occur. A decisive step in the abandonment of the doctrine as applied to animals took place as the result of experiments performed about 1665 by an Italian physician, Francesco Redi. He showed that the maggots that develop in putrefying meat are the larval stages of flies and will never appear if the meat is protected by placing it in a vessel closed with fine gauze so that flies are unable to deposit their eggs on it. By such experiments, Redi destroyed the myth that maggots develop spontaneously from meat. Consequently, the doctrine of spontaneous generation was already being weakened by studies on the development of plants and animals at the time when Leeuwenhoek discovered the microbial world. For technical reasons, it is far more difficult to show that microorganisms are not generated spontaneously, and as time went on the proponents of the doctrine came to center their claims more and more on the mysterious appearance of these simplest forms of life in organic infusions. Those who did not believe in the spontaneous generation of microorganisms were in the position, always difficult, of having to prove a negative point; in fact, it was not until the middle of the nineteenth century that the cumulative negative evidence became sufficiently abundant to lead to the general abandonment of this doctrine.

One of the first to provide strong evidence that microorganisms do not arise spontaneously in organic infusions was the Italian naturalist Lazzaro Spallanzani, who conducted a long series of experiments on this problem in the middle of the eighteenth century. He could show repeatedly that heating can prevent the appearance of animalcules in infusions, although the duration of the heating necessary is variable. Spallanzani concluded that animalcules can be carried into infusions by air and that this is the explanation for their supposed spontaneous generation in well-heated infusions. Earlier workers had closed their flasks with corks, but Spallanzani, who was not satisfied that any mechanical plug could completely exclude air, resorted to hermetic sealing. He observed that after sealed infusions had remained barren for a long time a tiny crack in the glass would be followed by the development of animalcules. His final conclusion was that to render an infusion *permanently* barren, it must be sealed hermetically and boiled. Animalcules could never appear unless new air somehow entered the flask and came in contact with the infusion.

Spallanzani's beautiful experiments showed clearly all the difficulties of work of this kind. However, faulty experiments continued to be performed, and the results continued to be brought forward as evidence for the occurrence of spontaneous generation. In the meantime, an interesting *practical* application of Spallanzani's discoveries had been made. Because his experiments had shown that even very perishable plant or animal infusions do not undergo putrefaction or fermentation when they have been rendered free of animalcules, it seemed probable that these chemical changes were in some way connected with the development of microbes. In the beginning of the nineteenth century François Appert found that one can preserve foods by enclosing them in airtight containers and heating the containers. He was able in this way to preserve highly perishable foodstuffs indefinitely, and "appertization," as this original canning process was called, came into extensive use for the preservation of foods long before the scientific issue had been finally settled.

In the late eighteenth century the work of Priestley, Cavendish, and Lavoisier laid the foundations of the chemistry of gases. One of the gases first discovered was oxygen, which soon was recognized to be essential for the life of animals. In the light of this knowledge, it seemed possible that the hermetic sealing recommended by Spallanzani and practiced by Appert was effective in preventing the appearance of microbes and the decomposition of organic matter, not because it excluded air carrying germs but because it excluded oxygen, required both for microbial growth and for the initiation of fermentation or putrefaction. Consequently, the influence of oxygen on these processes was a matter of much discussion in the early nineteenth century. It was finally shown that neither growth nor decomposition will occur in an infusion that has been properly heated, even when it is exposed to air, provided that the air entering the infusion has been previously treated so as to remove any germs that it contains.

## The Experiments of Pasteur

By 1860 some scientists had begun to realize that there is a *causal relationship* between the development of microorganisms in organic infusions and the chemical changes that take place in these infusions; *microorganisms are the agents that bring about the chemical changes.* The great pioneer in these studies was Louis Pasteur (Fig 1.3). However, the acceptance of this concept was conditional on the demonstration that spontaneous generation does not occur. Stung by the continued claims of adherents to the doctrine of spontaneous generation, Pasteur finally turned his attention to this problem. His work on the subject was published in 1861 as a *Memoir on the Organized Bodies Which Exist in the Atmosphere.*

Pasteur first demonstrated that air does contain microscopically observable "organized bodies." He aspirated large quantities of air through a tube that contained a plug of guncotton to serve as a filter. The guncotton was then removed and dissolved in a mixture of alcohol and ether, and the sediment was examined microscopically. In addi-

**FIGURE 1.3**
Louis Pasteur (1822–1895). Courtesy of the Institut Pasteur, Paris.

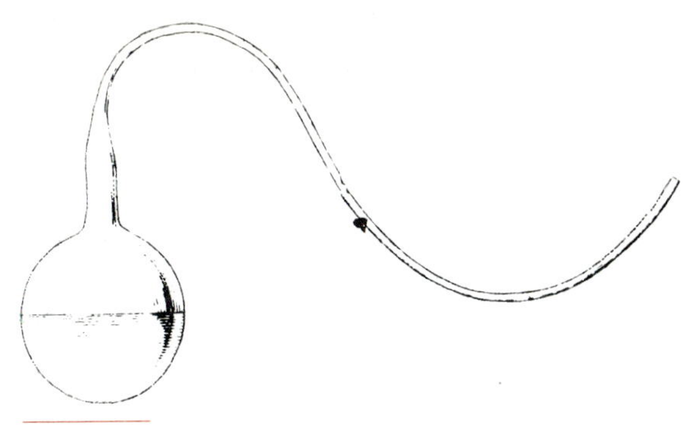

**FIGURE 1.4**
The swan-necked flask used by Pasteur during his studies on spontaneous generation. The construction of the neck permitted free access of air to the flask contents but prevented entry of microorganisms present in the air.

tion to inorganic matter, it contained considerable numbers of small round or oval bodies, indistinguishable from microorganisms. Pasteur next confirmed the fact that heated air can be supplied to a boiled infusion without giving rise to microbial development. Having established this point, he went on to show that in a closed system the addition of a piece of germ-laden guncotton to a sterile infusion invariably provoked microbial growth. These experiments showed Pasteur how germs can enter infusions and led him to what was perhaps his most elegant experiment on the subject. This was the demonstration that infusions will remain sterile indefinitely in open flasks, provided that the neck of the flask is drawn out and bent down in such a way that the germs from the air cannot ascend it. Pasteur's swan-necked flasks are illustrated in Figure 1.4. If the neck of such a flask was broken off, the infusion rapidly became populated with microbes. The same thing happened if the sterile liquid in the flask was poured into the exposed portion of the bent neck and then poured back.

Pasteur rounded out his study by determining in semiquantitative fashion the distribution of microorganisms in the air and by showing that these living organisms are by no means evenly distributed through the atmosphere.

## The Experiments of Tyndall

The last proponents of spontaneous generation maintained a stubborn rear-guard action for some years. The English physicist, John Tyndall, an ardent partisan of Pasteur, undertook a series of experiments designed to refute their claims; in the course of them, he established an important fact that had been overlooked by Pasteur, and in part accounted for the conflicting claims of the spontaneous generationists.

In a long series of experiments with infusions prepared from meat and fresh vegetables Tyndall obtained satisfactory sterilization by placing tubes of these infusions for five minutes in a bath of boiling brine. However, when he undertook similar experiments with infusions prepared from dried hay, this sterilization procedure proved completely inadequate. Worse still, when he then attempted to repeat his earlier experiments with other types of infusions, he found that they could no longer be sterilized by immersion in boiling brine, even for periods of as long as an hour. After many experiments, Tyndall finally realized what had happened. Dried hay contained spores of bacteria that were many times more resistant to heat than any microbes with which he had previously dealt, and, as a result of the presence of the hay in his laboratory, the air had become thoroughly infected with these spores. Once he had grasped this point, he proceeded to test the actual limits of heat resistance of the spores of hay bacteria and found that boiling infusions for even as long as $5\frac{1}{2}$ hours would not render them sterile with certainty. From these results he concluded that bacteria have phases, one relatively thermolabile (destroyed by boiling for five minutes) and one thermoresistant to an almost incredible extent. These conclusions were almost immediately confirmed by a German botanist, Ferdinand Cohn, who demonstrated that the hay bacteria can produce microscopically distinguishable resting bodies (*endospores*), which are highly resistant to heat.

Tyndall then proceeded to develop a method of sterilization by *discontinuous heating*, later called *tyndallization*, which could be used to kill *all* bacteria in infusions. Since growing bacteria are easily killed by brief boiling, all that is necessary is to allow the infusion to stand for a certain period to permit germination of the spores with a consequent loss of their heat resistance. A very brief period of boiling can then be used, and repeated, if need be, several times at intervals to catch any spores late in germination. Tyndall found that discontinuous boiling for 1 minute on five successive occasions would make an infusion sterile whereas a single continuous boiling for one hour would not. Recognition of the tremendous heat resistance of bacterial spores was essential to the development of adequate procedures for sterilization.

It has been stated that the work of Pasteur and Tyndall "disproved" the possibility of spontaneous generation, and their experimental findings have been used to support the contention that spontaneous generation has never occurred. This is an unjustifiable extension of their actual findings. The conclusion that we may safely draw is a much more limited one: that at the present time microorganisms do not arise spontaneously in properly sterilized organic infusions. It is probable that the primary origin of life on earth did involve a kind of spontaneous generation, although a far more gradual and subtle one than that envisaged by the proponents of the doctrine during the eighteenth and nineteenth centuries.

## THE DISCOVERY OF THE ROLE OF MICROORGANISMS IN TRANSFORMATION OF ORGANIC MATTER

During the long controversy over spontaneous generation, a correlation between the growth of microorganisms in organic infusions and the onset of chemical changes in the infusion itself was frequently observed. These chemical changes were designated as "fermentation" and "putrefaction." Putrefaction, a process of decomposition that results in the formation of ill-smelling products, occurs characteristically in meat and is a consequence of the breakdown of proteins, the principal organic constituents in such natural materials. Fermentation, a process that results in the formation of alcohols or organic acids, occurs characteristically in plant materials as a consequence of the breakdown of carbohydrates, the predominant organic compounds in plant tissues.

### Fermentation as a Biological Process

In 1837 three men, C. Cagniard-Latour, Th. Schwann, and F. Kützing, independently proposed that the yeast that appears during alcoholic fermentation is a microscopic plant and that the conversion of sugars to ethyl alcohol and carbon dioxide characteristic of the alcoholic fermentation is a physiological function of the yeast cell. This theory was bitterly attacked by such leading chemists of the time as J. J. Berzelius, J. Liebig, and F. Wohler, who held the view that fermentation and putrefaction are purely chemical processes. The science of chemistry had made great advances during the first decades of the nineteenth century and in 1828 the whole field of synthetic organic chemistry had been opened up by the first synthesis of an organic compound, urea, from inorganic materials. With the demonstration that organic compounds, until that time known exclusively as products of living activity, could be made in the laboratory, the chemists rightly felt that a large body of natural phenomena had now become amenable to analysis in physicochemical terms. The conversion of sugars to alcohol and carbon dioxide appeared to be a relatively simple chemical process. Accordingly, the chemists did not look with favor on the attempt to interpret this process as the result of the action of a living organism.

Ironically enough it was Pasteur, himself a chemist by training, who eventually convinced the scientific world that *all fermentative processes are the results of microbial activity.* Pasteur's work on fermentation extended with minor interruptions from 1857 to 1876. This work had a practical origin. The distillers of Lille, where the manufacture of alcohol from beet sugar was an important local industry, had encountered difficulties and called on Pasteur for assistance. Pasteur found that their troubles were caused by the fact that the alcoholic fermentation had been in part replaced by another kind of fermentative process, which resulted in the conversion of the sugar to lactic acid. When he examined microscopically the contents of fermentation vats in which lactic acid was being formed, he found that the cells of yeast characteristic of the alcoholic fermentation had been replaced by much smaller rods and spheres. If a trace of this material was placed in a sugar solution containing some chalk, a vigorous lactic fermentation ensued, and eventually a grayish deposit was formed, which again proved on microscopic examination to consist of the small spherical and rod-shaped organisms. Successive transfers of minute amounts of material to fresh flasks of the same medium always resulted in the production of a lactic fermentation and an increase in the amount of the formed bodies. Pasteur argued that the active agent, or new "yeast," was a microorganism that specifically converted sugar to lactic acid during its growth.*

Using similar methods, Pasteur studied a considerable number of fermentative processes during the following 20 years. He was able to show that fermentation is invariably accompanied by the development of microorganisms. Furthermore, he showed that each particular chemical type of fermentation, as defined by its principal organic end products (for example, the lactic, the alcoholic, and the butyric fermentations), is accompanied by the development of a *specific type of microorganism.* Many of these specific microbial types could be

* The agents of the lactic acid fermentation are in fact bacteria, but in Pasteur's time, the different taxonomic groups of microorganisms had not yet been clearly distinguished.

recognized and differentiated microscopically by their characteristic size and shape. In addition, they could be distinguished by the specific environmental conditions that favored their development. To cite one example of such *physiological specificity*, Pasteur observed very early that whereas the agent of alcoholic fermentation can flourish in an acid medium, the agents of the lactic fermentation grow best in a neutral medium. It was for this reason that he added chalk (calcium carbonate) to his medium for the cultivation of the lactic organisms; this substance serves as a neutralizing agent and prevents too strong an acidification of the medium that would otherwise occur as a result of the formation of lactic acid.

## The Discovery of Anaerobic Life

During his studies on the butyric fermentation Pasteur discovered another fundamental biological phenomenon: the *existence of forms of life that can live only in the absence of free oxygen.* While examining microscopically fluids that were undergoing a butyric fermentation, Pasteur observed that the bacteria at the margin of a flattened drop, in close contact with the air, became immotile, whereas those in the center of the drop remained motile. This observation suggested that air had an inhibitory effect on the microorganisms in question, an inference that Pasteur quickly confirmed by showing that passage of a current of air through the fermenting fluid could retard, and sometimes completely arrest, the butyric fermentation. He thus concluded that some microorganisms can live only in the absence of oxygen, a gas previously considered essential for the maintenance of all life. He introduced the terms *aerobic* and *anaerobic* to designate, respectively, life in the presence of and in the absence of oxygen.

## The Physiological Significance of Fermentation

The discovery of the anaerobic nature of the butyric fermentation provided Pasteur with an important clue for the understanding of the role that fermentations play in the life of the microorganisms that bring them about. Free oxygen is essential for most organisms as an agent for the oxidation of organic compounds to carbon dioxide. Such oxygen-linked biological oxidations, known collectively as *aerobic respirations*, provide the energy that is required for maintenance and growth.

Pasteur was the first to realize that the breakdown of organic compounds in the absence of oxygen can also be used by some organisms as a means of obtaining energy; as he put it, *fermentation is life without air*. Some strictly anaerobic microorganisms, such as the butyric acid bacteria, are dependent on fermentative mechanisms to obtain energy. Many other microorganisms, including certain yeasts, are *facultative anaerobes*, which have two alternative energy-yielding mechanisms at their disposal. In the presence of oxygen they employ aerobic respiration, but they can employ fermentation if no free oxygen is present in their environment. This was beautifully demonstrated by Pasteur, who showed that sugar is converted to alcohol and carbon dioxide by yeast in the absence of air but that in the presence of air little or no alcohol is formed; carbon dioxide is the principal end product of this aerobic reaction.

The amount of growth that can occur at the expense of an organic compound is determined primarily by the amount of energy that can be obtained by the breakdown of that compound. Fermentation is a less efficient energy-yielding process than aerobic respiration because part of the energy present in the substance decomposed is still present in the organic end products (for example, alcohol or lactic acid) characteristically formed by fermentative processes. As Pasteur was the first to show, the breakdown of a given weight of sugar results in substantially less growth of yeast under anaerobic conditions than under aerobic ones, thus establishing the relative inefficiency of fermentation as an energy source.

Pasteur's work showed that fermentations are "vital processes," which play a role of basic physiological importance in the life of many cells. Further development of knowledge about the nature of fermentation resulted from an accidental observation made in 1897 by H. Buchner. In attempting to preserve an extract of yeast, prepared by grinding yeast cells with sand, Buchner added a large quantity of sugar to it and was surprised to observe an evolution of carbon dioxide accompanied by the formation of alcohol. A soluble enzymatic preparation, able to carry out alcoholic fermentation, was thus discovered. Buchner's discovery inaugurated the development of modern biochemistry; the detailed analysis of the mechanism of cell-free alcoholic fermentation was eventually to show that this complex metabolic process can be interpreted as resulting from a succession of chemically intelligible reactions, each catalyzed by a specific enzyme. Today, the belief that even the most complex physiological process can be similarly understood in physicochemical terms is accepted as a matter of course by all biologists. In this sense, the intuition of the nineteenth-century chemists who battled against the biological theory of fermentation has proved to be a correct one.

## THE DISCOVERY OF THE ROLE OF MICROORGANISMS IN THE CAUSATION OF DISEASE

During his studies on fermentation, Pasteur, ever conscious of the practical applications of his scientific work, devoted considerable attention to the spoilage of beer and wine, which he showed to be caused by the growth of undesirable microorganisms. Pasteur used a peculiar and significant term to describe these microbially induced spoilage processes; he called them "diseases" of beer and wine. In fact, he was already considering the possibility that microorganisms may act as agents of infectious disease in higher organisms. Some evidence in support of this hypothesis already existed. It had been shown in 1813 that specific fungi can cause diseases of wheat and rye, and in 1845 M. J. Berkeley had proved that the great Potato Blight of Ireland, a natural disaster that deeply influenced Irish history, was caused by a fungus. The first recognition that fungi may be specifically associated with a disease of animals came in 1836 through the work of A. Bassi in Italy on a fungal disease of silkworms. A few years later J. L. Schönlein showed that certain skin diseases of man are caused by fungal infections. Despite these indications, very few medical scientists were willing to entertain the notion that the major infectious diseases of man could be caused by microorganisms, and fewer still believed that organisms as small and apparently simple as the bacteria could act as agents of disease.

### Surgical Antisepsis

The introduction of anesthesia about 1840 made possible a very rapid development of surgical methods. Speed was no longer a primary consideration, and the surgeon was able to undertake operations of a length and complexity that would have been unthinkable previously. However, with the elaboration of surgical technique, a problem that had always existed became more and more serious: *surgical sepsis*, or the infections that followed surgical intervention and often resulted in the death of the patient. Pasteur's studies on the problem of spontaneous generation had shown the presence of microorganisms in the air and at the same time indicated various ways in which their access to and development in organic infusions could be prevented. A young British surgeon, Joseph Lister, who was deeply impressed by Pasteur's work, reasoned that surgical sepsis might well result from microbial infection of the tissues exposed during operation. He decided to develop methods for preventing the access of microorganisms to surgical wounds. By the scrupulous sterilization of surgical instruments, by the use of disinfectant dressings, and by the conduct of surgery under a spray of disinfectant to prevent airborne infection, he succeeded in greatly reducing the incidence of surgical sepsis. Lister's procedures of antiseptic surgery, developed about 1864, were initially greeted with considerable skepticism but, as their striking success in the prevention of surgical sepsis was recognized, gradually became common practice. This work provided powerful *indirect* evidence for the germ theory of disease, even though it did not cast any light on the possible microbial causation of specific human diseases. Just as in the case of Appert's development of canning as a means of food preservation half a century before, so with Lister's introduction of surgical antisepsis: practice had run ahead of theory.

### The Bacterial Etiology of Anthrax

The discovery that bacteria can act as specific agents of infectious disease in animals was made through the study of anthrax, a serious infection of domestic animals that is transmissible to humans. In the terminal stages of a generalized anthrax infection, the rod-shaped bacteria responsible for the disease occur in enormous numbers in the bloodstream. These objects were first observed as early as 1850, and their presence in the blood of infected animals was reported by a series of investigators during the following 15 years. Particularly careful and detailed studies were carried out between 1863 and 1868 by C. J. Davaine, who showed that the rods are invariably present in diseased animals but are undetectable in healthy ones and that the disease can be transmitted to healthy animals by inoculation with blood containing these rod-shaped elements.

The conclusive demonstration of the bacterial causation, or *etiology*, of anthrax was provided in 1876 by Robert Koch (Figure 1.5), a German country doctor. He had no laboratory, and his experiments were conducted in his home, using very primitive improvised equipment and small experimental animals. He showed that mice could be infected with material from a diseased domestic animal. He transmitted the disease through a series of 20 mice by successive inoculation; at each transfer, the characteristic symptoms were observed. He then proceeded to cultivate the causative bacterium by introducing minute, heavily infected particles of spleen from a diseased animal into drops of sterile serum. Observing hour after hour the growth of the organisms in this culture medium, he saw the rods

**FIGURE 1.5**
Robert Koch (1843–1910). Courtesy of VEB George Thieme, Leipzig.

change into long filaments within which ovoid, refractile bodies eventually appeared. He showed that these bodies were spores, which had not been seen by previous workers (Figure 1.6). When spore-containing material was transferred to a fresh drop of sterile serum, the spores germinated and gave rise once more to typical rods. In this fashion, he transferred cultures of the bacterium eight successive times. The final culture of the series, injected into a healthy animal, again produced the characteristic disease, and from this animal the organisms could again be isolated in culture.

This series of experiments fulfilled the criteria which had been laid down 36 years before by J. Henle as logically necessary to establish the causal relationship between a specific microorganism and a specific disease. In generalized form, these criteria are (1) the microorganism must be present in every case of the disease; (2) the microorganism must be isolated from the diseased host and grown in pure culture; (3) the specific disease must be reproduced when a pure culture of the microorganism is inoculated into a healthy susceptible host; and (4) the microorganism must be recoverable once again from the experimentally infected host. Since Koch was the first to apply these criteria experimentally, they are now generally known as *Koch's postulates.*

Koch carried out another series of experiments that demonstrated the *biological specificity* of disease agents. He showed that another spore-forming bacterium, the hay bacillus, does not cause anthrax upon injection, and he also differentiated bacteria that cause other infections from the anthrax organism. From these studies he concluded that only one kind of bacillus is able to cause this specific disease process, while other bacteria either do not produce disease following inoculation, or give rise to other kinds of disease.

In the meantime, Pasteur had found a collaborator, J. Joubert, with a knowledge of medical problems. Unaware of Koch's work, Pasteur and Joubert undertook the study of anthrax. They did not add anything new to the conclusions reached by Koch, but they confirmed his work and provided additional demonstrations that the bacillus, and not some other agent, was the specific cause of the disease.

## The Rise of Medical Bacteriology

This work on anthrax abruptly ushered in the golden age of medical bacteriology, during which newly established institutes, created in Paris and in Berlin for Pasteur and Koch, respectively, became the world centers of bacteriological science. The

**FIGURE 1.6**
The first photomicrographs of bacteria, taken by Robert Koch in 1877. (a) Unstained chains of vegetative cells of *Bacillus anthracis*. (b) Unstained chains of *B. anthracis:* the cells contain refractile spores. (c) A stained smear of *B. anthracis* from the spleen of an infected animal. Note the rod-shaped bacilli and the larger tissue cells.

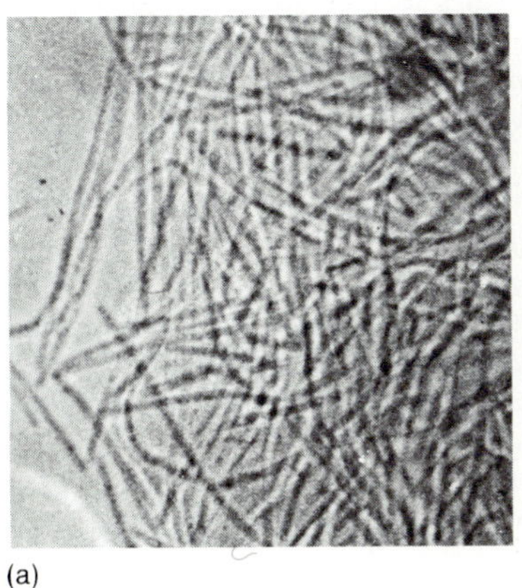

(a)

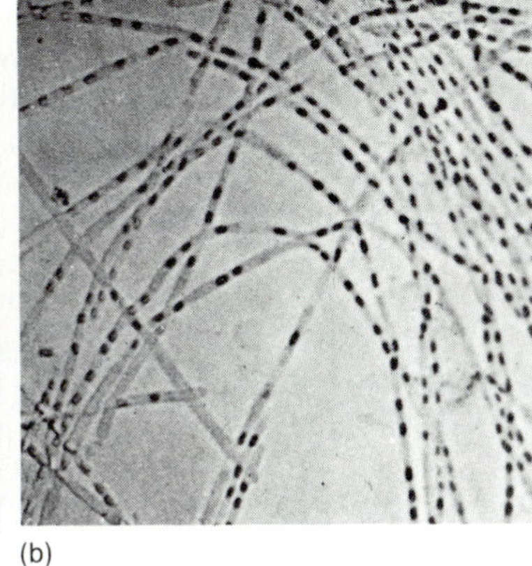

(b)

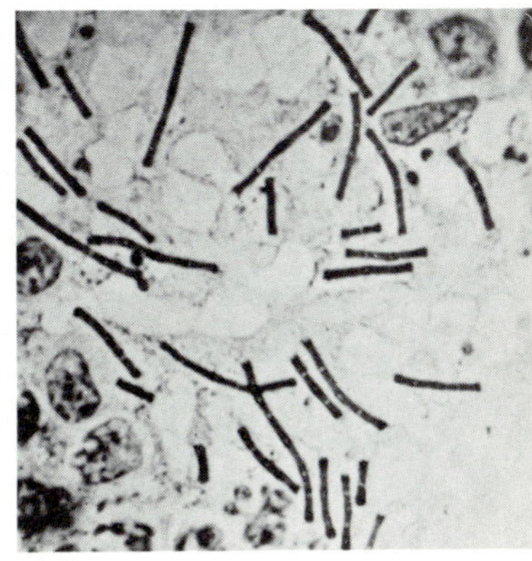

(c)

German school, led by Koch, concentrated primarily on the isolation, cultivation, and characterization of the causative agents for the major infectious diseases of man. The French school, under the leadership of Pasteur, turned almost immediately to a more subtle and complex problem: the experimental analysis of how infectious disease takes place in the animal body and how recovery and immunity are brought about. Within 25 years most of the major bacterial agents of human disease had been discovered and described, and methods for the prevention of many of these diseases, either by artificial immunization or by the application of hygienic measures, had been developed. It was by far the greatest medical revolution in all human history.

### The Discovery of Filterable Viruses

One of the early technical contributions from Pasteur's new institute was the development of filters able to retain bacterial cells and thus to yield bacteria-free filtrates. Infectious fluids were often tested for the presence of disease-producing bacteria by passing them through such filters; if the filtrate was no longer able to produce infection, the presence of a bacterial agent in the original fluid was indicated. In 1892 a Russian scientist, D. Iwanowsky, applied this test using an infectious extract from tobacco plants infected with mosaic disease. He found to his surprise that the filtrate was fully infectious when applied to healthy plants. His specific discovery was soon confirmed, and within a few years other workers found that many major plant and animal diseases are caused by similar, filter-passing, submicroscopic agents. A whole class of infectious entities, much smaller than any previously known organisms, was thus discovered. The true nature of these *viruses*, as they came to be known, remained obscure for many decades, but eventually it was established that they are a distinctive group of biological objects entirely different in structure and mode of development from all cellular organisms (see Chapter 9).

## THE DEVELOPMENT OF PURE CULTURE METHODS

Pasteur possessed an intuitive skill in the handling of microorganisms and was able to reach correct conclusions about the specificity of fermentative processes, even when working with cultures that contained a mixture of microbial forms. The classical studies of Koch and Pasteur on anthrax, which firmly established the germ theory of animal disease, were conducted under experimental conditions that did not really permit certainty that rigorously pure cultures of the causative organism had been obtained. There are pitfalls in working with mixed microbial populations, and not all the scientists who began to study microorganisms in the middle of the nineteenth century were as skillful as Pasteur and Koch. It was frequently claimed that microorganisms had a large capacity for variation with respect both to their *morphological form* and to their *physiological function*. This belief became known as the doctrine of *pleomorphism*, while the opposing belief, that microorganisms show constancy and specificity of form and function, became known as the doctrine of *monomorphism*.

### The Origin of the Belief in Pleomorphism

Let us consider what happens when a nutrient solution is inoculated with a mixed microbial population. The principle of natural selection at once begins to operate, and the microbe that can grow most rapidly under the conditions provided soon predominates. As a result of its growth and chemical activities, the composition of the medium changes; after some time, conditions no longer permit growth of the originally predominant form. The environment may now be favorable for the growth of a second kind of microorganism, also originally introduced into the medium but hitherto unable to develop, which gradually replaces the first as the predominant form in the culture. In this fashion one may obtain the *successive development of many different microbial types* in a single culture flask seeded with a mixed population. It is often possible to maintain the predominance of the form that first develops by repeated transfer of the mixed population at short intervals into a fresh medium of the same composition; this was essentially the device used by Pasteur in his studies on fermentation.

If one does not recognize the possibility of such microbial successions, it is easy to conclude that the chemical and morphological changes observable over the course of time in a single culture inoculated with a mixed population reflect *transformations undergone by a single kind of microorganism*. Between 1865 and 1885, claims for the extreme variability of microorganisms, based on such observations, were frequently made.

The term *pleomorphism* (derived from the Greek, meaning "doctrine of many shapes") implies that its proponents were concerned primarily with the possibilities of morphological variation. In fact, this was often not the case. Many pleomorphists insisted equally on the variability of function. For them, there was no such thing as a specific microbial agent for alcoholic fermentation or for a particular disease; they considered that it is the nature of the environment that determines both form and function. The widespread persistence of such beliefs represented a threat to the development of microbiology, and they were opposed by such leaders of the new discipline as Pasteur, Koch, and Cohn, who upheld the doctrine of monomorphism, insisting on the constancy of microbial form (and function).

Around 1870 it began to be realized that a sound understanding of the form and function of microorganisms could be obtained only if the complications inherent in the study of mixed microbial populations were avoided by the use of pure cultures. *A pure culture is one that contains only a single kind of microorganism.* The leading advocates of the use of pure cultures were two great mycologists (students of fungi), A. de Bary and O. Brefeld.

## The First Pure Cultures

Much of the pioneering work on pure culture techniques was done by Brefeld, working with fungi. He introduced the practice of isolating single cells, as well as the cultivation of fungi on solid media, for which purpose he added gelatin to his culture liquids. His methods of obtaining pure cultures worked admirably for the fungi but were found to be unsuitable when applied to the smaller bacteria. Other methods had, therefore, to be devised for bacteria. One of the first to be proposed was the *dilution method.* A fluid containing a mixture of bacteria was diluted with sterile medium in the hope that ultimately a growth could be obtained that took its origin from a single cell. In practice, the method is tedious, difficult, and uncertain; it also has the obvious disadvantage that one can only isolate in pure form the microorganisms that predominate in the original mixture.

Koch realized very early that the development of simple methods for obtaining pure cultures of bacteria was a vital requirement for the growth of the new science. The dilution method was obviously too tedious and uncertain for routine use. A more promising approach had already been suggested by the earlier observations of J. Schroeter, who had noted that on such solid substrates as potato, starch paste, bread, and egg albumen, isolated bacterial growths, or *colonies,* arose. The colonies differed from one another, but within each colony the bacteria were of one type. At first Koch experimented with the use of sterile, cut surfaces of potatoes, which he placed in sterile, covered glass vessels and then inoculated with bacteria. However, potatoes have obvious disadvantages: the cut surface is moist, which allows motile bacteria to spread freely over it; the substrate is opaque, and hence it is often difficult to see the colonies; and most important of all, the potato is not a good nutrient medium for many bacteria. Koch perceived that it would be far better if one could solidify a well-tried liquid medium with some clear substance. In this fashion, a translucent gel could be prepared on which developing bacterial colonies would be clearly visible. At the same time, the differing nutritional requirements of different bacteria could be met by modifying the composition of the liquid base. With this in mind, he added gelatin as a hardening agent. Once set, the gelatin surface was seeded by picking up a minute quantity of bacterial cells (*the inoculum*) on a platinum needle, previously sterilized by passage through a flame, and drawing it several times rapidly and lightly across the surface of the jelly. Different bacterial colonies soon appeared, each of which could be purified by a repetition of the streaking process. This became known as the *streak method* for isolating bacteria. The pure cultures were transferred to tubes containing sterile nutrient gelatin that had been plugged with cotton wool and set in a slanted position. Such cultures became known as *slant cultures.* Shortly thereafter, Koch discovered that instead of streaking the bacteria over the surface of the already solidified gelatin, he could mix them with the melted gelatin. When the gelatin set, the bacteria were immobilized in the jelly and there developed into isolated colonies. This became known as the *pour plate method* for isolating bacteria.

Gelatin, the first solidifying agent used by Koch, has several disadvantages. It is a protein highly susceptible to microbial digestion and liquefaction. Furthermore, it changes from a gel to a liquid at temperatures above 28° C. A new solidifying agent, *agar,* was soon introduced. Agar is a complex polysaccharide, extracted from red algae. A temperature of 100° C is required to melt an agar gel, so it remains solid throughout the entire temperature range over which bacteria are cultivated. However, once melted, it remains a liquid until the temperature falls to about 44° C, a fact that makes possible its use for the preparation of cultures by the pour plate method. It produces a stiff and transparent gel. Finally, it is a complex carbohydrate that is attacked by relatively few bacteria, so the problem of its liquefaction rarely arises. For these reasons, agar rapidly replaced gelatin as the hardening agent of choice for bacteriological work. Although no equally satisfactory synthetic substitute for agar has yet been discovered, certain bacterial polysaccharides have recently been shown to possess promising properties.

### The Development of Culture Media by Koch and His School

Pasteur had used simple, transparent liquid media of known chemical composition for the selective cultivation of fermentative microorganisms. For the isolation of the microbial agents of disease, different types of culture media were required, and this was the second major technical problem to which Koch and his collaborators devoted their attention. Disease-producing bacteria develop normally within the tissues of an infected host, so it seemed logical that their cultivation outside the animal body would succeed best if the medium resembled as much as possible the environment of the host tissues. This line of reasoning led Koch to adopt *meat infusions* and *meat extracts* as the basic ingredients in his culture media. *Nutrient broth* and its solid counterpart, *nutrient agar*, which are still the most widely used media in general bacteriological work, were the outcome of Koch's experiments along these lines. Nutrient broth contains 0.5 percent peptone, an enzymatic digest of meat; 0.3 percent meat extract, a concentrate of the water-soluble components of meat; and 0.8 percent NaCl, to provide roughly the same total salt concentration as that found in tissues. For the cultivation of more fastidious disease-producing organisms, this basal medium can be supplemented in various ways (e.g., with sugar, blood, or serum). Considering the specific purposes for which these media were designed, the choice of ingredients may be considered logical, although there is no evidence that the traditional inclusion of NaCl has any real value, for most bacteria are insensitive to changes in the salt concentration of their environment over a very wide range. As time went on, however, many bacteriologists came to consider that these media were universal ones, suitable for the cultivation of nearly all bacteria. This is untrue; bacteria vary greatly in their nutritional requirements, and no single medium is capable of supporting the growth of more than a very small fraction of the bacteria that exist in nature (see Chapter 2).

**FIGURE 1.7**
Sergius Winogradsky (1856–1953). Courtesy of Masson et Cie., Paris. Reprinted with the permission of the *Annales de l'Institut Pasteur*.

## MICROORGANISMS AS GEOCHEMICAL AGENTS

Although the role played by microorganisms as agents of infectious disease was the central microbiological interest in the last decades of the nineteenth century, some scientists carried forward the work initiated by Pasteur through his early investigations on the role of microorganisms in fermentation. This work had clearly shown that microorganisms can serve as specific agents for large-scale chemical transformations and indicated that the microbial world as a whole might well be responsible for a wide variety of other geochemical changes.

The establishment of the cardinal roles that microorganisms play in the biologically important cycles of matter on earth—the cycles of carbon, nitrogen, and sulfur—was largely the work of two men, S. Winogradsky (Figure 1.7) and M. W. Beijerinck (Figure 1.8). In contrast to plants and animals, microorganisms show an extraordinarily wide range of physiological diversity. Many groups are specialized for carrying out chemical transformations that cannot be performed at all by plants and animals, and thus play vital parts in the turnover of matter on earth.

One example of microbial physiological specialization is provided by the *chemoautotrophic bacteria*, discovered by Winogradsky. These bacteria can grow in completely inorganic environments, obtaining the energy necessary for their growth by the oxidation of reduced inorganic compounds, and use carbon dioxide as the source of their cellular carbon. Winogradsky found that there are several physiologically distinct groups among the autotrophic bacteria, each characterized by the ability to use a particular inorganic energy source; for example, the sulfur bacteria oxidize inorganic sulfur compounds, the nitrifying bacteria, inorganic nitrogen compounds.

FIGURE 1.8
Martinus Willem Beijerinck (1851–1931). Courtesy of Martinus Nijhoff, The Hague. Reprinted with permission.

Another discovery, to which both Winogradsky and Beijerinck contributed, was the role that microorganisms play in the fixation of atmospheric nitrogen, which cannot be used as a nitrogen source by most living organisms. They showed that certain bacteria, some symbiotic in higher plants and others free-living, can use gaseous nitrogen for the synthesis of their cell constituents. These microorganisms accordingly help to maintain the supply of combined nitrogen, upon which all other forms of life are dependent.

### Enrichment Culture Methods

For the isolation and study of the various physiological types of microorganisms that exist in nature, Winogradsky and Beijerinck developed a new and profoundly important technique: the technique of the *enrichment culture.* It is essentially an application on a microscale of the principle of natural selection. The investigator devises a culture medium of a particular defined chemical composition, inoculates it with a mixed microbial population, such as can be found in a small amount of soil, and then ascertains by examination what kinds of microorganisms come to predominate. Their predominance is caused by their ability to grow more rapidly than any of the other organisms present in the inoculum, hence the term *enrichment medium.* To take a specific example, if we wish to discover microorganisms that can use atmospheric nitrogen, $N_2$, as the only source of the element nitrogen, we prepare a medium that is *free of combined nitrogen* but which contains all the other nutrients—an energy source, a carbon source, minerals—necessary for growth. This is then inoculated with soil, placed in contact with $N_2$, and incubated under any desired set of physical conditions. Since nitrogen is an essential constituent of every living cell, the only organisms of all those present in the original inoculum that will be able to multiply in such a medium are those that can fix atmospheric nitrogen. Provided that such types are present in the soil sample, they will grow. Such experiments can be varied in innumerable ways by modifying such factors as the carbon source, the energy supply, the temperature, and the hydrogen ion concentration. For each particular set of conditions, a particular kind of microorganism will come to predominance, provided that there are any organisms existing in the inoculum that can grow under such conditions. The enrichment culture method is thus one of the most powerful experimental tools available to microbiologists; by its use they can isolate microorganisms with any desired set of nutrient requirements, provided that such organisms exist in nature.

## THE GROWTH OF MICROBIOLOGY IN THE TWENTIETH CENTURY

During the last decades of the nineteenth century microbiology became a solidly established discipline with a distinctive set of concepts and techniques, both in large measure outgrowths of the work of Pasteur. During the same period a science of general biology also emerged. It was the creation of Charles Darwin, who imposed a new order and coherence in the heretofore anecdotal materials of natural history by interpreting them in terms of the theory of evolution through natural selection. Logically, microbiology should have taken its place, alongside other specialized biological disciplines, in the framework of post-Darwinian general biology. In fact, however, this did not occur. For half a century after the death of Pasteur in 1895, microbiology and general biology developed in almost complete independence of one another. The major interests of microbiology in this period were the characterization of agents of infectious disease, the study of immunity and its functions in the prevention and cure of disease, the search for chemotherapeutic agents, and the analysis of the chemical activities of microorganisms. All these problems were both conceptually and experimentally remote from the dominant interests of biology in the early twentieth century: the organization of the cell and its role in reproduction and development; and the mechanisms of heredity and evolution in plants and animals. Even the distinctive and original technical innovations of microbiology were of little interest to contemporary biologists; their value became widely recognized only about 1950, when tissue and cell culture began to be applied extensively to plant and animal systems.

However, microbiology did contribute significantly to the development of the new discipline of biochemistry. The discovery of cell-free alcoholic fermentation by Buchner (see page 7) provided the key to the chemical analysis of energy-yielding metabolic processes. In the first two decades of the twentieth century parallel studies on the mechanisms of glycolysis by muscle and of alcoholic fermentation by yeast gradually revealed their fundamental similarity. Quite unexpectedly, vertebrate physiologists and microbial biochemists had found a common ground. A few years later the analysis of animal and microbial nutrition revealed another unexpected common denominator: the "vitamins" required in traces by animals proved chemically identical with the "growth factors" required by some bacteria and yeasts. The detailed study of the functions of these substances, conducted for reasons of facility in large measure with microorganisms, revealed that they are biosynthetic precursors of a variety of coenzymes, all of which play indispensable roles in the metabolism of the cell. These discoveries, spanning the period from 1920 to 1935, demonstrated the fundamental similarities of all living systems at the metabolic level—a doctrine proclaimed by biochemists and microbiologists under the slogan "the unity of biochemistry."

The second great advance of biology in the early twentieth century—the creation of the discipline of genetics, formed through the convergence of cytology and Mendelian analysis—had no immediate impact on microbiology. Indeed, it long seemed doubtful whether the mechanisms of inheritance operative in plants and animals likewise functioned in bacteria. The first important contact between genetics and microbiology occurred in 1941, when Beadle and Tatum succeeded in isolating a series of biochemical mutants from the fungus *Neurospora.* This opened the way to the analysis of the consequences of mutation in biochemical terms, and *Neurospora* joined the fruit fly and the maize plant as a material of choice for genetic research.

In 1943 an analysis by Delbrück and Luria of mutation in bacteria provided the technical and conceptual basis for genetic work on these microorganisms. Soon afterward several mechanisms of genetic transfer were shown to exist in bacteria, all significantly different from the mechanism of sexual recombination in plants and animals. In 1944 the work of Avery, McLeod and McCarty on the process of bacterial genetic transfer known as *transformation* revealed that it is mediated by free deoxyribonucleic acid (DNA). The chemical nature of the hereditary material was thus discovered.

The confluence of microbiology, genetics, and biochemistry between 1940 and 1945 brought to an end the long isolation of microbiology from the main currents of biological thought. It also set the stage for the second major revolution in biology, to which microbiologists made many contributions of fundamental importance: the advent of molecular biology.

The era of molecular biology began suddenly in 1953 with the publication of a brief paper by J. D. Watson and F. H. C. Crick in *Nature*, proposing that DNA was composed of two helical strands of alternating deoxyribose and phosphate residues held together by nucleic acid bases. The bases were covalently bonded to each of the deoxyribose residues and joined to each other at the core of the molecule by hydrogen bonds. The spatial constraints imposed by this molecular architecture permitted the sequence of the four types of bases attached to the strands to be completely variable, but necessitated specific hydrogen-bonded pairing between bases at the core. For example, adenine on one strand paired with thymine on the other; guanine on one strand with cytosine on the other. Thus the structure immediately suggested a rationale for the mechanisms of a number of fundamental biological processes, including replication and mutational alterations of DNA. However, the structure provided no good hints as to how genetic information might be encoded. The elucidation of the coding mechanism, or the "cracking of the genetic code" as it was called, required an additional decade and constituted a major accomplishment to which many scientists contributed. First the code was deduced to be *triplet* (three base pairs encode a single amino acid in a protein) and *commaless* (each triplet is read sequentially without intervening bases that might punctuate to maintain proper register of reading). Later the specific amino acid encoded by each of the 64 (i.e., $4^3$) possible triplets was determined (see Chapter 5). Deciphering the genetic code, which turned out to be common to all living organisms, constituted not only a major intellectual accomplishment but a major tool for the furtherance of molecular biology.

In the mid-1950s microbial biochemistry also set out in new directions. Until that time, it had concerned itself with understanding the chemistry of cellular components and reactions; then it turned to detailed studies on the coordination and regulation of cellular processes at the levels of enzyme action and controlled expression of specific genes. Largely through studies on microorganisms, an understanding began to develop of how the myriad of individual cellular reactions were efficiently and harmoniously controlled to accomplish biological functions.

In the mid-1970s a series of discoveries ushered in a new and powerful set of capabilities known as *recombinant DNA technology*. These allowed a remarkable merging of chemical and biological studies on genetic material. DNA can now be manipulated *in vitro* in a number of ways, including the controlled chemical joining of fragments of it derived from any biological source or from chemical synthesis, and the processed DNA can be introduced into a microbial cell to study the biological consequences of the manipulations. Recombinant DNA technology has already been successfully applied to a large number of fundamental biological questions, to practical issues like the production of medically useful proteins (e.g., insulin, human growth hormone, and Factor VIII), and to the improvement of industrial fermentations. It holds almost certain promise for the production of other therapeutically useful proteins and fermentation products as well as for crop improvement and the detection and treatment of genetic defects in humans.

Concurrent with the development of recombinant DNA technology came a major biological finding with profound evolutionary significance: the realization that a group of procaryotic microorganisms, termed Archaebacteria are as different from other bacteria as they are from plants and animals.

## FURTHER READING


**Books**

BROCK, T. D. ed. and trans., *Milestones in Microbiology.* Englewood Cliffs, N.J.: Prentice-Hall, Inc., 1961.

BULLOCH, W., *The History of Bacteriology.* New York: Oxford University Press, 1960; Republished: New York: Dover Publications, Inc., 1979.

DOBELL, C., *Antony van Leeuwenhoek and His "Little Animals."* London: Staples Press, 1932; Republished: New York: Dover Publications, Inc., 1960.

DUBOS, R., *Louis Pasteur, Free Lance of Science.* Boston: Little, Brown, 1950.

JACOB, F., *The Logic of Living Systems.* London: Allen Lane, 1974.

LARGE, E. C., *Advance of the Fungi.* London: Jonathan Cape, 1940; Republished: New York: Dover Publications, Inc., 1962.

STENT, G., ed., *Phage and the Origins of Molecular Biology.* Cold Spring Harbor, N. Y.: Laboratory of Quantitative Biology, 1966.

WATSON, J. D., *The Double Helix.* New York: Atheneum, 1968.

**Review**

VAN NIEL, C. B., "Natural Selection in the Microbial World," *J. Gen. Microbiol.* **13,** 201 (1955).

# Chapter 2
# The Methods of Microbiology

As a result of the small size of microorganisms, the amount of information that can be obtained about their properties from the examination of *individuals* is limited; for the most part, the microbiologist studies *populations*, containing millions or billions of individuals. Such populations are obtained by growing microorganisms, under more or less well-defined conditions, as *cultures*. A culture that contains only one kind of microorganism is known as a *pure* or *axenic culture*. A culture that contains more than one kind of microorganism is known as a *mixed culture*; if it contains only two kinds of microorganisms, deliberately maintained in association with one another, it is known as a *two-membered culture*.

At the heart of microbiology there accordingly lie two kinds of operations: *isolation*, the separation of a particular microorganism from the mixed populations that exist in nature; and *cultivation*, the growth of microbial populations in artificial environments (culture media) under laboratory conditions. These two operations come into play irrespective of the kind of microorganism with which the microbiologist deals; they are basic alike to the study of viruses, bacteria, fungi, algae, protozoa, and even small invertebrate animals. Furthermore, they have been extended in recent years to the study of cell or tissue lines derived from higher plants and animals (*tissue culture*). The unity of microbiology as a science, despite the biological diversity of the organisms with which it deals, is derived from this common operational base.

## PURE CULTURE TECHNIQUE

Microorganisms are ubiquitous, so the preparation of a pure culture involves not only the isolation of a given microorganism from a mixed natural microbial population, but also the maintenance of the isolated individual and its progeny in an artificial environment to which the access of other microorganisms is prevented. Microorganisms do not require much space for development; hence an artificial environment can be created within the confines of a test tube, a flask, or a petri dish, the three kinds of containers most commonly used to cultivate microorganisms. The culture vessel must be rendered initially *sterile* (free of any living microorganism) and, after the introduction of the desired type of microorganism, it must be protected from subsequent external contamination. The primary source of external contamination is the atmosphere, which always contains floating microorganisms. The form of a petri dish, with its overlapping lid, is specifically designed to prevent atmospheric contamination. Contamination of tubes and flasks is prevented by closure of their orifices with an appropriate stopper. This has traditionally been a plug of cotton wool, although metal caps or plastic screw caps are now often employed, particularly for test tubes.

The external surface of a culture vessel is, of course, subject to contamination, and the interior of a flask or tube can become contaminated when it is opened to introduce or withdraw material. This danger is minimized by passing the orifice through a flame, immediately after the stopper has been removed and again just before it is replaced.

The *inoculum* (i.e., the microbial material used to seed or *inoculate* a culture vessel) is commonly introduced on a metal wire or loop, which is rapidly sterilized just before its use by heating in a flame. Transfers of liquid cultures can also be made by pipette. For this purpose, the mouth end of the pipette may be plugged with cotton wool, and the pipette is sterilized in a paper wrapping or in a glass or metal container, which keeps both inner and outer surfaces free of contamination until the time of use.

The risks of accidental contamination may be further reduced by performing transfers in a hood or in a small closed room, the air of which has been specially treated to reduce its microbial content. Special hoods and other precautions may also be necessary to prevent accidental release of the organism if it causes disease, or if it has been constructed using recombinant DNA techniques (see Chapter 11).

### The Isolation of Pure Cultures by Plating Methods

Pure cultures of microorganisms that form discrete colonies on solid media (e.g., yeasts, most bacteria, many fungi and unicellular algae) may be most simply obtained by one of the modifications of the plating method. This method involves the separation and immobilization of individual organisms on or in a nutrient medium solidified with agar or some other appropriate gelling agent. Each viable organism gives rise, through growth, to a colony from which transfers can be readily made.

The *streaked plate* is in general the most useful plating method. A sterilized bent wire is dipped into a suitable diluted suspension of organisms and is then used to make a series of parallel, nonoverlapping streaks on the surface of an already solidified agar plate. The inoculum is progressively diluted with each successive streak, so that even if the initial streaks yield confluent growth, well-isolated colonies develop along the lines of later streaks (Figure 2.1). Alternatively, isolations can be made with *poured plates:* successive dilutions of the inoculum are placed in sterile petri dishes and mixed with the cooled but still molten agar medium, which is then allowed to solidify. Colonies subsequently develop embedded in the agar.

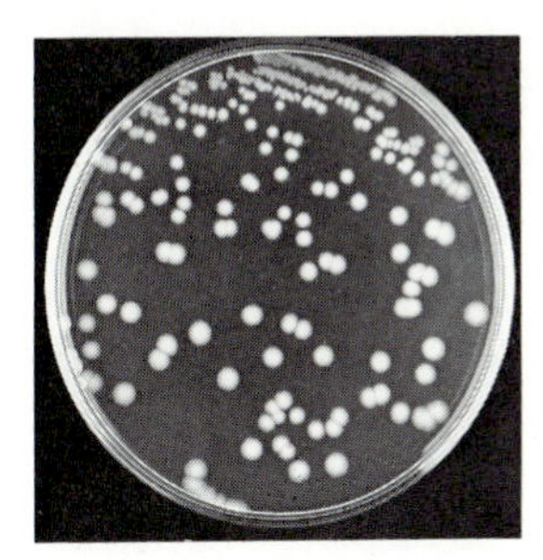

**FIGURE 2.1**
Isolation of a pure culture by the streak method. A petri dish containing nutrient agar was streaked with a suspension of bacterial cells. As a result of subsequent growth, each cell has given rise to a macroscopically visible colony.

The isolation of anaerobic bacteria by plating methods poses special problems. Provided that the desired organisms are not rapidly killed by exposure to oxygen, plates may be prepared in the usual manner and then incubated in closed containers, from which the oxygen is removed either by chemical absorption or evacuation. For more oxygen-sensitive anaerobes, a modification of the pour plate method, known as the *dilution shake culture*, is preferred. A tube of melted and cooled agar medium is inoculated and mixed, and approximately one-tenth of its contents is transferred to a second tube, which is then mixed and used to inoculate a third tube in a similar fashion. After 6 to 10 successive dilutions have been prepared, the

**FIGURE 2.2**
Isolation of a pure culture of anaerobic bacteria by the dilution shake method. A complete series of dilution shakes is shown. Note the confluent growth in the more densely seeded tubes (at right), and the well-isolated colonies in the two final tubes of the series (at left). After the agar had solidified, each tube was sealed with a mixture of sterile vaseline and paraffin to prevent the access of atmospheric oxygen, which inhibits the growth of anaerobic bacteria.

tubes are rapidly cooled and sealed, by pouring a layer of sterile petroleum jelly and paraffin on the surface, thus preventing access of air to the agar column. In shake culture the colonies develop deep in the agar column (Figure 2.2), and are thus not easily accessible for transfer. To make a transfer, the petroleum jelly-paraffin seal is removed with a sterile needle, and the agar column is extruded from the tube into a sterile petri dish by gently blowing a stream of gas through a capillary pipette inserted between the tube wall and the agar. The column is sectioned into discs with a sterile knife to permit examination and transfer of colonies.

Many bacteria are killed by even momentary exposure to air; thus, successful cultivation of these *strict anaerobes*, as they are called, requires extraordinary measures to exclude at all times even traces of oxygen. Two principal techniques are currently used to culture bacteria in the complete absence of oxygen; they are the *roll tube*, and *anaerobic glove box* techniques. The roll tube procedure, developed by R. E. Hungate, is used to obtain pure cultures of strict anaerobes by distributing cells in a thin layer of agar on the walls of a test tube, where they develop into isolated colonies. The test tube contains a few milliliters of molten agar medium that has been reduced chemically to remove dissolved oxygen and is tightly stoppered with a butyl rubber bung (small but lethal quantities of oxygen diffuse through ordinary rubber). The molten agar is inoculated with appropriate dilutions of the source of bacteria by inserting them through the rubber stoppers with a sterile syringe. The tubes are then laid on their sides in ice and rolled until the agar solidifies in a thin layer on the wall of the tube. After a period of incubation when colonies become visible, the bung is removed and isolated colonies picked from the agar with a needle or capillary tube. Whenever a tube is unstoppered, entry of air is prevented by continuously passing a stream of $O_2$-free gas (normally $CO_2$ or $N_2$) into the tube. To ensure that some entry of air has not inadvertently occurred, it is usual to include in the medium the redox dye *resazurine*, which changes from colorless to red at a midpoint $E_h$ of $-0.042$ V.*

Strict anaerobes may also be isolated using conventional streak plate techniques if all procedures are done within a *glove box*** that encloses a reducing atmosphere.

In isolating from a mixed natural population it is often possible, provided one's technique is good, to prepare a first plate, or dilution shake or roll tube series, in which many of the colonies that develop are well separated from one another. Can one then pick material from such a colony, transfer it to an appropriate medium, and call it a pure culture? Although this is often done, a culture so isolated may be far from pure. There is a significant probability that any particular colony was initiated

* $E_h$ is the effective reduction potential of a medium, resulting from the ratios of oxidized to reduced forms of all redox-active compounds. In the construction of media for the culture of strict anaerobes, it is lowered by the inclusion of strong reducing agents.

** A glove box is a sealed chamber in which objects are manipulated with hands inserted into rubber gloves that are attached to the front of the unit; a transparent panel allows the manipulations to be viewed from the outside.

by two similar microorganisms seeded together on the plate to produce a mixed colony. Furthermore, because microorganisms vary greatly in their nutritional requirements, no single medium and set of growth conditions will permit the growth of all the microorganisms present in a natural population. Indeed, it is probable that only a very small fraction of the microorganisms initially present will be able to form colonies on any given medium. Hence, for every visible colony on a first plate, there may be thousands of other microorganisms that were also deposited on the agar surface but that failed to give macroscopically visible growth, although they may still be viable. The probability is high that some of these organisms will be picked up and carried over when a transfer is made. *One should never pick from a first plate for the preparation of a pure culture.* Instead, a second plate should be streaked from a cell suspension prepared from a well-isolated colony. If all the colonies on this second plate appear identical, a well-isolated colony can be used to establish a pure culture.

Not all microorganisms able to grow on solid media necessarily give rise to well-isolated colonies. Certain motile flagellated bacteria (*Proteus*, *Pseudomonas*, for example) can rapidly spread over the slightly moist surface of a freshly poured plate. This can be prevented by the use of plates with well-dried surfaces, on which the cells are immobilized. Spirochetes and organisms that show gliding movement (e.g., myxobacteria, many cyanobacteria) can move over or through an agar gel, even when its surface is well dried. In such cases, the movement of the organisms in question may be an aid to their purification, since they can move away from other kinds of microorganisms immobilized on the agar. Thus, purification can often be achieved by allowing migration to occur and transferring repeatedly to fresh plates from the advancing edge of the migrating population.

The incorporation into the medium of selectively inhibitory substances is also sometimes helpful in making isolations from nature. Because of their biological specificity, certain antibiotics are particularly useful in this respect. Bacteria vary greatly in their sensitivity to the antibiotic penicillin, which can consequently be used at low concentrations to prevent the development of sensitive bacteria in the initial population. At higher concentrations, penicillin is generally toxic for procaryotic organisms but not for eucaryotic ones. It is thus a very useful agent for the purification of protozoa, fungi, and eucaryotic algae that are contaminated by bacteria. Conversely, procaryotic organisms are insensitive to polyene antibiotics such as nystatin, which are generally toxic for eucaryotic organisms. The incorporation of this kind of antibiotic into the isolation medium can sometimes be used to advantage in the purification of bacteria heavily contaminated by fungi or amebae. Many other variations on the theme of selective toxicity can be used to facilitate isolations by the plating method.

## The Isolation of Pure Cultures in Liquid Media

Plating methods are in general satisfactory for the isolation of bacteria and fungi because the great majority of the representatives of these groups can grow well on solid media. However, some of the larger-celled bacteria have not yet been successfully cultivated on solid media, and many protozoa and algae are also cultivable only in a liquid medium. Although plating methods for the isolation of viruses have been greatly extended in recent years, many of these organisms are most easily isolated by the use of liquid media. In the case of viruses, of course, a pure culture is never obtainable, since these organisms are obligate intracellular parasites; a two-membered culture, consisting of a specific virus and its biological host, represents the goal of purification for this microbial group.

The simplest procedure of isolation in liquid media is the *dilution method*. The inoculum is subjected to serial dilution in a sterile medium, and a large number of tubes of medium are inoculated with aliquots of each successive dilution. The goal of this operation is to inoculate a series of tubes with a microbial suspension so dilute that the probability of introducing even one individual into a given tube is very small: a probability of the order of 0.05. When a large number of tubes is seeded with an inoculum of this size, it can be calculated from probability theory that the fraction of tubes receiving one organism is 0.048; the fraction receiving two organisms is 0.0012; the fraction receiving three organisms is 0.00002. As a result, if a tube shows *any* subsequent growth, there is a very high probability that this growth has resulted from the introduction of a *single* organism. The probability that growth has originated from a single organism declines very rapidly as the mean number of organisms in the inoculum increases. It is therefore essential to isolate from a series of tubes the great majority of which show *no* growth.

The dilution method has, however, one major disadvantage: it can be used only to isolate the *numerically predominant* member of a mixed microbial population. It can almost never be effectively used for the isolation of larger microorganisms that are incapable of developing on solid media (e.g., protozoa, algae), because in nature these microorganisms are, as a rule, greatly outnumbered by bacteria. Hence, the usefulness of the dilution method is limited.

When neither plating nor dilution methods can be applied, the only alternative is to resort to the *microscopically controlled isolation of a single cell or organism from the mixed population*, a technique known by the name of *single-cell isolation*. The technical difficulty of single-cell isolation is inversely related to the *size* of the organism which one wishes to isolate: it is relatively easy to use with large-celled microorganisms, such as algae and protozoa, but becomes much harder with bacteria.

In the case of large microorganisms, purification involves the capture of a single individual in a fine capillary pipette and the subsequent transfer of this individual through several washings in relatively large volumes of sterile medium to eliminate microbial contaminants of smaller size. The successive operations can be performed manually, with control by direct microscopic observation at a relatively low magnification, such as that provided by a dissecting microscope.

The technique of the capillary pipette can no longer be applied if the organism that one wishes to isolate is so small that it cannot be readily observed at a magnification of 100 times or less, because one cannot achieve the necessary fineness of control to manipulate a capillary pipette directly at higher magnifications. In this event, a mechanical device known as a *micromanipulator* must be used in conjunction with specially prepared, very fine glass operating instruments. The essential purpose of a micromanipulator is to *gear down* manual control, so that very slight and precisely controlled movements of the operating instruments can be effected in a small operating area (a microdrop) under continuous microscopic observation at high magnifications (500 to 1,000 ×).

### Two-Membered Cultures

The goal of isolation is normally to obtain a pure culture. However, there are certain situations where this cannot be achieved or where achievement is so difficult as to be impractical. Under such circumstances, the alternative is to obtain the next best degree of purification, in the shape of a *two-membered culture*, which contains only two kinds of microorganisms. As already mentioned, a two-membered culture is in principle the only possible way to maintain viruses, since these organisms are all obligate intracellular parasites of cellular organisms. Obligate intracellular parasitism is also characteristic of several groups of cellular microorganisms. In all these instances a two-membered culture represents the nearest approach to cultivation under controlled laboratory conditions that can be achieved.

Many of the protozoa, which feed in nature on smaller microorganisms, are also most easily maintained in the laboratory as two-membered cultures in association with their smaller microbial prey. This is true, for example, of ciliates, amebaes, and slime molds. In such instances the association is probably never an *obligate* one, since careful nutritional studies on a few representatives of these groups have shown that they can be grown in pure culture; however, the nutritional requirements of protozoa are often extremely complex, so that the preparation of media for the maintenance of pure cultures is both difficult and laborious. For purposes of routine maintenance, and also for many experimental purposes, two-membered cultures are satisfactory.

The establishment of a two-membered culture is an operation that is conducted in two phases. First, it is necessary to establish a pure culture of the food organism (the host in the cases of obligate intracellular parasites, the prey in the case of protozoa). Once this has been achieved, the parasite or predator can be isolated by any one of a variety of methods (plating on solid media in the presence of the food organism, dilution in a liquid medium, single-cell isolation) and introduced into the pure culture of the food organism.

The successful maintenance of two-membered cultures requires considerable art because a reasonably stable biological balance between the two components is essential. The medium must be one that permits sufficient growth of the food organism to meet the needs of the parasite or predator but should not be so rich that the food organism can outgrow its associate or produce metabolic products that are deleterious to it.

## THE THEORY AND PRACTICE OF STERILIZATION

Sterilization is a treatment that *frees the treated object of all living organisms*. It can be achieved by exposure to lethal physical or chemical agents or, in the special case of solutions, by filtration.

To understand the basis of sterilization by lethal agents, it is necessary to describe briefly the

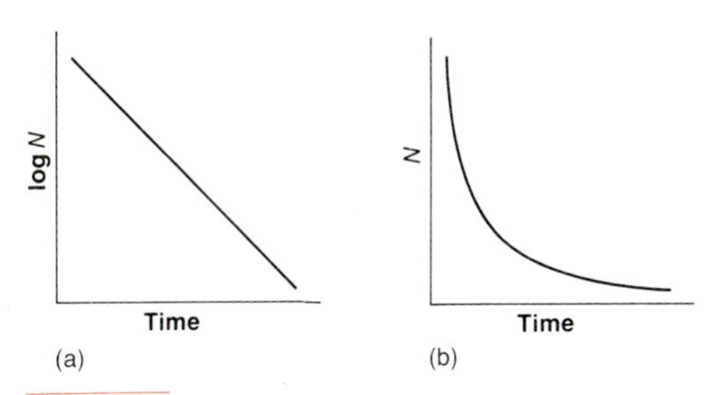


FIGURE 2.3

Exponential (logarithmic) order of death of bacteria. The same data are plotted semilogarithmically in (a) and arithmetically in (b); $N$ is the number of surviving bacteria.

kinetics of death in a microbial population. The only valid criterion of death in the case of a microorganism is *irreversible loss of the ability to reproduce*; this is usually determined by quantitative plating methods, survivors being detected by colony formation. When a pure microbial population is exposed to a lethal agent, the kinetics of death are nearly always *exponential:* the number of survivors decreases geometrically with time. This reflects the fact that all the members of the population are of similar sensitivity; probability alone determines the actual time of death of any given individual. If the logarithm of the number of survivors is plotted as a function of the time of exposure, a straight line is obtained (Figure 2.3); its negative slope defines the *death rate.*

The death rate tells one only what *fraction* of the initial population survives a given period of treatment. To determine the *actual number* of survivors, one must also know the *initial population size*, as illustrated graphically in Figure 2.4. Accordingly, for the establishment of procedures of sterilization, two factors have to be taken into account: the death rate and the initial population size.

In the practice of sterilization the microbial population to be destroyed is almost always a *mixed* one. Since microorganisms differ widely in their resistance to lethal agents, the significant factors become the initial population size and the death rate of the *most resistant* members of the mixed population. These are almost always the highly resistant endospores of certain bacteria. Consequently, spore suspensions of known resistance are the objects commonly used to assess the reliability of sterilization methods.

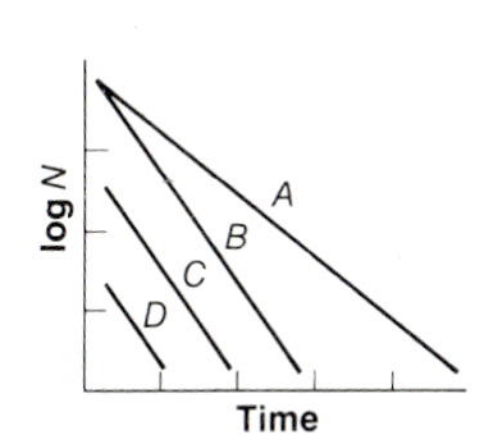


FIGURE 2.4

Relationship of death rate and population size to the time required for the destruction of bacterial cultures; $N$ is the number of surviving bacteria. Cultures $B$, $C$, and $D$ have identical death rates. Culture $A$ has a lower death rate.

Taking into account the kinetics of microbial death, we can formulate the practical goal of sterilization by a lethal agent in a slightly more refined way: *the probability that the object treated contains even one survivor should be infinitesimally small.* For example, if we wish to sterilize a liter of a culture medium, this goal will be achieved for all practical purposes if the treatment is one that will leave no more than one survivor in $10^6$ liters; under such circumstances, the probability of failure is very small indeed. Procedures of routine sterilization are always designed to provide a very wide margin of safety.

## Sterilization by Heat

Heat is the most widely used lethal agent for purposes of sterilization. Objects may be sterilized by dry heat, applied in an oven in an atmosphere of air, or by moist heat, provided by steam. Of the two methods, sterilization by dry heat requires a much greater duration and intensity because heat conduction is less rapid in dry than in moist air. In addition, bacteria can survive in a completely desiccated state and, in this state, the intrinsic heat resistance of vegetative bacterial cells is greatly increased, almost to the level characteristic of spores. Consequently, the death rate is much lower for dry cells than for fully hydrated ones.

Dry heat is used principally to sterilize glassware or other heat-stable solid materials. The objects are wrapped in paper or otherwise protected from subsequent contamination and exposed to a temperature of 170° C for 90 minutes in an oven.

Steam must be used for the heat sterilization of aqueous solutions. Treatment is usually carried out in a metal vessel known as an *autoclave*, which can be filled with steam at a pressure greater than atmospheric. Sterilization can thus be achieved at temperatures considerably above 100°C; laboratory autoclaves are commonly operated at a steam pressure of 1.06 kg/cm$^2$ (15 lb/in.$^2$) above atmospheric pressure, which corresponds to a temperature of 121° C. Even bacterial spores that survive several hours of boiling are rapidly killed at this temperature. Small volumes of liquid can be sterilized by exposure for 20 minutes; if larger volumes are to be sterilized, the time of treatment must be extended.

A temperature of 121° C within the autoclave

will be attained under a pressure of 1.06 $kg/cm^2$ *only if the atmosphere consists entirely of steam.* At the start of the operation, accordingly, all the dry air originally in the chamber must be expelled and replaced by steam; this is achieved by the use of a steam trap, which remains open as long as air is being passed through it but closes when the atmosphere is saturated. If some dry air remains in the sterilization chamber, the partial pressure of steam will be lower than that indicated on the pressure gauge, and the temperature will be correspondingly lower. For this reason an autoclave should always be equipped with both a temperature and a pressure gauge. The temperature within the sterilization chamber can be monitored by including, among the objects to be sterilized, special indicator papers, which change color if the heat treatment has been adequate.

### Sterilization by Chemical Treatment

Many of the substances used in preparing culture media are too heat-labile to be sterilized by autoclaving. For such substances, a reliable method of chemical sterilization would be extremely useful. The essential requirement for a chemical sterilizing agent is that it should be *volatile* as well as *toxic*, so that it can be readily eliminated from the object sterilized after treatment. The best available candidate is *ethylene oxide*, a liquid that boils at 10.7° C. It can be added to solutions in liquid form (final concentration of approximately 0.5 to 1.0 percent) at a temperature of 0 to 4° C, or used as a sterilizing gas at temperatures above the boiling point. However, it is chemically unstable, decomposing in aqueous solution to ethylene glycol, which is nonvolatile and may have undesirable effects. Furthermore, ethylene oxide is both explosive and toxic for humans, so special precautions must be taken in its handling. For these reasons, ethylene oxide sterilization has not become a routine laboratory procedure. It is, however, used industrially for the sterilization of plastic petri dishes and other plastic objects which melt at temperatures greater than 100° C.

### Sterilization by Filtration

The principal laboratory method used to sterilize solutions of heat-labile materials is *filtration* through filters capable of retaining microorganisms. The action of such filters is almost always complex. Microorganisms are retained in part by the small size of the filter pores and in part by adsorption on the pore walls during their passage through the filter. The importance of adsorption is indicated by the fact that a filter may effectively retain microorganisms even when the average diameter of its pores is somewhat greater than the mean size of the cells that are retained. Sterilization by filtration is subject to one major theoretical limitation. Since the viruses range down in size to the dimensions of large protein molecules, they are not necessarily retained by filters that can hold back even the smallest of cellular microorganisms. Consequently, it is never possible to be certain that filtration procedures that render a solution bacterium-free will also free it of viruses.

## THE PRINCIPLES OF MICROBIAL NUTRITION

To grow, organisms must draw from the environment all the substances they require for the synthesis of their cell materials and for the generation of energy. These substances are termed *nutrients.* A culture medium must therefore contain, in quantities appropriate to the specific requirements of the microorganism for which it is designed, all necessary nutrients. However, microorganisms are extraordinarily diverse in their specific physiological

**TABLE 2.1**

**Approximate Elementary Composition of the Microbial Cell**

| Element | Percentage of Dry Weight |
|---|---|
| Carbon | 50 |
| Oxygen | 20 |
| Nitrogen | 14 |
| Hydrogen | 8 |
| Phosphorus | 3 |
| Sulfur | 1 |
| Potassium | 1 |
| Sodium | 1 |
| Calcium | 0.5 |
| Magnesium | 0.5 |
| Chlorine | 0.5 |
| Iron | 0.2 |
| All others | ~0.3 |

Source: Data for a bacterium, *Escherichia coli*, assembled by S. E. Luria, in *The Bacteria*, ed. I. C. Gunsalus and R. Y. Stanier, Vol. 1, Chap. 1 (New York: Academic Press, 1960).

properties, and correspondingly in their specific nutrient requirements. Literally thousands of different media have been proposed for their cultivation, and in the descriptions of these media the reasons for the presence of the various components are often not clearly stated. Nevertheless, the design of a culture medium can and should be based on scientific principles, the *principles of nutrition*, which we shall outline as a preliminary to the description of culture media.

The chemical composition of cells, broadly constant throughout the living world, indicates the major material requirements for growth. *Water* accounts for some 80 to 90 percent of the total weight of cells and is always therefore the major essential nutrient, in quantitative terms. The solid matter of cells (Table 2.1) contains, in addition to *hydrogen* and *oxygen* (derivable metabolically from water), *carbon, nitrogen, phosphorus,* and *sulfur*, in order of decreasing abundance. These six elements account for about 95 percent of the cellular dry weight. Many other elements are included in the residual fraction. Nutritional studies show that *potassium, magnesium, calcium, iron, manganese, cobalt, copper, molybdenum,* and *zinc* are required by nearly all organisms. The known functions in the cell of these 15 elements are summarized in Table 2.2.

**TABLE 2.2**

**General Physiological Functions of the Principal Elements**

| Element | Physiological Functions |
|---|---|
| Hydrogen | Constituent of cellular water, organic cell materials |
| Oxygen | Constituent of cellular water, organic cell materials; as $O_2$, electron acceptor in respiration of aerobes |
| Carbon | Constituent of organic cell materials |
| Nitrogen | Constituent of proteins, nucleic acids, coenzymes |
| Sulfur | Constituent of proteins (as amino acids cysteine and methionine); of some coenzymes (e.g., CoA, cocarboxylase) |
| Phosphorus | Constituent of nucleic acids, phospholipids, coenzymes |
| Potassium | One of the principal inorganic cations in cells, cofactor for some enzymes |
| Magnesium | Important cellular cation; inorganic cofactor for very many enzymatic reactions, including those involving ATP; functions in binding enzymes to substrates; constituent of chlorophylls |
| Manganese | Inorganic cofactor for some enzymes, sometimes replacing Mg |
| Calcium | Important cellular cation; cofactor for some enzymes (for example, proteinases) |
| Iron | Constituent of cytochromes and other heme or nonheme proteins; cofactor for a number of enzymes |
| Cobalt | Constituent of vitamin $B_{12}$ and its coenzymes derivatives |
| Copper, zinc, nickel, molybdenum | Inorganic constituents of special enzymes |

All the required metallic elements can be supplied as nutrients in the form of the *cations of inorganic salts.* Potassium, magnesium, calcium, and iron are required in relatively large amounts and should always be included as salts in culture media. The quantitative requirements for manganese, cobalt, copper, nickel, molybdenum, and zinc are very small—so small, in fact, that it is often technically difficult to demonstrate their essentiality, since they are present in adequate amounts as contaminants of the major inorganic constituents of media. They are often referred to as *trace elements* or *micronutrients.* One nonmetallic element, phosphorus, can also be used as a nutrient when provided in inorganic form, as phosphate salts.

It should be noted that some biological groups have additional, specific mineral requirements; for example, diatoms and certain other algae synthesize cell walls that are heavily impregnated with silica and consequently have a specific *silicon* requirement, supplied as silicate. Although a requirement for *sodium* cannot be demonstrated for most microorganisms, it is required at relatively high concentrations by certain marine bacteria, by cyanobacteria, and by photosynthetic bacteria. In these groups it cannot be replaced by other monovalent cations.

The needs for *carbon, nitrogen, sulfur,* and *oxygen* cannot be so simply described because organisms differ with respect to the *specific chemical form* under which these elements must be provided as nutrients.

## The Requirements for Carbon

Organisms that perform photosynthesis and bacteria that obtain energy from the oxidation of inorganic compounds typically use the most oxidized form of carbon, $CO_2$, as the sole or principal source of cellular carbon. The conversion of $CO_2$ to organic cell constituents is a reductive process, which requires a net input of energy. In these physiological groups, accordingly, a considerable part of the energy derived from light or from the oxidation of reduced inorganic compounds must be expended for the reduction of $CO_2$ to the level of organic matter.

All other organisms obtain carbon largely from organic nutrients. Since most organic substrates are at the same general oxidation level as organic cell constituents, they usually do not have to undergo a primary reduction to serve as sources of cell carbon. In addition to meeting the biosynthetic needs of the cell for carbon, organic substrates must often supply the energetic requirements of the cell. Consequently, much of the carbon present in the organic substrate enters the pathways of energy-yielding metabolism and is eventually excreted again from the cell as $CO_2$ (the major product of energy-yielding respiratory metabolism) or as a mixture of $CO_2$ and organic compounds (the typical end products of fermentative metabolism). Organic substrates thus usually have a *dual nutritional role:* they serve at the same time as a source of carbon and as a source of energy. Many microorganisms can use *a single organic compound* to supply completely both these nutritional needs. Others, however, cannot grow when provided with only one organic compound, and they need a variable number of additional organic compounds as nutrients. These additional organic nutrients have a purely biosynthetic function, being required as precursors of certain organic cell constituents that the organism is unable to synthesize. They are termed *growth factors*, and their roles are described in greater detail below.

Microorganisms are extraordinarily diverse with respect to both the *kind* and the *number* of organic compounds that they can use as a principal source of carbon and energy. This diversity is shown by the fact that *there is no naturally produced organic compound that cannot be used as a source of carbon and energy by some microorganism.* Hence, it is impossible to describe concisely the chemical nature of organic carbon sources for microorganisms. This extraordinary variation with respect to carbon requirements is one of the fascinating physiological aspects of microbiology.

When the organic carbon requirements of *individual* microorganisms are examined, some show a high degree of versatility, whereas others are extremely specialized. Certain bacteria of the *Pseudomonas* group, for example, can use any one of over 90 different organic compounds as their sole carbon and energy source.

At the other end of the spectrum are some methane-oxidizing bacteria, which can use only two organic substrates, methane and methanol, and certain cellulose-decomposing bacteria, which can use only cellulose.

Most (and probably all) organisms that depend on organic carbon sources also require $CO_2$ as a nutrient in very small amounts, because this compound is utilized in a few biosynthetic reactions. However, as $CO_2$ is normally produced in large quantities by organisms that use organic compounds, the biosynthetic requirement can be met through the metabolism of the organic carbon and energy source. Nevertheless, the complete removal of $CO_2$ often either delays or prevents the growth of microorganisms in organic media, and a few bacteria and fungi require a relatively high concentration of $CO_2$ in the atmosphere (5 to 10 percent) for satisfactory growth in organic media.

## The Requirements for Nitrogen and Sulfur

Nitrogen and sulfur occur in the organic compounds of the cell principally in reduced form as amino and sulfhydryl groups, respectively. Most photosynthetic organisms, as well as many nonphotosynthetic bacteria and fungi, assimilate these two elements in the oxidized inorganic state, as nitrates and sulfates; their biosynthetic utilization thus involves a preliminary reduction. Some microorganisms are unable to bring about a reduction of one or both of these anions and must be supplied with the elements in a *reduced form.* The requirement for a reduced nitrogen source is relatively common and can be met by the provision of nitrogen as ammonium salts. A requirement for reduced sulfur is rare; it can be met by the provision of sulfide or of an organic compound that contains a sulfhydryl group (e.g., cysteine).

The nitrogen and sulfur requirements can often also be met by organic nutrients that contain these two elements in reduced organic combination (amino acids or more complex protein degradation products, such as peptones). Such compounds may also, of course, provide organic carbon and energy sources, meeting simultaneously the cellular requirements for carbon, nitrogen, sulfur, and energy.

Some bacteria can also utilize the most abundant natural nitrogen source, $N_2$. This process of nitrogen assimilation is termed *nitrogen fixation* and involves a preliminary reduction of $N_2$ to ammonia.

**TABLE 2.3**

**Relation of Some Water-Soluble Vitamins to Coenzymes**

| Vitamin | Coenzyme | Enzymatic Reactions Involving the Coenzyme Form |
|---|---|---|
| Nicotinic acid (niacin) | Pyridine nucleotide coenzymes ($NAD^+$ and $NADP^+$) | Dehydrogenations |
| Riboflavin (vitamin $B_2$) | Flavin nucleotides (FAD and FMN) | Some dehydrogenations, electron transport |
| Thiamin (vitamin $B_1$) | Thiamin pyrophosphate (cocarboxylase) | Decarboxylations and some group-transfer reactions |
| Pyridoxine (vitamin $B_6$) | Pyridoxal phosphate | Amino acid metabolism<br>Transamination<br>Deamination<br>Decarboxylation |
| Pantothenic acid | Coenzyme A | Keto-acid oxidation, fatty acid metabolism |
| Folic acid | Tetrahydrofolic acid | Transfer of one-carbon units |
| Biotin | Biotin | $CO_2$ fixation, carboxyl transfer |
| Cobalamin (vitamin $B_{12}$) | Various cobalamin derivatives | Molecular rearrangement reactions |

## Growth Factors

Any organic compound that an organism requires as a precursor or constituent of its organic cell material, but which it cannot synthesize from simpler carbon sources, must be provided as a nutrient. Organic nutrients of this type are known collectively as *growth factors*. Most growth factors fall into one of three classes, in terms of chemical structure and metabolic function:

**1.** *Amino acids*, required as constituents of proteins.

**2.** *Purines* and *pyrimidines*, required as constituents of nucleic acids.

**3.** *Vitamins*, a diverse collection of organic compounds that form parts of the prosthetic groups or active centers of certain enzymes (Table 2.3).

Because growth factors fulfill specific needs in biosynthesis, they are required in only small amounts relative to the principal cellular carbon source, which must serve as a general precursor of cell carbon. Some 20 different amino acids enter into the composition of proteins, so the need for any specific amino acid that the cell is unable to synthesize is obviously not large. The same applies to specific needs for a purine or a pyrimidine; five different compounds of these classes enter into the structure of nucleic acids. The quantitative requirements for vitamins are even smaller, since the various coenzymes of which they are precursors have catalytic roles and consequently are present at levels of a few parts per million in the cell, as shown in Table 2.4.

The biosynthesis of amino acids, purines, pyrimidines, and coenzymes typically involves com-

**TABLE 2.4**

**Concentrations of Several Water-Soluble Vitamins in the Cells of Bacteria (in parts per million of dry weight)**

| Vitamin | *Aerobacter aerogenes* | *Pseudomonas fluorescens* | *Clostridium butyricum* |
|---|---|---|---|
| Nicotinic acid | 240 | 210 | 250 |
| Riboflavin | 44 | 67 | 55 |
| Thiamin | 11 | 26 | 9 |
| Pyridoxine | 7 | 6 | 6 |
| Pantothenic acid | 140 | 91 | 93 |
| Folic acid | 14 | 9 | 3 |
| Biotin | 4 | 7 | — |

Note: In the cell these substances are present in the coenzyme form, but since their quantitative measurement after extraction is dependent on conversion to the corresponding vitamins, the data are presented in this form.

Source: Taken from R. C. Thompson, *Texas Univ. Publ.* **4237,** 87 (1942).

plex series of individual step reactions, which will be discussed in Chapter 5. The inability to perform *any one* of these step reactions makes an organism dependent on the provision of the end product as a growth factor. However, the growth factor itself may not be absolutely essential; if the blocked reaction occurs at an early stage in its biosynthesis, organic precursors that follow the blocked step may be able to satisfy the needs of the cell as specific nutrients. A close analysis of a particular growth-factor requirement shown by a number of different microorganisms usually reveals that they differ in the particular chemical form or forms of the growth factor that they require. This can be illustrated by considering the rather common requirement for vitamin $B_1$ (thiamin), which has the following structure:

```
         NH2              CH3
          |                |
          C               C=C—CH2—CH2OH
      N//   \C—CH2—N/        \
      |      ||      + \\CH—S
 CH3—C\\   /CH
        N

   pyrimidine              thiazole
```

Some microorganisms require the entire molecule as a growth factor. There are, however, some microorganisms that, given the two halves of the molecule as nutrients, can put them together. Others require only the pyrimidine portion because they can synthesize thiazole. Still others need only the thiazole portion because they can make and add the pyrimidine portion. For each type of organism described above, the *minimal* growth-factor requirement is different. Yet, in every case, what the organism must eventually have is the entire thiamin molecule, and if this compound is provided as nutrient, it can be used as a growth factor by all the types described. Even the entire thiamin molecule, however, is not the compound that organisms must eventually make as an essential component of their cells. The functional compound is the coenzyme cocarboxylase, which acts as a prosthetic group in several enzymatic reactions. This coenzyme is thiamin pyrophosphate and has the following structure:

```
         NH2              CH3                   O      O
          |                |                    ||     ||
          C               C=C—CH2—CH2O—P—O—P—OH  .
      N//   \C—CH2—N/        \                  |      |
      |      ||      + \\CH—S                   OH     OH
 CH3—C\\   /CH
        N
```

## The Roles of Oxygen in Nutrition

As an elemental constituent of water and of organic compounds, oxygen is a universal component of cells and is always provided in large amounts in the major nutrient, water. However, many organisms also require *molecular oxygen* ($O_2$). These are organisms that are dependent on aerobic respiration for the fulfillment of their energetic needs and for which molecular oxygen functions as a terminal oxidizing agent. Such organisms are termed *obligately aerobic.*

At the other physiological extreme are those microorganisms that obtain energy by means of reactions that do not involve the utilization of molecular oxygen and for which this chemical form of the element is not a nutrient. Indeed, for many of these physiological groups, molecular oxygen is a toxic substance, which either kills them or inhibits their growth. Such organisms are *obligately anaerobic.*

Some microorganisms are *facultative anaerobes,* able to grow either in the presence or in the absence of molecular oxygen. In metabolic terms, facultative anaerobes fall into two subgroups. Some, like the lactic acid bacteria, have an exclusively fermentative energy-yielding metabolism but are relatively insensitive to the presence of oxygen; such organisms are most accurately termed *aerotolerant anaerobes.* Others (e.g., many yeasts, enteric bacteria) can shift from a respiratory to a fermentative mode of metabolism. Such facultative anaerobes use $O_2$ as a terminal oxidizing agent when it is available but can also obtain energy in its absence by fermentative reactions.

Among microorganisms that are obligate aerobes, some grow best at partial pressures of oxygen considerably below that (0.2 atm) present in air. They are termed *microaerophilic.* This probably reflects the possession of enzymes that are inactivated under strongly oxidizing conditions and can thus be maintained in a functional state only at low partial pressures of $O_2$. Many bacteria that obtain energy by the oxidation of molecular hydrogen show this behavior, and it is known that hydrogenase, the enzyme involved in hydrogen utilization, is readily inactivated by oxygen.

## Nutritional Categories among Microorganisms

Originally, biologists recognized two principal nutritional classes among organisms: the *autotrophs*, exemplified by plants, which can use completely inorganic nutrients and the *heterotrophs*, exemplified by animals, which require organic nutrients. Today, these two simple categories are insufficient to encompass the variety of nutritional patterns known to exist in the living world, and various attempts to construct more elaborate systems of nutritional classification have been made. Perhaps the most useful, albeit relatively simple, nutritional classification is that based on two parameters: the nature of the *energy source* and the nature of the *principal carbon source*. With respect to energy source, there is a basic dichotomy between organisms that are able to use light as an energy source, termed *phototrophs*, and organisms that are dependent on a chemical energy source, termed *chemotrophs*. Organisms able to use $CO_2$ as a principal carbon source are termed *autotrophs*; organisms dependent on an organic carbon source are termed *heterotrophs*. By means of these criteria, four major nutritional categories can be distinguished.

**1.** *Photoautotrophs*, using light as the energy source and $CO_2$ as the principal carbon source. They include most photosynthetic organisms: higher plants, algae, and many photosynthetic bacteria.

**2.** *Photoheterotrophs*, using light as the energy source and an organic compound as the principal carbon source. This category includes certain of the purple and green bacteria.

**3.** *Chemoautotrophs*, using a chemical energy source and $CO_2$ as the principal carbon source. Energy is obtained by the oxidation of *reduced inorganic compounds*, such as $NH_4^+$, $NO_2^-$, $H_2$, reduced forms of sulfur (e.g., $H_2S$, S, $S_2O_3^{2-}$), CO, or ferrous iron. Only members of the bacteria belong to this nutritional category. As a result of their distinctive ability to grow in strictly mineral media in the absence of light, these organisms are sometimes termed *chemolithotrophs* (from the Greek word *lithos*, a rock).

**4.** *Chemoheterotrophs*, using a chemical energy source and an organic substance as the principal carbon source. The clear-cut distinction between energy source and carbon source, characteristic of the three preceding categories, loses its clarity in the context of chemoheterotrophy, where *both carbon and energy can usually be derived from the metabolism of a single organic compound*. The chemoheterotrophs include all metazoan animals, protozoa, fungi, and the great majority of bacteria. Certain further subdivisions within this very complex nutritional category can be made. One is based on the *physical state in which organic nutrients enter the cell*. The *osmotrophs* (for example, bacteria and fungi) take up all nutrients in dissolved form; the *phagotrophs* (for example, protozoa) can take up solid food particles by the mechanism termed *phagocytosis* (see p. 57).

It must be emphasized that the marked nutritional versatility of many microorganisms makes the application of this system of nutritional categories to some degree arbitrary. For example, many photoautotrophic algae can also grow in the dark, as chemoheterotrophs. Chemoheterotrophy is likewise an alternate nutritional mode for certain photoheterotrophs and chemoautotrophs. More or less by convention, such organisms are assigned to the category characterized by the simplest nutritional requirements: thus, phototrophy takes precedence over chemotrophy, autotrophy over heterotrophy. The qualifications *obligate* and *facultative* are often used to indicate the absence (or presence) of nutritional versatility. Thus, an *obligate photoautotroph* is strictly dependent on light for its energy source and on $CO_2$ for its principal carbon source, but a *facultative photoautotroph* is not.

In order to take into account the requirement for growth factors an additional pair of terms, *prototrophy* and *auxotrophy*, are sometimes employed. A prototroph can derive all carbon requirements from the principal carbon source. An auxotroph requires, in addition to the principal carbon source, one or more organic nutrients (growth factors). Both prototrophy and auxotrophy may occur among the organisms assigned to any one of the four major nutritional categories defined in terms of energy requirements and principal carbon sources. For example, auxotrophy, represented by an absolute requirement for one or more vitamins, is characteristic of many photoautotrophic algae and bacteria.

# THE CONSTRUCTION OF CULTURE MEDIA

In constructing a culture medium for any microorganism, the primary goal is to provide a balanced mixture of the required nutrients, at concentrations that will permit good growth. It might seem at first sight reasonable to make the medium as rich as possible, by providing all nutrients in great excess. However, this approach is not a wise one. In the first place, many nutrients become growth inhibitory or toxic as the concentration is raised. This is true of many organic substrates, such as salts of fatty acids (e.g., acetate) and even of sugars, if the concentration is high enough. Some inorganic constituents may also become inhibitory if provided in excess; many algae are very sensitive to the concentration of inorganic phosphate. Second, even if growth can occur in a concentrated medium, the metabolic activities of the growing microbial population will eventually change the nature of the environment to the point where it becomes highly unfavorable and the population becomes physiologically abnormal or dies. This may be brought about by a drastic change in the hydrogen ion concentration (pH), by the accumulation of toxic organic metabolites, or, in the case of strict aerobes, by the depletion of oxygen. Since the usual goal of the microbiologist is to study the properties and behavior of *healthy* microorganisms, it is wise to limit the total growth of cultures by providing a limiting quantity of one nutrient; in the case of chemoheterotrophs, the principal carbon source is usually selected for this purpose. Examples of the appropriate concentrations of nutrients will be provided in the various media described below.

The rational point of departure for the preparation of media is to compound a *mineral base*, which provides all those nutrients that can be supplied to any organism in inorganic form. This base can then be supplemented, as required, with a carbon source, an energy source, a nitrogen source, and any required growth factors; these supplements will, of course, vary with the nutritional properties of the particular organism that one wishes to grow. A medium composed entirely of chemically defined nutrients is termed a *synthetic medium*. One that contains ingredients of unknown chemical composition is termed a *complex medium*.

We may illustrate these principles by considering the composition of four media of increasing chemical complexity, each of which is suitable for the cultivation of certain kinds of chemotrophic bacteria (Table 2.5). All four media share a common mineral base. Medium 1 is supplemented with $NH_4Cl$ at a concentration of 1 g/liter but has no added source of carbon. However, if it is incubated aerobically, the $CO_2$ of the atmosphere will be available as a carbon source. In the dark the only organisms that can grow in this medium are chemoautotrophic nitrifying bacteria, such as *Nitrosomonas*, which obtain carbon from $CO_2$ and energy from the aerobic oxidation of ammonia; the ammonia also provides them with a nitrogen source.

Medium 2 is additionally supplemented with glucose at a concentration of 5 g/liter. Under aerobic conditions, it will support the growth of many bacteria and fungi, since glucose can commonly be used as a carbon and energy source for aerobic growth. If incubated in the absence of oxygen, it can also support the development of many facultatively or strictly anaerobic bacteria, able to derive carbon and energy from the fermentation of glucose. Note, however, that this medium is not a suitable one for any microorganism that requires growth

**TABLE 2.5**

**Four Media of Increasing Complexity**

| Common Ingredients | Additional Ingredients | | | |
|---|---|---|---|---|
| | MEDIUM 1 | MEDIUM 2 | MEDIUM 3 | MEDIUM 4 |
| Water, 1 liter | $NH_4Cl$, 1 g | Glucose,[a] 5 g | Glucose, 5 g | Glucose, 5 g |
| $K_2HPO_4$, 1 g | | $NH_4Cl$, 1 g | $NH_4Cl$, 1 g | Yeast extract, 5 g |
| $MgSO_4 \cdot 7H_2O$, 200 mg | | | Nicotinic acid, 0.1 mg | |
| $FeSO_4 \cdot 7H_2O$, 10 mg | | | | |
| $CaCl_2$, 10 mg | | | | |
| Trace elements (Mn, Mo, Cu, Co, Zn) as inorganic salts, 0.02–0.5 mg of each | | | | |

[a] If the media are sterilized by autoclaving, the glucose should be sterilized separately and added aseptically. When sugars are heated in the presence of other ingredients, especially phosphates, they are partially decomposed to substances that are very toxic to some microorganisms.

TABLE 2.6

**Medium for *Leuconostoc mesenteroides***

| | | | |
|---|---|---|---|
| WATER | 1 liter | | |
| ENERGY SOURCE | | | |
| Glucose | 25 g | | |
| NITROGEN SOURCE | | | |
| $NH_4Cl$ | 3 g | | |
| MINERALS | | | |
| $KH_2PO_4$ | 600 mg | $FeSO_4 \cdot 7H_2O$ | 10 mg |
| $K_2HPO_4$ | 600 mg | $MnSO_4 \cdot 4H_2O$ | 20 mg |
| $MgSO_4 \cdot 7H_2O$ | 200 mg | NaCl | 10 mg |
| ORGANIC ACID | | | |
| Sodium acetate | 20 g | | |
| AMINO ACIDS | | | |
| DL-α-Alanine | 200 mg | L-Lysine · HCl | 250 mg |
| L-Arginine | 242 mg | DL-Methionine | 100 mg |
| L-Asparagine | 400 mg | DL-Phenylalanine | 100 mg |
| L-Aspartic acid | 100 mg | L-Proline | 100 mg |
| L-Cysteine | 50 mg | DL-Serine | 50 mg |
| L-Glutamic acid | 300 mg | DL-Threonine | 200 mg |
| Glycine | 100 mg | DL-Tryptophan | 40 mg |
| L-Histidine · HCl | 62 mg | L-Tyrosine | 100 mg |
| DL-Isoleucine | 250 mg | DL-Valine | 250 mg |
| DL-Leucine | 250 mg | | |
| PURINES AND PYRIMIDINES | | | |
| Adenine sulfate · $H_2O$ | 10 mg | Uracil | 10 mg |
| Guanine · HCl · $2H_2O$ | 10 mg | Xanthine · HCl | 10 mg |
| VITAMINS | | | |
| Thiamine · HCl | 0.5 mg | Riboflavin | 0.5 mg |
| Pyridoxine · HCl | 1.0 mg | Nicotinic acid | 1.0 mg |
| Pyridoxamine · HCl | 0.3 mg | p-Aminobenzoic acid | 0.1 mg |
| Pyridoxal · HCl | 0.3 mg | Biotin | 0.001 mg |
| Calcium pantothenate | 0.5 mg | Folic acid | 0.01 mg |

Source: From H. E. Sauberlich and C. A. Baumann, "A Factor Required for the Growth of *Leuconostoc citrovorum*," *J. Biol. Chem.* **176,** 166 (1948).

factors; it contains only a single carbon compound.

Medium 3 is additionally supplemented with one vitamin, nicotinic acid. It can therefore support the growth of all those organisms able to develop in medium 2, together with others, such as the bacterium *Proteus vulgaris*, that require nicotinic acid as a growth factor.

For the three media so far described, the chemical nature of every ingredient is known; thus, they are good examples of synthetic media. Medium 4 is a complex medium, in which the $NH_4Cl$ and nicotinic acid of medium 3 have been replaced by a nutrient of unknown composition, yeast extract, at a concentration of 5 g/liter. It can support the growth of a great many chemoheterotrophic microorganisms, both aerobic and anaerobic, having no growth-factor requirements, relatively simple ones, or highly complex ones. The yeast extract provides a variety of organic nitrogenous constituents (partial breakdown products of proteins) which can fulfill the general nitrogen requirements, and it also contains most of the organic growth factors likely to be required by microorganisms.

Complex media are, accordingly, useful for the cultivation of a wide range of microorganisms, including ones whose precise growth-factor requirements are not known. Even when the growth-factor requirements of a microorganism have been precisely determined, it is often more convenient to grow the organism in a complex medium, particularly if the growth-factor requirements are numerous. This point is illustrated in Table 2.6, which describes the composition of a synthetic medium that will support growth of the lactic acid bacterium, *Leuconostoc mesenteroides*. This bacterium can also be cultivated satisfactorily in the complex medium 4. In this particular instance, accordingly, the yeast extract of medium 4 must furnish the following requirements: the organic acid, acetate; 19 amino acids; 4 purines and pyrimidines; and 10 vitamins.

The media described in Table 2.5 can support the development of microorganisms only if certain other requirements for growth are also met. These include a suitable temperature of incubation, favorable osmotic conditions, and a hydrogen ion concentration within the range tolerated by the organism in question. Suitable chemical adjustments may be required to accommodate the osmotic conditions and hydrogen ion concentration to the needs of some microorganisms for which these media are satisfactory with respect to their content of nutrients.

## The Control of pH

Although a given medium may be suitable for the *initiation* of growth, the subsequent development of a bacterial population may be severely limited by chemical changes that are brought about by the growth and metabolism of the organisms themselves. For example, in glucose-containing media, organic acids that may be produced as a result of fermentation may become inhibitory to growth.

In contrast, the microbial decomposition or utilization of anionic components of a medium tends to make the medium more alkaline. For example, the oxidation of a molecule of sodium succinate liberates two sodium ions in the form of the very alkaline salt, sodium carbonate. The decomposition of proteins and amino acids may also make a medium alkaline as a result of ammonia production.

To prevent excessive changes in hydrogen ion concentration either *buffers* or *insoluble carbonates* are often added to the medium.

The phosphate buffers, which consist of mixtures of monohydrogen and dihydrogen phosphates (e.g., $K_2HPO_4$ and $KH_2PO_4$), are the most useful ones. $KH_2PO_4$ is a weakly acidic salt, whereas $K_2HPO_4$ is slightly basic, so that an equimolar solution of the two is very nearly neutral, having a pH of 6.8. If a limited amount of strong acid is added to such a solution, part of the basic salt is converted to the weakly acidic one:

$$K_2HPO_4 + HCl \longrightarrow KH_2PO_4 + KCl$$

If, however, a strong base is added, the opposite conversion occurs:

$$KH_2PO_4 + KOH \longrightarrow K_2HPO_4 + H_2O$$

Thus, the solution acts as a buffer in that it resists radical changes in the hydrogen ion concentration when acid or alkali is produced in the medium. By using different ratios of acidic and basic phosphates, different pH values may be established, ranging from approximately 6.0 to 7.6. Good buffering action, however, is obtained only in the narrower range of pH 6.4 to 7.2 because the capacity of a buffer solution is limited by the amounts of its basic and acidic ingredients. Hence, the more acidic the initial buffer, the less is its ability to prevent increases in hydrogen ion concentration (decreases in pH) and the greater its capacity for reacting with alkali. Conversely, the more alkaline the initial buffer, the less is its ability to prevent increases in pH and the greater its ability to prevent acidification.

The phosphates are used widely in the preparation of media because they are the only inorganic agents that buffer in the physiologically important range around neutrality and that are relatively nontoxic to microorganisms. In addition, they provide a source of phosphorus, which is an essential element for growth. In high concentrations, phosphate becomes inhibitory, so the amount of phosphate buffer that can be used in a medium is limited by the tolerance of the particular organism being cultivated. Generally, about 5 g of potassium phosphates per liter of medium can be tolerated by bacteria and fungi.

When a great deal of acid is produced by a culture, the limited amounts of buffer that may be used become insufficient for the maintenance of a suitable pH. In such cases, carbonates may be added to media as "reserve alkali" to neutralize the acids as they are formed. In the presence of hydrogen ions, carbonate is transformed to bicarbonate, and bicarbonate is converted further to carbonic acid, which decomposes spontaneously to $CO_2$ and water. This sequence of reactions, all of which are freely reversible, can be summarized:

$$CO_3^{2-} \underset{-H^+}{\overset{+H^+}{\rightleftharpoons}} HCO_3^- \underset{-H^+}{\overset{+H^+}{\rightleftharpoons}} H_2CO_3 \rightleftharpoons CO_2 + H_2O$$

Because $H_2CO_3$ is an extremely weak acid and because it decomposes with the loss of $CO_2$ to the atmosphere, the addition of carbonates prevents the accumulation of hydrogen ions and hence of free acids in a medium. The soluble carbonates, such as $Na_2CO_3$, are strongly alkaline and are therefore not suitable for use in culture media. In contrast, *insoluble carbonates* are very useful ingre-

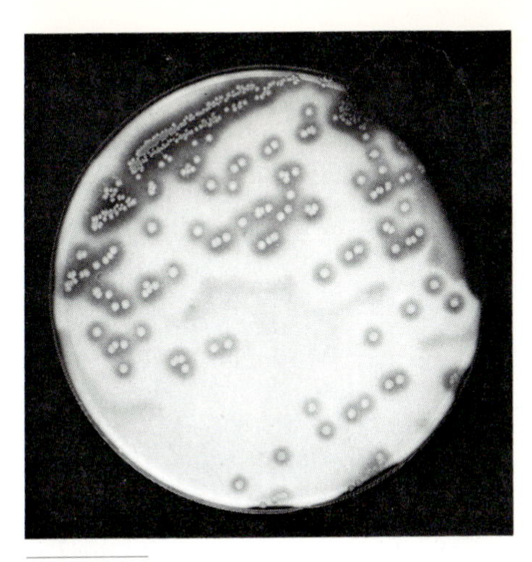

**FIGURE 2.5**

Colonies of *Streptococcus* growing on an agar medium containing a suspension of calcium carbonate. Lactic acid production has dissolved the calcium carbonate, producing clear zones around each colony.

dients for many culture media. Of these insoluble carbonates, finely powdered chalk ($CaCO_3$) is the most generally employed. Because of its insolubility, calcium carbonate does not create strongly alkaline conditions in the medium, especially if it is used in conjunction with other buffers. When, however, the pH of the liquid drops below approximately 7.0, the carbonate is decomposed with the evolution of $CO_2$. It thus acts as a neutralizing agent for any acids that may appear in a culture by converting them to their calcium salts.

The addition of $CaCO_3$ to agar media used for the isolation and cultivation of acid-forming bacteria helps to preserve neutral conditions. Furthermore, since the acid-forming colonies dissolve the precipitated chalk and become surrounded by clear zones, they can be easily recognized against the opaque background of the medium (Figure 2.5).

In some instances, neither buffers nor insoluble carbonates can be used to maintain a relatively constant pH in a culture medium. Special problems arise, for example, when very large amounts of acid are formed in media in which the presence of calcium carbonate is not desired. Even more serious difficulties are encountered in controlling the pH of slightly alkaline media in which basic substances are produced as a result of bacterial growth. This is due to the fact that phosphate buffers are not effective in the pH range 7.2 to 8.5, and few suitable buffers in this range are available. In certain cases, therefore, it is necessary to adjust the pH of the culture, either periodically or continuously, by the aseptic addition of strong acids or bases. In some laboratories and in industrial plants, elaborate mechanical devices are used for this purpose. With their aid, a continuous titration of the medium is feasible and the pH is kept nearly constant.

The media described in Table 2.5 are all slightly alkaline at the beginning because they contain the alkaline salt $K_2HPO_4$. Many organisms prefer neutral or slightly acidic conditions, which can be achieved by the use of appropriate buffers.

## The Avoidance of Mineral Precipitates: Chelating Agents

A troublesome problem often encountered in the preparation of synthetic media is the formation of a precipitate upon sterilization, particularly if the medium has a relatively high phosphate concentration. This results from the formation of insoluble complexes between phosphates and certain cations, particularly calcium and iron. Although it usually does not affect the nutrient value of the medium, it may make the observation or quantitation of microbial growth difficult. The problem can be avoided by sterilizing separately the calcium and iron salts in concentrated solution and adding them to the sterilized and cooled medium. Alternatively, one can incorporate in the medium a small amount of a chelating agent, which will form a soluble complex with these metals and thus prevent them from forming an insoluble complex with phosphates. The chelating agent most commonly used for this purpose is ethylenediaminetetraacetic acid (EDTA), at a concentration of approximately 0.01 percent.

## The Control of Oxygen Concentration

Oxygen is an essential nutrient for the obligately aerobic bacteria. Aerobic microorganisms can be grown easily on the surface of agar plates and in shallow layers of liquid medium. In unshaken liquid cultures, growth usually occurs at the surface. Below the surface, however, conditions become anaerobic, and growth is impossible. To obtain large populations in liquid cultures, *it is therefore necessary to aerate the medium.* Various types of shaking machines that constantly agitate, and thus aerate, the medium are available for laboratory use. Another method of aeration is the continuous passage of a stream of sterile air through a culture. To ensure a large surface of contact between gas and liquid, the air may be introduced through a porous "sparger," which delivers it in the form of very fine bubbles. However, even vigorous aeration is often insufficient to maintain an adequate concentration of dissolved oxygen, because the rate of $O_2$ utilization by the culture may exceed the rate of its diffusion into the culture medium.

These methods all have as their goal the maintenance of $O_2$ concentrations at the saturation point with respect to the partial pressure of $O_2$ in air. Maintaining constant dissolved $O_2$ concentrations less than saturation requires continuous adjustment of the rate of aeration in response to the dissolved oxygen concentrations sensed by an *oxygen electrode*; or alternatively, that the culture be aerated with a gas mixture in which the partial pressure of $O_2$ is less than that of air.

## Techniques for Cultivation of Obligate Anaerobes

Many of the more sensitive, strictly anaerobic microorganisms are rapidly killed by contact with molecular oxygen. The exposure of cultures to air should accordingly be minimized or avoided completely. Furthermore, many strict anaerobes can initiate growth only in media of low $E_h$ ($\sim 150$ mV, or even less). The use of media that are *prereduced* by the inclusion of such reducing agents as cysteine, thioglycollate, $Na_2S$, or sodium ascorbate is therefore a factor of cardinal importance for the cultivation of many anaerobes. Once prepared, such media must, of course, be protected from exposure to air during both storage and use. During use, this can be achieved by passage of a stream of $O_2$-free $CO_2$ or $N_2$ into the orifice of the opened culture vessel.

Liquid cultures of strict anaerobes are usually prepared in tubes or flasks completely filled with medium and closed by rubber stoppers or plastic screw caps. Isolation in solid media can be undertaken by several methods. Organisms able to tolerate a transient exposure to air can be isolated in shake tubes, or on streaked plates, which are placed after inoculation in sealed anaerobic jars. The atmosphere of the jars is then rendered $O_2$-free by evacuation and refilling with an inert gas (e.g., $N_2$); by chemical destruction of oxygen; or by a combination of both methods. Organisms unable to tolerate even momentary exposure to air are often isolated using roll tubes.

Although roll tubes allow the convenient isolation and subculturing of the most strict anaerobes, they are not well suited to growth of cultures for genetic or physiological study. Consequently, glove boxes containing a reducing atmosphere at slightly higher pressure than the surrounding air (to ensure that gas will not leak in through undetected holes) are being used increasingly. Within such a glove box the full range of microbiological and biochemical techniques may be performed.

## The Provision of Carbon Dioxide

A problem frequently encountered in the cultivation of photoautotrophs and chemoautotrophs is the provision of $CO_2$ in sufficient amounts. Although the diffusion of $CO_2$ from the atmosphere into the culture medium will permit growth to occur, the $CO_2$ concentration in the atmosphere is very low (0.03 percent in the open atmosphere, somewhat higher inside a building), and the growth rates of autotrophs are often limited by the availability of $CO_2$ under these conditions. The solution is to gas the cultures with air that has been artificially enriched with $CO_2$ and contains from 1 to 5 percent of this gas. The control of pH becomes a problem, for reasons already discussed (p. 30), and if this solution is adopted, care must be taken to modify the buffer composition of the medium. In the case of autotrophs that can be grown under anaerobic conditions in stoppered bottles (e.g., the purple and green sulfur bacteria) the requirement for $CO_2$ can be met by the incorporation of $NaHCO_3$ in the medium. Soluble carbonates cannot be used in media exposed to air because the rapid loss of $CO_2$ to the atmosphere causes the medium to become extremely alkaline.

## The Provision of Light

For the cultivation of phototrophic microorganisms (algae, photosynthetic bacteria), *light* is a requirement. The provision of adequate illumination combined with control of temperature is not a simple matter. In the cultivation of nonphotosynthetic organisms, temperature control is provided by the use of incubators, maintained by a thermostatic device at the desired value; however, most commercially available models are not designed with a system of internal illumination and cannot be used for the cultivation of phototrophic organisms.

A relatively uncontrolled and discontinuous illumination may be obtained by the exposure of cultures to daylight. Direct exposure to sunlight should be avoided, because the intensity may be too high, and the temperature may rise to a point where growth is prevented. Many phototrophic microorganisms can tolerate continuous illumination, and their growth is much more rapid under these conditions, so artificial light sources are advantageous. The *emission spectrum* of the lamp employed is important. Fluorescent light sources have the practical advantage of producing relatively little heat, so maintenance of a suitable temperature is not difficult. However, their emission spectra are deficient, compared to sunlight, in the longer wavelengths of the visible spectrum and the near infrared region. They are satisfactory for the cultivation of algae and cyanobacteria, which perform photosynthesis with light of wavelengths shorter than 700 nm, but provide little or no photosynthetically effective light for purple and green bacteria, which use wavelengths in the range 750 to 1,000 nm. The only suitable artificial light sources for the latter photosynthetic bacteria are incandescent lamps; however, if high intensities are used, the dissipation of heat becomes a problem. The easiest solution is to immerse culture vessels in a glass or plastic water bath that can be subjected to lateral illumination and maintained at the desired temperature by the circulation of water. The other solution is to construct a light cabinet with internal incandescent illumination, in which the temperature can be controlled by ventilation or refrigeration.

## SELECTIVE MEDIA

It is clear that no single medium or set of conditions will support the growth of all the different types of organisms that occur in nature. Conversely, any medium that is suitable for the growth of a specific organism is, to some extent, *selective* for it. In a medium inoculated with a variety of organisms, only those that can grow in it will reproduce, and all others will be suppressed. Further, if the growth requirements of an organism are known, it is possible to devise a set of conditions that will specifically favor the development of this particular organism, thus permitting its isolation from a mixed natural population, even when the organism in question is a minor component of the total population. Microorganisms can be selectively obtained from natural habitats (e.g., soil or water) either by *direct isolation* or by *enrichment*.

### Direct Isolation

If a mixed microbial population is spread over the surface of a selective medium solidified with agar (or some other gelling agent), every cell in the inoculum capable of development will grow and eventually form a colony. The spatial dispersion of the microbial population on a solid medium considerably reduces the competition for nutrients: under these circumstances, even organisms that grow relatively slowly will be able to produce colonies. Direct plating on a selective medium is the technique of choice when one wishes to isolate a considerable diversity of microorganisms, all able to grow under the conditions of culture employed.

### Enrichment

If a mixed microbial population is introduced into a liquid selective medium, there is a direct competition for nutrients among the members of the developing population. Liquid enrichment media therefore tend to select the microorganism of highest growth rate among all the members of the introduced population that are able to grow under the conditions provided.

The selectivity of an enrichment culture is not determined solely by the chemical composition of the medium used. The outcome of enrichment in a given medium can be significantly modified by variation of such other factors as temperature, pH, ionic strength, illumination, aeration, or source of inoculum. For example, medium 2 of Table 2.5 can be used to grow or to enrich enteric bacteria belonging to the genera *Escherichia* and *Enterobacter*. Members of the former genus are normal inhabitants of the intestinal tract of warm-blooded animals, while members of the latter genus are common soil inhabitants. *Escherichia* strains are naturally adapted to growth at somewhat higher temperatures than *Enterobacter* strains, and they are also less sensitive to the toxic effect of bile salts. Consequently, either incubation at an elevated temperature (45° C) or inclusion of bile salts will make medium 2 selective for *Escherichia*.

Temperature selection can also be used with great effectiveness for the isolation of cyanobacteria, which closely resemble many algae in nutritional and metabolic respects. Both algae and cyanobacteria can grow in a simple mineral medium, incubated in the light at 25° C. However, development of algae can be almost wholly prevented by incubation at a temperature of 35° C, since algae in general have lower temperature maxima than cyanobacteria.

An enrichment medium that is not initially highly selective may acquire greatly increased selectivity for a particular type of microorganism as a result of chemical changes produced by this organism during its development. Thus, fermentative bacteria and yeast are typically more tolerant than other organisms of the organic end products that they themselves produce from carbohydrates; in a carbohydrate-rich medium their development will therefore tend to suppress competing microorganisms.

In the isolation of endospore-forming bacteria (genera *Bacillus* and *Clostridium*), competition from nonsporulating bacteria can be largely eliminated by a pretreatment of the inoculum. Pasteurization of the inoculum, involving brief exposure to a high temperature (two to five minutes at 80° C) will

**TABLE 2.7**

**Primary Environmental Factors That Determine the Outcome of Enrichment Procedures for Chemoheterotrophic Bacteria with the Use of Synthetic Media**

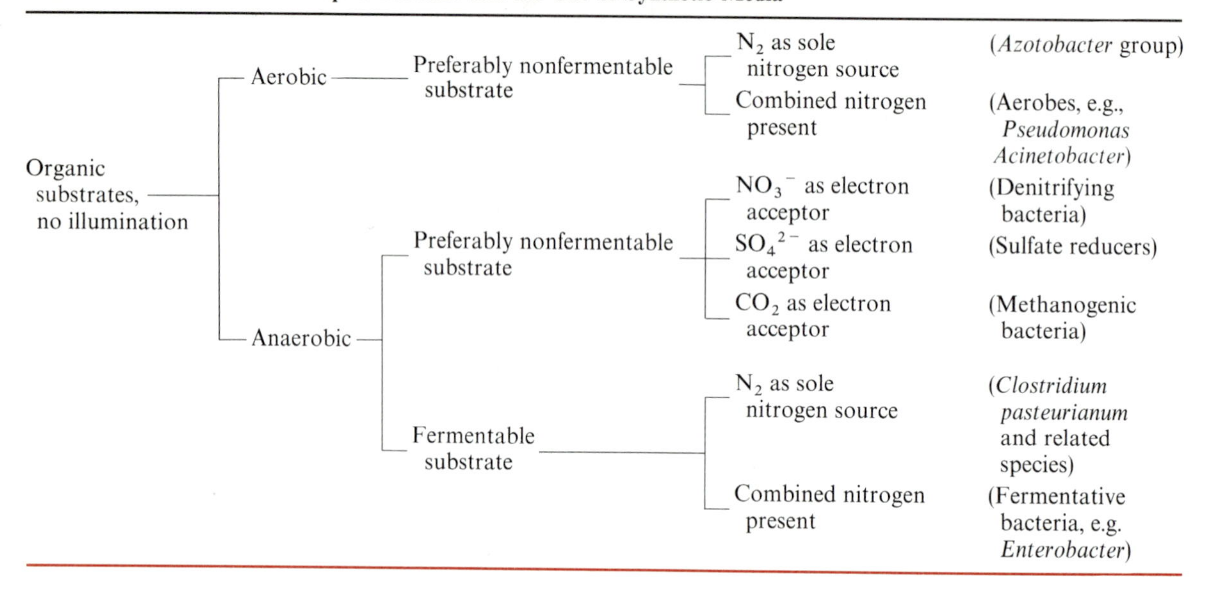

| | | | | |
|---|---|---|---|---|
| Organic substrates, no illumination | Aerobic | Preferably nonfermentable substrate | $N_2$ as sole nitrogen source | (*Azotobacter* group) |
| | | | Combined nitrogen present | (Aerobes, e.g., *Pseudomonas Acinetobacter*) |
| | Anaerobic | Preferably nonfermentable substrate | $NO_3^-$ as electron acceptor | (Denitrifying bacteria) |
| | | | $SO_4^{2-}$ as electron acceptor | (Sulfate reducers) |
| | | | $CO_2$ as electron acceptor | (Methanogenic bacteria) |
| | | Fermentable substrate | $N_2$ as sole nitrogen source | (*Clostridium pasteurianum* and related species) |
| | | | Combined nitrogen present | (Fermentative bacteria, e.g. *Enterobacter*) |

destroy most vegetative cells, leaving the much more heat-resistant spores relatively unaffected.

## Enrichment Methods for Some Specialized Physiological Groups

Enrichment culture is one of the most powerful techniques available to the microbiologist. An almost infinite number of permutations and combinations of the different environment variables, nutritional and physical, can be developed for the specific isolation of microorganisms from nature. Enrichment techniques provide a means for isolating known microbial types at will from nature, by taking advantage of their specific requirements, and can also be indefinitely elaborated as a means of obtaining hitherto undescribed organisms capable of growing in the environments devised by the scientist. Here we shall attempt to summarize a few of the enrichment procedures that can be used to isolate major physiological groups of microorganisms, principally bacteria, from nature.

### Synthetic Enrichment Media for Chemoheterotrophs

The nutritional and environmental conditions necessary for the enrichment of various groups of chemoheterotrophs in synthetic media are outlined in Table 2.7.

For the enrichment of fermentative organisms, the chemical nature of the organic substrate is important; to be attacked by fermentation, it must be neither too oxidized nor too reduced. Sugars are excellent fermentative substrates, but many other classes of organic compounds on the same approximate oxidation level can also be fermented. The enrichment cultures must be incubated anaerobically, not only because some fermentative organisms are obligate anaerobes, but also to prevent competition from aerobic forms. Nitrate should not be used as a nitrogen source, since it will also allow growth of denitrifying bacteria (see below). Calcium carbonate may be added if the organisms being enriched produce acid and are themselves sensitive to it.

Three special physiological groups of chemoheterotrophic bacteria, which can use inorganic compounds other than molecular oxygen as terminal oxidants for respiratory metabolism, may also be enriched in synthetic media under anaerobic conditions. In these cases, readily fermentable organic substrates such as sugars should be avoided. For the denitrifying bacteria, nitrate is included as a terminal oxidant, and acetate, butyrate, or ethyl alcohol as a carbon and energy source. For the sulfate-reducing bacteria, a relatively large amount of sulfate is included, since it is the specific terminal oxidant; lactate or malate provides the best source of carbon and energy. For the carbonate-reducing (methane-producing) bacteria, $CO_2$ must be present as an oxidant and such compounds as $H_2$, acetate, or formate as the source of energy. It should also be noted here that in enrichments for sulfate- and carbonate-reducing bacteria, the use of ammonia as a nitrogen source is desirable; addition of nitrate will favor development of the nitrate reducers. Similarly, in enrichments for carbonate and nitrate reducers, the concentration of sulfate should be kept to a minimum to prevent overgrowth by sulfate reducers.

Obviously, aerobic conditions are essential for the enrichment of microorganisms that obtain energy from aerobic respiration. Either fermentable or nonfermentable substrates are satisfactory organic nutrients. If a fermentable substrate is used, however, the culture must be aerated thoroughly and continuously, because the depletion of oxygen by respiration favors the enrichment of anaerobes. Readily fermentable compounds generally stimulate the growth of facultative anaerobes. Thus, when glucose or other sugars are included in enrichment media, *Enterobacter* appears as one of the principal organisms under aerobic as well as under anaerobic conditions.

The use of nonfermentable substrates usually results in the enrichment of organisms that are obligate aerobes, e.g., members of the genus *Pseudomonas*. For instance, benzoate-oxidizing strains of *P. putida* can be obtained by using sodium benzoate (1 g/liter) as the sole organic nutrient. If asparagine (2 g/liter) is used as the sole source of both carbon and nitrogen in the medium, asparagine-oxidizing strains of this or other related species become predominant.

Ammonium salts are generally used as a nitrogen source in synthetic enrichment media for aerobes. If no combined nitrogenous compounds are provided, cultures of the aerobic nitrogen-fixing bacteria of the *Azotobacter* group can be obtained. Members of this group can use a great variety of organic substrates, including alcohols, butyrate, benzoate, and glucose, as their only organic nutrients.

## The Enrichment of Chemoautotrophic and Photosynthetic Organisms

For the enrichment of chemoautotrophic and photoautotrophic organisms, organic compounds must be omitted from the medium, and $CO_2$ or bicarbonate must be used as the only source of carbon (Tables 2.8 and 2.9). Photosynthetic forms require light, whereas the chemoautotrophs should be cultivated in the dark to prevent the development of photosynthetic types. During incubation of cultures, either aerobic or anaerobic conditions must be maintained, depending on whether or not the organisms need oxygen. An exception to this rule is found in the selective cultivation of algae and cyanobacteria. Since these organisms *produce* oxygen in their metabolism, it makes virtually no difference whether the enrichment cultures are initially incubated under aerobic or anaerobic conditions.

**TABLE 2.8**

**Primary Environmental Factors That Determine the Outcome of Enrichment Procedures for Some Chemoautotrophic Bacteria**

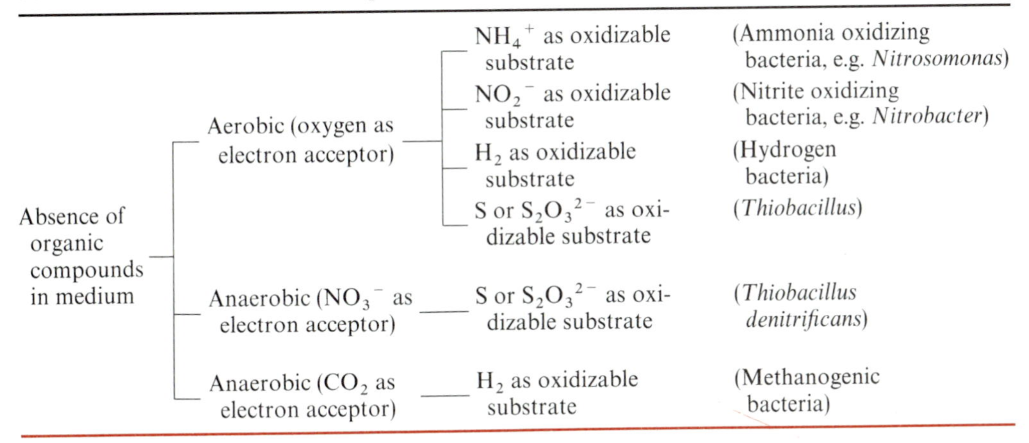

| | | | |
|---|---|---|---|
| Absence of organic compounds in medium | Aerobic (oxygen as electron acceptor) | $NH_4^+$ as oxidizable substrate | (Ammonia oxidizing bacteria, e.g. *Nitrosomonas*) |
| | | $NO_2^-$ as oxidizable substrate | (Nitrite oxidizing bacteria, e.g. *Nitrobacter*) |
| | | $H_2$ as oxidizable substrate | (Hydrogen bacteria) |
| | | S or $S_2O_3^{2-}$ as oxidizable substrate | (*Thiobacillus*) |
| | Anaerobic ($NO_3^-$ as electron acceptor) | S or $S_2O_3^{2-}$ as oxidizable substrate | (*Thiobacillus denitrificans*) |
| | Anaerobic ($CO_2$ as electron acceptor) | $H_2$ as oxidizable substrate | (Methanogenic bacteria) |

**TABLE 2.9**

**Primary Environmental Factors That Determine the Outcome of Enrichment Procedures for Photosynthetic Microorganisms**

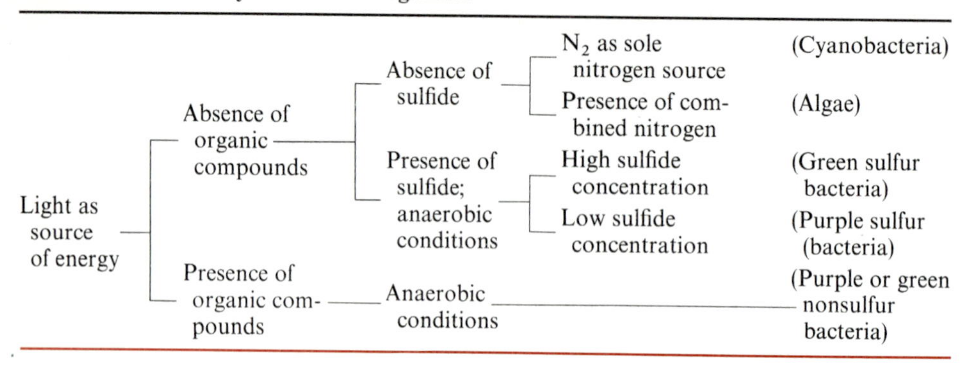

| | | | | |
|---|---|---|---|---|
| Light as source of energy | Absence of organic compounds | Absence of sulfide | $N_2$ as sole nitrogen source | (Cyanobacteria) |
| | | | Presence of combined nitrogen | (Algae) |
| | | Presence of sulfide; anaerobic conditions | High sulfide concentration | (Green sulfur bacteria) |
| | | | Low sulfide concentration | (Purple sulfur (bacteria) |
| | Presence of organic compounds | Anaerobic conditions | | (Purple or green nonsulfur bacteria) |

Media devised for the enrichment and propagation of photosynthetic organisms should contain sodium because this element is known to be required by photosynthetic bacteria.

For the enrichment of purple nonsulfur bacteria, the medium should include a suitable organic substrate and, in some instances, bicarbonate. The substrate should not be a readily fermentable one; acetate, butyrate, or malate is customarily used. Bicarbonate must be added if the substrate (e.g., butyrate) is more reduced than cell material because photosynthesis is accompanied by a net consumption of $CO_2$. With substrates such as malate, which are metabolized with a net production of $CO_2$, the addition of bicarbonate is unnecessary. Because the photoheterotrophic bacteria often require various growth factors, a small amount of yeast extract is generally added to the enrichment medium.

## The Use of Complex Media for Enrichment

Some bacteria cannot be enriched in defined media because they have extremely complex nutritional requirements. Nevertheless, such organisms may, in some cases, be obtained from nature by the use of specially designed complex media. The lactic acid bacteria illustrate this point. These organisms are characterized by their remarkable resistance to lactic acid, which they themselves produce in the fermentation of sugar. To enrich for lactic acid bacteria, a poorly buffered medium containing glucose and a rich source of growth factors is used (e.g., 20 g of glucose and 10 g of yeast extract per liter). After inoculation preferably with natural materials that are rich in lactic acid bacteria (e.g., vegetable matter, raw milk, sewage), the medium is incubated under anaerobic conditions. The first organisms to develop are usually bacteria such as *Enterobacter* and *Escherichia.* However, as lactic acid gradually accumulates, conditions become less and less favorable for these bacteria, whereas the lactic acid bacteria continue to grow. Eventually, the acidity of the medium becomes so high that the lactic acid bacteria predominate and most other organisms are destroyed.

Another good example of a complex medium that is quite selective for a specific group of organisms is one designed for the enrichment of the propionic acid bacteria. These organisms produce propionic acid, acetic acid, and $CO_2$ in fermentation. Although they can ferment glucose readily, they cannot compete either with *Enterobacter* or with the lactic acid bacteria in glucose media because they grow relatively slowly and do not tolerate acidic conditions. However, the propionic acid bacteria can also ferment lactic acid, which is not a suitable substrate for most other fermentative organisms. This capacity is the key to their enrichment. If a neutral medium containing 20 g of sodium lactate and 10 g of yeast extract per liter is inoculated with natural materials containing propionic acid bacteria and incubated at 30° C under anaerobic conditions, an enrichment of these organisms is obtained. Swiss cheese is a good inoculum for the cultures because propionic acid bacteria are the principal agents in its ripening.

Complex media can be used successfully for the selective cultivation of the acetic acid bacteria. These bacteria are especially adapted to environments that contain high concentrations of ethanol. They are also far less susceptible than other bacteria to inhibition by acetic acid, which they produce from ethanol in respiration. To enrich for them, a complex medium containing ethanol is inoculated with materials containing these bacteria and then incubated under aerobic conditions. Fruits, flowers, and unpasteurized (draught) beer are good sources of inoculum. A medium containing 40 ml of ethanol and 10 g of yeast extract per liter, adjusted to pH 6.0, can be used for enrichment. Beer and hard cider are also excellent enrichment media, for these fermented beverages closely resemble the natural environment in which acetic acid bacteria predominate. Although a large surface of contact with air must be ensured, the inoculated medium is usually not aerated vigorously because many acetic acid bacteria grow best in a pellicle that they form at the surface.

## LIGHT MICROSCOPY

In order to understand the indispensable role played by the microscope in the study of microorganisms, it is necessary to appreciate the intrinsic limitions of the eye as a magnifying instrument. The apparent size of an object viewed by the human eye is directly related to the angle that the object subtends at the eye: hence, if its distance from the eye is halved, its apparent size is doubled. However, the eye cannot focus on objects brought closer to it than approximately 25 cm; this is, accordingly, the distance of maximal effective magnification. In order to be seen at all, an object must subtend an angle at the eye of 1° or greater; and for a distance of 25 cm, this corresponds to a particle with a diameter of approximately 0.1 mm.

Most cells (and hence most unicellular microorganisms) are too small to be detected by the unaided human eye. In order to detect them, and to observe their form and structure, the use of a microscope is therefore essential. The function of the magnifying lens system of this instrument, interposed between the specimen and the eye, is to greatly increase the apparent angle subtended at the eye by objects within the microscopic field. In addition to this factor of *magnification*, two other factors, *contrast* and *resolution*, are of great importance. In order to be perceived through the microscope, an object must possess a certain *degree of contrast* with its surrounding medium; and in order to produce a clear magnified image, the microscope must possess a *resolving power* sufficient to permit the perception as separate objects of closely adjacent points in the image.

### The Light Microscope

As discussed in Chapter 1, Leeuwenhoek discovered the microbial world through the use of simple microscopes containing a single, biconvex lens of short focal length. The development and improvement of the more complex *compound microscopes* now employed required almost two centuries of research in applied optics.

A modern compound microscope contains three separate lens systems (Figure 2.6). The *condenser*, interposed between the light source and the specimen, collimates the light rays in the plane of the microscopic field. The *objective* produces a magnified image of the microscopic field within the microscope; and the *ocular* further enlarges this image and enables it to be perceived by the eye.

Single lenses have two inherent optical defects. They fail to bring the whole microscopic field into simultaneous focus (*spherical aberration*), and they produce colored fringes around objects in the field (*chromatic aberration*). These defects can be largely eliminated by placing additional, correcting lenses adjacent to a primary magnifying lens. Consequently, both the ocular and objective lenses of a modern compound microscope are multiple ones, designed to minimize these aberrations.

Correct adjustment of the condenser lens is of critical importance in providing a clear image. When a microscope is used at high magnifications, the condenser is usually positioned in order to provide *Koehler illumination*, in which the specimen is illuminated by parallel rays of light.

### Resolving Limit

The physical properties of light set a fixed limit to the effective magnification obtainable with a light microscope. Because of the wave nature of light, a very small object appears to be a disc, surrounded by a series of light and dark rings. Two adjacent points can be distinguished as separate, or *resolved*, only if the rings surrounding them do not overlap. The distance between two points that can just be distinguished from one another is known as the *resolving limit*, and it determines the maximal useful magnification of the light microscope.

The resolving limit ($d$) is defined by the equation

$$d = \frac{0.5\lambda}{N \sin \alpha} \qquad \textbf{(2.1)}$$

where $\lambda$ is the wavelength of the light source employed, $\alpha$ is the half angle of the objective lens, and $N$ is the refractive index of the medium between

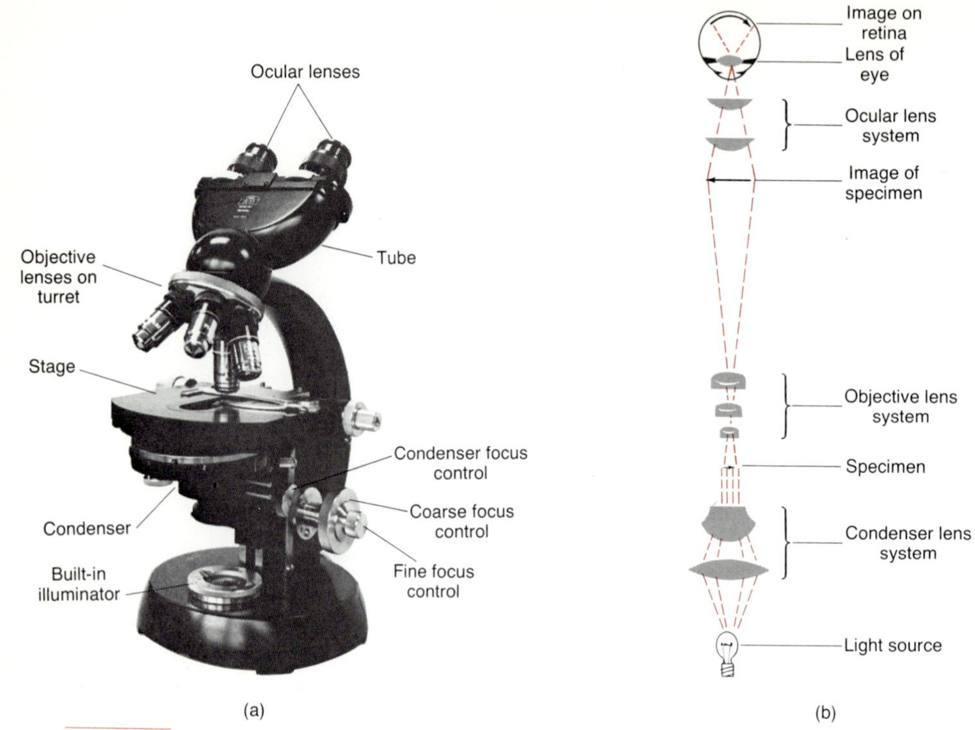


**FIGURE 2.6**

The modern compound microscope (a) with the principle parts identified. Schematic representation (b) of the optical system of a compound microscope. The light path shown is generalized; no attempt is made to show refractive events at individual lens elements. Light produced by the bulb is directed into the condenser lens system which either collimates it (*Köhler* illumination) or focuses it on the specimen (*critical* illumination). After passing through the specimen, the light traverses the objective lens system, which forms within the microscope tube an enlarged image of the specimen. This image is further enlarged by the ocular lens system which, acting in conjunction with the lens of the microscopist's eye, forms the final image on the retina.

the specimen and the front of the objective lens. The terms of the denominator ($N \sin \alpha$), commonly called the numerical aperture ($NA$), describe the properties of the objective lens. Up to a certain limit, an increase in the numerical aperture of the objective lens increases resolving power. If the medium between the specimen and the objective is air, a feasible diameter of the objective lens limits the $NA$ to approximately 0.65. The value of $N$ can be increased by filling the intervening space with oil, which has a higher refractive index than air. With *oil immersion lenses*, the $NA$ can be increased to a value as high as 1.4, although values of 1.25 are more common. Under the best obtainable conditions, the maximal resolution of the light microscope approaches 200 nm, with light of the shortest visible wavelength (approximately 426 nm); i.e., two adjacent points closer together than 200 nm cannot be resolved into separate images.

## Contrast and Its Enhancement in the Light Microscope

When a small biological object, such as a microbial cell, is observed in the living state, it is normally suspended in an aqueous medium, compressed into a thin layer between a slide and cover slip. The perception of such an object depends on the fact that it displays some degree of contrast with the surrounding aqueous medium as a result of the fact that less light is transmitted through it than through the medium. This decreased light transmission is caused by two factors: light absorbed by the cell and light refracted out of the optical path of the microscope, by a difference in refractive index between the cell and the surrounding medium. With the exception of intensely pigmented cell structures (e.g., chloroplasts) biological objects absorb very little light in the visible region of the spectrum; hence, the contrast which a living cell generates is attributable almost entirely to light refraction. However, its degree of contrast can be greatly increased by staining procedures: treatment with dyes that bind selectively either to the whole cell or to certain cell components, thus producing a much greater absorption of the incident light. Most staining treatments kill cells; and as a preliminary to staining, the cells are sometimes *fixed*, by treatments designed to minimize postmortem changes of structure. Commonly used chemical fixatives include osmic acid and aldehydes, notably glutaraldehyde. However, for light microscopy *heat* is the most commonly used fixative, despite the fact that it often leads to changes of cell shape or size.

With the wide availability of phase-contrast microscopes (see below), it is now rarely necessary to stain microbial cells in order to make them visible; the simplest, and for many purposes the most satisfactory, way to observe microorganisms with a light microscope is in the living state, as *wet mounts*. The principal value of staining procedures is to provide *specific information about the internal structure or the chemical properties of cells*. Thus, staining methods specific for deoxyribonucleic acid can reveal the structure and location of the nucleus; and a variety of special staining methods can be used to demonstrate intracellular deposits of such reserve materials as glycogen, polyphosphate, and poly-$\beta$-hydroxybutyrate. The Gram stain (see p. 145) and the acid-fast stain (see p. 510) are used to obtain information on the composition of the wall layers of bacterial cells. So-called *negative stains*, which do not enter the cell, are sometimes useful to reveal surface layers of very low refractive index, such as the capsules and slime layers that often surround microbial cells. These can be made visible by adding India ink to the suspending medium, since the carbon particles of the ink cannot penetrate the capsular layer, it is revealed as a clear zone surrounding the cell (Figure 2.7).

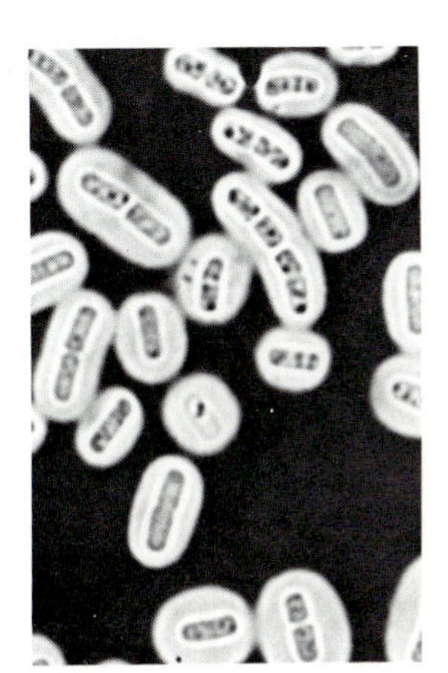

**FIGURE 2.7**

Bacterial capsules, demonstrated by dispersing the cells in India ink. The organism shown is *Bacillus megaterium* (× 2,160). Courtesy of C. F. Robinow.

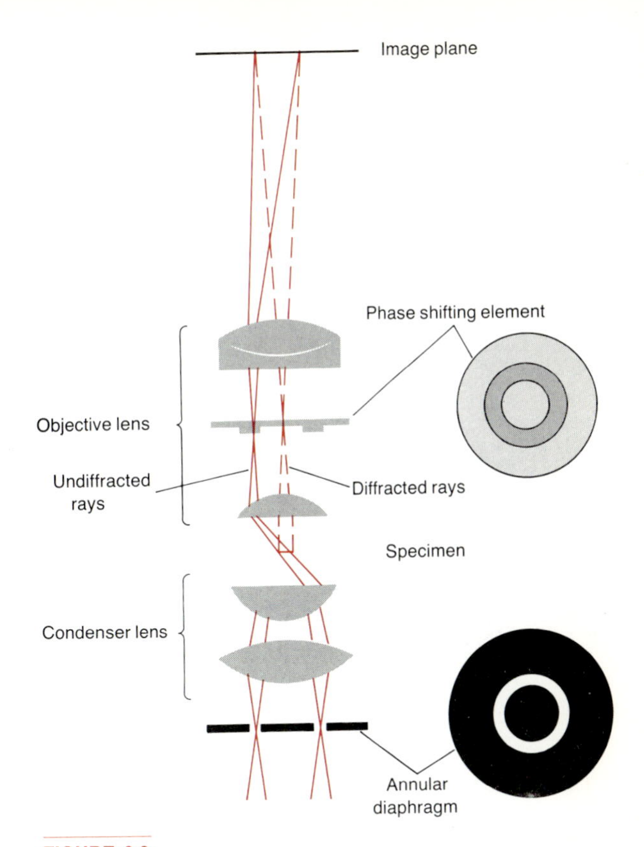


**FIGURE 2.8**

Schematic representation of the phase contrast microscope.

### Enhancement of Contrast by Phase Microscopy

The relatively low contrast of living cells as viewed with a conventional light microscope can be greatly increased by the use of an instrument with a modified optical system, known as the *phase contrast microscope* (Figure 2.8). Phase contrast microscopy is based on the fact that the rate at which light travels through objects is inversely related to their refractive indices. Since the frequency of light waves is independent of the medium through which they travel, the phase of a light ray passing through an object of higher refractive index than the surrounding medium will be relatively retarded. A system of rings in the condenser and the objective separate the rays diffracted from the specimen from those that are not; the two sets of rays are recombined, after the diffracted rays have been passed through a glass ring, which introduces an additional phase difference. The total phase shift is equal to one-half of a wavelength; hence, when the rays are recombined in the image plane, destructive interference greatly increases the contrast of cells or intracellular structures that differ slightly in refractive index from their surroundings.

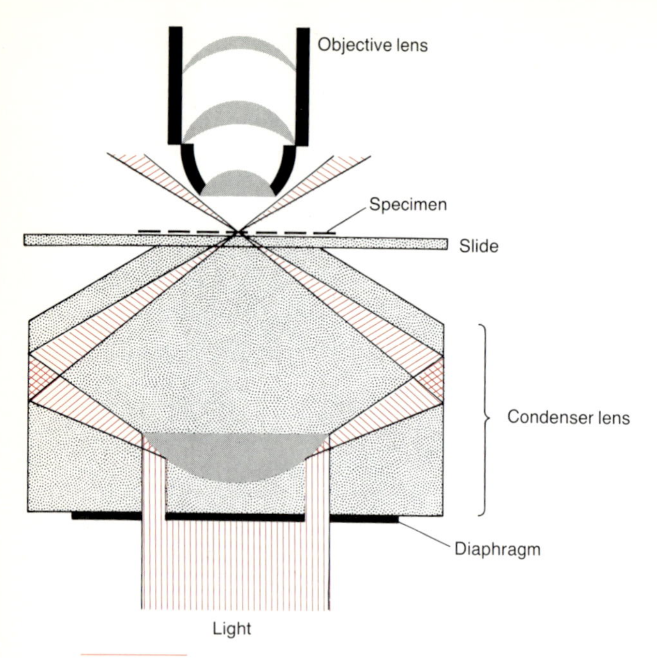


**FIGURE 2.9**
Schematic representation of dark field illumination.

INTERFERENCE CONTRAST MICROSCOPY An alternative optical technique for enhancing contrast is *interference contrast microscopy*. In an interference microscope, the specimen is illuminated with two coherent bundles of plane-polarized light, that differ by 90° in their planes of polarization. After passing through the objective, these two beams are displaced laterally relative to each other by a very slight amount (approximately equal to the limit of resolution of the microscope). Phase differences between the two bundles of light occur at boundaries between areas of different refractive index; as in the phase contrast microscope, these phase differences are manipulated to increase contrast. Although the interference contrast microscope does not produce halos around structures as are seen in the phase contrast microscope, it does not produce clear images if the object viewed is very thin, and it is much more costly. Phase contrast microscopy is hence the generally preferred method of observing wet mounts of most bacteria.

DARK-FIELD ILLUMINATION When light impinges on a small object, some light is scattered, making the object appear luminous and thus visible against a dark background. The technique of *dark-field illumination* exploits this phenomenon and permits the detection of objects so small as to otherwise provide insufficient contrast. Such illumination is achieved by the use of a special kind of condenser, which focuses on the specimen a hollow cone of light (Figure 2.9), the diverging rays of which do not enter the objective. Only light scattered by the specimen enters the objective and is observed.

### Ultraviolet and Fluorescence Microscopy

Since the resolving power of the light microscope is directly related to the wavelength of light employed, a slight improvement of resolution (about twofold) can be achieved by the use of an ultraviolet light source. Since glass is opaque to ultraviolet light of short wavelengths, the lens system must be composed of quartz, and a camera must be used to record the image, since the eye cannot perceive ultraviolet light. Its expense and complexity have limited the use of ultraviolet microscopy. However, a modification, known as *fluorescence microscopy*, has many important biological uses.

Certain chemical compounds that absorb ultraviolet light reemit part of the radiant energy as light of longer wavelength, situated in the visible region; this phenomenon is termed *fluorescence*. When exposed to ultraviolet light, a fluorescent object can thus be perceived as a brightly colored body against a black background. This is the principle of fluorescence microscopy. Only the condenser lens must be constructed of quartz, since it is the visible light emitted by the fluorescent specimen that is transmitted through the microscope. The major use of fluorescence microscopy in biology involves the techniques of *immunofluorescence*. When an animal is immunized with a specific antigen (e.g., a particular type of bacterium), its serum contains antibody proteins, which are capable of binding specifically to the immunizing antigen employed. The antibody proteins can be made intensely fluorescent by chemical conjugation with a fluorescent dye; and when they combine with the specific antigen, it also becomes fluorescent. By fluorescence microscopy, it is therefore possible to detect specifically the cells of a particular type of bacterium in a mixed microbial population treated with a fluorescent antiserum directed against this bacterium. This technique has also been used to study the mode of growth of the bacterial cell wall (Chapter 6).

## ELECTRON MICROSCOPY

The development of the electron microscope is one of the major achievements of applied physics in the twentieth century, and has revolutionized our knowledge of biological structure. It was based on the discovery that an electromagnetic field acts on a beam of electrons in a way analogous to the action of a glass lens on a beam of photons. An electron beam has the properties of an electromagnetic wave of very short wavelength; when accelerated through an electric field, its wavelength is inversely proportional to the square root of the accelerating voltage. A wavelength of only 0.04 nm, about 10,000 times shorter than that of visible light is routinely obtainable with accelerating voltages of about 100 kV. The resolving limit of the electron microscope is consequently several orders of magnitude lower than that of the light microscope (Eq. 2.1), and it thus permits the use of far higher effective magnifications. The path of electrons through a transmission electron microscope is directed in a manner analogous to the path of light rays through a light microscope. A beam of electrons projected from an electron gun is passed through a series of electromagnetic lenses. The condenser lens collimates the electron beam on the specimen, and an enlarged image is produced by a series of magnifying lenses. The image is rendered visible by allowing it to impinge on a phosphorescent screen (similar to the front of a cathode-ray tube in a television set). Since electrons can travel only in a high vacuum, the entire electron path through the instrument must be evacuated; consequently, specimens must be completely dehydrated prior to examination. Furthermore, only very thin specimens (with a thickness of 100 nm or less) can be observed in the conventional electron microscope, since the penetrating power of electrons through matter is weak. However, electron microscopes with accelerating voltages of a million kV have recently been developed; their electron beams can penetrate intact cells of more than a micrometer thickness. Such instruments are extremely large and costly, and there are only a few in existence.

In a transmission electron microscope, contrast results from the differential scattering of electrons by the specimen, the degree of scattering being a function of the number and mass of atoms that lie in the electron path. Since most of the constituent elements in biological materials are of low mass, the contrast of these materials is weak. It can be greatly enhanced by "staining" with the salts of various heavy metals (e.g., lead, tungsten, uranium). These may be either fixed on the specimen (positive staining) or used to increase the electron opacity of the surrounding field (negative staining). Negative staining is particularly valuable for the examination of very small structures such as virus particles, protein molecules, and bacterial flagella (e.g. see Figure 6.43). However, cells (even of very small microorganisms) are too thick to be examined satisfactorily in whole mounts. Observation of their internal fine structure requires that they be fixed, dehydrated, embedded in a plastic, and sectioned. Ultrathin sections (not more than 50 nm thick) are then positively stained with heavy metal salts, and are then mounted for examination.

Two other preparatory techniques, *metal shadowing* and *freeze-fracturing*, are frequently used for the observation of biological specimens with the transmission electron microscope. In metal shadowing the dried specimen is exposed at an acute angle to a directed stream of a heavy metal (platinum, palladium, or gold), thus producing an image that reveals the three-dimensional structure of the object (e.g. Figure 6.2). In the process of freeze-fracturing, the specimen is frozen and the frozen block is fractured with a knife, exposing various surfaces on and within the specimen. The fractured surface is shadowed at an acute angle with a heavy metal, and a supporting layer of carbon is evaporated onto the metal surface. The shadowed specimen is then destroyed by chemical treatment, and the replica is examined. The closely similar technique, *freeze-etching*, differs only in allowing a period of time for the sublimation of some of the surrounding ice before shadowing. Such etching of the surface provides more relief between structures and the surrounding aqueous milieu. These techniques have proved of particular value in the study of the wall and membrane structure of cells, since the fracture plane often follows the surface of a wall layer or the interior (hydrophobic) region of unit membranes (e.g. Figure 3.5).

## The Scanning Electron Microscope

The light microscope and the transmission electron microscope are fundamentally similar in operation: a broad beam of electromagnetic energy (consisting of photons or electrons) is passed through the specimen, which a magnified image of is formed by passing the emergent beam through lenses (glass or electromagnetic). In both instruments, the principle or refraction of electromagnetic energy by lenses is used to form a magnified image of the specimen. The recently developed *scanning electron microscope* utilizes a totally different principle of image formation, that of electronic amplification of signals generated by irradiating the surface of the specimen with a very narrow beam of electrons. Such irradiation causes low energy (secondary) electrons to be ejected from the specimen; these can be collected on a positively-charged plate (an *anode*) thereby generating an electric signal that is proportional to

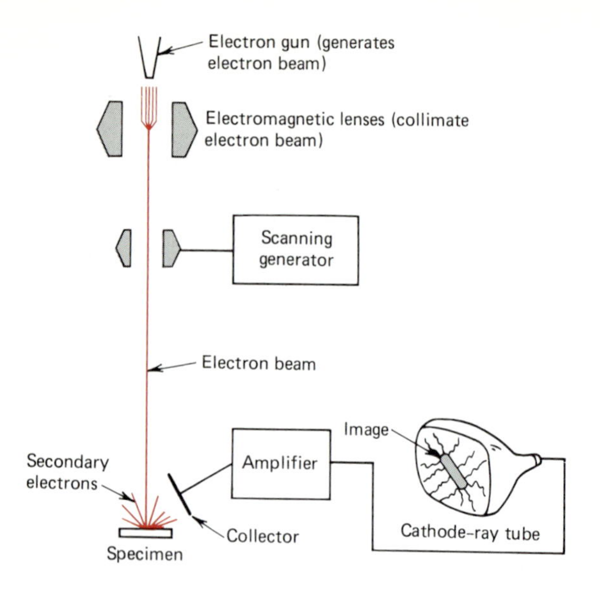


Schematic representation of a scanning electron microscope.

the number of electrons striking the anode. Since this number of electrons depends on the number ejected (in turn, a function of angle of a particular region of the surface with respect to the electron beam) and on the number reabsorbed by surrounding protuberances on the surface of the specimen, the electric signal can be used to generate an image of the specimen (Figure 2.10). By use of a scanning generator, the electron beam is caused to traverse the specimen in a raster pattern. The signal generated by secondary electrons striking the anode is amplified and used to modulate the intensity of a spot scanning a cathode ray tube (essentially the same as the picture tube of a television receiver) in precise register with the scanning pattern of the electron beam. Hence a magnified image of the surface topography of the specimen is presented on the cathode ray tube. The depth of focus of this instrument is several millimeters; and its range of effective magnification extends from about 20 × to more than 20,000 ×. An example of the type of image obtained is shown in Figure 18.5.

## FURTHER READING

### Books

Meynell, G. G., and E. Meynell, *Theory and Practice in Experimental Bacteriology.* New York: Cambridge University Press, 1965.

Norris, J. R., and D. W. Ribbins, eds., *Methods in Microbiology.* London and New York: Academic Press, 1969–present. A comprehensive reference work in many volumes.

Gerhardt, P., R. G. E. Murray, R. N. Costilow, E. W. Nester, W. A. Wood, N. R. Krieg, and G. B. Phillips, *Manual of Methods for General Bacteriology.* Washington: American Society for Microbiology, 1981.

Spencer, M., *Fundamentals of Light Microscopy.* New York: Cambridge University Press, 1982.

Starr, M. P., H. Stolp, H. G. Truper, A. Ballows, and H. G. Schlegel, *The Prokaryotes.* New York, Heidelberg, and Berlin: Springer-Verlag, 1981. A comprehensive account (in two volumes) of enrichment methods for most groups of bacteria.

# Chapter 3
# The Nature of the Microbial World

The term *microorganism* does not have the precise taxonomic significance of such terms as *vertebrate* or *angiosperm*, each of which defines a restricted biological group, all members of which share numerous common structural and functional properties. In contrast, any organism of microscopic dimensions is by definition a microorganism; and microorganisms occur in a wide diversity of taxonomic groups, some of which (e.g., the algae) also contain members far too large to be assigned to this category. In this chapter, the three major taxonomic groups that consist either in whole or in part of microorganisms—the eubacteria, the archaebacteria, and the protists—will be distinguished.

## THE COMMON PROPERTIES OF BIOLOGICAL SYSTEMS

Cellular organisms share a *common chemical composition*, their most distinctive chemical attribute being the presence of three classes of complex macromolecules: deoxyribonucleic acid (DNA), ribonucleic acid (RNA), and proteins. DNA is the constituent that carries in coded form all the genetic information necessary to determine the specific properties of the organism, known collectively as its *phenotype*. The genetic information is initially transcribed into complementary RNA sequences, in the form of molecules of RNA known as *messenger* RNA (mRNA). The mRNA molecules subsequently serve as templates for the synthesis of all the specific protein molecules characteristic of the organism; the translation of the transcribed

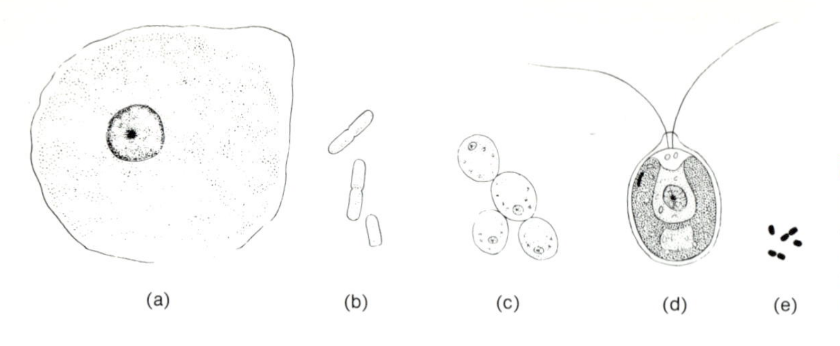


FIGURE 3.1

Drawings of several unicellular microorganisms on the same relative scale: (a) An amoeba; (b) a large bacterium; (c) a yeast; (d) a flagellate alga; (e) a small bacterium ( × 1,000).

genetic message is mediated by organelles known as *ribosomes*, composed of protein subunits and a special class of RNA, *ribosomal RNA* (rRNA). A third class of RNA molecule, the transfer RNAs (tRNAs) also participate in protein synthesis, as carriers of the amino acids that are assembled into linear sequence in the primary step of protein synthesis. The proteins of an organism include the enzymes that catalyze its activities, and the subunits from which many classes of proteinaceous cellular microstructures are assembled.

The chemical activities of an organism, catalyzed by its specific array of enzymes, are known collectively as *metabolic* activities. They include (1) the biosynthesis of the macromolecular constituents from the much simpler chemical substances (*nutrients*) derived from the external environment, and (2) the reactions necessary to generate the energy-rich substances that drive the processes of biosynthesis.

Most organisms share a common physical structure, being organized into microscopic subunits termed *cells*. All cells are enclosed by a thin membrane, the *cytoplasmic membrane*, which retains within its boundary the various molecules, large and small, necessary for the maintenance of biological function, and which at the same time regulates the passage of solutes between the interior of the cell and its external environment. Cells never arise *de novo:* they are always derived from preexisting cells by the process of growth and cell division.

These generalizations apply to all living objects with the exception of viruses. The general properties of viruses will be described at the end of this chapter.

## Patterns of Cellular Organization

The simplest cellular organisms consist of a single cell. Because cells are always of microscopic dimensions, such *unicellular organisms* are necessarily small, and thus fall in the general category of microorganisms. Unicellularity is widespread, though not universal, in the microbial groups known as *bacteria*, *protozoa*, and *algae*; it likewise occurs, though more rarely, in *fungi*. The very considerable differences that exist among the various groups of microorganisms are expressed solely in terms of differences with respect to the *size, form, metabolism*, and *internal structure* of the cell. Sketches of a few unicellular organisms, all drawn to the same scale, are shown in Figure 3.1.

A more complex mode of organization is *multicellularity*. Although a multicellular organism arises initially from a single cell, it consists in the mature state of many cells, attached to one another in a characteristic fashion that determines the gross external form of the organism. Multicellular organisms composed of a small number of cells may still be of microscopic dimensions; many examples exist among bacteria and algae. Such organisms are usually composed of similar cells, arranged in the form of a thread or filament. However, when the number of cells composing the organism is larger, the organism acquires a certain degree of structural complexity, simply from the manner in which the constituent cells are arranged. The best illustrations of such simple multicellular organization occur among the larger algae, which often have a characteristically plantlike form, even though there is little or no specialization of the component cells. Form is derived by the specific pattern in which the like structural units are arranged.

In metazoan animals and vascular plants, multicellular organization leads to a much higher degree of intrinsic structural complexity, as a consequence of the *differentiation of distinct cell types* during the development of the mature organism. This leads, through cell division, to the emergence of distinct *tissue regions*, each composed of a special type of cell; a further level of internal complexity may be attained by the association of different cell types into functional units known as *organs*. The structural complexity of a vascular plant or a metazoan animal thus proves upon microscopic analysis to be vastly greater than that of large but undifferentiated multicellular organisms such as the marine algae.

**TABLE 3.1**

**Some Major Differences Between Metazoan Animals and Vascular Plants**

| | | Vascular Plants | Metazoan Animals |
|---|---|---|---|
| Functional characters | Energy source | Light | Organic compounds |
| | Carbon source | $CO_2$ | Organic compounds |
| | Growth factor requirements | None | Complex |
| | Active movement | Absent | Present |
| Structural characters | Cells walls | Present | Absent |
| | Chloroplasts | Present | Absent |
| | Mode of growth[a] | Open | Closed |

[a] In animals the individual achieves a more or less fixed size and form as an adult. In most plants growth continues throughout the life of the individual, and the final size and form are much less rigidly fixed.

In a few groups, biological organization assumes a third form, known as *coenocytic structure.* A coenocytic organism is not composed of cellular subunits, separated from one another by their bounding membranes; instead, the multinucleate cytoplasm is continuous throughout the entire organism, which grows in size without undergoing cell division. This type of organization is characteristic of most fungi, and also occurs in many algae.

## The Problem of Primary Divisions among Organisms

It is a judgment of common sense, as old as humankind, that our planet is populated by two different kinds of organisms, plants and animals. Early in the history of biology this prescientific opinion became formalized in scientific terms: biologists recognized two primary kingdoms of organisms, the *Plantae* and the *Animalia.* The members of the two kingdoms appeared to be readily distinguishable by a whole series of characters, both structural and functional, some of which are summarized in Table 3.1. This traditional bipartite division was in fact a satisfactory one as long as biologists had to take into account only the more highly differentiated groups of multicellular organisms.

## The Place of Microorganisms

When exploration of the microbial world got under way in the eighteenth and nineteenth centuries, there seemed no reason to doubt that these simple organisms could be distributed between the plant and animal kingdoms. In practice, the assignment was usually made on the basis of the most easily determinable differences between plants and animals: the power of active movement and the ability to photosynthesize. Multicellular algae, which are immotile, photosynthetic, and in some cases plantlike in form, found a natural place in the plant kingdom. Although they are all nonphotosynthetic, the coenocytic fungi were also placed in the plant kingdom on the basis of their general immotility. Motile microscopic forms were lumped together as one group of animals, the Infusoria (Table 3.2).

Following the enunciation and acceptance of the cell theory (about 1840), biologists perceived that the Infusoria were a very heterogeneous group in terms of their cellular organization. Some of these microscopic forms (e.g., the rotifers) are invertebrate animals, with a body plan based on differentiation during multicellular development. Furthermore, the unicellular representatives can be subdivided into two groups: *protozoa,* with relatively large and complex cells, and *bacteria,* with much smaller and simpler cells. The old Infusoria was thus split three ways. Some of its component groups were classified as metazoan (multicellular) invertebrate animals. Others, the protozoa, were kept in the animal kingdom, but differentiated from all other animals on

**TABLE 3.2**

**Early Attempts (about 1800) to Allocate Microorganisms to the Plant and Animal Kingdoms**

| Plants | Animals |
|---|---|
| Algae (immotile, photosynthetic)<br>Fungi (immotile, nonphotosynthetic) | Infusoria (motile) |

the basis of their unicellular structure. Finally, the bacteria were transferred to the plant kingdom, despite their generally nonphotosynthetic nature, as a result of the discovery that the cyanobacteria, then considered to be algae, were characterized by cells with a comparably simple structure.

However, subsequent experience showed that the treatment of the protozoa (a large and complex microbial group) as unicellular animals led to considerable difficulties. Such protozoa as the ciliates and amebae, phagotrophic organisms devoid of cell walls, could be fitted quite satisfactorily into the confines of the animal kingdom, but other protozoa could not. On closer study, the flagellate protozoa proved to be a very odd assortment of creatures, in some of which motility by means of flagella was the only "animal-like" character. Some possessed cell walls, others did not. Some were phototrophs, others chemotrophs; and among the latter, both osmotrophic and phagotrophic representatives occurred. In short, this one microbial group shows all possible combinations of plantlike and animal-like characters. The problem of the placement of the flagellates became even more acute when it was recognized that in terms of cellular properties many of the phototrophic flagellates resembled very closely certain of the multicellular, immotile algae. Another protozoan group, the slime molds, also presented difficulties. In the vegetative state these organisms are phagotrophic and ameboid, but they can also form complex fruiting structures, similar in size and form to those characteristic of the true fungi. Should the slime molds be classified with the fungi, as plants, or with the protozoa, as animals?

Consequently, as knowledge of the properties of the various microbial groups deepened, it became apparent that at this biological level a division of the living world into two kingdoms cannot really be maintained on a logical and consistent basis. Some groups (notably the flagellates and the slime molds) were claimed both by botanists as plants and by zoologists as animals (Table 3.3). The problem is easy enough to understand in evolutionary terms. The major microbial groups can be regarded as the descendants of very ancient evolutionary lines that antedated the emergence of the two great lines that eventually led to the development of plants and animals. Hence, most microbial groups cannot be pigeonholed in terms of the properties that define these two more advanced evolutionary groups.

## The Concept of Protists

Dissatisfaction with existing classification, coupled with a clear understanding of the root of the trouble, led one of Darwin's disciples, E. Haeckel, to propose the obvious way out. In 1866 he suggested that logical difficulties could be avoided by the recognition of a *third* kingdom, the *protists*, to include protozoa, algae, fungi, and bacteria. The protists accordingly include both photosynthetic and nonphotosynthetic organisms, some plantlike, some animallike, some sharing properties of both the traditional kingdoms. What distinguished all protists from plants and animals was their *relatively simple biological organization*. Many protists are unicellular or coenocytic; and even the multicellular protists (e.g., the larger algae) lack the internal differentiation into separate cell types and tissue regions characteristic of plants and animals. A primary division

**TABLE 3.3**

**Final Effort (about 1860) to Allocate Microorganisms to the Plant and Animal Kingdoms**

| Plants | Contested Groups | Animals |
|---|---|---|
| | | Small metazoans |
| | | Rotifers |
| | | Nematodes (some) |
| | | Arthropods (some) |
| Algae (photosynthetic) | | Protozoa |
| | | Ciliates |
| Immotile forms ← | Photosynthetic flagellates → | Nonphotosynthetic flagellates |
| Fungi (nonphotosynthetic) | | |
| True fungi ← | Slime molds → | Ameboid protozoa |
| Bacteria | | |

**TABLE 3.4**

**Component Groups of the Three Kingdoms of Organisms Proposed by Haeckel (1866)**

| Properties | Plants | Animals |
|---|---|---|
| Multicellular; extensive differentiation of cells and tissues | Seed plants<br>Ferns<br>Mosses and liverworts | Vertebrates<br>Invertebrates |
| | Protists | |
| Unicellular, coenocytic, or multicellular, latter with little or no differentiation of cells and tissues | Algae<br>Protozoa<br>Fungi<br>Bacteria | |

in the biological world could accordingly be made in terms of the *degree of complexity of biological organization*; this could then be followed, for the more highly organized forms, by a secondary division on the basis of the properties long used to separate plants from animals (Table 3.4).

## EUCARYOTES AND PROCARYOTES

About 1950 the development of the electron microscope and of associated preparative techniques for biological materials made it possible to examine the structure of cells with a degree of resolution many times greater than that previously possible by the use of the light microscope. Within a few years many hitherto unperceived features of cellular fine structure were revealed. This led to the recognition of a profoundly important dichotomy among the various groups of organisms with respect to *the internal architecture of the cell:* two radically different kinds of cells exist in the contemporary living world. The more complex *eucaryotic cell* is the unit of structure in plants, metazoan animals, protozoa, fungi, and all save one of the groups that had traditionally been assigned to the algae. Despite the extraordinary diversity of the eucaryotic cell as a result of its evolutionary specialization in these groups, as well as the modifications that it can undergo during the differentiation of plants and animals, its basic architecture always has many common denominators. The less complex *procaryotic cell* is the unit of structure in two microbial groups: the *eubacteria* (including the *cyanobacteria*, formerly known as the "blue-green algae") and the *archaebacteria*, a heterogeneous group of microorganisms with procaryotic structure but with a cell chemistry that is strikingly different from that of the eubacteria. Indeed, the differences between the eubacteria and the archaebacteria are so profound that most microbiologists now believe that this distinction reflects an evolutionary separation as fundamental as that which divides the eucaryotes from either of the two groups of bacteria.

These newly recognized lines of demarcation run through Haeckel's proposed kingdom of protists. Protozoa, fungi, and algae (with the exception of the "blue-green algae") are eucaryotes, which share with plants and animals a common cell structure (eucaryotic) and many details of cell chemistry and function. The eubacteria include most bacterial groups (including the cyanobacteria). The archaebacteria include only three known groups indistinguishable from the eubacteria on structural grounds but profoundly different chemically. It is likely that a number of additional groups of archaebacteria will be recognized as details of the cell biology of poorly studied groups of bacteria accumulate.

We can thus distinguish on the basis of cell structure and function three major groups of cellular organisms (Table 3.5): the eucaryotes, the eubacteria, and the archaebacteria. The eucaryotes can be subdivided into three further groups: the plants, the animals, and the protists (a term that we shall restrict to the eucaryotic microorganisms). The eubacteria can be subdivided into Gram-negative eubacteria and Gram-positive eubacteria on the basis of the structure of the cell wall (although a third, relatively small group of eubacteria cannot be assigned to either of these groups because they lack a cell wall, the determining characteristics for assignment to either the Gram-negative or Gram-positive group). Our knowledge of the diversity of

**TABLE 3.5**

**Primary Subdivisions of Cellular Organisms That Are Now Recognized**

| Group | Cell Structure | Properties | Constituent Groups |
|---|---|---|---|
| Eucaryotes | Eucaryotic | Multicellular; extensive differentiation of cells and tissues | Plants (seed plants, ferns, mosses); Animals (vertebrates, invertebrates) |
| | | Unicellular, coenocytic or mycelial; little or no tissue differentiation | Protists (algae, fungi, protozoa) |
| Eubacteria | Procaryotic | Cell chemistry similar to eucaryotes | Most bacteria |
| Archaebacteria | Procaryotic | Distinctive cell chemistry | Methanogens, halophiles, thermoacidophiles |

the archaebacteria is still too rudimentary to attempt a systematic subdivision; the three provisionally recognized groups of archaebacteria are listed in Table 3.5.

In the following pages the principal features of cellular organization and function that distinguish these cell types are summarized.

## STRUCTURE OF THE CYTOPLASMIC MEMBRANE

All cells are bounded by a surface membrane known as the *cytoplasmic membrane*. Regardless of source, thin sections stained with heavy metals and viewed in the transmission electron microscope reveal a characteristic triple-layer appearance: two electron-dense layers separated by an electron-lucent zone, with a total width of approximately 8 nm. Membranes possessing this fine structure are termed *unit membranes*

In addition to their morphological similarity, the cytoplasmic membranes of all organisms have the same basic chemical structure: a lipid layer in which proteins are inserted (Figure 6.5). The lipids are *amphoteric* (they contain both hydrophobic and hydrophilic regions), and they orient themselves in such a way that the hydrophobic portions of the molecule lie within the membrane, from which water is excluded, while the hydrophilic portions are in contact with the water of the aqueous phase on both sides. The inserted proteins have hydrophobic amino acid residues buried within the membrane, and hydrophilic residues exposed on one or both sides.

The basic molecular architecture of unit membranes is the same in all cellular organisms. However, there is a fundamental dichotomy of chemical composition: the eucaryotes and the eubacteria contain lipids whose hydrophobic portion is a long, generally *unbranched* hydrocarbon chain joined in *ester* linkage to the hydrophilic portion; the archaebacterial lipids always contain *branched* hydrocarbon chains joined to the polar region by *ether* bonds (Figure 3.2).

The significance of these chemical differences is difficult to assess. Many archaebacteria occupy environments characterized by extremes of temperature (to nearly 100° C) or pH (to below 2.0), and it has been suggested that the unusual lipids have survival value under such extreme conditions. The ether linkage found in these lipids is more stable than the ester linkage to thermal breakage, and the branching of the hydrocarbon chains influences the fluidity of the lipid bilayer, regulation of which is essential to cell survival but which may be difficult at very high temperature. (Branched chain lipids decrease membrane fluidity and therefore are particularly suitable to high temperature environments; indeed many of the few eubacteria that possess branched chain lipids are thermophiles.) Nevertheless, eubacteria and archaebacteria are often found growing together in nature, so the special lipids of the archaebacteria are an alternate, not an essential, adaptation in most environments.

Other chemical differences distinguish the membranes of eucaryotes from those of most bac-

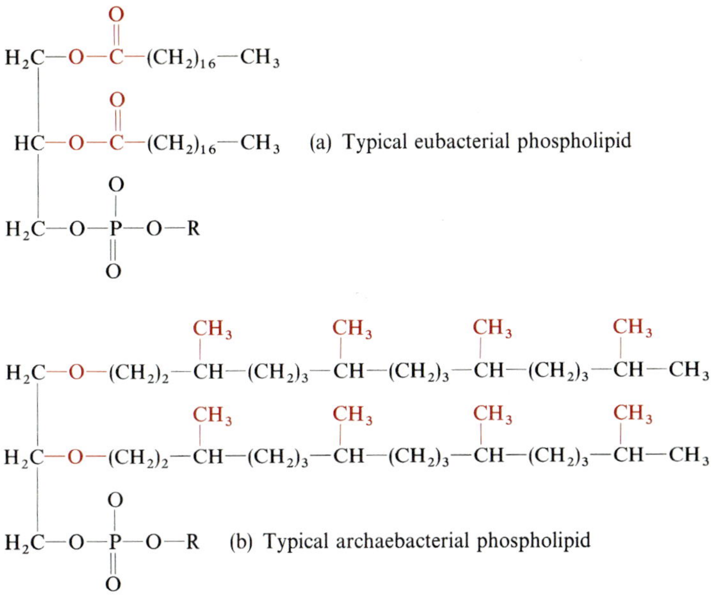

$$
\begin{array}{l}
H_2C{-}O{-}C({=}O){-}(CH_2)_{16}{-}CH_3 \\
\vert \\
HC{-}O{-}C({=}O){-}(CH_2)_{16}{-}CH_3 \\
\vert \\
H_2C{-}O{-}P(O)({=}O){-}O{-}R
\end{array}
$$

(a) Typical eubacterial phospholipid

$$
\begin{array}{l}
H_2C{-}O{-}(CH_2)_2{-}CH(CH_3){-}(CH_2)_3{-}CH(CH_3){-}(CH_2)_3{-}CH(CH_3){-}(CH_2)_3{-}CH(CH_3){-}CH_3 \\
\vert \\
H_2C{-}O{-}(CH_2)_2{-}CH(CH_3){-}(CH_2)_3{-}CH(CH_3){-}(CH_2)_3{-}CH(CH_3){-}(CH_2)_3{-}CH(CH_3){-}CH_3 \\
\vert \\
H_2C{-}O{-}P(O)({=}O){-}O{-}R
\end{array}
$$

(b) Typical archaebacterial phospholipid

FIGURE 3.2

Characteristic lipids of eubacteria and archaebacteria.

teria. The membrane lipids of nearly all eucaryotes include both *sterols* (Figure 3.3) and phospholipids whose constituent fatty acids are *polyunsaturated* (i.e., fatty acids that contain more than one double bond). Most bacteria lack significant amounts of sterols in their membranes; among these, many contain functionally equivalent *hopanoids*, which are rare in eucaryotes, or *squalene*, a biosynthetic precursor of sterols and hopanoids (Figure 3.3). The archaebacteria generally lack both these compounds. Most bacteria, with the exception of the cyanobacteria, lack polyunsaturated fatty acids: their lipids contain only saturated or monounsaturated fatty acids.

FIGURE 3.3

Polycyclic lipids found in microbial cell membranes.

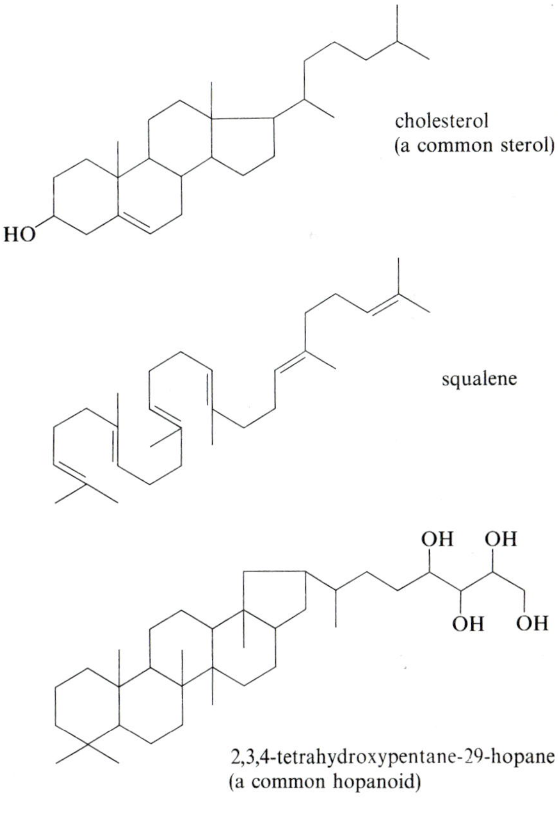

## STRUCTURE OF THE CYTOPLASM

The development of very high power electron microscopes (utilizing accelerating voltages up to a million volts), because of their greater penetrating power, has allowed the examination of very thick specimens, even whole cells. Examination of such thick sections at high magnification has revealed that the cytoplasm of eucaryotic cells, formerly thought to be composed of a relatively homogeneous gel in which the cytoplasmic organelles were embedded, has a marked fibrillar structure (Figure 3.4). It now appears that the bulk of the soluble protein of the cell is bound in this matrix; the space through which it penetrates is an aqueous solution mainly of small molecules. The fibrils interconnect all the organelles of the cell, including ribosomes

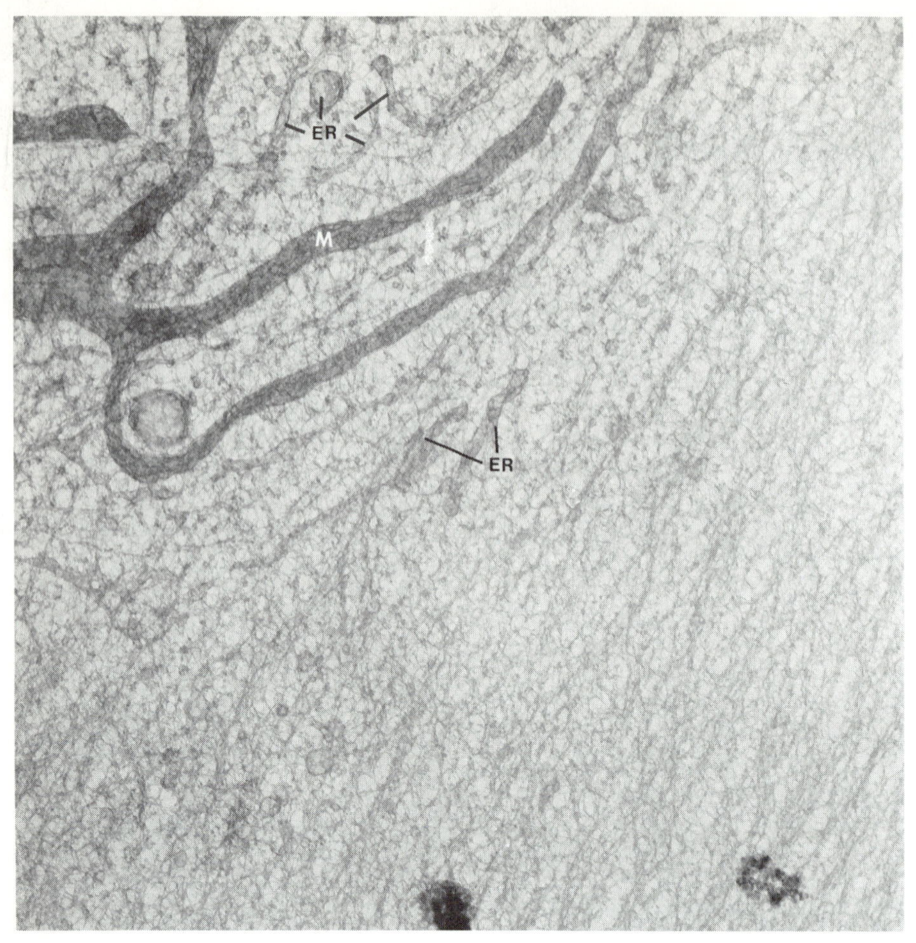


FIGURE 3.4
High voltage electron microscopic view of eucaryotic cytoplasm. The fibrillar nature of the cytoplasmic ground substance is clearly evident. M: Mitochondrion; ER: Endoplasmic reticulum. Reproduced with permission from: K. R. Porter and K. L. Anderson, "The structure of the cytoplasmic matrix preserved by freeze-drying and freeze-substitution." *Eur. J. Cell Biol.* **29,** 83–96 (1982).

previously thought to float freely in the cytoplasm. The number of different structural proteins that comprise the fibrils is unknown.

Equivalently detailed studies of the structure of the cytoplasm of procaryotic cells have not been done; preliminary examination of some eubacteria indicates an organization substantially similar to that of the eucaryotic cell.

## CYTOPLASMIC MEMBRANE SYSTEMS

The eucaryotic cell contains several unique membrane systems within the cytoplasm: the nuclear envelope, the endoplasmic reticulum, the Golgi apparatus, and one or two types of membrane-bounded organelles that house the electron transport machinery of the cell. Although these are composed of unit membranes, they are structurally and topologically distinct from the cytoplasmic membrane. They serve to segregate many of the functions of the eucaryotic cell into specialized and partly isolated regions, among which exchange of material is precisely regulated.

### The Nuclear Envelope

An important difference in the structure of the nucleus exists between eucaryotes and procaryotes. Indeed, this difference is the fundamental basis for

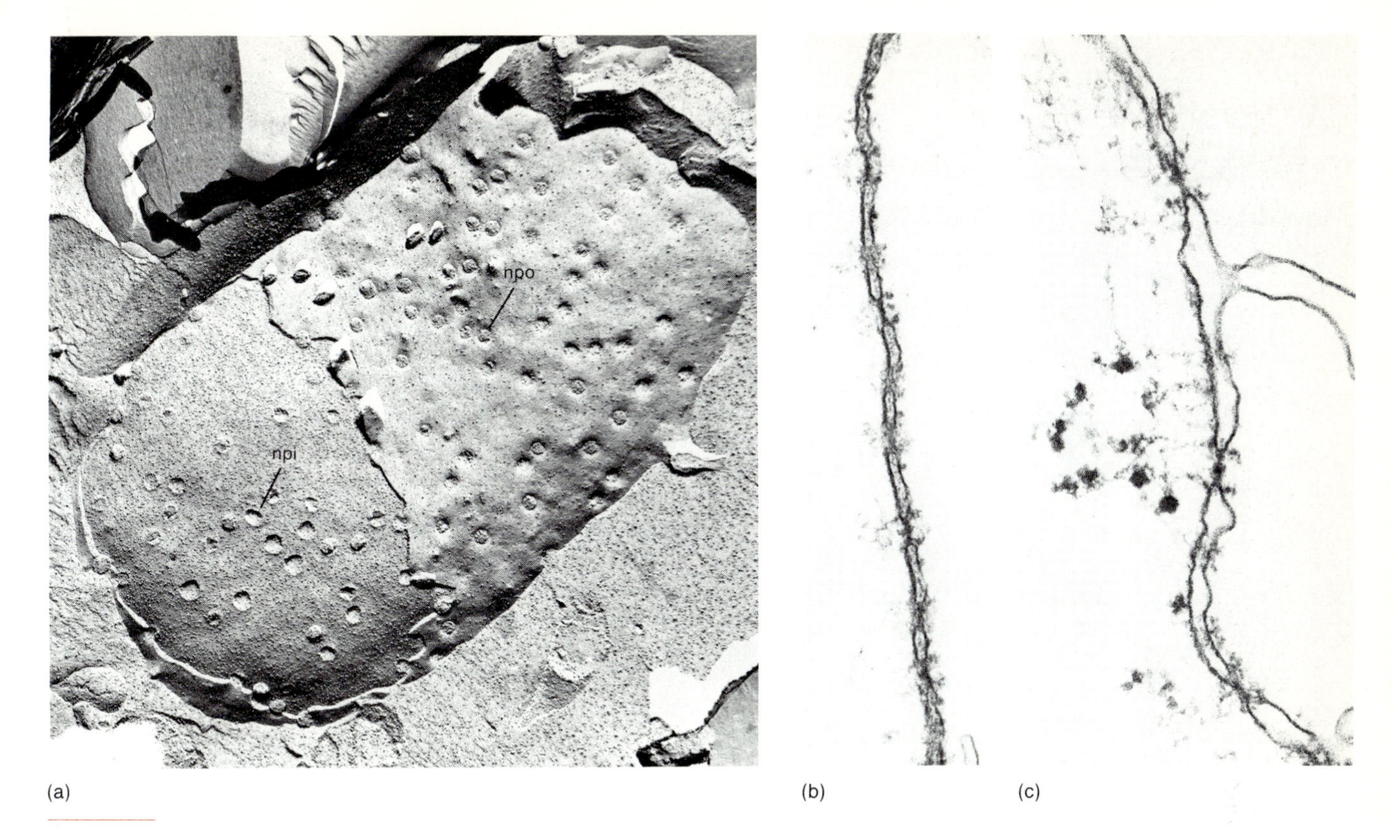


FIGURE 3.5
Electron micrographs of the nuclear envelope. (a) Freeze-fractured and etched preparation of the nucleus of a mouse cell. The cleavage plane has passed through the middle of the outer membrane of the envelope (upper right) and the inner membrane (lower left). Note the nuclear pores viewed from within the outer membrane of the envelope (npo) and from within the inner membrane of the envelope (npi). (b) Thin section of the nuclear envelope of *Xenopus laevis*, showing cross-sections of numerous pores. Pore complexes are visible as dark granules. (c) Thin section of the nuclear envelope of *Pleurodeles waltlii*, showing pore complexes and continuity of the nuclear envelope and the endoplasmic reticulum. (a) Courtesy of Dr. L. G. Chevance, Institute Pasteur; (b) and (c) reproduced with permission of the Rockefeller University Press from W. W. Franke, U. Scheer, G. Krohne, and E.-D. Jarasch, "The Nuclear Envelope and the Architecture of the Nuclear Periphery." *J. Cell. Biol.* **91,** 395–505 (1981).
(a) ( × 24,000); (b) ( × 49,550); (c) ( × 71,400).

distinguishing the groups: those cells with nuclei that are separated from the cytoplasm by a membrane are termed *eucaryotic*; those with nuclear regions not so separated are termed *procaryotic*.

The eucaryotic nucleus is enclosed within a *nuclear envelope* composed of two concentric unit membranes (Figure 3.5). These are fused together at regular intervals to produce a series of *nuclear pores* through which the nuclear and cytoplasmic compartments can exchange materials. The exchange is actively regulated; some small molecules may diffuse readily through the pores, whereas larger molecules, such as proteins and nucleic acids, are completely excluded even though they may be substantially smaller than the unobstructed pore (about 80 nm diameter). Their passage through the pore requires the active participation of a *pore complex* composed of nine subunits: eight around the periphery of the pore and a ninth in the center. These subunits, presumably composed principally of protein, are linked to other pore complexes by fine fibrils. Such interconnections are clearly apparent when the surrounding nuclear membrane is dissolved by detergent (Figure 3.6).

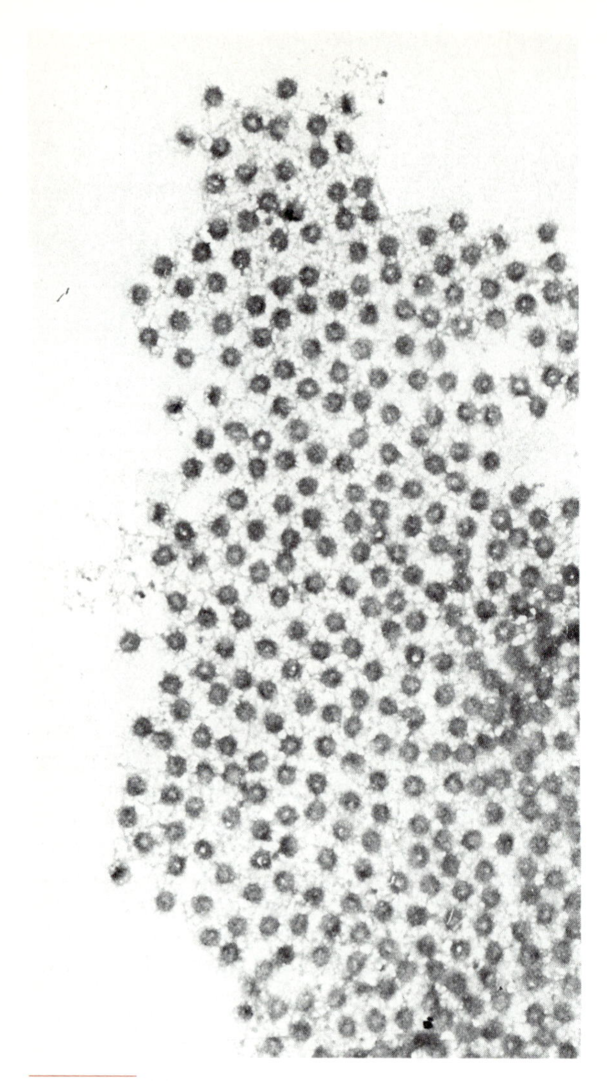

FIGURE 3.6 (above)

Nuclear pore complexes of *Triturus alpestris*, and their interconnecting fibrils. The nuclear membranes have been dissolved with detergent ( × 28,480). Reproduced with permission of the Rockefeller University Press from W. W. Franke, U. Scheer, G. Krohne, and E.-D. Jarasch, "The Nuclear Envelope and the Architecture of the Nuclear Periphery," *J. Cell. Biol.* **91,** 395–505 (1981).

## The Endoplasmic Reticulum and the Golgi Apparatus

The space between the two nuclear membranes is continuous with the lumen of the *endoplasmic reticulum* (ER), an extensive system of membranes that traverses the cytoplasm of the eucaryotic cell (Figure 3.7), enclosing a complex network of channels that ramifies throughout much of the cell. Much of the surface of the ER is coated with ribosomes (the *rough ER*, in contrast to the ribosome-free *smooth ER*). Proteins synthesized on these ribosomes pass into the lumen of the ER, from which they are transferred to the Golgi apparatus (see below). They include intrinsic membrane proteins, the hydrolytic enzymes of lysosomes, and proteins destined for secretion. The membranes of both rough and smooth ER are the sites of phospholipid synthesis. Thus one of the major roles of the ER is membrane synthesis; another is to act as a channel via which material made in one region of the cell can be conducted to another.

Most proteins synthesized on the rough ER are modified posttranscriptionally. They may be partially hydrolyzed to produce proteins with lowered molecular weight; glycosylated (to add covalently bound sugar residues); or sulfated (to add a covalently bound sulfate group). Lipids synthesized in the ER may also be glycosylated or sulfated. The contents of the lumen of the ER are sorted, concentrated, and transported to their appropriate destinations by the *Golgi apparatus*, a stack of flattened membrane vesicles (*cisternae*) normally located near the center of the cell (Figure 3.8).

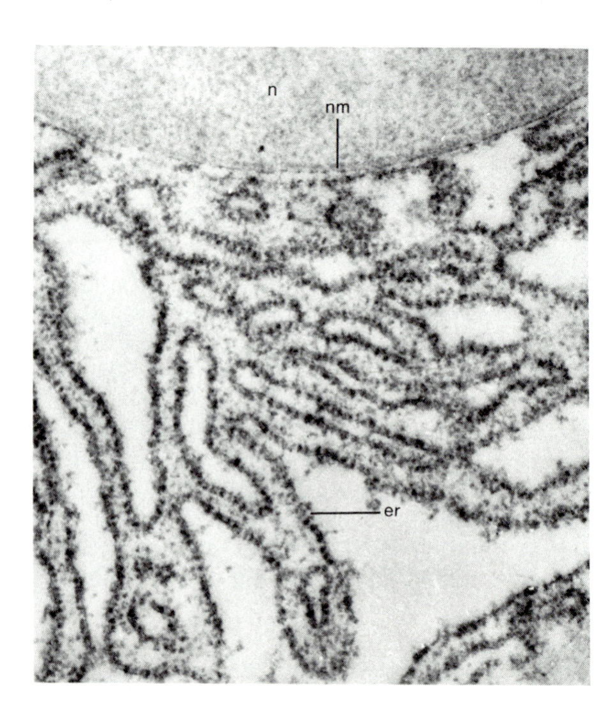


FIGURE 3.7 (right)

Electron micrograph of a thin section of a rabbit plasmocyte, showing a portion of the cytoplasm filled with rough endoplasmic reticulum, er; the field also includes a portion of the nucleus, n, surrounded by the nuclear membrane, nm ( × 35,000). Courtesy of Dr. L. G. Chevance, Institut Pasteur.

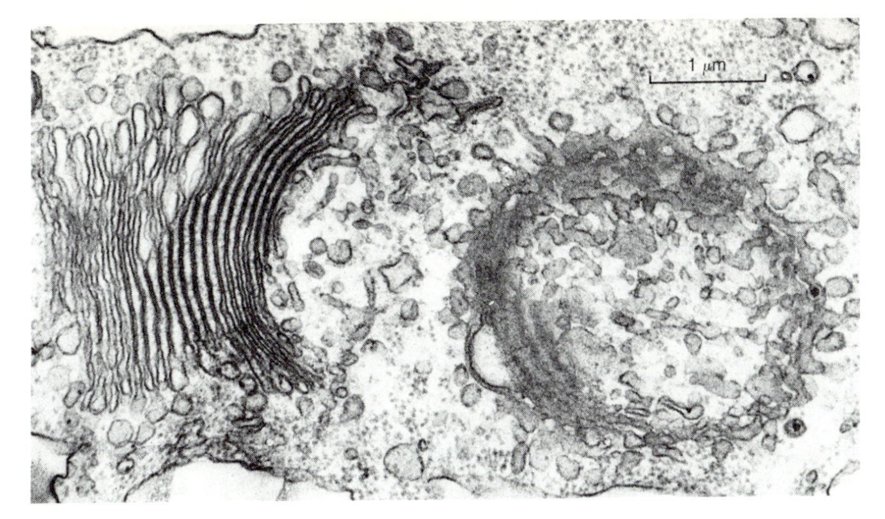


FIGURE 3.8

The Golgi apparatus as seen in an electron micrograph of a thin section of *Euglena gracilis* (×28,000). Two adjacent Golgi bodies have been sectioned in different planes. At left, vertical section through the stack of cisternae. At right, section parallel to the stack. Courtesy of Gordon F. Leedale.

Material is transferred from the ER to the lumen of the Golgi in membrane vesicles that bud from the ER and fuse with one of the Golgi cisternae. Material leaves the Golgi in the same way: the product of the ER, appropriately modified and concentrated by the Golgi, enters vesicles that bud from the cisterna. The surface properties of these vesicles vary according to their contents, thereby allowing them to fuse specifically with an appropriate target membrane, including the cell membrane (vesicles containing secretory proteins or material for insertion into the cell membrane), the phagosome membrane (lysosomes—see below), and the ER (vesicles containing ER membrane material acquired by the Golgi as a consequence of previous ER-to-Golgi transfer).

## Chloroplast and Mitochondrial Membranes

Electron transport systems that conserve energy and make it available for a variety of biological activities only function within a topologically closed membrane (Chapter 4). In eucaryotic organisms, photosynthetic and respiratory electron transport systems are incorporated into membrane-enclosed organelles termed *chloroplasts* and *mitochondria* respectively. Both have the same basic structure: an outer membrane enclosing the topologically more complex inner membrane system in which the components of electron transport are embedded. The inner membrane system of the chloroplast (Figure 3.9) is normally arranged into stacks of vesicles (often termed *thylakoids*); that of the mito-

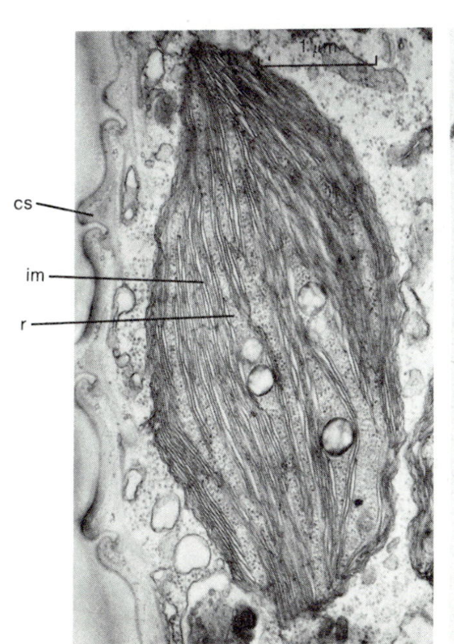


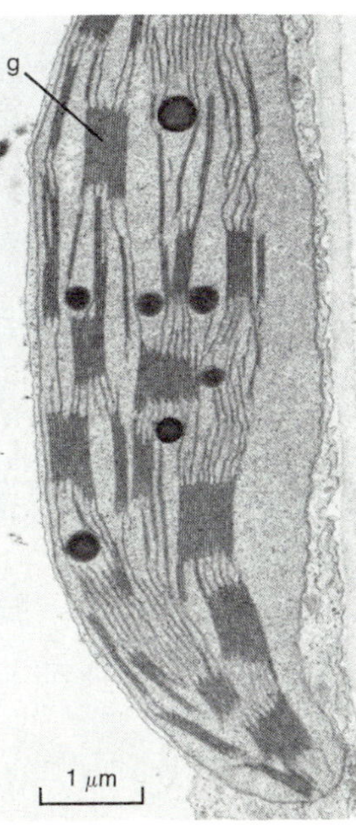


FIGURE 3.9

The structure of chloroplasts as revealed in electron micrographs of thin sections of eucaryotic cells. (a) Chloroplast of the unicellular alga *Euglena* (×21,200). The internal membranes, im, are arranged in irregular parallel groups and run in the long axis of the chloroplast. Ribosomes, r, are scattered between the lamellae. The chloroplast lies just below the sculptured cell surface, cs. From G. F. Leedale, B. J. D. Meeuse, and E. G. Pringsheim, "Structure and Physiology of *Euglena spirogyra*," *Arch. Mikrobiol.* **50,** 68 (1965). (b) Chloroplast of a sugar beet leaf (×14,840). The internal membranes tend to be arranged in dense, regular stacks, termed *grana* (g), in the chloroplasts of plants. Courtesy of W. M. Laetsch.

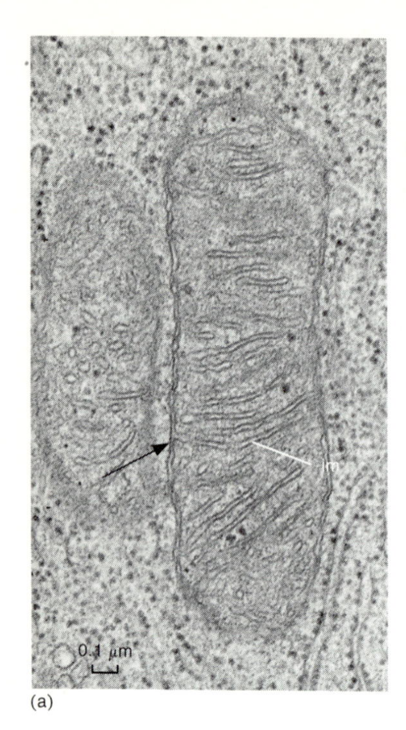

(a)

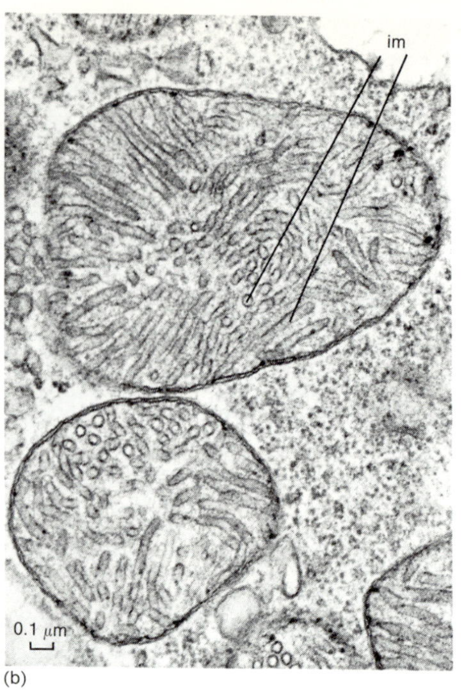

(b)

FIGURE 3.10

The structure of mitochondria as seen in electron micrographs of thin sections of eucaryotic cells: (a) Mitochondria in a mammary gland cell of the mouse (× 56,100). Numerous flattened internal membranes (im) arise by invagination from the inner enclosing membrane of the organelle (arrow). (b) Mitochondria of a ciliate, *Condylostoma* (× 56,100). The internal membranes (im) are tubular in cross section and are very abundant. Courtesy of Dorothy Pitelka.

chondrion (Figure 3.10) is concentric with the outer membrane but often extensively invaginated.

## Cytoplasmic Membrane Systems in Bacteria

One of the most striking structural features of the procaryotic cell is the *absence of internal compartmentalization by unit membrane systems* (Figure 3.11). The *cytoplasmic membrane is, in the great majority of procaryotes, the only unit membrane system of the cell.* However, the topology is often complex, with membranous infoldings penetrating deeply into the cytoplasm. The cyanobacteria provide the sole known exception to the rule that there is only one unit membrane system in the procaryotic cell. In these organisms the photosynthetic apparatus is located on a series of *thylakoids*, similar in structure and function to the thylakoids of a chloroplast. However, in cyanobacteria the thylakoids are not segregated within an organelle, but are dispersed throughout the cytoplasm (Figure 3.12).

A few other types of distinctive organelles do occur in some groups of bacteria, although none of these structures is of general distribution. All of these organelles are characterized by the *absence of unit membranes*; they are enclosed by a single-layered membrane of only some 2 to 3 nm in thickness. They include *chlorosomes*, which house the photosynthetic apparatus of the green bacteria; *gas vesicles*, which confer buoyancy on the cells of

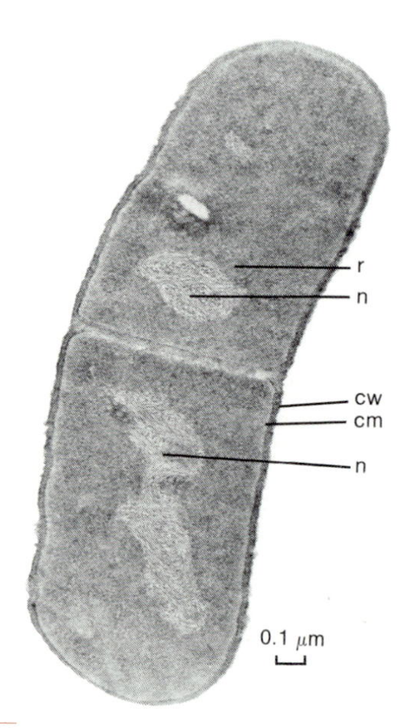


FIGURE 3.11

Electron micrograph of a thin section of a nonphotosynthetic unicellular microorganism, the bacterium *Bacillus subtilis*, which has a typical procaryotic cell structure. The dividing cell is surrounded by a relatively dense wall (cw), enclosing the cell membrane (cm). Within the cell, the nucleoplasm (n) is distinguishable by its fibrillar structure from the cytoplasm, densely filled with 70S ribosomes (r). Note the absence of internal unit membrane systems. Courtesy of C. F. Robinow.

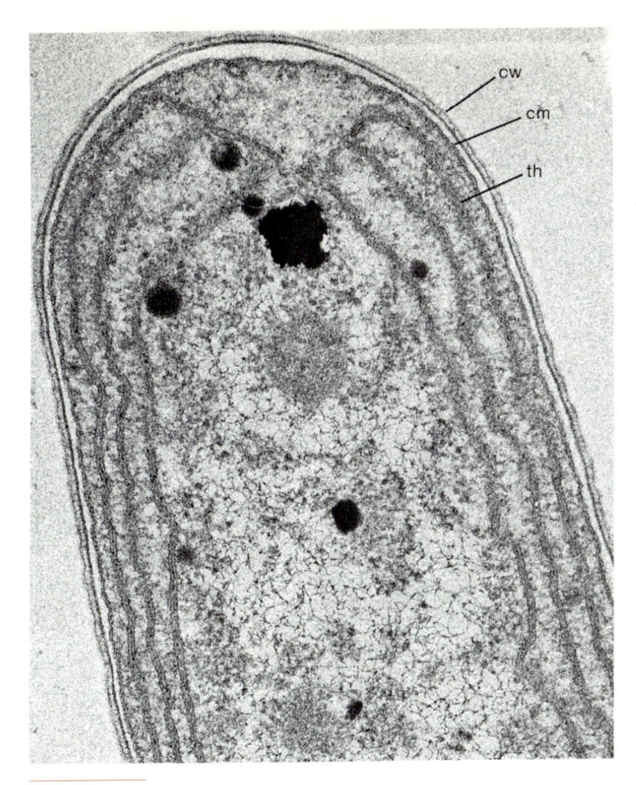


FIGURE 3.12

Electron micrograph of a thin section of a unicellular cyanobacterium. The cell is enclosed by a cell wall (cw) and cell membrane (cm). The thylacoids (th) which bear the photosynthetic apparatus are located in the cytoplasm, and are thicker than the cell membrane, since they are composed of two closely appressed membranes ( × 53,000). Courtesy of Dr. G. Cohen-Bazire.

a diversity of aquatic bacteria; and *carboxysomes*, which contain a key enzyme of reductive $CO_2$ assimilation in many autotrophic bacteria. The properties of these unique organelles will be discussed further in Chapter 6.

## CYTOSKELETAL ELEMENTS

The eucaryotic cell contains several classes of fibrous structures collectively termed *cytoskeletal* elements or structures. These are responsible for a variety of important activities. There are three major classes of such structures: *microtubules*, hollow fibers about 25 nm in diameter; *microfilaments*, proteinaceous fibers with a diameter of 6 nm; and *intermediate filaments*, a heterogeneous class of fibrous elements intermediate in diameter between microfilaments and microtubules.

### Microtubules

An element of structure that has many functions in the eucaryotic cell is the *microtubule*, an extremely thin cylinder, some 20 to 30 nm in diameter and of indefinite length. The walls of microtubules are composed of globular protein subunits (*tubulin*) with a relative molecular weight of 50,000 to 60,000.

Microtubules provide the structural framework of the mitotic spindle (see below); they also appear to play a role in the establishment and maintenance of the shape of many types of eucaryotic cells. In addition, a regular longitudinal array of microtubules occurs within the eucaryotic locomotor organelles known as *cilia* or *flagella*. The cilium or flagellum is enclosed within an extension of the cell membrane and contains a set of nine outer pairs of microtubules surrounding an inner central pair (Figure 3.13). The hydrolysis of ATP (see Chapter 4) by an enzyme associated with the microtubules causes them to move longitudinally relative to each other, hence moving the entire organelle.

The central two microtubules of cilia and flagella arise from a plate just within the cell, whereas the outer pairs originate from a cylindrical body or *centriole* (Figure 3.14), likewise composed of microtubules. In some eucaryotes (many animals) the centrioles are also associated with the formation of the microtubular system of the mitotic apparatus, being located at the two poles of the spindle.

### Microfilaments

One of the most abundant proteins in eucaryotic cells is *actin*. Much of it is soluble or, more likely, bound into the network of fibrils that forms the structure of the cytoplasm. The rest is polymerized into fibrils 6 nm in diameter (Figure 3.15) that bind another protein, *myosin*; when complexed with myosin, the filaments become *contractile*. Contraction requires energy, which is supplied by myosin-catalyzed hydrolysis of ATP.

Microfilaments are concentrated in the peripheral cytoplasm, where they appear to be anchored in the cytoplasmic membrane. These microfilaments probably mediate cytoplasmic streaming, ameboid movement (see below), and cell division, or *cytokinesis*. Other microfibrils are responsible for determining the viscosity of the cytoplasm: cross-linking of actin filaments into a network by special cross-linking proteins produces a high-viscosity gel; breakage of the cross-links and depolymerization of the microfilaments lowers the viscosity of the cytoplasm. The activity of

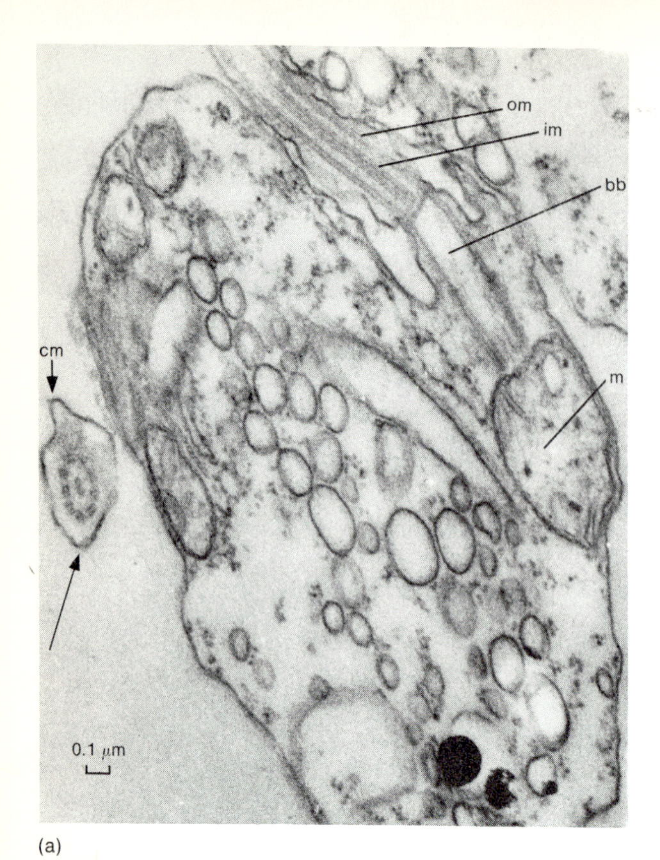


(a)

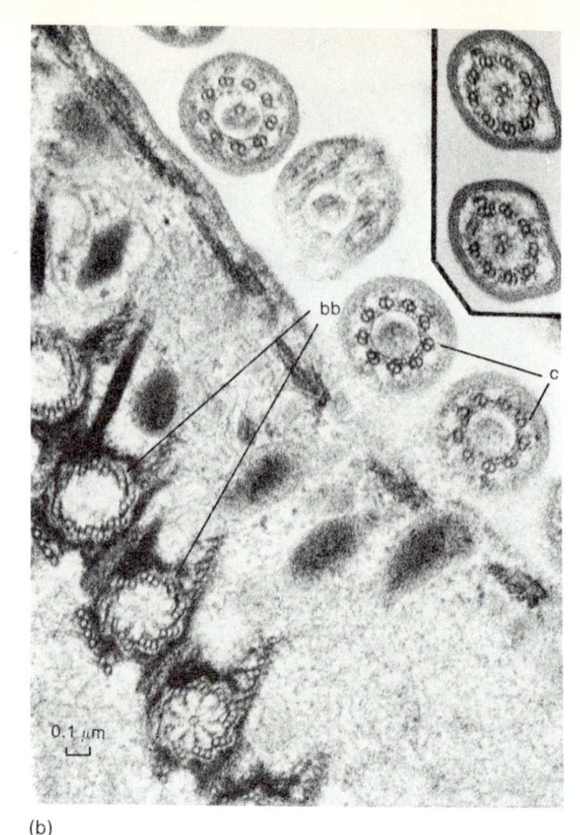


(b)

FIGURE 3.13

The fine structure of eucaryotic flagella and cilia, as revealed by electron micrographs of thin sections. (a) Longitudinal section through the cell of *Bodo*, a nonphotosynthetic flagellate (× 38,800): cylindrical basal body (bb); outer microtubules (om); inner microtubules (im). Underlying the basal body is a specialized mitochondrion (m). At left (arrow), transverse section of a flagellum external to the cell. Note enclosure by an extension of the cell membrane (cm). (b) Section through the body surface of a ciliate, *Didinium* (× 51,800). Within the cell (lower left) basal bodies (bb) have been sectioned transversely; their walls are composed of nine triple rows of microtubules. Just above the cell surface, several cilia (c) have been sectioned transversely; note the nine outer pairs of microtubules and the absence of the inner pair of microtubules. (c) Insert at upper right: section through two cilia at a point some distance from the cell surface. Note the inner pair of microtubules, the nine outer pairs, and the enclosing membrane. Courtesy of Dorothy Pitelka.

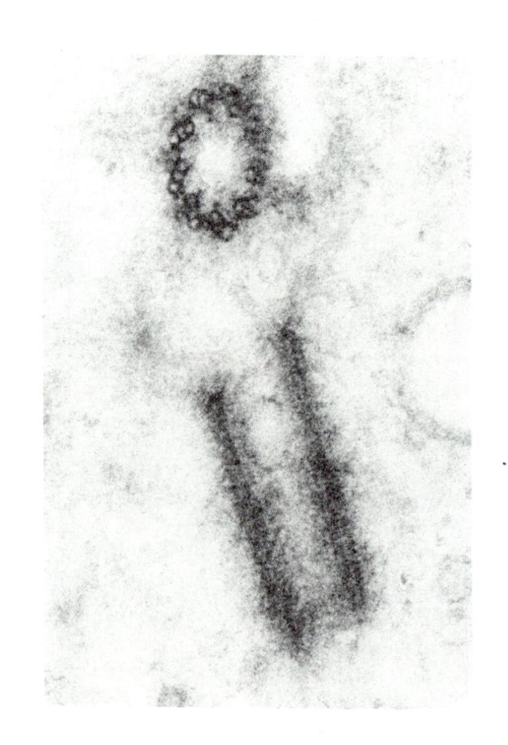

FIGURE 3.14 (right)

Electron micrograph of a pair of centrioles in a dividing human lymphosarcoma cell (× 104,000). One is sectioned transversely, the other longitudinally, revealing the typical hollow cylindrical structure of the organelle. Courtesy of G. Bernhard, Institut de Recherches sur le Cancer, Villejuif, France.

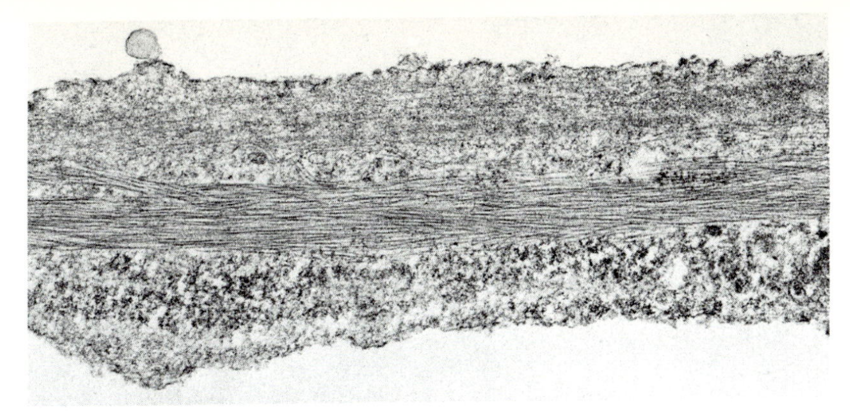

FIGURE 3.15

Electron micrograph of a thin section of preadipocytes from the mouse. The fine fibrils at the top are microfilaments; the thicker ones in the center are intermediate filaments (×25,500). From A. B. Novikoff, P. M. Novikoff, O. M. Rosen, and C. S. Rubin, "Organelle relationships in cultured 3T3-L1 preadipocytes," *J. Cell. Biol.* 87, 180–196. (1980). Reproduced with permission of the Rockefeller University Press.

both the contractile peripheral microfilaments and the viscosity-regulating microfilaments seems to be regulated by the concentration of calcium ions; gelation of the cytoplasm is inhibited by calcium, while contractility is stimulated. The basis for the calcium stimulation of contractility is beginning to emerge: ATP is hydrolyzed by myosin only if it is phosphorylated; phosphorylation of myosin is accomplished by a *protein kinase* that is activated by calcium.

### Intermediate Filaments

Both microtubules and microfilaments have substantially the same shape and size in all eucaryotic cells, and their constituent monomeric proteins are highly conserved in an evolutionary sense; that is, the amino acid sequences of these proteins (tubulin and actin, respectively) vary little over large taxonomic distances. In contrast, *intermediate filaments* (Figure 3.15) constitute a heterogeneous class of proteinaceous filaments, the only common characteristic of which is a size intermediate between that of microtubules and microfilaments: approximately 10 nm, although there is considerable variation. To date, intermediate filaments have been detected only in animal cells. Although their function is obscure, there is some evidence that one class serves to anchor cytoplasmic organelles to each other or to the cell membrane, thus playing a cytoskeletal role.

### Cytoskeletal Elements in Bacteria

There are no structures in bacteria homologous with any of the three classes of cytoskeletal elements of eucaryotes. In a number of instances short tubular structures with the approximate dimensions of microtubules have been seen in bacteria, but these are most probably fragments of bacterial viruses, many of which have a tubular structure. Besides lacking demonstrable cytoskeletal elements, the bacteria lack proteins sufficiently similar to tubulin or actin to cross-react immunologically (Chapter 30).

It has been suggested that the endoflagellum of the spirochetes is homologous with eucaryotic microtubules. Indeed, all bacterial flagella closely resemble microtubules (they are hollow tubes about 15 nm in diameter composed of globular protein subunits), but convincing evidence of a phylogenetic relationship is lacking.

## ENDOCYTOSIS AND EXOCYTOSIS

Although small molecules in solution can enter the eucaryotic cell by passage through the cytoplasmic membrane, the entry of other materials can occur by a second quite distinct mechanism: bulk transport of small droplets, enclosed by an infolding of the cytoplasmic membrane to form a membrane-enclosed vacuole (*phagosome*). The most familiar example of this phenomenon is the *phagocytosis* of bacteria or other small solid objects by phagotrophic protozoa, or by the phagocytic cells of metazoan animals. Droplets of liquid can enter the eucaryotic cell in a similar fashion, this process being termed *pinocytosis*. Phagocytosis and pinocytosis are known collectively as *endocytosis*. Endocytosis is a distinctively eucaryotic process of fundamental importance, which initiates both *intracellular digestion* (hydrolysis of biological macromolecules) and the *establishment of endosymbiosis* (see below).

One of the products formed in the Golgi apparatus is a membrane-bounded vesicle known as the *lysosome*. Lysosomes contain an extensive array of hydrolytic enzymes capable of breaking down most classes of biological macromolecules (e.g.,

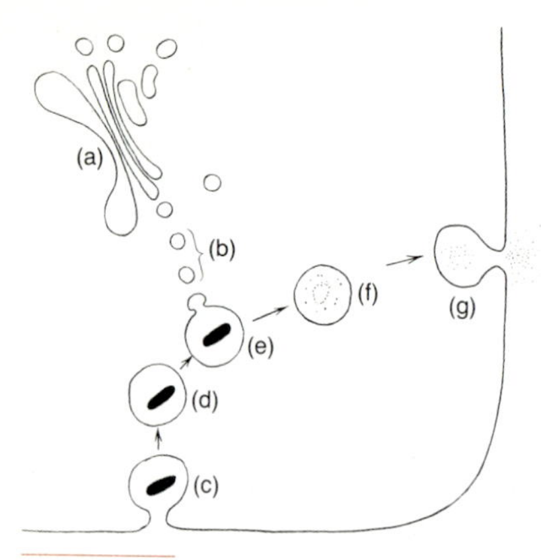


FIGURE 3.16
A diagrammatic representation of the events of intracellular digestion: (a) Golgi apparatus; (b) lysosomes produced from the Golgi apparatus; (c) phagocytic capture of a food particle (a bacterium) at the surface of the cell, during which the particle is almost completely surrounded by the cell membrane; (d) newly formed food vacuole; (e) coalescence of the food vacuole with a lysosome; (f) digestion of the vacuolar contents by hydrolytic enzymes released from the lysosome; (g) exocytosis of undigested waste. Modified from N. Novikoff, E. Essner, and N. Quintana, "Golgi Apparatus and Lysosomes." *Federation Proc.* **23,** 1010 (1964).

they contain a variety of nucleases, proteinases, lipases, etc.). These enzymes do not normally act upon the constituents of the cell in which they are formed, because they are segregated within the lysosomal membrane. However, the lysosomes can fuse with vacuoles formed through endocytosis, thus permitting hydrolysis of the materials (or cells) contained in these vacuoles; the soluble hydrolytic products are then taken into the surrounding cytoplasm (Figure 3.16). This process of intracellular digestion is the major feeding mechanism in many protozoa and primitive invertebrate animals. Throughout the animal world it plays a protective role by enabling the body to destroy potentially dangerous microorganisms that have entered its fluids or tissues. Indeed, this protective function has become the primary role of phagocytosis in vertebrates, where digestion takes place in the alimentary tract, outside of the body tissues.

The phenomenon of *intracellular symbiosis*, or *endosymbiosis*,* is widespread in all major groups of eucaryotes: plants, animals, and protists (Chapter 28). In order for the cells of future endosymbionts to gain access to the cytoplasm of the host cell, they must pass through the cytoplasmic membrane; this passage commonly occurs through phagocytic engulfment. In this situation, engulfment is not followed by lysosomal fusion.

Droplets or solid particles synthesized within the eucaryotic cell can also pass to the exterior by the converse mechanism, known as *exocytosis*. Here also the Golgi apparatus plays a key role, since materials destined for exocytosis are processed and packaged by the Golgi body. The secretion of enzymes and of hormones by specialized animal cells occurs in this manner, and in algae it has been shown that formation of the cell wall involves the exocytosis of small fragments of the wall fabric, synthesized endogenously and transported to the cell surface in Golgi vesicles.

Both endo- and exocytosis would change the surface area of the cytoplasmic membrane if no mechanism existed to ensure its remaining relatively constant. However, even cells that are actively phagocytosing or secreting maintain their surface area within quite close limits. A balancing mechanism exists in which the Golgi plays a major role: excess membrane added by exocytosis is removed by endocytosis and returned to the Golgi for recycling; similarly, excess membrane removed by endocytosis is recovered by the Golgi for reinsertion into the cell membrane, or else directly reinserted by exocytosis, as is often the case with phagosomes.

In bacteria, exocytosis is completely absent, and endocytosis extremely rare. The function of endocytosis, in the rare instances in which it occurs, is never to transport food material nor to establish an endosymbiosis. One form of endocytosis occurs in the early steps in the formation of the resting cell, termed an *endospore*, by some groups of eubacteria: following an asymmetric binary fission, unaccompanied by the deposition of cell wall material in the septum, the larger of the two daugther cells engulfs the smaller. Another form of endocytosis occurs in the uptake of DNA in the first step of genetic transformation in some bacteria (Chapter 11).

Even among those groups of bacteria that lack a rigid cell wall (*Thermoplasma* among the archaebacteria and the mollicutes among the eubacteria), phagocytosis is unknown. The biological properties that in eucaryotes depend on phagocytosis for their initiation (the abilities to perform intracellular digestion of particulate material and to establish endosymbiosis) are thus absent from the bacteria.

* An endosymbiosis is an association between two organisms in which *one organism lives entirely within the cytoplasm of the other.*

## OSMOREGULATION IN MICROORGANISMS

Most free-living organisms live in an environment with a water concentration considerably greater than that inside the cell. Since the cytoplasmic membrane is freely permeable to water, but not to many solutes, there is a tendency for water to enter the cell; unless this tendency is counterbalanced in some manner, the cell swells and eventually undergoes *osmotic lysis*. In many protists (algae, fungi) and most bacteria the danger of osmotic lysis is prevented mechanically by enclosure of the cell in a rigid wall of sufficient tensile strength to counterbalance water pressure and hence prevent lysis. The chemical nature of the cell wall varies among organisms. All but one small group of walled eubacteria possess a characteristic polymer *murein*, a form of *peptidoglycan* (a polymer characterized by short polypeptides attached to specific residues of a polysaccharide chain). The walls of archaebacteria and protists are more variable, several major types being found among each group; murein, however, is never found outside the eubacteria.

Many protozoa do not possess walls; in these protists a special type of vacuole, the *contractile vacuole*, functions as a cellular pump to collect water from within the cell and periodically discharge it to the exterior through coalescence with the cell membrane. The operation of the contractile vacuole is accordingly another mode of exocytosis that provides an active mechanism of osmoregulation.

Most bacteria that lack a rigid cell wall are osmotically sensitive and hence confined to environments of high osmolality; the few exceptions to this generalization (*Thermoplasma*, a few of the mollicutes) must have mechanisms (still poorly understood) to prevent osmotic lysis.

## STRUCTURE OF THE CHROMOSOME

In all cellular organisms the genetic information is stored as a linear series of bases in *deoxyribonucleic acid*, or *DNA*. The DNA of cells is a double-stranded helix in which the two strands wind around each other, making one complete turn about every 10 base pairs. The two strands are held together by hydrogen bonding between the bases adenine and thymine, and between guanine and cytosine. Each strand thus contains the information necessary to specify its complementary strand, a feature of central importance in both the replication and expression of DNA. This basic genomic structure is common to all cells; however, there are differences among the major groups with respect to the organization of their chromosomes.

### The Eucaryotic Chromosome

The genome of eucaryotic cells is always distributed over several chromosomes. Some of these chromosomes are circular and similar to those found in eubacteria; they are housed in the mitochondria and chloroplasts (see below). However, the bulk of the DNA is contained within the nucleus, and is always distributed among several chromosomes, each of which contains a single linear molecule of double-stranded DNA.

Within the nondividing (interphase) nucleus, the eucaryotic chromosomes are dispersed as long, threadlike strands with a distinct substructure: electron microscopy reveals a string of *nucleosomes* (Figure 3.17). Each nucleosome is composed of nine molecules of protein (termed *histones*) and

FIGURE 3.17

Nucleosomes (×61,650). From S. L. McKnight and O. L. Miller, "Ultrastructural Patterns of RNA Synthesis During Early Embryogenesis of *Drosophila melanogaster*." *Cell* **8,** 305–319 (1976). Copyright Massachusetts Institute of Technology.

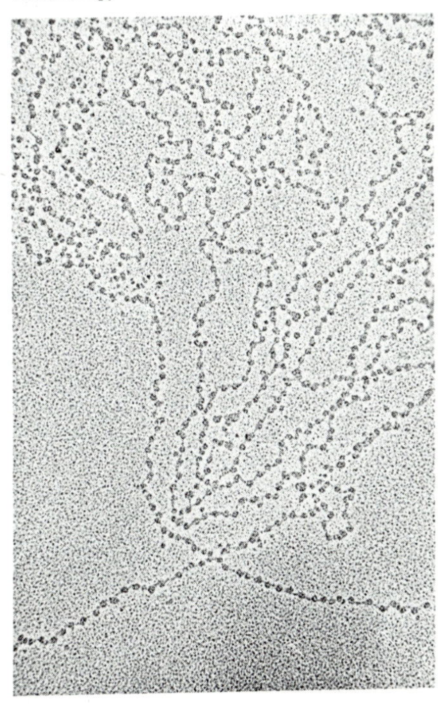

about 165 base pairs of DNA. Histones are DNA-binding proteins of relatively low molecular weight, with a high content of basic amino acids. Most eucaryotic cells have five different types of histone, of which four comprise the nucleosome core (two copies of each). The fifth histone (termed *H1*) is present in one copy per nucleosome, and apparently binds to the DNA where it enters and exits the nucleosome (Figure 3.18).

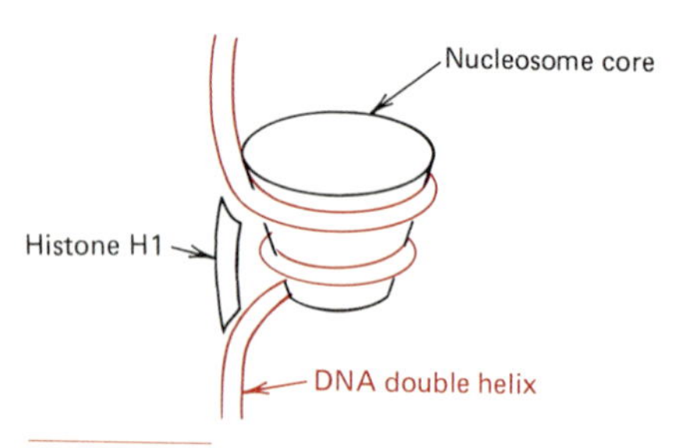


FIGURE 3.18
Schematic diagram of a nucleosome.

The formation and spacing of nucleosomes does not depend on specific signal sequences in the DNA; rather it is a consequence of the properties of the histone molecules themselves, and will occur spontaneously in a solution of DNA and histones.

During DNA replication, passage of the replication fork does not disrupt the nucleosome structure; nucleosomes are conserved and distributed *en masse* to one of the daughter strands of DNA. The other daughter strand associates with free histones to form new nucleosomes. Most probably nucleosomes are also conserved during cell division. Prior to mitosis or meiosis, the chromosomes shorten and thicken. The nature of this condensation is unknown; however, the nucleosome arrangement of DNA and histones is probably preserved within the coiled chromosomes.

In addition to being coiled into nucleosomes every few hundred base pairs, the eucaryotic chromosome is divided into loops or *domains* of about 50,000 to 100,000 base pairs. The DNA in each loop is *supercoiled*; i.e., the double helical DNA is itself coiled into a loose helix. The domains are divided from each other by the attachment of the DNA to a scaffolding of structural protein (Figure 3.19), the detailed structure of which is still obscure.

### The Eubacterial Chromosome

Unlike the eucaryotic genome, the eubacteria contain a single chromosome that is circular and covalently closed except during replication (the replication of a covalently closed circular molecule of nucleic acid imposes the topological requirement for at least one single-strand break). It is attached to the cell membrane during replication and segregation. In addition to the chromosome, a variable number of *plasmids* may be present. Like the chromosome, plasmids are covalently closed circular molecules of DNA, but they are not necessary for growth under all conditions and they are small compared to the chromosome.

The eubacteria contain a single type of histonelike protein (termed HU in *Escherischia coli*, the best-studied eubacterium). Although these proteins share with eucaryotic histones the properties of low molecular weight and a high content of basic amino acids, their ability to bind DNA at physiological ionic strengths is disputed. Nevertheless, the presence of such proteins, and the recent electron microscopic visualization of a nucleosome-like structure of the *Escherichia coli* chromosome (Figure 3.20), suggests that the eubacterial chromosome is condensed in a manner similar to that of eucaryotes. It also seems to be organized into supercoiled domains attached to a structural scaffold, as are eucaryotic chromosomes.

### The Archaebacterial Chromosome

The condensation of DNA by wrapping around a proteinaceous core appears also to characterize the archaebacterial chromosome. The archaebacterium *Thermoplasma* contains a single species of small, basic histonelike protein (termed HTa), that at physiological ionic strengths forms nucleosome-like complexes containing a core of four molecules of HTa around which a 40-base-pair length of DNA is wrapped. Other archaebacteria contain DNA-binding proteins presumably related to HTa; they have not been studied in equivalent detail.

## SEGREGATION OF THE CHROMOSOMES

All cells arise by division from preexisting ones. Thus mechanisms are needed to ensure that each daughter cell receives an appropriate number of each essential cell constituent. Because the genome is usually present in only a few copies per cell, its partition to daughter cells cannot be left to chance. Hence cells have evolved mechanisms for the orderly segregation of chromosomes.

FIGURE 3.19

The scaffold of eucaryotic metaphase chromosome. The histones have been removed, showing the DNA strands attached to an electron-dense core or scaffold. The chromatids remain attached to each other at the centromere. From U. K. Laemmli, S. M. Cheng, K. W. Adolph, J. R. Paulson, J. A. Brown, and W. R. Baumbach, "Metaphase Chromosome Structure: The Role of Non-Histone Proteins," *Symp. Soc. Quant. Biol.*, **XLII,** 351–360 (1977).

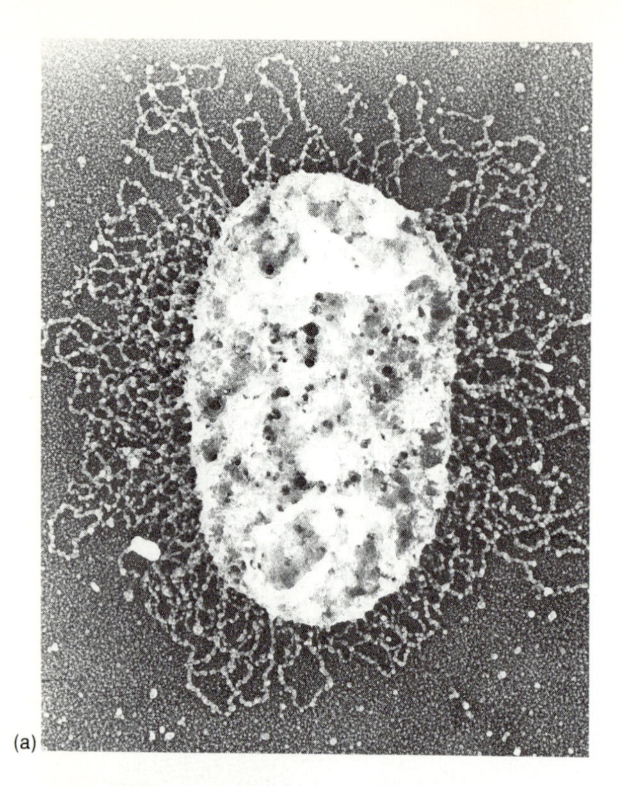
(a)

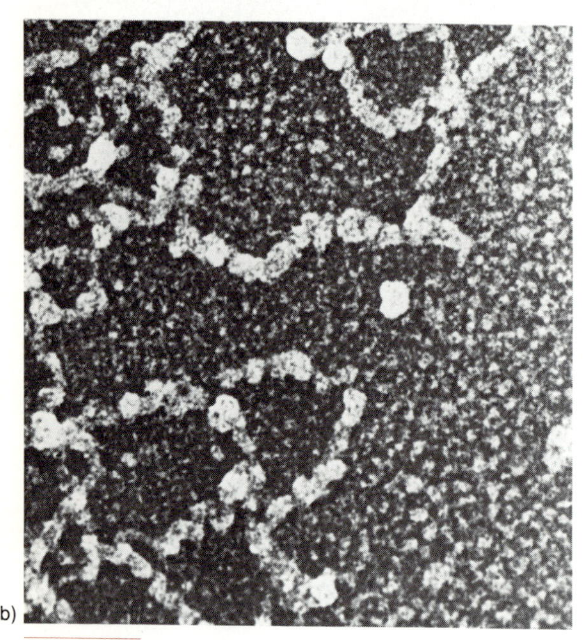
(b)

FIGURE 3.20

The eubacterial chromosome (a) Electron micrograph of osmotically lysed *E. coli*, showing the nucleosome-like organization of the chromosome released from the cell; (b) an enlarged view of the extruded chromosome. (a) (× 47,300); (b) (× 273,480). From J. D. Griffith, "Visualization of prokaryotic DNA in a regularly condensed chromatin-like fiber," *Proc. Natl. Acad. Sci.* (USA) **73,** 563–567 (1976).

## Chromosome Segregation in Eucaryotes

Prior to cell division, DNA replication in the interphase nucleus duplicates each chromosome, the copies of which (*chromatids*) remain attached at a point known as the *centromere*. Segregation of the copies is then achieved by a mechanism termed *mitosis*.

As mitosis begins, the chromosomes condense into compact units, daughter chromatids still remaining attached at the centromere. The complex topology of these condensed chromosomes is still a mystery, but the function of the condensation must be to prevent the tangling that would certainly occur during separation of the extended chromosomes. When condensation is complete, the chromosomes are individually visible in the light microscope, each with a characteristic length and centromere location. The number and appearance of the condensed chromosomes is a constant features of each eucaryotic species, and is termed its *karyotype*.

Simultaneously with condensation of the chromosomes, a *mitotic spindle* forms, which is a bipolar, spindle-shaped array of microtubules that spans the nucleus. Spindle formation is accompanied by dissolution of the nuclear membrane in most plants and animals; in many protists, however, the nuclear membrane remains intact and the spindle either forms entirely within the nucleus or the spindle penetrates through *polar fenestrae*, localized breaks in the nuclear membrane (Figure 3.21).

When chromosome condensation and spindle formation are complete, the chromosomes move to the equatorial region of the spindle where their centromeres attach to the spindle fibers; the two halves of the centromere apparently attach to microtubules originating from different poles of the spindle. Thus when the centromeres split and the chromosomes are pulled apart, one copy of each chromosome goes to each pole. The pulling mechanism is unknown; chromosomes may slide along the microtubules, being pulled by contractile microfilaments, or they may be pulled by progressive disassembly of the microtubules at the spindle pole, thus drawing the attachment point towards the pole (Figure 3.22).

In some protists, e.g., the dinoflagellate algae, chromosome separation appears to be accomplished by a different mechanism. Their spindle forms entirely outside the nucleus; individual spindle microtubules either penetrate through invaginations of the nuclear membrane, or the entire nucleus folds around the spindle (Figure 3.23). In either case the chromosomes attach to the nuclear membrane, which may, in turn, attach to the spin-

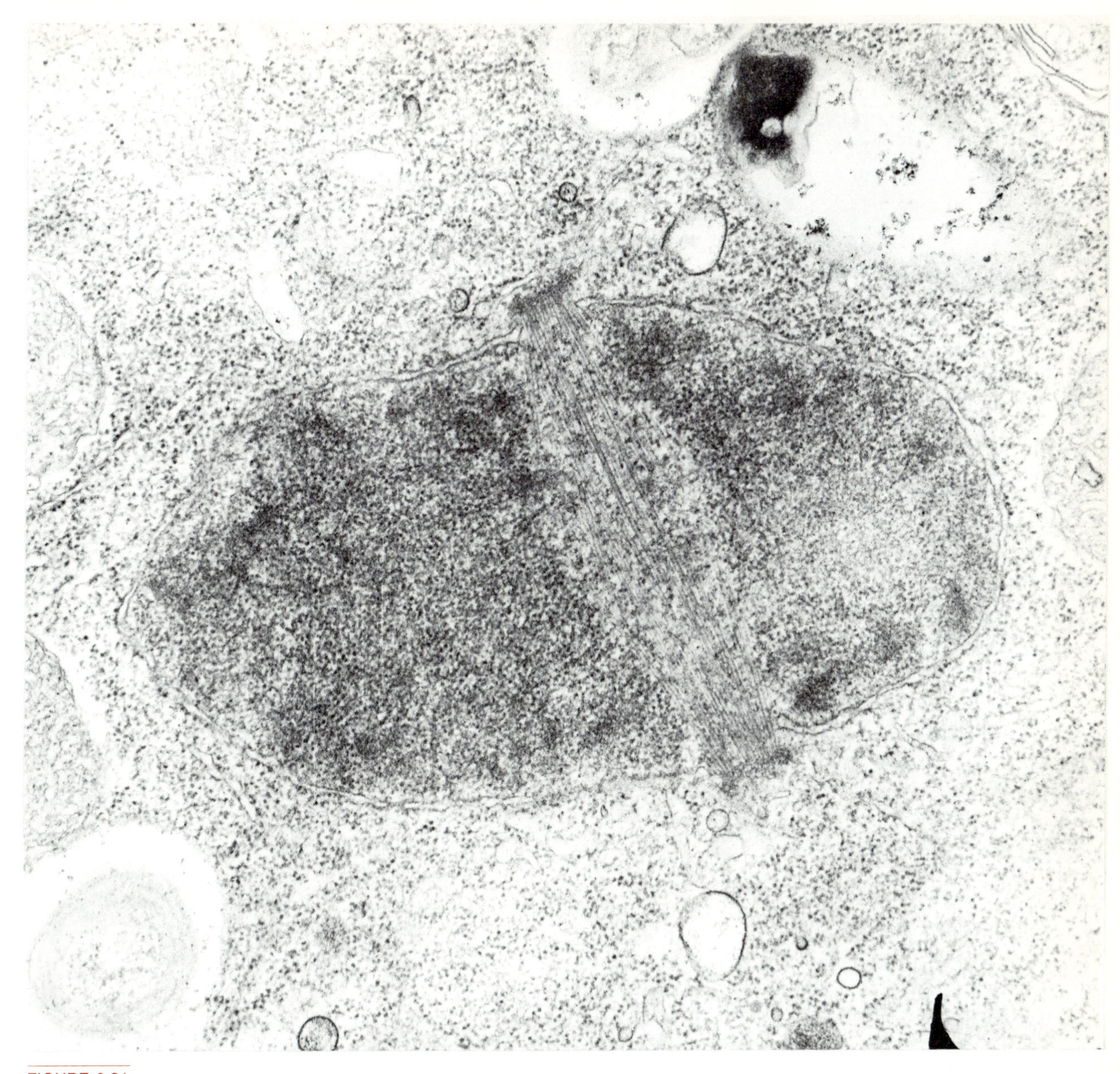

FIGURE 3.21

Electron micrograph of a thin section showing the mitotic spindle of the slime mold *Dictyostelium discoideum* pentrating polar fenestrae in the otherwise intact nuclear membrane, ×40,000. Courtesy of Dr. P. B. Moens.

dle. Chromosome separation is accomplished either by elongation of the spindle, pulling the attached nuclear membrane sites apart, or by localized membrane growth between the sites of attachment of daughter chromatids. In the latter case, the spindle microtubules would act as an armature on which the membrane-attached chromosomes slide.

When chromosome separation is complete, the spindle dissolves and nuclear membranes form around the two sets of chromosomes. Cell division normally occurs at this time, the plane of division coinciding with the equatorial plane of the mitotic spindle.

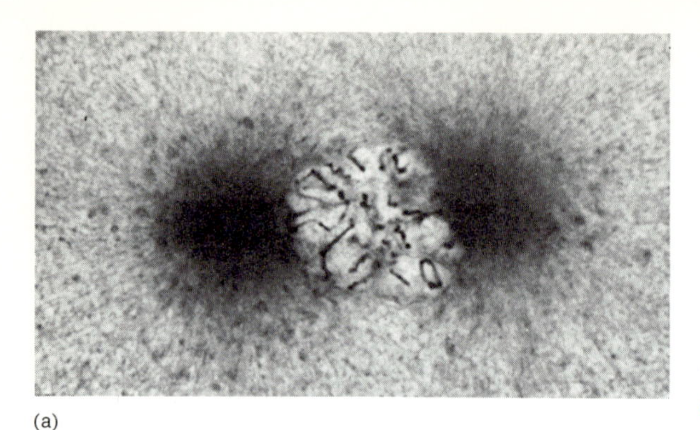

(a)

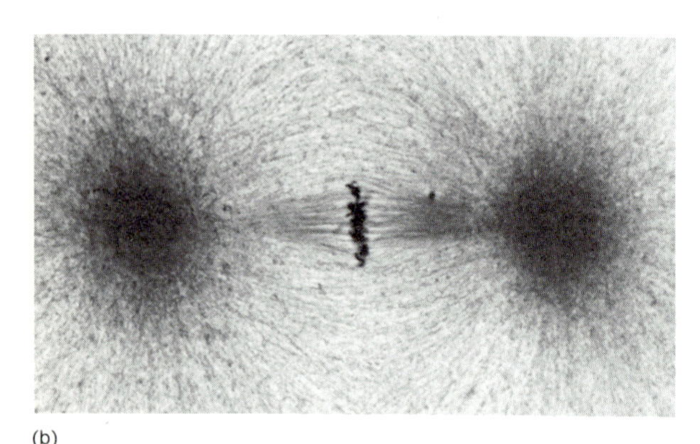

(b)

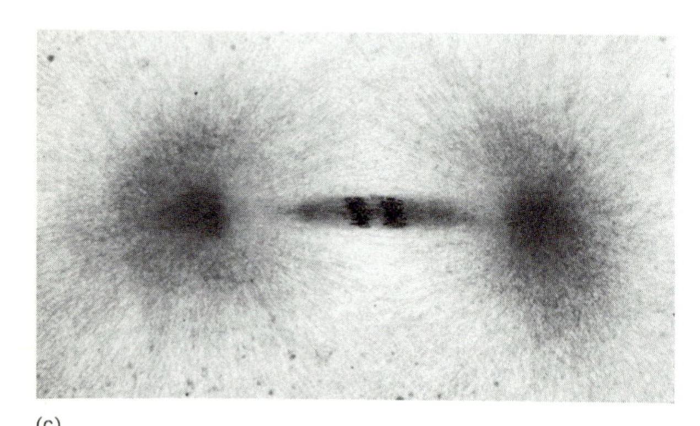

(c)

FIGURE 3.22

Photomicrographs illustrating three successive phases of a mitotic nuclear division. (a) Early organization of the spindle; the nuclear membrane has disappeared, and the chromosomes are already visible in the region of the organizing spindle. (b) The spindle is now fully developed, and the chromosomes are regularly aligned in its equatorial plane; this stage is often referred to as the *metaphase of mitosis*. (c) Separation of the two daughter sets of chromosomes has occurred, and each set is being withdrawn toward one pole of the spindle.

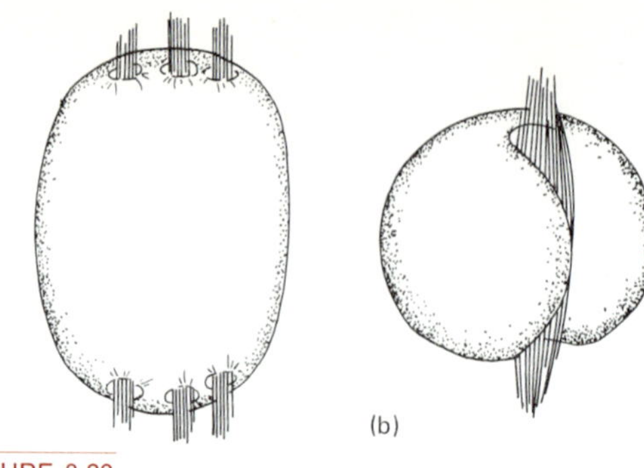


FIGURE 3.23

Dinoflagellate mitosis: (a) spindle fibers penetrating invaginations of the nuclear membrane; (b) the nucleus wrapped around the spindle fibers. Redrawn from H. Fuge, "Ultrastructure of the Mitotic Spindle," *Int. Rev. Cytol.* Supp. 6, 1–52 (1977).

## Chromosome Segregation in Eubacteria

Segregation of the genome following DNA replication in eubacteria is much simpler than in the nucleus of a eucaryote. It is achieved by an entirely different mechanism, which does not involve the complex sequence of structural events associated with mitosis. During the cell cycle the bacterial chromosome never undergoes changes of length and thickness by condensation, and the separation of daughter chromosomes is not mediated by a microtubular system. The mechanism of chromosomal segregation in eubacteria is not yet fully understood, but the available evidence suggests that it involves attachment to specific sites on the cytoplasmic membrane, with separation of daughter chromosomes being brought about by membrane growth between the sites. Plasmid segregation may occur by the same mechanism, each plasmid having its own membrane attachment site.

The separation of daughter genomes and the process of cell division are not as closely linked to one another in the eubacterial cell as they are in the eucaryotic cell, in which cell division typically begins in the terminal stages of mitosis. During rapid growth of unicellular eubacteria, nuclear division typically runs ahead of cell division (Figure 3.24). Each daughter cell thus contains two (or more) already separated chromosomes after the completion of cell division, the uninucleate state becoming reestablished only after the cessation of growth. Indeed, some bacteria contain so many copies of their chromosome (see page 238) that it may be feasible to depend on chance to ensure that each daughter cell receives at least one genome;

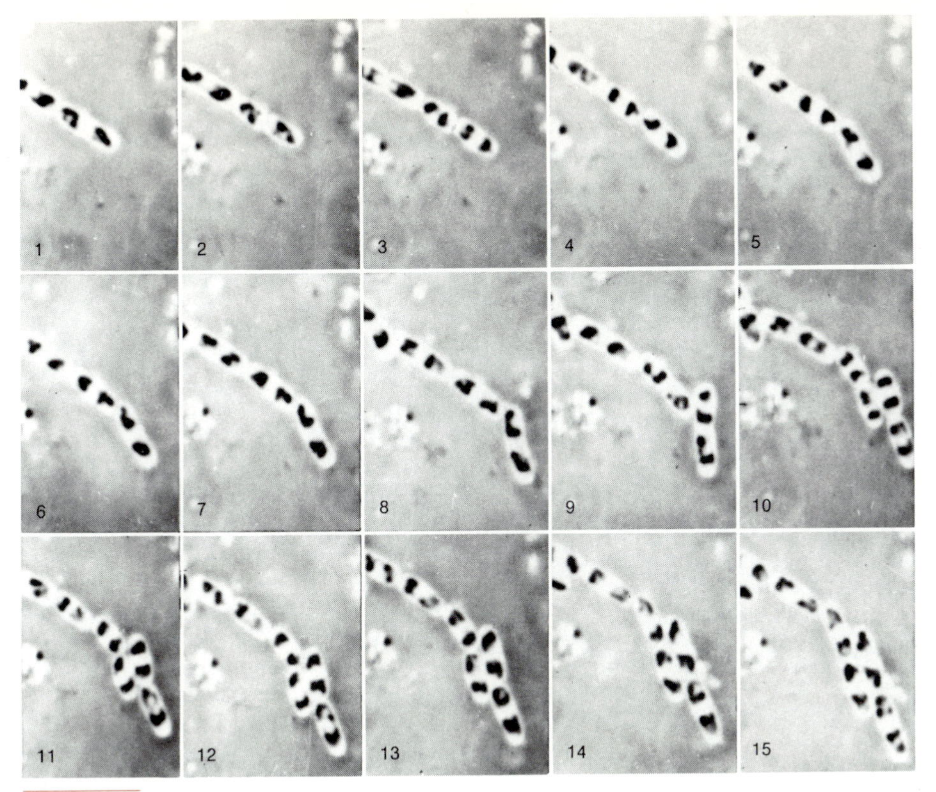


FIGURE 3.24

Successive photomicrographs of growth and nuclear division in a single group of *E. coli* cells suspended in a concentrated protein solution to enhance the contrast between nuclear and cytoplasmic regions (phase contrast, ×975). The sequence was taken over a total period of 78 minutes, equivalent to 2.5 bacterial divisions. Courtesy of D. J. Mason and D. Powelson.

consequently, even the simple mechanism of chromosome segregation via membrane attachment may not be universal among eubacteria.

### Chromosome Segregation in Archaebacteria

The mechanism of chromosome segregation has not been examined in any archaebacterial species; there are, however, no indications that they possess more than one chromosome, and a relatively simple segregation mechanism such as that characteristic of the eubacteria should suffice.

## TRANSCRIPTION AND TRANSLATION OF THE GENOME

In all cellular organisms, the first step in expression of the information in DNA is its transcription into RNA. This step is performed by the enzyme *RNA polymerase*. Subsequent steps vary with the organism and the function of the RNA. The RNA may be *processed* (enzymatically cut and reassembled, or otherwise chemically modified); processing is universal for all species of metabolically stable RNA (tRNA and rRNA), and is common for eucaryotic mRNA. Both tRNA and rRNA assume complex and characteristic secondary structures as a consequence of intramolecular base pairing; the metabolically active form of tRNA is the processed, folded molecule itself, whereas rRNA is complexed with a variety of specific protein molecules to form the *ribosome*.

The protein composition (subunit structure) of RNA polymerase is highly conserved in eubacteria. In all cases examined, the *core enzyme* is composed of four components: two identical "alpha" subunits and two very similar "beta" subunits. A fifth subunit, termed *sigma factor*, binds to the core enzyme and confers on it the specificity for accurate recognition of regions on the DNA at which transcription starts (*promotor regions*).

Eucaryotes normally have three different RNA polymerases. One of these (polymerase I) transcribes the genes that encode the precursors from which the 18S* and 28S rRNA are derived; it is specifically associated with the *nucleolus* (a region of the nucleus in which active transcription of the multiple copies of rRNA genes occurs). Polymerase II transcribes structural genes to produce mRNA, and polymerase III transcribes the genes that encode small stable RNA (5S rRNA and tRNA molecules). The subunit structure of these polymerases is more complex and more variable than the eubacterial polymerase; eight to ten different subunits appears to be characteristic. Typically three of the larger subunits of polymerase I show faint but detectable homology to the alpha and two beta subunits of the eubacterial RNA polymerase.

The archaebacteria, like the eubacteria, possess a single RNA polymerase; however, in most archaebacteria its subunit structure more closely resembles the eucaryotic polymerase in having multiple (5–11) different subunits.

## Sequence and Processing of Stable RNA

Analysis of sequences present in one of the three major classes of rRNA indicates that there are profound differences among the eucaryotes, the eubacteria, and the archaebacteria. In addition, although all three groups contain many modified nucleotides in their stable RNA, the modifications tend to be group specific. An example is the "common arm" sequence in tRNA; until the archaebacterial tRNA was analyzed, this sequence was thought to be universally characterized by a ribothymidine residue (created by methylation of a uridine residue). No archaebacterial tRNA has a ribothymidine in this position; rather they contain pseudouridine or the unique base 1-methylpseudouridine (Figure 3.25).

Another characteristic processing step in stable RNA synthesis is the cleavage of a single transcript into smaller RNA molecules by endonucleases. For instance, in many eubacteria, the genes for the three species of rRNA and one tRNA are clustered on the chromosome, and are transcribed as a single unit. Cleavage of this transcript releases the individual molecules of rRNA and tRNA. Similar endonucleolytic cleavages are necessary steps in the maturation of many other tRNAs. Eucaryotes also have clusters of rRNA and tRNA genes and, like the eubacteria, often produce a primary transcript that is cleaved to release individual stable RNA species. However, the arrangement of genes is quite different from that found in the eubacteria; in the case of rRNA genes, the 18S and 28S genes are transcribed as a unit (along with one tRNA gene) that does not include the gene for the 5S rRNA, which is transcribed separately.

Most of the archaebacteria, like the eubacteria, have the three rRNA genes clustered and transcribed together; however, these transcription units do not contain a tRNA gene.

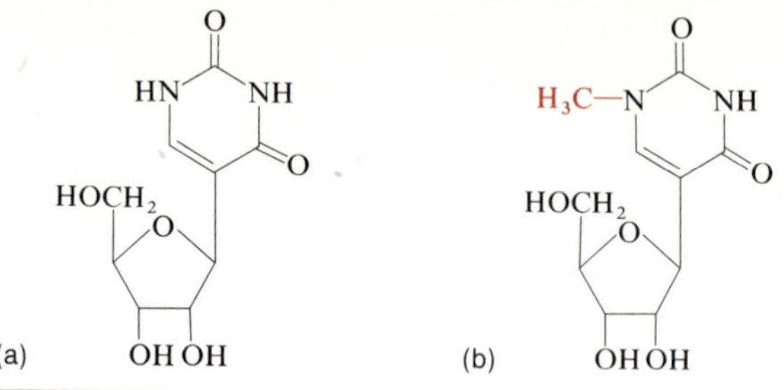


FIGURE 3.25

Bases that replace ribothymidine in archaebacterial tRNA (a) Pseudouridine; (b) 1-methylpseudouridine.

## Messenger RNA Processing

Three major classes of mRNA processing are characteristic of eucaryotes: *capping*, *tailing*, and *splicing*. Capping and tailing are universally performed; splicing is more common in the plants and animals than in the protists.

RNA is synthesized from the 5′ to the 3′ end. Since the substrates for polymerization are the 5′ nucleotide triphosphates, mRNA characteristically has a 5′ triphosphate and a 3′ hydroxyl (Figure 3.26). Capping and tailing modify mRNA at the 5′ and 3′ ends respectively.

*Capping* joins a modified guanosine residue in a 5′ to 5′ orientation to the beginning of the

FIGURE 3.26

Schematic diagram of newly synthesized mRNA. P = phosphate; R = ribose; B = base.

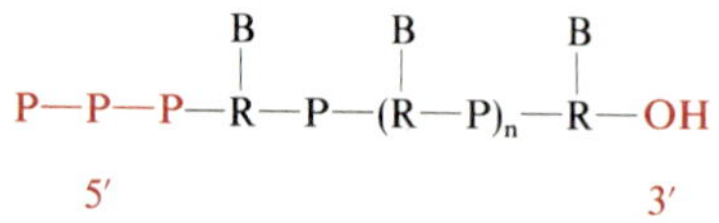

$$\begin{array}{c} \phantom{P-P-P-}B\phantom{-P-(}B\phantom{-P)_n-}B \\ P-P-P-R-P-(R-P)_n-R-OH \\ 5' \qquad\qquad\qquad\qquad 3' \end{array}$$

* S denotes "Svedberg unit," a measure of how fast the molecule or structure sediments in the gravitational field of an ultracentrifuge. These rates are influenced by size, shape, and mass of the sedimenting object.

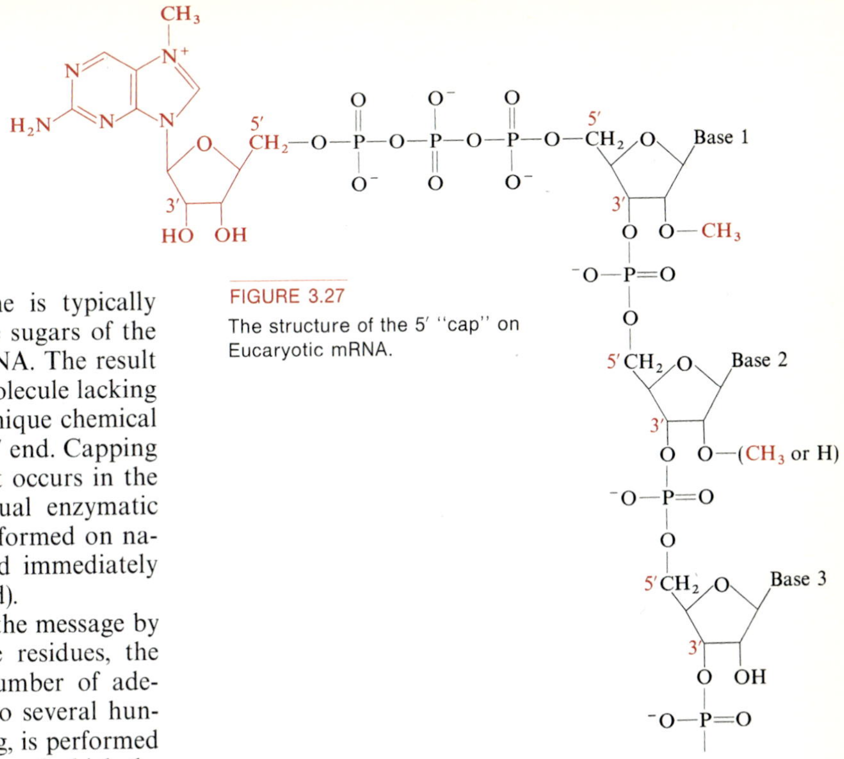

message (Figure 3.27). The guanine is typically modified by methylation, as are the sugars of the first one or two residues of the mRNA. The result of these modifications is an RNA molecule lacking a 5′ end and having in its place a unique chemical configuration at the newly created 3′ end. Capping is a complex series of reactions that occurs in the nucleus; approximately six individual enzymatic steps are involved, and they are performed on nascent mRNA (i.e., mRNA is capped immediately after transcription has been initiated).

*Tailing* modifies the 3′ end of the message by attaching a long string of adenine residues, the "poly-A tail" (Figure 3.28). The number of adenine residues is variable, often up to several hundred. This modification, like capping, is performed enzymatically in the nucleus; the gene of which the mRNA is a copy does not contain a string of A-T base pairs at its distal end. The tailing enzyme apparently recognizes the sequence AAUAAA near the 3′ end of an RNA transcript; RNA molecules without this sequence are not tailed.

*Splicing* is a processing step necessary for all messages that contain *introns*. Introns are noncoding sequences of variable length inserted within the surrounding coding sequences (sometimes termed *exons*) of a gene (Figure 3.29). The transcript of such a gene thus contains stretches of nonsense RNA interspersed with the sequences to be translated. Thus protein synthesis requires either that the introns be excised from the message or that the translation apparatus skip over them during the process of amino acid polymerization. In all known cases, introns are excised from the message before it participates in protein synthesis; the process of

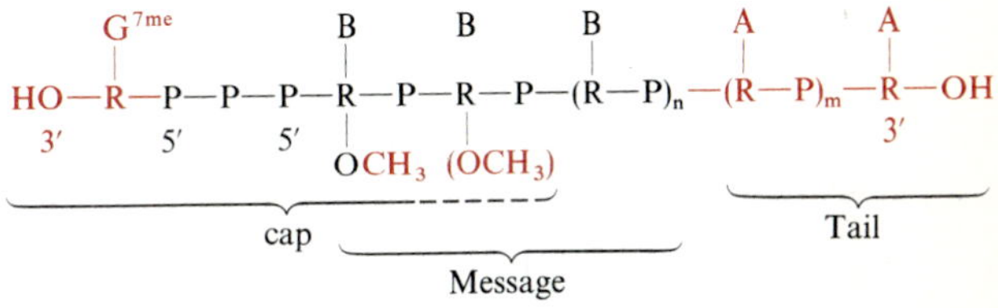

FIGURE 3.27
The structure of the 5′ "cap" on Eucaryotic mRNA.

$$\mathrm{HO{-}\underset{3'}{R}(G^{7me}){-}P{-}P{-}P{-}\underset{5'}{R}(B)(OCH_3){-}P{-}R(B)((OCH_3)){-}P{-}(R(B){-}P)_n{-}(R(A){-}P)_m{-}\underset{3'}{R}(A){-}OH}$$

cap — Message — Tail

FIGURE 3.28
Mature (processed) eucaryotic mRNA. R, ribose; P, phosphate; B, any base; A, adenine; $G^{7me}$, 7-methyl guanine.

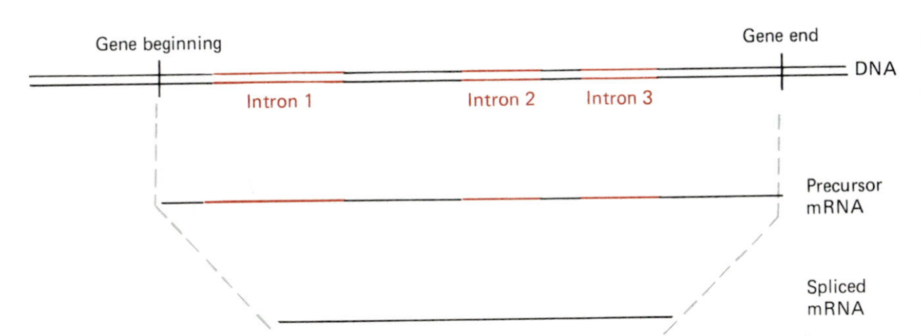


FIGURE 3.29
Schematic diagram of a typical eucaryotic gene with introns.

excision is known as splicing, because the introns are excised and the coding sequences ligated. Little is known about the enzymes that splice RNA; however, a cell-free system that will splice mRNA has recently been obtained, so we can expect rapid progress in this area. Clearly one of the most intriguing aspects of splicing is its accuracy; a single base mistake (either leaving one intron base behind or removing one base from an adjacent exon) would throw off the reading frame of the entire downstream portion of the message. Comparing the sequences of a number of intron-exon boundaries reveals some striking similarities but no single base sequence that uniquely signals the beginning or end of an intron. Most probably, a characteristic secondary structure of the mRNA, caused by intermolecular hydrogen bonding, signals splicing; while there is no unique intron boundary sequence, the sequences at each end of an intron show significant homology to the 5′ end of a particular species of small nuclear RNA termed U1; the function of this RNA, previously obscure, may well be to bring the two ends of an intron together and present them to the splicing enzyme(s).

All three processing steps are performed by enzymes in the nucleus, and are apparently required for message export from the nucleus. One possible function of capping and tailing is to distinguish RNA to be exported from RNA that is to be retained in the nucleus. If this is the case, there must be different export signals for stable RNA, which is neither capped nor tailed (and possibly for histone mRNA, which appears not to be tailed). It is clear, though, that processing is a prerequisite for export of messages, because mRNA that is not capped and tailed will not leave the nucleus. In some cases the splicing step is also necessary; using genetic engineering techniques (Chapter 11), it has been possible to produce genes that lack their normal introns; hence their transcripts should not need splicing. However, in certain cases these messages never leave the nucleus, even though they are capped and tailed.

Another possible function of message processing is to protect mRNA from exonucleolytic digestion. Capping results in a message with no free 5′ end and with a unique structure at the newly created 3′ end; it is unlikely that a 3′ exonuclease could use the capped end of mRNA as substrate. The other 3′ end, although structurally normal and hence a substrate for a 3′ exonuclease, has a long sequence of dispensable bases; exonucleolytic digestion of that end could proceed for several hundred residues before it affected the coding sequences. Indeed, it has been demonstrated that the half-life of uncapped mRNA is much less than that of the same mRNA when capped, when the two are either added to cell-free extracts or microinjected into eucaryotic cytoplasm.

Capping is also required for efficient translation; initiation of translation occurs at a much reduced rate if the mRNA is uncapped. Another function of capping may thus be to allow the cytoplasmic discrimination between RNA that is to be translated and RNA that is not.

A few mRNA species in eubacteria have been determined to have short poly-A tails; however, these are very rare, and the poly-A tails never approach the length of those in eucaryotes. Capping and splicing of eubacterial mRNA appear to be completely absent. Archaebacteria mRNA appears not to be capped or tailed; however, introns have been detected in archaebacteria.

### The Initiation of Translation

In all organisms, translation is initiated by the formation of a complex of mRNA, the small ribosomal subunit, and a specific *initiator tRNA* (Chapter 5). This complex forms at a methionine codon on the mRNA, and the initiator tRNA is charged with methionine. In the eubacteria, methionine bound to the initiator tRNA is modified by the formylation of its amino group; in the archaebacteria and in all eucaryotes the methionine is unmodified.

### Elongation Factors in Translation

The movement of the ribosome along a mRNA requires the participation of several *elongation factors* (Chapter 5). One of these, EF-2 of eucaryotes, has a characteristic posttranslational modification of histidine to form *diphthamide* (Figure 3.30), which renders it sensitive to diphtheria toxin (Chapter 31). This modification, universal among eucaryotes, appears to also be characteristic of archaebacteria; the analogous elongation factor of eubacteria, EF-G, does not contain diphthamide and is unaffected by diphtheria toxin.

FIGURE 3.30

Diphthamide. The portion in black is histidine; the portion in color is added posttranslationally.

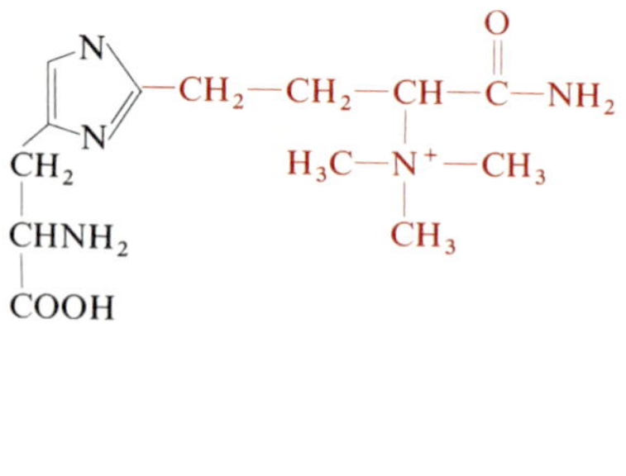

$$\text{(imidazole ring)}-CH_2-CH_2-CH(N^+(CH_3)_3)-C(=O)-NH_2$$

Imidazole ring attached to $CH_2-CHNH_2-COOH$; side chain: $-CH_2-CH_2-CH-C(=O)-NH_2$ with $H_3C-N^+-CH_3$ (and $CH_3$) on the CH.

**TABLE 3.6**

**Comparison of Procaryotic and Eucaryotic Ribosomes**

| Group | Ribosome Size | Subunit Sizes | Number of Proteins | rRNA |
|---|---|---|---|---|
| Procaryotes | 70S | 30S and 50S | 30S:21<br>50S:34 | 30S:16S rRNA (1,500 nucleotides)<br>50S:5S rRNA (120 nucleotides)<br>23S rRNA (3,000 nucleotides) |
| Eucaryotes | 80S | 40S and 60S | 40S:33<br>60S:45 | 40S:18S rRNA (2,000 nucleotides)<br>60S:5S rRNA (120 nucleotides)<br>5.8S rRNA (160 nucleotides)<br>28S rRNA (5,000 nucleotides) |

Note: All sizes and numbers are approximate; there is considerable variation within groups.

## Ribosome Structure

The ribosomes of procaryotic and eucaryotic cells are distinctive and characteristic in their shape, size, subunit size, and molecular composition (Table 3.6). In addition to the characteristic 80S ribosomes bound into the fibrillar matrix of the cytoplasm and to the surface of the rough ER (together often termed *cytoplasmic ribosomes*), eucaryotes have ribosomes in their mitochondria and chloroplasts that resemble 70S ribosomes more closely than they do 80S ribosomes (see below). Eucaryotes thus typically have two (or three) size classes of ribosome.

In recent years the use of *cross-linking* agents (compounds with reactive groups at each end of a molecule, that can react with adjacent macromolecules to bind them together), coupled with high-power electron microscopy, has allowed reconstruction of the three-dimensional conformation of the ribosome. Although this work has been done principally with *Escherichia coli*, a few comparative studies suggest that there are significant structural differences among the ribosomes of the three different cell types (Figure 3.31). Among the archaebacteria there appear to be two morphological types of ribosome, one characteristic of the thermoacidophiles, and one of *Halobacterium*.

The structural differences among ribosomes of the three groups is paralleled by differences in sensitivity to antibiotics that act on the ribosome (Chapter 33). For instance, the archaebacteria are resistant to a variety of antibiotics (e.g., the aminoglycosides, the macrolides, and chloramphenicol) that arrest protein synthesis on

FIGURE 3.31

Micrographs and schematic drawings of the large subunits (A and B) and the small subunits (C and D) of the ribosomes of (from left to right): the eubacterium *Synechocystis*; the archaebacterium *Halobacterium*; the archaebacterium *Thermoproteus*: and the eucaryotic protist Saccharomyces. All 250,000. From J. A. Lake, E. Henderson, M. Oakes, and M. W. Clark. "Eocytes: A new ribosome structure indicates a kingdom with a close relationship to eukaryotes." *Proc. Nat'l. Acad. Sci. USA* **81,** 3786–3790 (1984).

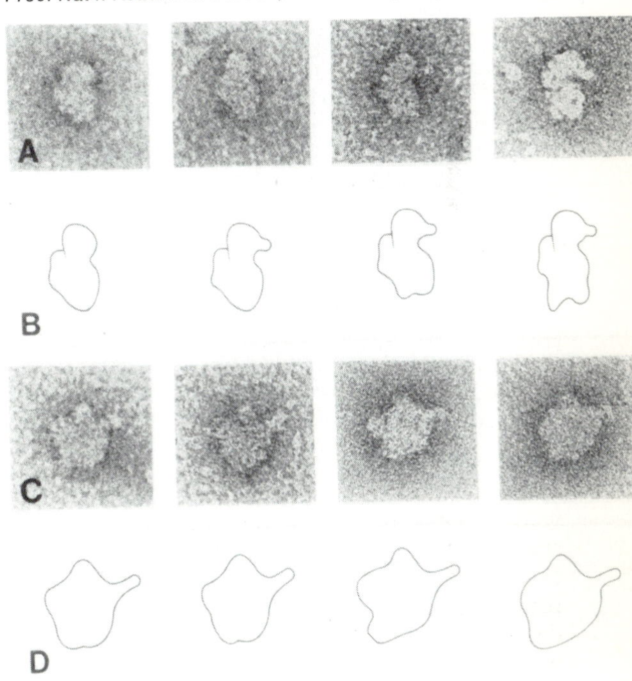

eubacterial, mitochondrial, and chloroplast ribosomes; they are also resistant to cyclohexamide, an inhibitor of 80S ribosomes.

### Coupling of Transcription and Translation

The processes of transcription and translation have been shown to be intimately coupled in eubacteria. Ribosomes bind the 5′ end of the nascent message and initiate the translation process immediately following the initiation of transcription. This coupling is of fundamental importance, as a common mechanism for the control of gene expression in eubacteria, namely *attenuation* (Chapter 12), depends on it.

Since in eucaryotes the processes of transcription and translation are physically separated by the nuclear membrane, such coupling cannot occur. A consequence of this is their inability to utilize attenuation as a genetic regulatory mechanism.

The archaebacteria, which do not have a nuclear membrane, are in principle capable of coupling transcription and translation. Whether they in fact do and, if so, whether they utilize such coupling in genetic regulation, are unknown.

## CHLOROPLAST AND MITOCHONDRIAL GENOMES

Both chloroplasts and mitochondria are organelles whose synthesis is directed in part by the nuclear genome and in part by the endogenous genome. In most cases the structure and expression of the endogenous genome has features of eubacterial rather than eucaryotic nuclear genomes, although some eucaryotic characteristics are found. Multiple copies per organelle of the endogenous genome are characteristic; 20 to 30 copies are common. The only extant eucaryotic organisms that do not have part of their total genomic content housed separately from the nucleus are certain obligately anaerobic, fermentative protozoa, which have most probably lost the ability to respire as a consequence of the loss of their mitochondrial genome. Partitioning of the genome thus appears to be a general characteristic of eucaryotes.

### Genome Structure in Chloroplasts and Mitochondria

The chloroplast genome is remarkably constant over wide taxonomic distances among the eucaryotic phototrophs (plants and algae). It is characteristically a covalently closed circular molecule of double-stranded DNA. Most chloroplast genomes are quite small; a relative molecular weight of about 100 million (i.e., about 5 percent the size of a typical eubacterial genome) is common.

Mitochondrial genomes are much more variable in terms of size and topology than those of chloroplasts. Most commonly, they are covalently closed circular double-stranded DNA, although linear double-stranded molecules have been described in a number of cases. The genome sizes of known mitochondrial DNAs vary over a range of several orders of magnitude (from ten million to over a billion relative molecular weight). Since the information encoded by the mitochondrial genome appears to be very much the same in all organisms, the size heterogeneity must reflect either the presence of noncoding DNA (present either as introns or as insertions between coding regions), or the presence of multiple copies of coding sequences.

### Expression of the Chloroplast and Mitochondrial Genomes

Both chloroplasts and mitochondria contain a complete transcription and translation apparatus, including RNA polymerase, ribosomes, and tRNAs. Many of the proteins involved are encoded in the nuclear genome, synthesized on cytoplasmic ribosomes, and transported into the organelles. The rRNAs, tRNAs, and some of the proteins needed for gene expression in these organelles are encoded by the organelle genome and synthesized within the organelle.

The chloroplast ribosomes are typically bacterial in terms of the sizes of the individual rRNA molecules (5S, 16S, and 23S), total ribosome size (about 70S), subunit size (30S and 50S), and total number of ribosomal proteins. Mitochondrial ribosomes are extraordinarily diverse in size, ranging from 55S to 80S, with corresponding variation in subunit size, rRNA size, and number of proteins. In all but those of higher plants, mitochondrial ribosomes lack the small rRNA molecule, equivalent to the 5S rRNA of bacteria and chloroplasts.

RNA polymerase from chloroplasts strongly resembles the eubacterial enzyme in possessing a core enzyme composed of four or five subunits, and an additional subunit that confers promotor specificity. Indeed, this additional subunit from at least one chloroplast enzyme is capable of replacing the bacterial sigma factor in conferring promotor specificity upon the bacterial RNA polymerase; therefore it must have substantial structural and sequence similarity to the bacterial sigma factor. The mitochondrial RNA polymerase is, however, unique among cellular RNA polymerases in being composed of a single relatively low molecular weight polypeptide.

Capping does not occur in either chloroplasts or mitochondria. Poly-A tailing of mRNA is, however, found in mitochondria, although it appears that the enzyme(s) responsible for tailing in mitochondria are qualitatively different from those that tail nuclear mRNA. Introns are also relatively common in mitochondrial DNA. Introns and poly-A tails are quite rare in chloroplasts.

### The Evolutionary Origins of Chloroplasts and Mitochondria

The striking similarities between the chloroplast of eucaryotic phototrophs and the entire cell of certain eubacterial phototrophs (the cyanobacteria and their relatives) suggests an evolutionary hypothesis: *the chloroplast and the eubacterial cell have a common evolutionary origin, different from that of the rest of the eucaryotic cell.* It is now generally believed that an endosymbiosis between a eubacterial phototroph (probably related to the ancestor of the contemporary cyanobacteria) and the progenitor of the bulk of the eucaryotic cell established the cell lines that led ultimately to the plants and algae. It is probable that the algae have polyphyletic origins; that is, the different groups of eucaryotic phototrophs probably originate from different ancestral endosymbioses.

This evolutionary hypothesis is supported by the discovery that certain protozoa (e.g., the flagellates *Cyanophora* and *Peliaina*, and the ameboid rhizopod *Paulinella*) have photosynthetic organelles termed *cyanellae.* Cyanellae have the fine structure typical of unicellular cyanobacteria and contain murein; however in genome size and copy number they resemble chloroplasts. They are most plausibly interpreted as relatively recently established endosymbioses, in which the progressive loss of bacterial characteristics (e.g., the capacity to synthesize murein) is not yet complete.

The origin of mitochondria is much more obscure; while mitochondria show more similarity to free-living respiratory eubacteria than to the eucaryotes, they display such a bewildering melange of eubacterial and eucaryotic characteristics that the endosymbiotic theory provides only a poorly satisfactory explanation of their origin. However, sequence analysis of their rRNA shows a specific relationship to the eubacteria; their diversity is thus probably due to rapid evolutionary divergence following the establishment of the endosymbiosis.

## SEXUAL PROCESSES IN MICROORGANISMS

### Sexual Processes in Eucaryotes

In all eucaryotic organisms, cellular fusion is the first step in the process of *sexual reproduction.* The two cells that participate are known are *gametes* and the resulting fusion cell as a *zygote.* Gametic fusion is followed by nuclear fusion, with the result that the zygote nucleus contains *two complete sets of genetic determinants,* one derived from each gametic nucleus.

Sexual reproduction is common in the life cycle of plants and animals. In vertebrates and many invertebrates, it is the *only* method for the production of a new individual. Plants can also be propagated asexually (e.g., by cuttings), and asexual modes of reproduction exist in many groups of invertebrates. Among the protists, sexual reproduction is rarely an obligatory event in the life cycle. Many of these organisms completely lack a sexual stage in their life cycles, and even in species in which sexuality does exist, sexual reproduction may occur infrequently, the formation of new individuals taking place principally by asexual means (for example, by binary fission or the formation of spores).

Sexual fusion results in a *doubling of the number of chromosomes,* since the nuclei of the gametes, each containing $N$ chromosomes, fuse to form the nucleus of the zygote, which consequently contains $2N$ chromosomes. Hence, in passing from one sexual generation to the next, there must at some stage be a *halving of the number of chromosomes,* if the chromosome content of the nucleus is not to increase indefinitely. In fact, the halving of the chromosome number is a universal accompaniment of sexuality. It is brought about by a special process of nuclear division termed *meiosis* (Figure 3.32). In animals, *meiosis takes place immediately prior to the formation of gametes.* In other words, each individual of the species has $2N$ chromosomes in its cells through most of the life cycle. Such an organism is termed *diploid.* This state of affairs is, however, by no means universal among sexually reproducing eucaryotic organisms. In many protists, *meiosis takes place immediately after zygote formation,* with the consequence that the organisms have $N$ chromosomes through most of the life cycle. Such organisms are termed *haploid.* In many algae and plants, as well as in some fungi and protozoa, there is a well-marked *alternation of haploid and diploid generations.* In this type of life cycle, the diploid zygote gives rise to a diploid individual, which forms, by meiosis, haploid *asexual* reproductive cells. Each such haploid cell gives rise to a haploid

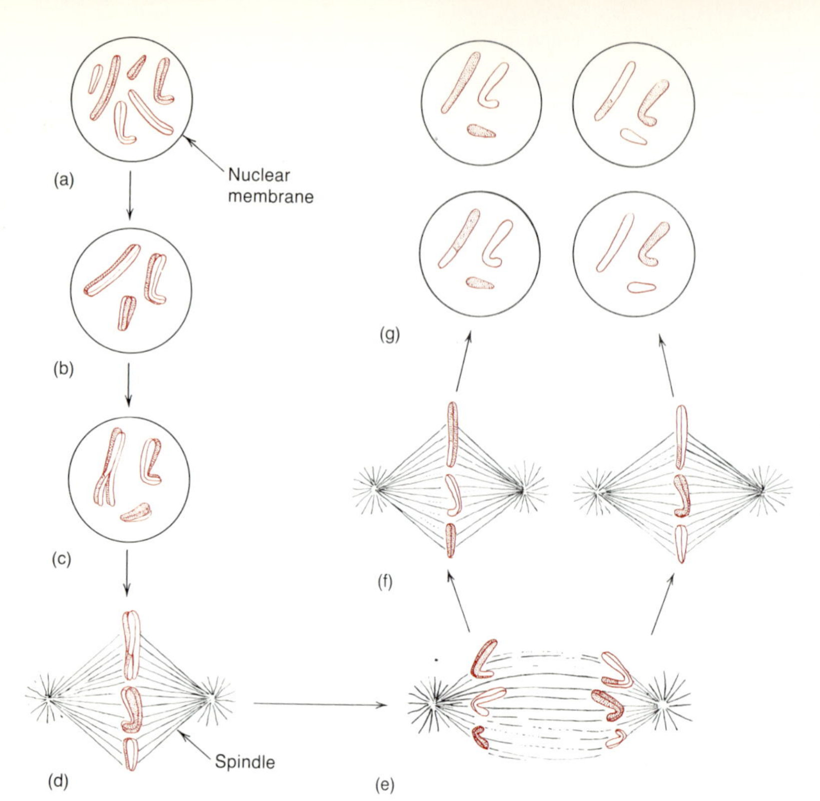


FIGURE 3.32

Meiosis in a hypothetical diploid plant cell (a) with three pairs of chromosomes. (b) Homologous chromosomes pair. (c) An exchange of segments (crossing over) takes place (shown for one chromosome only). (d) The chromosomes are shown at "first metaphase." (e) The chromosomes of each pair separate. (f) Two nuclei have formed, and within each a metaphase spindle forms. This time, however, the sister chromatids that make up each single chromosome separate. This phase, called the "second metaphase," is thus analogous to a mitotic division. (g) The four haploid nuclei that result from meiosis are shown.

individual, which eventually forms haploid gametes; gametic fusion, with the formation of a diploid zygote once again, completes the cycle.

## Sexual Processes in Bacteria

Most bacteria (probably all) normally exist and reproduce by asexual means in the haploid state. Consequently, persistent diploidy, which is characteristic of many groups of eucaryotes and which has had profound evolutionary consequences, plays no role in the evolution of bacteria. A diploid state can arise transiently in bacteria, as a result of genetic transfer, but full diploidy is rarely attained, as a consequence of the special mechanisms of genetic transfer characteristic of procaryotes.

As will be discussed in Chapter 11, genetic transfer among procaryotes always occurs by a unidirectional passage of DNA from a *donor cell* to a *recipient cell*. This may be mediated either by *conjugation*, involving direct cell-to-cell contact, or by the processes known as *transduction* and *transformation*. Transductional transfer to a recipient cell is mediated by certain bacterial viruses (bacteriophages) that incorporate fragments of the genome of the donor cell. Transformational transfer is mediated by free DNA fragments derived from the donor cell, which pass through the medium and are taken up by the recipient cell. As a rule, only small fragments of the donor genome are transferred by transduction and transformation. Although conjugational transfer can in principle permit transfer of the entire donor chromosome, this rarely occurs. Consequently, the recipient cell usually becomes a *partial diploid* (*merodiploid*), and subsequent genetic recombination involves exchanges between the complete haploid genome of the recipient and a fraction of the donor genome. The haploid state is usually rapidly restored after recombination, with elimination of supernumerary genes not incorporated into the recipient chromosome. Return to the haploid state does not, accordingly, involve a regular reduction division comparable to the eucaryotic process of meiosis.

Conjugational genetic transfer in bacteria does not necessarily involve the transfer of chromosomal determinants: the transferred material may be a plasmid. In this event, transfer is not followed by recombination; the plasmid, provided that it is capable of autonomous replication, may instead be maintained by the recipient cell independently of the chromosome. Hence, new genetic elements borne on plasmids, which possess few if any regions homologous with regions of the chromosome, can be introduced into and maintain themselves within the procaryotic cell. For this reason, plasmid transfer can occur between organisms of widely differing (chromosomal) genetic constitution.

## THE DIFFERENCES AMONG CELL TYPES: A SUMMARY

The numerous and profound differences of organization and function among the three cell types—eucaryotes, eubacteria, and archaebacteria—have been revealed gradually and are only now becoming fully recognized. It is, however, evident that they differ from each other with respect to the structures and mechanisms mediating a variety of fundamental cellular activities: transmission, transcription, and translation of the genetic material; the structure of their chromosomes; organization of electron transport systems; organization of the cytoplasm; structure of the cytoplasmic membrane; nutrient uptake; secretion; and movement. These differences are summarized in Table 3.7 and Figure 3.33.

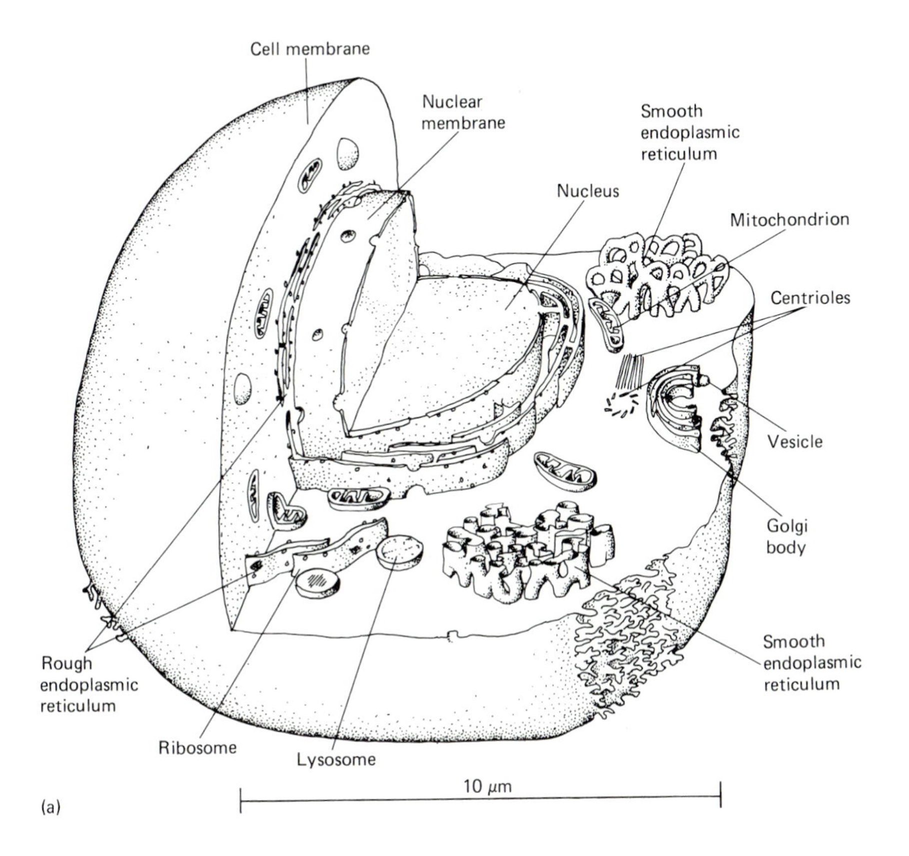


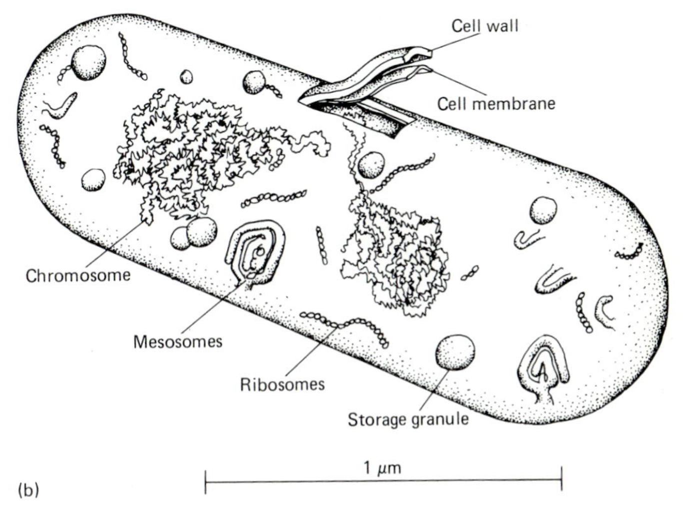


FIGURE 3.33

Schematic diagrams of (a) a eucaryotic cell and (b) a procaryotic cell. (a) Redrawn from H. Curtis, *Biology* (2nd edition), New York: Worth Publishers (1975). (b) Redrawn from E. J. Du Praw, *Cell and Molecular Biology*, New York and London: Academic Press (1968).

**TABLE 3.7**

**Basic Cell Types: A Summary**

| | Eucaryotes | Eubacteria | Archaebacteria |
|---|---|---|---|
| CELL STRUCTURE | Eucaryotic | Procaryotic | Procaryotic |
| CELL MEMBRANE | | | |
| Major lipids | Glycerol diesters | Glycerol diesters | Glycerol diethers or diglycerol tetraethers |
| Side chain | Fatty acids (usually unbranched, often polyunsaturated) | Fatty acids (usually unbranched, rarely polyunsaturated) | Polyisopranoid alcohols (branched, saturated) |
| Other | Steroids nearly universal | Steroids rare; hopanoids common | Steroids and hopanoids rare |
| Endo- and exocytosis | + | – | – |
| INTRACYTOPLASMIC MEMBRANE SYSTEMS | | | |
| Nuclear membrane | + | – | – |
| Endoplasmic reticulum | + | – | – |
| Golgi apparatus | + | – | – |
| Mitochondria | + | – | – |
| Chloroplasts | + (In phototrophs) | – | – |
| CYTOSKELETAL ELEMENTS | | | |
| Microtubules | + | – | – |
| Microfilaments | + | – | – |
| CELL WALL CONTAINS MUREIN | – | +[a] | – |
| GENETIC SYSTEMS | | | |
| Number of chromosomes | >1 | 1 | 1 |
| Chromosome topology | Linear | Circular | ? |
| Number of histones or histonelike proteins | 5 | 1 | 1 |
| Nucleosomes | Yes | Maybe | Probably |
| Chromosome segregation | Mitotic spindle | Cell membrane | ? |
| mRNA capping and tailing | + | – | ? |
| Introns | Usually present | Absent | Occasionally present |
| Genetic exchange initiated by: | Zygote fusion | Unidirectional or bidirectional DNA transfer | ? |
| Meiosis | + | – | – |
| Transcription and translation coupled | – | + | ? |
| Initiator tRNA | Methionyl- | Formylmethionyl- | Methionyl- |
| Diphthamide in elongation factor | + | – | + |
| Ribosome size | 80S (cytoplasmic) | 70S | 70S |

[a] With one exception (the Planctomyces group).

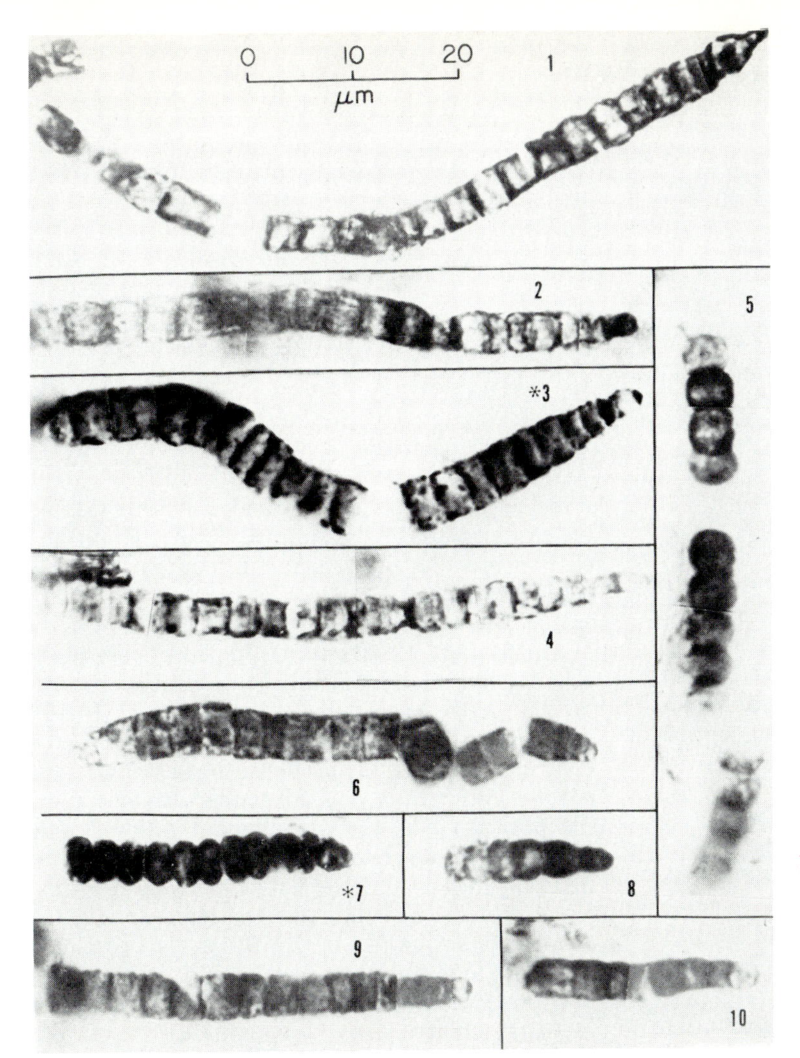


FIGURE 3.34
Precambrian microfossils of some filamentous procaryotic cells, from J. W. Schopf and J. M. Blacic, "New Microorganisms from the Bitter Springs Formation (Late Precambrian) of the North-Central Amadeus Basin, Australia." *J. Paleontol.* **45,** 925–960 (1971).

These profound differences among the extant cell types raise a number of fundamental questions concerning their evolution, questions as difficult to answer as they are important. There is fossil evidence of cells with procaryotic structure dating from well over 3 billion years ago, a striking finding since the earth probably cooled to physiological temperatures only about 4 billion years ago. Thus cells indistinguishable on structural grounds from modern eubacteria or archaebacteria developed very rapidly after conditions became permissive for life on earth, and have been continuously present for virtually the entire history of the planet (Figure 3.34). The precise relationship between the eubacteria and the archaebacteria is unclear; the most generally accepted speculation is that they evolved very early from a common ancestor, or "progenote," and have diverged ever since. The origin of the eucaryotic cell is much more obscure. It is generally accepted that the algal chloroplast has an evolutionary source different from that of the cell which houses it, and a similar origin for the mitochondrion seems probable. Whether the bulk of the cell, with the nucleus as its sole or principal site of storage of genetic information (the "urcaryote"), evolved from an archaebacterial ancestor or directly from the progenote is not clear. Although the fossil record indicates a fairly recent appearance of nucleated cells (1 to 2 billion years ago), the molecular evidence suggests that the line that has led to the modern eucaryotic cell has been genetically independent for nearly the entire course of cellular evolution. These possible relationships are shown schematically in Figure 3.35.

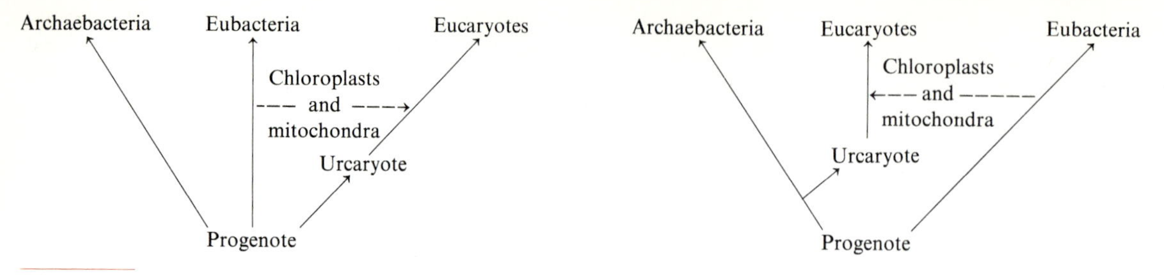


FIGURE 3.35
Possible evolutionary relationships among archaebacteria, eubacteria, and eucaryotes.

## THE GENERAL PROPERTIES OF VIRUSES

One class of microorganisms, the viruses, are acellular: they differ from cellular organisms in structure, chemical composition, and mode of growth.

The viruses are obligate parasites, capable of development only within the cells of susceptible host organisms. Viral hosts include almost all groups of cellular organisms, both procaryotes and eucaryotes. Viruses are transmitted from cell to cell in the form of small infectious particles, known as *virions*. Each virion consists of a core of nucleic acid, enclosed within a protein coat, or *capsid*, which is normally composed of a fixed number of identical protein subunits, the arrangement of which confers on the virion its external form. Certain virions possess additional structures. Many of those that infect animals are enclosed in lipoprotein membranes, usually derived from the host cell nuclear or cytoplasmic membrane. Certain of those that infect procaryotes have special proteinaceous tail structures attached to the capsid, which function in the attachment of the virion to the host cell, and the introduction of viral nucleic acid into the host.

The core of the virion contains only one kind of nucleic acid; depending on the virus, it may be double-stranded or single-stranded DNA or double-stranded or single-stranded RNA, but in all cases it provides the genetic information required for the synthesis of viral components and their assembly into new virions by the infected host cell.

Although all viruses are dependent on host cells for their development, the extent and nature of this dependence varies. The simplest viruses contain very little genetic information, sufficient to code at most for three proteins. In such cases, the genetic information and enzymatic machinery of the host cell play the predominant role in viral synthesis. The largest viruses contain genetic information sufficient to code for as many as 500 different proteins, including many enzymes specific to viral synthesis. In all cases, however, the provision of energy and of low molecular weight precursors of proteins and nucleic acids, together with much of the machinery of protein synthesis, is assured by the host cell. As a result of the transcription and translation of the information carried in the viral genome, the activities of the host cell are largely redirected towards the synthesis of viral components, which are then assembled into new virions within the infected cell. Intracellular maturation of the virions is followed by their release from the cell, which is generally killed. The properties of viruses are discussed in Chapter 9.

## FURTHER READING

### Books

ALBERTS, B., D. BRAY, J. LEWIS, M. RAFF, K. ROBERTS, and J. D. WATSON, *Molecular Biology of the Cell.* New York: Garland Publishing, Inc., 1983.

COLD SPRING HARBOR LABORATORY of QUANTITATIVE BIOLOGY, *Organization of the Cytoplasm* (*Cold Spring Harbor Symposia on Quantitative Biology*, Vol. 46). Cold Spring Harbor: Laboratory of Quantitative Biology, 1982.

**Reviews**

DEMEL, R. A. and B. DE KRUYFF, "The Function of Sterols in Membranes," *Biochem. Biophys. Acta* **457**, 109–132 (1976).

GREY, M. W., and W. F. DOOLITTLE, "Has the Endosymbiont Hypothesis Been Proven?" *Microbiol. Rev.* **46**, 1–42 (1982).

LIENHARD, G. E., "Regulation of Cellular Membrane Transport by the Exocytic Insertion and Endocytic Retrieval of Transporters," *Trends in Biochem. Sci.* **8**, 125–127 (1983).

PORTER, K. R., and J. B. TUCKER, "The Ground Substance of the Living Cell," *Scientific American* **244** (3), 57–67 (1981).

ROBINSON. D. G., and U. Kristen, "Membrane Flow via the Golgi Apparatus of Higher Plant Cells," *Int. Rev. Cytol.* **77**, 89–127 (1982).

ROHMER, M., P. BOUVIER, and G. OURISSON, "Molecular Evolution of Biomembranes: Structural Equivalents and Phylogenetic Precursors of Sterols," *Proc. Natl. Acad. Sci. USA* **76**, 847–851 (1979).

ROTHMAN, J. E., "The Golgi Apparatus: Two Organelles in Tandem," *Science* **213**, 1212–1219 (1981).

STANIER, R. Y., "Some Aspects of the Biology of Cells and Their Possible Evolutionary Significance," *Symp. Soc. Gen. Micro.* **20**, 1–38 (1970).

TAYLOR, R. F., "Bacterial Triterpenoids," *Microbiol. Rev.* **48**, 181–198 (1984).

WALLACE, D. C., "Structure and Evolution of Organelle Genomes," *Microbiol. Rev.* **46**, 208–240 (1982).

WEEDS, A., "Actin-Binding Proteins—Regulators of Cell Architecture and Motility," *Nature* **296**, 811–816 (1982).

WOESE, C. R., "Archaebacteria," *Scientific American* **244** (6), 98–122 (1981).

Chapter 4

# Microbial Metabolism: Fueling Reactions

The sum of all the chemical transformations that occur in cells is termed *metabolism*; the major net consequence of these transformations in microorganisms is the synthesis of a new cell. Although the number of individual metabolic reactions exceeds 1,000 and the interrelationships among them are complex, metabolic schemes in which the reactions are grouped by function can give a simplified overview of the process. Such a scheme of the metabolic activities of the well-studied chemoheterotroph *Escherichia coli* growing in a minimal salts medium with glucose as the carbon source is shown in Figure 4.1. By a set of reactions termed *fueling reactions*,* glucose and phosphate ions are metabolized to mobilize chemical energy in the form of the highly reactive compound ATP, to produce reducing power in the form of *pyridine nucleotides*, and to form 12 compounds termed *precursor metabolites* from which all cellular compounds are synthesized. Precursor metabolites along with sulfate and ammonium ions enter into a set of reactions, termed *biosynthetic*, which leads to synthesis of compounds termed *building blocks*. These are *polymerized* to form the macromolecules that are then *assembled* into the various cellular structures.

Although there are variations among microorganisms with respect to biosynthetic, polymerization, and assembly reactions, they are minor as compared with the vast diversity of fueling reactions. In some cases (e.g., chemoheterotrophs), ATP, reducing power, and precursor metabolites

* Fueling reactions, when they serve primarily to degrade a substrate and thereby generate ATP are called *catabolic*; when they produce biosynthetic building blocks, they are called *anabolic*; when they produce ATP and precursors of biosynthetic building blocks, they are called *amphibolic*.

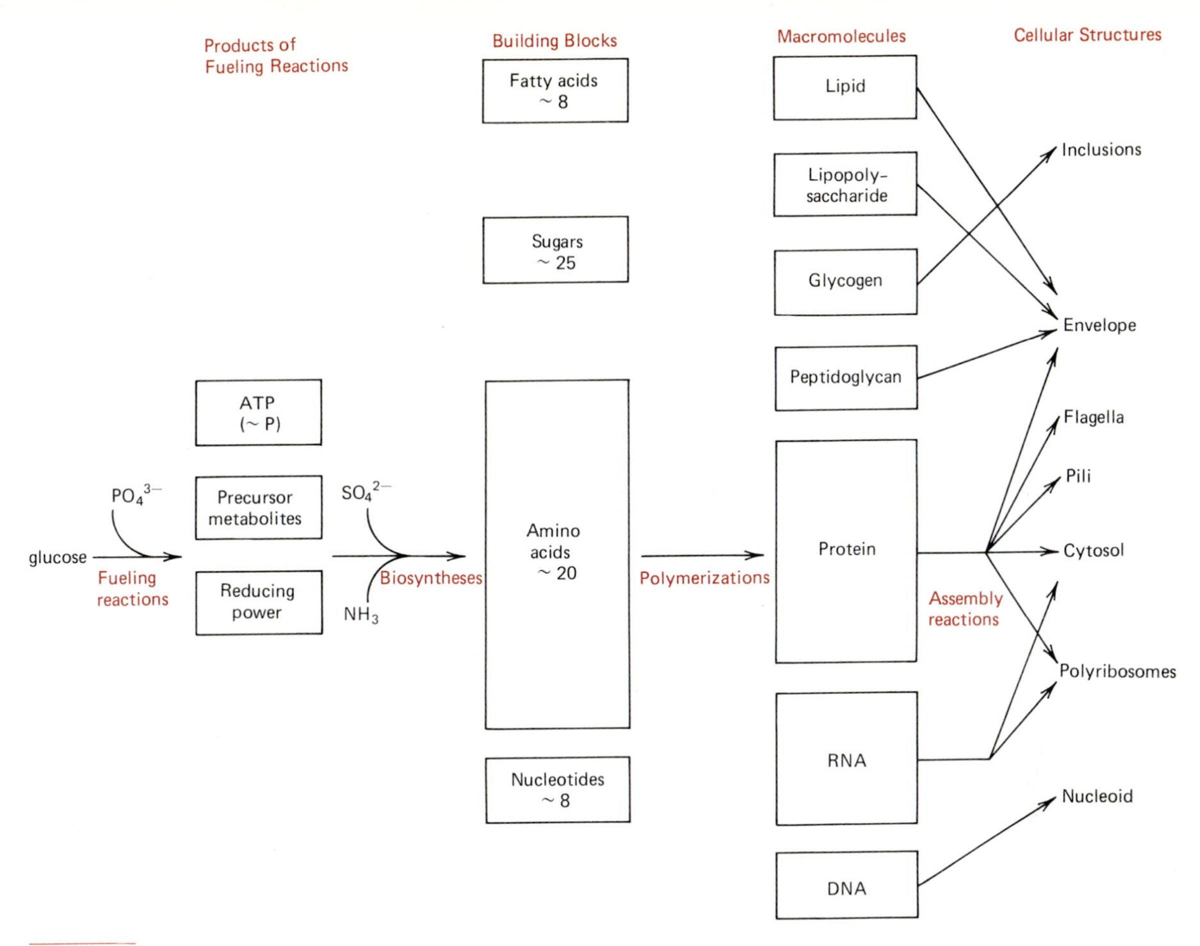


FIGURE 4.1

General pattern of metabolism leading to the synthesis of a cell of *E. coli* from glucose. Boxes indicating building blocks and macromolecules are proportional to their need in *E. coli*. The names and structures of precursor metabolites are shown in Figure 4.5. After J. L. Ingraham, O. Maaloe, and F. C. Neidhardt, *Growth of the Bacterial Cell* (Sunderland, Mass.: Sinauer Associates, Inc., 1983).

are generated by the same metabolic pathways. In other cases, e.g., chemoautotrophs, they are generated by distinct pathways. In this chapter we consider the diversity of microbial fueling reactions and the roles of their products, ATP, reducing power, and precursor metabolites in metabolism.

## THE ROLE OF ATP IN METABOLISM

All biosynthetic and polymerization pathways and possibly some assembly reactions require the participation of ATP, or other compounds containing particularly reactive phosphate groups. The chemical structure of ATP is shown in Figure 4.2. It is a derivative of AMP to which two additional

FIGURE 4.2

The structures of ATP (adenosine triphosphate), showing the various components of the molecule that can be obtained by hydrolysis.

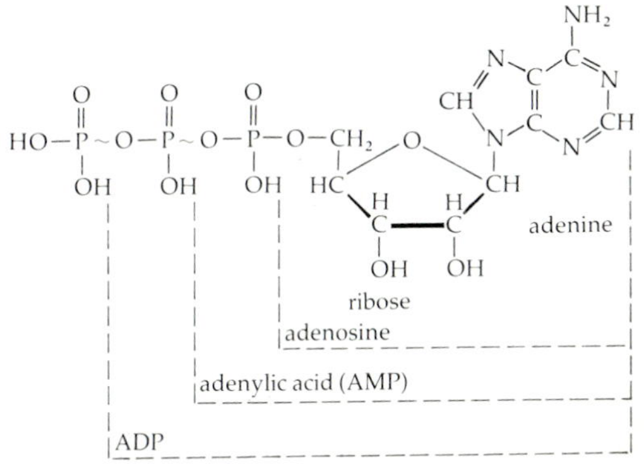

phosphate groups are attached through an *anhydride linkage*. The two bonds indicated by the symbol ~ are high-energy bonds* and thus are particularly reactive. Hence, ATP is able to donate phosphate groups to a number of metabolic intermediates, thereby converting them to activated forms. Their standard free energy is increased to a level that allows the phosphorylated intermediate to participate in biosynthetic reactions that are thermodynamically favorable ($\Delta G^\circ < 0$), while the comparable reaction with the unphosphorylated form as a reactant would be thermodynamically unfavorable ($\Delta G^\circ > 0$). Thus, ATP generation is required in order for biosynthetic pathways to function.

The special reactivity of the high-energy bonds of ATP is apparent when the $\Delta G^\circ$ (free energy) of their hydrolysis is compared with the $\Delta G^\circ$ of hydrolysis of the phosphate of AMP (attached to adenosine by an ester linkage, therefore less reactive and termed a *low-energy bond*); i.e.,

$$\underset{\text{(ATP)}}{\text{adenosine}}\text{—Ⓟ}\sim\text{Ⓟ}\sim\text{Ⓟ} + H_2O \longrightarrow \underset{\text{(ADP)}}{\text{adenosine}}\text{—Ⓟ}\sim\text{Ⓟ} + \text{Ⓟ} \qquad \Delta G^\circ = -7.3 \text{ Kcal}$$

$$\underset{\text{(ADP)}}{\text{adenosine}}\text{—Ⓟ}\sim\text{Ⓟ} + H_2O \longrightarrow \underset{\text{(AMP)}}{\text{adenosine}}\text{—Ⓟ} + \text{Ⓟ} \qquad \Delta G^\circ = -7.3 \text{ Kcal}$$

$$\underset{\text{(AMP)}}{\text{adenosine}}\text{—Ⓟ} + H_2O \longrightarrow \text{adenosine} + \text{Ⓟ} \qquad \Delta G^\circ = -3.4 \text{ Kcal}$$

### Other Compounds with High-Energy Bonds

ATP can be visualized as playing a role of trapping a portion of the free energy made available in fueling reactions and driving biosynthetic reactions by activating certain biosynthetic intermediates. Although ATP is directly involved in the majority of such activation reactions, a number of other highly reactive metabolites that also contain high-energy bonds enter into specific activation steps in certain pathways of biosynthesis. All these high-energy compounds can be formed at the expense of one or more of the high-energy bonds of ATP, but sometimes they are formed directly in catabolic reactions. These compounds, along with some representative activation steps in which they participate, are listed in Table 4.1.

* The term *high-energy bond* should not be confused with the term *bond energy* that is used by the physical chemist to denote the energy *required to break a bond between two atoms.*

**TABLE 4.1**

**High-Energy Compounds Other Than ATP That Activate Metabolic Intermediates and Thereby Drive Certain Reactions of Biosynthesis**

| High-Energy Compound[a] | Cause Activation in the Biosynthesis of: |
|---|---|
| Guanosine-Ⓟ ~ Ⓟ ~ Ⓟ<br>GTP | Proteins (ribosome function) |
| Uridine-Ⓟ ~ Ⓟ ~ Ⓟ<br>UTP | Peptidoglycan layer of the bacterial wall<br>Glycogen |
| Cytidine-Ⓟ ~ Ⓟ ~ Ⓟ<br>CTP | Phospholipids |
| Deoxythymidine-Ⓟ ~ Ⓟ ~ Ⓟ<br>dTTP | Lipopolysaccharide of bacterial wall |
| Acyl ~ SCoA<br>Acyl coenzyme A | Fatty acids |

[a] High-energy bonds are indicated by the symbol "~."

## THE ROLE OF REDUCING POWER IN METABOLISM

Like all oxidations, the biological oxidation of organic metabolites is the removal of electrons. In most cases, oxidation of a metabolite involves the removal of two electrons and thus the simultaneous loss of two protons; this is equivalent to the removal of two hydrogen atoms and is called *dehydrogenation.* Conversely, the *reduction* of a metabolite usually involves the addition of two electrons and two protons and can therefore be considered a *hydrogenation.* For example, the oxidation of lactic acid* to pyruvic acid and the reduction of pyruvic acid to lactic acid can be expressed as follows:

$$\underset{\text{lactic acid}}{\begin{array}{c} COOH \\ | \\ CHOH \\ | \\ CH_3 \end{array}} \underset{+2H}{\overset{-2H}{\rightleftharpoons}} \underset{\text{pyruvic acid}}{\begin{array}{c} COOH \\ | \\ C{=}O \\ | \\ CH_3 \end{array}}$$

The compounds that most often mediate biological oxidations and reductions (i.e., that serve as

* Most organic acids like lactic acid exist over the physiological range of pH partially in the protonated (acid) form and partially in the ionic (salt) form. Acid forms carry an *-ic* ending (e.g., lact*ic* acid), and ionic forms carry an *-ate* ending (e.g., lact*ate*). Conventionally, the actual mixture of the two is referred to by either designation alone. We do the same.

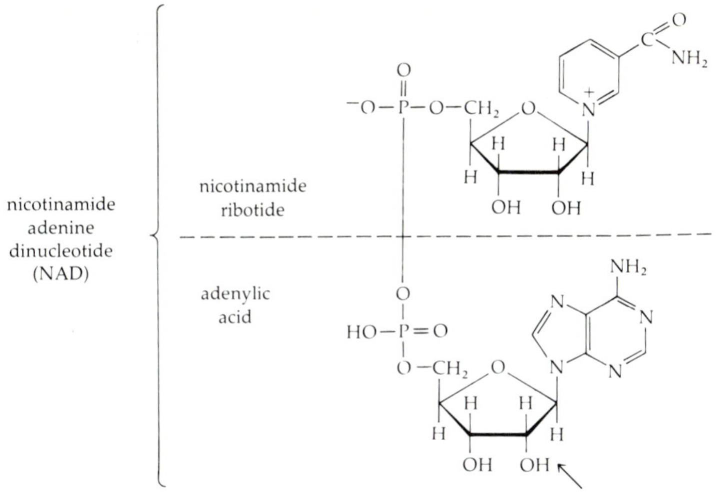

FIGURE 4.3
Structure of NAD (nicotinamide adenine dinucleotide). NADP (nicotinamide adenine dinucleotide phosphate) has an additional phosphate group which is esterified to the 2′ position of the ribose group in the adenyl moiety of the molecule (indicated by the arrow).

acceptors for hydrogen atoms released by dehydrogenation reactions, and as donors of hydrogen atoms required for hydrogenation reactions) are two pyridine nucleotides: *nicotinamide adenine dinucleotide* (NAD) and *nicotinamide adenine dinucleotide phosphate* (NADP), the structures of which are shown in Figure 4.3.

Both of these pyridine nucleotides can readily undergo reversible oxidation and reduction, the site of which is the nicotinamide group (Figure 4.4). It can be seen that the oxidized form of the pyridine nucleotides carries one hydrogen atom less than the reduced form; in addition, it has a positive charge on the nitrogen atom, which enables it to accept a second electron upon reduction. The reversible oxidation-reduction of NAD and NADP can thus be symbolized

$$NAD^+ + 2H \rightleftharpoons NADH + H^+$$

and

$$NADP^+ + 2H \rightleftharpoons NADPH + H^+$$

FIGURE 4.4
Oxidized and reduced forms of the nicotinamide moiety of pyridine nucleotides.

oxidized ⇌ (+2H / −2H) reduced + $H^+$

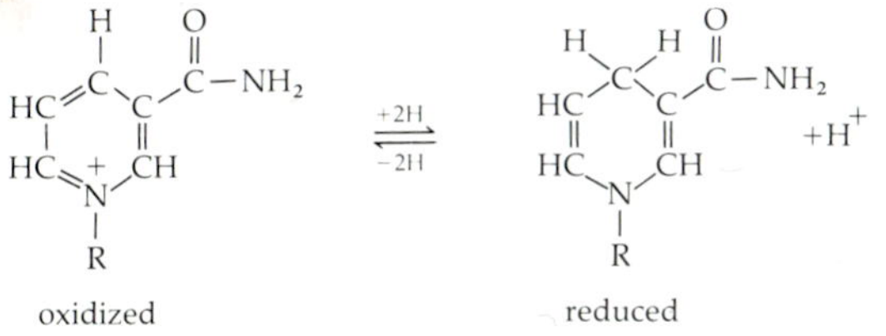

One may ask about the metabolic necessity of the existence of two types of pyridine nucleotides (NAD and NADP) that are so similar in function. The answer seems to lie in the fact that in catabolic reactions oxidized pyridine nucleotides are the usual reactant, and in biosynthetic reactions the reduced form is the usual reactant. Thus for catabolic reactions to proceed, pyridine nucleotides must be largely in oxidized form; for biosynthetic reactions, they must be largely in reduced form. Consequently, two kinds of pyridine nucleotides are required. Indeed, the intracellular pool of NAD is maintained largely in the oxidized state, and the pool of NADP is maintained largely in the reduced state. Certain fueling reactions reduce $NADP^+$ rather than $NAD^+$, but the details of how these differing states of oxidation are maintained is not entirely clear. However, an enzyme termed *transhydrogenase* undoubtedly plays an important role. It catalyzes the reaction

$$NADP^+ + NADH \rightleftharpoons NADPH + NAD^+$$

the equilibrium of which can be shifted by ATP.

## THE ROLE OF PRECURSOR METABOLITES IN METABOLISM

As stated earlier, in addition to providing ATP and reducing power, fueling reactions must supply 12 compounds, termed precursor metabolites, from which all biosyntheses begin. Their names and structures are shown in Figure 4.5.

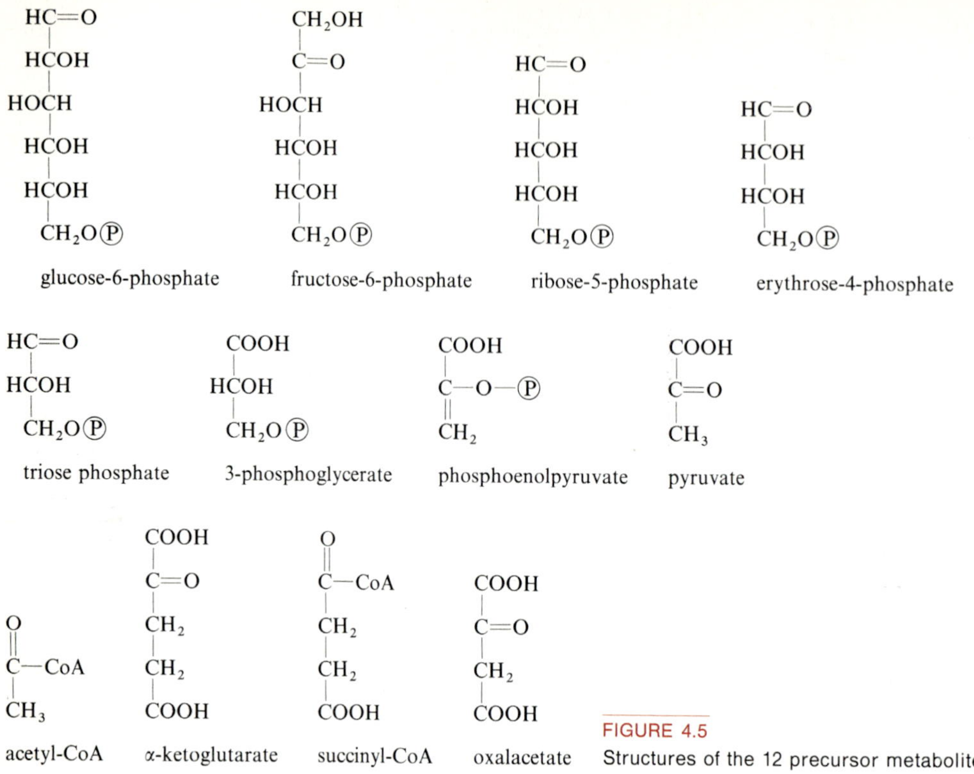

HC=O
HCOH
HOCH
HCOH
HCOH
$CH_2O$Ⓟ
glucose-6-phosphate

$CH_2OH$
C=O
HOCH
HCOH
HCOH
$CH_2O$Ⓟ
fructose-6-phosphate

HC=O
HCOH
HCOH
HCOH
$CH_2O$Ⓟ
ribose-5-phosphate

HC=O
HCOH
HCOH
$CH_2O$Ⓟ
erythrose-4-phosphate

HC=O
HCOH
$CH_2O$Ⓟ
triose phosphate

COOH
HCOH
$CH_2O$Ⓟ
3-phosphoglycerate

COOH
C—O—Ⓟ
‖
$CH_2$
phosphoenolpyruvate

COOH
C=O
$CH_3$
pyruvate

O
‖
C—CoA
$CH_3$
acetyl-CoA

COOH
C=O
$CH_2$
$CH_2$
COOH
α-ketoglutarate

O
‖
C—CoA
$CH_2$
$CH_2$
COOH
succinyl-CoA

COOH
C=O
$CH_2$
COOH
oxalacetate

FIGURE 4.5
Structures of the 12 precursor metabolites.

## BIOCHEMICAL MECHANISMS GENERATING ATP

In metabolism, ATP is generated by two fundamentally different biochemical mechanisms: *substrate level phosphorylation* and *electron transport.*

### Substrate Level Phosphorylation

In substrate level phosphorylation, ATP is formed from ADP by transfer of a high-energy phosphate group from an intermediate of a fueling pathway. The following reactions serve as an example:

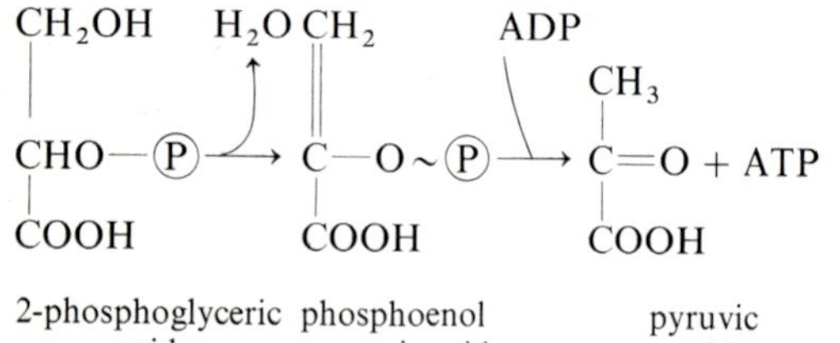

$$\begin{array}{c} CH_2OH \\ | \\ CHO\text{—}Ⓟ \\ | \\ COOH \end{array} \xrightarrow{\;H_2O\uparrow\;} \begin{array}{c} CH_2 \\ \| \\ C\text{—}O\sim Ⓟ \\ | \\ COOH \end{array} \xrightarrow{\;ADP \to ATP\;} \begin{array}{c} CH_3 \\ | \\ C=O \\ | \\ COOH \end{array} + ATP$$

2-phosphoglyceric acid — phosphoenol pyruvic acid — pyruvic acid

As a consequence of the removal of a molecule of water, the low-energy ester linkage of phosphate in 2-phosphoglyceric acid is converted to the high-energy enol linkage in phosphoenol pyruvic acid. This high-energy linked phosphate can then be transferred to ADP, the consequence of which is the generation of a molecule of ATP.

### Generation of ATP by Electron Transport

In a number of different modes of microbial metabolism including respiration and photosynthesis, ATP is generated by transporting electrons through a chain of carrier molecules with fixed orientation in the cell membrane.

Although the complexity and components of electron transport chains vary, they have certain common features: the components of the chain are carrier molecules capable of undergoing reversible oxidation and reduction; each member of the chain is capable of being reduced by reacting with the carrier molecule that precedes it and oxidized by the carrier that follows it (Figure 4.6).

In any specific example of an electron transport chain, certain members transport hydrogen atoms [an electron ($e^-$) plus a proton ($H^+$)] while others transport only an electron. The orientation of the carriers in the cell membrane is such that hydrogen carriers transport in the direction toward the outside of the cell and electron carriers transport

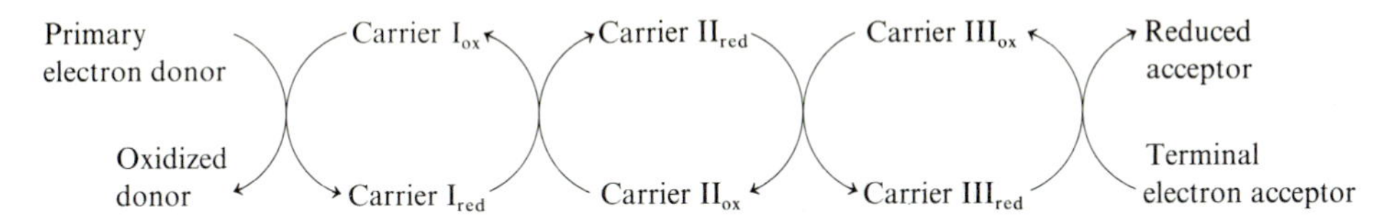


FIGURE 4.6
Schematic representation of an electron transport chain.

toward the inside. Thus, at each conjunction in the chain of a hydrogen carrier and an electron carrier, a proton is transported out of the cell (Figure 4.7). The cell membrane is otherwise impermeable to protons; as a consequence, electron transport traps a portion of the chemical energy released by the net reaction of the chain (oxidation of the primary electron donor by the terminal electron acceptor) in the form of a gradient across the membrane of protons and electrical charge. Such a gradient, termed a *protonmotive force* ($\Delta$p) is a form of potential energy capable of doing work: it drives certain permease systems that concentrate externally supplied substrates within the cell; it provides the energy for flagellar-mediated cell motility and it drives the energy-requiring synthesis of ATP from ADP.

The synthesis of ATP at the expense of protonmotive force is catalyzed by a complex membrane-bound enzyme, *ATP phosphohydrolase* (sometimes termed *ATPase*) composed, in all bacteria studied, of two multicomponent proteins $BF_0$ and $BF_1$. The subunit composition and membrane insertion of $BF_0$ and $BF_1$ are shown in Figure 4.8. The $\alpha$ and $\beta$ subunits of $BF_1$ are arranged alternately to form a hollow hexagon, the central hole of which contains the $\gamma$ subunit associated with other subunits $\delta$ and $\epsilon$. Thus $BF_1$ probably has the subunit structure $\alpha_3\beta_3\gamma\delta\,\epsilon$. The $\alpha$ and $\beta$ subunits form the catalytically active portion of the structure, the site where ATP is synthesized from ADP and inorganic phosphate; the $\gamma$, $\delta$, and $\epsilon$ subunits form a proton-translocating stalk and gate that bring to the active site at the proper rate the protons that drive the reaction. The peptides ($\zeta_3\eta_6$) form a proton channel through the membrane; they are highly hydrophobic, accounting for their intramembranal location.

ATP phosphohydrolase catalyzes a reversible reaction: ATP can be synthesized at the expense of a protonmotive force, or in certain cases a protonmotive force can be established at the expense of hydrolysis of intracellular ATP.

FIGURE 4.7
Schematic representation of a conjunction of a hydrogen carrier and an electron carrier in a transport chain illustrating the extrusion of a proton.

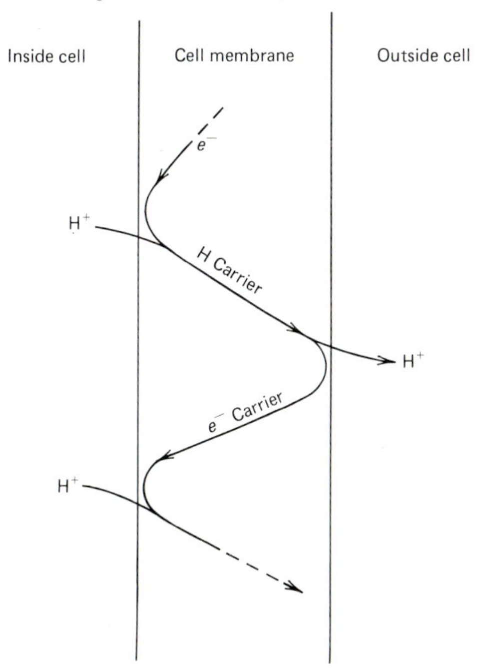


FIGURE 4.8
Schematic representation of the subunit composition and membrane insertion of ATP phosphohydrolase. After C. W. Jones, *Bacterial Respiration and Photosynthesis* (Washington, D.C.: American Society for Microbiology, 1982).

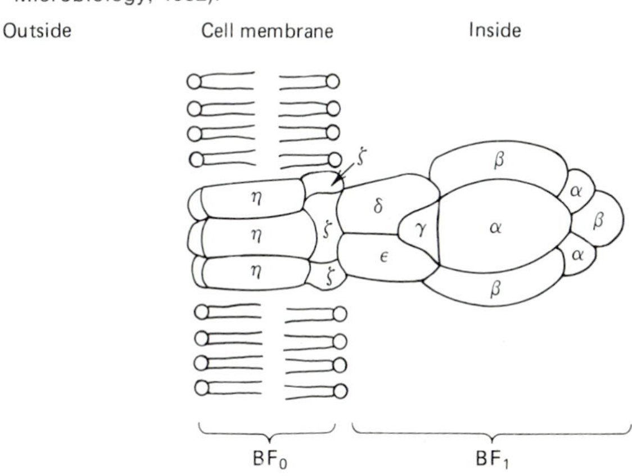

FIGURE 4.9

The $E'_0$ values of various half reactions of primary electron donors, electron carriers, and terminal electron acceptors. Certain electron carriers [flavins (FMN and FAD) and hemes] are bound to proteins as a consequence of which the $E'_0$ of their oxidation is substantially altered. The $E'_0$ values of the free $FMN/FMNH_2$ and $FAD/FADH_2$ couples (−0.21 and −0.21 V respectively are substantially modified when bound to proteins.

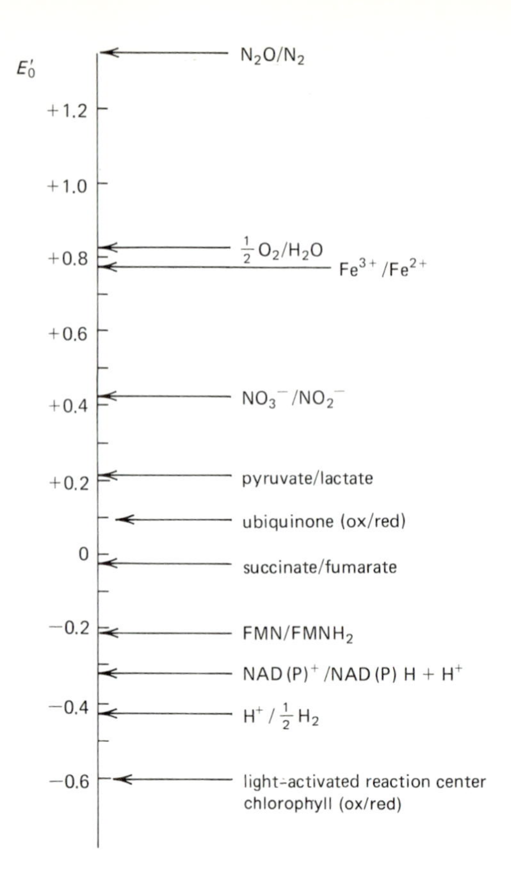


## Values of $E'_0$ for Components in Electron Transport Chains

In order for an electron transport chain to function, there must be a gradient of susceptibility to oxidation; i.e., each component must be capable of being reduced by the reduced form of the previous component and oxidized by the oxidized form of the subsequent component in the chain.

The relative susceptibility of a substance to oxidation or reduction can be described quantitatively in terms of its *electrode potential* or *reduction potential*; this is the relative voltage required to remove an electron from a given compound as compared with that required to remove an electron from $H_2$. Thus the *standard reduction potential* is that of the hydrogen electrode

$$\frac{1}{2}H_2 = H^+ + e^-$$

which is assigned an arbitrary value of 0.0 volts (V) under standard conditions (i.e., all reactants and products at 1 molar or 1 atm; pH 0.0). At pH 7.0, near which most biological reactions occur, the potential of the hydrogen electrode is −0.42 V. The symbol $E'_0$ designates electrode potentials measured under these conditions (i.e., 25°C, all reactants and products at 1.0 molar or 1 atm, pH 7.0).

Knowing $E'_0$ values of two *half reactions* (as reactions written with electrons as reactants or products are called), the free energy change of the coupled reaction can be calculated from the relationship

$$\Delta G'_0 = -nF\,\Delta E'_0$$

where $\Delta G'_0$ is the free energy change at pH 7.0; $n$ is the number of electrons transferred; $F$ is the faraday (a physical constant equal to 23,000 cal/v); and $\Delta E'_0$ is the algebraic difference between the potentials of the two half reactions. For example, the reduction of oxygen by hydrogen as ($H_2 + \frac{1}{2}O_2 \rightarrow H_2O$) can be divided into two half reactions.

$$H_2 = 2H^+ + 2e^- \; (E'_0 = -0.42\ \text{V})$$

and

$$H_2O = \tfrac{1}{2}O_2 + 2H^+ + 2e^- (E'_0 = +0.82\ \text{V})$$

The free energy change can be calculated to be

$$\begin{aligned}\Delta G'_0 &= -2 \times 23{,}000\,[0.82 - (-0.42)]\\ &= -57{,}040\ \text{cal}\end{aligned}$$

For a typical biological oxidation, for example, the oxygen-linked oxidation of NADH ($E'_0$ = −0.32 V), the analogous calculation shows a free energy change of −52,400 cal, not significantly different from that for oxidation of hydrogen.

The carriers in an electron transport chain participate in a series of reactions of increasing $E'_0$ values, between that of the primary electron donor and the terminal electron acceptor. The positions on the $E'_0$ scale of several typical electron carriers, primary electron donor, and terminal electron acceptors are shown in Figure 4.9.

## The Components of Electron Transport Chains

The electron transport chains of aerobic chemoheterotrophs are the most thoroughly studied.

Those involved in the oxidation of organic compounds always contain four different classes of molecules. Two classes, the *flavoproteins* and *quinones*, are hydrogen carriers; the others, the *iron-sulfur proteins* and *cytochromes* are electron carriers.

Flavoproteins have a yellow-colored prosthetic group derived biosynthetically from riboflavin (vitamin $B_2$) (Figure 4.10). The prosthetic group may be either flavin mononucleotide (FMN) or flavin adenine dinucleotide (FAD); both possess the same active site capable of undergoing reversible oxidation and reduction by donating or accepting two hydrogen atoms respectively (Figure 4.10). The flavoproteins are members of a large class and differ widely with respect to their $E_0'$ values. Some are active in the primary dehydrogenation of organic substrates (e.g., succinate); others participate as hydrogen carriers within an electron transport chain; and still others react directly with molecular oxygen with the formation of $H_2O_2$.

Most electron transport chains contain either ubiquinone, a substituted benzoquinone (Figure 4.11), or menaquinone, a substituted naphthoquinone. The former occur most frequently in Gram-negative bacteria and in the mitochondria of eucaryotes, and the latter most frequently in Gram-positive bacteria. A few facultative anaerobes, including *E. coli*, contain both quinones but tend to use ubiquinone for aerobic respiration. On reduction they accept two hydrogen atoms to form the corresponding quinol (Figure 4.11).

The *iron-sulfur proteins* contain two, four, or eight atoms of labile sulfur, so called because they are released as $H_2S$ by strong acids. The iron atoms in 2Fe-2S proteins are held in a lattice composed of four atoms of cysteine sulfur and the two labile sulfur atoms. The iron atoms in the 4Fe-4S proteins interact with the four labile sulfur atoms to form a cube that is held in place within the protein by four cysteine residues. The 8Fe-8S proteins contain two active centers identical to those found in 4Fe-4S protein, thus enabling them to accept on reduction, two electrons while 2Fe-2S and 4Fe-4S proteins can accept only one.

The cytochromes belong to the class of heme proteins, a class that also includes hemoglobin and catalase. All have one or more prosthetic groups derived from heme, a cyclic tetrapyrrole with an atom of iron chelated within the ring system (Figure 4.12). Electron transfer by the cytochromes involves a reversible oxidation of this iron atom

$$Fe^{2+} = Fe^{3+} + e^-$$

in an analogous way to the oxidation of an iron atom in the iron-sulfur proteins.

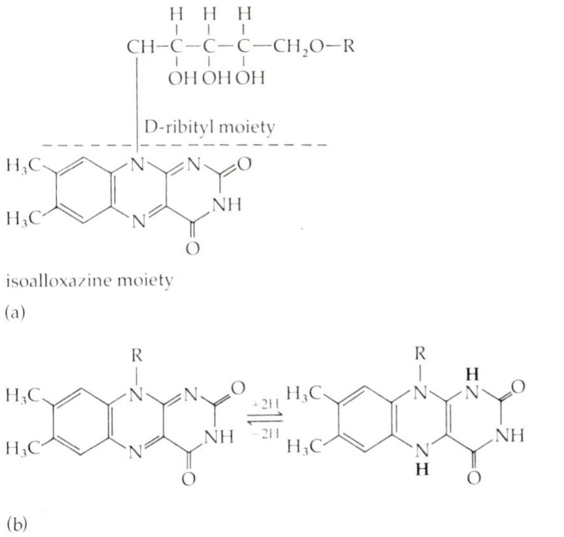

FIGURE 4.10

(a) Structures of the vitamin riboflavin (R is H) and of its two coenzyme derivatives, FMN (R is $PO_3H_2$) and FAD (R is ADP). (b) The reversible oxidation and reduction of the ring structure of FMN and FAD.

FIGURE 4.11

(a) Structure of ubiquinone. The number of isoprenoid units

$$-(CH_2-CH=\overset{CH_3}{\overset{|}{C}}-CH_2)-$$

in the side chain varies in different organisms from 6 to 10.
(b) Reversible reduction of a quinone to form a quinol.
(c) Structure of menaquinone.

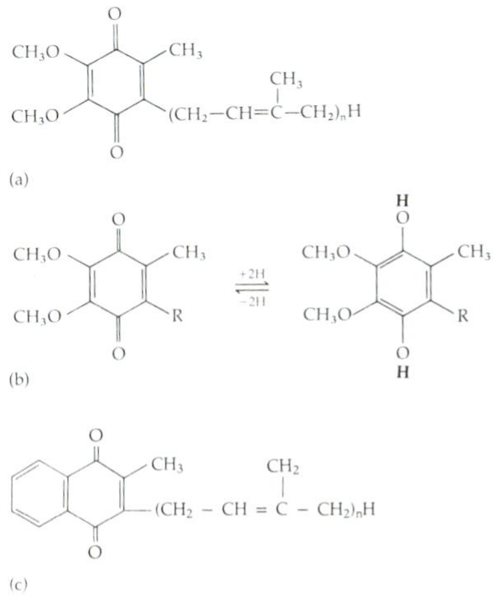

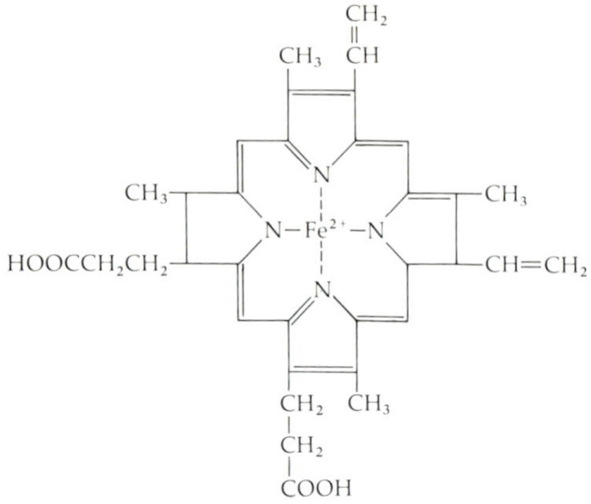

FIGURE 4.12
Structure of heme.

The cytochromes have characteristic absorption bands in the reduced state (Figure 4.13) that permit recognition of the several different members of the class, which are designated by a terminal letter (e.g., cytochrome *c*). There are four types of heme prosthetic groups known (*a*, *b*, *c*, and *d*) that differ principally in the nature of the substituent groups at the periphery of the heme. Individual cytochromes contain one or more of these groups in various combinations. The molecular weights and reduction potentials of cytochromes vary widely (Table 4.2).

Cytochrome content of bacteria is quite variable among species and environments. For example, an aerobically grown bacterium will have a different cytochrome composition from that of the same strain grown anaerobically or even grown with a restricted supply of oxygen. The presence of a *c*-type cytochrome in a particular bacterium is correlated with the outcome of an empirical procedure, the *oxidase test*, which has considerable diagnostic importance in the identification of aerobic bacteria. The test is performed by putting a small quantity of bacteria on a piece of filter paper soaked in a dye: either dichlorophenol indophenol or *N*,*N*-dimethyl-*p*-phenylene-diamine. These dyes, which are colorless in the reduced state, are rapidly oxidized to colored forms by "oxidase-positive" species (which contain a *c*-type cytochrome) but not by "oxidase-negative" species (which lack a *c*-type cytochrome).

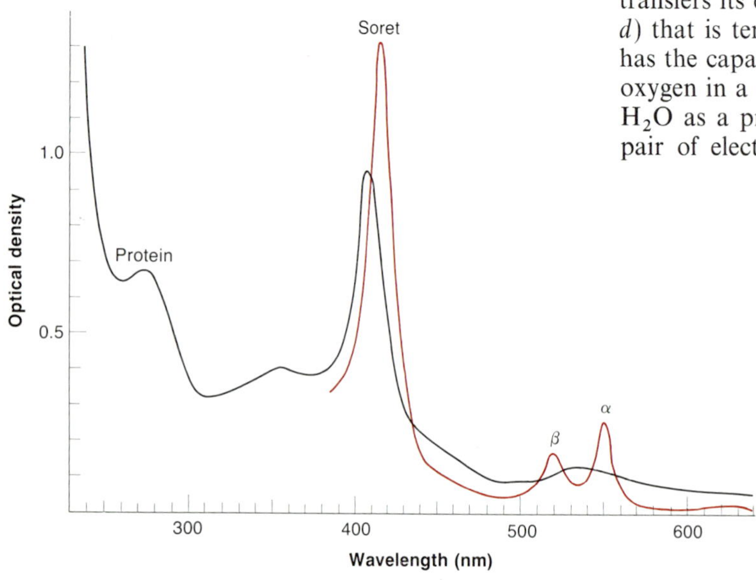


FIGURE 4.13
The absorption spectrum of cytochrome c. Solid line: oxidized cytochrome c; colored lines: reduced cytochrome *c*. The three principal bands in the reduced state (progressing from higher to lower wavelengths) are designated α, β, and Soret. The wavelength of their maximum absorption provides a means of identification. Their protein component absorbs light at shorter wavelengths.

## Arrangement of Electron Transport Chains in the Cell Membrane

As stated, bacteria are quite variable with respect to their cytochrome content; they are also variable with respect to other constituents of electron transport chains. A proposal for the aerobic respiratory chain in *Escherichia coli* is shown in Figure 4.14. NADH in the cytoplasm generated by a reaction in a fueling pathway transfers on the inner aspect of the membrane two hydrogen atoms to a flavoprotein embedded within the membrane. On the outer aspect of the membrane the flavoprotein transfers two electrons to an iron-sulfur protein, thereby releasing two protons outside the cell (into the periplasm). On the inner aspect the iron-sulfur protein reacts with two protons and a quinone, thereby reducing the quinone, which carries two more protons across the membrane and releases them outside when it transfers two electrons to a *b*-type cytochrome. Finally, the cytochrome *b* transfers its electrons to another cytochrome (*o* or *d*) that is termed a *cytochrome oxidase* because it has the capacity to transfer electrons to molecular oxygen in a reaction on the inner surface that has $H_2O$ as a product. It will be noted that for each pair of electrons transported through the chain,

TABLE 4.2

**Cytochromes of Aerobic Electron Transport Chain**

| Cytochrome | Prosthetic Group (number) | Molecular Weight | $E'_0$ value (V) |
|---|---|---|---|
| $a_1$ | 2 heme *a* | | +0.14 and +0.25 |
| $aa_3$ | heme *a* | 73,000 | +0.20 to +0.27 and +0.36 to +0.38 |
| *b* | heme *b* | 12,000 to 17,500 | −0.11 to +0.11 |
| *c* | heme *c* | 12,000 to 100,000 | +0.19 to +0.34 |
| *d* | 2 heme *d* | 350,000 | +0.28 |
| | 2 heme *b* | | +0.14 and +0.26 |
| *o* | 2 heme *b* | 28,000 | −0.12 to +0.42 |

Source: After C. W. Jones, *Bacterial Respiration and Photosynthesis, Aspects of Microbiology* (Washington, D.C.: American Society for Microbiology, 1982).

FIGURE 4.14

The electron transport chain of *Escherichia coli* showing the transfer of reducing power from NADH to $O_2$. The electron and hydrogen carriers embedded in the membrane are abbreviated as follows: flavoprotein (FMN), iron-sulfur protein (Fe-S), ubiquinone (Q), and cytochromes ($b_{556}$, $b_{558}$, *o* and *d*) There are several proposals for the actual composition of this electron transport chain; this is the one of B. A. Haddock and E. W. Jones, "Bacterial Respiration," *Microbiol. Rev.* **41,** 47 (1977).

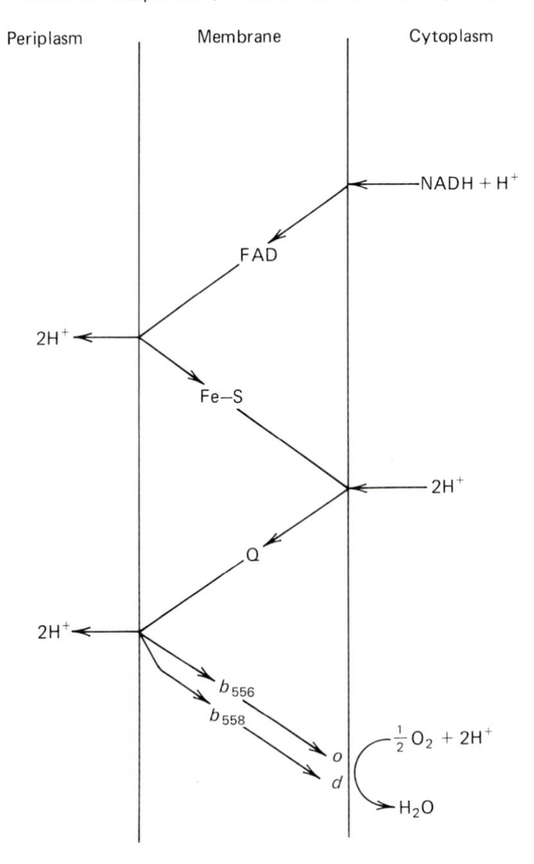


four protons are pumped out of the cell, thus creating a protonmotive force that can be used to do cellular work, for example ATP synthesis, through the membrane-bound ATP phosphohydrolase. It will also be noted that composition of the chain varies with the availability of oxygen. When fully aerobic, cytochrome $b_{550}$* serves as the terminal electron carrier and cytochrome *o* as the terminal oxidase. When the supply of oxygen is limited, cytochrome $b_{558}$ and *d* play the corresponding roles.

## THE BIOCHEMISTRY OF THE FUELING REACTIONS IN AEROBIC HETEROTROPHS

### Pathways of Formation of Pyruvate

Certain reaction pathways are common to both respiratory and fermentative metabolism. Among these are three pathways of conversion of sugars to the key intermediate pyruvic acid. They are the *Embden-Meyerhof pathway* (also called the *glycolytic pathway*), the *pentose phosphate pathway* (also called the *hexose monophosphate shunt*), and the *Entner-Doudoroff pathway*. The first two occur in many organisms including both eucaryotes and procaryotes. The third replaces the Embden-Meyerhof pathway in certain procaryotes (Table 4.3) as a means of metabolizing glucose. However, organisms like *E. coli* that metabolize glucose via the Embden-Meyerhof pathway have the capacity to synthesize enzymes of the Entner-Doudoroff

* Cytochromes are sometimes distinguished by a numerical subscript that corresponds to the wavelength in nm at which their α bands absorb light maximally.

TABLE 4.3

**Distribution of Pathways of Sugar Metabolism among Certain Bacteria**

| Bacterium | Embden-Meyerhof | Entner-Doudoroff |
|---|---|---|
| *Arthrobacter* spp. | + | − |
| *Azotobacter chroococcum* | + | − |
| *Alcaligenes eutrophus* | − | + |
| *Bacillus* spp. | + | − |
| *Escherichia coli* and other enteric bacteria | + | − |
| *Pseudomonas* spp. | − | + |
| *Rhizobium* spp. | − | + |
| *Thiobacillus* spp. | − | + |
| *Xanthomonas* spp. | − | + |

Source: After G. Gottshalk, *Bacterial Metabolism* (New York, Heidelberg, and Berlin: Springer-Verlag, 1979).

pathway; gluconic acid induces these enzymes to be synthesized.

In the Embden-Meyerhof pathway (Figure 4.15) two molecules of ATP are expended in the initial reactions to produce fructose-1,6-bisphosphate. This is cleaved to yield the triose phosphates, glyceraldehyde phosphate and dihydroxyacetone phosphate, which are freely interconvertible. The oxidation of glyceraldehyde phosphate, coupled with the production of two molecules of reduced pyridine nucleotide, is accompanied by an esterification of inorganic phosphate to yield 1,3-bisphosphoglyceric acid. The subsequent steps in the conversion of this compound to pyruvic acid are accompanied by the transfer of both phosphate groups to ADP (substrate level phosphorylation). Thus, the net yield of ATP by substrate level phosphorylation is two molecules; more are formed if the two molecules of NADH donate hydrogen atoms to an electron transport chain. In addition the Embden-Meyerhof pathway generates 6 of the 12 precursor metabolites (Figure 4.15).

As stated, the Embden-Meyerhof pathway is replaced in certain bacteria by a different pathway of conversion of glucose to pyruvic acid, the Entner-Doudoroff pathway. The first reaction, the formation of glucose-6-phosphate, is shared by the two pathways (Figure 4.16).

Following its oxidation to 6-phosphogluconic acid, the unique intermediate of the pathway, 2-keto-3-deoxy-6-phosphogluconic acid (KDPG), is formed by a dehydration step. KDPG is cleaved to one molecule of pyruvic acid and one molecule of glyceraldehyde-3-phosphate, which is metabolized by enzymes shared by the Embden-Meyerhof pathway to produce a second molecule of pyruvic acid. The net yield of the metabolism of one molecule of glucose through the Entner-Doudoroff pathway is one molecule of ATP (as contrasted with two from the Embden-Meyerhof pathway) and one molecule of NADPH. Five of the intermediates are precursor metabolites; a sixth, fructose-6-phosphate, is formed from glucose-6-phosphate by an enzyme shared by the Embden-Meyerhof pathway.

Regardless of whether a bacterium metabolizes glucose via the Embden-Meyerhof or the Entner-Doudoroff pathway, it has all or most of the enzymes of another sugar-metabolizing pathway, the pentose phosphate pathway (Figure 4.17). The explanation of this seeming metabolic redundancy apparently lies in the vital functions of the pentose phosphate pathway in supplying NADPH and two more precursor metabolites, ribose-5-phosphate and erythrose-4-phosphate.

The pentose phosphate pathway does not lead directly to pyruvate; rather it provides only for the oxidation of one of the carbon atoms of glucose. It involves the initial phosphorylation of glucose. The product enters into a series of reactions involving two NADP-linked oxidations and a decarboxylation to yield the pentose phosphate D-ribulose-5-phosphate. By epimerization, D-xylulose-5-phosphate and ribose-5-phosphate are formed. These two pentose phosphates are the starting point for a series of transketolase reactions (transfer of a 2-carbon glycoaldehyde group, ($CH_2OH$—CO—) and transaldolase reactions (transfer of a 3-carbon dihydroxyacetone group, $CH_2OH$—CO—CHOH—) leading eventually to the initial compound of the pathway, glucose-6-phosphate. The pathway is thus cyclic in nature. Passage of a molecule of glucose through the cycle

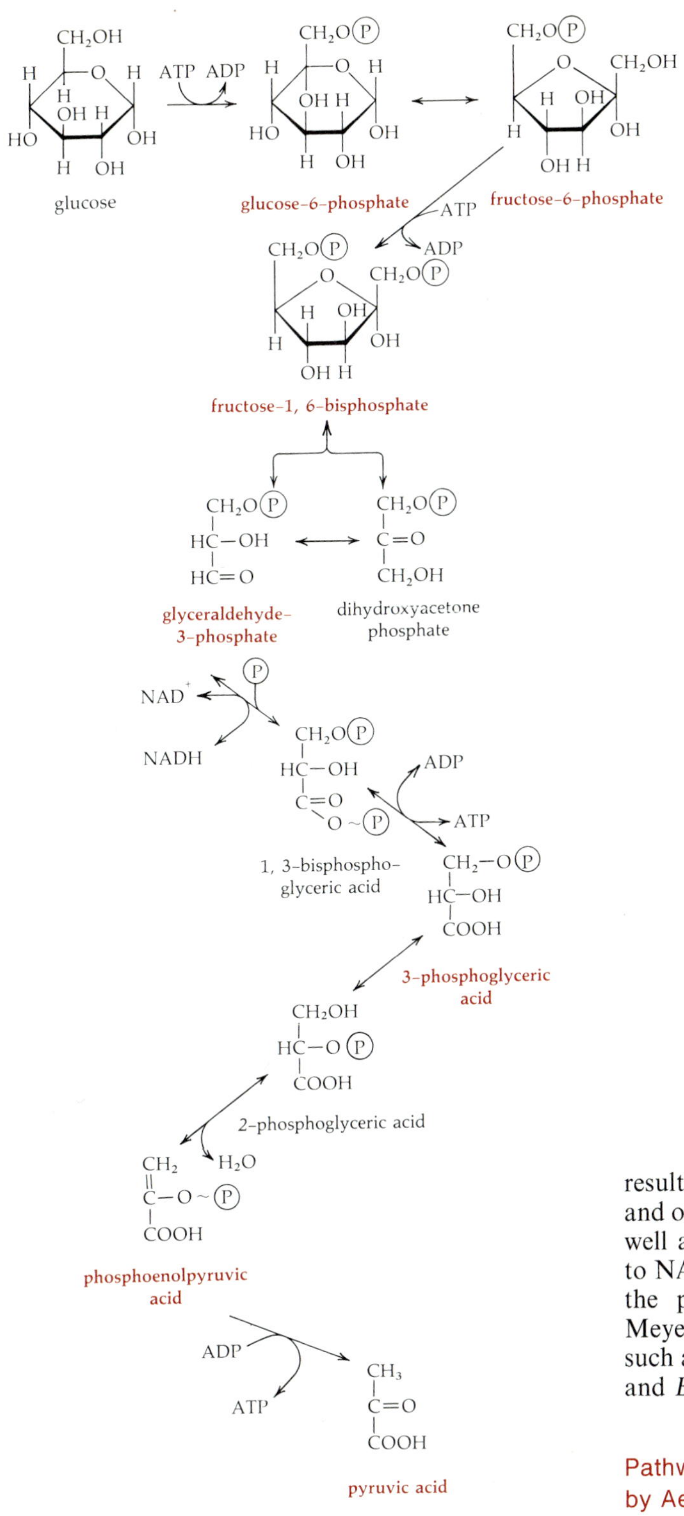

FIGURE 4.15

The Embden-Meyerhof (glycolytic) pathway of conversion of glucose to pyruvic acid. The six precursor metabolites formed as intermediates are shown in color.

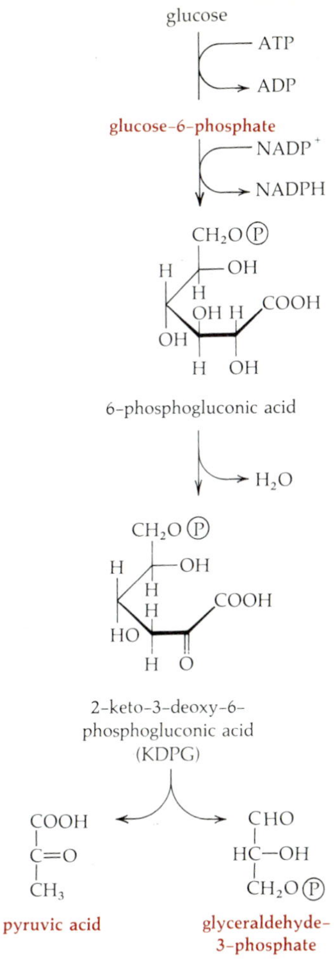

FIGURE 4.16

The Entner-Doudoroff pathway of conversion of glucose to pyruvic acid and glyceraldehyde-3-phosphate. Precursor metabolites are indicated in color.

results in the formation of three molecules of $CO_2$ and one molecule of glyceraldehyde-3-phosphate as well as the reduction of six molecules of $NADP^+$ to NADPH. Thus glucose can be oxidized without the participation of either an intact Embden-Meyerhof or Entner-Doudoroff pathway. Indeed, such appears to be the case in *Thiobacillus novellus* and *Brucella abortus*.

## Pathways of Utilization of Pyruvate by Aerobes

In most aerobes, pyruvate is oxidatively decarboxylated by an elaborate enzyme system termed the *pyruvate dehydrogenase complex* producing acetyl-coenzyme A (acetyl-CoA) as a product according

FIGURE 4.17

The pentose phosphate pathway of glucose oxidation. Precursor metabolites are shown in color.

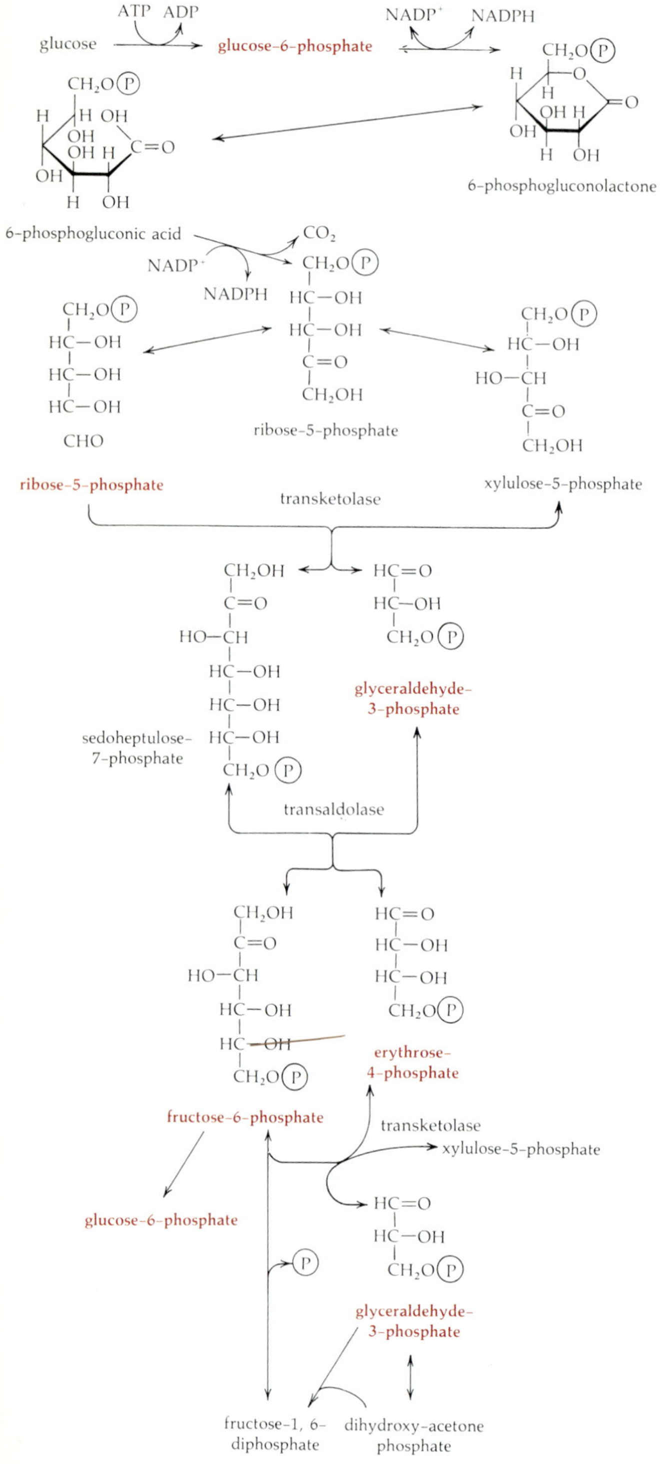

Net reaction:

glucose + 6 $NADP^+ \rightarrow$ glyceraldehyde-3-phosphate + 3 $CO_2$ + 6 NADPH

to the following reaction:

$$CH_3{-}CO{-}COOH + CoA + NAD^+ \longrightarrow CH_3{-}CO{-}CoA + CO_2 + NADH + H^+$$

Acetyl-CoA, being a precursor metabolite, enters into biosynthetic pathways; alternatively, it can be completely oxidized in a cyclic manner through a pathway known as the *tribcarboxylic acid* (*TCA*) cycle (Figure 4.18). This cycle is the major route of ATP generation in aerobic heterotrophs (by passage of electrons from reduced pyridine nucleotides through an electron transport chain). The TCA cycle also generates three of the precursor metabolites, α-ketoglutarate, succinyl-CoA, and oxaloacetate, so even strict anaerobes possess most of the enzymes of the cycle, lacking only the step between α-ketoglutarate and succinate (the α-ketoglutarate dehydrogenase complex). Thus, by reverse flow from oxaloacetate to succinyl-CoA and forward flow from citric acid to α-ketoglutarate, anaerobes are able to synthesize all precursor metabolites, even under anaerobic conditions. The TCA cycle effects the complete oxidation of one molecule of acetic acid to $CO_2$ and generates three molecules of reduced pyridine nucleotides, one molecule of ATP, and one molecule of reduced FAD which donates electrons to a transport chain independent of pyridine nucleotides.

If the TCA cycle operated exclusively for the terminal oxidation of acetyl CoA, it could be maintained without a net input of oxalacetic acid, the role of this compound being purely catalytic. However, as has been stated, the TCA cycle also generates precursor metabolites that are utilized in biosynthesis. Hence, in a growing organism, the cycle is never in fact closed and its maintenance requires a considerable net synthesis of oxalacetic acid by reactions termed *anaplerotic* because they function to resupply the TCA cycle with intermediates. They usually carboxylate either pyruvate or phosphoenolpyruvate:

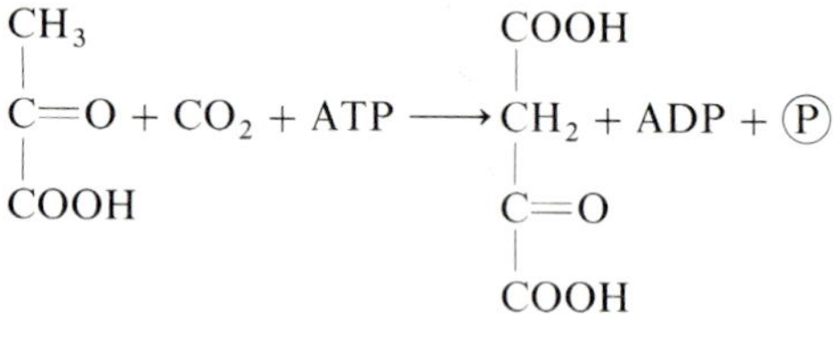

$$\underset{\text{pyruvate}}{CH_3{-}C({=}O){-}COOH} + CO_2 + ATP \longrightarrow \underset{\text{oxalacetate}}{COOH{-}CH_2{-}C({=}O){-}COOH} + ADP + Ⓟ$$

$$\underset{\text{phosphoenolpyruvate}}{CH_2{=}C(O{\sim}Ⓟ){-}COOH} + ADP + CO_2 \rightleftharpoons \underset{\text{oxalacetate}}{COOH{-}CH_2{-}C({=}O){-}COOH} + ATP$$

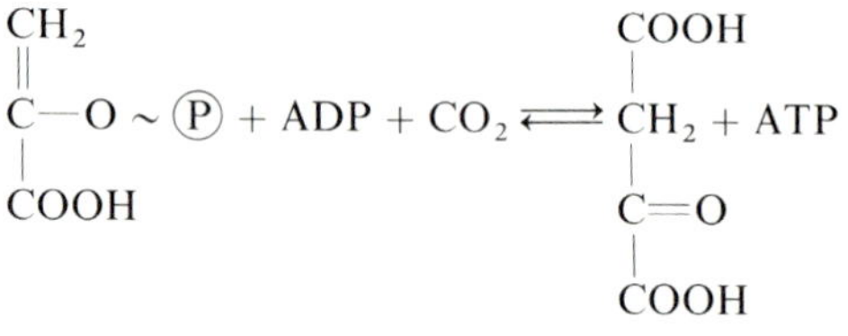

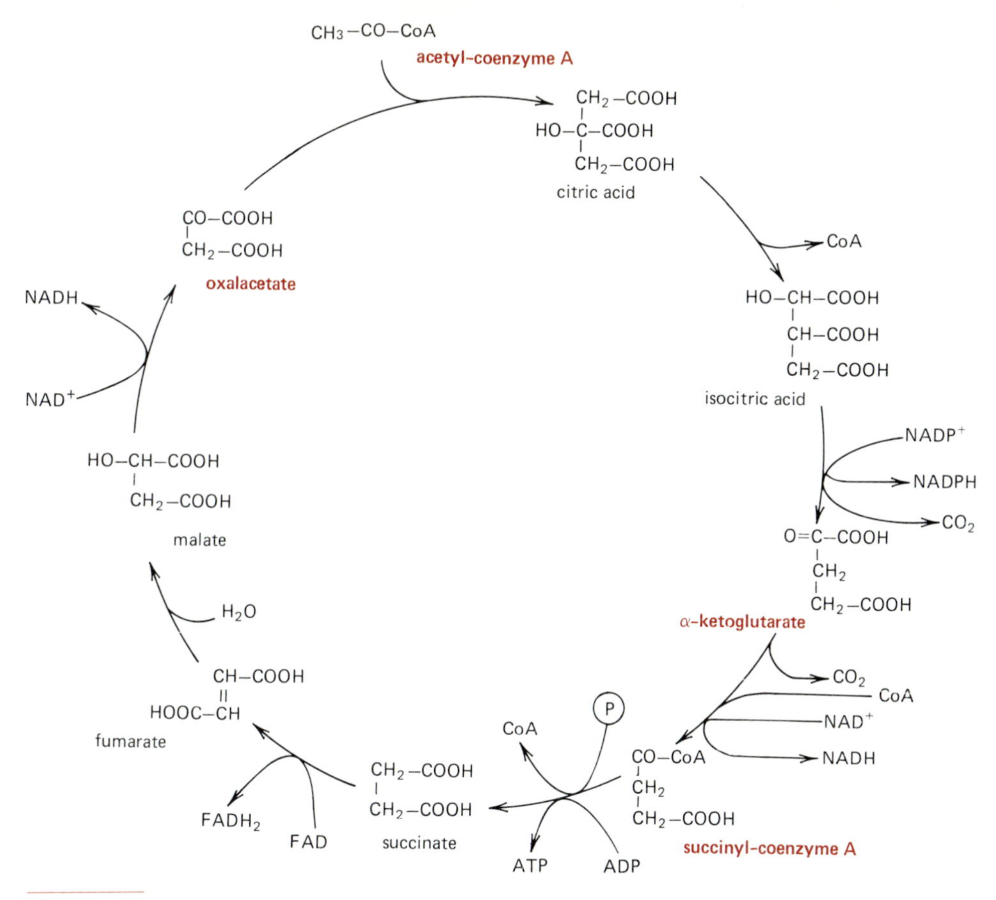


FIGURE 4.18
The tricarboxylic acid (TCA) cycle by which acetyl-CoA is oxidized. Precursor metabolites are shown in color.

As a result, carbon from pyruvate enters the cycle by two routes: via acetyl-CoA and via pyruvate or phosphoenolpyruvate.

## The Role of the Glyoxylate Cycle in Acetic Acid Oxidation

The special modification of the TCA cycle, known as the *glyoxylate cycle*, comes into play during oxidation of acetic acid or substrates (such as fatty acids) that are converted to acetyl-CoA without the intermediate formation of pyruvate. Under these circumstances, oxalacetate cannot be generated from pyruvate or phosphoenolpyruvate, because in aerobes there is no mechanism for synthesizing pyruvate from acetate: the oxidation of pyruvate by the pyruvate dehydrogenase complex is completely irreversible.

The supply of oxalacetate required for oxidation of acetate is replenished by the oxidation of succinate and malate, which are produced through a sequence of two reactions. In the first reaction, isocitrate, which is a normal intermediate of the TCA cycle, is cleaved to yield succinate and glyoxylate:

$$\begin{matrix} COOH \\ | \\ CHOH \\ | \\ CH{-}COOH \\ | \\ CH_2 \\ | \\ COOH \end{matrix} \longrightarrow \begin{matrix} COOH \\ | \\ CHO \end{matrix}\ \text{glyoxylate} + \begin{matrix} CH_2{-}COOH \\ | \\ CH_2{-}COOH \end{matrix}\ \text{succinate}$$

In the second reaction, acetyl-CoA is condensed with glyoxylate to yield malate:

$$\begin{matrix} COOH \\ | \\ CHO \end{matrix} + \text{acetyl-CoA} \longrightarrow HO{-}\begin{matrix} CH{-}COOH \\ | \\ CH_2{-}COOH \end{matrix} + CoA$$

In combination, these two reactions constitute a bypass whereby two carbon atoms lost from

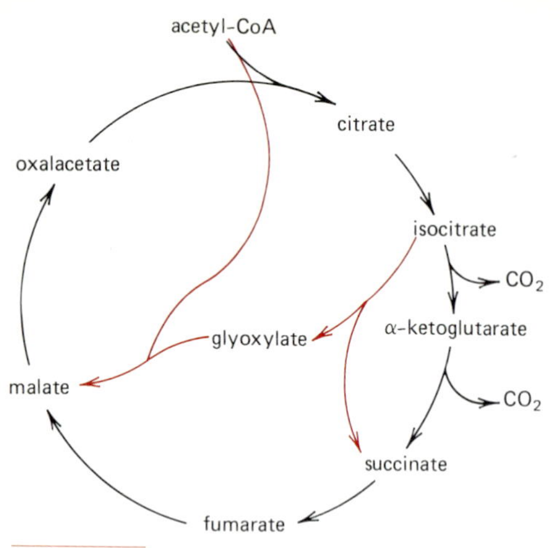


FIGURE 4.19

The glyoxylate bypass (colored) and its relation to the reactions of the citric acid cycle.

the TCA cycle as $CO_2$ are preserved in the glyoxylate cycle (Figure 4.19) as glyoxylate, which can then combine with acetyl-CoA to form malate, the immediate precursor of oxalacetate. Thus the glyoxylate cycle acts as an anaplerotic sequence allowing the normal TCA cycle to function.

## Special Pathways for Primary Attack on Organic Compounds by Microorganisms

In the preceding sections, we have emphasized the pathways by which glucose is metabolized. Sometimes these are termed *central metabolism*, because of the vital roles they play in the synthesis of precursor metabolites and in forming part of the pathways by which a variety of other substrates are metabolized. Owing to their essential nature, they are functionally reversible. For example, when an aerobe such as *E. coli* is grown in a medium that contains a compound such as succinate as a total source of carbon, the Embden-Meyerhof pathway must operate in a reverse direction in order to synthesize several precursor metabolites. It will be noted (Figure 4.15) that all reactions linking glucose-6-phosphate and pyruvate are reversible with the exception of the conversion of fructose-6-phosphate to fructose-1,6-bisphosphate. In order to overcome this apparent blockade, a second enzyme (fructose-1,6-bisphosphatase) is present that catalyzes an ATP-independent conversion of the diphosphate to the monophosphate form of fructose.

As stated earlier, there is probably no naturally occurring organic compound that cannot be used as a substrate for metabolism by some microorganism. However complex the structure of the substrate may be, its utilization always involves the same basic principle: a stepwise degradation to yield eventually one or more fragments capable of entering central metabolism. As examples of the many such specialized microbial pathways, we shall describe those involved in the utilization of the disaccharide lactose, the metabolic products of which enter the Embden-Meyerhof pathway, and certain aromatic compounds whose products enter the TCA cycle.

Lactose is hydrolytically split (Figure 4.20) yielding glucose, which enters central metabolism directly, and galactose, which is phosphorylated and then converted to another intermediate of central metabolism, glucose-1-phosphate, by a cyclic pathway. The galactose and glucose moieties of galactose-1-phosphate and UDP-glucose are exchanged, yielding glucose-1-phosphate and UDP-galactose, which can undergo epimerization to UDP-glucose. The catalytic amounts of UDP-glucose required for the cycle to function can be supplied by a reaction between UTP and glucose-1-phosphate, yielding pyrophosphate as the other product.

One route by which aromatic compounds are attacked by those bacteria that have this capacity is through one or another of two convergent

FIGURE 4.20

The pathway of utilization of the disaccharide lactose. Compounds that enter central metabolic pathways are colored.

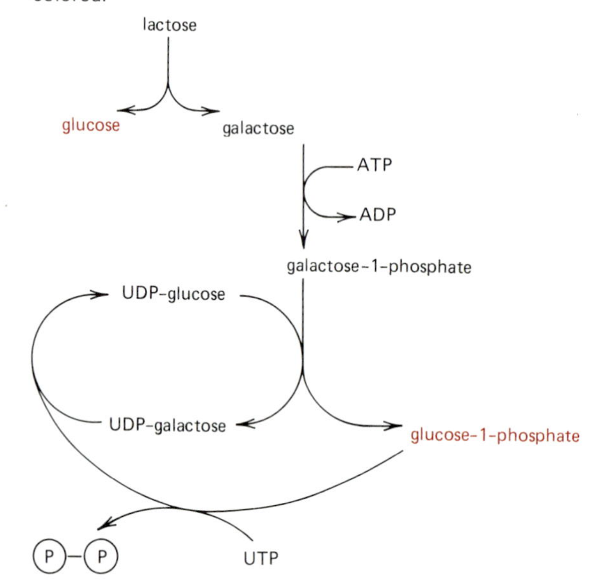

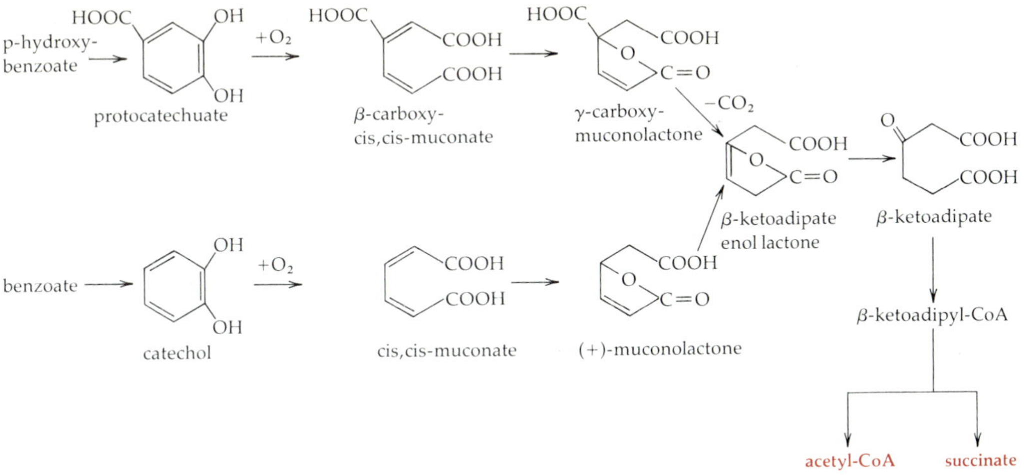

HOOC OH
p-hydroxy-
benzoate → $+O_2$ →
OH
protocatechuate

HOOC COOH COOH
β-carboxy-
cis,cis-muconate

HOOC COOH O C=O
γ-carboxy-
muconolactone

$-CO_2$

COOH O C=O
β-ketoadipate
enol lactone

O COOH COOH
β-ketoadipate

benzoate → OH OH $+O_2$ →
catechol

COOH COOH
cis,cis-muconate

COOH O C=O
(+)-muconolactone

β-ketoadipyl-CoA

acetyl-CoA
succinate

FIGURE 4.21
The chemistry of the β-ketoadipate pathway. Compounds that enter the TCA cycle are shown in color.

branches of the β-ketoadipate pathway (Figure 4.21). Through these reactions, the six carbon atoms of the aromatic nucleus are converted to those of an aliphatic acid; this is in turn cleaved to acetyl-CoA and succinate, both of which enter the TCA cycle.

A number of other structurally related compounds are metabolized to intermediates of the β-ketoadipate pathway and flow through it to intermediates of the TCA cycle. Some of these compounds and their points of convergence with the β-ketoadipate pathway are shown in Figure 4.22.

FIGURE 4.22
Certain compounds which are metabolized through the β-ketoadipate pathway: their structures and points of convergence with the pathway. Intermediates of the β-ketoadipate pathway (see Figure 4.21) are shown here in boldface type.

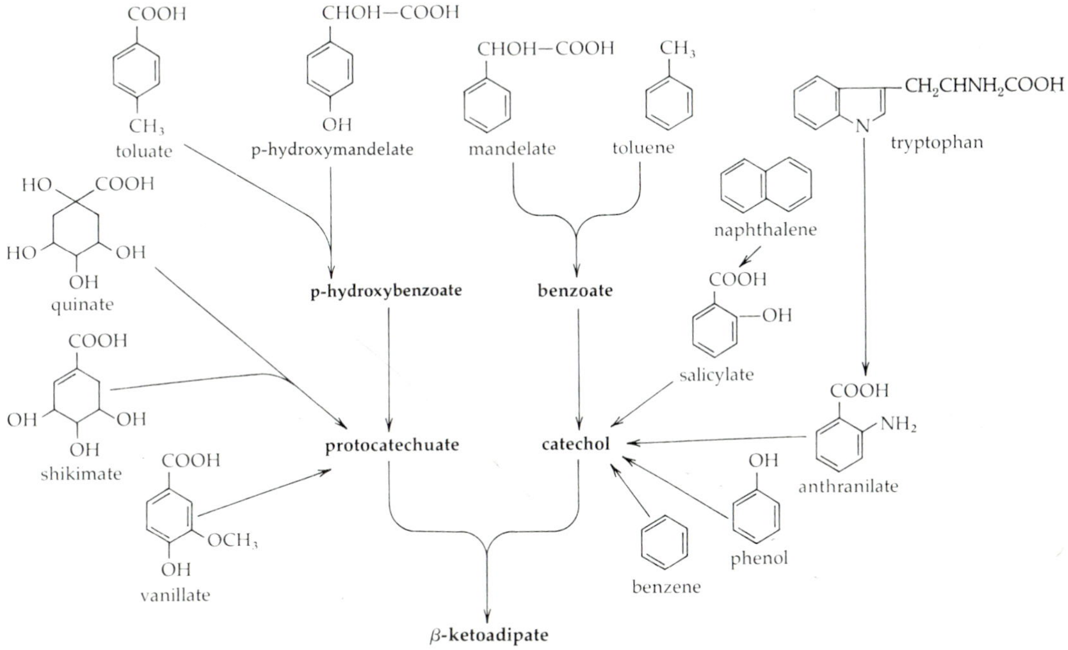

COOH $CH_3$
toluate

CHOH—COOH OH
p-hydroxymandelate

CHOH—COOH
mandelate

$CH_3$
toluene

$CH_2CHNH_2COOH$ N
tryptophan

HO COOH HO OH OH
quinate

naphthalene

COOH OH
salicylate

**p-hydroxybenzoate**

**benzoate**

COOH OH OH OH
shikimate

COOH $NH_2$
anthranilate

**protocatechuate**

**catechol**

COOH $OCH_3$ OH
vanillate

OH
phenol

benzene

**β-ketoadipate**

## THE FUELING REACTIONS OF ANAEROBIC CHEMOTROPHS

Anaerobic and facultatively anaerobic chemoheterotrophs have one or both of two patterns of fueling reactions, termed *anaerobic respiration* and *fermentation.*

### Anaerobic Respiration

Anaerobic respiration involves essentially the same biochemical pathways as the aerobic metabolism (or aerobic respiration) of heterotrophs, differing principally in the compound that serves as the terminal electron acceptor of the electron transport chain. Rather than molecular oxygen being the terminal acceptor, nitrate ($NO_3^-$), sulfate ($SO_4^{2-}$), fumarate, or trimethylamine oxide may serve the corresponding role. In the case of $SO_4^{2-}$ or $NO_3^-$, the products of their reduction are also capable of being terminal electron acceptors, thus forming a cascade of anaerobic respirations. In the case of the reduction of $NO_3^-$, some bacteria, by a set of three anaerobic respirations, produce dinitrogen as a final product; $NO_3^-$ is the terminal electron acceptor of the first chain; its reduction product, nitrite ($NO_2^-$), is the acceptor for the next; and its reduction product, nitrous oxide ($N_2O$), is the acceptor for the final one. This process is termed *denitrification* because it converts a nonvolatile form of nitrogen to a volatile one, thereby depleting an aqueous or terrestrial environment of fixed nitrogen which is essential for the growth and development of most organisms. The ways by which fixed nitrogen is replenished and the various molecular forms of nitrogen are cycled through the biosphere are discussed in Chapter 27.

### Fermentation

Fermentations are fueling reactions that do not require the participation of an electron transport chain, although these do play minor roles in certain cases. As a consequence of the lack of participation of an external electron acceptor, the organic substrate undergoes a balanced series of oxidative and reductive reactions; pyridine nucleotides reduced in one step of the process are oxidized in another. This general principle is illustrated by two fermentations: alcoholic fermentation (typical of the anaerobic metabolism of glucose by yeasts), and the homolactic acid fermentation (typical of the metabolism of certain lactic acid bacteria). Both of these fermentative processes (Figure 4.23) utilize the Embden-Meyerhof pathway: the two molecules of NAD reduced by this pathway are reoxidized in reactions involving the subsequent metabolism of pyruvate. In the case of the homolactic acid fermentation, this oxidation occurs as a direct conse-

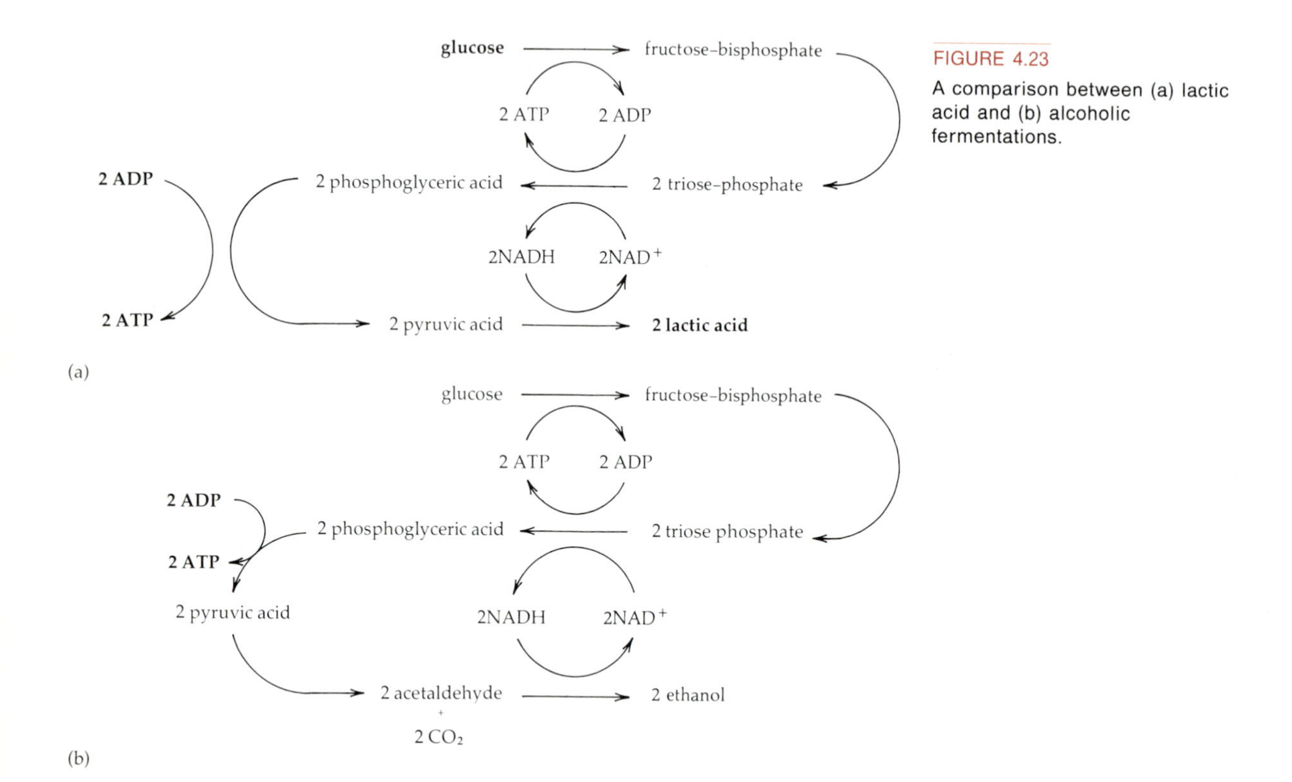


FIGURE 4.23
A comparison between (a) lactic acid and (b) alcoholic fermentations.

quence of the reduction of pyruvic acid to lactic acid. In the case of the alcoholic fermentation, pyruvic acid is first decarboxylated to form acetaldehyde; the reoxidation of NADH occurs concomitantly with the reduction of acetaldehyde to form ethanol.

The Embden-Meyerhof pathway is the most widespread one for the fermentative conversion of glucose to pyruvic acid, and is employed by many bacteria that produce fermentative end products other than lactic acid and ethanol. These differences exclusively reflect differences with respect to metabolism of pyruvic acid. The various patterns of pyruvic acid metabolism are summarized in Figure 4.24. Most bacterial fermentations may produce several end products; however, no single fermentation produces all of the end products shown in Figure 4.24.

Not all fermentative mechanisms follow the Embden-Meyerhof pathway. Certain fermentations of glucose follow the pentose phosphate pathway, and others follow the Entner-Doudoroff pathway. Fermentations of substrates other than sugars (e.g., amino acids) involve highly specific pathways.

The end products of fermentations and the pathway by which they are formed are group specific. These particular fermentations will be considered in subsequent chapters along with the other characteristics of the various physiological groups of bacteria.

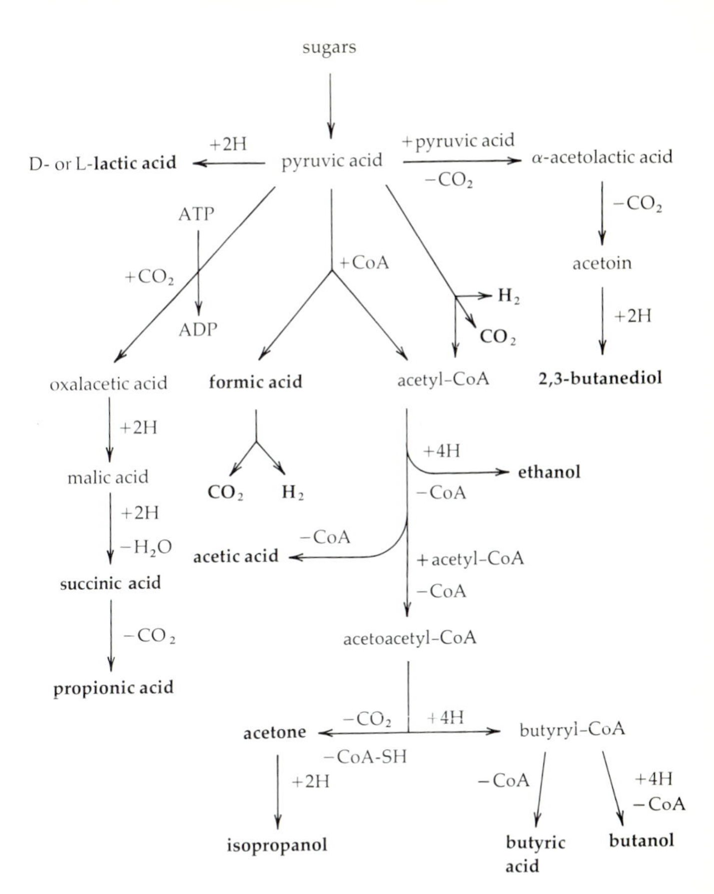


FIGURE 4.24

Derivations of some major end products of the bacterial fermentations of sugars from pyruvic acid. The end products are shown in boldface type.

## THE FUELING REACTIONS OF AUTOTROPHS

Unlike those of heterotrophs, the fueling reactions of autotrophs, which obtain all or nearly all of their cellular carbon from $CO_2$, occur in two biochemically distinct phases: (1) that leading to the synthesis of precursor metabolites, and (2) that leading to the synthesis of ATP and reduced pyridine nucleotides. One pathway of synthesis of precursor metabolites, the *Calvin-Benson cycle*, is shared by most photoautotrophs and chemoautotrophs.

### The Calvin-Benson Cycle: Synthesis of Precursor Metabolites

Most autotrophs capture (*fix*) $CO_2$ by a reaction catalyzed by *ribulose bisphosphate carboxylase* (Figure 4.25). Exceptions are the green bacteria, which utilize a pathway termed the *reverse TCA cycle* (discussed in Chapter 15), the sulfur-reducing bacteria, the autotrophic archaebacteria, and the acetogenic

FIGURE 4.25

The $CO_2$-fixing reaction of the Calvin-Benson cycle. The phosphorylated pentose, ribulose-1,5-diphosphate, accepts one molecule of $CO_2$ and is simultaneously cleaved yielding two molecules of glyceric acid-3-phosphate; the carboxyl group of one glyceric acid-3-phosphate molecule is thus derived from $CO_2$.

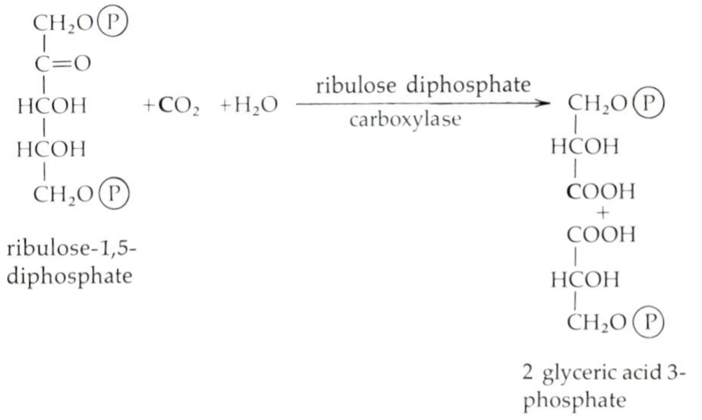

CH₂O(P) – C=O – HCOH – HCOH – CH₂O(P) $+ CO_2 + H_2O$ —(ribulose diphosphate carboxylase)→ CH₂O(P) – HCOH – COOH + COOH – HCOH – CH₂O(P)

ribulose-1,5-diphosphate → 2 glyceric acid 3-phosphate

bacteria which use the carbon monoxide pathway (discussed in Chapter 22). The primary product of $CO_2$ fixation via the Calvin-Benson cycle is glyceric acid-3-phosphate from which all precursor metabolites are synthesized. However, $CO_2$ fixation is dependent on a supply of ribulose bisphosphate. Consequently, most of the glyceric acid must be utilized to regenerate this $CO_2$ acceptor. Various intermediates of the cycle are precursor metabolites or can be converted to them. The reactions by which autotrophs synthesize precursor metabolites are thus complex, combining the Calvin-Benson cycle with certain reactions of the glycolytic and pentose phosphate pathways. For simplicity of analysis, the Calvin-Benson cycle can be divided into three phases: $CO_2$ fixation, reduction of fixed $CO_2$, and regeneration of the $CO_2$ acceptor (Figure 4.26). Only two reactions are specific to the Calvin-Benson cycle: the $CO_2$ fixation reaction itself and the regeneration of the $CO_2$ acceptor from its immediate precursor:

$$\text{ribulose-5-phosphate} + \text{ATP} \longrightarrow \text{ribulose-1,5-bisphosphate} + \text{ADP}$$

It will be noted that the Calvin-Benson cycle, rather than producing ATP and reducing pyridine nucleotides, utilizes these essential metabolites. In autotrophs they are synthesized by other mechanisms.

FIGURE 4.26

The schematic representation of the Calvin-Benson cycle illustrating its three phases: $CO_2$ fixation; reduction of fixed $CO_2$; and regeneration of the $CO_2$ acceptor.

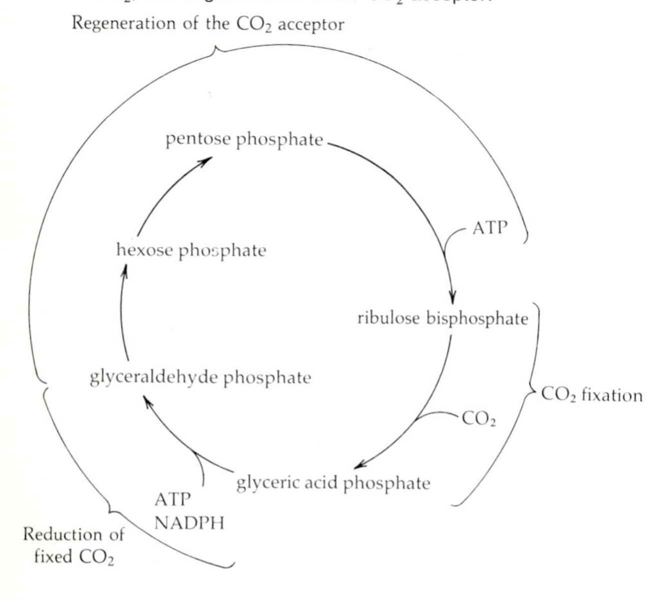


## Generation of ATP and Reduced Pyridine Nucleotides by Chemoautotrophs

Chemoautotrophs obtain ATP and reducing power by the oxidation of inorganic compounds. The substrates that can so serve as energy sources are $H_2$, CO, $NH_3$, $NO_2^-$, $Fe^{2+}$, and reduced sulfur compounds ($H_2S$, S, $S_2O_3^{2-}$). In this mode of respiratory metabolism, electrons from these compounds are passed through an electron transport chain that generates ATP by the mechanisms that function in heterotrophs: generation of a protonmotive force that drives ATP synthesis by a membrane-located ATP phosphohydrolase. The terminal electron acceptor of the chain of chemoautotrophs is usually $O_2$. Certain of these inorganic substrates ($H_2$ and CO) are sufficiently powerful reducing agents to be able to reduce pyridine nucleotides directly, but others are not. Those autotrophic bacteria that generate ATP by oxidizing these weaker reducing agents reduce pyridine nucleotides by a process termed *reverse electron transport*. A portion of the protonmotive force derived from the functioning of the normal electron transport chain is used to drive electrons in an otherwise thermodynamically unfavorable (reverse) direction through another chain linking the inorganic substrate to oxidized pyridine nucleotides, thereby reducing them.

An example of the coupling between forward and reverse electron transport is shown in Figure 4.27. In the oxidation of the nitrite ($NO_2^-$) to nitrate ($NO_3^-$), a reaction by which nitrifying bacteria obtain all of their ATP and reducing power, two electrons from each oxidized nitrite ion enter a branched electron transport chain. Some pairs flow in the thermodynamically favorable direction to $O_2$, thereby generating a protonmotive force. Others flow at the expense of this force in the thermodynamically unfavorable direction (reverse electron transport) to $NAD^+$, thereby reducing it and generating reducing power.

# PHOTOSYNTHESIS

Photosynthesis is the process by which plants and certain bacteria, termed *phototrophs*, convert radiant energy in the form of light into metabolic energy and reducing power. Other aspects of the metabolism of these organisms resemble the corresponding ones of chemotrophs. The unique reactions of photosynthesis are mediated by a class of molecules termed *chlorophylls*, representatives of which nearly all phototrophs contain. When these molecules absorb light energy they become activated to a state

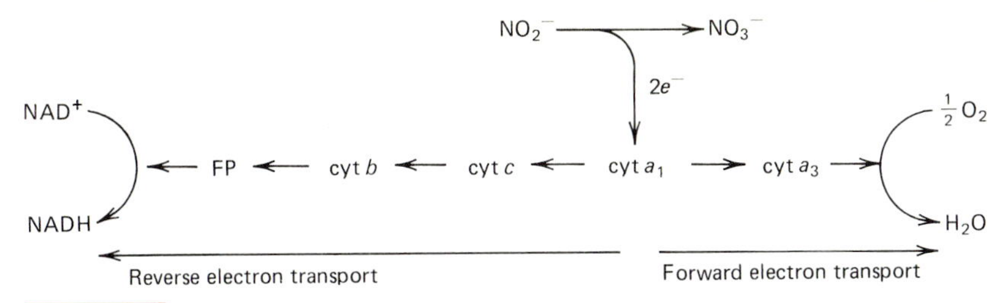


FIGURE 4.27

Forward and reverse electron transport chains of nitrifying bacteria. Electrons derived from the oxidation of nitrite ($NO_2^-$) to nitrate ($NO_3^-$) flow into a branched electron transport chain composed of various cytochromes (cyta$_1$, cyta$_3$, cyt*b* and cytc) and a flavoprotein (FP). Those that flow in the thermodynamically favorable (forward) direction generate a protonmotive force that is utilized in part to drive other electrons in the thermodynamically unfavorable (reverse) direction to reduce $NAD^+$. After G. Gottschalk, *Bacterial Metabolism* (New York, Heidelberg, and Berlin: Springer-Verlag, 1979).

that is readily oxidized by the transfer of an electron to a carrier molecule. In the oxidized state, chlorophyll is readily reduced by accepting an electron from another carrier molecule to regenerate the ground state. Thus through its photoactivation properties, chlorophyll is capable of serving both as primary electron donor and terminal electron acceptor of a membrane-located electron transport chain, through which the flow of electrons can generate a protonmotive force across the membrane in an analogous way to which a protonmotive force is generated in the process of respiration.

When chlorophyll serves both as donor and acceptor of electrons to a photosynthetic electron transport chain, no electrons can be withdrawn for the purpose of producing reducing power by reducing pyridine nucleotides. These electrons must be supplied from another source. In various phototrophs, these electrons are supplied by water, reduced sulfur compounds, hydrogen gas, or organic compounds.

In the following sections, the general biochemical principles of photosynthesis are discussed. Variations of the process among the various groups of phototrophs are discussed in greater detail in Chapter 15.

The photosynthetic apparatus of all organisms capable of carrying out photosynthesis consists of three essential components: an antenna of light-harvesting pigments, a photosynthetic reaction center, and an electron transport chain.

## Antenna of Light-Harvesting Pigments

Light-harvesting pigments can include chlorophylls, carotenoids, and phycobiliproteins; they function to absorb light energy and transmit it to the photosynthetic reaction center. The particular set of light-harvesting pigments that comprise an antenna system are group specific and their cumulative light-absorptive properties determine the range of wavelengths of light over which photosynthesis occurs, and therefore the habitat of phototroph. As a consequence of these variations in composition of antenna systems, phototrophs collectively are capable of utilizing as an energy source all radiant energy that falls in the wavelength of visible and near infrared light.

Radiant energy is always transferred in discrete packets known as *photons*, the energy content of which is inversely related to wavelength (Figure 4.28). When radiant energy is absorbed by matter, its possible effects are a function of the energy content of the photon and hence the wavelength of the radiation. Wavelengths longer than 1,200 nm, in the infrared region and beyond, have an energy content so small that the absorbed energy is immediately converted to heat; it cannot mediate chemical change. Wavelengths less than 200 nm, in the range of X rays, $\alpha$ particles, and cosmic rays, are termed *ionizing radiations* because their energy content is so high that molecules in their path are immediately ionized. Between these two extremes (200 to 1200 nm) lies the region of electromagnetic radiation that is capable of mediating biological reactions. It is this portion of the electromagnetic spectrum (ultraviolet, visible, and near-infrared light) that can serve for the performance of photosynthesis, and that constitutes the major portion of the energy content of solar radiation at the earth's surface. The shorter and longer wavelengths of the broad spectrum of electromagnetic radiation from the sun are filtered out during passage through the earth's atmosphere. Indeed, at sea level, three-quarters of the energy content of solar radiation is

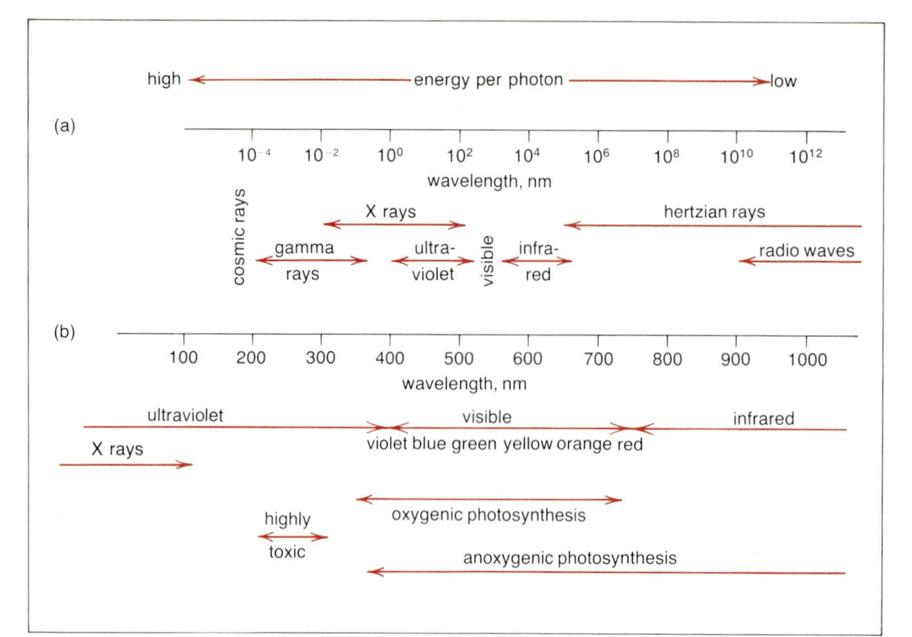


FIGURE 4.28

The electromagnetic spectrum. (a) The entire spectrum is plotted on an exponential scale. (b) The ultraviolet, visible, and near-infrared regions are greatly expanded and plotted on an arithmetic scale.

contained in light of wavelengths between 400 and 1,100 nm, and it is within these limits that pigments responsible for light capture in photosynthesis absorb maximally.

The chlorophylls, of which at least seven kinds occur in various groups of phototrophs, absorb light intensely in two regions of the spectrum: the violet around 400 nm, and the red or near-infrared, around 600 to 1100 nm. The region of peak absorption within the longer wavelengths varies with the particular chlorophyll species and the proteins with which they are associated.

Carotenoids have a single broad region of absorption between 450 and 550 nm. Phycobiliproteins absorb from 550 to 650 nm, between the major absorption regions of a chlorophyll.

All chlorophylls share the same molecular ground plan shown in Figure 4.29. These compounds are related structurally and biosynthetically to hemes (Figure 4.12), the prosthetic groups of cytochromes. Both types of molecules have a central tetrapyrrolic nucleus, within which a metal ion is chelated: iron in hemes, and magnesium in chlorophylls. Two additional properties distinguish chlorophylls chemically from hemes: chlorophylls contain a fifth ring, the pentanone ring, and one of the side chains of their tetrapyrrolic nucleus is esterified to an alcohol.

The carotenoids, of which a large number of different kinds occur in phototrophs, have the basic structure of long, unsaturated hydrocarbons with projecting methyl groups (Figure 4.30). In particular members of the class, this basic structure can be modified in several ways: by terminal ring closure to form six-membered alicyclic or aromatic rings, and by the addition of oxygenated substituents, notably hydroxyl, ethoxyl, or keto groups.

Phycobiliproteins are water-soluble chromoproteins containing linear tetrapyrroles.

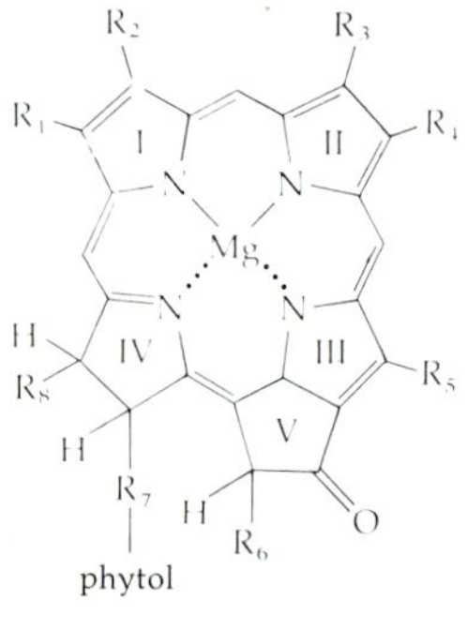

FIGURE 4.29

The molecular ground plan of the chlorophylls. The tetrapyrrolic nucleus (rings I, II, III, and IV) has the same derivation as that of the hemes, but it is chelated with magnesium. In chlorophylls, one or more of the pyrrole rings are reduced; in this diagram, ring IV is shown reduced, as is characteristic of chlorophyll a; $R_1$, $R_2$, and so forth, designate aliphatic side chains attached to the tetrapyrrolic nucleus. The presence of ring V, the pentanone ring, and the substitution of $R_7$ by a long-chain alcohol are characteristic features of the chlorophylls that do not occur in hemes.

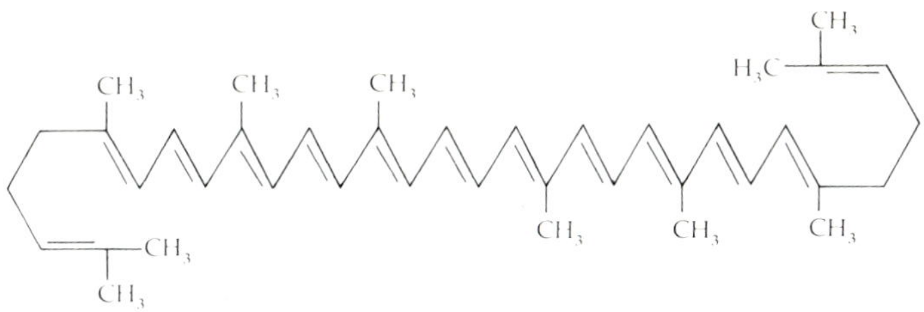

FIGURE 4.30

The molecular ground plan of the carotenoids, illustrated by an open-chain carotenoid which does not contain oxygen. This basic structure may be modified, in the different kinds of carotenoids, by terminal ring closure at one or both ends of the molecule and by the introduction of hydroxyl (—OH), methoxyl (—$OCH_3$), or ketone (=O) groups.

### Photochemical Reaction Centers

The photochemical reaction center contains the site where a molecule of chlorophyll becomes photoactivated and oxidized by donating an electron to a carrier molecule. Chlorophyll molecules in the reaction center differ from those in the antenna in two important respects: (1) they are associated with certain proteins that interact with them in a manner that decreases the energy required to raise them to the activated state, and (2) they are in close proximity with carrier molecules that can accept an electron from them when they are activated.

The more numerous chlorophyll molecules in the antenna (in the case of most purple bacteria there are 50 to 500 light-harvesting chlorophyll molecules per reaction center), being unassociated with carrier molecules, do not become oxidized upon activation; rather they transfer their absorbed energy by a process termed *inductive resonance* to an adjacent pigment and eventually to the reaction center. Since the reaction center chlorophylls are more easily activated, this transfer is associated with a slight loss of energy in the form of heat.

The energy required to activate a molecule of chlorophyll, designated P, in a reaction center can be reckoned by the maximum wavelength of a photon that can bring it about. Thus, reaction center bacteriochlorophyll (as bacterial chlorophylls are called) of a purple bacterium that is activated maximally by photons of wavelength 870 nm is designated, $P_{870}$.

One of the best-characterized reaction centers is that from the purple bacterium, *Rhodobacter sphaeroides*. It contains three polypeptides (21, 24, and 28 kilodaltons), four bacteriochlorophyll molecules, two bacteriopheophytin molecules (bacteriochlorophylls that lack magnesium), two ubiquinone molecules, and one iron molecule. It is presumed from a variety of experiments that two of the reaction centers of bacteriochlorophylls are in the $P_{870}$ state, owing to their association with the three polypeptides. When activated, each can donate an electron to an associated molecule of bacteriochlorophyll ($B_{800}$), which passes it to a molecule of bacteriopheophytin, which passes it sequentially through two molecules of ubiquinone. The second molecule of ubiquinone seems to act as a gate, accepting two electrons before passing them to the associated electron transport chain.

### Photosynthetic Electron Transport Chain

Photosynthetic electron transport chains located within the photosynthetic membrane are composed of the same sorts of carrier molecules—cytochromes, quinones, and iron-sulfur centers—that are found in respiratory electron transport chains, and they appear to function in the same manner. As electrons flow through the chain, a proton-motive force is generated that is used, in part, to synthesize ATP by a membrane-located ATP phosphohydrolase.

### Patterns of Electron Flow

The simplest pattern of electron flow is that alluded to at the beginning of this section, whereby reaction center chlorophyll in its photoactivated and oxidized states serves respectively as both an electron donor and acceptor for an electron transport chain [Figure 4.31 (a)]. Such a system is termed *cyclic photophosphorylation.* However, as stated previously, electrons cannot be withdrawn from it, so another source of electrons is required to generate reducing power in the form of reduced pyridine nucleotides. Fundamental differences in the patterns of electron flow exist, depending on whether or not the source of electrons is water, because the oxidation of (removal of electrons from) water occurs only at the relatively high potential of 0.8 V. The product of the oxidation is $O_2$ gas. Thus, those organisms (plants, algae, and cyanobacteria) that utilize water are said to carry out *oxygenic* or *plant photosynthesis.* Organisms (purple and green bacteria) that utilize other electron sources are said to carry out *anoxygenic* or *bacterial photosynthesis.*

In certain organisms that carry out anoxygenic photosynthesis, reducing power may be generated by reverse electron transport as it is in certain chemoautotrophs. In others, it is generated by

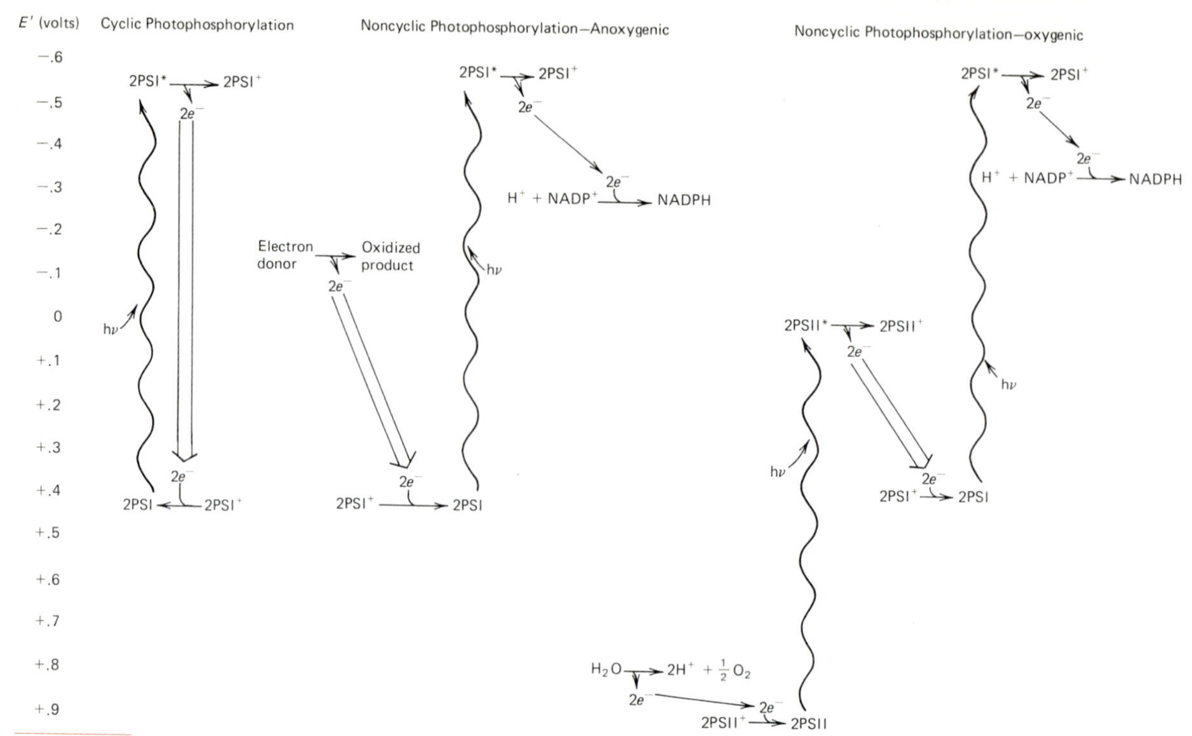


FIGURE 4.31

Patterns of electron flow in various forms of photophosphorylation: (a) cyclic photophosphorylation as it occurs in both anoxygenic and oxygenic photosynthesis; (b) noncyclic photophosphorylation as it occurs in anoxygenic or bacterial photosynthesis; (c) noncyclic photophosphorylation as it occurs in oxygenic or plant photosynthesis. The values of reduction potential ($E'$) at which the various electron transfers occur is indicated by their relative position to the scale at the left. Photosystems I and II (PSI and PSII) are indicated as being activated (∗) or oxidized (+). Wiggly lines indicate the range over which light (hv) activation occurs. Double arrows represent electron transport chains which have the potential of generating a cross-membrane protonmotive force.

a process termed, *noncyclic photophosphorylation* [Figure 4.31 (b)]. In this pattern of electron flow, oxidized reaction center bacteriochlorophyll accepts electrons from an electron transport chain fed by an alternate electron source, thus freeing electrons emanating from bacteriochlorophyll in the activated state to be used for the reduction of pyridine nucleotides.

The process of noncyclic photophosphorylation in organisms that carry out oxygenic photosynthesis is a more complicated one owing to the high potential required to remove electrons from water [Figure 4.31 (c)]. Unactivated chlorophyll molecules in the sorts of reaction centers we have discussed to this point, collectively termed *photosystem I*, do not constitute a sufficiently powerful oxidant to accept electrons from water. In oxygenic photosynthesis, photosystem *I* operates in tandem with another type of reaction center, *photosystem II*, that operates over a much higher range of reduction potential. Together photosystems I and II can mediate a change in reduction potential in excess of the amount required to oxidize water and reduce a pyridine nucleotide. The excess of reducing power is used by a photosynthetic electron transport chain that flows between the activated state of chlorophyll in photosystem II and the oxidized state of chlorophyll in photosystem I. Photosystem I alone mediates cyclic photophosphorylation [Figure 4.31 (a)].

## FURTHER READING


### Books

CLAYTON, R. K., *Photosynthesis: Physical Mechanisms and Chemical Patterns.* Cambridge: Cambridge University Press, 1980.

GOTTSCHALK, G., *Bacterial Metabolism.* New York, Heidelberg, and Berlin: Springer-Verlag, 1979.

INGRAHAM, J. L., O. MAALØE and F. C. NEIDHARDT, *Growth of the Bacterial Cell.* Sunderland, Mass.: Sinauer Associates, Inc., 1983.

JONES, C. W., *Bacterial Respiration and Photosynthesis.* Washington, D.C.: American Society for Microbiology, 1982.

LEHNINGER, A. L., *Principles of Biochemistry.* New York: Worth Publishers, 1982.

MANDESTAM, J., and K. MCQUILLEN, *Biochemistry of Bacterial Growth.* New York: John Wiley, 1982.

# Chapter 5
# Microbial Metabolism: Biosynthesis, Polymerization, Assembly

Despite their mechanistic diversity, all the metabolic pathways discussed in Chapter 4 have the same common function: the provision of ATP, reduced pyridine nucleotides, and the 12 precursor metabolites. In this sense, there is a fundamental unity underlying the superficial diversity of fueling reactions. This unity of biochemistry, a concept first emphasized by the microbiologist, A. J. Kluyver, in 1926, becomes even more evident when we analyze the ways in which building blocks are synthesized, and then polymerized into macromolecules that are assembled into cellular components. In all cells the principal macromolecules are proteins and nucleic acids, and the biochemical reactions leading to their formation show little variation among procaryotes and even between procaryotes and eucaryotes. There is, accordingly, a *central core of biosynthetic reactions that are similar in all organisms.* A greater degree of diversity occurs in the synthesis of certain other classes of cell constituents, in particular polysaccharides and lipids, since the chemical composition of these substances is often group specific.

In this chapter we will focus attention on the reactions of biosynthesis and polymerization that are common to most or all organisms. Some more specialized biosynthetic processes, distinctive of procaryotic organisms, will also be discussed.

## METHODS OF STUDYING BIOSYNTHESIS

Biochemistry was initially concerned with the elucidation of ATP-generating reactions, many of which (e.g., the fermentations) are chemically fairly simple. The principal technique employed was the direct study of the enzymes involved; from the reactants and products of the individual reactions, the complete reaction sequence was deduced.

The technique of *sequential induction* has also been used to advantage for the elucidation of inducible pathways. By comparing cells grown on the inducer substrate of the pathway under investigation with cells grown on a substrate that is metabolized through alternate pathways, deductions concerning probable intermediates of the inducible pathway can be made. Because enzymes of an inducible pathway are not synthesized in the absence of the primary inducer substrate and because, through sequential induction, all enzymes of a pathway are synthesized in its presence, probable intermediates of a pathway are identified as those that are immediately metabolized by cells grown on the primary inducer substrate. For example, benzoate is the primary inducer and catechol is an intermediate of the $\beta$-ketoadipate pathway (Chapter 4). If catechol is added to a suspension of cells of *Pseudomonas putida* that were grown on benzoate, immediate oxidation of catechol ensues; if it is added to cells of the same organism that were grown on asparagine, oxidation begins only after a lag period of about 40 minutes.

The elucidation of biosynthetic mechanisms has come more recently, largely through studies on bacteria. The information gained through these studies, however, was later shown to hold for other organisms. Work on this problem could not even be initiated until the role of ATP as an energetic coupling agent between catabolism and biosynthesis was established. Furthermore, the unraveling of biosynthetic pathways required the development of new techniques that, although helpful, are rarely essential for the analysis of catabolism. The most important of these is the *use of mutants* and the *use of isotopic labeling.*

### Use of Biochemical Mutants

Biochemical mutants (see Chapter 11) became an important tool for the study of biosynthesis after the demonstration in 1940 by G. Beadle and E. Tatum that it is possible to isolate so-called *auxotrophic* mutants. Such mutants require as growth factors biosynthetic intermediates that the parental strain can synthesize *de novo.* Such requirements are caused by the genetic loss of the ability to synthesize, in a functional form, one enzyme mediating a specific step in the affected pathway. The early studies with biochemical mutants led to the hypothesis that each individual enzyme is encoded by a specific gene, which became known as the *one-gene-one-enzyme* hypothesis. Now is it known that there are exceptions: certain genes play exclusively regulatory roles; others encode RNA that is not translated into protein; and some enzymes are composed of dissimilar subunits, each of which is encoded by a distinct gene. Still, the hypothesis remains a valid and useful generalization.

Biochemical mutants can be utilized in the following ways to determine the sequence of reactions in a biosynthetic pathway:

**1.** By determining the number of different genes that can undergo mutation resulting in a nutritional requirement for the same growth factor, the number of different enzymatically catalyzed reactions in the pathway of biosynthesis of that growth factor can be determined. For example, mutations in eight different genes lead to a requirement (auxotrophy) for the amino acid arginine; suggesting that hence, there are eight different enzymatically catalyzed reactions in the arginine pathway (Figure 5.1).

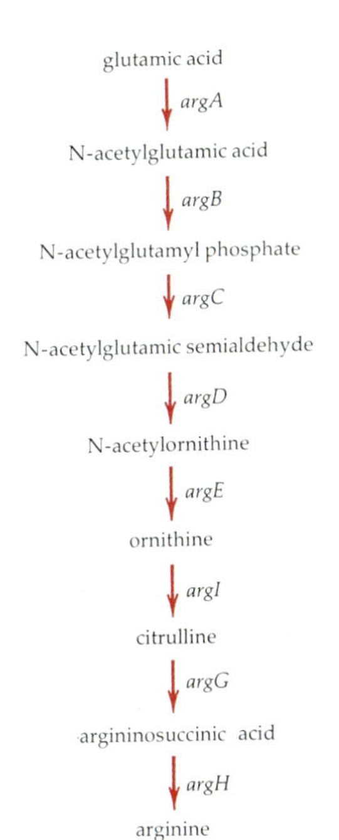


FIGURE 5.1

Reaction sequence leading to the biosynthesis of arginine in *Salmonella typhimurium.* The designations of the genes that encode the various enzymes are written to the right of the arrows.

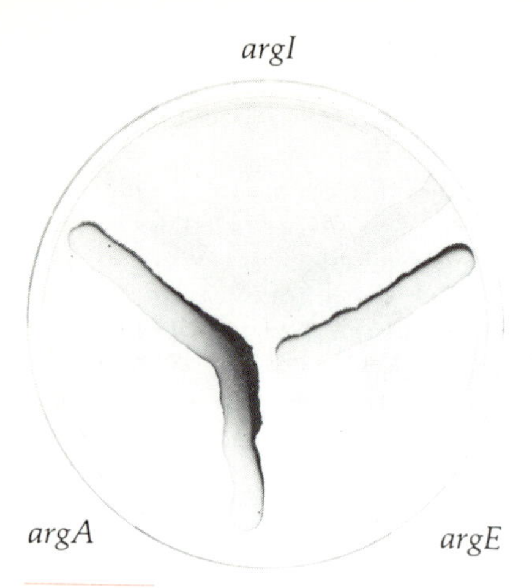


FIGURE 5.2

Strains of *Salmonella typhimurium* carrying mutations in genes *argA*, *argE* and *argI* (Figure 5.1) were streaked adjacent to one another on a plate lacking arginine. Since all three strains are genetically incapable of synthesizing arginine, they would be unable to grow if streaked alone on such a plate. However, the *argI* strain excretes ornithine into the medium, which allows the *argA* and *argE* strains to grow in that region. From such an experiment one can conclude that *argA* and *argE* encode enzymes which catalyze steps of the arginine pathway prior to that encoded by *argI*. Similarly, the *argE* strain excretes an intermediate allowing growth of the *argA* strain.

**2.** Genetic blockades in a pathway tend to cause the accumulation and excretion into the medium of metabolic intermediates prior to the blockade. These intermediates sometimes allow the growth of other mutant strains blocked in the same pathway at an earlier step; thus, the sequence of blockades in a series of mutant strains can be determined. For example, strains with mutations in the gene *argI* excrete ornithine (Figure 5.2), which can be utilized by strains blocked at the earlier steps under the control of genes *A* and *E*. In addition, the intermediates excreted by the mutant strains can be chemically isolated and identified.

**3.** Information on the sequence of reactions in a biosynthetic pathway can also be obtained by testing the growth response of mutant strains to suspected intermediates of the pathway being investigated. For example, an *argI* mutant strain will grow if arginine or citrulline is added to the medium; an *argA* strain will grow if arginine, citrulline, or ornithine is added. From such experiments citrulline and ornithine would appear to be intermediates of the arginine pathway, with ornithine being a biosynthetic precursor of citrulline.

### Use of Isotopic Labeling

When a biosynthetic building block (e.g., an amino acid) is added to a growing population of cells, it will often prevent its own endogenous synthesis (the mechanism by which this control is effected is discussed in Chapter 12). The exogenously furnished compound is, therefore, preferentially incorporated by the cell into biosynthetic end products. If the exogenously furnished compound is labeled with a radioisotope, chemical fractionation of the labeled cells can reveal the ultimate location of radioactivity in the various cell constituents. Such experiments show, for example, that $^{14}C$-labeled glutamic acid is incorporated into protein not only as glutamic acid residues but also as residues of two other amino acids, arginine and proline. This result demonstrates that glutamic acid is a biosynthetic precursor of arginine and proline.

Another valuable technique employing radioisotopes is *pulse labeling*. A growing culture is briefly exposed to a radioactive biosynthetic precursor. During this exposure a small quantity of the precursor enters the cell and starts to be distributed through the various pathways in which it participates. If samples of cell material are subjected to chemical fractionation at various times after pulse labeling, the sequence of chemical transformations in pathways leading from the radioactive precursor is revealed. The pathway for the conversion of $CO_2$ to organic compounds by photosynthetic organisms and chemoautotrophs was largely established by experiments of this kind.

Radioisotopic methods are also valuable for detecting the products of biosynthetic reactions catalyzed by extracts of cells, in which the reaction products are formed in quantities too small for ordinary chemical methods to be used. Such methods were indispensable in the early studies on the synthesis of protein.

## THE ASSIMILATION OF NITROGEN AND SULFUR

Of the six major bioelements (*carbon, nitrogen, sulfur, hydrogen, phosphorus,* and *oxygen*), precursor metabolites lack only two: nitrogen and sulfur. These become incorporated into cellular constituents as a consequence of certain reactions in biosynthetic pathways. Both elements enter biosynthetic metabolism in a reduced state: nitrogen as ammonia ($NH_3$) and sulfur as hydrogen sulfide ($H_2S$). But these elements are often available in the environment in other chemical forms: as constituents of organic compounds or in inorganic form at a different oxidation state.

## The Assimilation of Ammonia

The nitrogen atom of ammonia (valence of −3) is at the same oxidation level as the nitrogen atoms in the organic constituents of the cell. The assimilation of ammonia does not, therefore, necessitate oxidation or reduction. There are three $NH_3$ fixation reactions: one forming an amino group in glutamic acid

$$\underset{(\alpha\text{-ketoglutaric acid})}{HOOC{-}(CH_2)_2{-}\overset{O}{\overset{\|}{C}}{-}COOH} + NH_3 + NADPH + H^+$$

$$\xrightarrow{\text{glutamate dehydrogenase}} \underset{(\text{glutamic acid})}{HOOC{-}(CH_2)_2{-}\overset{NH_2}{\overset{|}{CH}}{-}COOH} + NADP^+ + H_2O$$

and two others forming amido groups in asparagine and glutamine

$$\underset{(\text{aspartic acid})}{HOOC{-}CH_2{-}CHNH_2{-}COOH} + NH_3 + ATP$$

$$\xrightarrow{\text{asparagine synthetase}} \underset{(\text{asparagine})}{(O{=})(NH_2{-})C{-}CH_2{-}CHNH_2{-}COOH} + AMP + Ⓟ{-}Ⓟ$$

and,

$$\underset{(\text{glutamic acid})}{HOOC{-}(CH_2)_2{-}CHNH_2{-}COOH} + ATP + NH_3$$

$$\xrightarrow{\text{glutamine synthetase}} \underset{(\text{glutamine})}{(O{=})(NH_2{-})C{-}(CH_2)_2{-}CHNH_2{-}COOH} + ADP + Ⓟ$$

All three products of $NH_3$ fixation (glutamic acid, asparagine, and glutamine) are direct precursors of proteins, and asparagine serves only in this role. However, both glutamic acid and glutamine play additional roles as agents for the transfer of amino and amido groups to all other nitrogenous precursors of cellular macromolecules. For example, the amino acids alanine, aspartic acid, and phenylalanine are formed by transamination between glutamic acid and nonnitrogenous metabolites, i.e.,

L-glutamic acid + pyruvic acid ⟶ α-ketoglutaric acid + L-alanine

L-glutamic acid + oxalacetic acid ⟶ α-ketoglutaric acid + L-aspartic acid

L-glutamic acid + phenylpyruvic acid ⟶ α-ketoglutaric acid + phenylalanine

and the amido group of glutamine is the source of the amino groups of cytidine triphosphate, carbamyl phosphate, NAD, and guanosine triphosphate, among others; e.g.,

uridine triphosphate + glutamine + ATP ⟶ cytidine triphosphate + glutamic acid + ADP + Ⓟ

The pathways of synthesis of glutamic acid and glutamine depend on the concentration of $NH_3$ available in the cell. At high concentrations of $NH_3$, the two sequential reactions, catalyzed by a dehydrogenase and glutamine synthetase, lead to the synthesis of these two compounds:

$$\alpha\text{-ketoglutaric acid} \xrightarrow{NH_3} \text{glutamic acid} \xrightarrow{NH_3} \text{glutamine}$$

However, the substrate affinity of α-ketoglutaric dehydrogenase for $NH_3$ is relatively low; consequently, this enzyme ceases to function effectively at low concentrations of $NH_3$, and the above pathway becomes inoperative. Under these conditions, a new enzyme, *glutamate synthase*, sometimes called GOGAT (an acronym for the alternate name glutamine-oxoglutarate amino transferase) is induced, which catalyzes the reaction:

$$\text{glutamine} + \alpha\text{-ketoglutaric acid} \rightleftharpoons 2\ \text{glutamic acid}$$

Under these conditions, the glutamine synthase reaction becomes the major route of $NH_3$ assimilation, i.e., instead of being synthesized by glutamate dehydrogenase, glutamic acid is synthesized by the reaction sequence:

$$\text{glutamic acid} + NH_3 + ATP \xrightarrow{\text{glutamine synthase}} \text{glutamine} + ADP + Ⓟ$$

$$\text{glutamine} + \alpha\text{-ketoglutaric acid} \xrightarrow{\text{glutamate synthase}} 2\ \text{glutamic acid}$$

NET REACTION: α-ketoglutaric acid + $NH_3$ + ATP ⟶ glutamic acid + ADP + Ⓟ

However, it will be noted that the glutamine synthase—GOGAT route of synthesizing glutamic acid utilizes ATP while the glutamate dehydrogenase route does not. Thus it is not surprising that, in most bacteria, regulatory mechanisms occur that assure that the glutamine synthase—GOGAT system is utilized only when the concentration of ammonia available to the cell is so low that growth rate would be depressed were it to be assimilated only via glutamate dehydrogenase.

## The Assimilation of Nitrate

Nitrate ion ($NO_3^-$) is used by many microorganisms as a source of nitrogen. The valence of the nitrogen atom in $NO_3^-$ is +5; consequently assimilation of nitrogen from this source involves a preliminary reduction to the oxidation level of ammonia, −3.

As discussed in Chapter 4, nitrate is also reduced when in certain bacteria it serves as a terminal electron acceptor for anaerobic respiration. Some microorganisms, including fungi and algae, that use nitrate as a nitrogen source cannot use it for anaerobic respiration; some bacteria that use it for anaerobic respiration cannot use it as a nitrogen source. A relatively small number of bacteria, including *Pseudomonas aeruginosa* use it for both purposes. However, these two processes that reduce nitrate are catalyzed by different enzyme systems.

The process of *assimilatory nitrate reduction* is mediated by two enzyme complexes called *assimilatory nitrate reductase* and *assimilatory nitrite reductase.*

$$NO_3^- \xrightarrow[\text{reductase}]{\substack{\text{assimilatory}\\ \text{nitrate}}} NO_2^- \xrightarrow[\text{reductase}]{\substack{\text{assimilatory}\\ \text{nitrite}}} NH_3$$

Although reduction of nitrate to ammonia by electrons derived from an organic substrate is thermodynamically capable of generating sufficient energy to phosphorylate ADP, in no case is ATP known to be generated as a consequence of nitrate assimilation. Indeed, the process can be viewed as costing ATP, because electrons utilized to reduce nitrate could otherwise have flowed through an ATP-generating electron transport chain.

Assimilatory nitrate reductase from eucaryotes utilize reduced pyridine nucleotides (either NADH or NADPH) as a source of electrons, and the electron transfer sequence is thought to be the same in all of them:

$$\text{NAD(P)H} \longrightarrow \text{Fe-S} \longrightarrow \text{FAD} \longrightarrow \text{cytochrome } b_{557} \longrightarrow NO_3^-$$

In contrast, the few assimilatory nitrate reductases examined in bacteria cannot accept electrons from reduced pyridine nucleotides. The actual donor remains unidentified, but is suspected to be a ferredoxin.

All nitrate reductases, whether assimilatory or dissimilatory, belong to a small group of enzymes that contain molybdenum. Current evidence suggests that, regardless of source, all these enzymes, with the single exception of nitrogenase (see following section), share a common molybdenum cofactor (Mo-co). Although the structure of Mo-co has not yet been fully elucidated, it has been shown to contain a pterin nucleus and in this respect is structurally related to the vitamin folic acid.

Assimilatory nitrite reductases from eucaryotes have been studied in some detail, but very little is known about the corresponding enzymes in bacteria. However, it seems clear in all cases that the complex 6-electron-transfer reaction is catalyzed by a single enzyme, and that hydroxylamine ($NH_2OH$) is formed as an intermediate of the process.

$$NO_2^- \xrightarrow[\text{reductase}]{\text{nitrite}} NH_2OH \xrightarrow[\text{reductase}]{\text{nitrite}} NH_3$$

## The Assimilation of Molecular Nitrogen

Gaseous nitrogen ($N_2$) with a valence of zero must also be reduced to ammonia prior to incorporation into nitrogenous components of the cell. This process, called *nitrogen fixation*, is limited to procaryotes.

Although the ability of certain bacteria, both free-living and symbiotic, to fix $N_2$ has been recognized for about 100 years, attempts to elucidate the biochemical mechanism of $N_2$ fixation were long frustrated by the difficulty of preparing active cell-free extracts. This was accomplished by L. Mortenson and his associates who first established the peculiar properties now known to be common to all $N_2$-fixing enzyme systems: (1) their extreme sensitivity to irreversible inactivation by low concentrations of $O_2$; and (2) their requirements for ATP, which must be supplied continuously by an ATP-generating system, because the enzyme is inhibited by high concentrations of ATP.

The enzyme system (termed *nitrogenase complex*) (Figure 5.3) is composed of two proteins termed *nitrogenase* (or component I or MoFe protein) and *nitrogenase reductase* (or component II or Fe protein). Electrons are transferred through a low-potential reductant, either ferredoxin or flavodoxin, to nitrogenase reductase. Then, concomitant with the burst of hydrolysis of molecules of ATP (more than 16 molecules of ATP are hydrolyzed for each molecule of $N_2$ that is reduced), electrons are transferred to nitrogenase where reduction of $N_2$ and $H^+$ to $NH_3$ and $H_2$ occurs. The active site at which reduction occurs is occupied by a special molybdenum cofactor (MoFe-co). Although the structure of MoFe-co has not been elucidated, it is clearly different from the molybdenum cofactor (Mo-co) shared by all other molybdenum-containing enzymes (see preceding section).

As indicated by the structure of the enzymes, the nitrogenase system is a complex one, and this complexity is further revealed by the fact that 15

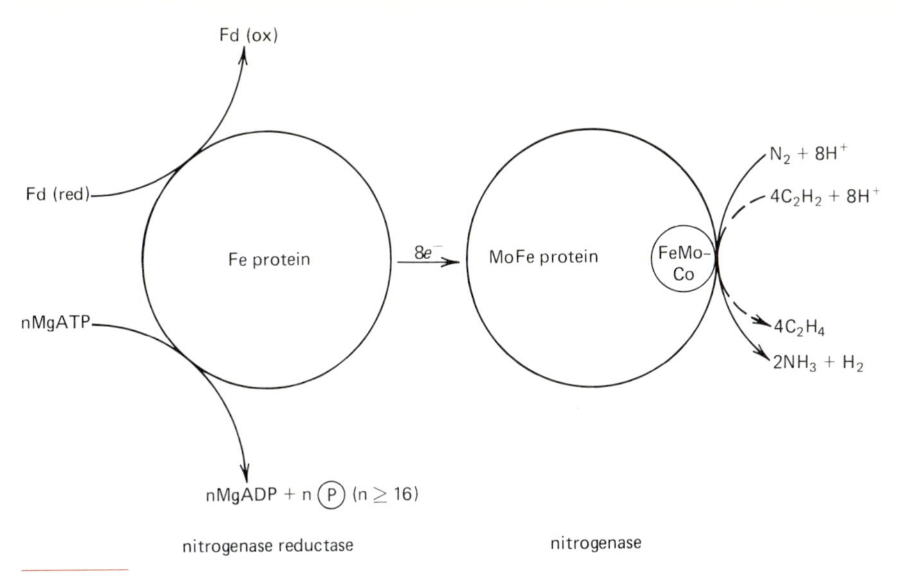


FIGURE 5.3
Structure and function of the nitrogenase complex. The nitrogenase complex is composed of two oxygen-sensitive proteins, nitrogenase reductase (also called Fe protein or Component II) and nitrogenase (also called MoFe protein or Component I). Electrons are transferred from reduced ferredoxin [Fd(red)] or in some cases flavodoxin to a magnesium-ATP (MgATP) complex of nitrogenase reductase, and then with the concomitant hydrolysis of at least 16 molecules of ATP to nitrogenase where reduction of dinitrogen ($N_2$) and $H^+$ to ammonia and hydrogen gas occurs at the active center occupied by the iron-molybdenum cofactor (FeMo-co). The complex is also capable (dotted lines) of reducing acetylene ($C_2H_2$) to ethylene ($C_2H_4$).

genes (*nif* genes) arranged in seven contiguous operons (see Chapter 11) encode it.

As indicated above, production of $H_2$ is an inevitable companion of $N_2$ reduction. Loss of this gas adds further to the energetic cost of nitrogen fixation. However, some, but not all, nitrogen-fixing bacteria possess a hydrogenase and therefore are able to gain some energy by oxidizing hydrogen.

The substrate specificity of nitrogenase is relatively low; a number of other compounds, including $N_3$, $N_2O$, HCN, $CH_3CN$, $CH_2CHCN$, and $C_2H_2$ are also reduced by it. Some of these reductions involve the transfer of only two electrons rather than the six required to reduce $N_2$. The proposed mechanism of the reaction suggests that such two-electron reductions should proceed at three times the rate of the reduction of $N_2$, and in most cases this is true.

The study of biological nitrogen fixation both in whole cells and in extracts has been greatly aided by the introduction of an assay method using the substrate acetylene, which is reduced to ethylene

$$CH{\equiv}CH \xrightarrow[2H^+]{2e^-} CH_2{=}CH_2$$

The product can be easily quantitated by gas chromotography, and the reaction is a highly specific one since no enzyme system other than nitrogenase can effect this reduction.

## The Assimilation of Sulfate

The great majority of microorganisms can fulfill their sulfur requirements from sulfate. Sulfate with a valence of $+6$ is reduced to sulfide (valence $-2$) prior to its incorporation into cellular organic compounds. Chemically, this is equivalent to the reduction of sulfate by the sulfate-reducing bacteria, which use it as the terminal electron acceptor in anaerobic respiration, as discussed in Chapter 4. The enzymatic mechanisms are different, however; the reduction of sulfate for use as a sulfur source is termed *assimilatory sulfate reduction* (by analogy with assimilatory nitrate reduction) to distinguish it from *dissimilatory sulfate reduction*, the use of sulfate as a terminal electron acceptor.

The pathway of assimilatory sulfate reduction to $H_2S$ is outlined in Figure 5.4. The initial two-electron reduction of sulfate occurs only after it has been converted to an activated form, adenylylsulfate, by a series of three enzymatic steps requiring the expenditure of three high-energy phosphate bonds. The final six-electron reduction is catalyzed by a huge, complex flavometallo-protein, *sulfite reductase*. Sulfite reductase from *E. coli* has a molecular weight of 750,000 and contains 4 FAD, 4 FMN, and 12 Fe prosthetic groups.

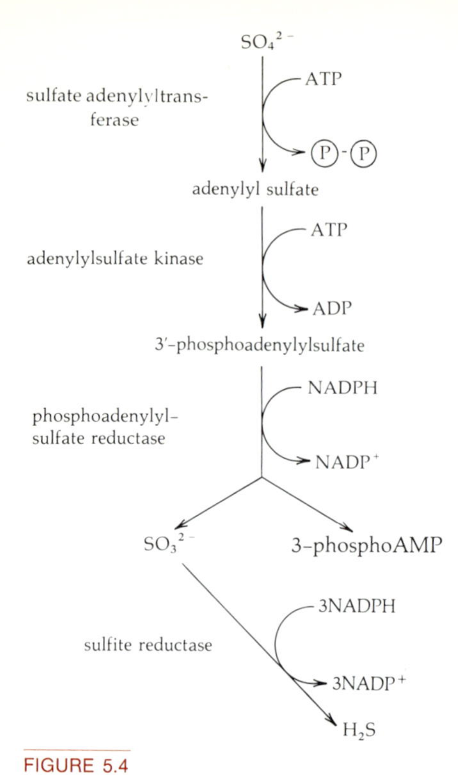


FIGURE 5.4
The assimilatory reduction of sulfate to produce $H_2S$ for use in biosynthetic reactions.

## THE STRATEGY OF BIOSYNTHESIS

On a weight basis most of the organic matter of the cell consists of macromolecules that belong to four classes: nucleic acids, proteins, polysaccharides, and complex lipids. These macromolecules are composed of lower molecular weight organic compounds termed building blocks. Each class of macromolecules is defined by the chemical type of the building blocks that are polymerized to form it: nucleotides in the case of nucleic acids, amino acids in the case of proteins, and simple sugars (monosaccharides) in the case of polysaccharides. Complex lipids are more variable and heterogeneous in composition; their precursors include fatty acids, polyalcohols, simple sugars, amines, and amino acids. As shown in Table 5.1, approximately 70 different kinds of building blocks are required to synthesize the four major classes of macromolecules.

In addition to the building blocks of macromolecules, the cell must synthesize a number of compounds that play catalytic roles. These include about 20 coenzymes and electron carriers.

In all, about 150 different small molecules are required to produce a new cell. These small molecules are, in turn synthesized from the 12 precursor metabolites formed in the course of catabolism by heterotrophs or of $CO_2$ assimilation by autotrophs (Chapter 4).

In the following pages we shall trace the pathways of biosynthesis of building blocks from precursor metabolites. In the concluding section of the chapter we shall discuss the processes by which they are polymerized into macromolecules, and how these are assembled into cellular structures.

**TABLE 5.1**

**Classes of Macromolecules of the Cell and Their Component Building Blocks**

| Macromolecule | Chemical Nature of Building Blocks | Number of Kinds of Building Blocks |
|---|---|---|
| Nucleic acids | | |
| RNA | Ribonucleotides | 4 |
| DNA | Deoxyribonucleotides | 4 |
| Proteins | Amino acids | 20 |
| Polysaccharides | Monosaccharides | ~15[a] |
| Complex lipids | Variable | ~20[a] |

[a] The number of building blocks in any particular representative of these macromolecules is usually much smaller.

## THE SYNTHESIS OF NUCLEOTIDES

The precursors of nucleic acids are purine and pyrimidine nucleoside triphosphates, all of which have the same general structure. A purine or pyrimidine base is attached through nitrogen atoms to a pentose; this combination is called a *nucleoside.* Phosphate groups are attached to the 5′ position of the nucleoside (to distinguish between the base and pentose moieties of a nucleoside, positions on the pentose are assigned a prime following the number). This combination is called a *nucleotide.* The general structure of nucleoside triphosphates is shown in Figure 5.5. The names and structures of specific nucleosides are shown in Figure 5.6. Nucleotides are symbolized by letters, A, G, U, C, or T, to indicate the purine or pyrimidine base they contain; MP, DP, or TP indicates whether they are mono-, di-, or triphosphates. Deoxynucleotides are indicated by a "d" (e.g., CDP symbolizes cytidine diphosphate, and dGTP symbolizes 2′-deoxyguanosine triphosphate). The two purine (dATP and dGTP) and two pyrimidine (dCTP and dTTP) nucleoside triphosphates containing deoxyribose are the specific precursors of DNA; the two purine (ATP and GTP) and two pyrimidine (CTP and UTP) nucleoside triphosphates containing ribose are specific precursors of RNA. Some of these nucleoside triphosphates also serve as activators (Chapter 4, Table 4.1) and thus play dual roles.

Deoxyribonucleotides are formed by the reduction of the corresponding ribonucleotides. The pathways of synthesis of ribonucleotides will, therefore, be considered first; later the manner by which ribonucleotides are converted to deoxyribonucleotides will be considered.

## Synthesis of Ribonucleotides

The ribose-phosphate moiety of all ribonucleotides is derived from 5-phosphoribosyl-1-pyrophosphate (PRPP) which, in turn, is synthesized from ribose-5-phosphate (a precursor metabolite generated in the pentose phosphate pathway) and ATP:

$$\text{ribose-5-phosphate} + \text{ATP} \xrightarrow[\text{synthetase}]{\text{PRPP}} \text{PRPP} + \text{AMP}$$

In the case of the purine ribonucleotides, PRPP is the starting point of the pathway. By suc-

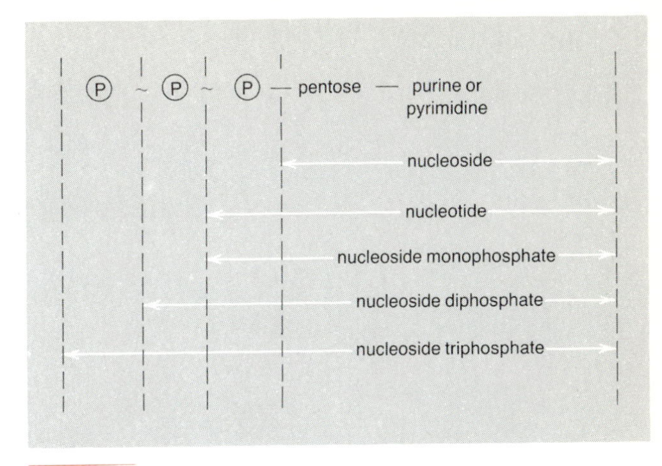


FIGURE 5.5
The general structure of nucleoside triphosphates. High-energy (anhydride) phosphate bonds are symbolized by a wavy line (~); low-energy (ester) phosphate bonds are symbolized by a straight line (—).

FIGURE 5.6
Names and composition of nucleoside triphosphates. Purines at the 9 position, and pyrimidines at the 3 position, are attached to the 1 position of pentoses to form nucleosides.

| BASE | | RIBONUCLEOSIDES | | 2′-DEOXYRIBONUCLEOSIDES | |
|---|---|---|---|---|---|
| Name | Base Structure | Pentose Structure | Name | Pentose Structure | Name |
| Purines | | | | | |
| adenine | 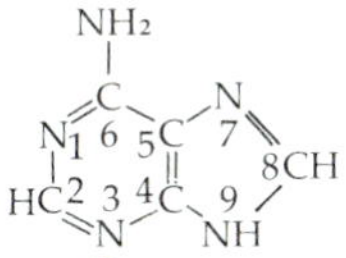 | | adenosine | | 2′-deoxyadenosine |
| guanine | 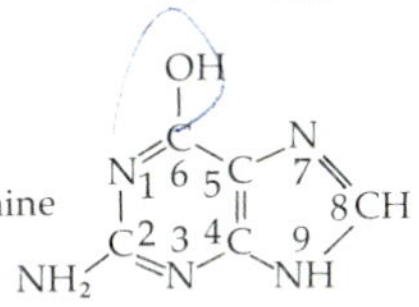 | 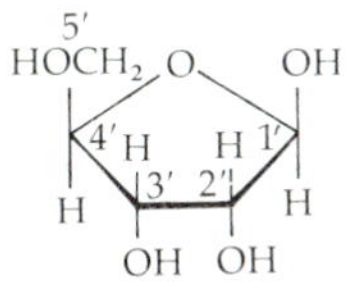 ribose | guanosine | 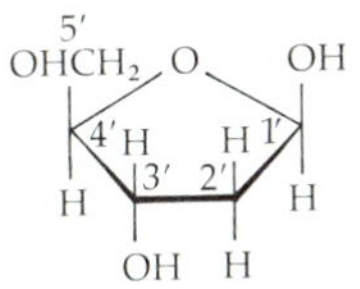 2′-deoxyribose | 2′-deoxyguanosine |
| Pyrimidines | | | | | |
| uracil | 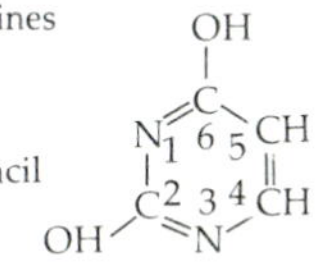 | | uridine | | 2′-deoxyuridine |
| cytosine | 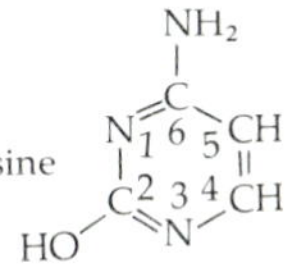 | | cytidine | | 2′-deoxycytidine |
| thymine | 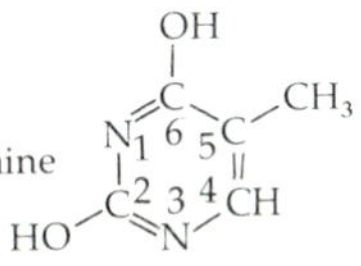 | | ribothymidine | | 2′-deoxythymidine |

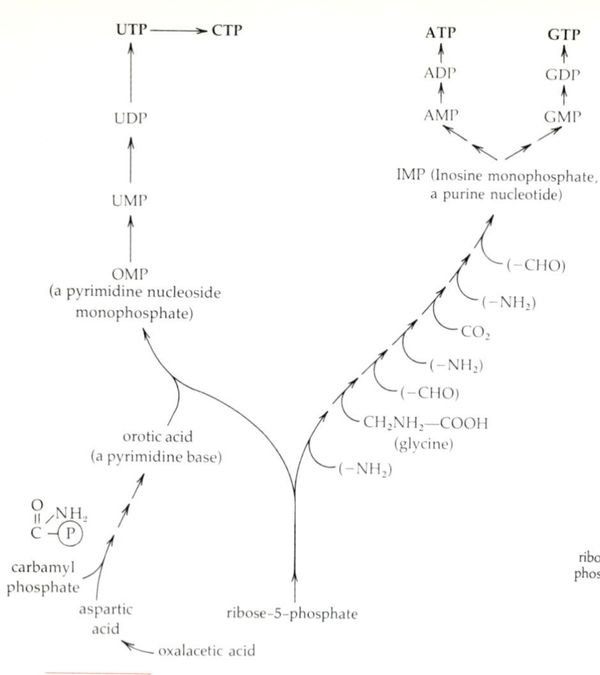


FIGURE 5.7

The general outlines of the pathways of synthesis of purine and pyrimidine ribonucleoside triphosphates.

cessive additions of amino groups and small carbon-containing groups, the nine-membered purine ring is synthesized, all intermediates of the pathway being ribonucleotides (Figure 5.7).

In contrast, the ribose-phosphate moiety of the pyrimidine ribonucleotides is added only after the six-membered pyrimidine ring has been completely synthesized by a condensation between aspartic acid and carbamyl phosphate.

With the single exception of CTP, all nucleoside triphosphates are synthesized from the corresponding nucleoside monophosphates. The general outlines of the pathways of synthesis of ribonucleotides are shown in Figure 5.7.

The detailed reactions by which the purine ribonucleoside monophosphates (AMP and GMP) are synthesized are shown in Figure 5.8.

Although the reactions leading from IMP to AMP and GMP (Figure 5.8) are irreversible, ancillary pathways exist that permit the interconversion of GMP and AMP through IMP (Figure 5.9). Thus, external sources of either guanine or adenine can satisfy the cell's requirement for both guanine- and adenine-containing nucleotides. The pathway between ATP and aminoimidazole carboxamide ribotide (AICAR) is also common to the pathway by which the amino acid, histidine, is synthesized (Figure 5.24).

The detailed reactions by which the pyrimidine ribonucleoside monophosphate, UMP, is synthesized are shown in Figure 5.10. The two purine ribonucleoside monophosphates, AMP and GMP, and the pyrimidine ribonucleoside monophosphate, UMP, are the precursors of the four essential ribonucleoside triphosphates (ATP, GTP, UTP, and CTP). The pathways of these conversions are shown in Figure 5.11.

FIGURE 5.8

The biosynthesis of the purine ribonucleotides, AMP and GMP.

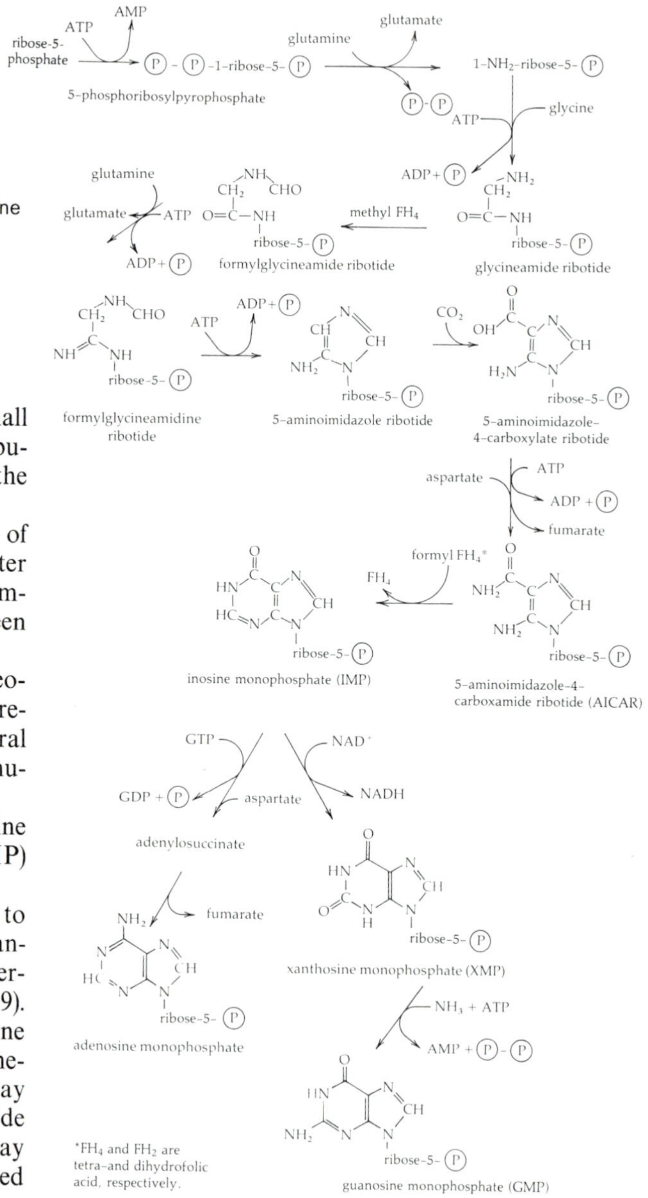

*$FH_4$ and $FH_2$ are tetra-and dihydrofolic acid, respectively.

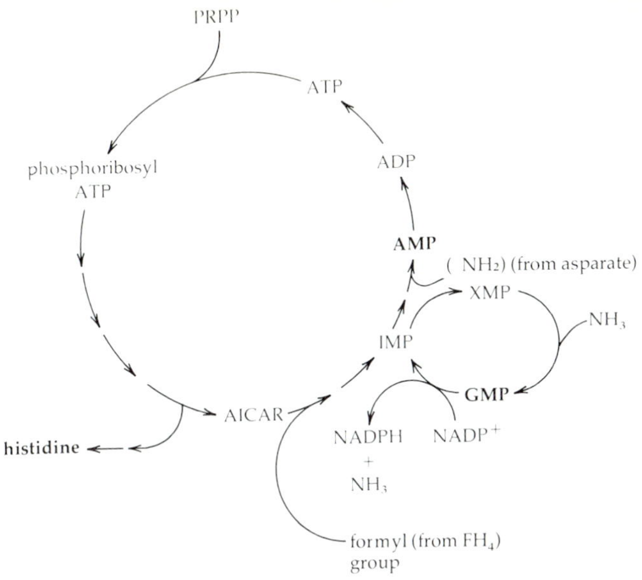


FIGURE 5.9

Interconversion pathways between GMP and AMP and the relation of one of these to the biosynthesis of the amino acid, histidine (see Figures 5.8 and 5.24).

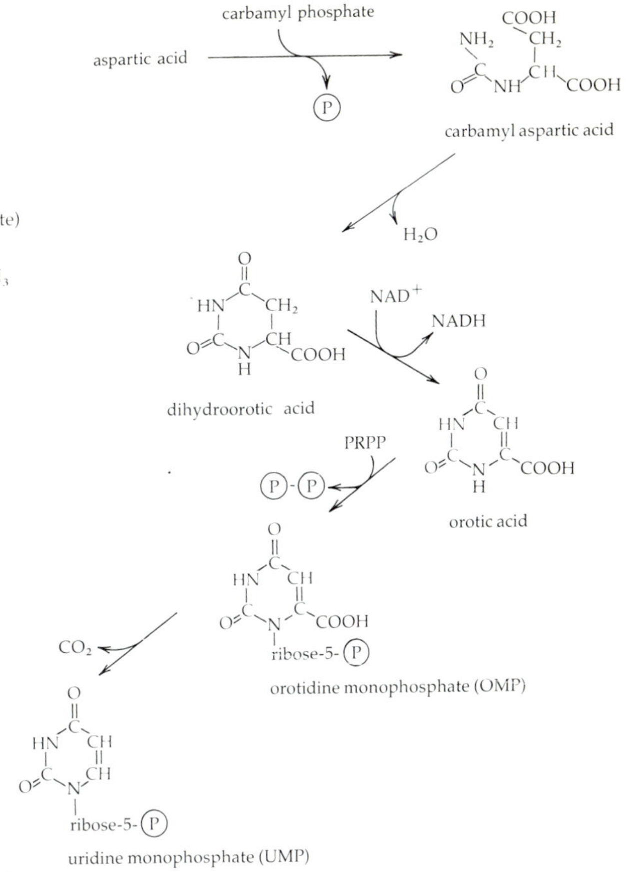

## Synthesis of the 2′-Deoxyribonucleotides

The four deoxyribonucleoside triphosphate precursors of DNA (dATP, dGTP, dCTP, and dTTP) are synthesized from ribonucleotides (Figure 5.12). Three of them (dATP, dGTP, and dCTP) are formed by reduction of the corresponding ribonucleotides by a single, highly regulated enzyme complex. In most bacteria, including *E. coli*, such reduction occurs at the level of the nucleoside diphosphate; however, in lactic acid bacteria it occurs at the level of the nucleoside triphosphates. In the former case, the products of reduction, the deoxynucleoside diphosphates (dADP, dGDP, and dCDP), are converted to triphosphates by a single enzyme, *nucleoside diphosphokinase*, the same enzyme that converts ribonucleoside diphosphates to triphosphates.

The fourth precursor of DNA, dTTP, is synthesized by a more circuitous route; dUTP, which is not normally a precursor of DNA, is an intermediate of this pathway. dUTP is formed both from dCTP by deamination and from dUDP by the action of nucleoside diphosphokinase. dUTP is then returned to the monophosphate level by the action of a specific pyrophosphatase before it is methylated to form dTMP and then returned to the triphosphate level by two kinase reactions. This curious pathway seems quite wasteful of ATP; nevertheless, it is apparently universal among procaryotes.

aspartic acid → (carbamyl phosphate; Ⓟ) → carbamyl aspartic acid

→ ($H_2O$) → dihydroorotic acid

→ ($NAD^+$ → NADH) → orotic acid

→ (PRPP; Ⓟ-Ⓟ) → orotidine monophosphate (OMP)

→ ($CO_2$) → uridine monophosphate (UMP)

FIGURE 5.10

Biosynthesis of the pyrimidine ribonucleotide, UMP.

FIGURE 5.11

Biosynthesis of ribonucleoside triphosphates from UMP, AMP, and GMP. Reactions a, b, and c are catalyzed by three specific kinases; reactions labeled d are catalyzed by a nonspecific kinase, nucleoside diphosphokinase. Reaction e symbolizes the many ATP-yielding reactions discussed in Chapter 4.

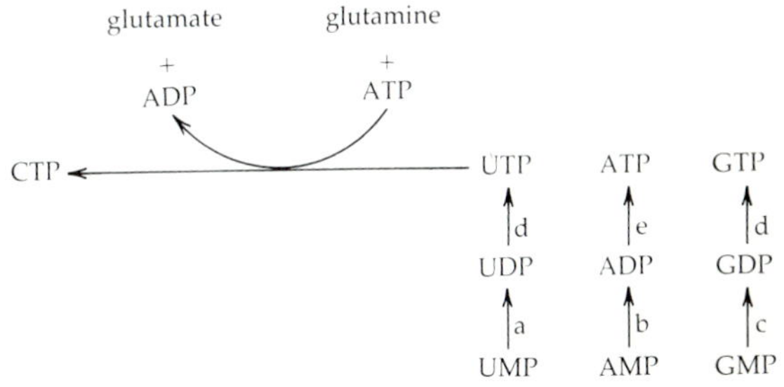

glutamate + ADP ; glutamine + ATP

CTP ← UTP ; ATP ; GTP

UDP →(d) UTP ; ADP →(e) ATP ; GDP →(d) GTP

UMP →(a) UDP ; AMP →(b) ADP ; GMP →(c) GDP

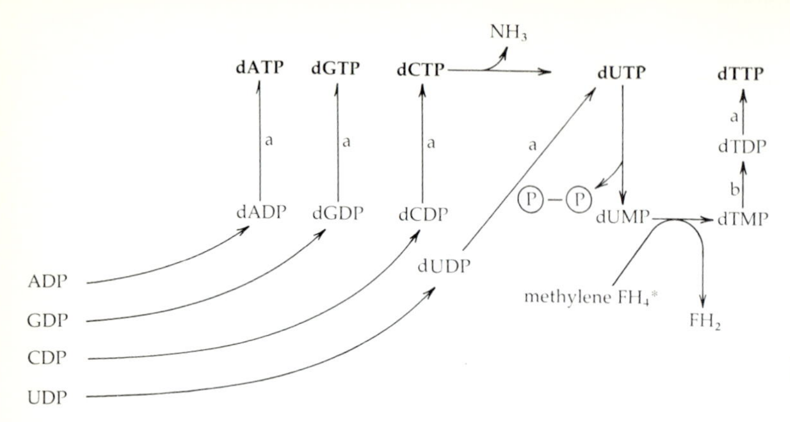


*$FH_4$ and $FH_2$ are tetra- and dihydrofolic acid, respectively.

FIGURE 5.12

Biosynthesis of deoxyribonucleoside triphosphates in *E. coli*. Reactions labeled a are all catalyzed by *nucleoside diphosphokinase;* reaction b is catalyzed by a specific kinase, *TMP kinase.*

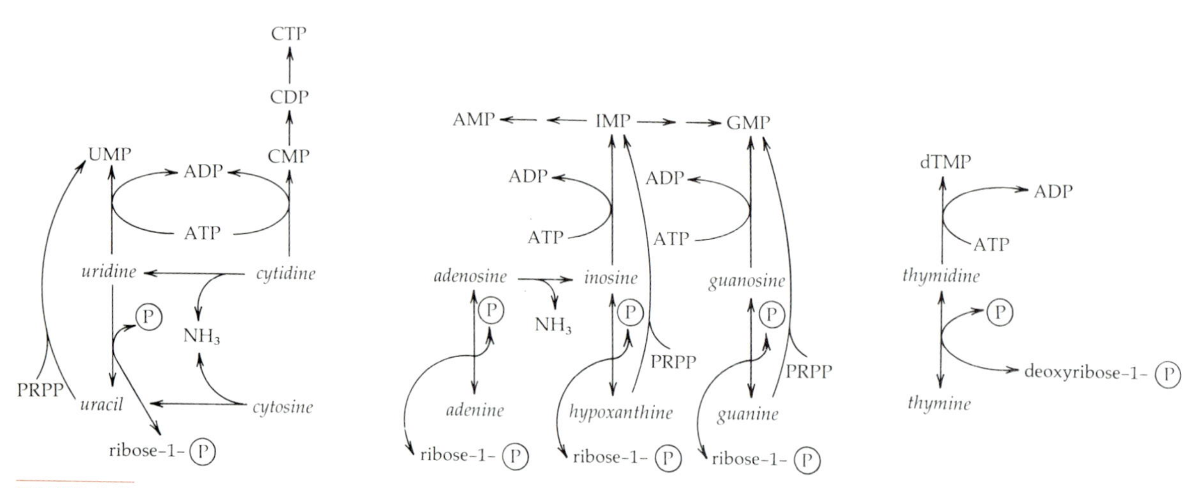


FIGURE 5.13

Pathways in enteric bacteria for the utilization of exogenous sources of purine and pyrimidine nucleotides.

## Utilization of Exogenous Purine and Pyrimidine Bases and Nucleosides

Most, but not all, bacteria are able to carry out the synthesis of all nucleoside triphosphates by the pathways outlined in Figures 5.8 through 5.12. They are also able to utilize purines and pyrimidines in the form of free bases as well as nucleosides, when these compounds are supplied in the medium. The pathways by which these compounds are utilized when supplied exogenously have been called *salvage pathways.* Although there are only minor variations among bacteria with respect to the *de novo* pathways of nucleotide biosynthesis, there are considerable variations with respect to the salvage pathways. The nucleotide salvage pathways found in enteric bacteria are shown in Figure 5.13.

The salvage pathway for thymine holds special significance to the microbial geneticist; since DNA is the only cellular constituent that contains thymine it provides a route by which DNA specifically can be made radioactive. The first reaction in the pathway (catalyzed by *thymidine phosphorylase*) by which thymine is converted to thymidine has an equilibrium constant near unity. As a result, exogenous thymine is not incorporated into DNA by enteric bacteria unless steps are taken to shift the equilibrium towards the direction of biosynthesis by increasing the intracellular concentration of the second substrate, deoxyribose-1-phosphate. Because this compound cannot penetrate the cell membrane, the steps taken to raise its intercellular concentration must be indirect: a deoxyriboside, which does penetrate the cell and is phosphorolytically cleaved to yield deoxyribose-1-phosphate, can be added to the medium; or a genetic blockade can be introduced in the step between dUMP and dTMP. The blockage causes dUMP to accumulate, which is then degraded intracellularly to

**TABLE 5.2**

**Biosynthetic Derivations of Amino Acids**

| Precursor Metabolite | | Amino Acid | Family |
|---|---|---|---|
| α-Ketoglutarate → | Glutamate → | Glutamine, Arginine, Proline | Glutamate |
| Oxaloacetate → | Aspartate → | Asparagine, Methionine, Threonine (→ Isoleucine), Lysine[a] | Aspartate |
| Phosphoenolpyruvate + erythrose-4-phosphate → | | Tryptophan, Phenylalanine, Tyrosine | Aromatic |
| 3-Phosphoglycerate → | Serine → | Glycine, Cysteine | Serine |
| Pyruvate → | | Alanine, Valine, Leucine | Pyruvate |
| Ribose-5-phosphate + ATP → | | Histidine | |

[a] In certain algae and fungi, lysine is synthesized from α-ketoglutarate (see text).

deoxyribose-1-phosphate. Many microorganisms including most pseudomonads lack completely the thymine salvage pathway.

## THE SYNTHESIS OF AMINO ACIDS AND OTHER NITROGENOUS CELL CONSTITUENTS

Twenty amino acids are required for the biosynthesis of proteins. Only one amino acid, histidine, has a completely isolated biosynthetic origin. The other 19 are derived through branched pathways from a relatively small number of precursor metabolites. They can be grouped, in terms of biosynthetic origin, into a total of five "families," as shown in Table 5.2. In addition, certain other nitrogenous cell constituents that do not enter into the synthesis of protein are also derived from these pathways (Table 5.3). We shall describe in a summary manner the pathways involved.

### The Glutamate Family

We have already discussed the origin of two members of the glutamate family (glutamic acid and glutamine) in the context of ammonia assimilation. The other two members of the glutamate family,

**TABLE 5.3**

**Derivation of Other Nitrogenous Constituents from the Pathways of Amino Acid Biosynthesis**

| Pathway (Family) | Other Nitrogenous Products |
|---|---|
| Glutamate[a] | Polyamines |
| Aspartate[a] | Diaminopimelic acid, dipicolinic acid |
| Aromatic | *p*-Hydroxybenzoic acid, *p*-aminobenzoic acid |
| Serine | Purines, porphyrins |
| Pyruvate | Pantothenic acid |

[a] In addition, glutamate, glutamine, and aspartate serve as amino donors in a number of biosynthetic pathways.

proline and arginine, are synthesized from glutamic acid by separate pathways (Figure 5.14).

### The Aspartate Family

The parent amino acid of the aspartate family, aspartic acid, arises by transamination of oxaloacetate and can be further amidated to yield the amide asparagine, in a reaction analogous to the formation of glutamine from glutamate. The other amino acids belonging to this family are formed through a branched pathway. The pathway leading to the synthesis of threonine, and the location of branch points leading to lysine, methionine, and isoleucine, are shown in Figure 5.15.

The lysine branch of the pathway is shown in Figure 5.16. This pathway of biosynthesis of lysine (sometimes called the *diaminopimelic acid* or *DAP pathway*) is characteristic of all procaryotes, higher plants, and most algae. Lysine is synthesized through a different pathway called the *α-aminoadipic acid* or *AAA pathway* (Figure 5.17) by euglenoid algae and higher fungi. Among the phycomycetes, some groups synthesize lysine through the AAA pathway, others through the DAP pathway. Metazoans are unable to synthesize lysine; they acquire it from dietary sources.

Two intermediates of the DAP pathway for the synthesis of lysine have special functions in procaryotes. Diaminopimelic acid is a component of the peptidoglycan of the cell wall of many eubacteria, and dihydrodipicolinic acid is the immediate precursor of dipicolinic acid, a major chemical constituent of endospores that contributes to their heat stability (see Chapter 22).

The methionine branch of the aspartate path-

FIGURE 5.14
Biosynthesis of proline and arginine.

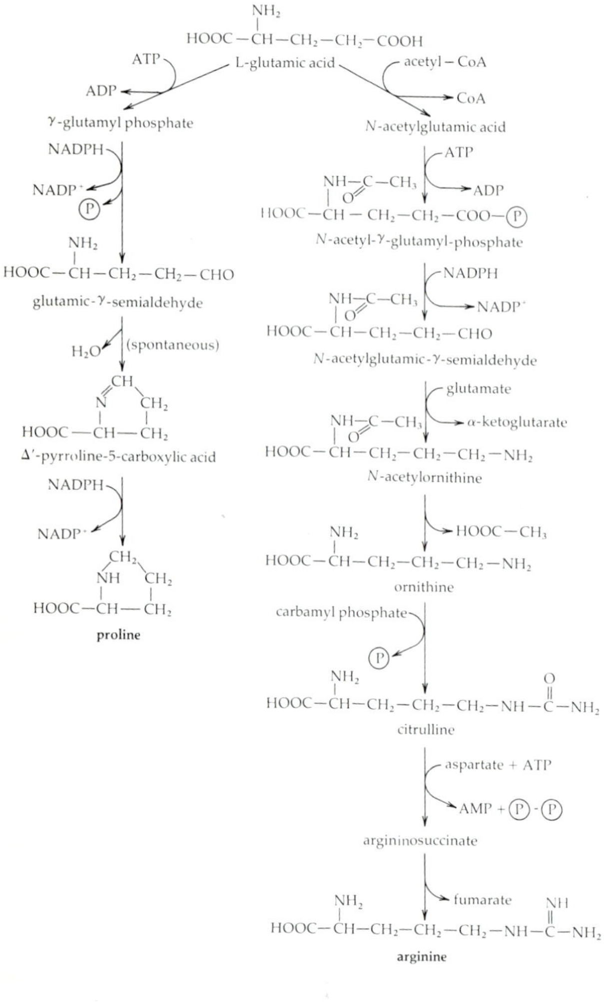

HOOC–CH(NH$_2$)–CH$_2$–CH$_2$–COOH
L-glutamic acid

ATP → ADP
γ-glutamyl phosphate
NADPH → NADP$^+$ + Ⓟ
HOOC–CH(NH$_2$)–CH$_2$–CH$_2$–CHO
glutamic-γ-semialdehyde
H$_2$O (spontaneous)
Δ'-pyrroline-5-carboxylic acid
NADPH → NADP$^+$
proline

acetyl–CoA → CoA
N-acetylglutamic acid
ATP → ADP
N-acetyl-γ-glutamyl-phosphate
NADPH → NADP$^+$
N-acetylglutamic-γ-semialdehyde
glutamate → α-ketoglutarate
N-acetylornithine
→ HOOC–CH$_3$
HOOC–CH(NH$_2$)–CH$_2$–CH$_2$–CH$_2$–NH$_2$
ornithine
carbamyl phosphate → Ⓟ
HOOC–CH(NH$_2$)–CH$_2$–CH$_2$–CH$_2$–NH–C(=O)–NH$_2$
citrulline
aspartate + ATP → AMP + Ⓟ-Ⓟ
argininosuccinate
→ fumarate
HOOC–CH(NH$_2$)–CH$_2$–CH$_2$–CH$_2$–NH–C(=NH)–NH$_2$
arginine

FIGURE 5.15
Biosynthesis of amino acids of the aspartate family.

HOOC–CH$_2$–C(=O)–COOH
oxalacetic acid
glutamate → α-ketoglutarate
HOOC–CH$_2$–CH(NH$_2$)–COOH
aspartic acid
ATP, +NH$_3$ → AMP + Ⓟ–Ⓟ
NH$_2$–C(=O)–CH$_2$–CH(NH$_2$)–COOH
asparagine
ATP → ADP
β-aspartyl phosphate
NADPH → NADP$^+$ + Ⓟ
OHC–CH$_2$–CH(NH$_2$)–COOH
aspartic β-semialdehyde → → → → → lysine (Figure 5.16)
NADPH → NADP$^+$
HOCH$_2$–CH$_2$–CH(NH$_2$)–COOH
homoserine → → → → methionine (Figure 5.18)
ATP → ADP
Ⓟ–O–CH$_2$–CH$_2$–CH(NH$_2$)–COOH
homoserine O-phosphate
→ Ⓟ
CH$_3$–CH(OH)–CH(NH$_2$)–COOH
threonine → → → → isoleucine (Figure 5.23)

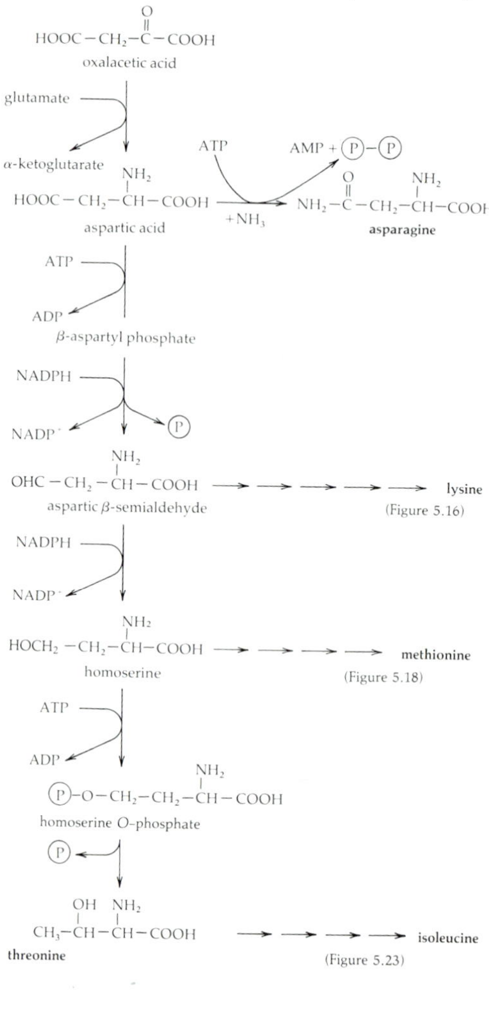

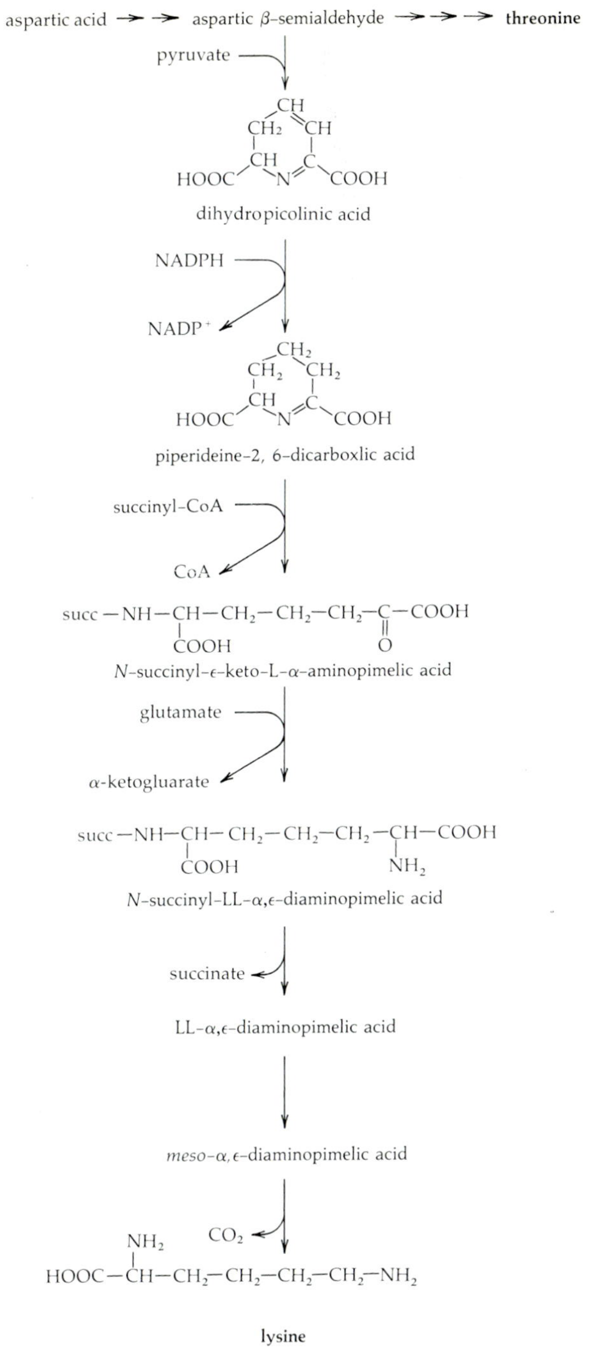

aspartic acid → → aspartic β-semialdehyde → → → **threonine**

pyruvate

dihydropicolinic acid

NADPH

$NADP^+$

piperideine-2, 6-dicarboxlic acid

succinyl-CoA

CoA

succ—NH—CH(COOH)—$CH_2$—$CH_2$—$CH_2$—C(=O)—COOH

*N*-succinyl-ε-keto-L-α-aminopimelic acid

glutamate

α-ketogluarate

succ—NH—CH(COOH)—$CH_2$—$CH_2$—$CH_2$—CH($NH_2$)—COOH

*N*-succinyl-LL-α,ε-diaminopimelic acid

succinate

LL-α,ε-diaminopimelic acid

*meso*-α,ε-diaminopimelic acid

$CO_2$

HOOC—CH($NH_2$)—$CH_2$—$CH_2$—$CH_2$—$CH_2$—$NH_2$

**lysine**

FIGURE 5.16

The lysine branch of the aspartate pathway (the DAP pathway).

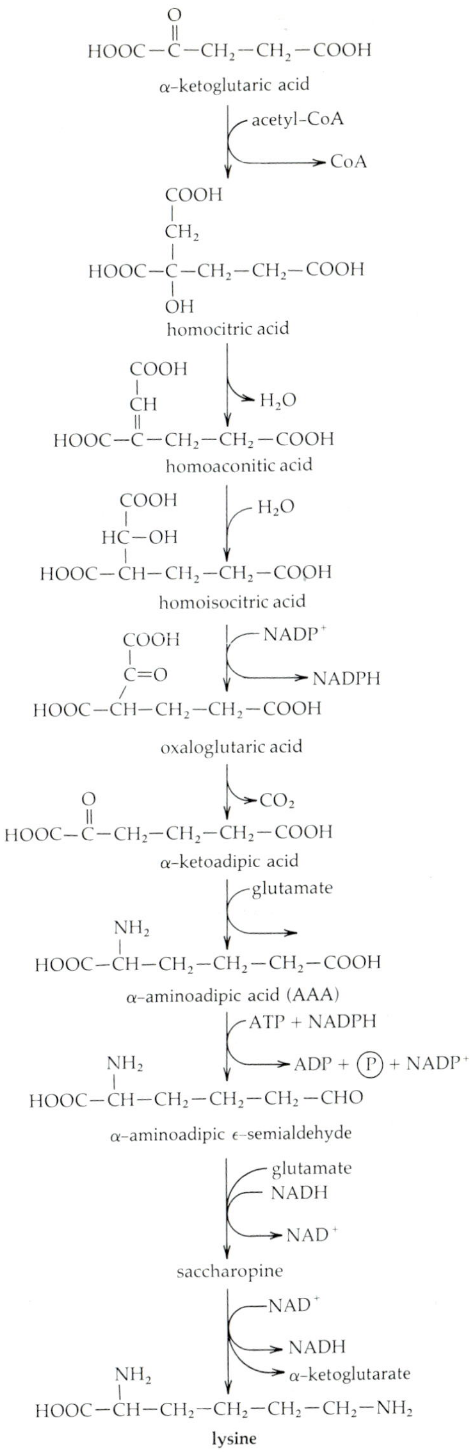

HOOC—C(=O)—$CH_2$—$CH_2$—COOH

α-ketoglutaric acid

acetyl-CoA

CoA

HOOC—C(OH)($CH_2$COOH)—$CH_2$—$CH_2$—COOH

homocitric acid

$H_2O$

HOOC—C(=CH—COOH)—$CH_2$—$CH_2$—COOH

homoaconitic acid

$H_2O$

HOOC—CH(HC(OH)—COOH)—$CH_2$—$CH_2$—COOH

homoisocitric acid

$NADP^+$

NADPH

HOOC—CH(C(=O)—COOH)—$CH_2$—$CH_2$—COOH

oxaloglutaric acid

$CO_2$

HOOC—C(=O)—$CH_2$—$CH_2$—$CH_2$—COOH

α-ketoadipic acid

glutamate

HOOC—CH($NH_2$)—$CH_2$—$CH_2$—$CH_2$—COOH

α-aminoadipic acid (AAA)

ATP + NADPH

ADP + (P) + $NADP^+$

HOOC—CH($NH_2$)—$CH_2$—$CH_2$—$CH_2$—CHO

α-aminoadipic ε-semialdehyde

glutamate

NADH

$NAD^+$

saccharopine

$NAD^+$

NADH

α-ketoglutarate

HOOC—CH($NH_2$)—$CH_2$—$CH_2$—$CH_2$—$CH_2$—$NH_2$

**lysine**

FIGURE 5.17

The AAA pathway of lysine biosynthesis.

way is shown in Figure 5.18. In certain bacteria, the final step of the pathway (methylation) can be catalyzed by two distinct enzymes. One requires folic acid as a cofactor; the other also requires vitamin $B_{12}$. Some bacteria, for example, *E. coli*, can synthesize folic acid, but they are unable to synthesize vitamin $B_{12}$. Thus, when growing in media that lack vitamin $B_{12}$, they synthesize methionine via the folic acid–dependent reaction. In media that contain vitamin $B_{12}$, the $B_{12}$-dependent reaction predominates.

Recently the interesting observation has been made that *Salmonella typhimurium* possesses genes that encode the ability to synthesize vitamin $B_{12}$, but these genes are expressed only when the bacterium grows anaerobically. Thus, probably the $B_{12}$-dependent route of methionine biosynthesis also predominates during anaerobic growth of this bacterium. Owing to the close metabolic similarity between *S. typhimurium* and *E. coli*, the same capability of $B_{12}$ synthesis might also be found in the latter bacterium.

The terminal steps in the synthesis of the fifth member of the aspartate family, isoleucine, are catalyzed by a series of enzymes that also catalyze analogous steps in the biosynthesis of a member of the pyruvate family, valine. Isoleucine biosynthesis will, accordingly, be discussed in the context of valine biosynthesis.

FIGURE 5.18

The methionine branch of the aspartic acid pathway.

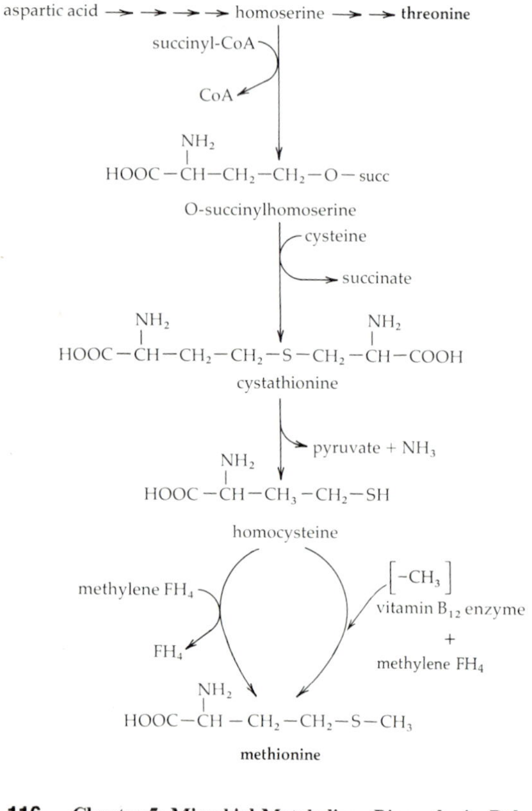

aspartic acid → → → → homoserine → → **threonine**

succinyl-CoA

CoA

$HOOC-CH(NH_2)-CH_2-CH_2-O-succ$

O-succinylhomoserine

cysteine

succinate

$HOOC-CH(NH_2)-CH_2-CH_2-S-CH_2-CH(NH_2)-COOH$

cystathionine

pyruvate + $NH_3$

$HOOC-CH(NH_2)-CH_2-CH_2-SH$

homocysteine

methylene $FH_4$

$FH_4$

$[-CH_3]$

vitamin $B_{12}$ enzyme

+

methylene $FH_4$

$HOOC-CH(NH_2)-CH_2-CH_2-S-CH_3$

**methionine**

## The Aromatic Family

The products of the aromatic pathway include the three amino acids: tyrosine, phenylalanine, and tryptophan. The first reaction of this pathway is a condensation between a precursor metabolite from the pentose-phosphate cycle, erythrose-4-phosphate, and one from the glycolytic pathway, phosphoenolpyruvate. Early steps of this pathway leading to the formation of chorismic acid and prephenic acid, both situated at major metabolic branch points, are shown in Figure 5.19. The tryptophan branch of the pathway is shown in Figure 5.20. The phenylalanine and tyrosine branches are shown in Figure 5.21. The aromatic pathway also furnishes, via chorismic acid, *p*-aminobenzoic acid (one precursor of folic acid), *p*-hydroxybenzoic acid (a precursor of the quinones, which are members of certain electron transport chains), and 2,3-dihydroxybenzoic acid (a component of certain siderophores, which participate in the entry of iron into the cell).

## The Serine and Pyruvate Families

The pathway for the formation of the amino acids of the serine family (serine, glycine, and cysteine) is shown in Figure 5.22.

The pathway for the formation of the amino acids of the pyruvate family (alanine, valine, and leucine), as well as isoleucine, which is synthesized by common enzymes, is shown in Figure 5.23. Pantothenate is synthesized from an intermediate in the biosynthesis of valine.

## Histidine Synthesis

The unbranched pathway of histidine biosynthesis is shown in Figure 5.24. The chain of five carbon atoms in the skeleton of this amino acid is derived from PRPP; two of these atoms contribute to the five-membered imidazole ring and the rest give rise to the three-carbon side chain. The remaining three atoms of the imidazole ring have a curious origin: a C-N fragment is contributed from the purine nucleus of ATP, and the other N atom from glutamine. This utilization of ATP as a donor of two atoms of the purine nucleus is unique. Its physiological rationale lies in the fact that cleavage of the purine nucleus of ATP leads to the formation

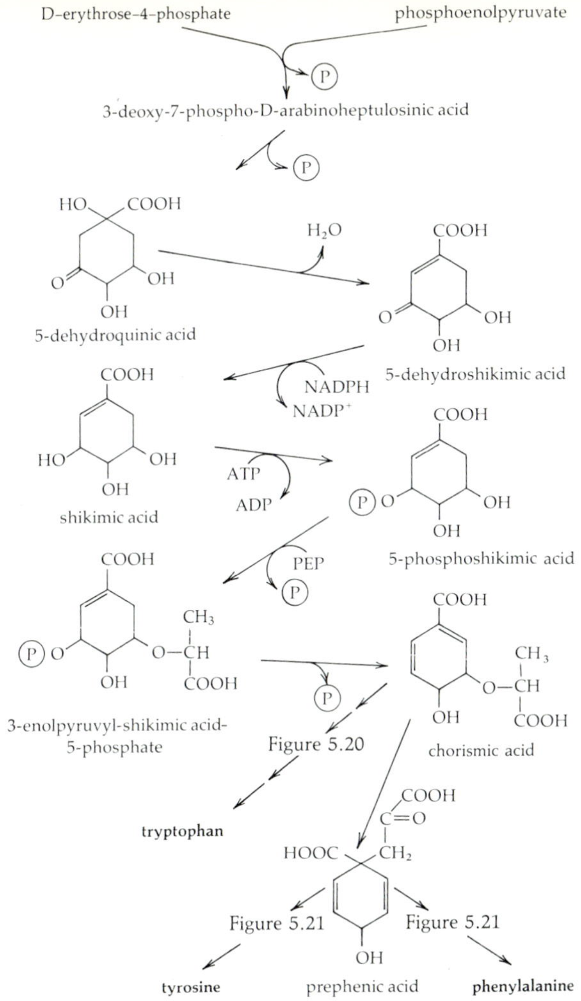

FIGURE 5.19
Biosynthesis of amino acids of the aromatic family.

of another biosynthetic intermediate, aminoimidazole carboxamide ribotide (AICAR), which is itself a precursor of purines (Figure 5.8). This intimate connection between the biosynthesis of histidine and purines has been discussed previously (Figure 5.9).

## Synthesis of Other Nitrogenous Compounds via Amino Acid Pathways

The pathways of amino acid biosynthesis also lead to the formation of intermediates that are converted to other essential cell constituents. Examples which have already been discussed are folic acid, ρ-hydroxybenzoic acid, ρ-aminobenzoic acid, 2,3 dihydroxybenzoate, diaminopimelic acid, dipicolinic acid, and purines. In quantitative terms, the most important class of nitrogenous compounds derived from a pathway of amino acid biosynthesis in pro-

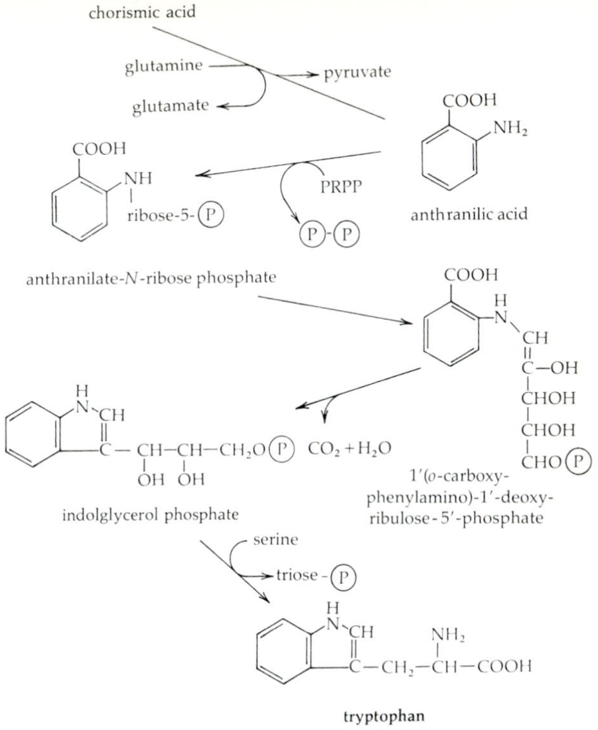

FIGURE 5.20
The tryptophan branch of the aromatic amino acid pathway.

FIGURE 5.21
The phenylalanine and tyrosine branches of the aromatic amino acid pathway.

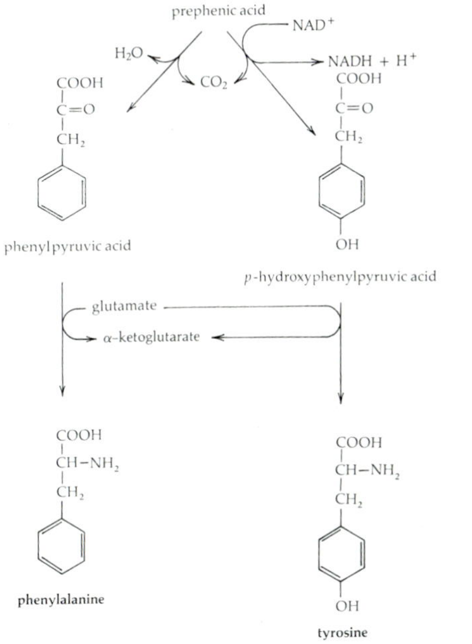

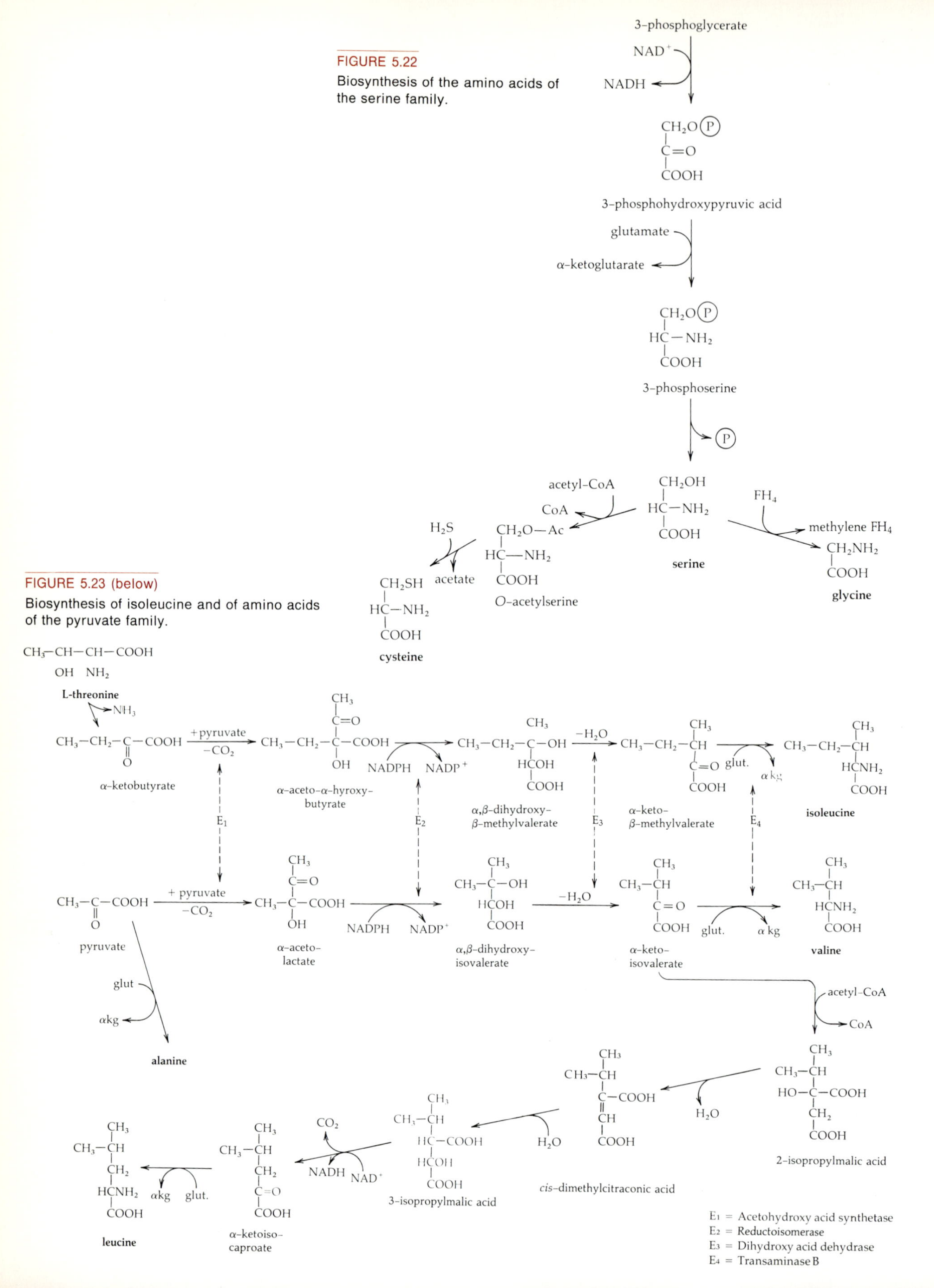


FIGURE 5.22
Biosynthesis of the amino acids of the serine family.

FIGURE 5.23 (below)
Biosynthesis of isoleucine and of amino acids of the pyruvate family.

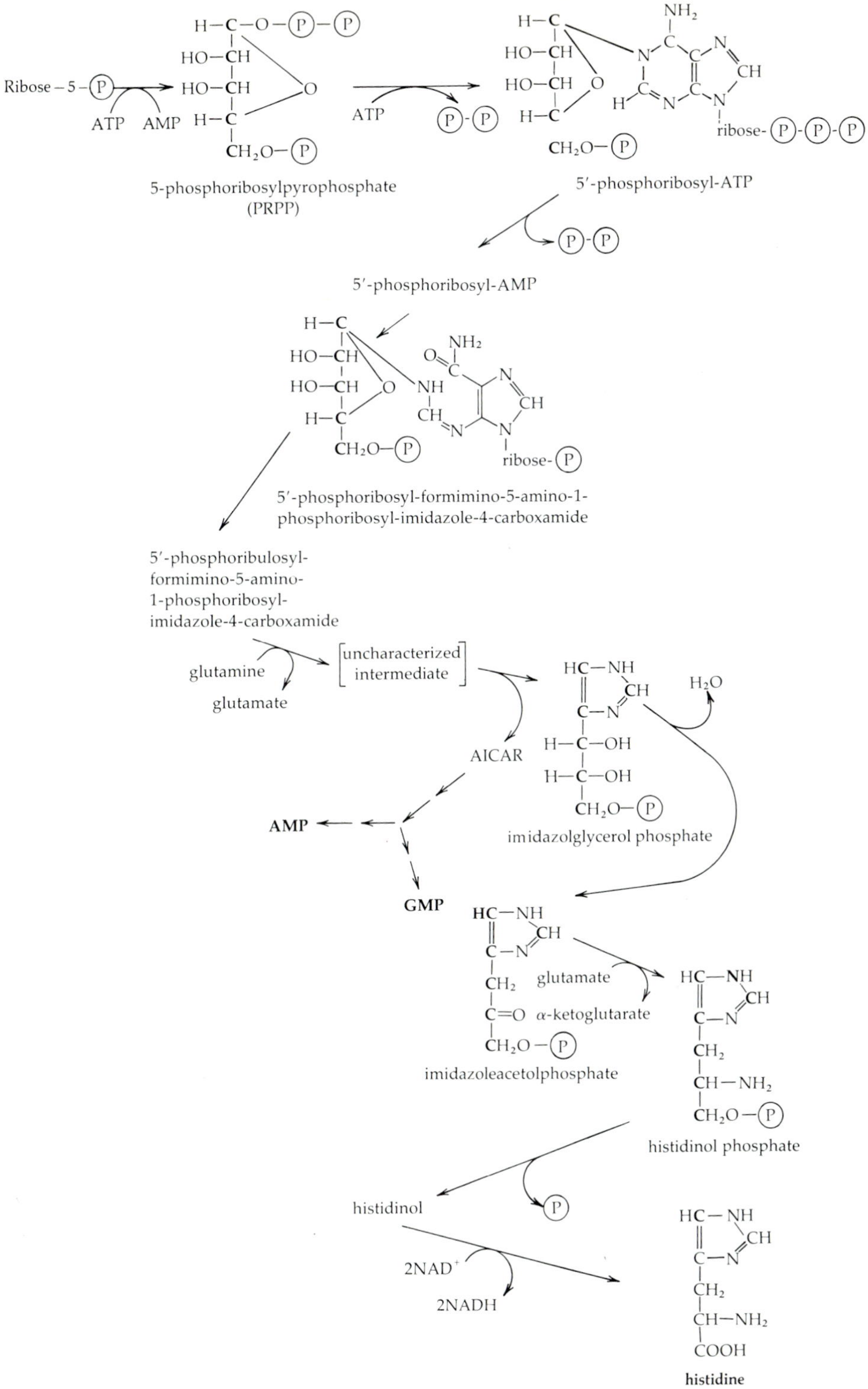

Ribose – 5 – (P)

ATP AMP

5-phosphoribosylpyrophosphate (PRPP)

ATP (P)-(P)

5′-phosphoribosyl-ATP

(P)-(P)

5′-phosphoribosyl-AMP

5′-phosphoribosyl-formimino-5-amino-1-phosphoribosyl-imidazole-4-carboxamide

5′-phosphoribulosyl-formimino-5-amino-1-phosphoribosyl-imidazole-4-carboxamide

glutamine glutamate

[uncharacterized intermediate]

AICAR

$H_2O$

imidazolglycerol phosphate

**AMP**

**GMP**

glutamate α-ketoglutarate

imidazoleacetolphosphate

histidinol phosphate

histidinol (P)

$2NAD^+$ 2NADH

**histidine**

FIGURE 5.24

The biosynthesis of histidine.

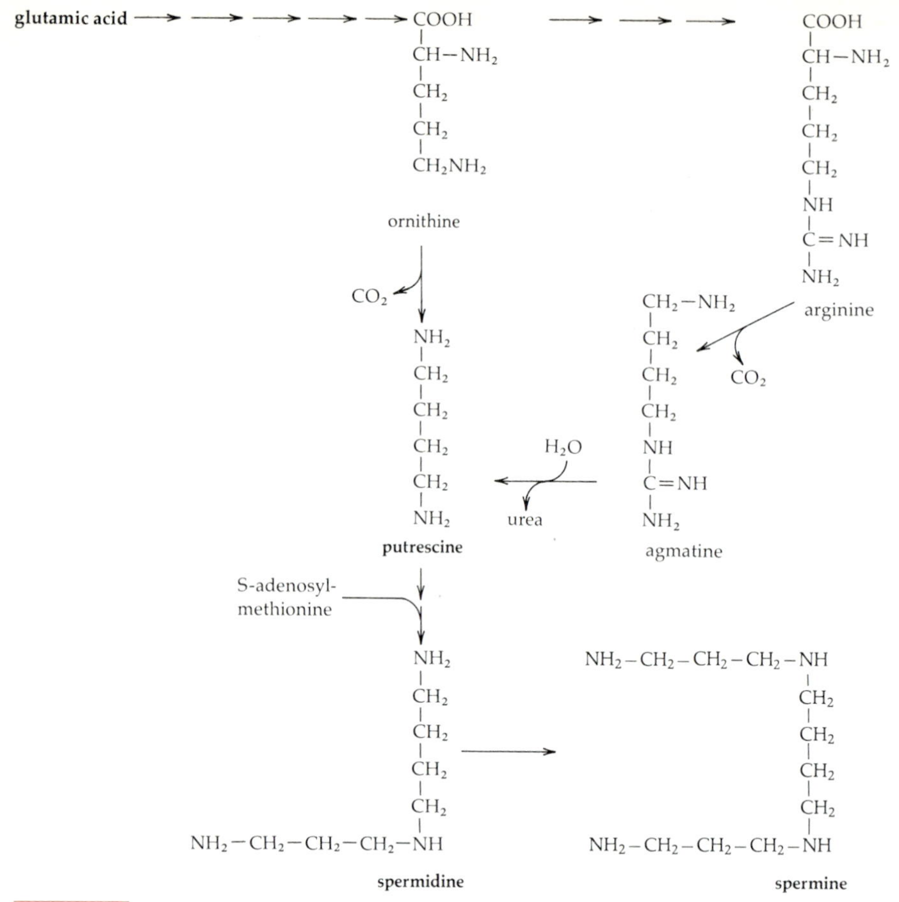


FIGURE 5.25
The pathway of synthesis of polyamines.

caryotes are the polyamines (putrescine, spermidine, and spermine), major cell constituents that arise from the arginine branch of the glutamate pathway (Figure 5.14).

During bacterial growth the flow through the arginine pathway produces roughly equal amounts of polyamines and of arginine. The pathway of synsynthesis of polyamines is shown in Figure 5.25.

Putrescine can be synthesized either from an intermediate of the arginine pathway, ornithine, or directly from arginine. The ornithine route predominates in cells growing in the absence of exogenous arginine. When arginine is supplied to the cells, the *de novo* arginine biosynthetic pathway ceases to function (Chapter 12). Under these conditions, the pathway from arginine comes into operation for the synthesis of polyamines.

Polyamines play a variety of physiological roles in cellular metabolism; important among these are neutralization of the negative charges of nucleic acids and the setting of intracellular ionic strength.

## THE SYNTHESIS OF LIPID CONSTITUENTS FROM ACETATE

Lipids are a class of cell constituents defined on the basis of their solubility properties instead of their chemical composition. They are insoluble in water and soluble in nonpolar solvents such as ether, chloroform, and benzene. They are chemically heterogeneous and include fats, phospholipids, steroids, isoprenoids, and poly-$\beta$-hydroxybutyrate. However, they can be grouped into two broad classes: those that contain esterified fatty acids and those that consist of repeating $C_5$ units with the structure of isoprene:

$$-CH_2-\overset{\displaystyle CH_3}{\overset{|}{C}}=CH-CH_2-$$

Certain lipids and their functions are listed in Table 5.4.

**TABLE 5.4**

**Lipids and Their Function**

| Lipid | Function: IN PROCARYOTES | Function: IN EUCARYOTES |
|---|---|---|
| I. LIPIDS CONTAINING ESTERIFIED FATTY ACIDS | | |
| A. Simple (a single monomeric unit) Poly-β-hydroxybutyrate | Reserve material | Absent |
| $HO-CH(CH_3)-CH_2-C(=O)-O-CH(CH_3)-CH_2-C(=O)-O-$ | | |
| B. Complex (fatty acids esterified to other compounds) | | |
| 1. Esterified to glycerol | | |
| a. Neutral fats 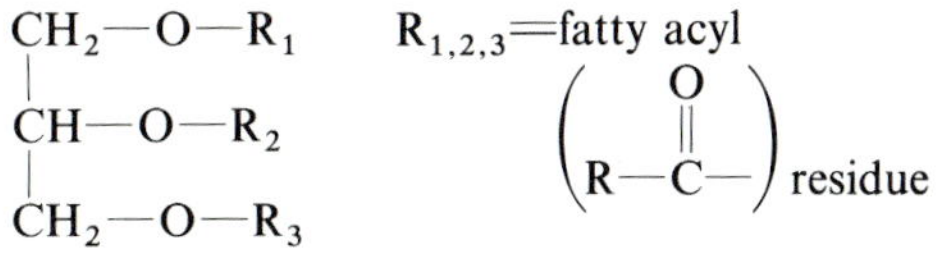 | Absent | Reserve material |
| $CH_2-O-R_1$ / $CH-O-R_2$ / $CH_2-O-R_3$; $R_{1,2,3}$=fatty acyl $(R-C(=O)-)$ residue | | |
| b. Phospholipid (see Figure 5.26) | Membrane constituent | Membrane constituent |
| c. Glycolipid 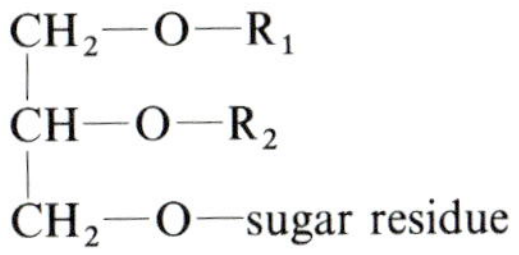 | Membrane constituent of cyanobacteria and green bacteria | Membrane constituent of chloroplasts |
| $CH_2-O-R_1$ / $CH-O-R_2$ / $CH_2-O-$sugar residue | | |
| 2. Esterified to an amino sugar Lipid A (See Figure 6.19) | Component of lipopolysaccharide wall layer | Absent |
| II. LIPIDS CONTAINING ISOPRENE UNITS | | |
| A. Polyisoprenoids 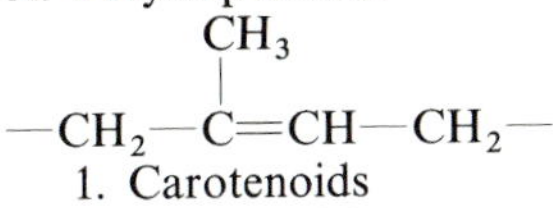 $-CH_2-C(CH_3)=CH-CH_2-$ | | |
| 1. Carotenoids $C_{40}(8 \times C_5)$ | Photoprotection and light-harvesting | Light-harvesting pigments |
| 2. Sterols $C_{30}(6 \times C_5)$ | Absent in most | Membrane constituent |
| 3. Bactoprenol $C_{55}(11 \times C_5)$ | Component upon which wall constituents are synthesized | Absent |
| B. Compounds with isoprenoid components | | |
| 1. Chlorophyll | Component of photosynthetic apparatus | Component of photosynthetic apparatus |
| 2. Quinones | Component of electron transport chains | Component of electron transport chains |

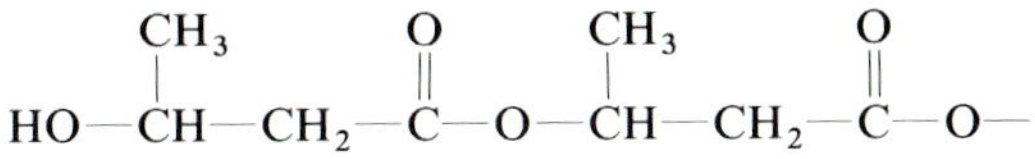

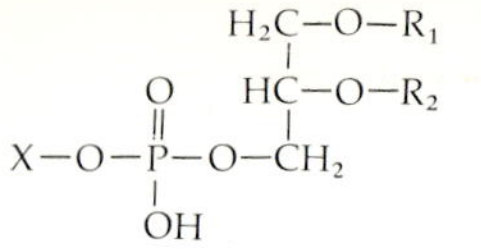

$$X-O-\underset{OH}{\overset{O}{\overset{\|}{P}}}-O-CH_2-CH(O-R_2)-CH_2-O-R_1$$

$R_1$ and $R_2$ represent acyl ($R-\overset{O}{\overset{\|}{C}}-$) residues

| X | Name of Phospholipid |
|---|---|
| $-CH_2-CH_2-NH_2$ | phosphatidylethanolamine |
| $-CH_2-CHNH_2-COOH$ | phosphatidylserine |
| $-CH_2-CHOH-CH_2OH$ | phosphatidyglycerol |
| $-CH_2-CHOH-CH_2-O-\underset{O}{\underset{\|}{\overset{O^-}{\overset{\vert}{P}}}}-OCH_2-CH(O-\overset{O}{\overset{\|}{C}}-R)-CH_2-O-\overset{O}{\overset{\|}{C}}-R$ | cardiolipin |

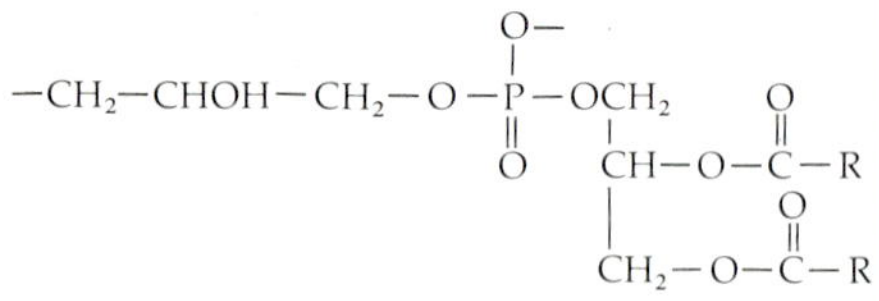

FIGURE 5.26
General structure of phospholipids.

The *phospholipids* are universal membrane components of bacteria and eucaryotes. Their general structure is shown in Figure 5.26. The chemical nature of the residue (X) attached to the phosphate group defines the class of phospholipid. In *E. coli* and *Salmonella typhimurium*, in which the most careful measurements have been made, the major phospholipid of the membrane is phosphatidylethanolamine (~75 percent). Lesser amounts of phosphatidylglycerol (~18 percent) and cardiolipin (~5 percent) and only traces of phosphatidylserine (~1 percent) are found.

## Synthesis of Fatty Acids

Fatty acids are synthesized separately and then esterified to form complex lipids. Scores of different kinds of fatty acids are found in bacteria: they contain different numbers of carbon atoms; they are straight chained or branched; they may or may not contain double bonds; they may or may not contain —OH groups; and they may or may not contain cyclopropane rings. In any particular bacterial

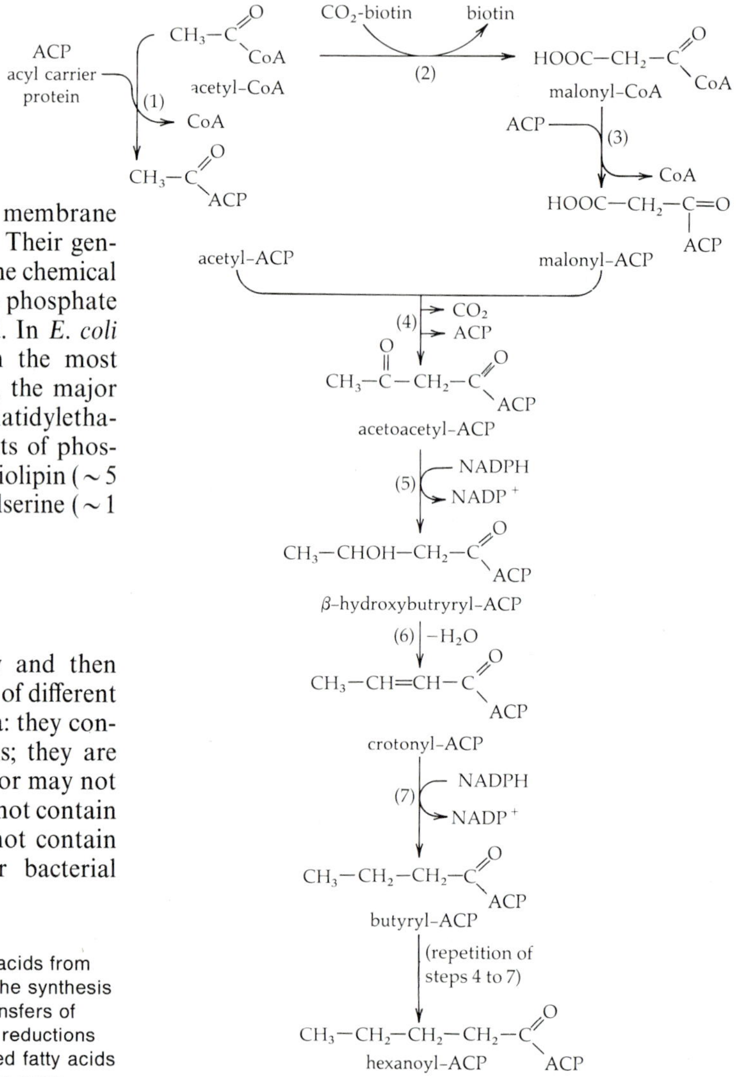

$$\underset{\text{acetyl-CoA}}{CH_3-C(=O)-CoA} \xrightarrow[(1)]{\text{ACP (acyl carrier protein)} \;\to\; CoA} \underset{\text{acetyl-ACP}}{CH_3-C(=O)-ACP}$$

$$\underset{\text{acetyl-CoA}}{CH_3-C(=O)-CoA} \xrightarrow[(2)]{CO_2\text{-biotin} \;\to\; \text{biotin}} \underset{\text{malonyl-CoA}}{HOOC-CH_2-C(=O)-CoA} \xrightarrow[(3)]{ACP \;\to\; CoA} \underset{\text{malonyl-ACP}}{HOOC-CH_2-C(=O)-ACP}$$

$$\text{acetyl-ACP} + \text{malonyl-ACP} \xrightarrow[(4)]{\to CO_2,\ \to ACP} \underset{\text{acetoacetyl-ACP}}{CH_3-\overset{O}{\overset{\|}{C}}-CH_2-C(=O)-ACP}$$

$$\xrightarrow[(5)]{NADPH \;\to\; NADP^+} \underset{\beta\text{-hydroxybutryryl-ACP}}{CH_3-CHOH-CH_2-C(=O)-ACP}$$

$$\xrightarrow[(6)]{-H_2O} \underset{\text{crotonyl-ACP}}{CH_3-CH{=}CH-C(=O)-ACP}$$

$$\xrightarrow[(7)]{NADPH \;\to\; NADP^+} \underset{\text{butyryl-ACP}}{CH_3-CH_2-CH_2-C(=O)-ACP}$$

$$\xrightarrow{\text{(repetition of steps 4 to 7)}} \underset{\text{hexanoyl-ACP}}{CH_3-CH_2-CH_2-CH_2-C(=O)-ACP}$$

FIGURE 5.27 (right)
Mechanism for the synthesis of saturated fatty acids from acetyl-CoA in *E. coli*. The reactions leading to the synthesis of hexanoyl-($C_6$)-ACP are shown. By further transfers of acetyl units from malonyl-ACP and subsequent reductions (repetitions of reaction steps 4 to 7), unbranched fatty acids of progressively greater chain length, containing even numbers of carbon atoms, are formed.

**TABLE 5.5**

**Fatty Acid Composition of Lipids in *Escherichia coli* and *Bacillus subtilis***

| | | | | | | |
|---|---|---|---|---|---|---|
| Number of carbon atoms | 14 | 14 | 14 | 15 | 15 | 16 |
| Number of double bonds | 0 | 0 | 0 | 0 | 0 | 0 |
| Number of hydroxyl groups | 0 | 1 | 0 | 0 | 0 | 0 |
| Structure[a] | Normal | Normal | Iso | Antiiso | Iso | Normal |
| Common name | Myristic | β-hydroxy myristic | | | | Palmitic |
| Percent in *E. coli*[b] | 6.1 | 4.8 | 0 | 0 | 0 | 37.1 |
| Percent in *B. subtilis*[c] | Trace | 0 | 3.9 | 36.6 | 12.1 | 6 |
| Number of carbon atoms | 16 | 16 | 17 | 17 | 17 | 18 |
| Number of double bonds | 1 | 0 | 0 | 0 | 0 | 1 |
| Number of hydroxyl groups | 0 | 0 | 0 | 0 | 0 | 0 |
| Structure[a] | Normal | Iso | Antiiso | Iso | Cyclo | Normal |
| Common name | Palmitoleic | | | | | *cis*-vaccenic |
| Percent in *E. coli*[b] | 28 | 0 | 0 | 0 | 3.2 | 20.8 |
| Percent in *B. subtilis*[c] | 0 | 11.1 | 14.4 | 15.9 | 0 | 0 |

[a] *Normal* indicates a straight chain fatty acid; *iso* indicates a methyl group branch at the penultimate carbon; *antiiso* indicates a methyl group branch at the antipenultimate carbon; *cyclo* indicates that the fatty acid contains an internal cyclopropane ring.
[b] From A. G. Marr and J. L. Ingraham, "Effect of Temperature on the Composition of Fatty Acids in *Escherichia coli*," *J. Bacteriol.* **84,** 1260 (1962).
[c] Recalculated from T. Kaneda, "Fatty Acids in the Genus *Bacillus*," *J. Bacteriol.* **93,** 894 (1967).

species the number of fatty acids is limited. *E coli*, for example, contains only six while *Bacillus subtilis* contains eight; only two fatty acids are common to both species (Table 5.5).

Certain generalizations can be drawn about the types of fatty acids found in bacteria. Like almost all fatty acids, most fatty acids found in bacteria contain an even number of carbon atoms. Although polyunsaturated fatty acids (more than one double bond) are common constituents of the lipids of eucaryotes, they are rare among procaryotes.

Saturated fatty acids are synthesized by the general pathway outlined in Figure 5.27. A special protein known as *acyl carrier protein* (ACP) plays a vital role. ACP is a small protein (MW 10,000) that is functionally and chemically analogous to CoA. Indeed, the clostridia synthesize short-chain fatty acids (butyrate and caproate) utilizing only CoA as a carrier of acyl groups, but synthesis of fatty acids usually requires that ACP be the carrier of acyl groups. The formation of long-chain fatty acids starts with the transfer of an acetyl group from CoA to ACP. This complex serves as an acceptor to which successive $C_2$ units are transferred. The $C_2$ donor is malonyl-ACP which is formed, in turn, by carboxylation of acetyl-CoA; during transfer of the $C_2$ unit, $CO_2$ is released and free ACP is regenerated. The product of $C_2$ transfer carries a terminal acetyl group, which in subsequent reactions is sequentially reduced, dehydrated, and reduced again, yielding an unsaturated acyl-ACP complex with two additional carbon atoms. Repetitions of this set of reactions progressively lengthen the fatty acid chain until the length characteristic of the particular bacterium (usually between $C_{14}$ and $C_{18}$) is reached.

Monounsaturated fatty acids are formed in various bacteria by one of two different pathways (Table 5.6), the *aerobic pathway* and the *anaero-*

**TABLE 5.6**

**Biological Distribution of Mechanisms for the Synthesis of Monounsaturated Fatty Acids**

| Anaerobic Pathway | Aerobic Pathway |
|---|---|
| *Clostridium* spp. | *Mycobacterium* spp. |
| *Lactobacillus* spp. | *Corynebacterium* spp. |
| *Escherichia coli* | *Micrococcus* spp. |
| *Pseudomonas* spp. | *Bacillus* spp. |
| Cyanobacteria | Fungi |
| Green bacteria | Protozoa |
| Purple bacteria | Animals |

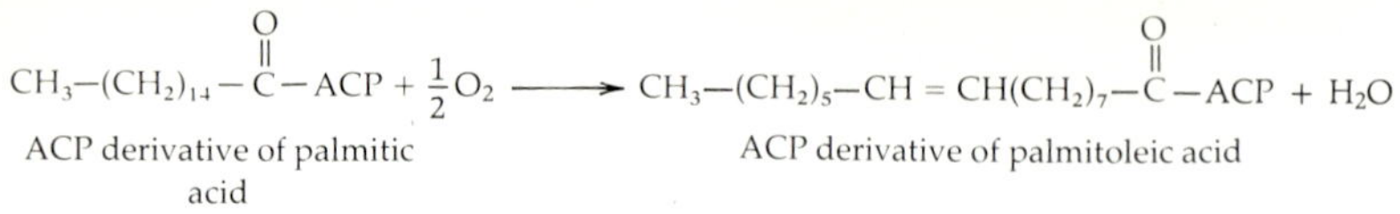

$$CH_3-(CH_2)_{14}-\overset{O}{\overset{\|}{C}}-ACP + \frac{1}{2}O_2 \longrightarrow CH_3-(CH_2)_5-CH=CH(CH_2)_7-\overset{O}{\overset{\|}{C}}-ACP + H_2O$$

ACP derivative of palmitic acid — ACP derivative of palmitoleic acid

FIGURE 5.28

The formation of the monounsaturated fatty acid, palmitoleic acid, from the corresponding saturated fatty acid, palmitic acid, by the aerobic pathway.

$CH_3-(CH_2)_5-CH_2^{\gamma}-CHOH^{\beta}-CH_2^{\alpha}-C(=O)-ACP$

β-hydroxydecanoyl-ACP

β,γ-dehydration → $CH_3-(CH_2)_5-CH^{\gamma}=CH^{\beta}-C^{\alpha}H_2-C(=O)-ACP$

↓ addition and reduction of three $C_2$ units from malonyl-ACP

$CH_3-(CH_2)_5-CH=CH-(CH_2)_7-C(=O)-ACP$

ACP derivative of palmitoleic acid ($C_{16}$, $\Delta^9$)

↓ addition and reduction of one $C_2$ unit from malonyl-ACP

$CH_3-(CH_2)_5-CH=CH-(CH_2)_9-C(=O)-ACP$

ACP derivative of *cis*-vaccenic acid ($C_{18}$, $\Delta^{11}$)

α,β-dehydration → $CH_3-(CH_2)_5-C^{\gamma}H_2-CH^{\beta}=CH^{\alpha}-C(=O)-ACP$

↓ reduction

decanoyl ($C_{10}$)-ACP

↓

saturated fatty acids of greater chain length

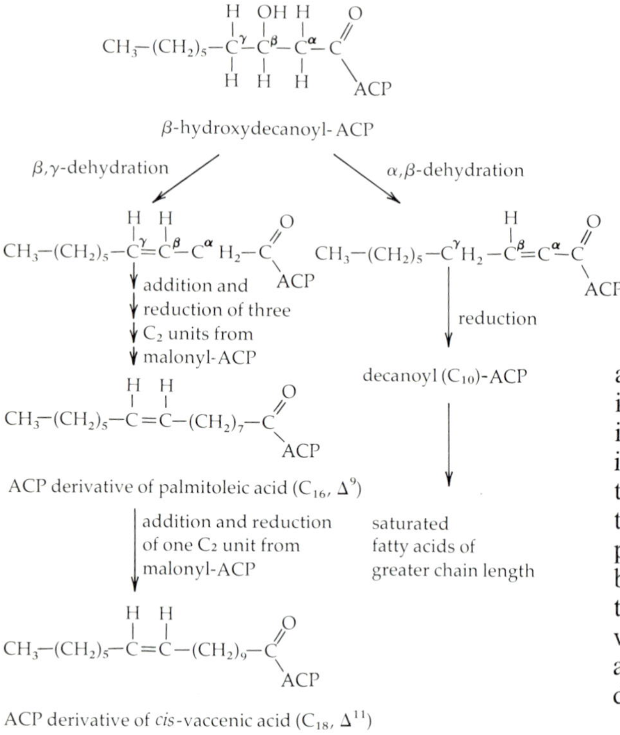

FIGURE 5.29

The anaerobic pathway to monounsaturated fatty acids, characteristic of many bacteria, showing its relationship to the pathway for saturated fatty acid synthesis.

*bic pathway* (which occurs in aerobes as well as anaerobes).

The aerobic pathway intervenes as a subsequent modification of fully synthesized (but still ACP-bound) saturated fatty acids, while in the anaerobic pathway unsaturation takes place during elongation of the fatty acid chain. The aerobic pathway requires the direct intervention of molecular oxygen (Figure 5.28).

The mechanisms of the anaerobic pathway are outlined in Figure 5.29. The $C_{10}$ hydroxyacyl intermediate, β-OH-decanoyl-ACP, can undergo normal α-β desaturation, leading to the formation of longer-chain saturated fatty acids, or it can undergo a β-γ dehydration, leading to the homologous monounsaturated fatty acids. Note that in the anaerobic pathway, the position of the double bond in the carbon chain of the eventual end products is determined by the point in biosynthesis at which it is introduced. Subsequent chain elongation leads to its location between carbon atoms 9 and 10 in the $C_{16}$ product (palmitoleic acid). In the $C_{18}$ product, however, the double bond becomes located between carbon atoms 11 and 12. Hence, bacteria that employ the anaerobic pathway contain *cis*-vaccenic acid as their monounsaturated $C_{18}$ fatty acid, rather than oleic acid, the product of direct desaturation of stearic acid by the aerobic pathway.

## Synthesis of Phospholipids

Phospholipids are synthesized from fatty acids and the precursor metabolite triose-phosphate by the pathway outlined in Figure 3.30. Dihydroxyacetone-phosphate is reduced to 3-glycerophosphate, which is subsequently esterified by two fatty acid residues. The resulting diglyceride, phosphatidic acid, is then activated by CTP to form CDP-diglyceride, which undergoes transfer reactions with serine and α-glycerophosphate, releasing CMP. The reaction product with serine, phosphatidylserine, itself constitutes a minor class of phospholipids. The major phospholipid class is the decarboxylation product, phosphatidylethanolamine. The reaction between CDP-diglyceride and α-glycerophosphate leads to the other phospholipid classes, phosphatidylglycerol and cardiolipin.

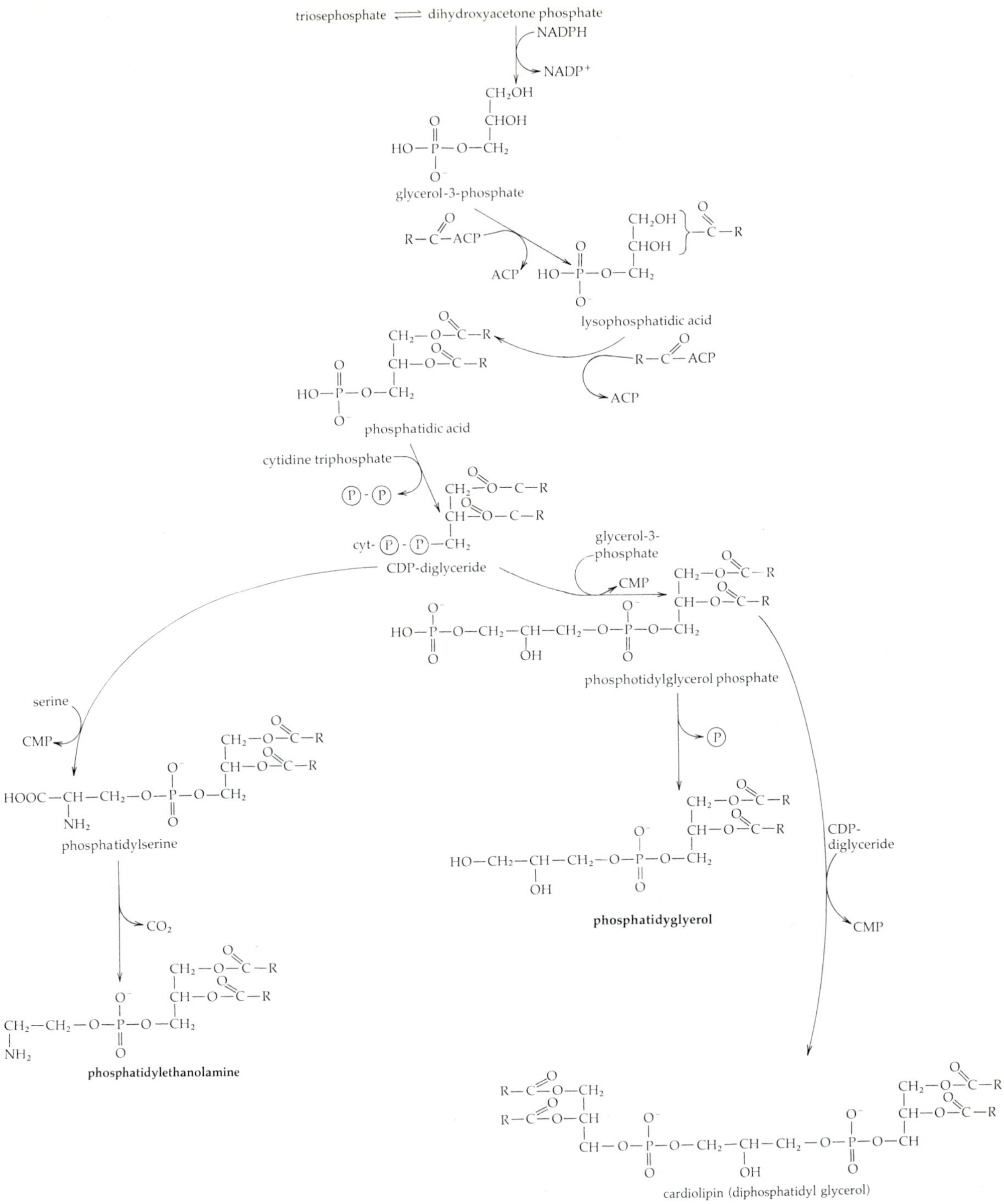

FIGURE 5.30

Pathway of formation of the major phospholipid classes found in *E. coli*. (The bracket (}) indicates that the acyl group (—C(=O)—R) may be esterified to either of these hydroxyl (—OH) groups.

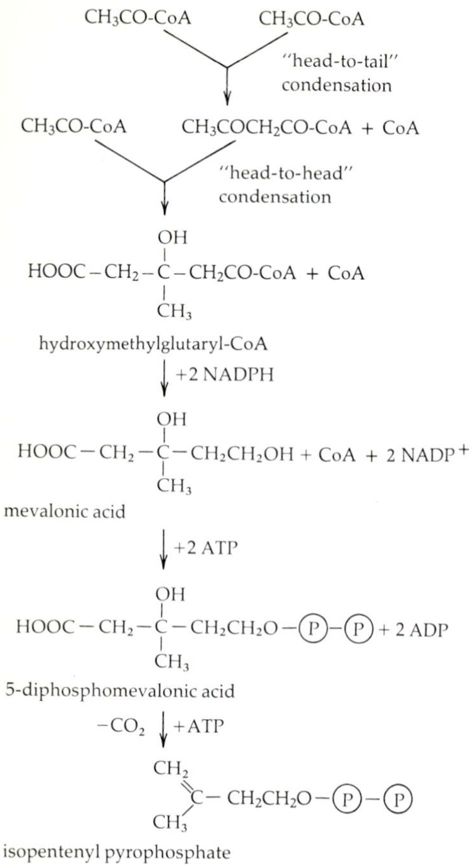

$CH_3CO\text{-}CoA$ + $CH_3CO\text{-}CoA$ → ("head-to-tail" condensation) → $CH_3COCH_2CO\text{-}CoA + CoA$

$CH_3CO\text{-}CoA$ + $CH_3COCH_2CO\text{-}CoA$ → ("head-to-head" condensation) → $HOOC-CH_2-C(OH)(CH_3)-CH_2CO\text{-}CoA + CoA$

hydroxymethylglutaryl-CoA

↓ +2 NADPH

$HOOC-CH_2-C(OH)(CH_3)-CH_2CH_2OH + CoA + 2\ NADP^+$

mevalonic acid

↓ +2 ATP

$HOOC-CH_2-C(OH)(CH_3)-CH_2CH_2O-Ⓟ-Ⓟ + 2\ ADP$

5-diphosphomevalonic acid

$-CO_2$ ↓ +ATP

$CH_2=C(CH_3)-CH_2CH_2O-Ⓟ-Ⓟ$

isopentenyl pyrophosphate

FIGURE 5.31

Synthesis of isopentenylpyrophosphate, the precursor of all polyisoprenoid compounds, from acetyl-CoA.

FIGURE 5.32

Chain elongation in polyisoprenoid biosynthesis.

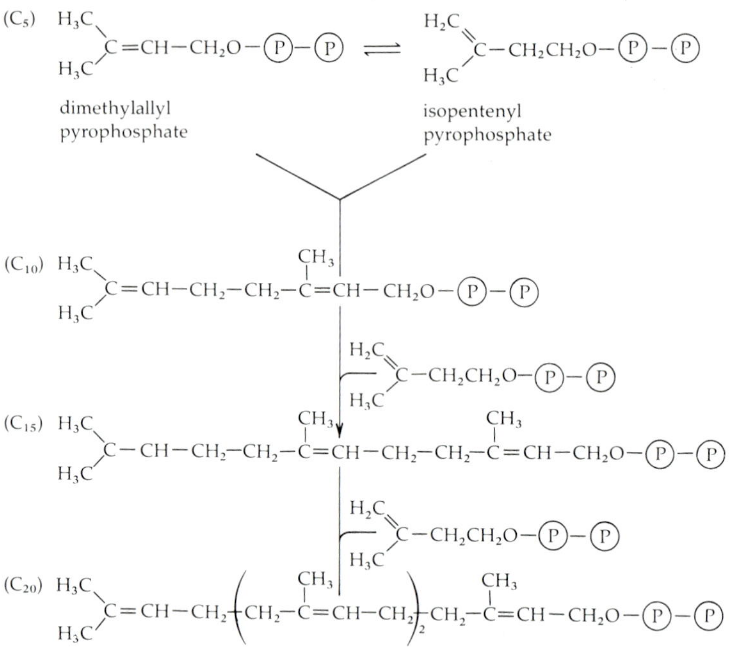

($C_5$) $(H_3C)_2C=CH-CH_2O-Ⓟ-Ⓟ \rightleftharpoons H_2C=C(H_3C)-CH_2CH_2O-Ⓟ-Ⓟ$

dimethylallyl pyrophosphate; isopentenyl pyrophosphate

($C_{10}$) $(H_3C)_2C=CH-CH_2-CH_2-C(CH_3)=CH-CH_2O-Ⓟ-Ⓟ$

+ $H_2C=C(H_3C)-CH_2CH_2O-Ⓟ-Ⓟ$

($C_{15}$) $(H_3C)_2C=CH-CH_2-CH_2-C(CH_3)=CH-CH_2-CH_2-C(CH_3)=CH-CH_2O-Ⓟ-Ⓟ$

+ $H_2C=C(H_3C)-CH_2CH_2O-Ⓟ-Ⓟ$

($C_{20}$) $(H_3C)_2C=CH-CH_2\left(CH_2-C(CH_3)=CH-CH_2\right)_2CH_2-C(CH_3)=CH-CH_2O-Ⓟ-Ⓟ$

## Synthesis of Polyisoprenoid Compounds

A large number of different cell constituents, including the membrane lipids of the archaebacteria, have carbon skeletons that consist of repeating $C_5$ units with the structure of isoprene. These *polyisoprenoid compounds* are synthesized exclusively from acetyl units; however, the mechanism of chain elongation differs markedly from that characteristic of fatty acid synthesis, diverging at the $C_4$ level (Figure 5.31). Acetoacetyl CoA undergoes a "head-to-head" condensation with acetyl CoA, to yield, after rearrangement, *mevalonic acid*, a branched $C_6$ acid. This is, in turn, converted, by two successive phosphorylations and decarboxylation, to *isopentenyl pyrophosphate*, the activated $C_5$ compound from which polyisoprenoid compounds are synthesized. The successive steps by which $C_{15}$ and $C_{20}$ derivatives are synthesized from this intermediate are shown in Figure 5.32. The tail-to-tail condensation of two molecules of the $C_{15}$ derivative, farnesyl pyrophosphate, yields squalene, a precursor of sterols. The analogous condensation of two molecules of the $C_{20}$ derivative yields phytoene, the precursor of carotenoids. The $C_{15}$ and $C_{20}$ polyisoprenoid alcohols, farnesol and phytol, are components of the chlorophylls. Further chain elongation by head-to-tail condensation yields polyisoprenoid compounds containing from 50 to 60 carbon atoms, as found in quinones.

## THE SYNTHESIS OF PORPHYRINS

Each of the many different organic molecules that serve as coenzymes or as prosthetic groups of enzymes is synthesized through a special pathway. As one illustration, we shall describe the synthesis of *porphyrins*. They fall into two major groups: the iron-containing *hemes*, which serve as prosthetic groups of cytochromes and many other enzymes, known collectively as *heme proteins*; and the magnesium-containing *chlorophylls*. Vitamin $B_{12}$, which is a precursor of the prosthetic group for certain enzymes that catalyze the transfer of $C_1$ groups, is synthesized from an intermediate in the biosynthetic pathway leading to the synthesis of porphyrins.

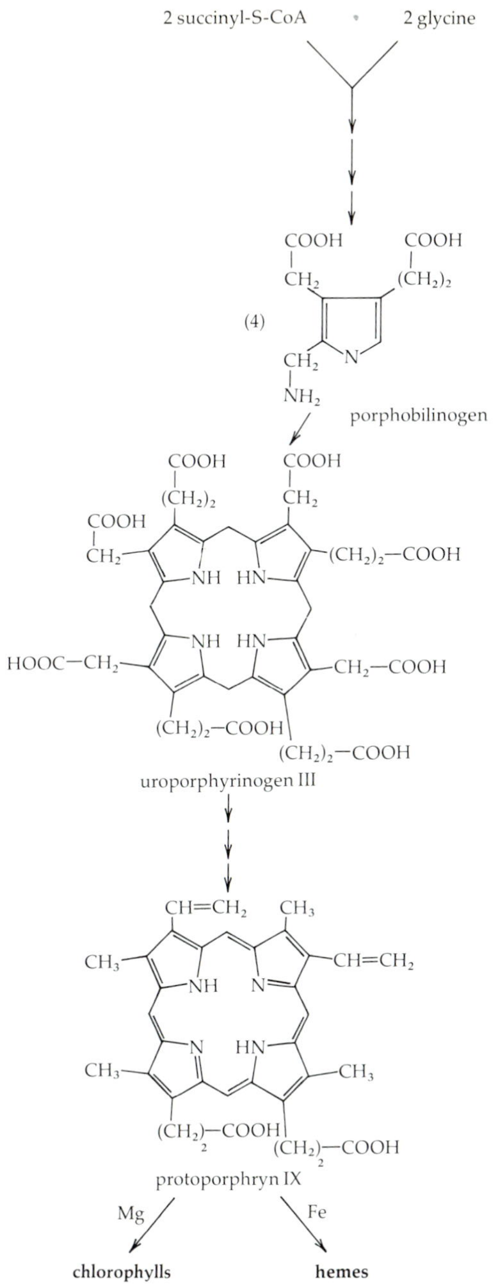

The synthesis of porphyrins is initiated by a condensation of the amino acid glycine with succinyl-CoA; this gives rise in three steps to porphobilinogen (Figure 5.33). The condensation of four molecules of this intermediate forms the tetrapyrrolic nucleus of uroporphyrinogen III; subsequent modifications and eventual oxidation yield protoporphyrin IX. The insertion of iron as a chelate in this molecule leads directly to the formation of a heme. Alternatively, if magnesium is chelated with the ring, a long series of subsequent steps leads to the formation of the chlorophylls characteristic of the various groups of photosynthetic organisms. Most of these steps in chlorophyll synthesis are common ones: the divergences that give rise to the various specific plant chlorophylls and the bacteriochlorophylls occur near the end of the biosynthetic sequence.

FIGURE 5.33

The general scheme of the synthesis of porphyrins. Green plants and cyanobacteria synthesize chlorophylls by a pathway that differs from this in the early steps (see Chapter 15).

## Variations of Biosynthetic Pathways among Bacteria

The various pathways of biosynthesis discussed in the previous sections of this chapter have been elucidated largely by studies on *Escherichia coli*. Remarkably, the same pathways have been found to occur in most other eubacteria and even in eucaryotes, with the major exception of the pathways by which lysine is synthesized (Figure 5.16 and 5.17), but certain other variations have also been found.

For example, the procaryotes, *Pseudomonas* spp. and *Corynebacterium glutamicum*, as well as the eucaryotes *Saccharomyces*, *Neurospora*, and *Chlamydomonas*, synthesize arginine by a slightly different pathway than do *E. coli* and *Salmonella typhimurium* (Figure 5.14). Rather than producing a *deacylase* that converts *N*-acetylornithine to ornithine, they produce a *transacylase* that transfers the acyl group to glutamate:

glutamate + *N*-acetylornithine ⟶ *N*-acetylglutamate + ornithine

thereby also effectively bypassing the first step of the pathway found in the enteric bacteria:

glutamate + acetyl-CoA ⟶ *N*-acetylglutamate + CoA

A variety of organisms including various species of *Pseudomonas*, cyanobacteria, and green plants can synthesize phenylalanine and tyrosine by a different pathway from the one found in enteric bacteria (Figure 5.21). Rather than the pathway branching at prephenic acid, this intermediate undergoes transamination to yield an intermediate named *pretyrosine*, which also lies on the common

| Type of Polymer | Designation of Bond | Structure of Bond |
|---|---|---|
| protein | peptide | 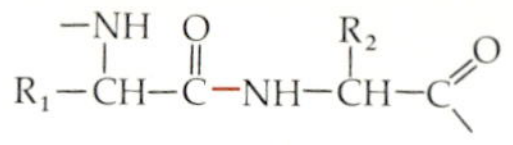 $R_1-CH(NH-)-C(=O)-NH-CH(R_2)-C(=O)-$ |
| polysaccharide | glycoside | 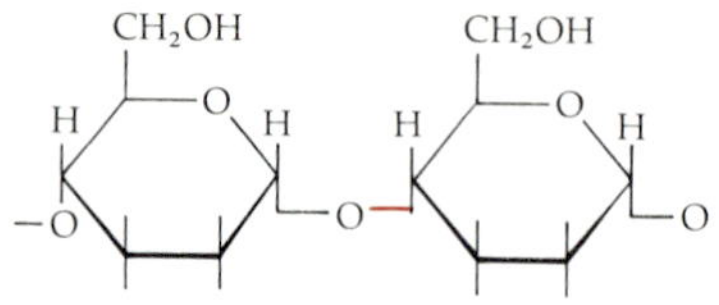 |
| nucleic acid | phosphodiester | 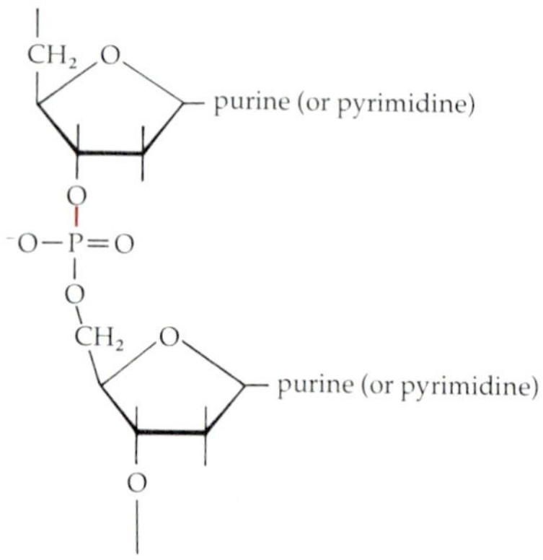 |

purine (or pyrimidine)

$^{-}O-P=O$

purine (or pyrimidine)

FIGURE 5.34

Nature of the bonds that link together the subunits in the major classes of biological polymers.

stem of the pathways of biosynthesis of the two amino acids. A dehydrase converts pretyrosine to phenylalanine and a dehydrogenase converts it to tyrosine.

There are a number of variations of the pathway by which isoleucine is synthesized (Figure 5.23). Most of these involve variations in the synthesis of the intermediate, α-ketobutyrate. Rather than synthesizing it from threonine as *E. coli* does, some bacteria can synthesize it from methionine, or acetate and pyruvate (*Leptospira*), or propionate (*Clostridrium sporogenes*). *C. sporogenes* also can synthesize another intermediate of the pathway, α-keto-β-methylvalerate by an alternate route, namely by carboxylation of α-methylbutyrate.

Other variations in biosynthetic pathways will almost certainly be revealed when still relatively unstudied groups like the strict anaerobes and the archaebacteria are more thoroughly investigated in this respect.

## THE POLYMERIZATION OF BUILDING BLOCKS: GENERAL PRINCIPLES

Proteins and nucleic acids are biopolymers composed of subunits (monomers) linked together by bonds that are characteristic of each class of macromolecule (Figure 5.34). The subunits of all biopolymers can be liberated in free form by hydrolysis. Thus, the biosynthesis of biopolymers involves the joining of subunits through reactions which are, in a formal chemical sense, the reverse of hydrolysis: namely, *dehydration.*

Biopolymers can be hydrolyzed to their subunits by either chemical or enzymatic means. Thus, their biosynthesis by simple dehydration is thermodynamically unfavorable. The net synthesis of all biopolymers is therefore accomplished by a preliminary *chemical activation* of the monomer.

**TABLE 5.7**

**Biopolymers and Their Monomeric Constituents, Showing the Activated Forms of the Monomers**

| Biopolymer | Constituent Monomer[a] | Activated Form of Monomer |
|---|---|---|
| Protein | Amino acids | Aminoacyl tRNAs |
| Nucleic acid | Nucleoside monophosphates | Nucleoside triphosphates |
| Polysaccharide | Sugars | Sugar-nucleoside diphosphates |

[a] Product formed by hydrolysis.

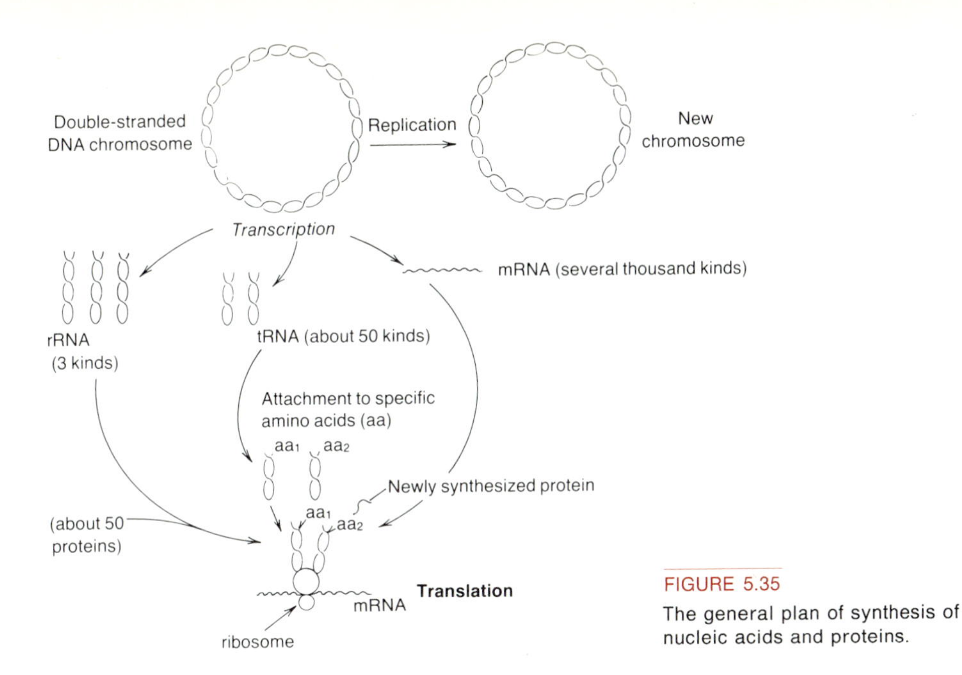


FIGURE 5.35
The general plan of synthesis of nucleic acids and proteins.

Such activation requires the expenditure of ATP, and involves the attachment of the monomer to a carrier molecule. Polymerization then occurs by transfer of the monomer from the carrier to the growing polymer chain, a thermodynamically favorable reaction. The activated forms of monomers of the major classes of biopolymers are shown in Table 5.7.

## The General Plan of Synthesis of Nucleic Acids and Proteins

A bacterial cell can synthesize several thousand different kinds of proteins, each containing, on the average, approximately 200 amino acid residues linked together in a definite sequence. The information required to direct the synthesis of these proteins is encoded by the sequence of nucleotides in the cell's complement of DNA, most of which is in the form of a double-stranded circular molecule, the bacterial chromosome (some bacteria also contain smaller circular molecules of DNA called plasmids, Chapter 10). By the process of *replication* the chromosome is precisely duplicated, thus assuring that progeny cells receive information enabling them to synthesize the same proteins.

The process by which the encoded information of the chromosome directs the order of polymerization of amino acids into proteins occurs in two steps: transcription and translation (Figure 5.35).

TRANSCRIPTION The information content of one of the strands of DNA is transcribed into RNA; i.e., the DNA strand serves as a template upon which a single strand of RNA is polymerized, the length of which corresponds to from one to several genes on the bacterial chromosome. One class of these RNA molecules, termed *messenger RNA* (mRNA), carries the information encoded in the DNA to the protein-synthesis machinery.

TRANSLATION Protein synthesis takes place on ribonucleoprotein particles called *ribosomes* [composed of ribosomal RNA (rRNA) and protein], which attach themselves to the molecule of mRNA. The information carried by the mRNA molecules is translated into protein molecules by a special class of RNA molecules called *transfer RNA* (tRNA). These molecules are multifunctional: they are able to bind to the ribosome, to be attached to specific amino acids, and to recognize specific nucleotide sequences of the mRNA. Each molecular species of tRNA recognizes a specific sequence of three nucleotides (a *codon*) on the mRNA molecule, and can be attached to a specific amino acid. Thus, the various amino acids are brought by their cognate tRNA molecules to the ribosome, where they are polymerized into protein in the sequence encoded by the mRNA.

The details of these processes will be discussed in subsequent sections.

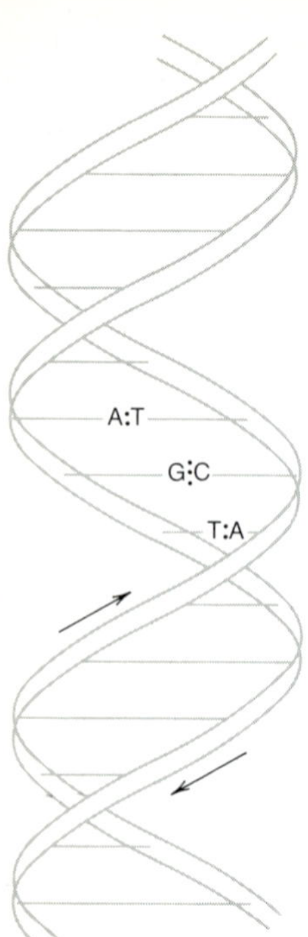


FIGURE 5.36 (left)

Schematic representation of the DNA double helix. The outer ribbons represent the two deoxyribosephosphate strands. The parallel lines between them represent the pairs of purine and pyrimidine bases held together by hydrogen bonds. Specific examples of such bonding is shown in the center section, each dot between the pairs of bases representing a single hydrogen bond. The direction of the arrows correspond to the 3′ to 5′ direction of the phosphodiester bonds between adjacent molecules of 2′ deoxyribose. After J. Mandelstam and K. McQuillen, *Biochemistry of Bacterial Growth*, 2nd ed. (New York: John Wiley, 1973).

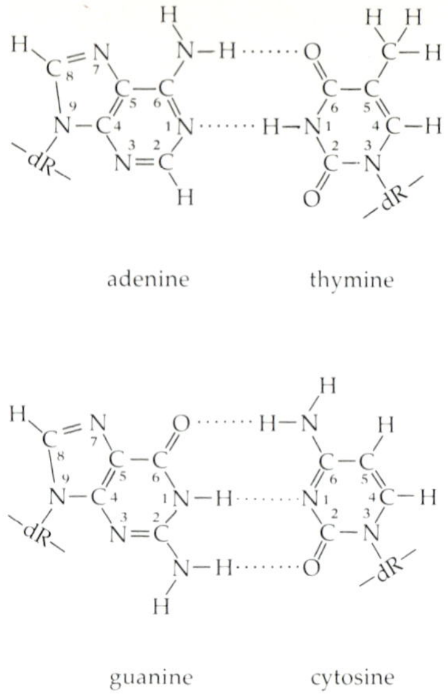

adenine thymine

guanine cytosine

FIGURE 5.37 (right)

The pairing of adenine with thymine and guanine with cytosine by hydrogren bonding. The symbol —dR— represents the deoxyribose moieties of the sugar-phosphate backbones of the double helix. Hydrogen bonds are shown as dotted lines.

FIGURE 5.38 (below)

The antiparallel nature of the double helix. Above, the complete structural formula of one base pair is shown. Below, a segment of duplex is shown diagrammatically. Note that the left-hand strand runs from 5′ to 3′, reading from top to bottom, while the right-hand strand runs from 3′ to 5′.

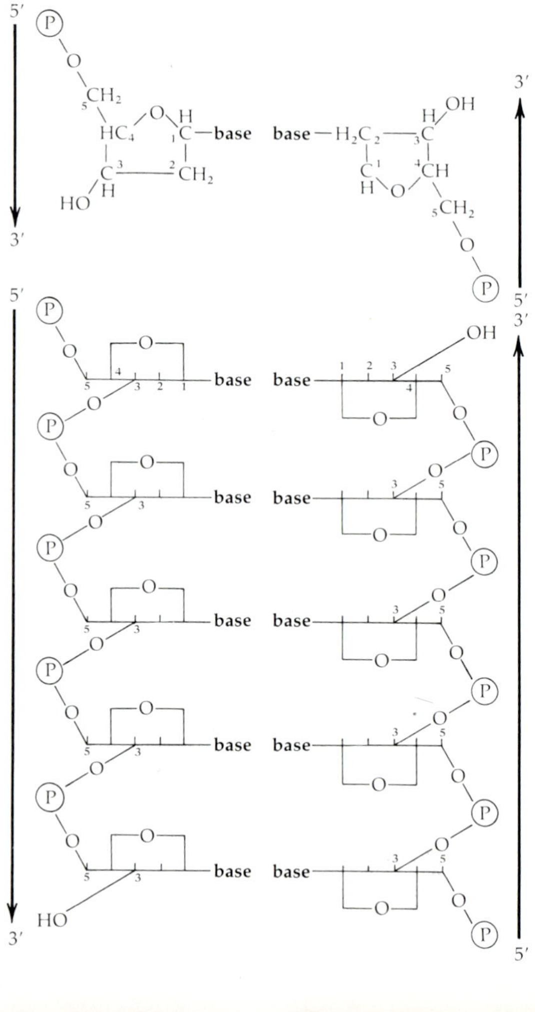

## THE POLYMERIZATION OF NUCLEOTIDES INTO DNA

The structure of the DNA molecule, elucidated by Watson and Crick in 1953, immediately suggests how it can be accurately replicated. It is a double helix, each strand of which consists of 2′-deoxyribose molecules linked together by phosphodiester bonds between the 3′-hydroxyl group of one and the 5′-phosphate group of the next. The purine and pyrimidine bases (attached to the 1 position of deoxyribose) project toward the center of the molecule, holding the two strands together by hydrogen bonding between specific purine-pyrimidine pairs. Guanine is paired with cytosine (G–C) and adenine is paired with thymine (A–T) (Figure 5.36). When the bases are present in their energetically most favorable forms (the keto, rather than the enol, form of the oxygenated bases, and the amino, rather than the imino, form of the aminated bases), only these pairs can fit within the hydrogen-bonding distances. The two hydrogen bonds that form between adenine and thymine, and the three hydrogen bonds that form between guanine and cytosine, are shown in Figure 5.37. The entire molecule can thus be described as a *linear sequence of nucleotide pairs*; the exact order of these pairs constitutes the genetic message, which contains all the information necessary to determine the specific structures and functions of the cell.

## The Antiparallel Structure of the DNA Double Helix

Each strand of the double helix is a polarized structure; its polarity results from the sequential linkage of polarized subunits, the deoxyribonucleotides. As shown in Figure 5.38, each nucleotide has a 5′-phosphate end and a 3′-hydroxyl end; when a series of nucleotides are connected by phosphodiester linkages, a polarized chain is formed which also has a 5′-phosphate end and a 3′-hydroxyl end.

The two strands of the double helix are *antiparallel* (i.e., they have *opposite polarity*). If we scan the diagram (Figure 5.38) from top to bottom, we see that the left-hand strand has 5′ → 3′ polarity, whereas the right-hand strand has 3′ → 5′ polarity. One consequence of the antiparallel structure of DNA will become apparent when we consider the process of DNA replication.

## DNA Polymerases

The polymerization of DNA is catalyzed by enzymes called *DNA polymerases.* In addition to the four deoxynucleoside triphosphates (dATP, dGTP, dCTP, and dTTP), which are substrates for the reaction, two molecules of nucleic acid are required: one is the DNA *template* to which the substrate deoxynucleoside triphosphate molecules pair according to the rules of hydrogen bonding (G with C and A with T); and a second is the primer, to which the nucleotides are attached as a consequence of the polymerization (Figure 5.39); the primer can be DNA or RNA. DNA synthesis proceeds in the 5′ to 3′ direction by the sequential formation of phosphodiester bonds between the α-phosphates of the 5′-deoxynucleoside triphosphates and the terminal 3′-hydroxyl group of the primer, with the release of one pyrophosphate molecule (Ⓟ—Ⓟ) for each deoxynucleotide added. The template DNA determines the sequence of addition of deoxynucleotides to the primer DNA molecule.

## Replication

Three different DNA polymerases are present in *E. coli:* polymerase I (Pol I), polymerase II (Pol II), and polymerase III (Pol III). Pol III catalyzes the addition of nucleotides to an RNA primer; Pol I can hydrolyze an RNA primer and duplicate single-stranded regions. The role of Pol II remains unclear.

Although the action of the DNA polymerases is simple and well understood, replication of the intact double-stranded bacterial chromosome is much more complicated and many questions remain to be answered. As previously stated, DNA polymerases require a single-stranded template; thus the double-stranded chromosome must be separated, at least locally, before replication can occur. Such separation forms a bubble in the chromosome at a specific site termed *oriC*, at which replication always initiates, but the bubble lacks a primer, which is required for DNA polymerase to function. Replication of a closed circular double helix of DNA like the chromosome would soon generate loops, termed *supercoiled twists*, in the molecule that would soon stop the process of replication. The required functions of strand separation, primer synthesis, and elimination of twists are mediated by a variety of proteins that, along with the

FIGURE 5.39

Schematic representation of a short fragment of double stranded DNA with a single-stranded region (the template) at the left end. The lower strand (the primer) is in the process of being lengthened by the action of DNA polymerase, which catalyzes a reaction between a deoxynucleoside triphosphate (in this case CTP) and the 3′-hydroxyl group of the primer strand. The arrow shows the direction of sequential additions of deoxyribonucleosides with the concomitant splitting off of pyrophosphate (Ⓟ—Ⓟ).

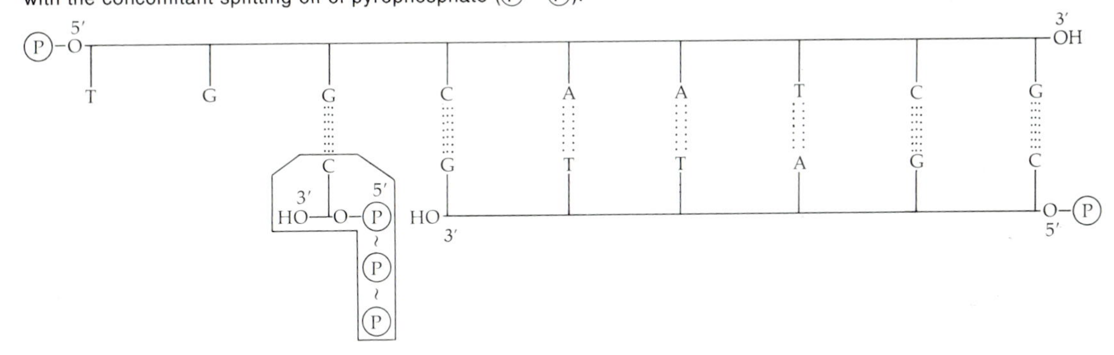

DNA polymerases, comprise a loose multienzyme complex termed the *replication apparatus.* This apparatus, which is composed of at least 13 different proteins, functions with remarkable speed and accuracy: approximately 3,000 nucleotides are polymerized per second (at 37° C) and only about one mistake (incorrect pairing) is made per $10^{10}$ nucleotides copied.

Strand separation, the first step in the process of replication requires the participation of several proteins and the hydrolysis of ATP because energy is required to break the hydrogen bonds between the complementary bases and to unwind the strands. The otherwise high energy cost of separation is reduced by the action of one of the proteins, the *helix-destabilizing protein* (HDP) that binds cooperatively to the separated single strands. Other proteins, including DNA-unwinding enzyme I (topoisomerase I) and *Rep protein*, collectively termed unwinding proteins, actively separate the DNA strands with the concomitant hydrolysis of ATP to ADP and inorganic phosphate. Another protein, *DNA gyrase*, prevents the formation of twists by periodically breaking a phosphodiester bond in one of the strands, thereby allowing free rotation of the opposite strand. Later this same enzyme reforms the same bond. The activity of DNA gyrase also requires the hydrolysis of ATP. The activities of the various proteins that mediate the processes of strand separation and elimination of twists are summarized in Figure 5.40.

Once the strands are separated, short pieces of RNA complementary to a portion of the single-stranded regions are synthesized on each of the strands. This RNA, which serves as a primer for DNA polymerase, is synthesized by a special RNA polymerase in the replication apparatus. Like other RNA polymerases it does not require a primer in order to function.

The opened chromosome with attached pieces of RNA forms two *replication forks* at which replication catalyzed by DNA polymerase proceeds in opposite directions around the circular chromosome until they meet at the *terminus:* bases of appropriate nucleoside triphosphates pair by hydrogen bonding with exposed bases in the single stranded region and phosphodiester bonds form between the terminal 3′-hydroxyl group of the primer molecule and the α-phosphate of the nucleoside triphosphate. Thus, replication on each strand always occurs in the 5′ to 3′ direction. Focusing on one of the replication forks (Figure 5.41) we note that, owing to the different polarities of the complementary single strands, their replication occurs in different directions. The strand that is replicated in the same direction as the movement of the

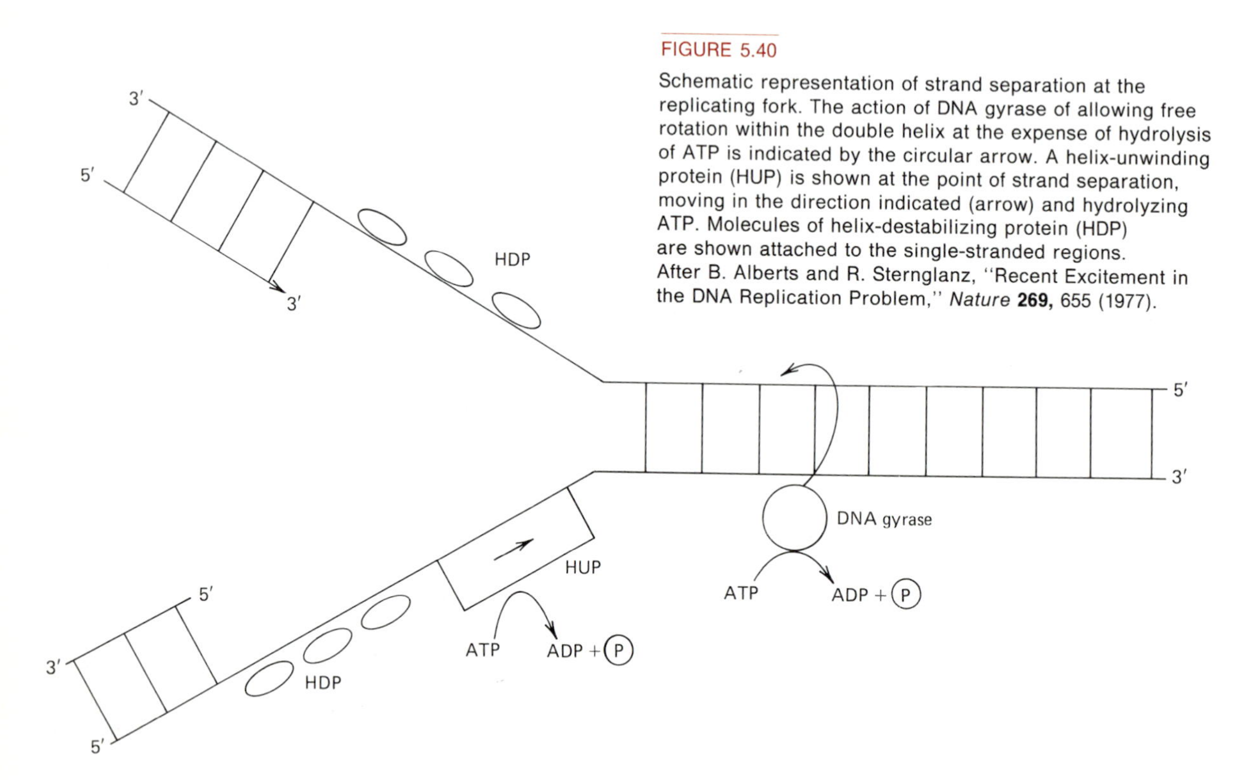


FIGURE 5.40

Schematic representation of strand separation at the replicating fork. The action of DNA gyrase of allowing free rotation within the double helix at the expense of hydrolysis of ATP is indicated by the circular arrow. A helix-unwinding protein (HUP) is shown at the point of strand separation, moving in the direction indicated (arrow) and hydrolyzing ATP. Molecules of helix-destabilizing protein (HDP) are shown attached to the single-stranded regions. After B. Alberts and R. Sternglanz, "Recent Excitement in the DNA Replication Problem," *Nature* **269,** 655 (1977).

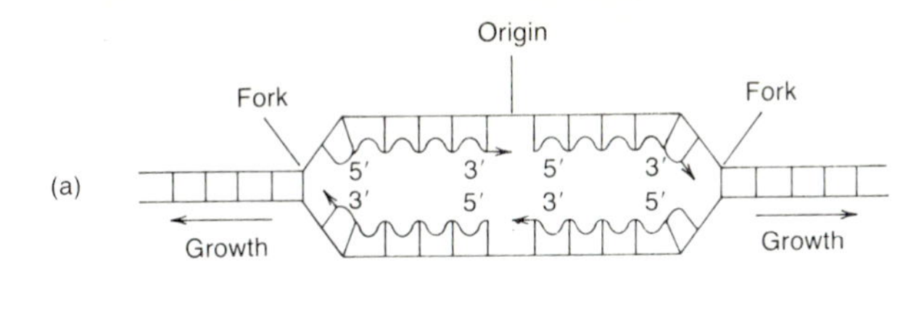


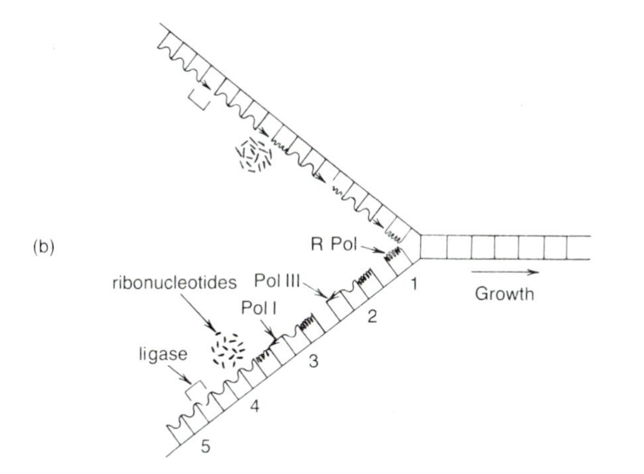


FIGURE 5.41

Schematic representation of the steps of replication of the bacterial chromosome. Part (a) represents a portion of a replicating bacterial chromosome at a stage shortly after replication has begun at the origin. The newly polymerized strands of DNA (wavy lines) are synthesized in the 5′ to 3′ direction (indicated by the arrows) using the preexisting DNA strands (solid lines) as a template. The process creates two replication forks which travel in opposite directions until they meet on the opposite side of the circular chromosome, completing the replication process. Part (b) represents a more detailed view of one of the replicating forks and shows the process by which short lengths of DNA are synthesized and eventually joined to produce a continuous new strand of DNA. For purposes of illustration, four short segments of nucleic acid are illustrated at various stages. In the first (1) primer RNA (thickened area) is being synthesized by an RNA polymerase (R Pol). Then, successively in (2) DNA is being polymerized to it by DNA polymerase III (Pol III); in (3) a preceding primer RNA is being hydrolyzed while DNA is being polymerized in its place by the exonuclease and polymerase activities of DNA polymerase I (Pol I); finally, the completed short segment of DNA (4) is joined to the continuous strand (5) by the action of DNA ligase (ligase).

replication fork is termed the *leading strand*; the other is termed the *lagging strand*. It will be noted that replication of the entire leading strand requires only a single molecule of primer RNA at *oriC* but replication of the lagging strand requires repeated synthesis of RNA primers as the replication fork moves. Reinitiation occurs on the lagging strand at intervals of about 1000 nucleotides, thus transiently creating short pieces of DNA attached to RNA (called *Okazaki fragments* for their discoverer, R. Okazaki). As synthesis of these fragments proceeds into the RNA primer of the previously synthesized one, the exonuclease activity of DNA polymerase I hydrolyzes the primer RNA as it replaces it with DNA. Finally the short pieces of DNA thus synthesized are joined to form a continuous complementary copy of the original strand by the action of an enzyme, *DNA ligase*, which catalyzes the two successive reactions:

$$\text{NAD}^+ + \text{enzyme} \longrightarrow \text{enzyme-AMP} + \text{nicotinic mononucleotide}$$

$$\text{enzyme-AMP} + 5'\text{-phosphate end of DNA} + 3'\text{-hydroxyl end of DNA} \longrightarrow \text{enzyme} + 5'\text{-}3'\text{phosphodiester bond} + \text{AMP}$$

It is interesting to speculate why DNA polymerases did not evolve (as RNA polymerases did) to be primer-independent, thus obviating the need for synthesis of an RNA primer on each Okazaki fragment. The answer might lie in the special requirements for accurate replication of DNA because of its role as a repository of the cell's genetic information. It has been estimated that the intrinsic mistake frequency of replication is about one incorrect base pairing per $10^5$ nucleotides. This is reduced to the observed mistake frequency (one per $10^{10}$) by an activity of DNA polymerase III, known as *proofreading*. This is accomplished at least in part by the enzyme's built-in 3′ to 5′ exonuclease activity, which removes a previous mismatch by moving backward. The enzyme does not catalyze replication in the forward direction unless it is followed by a properly matched base pair. Thus such a self-correcting polymerase cannot initiate chains *de novo*.

## THE SYNTHESIS OF RNA

Although the functional roles of the three major classes of RNA (mRNA, tRNA, and rRNA) differ markedly, their mechanism of polymerization from ribonucleoside triphosphates (termed *transcription*) are mechanistically identical. Indeed, the polymerization of all three major classes of RNA is catalyzed by the same complex enzyme *DNA-dependent RNA polymerase* (commonly called RNA polymerase). As stated earlier, polymerization of primer RNA that participates in replication is catalyzed by a different enzyme (sometimes called *primase*), but the mechanism of polymerization is similar whether it is catalyzed by RNA polymerase or primase. Primase utilizes a previously separated single strand of DNA as template; RNA polymerase participates in the separation of the strands as well as catalyzing polymerization. These enzymes utilize ribonucleoside triphosphates (ATP, CTP, GTP, of UTP) as substrate and a single strand of DNA as template. An example of a single step of the reaction can be written:

$$\text{RNA} + \text{ATP} \longrightarrow \text{RNA—AMP} + \textcircled{P} - \textcircled{P}$$

In this step the products are an RNA molecule lengthened by a single adenyl group and one molecule of pyrophosphate; C, G, and U moieties are added by analogous reactions when directed by their hydrogen bonding to the DNA template. The same hydrogen bondings occur in RNA polymerization as do in replication with the exception that uracil replaces thymine. Thus an A residue in the DNA directs incorporation of a U residue at the corresponding position in RNA. The phosphodiester bonds produced during polymerization form between a terminal 3′-hydroxyl group on the growing chain and the α-phosphate group of the nucleoside triphosphate: as with replication of DNA, chain growth proceeds in the 5′ to 3′ direction.

RNA polymerase is a complex enzyme composed, in most procaryotes, of four different types of subunits. One of these, sigma, is only loosely associated with the rest of the molecule (Table 5.8).

The structure of RNA polymerase is highly conserved among eubacteria; all so far studied have the same basic structure. But RNA polymerases in archaebacteria have been found to be quite different; *Sufolobus acidocaldarius*, for example, has an RNA polymerase composed of two large subunits and eight different small ones. The subunit structure of its RNA polymerase resembles RNA polymerase from yeasts more than those from eubacteria.

## SYNTHESIS OF PROTEINS

The products of transcription, mRNA, tRNA, and rRNA, all participate in the synthesis of proteins. Thus the rate and specificity of transcription determines, in large measure, the rate and the relative proportion of the various proteins that are synthesized. All control of transcription is affected by the frequency of its initiation and termination (a special form of termination, called *attenuation*, that plays an important role in regulating transcription is discussed in Chapter 12). In the following sections the process of protein synthesis by eubacteria is described; protein synthesis by archaebacteria and by eucaryotes differs in certain details.

### Initiation of Transcription

Initiation of transcription occurs at specific points on the bacterial chromosome that are defined by short sequences of bases of DNA called *promoters*. The sequences also determine which strand (termed the *sense strand*) will be transcribed and hence, because polymerization always occurs 5′ to 3′, the direction of transcription. The orientation of the various units of transcription (a gene or an operon; see Chapter 10) appears to vary without discernible pattern around the chromosome, i.e., with almost equal probability either DNA strand of the chromosome serves as the sense strand.

RNA polymerase binds to the promoter over a relatively long region extending from approximately 20 base pairs downstream (direction of transcription) to approximately 45 base pairs upstream from the point at which transcription is initiated. With the recently acquired capability of scientists to determine the exact sequence of bases in frag-

**TABLE 5.8**

**Subunit Structure of DNA-Dependent RNA Polymerase from *E. coli***

| Subunit | Molecular Weight | Number per Polymerase | | |
|---|---|---|---|---|
| Alpha | 41,000 | 2 | Core enzyme | Holoenzyme[a] |
| Beta | 155,000 | 1 | | |
| Beta-prime | 165,000 | 1 | | |
| Sigma | 86,000 | 1 | | |

[a] The holoenzyme initiates the transcription process; the core enzyme catalyzes polymerization.

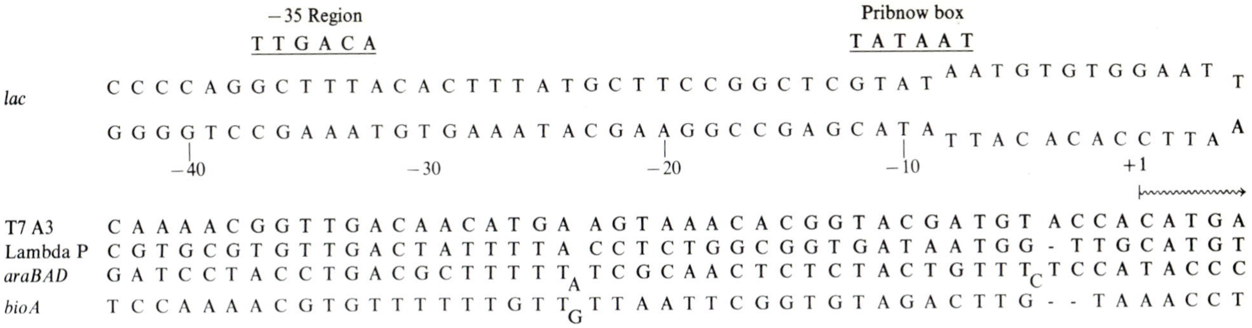

```
                  -35 Region                                        Pribnow box
                  T T G A C A                                       T A T A A T
                                                                              A A T G T G T G G A A T
lac       C C C C A G G C T T T A C A C T T T A T G C T T C C G G C T C G T A T                         T

          G G G G T C C G A A A T G T G A A A T A C G A A G G C C G A G C A T A                         A
              |                             |                     |     T T A C A C A C C T T A
             -40            -30            -20                   -10                 +1
                                                                                     ~~~~~~~~~~>

T7 A3     C A A A A C G G T T G A C A A C A T G A A G T A A A C A C G G T A C G A T G T A C C A C A T G A
Lambda P  C G T G C G T G T T G A C T A T T T T A C C T C T G G C G G T G A T A A T G G - T T G C A T G T
araBAD    G A T C C T A C C T G A C G C T T T T T   T C G C A A C T C T C T A C T G T T T   T C C A T A C C C
                                                 A                                     C
bioA      T C C A A A A C G T G T T T T T T G T T   T T A A T T C G G T G T A G A C T T G - - T A A A C C T
                                                 G
```

FIGURE 5.42

Base sequences of certain promoters in *E. coli* and its phages. The two upper rows (*lac*) show the sequence of bases in a mutant lactose promoter (*lacUV5*) that is transcribed at high frequency. The numbering follows the convention of assigning to base pairs increasing negative numbers from the point of transcription in the direction opposite to the one in which transcription occurs (wiggly arrow). Transcription initiates at +1. The separated bases are those that are unwound when RNA polymerase binds to the promoter. Above the *lac* promoter are shown the consensus sequences in the nontemplate strand of the two highly conserved regions: the −35 region and the Pribnow box. The four lower rows show the sequence of bases in the nontemplate strand of four other promoters: the A3 promoter of the coliphage T7 (T7 A3); the P promoter of the coliphage Lambda (Lambda P); the promoter of the arabinose operon (*araBAD*); and the promoter of the biotin A gene (*bioA*). Extra bases between the highly conserved regions are shown below the lines; lesser numbers of bases are indicated by dashes (–). From U. Siebenlist, R. B. Simpson, and W. Gilbert, 1980. "*E. coli* RNA Polymerase Interacts Homologously with Two Different Promoters," *Cell* **20,** 269 (1980).

ments of DNA (Chapter 13) the primary structure of a number of promoters (principally those from *E. coli*) have been determined. A comparison of these (Figure 5.42) reveals two highly conserved regions, suggesting that they play particularly important roles in promoter function. One of these sequences (TATAAT), termed the *Pribnow box,* lies 10 base pairs upstream from the point of initiation of transcription. The other (TTGACA), as its name (*−35 region*) implies, lies 35 base pairs upstream. These sequences, being the most frequently encountered, have been termed *consensus sequences.* Most promoters differ slightly from the consensus with respect to the precise sequence of the conserved region and/or the distance between them and the point of initiation. These variations are presumed to determine *promoter strength,* the frequency of initiation of transcription at a particular promoter. In *E. coli* growing in a rich medium at 37°C the strongest promoters initiate a transcription as frequently as every four seconds—the weakest ones as infrequently as a few times per hour.

An initiation begins when the RNA polymerase holoenzyme binds to the promoter, causing the strands of the DNA double helix to separate (melt); then polymerization begins with the sense strand as template. Soon after, the sigma ($\sigma$) subunit dissociates and the core enzyme alone catalyzes the remainder of the polymerization of the transcriptional unit. Thus the function of sigma seems primarily to be promoter recognition. It has been suggested that more than one type of sigma might occur, thereby further modulating promoter strength. However, additional sigma proteins have not been found in *E. coli,* although some proteins of unknown function (e.g., $\omega$) are sometimes associated with RNA polymerase and there is some evidence that a substitute sigma might modulate expression of certain proteins when *E. coli* is exposed to high temperature (Chapter 8). However, there is clear evidence for multiple sigma proteins in the aerobic spore-forming bacterium, *Bacillus subtilis.* It produces at least four sigma proteins (sigma$_{28}$, sigma$_{29}$, sigma$_{37}$, and sigma$_{55}$) designated by their molecular weight ($\times 10^{-3}$), and additional forms have been reported. Thus *B. subtilis* can produce at least four different forms of RNA polymerase. Strong evidence suggests that the variety of forms of RNA polymerase plays an important role in modulating the frequency of transcription of various units. The sporulating cell contains a complex mixture of forms that changes with the stage of sporulation, and even vegetative cells produce more than a single form. The particular forms of RNA polymerase are differentially effective in transcribing various promoters. Thus, as the proportion of the forms changes, expression of genes is modulated.

## Termination of Transcription

At the end of a transcriptional unit, RNA polymerase dissociates from the DNA template and polymerization ceases. Like initiation, termination of transcription is signaled by specific sequences of DNA called *terminators*. The precise nucleotide sequence of terminators varies. Some, termed *strong terminators*, stop transcription at high frequency; others, *weak terminators*, stop it at low frequency. In all cases the participation of an accessory protein, *rho*, which acts in an ATP-driven reaction to dissociate RNA polymerase from its template, raises the frequency of termination to near 100 percent. Thus strong promoters are sometimes called *rho-independent* and weak ones, *rho-dependent*.

Strong terminators share three features: they contain a set of A–T pairs within which termination occurs preceded by a set of G–C pairs. After a short interval upstream the sequence is repeated in reverse order (Figure 5.43). Thus, by hydrogen bonding between bases in this inverted repeat region, a stem-loop structure can form; it is presumed to facilitate release of RNA polymerase from the DNA template.

## Translation

The complex process collectively called translation, through which the information encoded by the sequence of bases in a molecule of mRNA directs the sequence in which amino acids are polymerized to form a protein, can be analyzed by considering the individual component reactions.

## Activation of Amino Acids

The activated forms of amino acids that are synthesized to form proteins are *aminoacyl-tRNAs*. They are synthesized in two steps by a group of enzymes, *aminoacyl-tRNA synthetases*. Each of these 20 enzymes is specific for a particular amino acid but some react with several different tRNA molecules, that is, several different types of tRNA molecules can accept the same amino acid. Various tRNAs are usually designated by a superscript of three capital letters indicating the amino acid that they accept and a numeric subscript (sometimes followed by a capital letter) to distinguish the various tRNAs that accept a particular amino acid.

FIGURE 5.43

(a) A hypothetical sequence of bases in a strong terminator. A set of A–T pairs (single underscore) and an inverted repeat of some of these (horizontal arrows). Termination occurs at the site of the vertical arrow. Noncritical nucleotides are designated N. (b) The terminal portion of the mRNA formed from the gene depicted in (a). (c) The stem-loop structure that can form in the mRNA by hydrogen bonding (·). After D. Pribnow, "Genetic Control Signals in DNA," in *Biological Regulation and Development*, ed. R. F. Goldberger (New York: Plenum Press, 1979).

(a)

(b) N N A A G C G C C G N N N N C C G G C G C U U U U U U—OH 3′ RNA

(c)

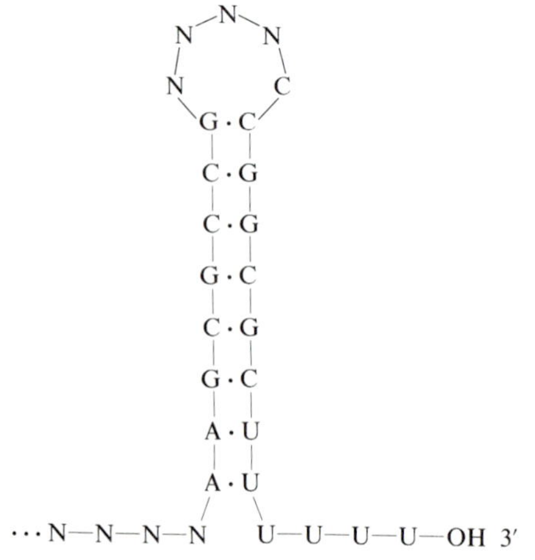

N N N
N C
G · C
C · G
C · G
G · C
C · G
G · C
A · U
A · U
···N—N—N—N U—U—U—U—OH 3′

RNA structure

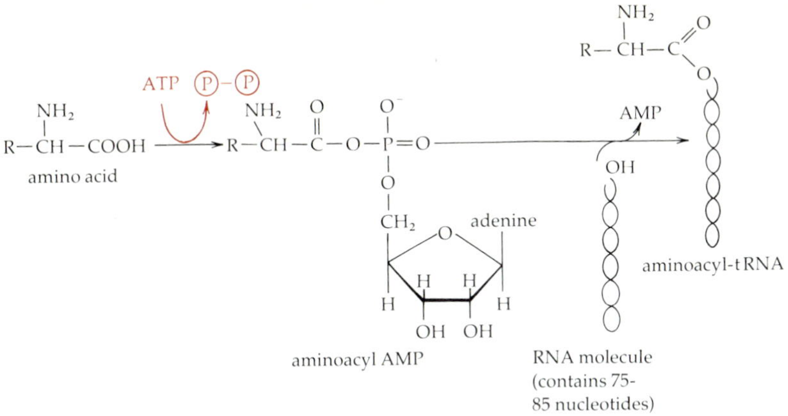

ATP Ⓟ–Ⓟ

amino acid

aminoacyl AMP

adenine

AMP

RNA molecule (contains 75-85 nucleotides)

aminoacyl-tRNA

FIGURE 5.44

The process of amino acid activation, in which an amino acid becomes attached to a specific tRNA molecule.

Thus $tRNA_2^{GLU}$ designates one of the species of tRNA that accepts glutamate.

In the synthetase reaction the amino acid reacts with ATP to form an enzyme bound intermediate, *aminoacyl adenylic acid*; then the aminoacyl group is transferred to the hydroxyl group of the terminal AMP residue that all tRNA molecules contain at their 3′ end (Figure 5.44).

## Synthesis of the Procaryotic Ribosome

Polymerization of activated amino acids into proteins is mediated by the 70S ribosome, a complex organelle that occurs in large numbers in procaryotic cells. A moderate sized *E. coli* cell will contain as many as 20,000 of these structures. Indeed, electron micrographs of thin sections of procaryotes reveal a cytoplasm that appears to consist almost entirely of closely packed ribosomes.

The 70S ribosome (Figure 5.45) dissociates reversibly into a 50S and 30S subunits *in vitro* if the concentration of $Mg^{2+}$ ions in the suspending buffer is lowered from $10^{-2}$ to $10^{-4}$ M. Each of these is an association of molecules of RNA and protein. The large one (50S) contains one molecule each of 23S and 5S RNA associated with 35 different proteins; the small one (30S) contains a single molecule of 16S RNA and 21 different proteins.

The large quantities of rRNA required for ribosome formation are supplied from seven redundant clusters of genes (operons, see Chapter 10) in *E. coli*. Each of these is composed of genes encoding one molecule of 16S, 23S, 5S and one or more molecules of tRNA in the sequence:

16S-spacer tRNA—23S—5S—distal tRNA

Not all clusters have a distal tRNA gene but each contains one or two spacer tRNA genes. The product of transcription of each gene cluster would

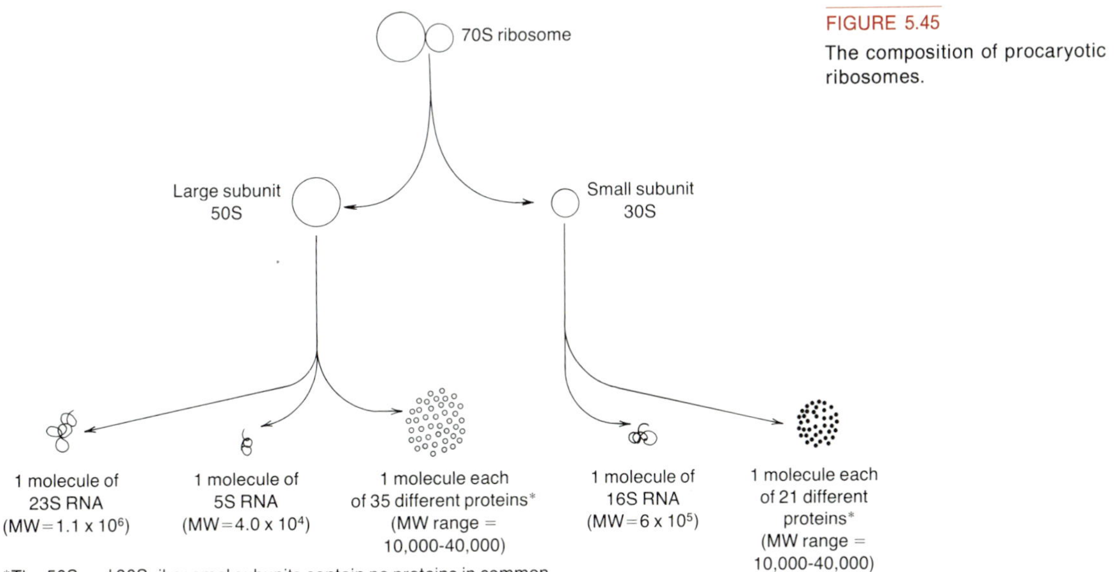


FIGURE 5.45

The composition of procaryotic ribosomes.

be a single large molecule of RNA. But as it is being polymerized it is cleaved by four different kinds of RNase (RNase III, RNase P, RNase F, and RNase E) into precursor forms of the constituent rRNAs and tRNAs. Precursor rRNAs immediately begin to associate in a precise pattern with appropriate molecules of ribosomal protein. During this process, at least three other RNases (RNase M16, RNase M23, and RNase M5) act on the precursor forms reducing them to the 16, 23, and 5S molecules found in the completely assembled ribosomal subunits. This entire process occurs with remarkable rapidity: each cell in a culture of *E. coli* growing at a moderate rate (doubling time, 40 minutes) at 37° C produces about eight completed ribosomes per second.

## Initiation of Translation

The first amino acid residue in a newly synthesized product of translation is always a modified methionine, N-formylmethionine. It is encoded by an AUG or GUG triplet on the mRNA. Thus initiation always begins at such an *initiation triplet*. But these triplets occur many places in most mRNA molecules so in themselves they are not adequate indicators of the site of initiation. A site on the mRNA that precedes the initiator codon by approximately 10 nucleotides contributes to the identification of a particular AUG or GUG codon as being an initiator triplet. This site, variously called a *ribosomal binding site* or a *Shine-Delgarno sequence* (for the two Australian biochemists who discovered its significance), is complementary to the 3′-hydroxyl end of the 16S RNA component of the 30S ribosomal subunit. Presumably hydrogen bonding occurs between these regions, placing the ribosome in proper register to begin polymerization at the initiator triplet.

Initiation of polymerization is preceded by the formation of an *initiation complex* composed of a 30S ribosomal subunit and a formylmethionyl-tRNA bound to the initiator region. Then the 50S subunit attaches producing a 70S ribosome attached to the mRNA. These reactions require the participation of three accessory proteins, called *initiation factors* (designated IF1, IF2, and IF3), and the hydrolysis of one molecule of GTP to GDP and inorganic phosphate.

## Elongation of the Peptide Chain

The 70S ribosome has two sites, designated the A (aminoacyl) and P (peptide) sites, that bind tRNAs and attached amino acids or peptides. The initiation complex forms with formylmethionyl tRNA at the P site; all other aminoacyl-tRNAs enter the A site. The addition of each amino acid to a growing peptide chain, including the one added to the formylmethionyl-tRNA in the initiation complex, includes the same series of three steps: recognition, peptidyl transfer, and translocation (Figure 5.46).

1. Recognition In the recognition step a molecule of aminoacyl-tRNA attaches to the A site with a sequence of three bases (a codon) on the mRNA molecule. The bases must be complementary to three bases (the *anticodon*) on the distal end of the aminoacyl-tRNA molecule in order for polymerization to proceed. Since the particular sequence of an anticodon of a tRNA molecule thereby determines that the amino acid attached to its 3′ OH terminus will enter into a peptide chain when its complementary codon is exposed in the A site, the sequence of bases of a codon is said to *encode* a particular amino acid. Collectively the relation between the 64 possible codons and amino acids is called the *genetic code* (Table 5.9). One notes that all but three codons, termed *nonsense codons*, correspond to a particular amino acid. The three nonsense codons signal chain termination.

Two protein factors, called *elongation factors* (EFTu and EFTs), participate in recognition; during this process one molecule of GTP is hydrolyzed to GDP and inorganic phosphate. One of the protein factors (EFTu) binds sequentially to GTP and aminoacyl-tRNA. This complex binds to the ribosome and, at recognition, GTP is hydrolyzed. GDP and EFTu are released as a complex which is dissociated by the action of EFTs. This cyclic series of reactions that participate in the recognition of all aminoacyl-tRNAs except formylmethionyl-tRNA can be summarized as follows:

$$\text{EFTu} \xrightarrow{\text{GTP}} \text{GTP—EFTu} \xrightarrow{\text{AA—tRNA}}$$

$$\text{GTP—EFTu—AA—tRNA} \xrightarrow{\text{ribosome}}$$

$$\text{ribosome-GTP-EFTu-AA-tRNA} \longrightarrow \text{ribosome-AA-tRNA} + (\text{P})$$

$$\text{GDP—EFTu} \xrightarrow{\text{EFTs}} \text{GDP} + \text{EFTu} + \text{EFTs}$$

2. Peptidyl Transfer After recognition, both ribosomal binding sites are occupied by aminoacylated tRNAs, so arranged as to facilitate a transfer reaction catalyzed by the 50S ribosomal subunit.

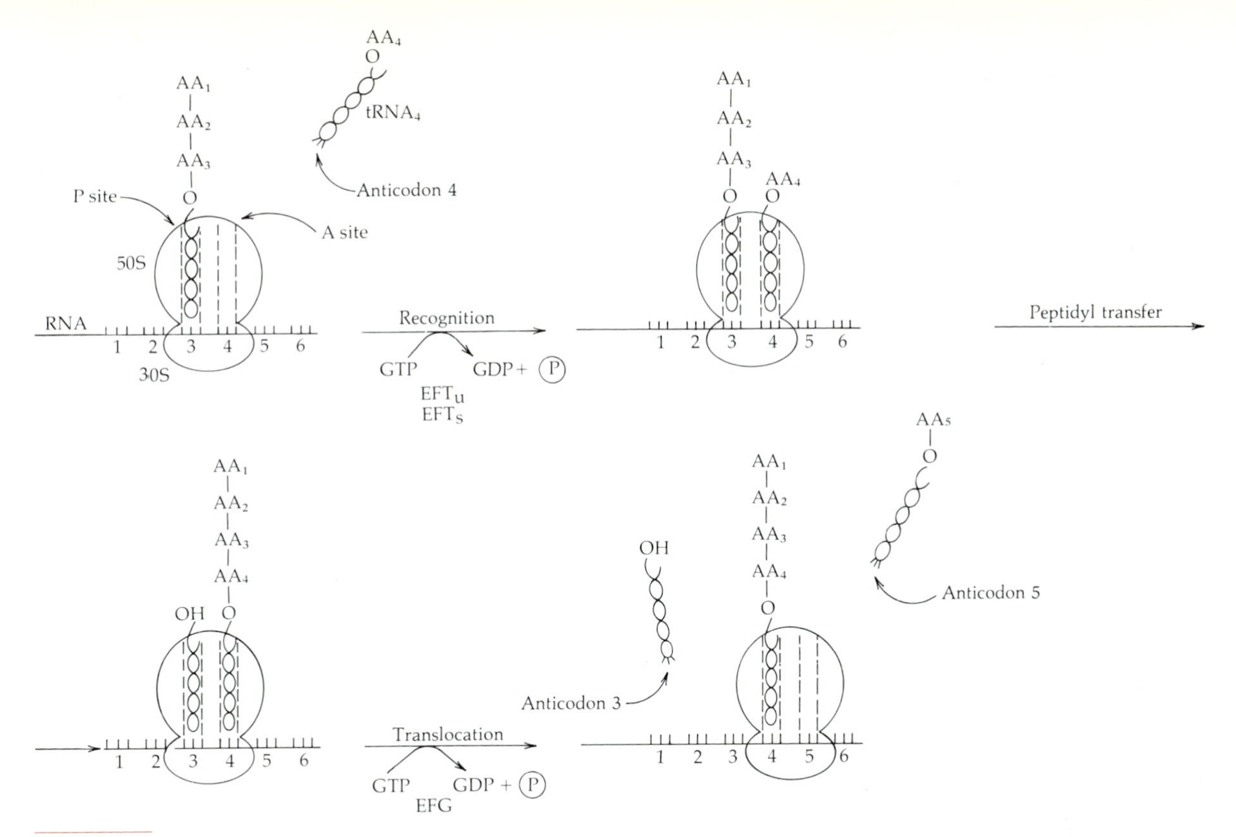


FIGURE 5.46

The sequence of events in the lengthening of the peptide chain. Amino acids (AA) are numbered by their order of addition to the peptide; numbered trios of lines symbolize the codons on the mRNA molecule.

**TABLE 5.9**

**The Genetic Code: Correspondence between Codons and Amino Acids**

| | Second bases | | | | | | | |
|---|---|---|---|---|---|---|---|---|
| First base | U | | C | | A | | G | |
| U | UUU | phe[a] | UCU | ser | UAU | tyr | UGU | cys |
| | UUC | phe | UCC | ser | UAC | tyr | UGC | cys |
| | UUA | leu | UCA | ser | UAA | (none)[b] | UGA | (none)[b] |
| | UUG | leu | UCG | ser | UAG | (none)[b] | UGG | try |
| C | CUU | leu | CCU | pro | CAU | his | CGU | arg |
| | CUC | leu | CCC | pro | CAC | his | CGC | arg |
| | CUA | leu | CCA | pro | CAA | glu-N | CGA | arg |
| | CUG | leu | CCG | pro | CAG | glu-N | CGG | arg |
| A | AUU | ileu | ACU | thr | AAU | asp-N | AGU | ser |
| | AUC | ileu | ACC | thr | AAC | asp-N | AGC | ser |
| | AUA | ileu | ACA | thr | AAA | lys | AGA | arg |
| | AUG | met | ACG | thr | AAG | lys | AGG | arg |
| G | GUU | val | GCU | ala | GAU | asp | GGU | gly |
| | GUC | val | GCC | ala | GAC | asp | GGC | gly |
| | GUA | val | GCA | ala | GAA | glu | GGA | gly |
| | GUG | val | GCG | ala | GAG | glu | GGG | gly |

[a] Amino acids are abbreviated as the first three letters in each case, except for glutamine (glu-N), asparagine (asp-N), and isoleucine (ileu).
[b] The codons UAA, UAG, and UGA are nonsense codons (see Chapter 10); UAA and UAG are called the ochre codon and the amber codon, respectively.

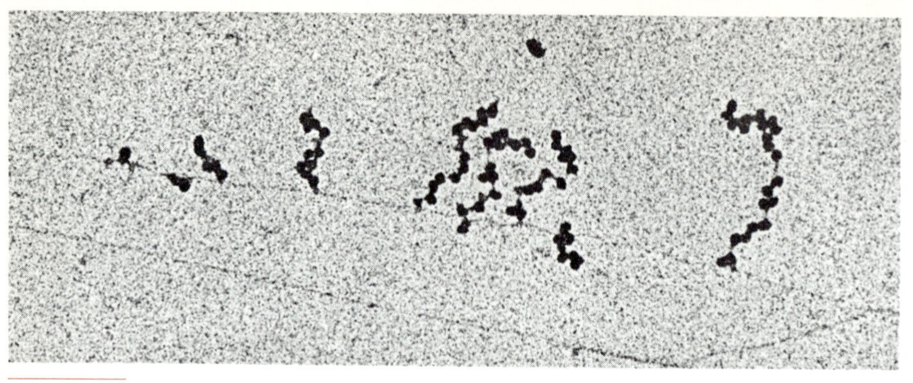

FIGURE 5.47

Photomicrograph of the simultaneous transcription and translation of a fragment of the chromosome of *E. coli*. The central horizontal line is DNA. The more wavy lines extending from it are molecules of mRNA to which a number of ribosomes are attached (polysomes). The gradual increase in length of the mRNA from left to right of the picture indicates that transcription by DNA-dependent RNA polymerase molecules (barely discernible at the junction of the DNA and mRNA) was proceeding in that direction, and that transcription began near the left edge of the picture. (×62,350) From B. Hamkalo and O. Miller, Jr., "Electronmicroscopy of Genetic Material," *Ann. Rev. Biochem.* **42,** 379 (1973).

The peptide bond forms between the terminal carboxyl group of the peptide and the alpha-amino group of the amino acid in the A site:

$$\text{peptide}-\text{CHR}_1-\overset{\overset{\displaystyle O}{\|}}{C}-\text{O-tRNA}_P +$$

$$\text{NH}_2-\text{CHR}_2-\overset{\overset{\displaystyle O}{\|}}{C}-\text{O-tRNA}_A \longrightarrow \text{tRNA}_P +$$

$$\text{peptide}-\text{CHR}_1-\overset{\overset{\displaystyle O}{\|}}{C}-\text{NH}-\text{CHR}_2-\overset{\overset{\displaystyle O}{\|}}{C}-\text{O-tRNA}_A$$

The peptidyl transfer reaction requires neither the participation of accessory protein factors nor the expenditure of energy in the form of hydrolysis of a nucleoside triphosphate.

3. TRANSLOCATION Peptidyl transfer is followed by a complex and incompletely understood series of reactions collectively called translocation: the free tRNA molecule in the P site is released and the ribosome moves three bases down the mRNA, thereby moving the peptide-bearing tRNA from the A site to the P site and putting in register in the empty A site the next codon to be read. Translocation requires one accessory protein (*elongation factor G*, EFG) and the hydrolysis of an additional molecule of GTP.

By repetitive recognition, peptidyl transfer, and translocation steps, successive amino acids are added to the peptide chain in the order encoded by the sequence of codons in the mRNA molecule. The process continues until a nonsense codon (UAG, UGA, or UAA) is reached, which causes the release of the completed protein from the 70S ribosome. The process requires the intervention of a protein *release factor* (RF). Then, through the action of IF3, the ribosome dissociates into its 30S and 50S subunits.

Within the cell, translation of a molecule of mRNA begins before its synthesis is complete. At any moment an average of about 20 ribosomes are translating the same molecule of mRNA. A molecule of mRNA to which a number of functioning ribosomes are attached is called a *polysome*. Excellent electron micrographs have been made which show the concurrent nature of transcription and translation as well as the formation of polysomes (Figure 5.47).

## The Secondary, Tertiary, and Quaternary Structure of Proteins

The sequence of amino acids in a protein, termed its *primary structure*, determines the complex shape that a protein will assume. Even before the nascent polypeptide chain detaches from the ribosome, it begins to fold on itself. First, parts of the polypeptide become coiled into a regular, helical structure called an alpha-helix, and other regions remain extended in a zigzag form, the so-called beta-conformation; these are designated as the proteins' *secondary structure*. In general, the presence of the amino acids, alanine, glutamic acid, leucine, and methionine foster the formation of the alpha-helix while proline, glycine, tyrosine, and serine inhibit it. Next the entire molecule assumes a specific three-dimensional shape, called the *tertiary structure* of the protein.

The fact that the secondary and tertiary structures of a protein are determined solely by primary structure can in certain cases be demonstrated experimentally. If a protein is *denatured* (caused to lose secondary and tertiary structure) under special conditions (e.g., in the presence of high concentrations of urea or guanidinium chloride), removal of the denaturating agent permits the protein to refold into its native state.

The fraction of the molecule that exists in the alpha-helical configuration or in the beta-conformation varies from 0 to 100 percent in different proteins. The alpha-helix is maintained by hydrogen bonds between the carboxyl oxygen of one peptide bond and the amino nitrogen of another three residues further along the chain. The resulting helix completes a 360° turn every 3.6 residues.

Tertiary structure of most proteins is maintained by several types of bonds of which the most important are *disulfide* and *hydrophobic*. Disulfide bonds occur by the formation of an oxidized linkage between the sulfhydryl moieties of two cysteine residues:

$$-SH + -SH = -S-S- + 2H$$

Hydrophobic bonds are derived from the lower energy state attained if hydrophobic (nonpolar) amino acids are closely packed together within the structure of the protein rather than being exposed to the external polar aqueous environment.

An important consequence of hydrogen bonding between different regions of the protein is the formation of beta-sheets, structures in which regions in the beta-conformation are held in precise alignment with one another at a distance of 0.7 nm.

The configuration of a protein is also maintained by *ionic bonds* that form between free carboxyl groups of acidic amino acids and free amino groups of basic amino acids. Examples of the secondary and tertiary structure of proteins is shown in Figure 5.48.

The quaternary structure of a protein is formed by the noncovalent association of two or more polypeptides termed *subunits*, or the covalent joining of them through —S—S— bonds. Glutamic dehydrogenase, for example, is composed of

FIGURE 5.48

Schematic representation of the secondary and tertiary structures of proteins with variable contents of alpha-helix and beta-conformation. The alpha-helical regions are represented as coiled ribbons and the beta-conformation regions as arrows. (a) Cytochrome *c*, a protein with only alpha-helical secondary structure. (b) Phosphoglycerate kinase, a protein with alpha-helical and beta-conformation, the latter arranged in sheets. (c) Triosephosphate isomerase, a protein also containing alpha-helical and beta-conformation regions.
J. S. Richardson, "The Anatomy and Taxonomy of Protein Structure," *Adv. Pro. Chem.* **34,** 168 (1981) by permission.

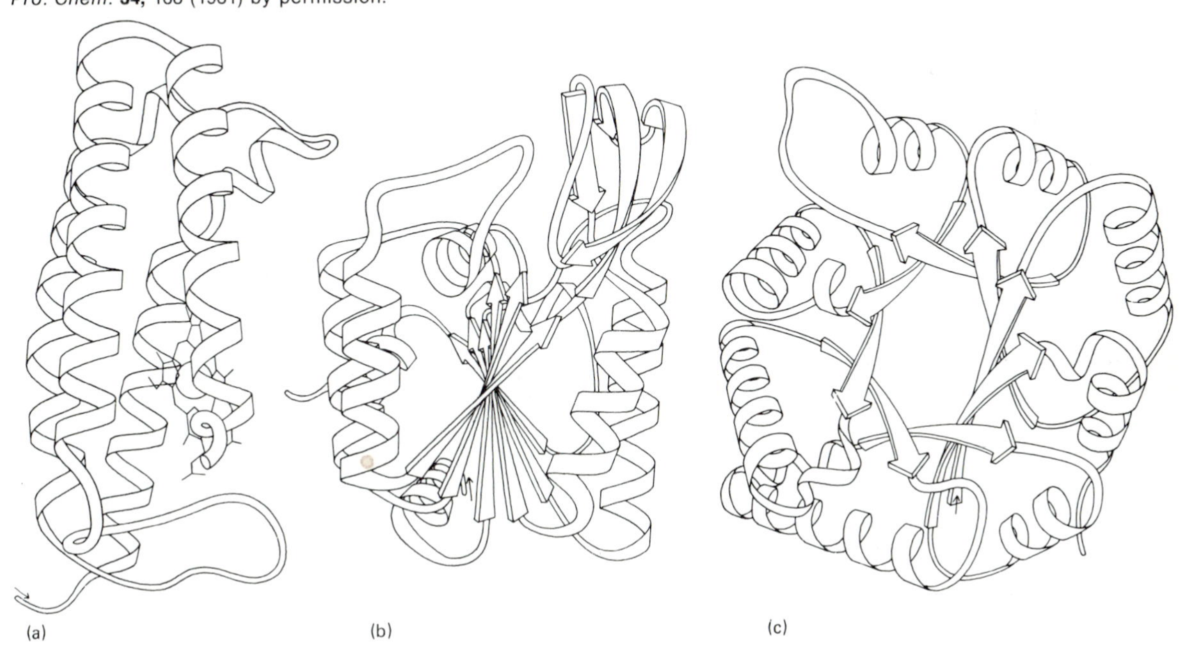

between 24 and 30 identical subunits, each with a molecular weight of about 40,000. However, most proteins have fewer subunits.

A fundamental principle of protein structure is that each higher-order structure is determined directly by lower-order structure; all the information needed to determine the structure of a protein is included in the sequence of the amino acids that comprise it.

## THE SYNTHESIS OF POLYSACCHARIDES

The properties of several systems for the synthesis of polysaccharides are described in Table 5.10. A characteristic feature of polysaccharide synthesis, like that of DNA synthesis, is the requirement for a primer. In the case of polysaccharide synthesis it is a short segment of the polysaccharide in question that acts as an acceptor of additional monomer units. In the synthesis of glycogen, where the function of the primer has been studied in detail, it has been found that the primer must contain more than four sugar units to function effectively (Figure 5.49). The molecular branching characteristic of glycogen is produced by a special *branching enzyme* that cleaves off small fragments from the end of the 1,4 linked linear polysaccharide chain, and inserts them in 1,6 linkage at another point.

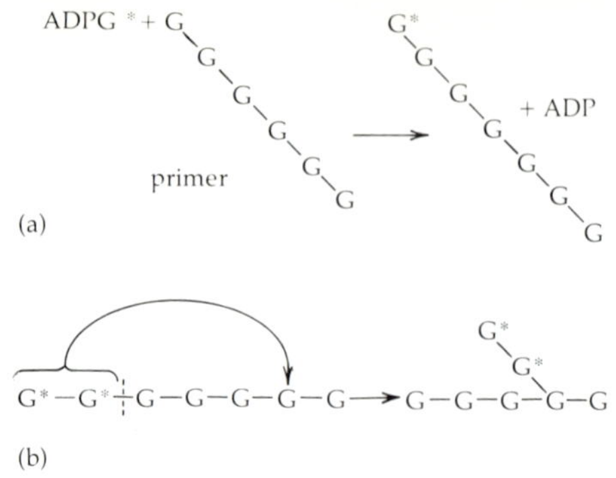

FIGURE 5.49
Chain elongation and branching in the enzymatic synthesis of the polysaccharide glycogen: G denotes glycosyl units. (a) Transfer of a glycosyl unit from ADPG to a primer molecule in glycogen synthesis. (b) Reaction catalyzed by the branching enzyme in glycogen synthesis.

## THE SYNTHESIS OF PEPTIDOGLYCAN

The repeating units of peptidoglycan are synthesized within the cytoplasm while bound to the nucleotide UDP; they are then transferred to a lipid carrier that facilitates their movement across the membrane. Finally, they are polymerized into peptidoglycan on the outside of the membrane by enzymes located on the membrane's outer surface.

Peptidoglycan differs from all biopolymers considered so far in that it is a two-dimensional network rather than a strand of molecules; it surrounds the cell like a sac. Therefore, its synthesis requires that the repeating units be chemically bonded in two dimensions (Figure 5.50). The chemical composition of the peptidoglycan sac is similar in all eubacteria, differing only in the amino acid composition of the tetrapeptide chain and the nature and frequency of bonding between tetrapeptide chains.

The steps of biosynthesis of peptidoglycan by *E. coli* and their cellular location are summarized in Figure 5.51. The N-acetylmuramic acid is synthesized stepwise in the cytoplasm while attached to UDP; it is then transferred to a $C_{55}$ isoprenoid carrier lipid (bactoprenol) in the membrane. In this form, an *N*-acetylglucosamine residue is added, completing the monomeric subunit of peptidoglycan. Attached to the long-chain lipid, the subunit can traverse the membrane to the outer face where it is added by formation of a

**TABLE 5.10**
**Polysaccharide-Synthesizing Systems**

| Polysaccharide | Repeating Unit | Precursor |
|---|---|---|
| Glycogen | α-D-Glucose (1 → 4) | UDP-glucose (animals), ADP-glucose (bacteria) |
| Cellulose | β-D-Glucose (1 → 4) | GDP-glucose |
| Xylan | β-D-Xylose (1 → 4) | UDP-xylose |
| Pneumococcus type III capsular polysaccharide | β-D-Glucuronic acid (1 → 4) β-D-Glucose (1 → 3) | UDP-glucuronic acid, UDP-glucose |

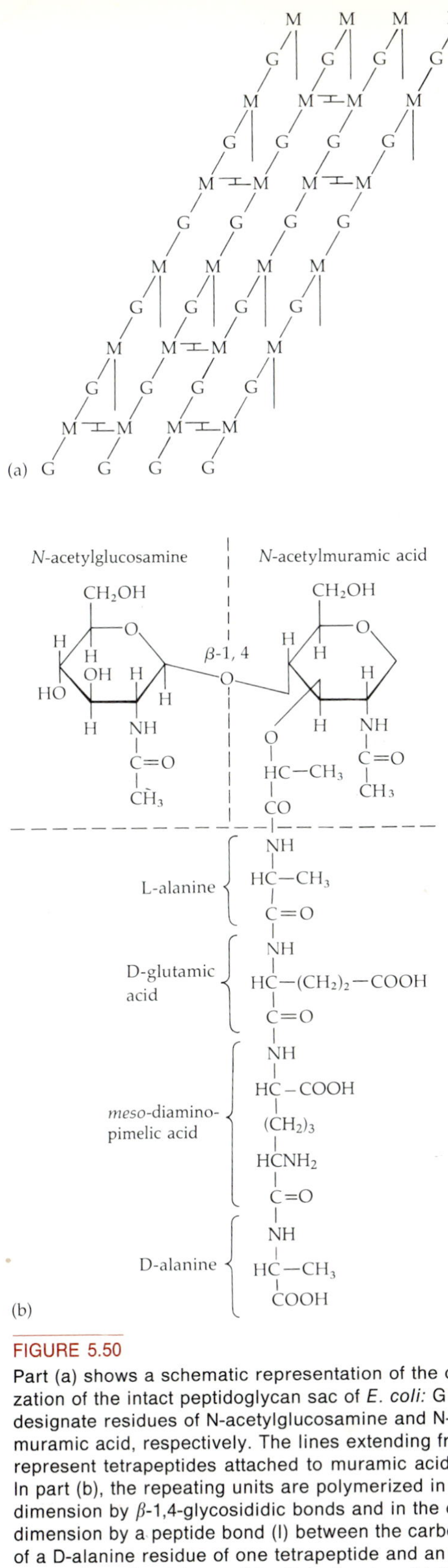

(a)

(b)

*N*-acetylglucosamine

*N*-acetylmuramic acid

β-1, 4

L-alanine

D-glutamic acid

*meso*-diaminopimelic acid

D-alanine

FIGURE 5.50

Part (a) shows a schematic representation of the organization of the intact peptidoglycan sac of *E. coli:* G and M designate residues of N-acetylglucosamine and N-acetylmuramic acid, respectively. The lines extending from M represent tetrapeptides attached to muramic acid residues. In part (b), the repeating units are polymerized in one dimension by β-1,4-glycosididic bonds and in the other dimension by a peptide bond (I) between the carboxyl group of a D-alanine residue of one tetrapeptide and an ϵ-amino group of another tetrapeptide. Not all pairs of tetrapeptides are so joined. From J. M. Ghuysen, "Bacteriolytic Enzymes in the Determination of Wall Structure," *Bacteriol. Rev.* **32,** 425 (1968).

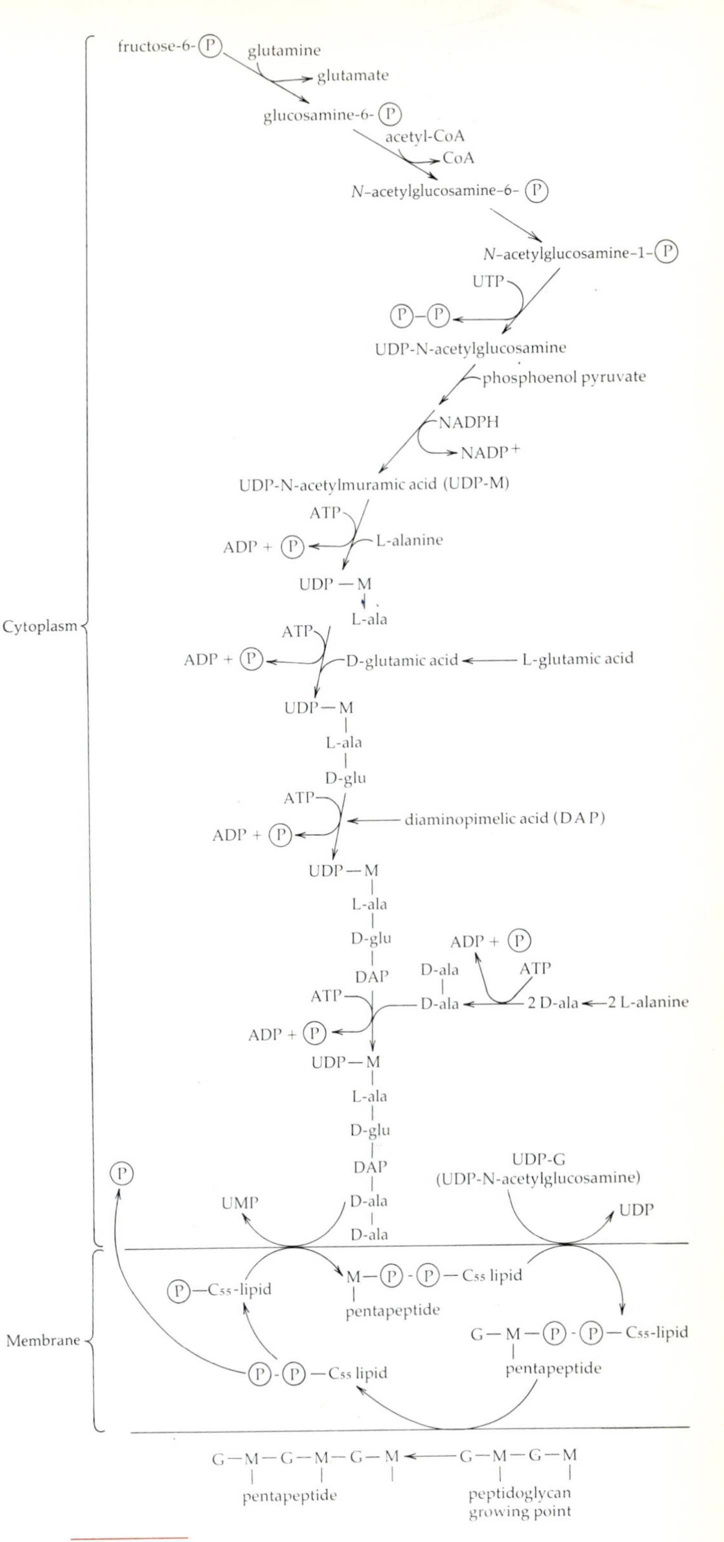


FIGURE 5.51

Pathway of peptidoglycan synthesis in *E. coli*.

1,4 glycosidic bond to a growth point in the peptidoglycan sac. Finally, if the new monomeric subunit is to be involved in a cross-link between peptide strands, an enzyme in the outer portion of the membrane catalyzes a *transpeptidation* reaction breaking the peptide bond between the two terminal D-alanine residues while forming a peptide bond between the subterminal D-alanine residue and the free amino group of diaminopimelic acid on an adjacent peptide strand. If the peptide is not to be involved in a cross link, the same enzyme removes the terminal D-alanine group without forming a new peptide bond.

Since the peptidoglycan layer is responsible for the structural strength that resists the internal osmotic pressure typical of bacteria growing in most environments, it must remain largely intact as the cell grows. However, the peptidoglycan layer can be likened to a mesh. Severing a mesh at one point does not reduce significantly its structural strength. During growth the peptidoglycan sac is opened at points by highly controlled autolytic enzymes, thus allowing enlargement through the insertion of new monomeric units.

In the case of *E. coli* it has been shown that the glycan strands of peptidoglycan are perpendicular to the long axis of the cell. The cell enlarges by breaking cross-linkages between strands, inserting new ones, and reforming cross-links. The pattern of synthesis provides an explanation for the well-established observation that during growth an *E. coli* cell increases in length but not girth.

## ASSEMBLY OF BIOPOLYMERS INTO CELLULAR COMPONENTS

Following the polymerization of macromolecules, some of them are chemically modified, transported to specific locations in the cell and associated to form cellular structures, e.g., the cell envelope, pili, flagella, the nucleoid, polysomes, and complexes of enzymes. Some of these processes are discussed in Chapter 6. The assembly of ribosomes and polysomes was discussed earlier in this chapter.

## FURTHER READING


**Books**

GOTTSCHALK, G., *Bacterial Metabolism.* New York, Heidelberg, and Berlin: Springer-Verlag, 1979.

INGRAHAM, J. L., O. MAALOE, and F. C. NEIDHARDT, *Growth of the Bacterial Cell.* Sunderland, Mass.: Sinauer Associates, Inc., 1983.

LEHNINGER, A. L., *Principles of Biochemistry.* New York: Worth Publishers, 1982.

**Review**

KORNBERG, A., "DNA Replication," *Trends in Biochem. Sci.* **9,** 122 (1984).

Chapter 6

# The Relations Between Structure and Function in Procaryotic Cells

The structure of the procaryotic cell has been described in Chapter 3. In this chapter we shall discuss in greater detail the organization of some of its component parts and the relationships between their structure and function.

## SURFACE STRUCTURES OF THE PROCARYOTIC CELL

### Taxonomic Significance

A differential stain of great practical value, which subsequently became known as the *Gram stain*, was discovered empirically in 1884 by Christian Gram. Many modifications exist, but they all embody the following essential steps: a heat-fixed smear of cells on a microscope slide is stained successively with a basic dye, crystal violet, and with a dilute iodine solution. The preparation is then briefly treated with a polar organic solvent (alcohol or acetone). *Gram-positive* bacteria resist decolorization with the solvent and remain stained a deep blue black. *Gram-negative* bacteria are rapidly and completely decolorized. In performing this staining procedure, it is essential to examine growing cells, because certain Gram-positive bacteria (for example, *Bacillus* spp.) lose the ability to retain the crystal violet–iodine complex after active growth ceases. Thus, the outcome of the Gram reaction is, to some extent, conditioned by the physiological state of the cell, but it also correlates with major differences in the chemical composition and ultrastructure of eubacterial cell walls. It distinguishes the Gram-positive from the Gram-negative bacteria. These two groups, along

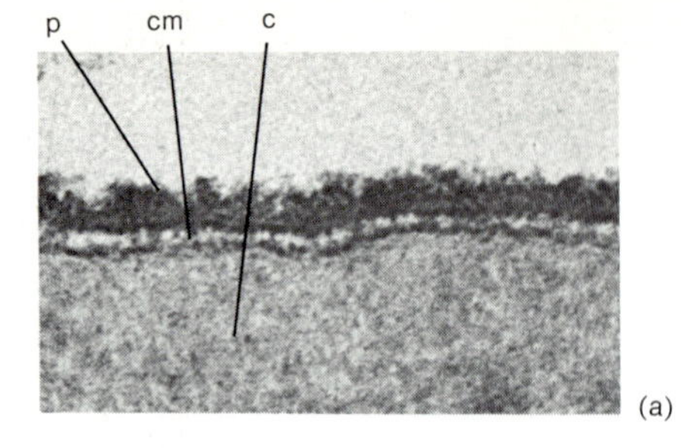


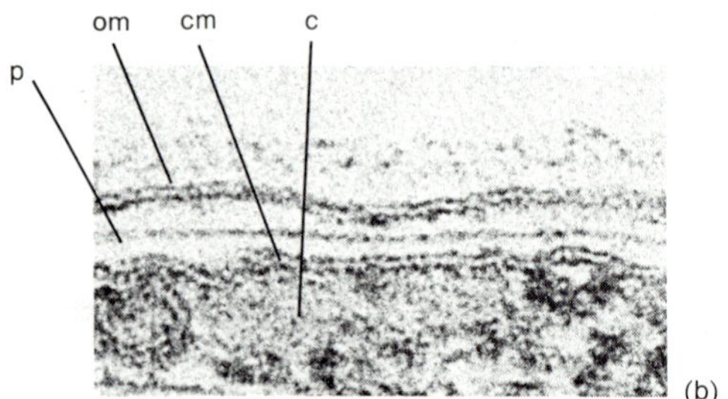


FIGURE 6.1
Electron micrographs of sections of the envelopes of (a) a Gram-positive bacterium and (b) a Gram-negative bacterium illustrating the differences in cell wall profiles: c, cytoplasm; cm, cell membrane; p, peptidoglycan layer; om, outer membrane. The region of Gram-negative envelope between cm and om is the periplasm.

with the mollicutes, which completely lack cell walls, constitute the eubacteria. The Gram-negative group is larger and more diverse than the Gram-positive group; the mollicutes group is quite small.

Electron micrographs of thin sections of the two types of cell wall reveal the wall of a Gram-positive bacterium to be a structure of almost uniform appearance, some 10 to 80 nm in width, whereas the wall of a Gram-negative bacterium is revealed to be composed of two readily distinguishable layers, both considerably thinner than the wall of a Gram-positive bacterium (Figure 6.1).

Chemical analyses show that the walls of Gram-positive bacteria contain peptidoglycan as a major component (generally accounting for 40 to 90 percent of the dry weight), with which are associated polysaccharides and a special class of polymers, the teichoic acids (see page 159).

The inner layer of the walls of Gram-negative bacteria is a very thin layer of peptidoglycan; the outer one is a membrane, termed the *outer membrane*.

In addition to the wall and the membrane, procaryotic cells may be enclosed by a loose outer layer known as a *capsule* or *slime layer*. Also, two classes of thread-shaped organelles, *flagella* and *pili*, occur on the cell surface of many bacteria.

**TABLE 6.1**

**Surface Structures of the Procaryotic Cell**

| Structure | Location | Structure and Dimensions | Chemical Composition |
|---|---|---|---|
| Membrane | Bounding layer of protoplast | Unit membrane, 7.5–8 nm wide | 20–30% phospholipid, remainder mostly protein |
| Wall | Layer immediately external to membrane | Gram-negative eubacteria: | |
| | | Inner single layer 2–3 nm wide | Peptidoglycan (murein) |
| | | Outer unit membrane layer 7–8 nm wide | Phospholipids, proteins, lipopolysaccharides |
| | | Gram-positive eubacteria: | |
| | | Homogeneous layer 10–80 nm wide | Peptidoglycan (murein); teichoic acids; polysaccharides |
| | | Archaebacteria: | |
| | | Variable | Variable |
| Capsule or slime layer | Diffuse layer external to wall | Homogeneous structure of low density and very variable width | Diverse; usually a polysaccharide, rarely a polypeptide |
| Flagella | Anchored in protoplast, traversing membrane and wall | Helical threads, 12–18 nm wide | Protein |
| Pili | Anchored in protoplast, traversing membrane and wall | Straight threads, 4–35 nm wide | Protein |

Table 6.1 summarizes the distinguishing properties of the various surface structures associated with the procaryotic cell.

## Early Studies on the Procaryotic Wall

Very little was known about the composition and functions of the outer layers of bacterial cells until 1952, when M. Salton developed methods for isolating and purifying cell walls (Figure 6.2). The first chemical analyses of such preparations revealed the complexity and diversity of bacterial wall composition, and notably the major compositional differences between the walls of Gram-positive and Gram-negative bacteria.

Salton showed that the hydrolytic enzyme lysozyme, previously known to lyse many Gram-positive bacteria, can completely destroy the isolated cell walls of such organisms as *Bacillus megaterium.* This observation enabled C. Weibull, in 1953, to perform a simple experiment that clearly revealed the respective functions of the bacterial wall and membrane (Figure 6.3). If cells of *B. megaterium* are suspended in an isotonic sucrose solution prior to lysozyme treatment, the enzymatic dissolution of the wall converts the initially rod-shaped cells into spherical *protoplasts*, which retain full respiratory activity and can synthesize protein and nucleic acids. In media that support good growth of intact cells, they are incapable of resynthesizing a cell wall after removal of the lysozyme, but they can increase in size and they have a limited capacity to reproduce. Under certain special conditions of cultivation, they can even be induced to regenerate cell walls and again assume a rod shape.

When a protoplast suspension is diluted, the protoplasts undergo immediate osmotic lysis. The only structural elements that remain after such lysis are the empty cell membranes or "ghosts," which can be readily isolated and purified by differential centrifugation. Such preparations, derived from lysozyme treatment of Gram-positive bacteria, have provided much information about the properties of the bacterial cell membrane.

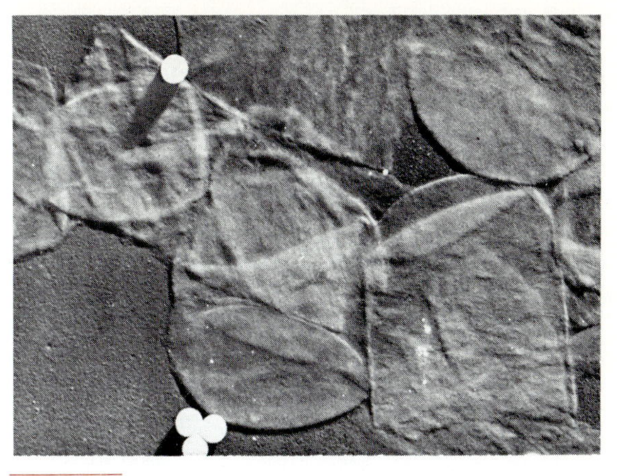

FIGURE 6.2
Isolated and purified cell walls of *Bacillus megaterium*. Electron micrograph. The white spheres are particles of latex exactly 0.25 μm in diameter, which were included in the preparation to show the scale of magnification. From R. Y. Stanier, "Some Singular Features of Bacteria as Dynamic Systems," in *Cellular Metabolism and Infection*, ed. E. Racker (New York: Academic Press, 1954).

Weibull's experiment showed that the cell wall has a vital mechanical function: it protects the cell from osmotic lysis in a hypotonic environment. In addition, it determines cell shape, and it appears to play a role both in movement and in division. However, the wall (at least in Gram-positive bacteria) does not contribute to metabolic activity or constitute part of the osmotic boundary.

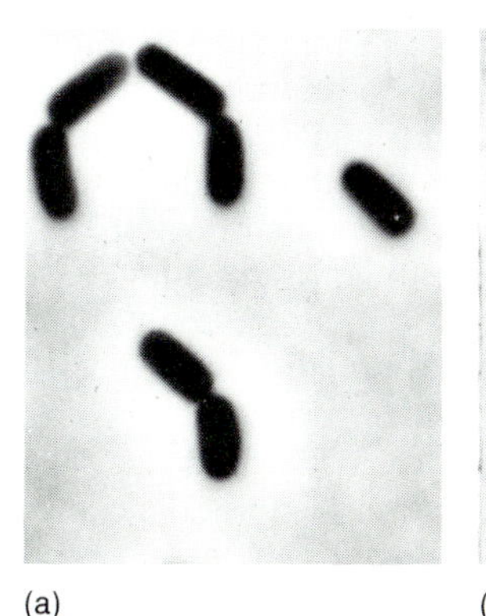

(a)

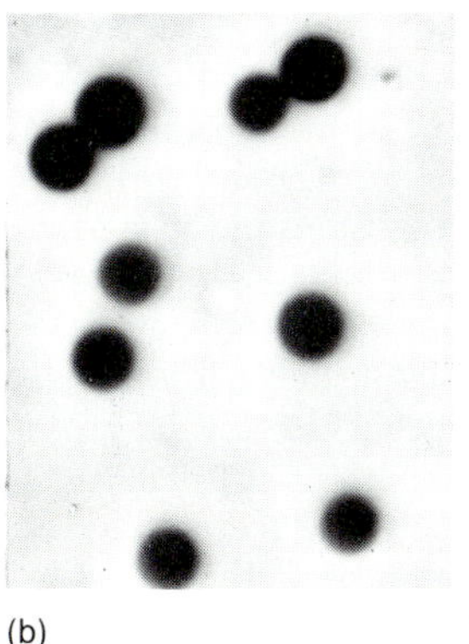

(b)

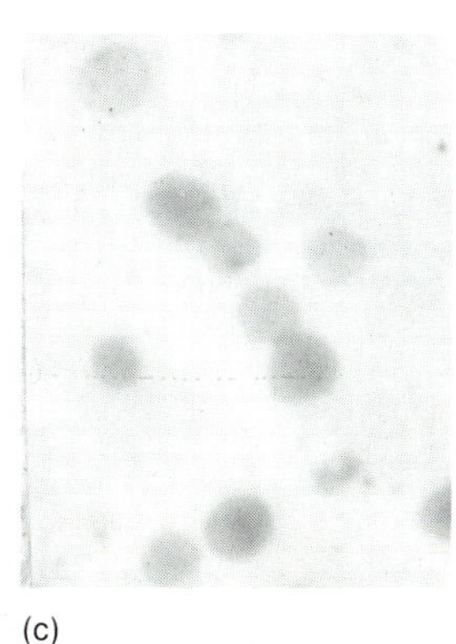

(c)

FIGURE 6.3
*Bacillus megaterium* (phase contrast, ×3,000). (a) The intact cells. (b) The spherical protoplasts, formed by enzymatic dissolution of the cell wall with lysosyme in an isotonic medium. (c) The ghosts (i.e., the empty cytoplasmic membranes), formed by osmotic rupture in a hypotonic medium. Courtesy of C. Weibull.

## The Surface Structures of Archaebacteria

The chemical composition of the surface structures of Archaebacteria is quite different from that of other bacteria. Their walls are composed of protein or a form of peptidoglycan termed *pseudomurein* rather than the typical *murein* that is found in all other bacterial walls. The lipids of their membranes are composed of glycerol ethers rather than the esters found in all other cellular organisms. These archaebacterial structures are discussed in Chapter 14, which deals exclusively with this newly discovered group. The discussions that follow here concern the nonarchaebacterial or eubacterial groups.

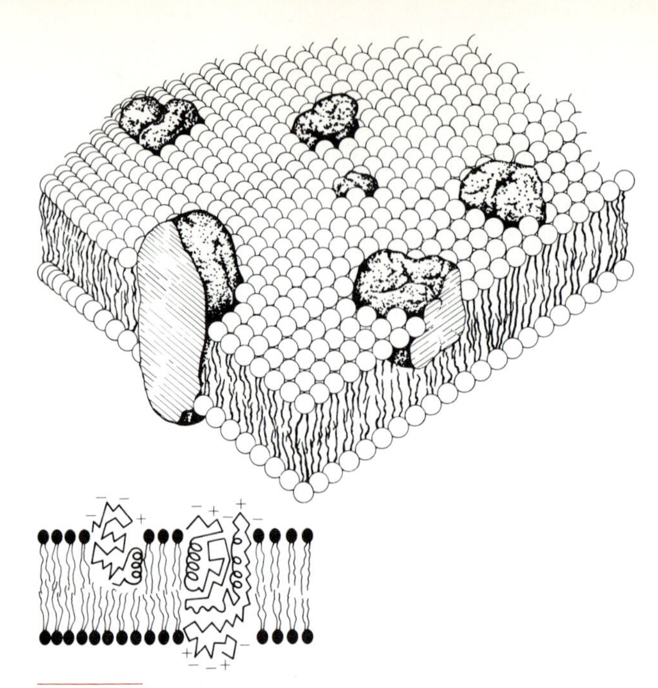

FIGURE 6.5

Schematic drawing to show the possible molecular organization of a unit membrane. Folded polypeptide molecules are visualized as being embedded in a phospholipid bilayer, with their hydrophobic regions extending beyond the bilayer of one or both of its surfaces. From S. J. Singer and A. L. Nicholson, "The Fluid Membrane Model of the Structure of Cell Membranes," *Science* **175,** 720 (1972).

## The Cell Membrane

The eubacterial cell membrane, which bounds the protoplast and is the cell's principal osmotic barrier, can be visualized in electron micrographs of thin sections of cells as a double line, about 8 nm wide: this is the typical fine structure of all so-called *unit membrane* systems. It is made up of a bilayer of phospholipids (Figure 6.4) which are the major lipids of bacterial cell membranes and which account for about 20 to 30 percent of their dry weight. The polar "head" regions of the phospholipids are located at the two outer surfaces of the bilayer, while the hydrophobic fatty acid chains extend into the center of the membrane, perpendicular to its plane (Figure 6.5). The membrane proteins, which account for more than one-half of the dry weight of the membrane, are intercalated into this phospholipid bilayer.

Isolated membranes are highly plastic structures. They can be disaggregated by treatment with detergents, and subsequently reassemble to form new, membrane-like structures. Furthermore, membrane fragments can reseal their edges to produce closed vesicles with permeability properties similar to those of the cells from which they are derived.

FIGURE 6.4

The general structure of phospholipids. These compounds are derivatives of glycerol-3-phosphate. The hydroxyl groups on carbon atoms 1 and 2 are esterified by long-chain fatty acids to yield acyl groups; these constitute the hydrophobic "tail" region of the molecule. The phosphate group esterified on carbon atom 3 can carry a variety of substituents (designated as —R), including glycerol, its acyl derivatives, and ethanolamine. This constitutes the hydrophilic "head" of the molecule.

$CH_2$—O—Acyl
| 
CH—O—Acyl } nonpolar, hydrophobic "tail"
| O ‖
$CH_2$—O—P—O—R polar, hydrophilic "head"
| 
$O^-$

The bacterial cell membrane is an important center of metabolic activity; it contains many different kinds of proteins, each of which probably has a specific catalytic function. Most of these proteins are tightly integrated into the hydrophobic region of the membrane, from which they can be separated only by methods (e.g., detergent treatment) that often destroy their activity. Major classes of proteins known to be localized in the membrane include (1) the permeases responsible for the transport of many organic and inorganic nutrients into the cell; (2) biosynthetic enzymes that mediate terminal steps in the synthesis of the membrane lipids, and of the various classes of macromolecules that compose the bacterial cell wall (peptidoglycans, teichoic acids, lipopolysaccharides and simple polysaccharides; (3) the proteins that participate in generation of ATP in those bacteria that do so by electron transport. In respiratory bacteria the components of the electron transport chain and the ATP phosphotransferase are located in the mem-

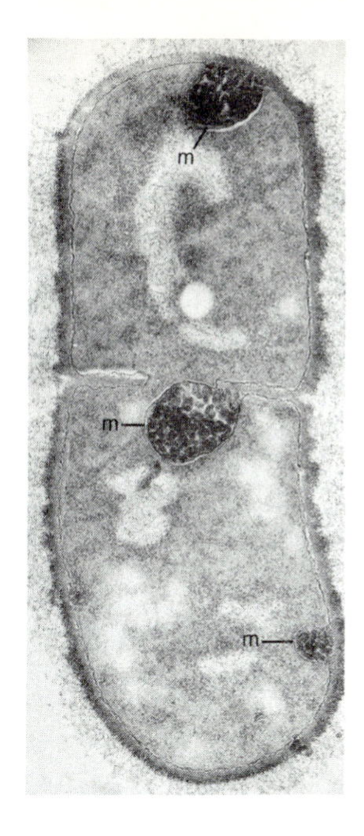


FIGURE 6.6

Electron micrograph of a thin section of a dividing cell of *Bacillus megaterium*, containing three mesosomes (m). One is located in association with the nearly formed transverse septum and wall. From D. J. Ellar, D. Lundgren, and R. A. Slepecky, "Fine Structure of *Bacillus megaterium* during Synchronous Growth," *J. Bacteriol.*, **94,** 1189 (1967).

brane. In purple bacteria these, as well as the other components of the complete photosynthetic apparatus (antenna pigments and reaction centers) are also located in the membrane. Lastly, much circumstantial evidence indicates that the procaryotic cell membrane contains specific attachment sites for the chromosome and for plasmids, and that it plays an active role in the partitioning of these genetic elements to daughter cells. In view of its numerous and varied function, it is not surprising that the bacterial membrane contains from 10 to 20 percent of the total cell protein.

Although the width of the cell membrane is fixed (being determined by the molecular configuration of the phospholipid bilayer), its area is not. In some bacteria, the membrane appears to have a simple contour, which closely follows that of the enclosing cell wall. In others, it is infolded, at one or more points, into the cytoplasmic region.

Complex, localized infoldings known as *mesosomes* occur in many bacteria, often at or near the site of cell division (Figure 6.6), and probably participate in the formation of the transverse septum. The continuity of the mesosome with the external surface of the membrane, not always evident in thin sections, is revealed in electron micrographs of whole cells negatively stained with a heavy metal salt that penetrates through the wall but does not enter the cytoplasm (Figure 6.7).

Membrane infoldings of a different type occur in purple bacteria (Figure 6.8) and in many nonphotosynthetic bacteria that possess a high level of respiratory activity, such as the nitrogen fixers of the *Azotobacter* group (Figure 6.9) and the nitrifying bacteria (Figure 6.10); the greatly enlarged total area of the membrane produced by such intrusions serves to accommodate more centers of respiratory (or photosynthetic) activity than could be housed in a membrane of simple contour. The most convincing evidence in support of this interpretation has come from studies on the membrane structure of certain purple bacteria, where the photosynthetic pigment content (and hence the photosynthetic activity) can vary widely in response to environmental factors (light intensity, presence or absence of oxygen). Here, the extent of the membrane intrusions is directly related to the pigment content and photosynthetic activity of the cells (Figure 6.11).

In most procaryotes, there is a physical continuity between membrane intrusions and the surface region of the membrane. This may not be true, however, of the cyanobacteria. In these organisms, the photosynthetic apparatus is contained in a system of flattened membranous sacs (thylakoids), which have been very rarely observed in connection with the cell membrane, and may be in large part physically distinct from it (Figure 6.12).

FIGURE 6.7

Mesosomes of *Caulobacter crescentus*: electron micrograph of whole cells negatively stained with phosphotungstate, which has penetrated the mesosomal involutions of the cell membrane, clearly outlining their positions (×22,100). Courtesy of Germaine Cohen-Bazire.

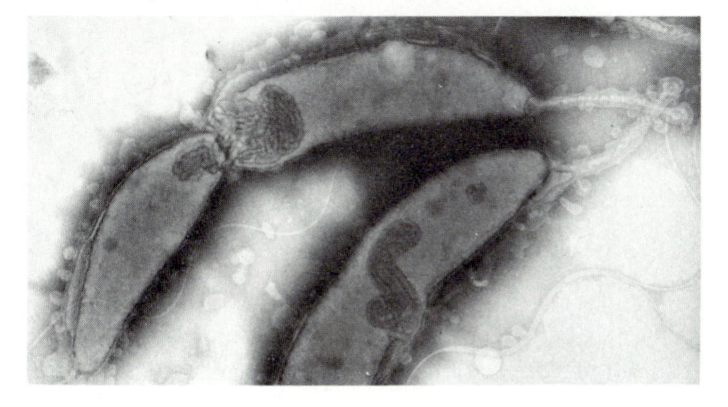

## The Bacterial Cell Wall: Its Peptidoglycan Component

Of the many classes of macromolecules that may be associated with procaryotic cell walls, only one—the *peptidoglycans*—is of well-nigh universal occurrence. The only procaryotes that possess walls devoid of peptidoglycans are certain archaebacteria (Chapter 14) and the *Planctomyces* group of eubacteria.

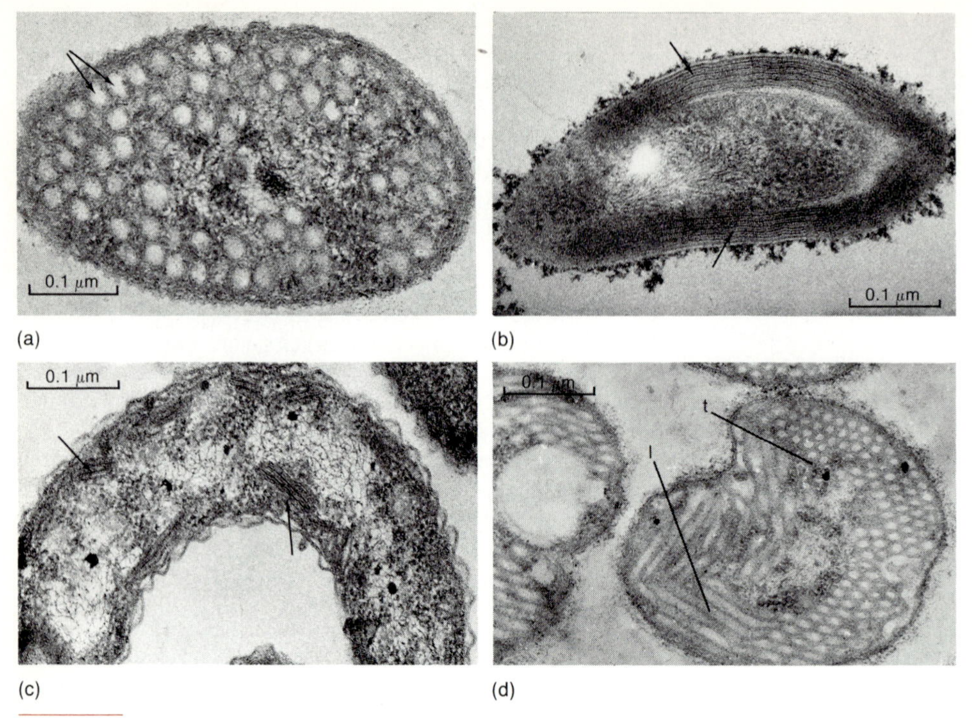


FIGURE 6.8

Electron micrographs of thin sections of several purple bacteria, illustrating variations in the structure of the internal membranes. (a) *Rhodobacter sphaeroides*, in which the membranes occur as hollow vesicles (arrows) ( × 5,640). (b) *Rhodopseudomonas palustris*, in which the membranes occur in regular parallel layers in the cortical region of the cell (arrows) ( × 5,640). (c) *Rhodospirillum fulvum*, in which the membranes occur in small, regular stacks (arrows) ( × 42,300). (d) *Thiocapsa* sp., in which the membranes are tubular; some of these tubes are sectioned longitudinally (l), others transversely (t) ( × 54,400). Micrographs (a), (b), and (c) courtesy of Germaine Cohen-Bazire; (d) courtesy of K. Eimhjellen.

FIGURE 6.9 (below)

Electron micrograph of a thin section of the nitrogen-fixing bacterium, *Azotobacter vinelandii*, showing vesicular intrusions of the membrane similar in structure to those of certain photosynthetic bacteria (see Figure 6.8). Courtesy of J. Pangborn and A. G. Marr.

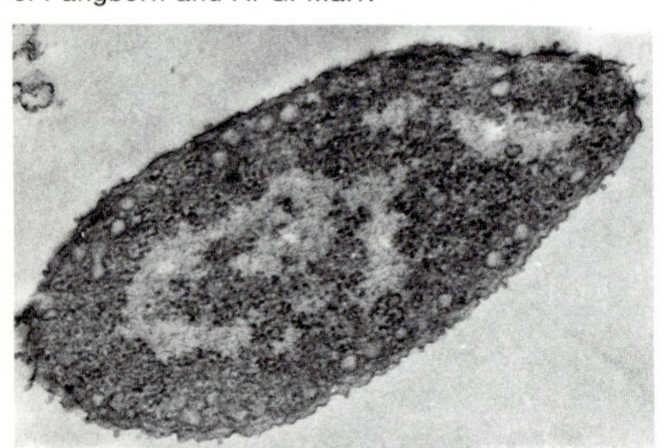

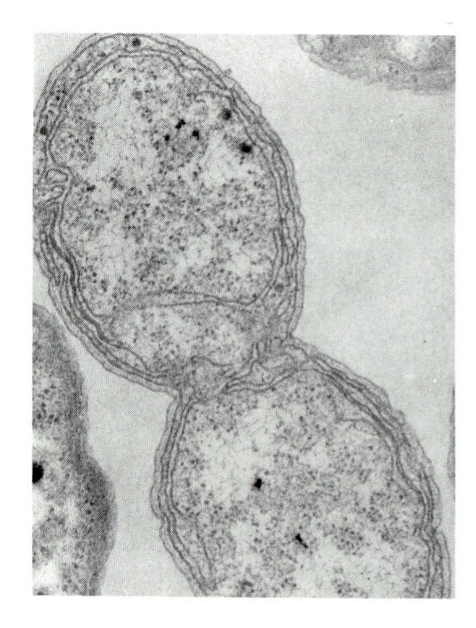

FIGURE 6.10 (left)

Electron micrograph of a thin section of *Nitrosomonas europaea*, an obligate chemoautotroph, showing membrane intrusions ( × 39,100). Courtesy of S. W. Watson.

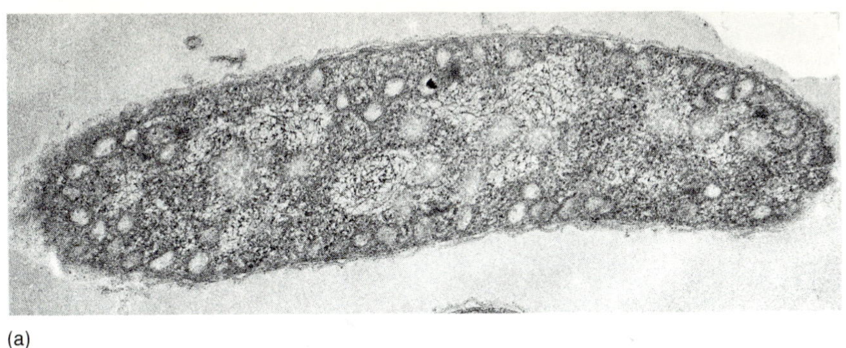

(a)

(b)

FIGURE 6.11

Electron micrographs of longitudinal sections of the photosynthetic bacterium, *Rhodospirillum rubrum* ($\times$ 52,700), showing the effect of the environment on the extent of the intrusion of the cytoplasmic membrane. (a) Cell from a culture grown in bright light (1,000 foot-candles) and having a relatively low chlorophyll content. (b) Cell from a culture grown in dim light (50 foot-candles) and having a high chlorophyll content. Courtesy of Germaine Cohen-Bazire.

FIGURE 6.12

Electron micrograph of a thin section of a unicellular cyanobacterium, showing the extensive array of internal membranes (thylakoids) characteristic of this procaryotic group ($\times$ 29,000). Courtesy of Germaine Cohen-Bazire.

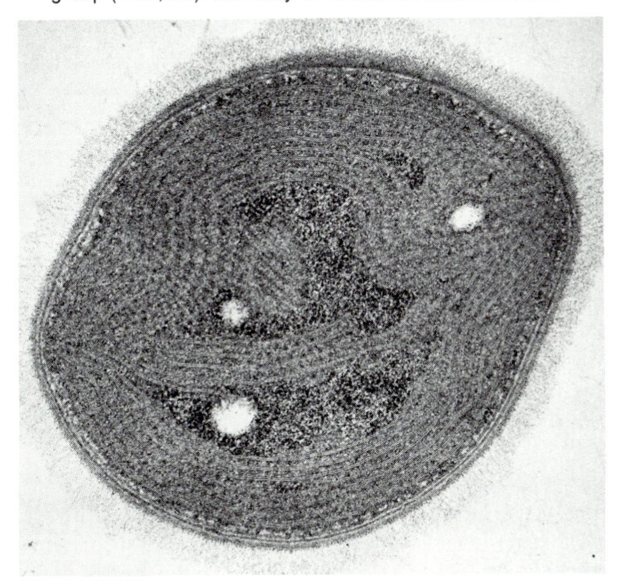

Peptidoglycans are heteropolymers of substituted sugars and amino acids synthesized uniquely by procaryotes. The form of peptidoglycan (termed *murein*) found in eubacteria is always composed of two acetylated amino sugars, N-acetylglucosamine, and N-acetylmuramic acid and a small number of amino acids, the particular representatives of which vary among different groups. Indeed, on the basis of these variations, over 100 different chemotypes of mureins are now known. Some of the amino acids in murein are "unnatural" in the sense that they never occur in proteins. The form of peptidoglycan, *pseudomurein*, found in archaebacteria differs from murein in two respects: N-acetylmuramic acid is never present (in its place is another amino sugar acid, *N-acetyltalosaminuronic acid*), nor are "unnatural" amino acids present. The structure of pseudomureins is discussed in Chapter 14.

The two amino sugars of murein form glycan strands composed of alternating residues of N-acetylglucosamine (G) and N-acetylmuramic acid (M) in beta-1,4 linkage (Figure 6.13). Each strand contains from 10 to 65 disaccharide units. Muramic acid, being a lactyl ether of glucosamine, provides a carboxyl group to which a peptide chain is attached. The particular sequence of amino acids shown in Figure 6.12 is the most common one found in bacteria, being typical of most Gram-negative and some Gram-positive bacteria.

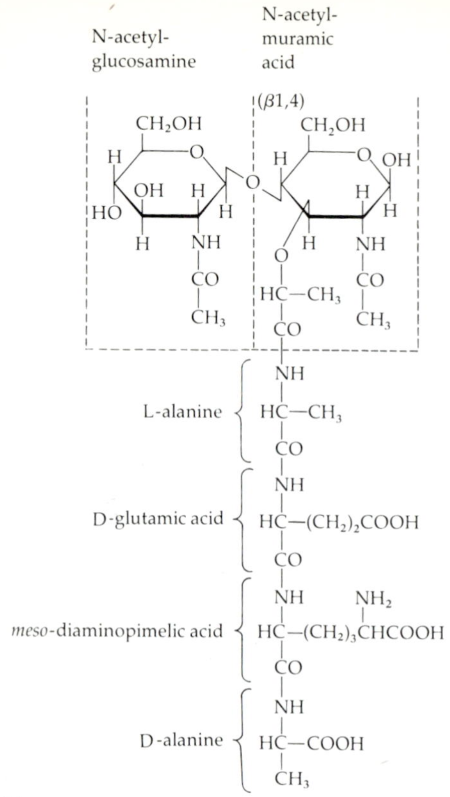

N-acetyl-glucosamine

N-acetyl-muramic acid

(β1,4)

L-alanine

D-glutamic acid

*meso*-diaminopimelic acid

D-alanine

(a)

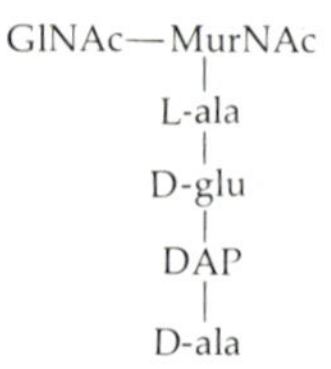

GlNAc—MurNAc
L-ala
D-glu
DAP
D-ala

(b)

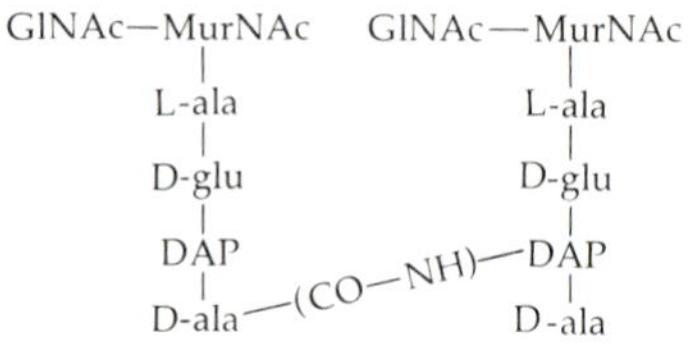

GlNAc—MurNAc   GlNAc—MurNAc
L-ala   L-ala
D-glu   D-glu
DAP   —(CO—NH)—DAP
D-ala   D-ala

(c)

FIGURE 6.13

General structure of a peptidoglycan (mDpm-direct type). (a) Complete structure of a single subunit, showing the linkage between the two amino sugars that make up the glycan strand, and between muramic acid and the four amino acids in the short peptide chain. (b) Schematic, simplified representation of structure shown in (a). (c) Representation of the mode of cross-linking between the terminal carboxyl group of D-alanine on one subunit and the free amino group of the diamino acid (diaminopimelic acid) on an adjacent subunit.

FIGURE 6.14

A schematic representation of the organization of the intact peptidoglycan sac of *E. coli:* G and M designate residues of N-acetylglucosamine and N-acetylmuramic acid, respectively, joined (diagonal lines) by β-1,4-glycosidic bonds. The vertical lines represent free tetrapeptide side chains, attached to muramic acid residues. The symbol ⌶ represents cross-linked tetrapeptide side chains. From J. M. Ghuysen, "Bacteriolytic Enzymes in the Determination of Wall Structure," *Bacteriol. Rev.* **32,** 425 (1968).

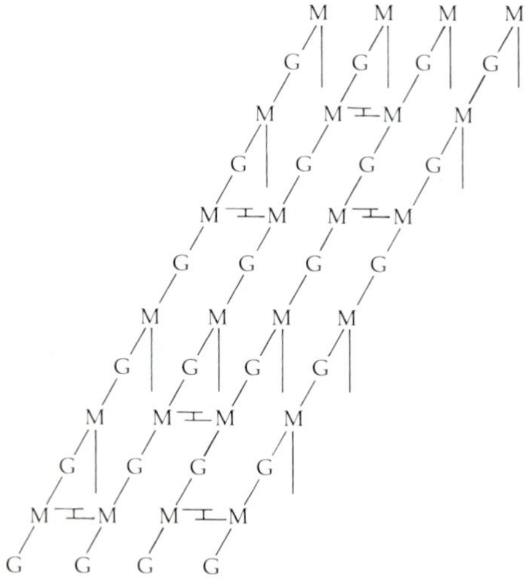

Adjacent peptide chains projecting from different glycan strands may be *cross-linked* by formation of a peptide bond between the carboxyl group of a terminal D-alanine in one chain and the free α-amino group of *meso*-diaminopimelic acid in another. Not all peptide chains participate in cross-linkage, but enough do to form a molecular mesh or fabric (Figure 6.14). In fact, the peptidoglycan layer of the bacterial cell wall is made up of a single giant macromolecule that completely encloses the protoplast. In Gram-negative bacteria it occurs as a single layer; in Gram-positive bacteria there are many layers.

The murein structure shown in Figure 6.13 sometimes termed *mDpm-direct* (Figure 6.15) is the most widespread one; it occurs in the walls of nearly all Gram-negative bacteria and many Gram-positive ones. But as stated earlier, there are many variations on this general pattern, some of which are illustrated in Figure 6.15.

In classifying types of mureins, the primary distinction depends on whether cross-linking occurs at the third (Group A) or second (Group B) amino acid in the peptide chain. In the case of the mDpm-direct type murein, cross-linkage depends on the availability in *meso*-diaminopimelic acid of a second amino group to form a peptide bond to the terminal

*Group A*

—G—M—G—
↓
ala
↓
D-glu
↓
m-dpm ← D-ala
↓ ↑
D-ala m-dpm←
↑

*m-Dpm-direct type*

—G—M—G—
↓
ala
↓
D-glu
↓
lys ← gly ← gly ← gly ← gly ← gly ← D-ala
↓ ↑
D-ala [← ser ← ala ← thr ← ala ←] lys ←
↑
[← D-asp ←]

*Lys-x-y type*

*Group B*

—G—M—G—
↓
gly
↓
d-glu → gly → lys ← D-ala
↓ ↑
lys [→ D-orn ←] lys
↓ ↑
D-ala

*[Lys]D-Glu-x-y type*

FIGURE 6.15

Variations in the molecular ground plan of murein. Group A mureins are cross-linked to the third amino acid residue in the peptide chain either directly to the terminal D-alanine residue of an adjacent chain (mDpm-direct type) or through a peptide bridge of variable length and composition (Lys-x-y type). Group B mureins are cross-linked to the carboxyl group to the second residue (D-glutamic acid) in the chain. G and M designate residues of N-acetyl glucosamine and N-acetylmuramic acid respectively. Amino acids are abbreviated according to the conventional three letter code: L-alanine, ala; D-alanine, D-ala; D-aspartic acid, D-asp; *meso*-diaminopimelic acid, m-Dpm; D-glutamic acid, D-glu; glycine, gly; L-lysine, lys; D-orńithine, D-orn; L-threonine, thr. Arrows between amino acids indicate polarization of the peptide bonds in the C to N direction. After O. Kandler, "Cell Wall Structures and Their Phylogenetic Implications," *Zbl. Bakt. Hyg.*, **I. Abt. Orig. C3,** 149 (1982).

alanine residue of an adjacent strand. Indeed, some diamino acid, be it *meso*-diaminopimelic acid, lysine, ornithine, or diaminobutyric acid, always participates in cross-linkage, the particular one differing among various murein chemotypes. In addition to direct linkage of peptide chains, some Group A mureins are joined through peptide bridges of variable length and composition.

Cross-links in Group B mureins are anchored to the alpha-carboxyl group of D-glutamic acid in the second position of the peptide chain. As a consequence the interpeptide bridge must contain a diamino acid to link the carboxyl termini of the two peptide side chains.

The Gram-negative bacteria with only a few exceptions contain the mDpm-direct type murein. All the other types including mDpm-direct are found among the Gram-positive bacteria.

## The Location of Peptidoglycan in the Walls of Gram-Negative Bacteria

The walls of Gram-negative bacteria have a comparatively low peptidoglycan content, seldom exceeding 5 to 10 percent of the weight of the wall. The location of the peptidoglycan layer in this type of wall was first established by W. Weidel and his collaborators, for walls of *Escherichia coli.* They showed that peptidoglycan constitutes the innermost layer of the multilayered wall and can be isolated as a very thin sac that retains the form and shape of the original cell, after other wall components have been stripped off it by appropriate treatments (see Figure 6.16).

The peptidoglycans of Gram-negative bacteria characteristically display a rather low degree of cross-linkage between the glycan strands: many of

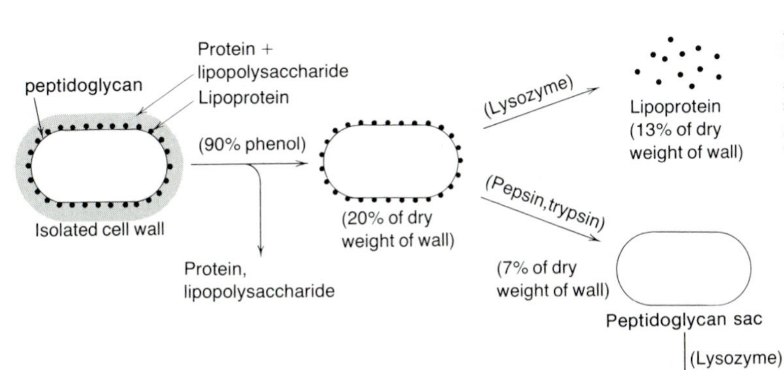


FIGURE 6.16

A diagrammatic representation of successive steps in the fractionation of the cell wall of *E. coli*. Reconstructed from experiments of W. Weidel, H. Frank, and H. H. Martin.

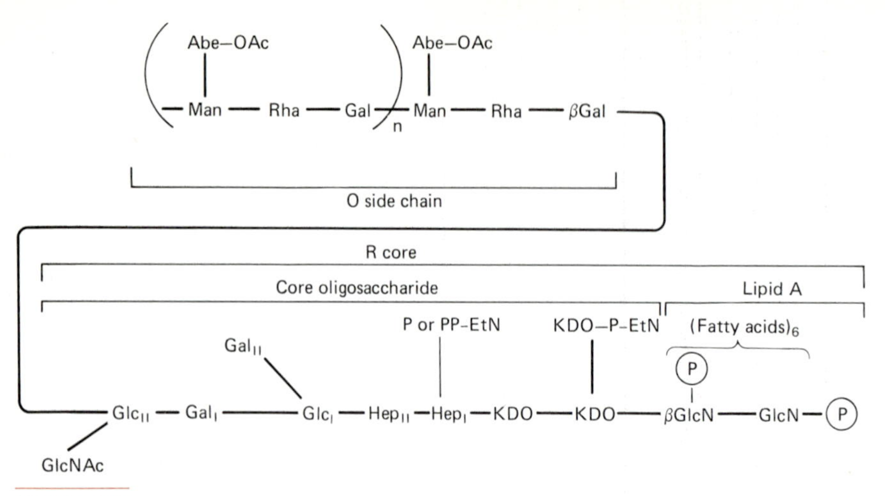


FIGURE 6.17
Structure of the lipopolysaccharide of *Salmonella typhimurium*. Abe, abequose; Màn, D-mannose; Rha, L-rhamnose; Gal, D-galactose; GlcNAc, N-acetyl-D-glucosamine; Glc, D-glucose; Hep, heptose; KDO, 2-keto-3-deoxyoctonic acid; EtN, ethanolamine; Ac, acetyl. Biosynthesis starts at the lipid A end, and the molecule is progressively elongated by the addition of sugar residues. After H. Nikaido, "Biosynthesis and Assembly of Lipopolysaccharide," in *Bacterial Membranes and Walls*, ed. L. Leive (New York: Marcel Dekker, 1973).

FIGURE 6.18
Schematic model of the outer membrane of *Escherichia coli* and *Salmonella typhimurium* showing the presumed arrangement of its components and its attachment to the murein layer. After H. Nikaido and M. Varra, "Molecular Basis of the Permeability of Bacterial Outer Membrane," *Microbiol. Rev.* **49,** 1 (1985).

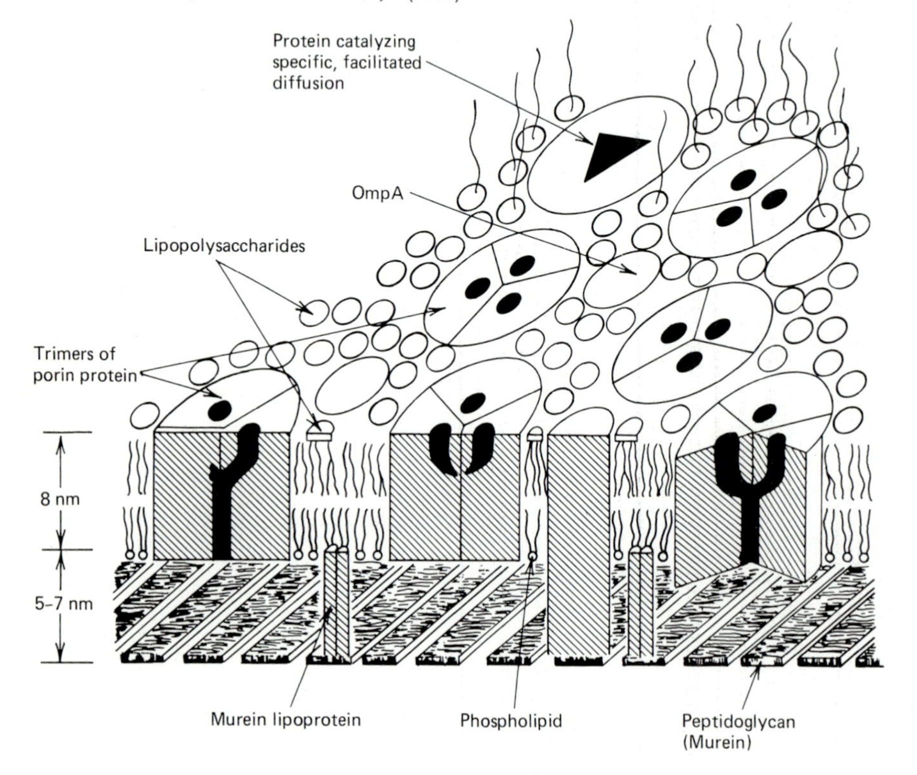

the peptide chains are not cross-linked (see Figure 6.14). The thickness of the peptidoglycan layer of the wall varies somewhat in different groups of Gram-negative bacteria. Calculations suggest that in many Gram-negative organisms it is a monomolecular (or at most bimolecular) layer.

## The Outer Membrane

Superimposed on the thin murein sac characteristic of Gram-negative bacteria is an outer layer that has the width and fine structure typical of a unit membrane. This layer, the *outer membrane*, has some chemical and physical properties in common with the cell membrane, and others that are quite different. Like the cell membrane, it is a lipid bilayer containing phospholipids and proteins, but in addition it contains large amounts of a unique lipid, *lipopolysaccharide* (*LPS*), which replaces, probably completely, phospholipids in the outer leaf of this unique structure. Although chemically quite different from a phospholipid, LPS has physical properties that are sufficiently similar so that it can participate in forming a membrane: one end of the molecule is hydrophobic and the other is hydrophilic; the hydrophobic end becomes inserted in the membrane's hydrophobic core and the hydrophilic end is on the outer surface. The structure of LPS is shown in Figure 6.17 and its orientation in the outer membrane is shown in Figure 6.18.

The lipopolysaccharides are complex molecules with molecular weights over 10,000 that vary widely in chemical composition, both within and between Gram-negative groups. Most work on their structure has been conducted on the forms present in the *Salmonella* group.

LPS is composed of three distinct regions: *lipid A*, the *R core region*, and the *O side chain*. Lipid A (Figure 6.19), the hydrophobic membrane-anchoring region of LPS, rather than carrying the two fatty acid residues typical of a phospholipid has six or seven attached to a phosphorylated glucosamine dimer. Unlike those in phospholipids, all the fatty acids in lipid A are saturated. Some are attached directly to the glucosamine dimer and others are esterified to the 3-hydroxy fatty acids that are characteristically present. Attached to the 6 position of one glucosamine residue in lipid A is the R core oligosaccharide: a short chain of sugars, which include two unusual ones, 2-keto-3-deoxyoctonoic acid (KDO) and heptose (Figure 6.19). The R core in turn bears the hydrophilic O side chain, likewise composed of sugars. It is much longer than the R core, being composed of many repeating tetra- or pentasaccharide units. The elucidation of this structure depended heavily on the availability of mutants, each blocked at a particular point in LPS biosynthesis (Figure 6.17). Biosynthesis of LPS is strictly sequential, starting with lipid A from which the oligosaccharide is built by successive sugar additions, the O side chain being added last. The innermost region, consisting of lipid A and three residues of KDO, appears to be essential, but the rest of the molecule is dispensable.

Lipid A

heptose

2-keto-3-deoxyoctonic acid (KDO)

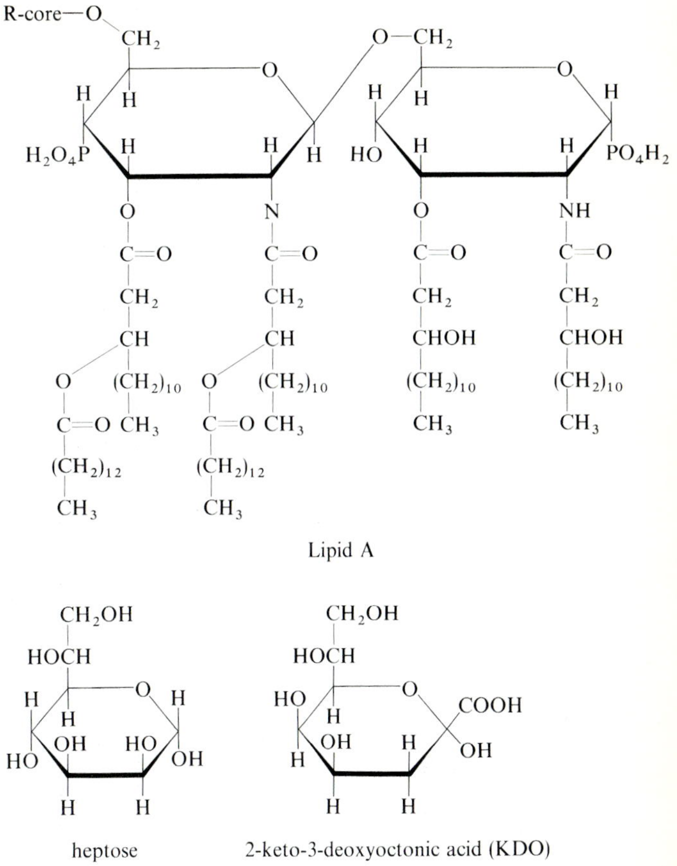

FIGURE 6.19

Structures of lipid A and certain constituents of the R core of lipopolysaccharide.

The peptidoglycan layer of the wall bears a small (MW = 7,200) specific type of lipoprotein, termed *murein lipoprotein*, which forms an anchoring bridge to the outer membrane. The C-terminus of this protein is a lysine residue which is peptide bonded to an amino group of a *meso*-diaminopimelic acid residue that is not cross-linked in the peptidoglycan layer. At the other end (the N-terminus) of the protein is a cysteine residue to which fatty acids are attached: one is attached in an amide linkage to the terminal amino group, and two more are esterified to a glycerol residue which

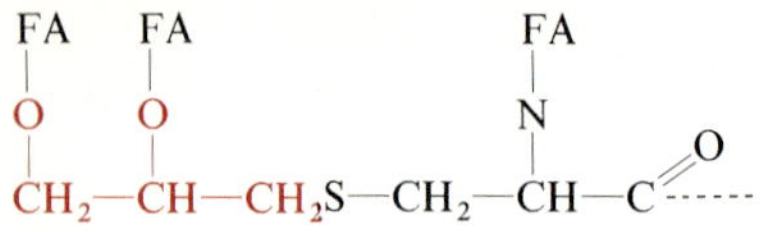

FA FA FA

O O N O

$CH_2-CH-CH_2S-CH_2-CH-C$······

FIGURE 6.20

Structure of the modification of the N-terminal cystein residue of murein lipoprotein showing the attachments of the glycerol residue (colored) and three fatty acids (FA). The dotted line indicates the peptide linkage to the penultimate amino acid residue.

is attached by sulfur ether linkage to the cysteine (Figure 6.20).

The resulting brushlike structure composed of the hydrophobic chains of three fatty acids becomes inserted into the inner leaf of the outer membrane, thereby anchoring it to the peptidoglycan layer.

Lipoprotein is quite abundant, there being $7 \times 10^5$ molecules of it in each cell, about half of which are bonded to peptidoglycan. In spite of its abundance and clear structural role, murein lipoprotein is apparently not essential to survival, at least under conditions of laboratory cultivation, because mutant strains have been isolated that grow well despite lacking this protein.

The physiological significance of the outer membrane is threefold: (1) it forms the outer limit of the *periplasm*, the region between the two membranes that contains a set of digestive enzymes (Chapter 8); (2) it presents an outer surface with strong negative charge which is important in evading phagocytosis and the action of complement (Chapters 29 and 30); and (3) it provides a permeability barrier and thereby increased resistance to a number of toxic agents. Among these are host defense agents like lysozyme, beta-lysin, and leucocyte protein, which are quite toxic to Gram-positive bacteria which lack an outer membrane. Destructive agents in the mammalian digestive tract, such as bile salts and digestive enzymes and a variety of antibiotics, are also excluded by the outer membrane.

Of course the outer membrane cannot present a barrier to all substances in the environment because all cellular nutrients must pass through it. Permeability of the outer membrane to nutrients is provided in part by proteins collectively termed *porins* which, in aggregates generally of three, form cross-membrane channels through which certain small molecules can diffuse (Figure 6.18). A variety of different porins are present in the outer membrane (Table 6.2). They vary with respect to the size of the channel they form and the environmental conditions that stimulate their synthesis. For example, in *E. coli* the pore formed by OmpF (*o*uter *m*embrane *p*rotein F) is slightly larger (1.2 nm diameter) than the one formed by OmpC (1.1 nm diameter). The synthesis of OmpF is repressed by elevated temperature (>37°C) and by growth in a medium of elevated osmotic pressure. The physiological rationale for this regulation of the synthesis of OmpF is presumed to be a mechanism of sensing whether the cell finds itself within a eucaryotic host or in an external environment. In the latter, usually cooler, environment, the concentration of substrate is typically quite low; this necessitates the presence of larger pores formed by OmpF to allow diffusion of substrate molecules to occur at a greater rate, because rate of diffusion is proportional to the product of concentration difference across the membrane and cross-sectional area of the pore. Within a host, where the concentration of substrates is typically much higher, the larger pore is unnecessary and even detrimental because antibacterial substances present in the host can enter more readily through these larger pores.

In addition to the nonspecific channels formed by porins, the outer membrane contains a variety of channels formed by other proteins that exhibit a remarkable specificity. For example, the channel sometimes called the *maltoporin*, formed by the inducible LamB protein, specifically allows the dif-

**TABLE 6.2**

**Channel Proteins in the Outer Membrane of *E. coli* and *S. typhimurium***

| Protein | Physiological Role |
|---|---|
| PORINS | |
| OmpC | Forms small (1.1 nm) pores |
| OmpD | Present in *S. typhimurium* only |
| OmpF | Forms larger pore (1.2 nm) than OmpC; repressed by elevated temperature and higher osmotic pressure |
| PhoE | Formed in response to restricted supply of phosphate |
| SPECIFIC CHANNEL PROTEINS | |
| LamB | Specific for the diffusional entrance of maltose and maltodextrins; induced by maltose; site of adsorption of phage lambda |
| Tsx | Specific for the diffusional entrance of nucleosides; site of adsorption of phage T6 |
| TonA | Specific for diffusional entrance of ferrichrome; site of adsorption of phages T1 and T6 |

fusional entrance of the disaccharide maltose and maltodextrans into the cell. Maltotriose diffuses through these channels at 100 times the rate of the similar-sized trisaccharide, raffinose. Presumably, proteins that bind tightly to a specific substrate are associated with these channels (see binding proteins, Chapter 8), thereby conferring specificity on them.

In addition to the channel-forming proteins, a protein termed OmpA is quite abundant in the outer membrane. Its specific role has not been clearly defined, but mutant strains that lack it produce a more fragile outer membrane, so we assume that OmpA contributes in some way to the membrane's structural integrity.

Although proteins constitute about half the mass of the outer membrane, until recently it was assumed that the number of different types of proteins located there was quite limited. Now it is clear that a large variety of different proteins are present in small quantities. With few exceptions, proteins in the outer membrane are not found in the cytoplasmic membrane.

The molecular basis of a remarkable property of the outer membrane that distinguishes it from other membranes, namely its impermeability to hydrophobic molecules, is not yet understood, but this property accounts for resistance to certain dyes (e.g., eosin, methylene blue, and brilliant green) that are used in certain selective media.

## The Periplasm

The region, termed *the periplasm*, between the cell membrane and the outer membrane of Gram-negative bacteria has been defined by ultrastructural and biochemical studies. Electron micrographs of the walls of Gram-negative bacteria (Figure 6.1) typically reveal an open region on either side of the

FIGURE 6.21

Electron micrograph of a section of *Escherichia coli* prepared using special techniques to avoid generation of artifacts: c, cell membrane; om, outer membrane. Note that no periplasmic space is seen, rather the intermembrane region seems to be completely filled. Bar, 0.1 $\mu$m. From J. A. Hobot, E. Carlemalm, W. Villiger, and E. Kellenberger, "Periplasmic Gel: New Concept Resulting from Reinvestigation of Bacterial Cell Envelope Ultrastructure by New Methods," *J. Bacteriol.* **160,** 143 (1984).

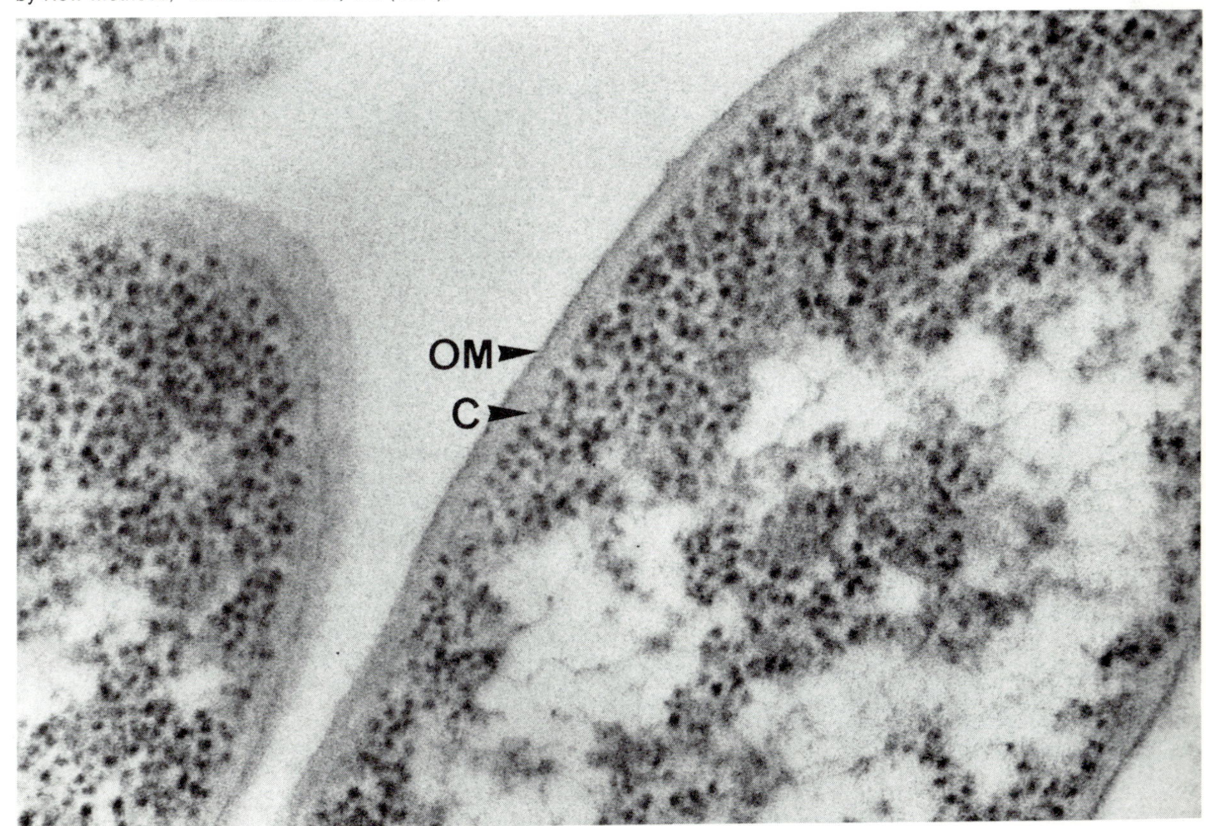

peptidoglycan layer that is identified as the periplasm. Biochemical studies (Chapter 8) have shown that a unique set of proteins (*periplasmic proteins*) are released from Gram-negative bacteria by treatments that disrupt the outer membrane while maintaining the cell membrane intact. Such studies establish the periplasmic location of this set of proteins.

The existence of a periplasm exterior to the cell membrane of Gram-negative bacteria can hardly be doubted, but its structure probably differs from that suggested by conventional electron microscopy. For example, one would expect that intracellular osmotic pressure would tightly press the cell membrane against the inner surface of the rigid peptidoglycan layer, thus eliminating the region of the periplasm that lies between these two layers of the envelope; one would also expect that the outer membrane would lie quite close to the outer surface of the peptidoglycan layer because these two layers are known to be linked by molecules of murein lipoprotein.

Quite recently, electron micrographs of Gram-negative walls (Figure 6.21) have appeared that indicate that the peptidoglycan layer is, indeed, in intimate contact with both the cell membrane and the outer membrane. These micrographs were made of material that was specially prepared to prevent damage to the cell envelope, and they suggest that the appearance of the periplasm in conventional micrographs (Figure 6.1) might be artifactual. The scientists who made these new micrographs suggest that the region between the two membranes of a Gram-negative bacterium is filled with highly hydrated peptidoglycan, only the outer portion of which is sufficiently cross-linked for it to be isolated as an intact structure; the inner portion of peptidoglycan consists of largely uncross-linked strands which detach from the intact saculus when the cell is disrupted. Thus, these authors believe that the intermembrane region ought more properly to be termed a *periplasmic gel.*

## Peptidoglycan in the Walls of Gram-Positive Bacteria

In most Gram-positive bacteria, peptidoglycan accounts for some 40 to 90 percent of the dry weight of the cell wall. The wall is homogeneous in fine structure, and considerably thicker (10 to 80 nm) than the peptidoglycan wall layer in Gram-negative groups. As already mentioned, there is considerable diversity with respect to peptidoglycan structure and composition among Gram-positive bacteria. The peptidoglycan matrix of the wall is covalently linked to other macromolecular wall constituents which may include a wide variety of polysaccharides and polyolphosphate polymers known as *teichoic acids.* The teichoic acids are water-soluble polymers, containing ribitol or glycerol residues joined through phosphodiester linkages (Figure 6.22). Their exact location in the cell envelope is not certain; most of the teichoic acid remains associated with cell wall material during cell fractionation, and covalent linkage to muramic acid has been demonstrated. However, a small percentage (consisting entirely of glycerol teichoic acids) remains associated with the cell membrane. This material, called *membrane teichoic acid* or *lipoteichoic acid,* has been found to be covalently linked to membrane glycolipid.

The teichoic acids constitute major surface antigens of those Gram-positive species that possess them, and their accessibility to antibodies has been taken as evidence that they lie on the outside surface of the peptidoglycan layer. Their activity is often increased, however, by partial digestion of the peptidoglycan; thus, much teichoic acid may lie within the peptidoglycan layer (Figure 6.23).

FIGURE 6.22

*Repeats units of some teichoic acids.* (a) Glycerol teichoic acid of *Lactobacillus casei* 7469 (R = D-alanine). (b) Glycerol teichoic acid of *Actinomyces antibioticus* (R = D-alanine). (c) Glycerol teichoic acid of *Staphylococcus lactis;* D-alanine occurs in the 6 position of N-acetylglucosamine. (d) Ribitol teichoic acids of *Bacillus subtilis* (*R* = glucose) and *Actinomyces streptomycini* (R = succinate). (The D-alanine is attached to position 3 or 4 ribitol). (e) Ribitol teichoic acid of the type 6 pneumococcal capsule.

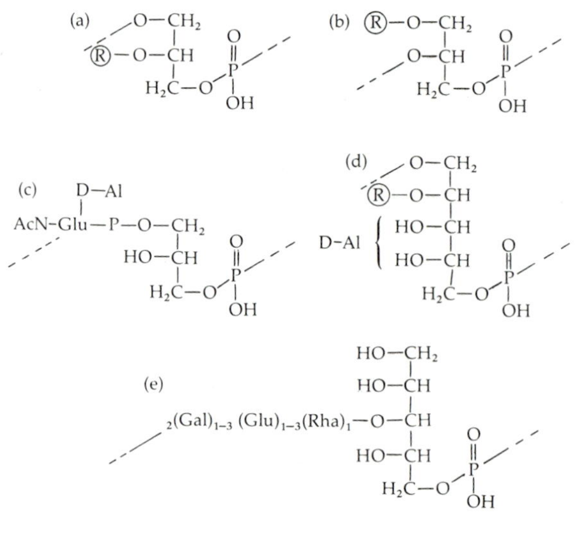

(a) O–$CH_2$; Ⓡ–O–CH; $H_2C$–O–P(=O)(OH)

(b) Ⓡ–O–$CH_2$; O–CH; $H_2C$–O–P(=O)(OH)

(c) D–Al; AcN-Glu–P–O–$CH_2$; HO–CH; $H_2C$–O–P(=O)(OH)

(d) O–$CH_2$; Ⓡ–O–CH; D-Al {HO–CH; HO–CH}; $H_2C$–O–P(=O)(OH)

(e) HO–$CH_2$; HO–CH; $_2(Gal)_{1-3}$ $(Glu)_{1-3}(Rha)_1$–O–CH; HO–CH; $H_2C$–O–P(=O)(OH)

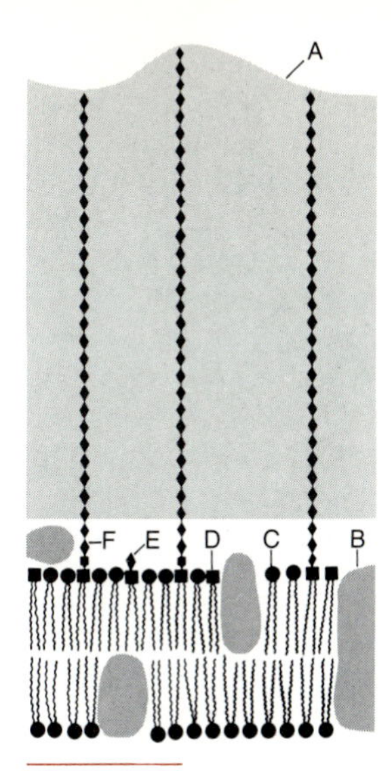


FIGURE 6.23

A model of the cell wall and membrane of a Gram-positive bacterium, showing lipoteichoic acid molecules extending through the cell wall. The wall teichoic acids, covalently linked to muramic acid residues of the peptidoglycan layer, are not shown. (A) Cell wall, (B) protein, (C) phospholipid, (D) glycolipid, (E) phosphatidyl glycolipid, (F) lipoteichoic acid. From D. van Driel et al., "Cellular Location of the Lipoteichoic Acids of *Lactobacillus fermenti* NCTC 6991 and *Lactobacillus casei* NCTC 6375," *J. Ultrastruct. Res.* **43,** 483 (1971).

The repeat units of some teichoic acids are shown in Figure 6.22. The repeat units may be glycerol, joined by 1,3- or 1,2- linkages; ribitol, joined by 1,5-linkages; or more complex units in which glycerol or ribitol is joined to a sugar residue such as glucose, galactose, or N-acetyl glucosamine. The chains may be 30 or more repeat units in length, although chain lengths of 10 or less are common.

Most teichoic acids contain large amounts of D-alanine, usually attached to position 2 or 3 of glycerol, or position 3 or 4 of ribitol. In some of the more complex teichoic acids, however, D-alanine is attached to one of the sugar residues. In addition to D-alanine, other substituents may be attached to the free hydroxyl groups of glycerol and ribitol: glucose, galactose, N-acetylglucosamine, *N*-acetylgalactosamine, or succinate. A given species may have more than one type of sugar substituent in addition to D-alanine; in such cases it is not certain whether the different sugars occur on the same or on separate teichoic acid molecules.

The function of the teichoic acids is unknown, but they do provide a high density of regularly oriented charges to the cell envelope, and these must certainly affect the passage of ions through the outer surface layers.

The walls of most Gram-positive bacteria contain no lipid; however, in corynebacteria, mycobacteria and nocardias, a variety of long-chain fatty acids are attached by ester bonds to the wall polysaccharides (Chapter 24). Proteins are absent from the walls of most Gram-positive bacteria. When present, they occur as a separate layer on the outer surface of the wall, often in a very regular ordered array.

## Function of the Peptidoglycan Layer

The well-nigh universal presence of a continuous peptidoglycan layer in the eubacterial cell wall, despite its many variations of chemical composition, indicates the functional importance of this particular wall polymer. The peptidoglycan layer appears to be the primary determinant of cell shape, as well as the wall component largely responsible for counteracting turgor pressure, and hence preventing osmotic lysis. The action on bacteria of agents that either attack peptidoglycan structure or inhibit peptidoglycan synthesis provide support for these conclusions. Bacterial lysis by lysozyme action, already discussed, is a result of the hydrolytic cleavage of the N-acetylmuramyl-N-acetylglucosamine linkage in the glycan strands, which destroys the integrity of the peptidoglycan fabric. The lethal effect of penicillins on growing populations of procaryotes containing peptidoglycans is a consequence of the fact that these antibiotics inhibit the terminal step of peptidoglycan synthesis: cross-linking of the peptide chains. The resultant weakening of the peptidoglycan fabric in the growing cell leads to osmotic lysis.

In this context, the properties of *bacterial L forms* are also highly significant. In 1935 it was observed that the bacterium *Streptobacillus moniliformis* can give rise on rich media to atypical colonies in which the normal rod-shaped cells are replaced by irregularly shaped, often globular growth forms. It was found that these so-called *L forms* could be propagated indefinitely on serum-enriched complex media. At first, it was believed that the L forms, so different in cell structure from *S. moniliformis*, were symbionts or parasites of the bacterium, but this interpretation had to be abandoned when it was found that they could, on rare occasions, revert to rods. Subsequent observations on L forms of other bacteria showed that the phenomenon was by no means confined to *S. moniliformis.*

After the discovery of penicillin, it was found that L forms are *penicillin resistant* and can in fact be selected by cultivation of bacteria on an osmo-

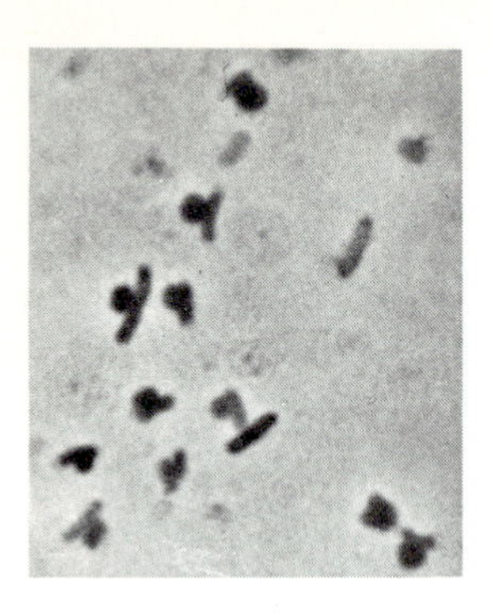

FIGURE 6.24

The genesis of L forms in *Proteus vulgaris:* conversion of the rods to spherical bodies during growth in the presence of penicillin (phase contrast, × 2,000). Courtesy of E. Kellenberger and K. Liebermeister.

tically buffered penicillin-containing medium (Figure 6.24). As a rule, removal of penicillin after short exposure leads to immediate reversion to the normal cell form. After longer periods of growth with penicillin, some bacteria may continue to grow in the L form when penicillin is removed. Both the duration of the penicillin exposure necessary to maintain L forms in the absence of penicillin, and the fraction of the population so converted, are variable. Furthermore, the stability of the resulting L forms, even those derived from a single bacterial species, is variable: some ("stable L forms") have never been observed to revert; others ("unstable L forms") do occasionally revert.

From the known mode of action of penicillin, L forms can be interpreted as bacteria in which the synthesis of the peptidoglycan layer has been severely deranged but which can continue to grow, even though with aberrant cell form, in media sufficiently concentrated to prevent their osmotic lysis.

In certain stable L forms, the components of the peptidoglycan layer are completely absent from the cells. All unstable L forms, and certain stable ones, still contain muramic acid, but the concentration is relatively low (about 10 to 15 percent of its concentration in normal cells). Furthermore, the muramic acid is in an unusual chemical state, being readily extractable with dilute acid, whereas the muramic acid in a normal cell wall is not. From these facts, the following general interpretation of the nature of L forms can be derived. *They are bacteria in which the primer* (see Chapter 5) *for peptidoglycan synthesis has been either eliminated or modified by penicillin treatment.* In the first case, no new peptidoglycan material can be deposited in the wall; in the second, new subunits are incorporated in an irregular and uncoordinated manner, so that the formation of the normal continuous peptidoglycan layer is prevented. On rare occasions, L forms of the latter type may succeed in resynthesizing a continuous peptidoglycan layer and will then revert to the normal bacterial form.

L forms can be obtained from both Gram-positive and Gram-negative bacteria. In the latter, the outer layer of the cell wall continues to be synthesized in an apparently normal fashion; this is indicated both by electron microscopic examination and by the fact that such L forms retain O antigens and are still susceptible to infection by phages for which the receptors are contained in the outer wall layer.

## The Topology of Wall and Membrane Synthesis

During the cell cycle of a bacterium the surface layers of the cell continuously change in form. The increase in the volume of the cell is accompanied by an extension of the area of both wall and membrane. With the onset of division, a vectorial change in wall deposition occurs: the transverse septum begins to grow inward, at right angles to the cell wall, until it forms a complete septum separating the two daughter protoplasts. Thereafter, the transverse septum peels apart into two layers, each of which becomes the newly formed pole of one of the daughter cells.

In the case of *E. coli* and many other rod-shaped bacteria, enlargement of the cell occurs largely in the longitudinal direction, the cross-sectional diameter remaining essentially constant during the cell cycle. Consistent with this pattern of growth, the glycan strands of the murein layer are oriented perpendicular to the long axis of the cell and the cross-linked peptide chains are parallel to it. As the cell elongates, cross-links are broken, new strands are inserted, and new cross-links are formed. This process is unusual in two respects: (1) the murein layer must remain at all times sufficiently intact to contain the cell's internal osmotic pressure, and (2) the process takes place outside the cell membrane.

In order to maintain the structural integrity of the murein sac, it is obvious that the breakage of cross-links (catalyzed in *E. coli* by two *murein peptidases*) must be highly regulated. If only a limited number of cross-links are broken at a time, the layer remains strong, much as a wire mesh does if only a few wires are cut.

The mechanism by which new peptidoglycan is assembled extracellularly is a topologically complicated one (Figure 6.25). The building blocks of a new chain are synthesized in the cytoplasm. On the inner surface of the cytoplasmic membrane, these are polymerized into peptidoglycan monomers and attached to a lipid carrier, *bactoprenol*

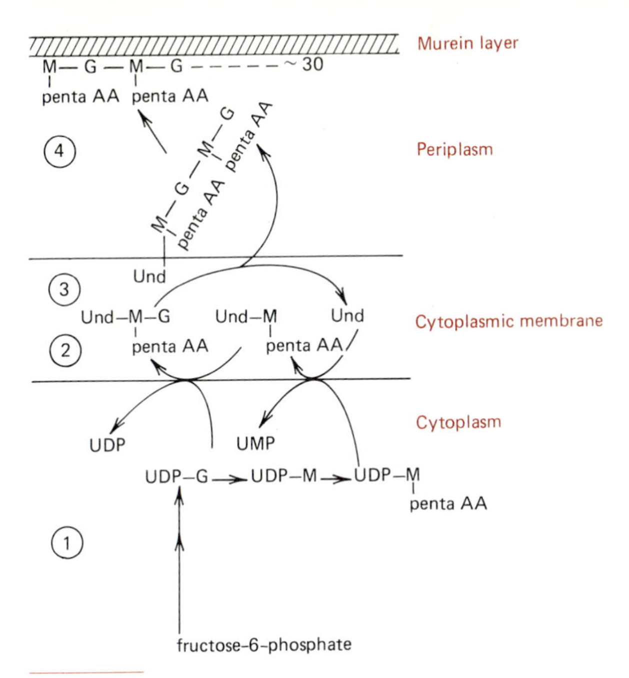


FIGURE 6.25

Pathway and topology of synthesis of a peptidoglycan strand. (1) In the cytoplasm N-acetylglucosamine (G) is synthesized and attached to uridine diphosphate (UDP). In this form it is converted to N-acetylmuramic acid M and five amino acids (penta AA) are added sequentially by enzymatic reactions that do not involve the participation of ribosomes. (2) On the inner aspect of the cytoplasmic membrane N-acetylmuramic acid with attached amino acids is transferred to undecaprenol (Und) with the release of uridine monophosphate (UMP). Then N-acetyl glucosamine is added with the release of UDP. (3) Attached to undecaprenol, the peptidoglycan monomer migrates through the membrane where on the outer aspect the peptidoglycan is polymerized. (4) When the length reaches about 30 disaccharide units, it is released from undecaprenol and migrates to the peptidoglycan and is integrated by cross-linkage. After J. L. Ingraham, O. Maaloe, and F. C. Neidhardt, *Growth of the Bacterial Cell (Sunderland, Mass.: Sinauer Associates, Inc., 1983).*

or *undecaprenol*, which mediates their transport through the cell membrane. On the outer surface the monomers, still attached to the membrane through the bactoprenol carrier, are polymerized into unit peptidoglycan strands that average about 30 disaccharides in length. These are released into the periplasm and then diffuse to a point in the wall where sufficient cross-links have been broken to permit entry of the new strand into peptidoglycan layer. There, they are incorporated by cross-linking the new strand to those already present in the wall. The energy necessary to form the cross-linking peptide bond comes from the simultaneous cleavage of the fifth amino acid, a D-alanine residue, in the peptide chain, a reaction catalyzed by the enzyme, *transpeptidase* (Figure 6.26). Transpeptidase also has a *carboxypeptidase* activity that removes terminal D-alanine residues from peptide strands that are not cross-linked. Thus the unit peptidoglycan chains when first synthesized carry pentapeptide strands, but after they have been incorporated into the murein layer they carry tetrapeptide strands. The beta-lactam class of antibiotics (Chapter 33), which includes penicillin, exerts its antibacterial action by inhibiting the activity of transpeptidase. Thus, the addition of penicillin to a culture causes the cells that are actively growing to lyse because crosslinks continue to be broken but new unit peptidoglycan strands cannot be inserted. After a few minutes, the murein layer becomes too weakened to contain the internal osmotic pressure and the cell lyses.

The increase in the area of the wall and membrane that accompanies cell growth might occur by the insertion of new material at specific growth points or by the intercalation of new material at numerous sites in the preexisting wall and membrane fabric. Many experiments designed to examine this question have been performed, often with conflicting results. It is important to keep in mind that secondary displacement of newly incorporated materials may occur. For example, the plasticity of the cell membrane and the outer wall layer of Gram-negative bacteria could cause an apparently random distribution of newly synthesized compo-

FIGURE 6.26

Cross-linkage reaction catalyzed by transpeptidase. The newly synthesized peptidoglycan strand (donor) carries a pentapeptide strand; the monomer in the previously incorporated strand (recipient) which was either not cross-linked or recently uncross-linked by the action of a murein peptidase carries a tetrapeptide strand. The transpeptidase reaction joins the terminal D-alanine (D-ala) residue to the free amino group of *meso*-diaminopimelic acid (m-DAP) in peptide linkage with release of the terminal D-alanine residue from the donor strand.

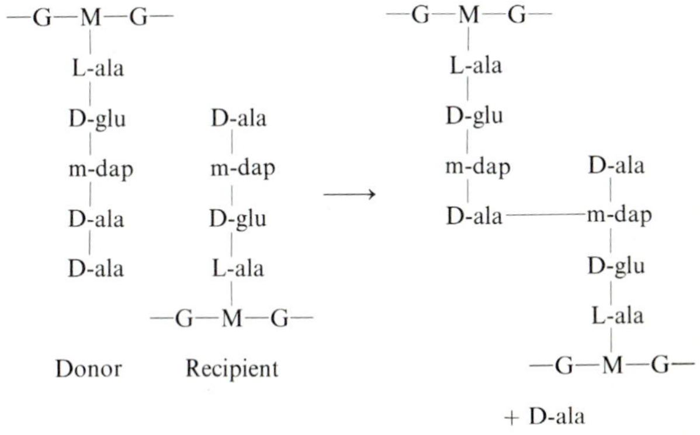

—G—M—G— | L-ala | D-glu | m-dap | D-ala | D-ala (Donor)

D-ala | m-dap | D-glu | L-ala | —G—M—G— (Recipient)

→

—G—M—G— | L-ala | D-glu | m-dap | D-ala———m-dap | D-glu | L-ala | —G—M—G— (with D-ala above m-dap)

Donor Recipient

+ D-ala

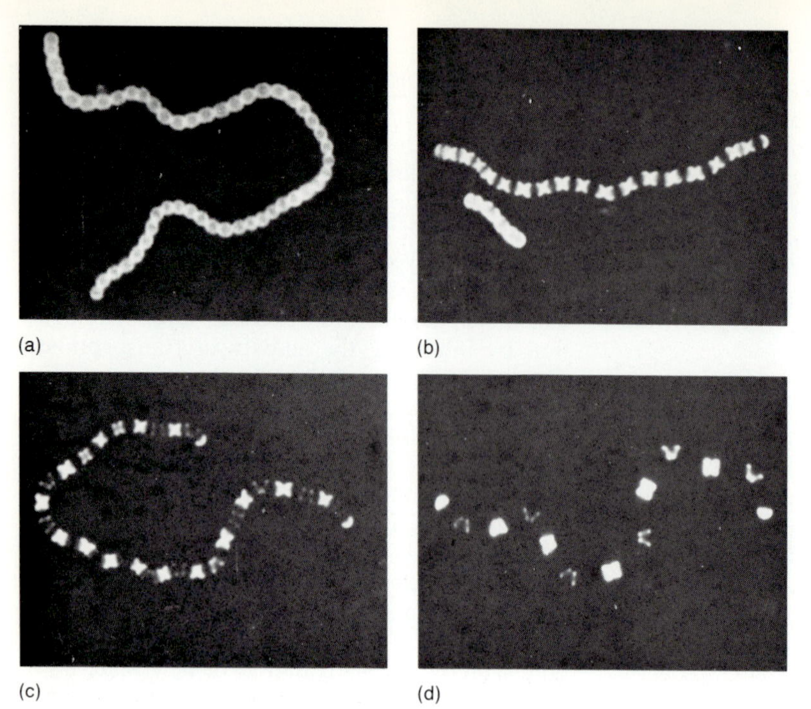


FIGURE 6.27

Growth of the wall of *Streptococcus pyogenes*, followed by ultraviolet photomicrography of growing chains of cells, in which the wall had been initially coated with a fluorescent antibody. (a) Immediately after antibody treatment; the cells are evenly fluorescent. (b) After 15 minutes of growth. New (nonfluorescent) wall material has been formed around the equator of each cell; the polar caps of the cells, previously labeled with fluorescent antibody, remain fluorescent. (c), (d) The appearance of cell chains after 30 and 60 minutes of growth, respectively. From R. M. Cole and J. J. Hahn, "Cell Wall Replication in *Streptococcus pyogenes*," *Science* **135,** 722 (1962).

FIGURE 6.28

(a) A diagrammatic drawing of the possible labeling patterns of stalked *Caulobacter* cells resulting from the experiment described in the text. Crosshatching indicates radioactive areas in the cell produced; arrows indicate the corresponding sites of elongation of the prostheca. (b) Radioautograph of a cell with elongated prostheca. From J. Schmidt and R. Y. Stanier, "The Development of Cellular Stalks in Caulobacteria," *J. Cell Biol.* **28,** 423 (1966).

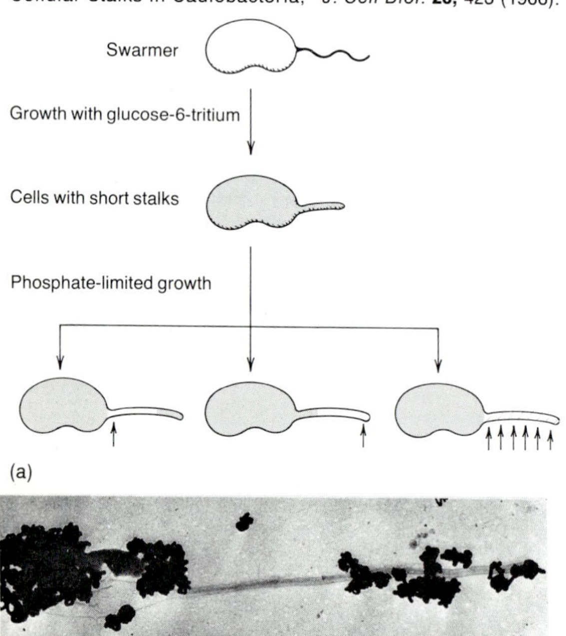


nents, even if they were initially incorporated at specific points. The same effect would result from a rapid turnover of wall or membrane constituents. Consequently, evidence for localized growth of walls or membranes is incontrovertible, whereas evidence that suggests random incorporation is often ambiguous.

In the streptococci, Gram-positive bacteria with spherical cells, clear evidence for a localized equatorial region of wall growth has been obtained by the use of antisera specifically directed against wall constituents. The antiserum, conjugated with a fluorescent dye, is used to coat the cells, making them intensely and uniformly fluorescent. During subsequent growth in the presence of non-fluorescent antiserum, the poles of the cells remain intensely fluorescent for several generations, new nonfluorescent wall areas being progressively inserted between the "old" wall material as growth proceeds (Figure 6.27). Thus, during exponential growth of streptococci, wall synthesis is highly localized, and the walls, once formed, are not secondarily modified.

Fluorescent antibody labeling experiments with growing cultures of Gram-negative bacteria show no indication of localized wall deposition; the intensity of fluorescence of the cells weakens uniformly and progressively as growth proceeds. In this case, however, the antibodies are directed against components of the outer wall (principally, the lipopolysaccharides), which could be redistributed by the plasticity of the lipid bilayer. One special case of highly localized growth, both of wall and membrane layers, has been demonstrated in a Gram-negative organism, *Caulobacter*. The filiform prostheca characteristic of caulobacters develops at each cell generation on the initially flagellated pole of the daughter swarmer cell, and normally attains a length of 2 to 3 $\mu$m. However, when cells are starved for inorganic phosphate, prosthecal elongation continues, the structure attaining a length of 10 to 15 $\mu$m. If swarmer cells are radioactively labeled in a uniform manner by growth with tritiated glucose, and then transferred to a medium containing nonradioactive glucose and a limiting amount of phosphate, the initially short prosthecae formed in the presence of tritiated glucose continue to elongate. Subsequent autoradiography of such cells shows that radioactive material is confined to the body of the cell and the distal end of the prostheca; the proximal part of the prostheca is not radioactive (Figure 6.28). It follows that elongation of the prostheca takes place through highly localized synthesis at the proximal end, close to its attachment to the body of the cell. Since the prostheca is made up of a membranous core, surrounded by both the inner and outer wall layers, it is evident that in this case the synthesis of *all* these structural elements is localized.

In view of the manifold functions of the cell membrane, in particular its postulated role in the segregation of the bacterial genome, the topology of membrane growth is of particular interest. Many studies of membrane growth have been conducted by isotopic or density labeling of precursors of membrane lipids, such as glycerol and fatty acids. This work provides no indication of localized membrane synthesis. However, certain experiments have used as markers of membrane growth certain inducibly synthesized membrane proteins, and the results obtained suggest that membrane growth is in fact highly localized. In *E. coli* the synthesis of $\beta$-galactoside permease, located in the membrane, is under the control of the lactose operon; approximately $10^4$ permease molecules are present in the membrane of a fully induced cell. Since the permease is required for lactose uptake, uninduced cells cannot grow with lactose, but induced cells can. When a mixture of induced and uninduced cells is exposed to penicillin in the presence of lactose as the sole carbon source, induced cells start to grow immediately, and undergo rapid lysis as a result of defective wall synthesis. Uninduced cells lyse only somewhat later, after permease synthesis has been again induced by lactose. The heterogeneity of a cell population with respect to permease function can consequently be determined by this method.

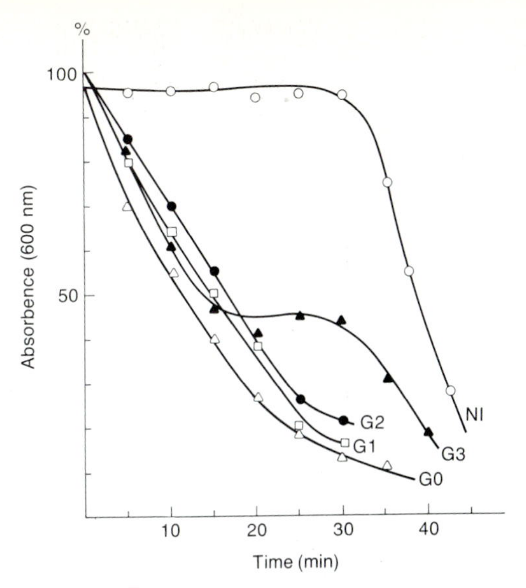


FIGURE 6.29
Time course of penicillin-induced lysis of *E. coli*, fully induced for galactoside permease (curve GO) and after growth for one, two, and three generations in absence of inducer (G1, G2, G3). Curve NI shows the time course of lysis of a noninduced control population. From A. Képès and F. Autissier, "Topology of Membrane Growth in Bacteria," *Biochem. Biophys. Acta* **265,** 443 (1972).

When a fully induced population of *E. coli* is transferred to a medium without inducer (glycerol as carbon source), all cells in the population remain subject to rapid penicillin lysis in the presence of lactose for two generations. By the third generation, however, only one-half the population undergoes immediate lysis (Figure 6.29). Thus, a large fraction of the cellular population suddenly becomes permease negative between the second and third generations. If one assumes that membrane synthesis by *E. coli* occurs in a central zone, separating the two "old" membrane poles by growth, it could be anticipated that one-half the population would contain no "old" membrane after two cell generations, starting from a newly divided cell. However, exponentially growing populations include cells at all stages of the growth cycle: if a cell halfway between two divisions is considered representative of such a nonsynchronized population, it will give rise to descendants with no "old" membrane only after its third division, which will occur after 2.5 generations of growth. These two situations are illustrated schematically in Figure 6.30. The fact that permease negative cells only begin to appear between the second and third generations after removal of the inducer is, therefore, fully compatible with the hypothesis that the membrane is extended by incorporation of newly synthesized material in the central region of the cell. The observations could also be satisfactorily explained on the assumption

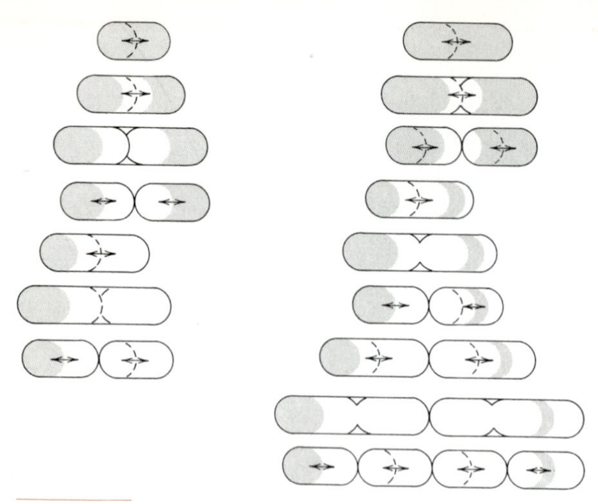

FIGURE 6.30

Distribution of parental membrane (shaded area) in descendants of a bacterium with a single median growing zone. Arrow indicates elongation of new membrane. At left, a newly formed daughter cell; at right, a cell midway between two divisions. From A. Képès and F. Autissier, "Topology of Membrane Growth in Bacteria," *Biochem. Biophys. Acta* **265,** 443 (1972).

that the membrane grows at some other localized site (e.g., the cell poles), but they are wholly incompatible with models that assume intercalation of material at many points over the surface of the membrane.

At first glance, synthesis of the outer membrane appears to present a dilemma because it occurs completely external to the cytoplasmic membrane. But this is not strictly correct, because at certain places on the cell surface, the cytoplasmic and outer membranes are joined at sites termed *Bayer's junctions.* In these regions (of which there are 200 to 400 in each cell) the outer surface of the cytoplasmic membrane is continuous with the inner surface of the outer membrane (Figure 6.31) creating pores that vary in diameter from 25 to 50 nm. At these sites no periplasm nor murein layer intervenes. The protein and lipid components of the outer membrane are synthesized on the inner aspect of the cytoplasmic membrane, translocated through Bayer's junctions, and are there incorporated into the outer membrane. This incorporation is fundamentally different from the incorporation of new peptidoglycan strands into the murein layer, because, with the exception of the joining of murein lipoprotein to the murein layer, no new covalent bonds need be formed. The physical properties of the protein and lipid components of the outer membrane determine their self-assembly into a membranous structure.

## Capsules and Slime Layers

Many procaryotes synthesize organic polymers that are deposited outside the cell wall, as a loose, more or less amorphous layer called a *capsule,* or *slime layer.* The term *capsule* is usually restricted to a layer that remains attached to the cell wall, as an outer investment of limited extent, clearly revealed by negative staining. However, these *exopolymers* often form much more widely dispersed accumulations, in part detached from the cells that produce them. Such variations in the location and extent of the layer are caused primarily by the abundance with which the exopolymer is formed, and by its degree of water solubility. This layer is clearly not essential to cellular functions, because many bacteria do not produce it and those that normally do so can lose the ability, as a result of mutation, without any effect on growth under laboratory conditions. But capsules do serve as virus receptors and as mediators of cell-cell interactions such as adherence to surfaces including specific mammalian tissues, and invasiveness and resistance to the action of phagocytes and other naturally occurring antibacterial agents (see Chapter 29).

Exopolymers vary widely in composition. A few *Bacillus* species produce exopolypeptides, made up of only one amino acid, glutamic acid, which is predominantly of D configuration. Glutamyl residues are linked through the $\gamma$ carboxyl group, as in the side chain of peptidoglycans. With this exception, the bacterial exopolymers are polysaccharides (Table 6.3). In terms of the mechanism of biosyn-

FIGURE 6.31

The molecular architecture of a Bayer's junction. Oblong structures depict lipopolysaccharides; circles depict phospholipids.

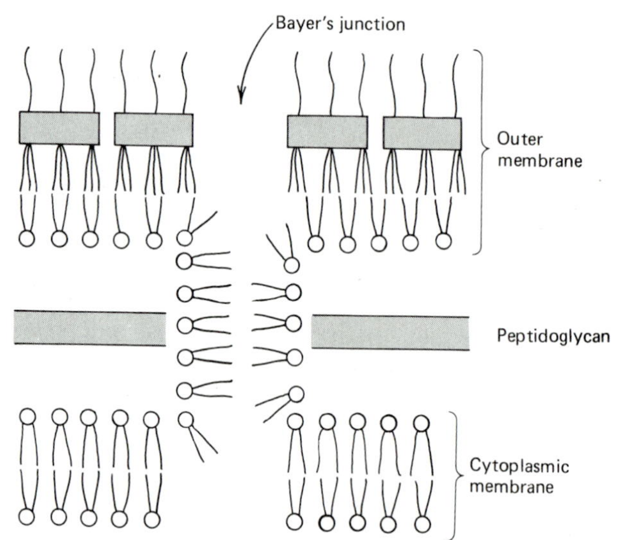

**TABLE 6.3**

**Bacterial Exopolymers**

| Polymer | Subunits | Structure (if known) | Organisms |
|---|---|---|---|
| A. EXOPOLYPEPTIDES | Glutamic acid (mainly D isomer) | $\gamma$-glutamyl-$\gamma$-glutamyl-Polyglutamic acid | *Bacillus anthracis, B. megaterium* |
| B. EXOPOLYSACCHARIDES SYNTHESIZED FROM SUGAR NUCLEOTIDES | | | |
| Cellulose | Glucose | $\beta$-glu 1 $\rightarrow$ 4 $\beta$-glu | *Acetobacter xylinum* |
| Glucan | Glucose | $\beta$-glu 1 $\rightarrow$ 2 $\beta$-glu | *Agrobacterium tumefaciens* |
| Colanic acid | Glucose, galactose, fucose, glucuronic acid, pyruvic acid | | Enteric bacteria |
| Polyuronides | Mannuronic acid, glucuronic acid | | *Pseudomonas aeruginosa, Azotobacter vinelandii* |
| Pneumococcal polysaccharides: | | | |
| Type II | Glucose, rhamnose, glucuronic acid | | |
| Type III | Glucose, glucuronic acid | -(-3-$\beta$-glucuronyl 1 $\rightarrow$ 4 $\beta$-glucosyl)- | *Streptococcus pneumoniae* |
| Type XIX | Glucose, rhamnose, N-acetyl-D-mannosamine, phosphate | | |
| Type XXXIV | Glucose, galactose, ribitol, phosphate | | |
| C. EXOPOLYSACCHARIDES SYNTHESIZED FROM SUCROSE | | | |
| Dextrans | Glucose (fructose) | $\alpha$-fru-$\beta$-glu 1 $\rightarrow$ 6 $\beta$-glu- | *Leuconostoc spp., Streptococcus spp.* |
| Levans | Fructose (glucose) | $\beta$-glu-$\alpha$-fru 2 $\rightarrow$ 6 $\alpha$-fru- | *Pseudomonas spp., Xanthomonas spp., Bacillus spp., Streptococcus salivarius* |

thesis, they fall into two main classes. The majority are synthesized via sugar nucleotide precursors (e.g., UDP-glucose, UDP-galactose, GDP-fucose, GDP-mannose), the formation of the polysaccharide chain involving successive transfers of the glycosyl residues, probably via a lipid carrier in the cell membrane, although this has been demonstrated in only a few cases. The biochemical mechanisms of synthesis thus closely resemble those that operate in the formation of the glycan strands of peptidoglycans, and of the polysaccharide moiety of lipopolysaccharides in Gram-negative bacteria. The sugars incorporated into this class of exopolysaccharides are synthesized by the cell, through the normal processes of intermediary metabolism. As a result, biosynthesis is little, if at all, influenced by the nature of the nutrients provided. A wide diversity of monosaccharides, amino sugars, and uronic acids occur in these compounds; some of the constituent sugars may bear acetyl or pyruvyl substituents. Apart from a few of the simpler, homopolymeric exopolysaccharides (for example, cellulose, synthesized by *Acetobacter xylinum*), the detailed structures have rarely been elucidated.

Two exopolysaccharides, *dextrans* and *levans*, have a different biosynthetic origin: they are formed directly from an exogenous substrate, the disaccharide sucrose, by successive addition of glycosyl

(a)

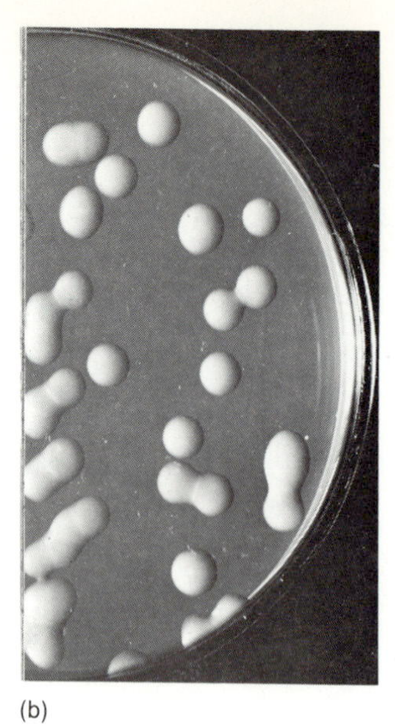
(b)

FIGURE 6.32

The formation of extracellular polysaccharides by bacteria. Two plates of *Leuconostoc mesenteroides*, streaked on glucose medium (a) and sucrose medium (b) The large size and mucoid appearance of the colonies on sucrose are caused by the massive synthesis and deposition around the cells of dextran.

units to an acceptor molecule of sucrose. Symbolizing the sucrose molecule as α-glu-β-fru, we can schematize the initial steps of levan synthesis as follows:

$$2\ \alpha\text{-glu-}\beta\text{-fru} \longrightarrow \alpha\text{-glu-}(\beta\text{-fru})_2 + \text{glu}$$
$$\alpha\text{-glu-}(\beta\text{-fru})_2 + \alpha\text{-glu-}\beta\text{-fru} \longrightarrow \alpha\text{-glu-}(\beta\text{-fru})_3 + \text{glu}$$

Thus, the levan molecule has a terminal glucosyl residue, to which is attached a chain of β-fructosyl residues in $2 \rightarrow 6$ linkage. The synthesis of dextrans occurs in an analogous fashion, by successive addition of α-glucosyl units to the fructosyl moiety of an acceptor molecule of sucrose. Therefore, dextrans contain a terminal β-fructosyl residue.

The formation of dextrans and levans without the ATP expenditure necessary to synthesize sugar nucleotide precursors is possible because the glycosidic bond energy in the disaccharide that serves as the substrate is conserved; chain elongation occurs by transglycosylation. For this reason, dextrans and levans cannot be formed at the expense of free monosaccharides: sucrose is the specific substrate for their synthesis. Consequently, dextran- and levan-forming bacteria produce these capsular materials only when they grow on a sucrose-containing medium. The colonies of such organisms thus have an appearance when grown with sucrose entirely different from that when grown with glucose or other sugars (Figure 6.32).

The capsules present on various isolates of *E. coli* have been extensively studied because they have been shown to play an important role in pathogenesis. All these capsules are made up of polymers composed of disaccharide *repeating units*, but the structures of these units vary from isolate to isolate on the basis of the monosaccharides they contain and the linkage between them and other groups (i.e., acetyl groups) which are sometimes attached to them. Indeed, over 50 chemically different types (*chemotypes*) of capsular material have been identified from various isolates of this single species.

The polysaccharide strands that comprise the capsule of *E. coli* have at one terminus a modified phospholipid group: rather than two fatty acid residues being attached to the glycerol unit, only one is attached, the other being replaced by the polysaccharide strand. This modified phospholipid confers a nonpolar character to the end of the polysaccharide strand, which can insert or associate with the outer membrane, thereby providing an anchor of capsular strands. Whether or not the ends of capsular strands from other Gram-negative bacteria are similarly modified is not yet known.

## THE MOLECULAR STRUCTURE OF FLAGELLA AND PILI

Although they differ both in function and in gross form, the two classes of filiform bacterial surface appendages, flagella and pili, share many common structural features. Both originate from the cell membrane and extend outward through the wall to a distance that may be as much as 10 times the diameter of the cell. The external part of these organelles can be detached from the cell by mechanical means (e.g., shearing in a blender), and subsequently isolated and purified. The filaments of flagella and pili are made up of specific proteins, known as *flagellins* and *pilins*. The protein subunits (monomers) can be prepared by treatment of the isolated flagella or pili with heat or acid; they are of relatively low molecular weight (17,000 to 40,000). Studies of isolated flagella and pili by electron microscopy and X-ray diffraction have shown that the protein monomers are assembled in helical chains, wound around a central hollow core. The structure of the filament is consequently a reflection of the properties of the specific type of protein subunit from which it is built; it is determined by the size

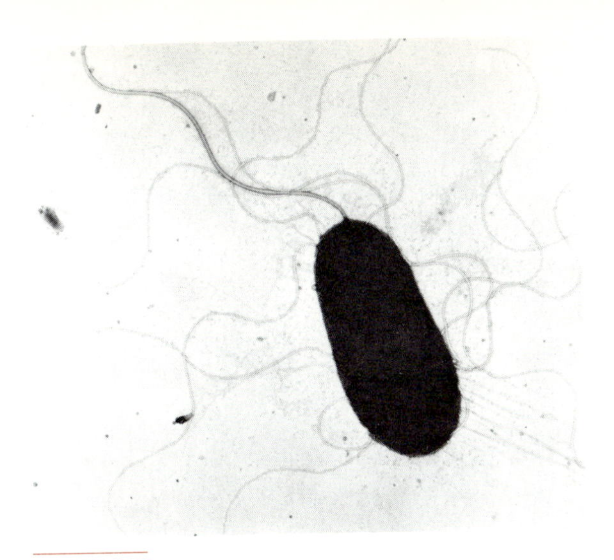

FIGURE 6.33
Electron micrograph of a *Vibrio* with mixed polar-peritrichous flagellation. The cell bears a single sheathed polar flagellum, together with numerous laterally inserted unsheathed flagella ( × 16,600). From R. D. Allen and P. Baumann, "Structure and Arrangement of Flagella in Species of the Genus *Beneckea* and *Photobacterium fischeri*, *J. Bacteriol.* **107,** 295 (1971).

of the subunit and by the number and pitch of the helical chains into which they aggregate. The flagella of different bacteria differ slightly both in diameter (12 to 18 nm), and in form (i.e., height and wavelength of the helical curvature). Different types of pili differ greatly in width (4 to 35 nm). These minor variations within each class of organelle evidently reflect differences in the assembly properties of different flagellins and pilins. It has been shown that a single mutational change in the amino acid sequence of a flagellin can cause a change in the height and wavelength of the flagellum formed from it.

Some bacteria bear flagella that are notably thickened. These *sheathed flagella* are surrounded by an extension of the cytoplasmic membrane (Figure 17.26). Under certain conditions of growth, some bacteria produce polar sheathed flagella along with many peritrichously arranged unsheathed flagella (Figure 6.33); such flagellation is termed *mixed flagellation.*

The probable ultrastructures of two filaments that have been studied in some detail, the flagellum of *Salmonella typhimurium* and the type I pilus of *Escherichia coli*, are shown schematically in Figure 6.34.

| | Flagellum<br>*(Salmonella typhimurium)* | Type I pilus<br>*(Escherichia coli)* |
|---|---|---|
| Form of organelle: | Helical filament,<br>14 nm in diameter | Straight filament<br>7 nm in diameter |
| Protein subunit: | Flagellin (40,000 daltons) | Pilin (17,000 daltons) |
| Suggested model for the assembly of the subunits (not drawn to same scale). Certain subunits are depicted in black in order to emphasize the helical structure. |  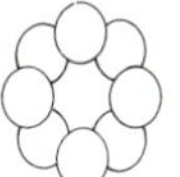 |  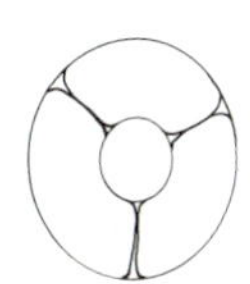 |

FIGURE 6.34
Models showing the probable helical arrangement of the protein subunits of a bacterial flagellum (left) and of a pilus (right).

One function of type I pili is adhesion to surfaces including those of eucaryotic cells, and hence these pili are sometimes essential to the first step in infection. In these cases the tip of the pilus attaches to specific receptor sites on the surface of the eucaryotic cell.

Another type of pili, termed *sex pili,* which are morphologically similar to type I pili but which are composed of different pilin monomers, are synthesized by cells that contain plasmids that determine the cell's capacity to carry out conjugative genetic exchange with other cells. These pili also play an attachment role, which in this case is the first step in the process of conjugation (Chapter 11).

## The Basal Structure of the Flagellum

The removal of flagella and pili by mechanical shearing breaks off these organelles near the cell surface; consequently, it does not reveal their basal structures. However, by more gentle methods of cell breakage (osmotic lysis, detergent treatment), it is possible to isolate flagella with their basal structures intact.

The entire flagellar apparatus is made up of three distinct regions. The outermost region is the helical flagellar filament, of constant width, made up of flagellin. Near the cell surface this is attached to a slightly wider *hook,* about 45 nm long, made up of a different kind of protein, which is in turn attached to a *basal body,* located entirely within the cell envelope (Figure 6.35).

The basal body consists of a small central rod, inserted into a system of rings. In Gram-negative bacteria the basal body typically bears two pairs of rings (Figure 6.36). The outer pair (L and P rings) are situated at the level of the outer membrane; apparently, their function is to serve as bushings for the insertion of the body through this layer. The inner pair (S and M rings) are located near the level of the cell membrane; the M ring is embedded either in it, or just below it, while the S ring lies just above, possibly attached to the inner surface of the peptidoglycan layer (Figure 6.37). On flagella of Gram-positive bacteria, only the lower (S and M) rings are present; apparently, the upper pair is not required to support the rod as it passes through the relatively thick and homogeneous Gram-positive wall. This difference is significant, since it implies that only the S and M rings are essential for flagellar function.

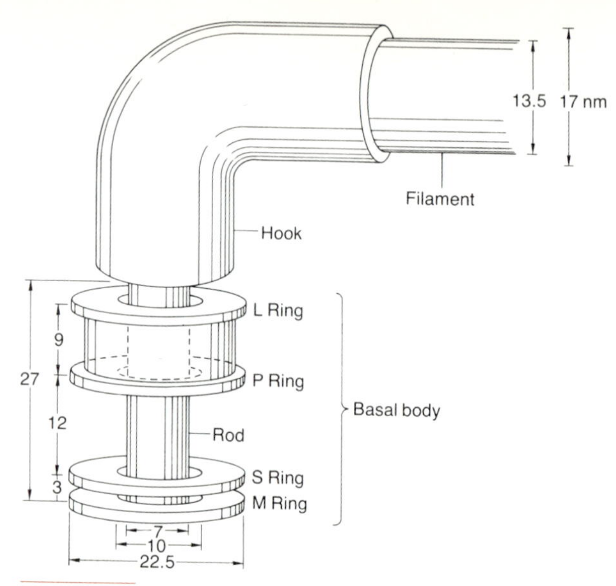


FIGURE 6.36

Diagrammatic model of the basal end of the flagellum of *E. coli,* based on electron micrographs of the isolated organelle. From M. L. De Pamphilis and J. Adler, "Fine Structure and Isolation of the Hook-Basal Body Complex of Flagella from *Escherichia coli* and *Bacillus subtilis,*" *J. Bacteriol.* **105,** 384 (1971).

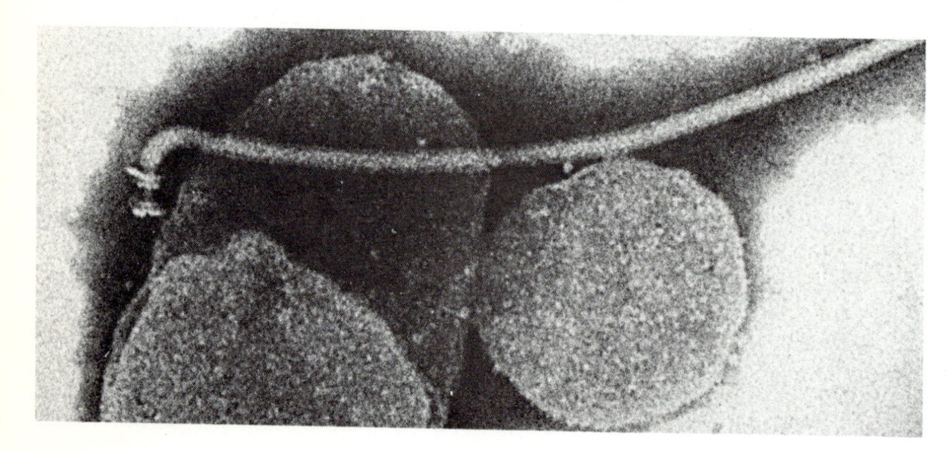

FIGURE 6.35

Electron micrograph of a negatively stained lysate of the purple bacterium, *Rhodospirillum molischianum,* showing the basal structure of an isolated flagellum (× 181,000). Note the basal hook and the attached paired discs. The other objects in the field are fragments of the photosynthetic membrane system. From G. Cohen-Bazire and J. London, "Basal Organelles of Bacterial Flagella," *J. Bacteriol.* **94,** 458 (1967).

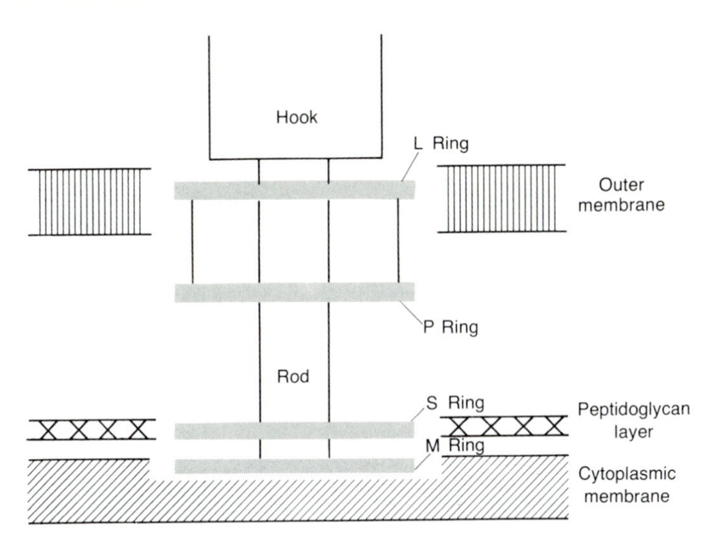


FIGURE 6.37
A model showing the possible topological relations between the basal structure of the flagellum and the outer cell layers of *E. coli*. Dimensions in nanometers. After M. L. De Pamphilis and J. Adler, "Attachment of Flagellar Basal Bodies to the Cell Envelope," *J. Bacteriol.* **105,** 396 (1971).

## Synthesis of the Flagellar Filament

Under appropriate physicochemical conditions, flagellin monomers can reassemble *in vitro* to produce filaments apparently identical with the flagella from which they are derived. This process of assembly requires the presence of seed structures: short flagellar fragments to which the flagellin molecules can attach. Such flagellar fragments possess a structural polarity; electron microscopy shows that one end is rounded and the other is indented. By marking the seed fragments with antiflagellar antibody, it can be demonstrated that flagellin subunits add only to the indented end. To which end of the intact flagellum, attached to the bacterial cell, does the indented end correspond? This question has been answered by an ingenious experiment on the growth of the flagella of *Salmonella typhimurium.* Addition of the amino acid analogue *p*-fluorophenylalanine to the growth medium causes *S. typhimurium* to synthesize abnormal ("curly") flagella of shorter wavelength than normal. When normal cells are cultivated for two to three hours in the presence of this analogue, and the distribution of curly waves on their flagella is then examined, it is found that these always occur on the distal parts of the flagella, the basal parts retaining the normal wavelenth. Hence, it follows that the elongation of the flagellar filament takes place by the addition of new flagellin subunits to its tip. It is most improbable that flagellins can attain the tip of a growing flagellum by excretion and passage through the medium; they are probably synthesized at the base of the flagellum and move outward through its hollow core to the site of incorporation.

## The Mechanism of Flagellar Movement

The role of flagella as agents of bacterial movement can be demonstrated by a very simple experiment. If cells are mechanically deflagellated by shearing, they become immotile. Regrowth of flagella is rapid, the normal number and length being restored in about one generation. During flagella regrowth the cells at first show only rotatory movement; translational movement begins after the flagella have attained a critical length.

The way by which bacterial flagella propel the cell has been discussed for many decades, but experimental evidence in support of a specific mechanism has been obtained only recently. This evidence, discussed below, indicates that flagella are semirigid helical rotors, each of which spins, either clockwise or counterclockwise, around its long axis. Movement is imparted to the organelle at its base by a flagellar "motor." It has been suggested that the motor operates by causing the S and M rings to rotate relative to each other. As already discussed, the M ring is situated in or just below the cell membrane. In order for relative rotation of the two rings to move the cell, it is essential that the S ring also be inserted into a structural fabric; most likely, it is attached to the inner surface of the cell wall. Such an attachment could account for the fact that complete removal of the bacterial cell wall by lysozyme treatment makes the protoplast immotile, even though it still bears intact flagella.

The cell evidently has the capacity to vary both the speed and the direction of rotation, as well as the frequency of stops and starts. Peritrichously flagellated bacteria swim in straight lines over moderate distances, these runs being interrupted periodically by abrupt, random changes of direction, known as *tumbles*. Recent observations suggest that smooth swimming of *E. coli* in a fixed direction is mediated by steady rotation of the flagella in a counterclockwise direction (viewing the flagellum along its rotational axis in the direction toward the cell). Tumbles are caused by a reversal of flagellar rotation, to the clockwise sense.

In spirilla, the multitrichous polar tufts of flagella are sufficiently thick to be visualized by phase contrast microscopy. During steady swimming the two polar tufts both rotate in the same sense. Spirilla never tumble; instead, changes in the direction of movement are brought about by a reversal of the sense of rotation of the flagellar tufts, and they result in an exact reversal, by 180°, of the path previously followed.

The mechanochemical basis for the operation of the flagellar motor is not known. However, there is good evidence that it is dependent on the transmembrane protonmotive force—not on the intracellular concentration of ATP.

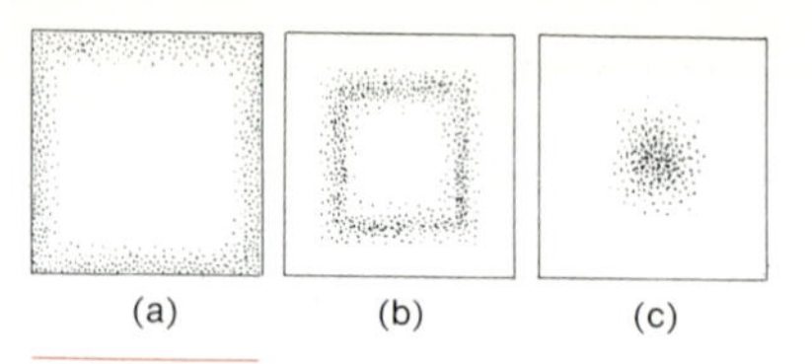


FIGURE 6.38
Aerotactic responses of motile bacteria (after Beijerinck). Suspensions of various bacteria were placed on slides under cover slips. (a) Aerobic bacteria accumulate near the edges of the cover slip, where oxygen concentration is greatest. (b) Microaerophilic bacteria accumulate at some distance from the edge. (c) Obligate anaerobes accumulate in the central, almost anaerobic region.

## THE CHEMOTACTIC BEHAVIOR OF MOTILE BACTERIA

The cells of a suspension of flagellated bacteria are normally in a state of continuous but random active movement. However, if certain chemical gradients are imposed on the population, the cells migrate to and accumulate in that part of the gradient that provides an optimal concentration of the chemical. Some of the substances to which they can respond (for the most part, nutrients) act as *attractants*, in the sense that the cells will accumulate in the more concentrated region of the gradient. Others (for the most part, toxic substances) act as *repellants*, in the sense that the cells avoid regions of high concentration, and accumulate in that part of the gradient in which the concentration is lowest. This behavior is known as *chemotaxis*. The specific substances that can elicit tactic responses differ for different bacteria; a particular chemical spectrum is characteristic for each species.

The system by which bacteria sense the concentration of attractants and repellants is far from completely understood but some key components have been discovered. Certain proteins, termed *binding proteins*, that participate in the entry of certain substrates into the cell by binding tightly and specifically to these substrates (Chapter 8) also serve as receptors for attractant and repellant signals. When an attractant or repellant binds to one of these proteins, a sequence of events involving, among other things, the methylation or demethylation of certain bacterial proteins transmits the signal to the flagellar motor.

*Molecular oxygen* elicits so-called *aerotactic* responses in most motile bacteria. Aerotactic patterns of accumulation can be readily observed in wet mounts in which an oxygen gradient is established by diffusion from the edges of the cover slip. Most strict aerobes accumulate close to the edge of the cover slip; however, the spirilla, which are microaerophils, accumulate in a narrow band some distance from the edge. For motile strict anaerobes, oxygen is a repellant; they accumulate in the center of the wet mount where the oxygen concentration is lowest (see Figure 6.38).

Detailed studies of the tactic responses of *Escherichia coli* to organic compounds (sugars and amino acids) have shown that by no means all compounds that can serve as nutrients (energy sources) are attractants. For example, the disaccharide maltose is an attractant; the disaccharide lactose (also a good substrate) is not, although its split products (glucose and galactose) are. Conversely, the amino acid serine is a powerful attractant, although pyruvate, the first product formed from it, is not. Compounds that act as attractants are not necessarily metabolizable; for example, D-fucose, a nonmetabolizable analogue of D-galactose, is almost as good an attractant as galactose. Systematic studies on the responses of *E. coli* to sugars and amino acids have involved the use of mutants blocked either in the metabolism of attractants or in the ability to detect them, and have also involved the examination of competition between different attractants. These studies have led to the identification of 11 different chemoreceptors, each capable of eliciting a tactic response to certain specific compounds

**TABLE 6.4**

**The Chemoreceptors of *Escherichia coli* for Sugars and Amino Acids, with their Specificities**

| Name of Chemoreceptor | Chemicals Detected (in order of decreasing effectiveness) | Chemicals Detected by More Than One Receptor |
|---|---|---|
| Aspartate | L-aspartate > L-glutamate > L-methionine | Asparagine, |
| Serine | L-serine > glycine > L-alanine | cysteine |
| Glucose | D-glucose, D-mannose | D-glucosamine, |
| Galactose | D-galactose, D-glucose > D-fucose > L-arabinose > D-xylose > L-sorbose | 2-deoxy D-glucose, Methyl-α-D-glucoside |
| Fructose | D-fructose | Methyl-β-D-glucoside |
| Mannitol | D-mannitol | |
| Ribose | D-ribose | |
| Sorbitol | D-sorbitol | |
| Trehalose | Trehalose | |
| N-acetylglucosamine | N-acetyl-D-glucosamine | |
| Maltose | Maltose | |

(Table 6.4). The nature of the receptor molecules for galactose and for maltose is known: they are the specific binding proteins for these two sugars, located in the periplasmic space of the cell. Mutants that lack one of these binding proteins lose their ability to respond to the sugar in question. The chemoreceptors serve purely as *gradient-sensing devices:* they are not directly associated with locomotion, since specific chemoreceptors can be lost by mutation without any impairment of motility.

Bacterial chemotaxis poses a special problem: how do such small organisms detect the concentration differences in chemical gradients over distances as short as the length of a single cell (2 to 3 $\mu$m)? Recent experiments show that the bacterium does not in fact make an instantaneous *spatial* comparison of the attractant concentrations at the two ends of the cell. Instead, it possesses a *temporal* gradient-sensing system, i.e., a kind of "memory" device that enables the cell to compare, over a short interval of time, present and past concentrations. The memory system has a decay time of many seconds. Thus, if a bacterium is swimming at 30 $\mu$m $\text{sec}^{-1}$ and if it possesses a memory with a decay time of 60 seconds, it can compare concentrations over a distance of about 1.8 mm, nearly 1,000 times its body length. The analytical accuracy required is thus several orders of magnitude less than that which would be required if it employed an instantaneous spatial sensing system.

Finally, we must consider how the bacterium uses the information derived from this time-dependent sensing process to migrate toward higher attractant concentrations. The mechanism, at least in *Escherichia coli* and related peritrichously flagellated bacteria, seems to be based on the frequency of tumbles: in other words, on the frequency with which the flagellar motor rotates in a clockwise or counterclockwise direction. Bacteria swimming up an attractant gradient sense a positive temporal gradient, and tumble less frequently than normal. However, bacteria swimming down an attractant gradient, and therefore sensing a negative temporal gradient, tumble more frequently than normal. Since each tumble causes a random change of the direction of movement, the net result is that cells placed in a gradient spend more time swimming up the gradient that down it: hence, their characteristic migration to the region of high attractant concentration.

## The Phototactic Behavior of Purple Bacteria

Many photosynthetic bacteria can respond to a *gradient of light intensity*, a phenomenon known as *phototaxis*. This behavior can be readily demonstrated by projecting a narrow spot of bright light onto an otherwise weakly illuminated suspension of motile purple bacteria, in which the cells are evenly dispersed and moving in a random fashion. Within 10 to 30 minutes, most of the population accumulates in the bright spot. The mechanism of this accumulation is shown in Figure 6.39. Swimming cells enter the light spot by random movement. Once within it, they are prevented from leaving again by a *shock movement* (i.e., an abrupt change

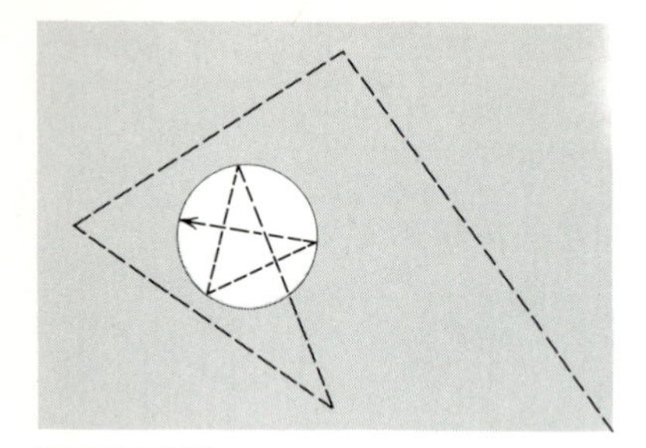

FIGURE 6.39

A diagrammatic illustration of the mechanism by which phototactic bacteria accumulate within an illuminated area. The dashed line indicates the trajectory of a purple bacterium in a darkened field that contains a single illuminated area (white circle).

in the direction of swimming), which occurs every time they penetrate the sharp gradient of light intensity that separates the brightly illuminated spot from the surrounding dim area. As this observation suggests, purple bacteria and indeed all polarly flagellated bacteria differ from peritrichously flagellated bacteria in the manner by which they change direction during a tactic response. Rather than changing their direction by tumbling, they abruptly reverse their direction of swimming.

If a wet mount of motile purple bacteria is illuminated not with white light but with a spectrum produced by focusing light that has passed through a prism on the preparation, the bacteria rapidly accumulate in a series of bands corresponding to the principal absorption bands of their photosynthetic pigment system (Figure 6.40). A careful quantitative study of the relative effectiveness of different wavelengths in mediating phototaxis has shown that the action spectrum for phototaxis by purple bacteria corresponds exactly to the action spectrum for the performance of photosynthesis.

## SPECIAL PROCARYOTIC ORGANELLES

As already mentioned, most procaryotes do not form intracellular organelles bounded by unit membranes; the only possible exceptions are the thylakoids, which house the photosynthetic apparatus of cyanobacteria. However, three classes of procaryotic organelles are bounded by *nonunit membranes*, made up of protein, at least in part, in all cases. These are *gas vesicles*, *chlorosomes*, and *carboxysomes* (*polyhedral bodies*).

### Gas Vesicles and Gas Vacuoles

Since most cells have a slightly higher density than water, they tend to sink in an aqueous medium at a rate that is a function of cell size. This effect can be counteracted by swimming against the gravitational pull. However, many aquatic procaryotes have developed another device to counteract gravitational pull: their cells contain gas-filled structures known as *gas vacuoles*. By light microscopy, gas vacuoles appear densely refractile, and have an irregular contour (Figure 6.41). If such cells are subjected to a sudden, sharp increase in hydrostatic pressure, the gas vacuoles collapse, and the cells simultaneously lose their buoyancy and become much less refractile. Electron microscopy shows that gas vacuoles are compound organelles made up of a variable number of individual *gas vesicles* (Figure 6.42). Each gas vesicle is a hollow cylinder, 75 nm in diameter with conical ends, and from 200 to 1000 nm in length. The vesicle is bounded by a layer of protein 2 nm thick and is banded by regular rows of subunits ("ribs") running at right angles to its long axis (Figure 6.43).

FIGURE 6.40

The pattern of phototactic accumulation of motile purple bacteria in a wet mount, which has been exposed to illumination in a spectrum. The cells accumulate massively at wavelengths corresponding to the absorption bands of their chlorophyll and carotenoids, which are both photosynthetically effective. The relatively weak accumulation around 500 nm corresponds to the positions of carotenoid absorption bands; the accumulations at 590, 800, 850, and 900 nm correspond to the positions of chlorophyll absorption bands. After J. Buder, "Zur Biologie des Bakteriopurpurins und der Purpurbakterien," *Jahrb. wiss. Botan.* **58,** 525 (1919).

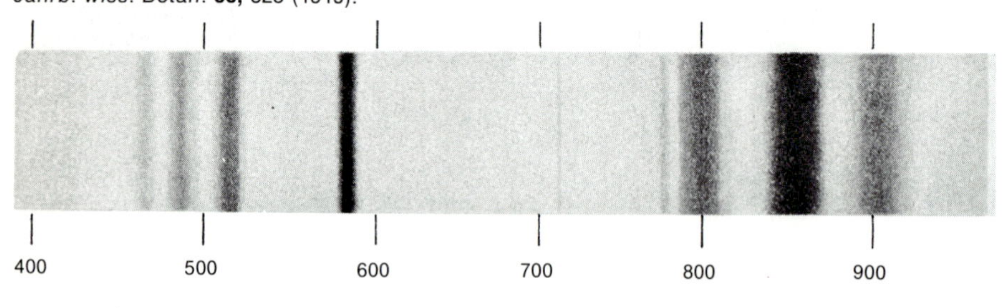

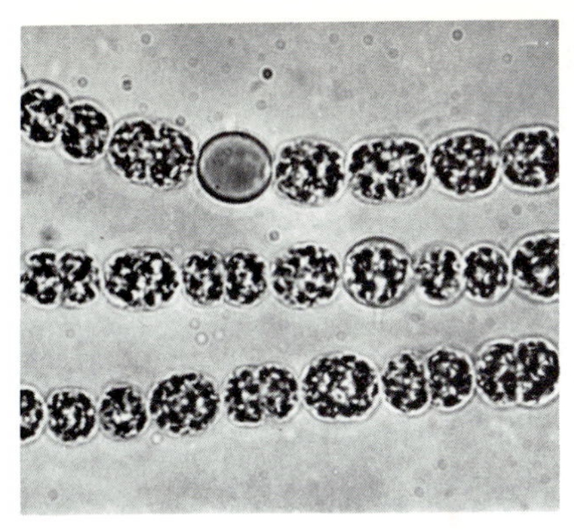

(a)

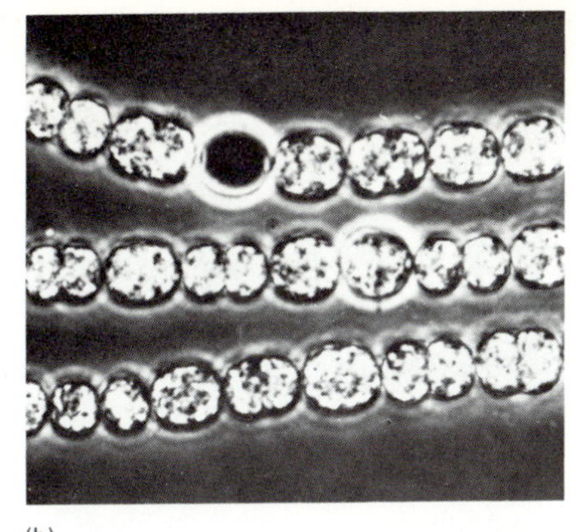

(b)

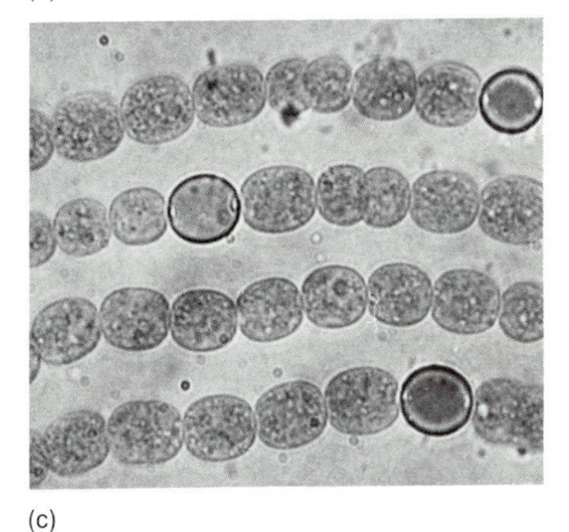

(c)

(d)

FIGURE 6.41

Filaments of a cyanobacterium containing gas vacuoles, as visualized by bright field (a) and phase-contrast (b) illumination. Filaments from the same culture, after collapse by pressure of the gas vacuoles, are visualized by bright field (c) and phase contrast (d) illumination. The clear cells are heterocysts, which never contain gas vacuoles. From A. E. Walsby, "Structure and Function of Gas Vacuoles," *Bacteriol. Rev.* **36,** 1 (1972).

FIGURE 6.42

Electron micrograph of a thin section of *Oscillatoria*, showing the intracellular arrangement of the cylindrical gas vesicles which compose gas vacuoles (×25,800). Courtesy of Germaine Cohen-Bazire.

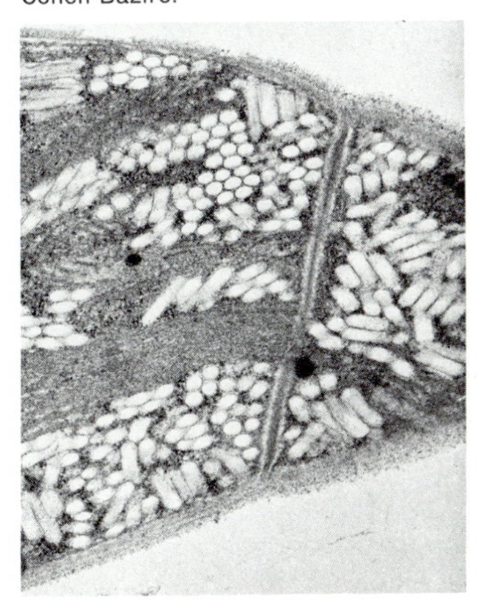

FIGURE 6.43

Electron micrograph of purified gas vesicles from *Oscillatoria*, negatively stained with uranyl acetate (×103,000). The vesicles are still inflated. Note the regular, banded fine structure of the vesicle wall. Courtesy of Germaine Cohen-Bazire.

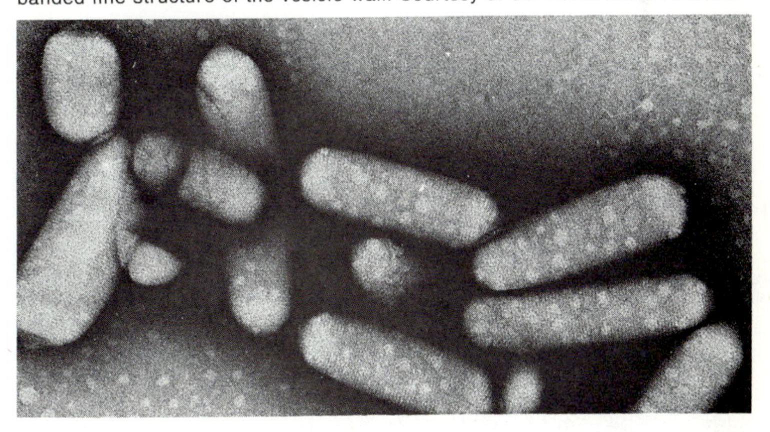

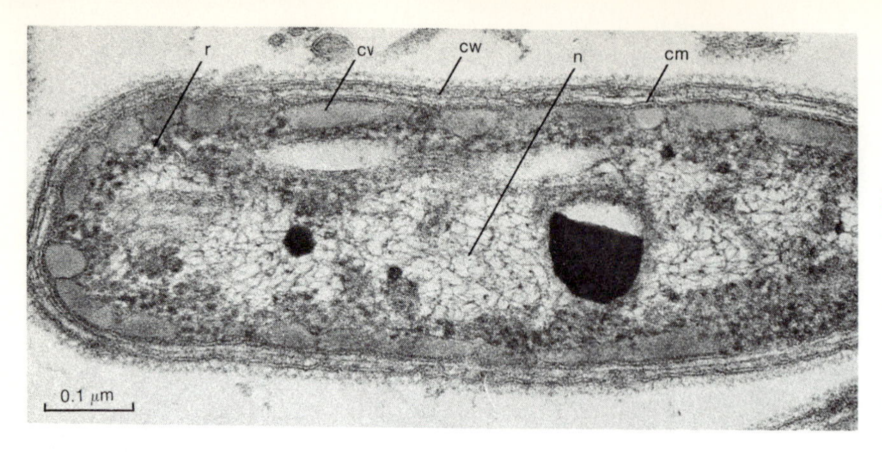


FIGURE 6.44

Electron micrograph of a thin section of the green bacterium *Pelodictyon*, showing the relationship of the chlorosomes (c) to other parts of the cell (×81,800): cw, cell wall; cm, cell membrane; r, ribosomes; n, nucleoplasm. Photo courtesy of Germaine Cohen-Bazire.

Because the membrane is freely permeable to all common gases, the vesicles can neither store nor accumulate gas. The composition and pressure of the gas in the vesicle are therefore functions of the dissolved gases in the surrounding medium. Water is excluded from the interior of the vesicles in the course of their formation and growth. This conclusion is confirmed by the observation that after pressure collapse, the vesicles do not recover; the cell can reacquire gas-filled vesicles only by *de novo* synthesis of these structures.

Two factors are probably important in maintaining a gas phase within vesicles. One is the structural rigidity of the enclosing protein membrane, which must be able to resist the various forms of pressure that normally act upon it. The other is the chemical character of the membrane protein. Its outer surface is clearly hydrophilic, being readily wettable, but the inner wall is strongly hydrophobic.

Gas vacuoles occur in a wide variety of procaryotes including archaebacteria, purple bacteria, cyanobacteria, green bacteria, and various groups of chemoheterotrophic eubacteria. The only common denominator of all these organisms is ecological: they occur in aquatic habitats. There can be little doubt, accordingly, that the function of gas vacuoles is to enable their possessors to regulate the buoyancy of the cell in order to occupy a position in the water column that is optimal for their metabolic activity with respect to light intensity, dissolved oxygen concentration, or the concentration of other nutrients.

## Chlorosomes

In one group of photosynthetic procaryotes, the green bacteria, part of the photosynthetic apparatus has a distinctive intracellular location: the antenna pigments are housed in a series of cigar-shaped vesicles, arranged in a cortical layer that immediately underlies, but is physically distinct from, the cell membrane (Figure 6.44). These structures, detectable only by electron microscopy, are 50 nm wide and 100 to 150 nm long, being enclosed by a single-layered membrane 3 to 5 nm thick. The photosynthetic pigments are entirely contained within them; these are probably the sites of the photosynthetic apparatus. The green bacteria are, accordingly, unique among photosynthetic organisms by virtue of the fact that the photosynthetic apparatus is not integrated into a unit membrane system.

## Carboxysomes (Polyhedral Bodies)

A number of photosynthetic bacteria (cyanobacteria, certain purple bacteria) and chemoautotrophic bacteria (nitrifying bacteria, thiobacilli) contain structures termed *polyhedral bodies*, 50 to 500 nm wide, with polygonal profiles, which are surrounded by a monolayer membrane about 3.5 nm wide and which have a granular content (Figure 6.45). These structures have been isolated (Figure 6.46); they contain most of the cellular content of ribulose bisphosphate carboxylase (carboxydismutase), the key enzyme in the fixation of $CO_2$ associated with the operation of the Calvin-Benson cycle. They have been termed *carboxysomes*, and evidently represent the principal site of $CO_2$ fixation in these autotrophic procaryotes.

## Magnetosomes

In 1975 a remarkable group of bacteria were described by R. P. Blakemore that are magnetotactic; i.e., when they are placed in a magnetic field as weak as 0.2 gauss they orient and swim towards one or another of the magnetic poles. The sensing organelles, termed *magnetosomes*, within these cells are uniformly shaped enveloped crystals of magnetite ($Fe_3O_4$), a ferrimagnetic mineral. These are often arranged in a string (Figure 6.47). The essen-

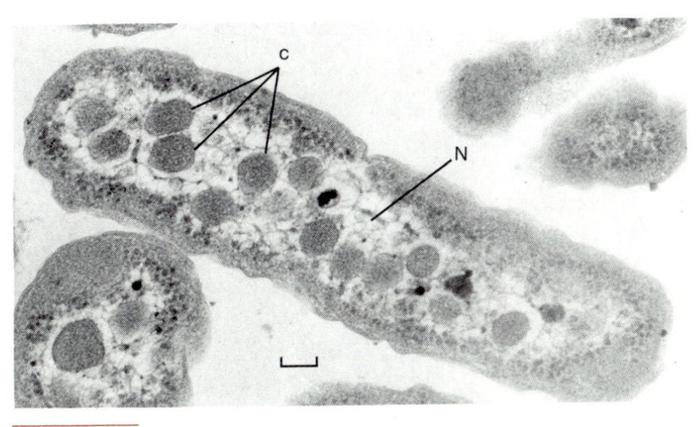


FIGURE 6.45

Electron micrograph of a section of *Thiobacillus*, containing numerous carboxysomes (c); N designates nucleoplasm. Bar indicates 0.1 μm. From J. M. Shiveley, F. L. Ball, and B. W. Kline, "Electron Microscopy of the Carboxysomes (Polyhedral Bodies) of *Thiobacillus neapolitanus*," *J. Bacteriol.* **116,** 1405 (1973).

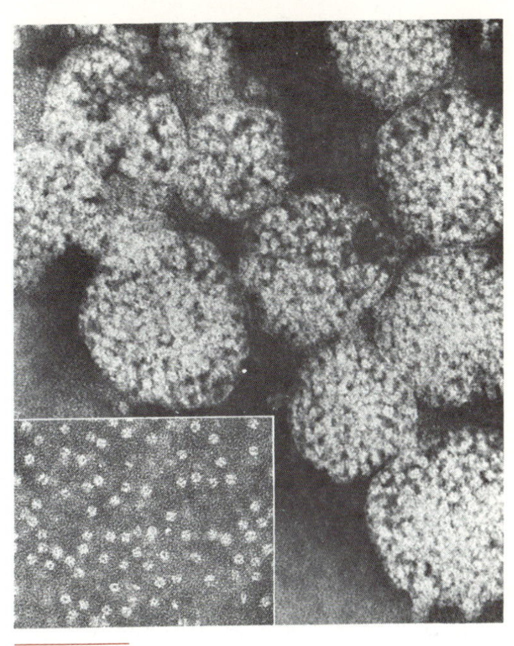

FIGURE 6.46

Electron micrographs of purified carboxysomes from *Thiobacillus*, and (inset) of the enzyme carboxydismutase, isolated from them. Negatively stained, × 108,000. From J. M. Shiveley, F. L. Ball, D. H. Brown, and R. E. Saunders, "Functional Organelles in Prokaryotes: Polyhedral Inclusions (Carboxysomes) of *Thiobacillus neapolitanus*," *Science* **182,** 584 (1973).

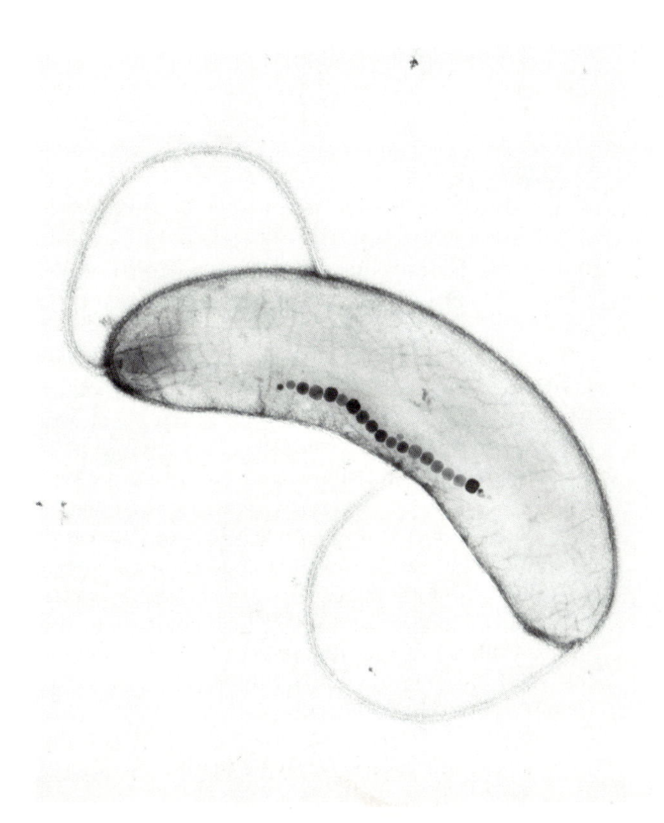

FIGURE 6.47

Transmission electron micrograph of a magnetotactic bacterium and the single string of magnetosomes that it contains. From R. Blakemore, "Magnetotactic Bacteria," *Science* **190,** 377 (1975).

tiality of the presence of magnetosomes to magnetotaxis can be clearly established by the simple experiment of growing these bacteria in a medium that contains a restricted concentration of iron: such cells do not contain magnetosomes and they are not magnetotactic.

Magnetosomes appear merely to orient the cells in a magnetic field and thereby determine the direction in which they swim; they are not an object that is pulled by a magnetic field as can be shown by the fact that dead cells are not magnetotactic.

The selective advantage of magnetotaxis seems to lie in the relationship of these bacteria to oxygen. All are aquatic organisms, found in either freshwater or marine environments and they are either microaerophiles or strict anaerobes. In northern latitudes, the earth's magnetic field is inclined downward; in southern latitudes it is directed upward. Thus, magnetotactic bacteria that happened to be suspended in the water column by some disturbance of the bottom layer could, by swimming along the magnetic field lines, return to the bottom where microaerophilic conditions are favorable for their survival. This hypothesis is materially strengthened by the observation that magnetotactic bacteria from the Northern Hemisphere are almost exclusively north-seeking and those from the Southern Hemisphere are almost exclusively south-seeking. In both cases, magnetotaxis directs cells in a downward direction. Curiously, populations of magnetotactic bacteria collected at the geomagnetic equator where the magnetic field has no vertical component are composed of equal mixtures of north-seeking and south-seeking cells.

**TABLE 6.5**

**The distribution of nonnitrogenous organic reserve materials among procaryotes**

A. GLUCANS
- Cyanobacteria (most representatives)
- Enteric bacteria (most genera, except those listed under B)
- Sporeformers: many *Bacillus* and *Clostridium* species

B. POLY-β-HYDROXYBUTYRATE
- Enteric bacteria: genera *Vibrio* and *Photobacterium*
- *Pseudomonas* (many species)
- Azotobacter group (*Azotobacter*, *Beijerinckia*, *Derxia*)
- *Rhizobium*
- *Moraxella* (some species)
- *Spirillum*
- *Sphaerotilus*
- *Bacillus* (some species)

C. BOTH GLUCANS AND POLY-β-HYDROXYBUTYRATE
- Cyanobacteria (a few species)
- Purple bacteria

D. NO DEMONSTRABLE RESERVE MATERIAL
- Green bacteria
- *Pseudomonas* (many species)
- *Acinetobacter*

Note: This list is partial and includes only groups in which the nature of reserve materials has been systematically investigated.

## THE PROCARYOTIC CELLULAR RESERVE MATERIALS

A variety of cellular reserve materials may occur in procaryotic organisms; they are frequently detectable as granular cytoplasmic inclusions.

### Nonnitrogenous Organic Reserve Materials

Two chemically different kinds of nonnitrogenous organic reserve materials, each of which can provide an intracellular store of carbon or energy, are widespread among procaryotic organisms (Table 6.5). They are glucose-containing polysaccharides (α-1,4-glucans) such as starch and glycogen; and a polyester of β-hydroxybutyric acid, poly-β-hydroxybutyric acid. The former class of substances also occurs as reserve materials in many eucaryotic organisms. Poly-β-hydroxybutyric acid is, however, uniquely found in procaryotic groups. Procaryotic organisms do not store neutral fats, which commonly occur as reserve materials in eucaryotic organisms.

As a general rule, only one kind of reserve material is formed by a given species. Thus, many bacteria of the enteric group and anaerobic sporeformers (*Clostridium*) synthesize only glycogen or starch as a reserve material, whereas many *Pseudomonas*, *Azotobacter*, *Spirillum*, and *Bacillus* species synthesize only poly-β-hydroxybutyrate. Certain bacteria can, however, synthesize both types of reserve material; this is characteristic of the purple bacteria. Finally, some bacteria (e.g., the fluorescent species of the genus *Pseudomonas*) do not synthesize any specific nonnitrogenous organic reserve material.

Bacterial polysaccharide reserves are usually deposited more or less evenly throughout the cytoplasm, in areas detectable with the electron microscope but not visible with the light microscope. The presence of large amounts of such reserves in cells can be revealed by treatment with a solution of iodine in potassium iodide, which stains unbranched polyglucoses such as starch dark blue and branched polyglucoses such as glycogen reddish brown. Deposits of poly-β-hydroxybutyrate, however, are readily visible with the light microscope, occurring as refractile granules of variable size scattered through the cell. They are specifically stainable with Sudan black, a property also shown, in other groups, by neutral fat deposits. For this reason, bacterial poly-β-hydroxybutyrate granules have sometimes been incorrectly identified as fat reserves.

As a general rule, the cellular content of these reserve materials is relatively low in actively growing cells: they accumulate massively when cells are limited in nitrogen but still have carbon and energy sources available. Under such circumstances, nucleic acid and protein synthesis are impeded and much of the assimilated carbon is converted to reserve materials, which may accumulate until they represent as much as 50 percent of the cellular dry weight. If such cells are then deprived of an external carbon source and furnished with an appropriate nitrogen source (e.g., $NH_4Cl$), the reserve materials can be used for the synthesis of nucleic acid and protein (Figure 6.48).

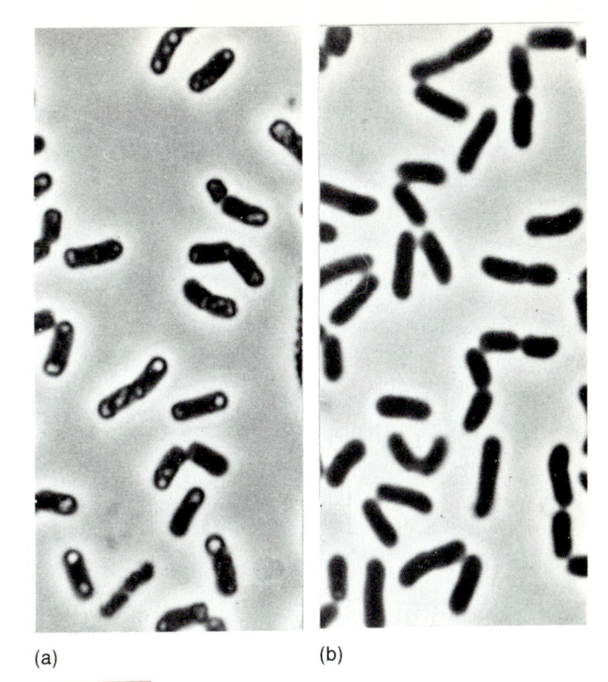


FIGURE 6.48
The formation and utilization of poly-β-hydroxybutyric acid in *Bacillus megaterium* (phase contrast, ×2,200). (a) Cells grown with a high concentration of glucose and acetate. All cells contain one or more granules of poly-β-hydroxybutyric acid (light areas). (b) Cells from the same culture after incubation for 24 hours with a nitrogen source, in the absence of an external carbon source. Almost all the polymer granules have disappeared. Courtesy of J. F. Wilkinson.

Essentially, the synthesis of polyglucoses or poly-β-hydroxybutyrate represents a device for accumulating a carbon store in a form that is *osmotically inert*. In the case of poly-β-hydroxybutyrate, polymer synthesis also represents a method of *neutralizing an acidic metabolite*, since the free carboxyl group of β-hydroxybutyric acid is eliminated through the formation of the ester bonds between the subunits of the polymer. The cell can thus accommodate a very large store of such materials, whereas an equivalent intracellular accumulation of free glucose or β-hydroxybutyric acid could have catastrophic physiological consequences.

The accumulation and subsequent reutilization of carbon reserves is mediated by special enzymatic machinery, under close regulatory control. Poly-β-hydroxybutyrate is formed through a branch of the metabolic route of fatty acid synthesis (Figure 6.49). The native polymer granules into which it is incorporated have associated with them a complex system for degradation of the polymer. The polymer in native granules cannot be attacked until the granules have been "activated" by an enzyme that requires $Ca^{2+}$ ions. This enzyme may be proteolytic, since its effect can be mimicked by trypsin. The activated granules become subject to the action of a depolymerase, which hydrolyzes the polymer to a dimeric ester; the ester is then converted to free β-hydroxybutyric acid by a specific dimerase. A remarkable feature of this system is that *the depolymerase cannot hydrolyze the chemically purified polymer*; the only substrate that it can attack is the activated polymer granule. Even relatively mild treatments of the native polymer granules (e.g., freezing and thawing) may render them unutilizable by the intracellular polymer-degrading system. Recently another lipid reserve material, poly-β-hydroxyoctanoic acid was found to accumulate in *Pseudomonas oleovoranas* when this organism was grown in a medium that contained *n*-octane as a source of carbon. This polymer, which is a homologue of poly-β-hydroxybutyrate (Figure 6.50) accumulates in granules that resemble poly-β-hydroxybutyrate granules and most probably serves a similar metabolic function.

The bacterial synthesis of glycogen is initiated by the formation of ADP-glucose from glucose-1-phosphate and ATP, through the action of the

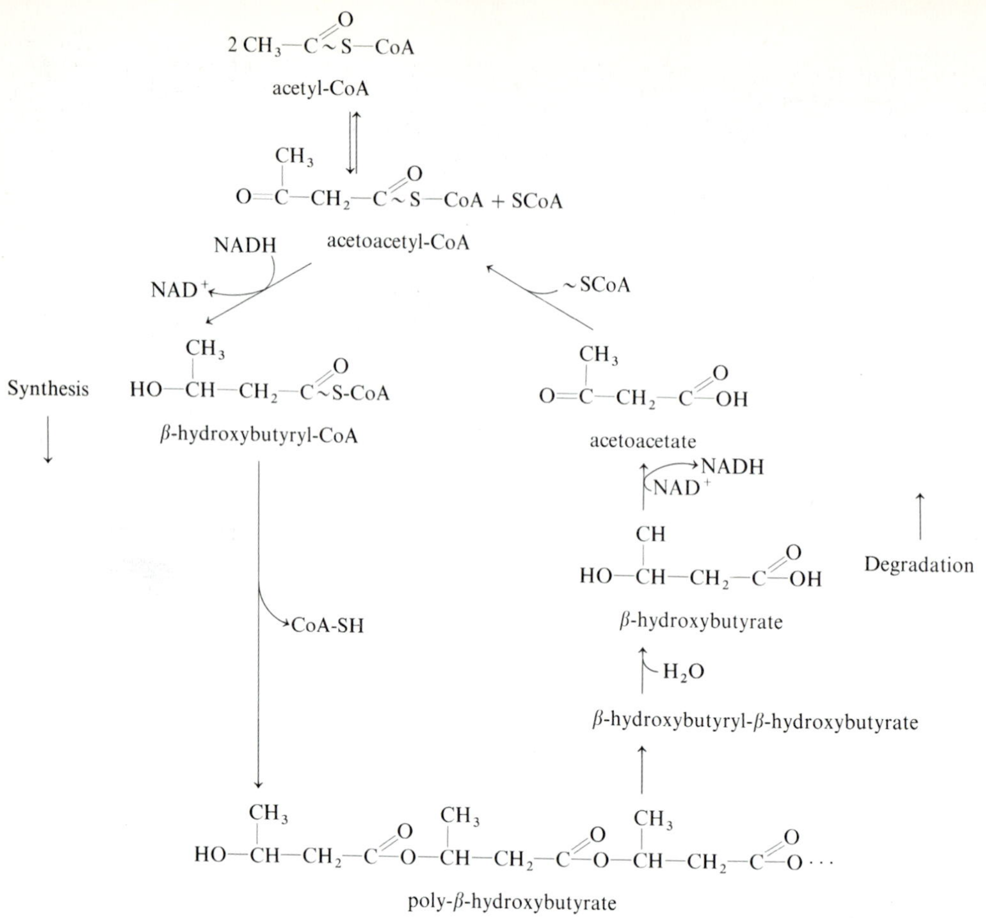

$2\ CH_3{-}C({=}O){\sim}S{-}CoA$

acetyl-CoA

$\rightleftharpoons$

$O{=}C(CH_3){-}CH_2{-}C({=}O){\sim}S{-}CoA + SCoA$

acetoacetyl-CoA

Synthesis ↓ (NADH → $NAD^+$)

$HO{-}CH(CH_3){-}CH_2{-}C({=}O){\sim}S\text{-}CoA$

$\beta$-hydroxybutyryl-CoA

↓ → CoA-SH

$HO{-}CH(CH_3){-}CH_2{-}C({=}O){-}O{-}CH(CH_3){-}CH_2{-}C({=}O){-}O{-}CH(CH_3){-}CH_2{-}C({=}O){-}O\cdots$

poly-$\beta$-hydroxybutyrate

Degradation ↑

$\beta$-hydroxybutyryl-$\beta$-hydroxybutyrate

↑ $H_2O$

$HO{-}CH(CH){-}CH_2{-}C({=}O){-}OH$

$\beta$-hydroxybutyrate

↑ ($NAD^+$ → NADH)

$O{=}C(CH_3){-}CH_2{-}C({=}O){-}OH$

acetoacetate

↑ (~SCoA) → acetoacetyl-CoA

FIGURE 6.49

The reactions involved in the synthesis and degradation of poly-$\beta$-hydroxybutyrate.

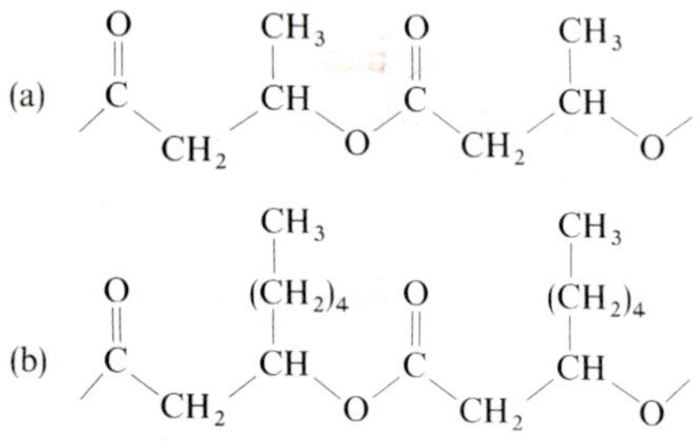

(a) $-C({=}O){-}CH_2{-}CH(CH_3){-}O{-}C({=}O){-}CH_2{-}CH(CH_3){-}O-$

(b) $-C({=}O){-}CH_2{-}CH((CH_2)_4CH_3){-}O{-}C({=}O){-}CH_2{-}CH((CH_2)_4CH_3){-}O-$

FIGURE 6.50

Comparison of the structures of (a) poly-$\beta$-hydroxybutyrate and (b) poly-$\beta$-hydroxyoctanoate.

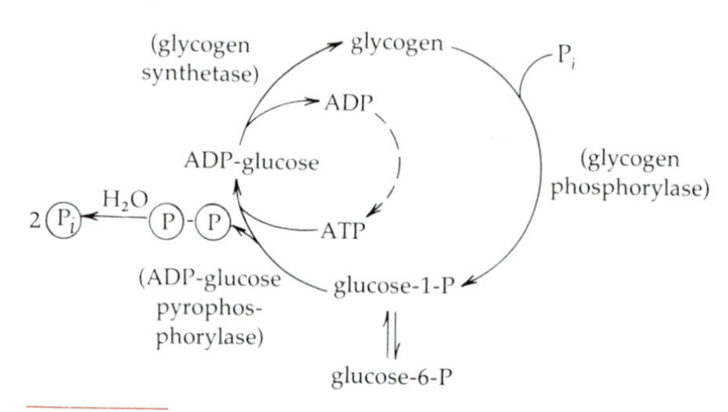


FIGURE 6.51

The reactions involved in the bacterial synthesis and degradation of glycogen.

enzyme ADP-glucose pyrophosphorylase:

$$\text{ATP} + \text{G-1-P} \rightleftharpoons \text{ADP-glucose} + \text{Ⓟ} - \text{Ⓟ}$$

followed by the transfer of a glucosyl unit to an acceptor molecule of $\alpha$-1,4-glucan, mediated by glycogen synthetase. The degradation of glycogen leads to the formation of glucose-1-phosphate, being mediated by glycogen phosphorylase (Figure 6.51).

## Nitrogenous Reserve Materials

As a rule, procaryotes do not produce intracellular nitrogenous organic reserve materials. However, many of the cyanobacteria accumulate a nitrogenous reserve material known as *cyanophycin* when cultures approach the stationary phase. Cyanophycin granules, which have a distinctive structured appearance in electron micrographs (Figure 6.52),

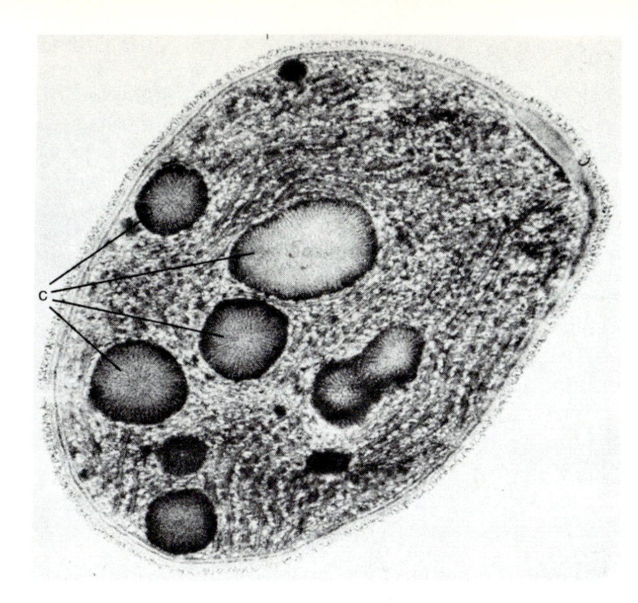


FIGURE 6.52
Electron micrograph of a section of a unicellular cyanobacterium, containing cyanophycin granules (c), (×28,800) Courtesy of Dr. Mary Mennes Allen.

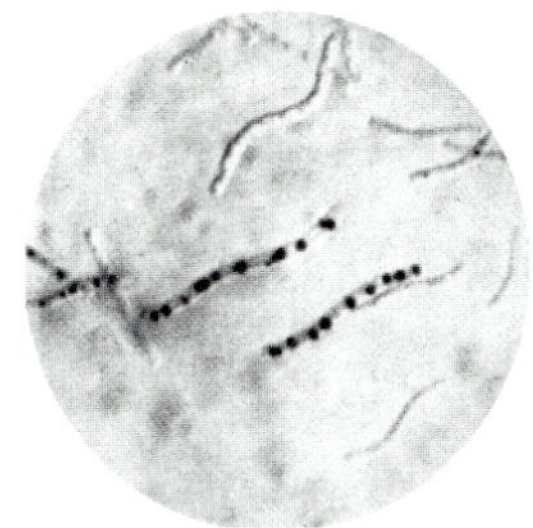

FIGURE 6.53
Polyphosphates in the cells of a *Spirillum*, demonstrated by staining with methylene blue (×850). From George Giesberger, *Beitrage zur Kenntnis der Gattung Spirillum Ehbg* (1936), p. 46. Dissertation, Univ. of Utrecht, Utrecht, Holland.

have recently been isolated and characterized as a copolymer of arginine and aspartic acid. This material, which can represent as much as 8 percent of the cellular dry weight, is rapidly degraded when growth is reinitiated. The formation of cyanophycin does not occur through the normal mechanism of protein synthesis, since it accumulates in cells when protein synthesis has been arrested by treatment with chloramphenicol.

### Polyphosphate Granules

Many microorganisms, both procaryotic and eucaryotic, may accumulate *polyphosphate granules*, which are stainable with basic dyes such as methylene blue (Figure 6.53). These bodies are also sometimes termed *volutin* or *metachromatic granules*, because they exhibit a *metachromatic effect*, appearing red when stained with a blue dye. In electron micrographs of bacteria they appear as extremely electron-dense bodies. The polyphosphates are linear polymers of orthophosphate, of varying chain lengths.

The conditions for polyphosphate accumulation in bacteria have been studied in some detail. In general, starvation of the cells for almost any nutrient other than phosphate leads to polyphosphate formation. Sulfate starvation is particularly effective and leads to a rapid and massive accumulation of polyphosphate. When cells that have built up a polyphosphate store are again furnished with sulfate, the polyphosphate rapidly disappears, and tracer experiments with $^{32}P$ show that the phosphate is incorporated into nucleic acids. The polyphosphate granules therefore appear to function primarily as an intracellular phosphate reserve, formed under a variety of conditions when nucleic acid synthesis is impeded. The formation of polyphosphate occurs by the sequential addition of phosphate residues to pyrophosphate, ATP serving as the donor:

$$\text{Ⓟ—Ⓟ} + \text{ATP} \longrightarrow \text{Ⓟ—Ⓟ—Ⓟ} + \text{ADP}$$
$$(\text{—Ⓟ—})_n + \text{ATP} \longrightarrow (\text{—Ⓟ—})_{n+1} + \text{ADP}$$

The degradation of polyphosphate, if it occurs by the reversal of this reaction, might also provide a source of ATP for the cell, although this function has not so far been firmly established.

### Sulfur Inclusions

Inclusions of inorganic sulfur may occur in two physiological groups: the purple sulfur bacteria, which use $H_2S$ as a photosynthetic electron donor, and the filamentous, nonphotosynthetic organisms, such as *Beggiatoa* and *Thiothrix*, which use $H_2S$ as an oxidizable energy source. In both these groups, the accumulation of sulfur is transitory and takes place when the medium contains sulfide; after the sulfide in the medium has been completely utilized, the stored sulfur is further oxidized to sulfate.

## THE NUCLEUS

### Recognition and Cytological Demonstration of Bacterial Nuclei

Basic dyes, which selectively stain the chromatin of the eucaryotic nucleus, stain most bacterial cells densely and evenly. The basophilic property of the bacterial cell is caused by the abundance of ribo-

tions of an extremely long circular molecule of DNA, highly folded to form a compact mass.

In 1963 J. Cairns succeeded in extracting DNA from *E. coli* under conditions that minimize its shearing. Cells were grown with $^{3}$H-labeled thymine, so that only their DNA would be radioactive, and were placed in a chamber sealed at one end with glass and at the other end with a membrane filter. The cells were then gently lysed and proteins were dissociated from the DNA; all the treatments were done by allowing the reagents to *diffuse* into and out of the chamber through the membrane filter, so that the material was never subjected to mechanical agitation. Finally, the membrane was punctured, the chamber was drained, and the membrane was mounted on a slide for radioautography.

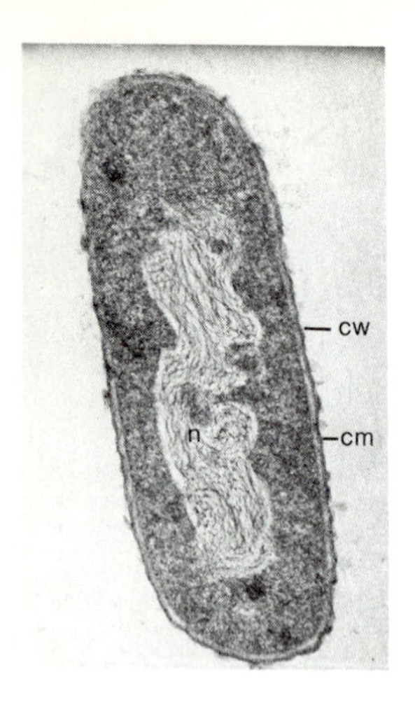


FIGURE 6.54

Thin section of a unicellular procaryotic organism, *Bacillus subtilis* (×20,000): n, nucleus; cm, cytoplasmic membrane; cw, cell wall. Courtesy of C. F. Robinow.

somes, which confers an unusually high nucleic acid content on the cytoplasmic region. Hence, in order to stain selectively the bacterial nucleus, the fixed cells must first be treated with ribonuclease or with dilute HCl, which hydrolyzes the ribosomal RNA. Subsequent staining with a basic dye then reveals the bacterial nuclei as dense, centrally located bodies of irregular outline; two to four are present in an exponentially growing cell. The growth and division of bacterial nuclei in living cells can be observed by phase-contrast microscopy, provided that the cells are suspended in a medium (e.g., a concentrated protein solution) that enhances the slight difference in contrast between nucleoplasm and cytoplasm.

The study of bacterial nuclear structure by electron microscopy was initially impeded by the difficulty of obtaining good fixation of the nuclear material. Once this problem had been solved, the nucleus was revealed in electron micrographs (Figure 6.54) as a region closely packed with fine fibrils of DNA. This region is not separated from the cytoplasm by a membrane, and contains no evident structures, apart from the DNA fibrils.

## The Bacterial Chromosome

By 1960 the cytological information about the structure of the bacterial nucleus had been complemented by genetic studies of *E. coli*, which suggested the presence of a single, circular linkage group. This in turn implied that each nucleus should contain a single, circular chromosome. If such were indeed the case, the fibrils of DNA revealed in the nuclear region by electron microscopy should represent sec-

FIGURE 6.55

Autoradiograph of the chromosome of *E. coli* strain K12, labeled with tritiated thymidine for two generations and extracted as described in the text. The scale at the bottom represents 100 μm. Inset: A diagram of the same structure, showing regions (a, b, c) in which both strands contain tritium (double solid lines) and in which only one strand is labeled (one solid and one dashed line); x and y indicate replication forks. From J. Cairns, "The Chromosome of *Escherichia coli*," *Cold Spring Harbor Symp. Quant. Biol.* **28**, 43 (1964).

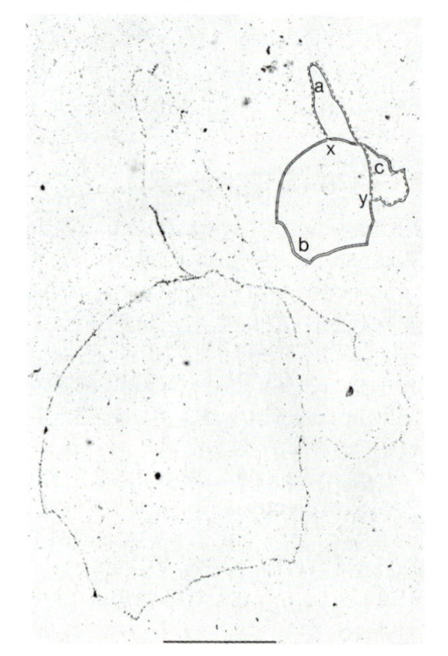

During the final draining of the chamber, the DNA of individual bacterial cells was spread out on the surface of the membrane filter. Examination of the developed radioautographs showed that the DNA was present as extremely long threads, the longest of which were slightly more than 1 mm in length. Furthermore, a few of the threads were *circular* (Figure 6.55). These threads are contained within cells which have an average length of approximately 2 $\mu$m.

The length of 1 mm for the DNA thread agrees well with the amount of DNA per nucleus as determined chemically, assuming that the radioactive structure in Cairn's pictures is an extended double helix. This amount of DNA represents approximately $4 \times 10^6$ base pairs, with a molecular weight of about $2.7 \times 10^9$.

DNA is a highly charged molecule, since adjacent bases are linked by phosphate groups, each with an ionized hydroxyl group. The resulting negative charges must therefore be balanced by an equivalent number of cationic groups. In eucaryotes this occurs through association of the chromosomal DNA with basic proteins (histones). Charge neutralization in bacteria is effected by polyamines, such as spermine and spermidine, and by $Mg^{2+}$, as well as by basic proteins.

Cairn's autoradiographic demonstration of the circular structure of the bacterial chromosome also revealed its mode of replication (see Figure 6.55). Replication begins at one point on the circumference of the chromosome, and the replication forks then move in opposite directions around the chromosome until they meet 180° away from origin of replication at a point termed the terminus.

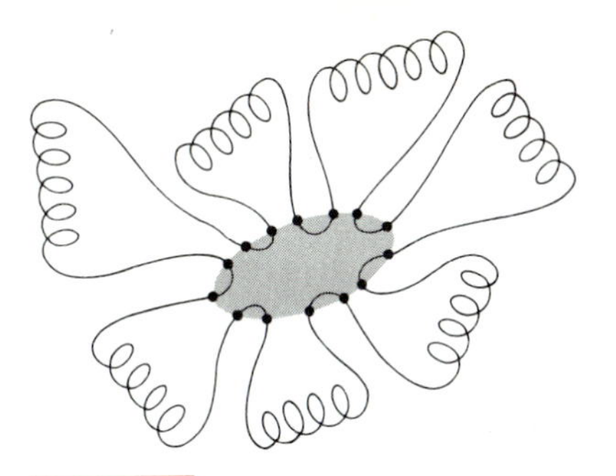

FIGURE 6.56
A schematic drawing illustrating the proposed structure of the folded chromosome of *E. coli*. The chromosome is shown as seven loops, each twisted into a superhelix (the actual number is much greater) and held together by a core of RNA (shaded area). After A. Worcel and E. Burgi, "On the Structure of the Folded Chromosome of *Escherichia coli*," *J. Mol. Biol.* **71**, 127 (1972).

## The Isolation of Bacterial Nuclei

If the bacterial chromosome consisted merely of a huge circular DNA molecule, folded in a random manner, its orderly replication and segregation could not possibly occur; some organization at a higher level must therefore be imposed on the chromosome. The recent isolation of structures that appear to be intact bacterial nuclei has provided a few clues to the nature of this organization. The structures in question have been obtained by gentle lysis of lysozyme-treated bacteria with nonionic detergents in 1.0 M NaCl. In addition to DNA, they contain a substantial amount of RNA and protein. They are rapidly sedimentable and are of low viscosity (in contrast to unfolded DNA of high molecular weight, which is extremely viscous).

Depending on the conditions of lysis, two structures of this type are obtainable. Lysis at 25° C releases structures with sedimentation coefficients of 1,300 to 2,200 S, which appear to be folded chromosomes, associated with a considerable amount of RNA and protein. The protein is largely RNA polymerase; and the RNA is newly transcribed single RNA strands. RNAse treatment causes a rapid increase in viscosity, which indicates that some of the associated RNA is responsible for holding the DNA into a compact form. The DNA in these structures is folded into a number (between 12 and 80) of supercoiled loops. In the light of these facts, the folded chromosome can be represented schematically as shown in Figure 6.56.

If lysis is conducted at a lower temperature (0 to 4° C), structures having considerably higher rates of sedimentation (3,000 to 4,000 S) are obtained. Electron microscopy (Figure 6.57) shows that they consist of folded chromosomes, attached to either one or two membrane fragments, from which they can be dissociated by gentle treatment. These observations show that binding of the chromosome to the membrane involves very weak forces. No folded chromosomes attached to membrane can be isolated from cells that have completed a round of DNA synthesis. Accordingly, it is possible that the resting chromosome may not be membrane-bound *in vivo*, membrane attachment taking place as a preliminary to the next round of replication.

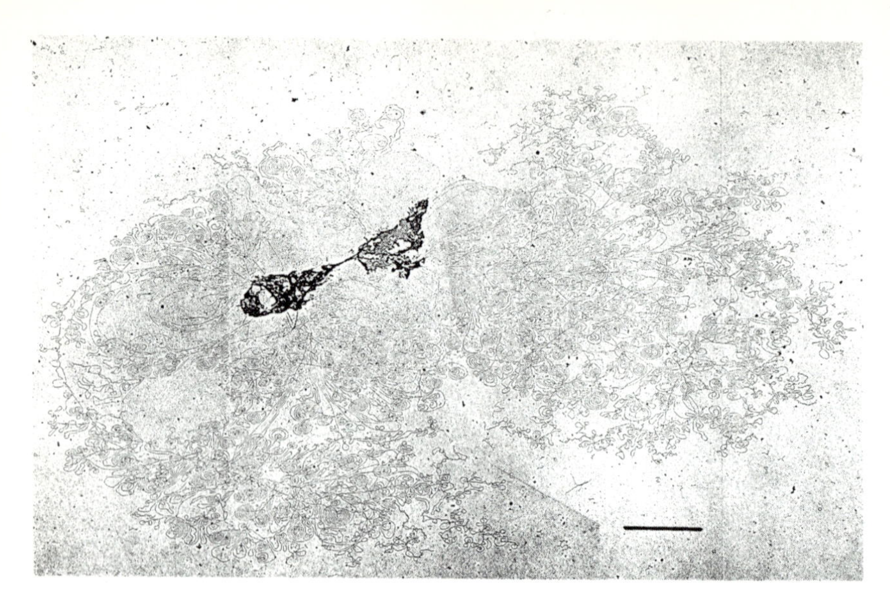

FIGURE 6.57

Electron micrograph of the isolated folded chromosome of *Escherichia coli*, attached to a fragment of the cell membrane (dark, irregular area in center of figure). The bar represents 2 μm. From H. Delius and A. Worcel, "Electron Microscopic Visualization of the Folded Chromosome of *Escherichia coli*," *J. Mol. Biol.* **82,** 107 (1974).

## FURTHER READING


### Book

LEIVE, L. ed., *Bacterial Membranes and Walls*. New York: Marcel Dekker, 1973.

### Reviews and original articles

BAYER, M. E., "Role of Adhesion Zones in Bacterial Cell-Surface Function and Biogenesis," in *Membrane Biogenesis*, ed. A. Tzagloff. New York: Plenum Press, 1975.

KANDLER, O., "Cell wall Structures and Their Phylogenetic Implications." *Zbl. Bakt. Hyg.*, **I. (Abt. Orig. C 3),** 149 (1981).

NIKAIDO H., and M. VARRA, "Molecular Basis of the Permeability of Bacterial Outer Membrane," *Microbiol. Rev.* **49,** 1 (1985).

OSBORN, M. J., and H. C. P. WU, "Proteins of the Outer Membrane of Gram-Negative Bacteria. *Ann. Rev. Microbiol.* **34,** 369 (1980).

PETTIJOHN, D. E., "Prokaryotic DNA in Nucleoid Structure," *CRC Cit. Rev. Biochem.* **4,** 175 (1976).

SHIVELY, J. M., "Inclusion Bodies of Procaryotes," *Ann. Rev. Microbiol.* **28,** 167 (1974).

SILVERMAN, M., and M. SIMON, "Bacterial Flagella," *Ann. Rev. Biochem.* **31,** 397 (1977).

SINDEN, R. R., and D. E. Pettijohn, "Chromosomes in Living *Escherichia coli* Cells Are Segregated into Domains of Supercoiling," *Proc. Natl. Acad. Sci. USA* **78,** 228 (1981).

TROY, F. A. II, "The Chemistry and Biosynthesis of Selected Bacterial Capsular Polymers," *Ann. Rev. Microbiol.* **33,** 519 (1979).

Chapter 7

# Microbial Growth

This chapter will describe the methods used for the measurement of microbial growth, together with various phases and modes of growth. The discussion will focus on the growth of unicellular bacteria, since they are ideal objects for study of the growth process and have largely served for the elucidation of its nature.

## THE DEFINITION OF GROWTH

In any biological system, growth can be defined as the orderly increase of all chemical components. Increase of mass might not really reflect growth because the cells could be simply increasing their content of storage products such as glycogen or poly-$\beta$-hydroxybutyrate. In an adequate medium to which they have become fully adapted, however, bacteria are in a state of *balanced growth*. During a period of balanced growth, an increase of the biomass is accompanied by a comparable increase of all other measurable properties of the population, e.g., protein, RNA, DNA, and intracellular water. In other words, cultures undergoing balanced growth maintain a constant chemical composition. The phenomenon of balanced growth simplifies the task of measuring the rate of growth of a bacterial culture; since the rate of increase of *all* components of the population is the same, measurements of *any* component suffice to determine the growth rate.

## THE MATHEMATICAL NATURE AND EXPRESSION OF GROWTH

A bacterial culture undergoing balanced growth mimics a first-order autocatalytic chemical reaction; i.e., the rate of increase in bacteria at any particular time is proportional to the number or mass of bacteria present at that time.

$$\text{rate of increase of cells} = k \text{ (number or mass of cells)} \tag{7.1}$$

The constant of proportionality, $k$, is an index of the rate of growth and is called the *growth rate constant*. Since we assume growth to be balanced, $k$ also relates the rate of increase of any given cellular component to the amount of that cellular component, or in mathematical terms,

$$\frac{dN}{dt} = kN, \qquad \frac{dX}{dt} = kX, \qquad \frac{dZ}{dt} = kZ \tag{7.2}$$

where $N$ is the number of cells/ml, $X$ is the mass of cells/ml, $Z$ is the amount of any cellular component/ml, $t$ is time, and $k$ is the growth rate constant. These equations, in fact, accurately describe the growth of most unicellular bacterial cultures. Other (nondifferential) forms of these equations are more useful in practice. Upon integration Eq. (7.2), yields

$$\ln Z - \ln Z_0 = k(t - t_0) \tag{7.3}$$

and on converting natural logarithms to logarithms to the base 10,

$$\log_{10} Z - \log_{10} Z_0 = \frac{k}{2.303}(t - t_0) \tag{7.4}$$

where the values of $Z$ and $Z_0$ correspond to the amount of any bacterial component of the culture at times $t$ and $t_0$, respectively. By measuring $Z$ and $Z_0$, one can compute the value of $k$, the growth rate constant of the culture. Thus, if the culture contains $10^4$ cells/ml at $t_0$ and $10^8$ cells/ml 4 hours later, the specific growth rate of the culture is

$$k = \frac{(8 - 4)2.303}{4} = 2.303 \text{ hours}^{-1} \tag{7.5}$$

The value of $k$ suffices to define the rate of growth of a culture. However, certain other parameters are also commonly used. One is the mean doubling time or generation time ($g$) defined as the time required for all components of the culture to increase by a factor of 2. The relationship between $g$ and $k$ can be derived from Eq. (7.3), since if the time interval considered $(t - t_0)$ is equal to $g$, then $Z$ will be twice $Z_0$. Making these substitutions, one obtains

$$k = \frac{\ln 2}{g} = \frac{0.693}{g} \tag{7.6}$$

In the case of the example we have chosen, the mean doubling time, $g$, of the culture is $g = 0.693/2.303 = 0.42$ hour or 25 minutes. This is a relatively high growth rate for a bacterium, as shown by the representative examples assembled in Table 7.1.

The above mathematical expressions for bacterial growth rate have been developed from the premise that the rate of increase is proportional to the number (or mass) present at any given time. From this premise, it was shown (Eq. 7.6) that doubling time ($g$) is constant during a period of balanced growth. The same equations can be derived from the premise that mean doubling time is constant and lead to the conclusion that the rate of increase of number (or mass) is proportional to the number (or mass) at any given time.

**TABLE 7.1**

**Maximal Recorded Growth Rates for Certain Bacteria, Measured at or Near Their Temperature Optimum, in Complex Media Unless Otherwise Noted**

| Organism | Temperature (°C) | Doubling Time (hours) |
|---|---|---|
| *Vibrio natriegens* | 37 | 0.16 |
| *Bacillus stearothermophilus* | 60 | 0.14 |
| *Escherichia coli* | 40 | 0.35 |
| *Bacillus subtilis* | 40 | 0.43 |
| *Pseudomonas putida* | 30 | 0.75[a] |
| *Vibrio marinus* | 15 | 1.35 |
| *Rhodobacter sphaeroides* | 30 | 2.2 |
| *Mycobacterium tuberculosis* | 37 | ~6 |
| *Nitrobacter agilis* | 27 | ~20[a] |

[a] Grown in synthetic medium.

### The Growth Curve

Equation (7.4) predicts a straight-line relationship between the logarithm of cell number (or any other measurable property of the population) and time [Figure 7.1(a)] with a slope equal to $k/2.303$ and

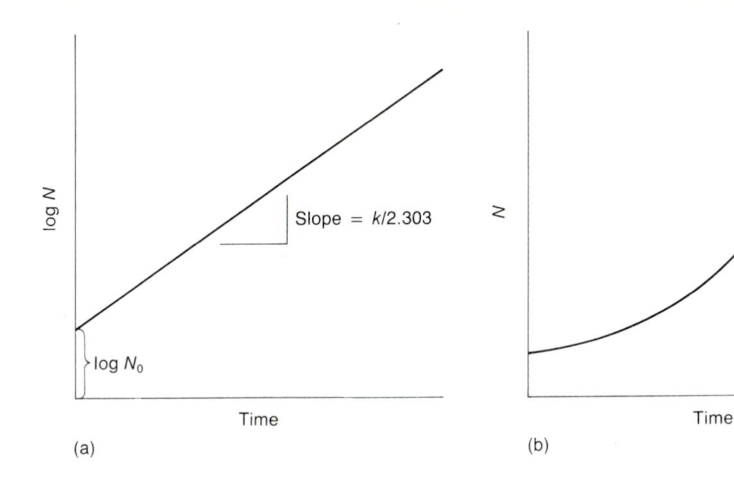


FIGURE 7.1
Comparison of methods of plotting growth data. Plotting the logarithm of cell density (number of cells/ml, $N$) of a culture undergoing balanced growth as a function of time yields a straight line (a); the slope of the line is the growth rate constant ($k$) divided by 2.303, and the intercept is $\log N_0$. Plotting the cell density directly as a function of time yields an exponential curve (b).

an ordinate intercept of $\log N_0$. By taking the antilogarithm, Eq. 7.4 can be written in the exponential form:

$$Z = Z_0 10^{k(t-t_0)/2.303} \quad (7.7)$$

which predicts an exponential relationship between the number of cells in the population (or any other measurable property) and time [Figure 7.1(b)]. Populations of bacteria growing in a manner that obeys these equations are said to be in the *exponential* or *logarithmic phase* of growth.

Microbial populations seldom maintain exponential growth at high rates for long. The reason is obvious if one considers the consequences of exponential growth. After 48 hours of exponential growth, a single bacterium weighing about $10^{-12}$ g with a doubling time of 20 minutes would produce a progeny of $2.2 \times 10^{31}$ g, or roughly 4,000 times the weight of the earth.

The growth of bacterial populations is normally limited either by the exhaustion of available nutrients or by the accumulation of toxic products of metabolism. As a consequence, the rate of growth declines and growth eventually stops. At this point a culture is said to be in the *stationary phase* (Figure 7.2). The transition between the exponential phase and the stationary phase involves a period of *unbalanced growth* during which the various cellular components are synthesized at unequal rates. Consequently, cells in the stationary phase have a chemical composition that is different from that of cells in the exponential phase. The cellular composition of cells in the stationary phase depends on the specific growth-limiting factor. Despite this, certain generalizations hold: cells in the stationary phase are small relative to cells in the exponential phase (since cell division continues after increase in mass has stopped), and they are more resistant to adverse physical (heat, cold, radiation) and chemical agents.

FIGURE 7.2
Generalized growth curve of a bacterial culture.

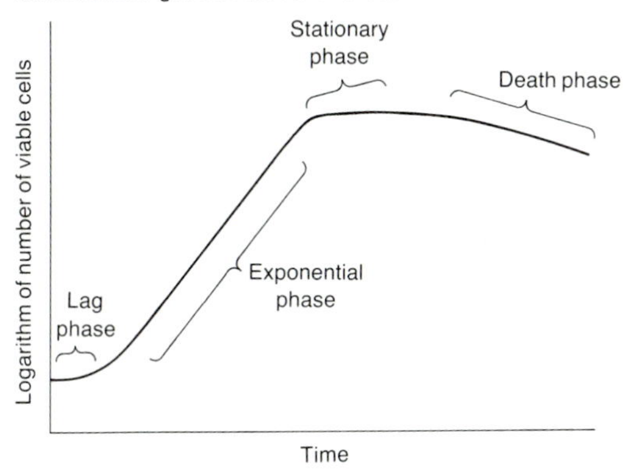


## The Death Phase

Bacterial cells held in a nongrowing state eventually die. Death results from a number of factors; an important one is depletion of the cellular reserves of energy. Like growth, death is an exponential function and hence in a logarithmic plot (Figure 7.2) the death phase is a linear decrease in number of *viable* cells with time. The death rate of bacteria is highly variable, being dependent on the environment as well as on the particular organism (e.g., enteric bacteria die very slowly, while vegetative cells of certain *Bacillus* spp. die rapidly).

## The Lag Phase

Cells transferred from a culture in the stationary phase to a fresh medium of the same composition undergo a change of chemical composition before they are capable of initiating growth. This period

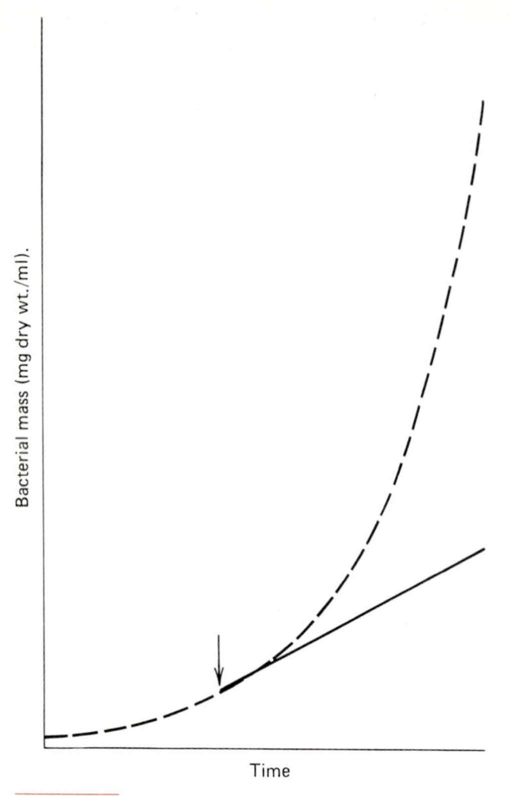


FIGURE 7.3

Linear growth of *E. coli* induced by the amino acid analogue *p*-fluorophenylalanine. The exponential growth of the culture (dotted line) becomes linear (solid line) after the time (indicated by vertical arrow) that the analogue was added.

of adjustment, called the *lag phase* (Figure 7.2), is extremely variable in duration; in general, its length is directly related to the duration of the preceding stationary phase.

### Arithmetic Growth

In certain abnormal situations, the kinetics of bacterial growth may become arithmetic rather than exponential. Under these circumstances, the rate of increase is a constant, i.e.,

$$\frac{dN}{dt} = C \qquad \textbf{(7.8)}$$

where $C$ is a constant; and on integration:

$$N - N_0 = C(t - t_0) \qquad \textbf{(7.9)}$$

The number of cells ($N$), rather than the logarithm of the number of cells, is a linear function of time ($t$).

A number of conditions can lead to arithmetic growth. For example, if a bacterium that requires nicotinic acid as a growth factor is deprived of this nutrient, it is unable to synthesize pyridine nucleotides, of which nicotinic acid is a specific biosynthetic precursor. As a result of the functions of pyridine nucleotides in electron transport, their levels in the cell determine the overall rate of metabolism and hence of growth. When no further synthesis of these compounds can occur, the growth rate of the population becomes directly proportional to the supply of pyridine nucleotides in the cells; the total catalytic power can no longer increase. Hence, the rate of increase of cells remains constant. The addition of *p*-fluorophenylalanine, an analogue of the natural amino acid, phenylalanine, to a culture also results in arithmetic growth (Figure 7.3). The analogue is sufficiently similar to the natural amino acid that it can become incorporated into newly synthesized proteins in place of phenylalanine. However, proteins containing *p*-fluorophenylalanine are largely nonfunctional, and hence they do not increase the catalytic capacity of the cell. As a consequence, the growth rate cannot increase beyond that determined by the cells' catalytic capacity at the time of addition of the analogue to the culture.

Certain mutations that preclude further synthesis of an essential protein in a particular environment lead to arithmetic growth when the culture is exposed to that environment.

## THE MEASUREMENT OF GROWTH

To follow the course of growth, it is necessary to make quantitative measurements. As discussed earlier, exponential growth is usually balanced so any property of the biomass can be measured to determine growth rate. As a matter of convenience, the properties measured are usually cell mass or cell number.

### Measurement of Cell Mass

The only direct way to measure cell mass is to determine the dry weight of cell material in a fixed volume of culture by removing the cells from the medium, drying them, and then weighing them. Such determinations are time consuming and relatively insensitive. With ordinary equipment it is difficult to weigh with accuracy less than 1 mg, yet this dry weight may represent as many as 5 billion bacteria.

The method of choice for measuring the cell mass of unicellular microorganisms is an optical one: the determination of the amount of light

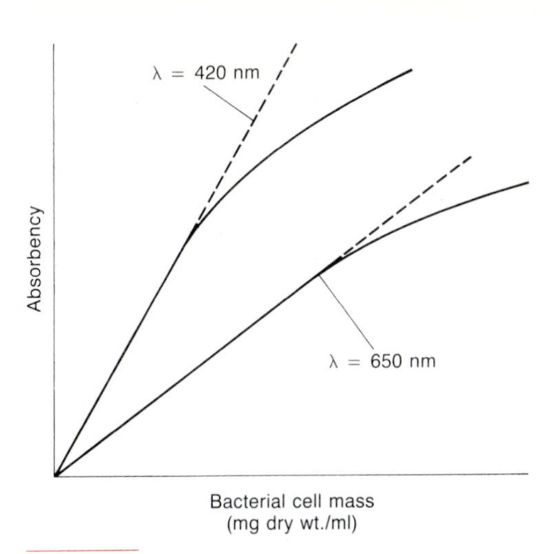


FIGURE 7.4

The relationship between absorbency of a suspension of bacteria and bacterial cell mass. Note that proportionality is strict only at low values of absorbency and deviates from strict proportionality (dashed line) at higher absorbency values; also note that the measurements are more sensitive with light of shorter wavelength ($\lambda$).

scattered by a suspension of cells. This technique is based on the fact that small particles scatter light proportionally, within certain limits, to their concentration. When a beam of light is passed through a suspension of bacteria, the reduction in the amount of light transmitted as a consequence of scattering is thus a measure of the bacterial mass present. Such measurements are usually made in a *photometer* or *spectrophotometer*. A spectrophotometer is provided with a prism or a diffraction grating thus allowing illumination of the sample with light of narrow but adjustable range of wavelengths; a photometer is provided with interchangeable filters that pass a relatively wide range of wavelengths. These instruments read in *absorbency* ($A$) units; absorbency is defined as the logarithm of the ratio of intensity of light striking the suspension ($I_0$) to that transmitted by the suspension ($I$):

$$A = \log \frac{I_0}{I} \quad (7.10)$$

These instruments are convenient for estimating cell concentration, and when calibrated against bacterial suspensions of known concentration (Figure 7.4), they provide an accurate and rapid way to estimate the dry weight of bacteria per unit volume of culture. It should be emphasized that such measurements are meaningful only when used in conjunction with a standard curve such as that shown in Figure 7.4. Since scattering is inversely proportional to the fourth power of the wavelength of light being scattered, the sensitivity of the measurements increases sharply if light of shorter wavelength is used; in general, however, the lower limit of sensitivity of the method is reached with bacterial suspensions that contain about 10 million per milliliter. More sensitive instruments for measuring scattering are called *nephelometers*. These have the light-sensing device arranged at right angles to the incident beam of light and hence directly measure the scattered light (Figure 7.5).

## Measurement of Cell Number

The number of unicellular organisms in a suspension can be determined microscopically by counting the individual cells in an accurately determined very

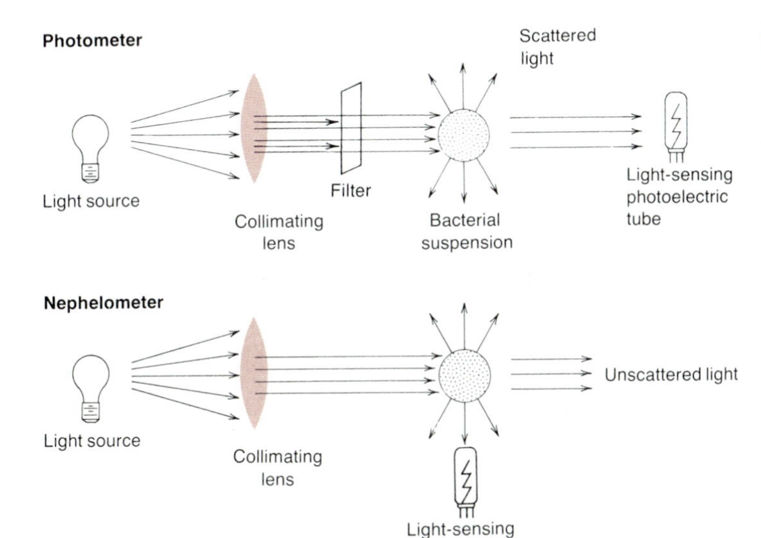


FIGURE 7.5

Comparison of arrangement of the optical components of a photometer and nephelometer. The greater sensitivity of the nephelometer depends on its measuring scattered light rather than residual unscattered light.

small volume. Such counting is usually done with special microscope slides known as *counting chambers.* These are ruled with squares of known area and are so constructed that a film of liquid of known depth can be introduced between the slide and the cover slip. Consequently, the volume of liquid overlying each square is accurately known. Such a direct count is known as the *total cell count.* It includes both viable and nonviable cells, since, at least in the case of bacteria, these cannot be distinguished by microscopic examination.

The principal limitation of the direct microscopic enumeration of bacterial populations is the relatively high concentrations of cells that must be present in the suspension. The high magnification required for seeing the bacteria limits the volume of liquid that can be examined carefully with the microscope; yet, a sufficient number of cells must be found in a known volume to make the count statistically significant. As a consequence, only suspensions that contain 10 million or more cells per milliliter can be counted with any degree of accuracy by this technique.

An electronic instrument, called the *Coulter counter* after its inventor, can also be used for the direct enumeration of cells in a suspension. A portion of the suspension is passed through a very fine orifice into a small glass tube. The orifice also serves to complete an electrical circuit through the suspending medium between electrodes on the interior and exterior of the tube. Detection depends on the difference in conductivity between the bacterium and the suspending liquid. Each time a bacterium passes through the orifice the conductivity drops; this event is detected and recorded electronically. The instrument can score the magnitude and duration of the changes in conductivity and thus register and record both the number and the distribution of size of a cellular population. The orifices commonly used to count bacteria are 30 micrometers in diameter. The suspending liquid must therefore be scrupulously free of inanimate particles (e.g., dust) since smaller ones will be counted as cells and larger ones will plug the orifice.

The enumeration of unicellular organisms can also be made by *plate count,* because single viable cells separated from one another in space by dispersion on or in an agar medium give rise through growth to separate, macroscopically visible colonies. Hence, by preparing appropriate dilutions of a bacterial population and using them to seed an appropriate medium, one can ascertain the number of viable cells in the initial population by counting the number of colonies that develop after incubation of the plates, and multiplying this figure by the dilution factor. This method of enumeration is often termed a *viable count:* in contrast to direct microscopic enumeration and electronic counting, it measures only those cells that are capable of growth on the plating medium used. The viable count is by far the most sensitive method of estimating bacterial number, since even a single viable cell in a suspension can be detected. Its accuracy depends on observing certain precautions. Significant numbers of colonies must be counted (the standard error is approximately equal to the square root of the number of colonies counted), preferably those on two or three plates with several hundred colonies per plate. Also, if one is determining the number of cells in a growing culture, it is important to realize that increase in cell numbers continues during the process of dilution, even if a medium inadequate for supporting continued growth is employed for the dilutions. Cell growth cannot be stopped without affecting viability. Rapid chilling prior to or during dilution can cause death of a significant portion of the population in certain cases, a phenomenon known as *cold shock.*

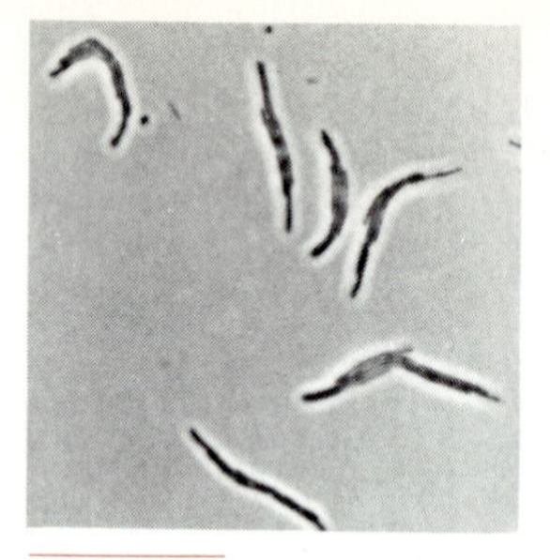

FIGURE 7.6

Determination of viability by a combination of growth and microscopic examination. A sample of a culture was spread on a thin layer of agar medium and was photographed after incubation for 3.5 hours. Viable cells have formed microcolonies. From J. R. Postgate, J. E. Crumpton, and J. R. Hunter, "The Measurement of Bacterial Viabilities by Slide Culture," *J. Gen. Microbiol.* **24,** 15 (1961).

A combination of *total cell count* and *viable count* can be used to determine the fraction of viable cells in the population. Alternatively, the viable fraction can be determined directly, by a technique that combines microscopic examination and growth. A sample of the population, appropriately diluted, is spread over a thin layer of agar medium on the surface of a sterile glass slide. The inoculated slide is incubated for a period sufficient to permit the occurrence of several cell divisions and is then examined microscopically with phase-contrast illumination. Under these conditions, the viable cells can be easily identified as a result of the fact that they have developed into microcolonies, and their number relative to nonviable cells, which remain single, can be precisely determined (Figure 7.6).

### Measurement of a Cell Constituent

Sometimes because of growth patterns (cells may grow as filaments or form clumps) or of complexity of the medium, it is impractical to measure cell mass or numbers. In these cases growth can be measured by determining the amount of a particular cell constituent (e.g., protein, peptidoglycan, DNA, RNA, or ATP) in the medium. Such measurements are often the most practical way to determine microbial mass and growth in a natural environment.

## THE EFFICIENCY OF GROWTH: GROWTH YIELDS

The net amount of growth of a bacterial culture is the difference between the cell mass (or number of cells) used as an inoculum and the cell mass (or number of cells) present in the culture when it enters the stationary phase. When growth is limited by a particular nutrient, there is a fixed linear relationship between the concentration of that limiting nutrient initially present in the medium and the net growth which results, as shown in Figure 7.7. The mass of cells produced per unit of limiting nutrient is, accordingly, a constant, the *growth yield* ($Y$). The value of $Y$ can be calculated from single measurements of total growth by the equation

$$Y = \frac{X - X_0}{C} \qquad (7.11)$$

where $X$ is the dry weight/ml of cells present when the culture enters the stationary phase, $X_0$ is the dry weight/ml of cells immediately after inoculation, and $C$ is the concentration of limiting nutrient.

The growth yield can be measured for any required nutrient and once determined can then be used to calculate the concentration of that nutrient in an unknown mixture simply by measuring how much growth a sample of the unknown mixture supports when added to a medium complete in all respects except for the limiting nutrient. Such a determination is called a *bioassay*. In the past, bioassays were extensively used for the determination of concentration of amino acids and vitamins in foodstuffs. Now, chemical and physical methods have come into more common use, but the principle of bioassay remains an important research tool for detecting and quantitating compounds with growth-promoting activities. To perform a bioassay one requires only a microbial strain for which the substance to be assayed is an essential nutrient.

In the case of a chemoheterotrophic bacterium, the growth yield measured in terms of organic substrate utilized becomes an index of the efficiency of conversion of substrate into bacterial mass. The data shown in Figure 7.7 were obtained with an obligately aerobic chemoheterotrophic pseudomonad growing in a synthetic medium with fructose as the sole source of carbon and energy. Inspection of the graph reveals a growth yield of approximately 0.4. Considering that the carbon content of fructose and cell material is 40 and 50 percent, respectively, the fraction of fructose carbon converted to cell carbon can be calculated to be about 0.5. Accordingly, this microorganism uses about half the carbon of fructose to make cells and oxidizes the other half to $CO_2$. Analogous experiments with other aerobic chemoheterotrophs utilizing sugars as the sole source of carbon reveal that the efficiency of conversion of carbon from the sugars to cellular carbon varies between about 20 and 50 percent.

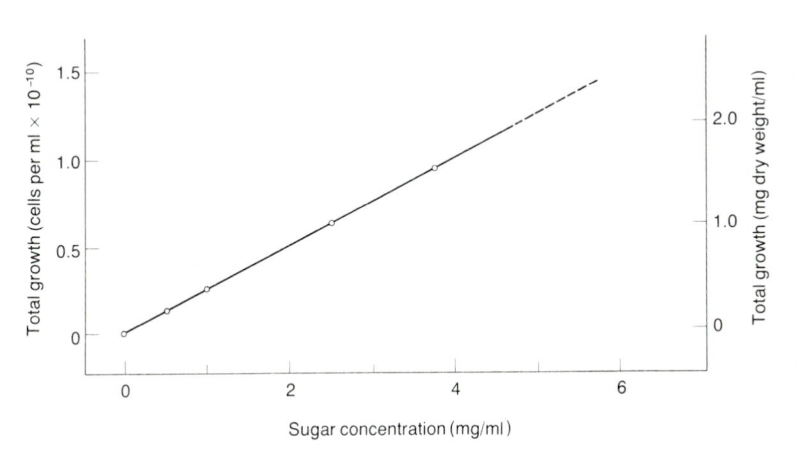


FIGURE 7.7
The relationship between total growth of an aerobic bacterium (*Pseudomonas* sp.) and the initial concentration of the limiting nutrient (fructose). The experiments were done in a synthetic medium with fructose as the sole source of carbon and energy. The slope of the line is the growth yield ($Y$) of the bacterium on fructose (see text).

**TABLE 7.2**

**Growth Yields of Fermentative Microorganisms, Measured in Terms of Glucose Fermented or ATP Produced**

| Organism | Fermentation and Pathway | Moles ATP Formed per Mole of Glucose Fermented | Molar Growth Yield Expressed as Grams of Cells Produced per Mole of | |
|---|---|---|---|---|
| | | | GLUCOSE FERMENTED | ATP PRODUCED |
| *Saccharomyces cerevisiae* (yeast) | Alcoholic, Embden-Meyerhof | 2 | 21 | 10.5 |
| *Streptococcus faecalis* | Homolactic, Embden-Meyerhof | 2 | 22 | 11.0 |
| *Lactobacillus delbruckii* | Homolactic, Embden-Meyerhof | 2 | 21 | 10.5 |
| *Zymomonas mobilis* | Alcoholic, Entner-Doudoroff | 1 | 8.6 | 8.6 |

These differences probably reflect differences in efficiency of generating ATP through catabolism of the substrate; evidence to support this inference is described below.

When certain obligately fermentative microorganisms are grown in rich media, radioactive tracer experiments show that little or no carbon from the fermentable substrate is converted into cellular material. Under these conditions, cellular material is derived from the other medium components (amino acids, purines, pyrimidines, etc.), and the fermentable substrate serves only as a source of energy. The ATP yield from many fermentations is known, and the yield of cells (grams dry weight) per mole of ATP produced ($Y_{ATP}$) approximates 10 g of cell material/mole of ATP, suggesting that the energy to polymerize carbon building blocks into macromolecules does not vary much among microorganisms. Such constancy is not seen in terms of yield of cells per *mole of substrate fermented* because different microorganisms may ferment a given substrate by pathways that differ in yields of ATP. For example, *Zymomonas mobilis* ferments glucose to produce ethyl alcohol via the Entner-Doudoroff pathway, with formation of 1 mole of ATP per mole of glucose; yeasts, in contrast, ferment glucose to ethyl alcohol via the Embden-Meyerhof pathway, yielding 2 moles of ATP per mole of glucose. Yeasts produce considerably more cells than *Zymomonas* per mole of glucose fermented but approximately the same amount per mole of ATP produced by the fermentation (Table 7.2). The approximate constancy of $Y_{ATP}$ can be used to deduce the ATP yield of an unknown dissimilatory pathway.

## SYNCHRONOUS GROWTH

So far, growth patterns of *populations* of bacteria have been described. Such studies permit no conclusions about the growth behavior of individual cells, because the distribution of cell size (and hence of cell age) in most bacterial cultures is completely random. Information about the growth behavior of individual bacteria can be obtained by the study of *synchronous cultures*, i.e., cultures composed of cells that are all at the same stage of the cell cycle. Measurements made on such cultures are equivalent to the measurements made on individual cells.

Synchronous cultures of bacteria can be obtained by a number of techniques. Synchrony can be *induced* by manipulations of environmental conditions, usually cyclic. In certain bacteria this is accomplished either by repetitive shifts of temperature or by furnishing fresh nutrients to cultures that have just entered the stationary phase. Alternatively, a synchronous population can be *selected* from a random population by physical separation of cells that are at the same stage of the cell cycle. This can be accomplished by differential filtration or by centrifugation. For physiological studies, techniques based on selection are preferable to those based on induction because the technique of induction may cause cyclic changes that are not typical of the normal cell cycle.

An excellent selective method for obtaining synchronous cultures is the Helmstetter-Cummings technique, which is based on the fact that certain bacteria stick tightly to cellulose nitrate (membrane) filters. The technique involves filtering an unsynchronized culture of bacteria through a (membrane) filter, then inverting the filter and allowing fresh medium to flow through it (Figure 7.8). After loosely associated bacteria have been washed from the filter, the only bacterial cells in the effluent stream of medium are those that arise through division. Hence, all cells in the effluent are newly formed and are therefore at the same stage of the cell cycle.

The growth of a culture of *E. coli* so synchronized is shown in Figure 7.9. The number of cells in the culture remains approximately constant for about one hour while the newly formed cells grow in size. Then, rather abruptly, the number of cells doubles. In the second division cycle, the plateau is less distinct and the population rise extends over a longer period, indicating that synchrony is already being lost. In the third division cycle, almost no indication of synchrony remains.

Synchronous cultures rapidly lose synchrony because various cells of a population do not all divide at the same size (age, or time following the previous division). From the rate of loss of synchrony one can calculate *the distribution of the age of cells at division.* Using electronic counting techniques, one can determine *the distribution of cell size in an exponentially growing culture.* These two distributions allow the calculation of growth rate as a function of cell size (Figure 7.10). The relationship is complex: very small cells grow slowly, cells of intermediate size grow more rapidly, and very large cells again grow slowly.

Similar computations permit an estimate of the distribution of cells of various ages (time elapsed since the division which produced them) in an exponential culture (Figure 7.11). Again, the relationship is somewhat complex, reflecting the relationship between growth rate and cell size. Figure 7.11 illustrates a general property of any expanding population: the numerical predominance of young individuals.

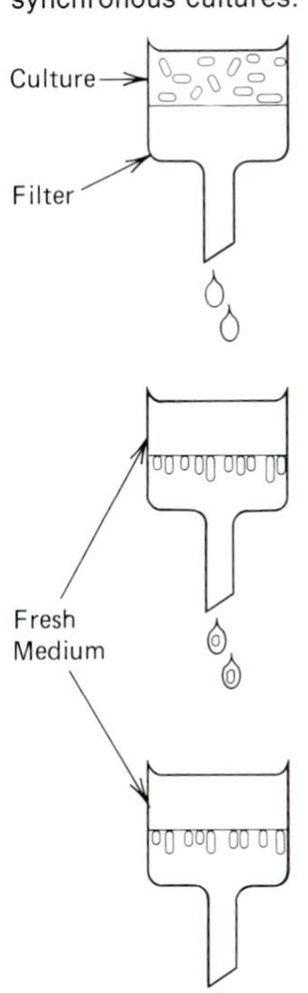


FIGURE 7.8
Helmstetter-Cummings technique of obtaining synchronous cultures.

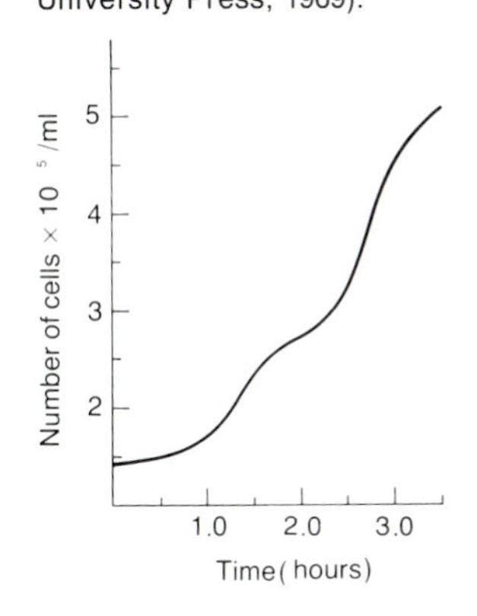


FIGURE 7.9
Synchronous growth of *E. coli* in glucose minimal medium. The effluent from a membrane culture (Helmstetter-Cummings technique) was collected for three minutes and incubated at 30° C. After A. G. Marr, P. R. Painter, and E. H. Nilson, "Growth and Division of Individual Bacteria," in *Microbial Growth*, 19th Symposium of the Society of General Microbiology (Cambridge: Cambridge University Press, 1969).

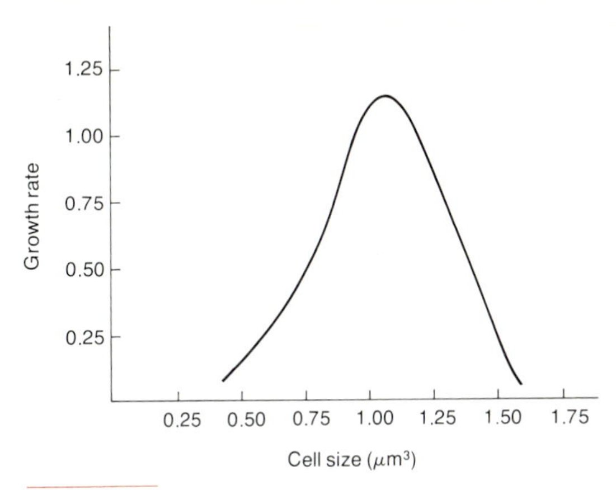


FIGURE 7.10
Growth rate ($\mu m^3$/hour) of individual cells of *E. coli* growing in a synthetic medium as a function of the size of the cell ($\mu m^3$). After A. G. Marr, P. R. Painter, and E. H. Nilson, "Growth and Division of Individual Bacteria," in *Microbial Growth*, 19th Symposium of the Society of General Microbiology (Cambridge: Cambridge University Press, 1969).

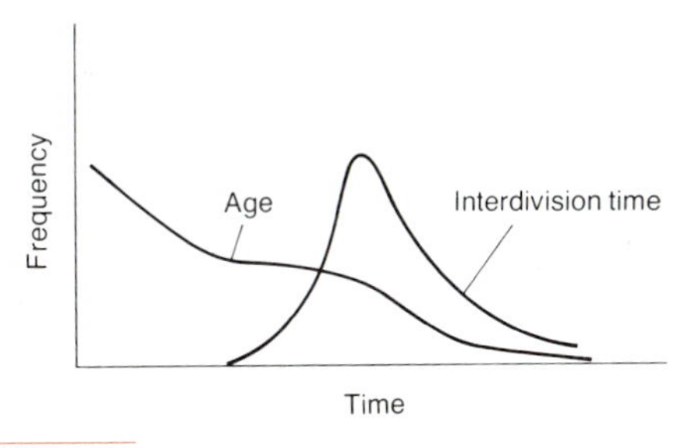


FIGURE 7.11
Comparison of the distribution of interdivision times and age of cells in an exponentially growing population of *E. coli*. After A. G. Marr, P. R. Painter and E. H. Nilson, "Growth and Division of Individual Bacteria," in *Microbial Growth*, 19th Symposium of the Society of General Microbiology (Cambridge: Cambridge University Press, 1969).

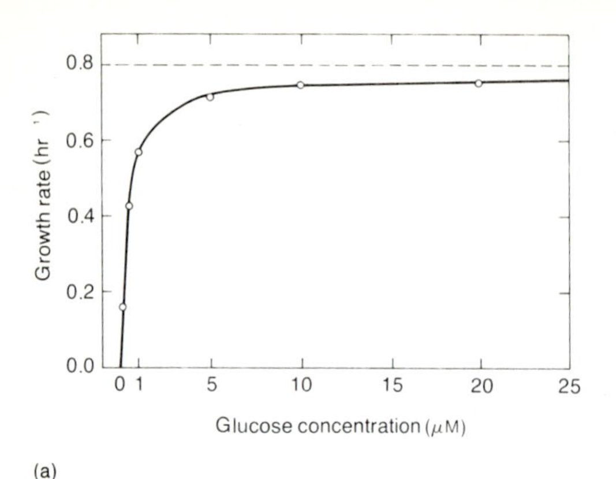


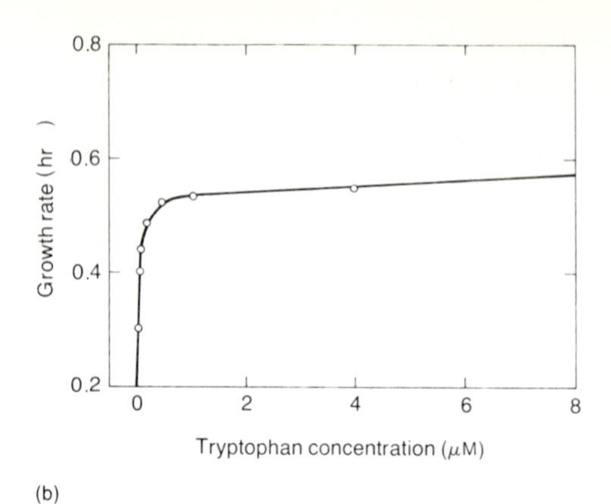


FIGURE 7.12

The effect of nutrient concentration on the specific growth rate of *E. coli:* (a) effect of glucose concentration; (b) effect of tryptophan concentration for a tryptophan-requiring mutant . From T. E. Shehata and A. G. Marr, "Effect of Nutrient Concentration on the Growth of *Escherichia coli*," *J. Bacteriol.* **107,** 210 (1971).

Perhaps the most clear illustration of the fact that the kinetics of growth of individual cells cannot be deduced from the kinetics of growth of the overall population is provided by the growth of *Caulobacter*. A population of this type of unicellular bacterium always consists of two structurally differentiated cell types, stalked cells and swarmer cells. The stalked cells always grow significantly faster than the swarmer cells (see page 415) even though the whole population grows exponentially at a constant rate.

## EFFECT OF NUTRIENT CONCENTRATION ON GROWTH RATE

In many respects, the bacterial growth process can be likened to a chemical reaction in which the components of the medium (the reactants) produce more cells (the product of the reaction), a process catalyzed by the bacterial population. The velocity of chemical reactions is determined by the concentration of reactants, but as we have seen, bacterial growth rate remains constant until the medium is almost exhausted of the limiting nutrient. This seeming paradox is explained by the action of permeases which are capable of maintaining saturating intracellular concentrations of nutrients over a wide range of external concentrations (Chapter 8). Nevertheless, at extremely low concentrations of external nutrients the permease systems are no longer able to maintain saturating intracellular concentrations, and the growth rate falls.

The curves relating growth rate to nutrient concentration are typically hyperbolic (Figure 7.12), and fit the equation

$$k = k_{\max} \frac{C}{K_s + C} \tag{7.12}$$

where $k$ is the specific growth rate at limiting nutrient concentration ($C$), $k_{\max}$ is the growth rate at saturating concentration of nutrient, and $K_s$ is a constant analogous to the Michaelis-Menten constant of enzyme kinetics, being numerically equal to the substrate concentration supporting a growth rate equal to $\frac{1}{2}k_{\max}$. Values of $K_s$ for glucose and tryptophan utilization by *E. coli* (Figure 7.12) are $1 \times 10^{-6}$ and $2 \times 10^{-7}$ *M*, respectively, or 0.18 and 0.03 microgram per milliliter. These very low values are attributable to the high affinities characteristic of many bacterial permeases, which can be construed as an evolutionary adaptation to growth in extremely dilute solutions. In this respect, conventional laboratory media are very different from many natural environments.

## CONTINUOUS CULTURE OF MICROORGANISMS

Cultures of the type so far discussed are called *batch cultures*; nutrients are not renewed and hence growth remains exponential for only a few generations. Microbial populations can be maintained in a state of exponential growth over a long period of time by using a system of continuous culture

(Figure 7.13). The growth chamber is connected to a reservoir of sterile medium. Once growth has been initiated, fresh medium is continuously supplied from the reservoir. The volume of liquid in the growth chamber is maintained constant by allowing the excess volume to be removed continuously through a siphon overflow.

If the fresh medium enters at a constant rate, the concentration of bacteria in the growth chamber remains constant after an initial period of adjustment. In other words, the bacteria in the growth chamber grow just fast enough to replace those lost through the siphon overflow. If the rate of entry of fresh medium is changed, another adjustment period occurs followed by maintenance of a constant population at a new density; the growth rate changes to match the new rate of loss of cells through the overflow. A continuous culture system responds in this manner to a wide variation in the rate of addition of fresh medium. However, no matter the rate of inflow of medium, bacteria cannot grow faster than they would in batch culture.

The question posed by the observation that culture densities in continuous culture systems remain constant is the following: how does the rate of addition of fresh medium to the culture vessel determine the growth rate of the culture? The explanation lies in the fact that the rate of growth of bacteria in continuous culture devices is always limited by the concentration of one nutrient. Consequently, the rate of addition of fresh medium determines the rate of growth of the culture: the system is self-regulating. Consider a continuous culture device that is operating at a constant rate of addition of fresh medium. After inoculation, the culture will at first grow at maximum rate ($k_{max}$). As the culture density increases, the rate of utilization of nutrients will increase until the depletion of one nutrient begins to limit the growth rate. As long as the growth rate exceeds the rate of loss through the siphon overflow, density will continue to increase, the steady-state concentration of limiting nutrient in the growth vessel will continue to decrease and, as a consequence, the growth rate will decrease until the rate of increase of cells through growth will just equal the rate of loss of cells through the overflow. Were the growth rate transiently to become lower than the rate of loss of cells, cell density would decrease, limiting nutrient concentration would increase, and the growth rate would increase until the balance between the growth rate and the loss of cells is again reached.

Using the fact that cell concentration in a continuous culture system remains constant and is self-regulating, we can describe the system in mathematical terms:

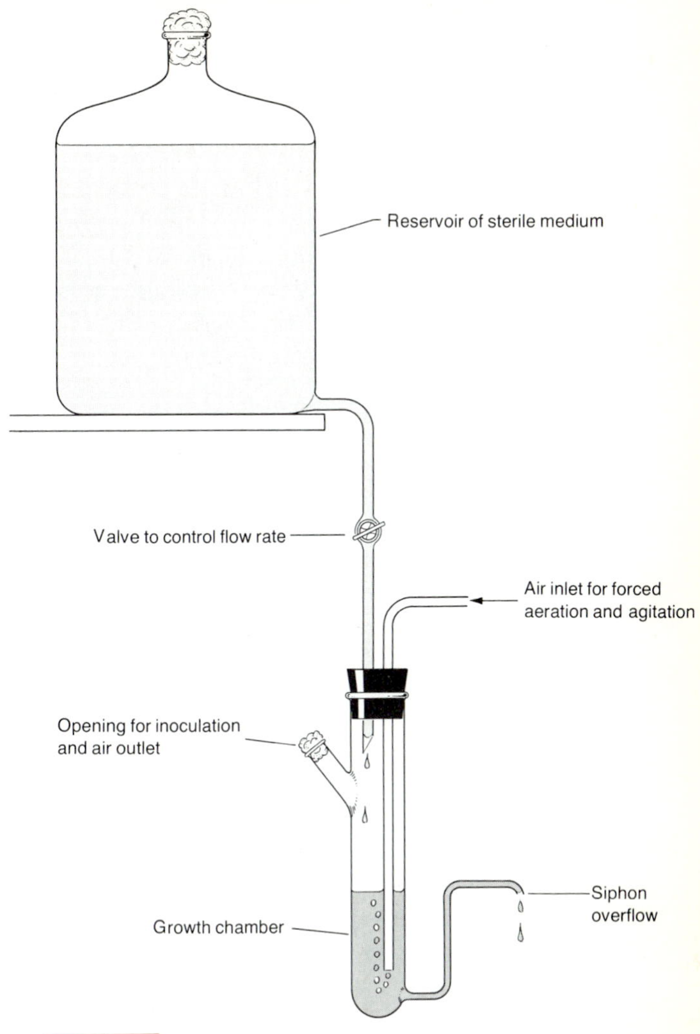


FIGURE 7.13
Simplified diagram of a continuous culture system.

$$\begin{pmatrix}\text{rate of production of} \\ \text{cells through growth}\end{pmatrix} = \begin{pmatrix}\text{rate of loss of cells} \\ \text{through the overflow}\end{pmatrix}$$

Previously we described the rate of production of bacterial mass ($dX/dt$) by Eq. 7.2:

$$\frac{dX}{dt} = kX$$

The rate of loss of cells through the overflow ($dX/dt$) can be stated as

$$\frac{dX}{dt} = \frac{f}{V} = DX \qquad \textbf{(7.13)}$$

where flow rate ($f$) is measured in culture volumes ($V$) per hour. The expression ($f/Vx$) is called the *dilution rate*, $D$. Thus,

$$kX = DX \tag{7.14}$$

or

$$k = D \tag{7.15}$$

which states that the growth rate equals the dilution rate in a stabilized continuous culture device. Substituting Eq. 7.12, we have

$$k_{max} \frac{C}{K_s + C} = D \tag{7.16}$$

and solving for $C$, we have

$$C = K_s \frac{D}{k_{max} - D} \tag{7.17}$$

which states the fundamental relationship between substrate concentration ($C$) in the growth vessel and dilution rate ($D$).

In steady-state operation of a continuous culture device, concentration of the limiting nutrient ($C$) also remains constant. Thus, the rate of addition of the nutrient must equal the rate at which it is utilized by the culture together with that lost through the overflow:

$$\begin{pmatrix}\text{substrate added}\\ \text{from reservoir}\end{pmatrix} = \begin{pmatrix}\text{substrate used}\\ \text{for growth}\end{pmatrix} + \begin{pmatrix}\text{substrate lost}\\ \text{through overflow}\end{pmatrix} \tag{7.18}$$

or

$$DC_r = \frac{dc}{dt} + DC$$

where $C_r$ is the concentration of limiting nutrient in the reservoir and $dc/dt$ is the rate of utilization of limiting nutrient for growth. Substituting $(dX/dt)(dc/dX)$ for $dc/dt$, and since $dX/dt = kX$ and $dc/dX = 1/Y$, we have

$$DC_r = \frac{kX}{Y} + DC \tag{7.19}$$

In the steady state, $D = k$; hence, solving for $X$ yields

$$X = Y(C_r - C) \tag{7.20}$$

the fundamental relation between cell concentration ($X$) and the concentration of limiting nutrient ($C$) in the growth vessel. Together, Eqs. 7.17 and 7.20 allow us to see the relationship between cell concentration, limiting nutrient concentration, and the dilution rate (Figure 7.14). Cell number and the concentration of limiting nutrient change little at low dilution rates. As the dilution rate approaches $k_{max}$, cell concentration drops rapidly to zero, and the concentration of the limiting nutrient approaches its concentration in the reservoir ($C_r$).

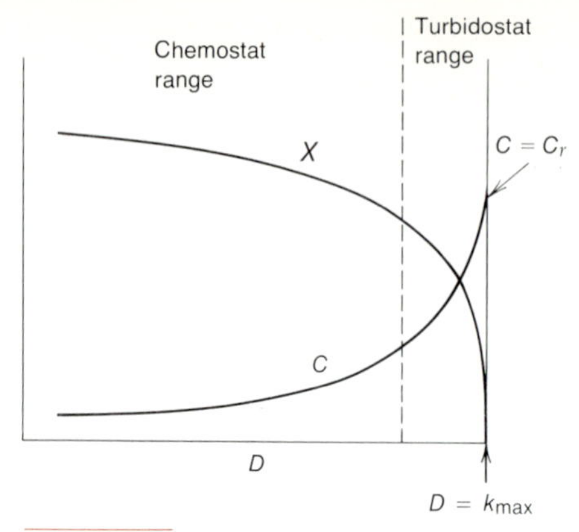


FIGURE 7.14

Relationship between cell density, $X$, limiting nutrient concentration, $C$, and dilution rate, $D$, in a continuous culture system.

## Chemostats and Turbidostats

Continuous culture systems can be operated as *chemostats* or as turbidostats. In a chemostat the flow rate is set at a particular value and the rate of growth of the culture adjusts to this flow rate. In a turbidostat the system includes an optical-sensing device which measures the absorbancy of the culture (culture density) in the growth vessel; the electrical signal from this device regulates the flow rate. Thus, the absorbancy of the culture controls the flow rate and the rate of growth of the culture adjusts to this flow rate.

As a practical matter, chemostats and turbidostats are usually operated at different dilution rates. In the chemostat, maximum stability is attained within a range of dilution rates over which cell concentration changes only slightly with changes in dilution rate, i.e., at low dilution rates. In contrast, in the turbidostat, maximum sensitivity and stability are achieved at high dilution rates, within a range over which culture biomass changes rapidly with dilution rate (Figure 7.14). A turbidostat can be operated at $k_{max}$ by setting an absorbance value that maintains a concentration of cells insufficient to cause any nutrient to fall to a growth rate–limiting concentration.

### Use of Continuous Culture Systems

Continuous culture systems offer two valuable features for the study of microorganisms. They provide a constant source of cells in an exponential phase of growth, and they allow cultures to be grown continuously at extremely low concentrations of substrate. The practical advantages of the former feature are obvious. Growth at low substrate concentrations is valuable in studies on regulation of synthesis or catabolism of the limiting substrate, in selection of various classes of mutants, and in ecological studies.

## MAINTENANCE ENERGY

It will be noted in Figure 7.14 that the curves for cell density and concentration of limiting nutrient are not extended into the range of very low dilution rates. This is because the response of a culture at very low dilution rates depends on the nature of the limiting nutrient. If the limiting nutrient is the energy source for the culture, growth ceases at very low dilution rates because a certain amount of energy, termed *maintenance energy*, is always used for purposes other than growth, including motility, DNA repair, and the bringing of nutrients into the cell against a concentration gradient. When the rate of entry of the source of energy is insufficient to supply more energy than that required for maintenance, growth cannot occur. Maintenance energy can be stated mathematically as follows:

$$\frac{dx}{dt} = Y\frac{dc}{dt} - ax \tag{7.21}$$

where $a$ is the *specific maintenance rate constant*. Since the term $Y\,dc/dt$ is $kx$, Eq. 7.21 is a variation of the growth equation (Eq. 7.2) that applies if the growth rate is limited by the supply of energy. Since values of $a$ are typically low (e.g., 0.018 $hr^{-1}$ for *E. coli* strain ML30, growing on glucose at 37° C), the effect of maintenance on growth rate is negligible in batch cultures but can be quite significant in nutrient-limited natural environments or chemostats operated at low values of $D$.

Equation 7.21 introduces a certain complication into the concept of $Y$. We previously defined $Y$ as the mass ratio of cells produced to substrate utilized. However, Eq. 7.21 excludes the amount of substrate used for maintenance. Rather than redefine $Y$ for the special case of a substrate that serves as an energy source, a new parameter can be used, $Y_G$, that fits the original definition of $Y$; the term, $Y$, is then reserved to describe substrate used for growth and maintenance. This leads to a new parameter *maintenance coefficient* ($m$) to describe maintenance, that is

$$\text{overall rate of substrate utilization} = \text{rate of substrate utilization for growth} + \text{rate of substrate utilization for maintenance} \tag{7.22}$$

which becomes

$$\frac{dC}{dt} = \left(\frac{dC}{dt}\right)_G + \left(\frac{dC}{dt}\right)_M \tag{7.23}$$

and,

$$\frac{1}{Y} = \frac{m}{k} + \frac{1}{Y_G} \tag{7.24}$$

The maintenance coefficient $m$ and specific maintenance rate $a$ are related by the equation

$$m = a/Y_G$$

## FURTHER READING


**Books**

KUBITSCHEK, H. E., *Introduction to Research with Continuous Cultures*. Englewood Cliffs, N. J.: Prentice-Hall, Inc., 1970.

MANDELSTAM, J., and K. MCQUILLEN, *Biochemistry of Bacterial Growth*, 3rd ed. Oxford: Blackwell, 1982.

MEYNELL, G. G., and E. MEYNELL, *Theory and Practice in Experimental Bacteriology*, 2nd ed. Cambridge: Cambridge University Press, 1970.

GERHARDT, P., ed. *Manual of Methods for General Bacteriology*, Washington: Am. Soc. Microbiol, 1981.

**Reviews**

MONOD, J., "The Growth of Bacterial Cultures," *Ann. Rev. Microbiol.* **3,** 371 (1949).

NOVICK, A., "Growth of Bacteria," *Ann. Rev. Microbiol.* **9,** 97 (1955).

# Chapter 8
# Effect of the Environment on Microbial Growth

This chapter will consider the interactions between the microbial cell and its environment.

## FUNCTIONS OF THE CELL MEMBRANE

Whether growing in an artificial medium or a natural environment, the concentration of solutes within a microbial cell is typically much higher than the concentration of solutes in the extracellular environment; and the composition of the intracellular solute mixture differs significantly from the extracellular one. These differences are maintained and, in the case of certain solutes, created by the cell membrane and some proteins that are associated with it. In general, the cell membrane is impermeable to large molecules and ions. Their passage is mediated by the activities of proteins associated with or imbedded within the cell membrane (Chapter 6).

In Gram-negative bacteria, the outer membrane also plays a role, although a qualitatively different one, in regulating the passage of solutes (see Chapter 6). In contrast to certain of the proteins associated with the cell membrane, no proteins in the outer membrane are known to pump solutes across it although some actively facilitate passage. The outer membrane has simple semipermeable properties: some solutes can pass through it or through pores in it; others cannot (Chapter 6). Thus the periplasm (the region between the two membranes of Gram-negative bacteria) has a solute composition that is distinct from that of cell interior and from that of the external environment. In general, solutes flowing in either direction between the cell interior and the external environment pass through the periplasm. However, as discussed in Chapter 6, in a number of regions on the cell surface, termed Bayer's junctions, the cell membrane and outer membrane are in direct contact. Evidence suggests that during genetic exchange and phage infection, DNA enters or leaves the cell through these regions without traversing the periplasm.

## ENTRY OF NUTRIENTS INTO THE CELL

Transport of various nutrients across the cell membrane occurs by a variety of mechanisms the simplest of which is *passive diffusion.*

### Passive Diffusion

Net flow of a solute by passive diffusion occurs only in response to a difference in its concentration across the cell membrane (a *concentration gradient*) and as a result of such flow the difference diminishes. The rate of flow is a direct function of the magnitude of the gradient and does not approach a limiting value even when the concentration difference is great. Passive diffusion occurs when there are regions of the membrane through which a particular solute can pass freely, much as small molecules can pass through the artificial membrane used for dialysis. Water and certain gases, such as oxygen and nitrogen, are the principal nutrients that cross the cell membrane by passive diffusion. All nutrients pass the outer membrane of Gram-negative bacteria, some through its pores, by passive diffusion.

### Facilitated Diffusion

The diffusion in or out of the cell of certain compounds to which the cell membrane is otherwise impermeable is mediated by specific membrane proteins, the presence of some of which are induced by their substrates. These proteins, collectively known as *permeases* or *carrier proteins,* bind to their substrate on the membrane's outer surface and, by mechanisms still largely unknown, mediate their passage through the membrane to the inner surface where the carrier-substrate complex dissociates, releasing the substrate into the cytosol. These proteins, therefore, are enzymes that catalyze the general reaction

substrate (outside the cell) $\longleftrightarrow$ substrate (inside the cell)

Facilitated diffusion is similar to passive diffusion in the sense that the substrate moves down a concentration gradient from a higher to a lower concentration; the process does not require the expenditure of metabolic energy. It differs from passive diffusion by its enzymatic nature: the process is rapid (more rapid than would be predicted from the laws governing simple diffusion); it exhibits considerable substrate specificity (optical enantiomorphs are often distinguished); the carrier proteins are often inducible; the rate of the reaction approaches a limiting value with increasing concentrations of substrate, i.e., it obeys normal enzyme (Michaelis-Menten) kinetics. The velocity of entry of substrate can be described as

$$v^{\text{entry}} = v_{\text{max}}^{\text{entry}} \frac{[\text{S}]_{\text{ex}}}{\text{K}_m^{\text{entry}} + [\text{S}]_{\text{ex}}} \qquad \textbf{(8.1)}$$

$$v^{\text{exit}} = v_{\text{max}}^{\text{exit}} \frac{[\text{S}]_{\text{in}}}{\text{K}_m^{\text{exit}} + [\text{S}]_{\text{in}}} \qquad \textbf{(8.2)}$$

where $[\text{S}]_{\text{ex}}$ and $[\text{S}]_{\text{in}}$ are concentrations of substrate outside and inside the cell respectively; $v_{\text{max}}^{\text{entry}}$ and $v_{\text{max}}^{\text{exit}}$ are the velocities of entry and exit at saturating concentrations of substrate; $\text{K}_m^{\text{entry}}$ and $\text{K}_m^{\text{exit}}$ are the Michaelis constants for the entry and exit processes.

One notes that as $[\text{S}]_{\text{ex}}$ increases, the fraction on the right side of Eq. 8.1 approaches a value of unity and $v^{\text{entry}}$ approaches $v_{\text{max}}^{\text{entry}}$ as a limit.

Since no metabolic energy is used in facilitated diffusion, the Michaelis-Menten parameters ($v_{\text{max}}$ and $\text{K}_m$) for entry and exit are equal. As a consequence, at equilibrium (when $v^{\text{entry}} = v^{\text{exit}}$) the internal and external concentrations of a nutrient transported by facilitated diffusion are equal.

Although facilitated diffusion is a common mechanism of transport in eucaryotic microorganisms, it is relatively rare among procaryotes. For example, sugars that characteristically enter eucaryotic microorganisms by facilitated diffusion, enter procaryotes by other transport mechanisms, *active transport* or *group translocation,* which are described below. One process of transport in a procaryote that is mediated by facilitated diffusion is the entry of glycerol into cells of the enteric group.

### Active Transport

The mechanisms of transport known collectively as active transport permit a solute to enter the cell against a thermodynamically unfavorable gradient of concentration; these mechanisms create concentrations of solutes within the cell that can be several hundred to a thousand times greater than those outside. The prevalence of active transport systems in bacteria can be correlated with the facts that bacteria frequently occur in dilute chemical environments but nevertheless exhibit rapid rates of metabolism. As we shall see, active transport systems appear to function as facilitated diffusion systems coupled to a source of metabolic energy, thereby permitting accomplishment of the chemical work necessary for creating and maintaining a concentration gradient across the cell membrane.* Several sources of metabolic energy drive the cell's various active transport systems—the electrostatic or pH gradient components of protonmotive force (Chapter 5), secondary gradients (e.g. of ions such as $Na^+$) derived from the protonmotive force by other active transport systems, and ATP.

## Binding Proteins

A variety of proteins, collectively termed *binding proteins,* occur in the periplasm of Gram-negative bacteria that have the property of binding with high affinity (with an affinity constant in the range of $10^{-7}$) to specific substrates. Not being located in the cell membrane, these proteins are not carrier proteins, but, as can be established by a simple experiment, they play an essential role in conjunction with a carrier protein in the transport of certain substrates. If Gram-negative bacteria are suspended in a buffered solution of 20 percent sucrose containing the chelating agent, ethylenediamine tetraacetic acid (EDTA), centrifuged, and rapidly resuspended in 0.5 mM $MgCl_2$ at 0°C, the outer membrane is damaged allowing proteins located in the periplasm, including binding proteins, to leak into the suspending solution. Cells treated by this procedure (termed *cold osmotic shock*) lose their ability to transport certain substrates but retain their ability to transport others. A direct correlation exists between the presence in the periplasm of a binding protein for a particular substrate and the loss, following cold osmotic shock, of the cell's ability to transport that substrate. Such transport systems are termed *shock-sensitive*; the others that remain in osmotically shocked cells are termed *shock-insensitive.* Binding proteins play two roles in transport: (1) by binding to the substrate they increase its effective concentration within the periplasm, providing a higher effective concentration to the carrier protein for transport into the cell; and (2) by interacting with the carrier protein, binding proteins stimulate transport activity.

Shock-sensitive and shock-insensitive systems appear to differ with respect to the source of metabolic energy that drives them: shock-sensitive active transport systems are usually driven by the hydrolysis of ATP or some other source of a high-energy phosphate bond; shock-insensitive systems are usually driven directly by protonmotive force.

## Secondary Active Transport

The establishment of a protonmotive force (Chapter 5) by proton extrusion associated with the passage of electrons through a membrane-bound transport chain or by hydrolysis of ATP by the membrane-bound ATPase is sometimes termed *primary active transport.* Protonmotive force, as discussed, is a source of energy that can drive the synthesis of ATP at the membrane-bound ATPase site; it can also drive molecules across the cell membrane. Such movement of a molecule across the cell membrane at the expense of a previously established gradient of another molecular species is termed *secondary active transport.*

There are three types of secondary active transport: *symport, antiport,* and *uniport* (Figure 8.1).

Symport is the simultaneous transport of two molecules by the same carrier; one molecule flows down its previously established gradient and the other flows with it [Figure 8.1 (a)]. Antiport is the simultaneous transport by the same carrier of two molecules in the opposite direction across the membrane, one of which flows down its concentration gradient, thus exchanging one gradient for another [Figure 8.1 (b)]. Active uniport is the flow of ions driven directly by an electrostatic gradient [Figure 8.1 (c)]. (Facilitated diffusion can be considered a passive uniport of an uncharged molecule.)

## Active Transport Linked to Phosphate Bond Energy

Experiments exploying mutant strains that lack membrane-bound ATPase have established that certain substrates can only be transported by these cells if they are also provided with an energy source,

* Although the concentration gradient usually runs from high in the cell interior to low in the exterior environment, the opposite direction of a gradient is established in certain cases. For example, such a reverse gradient is established by the active transport system encoded in certain tetracycline-resistance elements; they cause the active efflux of the antibiotic from the cell thereby protecting it from the toxic effects of the antibiotic.

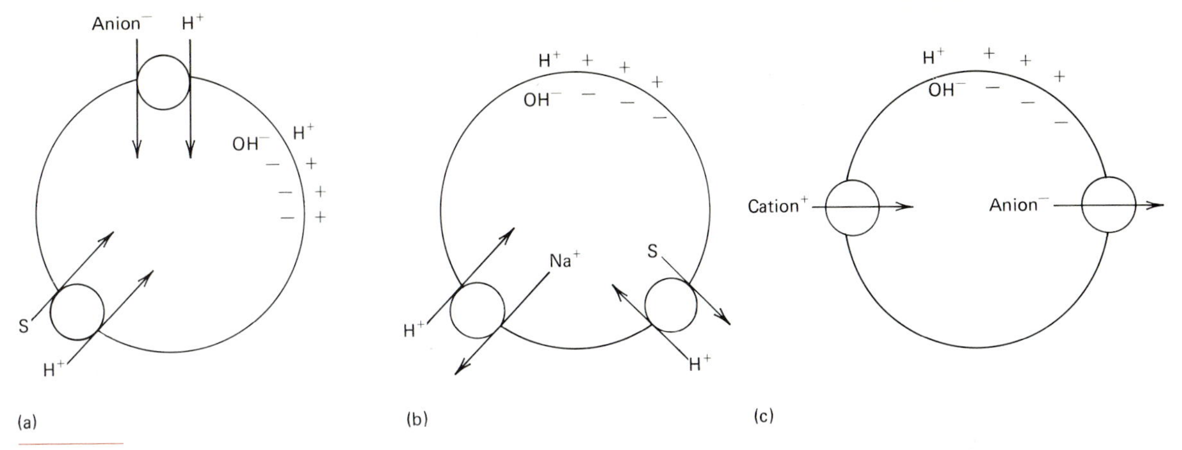


FIGURE 8.1

Schematic representation of secondary active transport system. The large circles represent cells; the small ones represent carrier proteins in the cell membrane. (a) Symport reactions. On the right is shown a protonmotive force created by primary active transport; the pH gradient drives (on the left) an electrogenic symport of an uncharged solute, S, with a proton, and (at the top) an electroneutral symport of an anion with a proton. (b) Antiport reactions. The pH gradient (at the top) drives (at the lower left) the electroneutral antiport of a cation and a proton, and (at the lower right) the electrogenic antiport of an uncharged solute, S, and a proton. (c) Uniport reactions. The protonmotive force (at the top) drives (at the left) the uniport of a cation into the cell, and (at the right) the uniport of an anion out of the cell. After B. P. Rosen and E. R. Kashket, "Energetics of Bacterial Transport," in *Bacterial Transport*, ed. B. P. Rosen (New York: Marcel Dekker, 1978).

the metabolism of which can generate ATP by substrate-level phosphorylation. Other substrates will enter even if the metabolized substrate only generates a protonmotive force. These experiments show that the transport systems by which the former class of substrates enter the cell are not capable of being driven directly by a protonmotive force; rather, their transport is dependent on phosphate bond energy. The mechanism by which phosphate bond energy is utilized to drive these transport systems remains unclear.

## Group Translocation

Certain substrates as they enter the cell are chemically modified to a form to which the membrane is impermeable. As a consequence, high concentration of the modified form can be generated and maintained within the cell at the expense of low concentrations of the unmodified form in the external environment. Thus a metabolic function equivalent to active transport is accomplished by a trapping reaction. Such mechanisms of transport are termed *group translocation.* Group translocation mechanisms differ fundamentally from true active transport because they do not establish a concentration gradient of a molecular species across the cell membrane. One compound is found in the external environment; a chemically modified form of it is found inside the cell. Group translocation mechanisms are particularly conserving of metabolic energy; the chemical change of the substrate that occurs on its entry into the cell requires the expenditure of energy in the form of a high-energy phosphate bond, but this change is also required for the substrate's further metabolism. Transport is thereby accomplished by a reaction that would also occur intracellularly even if the substrate were brought into the cell by another (energy-requiring) mechanism.

The most thoroughly studied of the group translocation systems is the *phosphotransferase system* (PTS) by which certain sugars are phosphorylated at the expense of phosphoenolpyruvate (PEP) as they enter the cell. Thus a PTS catalyzes the general reaction

$$\text{sugar}_{(\text{outside})} + \text{PEP}_{(\text{inside})} \longrightarrow \text{sugar-}\textcircled{P}_{(\text{inside})} + \text{pyruvate}_{(\text{inside})}$$

Each PTS is reasonably complex, involving the sequential action of four distinct phosphate-carrying proteins, termed HPr, Enzyme I, Enzyme III, and Enzyme II (Figure 8.2). The last member of the chain, Enzyme II (E II), is located within the membrane and serves as a carrier protein for the sugar

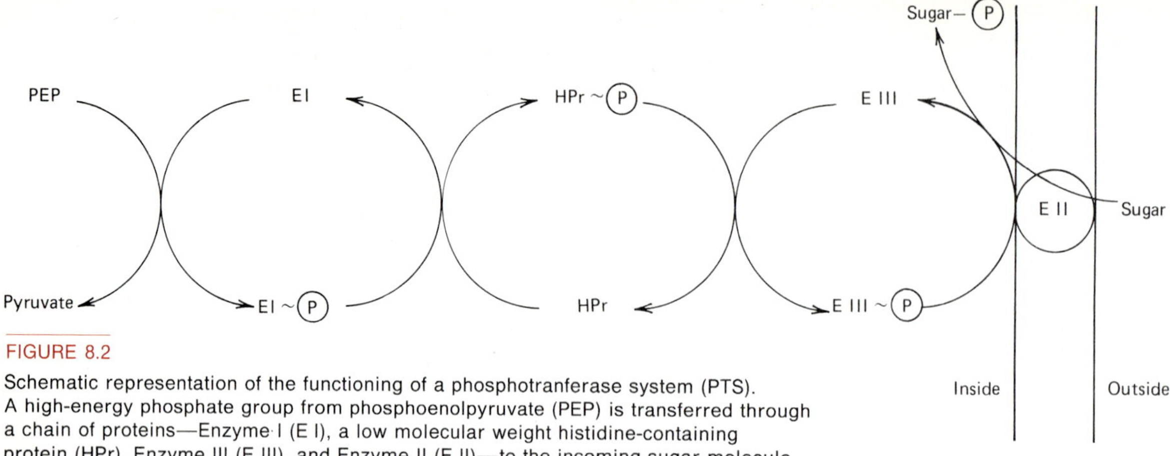


FIGURE 8.2

Schematic representation of the functioning of a phosphotranferase system (PTS). A high-energy phosphate group from phosphoenolpyruvate (PEP) is transferred through a chain of proteins—Enzyme I (E I), a low molecular weight histidine-containing protein (HPr), Enzyme III (E III), and Enzyme II (E II)—to the incoming sugar molecule. Enzyme II also serves as the membrane carrier protein. After S. S. Dills, M. R. Apperson, M. R. Schmidt, and M. R. Saier, "Carbohydrate Transport in Bacteria," *Microbiol. Rev.* **44,** 385 (1980).

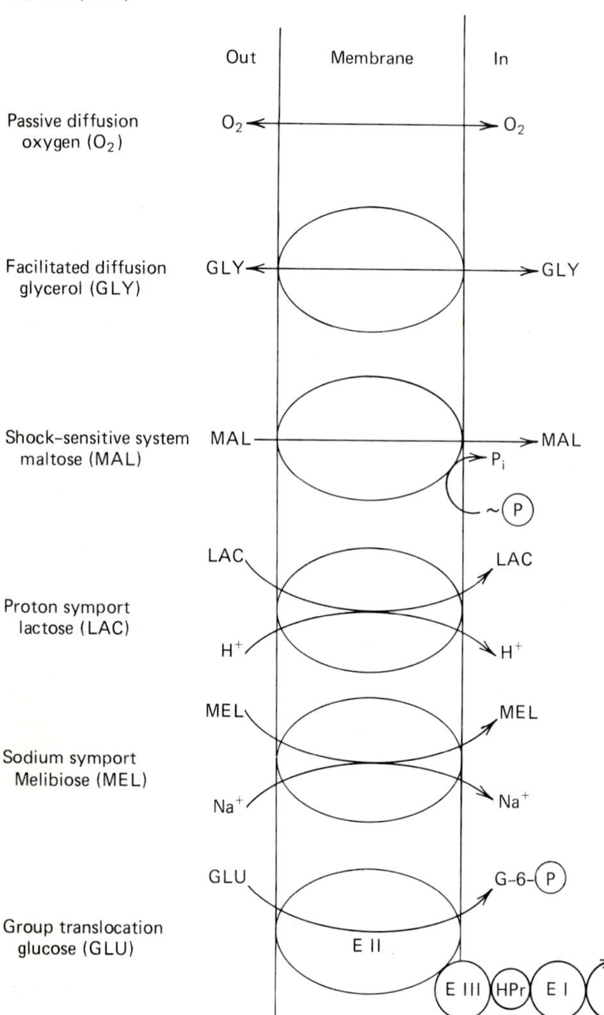


FIGURE 8.3

Schematic representation of examples of the various classes of membrane transport that occur in *E. coli*. Oxygen enters by passive diffusion; no carrier protein participates. Glycerol (GLY) enters by carrier-mediated facilitated diffusion; the intracellular concentration of this substrate never exceeds the extracellular concentration. Maltose (MAL) enters by a shock-sensitive system of active transport; high intracellular concentrations of this substrate are created by the expenditure of high-energy phosphate bonds. Lactose (LAC) enters by proton symport at the expense of the pH gradient previously established by primary active transport. Melibiose (MEL) enters by a sodium symport driven by the membrane potential component of protonmotive force previously established by primary active transport. Glucose (GLU) enters by a PTS through the activities of Enzyme I (E I), the histidine protein (HPr), Enzyme II (E II), and Enzyme III (E III); the intracellular product of the process is glucose-6-phosphate (G-6-Ⓟ). After S. S. Dills, A. Apperson, M. R. Schmidt, and M. H. Saier, "Carbohydrate Transport in Bacteria," *Microbiol. Rev.* **44,** 385 (1980).

substrate; the penultimate member of the chain is a membrane-associated protein, Enzyme III (E III), that catalyzes the transfer of a phosphate group to the entering sugar. The first two components, Enzyme I (E I) and HPr, carry the high-energy phosphate group to E III. Two components (EI and HPr) are nonspecific; they participate in all the PTSs of a particular cell. The other two, E III and EII, are specific for a particular sugar substrate and their synthesis is usually induced by the presence of that sugar in the cell's external environment.

In addition to the well-established PTSs, other examples of group translocation have been proposed. Among these is a coenzyme A transfer system mediated by acetyl-CoA-synthetase is presumed to play a role in the transport of fatty acids: as the fatty acid enters the cell it is converted to a CoA derivative. Similarly a group of enzymes, termed *phosphoribosyl transferases*, which catalyze the class of reactions

purine or pyrimidine base
+ phosphoribosyl-pyrophosphate
$\longrightarrow$ nucleoside monophosphate + Ⓟ − Ⓟ

are thought to constitute group translocation mechanisms for the uptake of adenine, guanine, hypoxanthine, xanthine, and uracil by certain bacteria.

### Summary of Membrane Transport Mechanisms

Specific examples of the various types of transport systems found in *Escherichia coli* are shown in Figure 8.3. Oxygen enters the cell by simple diffusion through the cell membrane; no specific membrane protein is associated with the process. Glycerol enters by facilitated diffusion; an inducible carrier protein mediates its passage in or out of the cell at the same concentration-dependent rates.* Maltose is transported by a shock-sensitive active transport system; a specific binding protein and a carrier protein are essential components, and a high-energy phosphate bond is expended. Lactose enters by proton symport; melibiose by sodium ion symport, and glucose by a PTS.

* An interesting consequence of glycerol's entering the cell by facilitated diffusion rather than active transport (as most substrates do) is the unusual dependence of growth rate on the concentration of glycerol in the medium when that substrate is the only carbon source provided. Unlike the apparent independence between growth rate and substrate concentration (except at very low concentration) seen with most substrates (Chapter 7), growth rate varies with glycerol concentrations over a broad range of concentrations. To obviate a constantly changing growth rate as glycerol is utilized, experimenters usually employ a higher (saturating concentration, ~0.4 percent) concentration of glycerol than of other substrates.

## UTILIZATION OF SUBSTRATES THAT CANNOT PASS THE CELL MEMBRANE

A number of macromolecules (e.g., starch, cellulose, or RNA) and highly charged small molecules (e.g., nucleotides) that cannot pass the cell membrane are nevertheless utilizable as substrates for growth. These substrates are enzymatically hydrolyzed (*degraded*) in the external medium by enzymes secreted by cells. The hydrolytic products of these substrates then enter the cell by specific transport system. This process of *extracellular digestion* is in many respects similar to the ones by which complex foods are broken down in the stomachs and intestines of animals; the enzymes that mediate extracellular digestion are termed *exoenzymes*. When microbial colonies develop on an agar medium containing particles of an insoluble macromolecule that is digestible, each colony is surrounded by an expanding clear area in which the insoluble substrate has been hydrolyzed by the action of exoenzymes. Table 8.1 lists some substrates of exoenzymes and the products formed by their action.

Most exoenzymes that have a digestive function are subject to catabolite repression (Chapter 12); i.e., their synthesis is repressed if more rapidly metabolizable sources of carbon and energy are present in the environment. In addition, many exoenzymes are inducible; their synthesis is triggered by the hydrolytic products, which can enter the cell. Since exoenzymes are synthesized at a low rate, even in the absence of the specific substrate, the presence of a macromolecular substrate in the environment always leads to the formation of small amounts of hydrolytic products which cause induction.

Not all exoenzymes function exclusively to provide the cell with utilizable carbon and energy sources. The 5′-nucleotidases, for example, convert nucleotides to which the membrane is impermeable into nucleosides that can enter the cell and be used for biosynthesis or be degraded as a carbon and energy source. Exoenzymes also include certain enzymes that destroy antibiotics: e.g., the penicillinases of many bacteria, which hydrolyze the beta-lactam ring of some penicillins and thereby detoxify them.

Many enzyme activities that occur as exoenzymes of Gram-positive bacteria, occur in the periplasms of Gram-negative bacteria (Table 8.2).

In addition to true exoenzymes, bacteria also produce certain incompletely excreted proteins; some (*membrane proteins*) enter the cell membrane and remain there rather than passing through, and, in Gram-negative bacteria, some (*periplasmic pro-*

**TABLE 8.1**

**Examples of Exoenzymes and Organisms That Produce Them**

| Exoenzyme | Macromolecular Substrate | Molecule that Enters the Cell | Example of Producing Microorganism |
|---|---|---|---|
| POLYSACCHARIDE-SPLITTING ENZYMES | | | |
| Amylase | Starch | Glucose, maltose oligoglycosides | *Bacillus subtilis* |
| Pectinase | Pectin | Galacturonic acid | *Bacillus polymyxa* |
| Cellulase | Cellulose | Glucose, cellobiose | *Clostridium thermocellum* |
| Lysozyme | Peptidoglycans | | *Staphylococcus aureus* |
| Chitinase | Chitin | Chitobiose | |
| PROTEINASES | | | |
| Peptidase | Peptides | Amino acids | *Bacillus megaterium* |
| NUCLEASES | | | |
| Deoxyribonuclease | DNA | Deoxribonucleosides[a] | *Streptococcus haemolyticus* |
| Ribonuclease | RNA | Ribonucleosides | *Bacillus subtilis* |
| ESTERASES | | | |
| Lipases | Lipids | Glycerol + fatty acids | *Clostridium welchii* |
| Poly $\beta$-hydroxybutyrate depolymerase | Poly $\beta$-hydroxybutyrate | $\beta$-hydroxybutyryl-$\beta$-hydroxybutyrate | *Pseudomonas* spp. |

[a] Although nucleotides are the primary product of hydrolysis, they are further hydrolyzed to nucleosides before entering the cell.

**TABLE 8.2**

**Certain Enzymes Located in the Periplasmic Space**

| Enzyme | Reactions Catalyzed |
|---|---|
| Ribonuclease I | Hydrolyzes RNA |
| DNA Endonuclease I | Internally cleaves DNA |
| Alkaline phosphatase | Removes phosphate groups from a number of organic compounds |
| 5′-nucleotidase | Converts a number of nucleotides to nucleosides |
| Acid hexose phosphotase | Removes phosphate groups from a number of sugar phosphates |
| Acid phosphatase | Removes phosphate groups from organic compounds |
| Cyclic phosphodiesterase | Converts ribonucleoside-2′, 3′-cyclic phosphates to the ribonucleoside-3′-phosphates, and further hydrolyzes the ribonucleoside-3′-phosphates to nucleosides |
| Penicillinase | Hydrolyzes and thereby inactivates penicillin |

| Protein | Charged Segment | Hydrophobic Segment |
|---|---|---|
| f1 | MET LYS LYS SER LEU VAL LEU LYS | ALA SER VAL ALA VAL ALA THR LEU VAL PRO MET LEU SER PHE ALA ALA↓GLU GLY··· |
| PhoA | MET LYS GLN | SER THR ILE ALA LEU ALA LEU LEU PRO LEU LEU PHE THR PRO VAL THR LYS ALA↓ARG··· |
| MalE | MET LYS ILE LYS THR GLY ALA ARG | ILE LEU ALA LEU SER ALA LEU THR THR MET MET PHE SER ALA SER ALA LEU ALA LYS··· |
| Bla | MET SER ILE GLN HIS PHE ARG | VAL ALA LEU ILE PRO PHE PHE ALA ALA PHE CYS LEU PRO VAL PHE ALA↓HIS PRO··· |
| Lpp | MET LYS ALA THR LYS | LEU VAL LEU GLY ALA VAL ILE LEU GLY SER THR LEU LEU ALA GLY↓CYS SER··· |
| LamB | MET MET ILE THR LEU ARG LYS | LEU PRO LEU ALA VAL ALA VAL ALA ALA GLY VAL MET SER ALA GLN ALA↓MET ALA VAL ASP. |

FIGURE 8.4
Amino acid composition of the charged and hydrophobic segments of the signal sequences of various proteins exported by *E. coli*. The major coat protein of f1 phage (f1) is located in the cell membrane; alkaline phosphatase (PhoA), the maltose-binding protein (MalE), and beta-lactamase (Bla) are located within the periplasm; lipoprotein (Lpp) and the receptor protein for phage lambda (LamB) are located in the outer membrane. The amino acids are indicated by their conventional abbreviations: arginine, ARG; aspartic acid, ASP; glycine, GLY; glutamic acid, GLU; glutamine, GLN; histidine, HIS; isoleucine, ILE; leucine, LEU; lysine, LYS; methionine, MET; proline, PRO; serine, SER; threonine, THR; and valine, VAL. The points at which the signal sequences are cleaved following their proper location is indicated by an arrow (↓). Note the presence of charged amino acids (ARG and LYS) in the charged segments, and the predominance of the highly hydrophobic amino acids (ALA, GLY, ILE, LEU, VAL) in the hydrophobic segments. After T. J. Silhavy, S. A. Benson, and S. D. Emr. "Mechanisms of Protein Localization," *Microbiol. Rev.* **47,** 313 (1983).

*teins*) pass through the cell membrane but not the outer membrane. Still others (*outer membrane proteins*) enter and remain in the outer membrane. Although not completely understood, the mechanisms by which exoenzymes are excreted and incompletely excreted proteins are properly located, share common features. However, fundamental to any such mechanism is the fact that the information that determines the eventual location of proteins is encoded in the primary structure (amino acid sequence) of the protein. Also, these proteins when first synthesized are larger (in a *precursor form*) than they are (in *mature form*) after they have entered their proper location: a portion of the amino-terminal end of the protein (termed the *signal sequence*) is removed after the protein is properly located. Comparison of the signal sequences from a variety of exported proteins shows that they have certain common features, thought to be related to their ability to enter or to cross membranes (Figure 8.4). Signal sequences are composed of two segments: at the immediate amino-terminal end lies a short segment that contains charged amino acids; next to this lies the somewhat longer *hydrophobic segment*, so called because it is totally composed of amino acids that carry hydrophobic side chains. It is this latter region of the signal sequence that is thought to play the pivotal role in enzyme localization. The hydrophobic nature of the region makes thermodynamically feasible its entry (probably in a form folded back on itself, thus making the hydrophobic region terminal) into the hydrophobic interior of a membrane. Thus the region is thought to lead the protein to its proper location; once there, the signal sequence is cleaved from the protein by the action of a membrane associated endopeptidase, termed the *signal peptidase*.

The temporal relationship between synthesis and export of a protein has also been the topic of recent investigations. In certain cases, termed *cotranslational export*, the signal sequence enters the membrane while more distal regions are still being synthesized. Indeed, cotranslational export binds the corresponding polysome to the membrane; one end of the growing protein that is being exported is attached to the polysome at the point where peptide elongation is occurring and the other is attached to the membrane by the entry into it, led by the signal sequence. However, in other cases (termed *posttranslational export*) excretion of a protein begins only after its synthesis has been completed.

The broad outlines of the mechanisms of protein export by procaryotes have been drawn, but many important details of how specific proteins are exported to specific sites within or beyond the cell membrane remain to be examined.

## EFFECTS OF SOLUTES ON GROWTH AND METABOLISM

Transport mechanisms play two essential roles in cellular function. First, they maintain the intracellular concentrations of all metabolites at levels sufficiently high to ensure operation of both catabolic and biosynthetic pathways at near-maximal rates, even when nutrient concentrations in the external medium are low. This is evidenced by the fact that the exponential growth rates of microbial populations remain constant until one essential nutrient in the medium falls to a very low value, approaching exhaustion. At this limiting nutrient concentration, the growth rate of the population rapidly falls to zero (Chapter 7). Second, transport mechanisms function in *osmoregulation*, maintaining the solutes (principally small molecules and ions) at levels optimal for metabolic activity, even when the osmolarity of the environment varies over a relatively wide range.*

Most bacteria do not need to regulate their internal osmolarity with precision because they are enclosed by a cell wall capable of withstanding a considerable internal osmotic pressure. Bacteria always maintain their osmolarity well above that of the medium. If the internal osmotic pressure of the cell falls below the external osmotic pressure, water leaves the cell and the volume of the cytoplasm decreases with accompanying damage to the membrane. In Gram-positive bacteria, this causes the cell membrane to pull away from the wall; the cell is said to be *plasmolyzed*. In Gram-negative bacteria the wall retracts with the membrane; this also damages the membrane.

Bacteria vary widely in their osmotic requirements. Some are able to grow in very dilute solutions, and some in solutions saturated with sodium chloride. Microorganisms that can grow in solutions of high osmolarity are called *osmophiles*. Most natural environments of high osmolarity contain high concentrations of salts, particularly sodium chloride. Microorganisms that grow in this type of environment are called *halophiles*. Bacteria can be divided into four broad categories in terms of their salt tolerance: *nonhalophiles*, *marine organisms*, *moderate halophiles*, and *extreme halophiles* (Table 8.3). Some halophiles, for example *Pediococcus halophilus*, can tolerate high concentrations of salt in the growth medium, but they can also grow in media without added NaCl. Other bacteria, including marine bacteria and certain moderate halophiles, as well as all extreme halophiles, require NaCl for growth. The tolerance of high osmolarity and the specific requirement for NaCl are distinct phenomena, each of which has a specific biochemical basis.

### Osmotic Tolerance

Osmotic tolerance—the ability of an organism to grow in media with widely varying osmolarities—is accomplished in bacteria by an adjustment of the internal osmolarity so that it always exceeds that of the medium. Intracellular accumulation of potassium ions ($K^+$) seems to play a major role in this adjustment. Many bacteria have been shown to concentrate $K^+$ to a much greater extent than $Na^+$ (Table 8.4). Moreover, there is an excellent correlation between the osmotic tolerance of bacteria and their $K^+$ content. For bacteria as metabolically diverse as Gram-positive cocci, bacilli, and Gram-negative rods, relative osmotic tolerance can be deduced from their relative $K^+$ contents after growth in a medium of fixed ionic strength and composition. Studies on *E. coli* have shown that the intracellular $K^+$ concentration increases progressively with increasing osmolarity of the growth medium. Consequently, both the osmolarity and the internal ionic strength of the cell increase.*

The maintenance of a relatively constant ionic strength within the cell is of critical physiological importance, because the stability and behavior of enzymes and other biological macromolecules are strongly dependent on this factor. In bacteria, the diamine putrescine (Chapter 5) probably always plays an important role in assuring the approxi-

* When a *solution* of any substance (*solute*) is separated from a *solute-free solvent* by a membrane that is freely permeable to solvent molecules, but not to molecules of the solute, the solvent tends to be drawn through the membrane into the solution, thus diluting it. Movement of the solvent across the membrane can be prevented by applying a certain hydrostatic pressure to the solution. This pressure is defined as *osmotic pressure*. A difference in osmotic pressure also exists between two solutions containing different concentrations of any solute.

The osmotic pressure exerted by any solution can be defined in terms of *osmolarity*. An osmolar solution is one that contains one *osmole* per liter of solutes, i.e., a 1.0 molal solution of an ideal nonelectrolyte. An osmolar solution exerts an osmotic pressure of 22.4 atmospheres at 0° C, and depresses the freezing point of the solvent (water) by 1.86° C. If the solute is an electrolyte, its osmolarity is dependent on the degree of its dissociation, since both ions and undissociated molecules contribute to osmolarity. Consequently, the osmolarity and the molarity of a solution of an electrolyte may be grossly different. If both the molarity and the dissociation constant of a solution of an electrolyte are known, its osmolarity can be calculated with some degree of approximation, as the sum of the moles of undissociated solute and the mole equivalents of ions. Such a calculation is accurate only if the solution is an ideal one, and if it is extremely dilute. Therefore, it is preferable to determine the osmolarity of a solution experimentally, e.g., by freezing-point depression.

* The ionic strength of a solution is defined by the equation $I = \frac{1}{2} \Sigma M_i Z^2$, where $M_i$ is the molarity of a given ion and $Z$ is the charge, regardless of sign. Since the $Z$ term is squared, the ionic strength of an ion increases exponentially with the magnitude of its charge either positive or negative. The magnitude of ionic charge, however, does not affect osmolarity.

**TABLE 8.3**

**Osmotic Tolerance of Certain Bacteria**

| Physiological Class | Representative Organisms | Approximate Range of NaCl Concentration Tolerated for Growth (%, g/100 ml) |
|---|---|---|
| Nonhalophiles | *Aquaspirillum serpens* | 0.0–1 |
| | *Escherichia coli* | 0.0–4 |
| Marine forms | *Alteromonas haloplanktis* | 0.2–5 |
| | *Pseudomonas marina* | 0.1–5 |
| Moderate halophiles | *Paracoccus halodenitrificans* | 2.3–20.5 |
| | *Vibrio costicolus* | 2.3–20.5 |
| | *Pediococcus halophilus* | 0.0–20 |
| Extreme halophiles | *Halobacterium salinarium* | 12–36 (saturated) |
| | *Halococcus morrhuae* | 5–36 (saturated) |

Note: Ranges of tolerated salt concentrations are only approximate; they vary with the strain and with the presence of other ions in the medium.

mate constancy of internal ionic strength. This has been shown through studies on *E. coli.* The concentration of intracellular putrescine varies inversely with the osmolarity of the medium; increases of osmolarity cause rapid excretion of putrescine. An increase in the osmolarity of the medium causes an increase in the internal osmolarity of the cell as a result of uptake of $K^+$; ionic strength is maintained approximately constant as a result of the excretion of putrescine. This is a consequence of the differing contributions that a multiply charged ion makes to ionic strength and osmotic strength of a solution; a change of putrescine$^{2+}$ concentration that alters ionic strength by 58 percent alters osmotic strength by only 14 percent.

Changes in osmotic strength or ionic strength of the growth medium also trigger a cellular response that changes the proportions in the outer membrane of *E. coli* of the two major protein constituents, OmpC and OmpF. These changes are thought to be adaptive, but the mechanism by which they alter the cell's ionic or osmotic tolerance remains unclear.

The presence in the medium of the amino acid proline dramatically increases a bacterium's ability to grow in a medium of high osmotic strength. The mechanism by which proline in-

**TABLE 8.4**

**Intracellular Concentrations of Solutes in Various Bacteria**

| Organism | Concentration (%, w/v) in Growth Medium of: | | Ratio of Intracellular to Extracellular Concentration of: | |
|---|---|---|---|---|
| | NaCl | KCl | $Na^+$ | $K^+$ |
| Nonhalophiles | | | | |
| *Staphylococcus aureus* | 0.9 | 0.19 | 0.7 | 27 |
| *Salmonella oranienburg* | 0.9 | 0.19 | 0.9 | 10 |
| Moderate halophiles | | | | |
| *Micrococcus halodenitrificans* | 5.9 | 0.02 | 0.3 | 120 |
| *Vibrio costicolus* | 5.9 | 0.02 | 0.7 | 55 |
| Extreme halophiles | | | | |
| *Sarcina morrhuae* | 23.4 | 0.24 | 0.8 | 64 |
| *Halobacterium salinarium* | 23.4 | 0.24 | 0.3 | 140 |

Source: Data from J. H. B. Christian and J. A. Waltho, "Solute Concentrations within Cells of Halophilic and Nonhalophilic Bacteria," *Biochem. Biophys. Acta* **65,** 506 (1962).

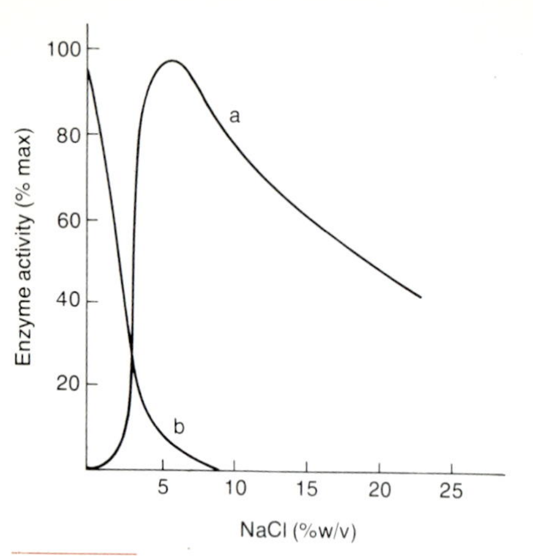


FIGURE 8.5

Effect of NaCl on the activity of the enzyme, malic dehydrogenase, from an extreme halophile (a) and from liver (b). Like most enzymes, the enzyme from liver becomes inactive in high concentrations of NaCl. The enzyme from the extreme halophile requires NaCl for activity. Redrawn from H. Larsen, "Biochemical Aspects of Extreme Halophism," *Advan. Microbiol. Physiol.* **1,** 97 (1967).

FIGURE 8.6

The influence of NaCl concentration on the growth rate of a typical marine bacterium *Pseudomonas marina*, growing (a) in a medium with concentrations of $Mg^{2+}$ and $Ca^{2+}$ (2 mM $MgSO_4$ and 0.55 mM $CaCl_2$) typical of a terrestrial environment, and (b) in a medium with concentrations of these ions (50 mM $MgSO_4$ and 10 mM $CaCl_2$) typical of a marine environment; in other respects, the media are the same. It will be noted that the higher concentration of $Mg^{2+}$ and $Ca^{2+}$ found in the marine environment spares the requirement for NaCl. After J. L. Reichelt and P. Baumann, "Effect of Sodium Chloride on Growth of Heterotrophic Marine Bacteria," *Arch. Microbiol.* **97,** 239 (1974).

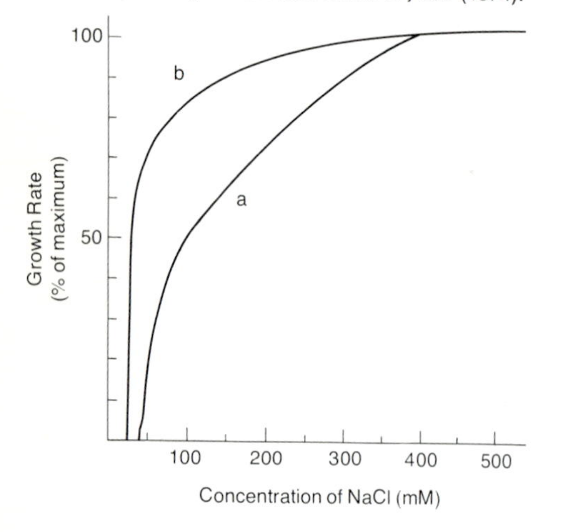


creases osmotic tolerance is also unclear, but its action is clearly intracellular because intracellular concentrations of proline correlate with osmotic tolerance regardless of whether it is raised by exogenous additions or by high rates of intracellular synthesis in certain mutant strains. Possibly proline directly protects intracellular proteins from high osmolarities. In certain organisms betaine mimics the effect of proline in osmotic protection.

## The Requirement for $Na^+$ in Bacteria

In most nonhalophilic bacteria it has not been possible to demonstrate a specific $Na^+$ requirement. In view of the extreme experimental difficulty of preparing a medium that is rigorously free of this very abundant ion, the possibility that nonhalophiles might require a very low concentration of $Na^+$ cannot be excluded. Among nonhalophiles a $Na^+$ requirement has been detected only for growth at the expense of certain specific carbon and energy sources: e.g., *E. coli* requires $Na^+$ for growth at maximal rate with L-glutamate, and *Enterobacter aerogenes* requires $Na^+$ for growth on citrate. In both cases, the quantitative requirement is small, and absolute growth dependence on $Na^+$ has not been demonstrated.

In contrast, bacteria of marine origin, most moderate halophiles, and extreme halophiles always require $Na^+$ for growth, at concentrations so high that their absolute dependence on this cation can be demonstrated experimentally without difficulty, even if the basal medium employed has not been prepared from specially purified ($Na^+$-free) ingredients.

In all these organisms, $Na^+$ probably plays a number of different roles, all indispensable to the maintenance of cellular function. In marine bacteria, there is good evidence that it assures the correct function of transport mechanisms. In the extreme halophiles, a high concentration of NaCl is essential in order to maintain both the stability and the catalytic activity of enzymes (Figure 8.5).

The walls of extreme halophiles of the genus *Halobacterium* lack a peptidoglycan layer, and are comprised exclusively of glycoprotein. Nevertheless, in very concentrated salt solutions (25 to 35 percent w/v, i.e., gm of solute per 100 ml solvent), similar to the natural environments in which these organisms live, the proteinaceous wall is sufficiently rigid to confer a cylindrical form on the cell. If the suspending medium is diluted to approximately 15 percent w/v, the cells become round, but they do not lyse. At lower salt concentration, the wall disaggregates into protein monomers, and lysis occurs. The cell wall of *Halobacterium* is therefore unique, by virtue of the fact that its structural integrity is assured by ionic bonds, a very high concentration of $Na^+$ (largely irreplaceable by other monovalent cations) being necessary to maintain the intermolecular association between the protein subunits of the wall.

For marine bacteria and other halophiles, the magnitude of the $Na^+$ requirement can be substantially reduced (by a factor as great as 2) by increasing the concentrations in the medium of two divalent cations, $Mg^{2+}$ and $Ca^{2+}$. The quantitative requirements of many halophiles for $Mg^{2+}$ and $Ca^{2+}$ also appear to be much greater than those of nonhalophiles. The influence of the NaCl content of two different media on the growth of a typical marine bacterium is shown in Figure 8.6.

## EFFECT OF TEMPERATURE ON MICROBIAL GROWTH

The rate of chemical reactions is a direct function of temperature and obeys the relationship originally described by Arrhenius:

$$\log_{10} v = \frac{-\Delta H^*}{2.303\,RT} + C$$

where $v$ represents the reaction velocity and $\Delta H^*$ the activation energy of the reaction, R the gas constant, and $T$ the temperature in degrees Kelvin. Hence, a plot of the logarithm of the velocity of chemical reaction as a function of $T^{-1}$ yields a straight line with a negative slope (Figure 8.7). Figure 8.8 shows a comparable plot of the rate of growth of *E. coli* as a function of $T^{-1}$. The curve is linear only over a portion of the temperature range for growth, since the growth rate falls abruptly at both the upper and the lower limits of the temperature range. The abrupt fall in growth rate at high temperatures is caused by the thermal inactivation of proteins and possibly of such cell structures as membranes. The *maximum* temperature for growth is the temperature at which these destructive reactions become overwhelming. This temperature is usually only a few degrees higher than the temperature at which growth rate is maximal, called the *optimum temperature.* Recently the very interesting observation has been made that if the square root of growth rate is plotted as a function of growth temperature a straight-line relationship is obtained over the linear portion of the Arrhenius plot and the low temperature range (Figure 8.9). Although this relationship has no obvious theoretical basis and fails by extrapolation to predict accurately the minimum temperature for growth, it should prove useful in predicting intermediate growth rates from limited data.

From the effect of temperature on the rate of a chemical reaction one would predict that all bacteria would continue to grow (although at progressively lower rates) as the temperature is reduced, until the system freezes. However, most bacteria stop growing at a temperature (the *minimum temperature of growth*) well above the freezing point of water. Every microorganism has a precise minimum temperature of growth, below which growth will not occur however long the period of incubation.

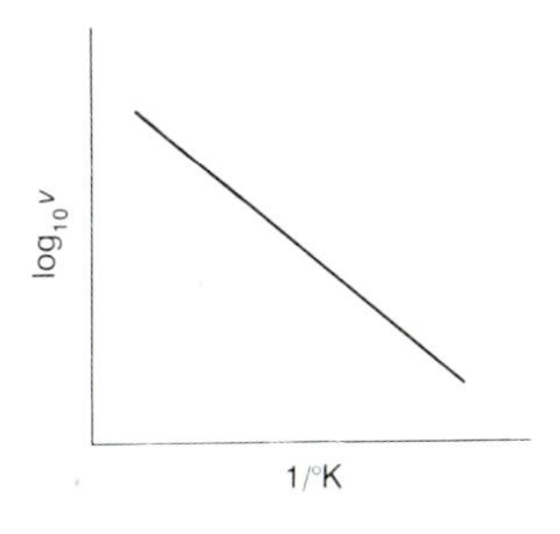


FIGURE 8.7
A generalized Arrhenius plot of the relationship between the velocity of a chemical reaction and temperature.

FIGURE 8.8
Arrhenius plot of growth rate of *E. coli* B/r. Individual data points are marked with corresponding degrees Celsius. *E. coli* B/r was grown in a rich complex medium (•) and a glucose—mineral salts medium (o). After S. L. Herendeen, R. A. VanBogelen, and F. C. Neidhardt, "Levels of Major Proteins of *Escherichia coli* during Growth at Different Temperatures, *J. Bacteriol.* **138,** 185 (1979).

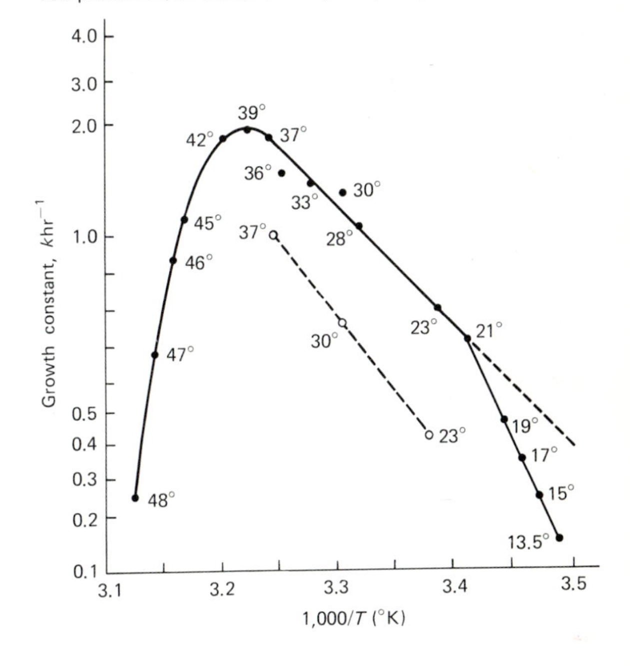

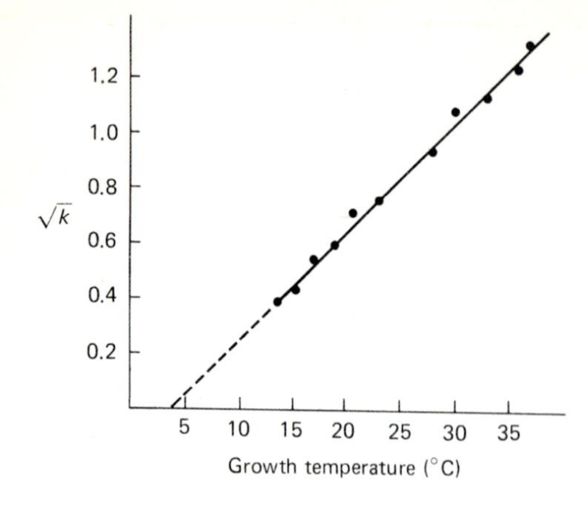


FIGURE 8.9

Plot of the data from Figure 8.8 according to the method of Ratkowski [D. A. Ratkowski, A. J. Olley, T. A. Meekin, and A. Ball, "Relationship between Temperature and the Growth Rate of Bacterial Cultures," *J. Bacteriol.* **149,** 1 (1982)]. The plot of the square root of the growth rate, $k$, vs. the growth temperature closely fit a straight line, but its extrapolation (dotted line) to zero (at 3.5° C) does not accurately predict the actual minimum temperature for growth (8° C).

**TABLE 8.5**

**Temperature Range of Growth of Certain Procaryotes**

| Organism | Temperature (°C): −10 0 10 20 30 40 50 60 70 80 90 100 | |
|---|---|---|
| *Bacillus globisporus* | ←——→→ | Psychrophiles |
| *Micrococcus cryophilus* | ←——→ | Psychrophiles |
| *Pseudomonas* sp | ..——→ | Psychrophiles |
| *Vibrio marinus* | ..——→ | Psychrophiles |
| *Xanthomonas pharmicola* | ←——→ | Psychrophiles |
| *Pseudomonas avanae* | ←——→ | Psychrophiles |
| *Xanthomonas rinicola* | ←——→ | Mesophiles |
| *Gaffkya homari* | ←——→ | Mesophiles |
| *Neisseria gonorrhoeae* | ←——→ | Mesophiles |
| *Escherichia coli* | ←——→ | Mesophiles |
| *Acholeplasma blastoclosticum* | ←——→ | Mesophiles |
| *Vibrio cholerae* | ←——→ | Mesophiles |
| *Fusobacterium polymorphum* | ←——→ | Mesophiles |
| *Haemophilus influenzae* | ←——→ | Mesophiles |
| *Lactobacillus lactis* | ←——→ | Mesophiles |
| *Bacillus subtilis* | ←——→ | Mesophiles |
| *Lactobacillus delbrueckii* | ←——→ | Mesophiles |
| *Mastigocladus laminosus* | ...——→ | Thermophiles |
| *Bacillus coagulans* | ←——→ | Thermophiles |
| *Synechococcus lividus* | ...——→ | Thermophiles |
| *Bacillus stearothermophilus* | ←——→ | Thermophiles |
| *Thermus aquaticus* | ....——→ | |

Note: Lines terminating in single arrows indicate established temperature limits of growth for at least one strain of the indicated species; variations exist among different strains of some species. Double-headed arrows indicate that the actual temperature limit lies between the arrow points. Solid lines terminating in dotted lines indicate that the minimum growth temperature is not established.

The numerical values of the *cardinal temperatures* (minimum, optimum, and maximum), and the range of temperature over which growth is possible vary widely among bacteria. Some bacteria isolated from hot springs are capable of growth at temperatures as high as 110°C; others, isolated from cold environments, can grow at temperatures as low as −10° C if high solute concentrations prevent the medium from freezing. On the basis of the temperature range of growth, bacteria are frequently divided into three broad groups: *thermophiles*, which grow at elevated temperature (above 55° C); *mesophiles*, which grow well in the midrange of temperature (20 to 45° C); and *psychrophiles*, which grow well at 0° C.

As is often true of systems of biological classification, this terminology implies a clearer distinction among types than is found in nature. The tripartite classification of temperature response does not take fully into account the variation among bacteria with respect to the extent of the temperature range over which growth is possible.*

The data describing the temperature ranges of growth of many different bacteria (Table 8.5) show the somewhat arbitrary nature of the designations *thermophile*, *mesophile*, and *psychrophile*. The range of temperature over which growth is possible is less variable than are the maxima and minima. Although some bacteria, like *Neisseria gonorrhoeae*, grow only within a very restricted range of temperature, most grow over a range of about 35° C and very few grow over a more extended range.

### Factors that Determine Temperature Limits for Growth

The factors that determine the temperature limits for growth have been revealed by two types of investigations: comparisons of the properties of organisms with widely different temperature ranges; and analyses of the properties of *temperature-sensitive* mutants, the temperature range of which has been decreased by a single mutational change. Temperature-sensitive mutants are of two types: *heat-sensitive mutants*, with decreased maximum growth temperatures; and *cold-sensitive mutants*, with increased minimum growth temperatures.

Studies on the kinetics of thermal denaturation both of enzymes and of cell structures that contain proteins (e.g., flagella, ribosomes) have shown that many specific proteins of thermophilic bacteria are considerably more heat-stable than their homologues from mesophilic bacteria. It is also possible to make an approximate determination of the overall thermal stability of soluble cell proteins by measuring the rates at which the protein in a cell-free bacterial extract becomes insoluble as a result of heat denaturation at several different temperatures. Experiments like this (Table 8.6) clearly demonstrate that virtually all the proteins of a thermophilic bacterium remain in the native state after a heat treatment that denatures virtually all the proteins of a related mesophile. It therefore follows that the adaptation of a thermophilic microorganism to its thermal environment can be achieved only through mutational changes affecting the primary structures of most (if not all) proteins of the cell.

Although the evolutionary adaptations that have produced thermophiles must have involved mutations that *increased* the thermal stability of their proteins, most of the mutations that affect the primary structure of a specific protein (e.g., an enzyme) decrease the thermal stability of that protein, even though many of these mutations may have little or no effect on its catalytic properties. Consequently, in the absence of counterselection by a thermal challenge, the maximal temperature for growth of *any* microorganism should decline progressively as a result of random mutations that

**TABLE 8.6**

**Stability of Cytoplasmic Proteins from Mesophilic and Thermophilic Bacteria at 60° C**

| Organism | Temperature Class | Percent of Proteins Denatured[a] |
|---|---|---|
| *Proteus vulgaris* | Mesophile | 55 |
| *Escherichia coli* | Mesophile | 55 |
| *Bacillus megaterium* | Mesophile | 58 |
| *Bacillus subtilis* | Mesophile | 57 |
| *Bacillus stearothermophilus* | Thermophile | 3 |
| *Bacillus* sp. (Purdue CD) | Thermophile | 0 |
| *Bacillus* sp. (Texas 11330) | Thermophile | 4 |
| *Bacillus* sp. (Nebraska 1492) | Thermophile | 0 |

Source: Data from H. Koffler and G. O. Gale, "The Relative Thermostability of Cytoplasmic Proteins from Thermophilic Bacteria," *Arch. Biochem. Biophys.* **66,** 249(1957).

[a] Percent of total trichloracetic acid—precipitable material from a sonic extract of cells which is coagulated by an 8-minute heat treatment at 60° C.

* Differences in temperature range among thermophiles are sometimes indicated by the terms *stenothermophile* (an organism which cannot grow below 37°C), and *eurithermophile* (an organism which can do so). Psychrophiles with temperature ranges that extend above 20°C are termed *facultative psychrophiles*; ones which cannot grow above 20°C are termed *obligate psychrophiles*.

affect the primary structure of its proteins. This inference is supported by the observation that psychrophilic bacteria isolated from antarctic waters contain a large number of exceptionally heat-labile proteins.

At low temperature, all proteins undergo slight conformational changes, attributable to the weakening of their hydrophobic bonds, which play an important role in determining tertiary (three-dimensional) structure. All other types of bonds in proteins become stronger as the temperature is lowered. The importance of precise conformation to the proper function of allosteric proteins (Chapter 12) and to the self-assembly of ribosomal proteins makes these two classes of proteins particularly sensitive to cold inactivation. Therefore, it is not surprising that mutations that increase the minimum temperature for growth usually occurs in genes that encode these proteins.

### Effect of Growth Temperature on Lipid Composition

The lipid composition of almost all organisms, both procaryotes and eucaryotes, alters with the growth temperature. As the temperature decreases, the relative content of unsaturated fatty acids in the cellular lipids increases.

An illustration of this phenomenon in *E. coli* is shown in Table 8.7. This change in fatty acid composition is a significant component of temperature adaptation in bacteria. The melting point of lipids is directly related to their content of saturated fatty acids. Consequently, the degree of saturation of the fatty acids in membrane lipids determines their degree of fluidity at a given temperature. Since membrane function depends on the fluidity of the lipid components, it is understandable that growth at low temperature should be accompanied by an increase in the degree of unsaturation of fatty acids.

**TABLE 8.7**

**Effect of Growth Temperature on the Amounts of Major Fatty Acids of *E. coli* (as percent of total fatty acids)**

| Fatty Acid | Temperature of Growth 10°C | 43°C |
|---|---|---|
| SATURATED FATTY ACIDS | | |
| Myristic | 3.9 | 7.7 |
| Palmitic | 18.2 | 48.0 |
| UNSATURATED FATTY ACIDS | | |
| Hexadecenoic | 26.0 | 9.2 |
| Octadecenoic | 37.9 | 12.2 |

Source: Data from A. G. Marr and J. L. Ingraham, "Effect of Temperature on the Composition of Fatty Acids in *E. coli*," *J. Bacteriol.* **8.4,** 1260 (1962).

## OXYGEN RELATIONS

The present atmosphere of the earth contains about 20 percent (v/v) of the highly reactive gas, oxygen. With the exception of many bacteria and a few protozoa, worms and molluscs, all organisms are dependent on the availability of molecular oxygen as a nutrient. The responses to $O_2$ among bacteria are remarkably variable, and this is an important factor in their cultivation (see Chapter 2). The aerobes are dependent on $O_2$; the facultative anaerobes use $O_2$ if it is available, but they also can grow in its absence; the anaerobes cannot utilize $O_2$. Anaerobes are of two types: the *obligate anaerobes*, for which $O_2$ is toxic, and the *aerotolerant anaerobes*, which are not killed by exposure to $O_2$. Although the toxicity of $O_2$ is most strikingly revealed by its effect in the obligate anaerobes, it is, in fact, toxic even for aerobic organisms at high concentration. Many obligate aerobes cannot grow in $O_2$ concentrations greater than atmospheric (i.e., $>20$ percent v/v). Indeed, some obligate aerobes require $O_2$ concentrations considerably lower than atmospheric (2 to 10 percent v/v) in order to grow. Aerobic bacteria that show such $O_2$ sensitivity under *all* growth conditions are called *microaerophiles.* Frequently, however, the requirement of a strict aerobe for a reduced $O_2$ concentration is conditional, being greatly influenced by either the energy source or the nitrogen source. For example, many hydrogen-oxidizing bacteria, which tolerate an atmospheric $O_2$ concentration when growing with organic substrates, require considerably lower $O_2$ concentrations when using $H_2$ as an energy source. Some aerobic nitrogen-fixing bacteria can fix $N_2$ only in the virtual absence of $O_2$ (see p. 106).

### The Toxicity of Oxygen: Chemical Mechanisms

All bacteria contain certain enzymes capable of reacting with $O_2$; the number and variety of these enzymes determine the physiological relations of the organism to oxygen. The oxidations of flavoproteins by $O_2$ invariably result in the formation of a toxic compound, $H_2O_2$, as one major product. In addition, these oxidations (and possibly other enzyme-catalyzed oxidations or oxygenations) produce small quantities of an even more toxic free radical,* superoxide or $O_2^{\cdot -}$.

In aerobes and aerotolerant anaerobes, the potentially lethal accumulation of superoxide ($O_2^{\cdot -}$) is prevented by the enzyme *superoxide dismutase*, which catalyzes its conversion to oxygen and hydrogen peroxide:

$$2O_2^{\cdot -} + 2H^+ \xrightarrow[\text{dismutase}]{\text{superoxide}} O_2 + H_2O_2$$

Nearly all these organisms also contain the enzyme *catalase*, which decomposes hydrogen peroxide to oxygen and water:

$$2H_2O_2 \xrightarrow[\text{catalase}]{} 2H_2O + O_2$$

One bacterial group able to grow in the presence of air (lactic acid bacteria, see Chapter 24) does not contain catalase. However, these organisms do not accumulate significant quantities of $H_2O_2$. Some of them decompose it by means of *peroxidases*, enzymes that catalyze the oxidation of organic compounds by $H_2O_2$, which is reduced to water; others employ a catalase-like reaction dependent on high intracellular concentrations of $Mn^{+2}$ (Chapter 23).

Superoxide dismutase and catalase, or peroxidase therefore play roles in protecting the cell from the toxic consequences of oxygen metabolism. The distribution of superoxide dismutase and catalase in bacteria with differing physiological responses to $O_2$ is shown in Table 8.8. Organisms that can tolerate an exposure to $O_2$ almost always contain superoxide dismutase, although not all necessarily contain catalase. However, *all strict anaerobes so far examined lack both superoxide dismutase and catalase.*

TABLE 8.8

**The Distribution of Superoxide Dismutase and Catalase in Bacteria with Differing Physiological Responses to $O_2$**

| Bacterium | Contains: Superoxide Dismutase | Contains: Catalase |
|---|---|---|
| Aerobes or facultative anaerobes | | |
| *Escherichia coli* | + | + |
| *Pseudomonas* spp. | + | + |
| *Deinococcus radiodurans* | + | + |
| Aerotolerant bacteria | | |
| *Butyribacterium rettgeri* | + | − |
| *Streptococcus faecalis* | + | − |
| *Streptococcus lactis* | + | − |
| Strict anaerobes | | |
| *Clostridium pasteurianum* | − | − |
| *Clostridium acetobutylicum* | − | − |

## The Photooxidative Effect

The toxicity of $O_2$ for living organisms can be greatly enhanced if the cells are exposed to light in the presence of air and of certain pigments, known as *photosensitizers*. Light converts the photosensitizer ($P$) to a highly reactive form, known as the triplet state ($P^*$):

$$P + h\nu \longrightarrow P^*$$

A secondary reaction between $P^*$ and $O_2$ produces singlet-state oxygen ($^1O_2$):

$$P^* + O_2 \longrightarrow P + {}^1O_2$$

Like the superoxide radical, singlet-state oxygen is a very powerful oxidant, and its formation within the cell is rapidly lethal.

One of the principal biological functions of carotenoid pigments is to act as *quenchers of singlet-state oxygen*, and thus protect the cell from photooxidative death. This function is of particular importance in phototrophs, since chlorophylls are powerful photosensitizers; the photosynthetic apparatus invariably contains carotenoid pigments. Their role in the prevention of lethal chlorophyll-mediated photooxidations was first shown in the purple bacterium *Rhodobacter sphaeroides*. *R. sphaeroides* grows photosynthetically under strictly anaerobic conditions and can also grow aerobically, either in the light or in the dark. Blue-green mutants, which have lost the ability to synthesize colored carotenoids, can still grow normally either anaerobically in the light or aerobically in the dark, but they are rapidly killed by simultaneous exposure to light and air. In phototrophic organisms that produce $O_2$ as a photosynthetic product, the loss of colored carotenoids totally abolishes photosynthetic function, since the accumulation of singlet-state oxygen, formed from metabolically generated $O_2$, occurs as soon as the cells are illuminated.

Many aerobic, nonphotosynthetic microorganisms also synthesize carotenoid pigments, which

* A free radical is a compound with an unpaired electron, indicated by a single dot in the structural formula. Having gained an extra electron, superoxide carries a negative charge.

are incorporated into the cell membrane and function as quenchers of singlet-state oxygen produced by such photosensitizing cellular pigments as the cytochromes. The role of carotenoids in protecting aerobic bacteria against the photodynamic action of sunlight has been demonstrated by studies on the light sensitivity of nonpigmented mutants of *Micrococcus luteus* and *Halobacterium salinarium*. This protection is probably of ecological importance in all aerobic bacteria that live in environments exposed to high light intensities.

## Oxygen-Sensitive Enzymes

Many enzymes, particularly enzymes of strict anaerobes, are rapidly and irreversibly denatured by exposure to $O_2$. Their purification and study must therefore be conducted under rigorously anaerobic conditions. A notable example is *nitrogenase*, the enzyme responsible for nitrogen fixation, which catalyzes the reaction

$$N_2 + 8H^+ + 8e^- \longrightarrow 2NH_3 + H_2$$

Even the nitrogenases from obligately aerobic nitrogen-fixing bacteria, such as the *Azotobacter* group, exhibit extreme oxygen sensitivity after extraction from the cell. In intact cells of *Azotobacter*, nitrogenase is evidently protected from inactivation by the high rate of utilization of $O_2$ by this organism. The nitrogenases of facultatively anaerobic nitrogen-fixing bacteria (*Enterobacter*, *Bacillus polymyxa*) are not so protected in the intact cell; consequently, these bacteria can fix nitrogen effectively only under anaerobic growth conditions. Most filamentous nitrogen-fixing cyanobacteria produce specialized cells (heterocysts) lacking photosystem II, in which nitrogenase is protected from oxygen inactivation (see Chapter 15).

## The Role of Oxygenases in Aerobic Microorganisms

Although the primary metabolic function of $O_2$ in strict aerobes is to serve as a terminal electron acceptor, it also serves as a *cosubstrate* for enzymes that catalyze some steps in the dissimilation of aromatic compounds and alkanes. These enzymes are termed *oxygenases*; they mediate a direct addition of either one or two oxygen atoms to the organic substrate. An example is the oxygenative ring cleavage of catechol, an intermediate in the dissimilation of many aromatic compounds:

$$\text{(catechol: ring with OH, OH)} + O_2 \xrightarrow[\text{oxygenase}]{\text{catechol}} \text{(ring-cleaved product: COOH, COOH)}$$

Many aerobic pseudomonads that are able to use aromatic compounds or alkanes as sole sources of carbon and energy are denitrifiers, and hence can grow anaerobically, using nitrate in place of $O_2$ as a terminal electron acceptor. However, this metabolic option can be exercised only with oxidizable substrates that are catabolized by dehydrogenases. Substrates in the dissimilation of which one or more steps are mediated by oxygenases cannot support anaerobic growth, since nitrate is unable to replace $O_2$ as a cosubstrate for oxygenases.

In eucaryotes and some procaryotes, the biosynthesis of sterols and unsaturated fatty acids involves steps mediated by oxygenases. Consequently, yeasts require sterols and unsaturated fatty acids as growth factors when growing fermentatively under anaerobic conditions, even though they can synthesize these cell components when they are growing aerobically.

## FURTHER READING

### Books


BROCK, T. D., *Thermophilic Microorganisms and Life at High Temperatures*. Heidelberg: Springer-Verlag, 1978.

ROSEN, B. P., ed. *Bacterial Transport*. New York: Marcel Dekker, 1978.


### Review


SILHAVY, T. J., S. A. BENSON, and S. D. EMR. "Mechanisms of Protein Localization," *Microbiol. Rev.* **47**, 313 (1983).

# Chapter 9
# The Viruses

Before the discovery in the nineteenth century that bacteria cause disease, the term *virus*, a Latin word for poisonous substance, was used to describe any disease-causing substance. Following that, this term was often used to describe microbial disease-causing agents. In the twentieth century, the term virus acquired a new and more specific meaning. Now it describes a large group of disease-causing agents that are fundamentally different from all cellular forms of life: viruses are infectious nucleic acid encapsulated in a protein coat. They may possess membranes but do not have any cytoplasm or metabolism of their own. Therefore, they must penetrate host cells where their nucleic acid directs the replication of viral macromolecular components which are then *assembled* into new viruses.

## THE DISCOVERY OF VIRUSES

Viruses were discovered in the nineteenth century, some 50 years before the development of the electron microscope, the only instrument capable of forming a visible image of such small objects. Consequently, viruses were not discovered in the sense that bacteria were when Leeuwenhoek first saw them. Rather, their existence and properties were deduced from the results of experiments done by a small number of scientists: no individual deserves exclusive credit for their discovery. The first demonstrations that viral diseases could be transmitted from one host to another under controlled laboratory conditions were published by L. Pasteur in 1884 in studies on rabies, and by A. Mayer in 1886 in studies on mosaic disease of tobacco plants. However, the nature of the causative agents of these infectious diseases was not clearly established by their experiments because neither of these viruses could be grown in pure culture by the methods available at that time. In 1892, D. Iwanowsky established by a simple experiment that the causative agent of tobacco mosaic disease is smaller than any bacterium then known; he passed an extract of diseased leaves through a porcelain filter with pores fine enough to block the passage of most bacteria and demonstrated that the filtrate remained highly infectious. As a result of this experiment, infectious agents that could pass through fine filters became known as *filterable viruses*.

In 1898 M. Beijerinck established that viruses possess the property of replication that is common to all living things by demonstrating that tobacco mosaic virus (TMV), the filterable virus that causes tobacco mosaic disease, proliferates in infected tissue; in that same year F. Loeffler and P. Frosch showed that the filterable agent of foot-and-mouth disease proliferates in cattle. Beijerinck further established that TMV proliferates only in growing plant tissue, an observation that led him to the correct conclusion that virus proliferation occurs intracellularly and is dependent on the active metabolism of host cells. In 1915 and 1917, respectively, F. Twort and F. d'Herelle independently discovered that some viruses, termed *bacteriophages* (i.e., eaters of bacteria) or simply *phages*, infect bacteria. Thus the three major biological groups—animals, plants, and bacteria—are all susceptible to viral disease.

For several decades after their discovery, viruses were distinguished by three properties: (1) they are infectious agents of disease, (2) they are quite small and hence are invisible in the light microscope and able to pass through filters that retain most bacteria, and (3) they do not proliferate in culture media designed to support growth of bacteria. Although these properties separate viruses from most bacteria, they are insufficient criteria for distinguishing between all bacteria and viruses. For example, the chlamydiae (Chapter 21), a group of eubacteria, pass through porcelain filters and can grow only inside host cells. Not surprisingly, the chlamydiae were initially classified as viruses.

In 1935, W. Stanley demonstrated the remarkable chemical simplicity of the plant virus TMV: he crystallized the virus and then showed that it is composed largely of protein. Later, other scientists showed that these crystals also contain a small but constant fraction of RNA. Chemical studies of other viruses revealed that some contain DNA in addition to protein, but no virus has been found that contains both DNA and RNA. In addition to nucleic acid, some viruses contain lipid, and some contain small amounts of carbohydrate conjugated to their protein components. The largest and most complex viruses, the poxviruses (Chapter 32), are composed of nucleic acid and several internal components surrounded by membranes, but even these relatively complex structures do not approach the chemical complexity of the simplest cells.

Initially, the significance of the finding that viruses possess either RNA or DNA, but not both, was not appreciated because the functions of nucleic acids were then unknown. Now it is clear that the nucleic acid, either DNA or RNA, functions as the viral genome. Those that contain RNA exhibit the highly unusual biological property of having genetic information permanently encoded in RNA. In some cases, the genomic RNA also functions as mRNA.

## VIRAL STRUCTURE

The first photograph of a virus was obtained in 1942 using an electron microscope that was primitive by modern standards. Detailed micrographs of virus structure were only obtained in the 1950s when better instruments became available. Viral particles, termed *virions*, exhibit a great variety of shapes and sizes (Figures 9.1 and 9.2). Some appear to be rods which on careful examination prove to be helical; some appear almost spherical; and others appear to be a combination of helical and nearly spherical structures. A helical virion is usually formed by association of many identical protein subunits clustered around a single molecule of nucleic acid, termed the *viral chromosome*, which occupies either the core or an internal groove of the helix (Figure 9.3). The protein subunits of the helix are termed *capsid* proteins or *capsomers*, and the entire helical structure, like all protein and nucleic acid complexes, is termed a *nucleoprotein*.

Virions that appear approximately spherical in electron micrographs are termed *polyhedral* because their protein building blocks (also termed *capsomers*) are arranged to form a polyhedral shell, usually composed of 20 triangular faces (an *icosahedron*), that surrounds viral nucleic acid (chromosome) and, in some cases, protein. The protein shell of a polyhedral virus is termed the *capsid* and, unlike the capsid of helical virions, can often be separated as an intact structure from the viral nucleic acid. The capsomers of polyhedral viruses typically are multimeric proteins composed of five, six, or more subunits. In the simplest cases, capsid proteins

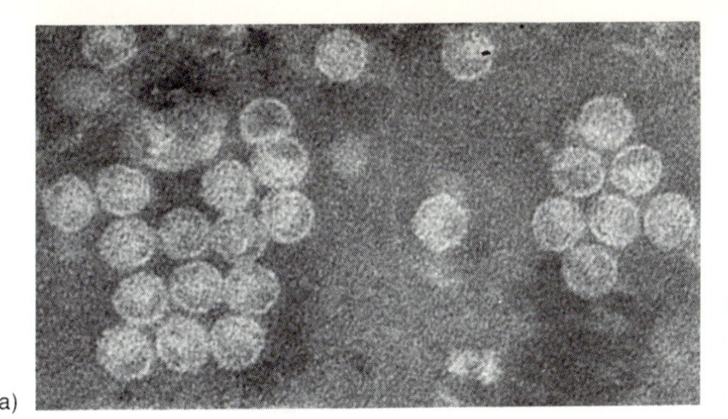

(a)

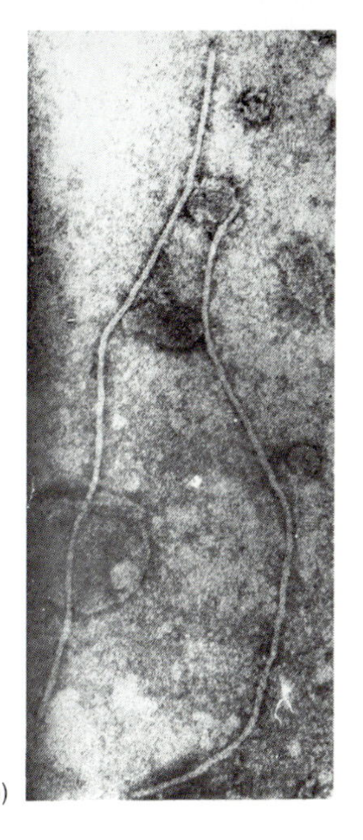

(b)

FIGURE 9.1 (left)

Bacteriophage virions: (a) icosahedral; (b) filamentous (helical symmetry). From D. E. Bradley, "The Structure of Some Bacteriophages Associated with Male Strains of *Escherichia coli*," *J. Gen. Microbiol.* **35,** 471 (1964).

FIGURE 9.2 (below)

Electron micrographs of a bacteriophage of *Bacillus subtilis* which adsorbs on the bacterial flagellum. (a) A free phage particle with helical tail fibers (× 120,000). (b) A phage particle that has adsorbed to a bacterial flagellum, around which the tail fibers are wrapped (× 120,000). (c) A group of phage particles attached to several flagella (× 61,000). From L. M. Raimondo, N. P. Lundh, and R. J. Martinez, "Primary Adsorption Site of Phage PBSI: the Flagellum of *Bacillus subtilis*," *J. Virol.* **2,** 256 (1968).

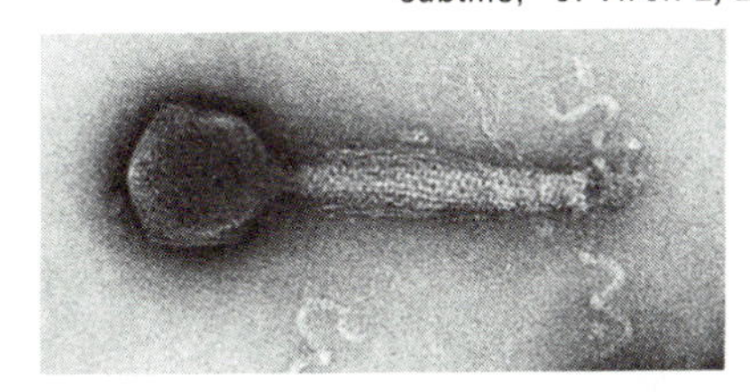

(a)

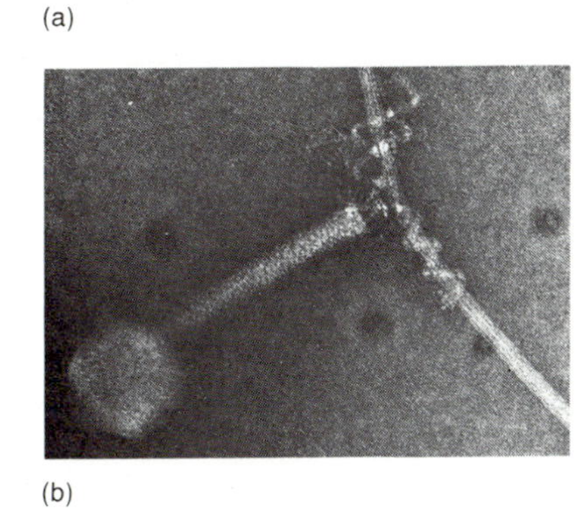

(b)

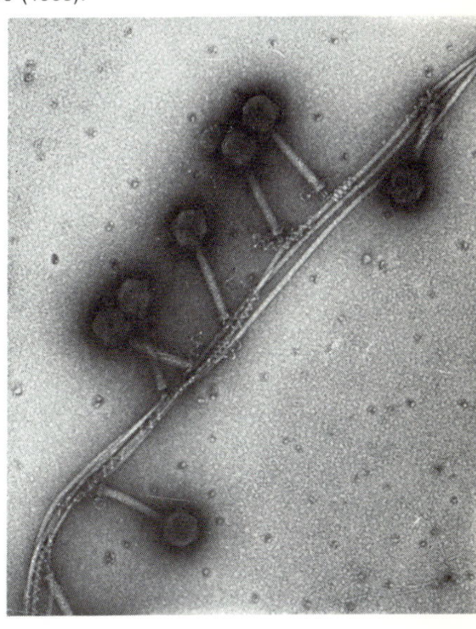

(c)

FIGURE 9.3 (below)

(a) A drawing of the structure of tobacco mosaic virus. For clarity, part of the ribonucleic acid chain is shown without its supporting framework of protein. From A. Klug, and D. C. D. Caspar, "The Structure of Small Viruses," *Adv. Virus Res.* **1,** 225 (1960).
(b) Electron micrograph of tobacco mosaic virus particles in phosphotungstic acid. From S. Brenner and R. W. Horne, "A Negative Staining Method for High Resolution Electron Microscopy of Viruses," *Biochem. Biophys. Acta* **34,** 103 (1959). Courtesy of R. W. Horne.

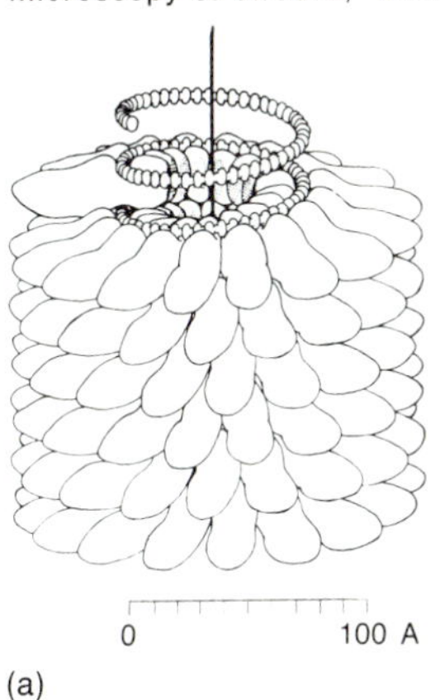


(a)

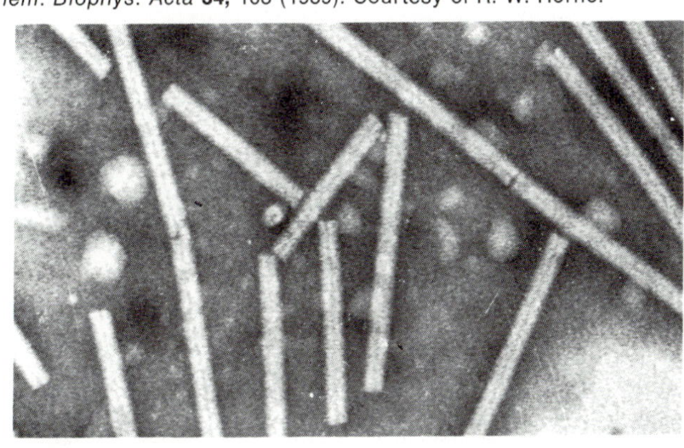

(b)

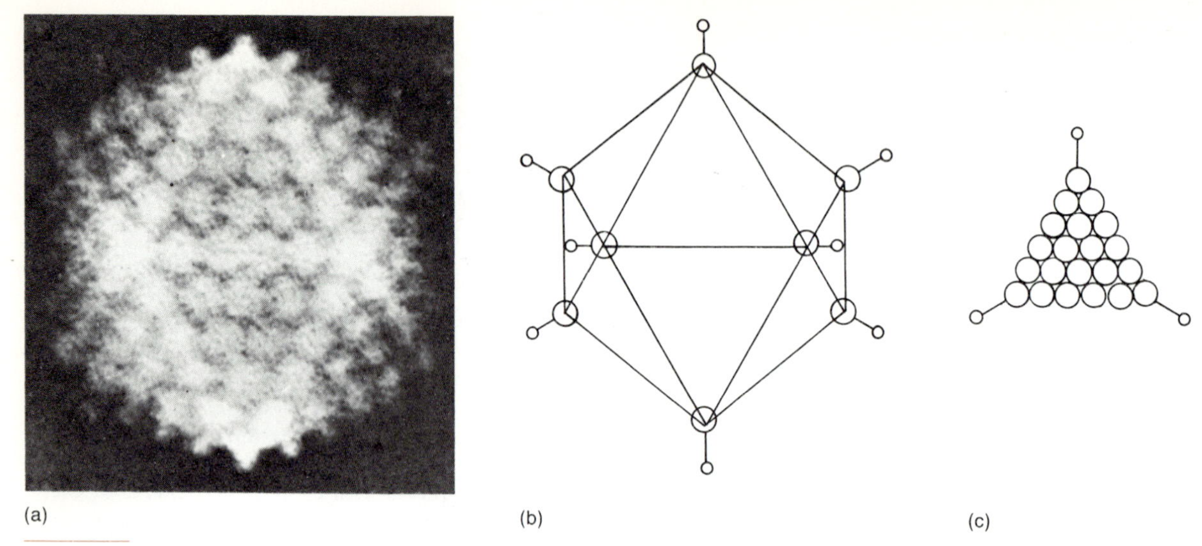


FIGURE 9.4

(a) Electron micrograph of an icosahedral virion (adenovirus). From R. C. Valentine and H. G. Pereira, "Antigens and Structure of the Adenovirus," *J. Mol. Biol.* **13,** 13 (1965). (b) Diagram of the edges of the icosahedron and the spikes at the vertices of the virion. Diagram of the capsomers arranged to form one face of the virion.

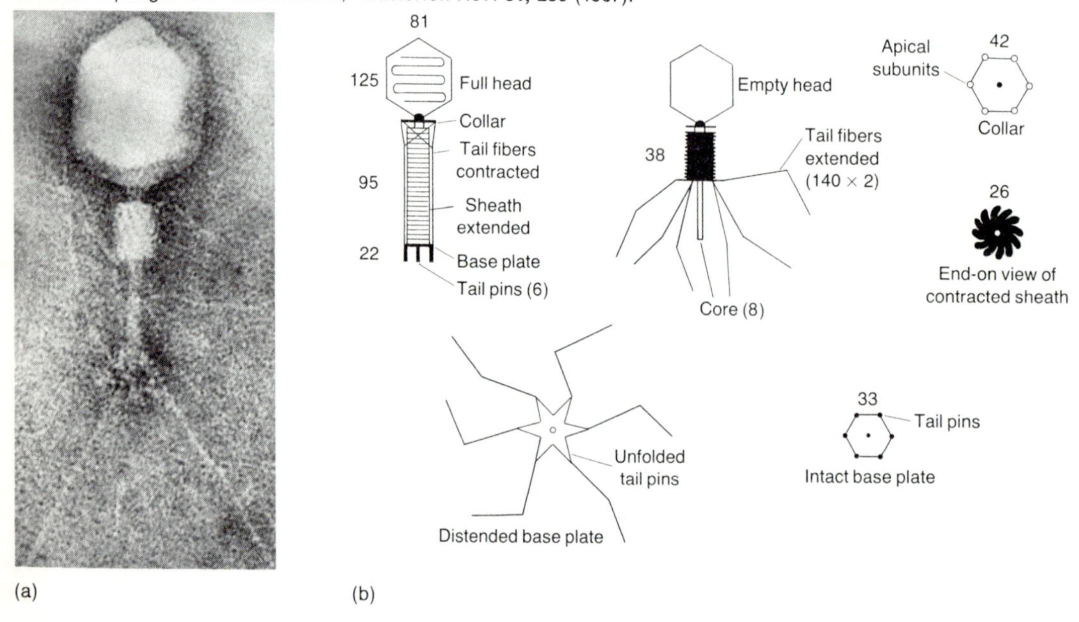


FIGURE 9.5

(a) An electron micrograph of a T-even bacterlophage negatively stained with phosphotungstic acid. Note the filled head, contracted sheath, core, and tail fibers. From S. Brenner et al., "Structural Components of Bacteriophage." *J. Mol. Biol.* **1,** 281 (1959). (b) T-even phage components, with dimensions indicated in nm. From D. E. Bradley, "Ultrastructure of Bacteriophages and Bacteriocins," *Bacteriol. Rev.* **31,** 230 (1967).

aggregate into pentamers, 12 of which form the capsid. More complex capsids are formed from 12 pentamers together with a number of hexameric capsomers. In all cases, the number of capsomers in a virion is a characteristic property of the class to which the polyhedral virus belongs. For example, adenoviruses have four hexamers on each edge, six hexamers on each face, and one pentamer at each vertex (Figure 9.4). As previously stated, an icosahedron has 20 faces, and thus it has 30 edges and 12 vertices so that adenoviruses have 240 ($6 \times 20 + 4 \times 30$) hexamers in addition to 12 pentamers.

Some phages are helical; some are polyhedral (Figure 9.1); and others are complex (Figure 9.2), being composed of a polyhedral structure (termed a *head*) connected to a helical structure (termed a *tail*). This combination, termed *binal* structure, is relatively common among phages, but is never found in animal or plant viruses. All the nucleic acid of binal phages is located within the head, the tail serving as an organ of attachment to host cells. The tails of most phages have the structure of a single protein tube, but some also have a thick outer tube, termed a *sheath*, surrounding an inner tube, termed a *core* (Figure 9.2). In some bacteriophages, the tail structure is remarkably complex. For example, in the *T-even phages* (phages T2, T4, and T6) the sheath contains 24 annular rings, which form a tube that is joined to the phage head by a thin *collar* (Figure 9.5). The tube's distal end is attached to a hexagonal *base plate*, which has a protein spike termed a *tail pin* at each of its corners. Six *tail fibers*, which are required for adsorption to bacteria, are also attached to the tail plate.

Viruses enclosed with a unit membrane envelope (Figure 9.6) are termed *enveloped* viruses; in contrast, those that lack an envelope are termed *naked viruses*. Most animal viruses are enveloped, but nearly all plant viruses and phages are naked. Some enveloped viruses, e.g., the herpesviruses, contain a polyhedral nucleoprotein core; others, e.g., the influenza viruses, contain a helical core. Viral membranes contain proteins encoded by viral genes, but in almost all cases their phospholipids are derived from host cell membranes; the poxviruses are an exception in this respect.

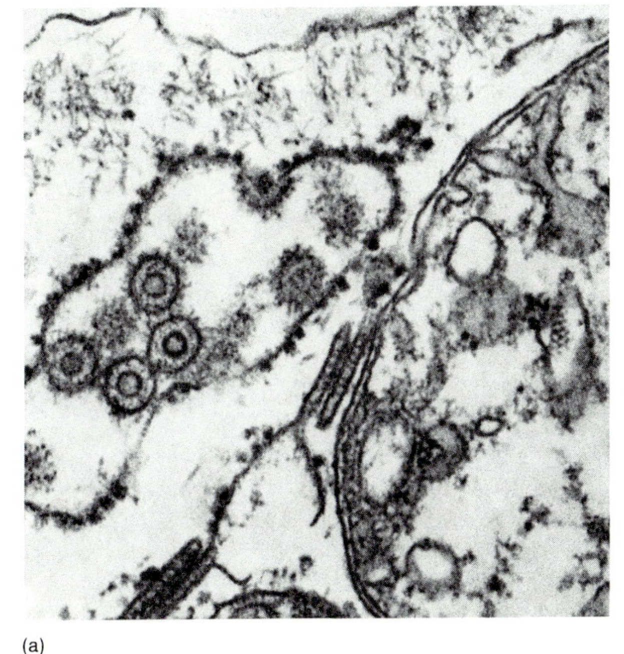

(a)

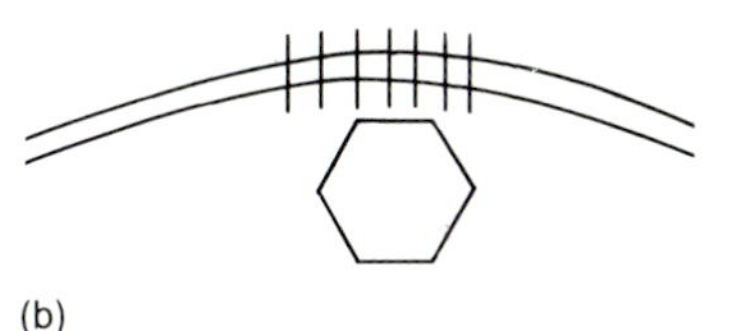

(b)

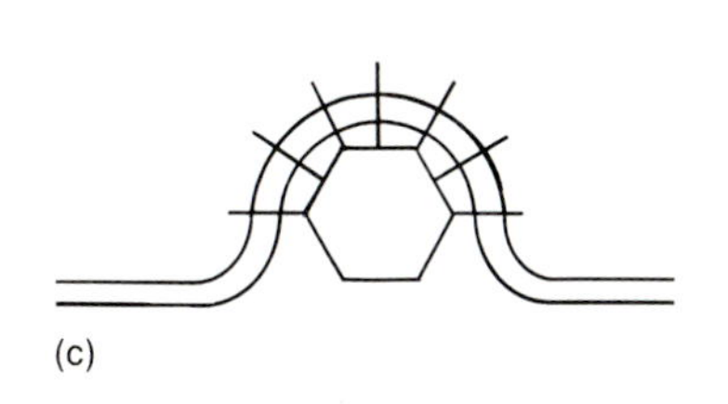

(c)

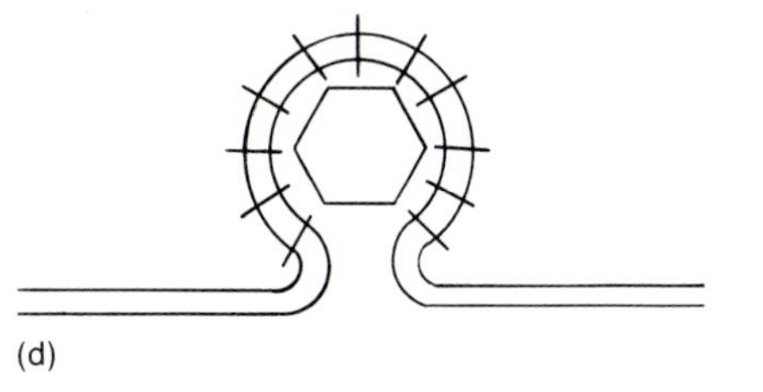

(d)

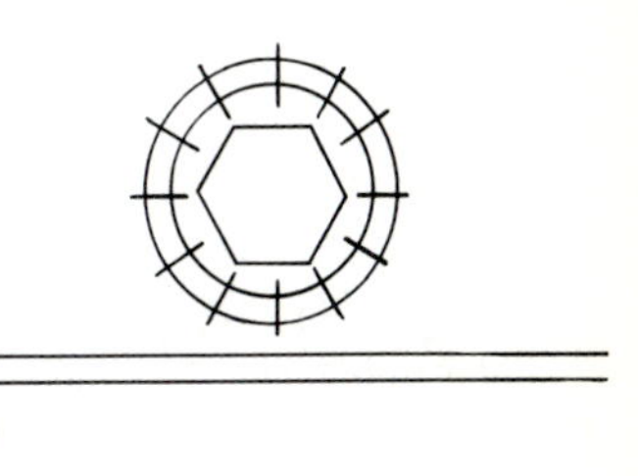

(e)

FIGURE 9.6

(a) Budding of an enveloped virus (feline syncytial virus). Five virions, each with a nucleoprotein core surrounded by a membrane envelope, are visible in the upper left region. Nearby, a virion can be seen in the process of acquiring its envelope as it buds jnto a pocket form by the cell membrane ($\times$ 80,000). Courtesy of J. S. Manning. (b), (c), (d) Diagram of the budding process as virion nucleoprotein associates with a region of membrane embedded with viral envelope protein. The viral core is wrapped entirely in the membrane, which pinches off to release the mature virion.

**TABLE 9.1**

**Some Families of Animal Viruses**

| Family Name | Morphology[a] | Enveloped (E) or Naked (N) | Approximate Size (nm) | Nucleic Acid[b] |
|---|---|---|---|---|
| Poxviridae (poxviruses) | 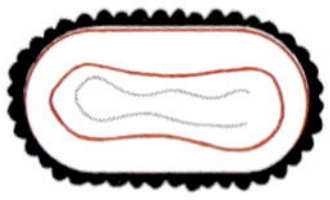 | E | 350 × 250 | Linear ds DNA |
| Herpesviridae (herpesviruses) |  | E | 200 | Linear ds DNA |
| Adenoviridae (adenoviruses) | 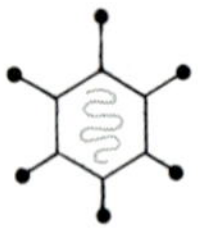 | N | 75 | Linear ds DNA |
| Parvoviridae (parvoviruses) | | N | 20 | Linear ss DNA |
| Papovaviridae (papovaviruses) | | N | 50 | Circular ds DNA |
| Baculoviridae (baculoviruses) | 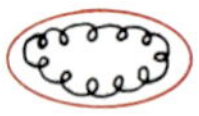 | E | 300 × 40 | Circular ds DNA |
| Picornaviridae (picornaviruses) | | N | 27 | Plus-strand RNA |
| Togaviridae (togaviruses) |  | E | 50 | Plus-strand RNA |
| Retroviridae (retroviruses) |  | E | 50 | Plus-strand RNA |
| Orthomyxoviridae (orthomyxoviruses) |  | E | 110 | Segmented: 8 minus-strand RNA molecules |
| Paramyxoviridae (paramyxoviruses) |  | E | 200 | Minus-strand RNA |
| Rhabdoviridae (rhabdoviruses) | 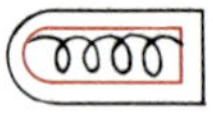 | E | 170 × 70 | Minus-strand RNA |
| Reoviridae (reoviruses) | | N | 65 | Segmented: 10–13 ds RNA molecules |

[a] Protein and nucleoprotein components are diagramed in black, nucleic acid in grey, and membranes (envelopes) in color.
[b] Single-stranded (ss) or double-stranded (ds) nucleic acid.

## CLASSIFICATION OF VIRUSES

In all viral taxonomies, individual viruses are grouped by the nucleic acid they contain (DNA or RNA), their size, and the architecture of their capsid. Presence of an envelope and chromosome structure (circular versus linear, and single-stranded versus double-stranded molecules) are also important properties in viral taxonomies. In some groups of RNA viruses, a virion contains several chromosomes, each encoding one or two viral proteins. Virions that exhibit this highly unusual genetic organization are said to possess a *segmented genome*. In some of the single-stranded RNA viruses, the chromosomes also serve as viral mRNA; in others, transcribed RNA, complementary to the RNA chromosome, serves as mRNA. Virion RNA that can function as mRNA is termed *plus-strand* RNA, while RNA that is complementary to virion mRNA is termed *minus-strand* RNA.

Animal, plant, and bacterial viruses are usually classified separately. Animal viruses are often grouped into families, genera, and species which are given Latin names, but English names are also used. Selected families of animal viruses together with their defining properties are listed in Table 9.1. A taxonomy employing Latin family names also has been proposed for bacterial viruses (Table 9.2), but this system is rarely used. Unlike the Latin systems, which resemble Linnaeus' classical taxonomy, the current English-named system for classifying plant viruses does not separate them into families, genera, or species. Instead, plant viruses are grouped on the basis of (1) the structure of the virion, (2) whether it contains DNA or RNA, and (3) its mode of transmission. In most cases, the groups of plant viruses are named after a prominent representative. For example, a group of plant viruses closely related to tobacco mosaic virus is termed the *tobacco group*. Selected groups of plant viruses together with their defining properties are listed in Table 9.3.

## THE VIRAL REPLICATION CYCLE

The essential features common to the replication cycles of all viruses include entry into the cytoplasm of a susceptible host cell, intracellular reproduction to produce progeny virions, escape of these into the environment, and survival there. All bacteria that are obligate intracellular parasites pass through a superficially similar cycle, but there are fundamental differences: viruses never reproduce by division. Rather, they are *replicated* by a process in which all molecular components are synthesized separately; then these are assembled into intact virions.

**TABLE 9.2**

**Some Families of Bacteriophages**

| Family Name | Morphology | Nucleic Acid[a] | Examples |
|---|---|---|---|
| Myoviridae | | Linear ds DNA | T2, T4, T6, P2 |
| Styloviridae | | Linear ds DNA | T5, $\lambda$ |
| Pedoviridae | | Linear ds DNA | T3, T7 |
| Microviridae | | Circular ss DNA | $\phi$X174 |
| Inoviridae | | Circular ss DNA | m13, fd |
| Leviviridae | | Linear ss RNA | Q$\beta$, R17, MS2 |
| Cystoviridae | | Linear ds RNA | $\phi$6 |

[a] Single-stranded (ss) or double-stranded (ds) nucleic acid.

**TABLE 9.3**

**Some Groups of Plant Viruses**

| Group (Representative) | Morphology and Approximate Size | Nucleic Acid[a] |
|---|---|---|
| Tobamoviruses (Tobacco mosaic virus) | Rigid helices 18 × 300 nm | Linear ss RNA |
| Potyviruses (Potato Y virus) | Flexible helices 11 × 700 nm | Linear ss RNA |
| Comoviruses (Cowpea mosaic) | Icosahedral with 30 nm diameter | Linear ss RNA |
| Caulimoviruses (Cauliflower mosaic) | Icosahedral with 50 nm diameter | Circular ds DNA |

[a] Single-stranded (ss) or double-stranded (ds) nucleic acid.

### Entry of Viruses into Host Cells

Plant viruses enter their hosts through breaches in the cell wall, often when insects with mouth parts contaminated by viruses feed on plant tissues. In contrast, the replication cycle of animal and bacterial viruses begins with a collision between a virion and a susceptible host cell: a structure on the virion surface binds to a specific molecular component, termed a *receptor*, on the cell surface. The process of attachment is termed *adsorption*.

In the case of polyhedral viruses that lack an envelope, one of the capsid proteins is responsible for adsorption. For example, capsomers at the vertices of an adenovirus particle (Figure 9.4) possess a protein spike that projects from the virion surface and interacts with specific components of the host cell membrane. Viruses that possess a membrane envelope have glycoproteins embedded in it, the carbohydrate residues of which adsorb to receptors in host cell membranes. In the case of binal phages, their tails adsorb to specific receptors on the cell envelope.

A variety of molecular components serve as viral receptors including, in the case of bacteriophages, teichoic acids, outer membrane proteins, lipopolysaccharides, flagella, and pili (see Table 9.4). This specificity of adsorption can be exploited to determine whether or not a cell carries a particular structure. For example, strains of enteric bacteria that possess the F plasmid (Chapter 11) carry F pili on their surfaces. Thus, a phage that adsorbs to this receptor can be used to determine whether or not a bacterial strain possesses the F factor. Similarly, phages that adsorb to a surface structure recognized by a specific antibody (Chapter 30) can be used as an alternative to the antibody in determining the serological type (Chapter 30) of a bacterial isolate. This procedure, termed *phage typing*, is widely used in epidemiological studies of food poisonings and other diseases.

Following adsorption of an animal or bacterial virus, its nucleic acid, sometimes accompanied by other virion components, enters the cell. Penetration by animal viruses that lack an envelope occurs by one of two processes: endocytosis or direct passage through the cell membrane. Virions entering by endocytosis are ingested into phagolysosomes (Chapter 29) from which some later escape into the cytoplasm. Enveloped viruses sometimes enter the cell by fusion between the viral membrane and the cell membrane, but recent studies suggest that endocytosis is the more common mechanism of entry. Escape of an enveloped virus from the phagolysosome occurs rapidly by fusion between the viral membrane and the membrane bounding the phagolysosome, thereby expelling the nucleoprotein viral core into the cytoplasm (Figure 9.7). Apparently, such fusion is triggered by the low pH of lysosomal contents.

**TABLE 9.4**

**Receptors for Bacteriophage Adsorption**

| Receptor | Phage | Host |
|---|---|---|
| Flagellum | PBS1 | *Bacillus subtilis* |
| F-pilus | M13, R17, fd | *Escherichia coli* |
| Lipopolysaccharide | T4, T7 | *E. coli* |
| Outer membrane proteins for: | | |
| Iron transport | T1 | *E. coli* |
| Nucleoside transport | T6 | *E. coli* |
| Maltose transport | $\lambda$ | *E. coli* |
| Teichoic acid | $\phi$29 | *B. subtilis* |

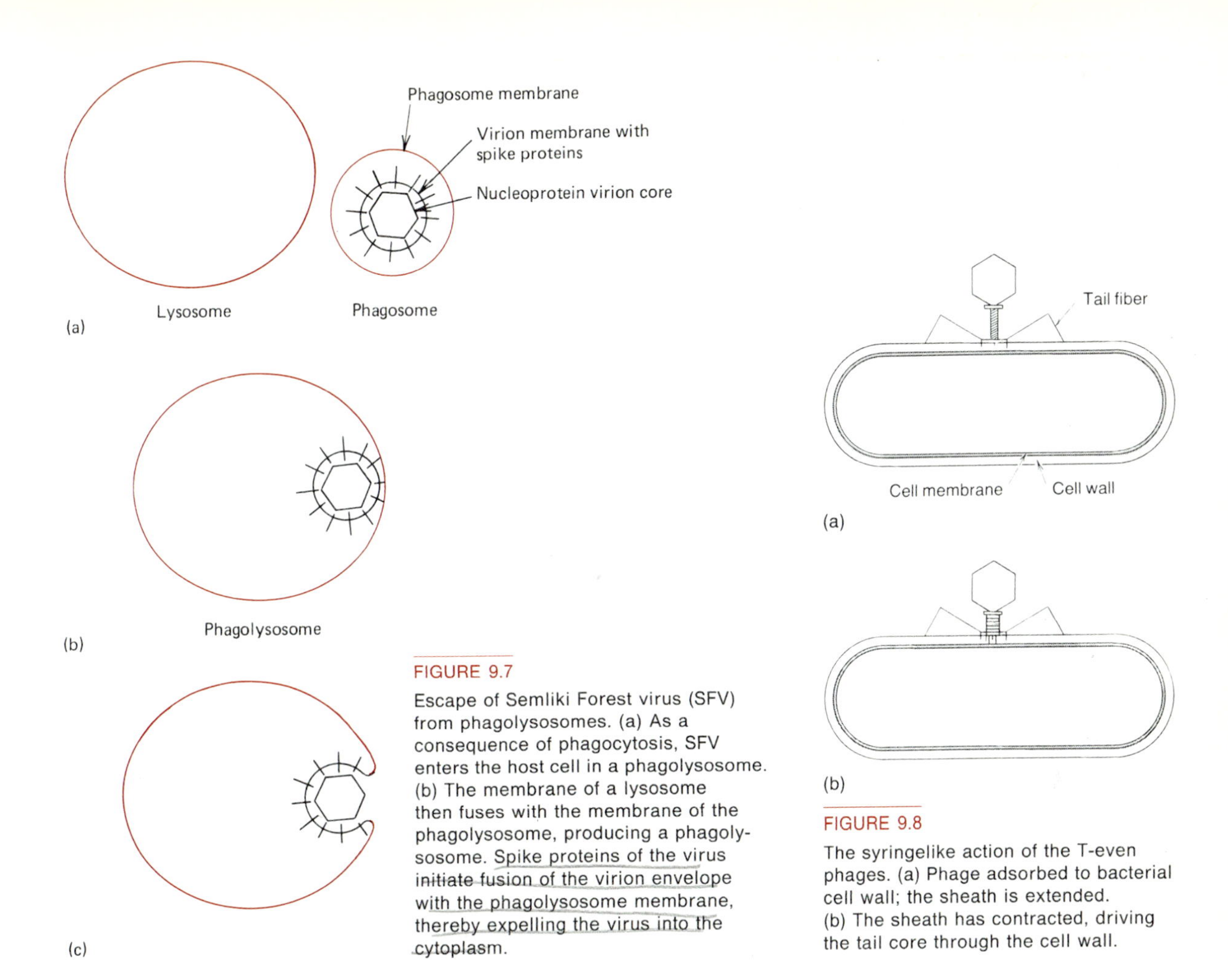


FIGURE 9.7

Escape of Semliki Forest virus (SFV) from phagolysosomes. (a) As a consequence of phagocytosis, SFV enters the host cell in a phagolysosome. (b) The membrane of a lysosome then fuses with the membrane of the phagolysosome, producing a phagolysosome. Spike proteins of the virus initiate fusion of the virion envelope with the phagolysosome membrane, thereby expelling the virus into the cytoplasm.

FIGURE 9.8

The syringelike action of the T-even phages. (a) Phage adsorbed to bacterial cell wall; the sheath is extended. (b) The sheath has contracted, driving the tail core through the cell wall.

Unlike animal and plant viruses, the capsid of most bacteriophages remains attached to a component of the cell surface, while the viral nucleic acid, often accompanied by internal proteins, penetrates into the cytoplasm. Filamentous (helical) phages are an exception: they adsorb to specific types of pili, and both phage DNA and capsid protein enter the cell, presumably as these pili are retracted. Few details of the penetration process of polyhedral phages are known, but the process of penetration of binal phages is understood in some detail. After adsorption of the tip of the tail to a component of the cell envelope, the DNA migrates through the hollow interior of the tail and penetrates the cell envelope by an unknown mechanism; in more complex T-even phages (Figure 9.5), the phage adsorbs to the cell surface by the tail fibers and pins; the sheath then contracts, driving the tail core through the outer membrane and the peptidoglycan layer (Figure 9.8); and phage DNA migrates into the cytoplasm.

## Uncoating

With the exception of a group of animal viruses, the *picornaviruses* (Chapter 32), nucleic acid of plant and animal viruses enters the cell still enclosed in its capsid and is liberated into the cytoplasm or nucleus by a process termed *uncoating*. Limited cleavage of capsid proteins by host proteases located on the cell surface or within lysosomes may initiate uncoating, but details of this process are poorly understood.

## Replication of Chromosomes of DNA Viruses

As stated before, viruses contain either DNA or RNA. In those with DNA, it is present within the virion either as a circular or a linear molecule, which is sometimes termed the *viral chromosome*. In some groups of viruses, chromosomes are single-stranded DNA; in others, they are double-stranded DNA.

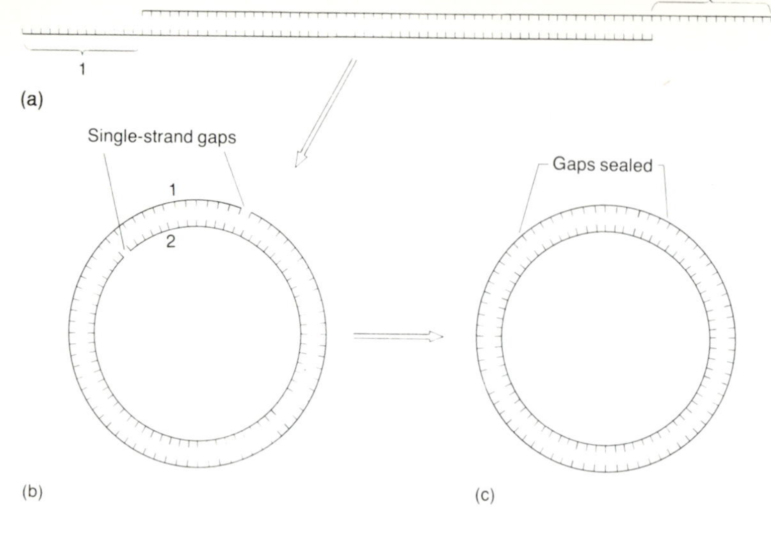


FIGURE 9.9

The conversion of the $\lambda$ genome from the linear form to the circular form. (a) The linear genome has single-stranded regions at each end that are complementary to each other. (b) The single-stranded regions are joined by hydrogen bonds between complementary bases. (c) The sugar-phosphate backbones are joined by polynucleotide ligase, forming a fully covalent circle.

Most DNA viruses have double-stranded chromosomes, but one group of animal viruses, the *parvoviruses*, possesses single-stranded linear DNA (Table 9.1), and several phages possess single-stranded circular DNA (Table 9.2). In most cases, linear DNA chromosomes are circularized within the host cell by one of two distinct mechanisms depending on whether the linear molecule has *cohesive ends* (Figure 9.9) or *terminal redundancy* (Figure 9.10). Cohesive ends are short, single-stranded regions that are complementary to each other; within host cells these anneal to form a circular molecule with a single nick in each DNA strand (Figures 9.9 and 9.11). Nicks are then sealed by the action of DNA ligase. Linear DNA chromosomes with terminal redundancy circularize by recombination within their homologous terminal regions (Figure 9.10).

Initial replication of the circular viral chromosome begins at a specific site and proceeds in both directions around the molecule by a process similar to that by which the bacterial chromosome is replicated (Chapter 5). Subsequent replications, in many cases, occur by a process termed *rolling circle replication* (Figure 9.12). A nick in one of the DNA strands is made by a specific endonuclease and the 3′OH end of the nicked strand (the primer) is extended by addition of nucleotides; the intact complementary strand serves as the template. The 5′ end is thus displaced, and later duplicated. In this manner, a double-stranded molecule is produced that can be much longer than the circumference of the viral chromosome. Such molecules, termed *concatemers*, are cleaved later to produce the chromosomes of progeny virions.

FIGURE 9.10

Alternative strategies for replication of linear double-stranded DNA chromosomes that possess terminal redundancy. In pathway (a), recombination occurs within the region of redundancy. In pathway (b), the chromosome is first replicated, producing molecules that can recombine to generate concatemers.

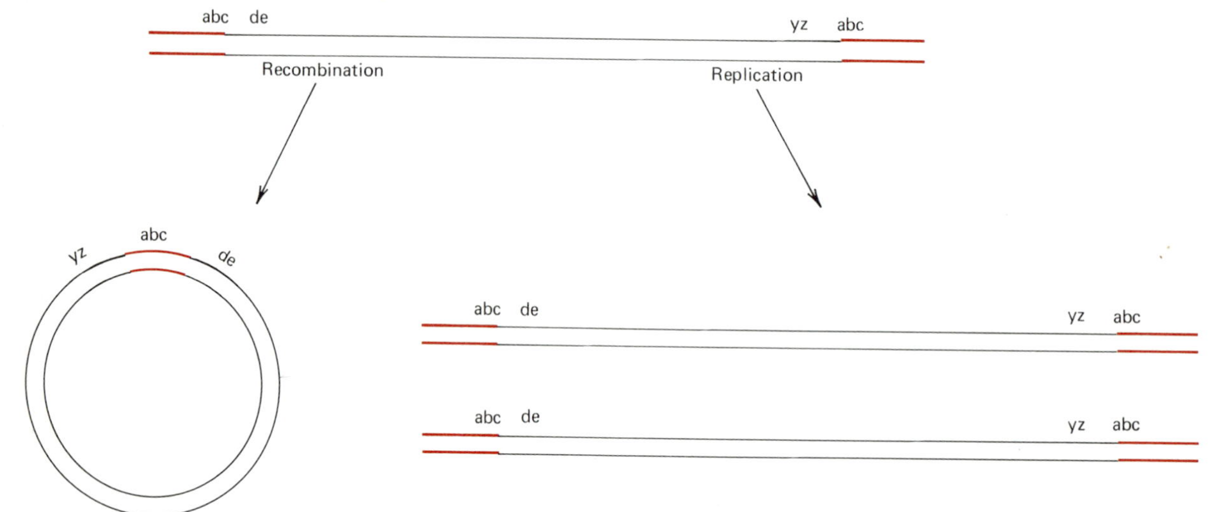

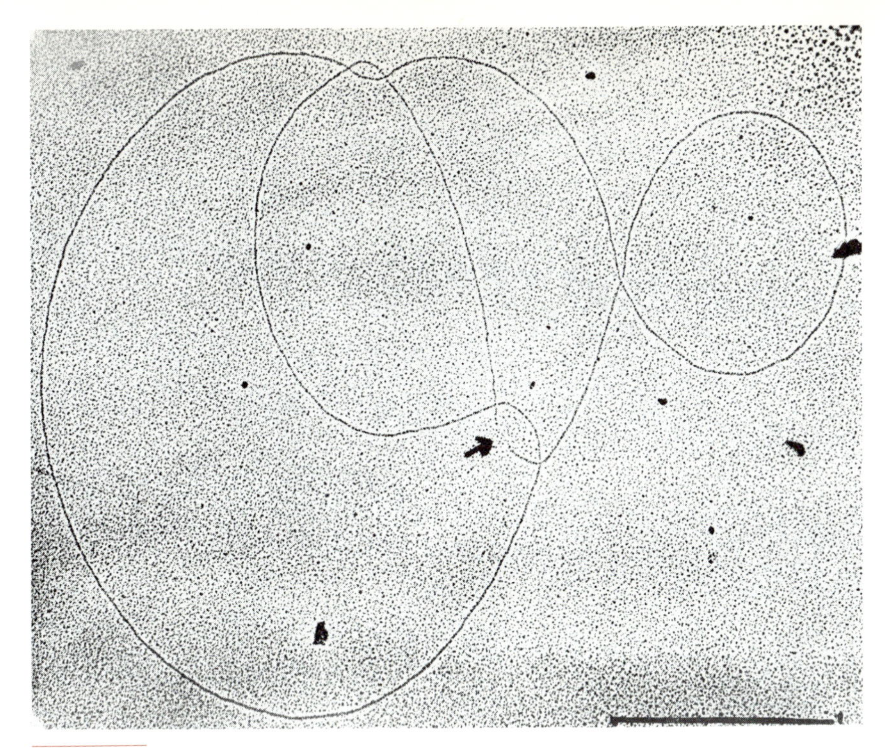

FIGURE 9.11

Lambda phage DNA. The scale marker represents 1 $\mu$m; the length of the DNA molecule is 16.3 $\mu$m. The arrow points to a region of discontinuity, believed to contain the cohesive ends described in the text. From H. Ris and B. L. Chandler. "The Ultrastructure of Genetic Systems in Prokaryotes and Eukaryotes," *Cold Spring Harbor Symp. Quant. Biol.* **28,** 1 (1963).

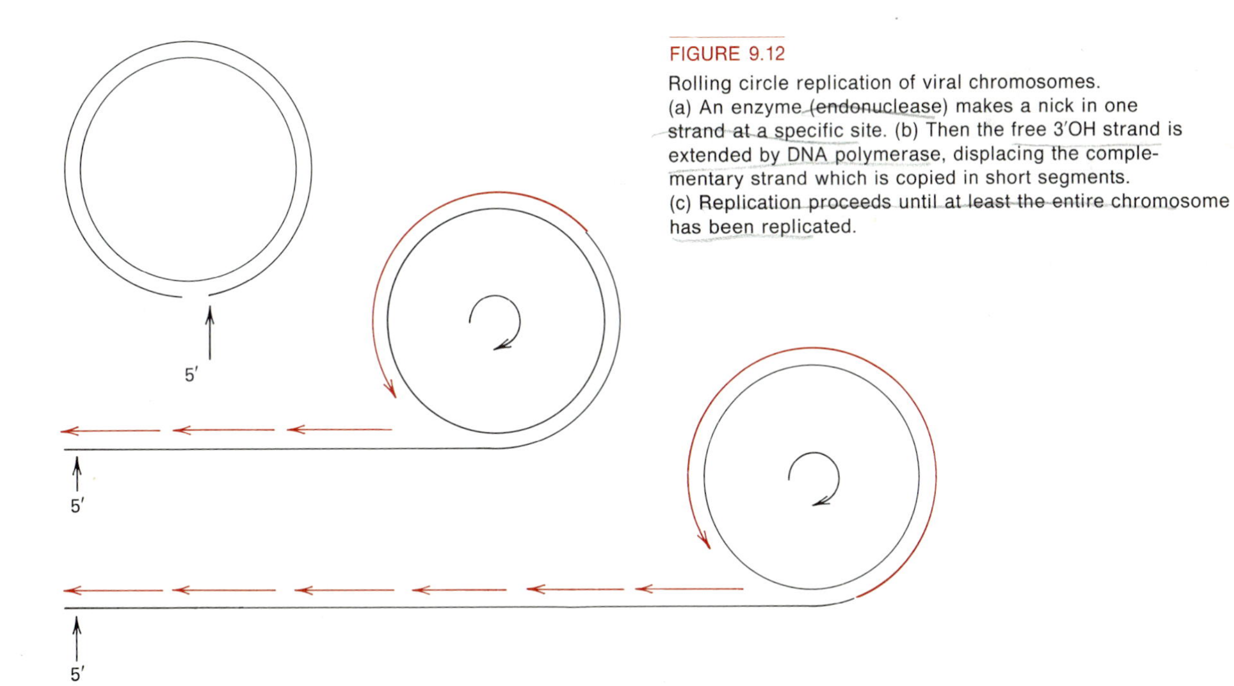


FIGURE 9.12

Rolling circle replication of viral chromosomes.
(a) An enzyme (endonuclease) makes a nick in one strand at a specific site. (b) Then the free 3′OH strand is extended by DNA polymerase, displacing the complementary strand which is copied in short segments.
(c) Replication proceeds until at least the entire chromosome has been replicated.

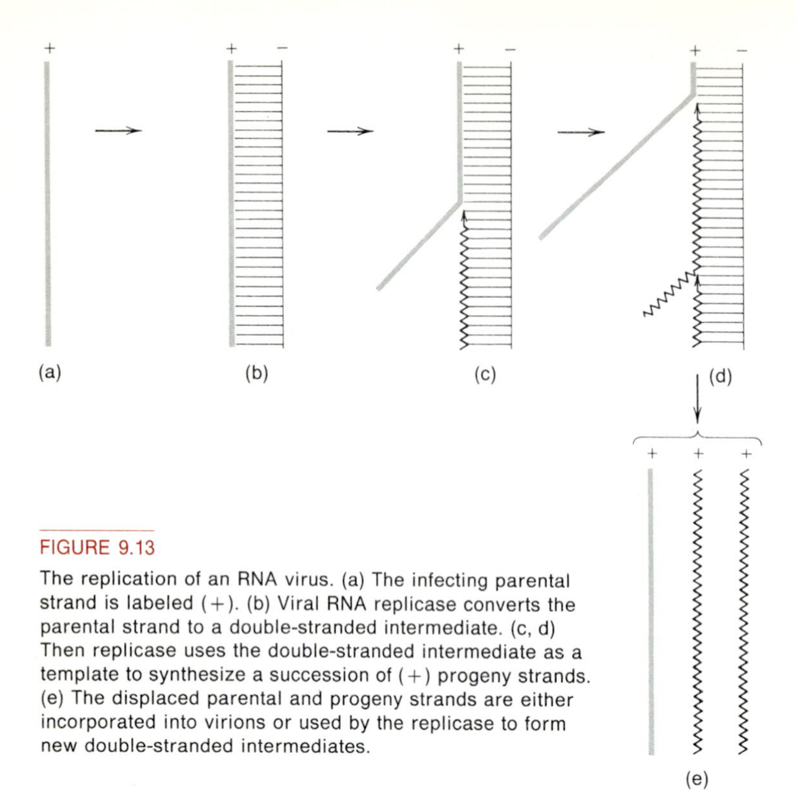


FIGURE 9.13

The replication of an RNA virus. (a) The infecting parental strand is labeled (+). (b) Viral RNA replicase converts the parental strand to a double-stranded intermediate. (c, d) Then replicase uses the double-stranded intermediate as a template to synthesize a succession of (+) progeny strands. (e) The displaced parental and progeny strands are either incorporated into virions or used by the replicase to form new double-stranded intermediates.

In some cases, circularization does not occur prior to replication. The linear molecule is repetitively replicated to produce a number of identical molecules (Figure 9.10) which recombine to generate concatemers. These are cleaved by an endonuclease into terminally redundant chromosomes.

## Replication of Chromosomes of RNA Viruses

With the exception of a group termed *retroviruses*, the chromosomes of which are replicated through DNA intermediates (Chapter 32), RNA chromosomes are replicated through double-stranded RNA intermediates. In the case of virions containing plus-strand RNA, the chromosome is translated immediately after it enters the host cell; one of the proteins thus produced is a *replicase* (RNA-dependent RNA polymerase) which the uninfected host lacks. Replicase catalyzes the synthesis of an RNA molecule termed a *minus strand* that is complementary to the chromosome. The viral chromosome is thus replicated in two stages (Figure 9.13). First, the minus strand is synthesized, forming a double-stranded molecule termed the *replicative form* (RF). Then, using the minus strand as a template, replicase catalyzes the synthesis of new copies of the plus strand.

Viral chromosomes composed of minus-strand or double-stranded RNA cannot be translated because they lack ribosome binding sites. All virions that contain such chromosomes also contain replicase molecules, which enter the host cell along with the chromosome. These enzymes catalyze the replication of the viral chromosome through a double-stranded replicative intermediate. In these viruses, additional plus strands are synthesized from the replicative intermediate to serve as mRNA for synthesis of viral proteins.

## Functions of Viral Gene Products

RNA phages (R17, MS2, f2, and Q$\beta$) are among the simplest viruses from a genetic standpoint: their chromosomes possess genes that encode only four proteins—the major capsid protein, the minor capsid protein, replicase, and a lysis protein. More complex viruses possess many genes for capsid proteins and also for assembly. For example, more than 40 genes are required just to make the capsids of T-even phages.

The building blocks and biosynthetic machinery, including amino acids, ribosomes, and nucleoside triphosphates, required for viral replication are synthesized by the host cell. However, some

dCMP $\xrightarrow[1]{}$ 5HMdCMP $\xrightarrow[2]{}$ 5HMdCDP $\xrightarrow[3]{}$ 5HMdCTP $\longrightarrow$ phage DNA

FIGURE 9.14

The pathway leading to synthesis of 5-hydroxymethyldeoxycytidine triphosphate (5HMdCTP) in bacteria infected with T-even phages. In the first reaction, N-formyltetrahydrofolate is the hydroxymethyl donor for synthesis of 5-hydroxymethyldeoxycytidine monophosphate (5HMdCMP). Kinases convert this compound first to 5-hydroxymethyldeoxycytosine diphosphate (5HMdCDP) and 5HMdCTP, which is incorporated into phage DNA by a phage-encoded DNA polymerase.

viral chromosomes contain *modified nucleotides* that are not normal components of the host cell. In these cases, viral genes encode enzymes that catalyze at least some steps in the synthesis of these unusual nucleotides. Unusual bases found in phage DNA include 5-hydroxymethylcytosine, which replaces cytosine in T-even phages; 5-hydroxymethyluracil, which replaces thymine; and a variety of others (Table 9.5).

Biochemical pathways leading to synthesis of unusual bases begin in most cases with a normal nucleoside monophosphate. A virus-encoded enzyme modifies this nucleotide, and other viral enzymes convert it to the unusual nucleoside triphosphate which is incorporated into the viral chromosome by a virus-encoded polymerase. An example of such a pathway is shown in Figure 9.14. To prevent incorporation of the normal base into viral chromosomes, some viruses encode enzymes that destroy the normal nucleoside triphosphate by converting it to the corresponding nucleoside monophosphate, thereby also providing more substrate for synthesis of the unusual triphosphate. The selective advantage offered by nucleic acids containing unusual bases appears to be their resistance to degradation by nucleases present in the host cell, as has been demonstrated by experiments with T-even phages, which normally contain 5-hydroxymethylcytosine glucosylated at the 5′OH position. Mutant phages that lack the ability to glucosylate their DNA are inactivated by a nuclease present in some strains of *E. coli*.

## Regulation of Expression of Viral Genes

Synthesis of viral proteins during the replication cycle is highly regulated, thus ensuring that proper amounts of them are synthesized at the proper time. The temporal control of their synthesis is achieved by a number of mechanisms that vary among different viral groups.

In replication of RNA phages, in which control can occur only by modulating translation, two types of regulatory mechanisms are known: (1) formation of secondary structures in the viral RNA, and (2) binding of specific proteins to the RNA (Chapter 12). An example of the first type of control is the unavailability of the ribosome binding site adjacent to the gene for the minor capsid protein of phage MS2 because the site is folded into a secondary structure termed a *hairpin loop* as a consequence of base pairing (Figure 9.15). But for a brief period following synthesis of a new plus-strand copy of this binding site (before secondary structure forms), ribosomes can translate the gene.

**TABLE 9.5**

**Modified Bases Found in DNA of Some Phages**

| Modified Base | Base Replaced | Extent of Modification (percent) | Phage | Host |
|---|---|---|---|---|
| 5-Hydroxymethylcytosine | Cytosine | 100 | T2, T4, T6 | *Escherichia coli* |
| 5-Methylcytosine | Cytosine | 100 | χP12 | *Xanthomonas oryzae* |
| 5-Hydroxymethyluracil | Thymine | 100 | φe | *Bacillus subtilis* |
| Uracil | Thymine | 100 | PBS2 | *Bacillus subtilis* |
| α-Putrescenylthymine | Thymine | 50 | φW14 | *Pseudomonas acidovorans* |
| 5-Dihydroxypentyluracil | Thymine | 41 | SP15 | *Bacillus subtilis* |
| α-Glutamyladenine | Adenine | 20 | SP10 | *Bacillus subtilis* |
| 2-Aminoadenine | Adenine | 100 | S2L | *Synechococcus elongus* |

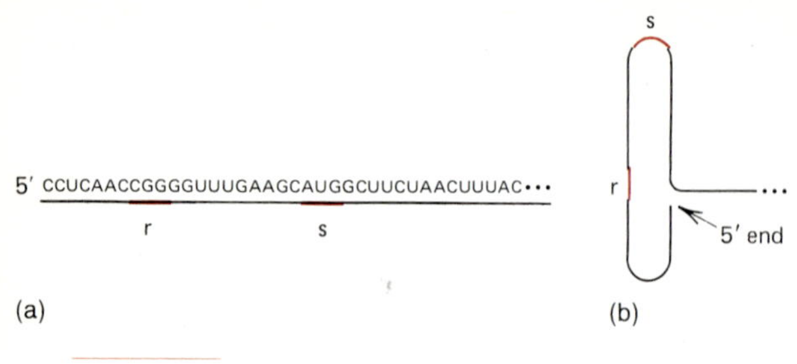


FIGURE 9.15

Regulation of minor coat protein gene (A gene) expression in phage R17. (a) The ribosome binding site and the start codon of the A gene are indicated by r and s, respectively, on the base sequence of the 5′ end of the R17 chromosome (and mRNA). In (b), a thermodynamically favorable hairpin structure, stabilized by hydrogen bonding between base pairs, is diagramed. Note that the ribosome binding site is involved in a double-stranded region of the hairpin and is therefore unavailable for initiation of translation. Redrawn from J. A. Steitz, "Discriminatory Ribosome Rebindings of Isolated Regions of Protein Synthesis Initiation from the Ribonucleic Acid of Bacteriophage R17," *Proc. Natl. Acad. Sci., USA,* **70,** 2605 (1973).

FIGURE 9.16

Genetic organization of the λ chromosome. Early transcription (←, →) starts at the left promoter (PL) and the right promoter (PR), and stops at termination sites (•) where N protein acts to prevent termination of transcription (Genes cI and *int* are also transcribed during this phase). After N protein reaches a sufficient concentration, delayed early transcription (←-, -→) commences, but terminates at a site (▪) where Q protein prevents termination. When Q protein reaches a sufficient concentration, late transcription (.....) proceeds through genes for phage capsid proteins and lysis.

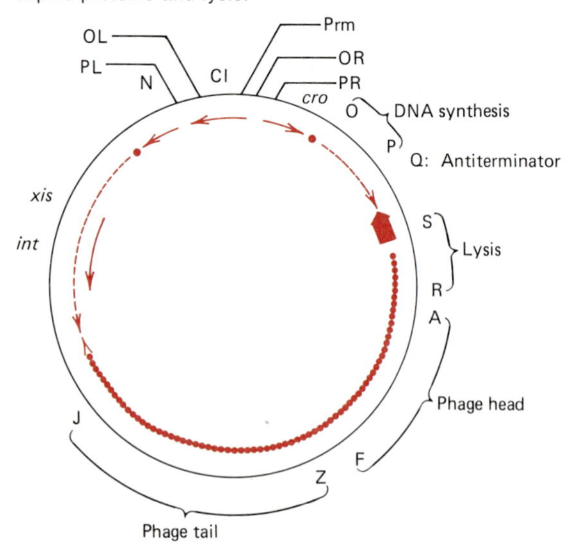


An example of the second type of control is the inhibition, during late stages of infection with phage MS2, of synthesis of viral replicase. This results from the accumulation of capsid protein, some molecules of which bind near the replicase gene, thereby preventing ribosome binding and subsequent translation of the gene.

In the case of DNA viruses, gene expression is regulated by a variety of different mechanisms, some of which appear to be unique to viral systems. One that is widespread among DNA phages is mediated by proteins termed *antiterminators* which inhibit *rho* factor–dependent termination of transcription (see Chapter 5). An example of their action is the temporal control of expression of genes by phage λ: some genes are expressed soon after infection (immediate-early phase), some are expressed slightly later (delayed-early phase), and others are expressed only during the final stages of infection (late phase). One of the phage genes expressed immediately following infection is gene N, which encodes an antiterminator, the N protein. Another set of genes that encodes proteins expressed during the delayed-early phase is not expressed immediately because they are preceded by *rho* factor–dependent termination sites (Figure 9.16). Only after sufficient N protein has been synthesized is termination at these sites inhibited; then delayed-early genes are expressed.

Expression of genes during the late phase is regulated in a similar manner. A second antiterminator, the Q gene product (Q protein), is synthesized during the delayed-early phase as a consequence of N protein control. When the intracellular concentration of Q protein reaches a certain level, it inhibits termination at a site preceding a third set of genes, permitting them to be expressed during the late phase. By these mechanisms, proteins are synthesized at the time they are required for viral synthesis. For example, protein components of the head and tail of the virion and lysis factors are required only during the final stages of assembly of the virions and during their release from the cell; genes encoding these proteins are expressed only during the late phase.

A variety of other mechanisms for temporal control of gene expression act during intracellular replication of other phages. In most cases of temporal control of gene expression, the same principle applies: expression of genes during later phases is dependent on prior expression of another gene (Table 9.6). As discussed in the case of phage λ, expression of later genes is dependent on prior expression of an antiterminator. In the case of phages T7 and PBS2, it is dependent on prior ex-

**TABLE 9.6**

**Some Phage Gene Products That Regulate Development**

| Protein | Phage | Function |
|---|---|---|
| N protein | λ | Antiterminator |
| Q protein | λ | Antiterminator |
| Gene I protein | SPO1 | Antiterminator |
| Transcriptase | T7, PBS2 | Transcription of late genes |
| Products of genes 33, 45, and 55 | T4 | Sigma-like factor |

pression of a gene encoding a special RNA polymerase, termed a *transcriptase*, that is required to transcribe late genes. In the case of T-even phages, it is dependent on expression of a gene that encodes an enzyme that modifies the RNA polymerase present in the host cell.

### Deleterious Effects of Viral Replication on Metabolism of Host Cells

Effects of intracellular growth of viruses on host cells differ widely, but they are always deleterious. Replication of some viruses, e.g., the filamentous DNA phages, only slightly inhibits the rate of growth of the host cell. At the other extreme, replication of T-even phages blocks expression of all host genes by producing several nucleases that destroy much of the host cell's DNA (Figure 9.17). Even genes that survive the action of these nucleases are not transcribed because viral-encoded enzymes adenylylate and thereby inactivate the host RNA polymerase, rendering it incapable of recognizing promoters in the host genome.

### Virion Assembly

Assembly of virions begins after capsid proteins and viral nucleic acid accumulate in the host cell. In TMV and related helical viruses, assembly is a relatively simple process in which capsid proteins associate with the viral chromosome and wind into a helix. No assistance from the host cell is required for TMV assembly as it will occur spontaneously *in vitro* when capsid proteins and nucleic acid are mixed. Assembly of the helical bacteriophages is more complex in that assembly and release of mature virions appear to be linked. Assembly occurs at the inner surface of the cytoplasmic membrane, and, as capsid proteins attach to viral DNA, the growing virion filament is extruded through the cell envelope (Figure 9.18).

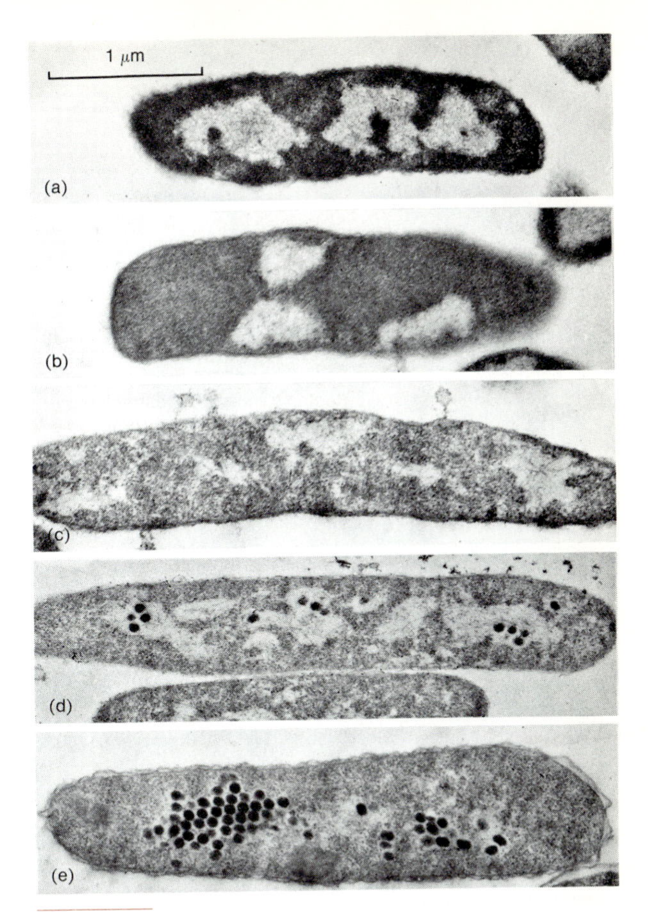


FIGURE 9.17

The course of phage infection in the bacterium *E. coli*. illustrated by electron micrographs of successive thin sections of the cells (×25,700). (a) Uninfected cell. (b) Four minutes after infection. Note change in structure of the DNA-containing region (light areas of cell). (c) Ten minutes after infection. (d) Twelve minutes after infection. The first new phage bodies, or DNA condensates (dark spots), are developing within the DNA-containing regions of the cell. (e) Thirty minutes after infection (shortly before lysis). Many new phage bodies are evident in the infected cell. Courtesy of E. Kellenberger, E. Boy de la Tour, J. Sechaud, and A. Ryter.

Assembly of icosahedral and binal viruses is somewhat different from that of helical virions: in most cases, capsid proteins assemble to form an empty structure, termed a *procapsid*, that has the shape and size of the capsid (or of the head in the case of binal phages). Then viral nucleic acid enters this structure and becomes condensed into a densely packed state. In the case of icosahedral viruses, the procapsid is then sealed, becoming impenetrable to large molecules. In the binal phages, tails are assembled separately and are then joined to filled heads.

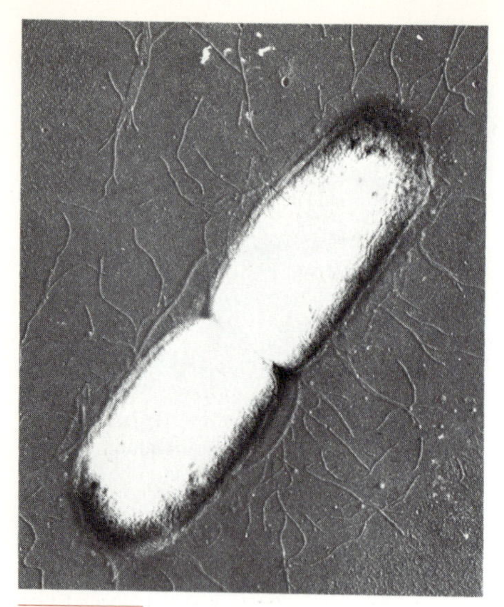

FIGURE 9.18

The liberation of filamentous phage by extrusion from living bacterial cells. The micrograph was made 30 minutes after the cells had been infected and washed free of unadsorbed phage particles. From P. H. Hofschneider and A. Preuss. "M13 Bacteriophage Liberation from Intact Bacteria as Revealed by Electron Microscopy," *J. Mol. Biol.* **7,** 450 (1963).

In most cases, the final step in assembly of an enveloped virus is the acquisition of a piece of host membrane that wraps around the viral nucleoprotein core as it passes through the intact host membrane. Before this step can occur, virus-encoded proteins aggregate in the cytoplasmic or the inner nuclear membrane. In the case of enveloped animal viruses, viral membrane proteins are synthesized by ribosomes bound to the rough endoplasmic reticulum and become embedded in the adjacent membrane as they are synthesized. These proteins are usually glycosylated by enzymes present within the endoplasmic reticulum. In the case of enveloped RNA viruses, viral membrane protein migrates to the cytoplasmic membrane by way of the Golgi apparatus. The nucleoprotein core of the virion then migrates to the inner surface of the membrane and becomes enveloped in membrane containing viral proteins as it leaves the cell by a process termed *budding* (Figure 9.6). In the case of enveloped DNA viruses, the membrane protein migrates to the inner nuclear membrane where nucleocapsids assembled in the nucleus associate with areas of inner nuclear membrane that are embedded with viral membrane proteins. This membrane envelops the nucleoprotein core, and the virion leaves the nucleus. The mature virion migrates into the endoplasmic reticulum.

### Escape

As discussed above, enveloped RNA viruses and filamentous bacteriophages escape from the cell as a part of the final assembly of the virion. Enveloped DNA viruses can migrate from the endoplasmic reticulum into vesicles from which they escape as a consequence of exocytosis. Some nonenveloped animal viruses, e.g., the adenoviruses, pass directly through the cytoplasmic membrane without apparent injury to the host cell. However, many animal and plant viruses kill the host cell and are thereby released by the autolysis that follows cell death. Most phages escape by causing the host bacteria to lyse. In some cases, a phage gene expressed during late phases encodes lysozyme, an enzyme that cleaves glycoside bonds in peptidoglycan. In some other cases, a phage gene encodes an enzyme that cleaves the peptide cross-links in peptidoglycan. But in many cases, the cause of lysis is not known.

### Infectious Viral Nucleic Acid

Purified nucleic acid of some viruses can penetrate, at low efficiency, susceptible cells and thereby establish an infection. This artificial process is termed *transfection.* Since transfection does not require specific interaction between a capsid structure and a cell receptor, a far wider range of host cell types or species can be transfected than can be infected by intact virions.

## DETECTION AND ENUMERATION OF VIRUSES

Plant virions are typically enumerated by the *plant leaf local lesion assay:* a suspension of virions is applied to the surface of a leaf along with an abrasive material that tears small holes in the walls of plant cells. Each virion that enters a host cell initiates a local infection that spreads to surrounding cells, creating a region of infection that becomes discolored and thus easily recognized.

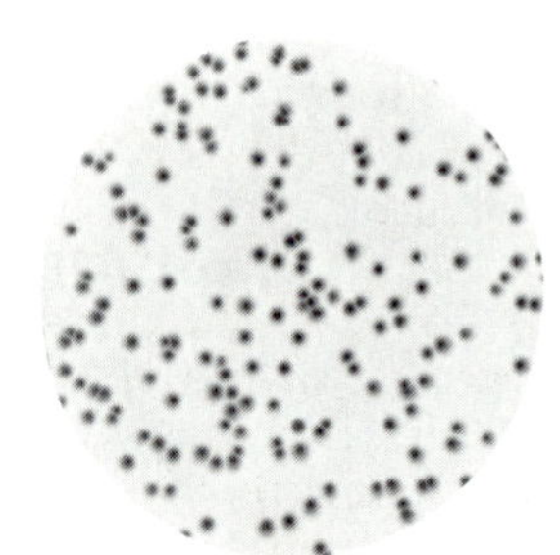

FIGURE 9.19

Plaques formed by bacteriophage T2. Courtesy of G. S. Stent.

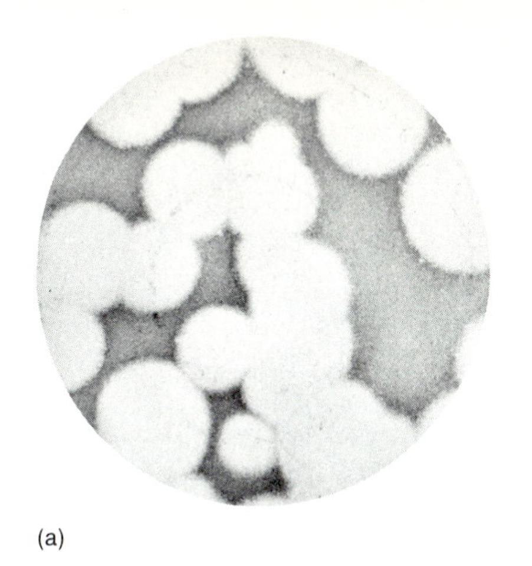

(a)

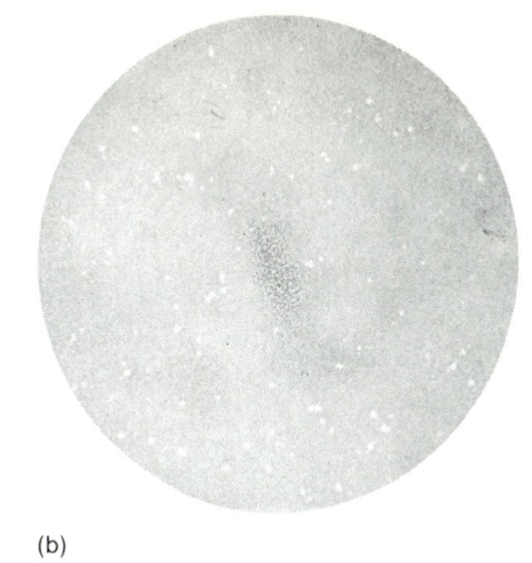

(b)

FIGURE 9.20

Plaques formed by encephalomyocarditis virus on a layer of animal cells. Two genetically distinct plaque types are shown; the large plaques (a) are 10 to 12 mm diameter, and the small plaques (b) are 0.5 mm diameter. Courtesy of H. Liebhaber.

Before the development of techniques for culturing animal cells *in vitro*, animal viruses were usually enumerated by the *pock assay* performed in developing chicken embryos: a sample containing virions is injected into a fluid compartment of the egg. These virions adsorb to cells in one of the internal membranes of the developing embryo, producing regions of infection termed *pocks* that can be recognized as being thickened and discolored.

### The Plaque Assay

Most commonly, animal and bacterial viruses are enumerated by infecting host cells that are growing in a thin layer on a medium partially solidified by agar. An infected cell in such a culture establishes a local, spreading infection where, depending on the infecting virus, cells either die or grow abnormally slowly. These infected zones, termed *plaques*, differ in appearance from the surrounding cell layer. Plaques formed by phages are usually clear, circular regions in a turbid layer of cells termed a *lawn* (Figure 9.19). Plaques formed by animal viruses (Figure 9.20) are sometimes visualized following application of a dye that stains live cells but not those killed by the viral infection.

## KINETICS OF VIRAL MULTIPLICATION

In 1940, M. Delbruck performed a type of experiment, termed the *one-step growth experiment*, with bacteriophage T2 that has proven extremely useful for analyzing the kinetics of replication of a variety of bacterial and animal viruses. The principle of the experiment is to infect simultaneously a large number of host cells; subsequent measurements on the cell culture then reflect the sequence of events that occur when a single susceptible host cell is infected by a virion. In the original experiment, a suspension of phage virions was mixed with a dense culture of susceptible bacteria at a ratio such that few cells were infected by more than one virion. Following a brief period during which most of the viruses adsorbed to host cells, the mixture was diluted about 1,000-fold, rendering further virion-cell collisions, and therefore further infections, unlikely. At intervals, samples were removed from the culture, and the number of virus particles and infected bacteria (collectively termed *infectious centers*) was determined by plaque counts. For a time (the *latent period*) the number of infectious centers remained constant. This number then suddenly increased (the *burst period*) as the infected cells lysed, each releasing numerous virus particles. When all infected cells had lysed, the number of infectious centers again remained relatively constant because further cycles of growth were largely prevented by the initial dilution of the culture. If one interprets such an experiment assuming that adsorption is efficient, the *burst size*, or number of virus particles released from a single infected cell, can be determined by dividing the number of infectious centers present after the burst by the number present before. In the example shown in Figure 9.21 (a), the burst size is about 300; in other cases it can be thousands. Additional information can be gained by exposing each sample briefly to chloroform before plating to determine the plaque count. Such treatment allows a distinction to be made between infected cells and phage virions because bacterial cells (including phage-infected cells) are lysed, whereas virions are unaffected. Plaque counts of chloroform-treated samples reflect only the

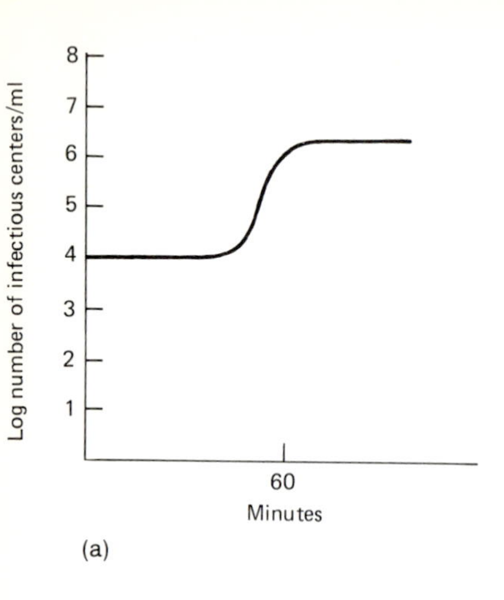


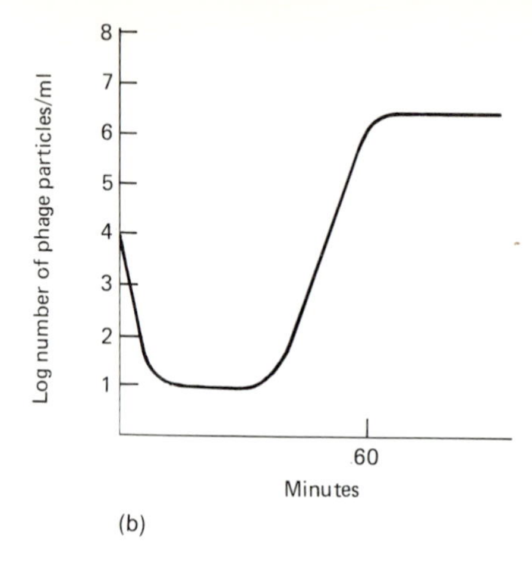


FIGURE 9.21

One-step growth of bacterial viruses. In (a) samples of the infected culture are removed and the number of infectious centers (free phage particles and infected cells) determined by the plaque method. In (b) each sample is treated with chloroform prior to plating, so that only free phage particles are determined.

number of phage particles present in the sample. If a one-step growth experiment is done in which samples are exposed to chloroform prior to performing the plaque count [Figure 9.21 (b)], a striking feature of growth emerges: virions, as stable infectious entities, disappear immediately after infection, and reappear in large numbers only late in the latent period. The period in which infectious particles are not recoverable from the cells is termed the *eclipse period.* (The few viruses detected during the eclipse period are those that did not adsorb to cells prior to the initial dilution.) We now know that the eclipse is the period during which viral components are being synthesized prior to assembly into virions.

One-step growth experiments can be used to follow the kinetics of synthesis of viral components as well as intact virions. For example, samples from the infected culture could be analyzed for their content of viral nucleic acid, viral protein, or other components.

FIGURE 9.22

Plaques formed by the wild type and by a virulent mutant of the bacteriophage, lambda. The wild-type particles form cloudy plaques as a result of the growth of lysogenized cells; the mutant particles, which are unable to lysogenize the bacterial host, form clear plaques with sharp edges. Courtesy of C. Radding.

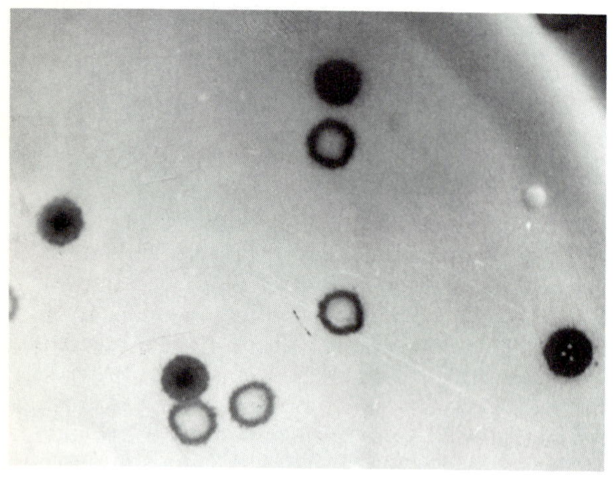

## LYSOGENY

Soon after the discovery of bacteriophages, some bacterial strains, termed *lysogenic strains* or *lysogens,* were found that produced phage virions spontaneously during growth of the culture, but the majority of cells in the culture were unaffected by these virions. Phages produced by lysogenic strains are termed *temperate* phages, and the relationship between the phage and the bacterium is termed *lysogeny.* Phages like T2 that do not cause lysogeny are termed *virulent.* When a temperate phage virion infects a susceptible nonlysogenic cell, one of two possible developmental cycles ensues: either phage replication occurs or the infected cell is converted into a lysogen. As stated, a temperate phage normally cannot replicate in the lysogenic strain that produces it, but it can in a closely related strain that is not lysogenic for that particular phage. Resistance of lysogens to the phage that they produce is termed *immunity to superinfection.* Therefore, when a temperate phage forms a plaque, some lysogenic bacteria are produced within the infected area and are able to grow, resulting in formation of a *turbid plaque* (Figure 9.22).

For many years, the basis for lysogeny and immunity to superinfection was unclear. However, in the 1950s, several basic properties of lysogens were discovered. First, A. Lwoff showed that the virions always present in a lysogenic culture are produced by lysis of a small fraction of the cells in

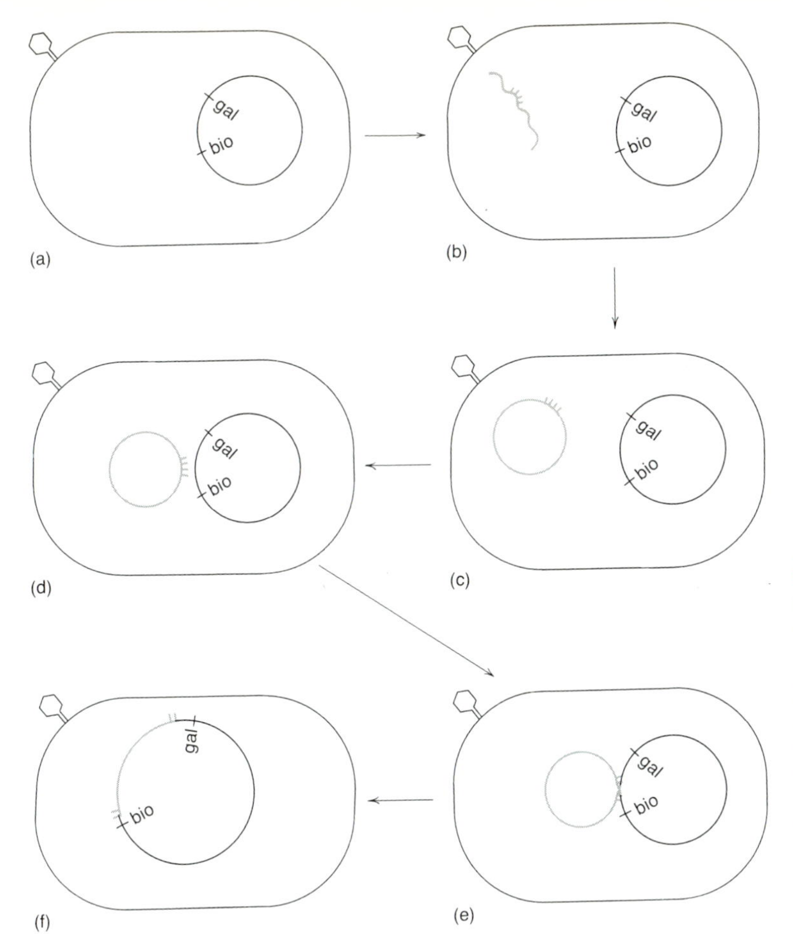


FIGURE 9.23
The formation of λ prophage.
(a) Adsorption of the virus. (b) Injection of viral DNA. (c) Circularization of the viral genome. (d) Pairing of homologous regions on the viral and bacterial genomes. (e) A crossover event occurs within the region of pairing. (f) The two genomes have been integrated, forming a single circle. Note that the attachment site for λ is at a specific location, between the loci *gal* and *bio*. The specific attachment site on λ is indicated by four short vertical lines on the DNA strand.

the culture. Next, electron micrographs of lysogenic bacteria revealed that most cells do not contain virions, and biochemical studies showed that lysogenic bacteria contain little, if any, capsid protein but do contain phage DNA, termed the *prophage*, within the bacterial cell. A temperate phage in the prophage state is replicated and segregated with the bacterial chromosome, and most of the viral genes it contains are not expressed.

### Lysogeny: Phage λ Type

The λ prophage integrates in the host chromosome between two sets of bacterial genes, the *gal* genes which encode enzymes that degrade the sugar galactose, and the *bio* genes which encode biotin synthesis. Most temperate phages resemble phage λ in that they integrate into the bacterial chromosome at specific sites, but there are exceptions. One temperate phage, mu, has the unusual ability to integrate anywhere in the bacterial chromosome.

Integration of the λ chromosome requires the action of a protein termed *integrase* encoded by a λ gene and occurs at a 13 base pair region of DNA sequence homology between the bacterial and phage chromosomes. Following alignment of these regions of homology, integrase catalyzes a reciprocal crossover between the chromosomes, resulting in an integrated prophage (Figure 9.23).

### Lysogeny: Phage P1 Type

A second group of temperate phages, of which phage P1 is a typical representative, differs from those of the λ-type with respect to the state in which the prophage is maintained. In lysogeny of the P1 type, viral DNA does not normally become integrated into the host chromosome. Instead, the prophage exists as a circular, self-replicating, double-stranded DNA element termed a *plasmid* (see Chapter 10).

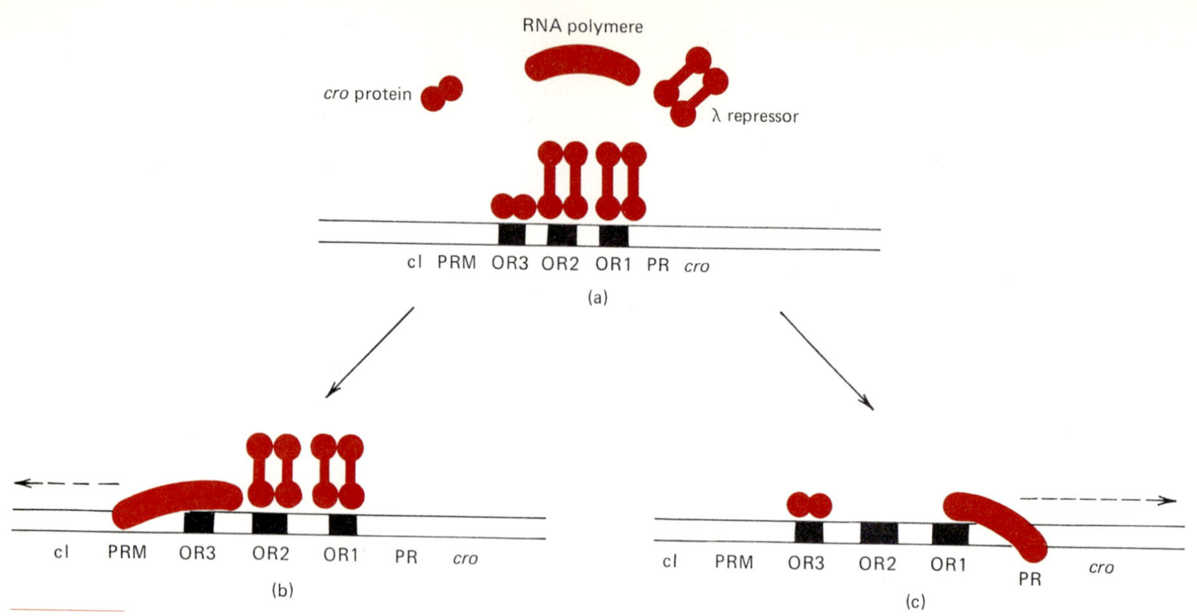


FIGURE 9.24

Regulation of λ by repressor and *cro* protein. (a) Repressor bound to the λ chromosome at site OR1 and OR2, and *cro* protein bound at OR3, prevent RNA polymerase from binding to this region of DNA. (b) If *cro* protein dissociates first, RNA polymerase binds to the promoter for repressor maintenance (PRM) and transcribes the repressor gene, cI. This results in an increased concentration of repressor and the establishment of lysogeny. (c) If repressor dissociates first, then RNA polymerase binds to the right promoter and begins transcription of genes *cro* through J (Figure 9.15), leading to the production of phages and lysis of the cell.

## Regulation of Lysogeny in Phage λ

Whether lysis or lysogeny follows infection of a susceptible cell by phage λ depends on the relative concentrations of two viral-encoded proteins, *λ repressor* and *cro protein*, produced during early phases of infection. Both proteins can bind to λ DNA at specific sites, termed the right operator (OR) and the left operator (OL), thereby interfering with binding of RNA polymerase to adjacent promoters. By this mechanism, both repressor and *cro* protein can stop early phases of gene expression (Figure 9.24).

Binding at one site within the right operator (OR1), prevents expression of a set of genes including the one (*cro*) that encodes the *cro* protein, and binding to another, OR3, prevents expression of the gene that encodes repressor. However, repressor binds preferentially to OR1, and *cro* protein binds preferentially to OR3. Thus, *cro* protein and repressor compete for their own biosynthesis; each inhibits synthesis of the other. Moreover, binding of a repressor to OR1 facilitates binding of a second repressor molecule to an adjacent site, OR2. When OR3 is unoccupied, repressor bound to OR2 facilitates expression of the repressor gene. Thus, in the absence of *cro* protein, repressor stimulates its own synthesis. In addition, it shuts off expression of genes required for phage replication.

Following the intracellular accumulation of repressor and *cro* protein during early gene expression, repression at OR temporarily stops further synthesis of both proteins. At this stage, repressor is usually bound to OR1 and OR2, but *cro* protein is bound to OR3. Meanwhile, phage DNA synthesis proceeds, and one or more phage chromosomes usually become integrated into the host chromosome by the action of λ integrase, which is produced during this period. As the host cell continues to enlarge, the intracellular concentrations of both *cro* protein and repressor are decreased. Thus, if *cro* protein is in relatively low abundance, it will dissociate from OR3 while repressor remains bound to OR1 and OR2, which causes more repressor to be synthesized and lysogeny to be established. On the other hand, if the repressor is in relatively low abundance, it dissociates from OR1, allowing expression of *cro* and all viral genes, the products of which catalyze viral replication and eventual lysis of the host cell.

### Induction

In many temperate phages including $\lambda$, lysogeny can be interrupted by briefly exposing a lysogenic strain to ultraviolet light or to a chemical that damages DNA, thereby creating single-stranded regions. The consequence is initiation of viral replication that results eventually in lysis of the cell and liberation of progeny virions. This process, termed *induction*, is mediated by the host cell's *recA* protein, which is activated when it binds to single-stranded regions of DNA, in this case, those formed by ultraviolet light or the chemical (Chapter 12). The activated *recA* protein is a protease that cleaves and thereby destroys the $\lambda$ repressor, thus allowing viral genes to be expressed. The products of two of these genes, *int* and *xis*, act together to excise the prophage from the host chromosome. The phage then undergoes normal vegetative replication; the cell lyses and virions are released. The low level of induction that occurs spontaneously in a few cells of a lysogenic culture accounts for the presence of extracellular virions that such a culture always contains.

### Lysogenic Conversion

When some temperate phages form lysogens, the phenotype of the host cell is altered; this process is termed *lysogenic conversion*. For example, the lipopolysaccharide structure (O-antigen) of *Salmonella typhimurium* is altered following lysogenization by phages termed epsilon phages, and some bacteria produce toxins when they carry certain prophages. Important human diseases that result from toxins under the control of lysogenic conversion include diphtheria and botulism (Chapter 31).

## VIROIDS

A class of extremely simple infectious agents, termed *viroids*, was discovered in 1967. They are merely infectious single-stranded RNA unassociated with any virion structure. To date, they have been found only to cause diseases of plants. They exist both intracellularly and extracellularly as circular single-stranded RNA molecules that contain fewer than 400 nucleotides. Therefore, they are smaller than any known viral chromosome. They are large enough to encode a single small protein, but it is doubtful that they do, because the RNA of viroids lacks the signals (ribosome-binding sites, start codons, and stop codons) that are needed for translation of RNA into a protein. It seems probable that a host cell RNA polymerase replicates the viroid chromosome.

## PRIONS

Two human diseases, *kuru* and *Jakob-Creutzfeldt* disease, and one disease of sheep, *scrapie*, are caused by macromolecules that are smaller than any known virus. These diseases are characterized by a progressive, fatal degeneration of brain tissue. Attempts to purify the causative agents have not been successful, but a preparation that is highly enriched for the causative agents of kuru and scrapie have been obtained: it is largely composed of proteins with molecular weights near 30,000. Infectious macromolecules causing these diseases have been called *prions*, and if they prove to be composed entirely of protein, they will constitute a fundamentally new group of infectious agents. How such an agent might reproduce is, of course, a mystery.

## FURTHER READING


### Books

DULBECCO, P., *Virology*. Hagerstown, Md.: Harper & Row, 1980.

FRAENKEL-CONRAT, H., and P. C. KIMBALL, *Virology*. Englewood Cliffs, N.J.: Prentice-Hall, Inc., 1982.

HAHON, N., ed., *Selected Papers on Virology*. Englewood Cliffs, N.J.: Prentice-Hall, Inc., 1964.

HENDRIX, R. W., W. ROBERTS, F. W. STAHL, and R. A. WEISBERG, eds., *LambdaII*. Cold Spring Harbor, N.Y.: Cold Spring Harbor Laboratory, 1983.

JOKLIK, W. K., *Principles of Animal Virology*. New York: Appleton-Century-Crofts, 1980.

LURIA, S. E., J. E. DANIELLI, JR., D. BALTIMORE, and A. CAMPBELL, *General Virology*, 3rd ed. New York: John Wiley, 1978.

MATHEW, R. E. F., *Plant Virology*, 2nd ed. New York: Academic Press, 1981.

### Reviews

DIENER, T. O., "Viroids: Structure and Function," *Science* **205,** 859 (1979).

FIERS, W., "Structure and Function of RNA Bacteriophages," in *Comprehensive Virology*, Vol. 13, ed. H. Fraenkel-Conrat and R. Wagner, Chap. 3. New York: Plenum Press, 1979.

HOWE, M., and E. G. BADE, "Molecular Biology of Bacteriophage Mu," *Science* **190,** 624 (1975).

PRUSINER, S. B., "Prions," *Scientific American* **251** (4), 50 (1984).

PTASHNE, M. A. D. JOHNSON, and C. O. PABO, "A Genetic Switch in a Bacterial Virus." *Scientific American* **247** (1), 128 (1982).

WARREN, R. A. J., "Modified Bases in Bacteriophage DNAs," *Ann. Rev. Microbiol.* **34,** 137 (1982).

Chapter 10

# Microbial Genetics: Gene Function and Mutation

The broad outlines of gene expression—the process by which information encoded in the DNA comprising a cell's genome directs the sequence in which constituent monomers (nucleotides and amino acids) are polymerized (into RNA and protein respectively)—are discussed in Chapter 5. With this background, a gene can be defined in biochemical terms: *a gene is a segment of DNA that determines, through the process of transcription, the sequence of nucleotides in an RNA molecule or, by the processes of transcription and translation, the sequence of amino acids in a protein.* The stable end product of most genes is a protein molecule; in such cases the RNA molecule, termed *messenger RNA (mRNA)*, is an unstable intermediate; (in the case of *E. coli* growing at 37° C mRNA turns over with an average half life of about 1.3 minutes.) However, the RNA product of transcription itself is the stable end product of a limited number of genes (about 75 out of a total of about 3,500 in the case of *E. coli*). The RNA molecules produced by these genes are termed *stable RNA*; they include transfer RNAs (tRNA) and the RNA components of ribosomes (rRNA). They do not serve a messenger role as templates for translation.

## THE BACTERIAL GENOME

Most bacterial genes are encoded in a single large circular molecule of DNA, the bacterial chromosome. This molecule in *E. coli* is about 1 millimeter in circumference with a molecular weight of $2.56 \times 10^9$; i.e., it contains about $3.8 \times 10^3$ kilobase pairs (kb). Most bacteria contain chromosomes that are about the same size as the one in *E. coli* but in special cases significant variations occur (Table 10.1). Usually the size of the procaryotic chromosome correlates with the cell's physiological or morphological complexity. For example, the chromosome of *Desulfovibrio*, a strict anaerobe with relatively limited metabolic capacity, is less than half the size of the *E. coli* chromosome; the chromosome of certain *Mycoplasma* strains (Chapter 25), which lack a cell wall and are metabolically simple, are five to six times smaller than the *E. coli* chromosome. On the other hand, the chromosome of *Myxococcus xanthus* (Chapter 18), which has a complex developmental cycle, is about one and a half times larger than the *E. coli* chromosome. However, the largest procaryotic chromosomes encountered so far are those of the cyanobacterium, *Calothrix*; they are over three times the size of the *E. coli* chromosome. *Calothrix* has a complex developmental cycle as compared with most procaryotes, but this complexity alone cannot account for the size of its chromosome, because certain strains of *Anabaena*, which are closely related to *Calothrix* and seemingly as complex, have chromosomes only slightly larger than the *E. coli* chromosome.

Assuming that the molecular weight of an average bacterial protein is about 45,000, one can calculate that the average gene that encodes it is about 1.1 kb long and hence that the *E. coli* chromosome has sufficient capacity to encode about 3,500 genes. However, it is far from clear that the *E. coli* chromosome contains this many genes. Only about 1000 have been identified thus far on the chromosome of this most thoroughly studied bacterium. These account for most of the enzymes required for the cell's known fueling, biosynthetic, polymerization, and assembly reactions as well as structural proteins, and those proteins associated with motility, tactic responses, and transport mechanisms. This discrepancy of about 2,500 between coding capacity and identified genes might be a true reflection of the number of bacterial genes, and therefore cellular functions, yet to be discovered, or the actual number of genes the chromosome contains might be significantly less than 3,500 because the chromosome might contain regions that don't contain genes. No such regions have as yet been identified, and their existence seems unlikely because if they were truly functionless they would constitute a selective disadvantage to the cell.

Not all bacterial genes are encoded within the chromosome. Many, but not all, bacteria contain one or more different small circular DNA molecules, termed *plasmids*. These elements, which vary in size from a few to several hundred kb in length, function in many respects as small chromosomes: they are self-replicating and they encode a variety of cellular functions. But in several respects, in addition to their markedly smaller size, they differ from chromosomes: they are dispensable, because the types of functions they encode only benefit the cell in a limited set of environments, and none is known to encode essential cellular functions. For example, some plasmids encode enzymes that inactivate antibiotics or other toxic compounds that are sometimes present in a cell's environment. Plasmids also differ from chromosomes by not being restricted to a single host. Some plasmids are transferred among and are stably replicated in only a small group of closely related host bacteria; others have a host range so broad as to include almost all Gram-negative species. The diversity of plasmids and their properties are discussed in Chapter 11.

The chromosome and such plasmids as a bacterium might contain constitute its *genome*.

**TABLE 10.1**

**Molecular Weight of the Chromosomes of Certain Procaryotes**

| Procaryote | Molecular Weight of Chromosome |
|---|---|
| *Mycoplasma* spp. | $0.4–0.5 \times 10^9$ |
| *Desulfovibrio* sp. | $1.1 \times 10^9$ |
| *Escherichia coli* | $2.56 \times 10^9$ |
| *Anabaena* sp. | $3.17 \times 10^9$ |
| *Myxococcus xanthus* | $3.79 \times 10^9$ |
| *Calothrix* sp. | $8.58 \times 10^9$ |

### Arrangement of Genes on the Chromosome

In some cases, genes occur individually on the chromosome; from such genes a single species of mRNA is transcribed and translated into a single type of protein. But frequently a small cluster of genes encoding related functions is transcribed from a single promoter to form a multigenic mRNA molecule that is translated to form as many proteins as there are genes in the cluster. Such a cluster of genes is termed an *operon*. The operonic arrangement of genes appears typical of many bacteria and to be selectively advantageous because it provides an economy of regulatory mechanisms. Since most mechanisms of regulating gene expression (see Chapter 12) function by modulating the frequency

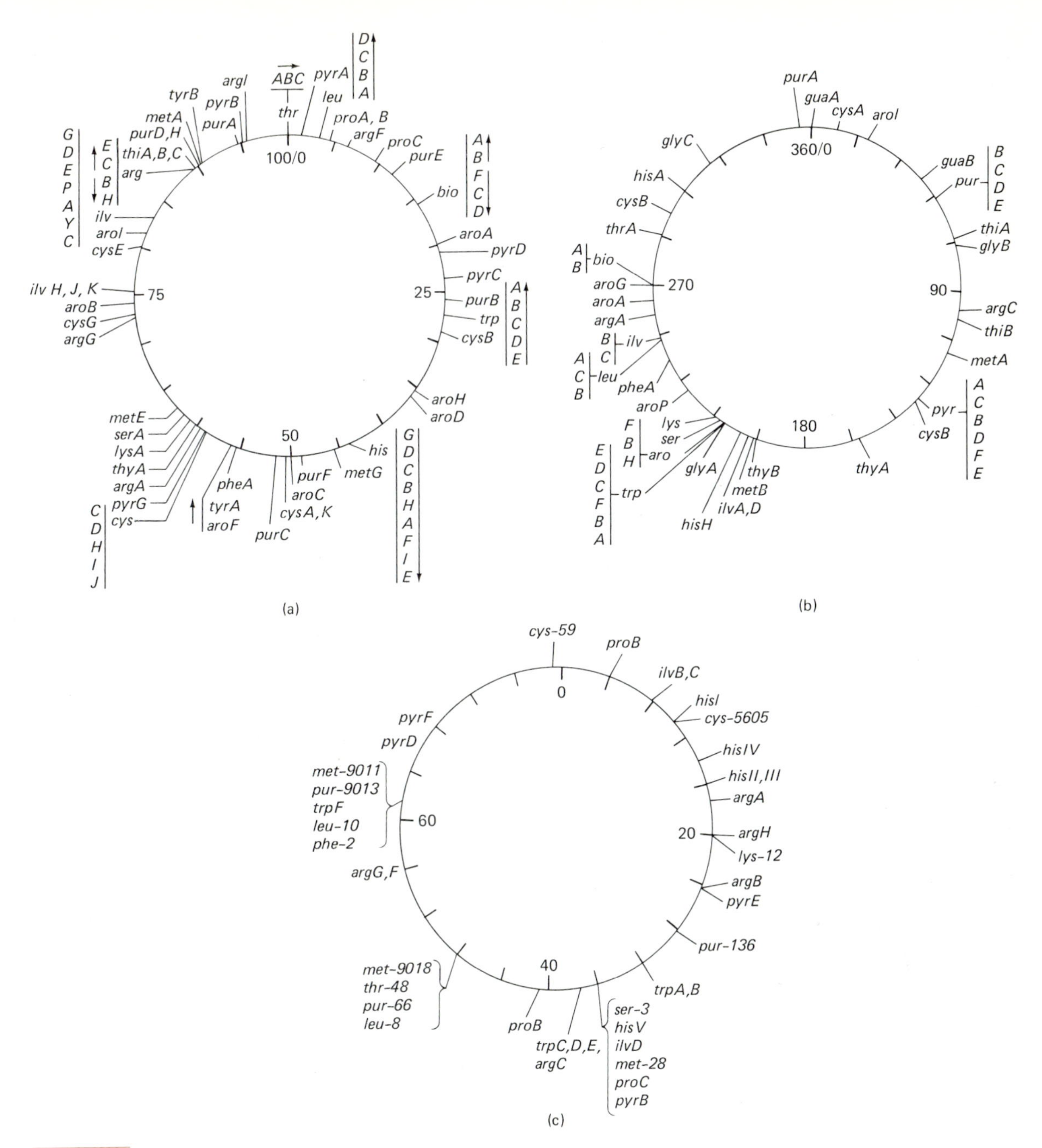

FIGURE 10.1
Relative location of genes encoding enzymes in biosynthetic pathways on the chromosomes of (a) *Escherichia coli* or *Salmonella typhimurium*, (b) *Bacillus subtilis*, and (c) *Pseudomonas aeruginosa*. Gene designations of biosynthetic pathways are: *arg*, arginine; *aro*, aromatic amino acids; *bio*, biotin; *cys*, cysteine; *gly*, glycine; *gua*, guanine; *his*, histidine; *ilv*, isoleucine-valine; *leu*, leucine; *lys*, lysine; *met*, methionine; *phe*, phenylalanine; *pro*, proline; *pur*, purine; *pyr*, pyrimidine; *ser*, serine; *thi*, thiamin; *thr*, threonine; *thy*, thymine; *trp*, tryptophan; *tyr*, tyrosine. Genes arranged in operons are shown joined by a horizontal or vertical bar; arrows indicate direction of transcription. After B. J. Bachmann, "Linkage Map of *Escherichia coli*, Edition 7," *Microbiol. Rev.* **47,** 180 (1983); D. J. Henner and J. A. Hoch, "The *Bacillus subtilis* Chromosome," *Microbiol. Rev.* **44,** 57 (1980); and P. L. Royle. H. Matsumoto, and B. W. Holloway, "Genetic Circularity of the *Pseudomonas aeruginosa* PAO chromosome," *J. Bacteriol.* **145,** 145 (1981).

with which transcription occurs, the clustering of genes into operons with related function, and therefore requiring similar regulation of expression, reduces the required number of regulatory systems.

As discussed in Chapter 5, the orientation of a promoter on the chromosome determines which strand (the sense strand) will be transcribed and thereby the direction in which transcription will occur. There seems to be no preference as to which strand of the chromosomal DNA serves as the sense strand; at different points on the chromosome one strand or the other serves with apparent equal probability.

The relative position of a number of genes on several bacterial chromosomes has been determined. The most completely studied bacterium in this respect (as well as in most others) is *Escherichia coli*. As stated, in this bacterium the relative location on the chromosome of over 900 different genes has been determined (*mapped*). A slightly lesser number of genes have been mapped on the chromosome of the closely related bacterium *Salmonella typhimurium*. On the *Bacillus subtilis* chromosome, over 360 genes have been mapped, and on the *Pseudomonas aeruginosa* chromosome about 100 genes have been mapped. Less complex genetic maps of a variety of other bacteria have been constructed.

Certain generalizations can be drawn by comparing the several available linkage maps of bacterial chromosomes (Figure 10.1). On a broad scale there is no discernible similarity between the genetic maps of distantly related bacteria such as *Escherichia coli*, *Bacillus subtilis*, and *Pseudomonas aeruginosa*, but often the arrangement of genes within homologous operons, e.g., the ones encoding enzymes of tryptophan biosynthesis, is strikingly similar. The arrangement of genes on the chromosome of closely related bacteria, e.g., those of the enteric group, is quite similar but the selective pressures that conserve gene order are unknown. Most probably frequent genetic exchange does not constitute one of these selective pressures because *E. coli* and *S. typhimurium*, which have diverged sufficiently with respect to the sequence of bases in homologous genes to render recombination extremely rare, still have an almost identical order of homologous genes on their chromosomes. However, close relatedness between bacteria is not always associated with a similar arrangement of genes on their chromosomes. For example, there is no apparent similarity in gene arrangement between *Pseudomonas aeruginosa* and *Pseudomonas putida* although both of these species belong to the same group of fluorescent pseudomonads. The arrangement of genes on the chromosome of the latter organism is striking with respect to *supraoperonic* clustering of genes encoding similar metabolic functions; i.e., genes encoding biosynthetic functions are largely restricted to one region of the chromosome and those encoding fueling reactions are largely restricted to another region. A similar pattern of supraoperonic clustering of functionally related genes (but a completely different gene order) occurs on the chromosome of the unrelated actinomycete, *Streptomyces coelicolor*.

All procaryotes so far studied have proven to be *haploid*; i.e., an individual cell contains only a single type of chromosome. However, they are often polycaryotic. That is, they usually contain several copies of the same chromosome. For example, *E. coli* cells growing in a rich medium contain about 4 genomes per cell; *Deinococcus radiodurans* cells contain 4 to 8; *Desulfovibrio gigaris* contain 9 to 17; *Azotobacter vinelandii* cells contain 20 to 40.

## MUTATIONS

As a consequence of normal chromosomal replication, or exposure to certain chemical or physical agents termed *mutagens*, the sequence of bases in the bacterial genome occasionally changes. Any such change is called a *mutation*. Under normal conditions of cultivation, mutations occur only rarely—a population typically contains about one cell in $10^8$ that carries a detectable mutation in any given gene, but exposure to certain powerful mutagens increases dramatically the frequency of cells that carry mutations. For example, exposure to the chemical mutagen termed *nitrosoguanidine* increases the frequency of cells in a population that carry a mutation in any given gene about $10^5$-fold; following treatment with this mutagen about one cell in $10^3$ will be found to carry a detectable mutation in any given gene. Since a cell has several thousand genes, each cell will carry about one detectable mutation somewhere on its chromosome.

Mutations may involve a change in only a single base pair (*microlesions*) of the cell's DNA or they may produce a change that extends over a number of base pairs or even a number of genes (*macrolesions*) (Table 10.2). If a microlesion involves the loss or gain of a base pair it is called a *frame-shift mutation* because it changes the reading frame of all codons of the gene or operon distal to the point of the mutation. If a microlesion involves the change of one base pair [e.g., an adenine-thymine (A—T) pair] to another [for example, a guanine-cytosine

**TABLE 10.2**

**Classification of Mutations Based on Genotype**

I. MACROLESIONS
   A. Deletions
   B. Duplications
   C. Inversions
   D. Insertions
   E. Translocations

II. MICROLESIONS
   A. Base-pair substitutions (transitions and transversions)
      1. Neutral
      2. Missense
      3. Nonsense
   B. Frame shift
      1. +1 or +2
      2. −1 or −2

(G–C) pair] it is called a *base-pair substitution mutation.* Because of the fundamental differences in the mechanism of their formation, base substitution mutations are subdivided into *transition* and *transversion mutations* depending on whether the purine base of a pair is changed to another purine base (transition) or to a pyrimidine base (transversion) (Figure 10.2).

Macrolesions include a variety of changes in DNA: the complete elimination of a segment of DNA (*deletion*), the tandem repetition of a segment of DNA (*duplication*), the inverting of a segment of DNA (*inversion*), the introduction of a new segment of DNA within an existing sequence (*insertion*), and the movement of a segment of DNA to another site in the genome (*translocation*).

The remarkably high frequency of the occurrence of genetic duplications has only recently been recognized, because unless they occur totally within a single gene or operon, they cause no loss of genetic function. As many as one cell in a thousand carries a duplicate copy of any given gene, because duplications form at a very high frequency, but they exist only transitorially because they are lost at a high frequency. (The mechanisms by which duplications are formed and lost are discussed in Chapter 11.) It is clear that duplications play an important role in evolution because they provide a mechanism by which additional DNA is added to a cell's genome.

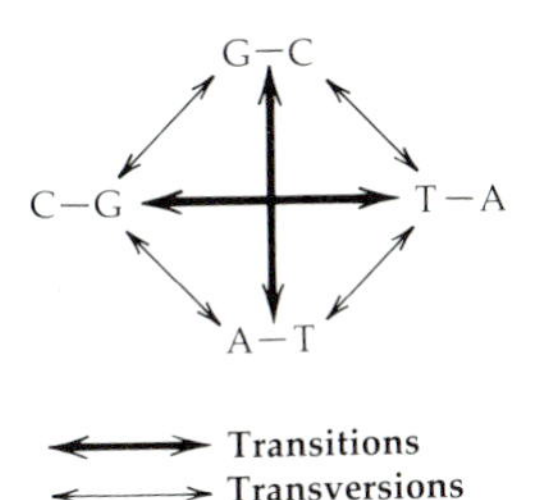


FIGURE 10.2
Base-pair substitutions. Those in which a purine is replaced by a different purine and a pyrimidine is replaced by a different pyrimidine are called *transitions*. Those in which a purine replaces a pyrimidine, or vice versa, are called *transversions*.

## The Consequences of Mutation

The complete set of genes encoded within a genome constitute a cell's *genotype*; the expressed products of these genes and their activities constitute a cell's *phenotype*. Mutations change the cell's genotype in ways discussed in the previous section; they can also affect the cell's phenotype in a variety of ways ranging from no detectable change to lethality.

Let us consider first the possible phenotypic consequences of base-pair substitution mutations on the protein product of a gene. It will be recalled (Chapter 5) that the genetic code is highly degenerate. Of the 20 amino acids occurring in proteins, only two (methionine and tryptophan) are designated by a single codon. Nine others are designated by two codons; one (isoleucine) is designated by three codons; five are designated by four codons; and three are designated by six codons. Thus, a base-pair substitution mutation that changes a codon to a redundant one encoding the same amino acid would not change the protein product of the mutated gene; one that changes a codon to one designating a different amino acid is called a *missense mutation.* The consequences of such a mutation on the activity of the protein product of the mutationally altered gene are quite varied. If the protein product is an enzyme (by far the most probable one of any given bacterial gene) its catalytic activity could be changed or eliminated, its heat stability could be altered, and its susceptibility to degradation by intracellular proteases (*turnover*) could be altered. Studies on the consequences of random missense mutations in the gene *lacZ* in *E. coli* (encoding the enzyme beta-galactosidase that catalyzes the hydrolysis of the disaccharide, lactose), have established that heat stability is the property that is most frequently affected by missense mutations. About 70 percent of such mutations caused detectable loss of heat stability of the protein, but only about 2 percent caused detectable loss of the enzyme's catalytic activity. Although the effect of missense mutations on susceptibility of a protein to turnover was not measured in these studies, it probably parallels loss of heat stability. Thus, the most probable consequence of a missense mutation is a lowering of the maximum temperature at which the affected bacterium can grow under conditions that require the functioning of the product of the affected gene. Considerably less probable

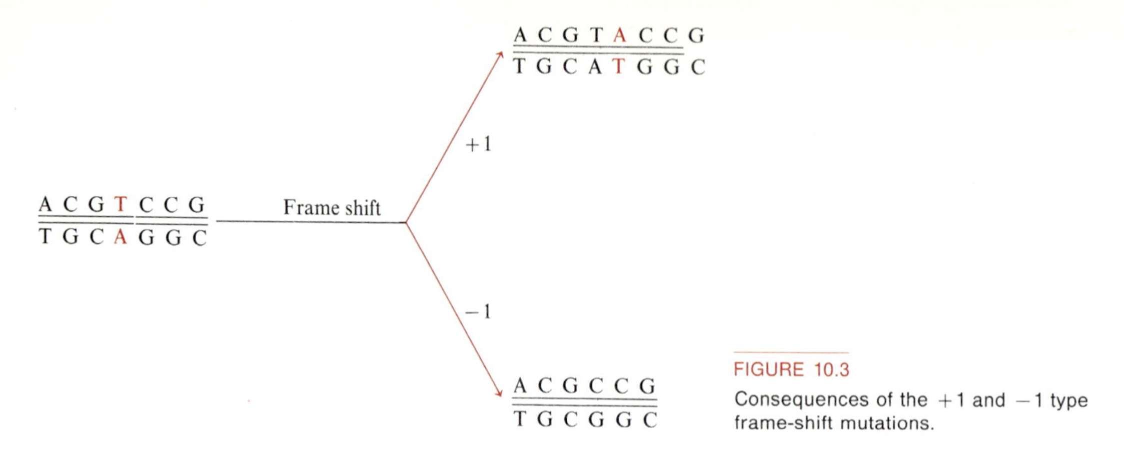


FIGURE 10.3
Consequences of the +1 and −1 type frame-shift mutations.

is the loss of the bacterium's ability to grow under these conditions at all temperatures.

A base substitution mutation might also cause one of the codons encoding an amino acid in the protein to become one of the nonsense codons. Such a base-pair substitution is called a *nonsense mutation.* As discussed in Chapter 5, nonsense codons designate the ends of genes. When a translating ribosome encounters one of these codons on the mRNA molecule, it disengages, terminating the attached polypeptide chain. Thus, generation of a nonsense codon within a gene causes premature termination of the growing polypeptide chain releasing a truncated protein that is usually catalytically inactive.

Frame-shift mutations of either the +1 type or the −1 type (addition or deletion of a single base, respectively) change the reading frame of all codons beyond the point of mutation, causing, with high probability, all the amino acids encoded beyond that point to be changed (Figure 10.3). Moreover, there is a high probability that one of these newly generated codons will be a nonsense codon which will cause termination of the polypeptide chain. Thus, frame-shift mutations, like nonsense mutations, cause, with high probability, an inactive product to be synthesized.

Macrolesions share the common property of generating one or two *improper junctions*, i.e., junctions between segments of DNA that do not occur in the unmutated form of DNA. Unless the macrolesion occurs totally within a single gene, the consequences of the mutation is the fusing of a portion of one gene to another. If the sense strand of the fused genes is the same, and if the reading frame of the promoter-distal gene is not changed, (which would be the case in one-third of all random fusions), the product of the fused genes will be a *chimeric protein*—a fusion of the amino terminal region of one protein with the carboxy terminal region of another. If the reading frame of the promoter-distal gene is changed, the consequences on the carboxy terminal region are the same as those that result from a frame-shift mutation.

In other respects, the consequences of various types of macrolesions are quite variable. Deletions that remove a gene, of course, always cause a complete loss of the function of that gene; and deletions that remove a portion of a gene almost always cause the loss of function of the gene. On the other hand, duplications cause only the relatively subtle effects deriving from genes being present in two copies: generally more gene product is produced. Inversions that occur within a gene cause loss of function of that gene with high probability. Inversions that cover several genes (not in the same operon) cause, with high probability, loss of function of both genes in which the inverted segment ends but the intervening genes remain functional. Insertions almost always cause loss of function of the gene where they occur. The consequence of a translocation is normally restricted to the gene into which the translocated fragment is inserted and is usually the same as an insertion. Most translocations occur as a consequence of *replicative translocation*, after which the translocated segment is found in its original site as well as the new one.

## Mutagens

Mutagens are chemical or physical agents that increase the frequency at which mutations occur during growth of a culture. These agents act in one or the other of two quite different ways: (1) some chemical mutagens become associated with DNA (*intercalating agents*) or become incorporated into it (*base analogues*); (2) a large variety of other mutagens react chemically with DNA—usually with one of its purine or pyrimidine bases. Commonly used chemical mutagens are shown in Table 10.3.

**TABLE 10.3**

**Examples of Types of Chemical Mutagens and Their Mode of Action**

| Name | Structure | Action |
|---|---|---|
| I. MUTAGENS THAT ASSOCIATE WITH OR BECOME INCORPORATED INTO DNA | | |
| A. Base analogue | | |
| 2-aminopurine | (purine ring structure with $NH_2$ at position 2) | Incorporates into DNA; causes transition mutations |
| B. Intercalating agents | | |
| ICR-191 | (acridine ring with $NH(CH_2)_3NH(CH_2)_2Cl$, $OCH_3$, Cl) | Causes frame-shift mutations |
| II. MUTAGENS THAT REACT WITH DNA | | |
| A. Alkylating agents | | |
| nitrosoguanidine[a] | $O{=}N{-}N(CH_3){-}C({=}NH){-}NH{-}NO_2$ | Alkylates purines; causes transitions, transversions, and −1 frame shifts |
| B. Other DNA modifiers hydroxylamine | $HONH_2$ | Hydroxylates 6 amino groups of cytosine; causes G-C to A-T transitions |

[a] Trivial name for N-methyl-N′-nitro-N-nitrosoguanidine.

As stated in Chapter 5, the bases of DNA are most energetically stable when their oxygenated substituents are in the keto form (=O) and their reduced nitrogen substituents are in the amino form ($-NH_2$). In these states adenine pairs with thymine and guanine with cytosine. However, at significant low frequency the bases undergo a tautomeric shift to their enol (—OH)-imino (=NH) forms. In this state their hydrogen bonding properties, and therefore the patterns of pair formation change: adenine now pairs with cytosine and guanine with thymine. If, when replication occurs, a base is in its enol-imino form, an inappropriate base will be introduced into the newly synthesized strand, and, unless it is removed by the proofreading properties of DNA polymerase, a transition mutation will be introduced into the genome at that point (Figure 10.4). Base analogues are effective mutagens because they become incorporated into DNA and undergo tautomeric shifts more frequently than natural bases. The mechanism of mutagenic action of base-analogue mutagens can be illustrated by considering a commonly used mutagen of this type, 2-aminopurine. In its usual amino form it pairs with thymine; in its less frequent imino form it pairs with cytosine. Thus, it usually has pairing properties like adenine, but the probability of its being in the imino form is greater. It is a sufficiently close analogue of adenine to be a substrate for the enzymes of the pathway that incorporates exogenous adenine into DNA, so if 2-aminopurine is added to a growth medium it passes through this pathway and becomes incorporated in DNA. If, at the moment of incorporation it is in amino form, it enters the double helix in place of an adenine residue, and if this base analogue later undergoes a tautomeric shift during replication, the eventual consequence is an A/T to G/C transition mutation. If 2-aminopurine enters in the imino form and later undergoes a tautomeric shift, the eventual consequence is G/C to A/T transition. Thus 2-aminopurine causes both A/T to G/C and G/C to A/T transition mutations (Figure 10.5).

Intercalating agents are planar molecules and can insert between the stacked pairs of bases in the core of the DNA molecule. Such incorporation

guanine

thymine (enol form)

(a)

adenine (imino form)

cytosine

(b)

FIGURE 10.4

Changes in base pairing as a result of tautomeric shifts. In the enol form (a) thymine forms hydrogen bonds with guanine instead of with adenine. In the imino form (b), adenine forms hydrogen bonds with cytosine instead of with thymine. Similar shifts in guanine and cytosine will also cause changes in base pairing.

distorts the backbone of the double helix in such a way that frame-shift mutations can occur when the distorted helix is replicated.

Mutagenic agents that react with DNA cause a variety of chemical changes, some of which are highly specific. For example, hydroxylamine reacts specifically with cytosine converting it to 6-hydroxylaminouracil which pairs with adenine, a process that causes G/C to A/T transitions when this chemically altered DNA is replicated.

Nitrous acid is somewhat less specific in its action in that it reacts with all bases (A, G, and C) that contain amino groups, thereby converting the amino groups to hydroxyl groups. As the altered bases have pairing properties different from those of the naturally occurring bases from which they were derived, treatment with this mutagen causes both A/T to G/C and G/C to A/T transition mutations.

The most powerful known chemical mutagens include the *alkylating agents*, which add methyl or ethyl groups to the heterocyclic nitrogen atoms of the bases. Although the mechanisms by which these alkylations become mutagenic are not entirely clear, a variety of different types of mutations result. For example, alkylating agents are known to cause transition, transversion, and −1 (but not +1) frameshift mutations.

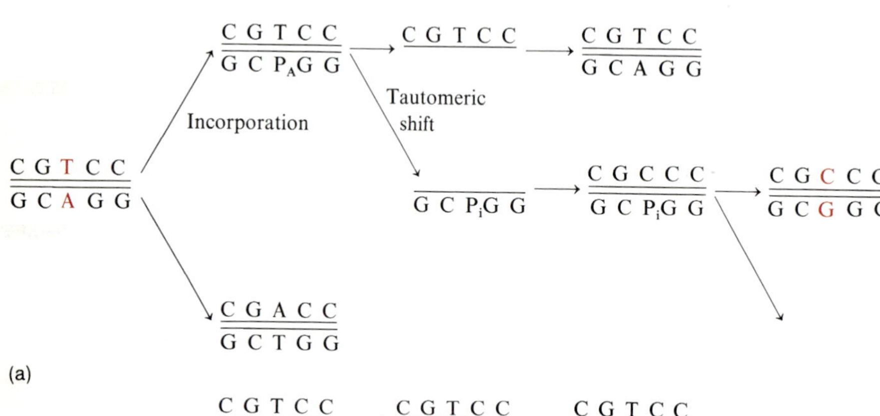


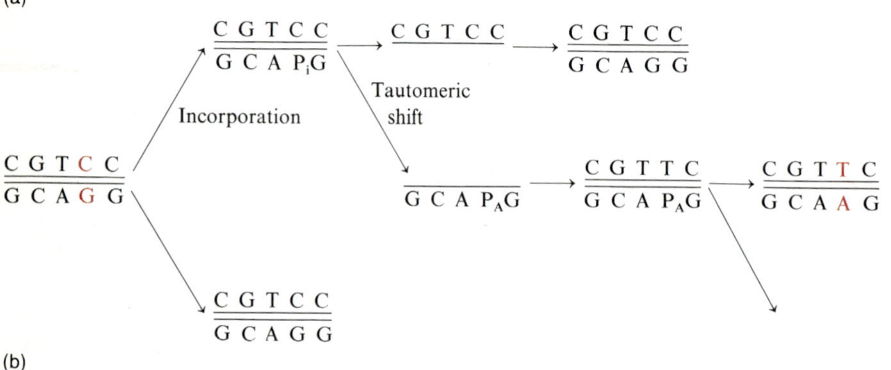


FIGURE 10.5

Mutagenic effect of base analogues. (a) When 2-aminopurine (P) is incorporated in its amino form ($P_A$), a subsequent tautomeric shift causes a transition of AT to GC. (b) When it is incorporated in its imino form ($P_I$), a subsequent tautomeric shift causes a transition from GC to AT.

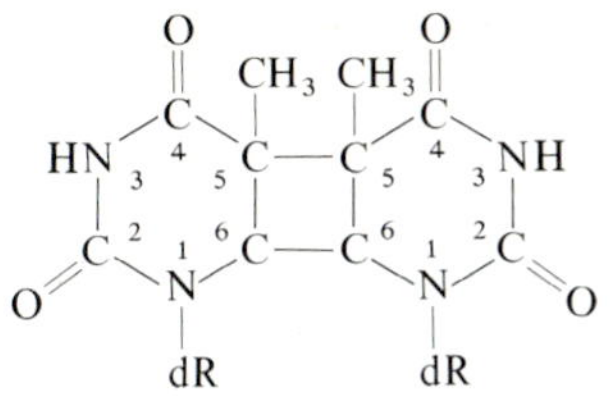

FIGURE 10.6
Structure of a thymine dimer, the most frequent type of pyrimidine dimers that form in DNA during ultraviolet irradiation.
The monomers of the dimer are attached to adjacent deoxyribose (dR) residues in one strand of duplex DNA.

**TABLE 10.4**

**Classification of Certain Mutations Based on Phenotype**

| Mutation | Target Gene | Mutant Phenotype |
| --- | --- | --- |
| Auxotroph | In biosynthetic pathway | Requires exogenous nutrient |
| Carbon source | In carbon dissimilatory pathway | Cannot utilize a particular carbon source |
| Nitrogen source | In nitrogen dissimilatory pathway | Cannot utilize a particular nitrogen source |
| Antibiotic resistance | Gene encoding target protein | Can grow in presence of antibiotic |
| Cryptic | Gene encoding a permease | Cannot mediate a particular cellular function, although all relevant intracellular enzymes are present |

Various types of radiation are also powerful mutagens. X-rays cause breaks in chromosomes that reform in a variety of ways, causing most types of macrolesions. Ultraviolet (UV) light is absorbed by DNA and the energy so released causes dimerization between adjacent pyrimidine residues on the same DNA strand. The occurrence of these pyrimidine dimers (Figure 10.6) triggers a repair mechanism that excises the pyrimidine dimers exposing a short region of single-stranded DNA on the opposite strand. Single-stranded regions are also formed on the opposite strand if a replication fork passes a pyrimidine dimer that has not been excised. The presence of these single-stranded regions induces the formation of a general DNA repair mechanism, termed the *SOS system.* Among the nine or more enzymes that are induced under SOS control to cope with DNA damage is a rapidly acting but relatively inaccurate DNA repair system that fills in the gaps opposite the single-stranded regions. This *error-prone repair*, as it is called, introduces the mutations that follow UV irradiation.

## Phenotypic Consequences of Mutations

In the preceding sections we have discussed the various effects of mutations on a cell's genotype. Depending on the genes in which they occur and their impact on the activity of the gene's protein product, mutations can change a cell's phenotype in a variety of ways. Many mutations inactivate indispensable gene products and therefore kill the cell. But many others inactivate gene products that are not essential under all conditions of growth; i.e., loss of these products is not lethal to the cell. Clones carrying the latter type of mutations can be maintained in cultures: they differ phenotypically in a variety of ways from their unmutated parents termed *wild-type strains* (Table 10.4).

A strain that carries a mutation that inactivates the product of a gene encoding an enzyme in a biosynthetic pathway loses the ability to synthesize the end product of that pathway. If the end product can enter the cell at an adequate rate (which is often the case) the mutant clone can grow in media that contain the end product. Such mutant strains are called *auxotrophs;* their parents are called *prototrophs.*

Strains that carry protein-inactivating mutations in genes encoding enzymes that participate in catabolic pathways lose the ability to grow at the expense of the primary substrate of the pathway, but such mutant strains can be maintained in media provided with other primary substrates.

Other mutations that alter the target protein of an antibiotic or other toxic chemical can render the cell resistant to the antimicrobial agent. Still other mutant strains might have lost some nonessential capacity, such as motility, a tactic response, and so forth.

## Conditionally Expressed Mutations

The phenotypic expression of certain mutations is conditionally dependent on the cell's environment; i.e., in certain environments the mutant clone expresses a wild-type phenotype; in others, a mutant phenotype is expressed. This class of mutations is particularly valuable for studies on microbial physiology, because mutant clones can be maintained with such mutations in any gene, even one that encodes an indispensable cellular activity the loss

**TABLE 10.5**

**Types of Conditionally Expressed Mutants**

| Type | Defect |
| --- | --- |
| Temperature-sensitive | Gene product, usually a protein but sometimes a tRNA, cannot function or be synthesized (*temperature-sensitive synthesis*) at the restrictive temperature; but functions or is synthesized at permissive temperature. |
| Heat-sensitive | High temperature is restrictive; low temperature is permissive (usually 42° and 30° C respectively in the case of enteric bacterial mutant strains). |
| Cold-sensitive | Low temperature is restrictive; high temperature is permissive (usually 20° and 37° C respectively in the case of enteric bacterial mutant strains). |
| Osmotically remedial | Permissive and restrictive conditions are determined by the osmotic strength of the growth medium. |
| Streptomycin remedial[a] | Permissive condition is growth in a medium that contains streptomycin; restrictive condition is growth in a medium that lacks the antibiotic. |

[a] Other aminoglycoside antibiotics, including neomycin and kanamycin can sometimes substitute for streptomycin.

FIGURE 10.7

Generation (a) and function (b) of a nonsense suppressor.

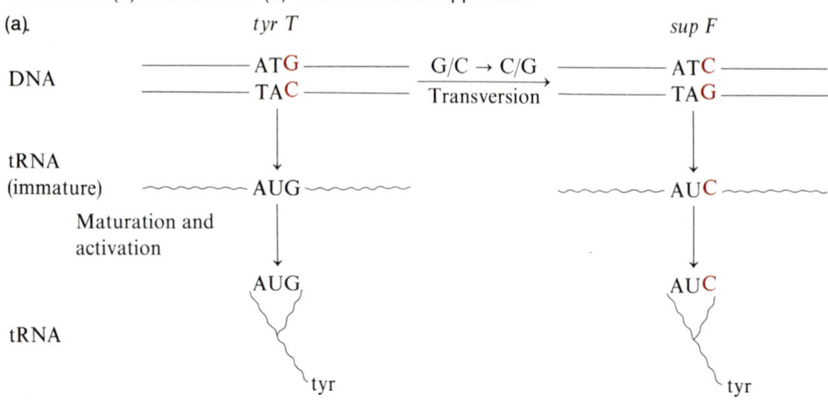


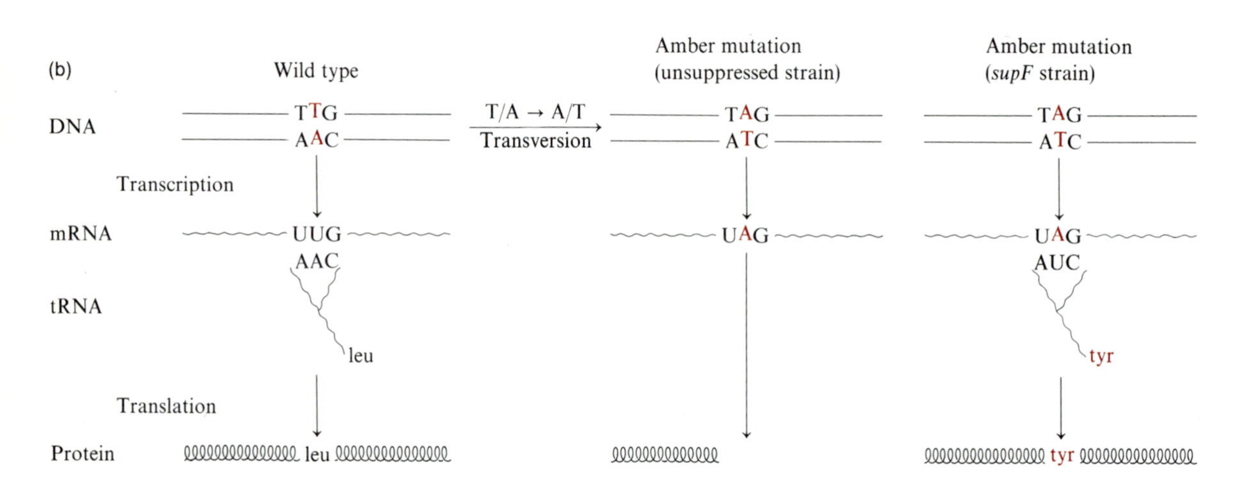

of which would be lethal. The mutant clone can be maintained in culture in the environment in which the wild-type phenotype is expressed (*permissive condition*), and the physiological consequence of the mutation can be evaluated in the environment in which the mutant phenotype is expressed (*restrictive condition*).

The various classes of conditionally expressed mutations along with their permissive and restrictive conditions are listed in Table 10.5. The biochemical bases of conditionally expressed mutations are varied. The *temperature-sensitive* type mutations cause the gene product to become nonfunctional at either high or low temperature (see Chapter 8); the former subclass of mutations are termed *heat-sensitive* and the latter *cold-sensitive*. Most genes with temperature-sensitive mutations encode products that are intrinsically unable to function at the restrictive temperature, but a few, termed *temperature-sensitive-synthesis mutations*, encode products that are nonfunctional only if they are synthesized at the restrictive temperature, at which the secondary or tertiary structure of the protein product forms incorrectly. However, if the protein is synthesized at the permissive temperature a properly folded product is synthesized that can function at the restrictive temperature as well as the permissive one.

Osmotically remedial mutations cause the protein product to be particularly sensitive to osmotic strength. Usually the protein is stable only in the presence of higher concentrations of solutes. Thus the permissive condition for a strain with such a mutation is a medium with a higher concentration of salts. Often such strains cannot grow in commonly-employed complex media but can grow in synthetic media because the latter typically contain higher concentrations of salts.

Streptomycin-remedial mutants express a near wild-type phenotype when low levels of an aminoglycoside antibiotic (streptomycin, neomycin, or kanamycin) are added to the culture medium. These antibiotics exert their remedial effect by altering the translation mechanism rather than the gene products. These antibiotics bind to the 30S ribosomal subunit, thereby increasing the error frequency of translation (the frequency with which an amino acid other than the encoded one is inserted into the growing peptide chain). Thus, in the presence of the antibiotic, incorrect forms of all the cell's proteins, including the mutant protein, are synthesized. Some of the mistranslated forms of the mutant protein are functional. By reversing the phenotypic consequences of a mutation, aminoglycoside antibiotics are said to *suppress* the defects of streptomycin-remedial mutants. Certain mutant forms of tRNA molecules also can suppress some mutations. These mutations in genes encoding tRNA molecules are called *suppressor mutations*, or *suppressors*. A suppressor mutation, like streptomycin, changes the translation mechanism, thereby producing some gene product that is functionally active.

The action of suppressor mutations can be illustrated by considering the action of a specific mutant allele (*supF*) in *E. coli* that suppresses the amber-type nonsense mutations (Figure 10.7). In the specific case considered, the amber codon (UAG) was generated by an A/T to T/A transversion mutation, thereby changing the wild-type codon (UUG), which encodes leucine to the amber nonsense codon (UAG), which signals chain termination. The suppressor mutation is a G/C to C/G transversion in a gene encoding one of the tyrosine tRNAs. The mutation changes the anticodon from AUG to AUC, thereby allowing it to recognize the amber codon and to insert a tyrosine residue at this site. Suppression occurs if the tyrosine-containing protein is functionally active. As is the case with exposure to streptomycin many faulty proteins are synthesized in cells that contain *supF* (or another suppressor mutation), because tyrosine residues are frequently inserted at the sites of other amber codons that should properly signal chain termination. But the codons designating tyrosine continue to be properly translated because the cells contain other (unmutated) species of tyrosine-recognizing tRNA molecules.

Analogous mutant forms of tRNA suppress other mutations as well as missense or even frameshift mutations.

## MUTANT METHODOLOGY

Much of the detailed information now available about the metabolism and activities of microorganisms has come from the study of mutant strains that have lost a specific cellular function. The rationale of the *mutant methodology*, as this set of procedures is sometimes called, is simple, direct, and powerful: *a mutation alters or eliminates the functioning of a particular gene product; by observing the effect of genotypic change on the cell's phenotype, one can deduce the cellular function of the gene product.* For example, strains with certain mutations in the *argI* gene do not synthesize a particular enzyme, ornithine carbamoyltransferase, and are able to grow only if their medium is supplemented with the amino acid arginine or with the intermediates of the pathway leading to biosynthesis of that amino acid, provided that the intermediates enter the pathway subsequent to the genetic blockage. From such a study one learns that ornitine carbamoyltransferase catalyzes an essential reaction in the biosynthesis of arginine. The use of the mutant methodology for determining the pathway of biosynthesis of metabolic intermediates is summarized in Chapter 5, p. 103, but the methodology can also address more subtle questions dealing with regulation and behaviorial responses like chemotaxis (Chapter 6).

## ISOLATION OF MUTANT STRAINS

The ways by which mutant strains are isolated are described in the following sections.

### Phenotypic Expression

As discussed, effective mutagenic treatments are able to increase the frequency of mutant cells in a population by as much as 100,000–fold. The periods of mutagenic treatment are typically quite brief; e.g., attaining a maximal mutagenic effect of nitrosoguanidine usually involves only a 15-minute exposure. But following mutagenesis, a period of growth, sometimes quite lengthy, termed *phenotypic lag* must occur before the genotypically mutant cell expresses a mutant phenotype. The reasons why phenotypic expression of a mutation lags the genetic change that caused it can be explained by considering the case of a mutation that causes a cell to become resistant to certain bacteriophages (Chapter 9) that infect bacteria by adsorbing to specific proteins on the cell's surface. Mutants that are resistant to these phages lack these receptor proteins or produce altered forms of them to which the phages cannot adsorb. As long as some of the wild-type form of the approximately 200 receptors are present on the cell surface, that cell remains sensitive to infection by the phage. Before the cell loses these receptors, it must become a homocaryotic mutant cell (i.e., contain a single type of nucleous) and preexisting wild-type receptors must be diluted from the cell surface by growth of new surface and cell division.

Since most bacteria have several haploid nucleoi at the time of mutation, the cell is a mutant heterocaryon because it contains two different forms (alleles) of at least one gene. If a cell contains four nuclei, two cell divisions must occur before a homocaryotic mutant cell appears (Figure 10.8). But even the newly formed homocaryotic cell would contain many active phage receptors on its surface, and, hence, would remain phage-sensitive. Only after a number of subsequent cell divisions (about 12) would a phenotypically phage-resistant cell first appear (Figure 10.9).

Because of the requirement for homocaryon formation and gene product dilution, the phenotypic lag before recessive mutations are expressed is extended. It is typically much shorter if the mutant phenotype is dominant because there is no need either to form a homocaryon or dilute the wild-type gene product; as soon as sufficient mutant gene product is produced, the phenotype of the mutant is expressed. For example, many chemicals stop the growth of bacteria by entering the cell and inhibiting the activity of specific target enzymes. Certain mutations in the gene encoding one of these target enzymes render it insensitive to the corresponding inhibitory chemical. A mutant phenotype of this class of mutations is usually dominant and the lag between the genotypic and phenotypic change is quite brief, because as soon as some mutant (resistant) enzyme is synthesized, the cell is able to grow in the presence of the inhibitory chemical.

### Enrichment of Mutant Cells in a Population

Even after effective mutagenesis, any particular mutant is relatively rare in a population and some with particular altered gene products can be quite rare. So, in order to obtain a pure culture of mutant cells, one or another of a number of schemes are employed to increase the frequency of mutant cells in the population (Table 10.6). Sometimes, when conditions can be designed to favor the growth of the mutant class, enrichment procedures can be quite simple. For example, phage-resistant clones can be isolated simply by plating the mutagenized, phenotypically expressed population on plates containing phage virions. Cells expressing the parental phenotype are killed; only those phage-resistant cells in the population develop into colonies. Similarly, mutant strains resistant to an antibiotic or a toxic chemical can be isolated by plating the population with the antibiotic or chemical. Such isolation or enrichment procedures are termed *direct enrichment*.

However, not all types of mutants can be isolated by direct enrichment. For example, it is not normally possible to devise conditions that favor the growth of auxotrophic mutant cells in a population composed principally of prototrophs, or con-

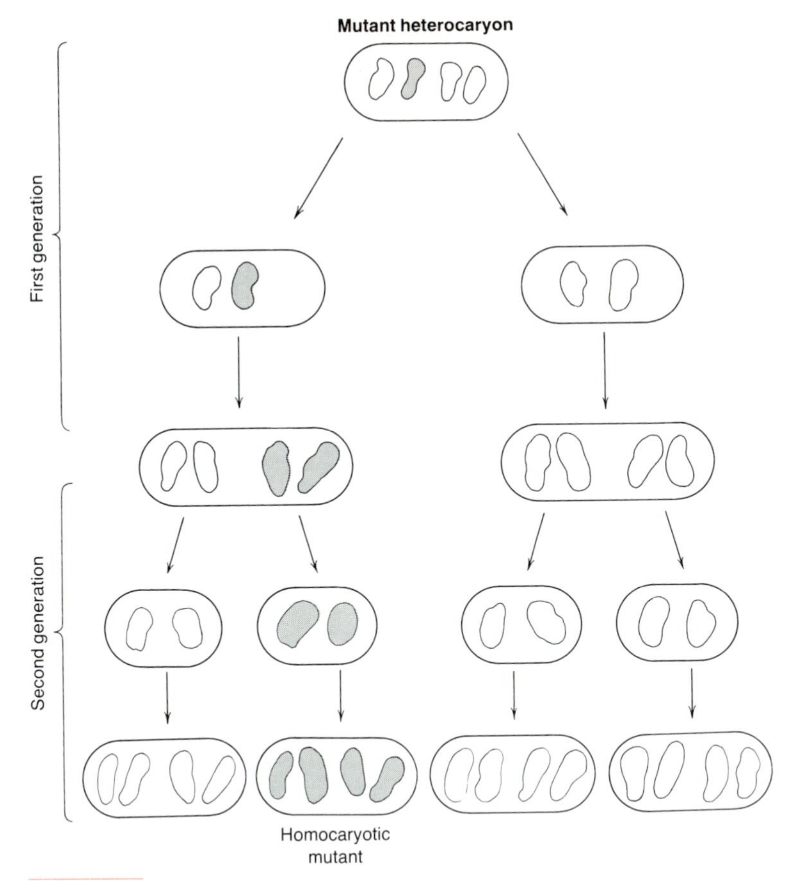


FIGURE 10.8

Bacterial cells with either two or four nuclei. If a mutation first occurs in a tetranucleate cell, two generations are required before a homocaryotic mutant cell appears. If the mutation is recessive, it cannot be expressed until the mutant nucleus has completely segregated from unmutated nuclei.

FIGURE 10.9

Delay in phenotypic expression of mutation. A suspension of phage-sensitive bacteria is treated with a mutagen and the survivors are plated on phage-coated agar. Only induced phage-resistant mutants appear. In this experiment the survivors were allowed to produce varying numbers of generations of growth before plating.

**TABLE 10.6**

**Examples of Schemes Suitable for Enriching a Mutagenized Culture for a Particular Mutant Type**

| Mutant Class | Treatment |
| --- | --- |
| I. DIRECT ENRICHMENT | |
| Phage resistance | Plate culture on a medium containing phage virions |
| Antibiotic or chemical resistance | Plate culture on a medium containing antibiotic or chemical |
| II. COUNTERSELECTION | |
| Auxotroph | Counterselect in absence of required nutrient |
| Carbon source | Counterselect in a medium containing only the carbon source that mutant strain cannot metabolize |
| Nitrogen source | Counterselect in a medium containing only the nitrogen source that the mutant strain cannot metabolize |

**TABLE 10.7**

**Counterselective Agents and the Mechanism by Which They Kill Growing Cells**

| Agent | Mechanism |
|---|---|
| Penicillin | Kills growing cells by inhibiting formation of cross-links in peptidoglycan. |
| 8-Azaguanine | Growing cells incorporate 8-Azaguanine into DNA, rendering it nonfunctional. |
| Radioactive nutrient | Growing cells incorporate the radioactive nutrient in cellular components. These cells die slowly (during subsequent storage) as radioactive decays occur. |
| Thymine deprivation | A culture of thymine auxotrophs die when deprived of thymine. Such a culture can be used to enrich a second mutation because they will not die if they lack an essential nutrient, or are unable to grow for some other reason. |

ditions that favor mutant cells unable to utilize a particular carbon or nitrogen source in a population of cells that can utilize it. The frequency of such mutants in a population can be increased by the use of one of a set of procedures, termed *counterselection*, so named because they result in killing cells that express a parental phenotype, thereby increasing the fraction of mutant cells in the surviving population. The efficacy of counterselection depends on employing a chemical or a condition of growth that kills cells only if they are growing.

The antibiotic penicillin is an effective agent for counterselection; it acts by inhibiting the formation of cross-links in peptidoglycan. As a consequence, cells that grow in the presence of penicillin synthesize a weakened peptidoglycan layer that is incapable of containing the intracellular osmotic pressure and such a cell bursts. But nongrowing cells survive in media that contain penicillin because the antibiotic does not affect preformed peptidoglycan. Thus, in a minimal medium containing penicillin but lacking the essential nutrient of an auxotrophic mutant, the mutant cell survives because it cannot grow, but the prototrophic parental-type cell is killed as it starts to grow, synthesizes weakened peptidoglycan, and bursts. Certain counterselective agents and their mechanisms of selectively killing growing cells are shown in Table 10.7.

## Detection of Mutant Clones

A variety of procedures have been devised to make colonies of the desired mutant type visually distinguishable from colonies of the wild type. For example, the colorless compound tetrazolium is reduced to the brilliantly red, insoluble product, formazan, only within a narrow pH range. Thus, in a complete medium supplemented with a high concentration of a fermentable sugar, cells able to ferment that sugar lower the pH to the point where the dye is not reduced, and form white colonies. Mutant cells unable to ferment the provided sugar, however, reduce the tetrazolium intracellularly to formazan and produce bright red colonies. By this technique it is possible to detect a single fermentation-deficient mutant colony among $10^3$ wild-type colonies on a petri dish.

In some cases, the only reagents that are able to stain mutant colonies differentially are also lethal. For example, one may wish to select mutants that form glycogen, which can only be detected by staining (and killing) the colonies with iodine. In such cases, the technique of *sib selection by replica plating* is used: a plate bearing thousands of colonies is replicated, as described below, and the replica plate is flooded with an iodine solution. If a glycogen-positive mutant colony is detected, an inoculum of *live* mutant cells can be recovered from the corresponding location on the original plate.

In replica plating, a piece of sterile velvet is stretched over a cylindrical block of wood or metal that is slightly smaller in diameter than a petri dish. The block is placed with the velvet surface facing upward; the petri dish with the lawn of bacterial colonies is inverted, and its surface is gently pressed against the velvet. The projecting fibers of the velvet, numbering thousands per square inch, act as inoculating needles, sampling every colony in the lawn. The petri dish is removed, and a fresh plate of agar is pressed against the velvet in order to receive an inoculum from each colony. The plates are identically oriented at each application

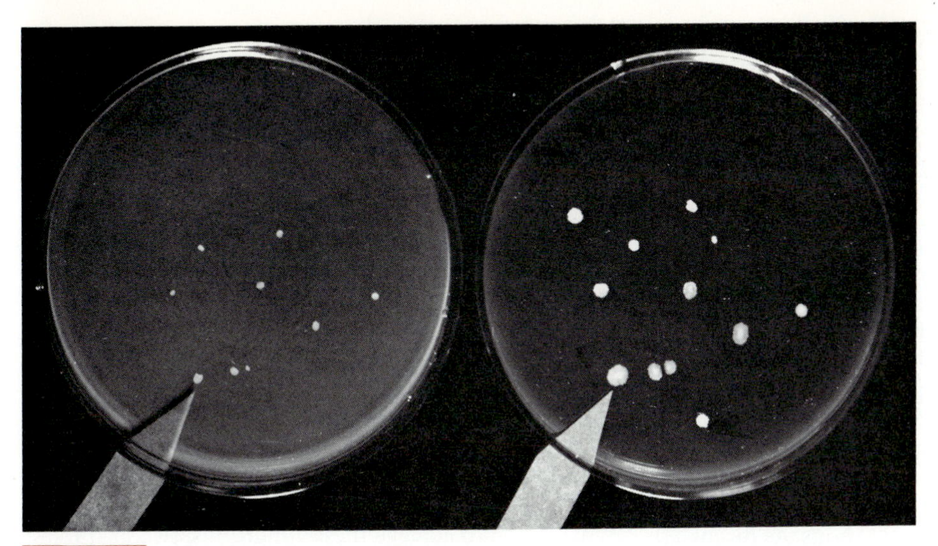

FIGURE 10.10

Replica plating. The master plate of nutrient agar (not shown) bore twelve colonies. One replica was prepared on nutrient agar (right) and one on a synthetic medium lacking growth factors (left). The two plates are similarly oriented, and the arrow points to sister replicas of one colony. Note that although twelve colonies developed on the complex medium, only nine were formed on the synthetic medium. The three colonies that failed to give replicas on the synthetic medium were made up of mutants that required growth factors for their development.

of the velvet with respect to marks placed on their rims, so that the colonies that appear on the replica plate after incubation occupy positions congruent with those of their siblings on the original plate (Figure 10.10).

The inoculum on the velvet surface is usually large enough to permit a series of different agar plates to be sequentially "printed" from it. Thus, replica plating can also be used to test inocula from a very large number of colonies on a "master plate" for their ability to grow on as many as eight or ten different selective media. This technique has made practicable the multiple analyses which are basic to microbial and molecular genetics.

The various steps involved in the isolation of a mutant are illustrated by the complete scheme suitable for isolating a heat-sensitive arginine auxotrophic mutant shown in Figure 10.11.

FIGURE 10.11

Isolation of a heat sensitive arginine auxotrophic mutant.

Culture

↓ Treat with suitable mutagen to increase probability of desired mutant being present.

Mutagenized cullture

↓ Grow for five or six doublings in a medium that permits desired mutant to replicate: either at permissive temperature (30° C) or at restrictive temperature (40° C) with arginine present.

Phenotypically expressed culture

↓ Initiate growth of culture under conditions that preclude growth of desired mutant, i.e. at 40° C in a medium lacking arginine. Add penicillin.

Culture in which the frequency of desired mutant cells is enriched

↓ Plate and incubate under conditions that support growth of mutant and parental clones e.g. at 30° C on a medium lacking arginine or at 40° C on a medium containing arginine; replicate to restrictive conditions, i.e. to 40° C on a medium containing arginine. Pick colony that grows on the first set of plates but not the second.

Mutant clone

## POPULATION DYNAMICS

A growing population of microbial cells is in a dynamic state with respect to the presence of mutant types. Two parameters are involved in this phenomenon: *mutation rate*, which can be assigned a rate constant, and *mutant frequency* (or mutant proportion), which is a variable parameter determined both by the rate of mutation and by the *rate of selection* of the mutant type.

## The Estimation of Mutation Rate

For a population of microbial cells, the *mutation rate* can be defined as the *probability that any one cell will mutate during a defined interval of time.* The first measurements of bacterial mutation rates were performed by Luria and Delbrück in 1943, at a time when the nature of the genetic material and the mechanism of mutation were unknown.They chose as their time parameter the *division cycle*, or the bacterial generation; their formula for mutation rate is thus *the number of mutations per cell per generation*, averaged over many generations. As discussed, most spontaneous mutations represent errors in template action that occur when the DNA double helix replicates. Since under normal conditions DNA replication and cell division are coordinated, the Luria-Delbrück formula is valid under normal physiological conditions of microbial growth.

In the Luria-Delbrück formula, mutation rate is the probability of a mutation occurring when one cell doubles in size and divides to form two cells. This series of events is called a *cell-generation*. The number of cell-generations can be simply determined for any culture, since each cell-generation increases the number of cells in the culture by one. Thus, the number of cell-generations equals the net increase in cells over the period of cultivation, and is given by the expression

$$n - n_0$$

where $n$ is the final number of cells and $n_0$ the number of cells at time zero. A small correction has to be applied to this expression, since at the moment that the culture is sampled to measure $n$, the cells are in varying stages of completion of their next division cycle. In an exponentially growing, nonsynchronized culture the average progress toward the next generation is such that the true number of cell-generations accomplished by the culture is

$$\frac{n - n_0}{\ln 2} = \frac{n - n_0}{0.69}$$

where ln 2 is the logarithm of 2 to the base $e$. The mutation rate is equal to the average number of mutations per cell-generation; we then have the equation

$$a = \frac{m}{\text{cell-generations}} = \frac{m}{(n - n_0)/0.69} = (0.69)\frac{m}{n - n_0}$$

where $a$ stands for the mutation rate and $m$ for the average number of mutations occurring when $n_0$ cells increase in number to $n$ cells.

To determine the mutation rate of a given culture, it is thus necessary to determine $m$. A simple way to achieve this is to allow the mutations to take place in a population of cells growing on a solid medium. Under such conditions, each mutation gives rise to a mutant clone that is fixed *in situ* and—with appropriate manipulations—can be detected as a single colony.

In practice, this means that a population of cells must be permitted to undergo a limited number of cell divisions on an agar plate, following which the conditions must be changed so that only the mutant clones can continue growing to form visible colonies. A variety of methods has been introduced to achieve these conditions; two examples will suffice.

**1.** Cells of an auxotrophic strain are spread on minimal agar containing a sufficient amount of the required growth factor to permit a limited number of divisions. Growth of the parental type then ceases; any prototrophic mutants, which no longer require the growth factor, are able to continue growth and to form visible colonies. The number of such mutants present in the inoculum must be subtracted; this number is determined by including one set of plates with no growth factor in the agar.

**2.** A population of streptomycin-sensitive cells is deposited on a membrane filter, and the membrane is placed on nutrient agar for a limited time. The membrane is then transferred to the surface of nutrient-streptomycin agar; the sensitive parents are killed, while resistant mutants that arose during growth on the nutrient agar form visible colonies. The number of resistant mutants present in the inoculum is determined by plating one set of membranes on nutrient-streptomycin agar at zero time.

In each of these examples the number of colonies per plate (corrected by subtraction of the number of mutants in the inoculum) equals the number of *mutations* per plate. It then remains only to determine the number of cell-generations per plate, which is accomplished by washing the cells off several plates in a known volume of liquid and performing a viable count. The number of cell-generations is, as stated above, equal to $(n - n_0)/0.69$; here $n$ is the average number of cells per plate at the time that the parental population is killed or inhibited and $n_0$ the average number of cells per plate in the original inoculum.

As an example, $1.0 \times 10^6$ streptomycin-sensitive cells are deposited on each of a series of membrane filters. One set of filters is put on streptomycin agar at zero time, and the average number of resistant mutants in the inoculum is found to be 1.2 per filter. Another set of membranes is placed on nutrient agar and incubated for 6 hours. At the end of that time, half the membranes are placed on streptomycin agar, and half are used to determine the viable count. The viable count is found to be $1.0 \times 10^9$ cells per filter; the streptomycin plates, after incubation, show an average of 4.2 colonies per filter. The mutation rate, $a$, is calculated as follows:

$$a = (0.69)\frac{m - m_0}{n - n_0}$$

where $m$ is the final number of mutant colonies and $m_0$ the number of mutants in the inoculum. Substituting the experimentally determined figures, we get

$$a = (0.69)\frac{4.2 - 1.2}{(1.0 \times 10^9) - (1.0 \times 10^6)} = \frac{2.1}{1.0 \times 10^9}$$

The mutation rate to streptomycin resistance was thus $2.1 \times 10^{-9}$ per cell-generation.

The number of mutations occurring in a *liquid* culture of microbial cells can be estimated by a statistical method, as shown by Luria and Delbrück in their original paper. A population of wild-type cells (e.g., streptomycin-sensitive) is used to inoculate a series of 20 or more cultures, each receiving an inoculum small enough to contain no streptomycin-mutants. The cultures are incubated until a high cell density has been reached, and then the entire contents of each tube are spread on a single plate of streptomycin agar. The plates are incubated until the colonies of streptomycin-resistant mutants are countable.

In most cases, the number of mutants found on the plate does not tell us the number of *mutations* that occurred in the corresponding tube; 16 mutants, for example, could have arisen from 16 mutations occurring during the last generation of the population, or from 1 mutation occurring four generations earlier. The exception is the plate with *zero mutants:* this represents a culture in which *zero mutations* took place.

Luria and Delbrück showed that the average number of mutations per culture is related to the fraction of cultures experiencing zero mutations by the zero term of the Poisson distribution:*

$$P_0 = e^{-\bar{m}}$$

where $P_0$ is the fraction of cultures with zero mutants (and hence zero mutations), and $\bar{m}$ is the average number of mutations per culture.

Solving for $\bar{m}$, we get

$$\bar{m} = -\ln(P_0)$$

For example, if 30 percent of the cultures had no mutations, $P_0$ equals 0.30, and $\bar{m} = -\ln(0.30)$, or 1.2. Suppose, in this example, that each tube grew from an inoculum of $1 \times 10^4$ cells to a final population size of $1 \times 10^9$ cells. The mutation rate is then calculated as follows:

$$a = (0.69)\frac{m}{n - n_0}$$

$$= \frac{(0.69)(1.2)}{(1 \times 10^9) - (1 \times 10^4)}$$

$$= \frac{0.83}{1 \times 10^9} = 0.83 \times 10^{-9}$$

In practice, a preliminary experiment is carried out to determine a suitable incubation period. If the cultures are allowed to grow too long, for example, every culture will have more than one mutant and the method cannot be applied.

## Mutational Equilibrium

There is a direct relationship between the mutation rate and the increase in the proportion of mutants in a culture at each generation, assuming that neither the mutant nor the parent type has selective advantage over the other. Suppose, for example, that a culture is started from a small inoculum and that the first two mutations occur when there are $1 \times 10^8$ cells in the culture. The proportion of mutants in the culture is then $2 \times 10^{-8}$. The culture continues to grow, and at the next generation there are $2 \times 10^8$ parent cells. The mutants also divide, however, and there are now four mutants in $2 \times 10^8$ cells; the proportion thus remains $2 \times 10^{-8}$. If no further mutations were to occur, and the mutant cells divided as often as the parent cells, the proportion would remain constant.

Let us assume that mutations do continue to occur, however, with a probability of $2 \times 10^{-8}$ per cell-generation. Then at each generation, for every $10^8$ parent cells, two new mutants will be added to the culture, and the proportion of mutants to parents is increased by just that amount. This is shown in Table 10.8 and in Figure 10.12. In Figure 10.12 the slope of the line $a/b$ directly represents the

* The Poisson distribution describes the proportion of subcultures of a population which will have experienced 0, 1, 2, 3, . . . , $n$ mutations, when the subculture size is very large compared to the number of mutations.

**TABLE 10.8**

**Hypothetical Increase in Proportion of Mutants in a Growing Culture as the Result of New Mutations**

| Generation | Average Number of Parent Cells During Generation | Number of Mutant Cells[a] NEW | OLD | | | | TOTAL | Proportion: Mutants/ Parents |
|---|---|---|---|---|---|---|---|---|
| $n$ | $1 \times 10^8$ | 2 | | | | | 2 | $2 \times 10^{-8}$ |
| $n + 1$ | $2 \times 10^8$ | 4 | 4 | | | | 8 | $4 \times 10^{-8}$ |
| $n + 2$ | $4 \times 10^8$ | 8 | 8 | 8 | | | 24 | $6 \times 10^{-8}$ |
| $n + 3$ | $8 \times 10^8$ | 16 | 16 | 16 | 16 | | 64 | $8 \times 10^{-8}$ |
| $n + 4$ | $16 \times 10^8$ | 32 | 32 | 32 | 32 | 32 | 160 | $10 \times 10^{-8}$ |

[a] The numbers enclosed in the dotted line show how many mutants there would be if no further mutations took place after the first two. Note that the *proportion* of mutants then would have remained constant at $2 \times 10^{-8}$.

mutation rate; the units in which the slope is expressed are "mutants per $10^8$ cells per generation."

What happens when such a population grows indefinitely? At first thought, one might expect the proportion of mutant cells to increase until it reaches 100 percent. This is prevented, however, by the phenomenon of *reverse mutation.* Many mutations are capable of mutating back to the original state, and this reverse mutation will have its own characteristic rate. When a population of bacteria has accumulated a high enough number of mutants, reverse mutations will become significant; the proportion of mutants will ultimately level off at the point where the forward mutations and reverse mutations just balance each other. Assume, for example, that in a population of cells of type X, the mutation X → Y occurs at a certain rate. As the population grows, the proportion of Y mutants will increase. When there are sufficient Y cells, mutation Y → X will have a chance to occur, and eventually the population will reach a true equilibrium state in which the number of forward mutations (X → Y) just equals the number of reverse mutations (Y → X) at each generation. From then on, the proportion of mutants remains constant.

Figure 10.13 illustrates the fact that the same equilibrium proportion of Y mutants is reached whether one starts with a pure culture of cells of type X or a pure culture of cells of type Y. The actual proportion attained is a function of the relative rates of forward and reverse mutation. For example, if these rates are equal, there will be an equal number of X and Y cells at equilibrium. The relationship is simply expressed as

$$\text{equilibrium proportion of Y cells} = \frac{\text{rate of mutation } (X \rightarrow Y)}{\text{rate of mutation } (Y \rightarrow X)}$$

Thus, in the absence of selection, an equilibrium proportion of mutants should eventually be achieved if the population is allowed to multiply indefinitely.

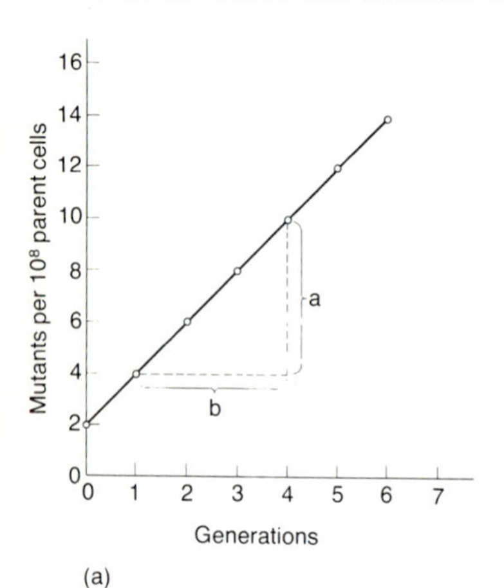


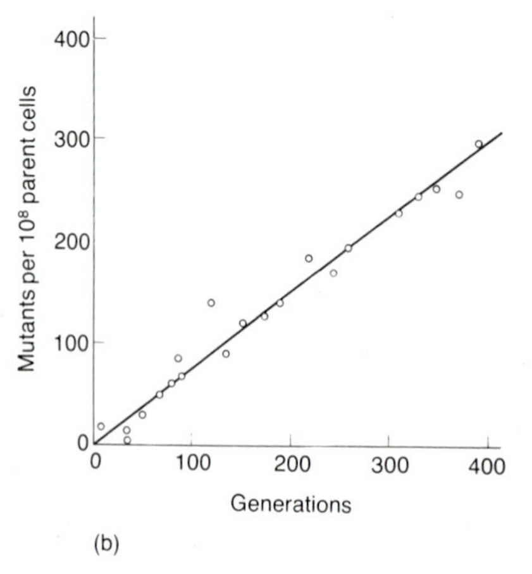


FIGURE 10.12
The increasing proportion of mutants in a culture as a result of spontaneous mutation. (a) The theoretical increase that results from exactly two new mutants per $10^8$ cells appearing at each generation. The mutation rate ($2.0 \times 10^{-8}$ per generation) is expressed by the slope of the line, which is *a/b*, or $(6 \times 10^8)/3$. (b) The results of an actual experiment in which the proportion of mutants in a culture has been determined by plating at successive times. The mutation rate is found to be $0.75 \times 10^8$ per generation since this is the slope of the plotted line.

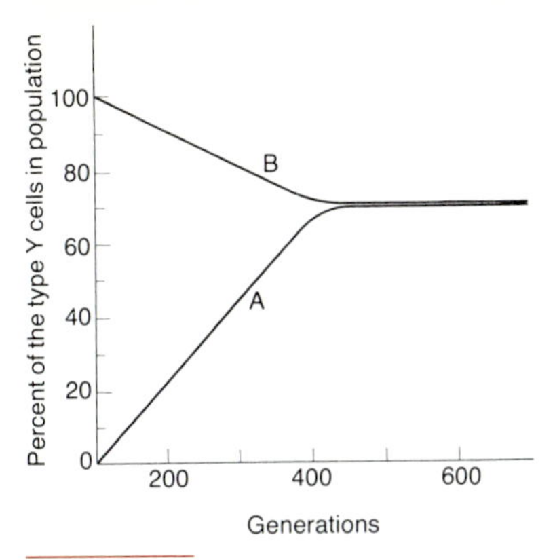


FIGURE 10.13

The attainment of an equilibrium proportion of mutants in a culture. In the case of curve A, the experiment was begun with a population having no cells of type Y. As a result of forward mutation, the proportion of type Y cells rose until they constituted about 70 percent of the population. At this point, back mutation and forward mutation just balanced each other. In the case of curve B, the experiment was begun with a pure culture of type Y cells. The proportion of Y cells decreased as the result of Y → X mutations, until again an equilibrium was reached at 70 percent.

### Effects of Selection on the Proportions of Mutants Types

In the section above on mutational equilibrium we saw that the proportion of a given mutant type in a microbial population increases, in the absence of any selective advantage, in proportion to the mutation rate. Suppose, for example, that we have produced by mutagenesis a strain of *E. coli* requiring the amino acid histidine for growth. A pure culture of this strain, here designated $h^-$, is put on a slant. The fully grown slant culture will probably contain one $h^+$ mutant (able to synthesize histidine and hence not requiring it for growth) for every million or so $h^-$ cells, and this proportion will increase with succeeding generations as the stock culture is transferred from slant to slant. The medium contains ample histidine, so there is negligible selective advantage for either type of cell. Assuming that about ten generations are accomplished on each slant and that the culture is transferred several times a year, one should expect that in a few years the culture would contain a greatly increased proportion of $h^+$ cells.

In practice, however, this rarely happens, even when calculations based on observed forward and reverse mutation rates predict that it will. Instead, we find that an apparent equilibrium is reached long before it should be, and always in favor of the genetic type with which the culture was started. The proportion of mutant cells in the culture increases only up to a very low value—perhaps $1 \times 10^{-6}$—and fluctuates around this value.

This puzzling observation has been found to result from the phenomenon of *periodic selection* (Figure 10.14). At fairly regular intervals in a population of bacteria, mutants arise that are better fitted to the environment and that eventually displace the parental type as a result of selection. We cannot always define the properties of the new mutant that give it this advantage. It might be an intrinsically faster growth rate, or it might be that the new type produces metabolic products that inhibit the parent type. In any case, the better-adapted mutant overgrows the culture, only to be replaced in turn by a mutant that is still better adapted. The process of replacement may be repeated many times, for new mutants that can displace the predominant type from the population continue to arise. *This periodic change in the population has a direct effect on the equilibrium proportion of all other mutants.* Let us consider the specific case of the $h^+$ mutants mentioned earlier. Suppose that a better-adapted mutant appears in the culture at the moment when the proportion of $h^+$ cells has risen to $1 \times 10^{-6}$. The better-adapted mutant could theoretically arise from either an $h^-$ cell or an $h^+$ cell, but since there are $10^6$ $h^-$ cells for every $h^+$ cell, the odds are a million to one that the new type will arise in the $h^-$ population.

FIGURE 10.14

Periodic selection. (a) The successive appearance and disappearance of different $h^+$ mutants. (b) The same data are replotted in terms of total $h^+$ mutants of all types. The resulting slightly fluctuating curve represents the pseudoequilibrium level that the proportion of $h^+$ mutants reaches.

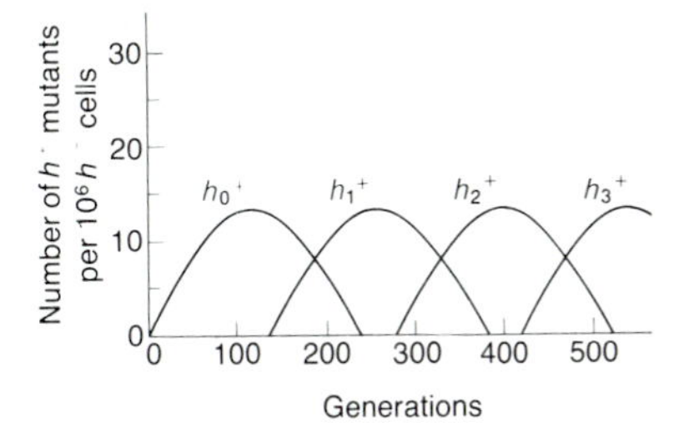


(a)

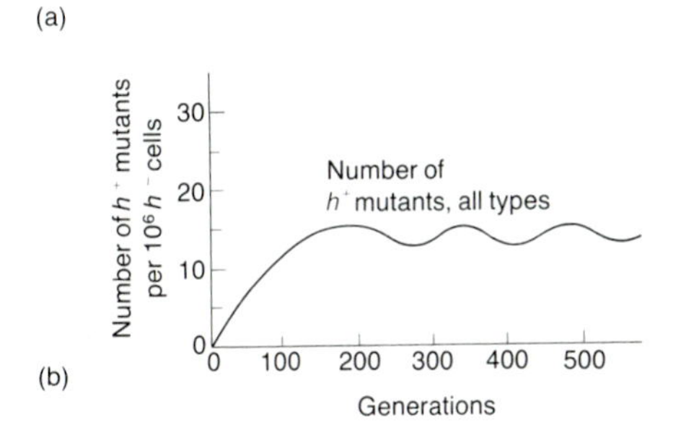


(b)

The better adapted mutant will thus have a selective advantage over all other cells in the population, which it will soon displace. Since the better adapted type is genetically $h^-$, all $h^+$ cells in the population should disappear as the result of selection.

The total disappearance of $h^+$ cells is prevented, however, by the occurrence in the better-adapted $h^-$ population of new mutations to $h^+$. Since the new $h^+$ cells are not at a selective disadvantage, they will increase in proportion until the cycle is started over again by the appearance of an even better adapted type.

This process can be expressed symbolically as follows. Let us call the original cells $h_0^-$ and $h_0^+$ and the first better-adapted mutant $h_1^-$. In the new $h_1^-$ population, mutations to $h_1^+$ will occur. As time goes on and the proportion of $h_0^+$ cells drops, there is a corresponding increase in the number of $h_1^+$ cells. The cycle is repeated again and again. The $h_1^-$ cells give rise to a still better adapted type, $h_2^-$, which displaces the $h_1^-$ and $h_1^+$ cells. The loss of $h_1^+$ cells is compensated for by the appearance of $h_2^+$ mutants. The mutational pattern can be diagrammed as

$$\begin{array}{ccccccc} h_0^- & \longrightarrow & h_1^- & \longrightarrow & h_2^- & \longrightarrow & h_3^- \\ \downarrow & & \downarrow & & \downarrow & & \downarrow \\ h_0^+ & & h_1^+ & & h_2^+ & & h_3^+ \end{array}$$

The left graph in Figure 10.14 shows the way that successive waves of $h^+$ mutants rise and fall in the population. The right graph in Figure 10.14 shows the apparent stability of the population with respect to the characters $h^+$ and $h^-$, when $h^+$ mutants are considered as a single class.

The level that the proportion of mutants reaches can be called a *pseudoequilibrium*, because it is really the composite result of a series of discrete, nonequilibrium events. With ordinary mutation rates, which are very low, the occurrence of periodic selection results in the attainment of such pseudoequilibria. It is only when both the forward and back mutation rates are extremely high that true equilibria, as illustrated in Figure 10.13, can be attained. In such cases, the proportion of mutant cells rises so rapidly that better-adapted mutants have an equal chance of appearing in the mutant or in the parent population.

Periodic selection is a subtle phenomenon, since the mutant type that is being experimentally observed (e.g., the $h^+$ mutant is the case described above) is not the one subject to selection. As the above example shows, the selection of one type of mutant in a population may prevent any other mutant type from increasing in proportion.

## SELECTION AND ADAPTATION

### The Genetic Variability of Pure Cultures

As a general rule, any one gene has only one chance in about 100 million of mutating at each cell division. At first sight, therefore, mutation might appear too rare to be of much significance. Suppose, however, that we have a "pure culture" of a bacterium in the form of 10 ml of a broth culture that has grown to the stationary phase. Such a culture will contain about 10 billion cells; for any given gene, there may well be several thousand mutant cells present in the culture. Even during the growth of a single bacterial colony, which may contain between $10^7$ and $10^8$ cells, a large number of mutants will arise (Figure 10.15).

Thus, a large population of bacteria is endowed with a high degree of potential variability, ready to come into play in direct response to changing environmental conditions. Because of their exceedingly short generation times and the consequent large sizes of their populations, these haploid organisms possess a store of latent variation despite the fact that they cannot accumulate recessive genes as can a population of diploid organisms. In practice, this means that no reasonably dense culture of bacteria is genetically pure; even a slight change in the medium may prove selective and bring about a complete change in the population within a few successive transfers. This explains, for example, why many "delicate" pathogenic bacteria, which prove difficult to cultivate when first isolated from their hosts, gradually become better adapted to the conditions of artificial media.

### Selective Pressures in Natural Environments

So far we have considered only the selective forces that may operate in artificial cultures. In nature, however, selection acts in an even more stringent fashion. A microbe in the soil, for example, must be able not only to survive under a given set of physicochemical conditions, but also to survive in competition with the numerous other microbial forms that occupy the same niche. Any mutation that decreases, even to the slightest extent, the ability of the organism to compete, will be selected against and quickly eliminated. Nature tolerates little variation within microbial populations, for the laws of competition demand that each type retain the array of genes that confers maximum fitness.

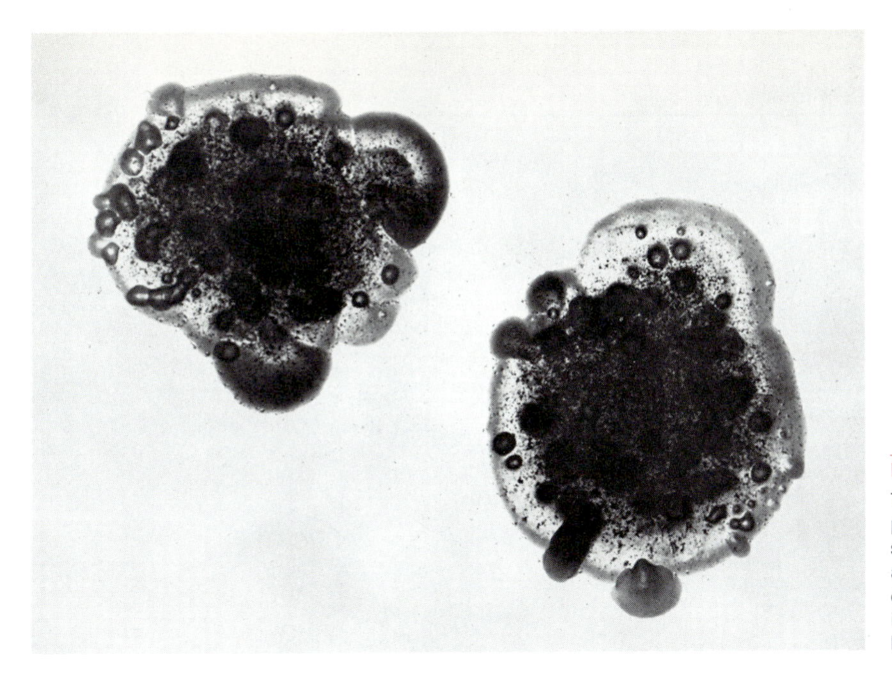

FIGURE 10.15

Two bacterial colonies showing papillae, which represent secondary growth of mutants that arose during the formation of the original colonies. From V. Bryson, in W. Braun, *Bacterial Genetics*. Philadelphia: Saunders, 1953.

As soon as an organism is isolated in pure culture, the selective pressures resulting from biological competition are removed. The isolated population becomes free to vary with respect to characters that are maintained stable in nature by selection. In adapting to existence in laboratory media, organisms may undergo genetic modifications that would lead to their speedy suppression in a competitive environment.

## THE CONSEQUENCES OF MUTATION IN CELLULAR ORGANELLES

Part of the genome of eucaryotic organisms is carried in the mitochondria and chloroplasts. Each kind of organelle contains and reproduces DNA that determines certain of its phenotypic properties. The properties of a chloroplast or a mitochondrion are thus controlled in part by nuclear and in part by organellar genes, both subject to change by mutation. Until recently, it has been difficult to select mutations that specifically affect organellar DNA. However, it has been found that mutations in yeast that confer resistance to certain antibiotics (those known to affect protein synthesis in bacteria) take place in the mitochondrial DNA. It has thus become possible to study experimentally the transmission of many different mitochondrial mutations. Each cell contains a population of mitochondria; hence, the relative growth rates of normal and mutated mitochondria determine the stability of such a mutation during vegetative growth. The situation is entirely comparable to that of a growing bacterial population which contains two genetically different kinds of cells, and the outcome can also be determined by environmental factors.

Organellar mutations have also been observed in the chloroplast of the unicellular alga, *Chlamydomonas*. R. Sager and her colleagues have induced mutations in chloroplast DNA affecting the ability of the cell to photosynthesize, as well as to resist the action of certain antibiotics.

## MUTANT TYPES OF BACTERIOPHAGES

In addition to the conditionally expressed lethal mutations described earlier, phages can undergo mutations that produce nonlethal changes in phenotype. The most readily observed nonlethal changes are those that produce alterations in *plaque morphology* and alterations in *host range*.

When a wild-type phage, such as T4, is plated on a sensitive host, such as *E. coli* strain B, under carefully standardized conditions, the plaques that appear are homogeneous and characteristic in appearance: they are small, with irregular fuzzy edges. When a large number of plaques is examined, how-

**TABLE 10.9**

**Some Mutant Types of T-even Bacteriophages**

| Type | Phenotype | Primary Effect of Mutation |
|---|---|---|
| Rapid lysis | Large plaques with sharp edges | Unknown |
| Minute | Very small plaques | Slow synthesis of phage, or precocious lysis of host cell |
| Host range | Adsorbs to bacteria that are resistant to wild-type phage | Altered polypeptides of the tail fibers |
| Cofactor-requiring | Requires a cofactor such as tryptophan for adsorption to host | Abnormal tail fibers bind to sheath, require cofactor to be released |
| Acriflavin-resistance | Forms plaques on agar containing concentrations of acriflavin that are lethal for wild-type phage | Causes host cell membrane to have reduced permeability for acriflavin |
| Osmotic shock | Survives rapid dilution from 3.0 M NaCl into distilled water | Alteration in head protein increases permeability of head |
| Lysozyme | Does not produce halo around plaque | Abnormal lysozyme synthesis |

Source: Modified from G. Stent, *Molecular Biology of Bacterial Viruses* (San Freeman, 1963).

ever, a few aberrant types are always observed; when particles from such plaques are picked and replated, the aberrant plaque type is found to breed true and thus to reflect a genetic mutation. Several mutant phenotypes are listed in Table 10.9.

Earlier in this chapter we described the occurrence of phage-resistant mutants in populations of phage-sensitive bacteria. These mutants owe their resistance to the production of altered surface receptors, such that they no longer adsorb wild-type phage particles; *E. coli* strain B, for example, can mutate to the state designated B/2, which does not adsorb phage T2. If $10^6$ or more particles of T2 are plated on a lawn of B/2 cells, however, a few plaques appear; when particles from these plaques are isolated and purified, they are found to be *host-range mutants*, which can now adsorb to cells of B/2 as well as to cells of *E. coli* strain B. The mutation in this case consists of a base-pair change in the gene governing the structure of the tail-fiber proteins, which are the adsorption organs of phage T2. The mutant phage is designated T2*h*.

By plating cells of B/2 with the mutant phage, one can select a new class of mutant bacteria that is resistant to phage T2*h*. The entire cycle can now be repeated: a second-step host-range mutant of the phage can be selected, which can adsorb to the new resistant bacterium. Apparently, any altered configuration of the bacterial surface receptor can be matched by an alteration in the adsorption organ of the phage. In nature the mutational capacities of cell and virus permit both to exist: at any given moment there are both susceptible hosts available to the virus as well as cells that can resist viral attack.

## FURTHER READING

### Books


Birge, E. A., *Bacterial and Bacteriophage Genetics.* New York, Heidelberg, and Berlin: Springer-Verlag, 1981.

Glass, R. E., *Gene Function.* Berkeley and Los Angeles: University of California Press, 1982.

Lewin, B., *Genes.* New York, Chichester, Brisbane, Toronto, and Singapore: John Wiley, 1983.

Chapter 11

# Microbial Genetics: Genetic Exchange and Recombination

The ability to exchange genes within a population is a nearly universal attribute of living things. Although the details of the process vary enormously, all systems of genetic exchange among eucaryotes involve the same cellular event: two haploid cells (*gametes*) fuse to form a diploid *zygote*; i.e., a complete complement of genes is contributed by each gamete. This almost never occurs during exchange of genes among procaryotes. In all those cases that have been studied sufficiently to reveal the molecular details of the process, only a small portion of the genome from one procaryotic cell (the *donor*) is transferred to another (the *recipient*), thus forming an incomplete zygote (termed a *merozygote*) that contains the complete genetic complement of the recipient (the *endogenote*) but, with very few exceptions, only a portion of the genetic complement of the donor (the *exogenote*) contained in the fragment of the chromosome that is transferred to the recipient.

In no known case is genetic exchange among procaryotes an obligatory step (as it often is among eucaryotes) in the completion of an organism's life cycle. Rather, genetic exchange seems to be an occasional process that occurs by three quite different mechanisms in various procaryotes. The three mechanisms of genetic exchange are called *transformation*, *transduction*, and *conjugation*.

In the case of transformation, DNA is released from cells into the surrounding medium, and recipient cells incorporate it into themselves from this solution.

In the case of transduction, DNA is transferred from one procaryotic cell to another as a consequence of a rare formation of an aberrant phage virion in which some or all of its normal complement of DNA is replaced by bacterial DNA (donor DNA). When such a phage virion attaches to and introduces this DNA into another bacterial cell (the recipient), genetic exchange is effected.

In the case of conjugation, genetic exchange occurs between cells in direct contact with one another by a process that is, in all known cases, encoded by plasmid-borne genes. Usually only the plasmid itself is transferred from donor to recipient by this process, but sometimes chromosomal genes are transferred as well.

## BACTERIAL TRANSFORMATION

Studies on bacterial transformation have had special impact on bacterial genetics in particular, and on biology in general. It was the first mechanism of bacterial genetic exchange to be discovered. In 1928 F. Griffith showed that injection of mice with an avirulent (not capable of causing disease, see Chapter 31) strain of *Streptococcus pneumoniae* (pneumococcus) together with heat-killed cells of a virulent strain killed the mice, although injection of mice with either culture alone caused no disease. On autopsy, these mice were found to contain live virulent cells of *S. pneumoniae*. These and subsequent experiments established that surviving cells were recombinant: they exhibited certain properties (including virulence) that were typical of the killed cells and others that were typical of the avirulent culture. Thus a genetic exchange had occurred between the dead cells and the live ones. Subsequent experiments by other investigators established that this type of genetic exchange could occur *in vitro*, and it was presumed that a particular substance, termed the *transforming principle*, mediated it.

In 1944, O. T. Avery, C. M. MacLeod, and M. McCarty purified the pneumococcal transforming principle and identified it as being DNA. Indeed, these experiments were the first to establish in any biological system that DNA is the macromolecule in which genetic information is encoded.

The word *transformation* then came to be used to describe genetic exchange among procaryotes that was mediated by DNA which at one point in the process of genetic exchange was dissolved in the external medium. As such, the definition of transformation is a very general one: included in it are a number of mechanistically distinct processes, the diversity of which has only recently been fully appreciated.

**TABLE 11.1**

**Bacteria Known to Encode a Capacity for Natural Transformation**

| |
|---|
| Gram-Positive Bacteria: |
| *Streptococcus pneumoniae, S. sanguis* |
| *Bacillus subtilis, B. cereus, B. lichiniformis, B. stearothermophilus* |
| *Thermoactinomyces vulgaris* |
| Gram-Negative Bacteria |
| *Neisseria gonorrheae* |
| *Acinetobacter calcoaceticus* |
| *Moraxella osloensis, M. urethalis* |
| *Psychrobacter* spp. |
| *Azotobacter agilus* |
| *Haemophilus influenzae, H. parainfluenzae* |
| *Pseudomonas stutzeri, P. alcaligenes, P. pseudoalcaligenes, P. mendocina* |

### Types of Transformation Mechanisms Found among Procaryotes

Cells that are in a state in which they can be transformed by DNA in their environment are said to be *competent*. In a significant number of bacteria (Table 11.1), entry into the competent state is encoded by chromosomal genes and signaled by certain environmental conditions. Such bacteria are said to be capable of undergoing *natural transformation*. Many other bacteria do not become competent under ordinary conditions of culture but they can be made competent by a variety of highly artificial treatments such as exposure of cells to high concentrations of divalent cations. Such systems of transformation have been termed *artificial transformation*.

Comparisons of the natural transformation systems of two procaryotes, *Streptococcus pneumoniae* and *Haemophilus influenzae*, that have been quite thoroughly studied illustrate the many variations that occur in this process (Table 11.2).

Until quite recently, the pattern of transformation exhibited by *Streptococcus pneumoniae* was considered to be typical of all naturally transformable Gram-positive bacteria, and that exhibited by *Haemophilus influenzae* to be typical of all naturally transformable Gram-negative bacteria. More recently, studies on the natural transformation of plasmids have shown that this is not the case; differences in mechanism of transformation do not necessarily correspond with the nature of the cell wall as revealed by the Gram reaction.

**TABLE 11.2**

**Differences between the Natural Transformation Systems Encoded by *Streptococcus pneumoniae* and *Haemophilus influenzae***

| Property | *Streptococcus* | *Haemophilus* |
|---|---|---|
| Competence factor triggers competence | Yes | No |
| Form in which DNA enters cell | Single-stranded | Double-stranded |
| Source of DNA that can enter cell | Any | Only homologous |
| Form of DNA bound to cell surface | Double-stranded | Double-stranded |
| Physical state of DNA within cell | Protein-bound | Transformasome-contained |
| Exhibits eclipse period | Yes | No |

## Natural Transformation Systems: *Streptococcus pneumoniae*

Cells in cultures of the Gram-positive bacterium, *Streptococcus pneumoniae*, rapidly become competent during the exponential phase of growth (Figure 11.1). This conversion of noncompetent cells into competent ones is mediated by a small protein termed the *competence factor*. The competence factor is constantly produced and excreted into the medium by cells of *S. pneumoniae*, but only when the density of the population of the cells in the culture, and hence the concentration of competence factor in the suspending medium, rises to a certain critical value does competence develop. A set of about 12 proteins is synthesized that mediate the process of transformation. With these proteins competent cells can absorb double-stranded DNA to their outer surface at several sites and cleave it through the action of surface-bound enzymes into smaller fragments. Then one strand of the fragment is digested by the nuclease and the other enters the cell while being bound to a competence-specific DNA-binding protein (Figure 11.2).

*Streptococcus pneumoniae* will take up and process DNA regardless of its source: for example, DNA from salmon sperm is taken up as readily as DNA from another *S. pneumoniae* cell. However, only if the DNA is homologous with the endogenote will it become integrated and thereby genetically alter the recipient cell. Fragments of nonhomologous DNA, not themselves constituting a replicon and not becoming part of one by integration into the endogenote, are not replicated and are eventually degraded; they cause no heritable change in the recipient cell.

Integration of homologous DNA occurs by a process of strand replacement (Figure 11.2) forming as an immediate product a *heteroduplex* region, one strand of which is the newly entered one and the other of which is the homologous region of the endogenote. Owing to their different origins, the two strands comprising the heteroduplex might not be identical (certainly the case if the consequence of transformation is a heritable change in the recip-

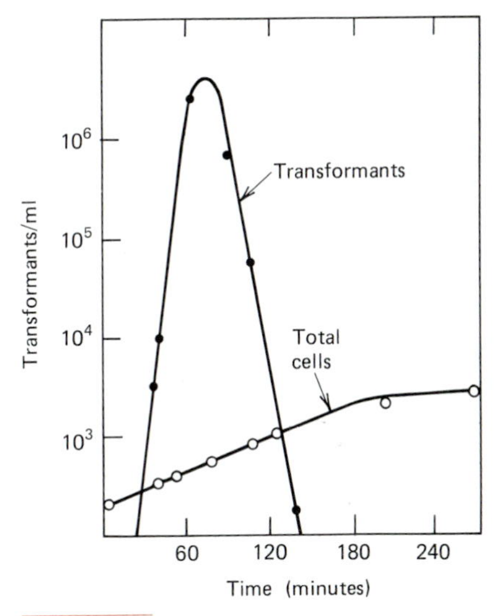


FIGURE 11.1
Course of development of competence during growth of a culture of *Streptococcus pneumoniae*. The number of competent cells in the culture was estimated from the number of cells that could be transformed (transformants) in samples taken at various times. The total number of cells in the culture (Total cells) is plotted on a different scale; the maximum number present at 240 minutes was $6.5 \times 10^8$/ml. After A. Tomasz, "Control of Competent State in *Pneumococcus* by a Hormone-like Cell Product: An Example of a New Regulatory Mechanism in Bacteria." Nature **208,** 155–159 (1965).

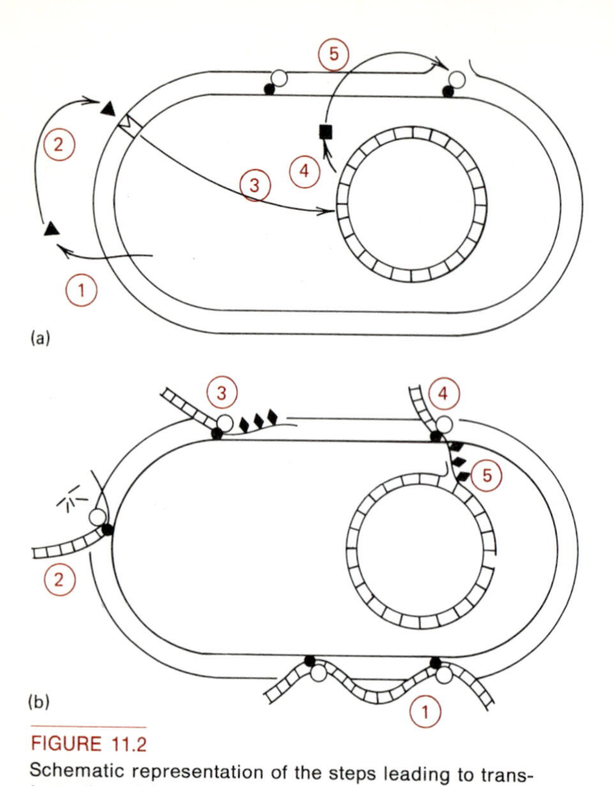


FIGURE 11.2
Schematic representation of the steps leading to transformation of *Streptococcus pneumoniae*. (a) Development of competence: (1) Cells in the culture produce a soluble protein termed competence factor (▲) that (2) adsorbs at a site on the cell surface, M, causing (3) certain genes to be expressed. Among these (4) is an autolysin (■) that exposes (5) a DNA-binding protein (○) and a nuclease (●). (b) Transformation: A long strand of double-stranded DNA is bound to the cell surface (1) where the nuclease (●) degrades (2) one of these strands. The remaining single strand is bound (3) to a DNA-binding protein (◆). In this form it enters the cell (4) and becomes integrated into the chromosome (5) by single-strand replacement. After H. O. Smith, D. B. Danner, and R. A. Deich, "Genetic Transformation," *Ann. Rev. Biochem.* **50,** 41 (1981).

ient), in which case there will be regions in which the heteroduplex is not held together by hydrogen bonding. The existence of such regions can trigger the operation of the recipient cell's DNA repair system through the action of which the mismatched regions of the exogenote sometimes are removed and replaced by bases complementary to the endogenote. (Curiously, it appears that the bases of the exogenous strand, rather than bases from the resident endogenote strand, are preferentially removed.) If the mismatched bases of the transformed strand are removed (a process sometimes termed *correction*) before that portion of the chromosome is replicated no heritable change in the recipient results. If, on the other hand, replication occurs first, two homoduplex copies are made, one of which is identical to the DNA taken from outside the cell; the consequence is genetic change of the recipient cell by transformation.

Early studies on transformation of *S. pneumoniae* were genetic, in that the progress of the process was followed by extracting DNA from the recipient and testing its ability to transform another recipient. For example (Figure 11.3), DNA from a streptomycin-resistant strain might be used to transform a sensitive strain and, at various times after the DNA encoding streptomycin resistance is added to the recipient; extracts from the recipient are used to transform another culture of the sensitive strain. The results of such experiments were at first quite puzzling: for a certain period of time after addition of the DNA encoding streptomycin resistance, extracts of the recipient failed to transform another recipient to streptomycin resistance. The period during which the transforming DNA seems to disappear as judged by transformability has been called the *eclipse period.* With our current state of knowledge of the process of transformation in *S. pneumoniae*, the reasons for the occurrence of an eclipse period are understood. Only double-stranded DNA is bound to the cell surface in the first step of the transformation process, so during that period in which the entering DNA is in a single-stranded form it is inactive when assayed for ability to transform another cell.

FIGURE 11.3
The eclipse phase of transforming DNA. At periods after DNA encoding resistance to streptomycin is added to a culture of streptomycin-sensitive cells, DNA from samples of these cells is extracted and used to transform other streptomycin-sensitive cells to resistance. At first no DNA is capable of mediating this transformation. Only after a period of about 10 minutes does the DNA extract gain the ability to transform maximally. After M. S. Fox, "Fate of Transforming DNA Following Fixation by Transformable Bacteria," *Nature* **187,** 1004–1006 (1960).

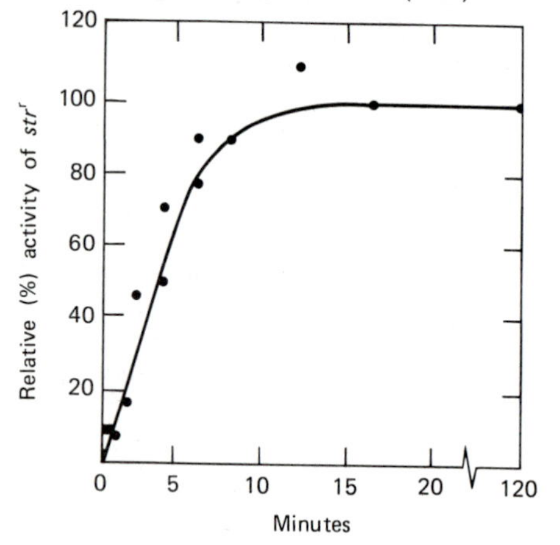

## Natural Transformation Systems: *Haemophilus influenzae*

The process by which the Gram-negative bacterium *Haemophilus influenzae* is transformed differs in many fundamental respects from the streptococcal system just discussed. No competence factor is produced that triggers development of the competent state. Rather, cells become competent as a consequence of growth in rich media. In laboratory studies, specially formulated media have been devised for this purpose that are remarkably effective, making it possible to obtain a culture in which all cells are competent.

The *Haemophilus* system of transformation differs from the *Streptococcus* system in two additional ways: (1) only homologous DNA (DNA from the same or a closely related species of *Haemophilus*) is bound to and taken into cells with any significant efficiency, and (2) transforming DNA enters the recipient in double-stranded form, remaining that way up to the time of its integration into the endogenote.

The past few years have seen the clarification of the general mechanisms by which double-stranded DNA is bound to the cell and taken into it and by which this binding and uptake of DNA is restricted to homologous DNA. The outer membrane of competent cells (but not noncompetent ones) contains, on the average, about 10 vesicular structures that appear to be localized extensions of that membrane or *blebs* (Figure 11.4), at the base of which are small pores (Figure 11.5). Imbedded in this membranous region is a protein that binds specifically with an 11–base pair sequence of DNA (5′-AAGTGCGGTCA-3′) that occurs at about 600 sites on the *Haemophilus* chromosome or about one site per 4,000 base pairs. Since at random, a sequence this long would be expected to occur only once in $4 \times 10^6$ (or $4^{11}$) base pairs, there can be little doubt that the frequent occurrence of this sequence on the chromosome has been selected as a means of restricting the uptake of extracellular DNA to homologous DNA.

Shortly after homologous DNA is added to a competent culture of *Haemophilus* the outwardly

FIGURE 11.4

Electron micrograph of thin section of competent *Haemophilus influenzae* cells. Arrow indicates blebs on the outer membrane that form as competence develops. From M. E. Kahn, G. Maul, and S. H. Goodgal, "Possible Mechanism for Donor DNA Binding and Transport in *Haemophilus*," *Proc. Natl. Acad. Sci. USA* **79,** 6370–6374 (1982).

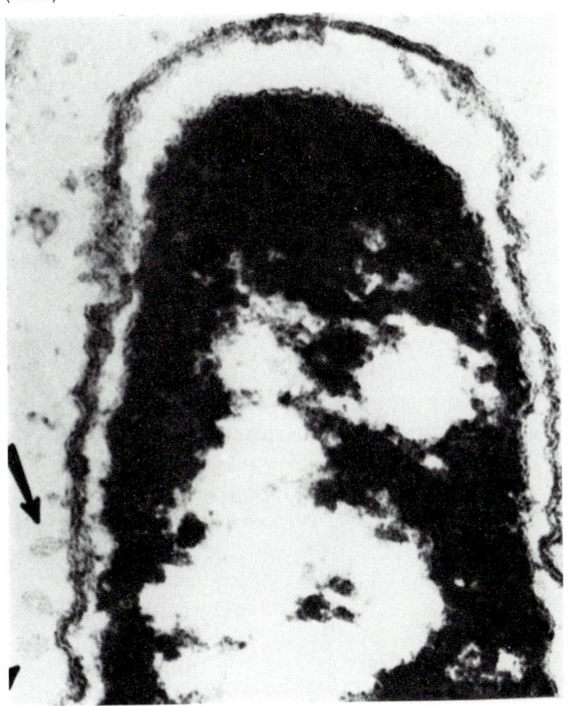

FIGURE 11.5

Pore structure on the inner face of the outer membrane at the base of blebs associated with competence of *Haemophilus parainfluenzae*. From M. E. Kahn, G. Maul, and S. H. Goodgal, "Possible Mechanisms for Donor DNA Binding and Transport in *Haemophilus*," *Proc. Natl. Acad. Sci. USA* **79,** 6370–6374 (1982).

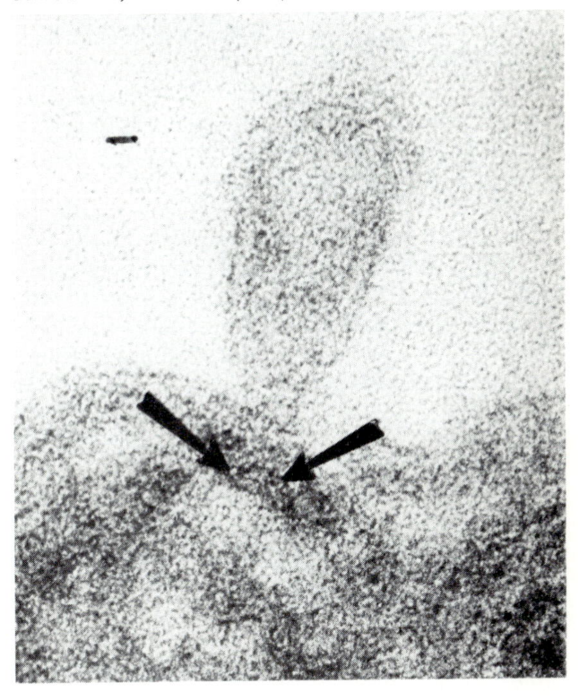

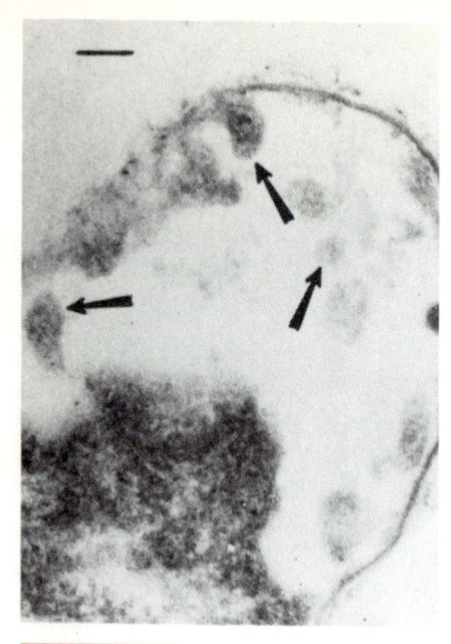

FIGURE 11.6

Thin section of competent cells of *Haemophilus parainfluenzae* after exposure to homologous DNA. Invaginated (periplasmic) vesicles are indicated by arrows. Bar = 0.1 μm. From M. E. Kah, G. Maul, and S. H. Goodgal, "Possible Mechanism for Donor DNA Binding and Transport in *Haemophilus*," *Proc. Natl. Acad. Sci. USA.* **79,** 6370–6374 (1982).

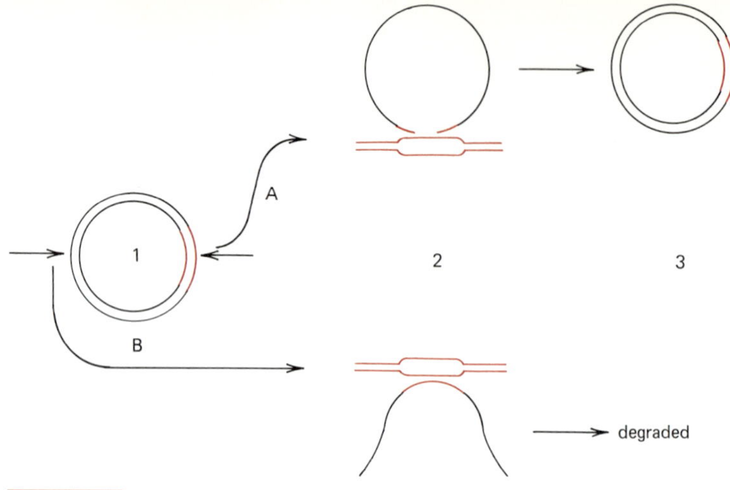


FIGURE 11.7

Model explaining the dependency of the transformability of *Bacillus subtilis* by plasmids on homology between a region of the plasmid and the endogenote. At the surface (1) of the cell the plasmid is cleaved by a nuclease. If the cleavage occurs within the region of homology, A, the single-stranded form of the linear molecule that enters the cell can pair with the homologous region of the endogenote (2) thereby holding the cut ends in proper position to be joined by the action of DNA ligase. By duplication (3) a double-stranded molecule is generated that can be stably replicated in the recipient. However, if the cleavage occurs within a region, B, that is not homologous with the endogenote, pairing (2) will not promote ligation. This linear molecule cannot be replicated and is eventually destroyed by intracellular nucleases (3). After U. Canosi, A. Iglesias, and T. A. Trautner, "Plasmid Transformation in *Bacillus subtilis*' DNA into Plasmid pC1974:," *Mol. Gen. Genet.* **181** 434–440 (1981).

extending vesicles disappear and internal ones appear (Figure 11.6). One might presume that the binding of homologous DNA on the surface of the vesicle causes it to invaginate, trapping the molecule of DNA within it. Regardless of mechanism, strong biochemical evidence in addition to the morphological evidence supports the existence within *Haemophilus* cells undergoing transformation of membranous vesicles that contain DNA. They have been termed *transformasomes.* DNA within a transformasome is in a protected state, resistant to the action of DNase, restriction enzymes, and modification enzymes. The DNA remains within the transformasome as it traverses the cytosol exiting in single-stranded form only immediately before it recombines with the endogenote.

## Natural Transformation by Plasmids

Intact plasmids are taken up by competent *Haemophilus influenzae* cells, if the plasmid contains the proper 11–base pair sequence found on the chromosome. However, competent cells of *Bacillus subtilis,* which are transformed by chromosomal DNA much as *S. pneumoniae* is, always cleave plasmids as they enter the cells and reduce them to single-stranded form (Figure 11.7). Linearized plasmids are incapable of being replicated; they can become a heritable part of the recipient cell's genome only if they are recircularized within the cell, and recircularization is dependent on there being homology between the plasmid and the endogenote (Figure 11.7). If the cleavage occurs within this region of homology, pairing of it with the endogenote brings the ends in juxtaposition so that a double-stranded circular plasmid can be reconstructed by the action of DNA polymerase and DNA ligase. If the cleavage occurs within the nonhomologous region of the plasmid, recircularization is not possible, and, as a consequence, the plasmid cannot be replicated. The homology between endogenote and plasmid need not be preexisting; the transformed plasmid itself can introduce the required homology if it exists as a tandem dimer or if two copies of the plasmid (cut at different sites) are introduced into the same cell. There is considerable evidence that plasmid transformation of the other well-studied transformable Gram-positive bacterium, *S. pneumoniae,* occurs by the same sequence of events as are shown in Figure 11.7. And this mechanism also applies to transformation of plasmids of the Gram-negative bacterium, *Pseudomonas stutzeri.*

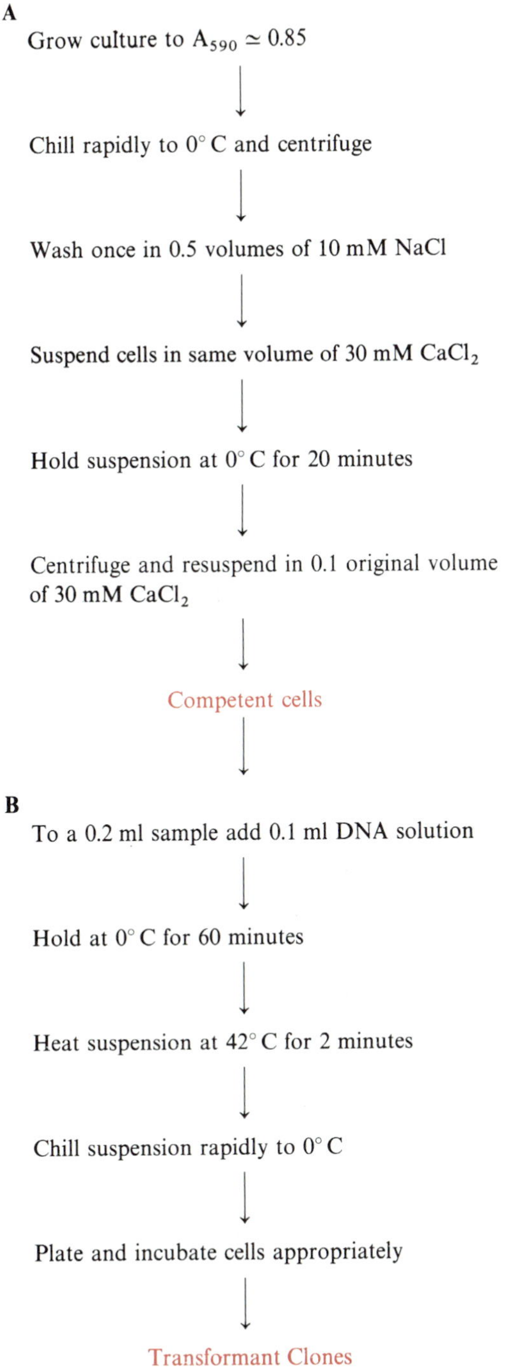
A
Grow culture to $A_{590} \simeq 0.85$
Chill rapidly to 0° C and centrifuge
Wash once in 0.5 volumes of 10 mM NaCl
Suspend cells in same volume of 30 mM $CaCl_2$
Hold suspension at 0° C for 20 minutes
Centrifuge and resuspend in 0.1 original volume of 30 mM $CaCl_2$
Competent cells
B
To a 0.2 ml sample add 0.1 ml DNA solution
Hold at 0° C for 60 minutes
Heat suspension at 42° C for 2 minutes
Chill suspension rapidly to 0° C
Plate and incubate cells appropriately
Transformant Clones

FIGURE 11.8
Scheme for artificially transforming cells of *Escherichia coli.* A series of manipulations (A) make the cells competent and another series (B) cause the transforming DNA to enter them. After S. N. Cohen, A. C. Chang, and L. Hsu, "Nonchromosomal Antibiotic Resistance in Bacteria. Genetic Transformation of *E. coli* by R-factor DNA," *Proc. Natl. Acad. Sci. USA* **69,** 2110 (1972).

### Artificial Transformation

The development of competence and uptake of DNA by naturally transformable bacteria is chromosomally encoded. In *Streptococcus pneumoniae* at least 12 genes govern the process. Interestingly, as we shall see, transformation is the only mechanism of genetic exchange among procaryotes that is chromosomally encoded (transduction occurs as a consequence of aberrant phage development; conjugation is plasmid encoded) and it must have evolved as a mechanism of genetic exchange, a biological function that occurs in all major groups of organisms and one that is highly selected. In spite of this, many bacteria, including *Escherichia coli,* do not possess a system of natural transformation. But almost all that have been tested can be made competent for transformation by plasmids by a variety of highly artificial treatments of the culture (as might be expected, these methods are ineffective for naturally transformable bacteria that cleave the incoming plasmid). A scheme for making *Escherichia coli* competent is shown in Figure 11.8.

Plasmids enter such cells as intact double-stranded molecules. Double-stranded linear molecules also enter but in many cases these are rapidly degraded by intracellular nucleases. In the case of *Escherichia coli* artificial transformation by linear DNA is possible if a mutant strain that lacks two DNA-cleaving nucleases is used as a recipient.

### The Role of the Donor Cell in Transformation

In the laboratory, purified solutions of DNA are usually employed in studies on transformation. This begs the question of how DNA becomes available for transformation in nature. Curiously, the role of the donor cell has received very little study. It has been assumed by many bacterial geneticists that the role of the donor cell is completely passive: that donation of DNA depends on the occasional and random lysis of certain cells in the population. However, recent experiments suggest that DNA might be actively extruded from certain competent cells by a genetically encoded pathway.

## BACTERIAL CONJUGATION

In 1946 J. Lederberg and E. L. Tatum discovered a genetic exchange occurring between certain strains of *E. coli* that eventually proved to be different from transformation in a number of respects: exchange was dependent on direct contact between

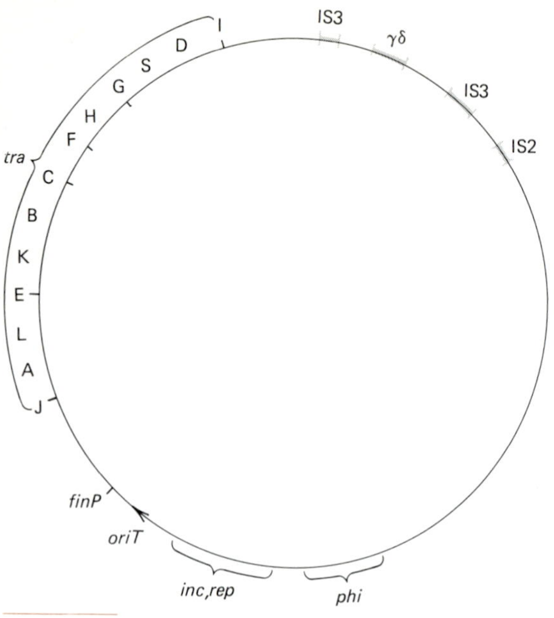


FIGURE 11.9

Genetic map of the F plasmid showing the relative position of genes encoding transfer functions (*tra*), fertility inhibition (*finP*), origin of transfer replication (*oriT*), incompatibility (*inc*), replication (*rep*), and phage inhibition (*phi*). The positions of insertion sequences IS2, IS3, and $\gamma\delta$, are also shown. The length of the genome is 94.5 kilobases. After J. A. Shapiro, "F, the *E. coli* Sex Factor," in *DNA, Insertion Elements, Plasmids and Episomes*, A. I. Bukhari, J. A. Shapiro and S. L. Adhya (Cold Spring Harbor, N.Y.: Cold Spring Harbor Laboratory, 1977).

cells; it occurred even if DNase was present in the medium; and it was polarized—i.e., certain strains designated as $F^+$ (fertility plus) always acted as donors (*males*) and others designated as $F^-$ (fertility minus) always acted as recipients (*females*). On the basis of these observations it was assumed correctly that genetic material was transferred directly from male to female cells without passing through the suspending medium. The process was termed *conjugation*. In subsequent years it was discovered that $F^+$ strains all contain a plasmid (termed the *F plasmid*) that carries all the genes that encode conjugative genetic transfer. Indeed, the F plasmid is not known to encode any additional function other than this one and its own replication (Figure 11.9).

## Properties of the F Plasmid

The F plasmid encodes transfer of itself to other cells that lack an F plasmid. Thus, if $F^-$ cells are added to an $F^+$ culture, all of them rapidly become $F^+$; i.e., transfer of the F plasmid occurs at a high frequency. At a considerably lower frequency, chromosomal genes are transferred to the $F^-$ cell along with the F plasmid.

The F plasmid shares certain genetic properties with all other plasmids. It carries certain genes (*rep*) that allow it to be replicated by the host cell and, as directed by the genetic region *inc*, it exhibits the phenomenon of incompatibility; i.e., if a certain plasmid is present in a cell, replication of closely related plasmids is inhibited. Thus, closely related plasmids are said to be incompatible because only one member of such a group of plasmids (termed an *incompatibility group*) can be stably replicated in the same cell. The biochemical mechanism by which incompatibility is expressed is not yet understood, but the phenomenon has proven useful for classifying plasmids as is discussed later in this chapter. The F plasmid belongs to an incompatability group termed IncF1.

Not all plasmids encode self-transfer, as do the F plasmid and a number of others, collectively termed *conjugative plasmids*. But of those that do and that occur in Gram-negative bacteria, the mechanism of gene transfer appears to be similar. In the case of the F plasmid, transfer is encoded by 13 genes, *traA* through *traL* and *traS*, that form an operon. Among their various functions, some encode the synthesis of special pili, termed *F pili* or *sex pili*. Others encode a special type of replication termed *transfer replication* of the F plasmid that occurs during transfer and mediates it.

Unlike certain other conjugative plasmids in which expression of the *tra* genes is repressed, the *tra* genes of F plasmids are always derepressed. Thus, $F^+$ cells always have F pili on their outer surface (Figure 11.10) unless they have been subjected to vigorous shaking, an action that breaks off these long (several $\mu$m) brittle appendages (Chapter 6). An F pilus binds specifically to a protein in the outer membrane (OmpA) of $F^-$ cells, thereby initiating transfer replication and the process of gene transfer by conjugation (Figure 11.11). A nick (the cutting of a single strand of DNA) is made in the F plasmid at the site termed *oriT* (*ori*gin of *t*ransfer), then a type of replication sometimes termed the *rolling circle* mechanism (see Chapter 9) ensues in which the intact strand is used as template and the 3′ end generated by the nick is used as primer. By this action the 5′ end of the nicked single strand is displaced and is transferred to the $F^-$ cell. It is not clear how this transfer occurs. The single-stranded molecule might pass through the hollow core of the F pilus or it might pass from the $F^+$ cell to the $F^-$ at some other point of contact between them. If the latter case is

FIGURE 11.10

F pilus of *E. coli*. The donor bacterial cell covered by numerous appendages termed Type I pili (which play no role in conjugation) is connected to the recipient cell (without appendages) by an F pilus. The F pilus has absorbed along its length numerous phages that attached specifically to this organelle. Courtesy of Charles C. Binton, Jr.

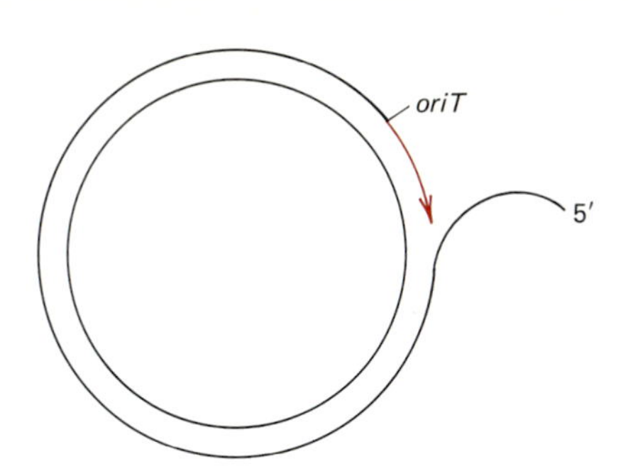


FIGURE 11.11

Transfer replication of an F plasmid. An F-encoded nuclease cleaves one strand of the plasmid at *oriT*. Then replication at arrow head occurs by a rolling circle mechanism whereby the newly synthesized DNA (colored) displaces a preexisting single strand, the 5′ end of which enters the $F^-$ cell.

true, the function of the F pilus would only be to hold the two cells together so that another direct *conjugative bridge* can form. Within the $F^-$ cell the single-stranded molecule of DNA is duplicated by a chromosomally encoded DNA polymerase and recircularized.

The question of how recircularization occurs is also an open one. It has been presumed for some time that the length of single-stranded DNA transferred to the recipient exceeded that of the intact plasmid, and recircularization of the transferred DNA occurred as a consequence of a recombinational crossover event (see the section on recombination later in this chapter) between its redundant ends. However, recent studies have shown that recircularization can occur in cells that are genetically incapable of mediating recombination, and there is no evidence for more than a genome's worth of plasmid DNA being transferred to the recipient. Possibly the ends of the transferred piece of DNA become attached to a protein located in the cytoplasmic membrane, thereby holding the ends in appropriate juxtaposition to be joined by the action of DNA ligase. Once circularized, the transferred F plasmid can be replicated in the recipient cell, the genes it carries are expressed, and the cell then becomes $F^+$.

## Hfr Strains

As stated above, when $F^+$ and $F^-$ cells are mixed, F plasmid is transferred at a high frequency, but at a much lower frequency chromosomal genes are also transferred. This occurs in part as a consequence of the presence in a $F^+$ culture of cells termed *Hfr* (*h*igh *f*requency *r*ecombination) in which the F plasmid and the bacterial chromosome have become integrated into a single large circular molecule. When such a cell comes in contact with an $F^-$ cell, replicative transfer begins within the F plasmid region at *oriT* and continues into the chromosomal region of the large circular molecule. Thus, chromosomal genes as well as F plasmid genes are transferred to the recipient $F^-$ cell. As the bridge between the pair of mating cells is a somewhat fragile one, they frequently break apart spontaneously, thereby interrupting the genetic transfer. A considerable time (about 100 minutes at 37° C) is required for the entire chromosome to be transferred. The pair of cells rarely stays together long enough for this to happen, so usually the major part of the chromosome is not transferred and, therefore, neither is the portion of the F plasmid at the distal end of the chromosome.

Hfr cells form as a result of a genetic crossover between regions of homology shared by the F

FIGURE 11.12

Mechanism of formation of an Hfr cell from the $F^+$ cell. Homologous insertion sequences (colored) pair and by a crossover event the plasmid becomes integrated. The arrow in the F plasmid and the Hfr chromosome indicates the position of *ori* and the 5′ end of the single strand of DNA that enters an $F^-$ cell.

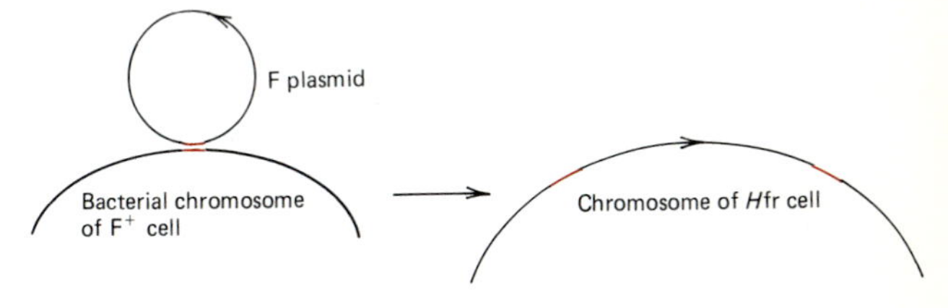

plasmid and the bacterial chromosome. These regions are, in all cases studied, identical DNA sequences called *insertion sequences* (see later in this chapter) that also occur at various places in the bacterial chromosome. The F plasmid carries four insertion sequences (Figure 11.9). The mechanism of formation of an Hfr cell is shown in Figure 11.12.

### Properties of Clones of Hfr Cells

Pure cultures of Hfr cells can be isolated and studied. These strains exhibit a number of interesting properties. Since all cells in a culture of an Hfr strain are the product of the same event of integration between an F plasmid and the chromosome, all cells contain chromosomes carrying an F plasmid at the same site with the origin of genetic transfer oriented in the same direction. Mixing an Hfr and an $F^-$ culture causes a transfer of chromosomal genes beginning at the site of insertion of the F plasmid. Since the rate of transfer of genes is set by the rate of replication by the rolling circle mechanism and since this rate is relatively constant (about $10^4$ nucleotides or 1 percent of the chromosome per minute at 37° C) the distance of a particular gene from the site of insertion of F (*Hfr origin*) can be judged by the time between initiation of conjugation and the time at which that gene enters the $F^-$ cell. As a result, distances on the chromosome of *E. coli* are reckoned in minutes. This time can be measured by doing an experiment in which, at various times, samples of the mating mixture are taken, shaken vigorously to break apart the mating pairs, and plated on selective media in order to determine how many $F^-$ cells have received and have integrated into their chromosome, genes from the Hfr cells, i.e., the number of recombinant cells. An example of the genetic composition of Hfr and $F^-$ strains and the media used to score recombinants is shown in Table 11.3; the results of such an experiment is shown in Figure 11.13. It will be noted that $Leu^+$, $Lac^+$ and $Gal^+$ recombinants begin to occur at 9, 15, and 24 minutes after conjugation was initiated by mixing the cultures. Thus their relative location on the chromosome is established and, if the site of insertion of the F plasmid is known, the absolute location of these genes is established. One also notes that the maximum number of recombinants that develop decreases as the distance between the F plasmid and the particular gene increases, because mating pairs spontaneously break apart.

Because in an Hfr × $F^-$ mating distal genes are inherited at such a low frequency, a particular Hfr strain can be used effectively to locate (*map*) only those genes that lie in approximately the first third of the chromosome to be transferred to the $F^-$ cell. A variety of different Hfr strains, i.e., strains in which the F plasmid is inserted at different locations on the chromosome in either the clockwise or counterclockwise orientation, have been isolated. Together these can be used to map any gene (Figure 11.14).

**TABLE 11.3**

**Mutations (Genotype) and Their Phenotypes in an $F^-$ strain of *Escherichia coli* and Media that Could Be Used to Select Recombinant Clones Produced in a Cross between It and an Hfr Strain That Carries Wild-Type Alleles of Those Genes**

| Mutation | Phenotype | Contents of Media to Select Wild-Type Recombinants[a] |
|---|---|---|
| *leu-87* | Leucine auxotrophy | Glucose, streptomycin |
| *lac-89* | Inability to use lactose as carbon source | Lactose, leucine, streptomycin |
| *gal-91* | Inability to use galactose as carbon source | Galactose, leucine, streptomycin |
| *rpsL93* | Resistance to the antibiotic streptomycin | —[b] |

[a] Medium contains inorganic salts in addition to the carbon sources (glucose, lactose, or galactose), growth factors (leucine), and antibiotic (streptomycin) indicated. Each medium is devised to select only the indicated class of recombinants. For example, recombinants that carry a wild-type allele in place of *lac-89* can be selected in a medium containing lactose, leucine, and streptomycin. The Hfr strain cannot grow because streptomycin is present; the $F^-$ strain cannot grow because it cannot utilize lactose as carbon source. Leucine is present because the recombinant may carry the *leu-87* allele that confers leucine auxotrophy.
[b] The wild-type allele confers sensitivity to streptomycin and therefore cannot be selected directly.

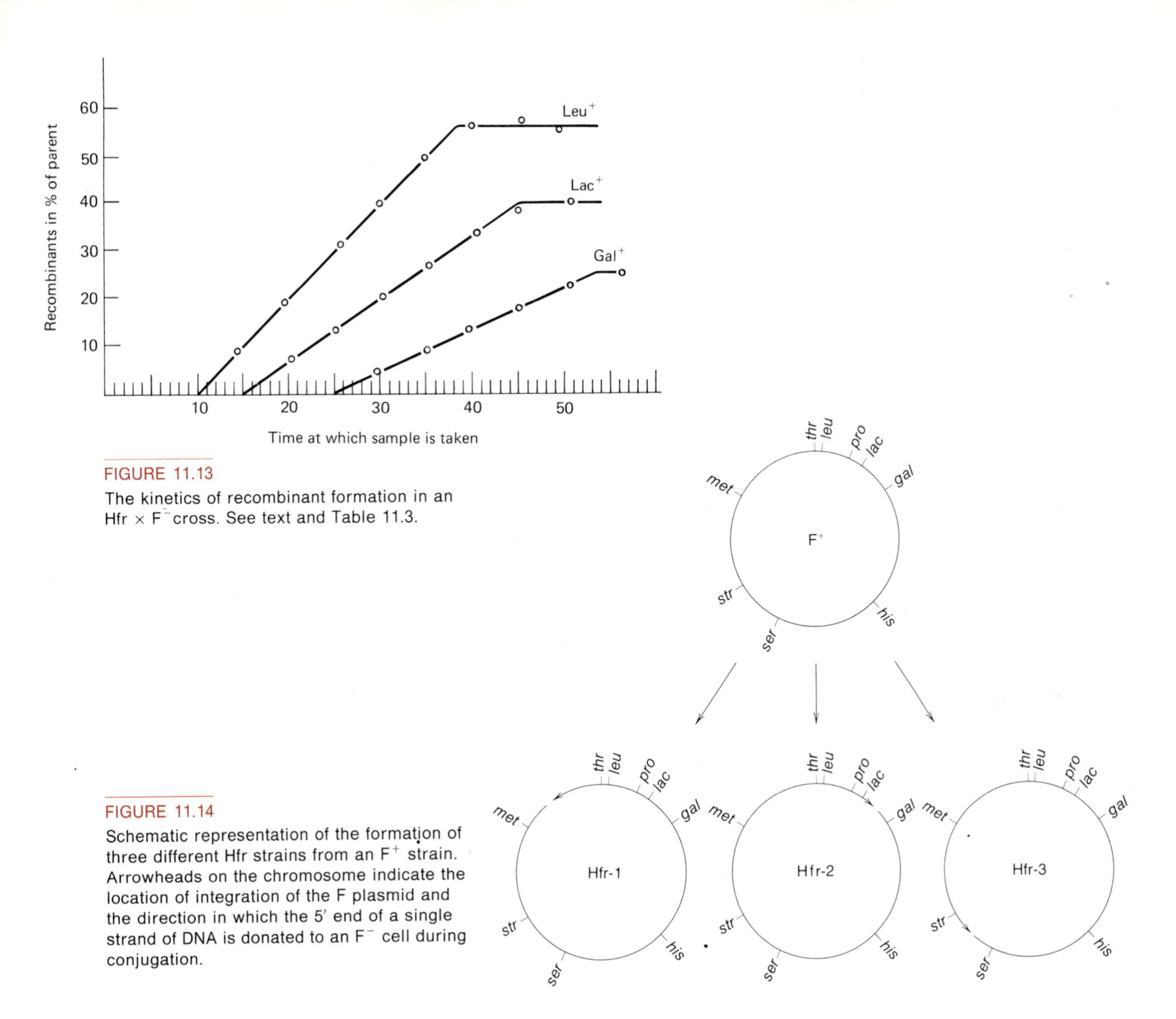


FIGURE 11.13
The kinetics of recombinant formation in an Hfr × F$^-$ cross. See text and Table 11.3.

FIGURE 11.14
Schematic representation of the formation of three different Hfr strains from an F$^+$ strain. Arrowheads on the chromosome indicate the location of integration of the F plasmid and the direction in which the 5′ end of a single strand of DNA is donated to an F$^-$ cell during conjugation.

## F-Mediated Transfer of Other Plasmids

To this point we have considered the ability of the F plasmid to mediate its own transfer to a recipient cell and, under certain conditions, to mediate the transfer of chromosomal DNA. The F plasmid is also capable of mediating the transfer of other plasmids not in themselves capable of self-transfer. Plasmid ColEl (see later in this chapter) is such a plasmid. If a culture containing ColEl is mixed with one that lacks it, no transfer occurs, but if the ColEl-containing culture also contains the F plasmid, ColEl is transferred to the recipient at almost as high a frequency as F itself. The mechanism of the F-mediated transfer of ColEl is analogous to, but not mechanistically the same as, the transfer of chromosomal DNA by an Hfr strain because ColEl and F occur in the recipient cell as separate circular molecules of DNA; some more subtle interaction must occur between the F plasmid and the other plasmid. The cotransferred plasmid is said to be *mobilized* by F, and the ability of a plasmid to be mobilized depends on its containing a specific region of DNA, termed *mob.* Loss of this region is associated with loss of F-mediated transfer of the plasmid.

Indeed the poorly understood phenomenon of mobilization probably also accounts for some of the transfer of chromosomal DNA in an F$^+$ × F$^-$ cross, because the frequency of such transfers is higher than that which would be expected to result from the number of Hfr cells that an F$^+$ culture always contains.

## Other Systems of Conjugation in Gram-Negative Bacteria

As we have seen in the preceding sections of this chapter, the capacity of *E. coli* to exchange genetic material by conjugation and the way in which this occurs are determined by the presence of the F plasmid in certain strains. This plasmid has the capacity to transfer itself from one cell to another. Occasionally it interacts with another element of the bacterial genome (the chromosome or another plasmid) causing some or all of this other element also to be transferred.

Therefore, in inquiring about the distribution of conjugative genetic exchange among bacteria, three question arise: (1) in which bacteria can the F plasmid be stably replicated? (2) which other plasmids are capable of self-transfer? (3) do other plasmids that are capable of self-transfer interact with the bacterial genome in such a way as to bring about its transfer? These questions are addressed in the following sections.

The F plasmid can replicate only in enteric bacteria (Chapter 19) so F plasmid–mediated conjugation can occur only within this group. But a vast number of other plasmids that are capable of self-transfer exist in Gram-negative bacteria. These plasmids are called *conjugative plasmids*. However not all conjugative plasmids readily mobilize the transfer of chromosomal DNA. In certain cases, those that lack this ability have been shown to gain it when they gain a small bit of DNA shown to be an *insertion sequence*. An example of this phenomenon is seen in the case of plasmid R68, termed a *broad host range* plasmid because it has the capacity to replicate in a large number of, if not all, Gram-negative bacteria. Although plasmid R68 transfers itself at a high frequency from one cell to another, mobilization of the chromosome or another plasmid occurs at only a very low frequency. But derivatives of plasmid R68 have been isolated that mobilize chromosomal and plasmid DNA from *Pseudomonas aeruginosa* and a variety of other Gram-negative bacteria at a high frequency. One of these, plasmid R68.45 mobilizes chromosomal DNA from *P. aeruginosa* at about a $10^5$-fold higher frequency than plasmid R68 does. It is also able to mobilize efficiently the chromosomes of many other Gram-negative bacteria and has been used to study the chromosomal genetics of a wide range of genera. Plasmid R68.45 differs from plasmid R68 by containing a 2.10 kb insertion sequence designated IS21.

As discussed earlier, the F plasmid also owes its ability to mobilize chromosomal DNA to its complement of insertion sequences. However, the mechanisms of F-mediated and R68.45 mediated mobilizations appear to differ fundamentally. The insertion sequences in the F plasmid also occur in the chromosome, and mobilization appears to depend on homologous recombination events between these two regions. But IS21 sequences do not occur in the chromosomes of those bacteria that have been investigated in this respect. Thus homologous recombination cannot be the basis for its ability to mobilize. Rather, mobilization in this case appears to be a consequence of the ability of the IS21 sequence to undergo transposition, sometimes causing the R68.45 plasmid to become integrated into the chromosome, at least transiently. Possibly as a consequence of the transient nature of this integration, strains that contain the R68.45 plasmid donate to recipient cells segments of chromosome from multiple origins.

Hfr-like strains derived from plasmids other than F are also encountered in certain Gram-negative bacteria. For example, certain Hfr-like strains of *Pseudomonas aeruginosa* that carry plasmids termed FP have played a major role in mapping the genes on the chromosome of this organism.

## Genetic Exchange by Conjugation among Gram-Positive Bacteria

Conjugational genetic exchange has been less frequently encountered among Gram-positive bacteria than among Gram-negative bacteria, and judging from one of the best-studied examples of the former class, that of conjugal transfer of plasmids among strains of *Streptococcus faecalis*, certain aspects of the process differ fundamentally (Figure 11.15).

Pili do not play a role in this process, rather plasmid-containing cells form clumps with cells that lack the plasmid, and plasmid transfer occurs within these clumps. Clumping results from the interaction between an *aggregation substance* on the surface of the plasmid-containing cell and a *binding substance* on the surface of a plasmid-lacking recipient. Binding substance is always present on the surface of cells of *S. faecalis.* Both plasmid-containing and plasmid-lacking cells produce it, but aggregation substance is produced only when a plasmid-containing (donor) cell is in close proximity of a cell that lacks that particular plasmid (recipient cell). The recipient cell produces a chromosomally encoded small molecule (a *pheromone*) that diffuses through the medium, enters the donor cell, and induces a plasmid-encoded gene to

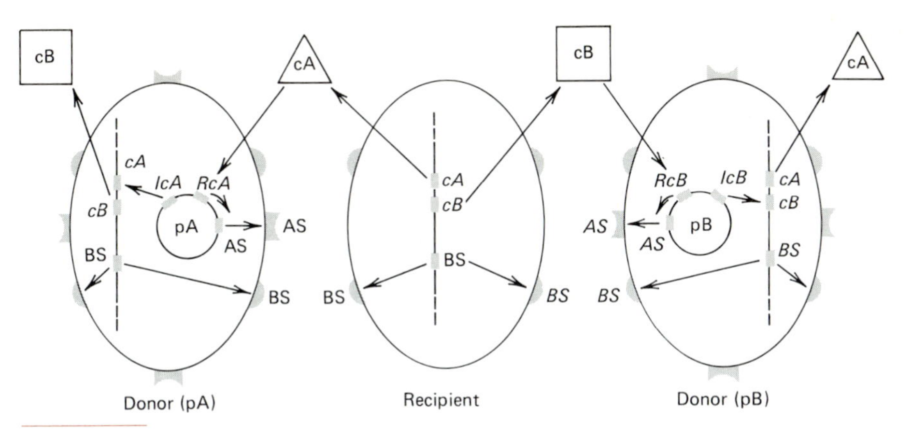


FIGURE 11.15

Conjugation of *Streptococcus faecalis*. The scheme shows the response of two donor cells (pA and pB) to pheromones (cA and cB) produced by the recipient. These react to chromosomal genes (RcA and RcB) which cause another gene (AS) to produce an aggregation substance (AS) which causes the cells to bind to donors at binding substance (BS) site. Cells that carry a particular plasmid (e.g., pA) do not produce the corresponding pheromone (cA) because the product of a plasmid gene (IcA) inhibits expression of the structural gene (cA). After D. B. Clewell, "Plasmids, Drug Resistance and Gene Transfer in the Genus *Streptoccocus*." *Microbiol. Rev.* **45,** 409 (1981).

synthesize aggregation substance. Although donor cells also carry chromosomal genes that encode synthesis of the pheromone, they do not produce it, and therefore do not produce aggregation substance in the absence of a recipient cell, because the product of a plasmid-encoded gene (*Ica*) represses synthesis of the pheromone.

## TRANSDUCTION

Occasionally during the intracellular development of certain bacteriophages, mistakes occur and a defective virion termed a *transducing particle* is produced. It contains DNA from the bacterial genome replacing part or all of the normal complement of phage DNA. The protein capsid of such transducing particles does not differ from the capsid of a normal phage virion. Since it is the capsid that determines a phage's ability to attach to a sensitive bacterial cell and to inject its complement of DNA into the cell, the transducing particle can introduce bacterial DNA derived from the cell in which it developed into another sensitive cell. The result is a transfer of genetic material between these two cells.

Phages that occasionally produce transducing particles and therefore are capable of effecting transductional genetic exchange are relatively commonly encountered in the microbial world. The host range of bacteriophages, including those capable of mediating transduction, is typically quite narrow. Therefore, in order to make transductional crosses, an investigator must, in general, have a different transducing phage for each bacterial species in which crosses are to be made. Nevertheless, it has usually proven possible to isolate transducing phages for most bacteria when serious attempts have been made to do so.

Two types of transducing particles and therefore two types of transduction exist. One of these is termed *generalized* or *nonspecialized* transduction, because it mediates the exchange of any bacterial gene. The other is termed *restricted* or *specialized transduction* because it mediates the exchange of only a limited number of specific genes. Most transducing phages are capable of mediating only one of these two types of transductions but a limited number are capable of mediating both. Regardless of the particular phage that mediates either of these types of transductions, the general developmental patterns leading to either type of transducing particle is similar. As an example of the developmental pattern leading to the generalized transducing particles we shall consider phage P22 (a phage that also produces specialized transducing particles) that infects *Salmonella typhimurium*. As an example of the pattern leading to specialized transducing particles we shall consider phage lambda ($\lambda$) that infects *Escherichia coli*.

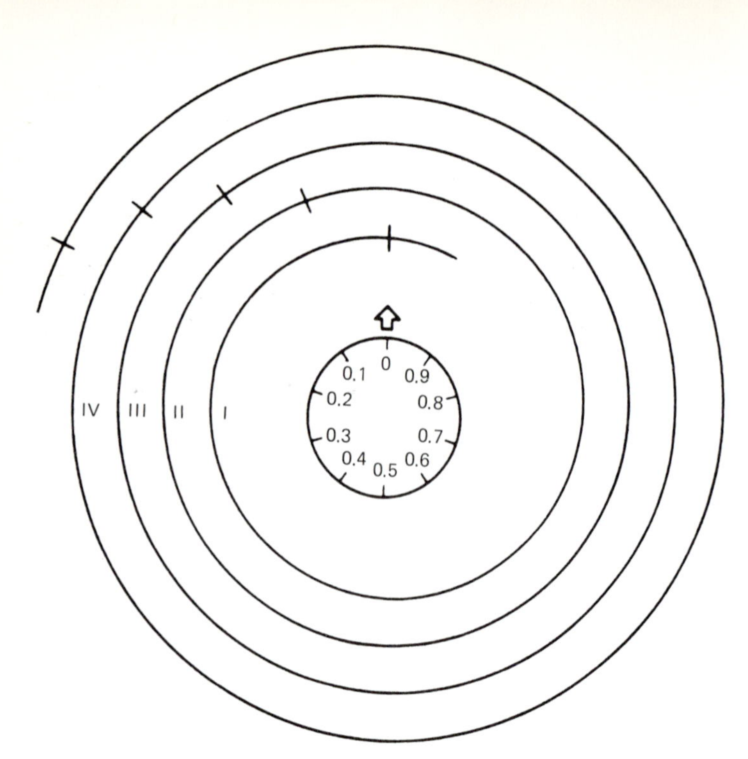


FIGURE 11.16

Mechanism of packaging DNA from a concatemer into heads of phages that mediate generalized transduction. The inner circle represents the genetic map divided into arbitrary coordinates, the *pac* site of which is indicated by the arrow. The outer spiral represents a concatemer generated from it. The phage headful amounts generated from the concatemer are indicated by roman numerals and delineated by the cross-hatches. This amount exceeds the size of the genome by approximately 2 percent. Thus, each successive fragment (I through IV) ends at a different site on the genome. After E. N. Jackson, F. Laski, and C. Andres, "Bacteriophage P22 Mutants That Alter the Specificity of DNA Packaging," *J. Mol. Biol.* **154,** 551 (1982).

## Generalized Transduction Mediated by Phage P22

The presence of a few generalized transducing particles in all phage P22 lysates is a seemingly inevitable consequence of the way that this phage's DNA is normally packaged into the head of the phage capsid (Figure 11.16). Within the infected cell, probably by a rolling circle mechanism, long stretches of phage DNA composed of tandemly repeated phage genomes (*concatemers*) are synthesized. In preparation for packaging into the phage head, the concatemer is initially cleaved by the action of a phage-encoded endonuclease at a specific site termed a *pac* (*pac*kaging) site. Then, starting from the point of cleavage, headful amounts of phage DNA are sequentially packaged into developing phage heads. This *headful mechanism* of packaging is somewhat inaccurate with respect to the amount of DNA inserted into the phage head, but the headful amount exceeds the length of the phage genome by about 2 percent. Thus each virion receives a length of DNA that is a complete phage genome plus a small amount of additional DNA at one end that is the same as the sequence at the other end of the molecule. The packaged DNA is therefore *terminally redundant*. As a consequence of packaging more than a complete genome in each phage head, sequentially packaged lengths of DNA terminate at different sites within the phage genome; they are said to be *circularly permutated*.

It will be noted that the headful mechanism of packaging phage P22 is specific for (or recognizes) phage DNA only at the times when the endonucleolytic act occurs at the *pac* site. All other steps in the packaging process proceed in the same manner regardless of the source of DNA. Transducing particles are made when cleavages occur in the bacterial genome at sites that resemble the *pac* site sufficiently to allow the phage-encoded endonuclease to act. Then from the cut end, successive phage headfuls of the bacterial genome (chromosome or a plasmid) are packaged, thereby forming transducing particles.

Phage P22 is a temperate phage (see Chapter 9) but not all phages capable of mediating generalized transduction are, nor is the capacity of the formation of generalized transducing particles. Indeed, in laboratory practice, mutants ($int^-$) incapable of entering the prophage state are usually employed to produce phage lysates to be used for transduction. A second type of phage mutation, termed *HT*, is also commonly employed. These decrease the specificity of the endonuclease that acts at the *pac* site so that *pac*-like sites in the bacterial genome are cleaved with greater frequency. When bacteria are infected with HT mutants of phage P22 the resulting yield of phage contains approximately equal numbers of transducing particles and normal virions.

Since phage lysates used to mediate generalized transduction contain infectious virions, precautions must be taken to prevent the killing of transduced cells (*transductants*) by these virions. Usually this is done by adding the transducing lysate to the recipient culture at a low multiplicity of infection, thereby rendering unlikely the infection of a potential transductant by an infectious virion. Occasionally heat-sensitive phage mutants are employed: lysates are prepared at a permissive temperature and transductions are done at a restrictive temperature, at which the virions in the lysate are incapable of killing potential transductant cells. Occasionally lysates are irradiated with ultraviolet light before they are added to the recipient bacterial culture. Such treatment inactivates the virions but has little effect on the transducing activity of transducing particles in the lysate.

### Laboratory Exploitation of Generalized Transduction

Transductional crosses provide a convenient way to construct bacterial strains with new combinations of mutations. They also provide a variety of ways to gain information about the distance between genes (*linkage*) on the bacterial genome. Among these are the determination of the possibility and frequency with which different genes are included within the same generalized transducing particle. Owing to the small size of phage genomes (that of phage P22 is 39 kb and that of phage P1 is 90 kb—about 1 percent and 2 percent, respectively, of the size of an enterobacterial chromosome), and the fact that only an equivalent amount of DNA is included in a transducing particle, only relatively closely linked genes can be incorporated into the same transducing particle, or as it is termed, be *cotransduced*. The frequency at which two genes are cotransduced by a population of phages can be used to estimate their linkage, i.e., their relative distance from each other on the chromosome. Assuming that all regions of the genome are transduced with the same frequency, (this approaches the actual situation when HT mutants are used), the physical distance between genetic markers can be calculated from the formula, $C = 1 - t + t \ln t$, in which $C$ is the frequency of cotransduction and $t$ is the linear distance between markers expressed as a fraction of the phage genome. For example, if the frequency of cotransduction of two markers by phage P22 is 0.43 (43 percent), the linear distance between them is 9.0 kb.

### Specialized Transduction Mediated by Phage Lambda

In contrast to generalized transduction, specialized transduction is only mediated by temperate phages, because specialized transducing particles are formed by inaccurate excision of prophages when they are induced to lytic growth (Figure 11.17). Rather than only the phage genome being excised by a recombinant event between the ends

FIGURE 11.17

Mechanism of formation of specialized transducing particles by phage lambda. (A) On infection phage lambda DNA enters the cell as a linear molecule. (B) Its complementary single-stranded ends (m and m′) anneal circularizing the molecule. (C) A crossover mediated by a phage gene (int) occurs between complementary regions on the phage (*attλP*) and the chromosome (*attλB*) integrating the two molecules and creating *attL* and *attR* ($C_1$). By the action of *int* and *xis* this process can be reversed ($C_2$). (D) But sometimes at a lower frequency($10^{-5}$) the crossover occurs between a region within the prophage and the chromosome creating a transducing particle (*λdgal*) that carries a chromosomal gene (*gal*).

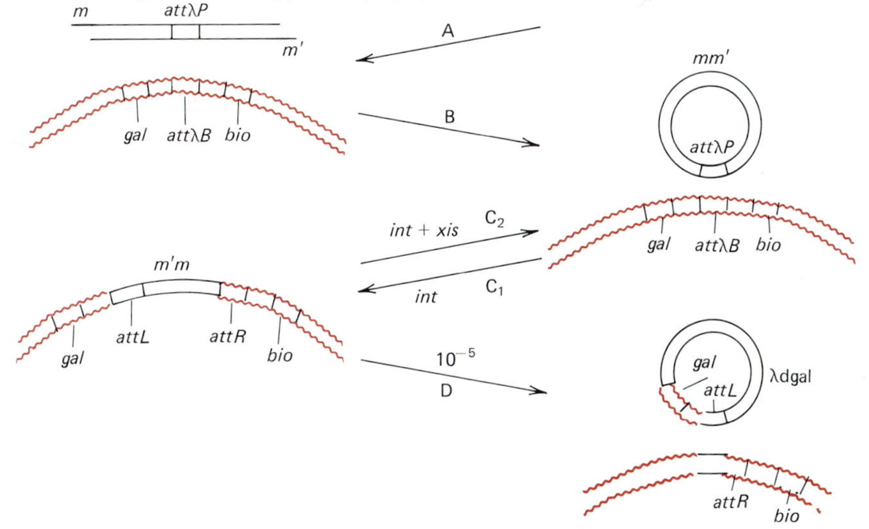

of the prophage region, a segment of DNA is excised that includes a portion of the bacterial chromosome and a portion of the prophage. If a significant portion of the bacterial chromosome is excised, the entire prophage cannot be included in the specialized transducing phage because the phages are viable only if they contain an amount of DNA greater than 73 percent and less than 110 percent of the phage genome. In a sense, a specialized transducing phage is a new organism having gained certain genes from the bacterial chromosome and lost others from its own genome. These phages are designated by the bacterial genes that they carry and by the degree of deficiency caused by the loss of phage genes. For example, *λdgal* designates a specialized transducing lambda phage that carries *gal* genes (encoding the dissimilation of the hexose *gal*actose) and is *d*efective because it has lost phage genes rendering it incapable of development except in a cell that contains another lambda phage (termed a *helper phage*) that contains the missing genes and thereby complements the lost phage functions of the specialized transducing phage; *λplac* designates a specialized transducing phage that carries *lac* genes (encoding the dissimilation of the disaccharide *lac*tose) but retains sufficient phage functions to permit it, on its own, to form *p*laques.

Aberrant excisions that generate specialized transducing phages are rare, occurring among total excisions at a frequency of about $10^{-6}$, so lysates obtained by inducing prophages contain very few specialized transducing phages and therefore are capable of causing specialized transduction only at a low frequency. Such lysates are termed *low frequency transducing* (LFT) lysates. But the transductant clones derived by infecting a culture with a LFT lysate all contain specialized transducing phages in the prophage state. Induction of a culture grown from such a transductant yields a *high frequency transducing* (HFT) lysate. If it is derived from a culture in which the specialized transducing phage is capable of independent growth, the lysate will consist almost exclusively of specialized transducing phage; if it is derived from a culture with a defective specialized transducing phage, only about 50 percent of the phages in the lysate will be transducing; the others will be progeny of the helper prophage that must be present in tandem with the transducing prophage to complement it.

As we have seen, specialized transducing phages contain only those genes that are immediately adjacent to the site of integration of the prophage. Phage lambda usually integrates at a site, *attλ*, located between two operons, *bio* (encoding the biosynthesis of the vitamin, *bio*tin) and *gal* (encoding the catabolism of the sugar, *gal*actose), so specialized transducing phages carrying either *gal* or *bio* genes can be generated by inducing a normal lambda lysogen. However, at a lower frequency, phage lambda will lysogenize strains of *E. coli*, from which *attλ* has been deleted. In such strains, the phage lambda integrates at several secondary sites, allowing the generation of specialized transducing phages that carry genes adjacent to these several secondary sites. In addition, by employing a variety of specialized genetic techniques, *attλ* can be transposed to other locations on the chromosome. As a consequence it has become possible to construct a variety of specialized transducing phages that carry many different genes.

The experimental utility of specialized transducing phages is considerable because they provide a remarkable means of obtaining high concentrations of specific genes, the condition that occurs in an HFT lysate.

## GENETIC ANALYSIS OF THE ACTINOMYCETES

A conjugational mechanism of genetic exchange occurs in certain members of the actinomycete group that seemingly differs markedly from the systems encountered among Gram-positive bacteria. Particular interest in the genetic system of these actinomycetes derives from certain of its unique properties and the industrial importance of the organisms involved. Crosses are effected by growing together two genetically different strains on nonselective media and then transferring the resulting cell mass to selective plates. Cultures to be crossed can be initiated either from spores or from an inoculum of fragmented mycelia. Most experiments on genetic exchange among actinomycetes have been done by D. A. Hopwood and his coworkers using *Streptomyces coelicolor* as the principle object of study. The fertility of *S. coelicolor* as isolated from nature, termed IF (*i*nitial *f*ertility) is relatively low, but two other fertility types, NF (*n*ormal *f*ertility) and UF (*u*ltra *f*ertility) with markedly higher fertility, have been obtained (Table 11.4, Figure 11.18). It will be noted that all fertility types are self- and inter-fertile, although the level of fertility varies over five orders of magnitude. In all such crosses, large portions of the chromosome are exchanged. Distances between genes on the chromosome are measured in terms of the probability of a crossover occurring between them and expressed in centimorgans (cm). One centimorgan is a unit of map distance corresponding to 1 percent recombination when corrected for

(a)

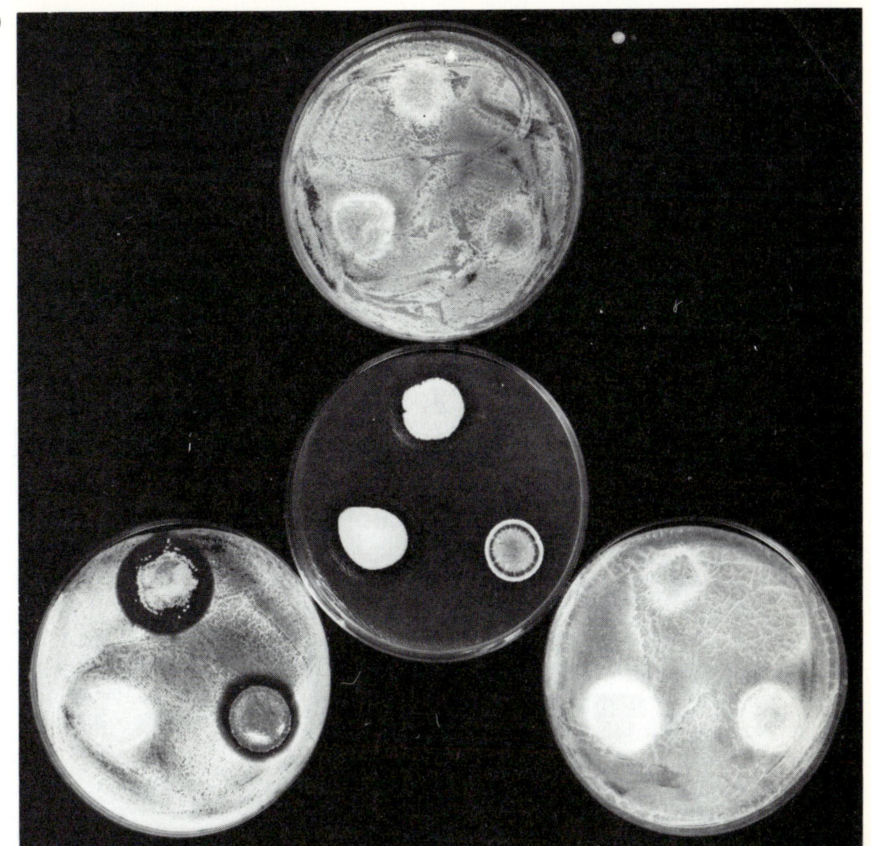

FIGURE 11.18

Interactions between fertility types (on all plates: top, IF; left, UF; right, NF) of *Streptomyces coelicolor* on the central plate (a) are of IF, NF, and UF strains, all carrying the marker *pheA1*. This plate was replicated to lawns of tester strains of each fertility type, carrying the marker *uraA1*, on nonselective medium to make the three crossing plates illustrated. Note inhibition of aerial mycelium (dark zones) of UF background by NF and IF patches. (b) The crossing plates in the upper picture were replicated to selective minimal medium to give the three recombinants, which appear white against the dark background. From D. A. Hopwood, "Genetics of the Actinomycetales," in *Actinomycetales*, ed. G. Sykes and F. A. Skinner (London and New York: Academic Press, 1973).

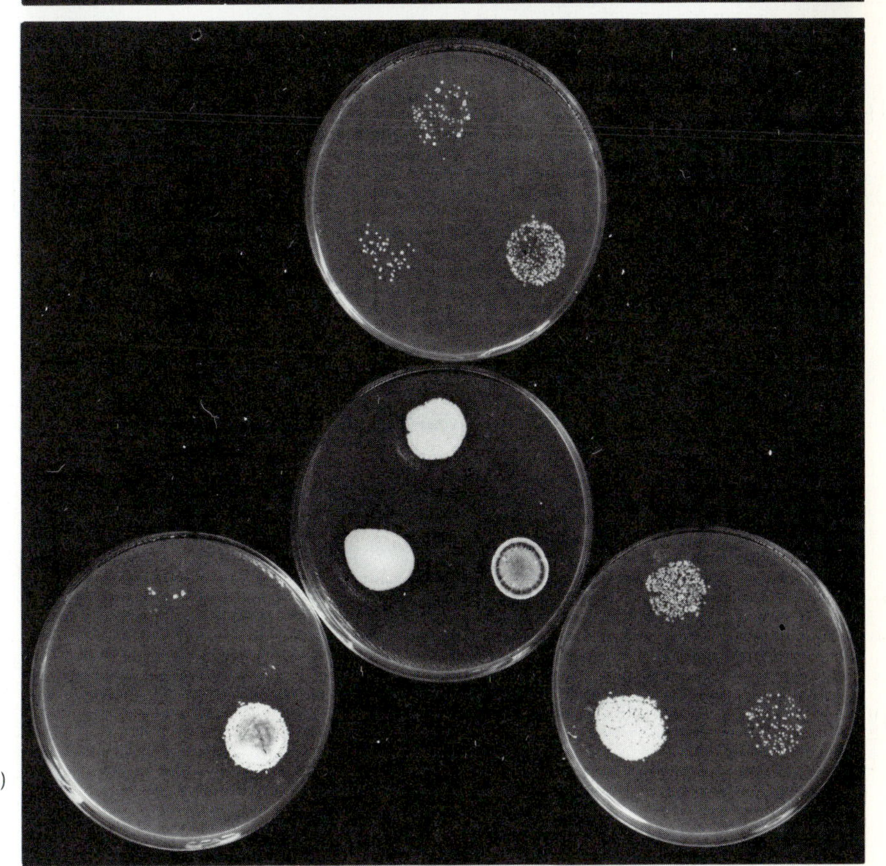

(b)

**TABLE 11.4**

**Frequency of Recombinants in Mixed Cultures of Different Fertility Types of *S. coelicolor* A3(2)**

| Fertility Type | Fertility Type IF | NF | UF |
|---|---|---|---|
| IF | $10^{-4}$ | $10^{-1}$ | $10^{-4}$ |
| NF | | $10^{-2}$ | 1 |
| UF | | | $10^{-5}$ |

Source: From D. A. Hopwood, "Genetics of the Actinomycetales," in *Actinomycetales: Characteristics and Practical Importance*, ed. G. Sykes and F. A. Skinner (London and New York: Academic Press, 1972).

the effects of multiple crossovers. Usually multiply marked strains are crossed; to assure that one is dealing with recombinants, strains with genetic characters from each parent are selected and the number of recombinations between these and other markers are scored. From these data, distances in centimorgans are calculated. The feasibility of this method depends in part on the relative infrequency at which crossovers occur in *Streptomyces* species. The chromosome of *S. coelicolor* is composed of only about 260 cm. A similar measurement of the chromosome of *E. coli* (which is physically two to three times smaller than the *S. coelicolor* chromosome) would exceed 2,000 cm, rendering infeasible measurements by this method of distances greater than only a few percent of the length of the chromosome.

Less well studied systems of conjugation have been found to occur in certain species of *Nocardia* (*canicruria*, *erythropolis*, and *mediterranei*) and *Micromonospora* (*chalcea*, *purpurea*, and *ichinospora*).

## THE MAJOR GROUPS OF PLASMIDS

Although the existence of plasmids was inferred from genetic studies in the 1950s, it has been only during the past decade, with the advent of simple and accurate techniques of detection, that the impact of plasmids on the biology and ecology of procaryotes has been fully appreciated. Recent studies have shown that a high percentage of bacteria contain one or more types of these dispensable genetic elements, and studies on them have distinguished and characterized many hundred different types. This complex array has been grouped or classified on the basis of a number of their properties (Table 11.5). All plasmids share the properties of being dispensable, self-replicating, circular molecules of double-stranded DNA, but they differ in many other respects, including among others their size, the number of copies of them per cell, their compatibility with other plasmids, the host bacterium in which they can replicate, and the types of functions they encode. Some of these properties are discussed in the following sections.

### Detection and Isolation of Plasmids

Often the mean G + C (guanine + cytosine) content (see Chapter 13) and therefore the density of the DNA of a plasmid differs sufficiently from that of the chromosome of the host cell so that these two components of a cell's genome can be separated and subsequently purified by submitting the DNA fraction of a cell extract to ultracentrifugation in a density gradient of cesium chloride.

**TABLE 11.5**

**Properties of Certain Plasmids Studied in Enteric Bacteria and Pseudomonads**

| Plasmid[a] | Incompatibility[b] Group | Host Range | Phenotype[c] | Size (kb) |
|---|---|---|---|---|
| F | IncF1 | Many enteric | $Tra^+$ | 94 |
| R1097 | IncH | *Salmonella* spp., *E. coli* | $Tra^+$, Ap, Cm, Sm, Su, Tc | 180 |
| RP1 | Inc P1 | Almost all Gram-negative bacteria | $Tra^+$, Ap, Km, Nm, Tc | 60 |
| pBR322 | [d] | Some enteric bacteria | $Tra^-$, Ap, Tc | 3.9 |

[a] In the past, no system existed to designate plasmids; a variety of letters and numbers were used. Now plasmids are designated by a lowercase "p" followed by two capital letters and a number, e.g., pBR322.
[b] Over 23 incompatibility groups are known for plasmids found in enteric bacteria.
[c] $Tra^+$, encodes self-transfer; Ap, Cm, Km, Nm, Sm, and Tc encode resistance respectively to the antibiotics ampicillin, chloramphenicol, kanamycin, neomycin, streptomycin, and tetracycline. Su encodes resistance to sulfa drugs.
[d] Belongs to the same group as ColE1 which has not yet been designated.

However the apparent differences in density can be amplified if certain dyes, e.g., ethidium bromide, are added to the cesium chloride–DNA mixture. These dyes intercalate between the stacks of base pairs at the core of a DNA molecule, unwinding the helix slightly and thereby decreasing its density. If the DNA is a closed circular molecule, positive superhelical turns are generated that cause increasing tension within it as increasing numbers of ethidium bromide molecules become intercalated* eventually preventing further incorporation of the dye.

But no such restriction applies to linear DNA molecules because they can rotate freely in response to intercalation, thus relieving the tension. As a consequence, linear DNA will intercalate a greater quantity of ethidium bromide than an equivalent amount of circular DNA, and its density will be decreased to a greater extent. The separation of plasmid from chromosomal DNA by density gradient ultracentrifugation in the presence of ethidium bromide therefore depends on the plasmid DNA remaining circular and chromosomal DNA being fragmented into linear pieces during the process of their extraction from the intact cell. Using conventional procedures of breaking the cells and extraction, the shearing forces generated inevitably fragment the large, and therefore fragile, chromosome, but the smaller plasmid molecules are more resistant to breakage, and they can be extracted largely intact if proper precautions are taken to minimize shearing forces. Even breaks, or nicks, as they are called, in one of the strands of DNA in a plasmid allow free rotation and destroy superhelicity. The resulting structure, termed an *open circle*, intercalates as much ethidium bromide as a linear fragment, and hence such structures migrate with chromosomal DNA during ultracentrifugation in an ethidium bromide–containing cesium chloride gradient.

An electron micrograph of linear fragments, open circular and supercoiled plasmid DNA is shown in Figure 11.19.

Plasmids can also be detected and isolated by subjecting cell extracts to *gel electrophoresis*. The extract is placed on the edge of gel slab (the usual gelling agent is a modified form of agar, termed *agarose*) and subjected to an electric field that causes the negatively charged DNA molecules to migrate towards the positive pole. The rate of migration, and therefore the position in the gel slab at the

* DNA structures naturally have negative superhelical turns. Intercalation of increasing amounts of ethidium bromide progressively removes these, and then generates increasing numbers of positive superhelical turns.

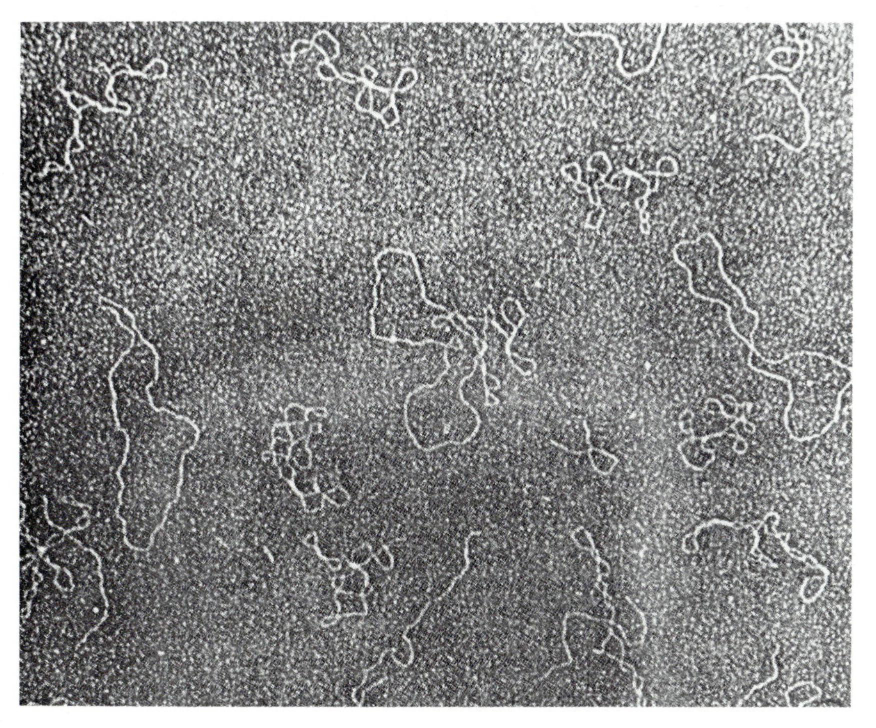

FIGURE 11.19

Electron micrographs of purfied DNA of plasmid ColEI showing open circular, supercoiled, and linear forms of the molecule. From T. F. Roth, and D. R. Helinski, "Evidence for Circular Forms of a Bacterial Plasmid," *Proc. Natl. Acad. Sci. USA* **58**, 650 (1967).

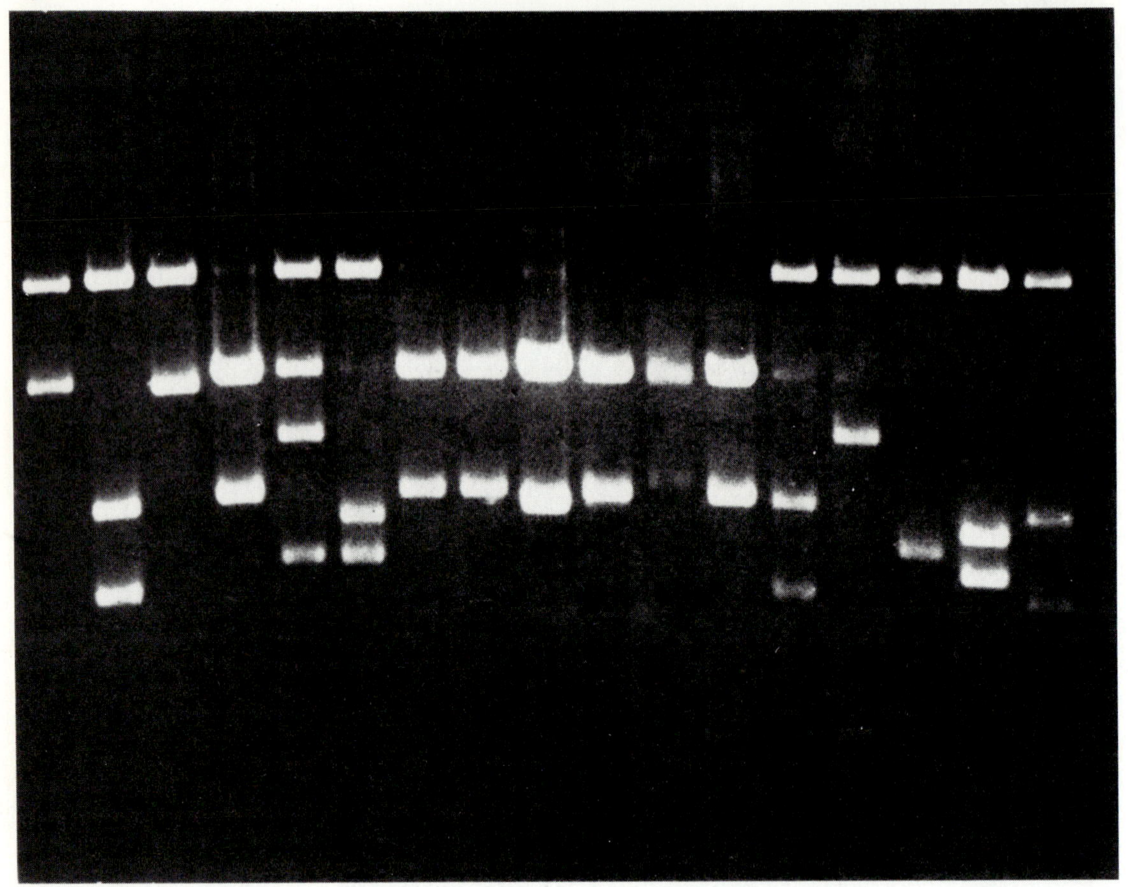


FIGURE 11.20

Photograph of an agarose gel on which various molecular species of DNA were separated by electrophoresis. Various samples of DNA were added to the wells (numbered wt through 16) at the top of the gel and the gel was exposed to an electric field causing the various classes of DNA molecules to migrate toward the anode (bottom). The location of the classes of DNA molecules was revealed by staining with ethidium bromide and viewed with ultraviolet light. All lanes contain DNA fragments generated by digesting a set of plasmids with restriction endonucleases, *EcoR1*, *Sal* and *HpaI*. Lane wt contains the digest of a plasmid that is not attacked by *HpaI*; two plasmids are generated [upper: 3.6 Kilobase pairs (Kb), lower: 2.3Kb]. Subsequent lanes are digests of derivatives of the same plasmid into which sites sensitive to digestion by *HpaI* have been inserted. By analyzing the gel patterns, the location of the insertion can be deduced. For example, in lane 1 an insertion was inserted in the 2.3Kb fragment because this fragment is replaced by two smaller ones. (Courtesy of Martin Privalsky).

completion of the process, depends on the size of the molecule and whether it is in linear, open circular, or superhelical form. The location of the separated bands of DNA in the gel can be visualized under ultraviolet light when stained by ethidium bromide. Ethidium bromide by its capacity to intercalate becomes concentrated in DNA-containing regions of the gel and fluoresces, emitting visible light, when irradiated with ultraviolet light (Figure 11.20).

## R Factors

Many plasmids carry genes that confer on the host cell resistance to antibiotics and other toxic agents. Plasmids of this type are sometimes called *R factors*. The rapid spread in recent years of R factors among pathogenic bacteria has had profound effects on medical practice. Among many types of pathogenic bacteria once controllable by a variety of antibiotics and other chemotherapeutic agents, strains are now frequently encountered that are highly

**TABLE 11.6**

**Products and Their Mode Action of Certain Plasmid-Borne Genes that Encode Resistance to Antibiotics**

| Gene | Resistance Conferred to: | Encoded Protein | Action[a] |
|---|---|---|---|
| Cm | Chloramphenicol | Acetyltransferase | Acetylates drug |
| Sm | Streptomycin | Adenyltransferase; Phosphotransferase | Adenylates drug Phosphorylates drug |
| Sp | Spectinomycin | Adenyltransferase | Adenylylates drug |
| Tc | Tetracycline | Tet proteins | Excludes drug from cell[b] |
| Ap | Ampicillin | β-Lactamase | Hydrolyzes four-membered ring of penicillins |

[a] The various chemical modifications of these antibiotics destroy their antimicrobial activity.
[b] These proteins function to transport the drug out of the cell.

resistant to a variety of these drugs because they carry R factors.

The existence of R factors was discovered in Japan in 1955. R factors are now encountered in certain strains of almost all pathogenic bacteria. They spread rapidly by conjugative transfer through a bacterial population and persist under the selective pressure of drug treatment. Some are self-transferable and others are mobilized by other conjugative plasmids.

The number of inhibitory substances for which resistance may be mediated by R factors has grown to include almost all antibiotics, many other chemotherapeutic agents, and a variety of heavy metals including mercury, cadmium, nickel, and cobalt. Different R factors carry different combinations of resistance genes, ranging from one to as many as seven or eight. The mechanisms of resistance conferred by these genes tend to be different from those that are chromosomally determined. The plasmid genes often encode enzymes that chemically inactivate the drug, or by active export eliminate it from the cell. Rarely, plasmid genes encode redundant resistant forms of the drug's target enzyme. In contrast, chromosomal mutations to drug resistance usually modify the cellular target of the drug so as to render it resistant to the action of the drug. The action of products of some resistance genes is summarized in Table 11.6.

## Other Plasmid-Encoded Characters

Many bacterial strains liberate plasmid-encoded proteinaceous toxins called *bacteriocins*, which are active only against closely related strains of bacteria. Toxins of this type that are liberated by strains of *E. coli* are called *colicins*. Most are simple proteins; several different types have been isolated which kill sensitive cells by different mechanisms. The producing strains are immune to their action because the plasmids that encode these agents also carry genes that protect the host cell. The principal ones that have been studied are ColB, ColE1, ColE2, ColI, and ColV; the colicins that they produce are similarly designated.

Plasmids also encode a variety of other functions. Some encode toxins, the production of which causes certain bacteria to be pathogenic to mammals (see Chapter 31). Others encode functions that make certain bacteria pathogenic to plants, or able to synthesize pigments or antibiotics, or to degrade certain unusual carbon sources. The ability of *Rhizobium* to invade root hairs of plants as the initial step in forming a nitrogen-fixing nodule is also encoded by plasmids. Examples of some of the many bacterial functions known to be encoded on plasmids are listed in Table 11.7. In spite of the many recent advances made in discovering which cellular functions are encoded by plasmids, some plasmids have been detected that have no known function other than the ability to encode their own replication; these are called *cryptic* plasmids.

## Incompatibility among Plasmids

It has been known for some time that certain pairs of plasmids (for example, an F plasmid and an F′ plasmid, or two different kinds of F′ plasmids) cannot be stably replicated in the same bacterial cell. Two such plasmids are said to be *incompatible*, and a collection of incompatible plasmids constitutes an

## TABLE 11.7

**Some Cellular Functions Known to be Encoded by Plasmids**

A. BY ALL PLASMIDS
  1. Self-replication

B. BY SOME PLASMIDS
  1. Self-transfer
  2. Resistance to antimicrobial agents
     a. Antibiotics: e.g., streptomycin, penicillin, kanamycin, neomycin, chloramphenicol, tetracycline
     b. Synthetic chemotherapeutic agents: e.g., sulfa drugs and trimethoprin
     c. Heavy metals: e.g., Ag, Hg, and Cd
  3. Pigment production
  4. Toxin production
  5. Catabolic functions: e.g., lactose, octane, and camphor degradation
  6. Phage-sensitivity and resistance
  7. Antibiotic production
  8. Bacteriocin production
  9. Induction of plant tumors
  10. $H_2S$ production
  11. Host controlled restriction and modification

*incompatibility group*. A variety of evidence suggests that plasmid members of the same incompatibility group (designated Inc followed by a capital letter and sometimes also a number), are closely related, so this property is often used as a system of classification. Knowing to which incompatibility group a particular plasmid belongs often reveals other facts about it. For example, members of the IncP1 incompatibility group all have the property of being able to replicate stably in a broad range of bacterial host cells.

## RECOMBINATION

Recombination occurs in procaryotes, as it does in eucaryotes, between homologous regions of DNA, but unlike the situation in eucaryotes, recombination is almost always an essential component of genetic exchange of chromosomal genes among procaryotes because the fragments of DNA transferred by transformation, transduction, or conjugation are rarely capable of self-replication (they do not constitute a replicon). For the genes carried on the transferred fragment to be stably inherited by the recipient clone, the portion of the fragment that contains them must be incorporated into a replicon (the chromosome or a plasmid) within the recipient cell by two recombinational events. A single recombination (or crossover) between a linear fragment and the chromosome does not yield a circular molecule, rather it yields a linear structure which cannot be replicated. In contrast a recombination between a circular element and the chromosome yields an integrated circular structure (Figure 11.21).

Four different types of recombination can be distinguished: *homologous or general recombination*, *illegitimate recombination*, *site-specific recombination*, and *replicative recombination* (Table 11.8).

Most recombinations, including those associated with genetic exchange between cells, occur

FIGURE 11.21

Consequences of crossovers between various genetic elements. (a) A single crossover between two eucaryotic chromosomes produces two recombinant chromosomes both of which are replicons. (b) A single crossover between a linear fragment and a bacterial chromosome produces a single linear molecule that is not a replicon. (c) Two crossovers between a linear fragment and a bacterial chromosome integrates a portion of the fragment into chromosome and produces a fragment that is not a replicon. (d) A single crossover between a plasmid and a bacterial chromosome produces a slightly larger chromosome with the plasmid integrated into it.

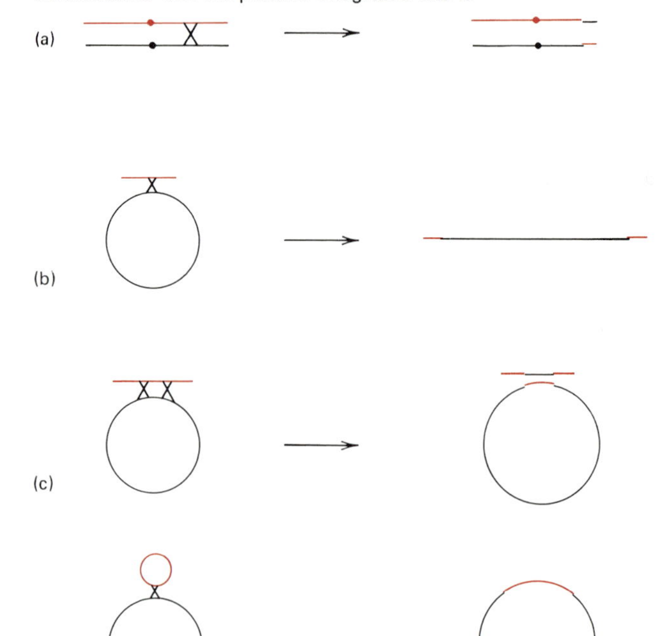

**TABLE 11.8**

**Types of Recombination**

I. General recombination: can occur between any pair of homologous sequences of DNA
II. Illegitimate recombination: occurs between nonhomologous regions of DNA
III. Site-specific recombination: occurs between certain homologous sequences of DNA, e.g., between a temperate phage and the bacterial chromosome
IV. Replicative recombination: occurs between certain genetic elements (transposable elements) and many other DNA sequences

between homologous regions of DNA by general recombination; these are catalyzed by products of a set of chromosomal genes, *recA* through *recF*. The mutational loss of activity of some of these causes only a decrease in the frequency of general recombination, suggesting that general recombination can occur by at least two parallel pathways. But loss of *recA* function completely eliminates general recombination, implying that both pathways require the participation of the RecA protein as the product of the *recA* gene is called.

Illegitimate recombination is a rare event that occurs between nonhomologous regions of DNA. Loss of RecA does not affect the frequency of illegitimate recombination; indeed, recombinations that occur in the absence of RecA can be assumed to have occurred as a consequence of illegitimate recombination.

Site-specific recombination is typified by the event that leads to *reduction* of lambda and certain other temperate phages to the prophage state. Recombination occurs within homologous segments shared by the phage and the chromosome (the *att* site), and is catalyzed by phage-encoded enzymes. As a consequence, site specific recombination does not require the participation of the RecA protein.

The transposition of insertion sequences and transposons is mediated by replicative recombination. Like site-specific recombination, replicative transposition is catalyzed by proteins encoded in genes on the genetic element undergoing recombination; it also occurs in the absence of RecA protein.

## Molecular Mechanism of General Recombination

The final consequence of general recombination is the same as breaking two double-stranded molecules of DNA and rejoining each to the other. Several plausible models by which this might occur have been proposed, one of which is shown in Figure 11.22. The two DNA duplexes pair, nicks are made in homologous regions of two strands with the same polarity, thus allowing them to pair with the complementary strand of the other duplex. It is at this point that the RecA protein functions in an ATP-driven reaction. On resealing the nick, a single-strand crossover has been effected. In this form the crossover point is free to migrate in either direction, a process termed *branch migrations*, creating heteroduplex regions composed of one strand from each of the original DNA duplexes. Rotating this structure enables a planar molecule to be visualized. At this point paired nicks are again made in two strands with the same polarity, and these are resealed to each other. If these nicks are not made in the strands not involved in the original crossover, general recombination is accomplished, the net result of which is the joining of one of the original strands to the other at a point composed of a short heteroduplex region. If, on the other hand, the second pair of nicks and subsequent exchange is made in the strands that were involved in the original crossover, no general recombination occurs; only a short heteroduplex region is created.

## Insertion Sequences, Transposons, and Replicative Recombination

In this and the previous chapter, references have been made to transposable genetic elements, termed *insertion sequences* and *transposons*, that have the ability to transpose to various sites on the bacterial genome. When such a transposition occurs, two copies of the element are generated, one at its original site and a second at the new one. Because of this, the mechanism by which these transposable elements are inserted at the new site is termed *replicative recombination*.

Insertion sequences and transposons differ in two respects: size and gene content. Insertion sequences are small elements (slightly smaller or larger than 1 kb) that encode only their capacity for replicative recombination. The latter are larger apparently composite elements (up to about 10 kb in length) terminated by insertion sequences and containing genes within the central region, termed the *core*, that encode various functions, typically resistance to an antibiotic, but sometimes the capacity to degrade a particular compound as a source of carbon. The capacity of transposons for replica-

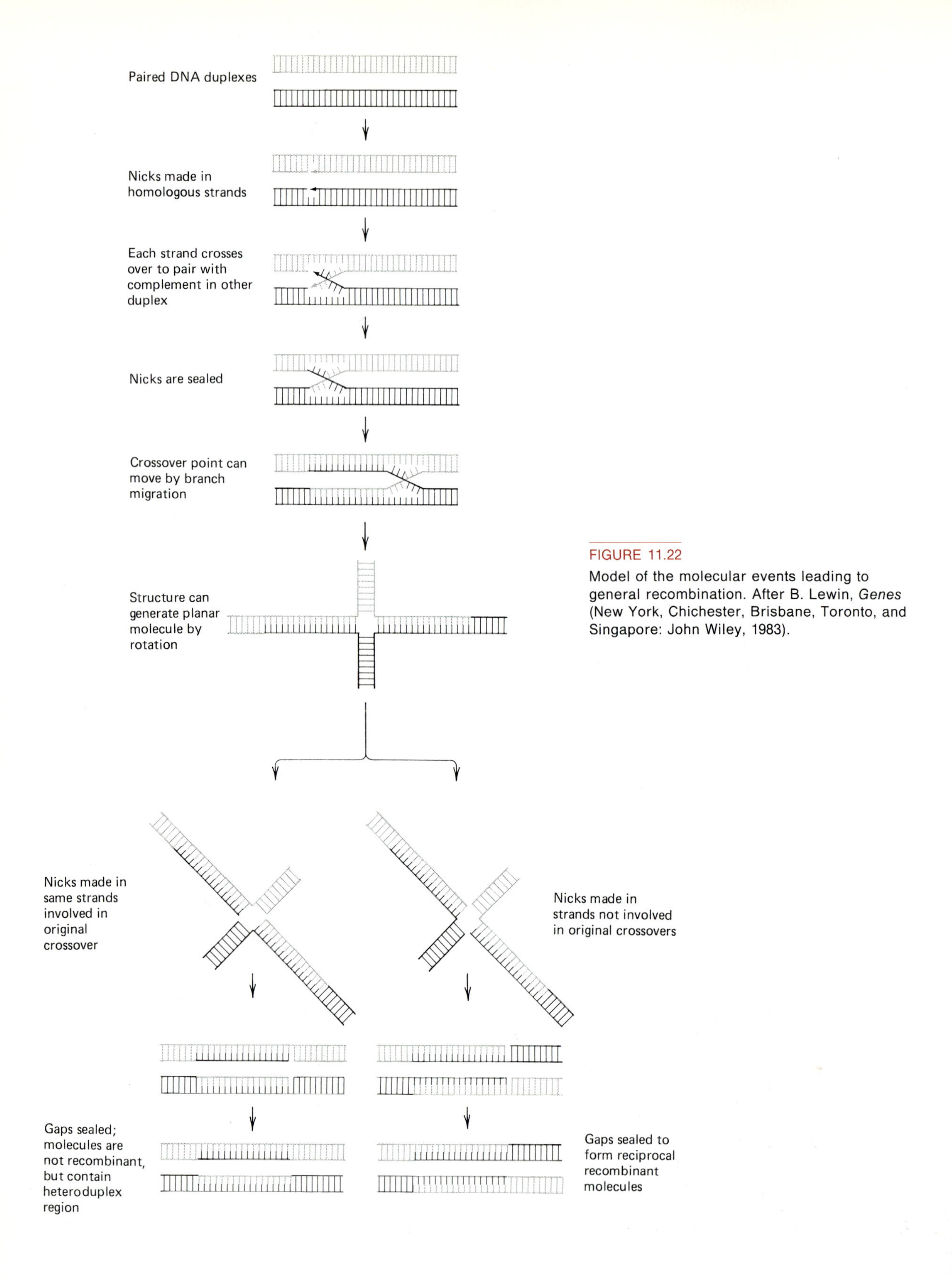
Paired DNA duplexes
Nicks made in homologous strands
Each strand crosses over to pair with complement in other duplex
Nicks are sealed
Crossover point can move by branch migration
Structure can generate planar molecule by rotation
Nicks made in same strands involved in original crossover
Nicks made in strands not involved in original crossovers
Gaps sealed; molecules are not recombinant, but contain heteroduplex region
Gaps sealed to form reciprocal recombinant molecules

FIGURE 11.22

Model of the molecular events leading to general recombination. After B. Lewin, *Genes* (New York, Chichester, Brisbane, Toronto, and Singapore: John Wiley, 1983).

**Table 11.9**

**Properties of Selected Insertion Sequences and Transposons**

| Element (designation) | Size (bp) | Polarity[a] (orientation) | Properties Encoded[b] |
|---|---|---|---|
| Insertion Sequences | | | |
| IS1 | 768 | Both | None |
| IS2 | 1,327 | One | None |
| IS3 | 1,400 | One | None |
| Transposons | | | |
| Tn1 | 4,957 | One | Ampicillin |
| Tn5 | 5,400 | Both | Kanamycin[c] |
| Tn9 | 2,638 | Both | Chloramphenicol |
| Tn10 | 9,300 | Both | Tetracycline |

[a] Indicates that transcription and therefore expression of distal sequences in a gene or genes in an operon is inhibited if the element is inserted in one or the other orientation.
[b] All these elements encode proteins mediating their own transposition.
[c] In *E. coli*, Tn5 expresses only a protein that confers resistance to kanamycin. In most other bacteria it expresses a second protein conferring resistance to streptomycin.

tive recombination resides in their terminal insertion sequences. The properties of some insertion sequences and transposons are summarized in Table 11.9.

Certain of the properties of transposable genetic elements are revealed by experiments done by E. Lederberg that led to the discovery of insertion sequences. Among a group of mutant clones that had lost the capacity to metabolize galactose as a carbon source because they no longer synthesized galactokinase, the enzyme encoded by *galK*, she found that some were found to carry mutations with unusual properties. Although *galK* function had been lost, the mutational events had occurred in a more promotor proximal gene, *galT* or *galE*, of the same operon; these mutations eliminated the activity of the product of the gene in which they occurred and in all downstream genes of the operon. Thus these mutations fell into a class of mutations, termed *polar*. Some nonsense and deletion mutations exhibit polar effects, but the one under investigation had properties that differed from nonsense and delection mutations. They were more polar (they produced lower levels of the products of the downstream genes); they were not suppressed by nonsense suppressor mutations (as nonsense mutations are); and they reverted to wild type at a low but significant rate (as deletion mutations never do). Subsequent experiments established that the mutations were not caused by microlesions in the base sequence of the DNA of the affected gene, nor by a loss of nucleotide pairs, but rather by the gain of new base sequences. Further, it was shown that a number of different mutations were caused by the gain of the same sequence. This class of mutation-causing sequences was termed *insertion sequences.* Later it was shown that these elements encoded their own transposition and were responsible, as well, for the transposing capacity of transposons. The insertion sequences at the ends of transposons occur in either possible orientation with respect to one another—as the same sequence of bases reading in the same direction (*direct repeats*), or the reverse (*inverted repeats*). A plausible mechanism of replicative recombination is shown in Figure 11.23.

The discovery of transposons that undergo replicative recombination at relatively high frequency (Tn5 on introduction into a new cell undergoes replicative recombination at a frequency of about $10^{-5}$) provided an explanation for the spread throughout a bacterial population of certain genes such as those that encode resistance to antibiotics. Such genes on transposons can transpose to plasmids, thereby forming multiple drug resistance plasmids. These plasmids can then be transferred to other strains, and if the plasmids have broad host range properties, the resistance genes can be spread to other species and genera.

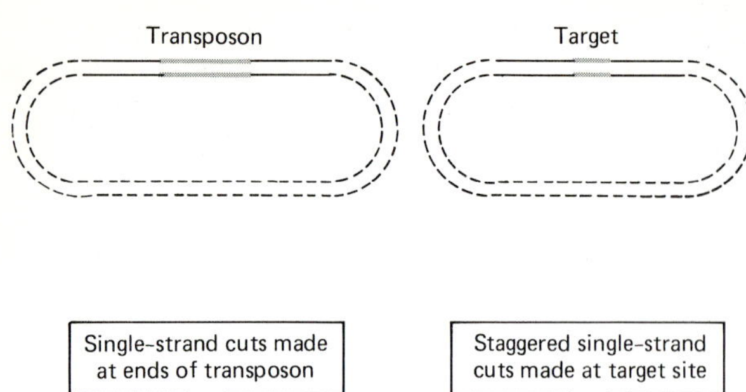
Transposon
Target

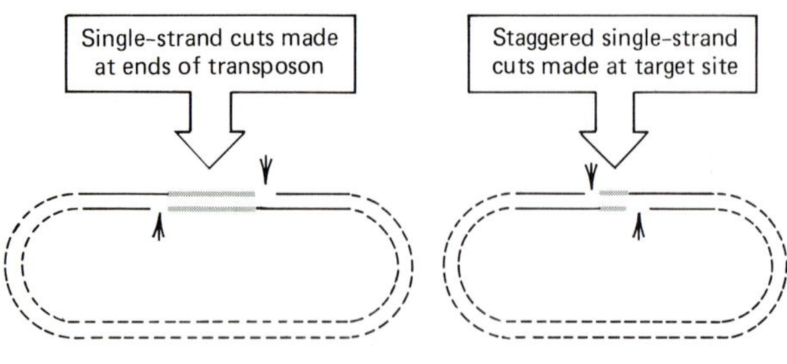
Single-strand cuts made at ends of transposon
Staggered single-strand cuts made at target site

## GENETIC ENGINEERING

The extremely rapid increase over the past decade of knowledge about the enzymes that alter DNA, combined with feasible methods of synthesizing DNA with known sequences, has led to a remarkable capacity to manipulate DNA in vitro. By transformation these artificial DNA constructs can be introduced into an appropriate cell, and if they are, or become, associated with a replicon, they will become a stable addition to the cell's genome. If they are associated with appropriate promoters and contain appropriate translational start signals, they will be expressed at high levels. Collectively this set of procedures, which continues to expand and to be improved, has been termed *recombinant DNA technology* or, because the technology offers the capacity to plan and synthesize new DNA constructs in a series of steps, *genetic engineering*. The technology, along with the recently developed methods of determining the sequence of bases in a length of DNA (Chapter 13), has become an extremely powerful set of tools for analyzing gene structure and function. The technology also promises a wide variety of practical applications to medicine, agriculture, and industry. Indeed, a large number of companies have been organized to exploit the technology. Already commercial products produced by use of these procedures (see Chapter 33) are beginning to appear in the marketplace. Among these products are human insulin and human growth hormone. Both are produced by strains of *E. coli* into which the appropriate human genes were introduced by techniques of recombinant DNA technology.

Even a brief consideration of the myriad of procedures that comprise this technology is beyond the scope of this book, but excellent books are available that are devoted exclusively to describing these procedures and their applications (see the partial listing at the end of this chapter).

Fundamental to the technology, as it is to any type of construction, is the cutting, shaping, and rejoining of component parts (in this case DNA) of the final product (in this case new genes arranged in a replicon). Examples of these techniques and

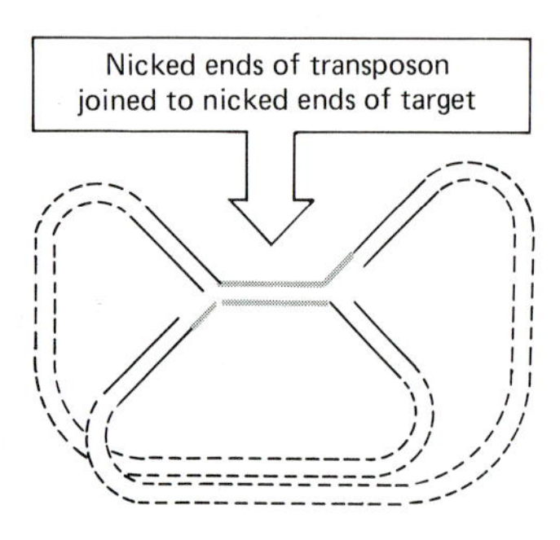
Nicked ends of transposon joined to nicked ends of target

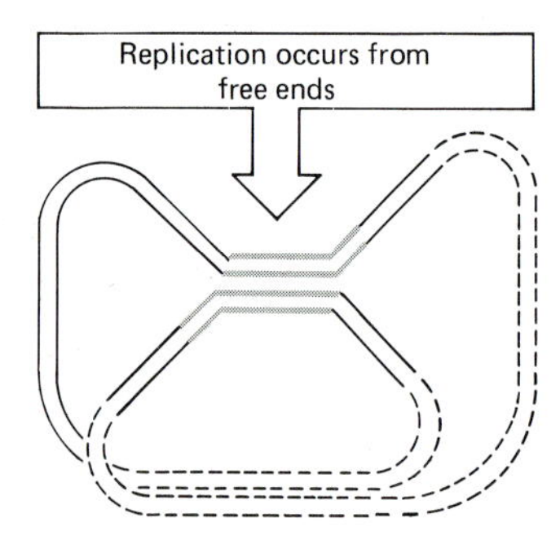
Replication occurs from free ends

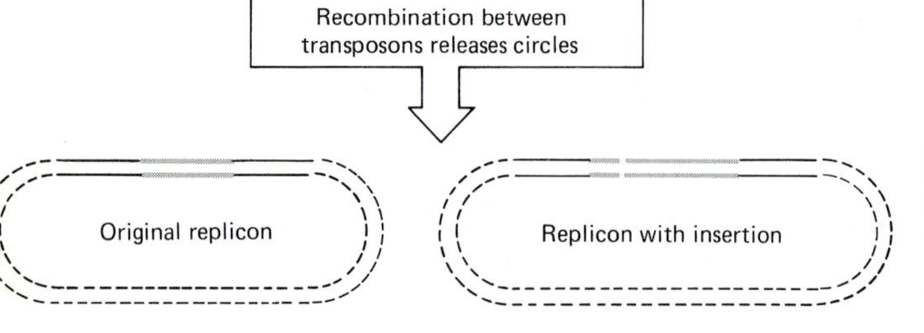
Recombination between transposons releases circles
Original replicon
Replicon with insertion

FIGURE 11.23
Model of the molecular events leading to replicative recombination. After B. Lewin, *Genes* (New York, Chichester, Brisbane, Toronto, and Singapore: John Wiley, 1983).

**TABLE 11.10**

**Examples of Some Molecular Vectors Used in Genetic Engineering**

| Vehicles | Advantages |
|---|---|
| Plasmids | |
| pBR322 | Small size; occur as multiple copies in a cell |
| pRK290 | Can replicate in a variety of Gram-negative bacteria |
| pHV14 | Can replicate in *E. coli* and *B. subtilis* |
| Cosmids[a] | Ensures that a large DNA sequence is cloned |
| Phages | |
| Lambda | The cloned DNA can be manipulated as the phage |
| M13 | Cloned DNA can be obtained in single-stranded form and is therefore more easily sequenced |

[a] Plasmids that contain a gene (*cos*) from phage λ, thereby allowing it to be packaged, *in vitro*, in a phage λ head.

replicons that have been used (sometimes termed *vectors*) are listed in Table 11.10.

The insertion of a gene into a vector and the introduction of this new genetic construct into a host cell is referred to as the *cloning* of the gene of interest because the cell clone that develops constitutes, as well, a clone of that gene.

### The Cutting and Rejoining of DNA

A class of enzymes that is frequently used in recombinant DNA technology to cleave DNA are certain *restriction endonucleases*. Most bacteria produce one or more representatives of this class of enzymes, probably because they offer the selective advantage of providing some protection from infection by phages that contain double-stranded DNA. These enzymes recognize specific sequences of bases, usually four or six base pairs in length, and cleave the two strands. If the entering phage DNA is cleaved in such a manner, the molecule becomes susceptible to further degradation by the nonspecific exodeoxynucleases that all bacteria contain. Of course any sequence of four or six base pairs occurs at significant frequency [at a probability of 1 in every 256 ($4^4$) or 4,069 ($4^6$) base pairs respectively]. Thus, the resident bacterial genome also contains large numbers of such sites; however, these are protected from cleavage by an enzymatic activity, termed *modification*, that methylates one of the bases within the recognition site, sometimes called a *restriction site*, rendering it resistant to the action of the endonuclease. Sometimes the restriction endonuclease itself catalyzes modification; in other cases it is catalyzed by a separate enzyme. Thus only DNA that has recently entered a bacterial cell is cleaved by the cell's restriction endonucleases.

Unsurprisingly in view of their presumed function, restriction endonucleases from different bacteria exhibit quite different specificities. Over 200 types of them have now been characterized. They are designated by the bacterium in which they occur by a capitalized initial of the genus and followed by the first two lowercase letters of the species. Sometimes additional letters or numbers are added to distinguish between the several enzymes produced by some species. For example *Eco*R1 and *Hin*dIII are restriction endonucleases found in strains of *Escherichia coli* and *Haemophilus influenzae* respectively. Some restriction endonucleases cut asymmetrically producing short single stranded regions at their ends; others cut symmetrically producing so called *blunt ends*.

The specificities and sources of some representative commonly used restriction endonucleases are shown in Table 11.11.

The special utility of restriction endonucleases in genetic engineering rests on their remarkable specificity. DNA from any source, except from the strain from which the endonuclease is isolated is cleaved at the restriction sites it contains, and in the case of enzymes that cut asymmetrically, the single-stranded ends are complementary. As a con-

**TABLE 11.11**

**Specificities and Sources of Certain Restriction Endonucleases Used in Recombinant DNA Technology**

| Enzyme | Source | Specificity |
|---|---|---|
| *Eco*RI | *Escherichia coli*[a] | ↓<br>G AATTC<br>CTTAA G<br>↑ |
| *Hae*III | *Haemophilus aegyptus* | ↓<br>GG CC<br>CC GG<br>↑ |
| *HindIII* | *Haemophilus influenzae* | ↓<br>A AGCTT<br>TTCGA A<br>↑ |

[a] Enzyme encoded in a plasmid.

sequence, DNA from different sources can pair (*anneal*) stably at reduced temperature, and the gaps can be resealed by the action of a DNA ligase. For this purpose a ligase from phage T4-infected bacteria (termed *T4 ligase*) is commonly used. It utilizes ATP rather than NAD as cofactor (Chapter 5). Fragments of DNA with blunt ends can also be ligated but at a low rate owing to the highly unstable and transitory nature of the end-to-end association of blunt ends of DNA. A scheme for cloning a gene into a commonly employed plasmid vehicle, pBR322 is outlined in Figure 11.24. Plasmids are commonly used cloning vehicles but phages are also used.

It will be noted that these techniques are nonspecific with respect to the genes that the cloned fragment contains. If the source of DNA to be cloned is contained in a total extract of the DNA in a bacterial cell, the products of the initial endonuclease digestion will contain as many as 1,000 different fragments (as we have seen, averaging about 4 kb in length if the restriction site contains six base pairs), only one or two of which contains a particular gene of interest. If the gene can complement a growth-promoting bacterial function, following transformation of the ligated mixture into an appropriate bacterial host, conventional selection procedures can be employed to detect the bacterial clone that contains the desired cloned gene. For example, in order to isolate a cloned gene encoding a step in the biosynthesis of a particular amino acid, the ligated mixture is introduced into a bacterial strain that carries a mutational block in that gene. Those transformed clones that can develop on a medium that lacks the amino acid most probably carry a clone of the desired gene.

However, different procedures must be employed to clone a eucaryotic gene, the product of which (for example, insulin) does not serve a growth-promoting function in bacteria. Rather than selecting among a large variety of cloned genes in the ligated mixture, the gene is purified prior to cloning. Usually purification is accomplished (Figure 11.24) by isolating the mRNA product of the desired gene. Using this as a template, appropriate enzymes as catalysts, and deoxynucleoside triphosphates as substrates, the gene can be synthesized *in vitro*. Then the synthetic gene is cloned. This procedure offers the added advantage of eliminating introns which commonly occur in eucaryotic genes but which cannot be eliminated during processing by the bacterial cell. The desired mRNA can some-

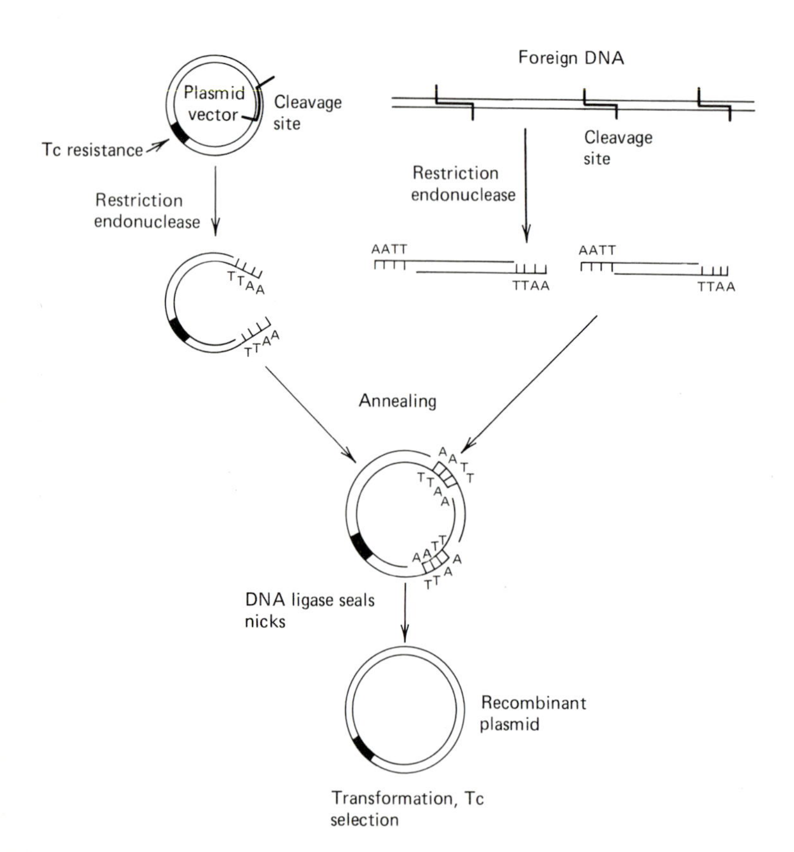


FIGURE 11.24
Scheme for cloning a gene using a plasmid as a vehicle and the restriction enzyme *Eco*R1.

times be purified simply by exploiting the property of certain tissues of producing high levels of specific messages (for example, insulin message by the pancreas). If this is not possible, the proper mRNA can be labeled and thereby purified by hybridizing it to a short complementary segment of radioactive single-stranded DNA. Appropriate segments of DNA, or *probes* as they are called, can be synthesized *in vitro* if a portion of the amino acid sequence of the gene's protein product has been determined. From this information, the genetic code can be used to deduce coding sequences in DNA. Since the genetic code is redundant, a number of different nucleotide sequences can encode the same sequence of amino acids. Accordingly the complete set of labeled probes must be synthesized to ensure detection of the desired mRNA.

The armamentarium of recombinant DNA technology continues to grow at a rapid rate, as does its possible applications. It is difficult to foresee all the directions this new technology will take but it is obvious that its impact will be profound.

## FURTHER READING


**Books**

GLASS, R. E., *Gene Functions.* Berkeley and Los Angeles: University of California Press, 1982.

LEWIN, B., *Genes.* New York: John Wiley and Sons, 1983.

OLD, R. W. and S. B. PRIMROSE, *Principles of Gene Manipulation*, 2*nd* Ed. Berkeley and Los Angeles: University of California Press, 1981.

**Reviews**

CLEWELL, D. B., "Plasmids, Drug Resistance and Gene Transfer in the Genus *Streptococcus*", *Microbiol. Rev.* **45,** 409 (1981).

SMITH, H. O. and D. B. DANNER, "Genetic Transformation", *Ann. Rev. Biochem.* **50,** 41 (1981).

WILLETS, N. and B. WILKINS, "Processing of Plasmid DNA during Bacterial Conjugation," *Microbiol. Rev.* **48,** 24 (1984).

# Chapter 12 Regulation

In Chapter 5 we considered the individual biochemical reactions that lead to the duplication of a bacterial cell and permit it to carry out certain essential functions. These can be likened to a highly complex pattern of converging assembly lines in a factory that manufactures at changing rates a particularly elaborate but variable product. Of course, the existence of all these assembly lines is essential to the mission of the factory, and independent control of their rate of functioning is essential in order to avoid waste and complete chaos. This factory analogy is not completely adequate to explain control of bacterial metabolism unless we also assume that the composition and even the existence of particular assembly lines varies in response to changes in demand for a product.

## TYPE OF CONTROL MECHANISMS

That bacterial cells possess elaborate control mechanisms can be appreciated from a variety of simple observations. Among these are the following: (1) bacteria synthesize building blocks (amino acids, nucleotides, etc.) at an optimal rate to supply polymerization reactions regardless of whether these reactions are occurring slowly or rapidly; (2) bacterial cells produce the enzymes to degrade seldomly encountered sources of carbon, nitrogen, and phosphorous only when one of these sources is present and a more readily metabolizable source is lacking in the medium; and (3) bacteria can modulate their complement of ribosomes in response to changes in

**TABLE 12.1**

**Comparison of Ribosome Content and Rate at Which Individual Ribosomes Function in Cultures of *Salmonella typhimurium* Growing Rapidly and Slowly**

| Medium[a] | Growth Rate (doublings/hr) | Ribosomes per Cell | Amino Acids Polymerized per Second | |
|---|---|---|---|---|
| | | | PER CELL | PER RIBOSOME |
| Nutrient broth | 2.4 | 69,750 | 1,116,000 | 16 |
| Glycerol-salts | 0.6 | 7,140 | 99,960 | 14 |

[a] Nutrient broth is an extract of beef, yeast, and casein that contains almost all monomers contained in macromolecules. Glycerol-salts contains glycerol as the sole organic constituent; all other required nutrients are supplied in the form of inorganic salts.

composition of the growth medium: if the medium is rich and therefore adequate to support growth at a high rate, thereby necessitating a rapid rate of protein synthesis, the complement of ribosomes is large; in poor media that support growth at only a low rate, the complement of ribosomes is correspondingly low. Indeed, at any growth rate (except at very low ones), the cell's complement of ribosomes is precisely set so that nearly all of them constantly function at maximum efficiency (Table 12.1). The rate of protein synthesis by a culture growing slowly in a poor medium is much less than the rate of protein synthesis by a rapidly growing culture. Correspondingly, cells in the slowly growing culture contain fewer ribosomes, but individual ribosomes in both cultures function at the same rate.

Cellular control mechanisms act at all levels of metabolism. Examples of some of them are shown in Figure 12.1. A few operate by changing the actual structure of the DNA. Many operate by modulating the rate of transcription of specific genes or operons. Some modulate the rate of translation of specific mRNA molecules, and many modulate the activity of specific enzymes.

Control mechanisms can also be classified by the type and specificity of the metabolic targets of their action (Table 12.2). Dissimilatory enzymes

FIGURE 12.1

Examples of cellular control mechanisms that act at various levels of metabolism.

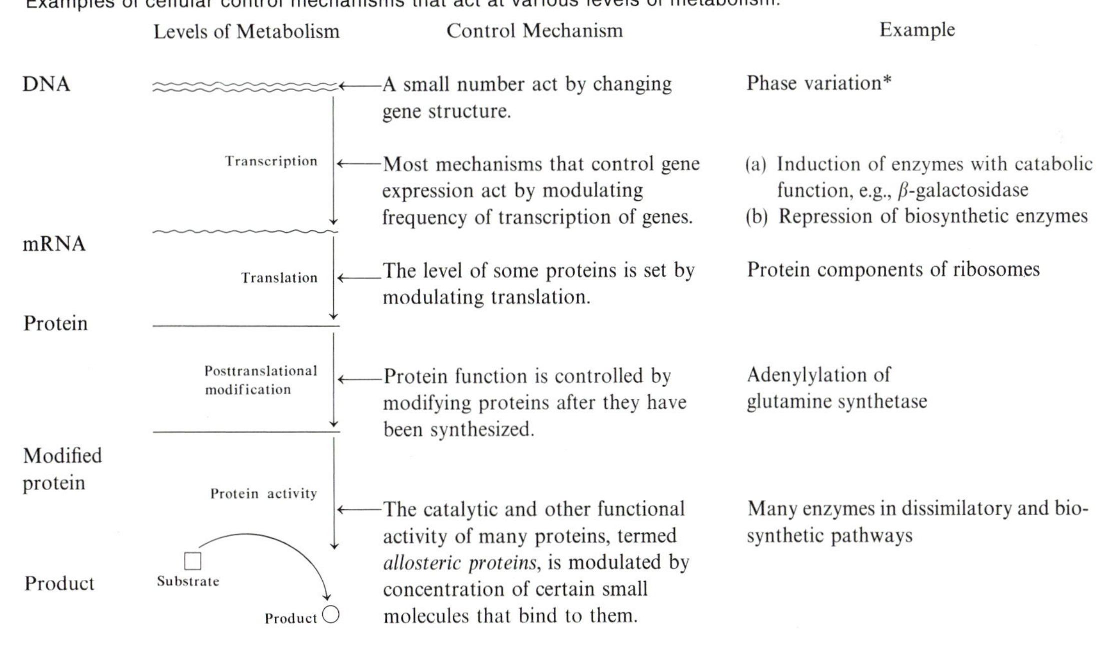

Levels of Metabolism
Control Mechanism
Example
DNA
A small number act by changing gene structure.
Phase variation*
Transcription
Most mechanisms that control gene expression act by modulating frequency of transcription of genes.
(a) Induction of enzymes with catabolic function, e.g., β-galactosidase
(b) Repression of biosynthetic enzymes
mRNA
Translation
The level of some proteins is set by modulating translation.
Protein components of ribosomes
Protein
Posttranslational modification
Protein function is controlled by modifying proteins after they have been synthesized.
Adenylylation of glutamine synthetase
Modified protein
Protein activity
The catalytic and other functional activity of many proteins, termed allosteric proteins, is modulated by concentration of certain small molecules that bind to them.
Many enzymes in dissimilatory and biosynthetic pathways
Product
Substrate
Product


***Salmonella typhimurium* possesses the genetic capacity to produce two different types of flagella. Inversion of a small segment of DNA switches production from one type to the other. The change is termed *phase variation*.

TABLE 12.2

**Metabolic Targets of Various Cellular Control Mechanisms**

| Target | Mechanism | Action |
|---|---|---|
| I. Biosynthetic pathway (for an amino acid) | 1. Repression-Derepression | When the end product is present in excess, further synthesis of component enzymes is stopped. When intracellular concentrations of the end product are inadequate, component enzymes are synthesized at elevated rates. |
| | 2. End-product inhibition | The activity of one of the enzymes in the pathway (usually the first one) is inhibited by the end product of the pathway. |
| II. Pathway of dissimilation of carbon sources other than glucose (e.g., lactose) | 1. Induction | Synthesis of the enzymes is induced by the presence in the medium of the primary substrate. |
| | 2. Catabolite repression | Synthesis of the enzymes is inhibited by the presence of other readily metabolized substrates even if the primary substrate is present. |
| III. Pathway of dissimilation of a nitrogen source | 1. Induction | See II.1 above. |
| | 2. Nitrogen control | Synthesis of the enzymes is inhibited if the most readily utilizable source, $NH_4^+$, is present. |
| IV. Biosynthesis of ribosomes | 1. Ribosome feedback | Synthesis of ribosomal RNA is inhibited by an excess in the cell of mature ribosomes. |
| | 2. Translation control | Synthesis of ribosomal proteins is inhibited by the free proteins (the excess that is not assembled into ribosomes). |

FIGURE 12.2

Schematic representation of control mechanisms that function to maintain the intracellular concentration of an amino acid at an optimal level.

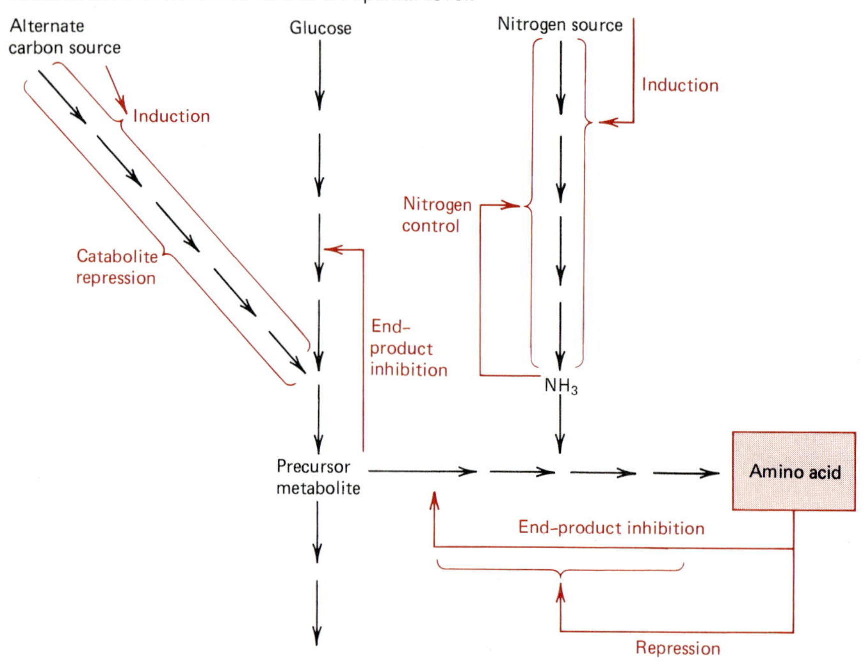

that are synthesized only when the cognate substrate is available are said to be *inducible* by that substrate. Biosynthetic enzymes that are synthesized in lesser amounts when the concentration of end product of the pathway rises are said to be *repressible*. Enzymes (or other proteins), the activity of which is modulated by the concentration of small molecules (other than their substrates or products), are termed *allosteric enzymes* or *allosteric proteins*.

Most cellular control mechanisms are activated in response to changes in the concentration of specific small molecules termed *effectors*. In the case of relatively specific control mechanisms such as those that regulate the functioning of a particular biosynthetic or dissimilatory pathway, the effector is usually part of that pathway: the substrate, an intermediate, or the end product. But special compounds that are not metabolic intermediates trigger control mechanisms with coordinated impact on certain sets of pathways such as those that repress alternate pathways of utilization of carbon, nitrogen, or phosphorous when a more readily metabolizable source of these essential nutrients is available. Sometimes these signal compounds are called *alarmones*, because variations in their intracellular concentration reflect a general cellular insufficiency. Two such compounds have been identified: cyclic-3′,5′-adenosine monophosphate (cAMP, or cyclic AMP) which signals catabolite repression in response to a sufficiency of carbon source, and 3′-diphospho-5′-diphosphoguanosine (ppGpp or guanosine tetraphosphate) the intracellular concentration of which rises in response to the insufficiency of any amino acid.

## Coordination of Control Mechanisms: Synthesis of an Amino Acid

The general patterns of regulation of the biosynthesis of a biosynthetic building block (in this case an amino acid) are outlined in Figure 12.2. The net consequence of these controls is homeostasis with maximal metabolic economy of the intracellular concentration of the amino acid in spite of changes in the extracellular environment.

If the intracellular concentration of the amino acid were to rise because of its appearance in the growth medium or a decreased intracellular demand for it, end-product inhibition of the first enzyme in the pathway would immediately stop or appropriately decrease the flow through the pathway. Thus, end-product inhibition alone is sufficient to maintain homeostasis of end products; repression effects the economy of their synthesis. If the pathway is functioning at a decreased rate or not at all, further synthesis of its component enzymes would constitute a waste of metabolic capacity. Although the onset of repression, inhibiting further enzyme synthesis and therefore effecting metabolic economy, is very rapid, the impact on the intracellular concentration of the repressed enzyme is delayed; its concentration declines principally by dilution during subsequent growth, because growing cells degrade proteins very slowly (at a rate of about 3 percent per hour; Chapter 10). Even if further synthesis of an enzyme is stopped completely, its intracellular concentration would decrease by a factor of two only after the number of cells in the population had doubled.

Other controls regulate the flow of metabolites into the biosynthetic pathway. End-product feedback loops exist within the pathway to maintain the concentration of precursor metabolites at optimum levels. If an alternate carbon source is the only one present, enzyme induction assures that the enzymes necessary to supply precursor metabolites are present. But if glucose is also present, these enzymes are unnecessary; under these conditions, catabolite repression acts to prevent their induction.

The dissimilatory pathways that supply nitrogen (in the form of ammonia) to the biosynthetic pathway are also typically induced by the primary substrate, a biosynthesis that is unnecessary if ammonia is available directly from the medium or from another more readily metabolizable source of nitrogen. Nitrogen control triggered by the elevated concentrations of ammonia (although another unknown compound is the actual effector) acts to prevent induction by the primary substrate. Many nitrogen sources (being organic compounds) also supply intermediates to the pathway of central metabolism. Accordingly, many pathways that supply nitrogen are subject to catabolite repression as well as nitrogen control.

## Coordination of Control Mechanisms: Synthesis of Ribosomes

As mentioned earlier, the bacterial cell regulates the biosynthesis of ribosomes to match its need for protein synthesis (Table 12.1). The patterns by which this very important control is effected is only now becoming understood (Figure 12.3). By still unknown mechanisms, an excess of ribosomes (more than those necessary to occupy fully all intracellular mRNA molecules) inhibits transcrip-

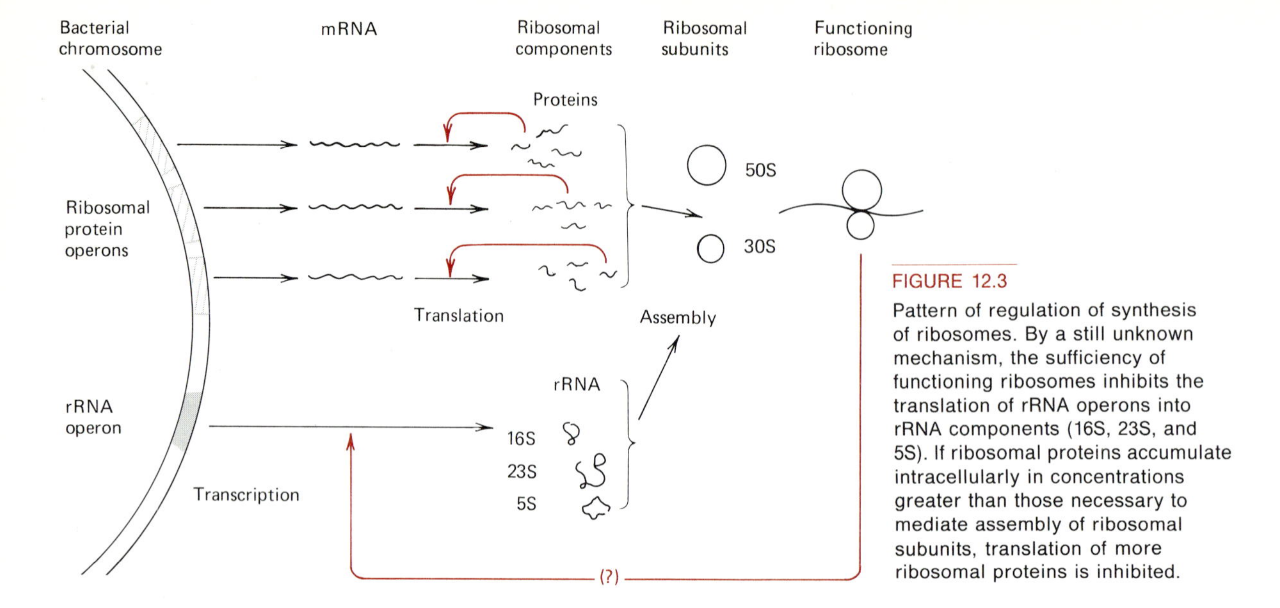


FIGURE 12.3

Pattern of regulation of synthesis of ribosomes. By a still unknown mechanism, the sufficiency of functioning ribosomes inhibits the translation of rRNA operons into rRNA components (16S, 23S, and 5S). If ribosomal proteins accumulate intracellularly in concentrations greater than those necessary to mediate assembly of ribosomal subunits, translation of more ribosomal proteins is inhibited.

tion of rRNA genes. Ribosomal protein molecules bind to rRNA in the process of ribosomal self-assembly. If the proteins are synthesized more rapidly than they are utilized in assembly, they inhibit translation of their own encoding mRNA molecules. Thus, under the regulation of these controls, ribosomes are synthesized in response to cellular requirements for their use, and the synthesis of their components is coordinated. An additional mechanism prevents over-synthesis of rRNA following the arresting of protein synthesis by a sudden deprivation of an amino acid. When this occurs, functioning ribosomes that lack a cognate amino acyl-tRNA at their A site, and are therefore unable to catalyze protein elongation, catalyze the *idling reaction*, the eventual product of which is guanosine tetraphosphate, ppGpp.

$$\text{pppG(GTP)} + \text{pppA(ATP)} \xrightarrow[\text{reaction}]{\text{idling}} \text{pppGpp} + \text{AMP}$$

$$\text{pppGpp} \longrightarrow \text{ppGpp} + \textcircled{P}$$

The idling reaction requires the participation of an accessory protein factor, the *stringent factor* encoded by a gene designated *relA*; the second reaction is catalyzed by various ribosomal proteins.

Accumulation of ppGpp, which occurs in response to a deprivation of any amino acid, triggers a number of cellular controls collectively termed the *stringent response*. Among these are a general stimulation of synthesis of amino acids, and an inhibition of synthesis of phospholipids and rRNA. Thus rRNA does not accumulate in normal cells (although it does in cells that lack stringent factor owing to a mutation in the *relA* gene) following starvation for any amino acid. If rRNA were to accumulate in the absence of protein synthesis, a generalized cellular toxicity would result because in the absence of free ribosomal proteins a variety of other basic proteins bind to rRNA, thereby becoming unable to perform their normal function. Indeed, *relA* mutants that have accumulated rRNA in response to starvation for an amino acid undergo a prolonged lag following replenishment of the amino acid before growth reinitiates.

## Mechanisms of End-Product Inhibition: Allosteric Proteins

Allosteric proteins* are proteins whose *properties change* if certain specific small molecules, *effectors*, are bound to them. Hence, *allosteric proteins are mediators of metabolic change which is directed by changes in concentration of the small effector molecules.*

There are two classes of allosteric proteins: *allosteric enzymes*, whose activities are either enhanced or diminished when combined with their effectors, and *regulatory allosteric proteins*, devoid of catalytic activity, which bind to DNA and modulate the synthesis of specific enzymes.

Regulatory allosteric proteins attach to the bacterial chromosome near the specific structural genes whose repression they control. This attachment can be modified by the binding of small effector molecules to the regulatory proteins, thereby

* The word allosteric means *differently shaped*, and it alludes to the fact that the effectors that regulate the activity of an allosteric enzyme have a structure different from that of the substrate or product of the enzyme.

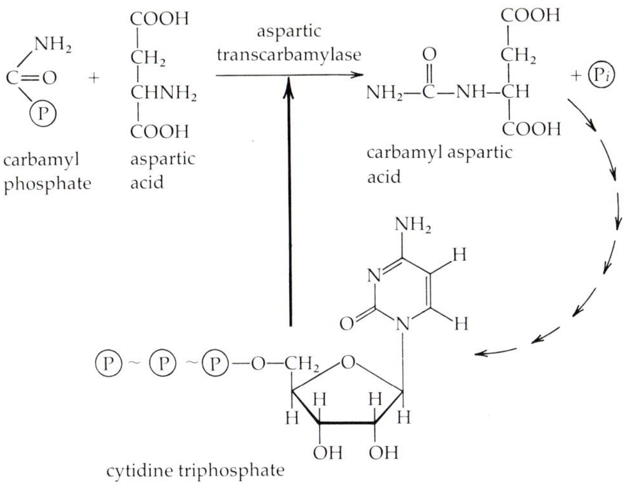

FIGURE 12.4

The allosteric control of the first step in pyrimidine biosynthesis (condensation of carbamyl phosphate and aspartic acid to form carbamyl aspartic acid). The enzyme responsible, aspartic transcarbamylase, is allosterically inhibited (bold arrow) by cytidine triphosphate, the eventual product of the biosynthetic sequence.

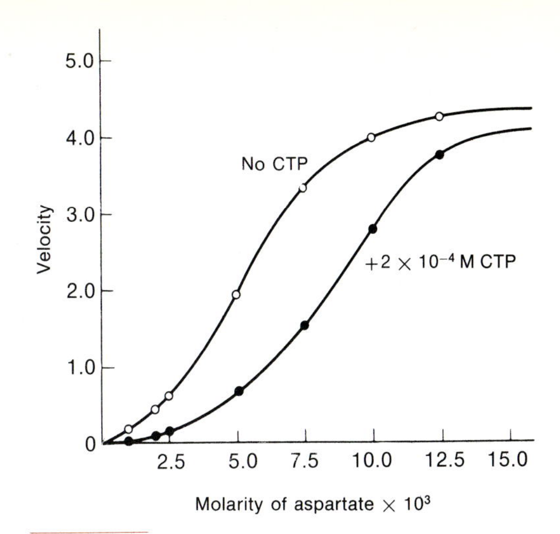


FIGURE 12.5

The rate of reaction of aspartic transcarbamylase as a function of the concentration of one of its substrates, aspartic acid. Note the sigmoid nature of the curve. The effect of the allosteric inhibitor, CTP, on aspartic transcarbamylase activity is also shown. Redrawn from J. C. Gerhart and A. B. Pardee, "The Enzymology of Control by Feedback Inhibition," *J. Biol. Chem.* **237,** 891 (1962).

changing the rate at which specific messenger RNAs are synthesized.

The most thoroughly studied allosteric proteins are the allosteric enzymes, exemplified by aspartic transcarbamylase (ATCase). ATCase catalyzes the first reaction in the pathway of biosynthesis of pyrimidines (Figure 12.4). Its activity is inhibited by an end product of the pathway, cytidine triphosphate (CTP). Thus, elevated intracellular concentrations of CTP inhibit the functioning of ATCase and consequently the formation of more CTP until its concentration decreases to an optimal level. The purine nucleotide, ATP, a second effector of ATCase, *activates* the enzyme, and thus serves to coordinate the synthesis of purine and pyrimidine nucleotides.

The rate of the reaction catalyzed by ATCase is a sigmoid function of substrate concentration (Figure 12.5), rather than a hyperbolic function which is typical of nonallosteric enzymes (Figure 12.6). Such kinetics are frequently associated with allosteric enzymes. The specific action of the allosteric inhibitor (also shown in Figure 12.5) is to increase the sigmoid nature of the curve, and hence to reduce the rate of reaction at low concentrations of substrate; such inhibition is largely reversed by increasing the concentration of the substrate. The sigmoid nature of the curve relating activity to substrate concentration also shows that the enzyme has more than one site able to bind substrate molecules (*catalytic sites*). The attachment of a substrate molecule to one of these sites increases the ability of the enzyme to bind additional molecules of substrate at other catalytic sites; i.e., there is a *cooperative interaction* of substrate molecules with the enzyme. Thus, as substrate concentrations are increased, the *rate of acceleration* of the velocity of the reaction increases. The same relationship obtains for effector molecules; they too interact cooperatively with the enzyme at the specific *effector sites.*

FIGURE 12.6

The rate of reaction of a typical nonallosteric enzyme (nucleoside diphosphokinase) as a function of the concentration of one of its substrates, ATP. Note the hyperbolic nature of the curve. Redrawn from C. L. Ginther and J. L. Ingraham, "Nucleoside Diphosphokinase of *Salmonella typhimurium*," *J. Biol. Chem.* **249,** 3406 (1974).

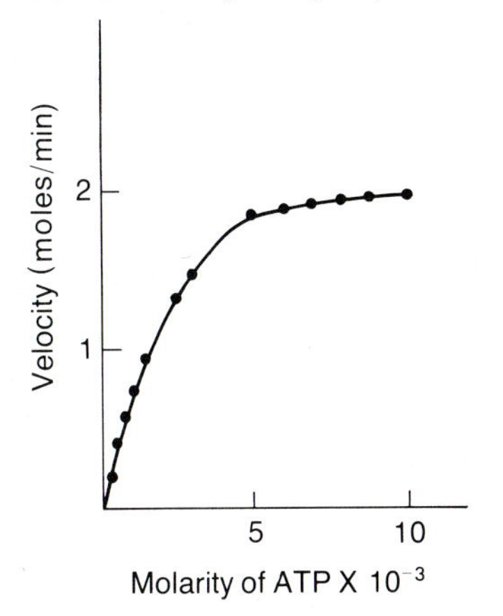

As might be suggested by the complexity of the curves shown in Figure 12.5, allosteric enzymes are almost always proteins of relatively high molecular weight composed of multiple subunits. As a rule, these subunits are identical, each possessing both a catalytic and an allosteric site. However, ATCase is composed of two different kinds of subunits, one with catalytic function and the other with regulatory functions. This fact makes it particularly easy to show that the allosteric and catalytic sites are topologically distinct. Upon mild chemical treatment (e.g., with *p*-Cl-mercuribenzoate), ATCase dissociates into subunits. One (the catalytic subunit) possesses all the catalytic activity of the intact enzyme, but is insensitive to allosteric inhibition by CTP or to allosteric activation by ATP. The other (the regulatory subunit) has no catalytic activity, but has the capacity to bind either CTP or ATP (Figure 12.7). Thus, binding CTP to one subunit inhibits the specific, enzyme-catalyzed reaction, which occurs on the other. The precise mechanism of such allosteric inhibition remains unexplained, but it evidently involves a conformational change in the enzyme. When the concentration in the cell of the end product (effector) of a given biosynthetic pathway rises, the catalytic activity of the allosteric enzyme with which it combines is reduced. Since the activity of this enzyme in turn controls the rate of biosynthesis of the end product (effector), the formation of the latter is also reduced and its intracellular concentration begins to fall. Therefore, allosteric inhibition also decreases. Through this device of *feedback regulation*, termed *end-product inhibition*, the intracellular concentrations of biosynthetic intermediates are very closely controlled. Typically, the enzyme that mediates the first reactions of a given biosynthetic pathway is the specific target of inhibition by the end product (or products) of that pathway. It is evident that when the first enzyme of a specific pathway is the target of regulation, neither the end product nor the intervening intermediates in its formation can accumulate in the cell. By such regulation, metabolic intermediates also regulate the rates of operation of catabolic pathways.

## MECHANISMS OF CONTROL OF TRANSCRIPTION

In procaryotes, frequency of transcription is controlled by at least three different mechanisms: (1) by DNA-binding proteins, (2) by *attenuation*, and (3) by exchange of the sigma factor components of RNA polymerase. These mechanisms are discussed in the following sections.

### Transcription Control: DNA-Binding Proteins

Two proteins participate in the dissimilation of the disaccharide, lactose (glucose-$\beta$-*D*-galactoside): galactoside permease, which mediates its entry into the cell, and $\beta$-galactosidase, which catalyzes its

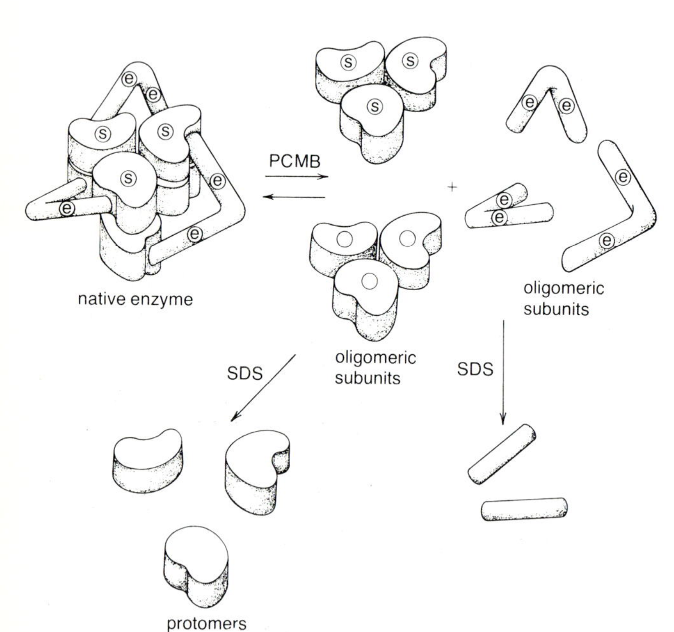


FIGURE 12.7

Dissociation of native aspartic transcarbamylase into two catalytic and three regulatory subunits (oligomers) by mild chemical treatment such as with *p*-Cl-mercuribenzoate (PCMB). Stronger chemical treatment such as with sodium dodecylsulfate (SDS) reveals that each catalytic subunit is comprised of three, and each regulatory subunit of two, polypeptide chains (protomers). Each catalytic protomer contains a single catalytic site, s, at which the substrates are bound, and each regulatory protomer contains a single regulator site, e, at which the effectors are bound. Catalytic subunits are catalytically active but insensitive to allosteric inhibition or activation. Regulatory subunits are catalytically inactive, but they retain the capacity to bind the allosteric effectors. Enzyme model is that of J. A. Cohlberg, V. R. Pigiet, and H. K. Schachman, "Structure and Arrangement of the Regulatory Subunits in Aspartate Transcarbamylase," *Biochemistry* **11**, 3393 (1972).

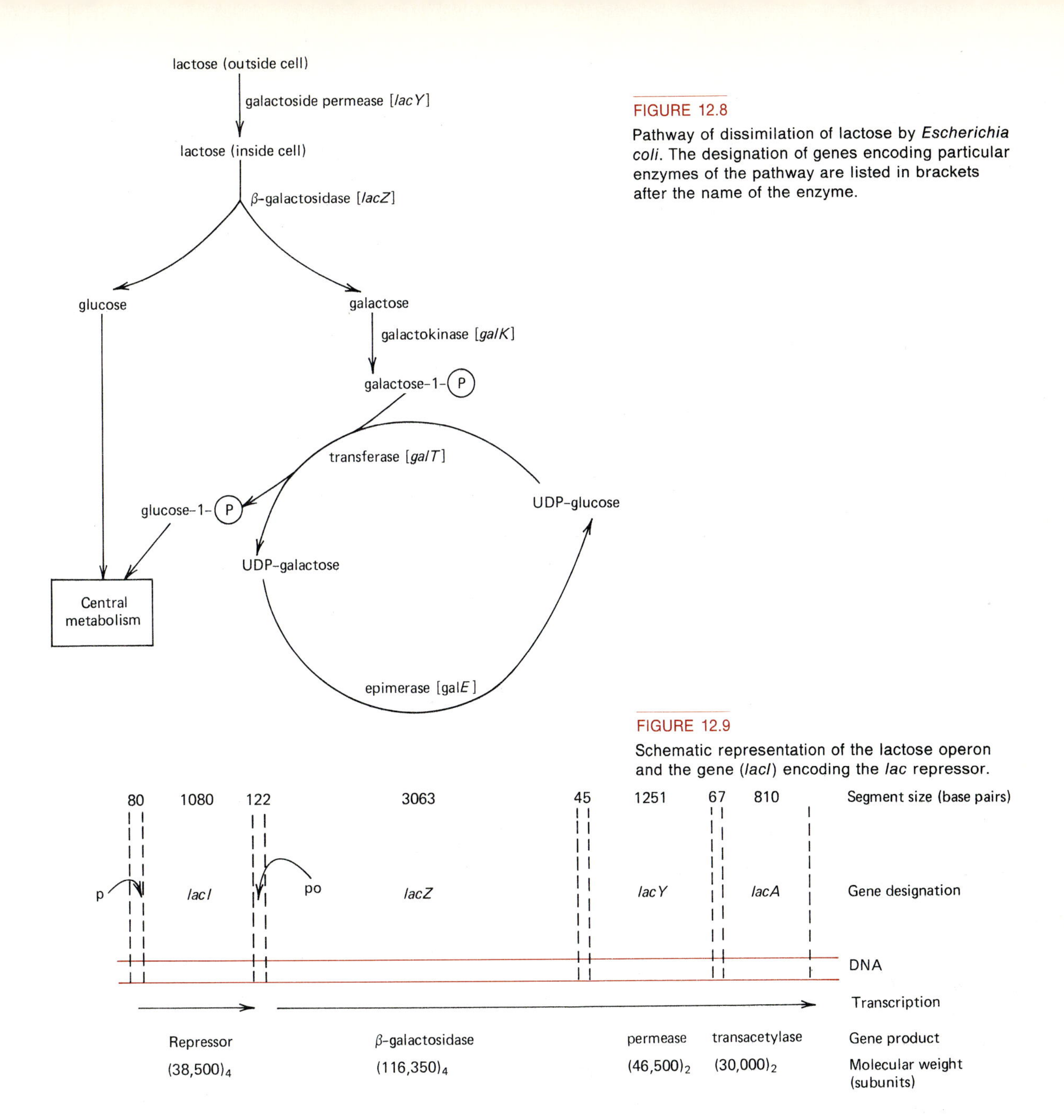


FIGURE 12.8

Pathway of dissimilation of lactose by *Escherichia coli*. The designation of genes encoding particular enzymes of the pathway are listed in brackets after the name of the enzyme.

FIGURE 12.9

Schematic representation of the lactose operon and the gene (*lacI*) encoding the *lac* repressor.

hydrolytic cleavage into two hexoses, glucose and galactose. These are metabolized further by other enzyme systems (Figure 12.8). The *structural genes* encoding these two proteins, along with another that encodes galactoside transacetylase (an activity with unknown cellular function) constitute an operon (Chapter 10, Figure 12.9), the expression of which has been studied since the 1930s. The intracellular levels of β-galactosidase rise about 1,000-fold when the primary substrate, lactose, is added to the external medium, but if another readily metabolizable substrate is also present, enzyme levels remain low (Figure 12.10). Thus, the lactose operon is inducible and subject to catabolite repression. Both of these regulations are affected by DNA-binding proteins: the *lac repressor*, encoded by a gene, *lacI* which is closely linked to the *lac* operon but is not a part of it, has the sole metabolic function of mediating control by induction; CAP (*c*atabolite *a*ctivating *p*rotein)* encoded by a gene (*crp*) located on the chromosome at a considerable distance from the *lac* operon mediates catabolite repression of many inducible operons, one of which is *lac*.

* Sometimes CAP is termed CRP (*c*yclic AMP *r*eceptor *p*rotein).

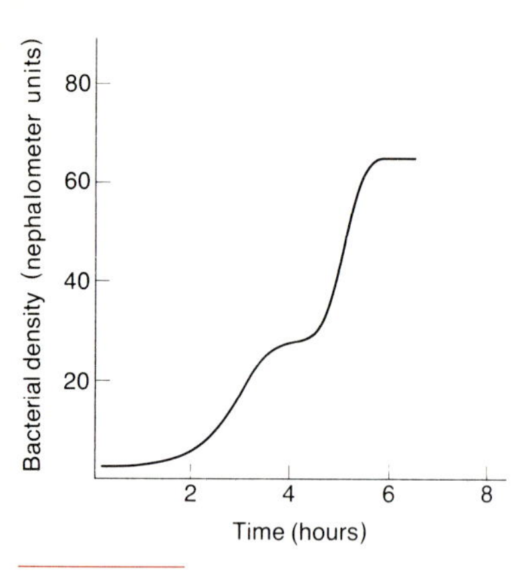


FIGURE 12.10

Growth of *E. coli* in a mineral salts medium initially containing equal quantities of glucose and lactose as sources of carbon. The transient cessation of growth after about four hours reflects the complete utilization of glucose. Synthesis of β-galactosidase and galactoside permease begins during the lag period and the second phase of growth occurs at the expense of the utilization of lactose. After J. Monod, *La Croissance des Cultures Bacteriennes* (Paris: Herman, 1942).

The immediate product of *lacI* aggregates to form a tetramer and in this form it has the capacity of binding with high affinity and specificity to a site, called the operator (*lacO*), which lies between the promoter and the structural genes of the operon (Figure 12.11). It will be noted that *lacO* overlaps the region of *lacP* to which RNA polymerase binds. Thus, when the repressor is bound at *lacO*, RNA polymerase cannot bind to *lacP* and transcription of the operon is inhibited. Direct competition between the *lac* repressor and RNA polymerase for DNA binding sites has been established by *in vitro* experiments; each protein inhibits binding of the other. But transcription initiates rapidly when the inducer is present. It binds to the repressor, thereby altering its conformation and its affinity for *lacO*. In this form the repressor dissociates from *lacO* allowing transcription. Within minutes, synthesis of β-galactosidase occurs at maximal rate.

Although lactose brings about induction of the *lac* operon, it is not the actual inducer that binds to the repressor. Rather, by a secondary activity of the low levels (*basal levels*) of β-galactosidase that are present even in noninduced cells, some lactose is converted to the actual inducer, *allolactose*.

It will be noted that the action of the *lac*-repressor is exclusively inhibitory or, as is commonly stated, *negative* in action. The free repressor inhibits transcription (and thereby enzyme synthesis); when bound to the inducer, its inhibitory capacity is virtually eliminated. Other DNA-binding proteins, termed *activators*, actively stimulate transcription; they are often described as being *positive* in action. CAP, the mediator of catabolite repression is an example of a DNA-binding protein that is positive in its action. This dimeric product of the *crp* gene is inactive unless bound to the small molecule effector, cyclic AMP (cAMP), but in association with it, CAP binds to a region of *lacP* (and the promoters of other genes or operons subject to catabolite repression) with the consequence that initiations of transcription are stimulated. It is not completely clear how the stimulatory effect is achieved, but the result is the conversion of an otherwise extremely weak promoter into quite a strong one. The intracellular levels of CAP remain quite constant under various conditions of growth. However, levels of cAMP are responsive to carbon nutrition. In the presence of a rapidly metabolizable source of carbon, like glucose, intracellular

FIGURE 12.11

Structure of promoter-operator region of the *lac* operon. After G. S. Stent and R. Calendar, *Molecular Genetics* (San Francisco: W. H. Freeman and Co., 1978).

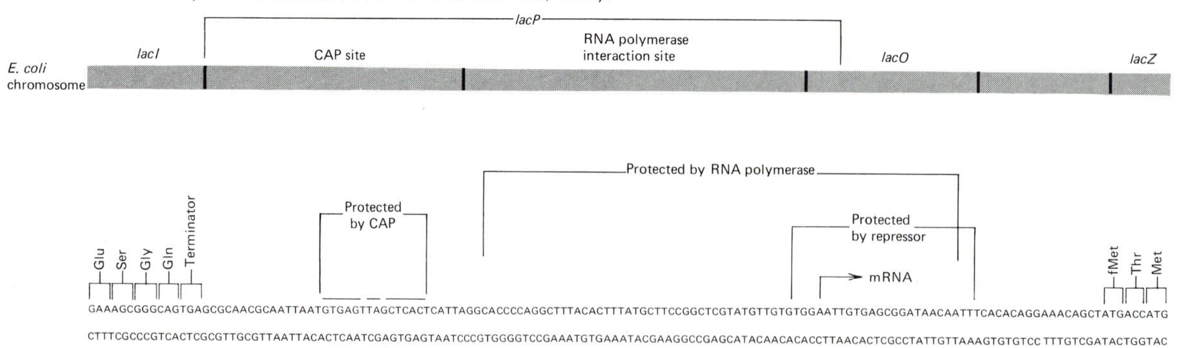

**TABLE 12.3**

**Comparison of the Mechanisms by Which DNA-Binding Proteins Mediate Control of the *lac* Operon and one of the *arg* Operons**

| | *lac* | *arg* |
|---|---|---|
| Function of encoded proteins | Catabolism of lactose, a carbon source | Biosynthesis of arginine, an amino acid |
| Regulatory gene | *lacI* | *argR* |
| Effector | Allolactose | Arginine |
| Activity of: | | |
| Free repressor | Inhibits transcription | No effect |
| Repressor-effector complex | No effect | Inhibits transcription |
| Consequence | Enzymes produced (induced) when substrate (lactose) is present in medium | Enzymes not produced (repressed) when end product (arginine) is present in medium |

levels of cAMP are low; they rise in cells growing at the expense of a slowly metabolizable source, like succinate, or in cells that are starved for a source of carbon.* Before the mechanism was understood, the term *catabolite repression* was adopted because rapid catabolism appeared to repress synthesis of certain enzymes. Now it is clear that slow catabolism stimulates their synthesis. Nevertheless, the term catabolite repression remains in current use.

The key linkage between the rate of catabolism and the intracellular levels of cAMP remains unknown. Synthesis of cAMP is catalyzed by *adenylate cyclase* via the reaction

$$\text{ATP} \longrightarrow 3',5'\text{-cyclic AMP} + \text{Ⓟ—Ⓟ}$$

and its hydrolytic degradation,

$$\text{cAMP} + H_2O \longrightarrow \text{AMP}$$

by *cyclic phosphodiesterase*. It is not yet known whether the intracellular concentrations of the effector are set by modulating its rate of synthesis or its rate of degradation, but, at present, the latter seems more probable. However, as expected, inactivating mutations in either the genes (*cya*) that encodes adenylate cyclase or *crp* that encodes CAP, exert the same metabolic consequence: genes and operons under catabolite repression are expressed only at very low levels regardless of the type of carbon source being metabolized and of the presence or absence of inducers.

For some time, DNA-binding proteins were considered to be the exclusive route by which transcriptional control of gene expression was effected. This conviction gained credence as increasing numbers of genes were discovered that encode DNA-binding proteins. The mechanism of action of binding proteins has remarkable flexibility. When bound to DNA the protein can either stimulate or inhibit transcription, and the small-molecule effector can either stimulate or inhibit binding of the protein to DNA.

This flexibility can be illustrated by comparing the regulation of the *lac* and *arg* operons (Table 12.3). Both are under negative control of a repressor protein, but the former is expressed at high levels when the effector (the inducer, allolactose) is present and the latter when the effector (the *corepressor* arginine) is absent or present in low concentrations. This seemingly opposite effect of the repressor is a simple consequence of the fact that, in the case of the lactose operon, the free repressor binds to the operator and, in the case of an arginine operon, the repressor-effector complex binds to the operator.

Although the genes encoding enzymes of a pathway controlled by a DNA-binding protein are quite commonly clustered in a single operon, this is not always the case. The nine genes of the arginine biosynthetic pathway are widely scattered on the chromosome; four are organized into two operons and the other five occur singly. Each of these genes or operons is preceded by an operator region, and all are controlled by the same protein repressor (product of the *argR* gene). The repressor-arginine complex binds to all of them, thereby inhibiting transcription. Such dispersed genes subject to control by the same regulatory mechanism are said to constitute a *regulon.*

* The situation described here applies to *E. coli* and many other but not all bacteria. For example, certain species of *Pseudomonas* metabolize succinate more rapidly than glucose; in these organisms succinate exerts a more powerful catabolite repression.

## Transcription Control: Attenuation

The gradual realization that bacteria might modulate transcription by mechanisms not involving the participation of DNA-binding proteins came from the accumulated evidence that DNA-binding proteins were not involved in the transcriptional regulation of a number of biosynthetic pathways because mutant strains lacking these proteins could not be isolated. DNA-binding proteins regulate a number of biosynthetic pathways, the most thoroughly studied of which is the one leading to the biosynthesis of the amino acid tryptophan. In one important respect the regulation of this pathway is atypical because it is under dual control—by a repressive DNA-binding protein (encoded by *trpR*) and by a mechanism termed *attenuation*. Other pathways are regulated only by one or the other mechanism.

A clue to the mechanism of attenuation came from the fact that certain mutations in the region between the operator (site of binding of the *trpR* product tryptophan complex) and the first structural gene (*trpE*) of the operon caused increased transcription of the operon. The region was termed an *attenuator* (of transcription), and was subsequently shown to be the site at which control of expression of the structural genes can occur. Transcription that initiates at the promoter proceeds through 141 nucleotides of the region termed the *leader sequence* (*trpL*), to a point at the end of the attenuator sequence where termination of transcription might occur (Figure 12.12). If it does, the *trp* structural genes are not transcribed, so expression of the operon is repressed, an action that occurs with high frequency if the intracellular concentration of tryptophan is elevated.

Whether or not termination occurs is determined by the possible secondary structures that can form in the RNA molecule synthesized to that point. Two distinct mutually exclusive structures can form (Figure 12.13) only one of which, termed the *terminator hairpin*, causes termination. Thus expression of the *trp* genes depends on which of the two structures form, and this in turn is determined by the rate of translation of a previous region of the leader RNA into a short (14 amino acids) peptide, termed the *leader peptide*. The completed leader peptide serves no cellular function (indeed, it is rapidly degraded by intracellular proteases); only its synthesis serves a regulatory role. The leader peptide contains two sequential tryptophan residues that are central to this role. If

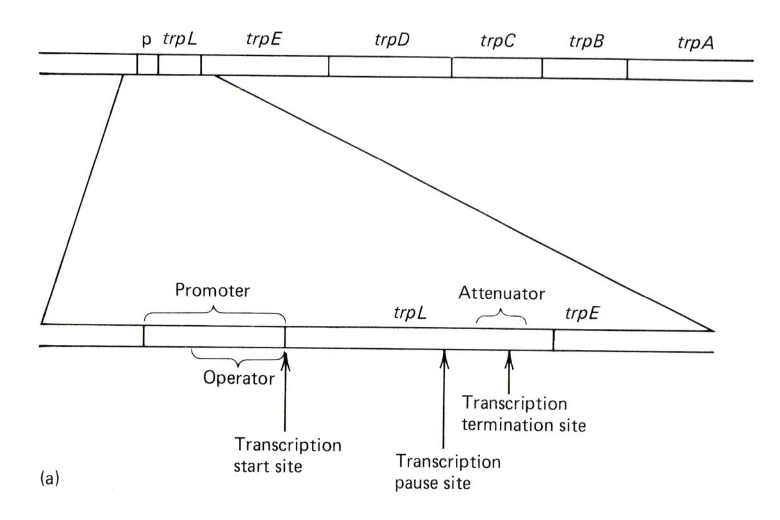


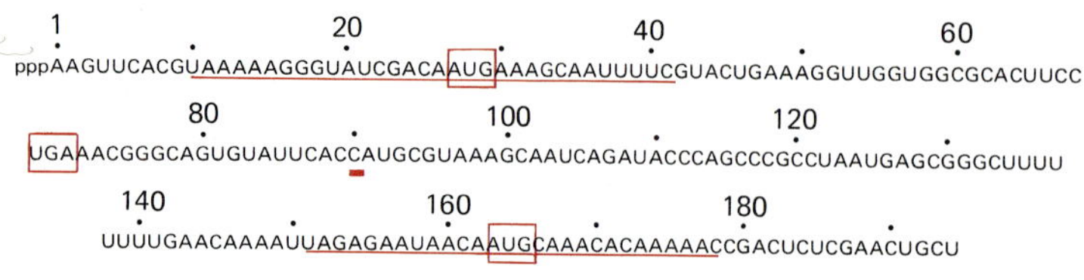


FIGURE 12.12

Structure of the *trp* operon. (a) The relationship between the regulatory elements, p and *trpL* to the structural genes *trpE*, *D*, *C*, *B*, and *A* that encode the enzymes of the pathway. (b) Sequence of the mRNA product of the regulatory region. Boxed triplets indicate start (AUG) codons for the *trpL* and *trpE* products and the stop (UGA) for *trpL*. The amino acid sequence for *trpL* and the first portion of *trpE* is shown in register with the mRNA. After C. Yanofsky, "Attenuation in the Control of Expressions of Bacterial Operons," *Nature* **289**, 751 (1981).

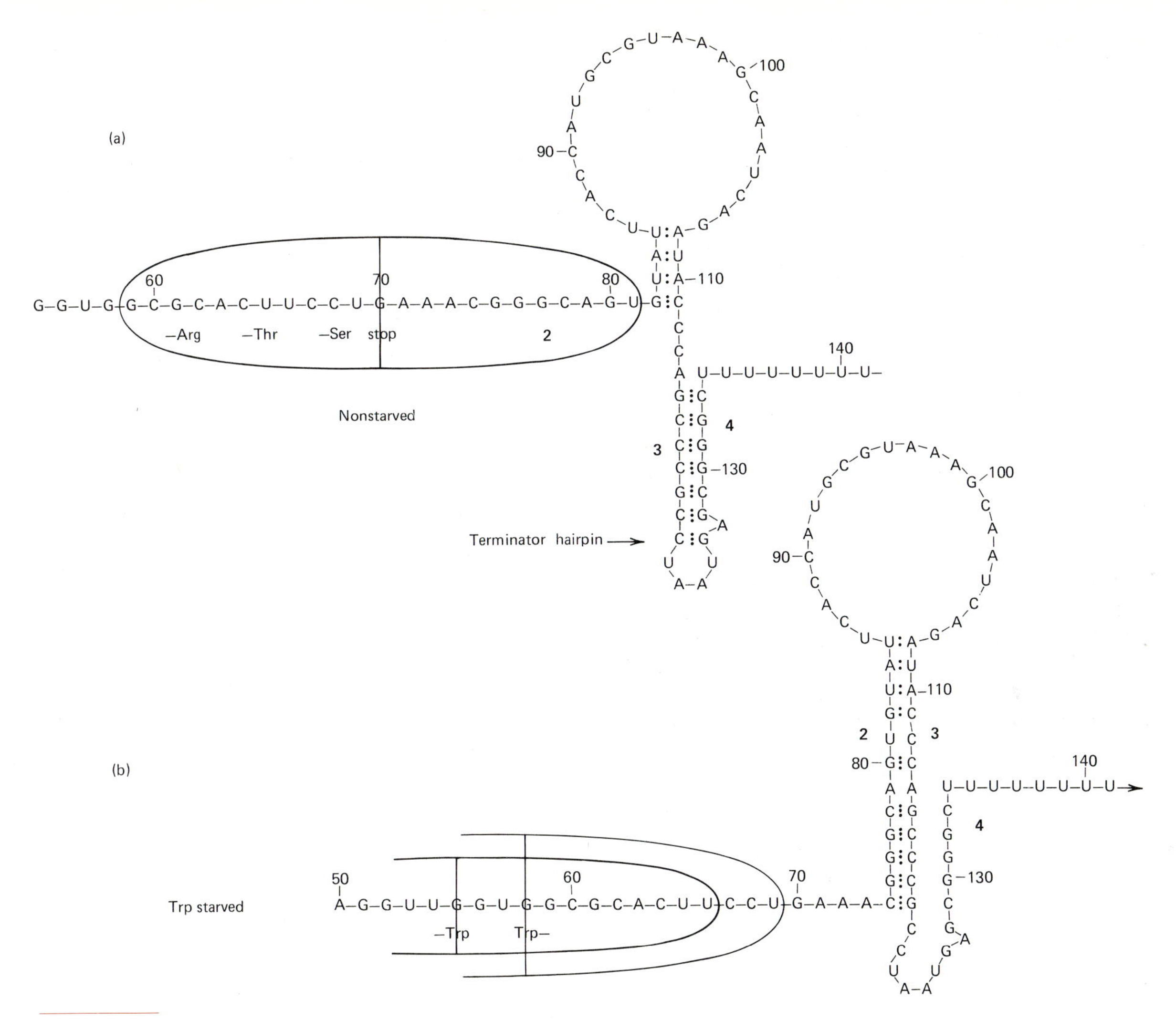


FIGURE 12.13

Secondary structures that form in trp mRNA. (a) If amino acids are present in adequate amounts (nonstarved), the ribosome (large oval) moves quickly to the stop codon of *trpL* preventing pairing between regions 2 and 3 and thereby allowing pairing between regions 3 and 4 (the terminator hairpin). (b) If trptophan is present in inadequate amounts (Trp starved) the ribosome pauses at one of the *trp* codons (UGG) allowing regions 2 and 3 to pair and thereby preventing formation of the terminator hairpin. After B. Lewin, *Genes*, (New York, Chichester, Brisbane, Toronto, and Singapore: John Wiley, 1983).

tryptophan, and therefore tryptophanyl-tRNA is present in low concentrations, translocation will pause at the UGG codons. In this position the ribosome overlaps region 1, preventing its pairing with region 2 so that, as transcription proceeds, region 2 is free to pair with region 3, thus preventing formation of the terminator hairpin (region 3–region 4) and allowing transcription of the structural genes to proceed. However, if tryptophan is present in adequate amounts the ribosome proceeds to the end of the coding region (the UGA nonsense codon) where it pauses before release. In this position it overlaps region 2, preventing its pairing with region 3 and thereby allowing region 3 to pair with region 4 forming the terminator hairpin that prevents continued transcription of the structural genes.

The two controls of tryptophan biosynthesis (repressor control and attenuation) function at different levels of deprivation of tryptophan. Because the rate of translation of the leader peptide is constant (and therefore the frequency of formation of the hairpin terminator is constant) over the range of adequate to high concentrations of tryptophan,

| Operon | Amino Acids in Leader Sequence | | | | | | | | | | | | | | | | | | | | | | |
|---|---|---|---|---|---|---|---|---|---|---|---|---|---|---|---|---|---|---|---|---|---|---|---|
| *pheA* | Met | Lys | His | Ile | Pro | PHE | PHE | PHE | Ala | PHE | PHE | PHE | Thr | PHE | Pro | | | | | | | | |
| *his* | Met | Thr | Arg | Val | Gln | Phe | Lys | HIS | HIS | HIS | HIS | HIS | HIS | HIS | Pro | Asp | | | | | | | |
| *leu* | Met | Ser | His | Ile | Val | Arg | Phe | Thr | Gly | LEU | LEU | LEU | LEU | Asn | Ala | Phe | Ile | Val | Arg | Gly | Arg | Pro | Val |
| | | | Gly | Gly | Ile | Gln | His | | | | | | | | | | | | | | | | |
| *thr* | Met | Lys | Arg | ILE | Ser | THR | THR | ILE | THR | THR | THR | ILE | THR | ILE | THR | ILE | THR | THR | Gly | Asn | Gly | Ala | Gly |
| | | | Gly | Ala | Ala | Leu | Gly | Arg | Gly | Lys | Ala | | | | | | | | | | | | |
| *ilv* | Met | Thr | Ala | LEU | LEU | Arg | VAL | ILE | Ser | LEU | VAL | VAL | ILE | Ser | VAL | VAL | VAL | ILE | ILE | ILE | Pro | Pro | Cys |

FIGURE 12.14

Amino acid sequences of the leader peptides of various amino acid operons. In all cases, they contain a central cluster of the amino acids (caps) that are the end product of the pathway. In the threonine (*thr*) pathway that leads also to synthesis of isoleucine (ILE) (Chapter 5), both amino acids are present in the cluster. The isoleucine-valine pathway (*ilv*) leads to the synthesis of three amino acids, leucine (LEU), isoleucine (ILE), and valine (VAL), all of which are present in the cluster. After C. Yanofsky, "Attenuation in the Control of Expression of Bacterial Operons," *Nature* **289,** 751 (1981).

attenuator control (causing derepression of the operon) functions only under conditions of virtual starvation for the amino acid. In the adequate to high range of tryptophan concentrations expression of the operon is appropriately set by the binding of the repressor-tryptophan complex to the operator region.

The amino acid composition of leader sequences of several other operons controlled by attenuation is shown in Figure 12.14. Several interesting variations are apparent. Rather than containing only two residues of the product amino acid, as is the case with tryptophan, most leader peptides contain considerably more residues. For example, the histidine leader contains seven histidine residues. The greater number probably reflects the fact that attenuation is the only mechanism that operates to control the histidine pathway and the greater number of residues allows the mechanism to operate over a greater range of concentrations of the end product.

The threonine leader peptide contains isoleucine residues as well as threonine residues allowing control by both amino acids, a mechanism that is advantageous to the cell because isoleucine is synthesized from threonine (Chapter 5). Similarly, the leader sequence of the isoleucine-valine operons contains isoleucine, valine, and leucine, presumably because the encoded enzymes participate in the biosynthesis of all three amino acids.

### Transcription Control: Multiple Sigma Factors

As discussed in Chapter 5, the sigma factor, which associates loosely with RNA polymerase, is largely responsible for the specificity of its binding to promoters. Unlike *E. coli*, which possess a single sigma factor,* *Bacillus subtilis*, which undergoes morphogenesis late in the growth cycle when endospores develop within many vegetative cells, has the capacity to produce at least four sigma factors (sigma$^{28}$, sigma$^{29}$, sigma$^{37}$, and sigma$^{55}$ (designated by a superscript that indicates their molecular weight $\times 10^{-3}$); the complement of these changes at various stages of growth. Thus *Bacillus subtilis* produces RNA polymerases with differing promotor specificities, the proportion of which modulates gene expression.

## CONTROL OF TRANSLATION

Synthesis of the 52 different proteins contained in the 70S ribosome is regulated at the level of translation. They are encoded in a number of different operons, each of which is regulated by one of its protein products (Figure 12.15). That product binds, at a slightly lesser affinity than it binds to rRNA in the process of ribosomal assembly, to the polygenic message at a site near the ribosome binding site. As a consequence, a ribosome cannot bind to the message and translation of the message cannot occur.

This mechanism assures that the synthesis of ribosomal proteins is precisely coordinated with the need for them to be assembled into ribosomes. If a site is available on a partially assembled ribo-

* Preliminary evidence suggests that there might be exceptions to this long-held generalization. When *E. coli* is exposed to elevated temperatures, a set of proteins, termed *heat-shock proteins*, is rapidly synthesized. A substitute sigma factor synthesized under these conditions may account for the high differential rate of synthesis of these proteins. Also, it appears that *nitrogen control* (synthesis of certain enzymes at high differential rates when $NH_4^+$ is in short supply) might be effected by a substitute sigma factor.

some, the protein binds there ; if it is not, the excess protein molecules bind to the mRNA, preventing the further synthesis of the protein. Since the protein's affinity for rRNA exceeds that for mRNA, synthesis of ribosomal proteins reinitiates as soon as there is a demand for them.

Thus, there are some similarities and some differences between control of transcription and translation by binding proteins. In both cases, control depends on binding of a protein to a specific site on a nucleic acid—DNA in transcriptional control and mRNA in translational control. In both cases the bound protein prevents attachment of a catalyst to its functional site—RNA polymerase to the promoter in the case of transcription control and ribosome to the ribosomal binding site in the case of translation control. But the determination of whether or not the protein binds is quite different. In cases of transcription control, the binding protein is an allosteric protein, the activity of which with respect to DNA binding is regulated by a small molecule, the inducer or corepressor. In contrast, ribosomal proteins have no allosteric properties; they always have the capacity to bind to mRNA and do so whenever their intracellular concentration exceeds the amount needed for assembly of ribosomes.

## POSTTRANSLATIONAL CONTROL

Posttranslational control involves the enzymic modification of a protein that alters its activity. Examples are common among eucaryotes. Some eucaryotic enzymes are synthesized as inactive proenzymes that become activated, when required, by proteases that cleave them at specific sites. The activity of many other eucaryotic proteins is altered by the action of *protein kinases* that catalyze the transfer of a $\gamma$-phosphate group from ATP to the protein.

Examples of posttranscriptional control in procaryotes are much less common, but the best-studied one is the control of the activity of glutamine synthetase, an enzyme that plays a pivotal role in the nitrogen metabolism of many bacteria. It catalyzes the reaction

$$\text{glutamate} + \text{ATP} + \text{NH}_4^+ \xrightarrow[\text{synthetase}]{\text{glutamine}} \text{glutamine} + \text{ADP} + \text{Ⓟ}$$

which is the only one by which glutamine can be synthesized. If this amino acid is not present in the medium, the reaction must function to satisfy the requirement of protein synthesis and a number of other biosynthetic reactions in which glutamine

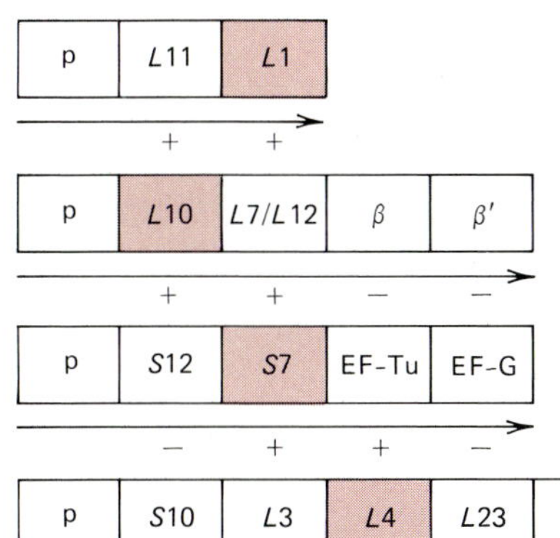


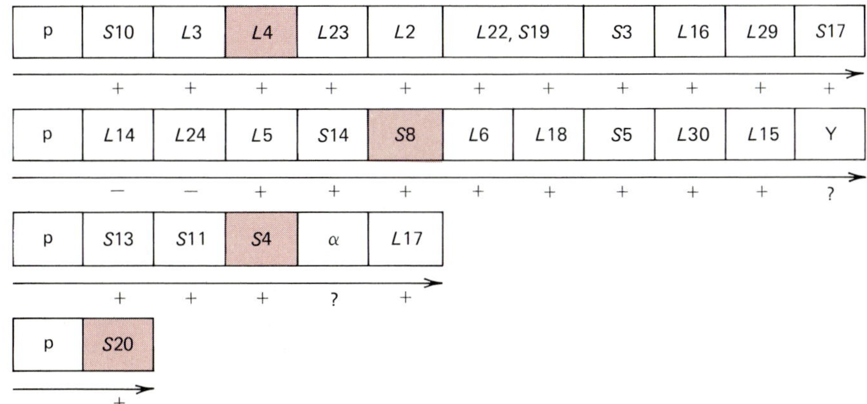


FIGURE 12.15

Translational control of expression of ribosomal RNA operons. Horizontal rows of boxes represent individual operons read from the promoter, p, at the left. In each case one of the encoded proteins (colored box) in each operon acts as a translational repressor of the operon. Proteins under or not under this control are designated by a "+" or "−" below the box. Protein components of the 30S ribosomal subunit are designated by "S" (small) followed by a number; those in the 50S subunit are designated by "L" (large) followed by a number. Other proteins that participate in protein synthesis are also encoded in those operons: $\alpha$, $\beta$, and $\beta'$, subunits of RNA polymerase; EF-Tu, EF-G, elongation factors; and Y, a protein that participates in protein secretion. After M. Nomura, "The Control of Ribosome Synthesis," *Scientific American* **250,** 102 (1984).

participates as an obligatory nitrogen donor. But the majority of organic nitrogen in a cell is derived from glutamate via transamination; hence, the major flow from ammonia to organic nitrogen passes through glutamate, which can be synthesized either from glutamine by the glutamate synthase reaction

$$\text{glutamine} + \alpha\text{-ketoglutarate} \xrightarrow[\text{synthase}]{\text{glutamate}} 2\ \text{glutamate}$$

or by the ammonia-incorporating glutamate dehydrogenase reaction

$$\alpha\text{-ketoglutarate} + \text{NADH} + NH_4^+ \xrightarrow[\text{dehydrogenase}]{\text{glutamate}} \text{glutamate} + NAD^+ + H_2O$$

Thus the glutamine synthase reaction, in addition to being the sole route of synthesizing glutamine, also serves as part of an alternate route by which glutamate can be synthesized, and by which the major flow of ammonia into nitrogenous constituents of the cell can occur. However, the two major ammonia-incorporation reactions differ in important respects. The glutamine synthase reaction requires ATP and functions well at low ammonia concentrations; the glutamate dehydrogenase reaction does not require ATP but has only a relatively low affinity for ammonia. Therefore, the former must function as the major route of ammonia incorporation when the supply of ammonia is restricted, but the latter is a more economic one when the supply of ammonia is adequate.

Part of the complex set of mechanisms by which the activity of glutamine synthase is controlled is one mediated by posttranscriptional modification of glutamine synthase (Figure 12.16) by means of an enzyme that adds an adenyl group to it, rendering it less active when the supply of ammonia is high. Another enzyme hydrolytically removes this group when the concentration of ammonia decreases. The adenylylating and deadenylylating enzymes themselves are allosterically controlled.

FIGURE 12.16

Covalent modification of glutamine synthetase through adenylation is stimulated by high concentrations of ammonia and results in a reduction of activity of the enzyme. The adenylated form of glutamine synthetase is sensitive to inhibition by the end products of glutamine metabolism, whereas the unsubstituted form is considerably less sensitive.

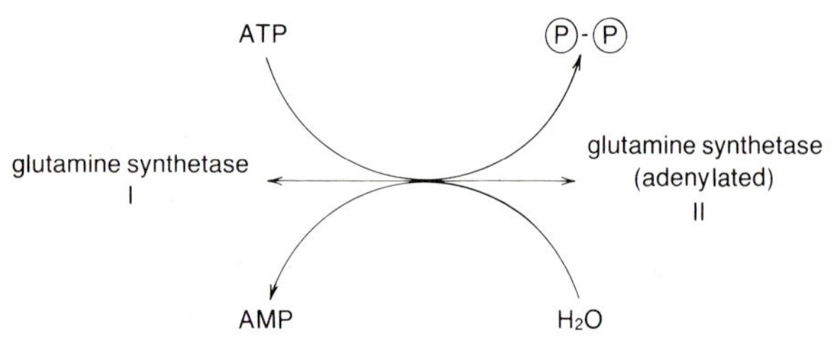


## ALTERATION OF GENE STRUCTURE

One example of a mechanism that alters gene structure in a manner that increases the ability of bacteria to survive in natural environments has been discovered and preliminary evidence suggests that a number of other similar mechanisms probably operate in procaryotes. The well-studied example changes the type of flagella synthesized by a *Salmonella typhimurium* cell. This mechanism is not a control mechanism in the sense that the alteration occurs in response to a change in the environment (although preliminary evidence suggests that some mechanisms that change gene structure might respond to specific environmental signals). Rather the change occurs randomly, but at a frequency that assures some cells will always be present that can survive in a particular hostile environment.

*Salmonella typhimurium* possess the genetic capacity to produce two types of flagella from nonallelic genes, but an individual cell produces only one type or the other: either the so-called H1 type or the H2 type. Cells producing the former are said to be in phase 1 and those producing the latter in phase 2. *Phase variation*, as the phenomenon of abrupt change from one phase to the other is termed, occurs about once in every 1,000 bacterial divisions. It is selectively advantageous to the organism in its parasitic state because flagella are actively antigenic, stimulating the host to produce antibodies that mediate the destruction of cells that produce the corresponding antigen. But not all the infecting bacteria will be eliminated, if a single type of antibody is produced by the host. Owing to phase variation, some cells will be in the phase that produces the other flagellar type thereby rendering these cells resistant to the action of the antibody.

The genes encoding H1- and H2-type flagellins (residing at different locations on the chromosome) are transcribed from different promoters (Figure 12.17). The H2-encoding gene lies in an operon that also contains a repressor that acts at the promoter of the H1-gene. Thus H1 is expressed only if H2 is not.

Expression of the H2-repressor operon is determined by the orientation of a 995–base pair segment that lies immediately adjacent to it (Figure 12.18) because the promoter lies within it. In the orientation encoding the H2 phase, transcription

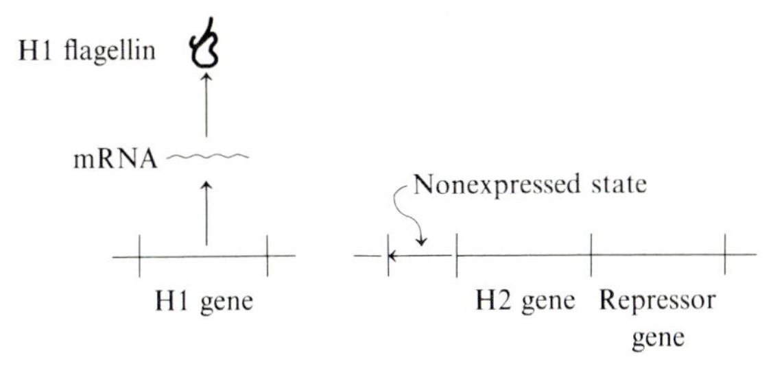


Phase 2

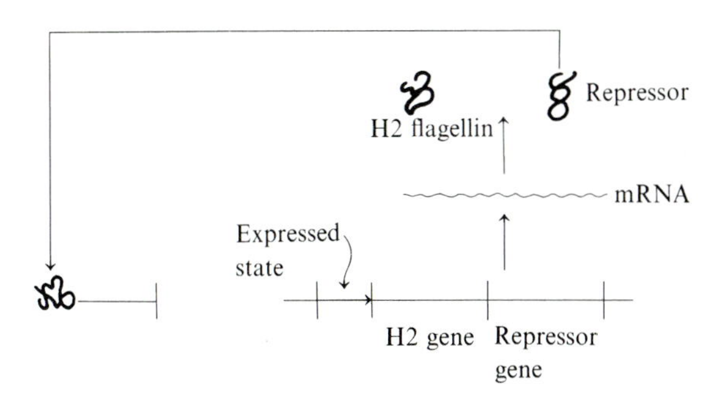


FIGURE 12.17 (left)

Regulation of phase variation. In phase 1, genes of the H2 operon are not expressed, and the H1 gene is. In phase 2 the H2 operon is expressed producing H2 flagellin and a repressor that prevents expression of the H1 gene.

Phase 2

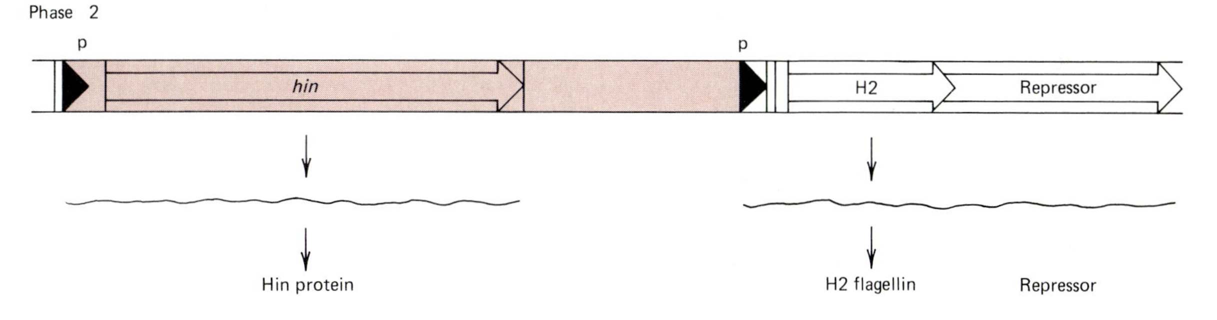


Phase 1

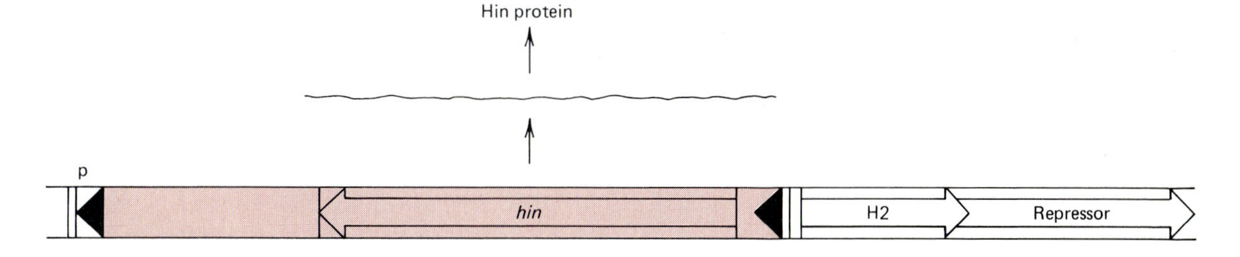


FIGURE 12.18 (below)

Regulation of expression of the H2 operon. In one orientation (phase 2) of the 995 base pair segment (colored) a promoter (p) is in the proper position for transcription of the H2 and repressor genes. In the other (phase 1) they are not transcribed. In either orientation, the *hin* gene is transcribed from its own promoter, p.

proceeds in the direction of structural genes; in the other, it probably proceeds in the opposite direction; H2 and repressor are not synthesized, so H1 is.

Inversion of the controlling segment is mediated by a protein, Hin, encoded by a gene, *hin*, that lies within the segment. Hin acts to invert the segment by catalyzing a recombination between the 14–base pair repeated regions that occur in inverted order at the ends of the sequence.

## PATTERNS OF REGULATION

Certain generalizations can be made about the regulation of metabolic pathways. In unbranched biosynthetic pathways, the first enzyme of the pathway is regulated by end-product inhibition and the biosynthesis of all enzymes of the pathway are subject to end-product repression. In catabolic pathways involving the metabolism of carbon sources that are commonly present in the environment of the organism, flow through the pathway is regulated largely by allosteric enzymes; the enzymes of the pathway are constitutively synthesized and are subject to minimal catabolite repression. Pathways of catabolism of substrates that occur more rarely in the environment of the organism are regulated primarily by enzyme induction and by catabolite repression, rather than by allosteric enzymes.

## End-Product Inhibition in Branched Pathways

Many biosynthetic pathways have two or more end products. End-product inhibition in such pathways is more complex than it is in simple, unbranched pathways. In a branched pathway leading, for example, to two different amino acids (Figure 12.19), it is obvious that feedback inhibition exerted by an end product (e.g., amino acid I) on the enzyme catalyzing the first step of the pathway (enzyme **a**) would likewise prevent synthesis of other end products (i.e., amino acid II). Consequently, the presence of amino acid I in the medium would effectively prevent endogenous synthesis of amino acid II, and growth would cease. In fact, feedback inhibition by the end product of a branched biosynthetic pathway is often exerted specifically on the enzyme that catalyzes the initial step following a metabolic branch point in the chain of reactions that leads specifically to its synthesis. Thus, amino acid I exerts feedback control on enzyme **d,** and amino acid II on enzyme **g.** The effective regulation of a branched biosynthetic pathway nevertheless requires feedback control of the initial enzyme, **a,** which catalyzes the first step of the common pathway (e.g., A → B → C).

FIGURE 12.19

Generalized scheme of a branched biosynthetic pathway leading to two essential metabolites (in this case, amino acids). Arrows indicate reactions catalyzed by enzymes (lowercase letters) producing biosynthetic intermediates (capital letters). Colored lines lead from end-product inhibitors to susceptible allosteric enzymes.

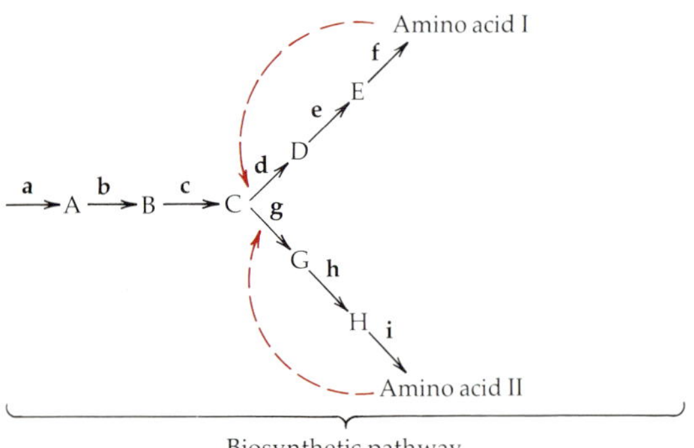


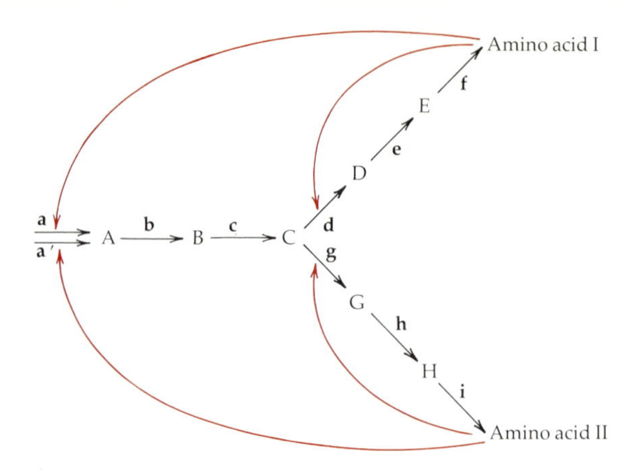


FIGURE 12.20

Scheme of regulation of a branched biosynthetic pathway by end-product inhibition of isofunctional enzymes. (Symbols are the same as in Figure 12.19.)

A number of different feedback mechanisms that permit this are known.

**1.** *Isofunctional enzymes.* The cell synthesizes two enzymes (**a** and **a′**) which have the same catalytic activity but are subject to feedback inhibition by different end products (Figure 12.20). When neither end product is present in the environment, the combined activities of enzymes **a** and **a′** produce a sufficient quantity of the intermediate A to meet the cellular demands for both end products. If one end product is present in the environment, synthesis of A is reduced as a result of the specific feedback inhibition of either **a** or **a′.** If both end products are present in the environment, the pathway ceases to function, since both **a** and **a′** are inhibited.

**2.** *Concerted feedback inhibition.* The reaction subject to control is catalyzed by a single enzyme, **a,** with two different allosteric sites, each of which binds one of the specific end products of the pathway (Figure 12.21). When only one of these sites is occupied by an effector, activity of the enzyme is not affected. However, when both effectors are bound to the enzyme, it becomes inactive. Concerted feedback inhibition exerts a somewhat imprecise control, since the rate of the reaction is unchanged if only one end product is present. However, it does prevent operation of the pathway when both end products are present.

**3.** *Sequential feedback inhibition.* The reaction subject to control is catalyzed by a single enzyme, **a,** for which the effector is not an end product of the pathway, but the intermediate (C) immediately preceding a metabolic branch point. Elevated concentrations of the end product (amino

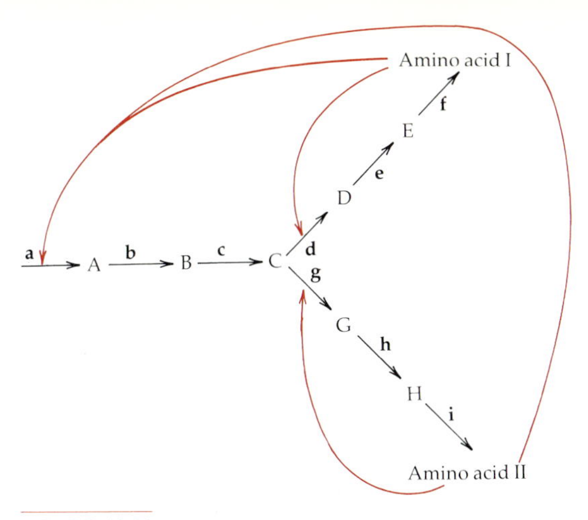


FIGURE 12.21
Scheme of regulation of a branched biosynthetic pathway by concerted feedback inhibition. (Symbols are the same as in Figure 12.19.)

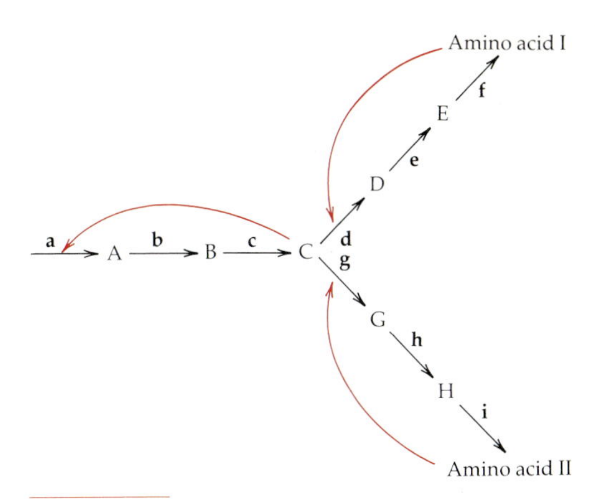


FIGURE 12.22
Scheme of regulation of a branched biosynthetic pathway by sequential feedback inhibition. (Symbols are the same as in Figure 12.19.)

acids I or II) inhibit enzymes **d** and **g,** and thus they cause a rise in the intracellular concentration of C. This, in turn, inhibits the activity of enzyme **a** (Figure 12.22).

**4.** *Cumulative feedback inhibition.* In certain branched pathways leading to multiple end products, a single allosteric enzyme has effector sites for all end products. Each end product (even at a high concentration) causes only partial inhibition of the enzyme, and inhibitory effects of the different end products are additive. Thus, the rate of the reaction mediated by enzyme **c** is determined by the number (and concentration) of different end products of the pathway that are present in the environment. This control is known as *cumulative feedback inhibition.*

**5.** *Combined activation and inhibition.* In certain cases, a biosynthetic intermediate formed by a specific reaction sequence enters two completely independent pathways, a situation illustrated by the role of carbamyl phosphate as a common intermediate in the synthesis both of arginine and of pyrimidines (Figure 12.23). In enteric bacteria the enzyme responsible for the synthesis of this intermediate, carbamyl phosphate synthetase, is allosterically inhibited by a metabolite of the pyrimidine pathway, UMP, and is allosterically activated by an intermediate of the arginine pathway, ornithine. If pyrimidines are available from the medium, the intracellular pool of UMP rises and carbamyl phosphate synthetase is inhibited. The resulting deple-

FIGURE 12.23
Regulation of carbamyl phosphate synthetase activity in enteric bacteria. Colored lines lead from inhibitors to susceptible allosteric enzymes. Jagged line (ornithine to carbamyl phosphate synthetase) indicates activation. The dual control of the activity of carbamyl phosphate synthetase through inhibition of uridylic acid (UMP) and activation by ornithine assures that proper amounts of carbamyl phosphate are synthesized under all growth conditions.

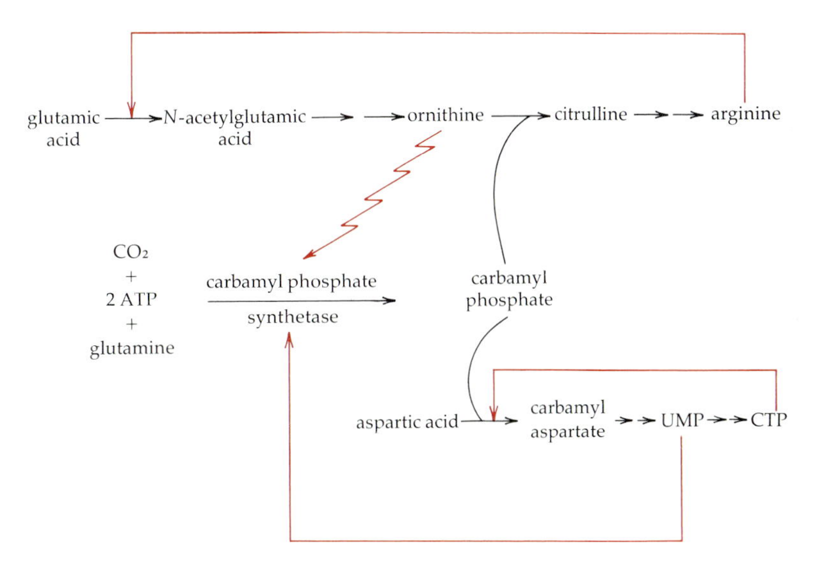

tion of carbamyl phosphate causes ornithine to accumulate which, in turn, activates the enzyme, thus making sufficient carbamyl phosphate available for the synthesis of the alternate end product, arginine. Alternatively, if arginine is available from the medium, the biosynthesis of ornithine is arrested as a consequence of the feedback inhibition by arginine of *N*-acetylglutamic acid synthetase. As a result, the intracellular concentration of ornithine falls, and the activity of carbamyl phosphate synthetase is decreased.

## Enzyme Repression in Branched Biosynthetic Pathways

The repression of enzyme synthesis in branched biosynthetic pathways is, like feedback inhibition, both complex and variable in mechanism. For instance, the synthesis of carbamyl phosphate synthetase by *E. coli* is partially repressed either by arginine or by cytidine triphosphate (CTP), and is fully repressed by both together. Thus, the synthesis of this key allosteric enzyme is regulated independently by two end products, neither of which affects the activity of the enzyme:

In the case of isofunctional enzymes subject to independent allosteric control by different end products, the synthesis of each enzyme is frequently controlled by the end product that inhibits its activity; an illustration of this will be described below.

## Examples of Regulation of Complex Pathways

In *E. coli* the conversion of aspartic acid to aspartyl phosphate is mediated by three isofunctional enzymes of which two (designed as **a** and **c** in Figure 12.24) also mediate the conversion of aspartic acid semialdehyde to homoserine. Enzyme **a,** possessing both these functions, is feedback inhibited and its synthesis is repressed by threonine. Enzyme **c,** which similarly possesses both functions, is inhibited and repressed by lysine. The third aspartokinase (enzyme **b**), is not subject to end-product inhibition, but its synthesis is repressed by methionine (Table 12.4).

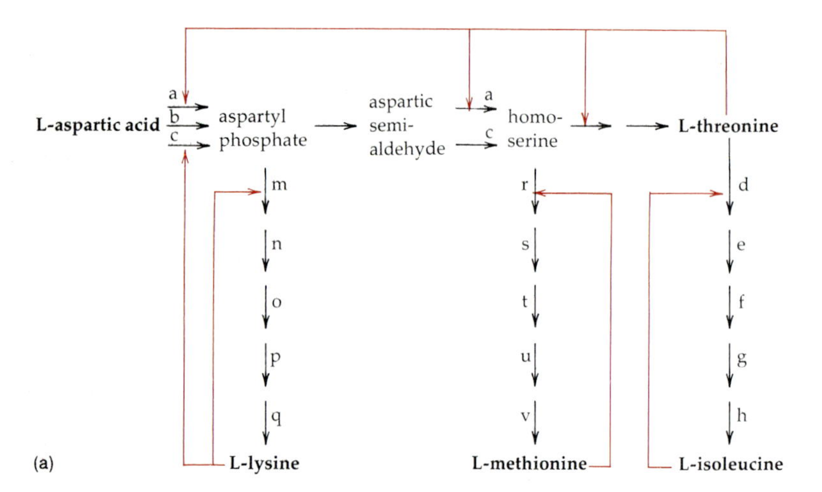


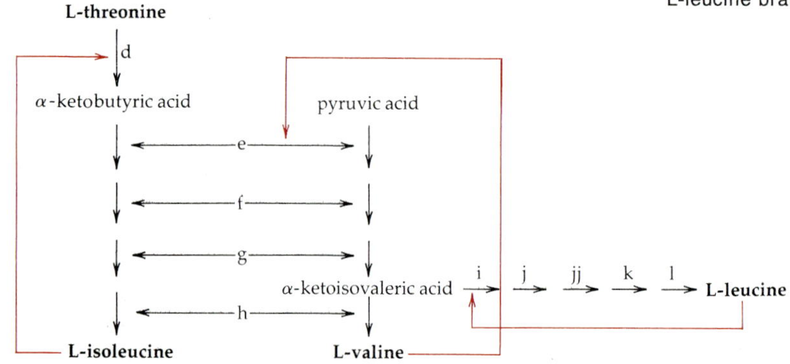


FIGURE 12.24

A simplified diagram of the aspartate pathway in *E. coli*. Each solid arrow designates a reaction catalyzed by one enzyme. The biosynthetic products of the pathway (in boldface) are all allosteric inhibitors of one or more reactions. Allosteric inhibitions are indicated by colored arrows. Careful study of this diagram reveals that with a single exception (the inhibition exerted by valine, see text) the inhibition imposed by one amino acid does not cause starvation for a different amino acid. Part (a) shows the regulatory interrelationships of the L-lysine, L-methionine, and L-isoleucine branches of the pathway. Part (b) shows the regulatory interrelationships of the L-isoleucine, L-valine, and L-leucine branches.

**TABLE 12.4**

**Control of the First Step of the Aspartate Pathway, Mediated by Three Different Aspartokinases, in the Bacterium *Escherichia coli***

| Enzyme | Corepressor | Allosteric Inhibitor |
|---|---|---|
| Aspartokinase I | Threonine and Isoleucine | Threonine |
| Aspartokinase II | Methionine | No allosteric control |
| Aspartokinase III | Lysine | Lysine |

The enzymes of the L-lysine branch (**m**–**q**) and the L-methionine branch (**r**–**v**) catalyze reactions leading in each case to a single end product and are subject to specific repression by that end product (L-lysine and L-methionine, respectively).

The third branch of the aspartate pathway is subject to much more complex regulation, for two reasons. First, L-threonine, formed through this branch, is both a component of proteins and an intermediate in the synthesis of another amino acid, L-isoleucine. Second, four of the five enzymes (**e**–**h**) which catalyze L-isoleucine synthesis from L-threonine, also catalyze analogous steps in the completely separate biosynthetic pathway by which L-valine is synthesized from pyruvic acid. The intermediate of this latter pathway, α-ketoisovaleric acid, is also a precursor of the amino acid L-leucine. These interrelationships are shown in Figure 12.24(b).

L-isoleucine is an end-product inhibitor of the enzyme, **d,** catalyzing the first step in its synthesis from L-threonine; this enzyme has no other biosynthetic role. L-valine is an end-product inhibitor of an enzyme (**e**) which has a dual metabolic role, since it catalyzes steps in both isoleucine and valine biosynthesis. In certain strains of *E. coli*, this enzyme is extremely sensitive to valine inhibition, with the result that exogenous valine prevents growth, an effect which can be reversed by the simultaneous provision of exogenous isoleucine.

**TABLE 12.5**

**Repressive Control of the Enzymes of the Isoleucine-Valine-Leucine Pathway (See Figure 12.24)**

| Enzymes | Corepressor |
|---|---|
| **d, e, f, g, h** | Isoleucine + valine + leucine |
| **i, j, jj, k, l** | Leucine |

The L-leucine branch of the valine pathway is regulated by L-leucine, which is an end-product inhibitor of the first enzyme, **i,** specific to this branch. These interrelationships are shown in Figure 12.24 (b).

As shown in Table 12.5, many of the enzymes that catalyze steps in the synthesis of L-isoleucine, L-valine, and L-leucine are subject to repression only by a mixture of the three end products, a phenomenon known as *multivalent repression.* However, the five enzymes specific to L-leucine synthesis are specifically repressed by this amino acid alone.

As previously mentioned, catabolic pathways mediated by constitutive enzymes are regulated exclusively by allosteric modulation of enzyme activity. This form of pathway control is schematized in Figure 12.25 for the pathway of glucose metabolism and glycogen synthesis in *E. coli.* Excess concentrations of the catabolic intermediates, fructose-1,6-diphosphate and phosphenolpyruvate, signal the diversion of carbon flow into glycogen.

## The Diversity of Bacterial Regulatory Mechanisms

With very few exceptions, biosynthetic pathways are biochemically identical in all microorganisms. However, a given pathway may be subject, in different organisms, to markedly different modes of regulation which tend to be group specific and probably indicate evolutionary relatedness. Certain examples of regulatory diversity in end-product inhibition are summarized in Table 12.6.

Although the members of a given bacterial group may all possess the same mechanism for the regulation of a particular enzyme (e.g., the isofunctional aspartokinases characteristic of the bacteria of the enteric group), a different kind of regulation may operate in the same group with respect to the key enzyme of another pathway (e.g., the regulation by multiple allosteric controls of a single carbamyl phosphate synthetase in the enteric bacteria).

Many catabolic pathways are likewise biochemically identical in a wide diversity of bacterial groups. Here, also, the patterns of regulation are both diverse and group-specific. The β-ketoadipate pathway, for example, serves for the oxidation of aromatic substrates in several groups of bacteria. It is mediated by inducible enzymes, the synthesis of which is mediated in a markedly different way in the *Pseudomonas* and *Acinetobacter* groups (Figure 12.26).

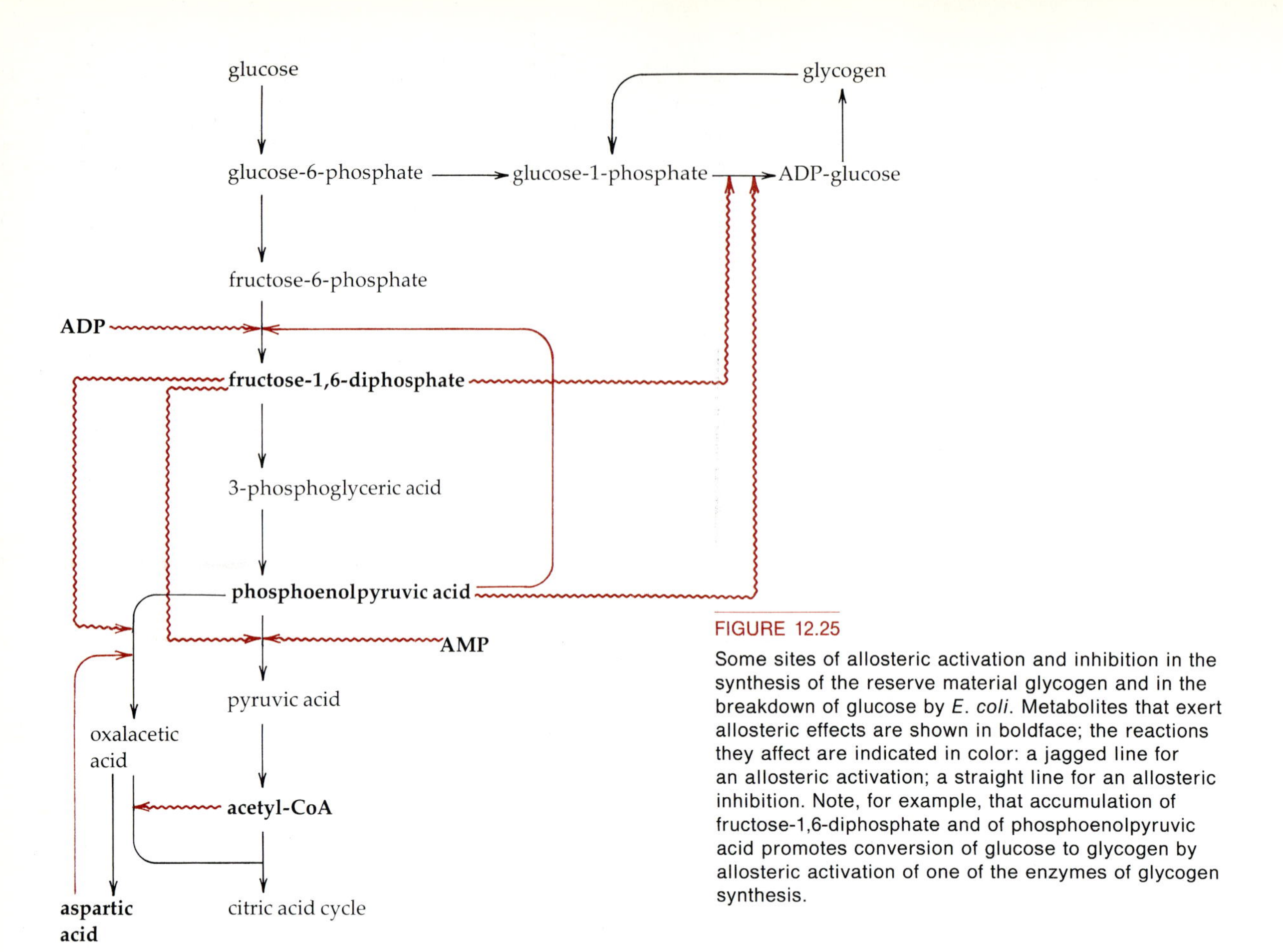


FIGURE 12.25

Some sites of allosteric activation and inhibition in the synthesis of the reserve material glycogen and in the breakdown of glucose by *E. coli*. Metabolites that exert allosteric effects are shown in boldface; the reactions they affect are indicated in color: a jagged line for an allosteric activation; a straight line for an allosteric inhibition. Note, for example, that accumulation of fructose-1,6-diphosphate and of phosphoenolpyruvic acid promotes conversion of glucose to glycogen by allosteric activation of one of the enzymes of glycogen synthesis.

**TABLE 12.6**

**Examples of Variation among Microorganisms in Feedback Regulation of Certain Key Enzymes in Branched Pathways**

| Enzyme | Organism | Type of Control |
|---|---|---|
| Aspartokinase | *E. coli* and other enteric bacteria | Isofunctional enzymes |
| | *Pseudomonas* spp. | Concerted feedback inhibition |
| 3-Deoxyarabinoheptulosonic acid-7-phosphate synthetase (first enzyme of aromatic amino acid pathway) | *E. coli* and other enteric bacteria | Isofunctional enzymes |
| | *Neurospora* | Isofunctional enzymes |
| | *Bacillus* spp. | Sequential feedback inhibition |
| | *Pseudomonas* spp. | { Concerted feedback inhibition; Cumulative feedback inhibition } |
| Carbamyl phosphate synthetase | *E. coli* and other enteric bacteria | Inhibition and activation |
| | *Pseudomonas putida* | Inhibition and activation |

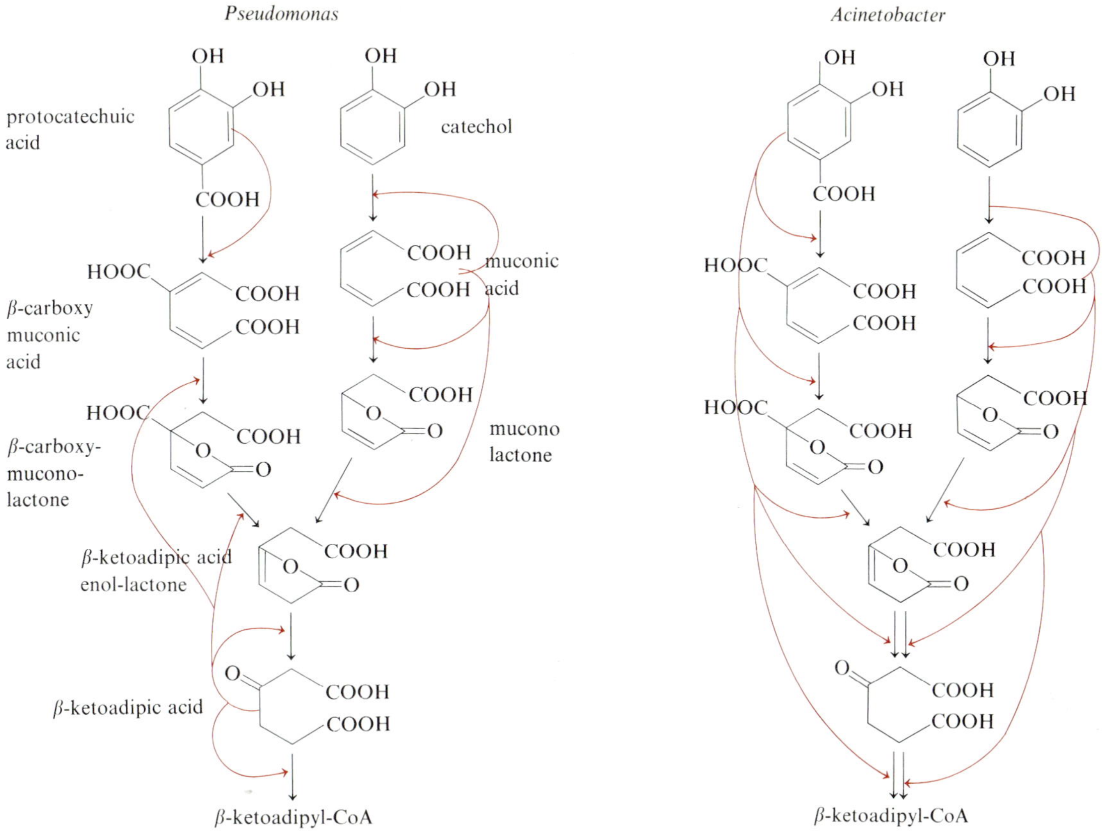

FIGURE 12.26

Comparison of the patterns of induction in *Pseudomonas* and *Acinetobacter* of the β-ketoadipate pathways of oxidation of protocatechuic acid and catechol. Colored arrows lead from inducer to induced enzyme(s). Parallel arrows indicate isofunctional enzymes.

## REGULATION OF DNA SYNTHESIS AND CELL DIVISION

In Chapter 5 the synthesis of DNA was described as a process in which the two strands of the double helix separate to produce two single strands, each of which serves as a template for the polymerization of deoxyribonucleotides. It will be recalled that the replication process creates a *fork* in the DNA molecule; as the DNA separates, the fork progresses, accompanied by replication of the two branches (Figure 5.38).

Regulation of DNA synthesis must be precisely controlled since at division each daughter cell receives at least one full complement of genetic material. Evidence that DNA synthesis and cell division are intimately interconnected comes from the observation that a wide variety of chemical treatments or mutational changes that inhibit DNA synthesis also inhibit cell division. Cells in which DNA synthesis has been inhibited elongate without division, eventually forming very long cells (Figure 12.27).

Data obtained largely through studies on *E. coli* indicate that segregation of DNA into daughter cells is a simple consequence of cell growth. The circular chromosome is attached to the cytoplasmic membrane. Following replication, the newly formed chromosome becomes attached to an adjacent site on the membrane. These sites are separated by interstitial membrane growth, and a cross wall is then formed between the two chromosomes (Figure 12.28). If, for any reason, chromosomal replication is not completed, cell division does not occur. The relationship between completion of replication of the chromosome and subsequent cell division appears to be causal; when chromosomal replication is completed, a series of metabolic events are initiated which eventually result in cell division. *Re-*

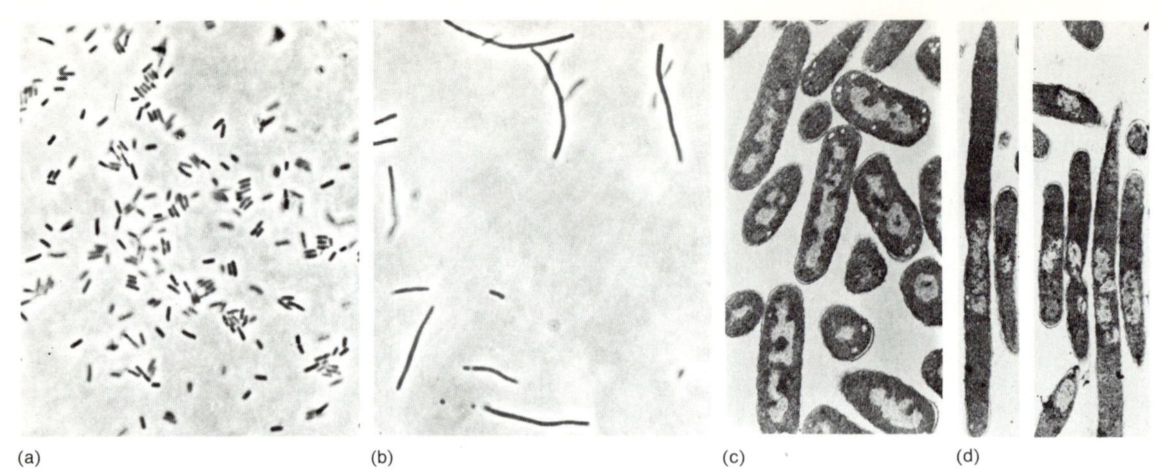


FIGURE 12.27 (above)

Phase (a), (b) and electron (c), (d) micrographs of *E. coli* B illustrating the consequences of inhibition of DNA synthesis for cell division. Specific inhibition of DNA synthesis (in this case effected by addition of mitomycin C to an exponentially growing culture) allows cell growth to continue, but stops further cell division. Thus, the normal short rods (a) become highly elongated forms. Electron micrographs (c) before and (d) 3 hours after addition of mitomycin C show in fact that little or no increase of the nuclear material (light central regions of the cells) occurs in the presence of the drug. From H. Suzuki, J. Pangborn, and W. W. Kilgore, "Filamentous Cells of *Escherichia coli* Formed in the Presence of Mitomycin," *J. Bacteriol.* **93,** 684 (1967).

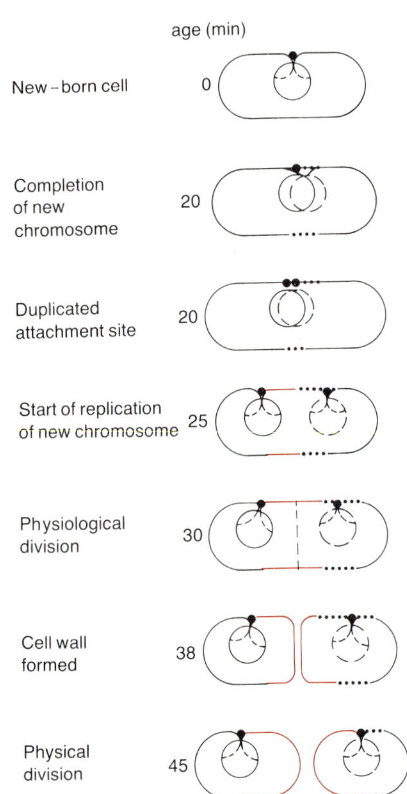


FIGURE 12.28 (right)

Diagram showing relationship between DNA replication, nuclear segregation, physiological division, and cell separation in *E. coli* growing with a doubling time of 45 minutes. Dotted lines (.....) indicate area of wall and membrane growth prior to duplication of attachment points. Dashed lines (---) indicated newly replicated portion of the chromosome. Colored lines indicate area of wall and membrane growth after formation of new attachment. Septum formation is indicated by a vertical dashed line (¦) at 30 minutes. After D. J. Clark, "The Regulation of DNA Replication and Cell Division in *E. coli* B/r," *Cold Spring Harbor Symp. Quant. Biol.* **33,** 825 (1968).

*gardless of growth rate, cell division in* E. coli *always occurs about 20 minutes after completion of chromosomal replication* (at 37° C).

Figure 12.28 also shows that chromosomal replication is bidirectional (i.e., two replication forks are formed which travel in opposite directions around the chromosome; when they meet, chromosomal replication is completed). In *E. coli,* the time required for a complete doubling of DNA (i.e., the time required for one of the pairs of forks to travel one-half the length of the molecule) is approximately 40 minutes at 37° C. If the organism grows in a medium in which the cycle of growth and cell division takes longer than 40 minutes, the cycle of DNA synthesis still occupies only 40 minutes; during the remainder of the division cycle, DNA synthesis stops. Therefore, DNA synthesis in slowly growing cells is discontinuous. In nonsynchronous cultures, however, the quantity of DNA in the total population increases continuously because the cell cycles are not in phase.

Thus, two events of fixed duration make up the division cycle: the time required to replicate the chromosome, which we will call *C*, and the time which elapses between completion of replication and cell division, which we will call D. In the case of *E. coli* growing at 37°C, *C* and *D* are about 40 and 20 minutes, respectively (except at very low growth rates). If the culture is growing with a doubling time approximately equal to *C* (Figure 12.29), a second round of chromosomal replication is initiated, and it continues during the *D* period of 20 minutes which precedes cell division. Under these conditions, each daughter cell receives a chromosome that contains two replication forks. At doubling times greater than 40 minutes, the portion of the *D* period during which no chromosome replication occurs increases. As a consequence, at the time of division the daughter cells receive chromosomes with smaller and smaller replication forks. At generation times greater than $C + D$ (viz., greater than 60 minutes), daughter cells receive chromosomes without replication forks, because reinitiation of chromosome replication always begins 60 minutes prior to the next cell division.

Thus, in cells growing with doubling times equal to or greater than *C*, the coordination of DNA synthesis with cell division is achieved by two means: (1) chromosomal replication is initiated at time intervals equal to the doubling time of the culture, and (2) initiation occurs 60 minutes prior to cell division. The only variable in this process is the time at which initiation of chromosomal replication occurs; *C* and *D* are invariant. In rich media in which *E. coli* is capable of growing at doubling times less than *C*, conditions 1 and 2 still hold. Because reinitiation begins at intervals equal to doubling time which is less than *C*, *multiple forks* are generated, thereby allowing a rate of DNA synthesis that is commensurate with the higher growth rate.

The relation between initiation of chromosomal replication and cell division at various growth rates is summarized in Figure 12.29. It can be seen that the critical factor is the *time* and *frequency* of initiation of a new round of DNA replication.

The initiation process requires that protein be synthesized; initiated cycles of DNA synthesis can be completed if protein synthesis is blocked, but *a new cycle cannot be initiated.* This has led to the hypothesis that initiation is under the positive control of a specific regulatory protein, the *initiator.* When the concentration of this initiator rises to a certain threshold level, initiation occurs, after which the initiator is destroyed. The time required for effective concentrations of the initiator to be synthesized is precisely equal to the doubling time of the culture. The initiator hypothesis is consistent with all known facts, but so far an initiator protein has not been identified.

FIGURE 12.29

Schematic diagram of the coupling between chromosome replication and cell division of *E. coli* growing at various doubling times. The circular chromosome (here represented as a linear molecule broken at the point of initiation of replication) is always replicated in 40 minutes (*C* time); 20 minutes later (*D* time) cell division occurs (vertical arrow). Initiation of replication (indicated by the terminal circles) always occurs 60 minutes ($C + D$) prior to division.

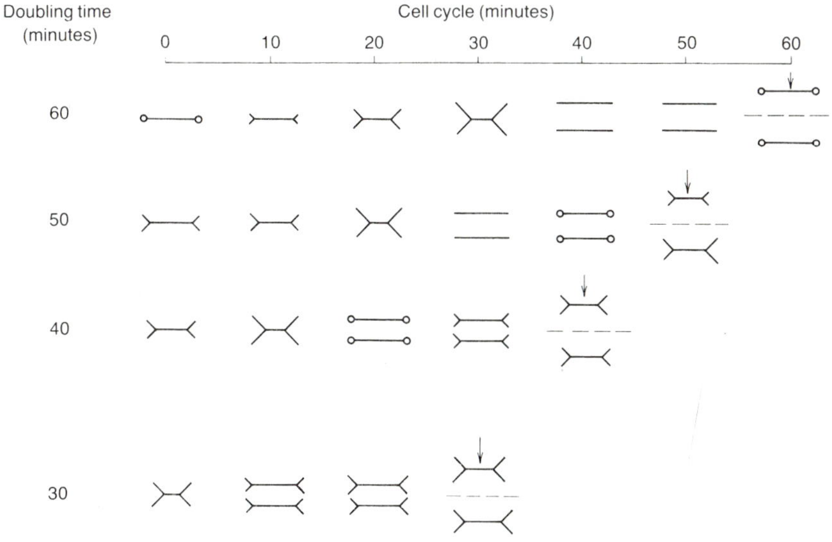

## FURTHER READING


**Books**

CLARK, B. F. C., and H. U. PETERSON, eds., *Gene Expression.* Copenhagen: Munksgaard, 1984.

LOSICK, R., and L. SHAPIRO, eds., *Microbial Development.* Cold Spring Harbor Laboratory, 1984.

MILLER, J. H. and W. S. REZNIKOFF, eds., *The Operon.* Cold Spring Harbor Laboratory, 1978.

**Reviews**

GALLANT, J., "Stringent Control in *E. coli*", *Ann. Rev. Genetics.* **13,** 393 (1979).

NOMURA, M., D. DEAN and J. L. YATES, "Feedback Regulation of Ribosomal Protein Synthesis in *Escherichia coli*", *Trends in Biochem. Sci.* **7,** 92 (1982).

ULLMANN, A. and A. DANCHIN, "Role of Cyclic AMP in Bacteria", *Adv. in Cyclic Nucleotide Res.* **15,** 1 (1983).

YANOFSKY, C. "Attenuation in the Control of Expression of Bacterial Operons," *Nature* **289,** 751 (1981).

# Chapter 13
# The Classification and Phylogeny of Bacteria

The art of biological classification is known as *taxonomy*. It has two functions: the first is to identify and describe as completely as possible the basic taxonomic units, or *species*; the second, to devise an appropriate way of arranging and cataloging these units.

## SPECIES: THE UNITS OF CLASSIFICATION

The notion of a species is complex. Speaking broadly, a species consists of an assemblage of individuals (or, in microorganisms, of clonal populations) that share a high degree of phenotypic similarity, coupled with an appreciable dissimilarity from other assemblages of the same general kind. The recognition of species would not be possible if natural variation were continuous, so that an intergrading series spanned the gap between two assemblages of markedly different phenotype. However, it became evident early in the development of biology that, among most groups of plants and animals, reasonably sharp discontinuities do separate the members of a group into distinguishable assemblages. Hence, the notion of the species as the base of taxonomic operation proved workable.

Every assemblage of individuals shows some degree of internal phenotypic diversity, because genetic variation is always at work. Hence, it becomes a matter of scientific tact to decide what *degree* of phenotypic dissimilarity justifies the breaking up of an assemblage into two or more species; or, to put the matter another way, how much internal diversity is permissible in a species. Opinions on this question vary. Taxonomists themselves can be broadly divided into two groups: "lumpers," who set wide limits to a species, and "splitters," who differentiate species on more slender grounds.

For plants and animals that reproduce sexually, a species can be defined in genetic and evolutionary terms. As long as a sexually reproducing population is free to interbreed at random, its total gene pool undergoes continuous redistribution, and new mutations, the source of phenotypic variation, are dispersed throughout the population. Such an interbreeding population may evolve in response to environmental changes, but it will evolve with reasonable uniformity. *Divergent evolution*, eventually leading to the emergence of new species, can occur only if a segment of the population becomes reproductively isolated in an evironment that is different from that occupied by the rest of the population. Reproductive isolation is probably usually geographic in the first instance; a physical barrier of some sort (for example, a mountain range or a body of water) is interposed between two parts of the initially continuous population. Within each of these subpopulations, a common gene pool is maintained by interbreeding, but through chance mutation and selection, the two subpopulations are now free to evolve along different lines. They will continue to diverge, as long as the geographical barrier persists. Eventually, the cumulative differences become so great that *physiological* isolation is superimposed on geographic isolation; members of the two populations are no longer capable of interbreeding if they are brought together. Hence, even if the two populations subsequently commingle once more, their gene pools remain permanently separated; a point of no return has been reached. These evolutionary considerations lead to a dynamic definition of the species as a stage in evolution at which actually or potentially interbreeding arrays have become separated into two or more arrays physiologically incapable of interbreeding. This definition is, in fact, an *explanation* of the origin of specific discontinuities in nature. At the same time, it provides an experimental criterion for the recognition of species differences: inability to interbreed.

Because most microorganisms are haploid, and reproduce predominantly by asexual means, the concept of the species that has emerged from work with plants and animals is evidently inapplicable to them. A microbial species cannot be considered an interbreeding population: the two offspring produced by the division of a bacterial cell are reproductively isolated from one another, and, in principle, they are free to evolve in a divergent manner. Genetic isolation is to some degree reduced by sexual or parasexual recombination in eucaryotic microorganisms and by the special mechanisms of recombination distinctive of bacteria. However, it is very difficult to assess the evolutionary effect of these recombinational processes, because the frequencies with which they occur in nature are unknown. In bacteria the problem is further complicated by plasmid transfer, which is relatively nonspecific, and permits exchanges of genetic material among bacteria of markedly different genetic constitution.

Since the dynamics of microbial evolution are so unlike the dynamics of evolution of plants and animals, there is no theoretical basis for the assumption that microbial evolution has led to phenotypic discontinuities that would justify the recognition of species. However, the experience of microbial taxonomists has shown that when many strains of a given microbial group are thoroughly analyzed, they can usually be divided into a series of discontinuous clusters: it is such *clusters of strains* that the microbial taxonomist recognizes empirically as species. Further insights into the dynamics of microbial evolution may eventually permit a formal definition of the microbial species; if so, this will most likely be different from the species definition applicable to plants and animals.

In bacterial populations, genetic change can occur so rapidly by mutation that it would be unwise to distinguish species on the basis of differences in a small number of characters, governed by single genes. Accordingly, the best working definition of a bacterial species is *a group of strains that show a high degree of overall phenotypic similarity and that differ from related strain groups with respect to many independent characteristics.*

## The Characterization of Species

Ideally, species should be characterized by complete descriptions of their phenotypes or—even better—of their genotypes. Taxonomic practice falls far short of these ideals; in most biological groups, even the phenotypes are only fragmentarily described, and genotypic characterizations are incomplete.

As a general rule, the phenotypic characters that can be most easily determined are structural or anatomical ones that can be directly observed. For this reason, biological classification is still based, at most levels, almost entirely on structural properties. Virtually the only exception is the classification of bacteria. The extreme structural simplicity of bacteria offers the taxonomist too small a range of characters upon which to base adequate characterizations. Hence, the bacterial taxonomist has always been forced to seek other kinds of characters—biochemical, physiological, ecological—with which to supplement structural data. The classification of bacteria is based, to a far greater extent than that of any other biological groups, on *functional* attributes. Most bacteria can be identified only by finding out what they can do, not simply how they look.

This confronts bacterial taxonomists with an additional problem. To find out what a bacterium can do, they have to perform experiments with it. The number of possible experiments that can be performed is extremely large, and although all will reveal facts, the facts so revealed will not necessarily be taxonomically significant ones, in the sense of contributing to a differentiation of the organism under study from related assemblages. Consequently, bacterial taxonomists can never be sure that they have performed the right experiments for taxonomic purposes; they may well have failed to perform certain experiments that would have shown them significant clustering in a collection of strains, and therefore erroneously conclude that they are dealing with a continuous series. There is no obvious way to get around this difficulty, except to make phenotypic characterizations as exhaustive as possible. However, an emerging alternative may soon resolve this dilemma; the molecular techniques for characterizing bacterial *genotypes* provide a possible objective basis for defining a bacterial species. These techniques are discussed later in this chapter.

## The Naming of Species

According to a convention known as the *binomial system of nomenclature*, every biological species bears a latinized name that consists of two words. The first word indicates the taxonomic group of immediately higher order, or *genus* (plural, *genera*) to which the species belongs, and the second word identifies it as a particular species of that genus. The first letter of the genetic (but not of the specific) name is capitalized, and the whole phrase is italicized: *Escherichia* (generic name) *coli* (specific name). In contexts in which no confusion is possible, the generic name is often abbreviated to its initial letter: *E. coli*.

A rigid and complex set of rules governs biological nomenclature; the rules are designed to keep nomenclature as stable as possible. The specific name given to a newly recognized species cannot be changed unless it can be shown that the organism has previously been described under another specific name, in which case the older name is used because it has priority. Unfortunately, the same stability does not govern the generic half of the name, since the arrangement of related species into genera is an operation that can be carried out in different ways and that often changes in the course of time as new information becomes available. For example, *E. coli* has in the past been placed in the genus *Bacterium*, as *Bacterium coli* and in the genus *Bacillus*, as *Bacillus coli*. These three names are synonyms, since they all refer to one and the same species. This consequence of the binomial system can be very confusing, and taxonomic descriptions usually list all such synonyms in order to minimize the confusion. Binomial nomenclature is used for all biological groups except viruses. The virologists are currently divided over the best way to designate members of this group; some wish to extend the binomial system to the viruses, whereas others would prefer another system, which gives in coded form information about the properties of the organism.

In bacterial taxonomy, when a new species is named, a particular strain is designated as the *type strain*. Type strains are preserved in culture collections; if one is lost, a *neotype strain*, which resembles as closely as possible the description of the type strain, is chosen. The type strain is important for nomenclatural purposes, since the specific name is attached to it. If other strains, originally included in the same species, prove on subsequent study to deserve recognition as separate species, they must receive new names, the old specific name resting with the type strain and related strains.

In the taxonomic treatment of a biological group, the individual species are usually grouped in a series of categories of successively higher order: genus, family, order, class, and division (or phylum). Such an arrangement is known as a *hierarchical* one, because each category in the ascending series unites a progressively larger number of taxonomic units in terms of a progressively smaller number of shared properties. It should be noted that the genus has a position of special importance, since according to the rules of nomenclature a species cannot be named unless it is assigned to a genus. The allocation of a species to a taxonomic category higher than the genus does not carry any essential *nomenclatural* information; it is merely indicative of the position of an organism, relative to other organisms, in the system of arrangement adopted.

## THE PROBLEMS OF TAXONOMIC ARRANGEMENT

In dealing with a large number of different objects, some system of orderly arrangement is essential for purposes of data storage and retrieval. It does not matter what criteria for making the arrangement are adopted, provided that they are unambiguous and convenient. Books can be arranged in different ways: for example, by subject, by author, or by title. Different individuals tend to adopt different systems, depending on their particular needs and tastes. Such a system of classification, based on arbitrarily chosen criteria, is termed an *artificial* one.

The earliest systems of biological classification were largely artificial in design. However, as knowledge about the anatomy of plants and animals increased, it became evident that these organisms conform to a number of *major patterns* or *types*, each of which shares many common properties, including ones that are not necessarily obvious upon superficial examination. Examples of such types are the mammalian, avian, and reptilian types among vertebrate animals. The first system of biological classification that attempted to group organisms in terms of such typological resemblances and differences was developed in the middle of the eighteenth century by Linnaeus. The Linnaean arrangement was more useful than previous artificial arrangements, since the taxonomic position of an organism furnished a large body of information about its properties: to say that an animal belongs to the vertebrate class Mammalia immediately tells one that it possesses all those properties which distinguish mammals collectively from other vertebrates. Because Linnaean classification expressed the *biological nature* of the objects that it classified, it became known as a *natural system* of classification, in contrast to preceding artificial systems.

### The Phylogenetic Approach to Taxonomy

When the fact of biological evolution was recognized, another dimension was immediately added to the concept of a natural classification. For biologists of the eighteenth century, the typological groupings merely expressed *resemblances*; but for post-Darwinian biologists, they revealed *relationships*. In the nineteenth century the concept of a "natural" system accordingly changed: it became one that grouped organisms in terms of their *evolutionary affinities*. The taxonomic hierarchy became in a certain sense the reflection of a family tree, and taxonomy suddenly acquired a new goal: the restructuring of hierarchies to mirror evolutionary relationships. Such a taxonomic system is known as a *phylogenetic system*.

### Numerical Taxonomy

An alternative approach is an empirical one: the attempt to base taxonomic arrangement upon *quantification* of the similarities and differences among organisms. This was first suggested by Michel Adanson, a contemporary of Linnaeus, and is known as *Adansonian* (or *numerical*) *taxonomy*. The underlying assumption is that, provided each phenotypic character is given explicit weighting, it should be possible to express numerically the taxonomic distances between organisms, in terms of the number of characters they share, relative to the total number of characters examined. The significance of the numerical relationships so determined is greatly influenced by the number of characters examined; these should be as numerous and as varied as possible, to obtain a representative sampling of phenotype.

Until recently, the Adansonian approach appeared impractical because of the magnitude of the numerical operations involved. This difficulty has been obviated by the advent of computers, which can be programmed to compare data for a large number of characters and organisms and to compute the degrees of similarity. For any pair of organisms, the calculation of similarity can be made in two slightly different ways (Table 13.1). The *similarity coefficient* $S_J$ does not take into account characters negative for both organisms, being based

**TABLE 13.1**

**The Determination of Similarity Coefficient and Matching Coefficient for Two Bacterial Strains, Both Characterized with Respect to Many Different Characters**

Number of characters positive in both strains: $a$

Number of characters positive in strain 1 and negative in strain 2: $b$

Number of characters negative in strain 1 and positive in strain 2: $c$

Number of characters negative in both strains: $d$

$$\text{Similarity coefficient } (S_J) = \frac{a}{a + b + c}$$

$$\text{Matching coefficient } (S_S) = \frac{a + d}{a + b + c + d}$$

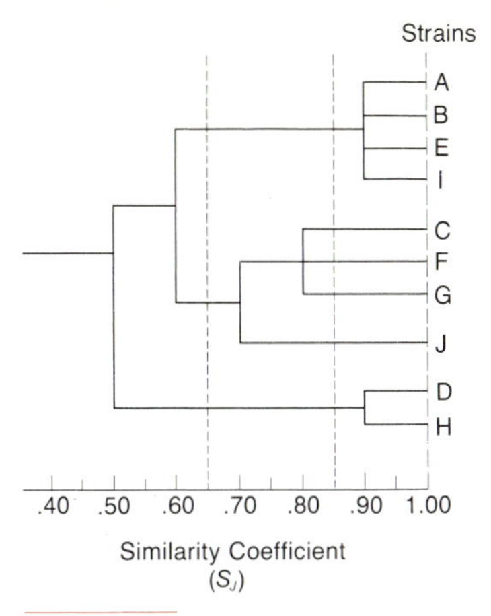


FIGURE 13.1

A dendrogram showing similarity relationships among 10 bacterial strains. The two dotted vertical lines indicate possible similarity levels at which successive ranks (e.g., genus and species) in the taxonomic hierarchy might be established. After P. H. A. Sneath, "The Construction of Taxonomic Groups," in *Microbial Classification*, ed. G. C. Ainsworth and P. H. A. Sneath (New York: Cambridge University Press, 1962).

only on positive matches; the *matching coefficient* $S_S$ includes both positive and negative matches in the calculation.

The data can then be transposed into a *dendrogram* (Figure 13.1), as a basis for determining taxonomic arrangement in terms of numerical relationships. The two dotted vertical lines in Figure 13.1 indicate similarity levels that might be considered appropriate for recognizing two different taxonomic ranks (e.g., genus and species).

Numerical taxonomy does not have the evolutionary connotations of phylogenetic taxonomy, but it provides an objective and stable basis for the construction of taxonomic groupings. Perhaps its greatest advantage is that it cannot be applied at all until a relatively large number of characters have been determined, so that its use encourages a thorough examination of phenotypes. The analyses are open to continuous revision and refinement as more characters in a given group are determined.

## NEW APPROACHES TO BACTERIAL TAXONOMY

The growth of molecular biology has opened up a number of new approaches to the characterization of organisms, which have had a profound impact on the taxonomy of bacteria. Of particular value are certain techniques that give insights into genotypic properties and thus complement the hitherto exclusively phenotypic characterizations of these organisms. Several kinds of analysis performed upon isolated nucleic acids furnish information about genotype: the analysis of the base composition of DNA, the study of chemical hybridization between nucleic acids isolated from different organisms, and the sequencing of nucleic acids.

### The Base Composition of DNA: Its Determination and Significance

DNA contains four bases: adenine (A), thymine (T), guanine (G), and cytosine (C). For double-stranded DNA, the base-pairing rules (see Chapter 5) require that A = T and G = C. However, there is no chemical restriction on the molar ratio (G + C):(A + T). Early in the chemical study of DNA, analyses showed that this ratio in fact varies over a rather wide range in DNA preparations from different organisms, and subsequent work has revealed that the base composition of DNA is a character of profound taxonomic importance, particularly among microorganisms.

Although DNA base composition may be determined chemically, after hydrolysis of a DNA sample and separation of the free bases, it can be determined more easily by physical methods, and these are now the ones principally used. The "melting temperature" of DNA (i.e., the temperature at which it becomes denatured, by breakage of the hydrogen bonds that hold together the two strands) is directly related to G + C content, because hydrogen bonding between GC pairs is stronger than that between AT pairs (a consequence of the fact that GC pairs form three hydrogen bonds, while AT pairs form only two). Strand separation is accompanied by a marked increase in absorbance at 260 nm, the absorption maximum of DNA, and this can be easily measured in a spectrophotometer. When a DNA sample is gradually heated, the absorbence increases as the hydrogen bonds are broken and reaches a plateau at a temperature at which the DNA has all become single-stranded (Figure 13.2). The midpoint of this rise, the melting temperature ($T_m$), is a measure of the G + C content. The G + C content of DNA may also be determined by subjecting a DNA sample to centrifugation in a CsCl gradient, and determining optically the position at which the DNA bands in the gradient, which affords a precise measure of its density (Figure 13.3). This method can be used because the density of DNA is also a function of the (G + C):(A + T) ratio.

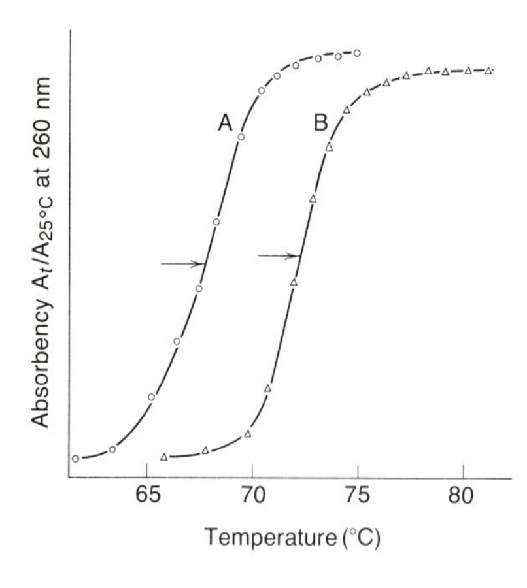


FIGURE 13.2

Melting curves determined optically for two samples of bacterial DNA. Curve A: DNA of *Lactobacillus acidophilus* ($T_m$ 67.7°C). Curve B: DNA of *Leptospira* sp. ($T_m$ 72.1°C). The ordinate expresses the absorbency at 260 nm of the DNA sample at each temperature on the abscissa, relative to its absorbence at 25°C. The midpoint of the absorbency increase (arrows) is the temperature ($T_m$) at which approximately half of the hydrogen bonds holding together the DNA double helices have been broken. This temperature is directly related to the G + C content of the DNA sample; the higher the G + C content, the higher the $T_m$. Data courtesy of M. Mandel.

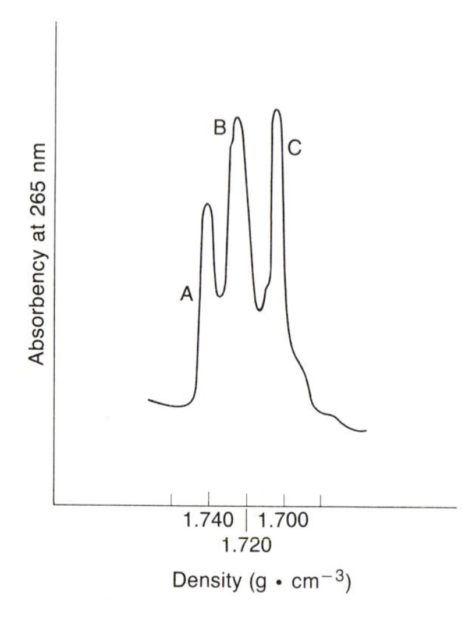


FIGURE 13.3

The positions in a CsCl density gradient assumed after centrifugation by three different DNAs of differing G + C content. (A) DNA of a bacteriophage of *Bacillus subtilis*, (B) DNA of *Thiobacillus novellus*. (C) DNA of *Leptospira* sp. Note that centrifugation sharply separates the three DNAs, each of which bands to a position in the CsCl density gradient that corresponds to its G + C content: the lower the G + C content of a given DNA, the lower the density at which it forms a band. The order of G + C content for the three DNAs is *B. subtilis* phage > *T. novellus* > *Leptospira* sp. Data courtesy of M. Mandel.

Physical methods of analysis also provide an indication of the *molecular heterogeneity* of a DNA sample. If every molecule of DNA had the same G + C content, both the thermal transition in a melting curve and the band position in a CsCl gradient would be extremely sharp. The steepness of the curve for thermal transition and the narrowness of the band in a gradient are therefore directly related to the homogeneity of G + C content in a population of DNA molecules. Even when DNA has been considerably fragmented by shearing (an unavoidable consequence of normal handling of large DNA molecules like the bacterial chromosome), preparations from most organisms remain relatively homogeneous by these criteria, which indicates that the mean G + C content varies little in different parts of the genome. The only major exceptions are preparations from organisms that contain two genetic elements of different G + C content. Thus, in preparations from certain eucaryotic organisms, DNA of mitochondrial or chloroplast origin may differ appreciably in G + C content from the nuclear DNA, and there is sometimes a marked molecular heterogeneity in the DNA of a bacterium that harbors a plasmid. In such cases, the minor constituent may form a distinct *satellite band* in a CsCl gradient; this phenomenon provided one of the clues that led to the discovery of DNA in mitochondria and chloroplasts.

Since no DNA preparation shows *absolute* molecular homogeneity, the G + C content is always a *mean* value and represents the peak in a normal distribution curve.

## The Taxonomic Implications of DNA Base Composition

The mean DNA base compositions characteristic of the nuclear DNA in major groups of organisms are shown in Figure 13.4. In both plants and animals the ranges are relatively narrow and quite similar, centering about a value of 35 to 40 percent G + C. Among the protists the ranges are much

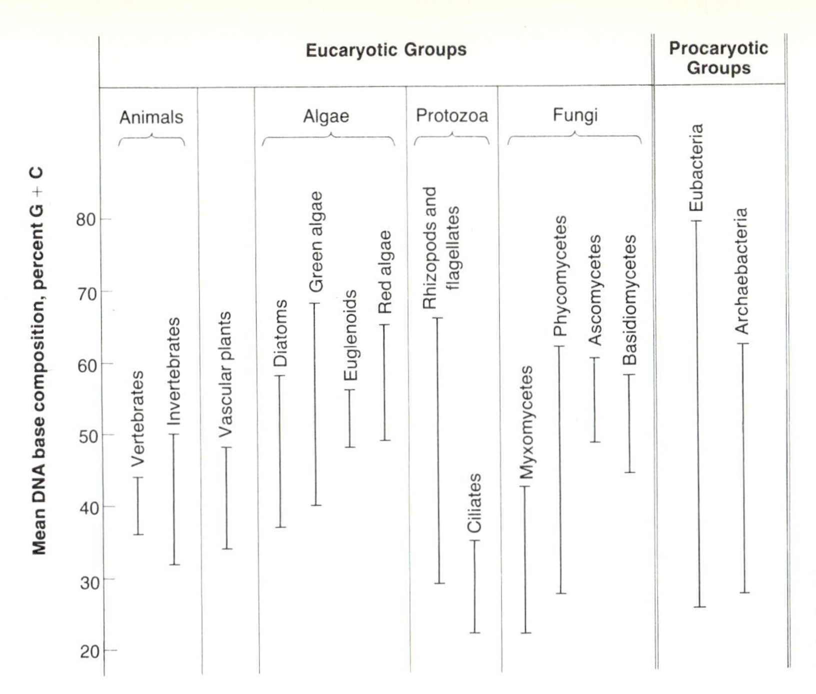


FIGURE 13.4
The ranges of mean DNA base composition (percent G + C) characteristic of major biological groups.

wider. The widest range of all occurs among the procaryotes, in which the range extends from about 25 to nearly 80 percent G + C. If, however, one examines the mean G + C content of many different strains that are assigned to a *single microbial species*, the values are closely similar or identical, as shown by the data for several *Pseudomonas* species assembled in Table 13.2. Each bacterial species, accordingly, has DNA with a characteristic mean G + C content; this can be considered one of its important specific characters. Furthermore, *a substantial divergence between two organisms with respect to mean DNA base composition reflects a large number of individual differences between the specific base sequences of their respective DNAs.* It is *prima facie* evidence for a major genetic divergence and hence for a wide evolutionary separation. The very broad span of values characteristic of the procaryotes, accordingly, reveals the great evolutionary diversity of this particular biological group, and it also suggests its evolutionary antiquity.

However, two organisms with identical mean DNA base compositions may differ greatly in genetic constitution. This is evident from the very similar base ratio values for DNA from all plants and animals. Hence, major evolutionary divergence is not *necessarily* expressed by a divergence of mean base composition. When two organisms are closely similar in their DNA base composition, this fact can be construed as indicative of genetic and evolutionary relatedness only if the organisms *also* share a large number of phenotypic properties in common or are known to resemble one another in genetic constitution (e.g., different strains that belong to a

**TABLE 13.2**

**Constancy of G + C Content in the DNA of Bacterial Strains Belonging to a Given Species**

| *Pseudomonas* spp. | Number of Strains Examined | Percent G + C (mean value ± standard deviation) |
|---|---|---|
| *P. aeruginosa* | 11 | 67.2 ± 1.1 |
| *P. acidovorans* | 15 | 66.8 ± 1.0 |
| *P. testosteroni* | 9 | 61.8 ± 1.0 |
| *P. cepacia* | 12 | 67.6 ± 0.8 |
| *P. pseudomallei* | 6 | 69.5 ± 0.7 |
| *P. putida* | 6 | 62.5 ± 0.9 |

Source: From M. Mandel, *J. Gen. Microbiol.* **43,** 273 (1966).

single bacterial species). In such a case, near-identity of DNA base composition provides supporting evidence for their genetic and evolutionary relatedness.

Mean DNA base composition is a character of particular taxonomic value among bacteria, since the range for the group as a whole is so wide. Values have now been determined for a large number of representative strains and species belonging to every major subgroup of procaryotes. Although the values for the constituent species in a genus differ somewhat (see Table 13.2), the total range within a bacterial genus is in general fairly narrow (rarely greater than 10 to 15 percent), and can indeed be considered as an important character for the definition of a genus. Furthermore, the members of multigeneric clusters that had been recognized as similar on phenotypic grounds have proved in some cases to be closely similar in DNA base composition. Thus, the fruiting myxobacteria and the euactinomycetes, two large bacterial multigeneric clusters, are both characterized by narrow spans of DNA base composition that lie near the high end of the G + C scale.

It has become evident, however, that some bacterial groups that had been closely associated by virtue of phenotypic resemblances are in fact unrelated genetically, as revealed by wide differences in DNA base composition. The gliding, unicellular nonphotosynthetic bacteria of the *Cytophaga* group, similar to fruiting myxobacteria with respect to vegetative cell structure, lie almost at the opposite end of the G + C scale; and the same situation occurs among the Gram-positive cocci of the *Staphylococcus* and *Micrococcus* groups (Figure 13.5).

FIGURE 13.5

Ranges of mean DNA base composition of certain representatives in two bacterial assemblages defined by phenotypic properties: gliding bacteria and Gram-positive cocci. The wide separations with respect to DNA base composition between the *Cytophaga* group and the fruiting myxobacteria and between the genera *Staphylococcus* and *Micrococcus* reveal that they are genetically unrelated.

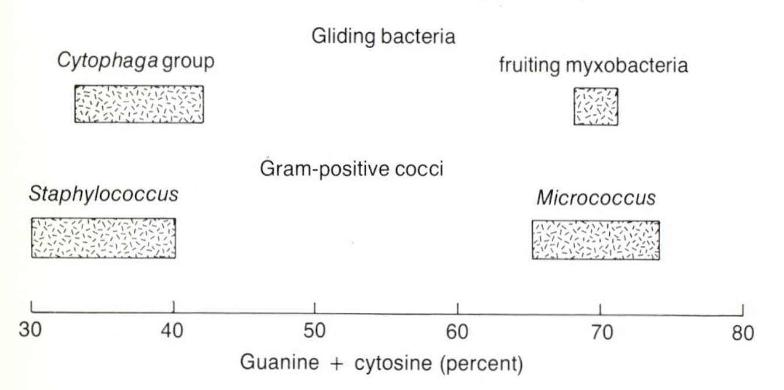


## Nucleic Acid Hybridization

Upon rapid cooling of a solution of thermally denatured DNA, the single strands remain separated. However, if the solution is held at a temperature from 10°C to 30°C below the $T_m$ value, specific reassociation ("annealing") of complementary strands to form double-stranded molecules occurs. There is always some random pairing, but since a randomly matched duplex contains many mismatched base pairs, its thermal stability is low and its strands separate very rapidly at temperatures near the $T_m$. In contrast, pairing of complementary strands forms duplexes that are quite stable because each base participates in interstrand hydrogen bonding. Thus at temperatures near the $T_m$, only duplexes between strands with a high degree of complementarity persist; the closer that the temperature of incubation is to the $T_m$, the more stringent is the requirement for precise base pairing.

Shortly after discovery of this phenomenon, it was shown that when DNA preparations from two related strains of bacteria are mixed and treated in this manner, *hybrid DNA molecules are formed.* One bacterial strain was grown in a medium containing $D_2O$, so that its DNA was "heavy" as a result of deuterium incorporation. After the two DNA samples had been mixed, denatured, and annealed, hybrid molecules could be detected by centrifugation in a CsCl gradient, where they formed a band intermediate in position between those of the "light" and "heavy" duplexes (Figure 13.6). In similar experiments, conducted with DNA preparations from two unrelated bacteria, no hybridization could be detected; upon annealing, duplexes were formed only by specific pairing between single strands originally derived from the same DNA.

The discovery of the reassociation of single-stranded DNA molecules from different biological sources to form hybrid duplexes laid the foundations of an entirely new approach to the study of genetic relatedness in bacteria. *In vitro* experiments on DNA-DNA reassociation permit an assessment of the overall degree of genetic homology between two bacteria. Furthermore, since duplexes can also be formed between single-stranded DNA and complementary RNA strands, analogous DNA-RNA reassociations can be performed. If the RNA preparations consist of either tRNAs or rRNAs, such experiments permit an assessment of the genetic homology between two bacteria *with respect to specific, relatively small segments of the chromosome:* those that code the base sequences either of the transfer RNAs or of the ribosomal RNAs. As will be discussed later, DNA-rRNA similarities are of particular taxonomic interest.

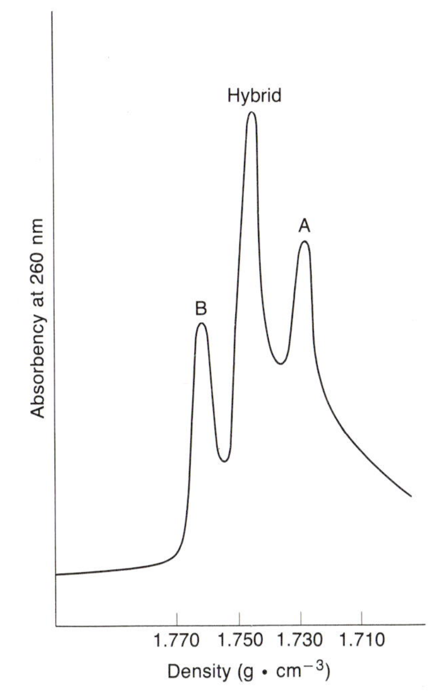


FIGURE 13.6

An experiment demonstrating the formation of hybrid DNA molecules through reassociation of single-stranded (denatured) DNA molecules to form double helical DNA. DNA was prepared from two different strains of *Pseudomonas aeruginosa;* strain A was grown in a normal medium and strain B was grown in a medium containing "heavy water" ($D_2O$) and $^{15}NH_4Cl$. Consequently, the DNA of strain B, although identical in base composition to the DNA of strain A, had a higher density as a result of its content of heavy atoms. The two DNAs were isolated, denatured by heating, mixed, and then annealed, after which residual single-stranded molecules were eliminated by treatment with a specific nuclease. The preparation was then centrifuged in a CsCl gradient. Three peaks of double-stranded DNA are apparent in the density gradient. Peak A corresponds to "light" double-helical DNA, formed by reassociation of single-stranded DNA from strain A; peak B corresponds to "heavy" double-stranded DNA, formed by reassociation of single-stranded DNA from strain B. Between these two peaks, a third peak of intermediate density occurred; this corresponds to hybrid double-stranded DNA, formed by specific reassociation of single strands, one of which was derived from strain A, and one from strain B. Data courtesy of M. Mandel.

## The Techniques and Interpretations of Reassociation Experiments

The density gradient method (Figure 13.6) is too cumbersome for routine use, and a variety of simpler methods for measuring nucleic acid reassociation have been developed. All are based on the same general procedure: formation of duplexes between two denatured DNA samples, one of which is labeled with a radioisotope; separation of the duplexes from residual single-stranded nucleic acid; and measurement of their radioactivity. A point of reference is always required: it is provided by the DNA of a *reference strain,* which is prepared both labeled and unlabeled. The amount of reassociation between these two homologous DNAs is determined, and is assigned an arbitrary value of 100. The amount of reassociation between the reference DNA and DNAs from heterologous strains can then be measured and expressed as a percentage of the normalized value for the homologous DNA-DNA reassociation. Depending on the particular technique used, either the reference DNA or the heterologous DNA may be labeled. The same principles apply to DNA-RNA reassociation experiments.

In every type of reassociation experiment the temperature of reassociation, the ionic strength of the solvent, and the mean length of the DNA fragments must be standardized, since these factors all affect the rate of duplex formation. When they are held constant, reassociation is determined solely by the *DNA concentration* and the time of incubation. Provided that these two parameters are correctly chosen, nearly complete reassociation of complementary strands will occur.

The temperature at which reassociation is conducted is a factor of paramount importance. For obvious reasons, this temperature must always be below the $T_m$ value of the reference DNA. At so-called *stringent* temperatures, 10° to 15° C below $T_m$, only complementary strands with a very high degree of base sequence homology can form stable duplexes. At somewhat lower temperatures, stable duplexes can also be formed between complementary strands in which base pairing is less perfect and hydrogen bonding therefore somewhat weaker.

If reassociation experiments are conducted at two different temperatures within the specific reassociation range (e.g., $T_m$ − 15° C and $T_m$ − 30° C), the *degree of base sequence homology* between a series of heterologous DNAs and the reference DNA can be roughly assessed. If homology is very high, reassociation is little affected by temperature; if base pairing is less perfect, reassociation will be markedly reduced at the higher temperature (Table 13.3).

Carefully controlled DNA-DNA reassociation experiments can thus provide much semiquantitative information about the degree of genetic homology between related strains or species of bacteria. However, if evolutionary divergence has led to numerous differences of base sequence between the two genomes, specific DNA-DNA reassociation

## TABLE 13.3

**The Effect of Incubation Temperature on DNA Duplex Formation between Radioactive DNA Prepared from *Escherichia coli* and Unlabeled DNAs from Other Bacteria Belonging to the Enteric Group**

| | Related Duplex Formation at | |
|---|---|---|
| | 60° C ($T_m$ − 30° C) | 75° C ($T_m$ − 15° C) |
| *E. coli/E. coli*[a] | 100 | 100 |
| *E. coli/Shigella boydii* | 89 | 85 |
| *E. coli/Salmonella typhimurium* | 45 | 11 |
| *E. coli/Enterobacter hafniae* | 21 | 4 |

[a] Control, using labeled and unlabeled DNA from the reference strain of *E. coli:* values normalized to 100.

Source: Data from D. J. Brenner and S. Falkow, *Adv. Genetics* **16,** 81 (1971).

## TABLE 13.4

**Nucleic Acid Homologies between *Pseudomonas acidovorans* and Four Other *Pseudomonas* Species, as Revealed by Parallel DNA-DNA and DNA-rRNA Hybridization Experiments**

| | Relative Duplex Formation | |
|---|---|---|
| | DNA-DNA | DNA-rRNA |
| *P. acidovorans/P. acidovorans* | 100 | 100 |
| *P. acidovorans/P. testosteroni* | 33 | 92 |
| *P. acidovorans/P. delafieldii* | 0 | 89 |
| *P. acidovorans/P. facilis* | 0 | 87 |
| *P. acidovorans/P. saccharophila* | 0 | 79 |

Source: Data from N. J. Palleroni, R. Kunisawa, R. Contopoulou, and M. Doudoroff, *Int. J. Syst. Bacteriol.* **23,** 333 (1973).

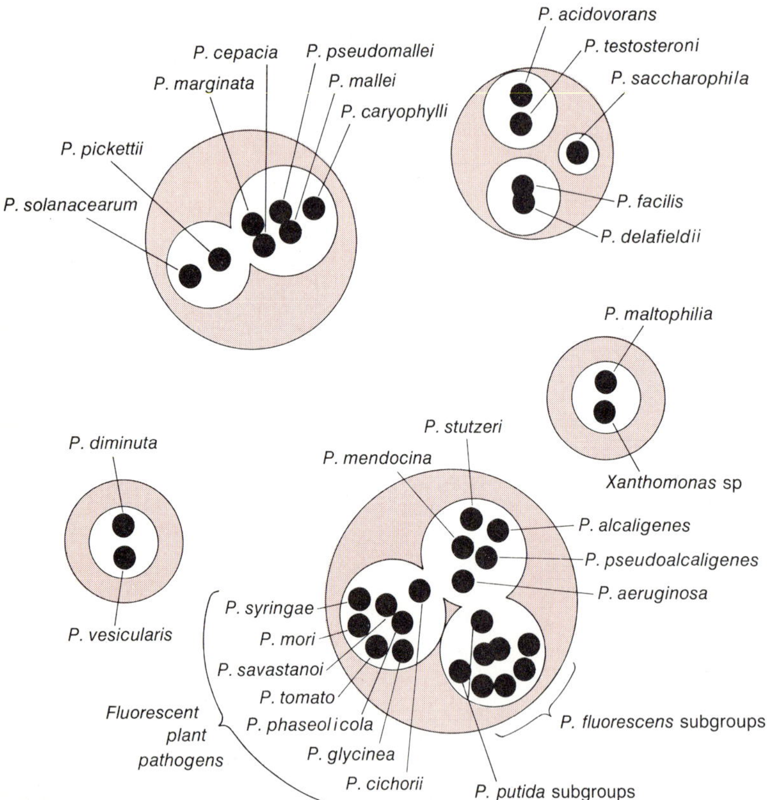


FIGURE 13.7

A schematic diagram of genetic relationships among aerobic pseudomonads, as revealed by nucleic acid reassociation *in vitro*. Each large, shaded circle represents an rRNA homology group, determined by DNA-rRNA hybridizations. The smaller white circles define DNA homology groups. The black dots within each white circle represent species (or biotypes) among which relatedness can be shown by DNA-DNA hybridization: the distance between any two black dots is a rough measure of the degree of shared genetic homology. After N. J. Palleroni et al., *Int. J. Syst. Bacteriol.* **23,** 333 (1973).

becomes too weak to be measured. The range of organisms among which genetic homology is detectable can be greatly extended by parallel studies on DNA-rRNA reassociation, because the relatively small portion of the bacterial genome that codes for ribosomal RNAs has a much more highly conserved base sequence than the bulk of the chromosomal DNA. As a result, it is frequently possible to detect by DNA-rRNA reassociation relatively high homology between the genomes of two bacteria which show no significant homology by DNA-DNA reassociation (Table 13.4).

In any given bacterial group the value of nucleic acid reassociation studies is directly related to the number of strains and species that have been compared. Extensive comparative data are now available for several major bacterial groups. The insights that have been obtained into the genetic relationships among about 30 species of aerobic pseudomonads, currently classified in the genera *Pseudomonas* and *Xanthomonas*, are shown schematically in Figure 13.7. Each large, shaded circle in this diagram embraces a single rRNA homology group. The species included in each group all show a much higher degree of relatedness with one another than with any species belonging to other rRNA homology groups. In fact, the *intergroup* level of rRNA homology among aerobic pseudomonads are no greater than those with the enteric bacterium, *Escherichia coli.*

Within each rRNA homology group, the smaller, unshaded circles define the limits of interspecific genetic homology detectable by DNA-DNA hybridization. It can be seen that one of the rRNA homology group includes three isolated DNA homology groups (*Pseudomonas acidovorans* + *P. testosteroni*; *P. saccharophila*; *P. facilis* + *P. delafieldii*). In the four other rRNA homology groups, all the constituent species can be shown to be genetically interrelated by pairwise DNA-DNA hybridization experiments, even though some pairs of species within a given DNA homology group may be so distantly related that no DNA homology can be detected between them. This is true, for example, of the two species of fluorescent pseudomonads, *P. aeruginosa* and *P. syringae*; however, both show some degree of relatedness in terms of DNA-DNA reassociation with a third species, *P. fluorescens.* Another interesting point concerns the yellow-pigmented plant pathogenic pseudomonads, currently classified in a separate genus, *Xanthomonas.* Both DNA-rRNA and DNA-DNA hybridization studies show that they form a homology group with one otherwise genetically isolated *Pseudomonas* species, *P. maltophilia.*

## Nucleic Acid Sequencing

Reassociation studies utilizing heterologous DNAs or RNAs can yield substantial information about the similarity of their base sequences; however, such studies cannot establish the actual order of bases in the molecule. Knowledge of the sequence of bases in nucleic acid would not only establish the similarity between two organisms, but would provide much additional information as well; for instance, the amino acid sequence of any protein could be read directly from the base sequence, and regions with regulatory function could be identified. Recently, methods have been developed for rapidly sequencing both DNA and RNA.

### DNA Sequencing

Either of two different procedures can be used to sequence substantial lengths of DNA: the *Sanger* method or the *Maxam and Gilbert* method. Although they differ in chemical detail, both rest on the same basic principle: a series of DNA fragments are generated which have a common starting point, but variable termini ("nested" fragments); by determining their exact length and the terminal base, the sequence can be inferred with great accuracy. As an example, the Sanger method will be described (Figure 13.8).

In order to determine the sequence of bases in a segment of DNA, a large number of identical copies of it must be obtained. If the molecule that contains the sequence is relatively small (e.g., a viral genome or a plasmid), one can feasibly sequence the entire molecule. In such a case one need only separate the molecule from contaminating nucleic acid (e.g., fragments of the bacterial chromosome). If the sequence is part of a much larger molecule, such as the bacterial chromosome, the usual approach is to *clone* the desired segment, and sequence a restriction fragment purified from the cloned DNA (see Chapter 11). Since the Sanger method uses single-stranded DNA, the segment must either be digested with a nuclease to convert it to single-stranded DNA or, preferably, cloned in a single-stranded vector to begin with (such as the bacteriophage M13).

Sequencing of a purified single-stranded fragment is accomplished by generating a series of complementary nested fragments of DNA by incubating the fragment with DNA polymerase, a short primer sequence complementary to a region of the fragment (often a short restriction fragment itself), the four deoxy nucleoside triphosphates (one or more of which are radioactive), and a small

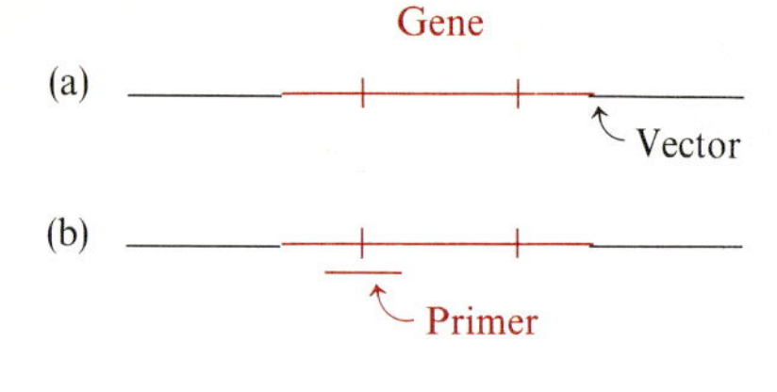

(d) Mix 1: dATP, dCTP, dGTP, dTTP, ddATP

--—— TGCTTddA
--—— TGCTTATGCTddA
--—— TGCTTATGCTAGGTddA
etc.

Mix 2: dATP, dCTP, dGTP, dTTP, ddCTP

--—— TGddC
--—— TGCTTATGddC
--—— TGCTTATGCTAGGTAddC
--—— TGCTTATGCTAGGTACGddC
etc.

Mix 3: dATP, dCTP, dGTP, dTTP, ddGTP

--—— TddG
--—— TGCTTATddG
--—— TGCTTATGCTAddG
--—— TGCTTATGCTAGddG
--—— TGCTTATGCTAGGTACddG
etc.

Mix 4: dATP, dCTP, dGTP, dTTP, ddTTP

--—— TGCddT
--—— TGCTddT
--—— TGCTTAddT
--—— TGCTTATGCddT
--—— TGCTTATGCTAGGddT
etc.

FIGURE 13.8

DNA sequencing by the Sanger method. (a) the single-stranded vector containing the cloned fragment to be sequenced (b) a purified restriction fragment complementary to the 5′ end of the gene is added as primer (c) the sequence near the 5′ end of the gene, showing the 5′ end of the primer (d) the primed template is added to four different polymerization mixtures, each containing DNA polymerase, the four deoxynucleoside triphosphates, and each a different dideoxynucleoside triphosphate. The families of new complementary strands synthesized in each mixture are shown.

amount of one *dideoxy* nucleoside triphosphate (e.g., dideoxythymidine triphosphate). Chance incorporation of dideoxythymidine, which lacks a 3′-hydroxyl group, terminates polymerization at that point. Thus, after a suitable incubation period, the mixture will contain radioactive DNA strands of variable length, all of which terminate in a thymidine residue. The number of classes of DNA molecules, differing from each other in length, depends on the number of thymidine residues, since for every thymidine residue in the new strand, there will be a family of radioactive DNA molecules that terminates at that point.

In practice four parallel incubations are performed, identical except that each contains a different dideoxy nucleoside triphosphate, and hence will terminate at a different base. The mixture is then denatured and electrophoresed* to separate the newly synthesized strands by size. The position of the radioactive bands on the gel are visualized by appressing the gel tightly to a sheet of X-ray film, which is exposed by the localized decay of the radioisotope incorporated into the DNA. Upon development, the film displays a series of exposed bands, each of which corresponds to a size class of single-stranded DNA (Figure 13.9). The sequence can then be inferred directly from this *autoradiogram*; the shortest fragment will be found in the incubation mixture that contains the dideoxy analogue of the first base after the primer; the next shortest fragment will be found in the mixture that contains the dideoxy analogue of the second base after the primer; and so on for up to hundreds of bases.

DNA sequencing techniques were devised by molecular biologists and have been applied primarily to genetic problems with dramatic success, but to date the application of these techniques to taxonomic problems has been minimal. However, as the techniques are more widely applied, and a rapidly expanding library of sequences begins to include a number of homologous sequences from different organisms, taxonomists will undoubtedly come to depend increasingly on comparison of DNA base sequence to determine organismic relationships. Indeed it is probably not unrealistic to anticipate that in the near future the complete base sequence for the genomes of a number of microorganisms will be available, and microbial taxonomy may become an exact, quantitative science.

* *Electrophoresis* is an analytic technique that separates compounds on the basis of their mobility in an electrical field. Mobility is a function of the *charge density*, the total electrical charge per unit molecular weight.

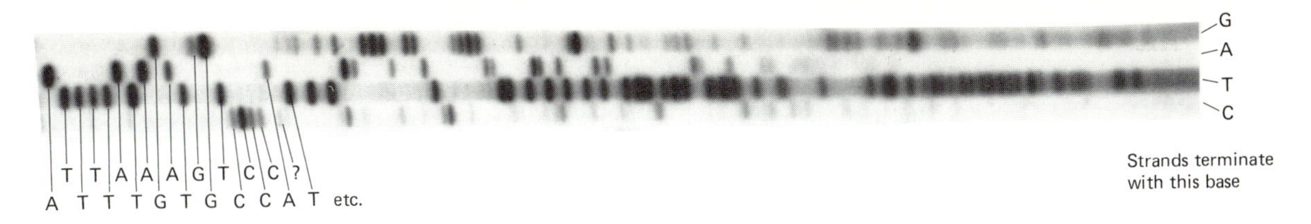


FIGURE 13.9

A dideoxy sequencing gel. Four parallel reaction mixtures were incubated in the presence of one of the four dideoxynucleoside triphosphates (indicated at the right). Electrophoresis was from right to left, with the shortest strands migrating fastest. The sequence of the newly synthesized DNA is shown below the gel. Courtesy of M. L. Privalsky.

## RNA Fingerprinting and Sequencing

Methods for sequencing RNA, comparable to those for sequencing DNA, have also been developed. They are, however, only beginning to receive wide application in microbial taxonomy. A number of tRNAs and 5S rRNAs have been sequenced but, because these are relatively small molecules, the amount of information they contain is fairly low. More useful for taxonomic purposes are sequences of larger molecules, such as the 16S or 18S rRNAs. Since sequencing these longer molecules was not feasible several years ago when Carl Woese initiated studies on the phylogeny of 16S RNA, alternative methods were devised. Since it was easy to sequence short RNA molecules, Woese digested 16S RNA with a specific endonuclease producing a variety of short oligonucleotides, which were separated and sequenced. This economical approach has allowed him and others to assemble catalogues of oligonucleotide sequences from hundreds of microorganisms; indeed it was the finding that the 16S rRNA sequences of methanogens, halophiles, and thermoacidophiles were unexpectedly divergent from those of other bacteria that allowed the recognition of the archaebacteria as a distinctive assemblage of bacteria.

The method devised by Woese and his collaborators is as follows: radioactive 16S rRNA (obtained by growing organisms in the presence of phosphate containing the radioactive isotope $^{32}P$) is purified and digested with the endonuclease $T_1$. This endonuclease cleaves the RNA on the 3′ side of every guanosine residue; hence, a mixture of oligonucleotides is obtained that range in size from a single nucleotide to a dozen or more, all containing a single guanosine residue at the 3′ terminus. These are separated by two dimensional electrophoresis (the mixture is electrophoresed in one dimension in one buffer system, then in a second dimension in another buffer system) and their position determined by autoradiography. The labeled oligonucleotides can then be removed and sequenced. In practice, only those with a chain length of five residues or more contain sufficient taxonomic information to be worth cataloguing. Oligonucleotide catalogues of a number of organisms are analyzed with the aid of a computer program, which compares them in a pairwise fashion; results may be displayed as a dendrogram. Much of our current understanding of the broad outlines of bacterial phylogeny is based on the results obtained by this method (see below).

In addition to the calculation of similarity coefficients among organisms, a detailed analysis of 16S rRNA oligonucleotide catalogues or complete sequences reveals that there are both quite variable and highly conserved regions of the molecule, presumably reflecting different degrees of functional constraint on different parts of the molecule. Separate comparison of the variable and conserved regions allows determination of both close and distant relationships. In addition, careful inspection of the conserved regions has revealed that there are sequences that are characteristic of major groups of bacteria, termed *signature sequences*. At great phylogenetic distances (very low $S_J$ values) these signature sequences are important in allowing the determination of relationships.

More recently, rapid techniques for sequencing long molecules of RNA have been devised, and these are now used in preference to fingerprinting. To sequence a molecule of rRNA, bulk RNA is isolated from a culture (there is no need to purify it), and a short RNA primer is added that is homologous to one of the conserved regions of rRNA. Dideoxy sequencing can then be performed directly on the RNA template using the enzyme *reverse transcriptase*, a viral enzyme that produces a DNA complement to a RNA molecule (Chapter 9).

Comparison of complete sequences of 16S rRNA among organisms gives results identical to those obtained by comparing oligonucleotide catalogues, but with the advantage that more distant relationships can be determined with greater accuracy.

## BACTERIAL PHYLOGENY

Complete sequences or oligonucleotide catalogues of 16S rRNA have now been accumulated for more than 450 bacteria; analysis of these catalogues has allowed the recognition that the bacteria can be divided into several major groups based on the similarity of their ribosomal RNA. It is now generally agreed that the degree of sequence similarity of this molecule is a reflection of the phylogenetic distance among organisms. Because there are very tight constraints on the structure of rRNA (a consequence of the necessity for it to assume a precise secondary structure, and to interact with a variety of proteins to form a functional ribosome), the rate of change of the sequence of genes encoding it is much less than that of the bulk of the genome, and it is accordingly possible to determine relationships over vast evolutionary distances. Indeed, comparison of complete sequences of eubacterial and archaebacterial 16S and eucaryotic 18S rRNA shows that even at the greatest phylogenetic distances known among cellular organisms about 50 percent homology remains. Accordingly, this *rRNA functions as a molecular clock, and allows accurate determination of phylogenetic distance.*

It is, of course, possible that the phylogeny of the rRNA is not the same as the phylogeny of the rest of the genome; such a situation could occur as the result of genetic transfer among different organisms. If transfer of chromosomal material across large phylogenetic distances were common, the bacterial genome would be polyphyletic, and no single organismic phylogeny could be drawn. Although genetic transfer of plasmids across large phylogenetic distances is known, there is no evidence that transfer and integration of chromosomal material is common except among closely related organisms. Indeed, such genetic exchange would not be expected because homology between the transferred fragment and the endogenote would most probably be too low to permit recombination. The phylogeny of the 16S rRNA can thus be provisionally regarded as an organismic phylogeny. A critical test of this idea will come with the comparison of other sequences distributed over comparable phylogenetic distances and comparably conserved; possibilities include the genes for the proton-translocating ATPase, DNA polymerase, RNA polymerase, histonelike proteins, and ferredoxin.

**TABLE 13.5**

**Similarity Coefficients ($S_J$) among Selected Representatives of Cellular Organisms**

| | 1 | 2 | 3 | 4 | 5 | 6 | 7 | 8 | 9 |
|---|---|---|---|---|---|---|---|---|---|
| Eucaryotes | | | | | | | | | |
| 1. *Saccharomyces cerevisiae* (yeast) | 1.0 | | | | | | | | |
| 2. *Lemna minor* (alga) | *0.29* | 1.0 | | | | | | | |
| 3. *L-cell* (animal) | *0.33* | *0.36* | 1.0 | | | | | | |
| Eubacteria | | | | | | | | | |
| 4. *Escherichia coli* | 0.05 | 0.10 | 0.06 | 1.0 | | | | | |
| 5. *Bacillus firmus* | 0.08 | 0.06 | 0.07 | *0.25* | 1.0 | | | | |
| 6. *Aphanocapsa* sp. | 0.11 | 0.09 | 0.09 | *0.26* | *0.26* | 1.0 | | | |
| Archaebacteria | | | | | | | | | |
| 7. *Methanosarcina barkeri* | 0.08 | 0.07 | 0.07 | 0.12 | 0.12 | 0.10 | 1.0 | | |
| 8. *Halobacterium halobium* | 0.10 | 0.09 | 0.11 | 0.07 | 0.10 | 0.13 | *0.28* | 1.0 | |
| 9. *Thermoplasma acidophilum* | 0.08 | 0.09 | 0.07 | 0.09 | 0.09 | 0.10 | *0.23* | *0.23* | 1.0 |

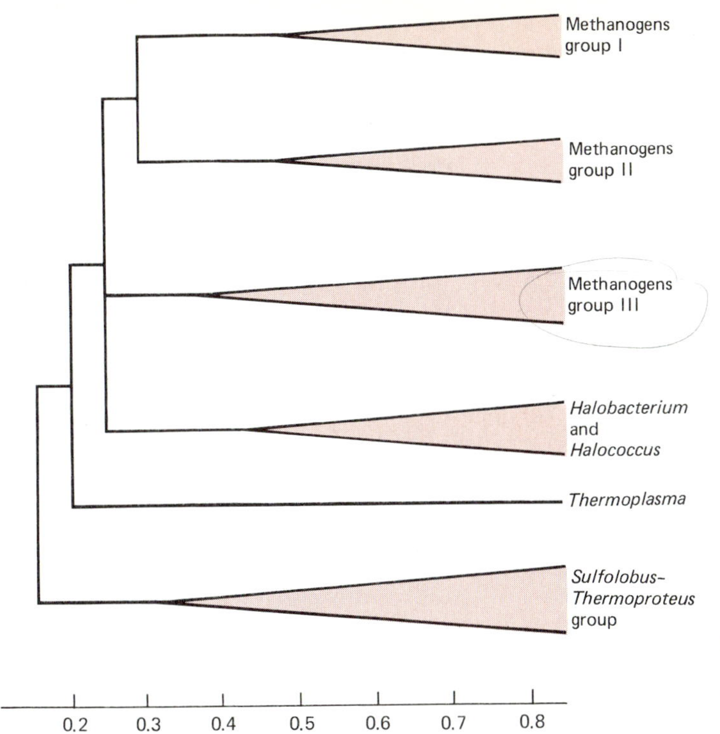


FIGURE 13.10
16S rRNA similarities among the archaebacteria.

## The Primary Divisions of Cellular Organisms

Comparison of oligonucleotide catalogues of eucaryotes, eubacteria, and archaebacteria shows very little detectable homology; $S_J$ values across the boundaries of these groups are typically around 0.1, while within a group they are typically above 0.2 (Table 13.5). Analysis of complete sequences of 16S rRNA shows that there is less than 60 percent homology across group boundaries, but greater than 70 percent within a group.

## Constituent Groups of Archaebacteria

Three major subgroups of archaebacteria can be distinguished on the basis of $S_J$ values: the methanogen-halophile group; *Thermoplasma*; and the *Sulfolobus-Thermoproteus* group (Figure 13.10). The *Thermoplasma* group contains a single organism, while the other two groups are quite heterogeneous.

## Constituent Groups of Eubacteria

Ten major groups of eubacteria are currently recognized (Table 13.6). $S_J$ values between groups are in most cases approximately 0.2–0.3, with the exception of the *Planctomyces* group, which is almost as distant from other eubacteria as it is from the archaebacteria.

THE GRAM-POSITIVE BACTERIA There are two major subgroups of the Gram-positive eubacteria: the *actinomycete group* or *high G + C group*, and the *low G + C group* (Figure 13.11). The low G + C group is a very heterogeneous one. It includes all the endospore-forming bacteria, the lactic acid bacteria, *Staphylococcus*, and the mollicutes. The genus

**TABLE 13.6**

**Constituent Groups of Eubacteria Based on rRNA Sequence Analysis**

1. Gram-positive eubacteria
2. Purple bacteria—pseudomonad group
3. Spirochetes
4. *Bacteroides-Cytophaga* Group
5. Cyanobacteria
6. Green sulfur bacteria
7. Green nonsulfur bacteria
8. Sulfur-reducers and myxobacteria
9. Radioresistant micrococci
10. *Planctomyces* group

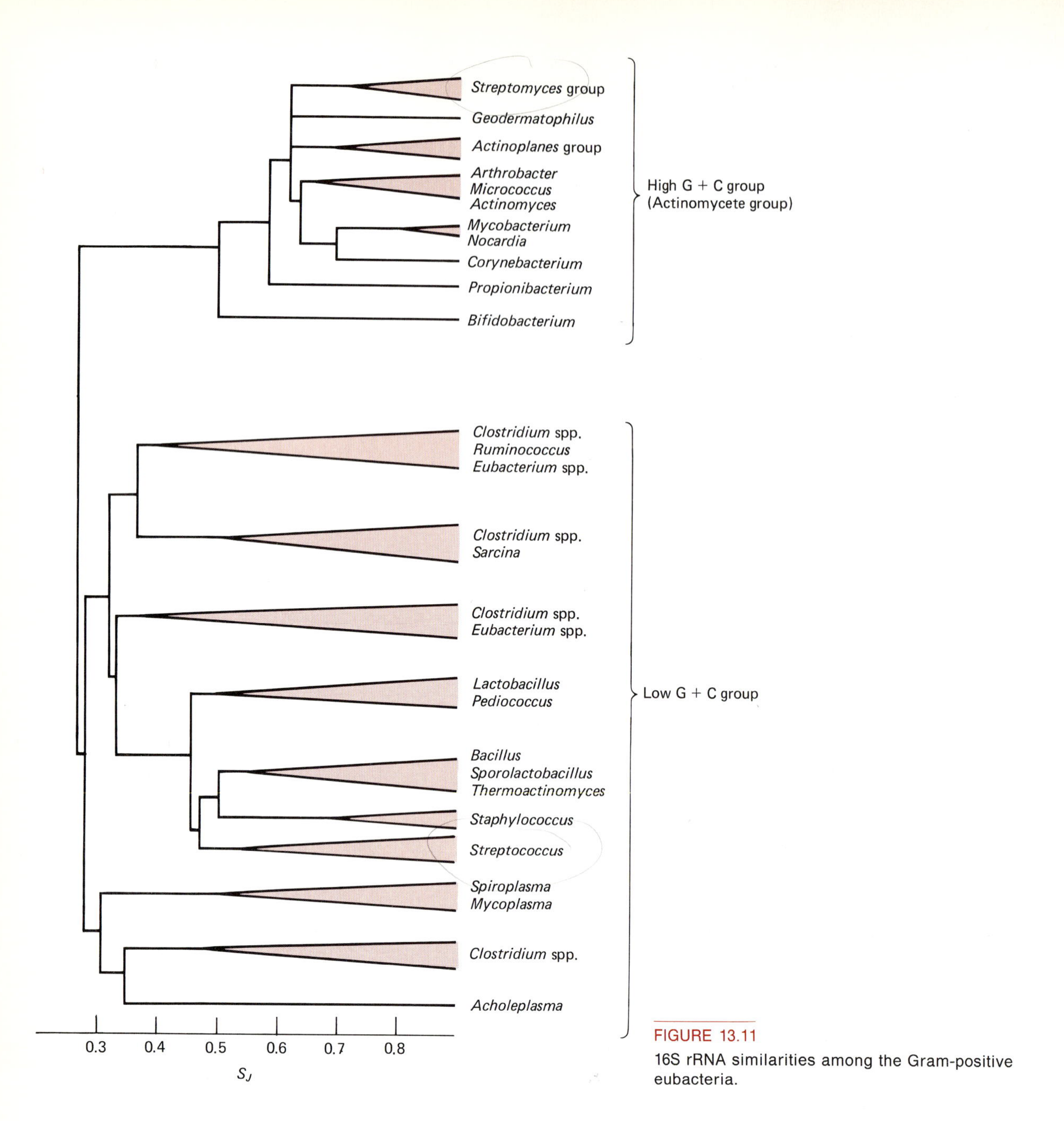


FIGURE 13.11

16S rRNA similarities among the Gram-positive eubacteria.

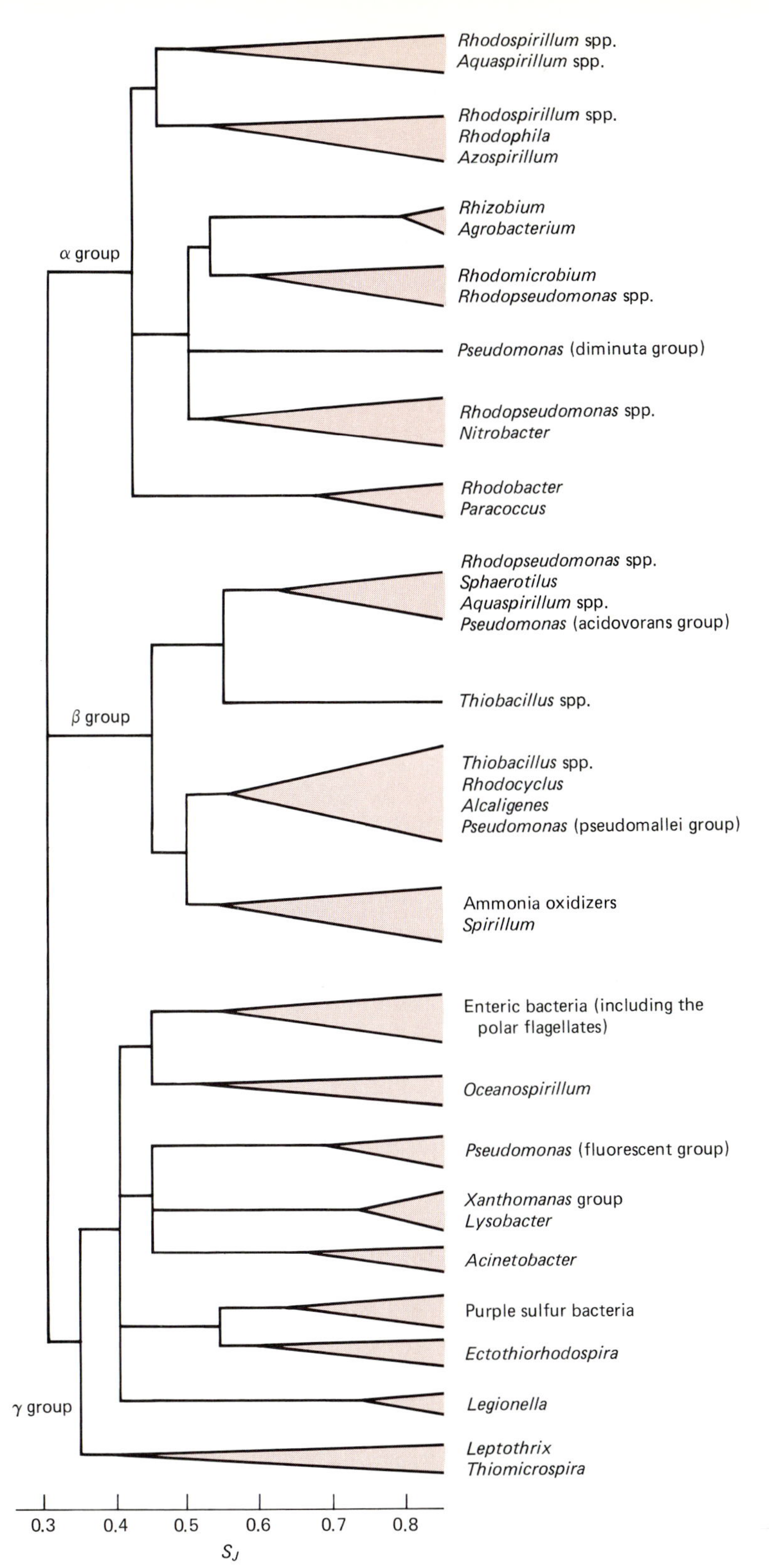


FIGURE 13.12
16S rRNA similarities within the purple bacteria–pseudomonad group.

*Clostridium*, defined on the basis of endospore formation and an anaerobic metabolism, is clearly polyphyletic.

THE PURPLE BACTERIA AND PSEUDOMONADS Most of the Gram-negative bacteria fall within the confines of this very large and heterogeneous assemblage. Three subgroups can be discerned (Figure 13.12): the alpha and beta subgroups contain the purple nonsulfur bacteria, most of the spiral bacteria, and several rRNA homology groups of *Pseudomonas*; the gamma subgroup contains the purple sulfur bacteria, pseudomonads, spiral bacteria, and the enteric group.

THE SPIROCHETES All the helical bacteria motile by means of endoflagella fall into this phylogenetically coherent group. Two main groups of *Spirochaeta* are evident, one of which contains the treponemes. *Borrelia* and *Leptospira* are more distantly related ($S_J$ values of 0.25 and 0.2, respectively, with the *Spirochaeta* groups).

THE BACTEROIDES-CYTOPHAGA GROUP The Gram-negative anaerobic rods (*Bacteroides*) form a coherent phylogenetic group, with intragroup $S_J$ values above 0.45. These bacteria are distantly related to a heterogeneous group that contains the sheathed organism *Haliscomenobacter*, *Flavobacterium*, and the gliding organisms *Cytophaga*, *Flexibacter*, *Sporocytophaga*, and *Saprospira*.

THE CYANOBACTERIA The cyanobacteria form a coherent phylogenetic group that contains *Prochloron* and, at greater distances ($S_J$ values of 0.25–0.3) the chloroplasts of red and green algae, and green plants.

THE GREEN SULFUR BACTERIA AND THE GREEN NONSULFUR BACTERIA The green photosynthetic bacteria comprise two distinct groups, very distantly related to each other ($S_J$ values of about 0.18).

THE SULFUR REDUCERS AND MYXOBACTERIA The anaerobic sulfur- and sulfate-reducing bacteria, the fruiting myxobacteria, and the bdellovibrios form a distinct cluster, although the relationships among the three subgroups are quite distant.

THE RADIORESISTANT MICROCOCCI Several species of the Gram-positive genus *Micrococcus* have long been known to be atypical: they are characterized by a high level of resistance to radiation, and an unusual cell wall. The finding that they are phylogenetically very distant from the other micrococci, constituting an isolated group, has led to the suggestion that they be renamed *Deinococcus*.

THE PLANCTOMYCES GROUP These budding bacteria constitute a heterogeneous group that shows very little detectable homology to either the eubacteria or the archaebacteria, although they are slightly closer to the eubacteria. Their lipids and their insensitivity to diphtheria toxin also suggest that they are a group of eubacteria. However, they are the only walled eubacteria to lack peptidoglycan.

## Taxonomic Implications of Bacterial Phylogeny

Until very recently, firm phylogenetic conclusions could not be drawn regarding bacteria. The extent of morphological variation among the bacteria is quite narrow, and consequently the fossil record, in addition to being fragmentary, is uninformative about the details of bacterial evolution. Conclusions based on phenotypic similarity rested on assumptions less reasonable than those that underlie phylogenies based on molecular evidence; furthermore, different conclusions resulted from approaches based on different phenotypic characteristics. Consequently, bacterial taxonomists largely abandoned the goal of a phylogenetic taxonomy and concentrated on devising systems that were useful and stable, based on phenotypic characteristics that often had no intrinsic importance.

The rapid expansion in recent years of our knowledge of the molecular homologies among bacteria, based on the techniques of nucleic acid reassociation, fingerprinting, and sequencing, has for the first time offered the opportunity for a reliable and objective phylogenetic taxonomy of the bacteria (provided, of course, that additional comparative studies confirm that the bacterial genome is largely monophyletic). Bacterial taxonomists are thus faced with a difficult choice. A decision to attempt to formulate bacterial systematics in phylogenetic terms would be in accord with the practice of other organismic taxonomies, and would establish a biological rationale for bacteriological systematics. However, extensive, confusing, and divisive nomenclatorial changes would be necessary; for instance, inspection of Figures 13.11 and 13.12 shows that many genera would have to be subdivided, some into as many as six genera. In many cases, the criteria for generic assignment of an organism include no easily determinable phenotypic traits, and depend solely on the molecular evidence, the routine determination of which is common in only a few laboratories in the world.

For these reasons, a change to a phylogenetic system of taxonomy will occur slowly, if at all. The taxonomy presented in the following chapters is accordingly in marked variance with the phylogenies presented in this chapter. Also, above the generic level, the practice established in previous editions of this book is continued: groups of genera united by common physiological or morphological traits are grouped together, regardless of whether the molecular evidence indicates relationship. For example, the pseudomonads and the spiral bacteria are discussed as separate groups of aerobic chemoheterotrophs in Chapter 17 and the purple bacteria are discussed as phototrophs in Chapter 15, despite the evidence that the three groups are phylogenetically intermixed.

## FURTHER READING

### Books

KRIEG, N. R., and J. G. Holt, eds., *Bergey's Manual of Systematic Bacteriology*, Vol. 1. Baltimore: Williams and Wilkins (1984). This is the first of a four-volume continuation of the classic *Bergey's Manual of Determinative Bacteriology*, which went through eight editions. The new version will have considerably more information than the old, which was aimed primarily at aiding identification of bacteria. Publication of the remaining volumes is anticipated on the following schedule: Vol. 2 (P. H. A. Sneath, and J. G. Holt, eds.), 1985; Vol. 3 (J. T. Staley, and J. G. Holt, eds.), 1986; Vol. 4 (S. T. Williams, and J. G. Holt, eds.), 1986. This work has since the first edition in 1923 been widely regarded as the definitive work in bacterial taxonomy.

STARR, M. P., H. STOLP, H. G. TRUPER, A. BALLOWS, and H. G. SCHLEGEL, *The Prokaryotes*. New York: Springer-Verlag (1981).

Both of the above books contain an enormous amount of information that is relevant to the individual organisms discussed in the following 12 chapters; they should be considered essential resources for anyone interested in the diversity of bacteria.

### Reviews

BAUMANN, P., L. BAUMANN, M. J. WOOLKALIS, and S. S. BANG, "Evolutionary Relationships in *Vibrio* and Photobacterium: A Basis for a Natural Classification," *Ann. Rev. Microbiol.* **37,** 369–398 (1983).

SCHLEIFER, K. H., and E. STACKEBRANDT, "Molecular Systematics of Prokaryotes," *Ann. Rev. Microbiol.* **37,** 143–187 (1983).

# Chapter 14
# The Archaebacteria

The archaebacteria are a heterogeneous group that are phylogenetically very distant from the eubacteria. Their distinguishing properties, some of which were discussed in Chapter 3 (see Table 3.7), include membrane lipids that contain ether-linked isoprenoid side chains, in contrast to the ester-linked hydrocarbons in all other biological systems; substantial differences between them and the eubacteria with respect to the subunit structure of their RNA polymerase; their universal lack of muramic acid as a constituent of the cell wall peptidoglycan in contrast to its nearly universal presence among walled eubacteria; their lack of ribothymine in tRNA; their use of methionyl-$tRNA^{met}$ rather than formylmethionyl-$tRNA_f^{met}$ as initiator tRNA; and their sensitivity to diphtheria toxin. Additional differences between the translational and transcriptional apparatus of archaebacteria and eubacteria undoubtedly exist.

Comparisons of 16S rRNA sequences from eubacteria and archaebacteria confirm the phylogenetic distance between the two groups that is suggested by their chemical differences (Chapter 13) similarity coefficients ($S_J$) among the most distantly related eubacteria are typically 0.2 or larger, those among archaebacteria slightly less, but $S_J$ values between the two groups are characteristically 0.1 or less. Complete sequence analysis of 16S rRNA has confirmed that inter-group similarities (less than 60 percent) are substantially less than those within any group (more than 70 percent).

## CONSTITUENT GROUPS OF ARCHAEBACTERIA

Three major groups of archaebacteria can be discerned, based on metabolic or ecological properties: the methanogens, the halophiles, and the thermoacidophiles. The methanogens are distinguished by their unique energy metabolism, in which methane is a prominent end product. The halophiles and the thermoacidophiles are distinguished by their habitats: highly saline environments for the former, high temperature and low pH for the latter. However, two of the groups are internally heterogeneous. The methanogens, although physiologically homogeneous, are, on the basis of 16S rRNA fingerprints, composed of at least three different subgroups (Chapter 13). Even more substantial phylogenetic distances separate the constituent groups of thermoacidophiles.

The extreme genetic heterogeneity among the archaebacteria, reflected in the low $S_J$ values, is presumably a reflection of considerable antiquity; it is commonly believed that the archaebacteria are evolutionary relics, survivors of a group whose phylogenetic roots go at least as deep as and probably deeper than those of the eubacteria. This belief is reflected in the name *archae*bacteria.

## ARCHAEBACTERIAL LIPIDS

Two types of lipid structure are found among the archaebacteria: glycerol diethers and diglycerol tetraethers (Figure 14.1). Their hydrocarbon chains are normally the $C_{20}$ *phytane* or the $C_{40}$ *biphytane*, respectively. However, small amounts of $C_{25}$, $C_{30}$, and $C_{35}$ isoprenoid hydrocarbons are occasionally found, and in the thermoacidophiles, one or two cyclopentane rings also commonly occur in the $C_{40}$ chains. Two adjacent hydroxyl groups on the glycerol moiety are ether-linked to these hydrocarbon chains; the third may remain free, or be ether- or ester-bonded to a phosphate group, a sugar, or a sugar alcohol.

Membranes that contain diglycerol dibiphytanyl tetraether probably consist of a monolayer rather than a bilayer, with each lipid molecule spanning the entire membrane (Figure 14.1). Such an arrangement may increase the membrane's mechanical strength and resistance to chemical agents.

In general, the thermoacidophiles contain mainly dibiphytanyl tetraethers; the halophiles, mainly diphytanyl diethers. Two different patterns of distribution occur in the methanogens: coccoid cells contain only diphytanyl diethers, while the rest contain both diphytanyl diethers and dibiphytanyl tetraethers.

## THE METHANOGENS

Biological methane ($CH_4$) formation is a geologically important process that occurs in most anaerobic environments where organic matter undergoes decomposition: swamps, lake sediments, the intestinal tract of animals, and anaerobic sewage digestors. It results from the activities of a highly specialized group of bacteria that convert fermentation products formed by other anaerobes (notably $CO_2$, $H_2$, formate, and acetate) to methane or methane and $CO_2$. Since methane is a gas that is sparingly soluble in water, it escapes from the anaerobic environment and hence may be aerobically oxidized by the members of another bacterial group, the *methophiles* (Chapter 16), usually at the interface between anaerobic and aerobic conditions. The methanogens are consequently terminal members of the anaerobic food chain, whose metabolic activity prevents the sequestering of large amounts of organic material in anaerobic ecosystems. In some anaerobic environments, e.g., sewage digestors and some sanitary landfills, methanogenesis occurs at a rate that makes collection and compression of this principal component of natural gas economically feasible.

In those anaerobic environments in which methanogenesis is not quantitatively important, a comparable ecological role is played by the sulfur-reducing eubacteria (Chapter 20). The factors that determine which of the two groups will predominate are not well understood; an affinity of uptake systems for acetate and $H_2$ that is higher in the sulfur-reducing bacteria than in methanogens is thought to contribute. The recent demonstration that methanogens are capable of sulfur reduction (see below) also suggests that perhaps they contribute to mineralizations even in the absence of methanogenesis.

### Diversity of the Methanogens

Analysis of fingerprints of their 16S rRNA suggests that methanogens comprise at least three major groups (Chapter 13): Group I contains *Methanobacterium* and *Methanobrevibacter*; Group II contains *Methanococcus*; and Group III contains

(a)

$CH_2-O-R_1$
$CH-O-R_1$
$CH_2-O-X$

$CH_2-O-X$
$CH_2-O-R_2-O-CH$
$CH-O-R_2-O-CH_2$
$CH_2-O-X$

(b) phytane

biphytane

pentacyclic derivatives of biphytane

(c)

FIGURE 14.1

Archaebacterial lipids: (a) general structure of diethers (left) and tetraethers (right); (b) typical hydrocarbon chains: phytane (in diethers) and biphytane and its pentacyclic derivatives (in tetraethers); (c) postulated arrangement of tetraethers to form a monolayer membrane. *X* may be H, or any of a variety of saccharides or phosphate derivatives.

**TABLE 14.1**

**The Methanogens**

| Group | Representative Genera | Percent G + C | Characteristic Lipids | Cell Wall Structure | Gram Reaction | Motility | Substrates Used |
|---|---|---|---|---|---|---|---|
| I | *Methanobacterium* | 32–50 | $C_{20}$ diethers and $C_{40}$ tetraethers | Pseudomurein | + | – | $H_2$; some use formate |
| | *Methanobrevibacter* | 27–32 | | | | | |
| II | *Methanococcus* | 30–32 | $C_{20}$ diethers | Protein (trace of glucosamine) | – | + | $H_2$; formate |
| III | *Methanospirillum* | 45–47 | $C_{20}$ diethers and $C_{40}$ tetraethers | Proteinaceous sheath; unknown structural material | – | + | $H_2$; formate |
| | *Methanosarcina* | 38–51 | $C_{20}$ diethers | Heteropolysaccharide | + | – | $H_2$; formate; methanol; methylamine; acetate |

several genera, including *Methanospirillum* and *Methanosarcina* (Table 14.1). Immunological comparisons are in substantial accord with this division, but the pattern of distribution of lipid type is not. Group II contains exclusively diphytanyl diethers, Group I strains all contain *both* diphytanyl diethers and dibiphytanyl tetraethers, while group III contains some strains whose lipid profiles resemble that of Group I and some that resemble Group II. The heterogeneity of membrane lipids of Group III is paralleled by heterogeneity of cell wall chemistry (see below): *Methanospirillum* has a wall of still undetermined chemical composition overlaid by a proteinaceous sheath; *Methanosarcina* has a wall composed principally of an unusual acidic heteropolysaccharide. Possibly new groups of methanogens are yet to be discovered; for instance, a methane-producing strain apparently lacking a cell wall, *Methanoplasma*, has recently been isolated in pure culture. Thus the present taxonomy of methanogens must be considered tentative.

## The Cell Walls of Methanogens

At least three different types of cell wall are found among the methanogens. The most chemically complex is that of the Group I methanogens, which is rigid and composed principally of *pseudomurein*, a peptidoglycan similar to the murein of eubacteria [Figure 14.2 (a)]. Pseudomurein contains N-acetyltalosaminuronic acid instead of N-acetylmuramic acid, and lacks D-amino acids. Preliminary evidence indicates that the peptide moiety of pseudomurein is as variable as that of eubacterial murein. In appearance the wall of Group I methanogens resembles those of Gram-positive eubacteria [Figure 14.2 (b)]; in *Methanobacterium* species it is thick and homogeneous; in *Methanobrevibacter* species it is triple-layered [Figure 14.2 (c)]. The inner layer of the *Methanobrevibacterium* wall probably contains the pseudomurein; the chemical composition of the outer two layers has not yet been determined.

*Methanospirillum*, one of the genera of Group III, has the most complex cell wall among the methanogens (Figure 14.3). Its wall is a flexible envelope composed of at least two layers: an inner, electron-dense one of unknown chemical composition, and an outer one which appears membranelike in cross section but is composed entirely of protein. This protein is resistant to hydrolysis by proteinases (e.g., trypsin), and to solubilization by detergents (e.g., sodium dodecyl sulfate, SDS). The outer layer does not participate in septum formation; rather, it is a sheath that envelopes the individual cells forming a helical trichome. Within this trichome, individual cells are separated by spaces, some of which contain structural material of unknown composition. "End caps" similar to these intercalary spaces are found on the ends of the trichomes.

*Methanosarcina*, the other member of Group III, contains a thick, rigid, lamellar wall of moderate electron density (Figure 14.4) composed of an acidic heteropolysaccharide, the principal constituents of which are galactosamine, neutral sugars, and uronic acids. It has a high ash content, but the nature of the mineral(s) is unknown (the sulfate and phosphate content are low).

*Methanococcus*, the sole representative of Group II, has a flexible cell wall composed principally of protein, with traces of glucosamine.

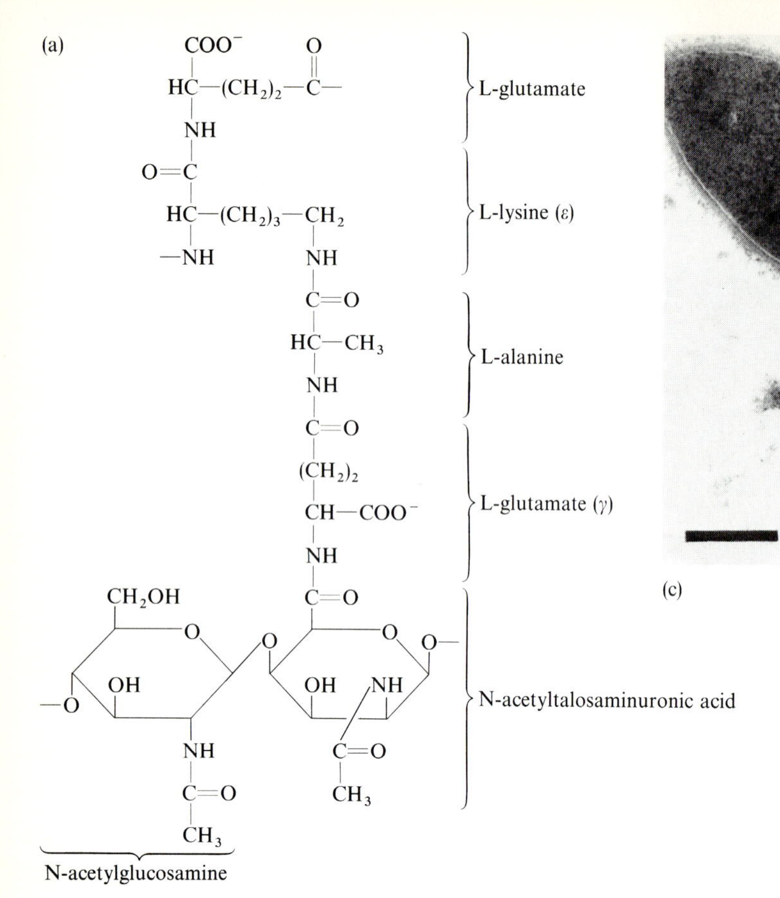


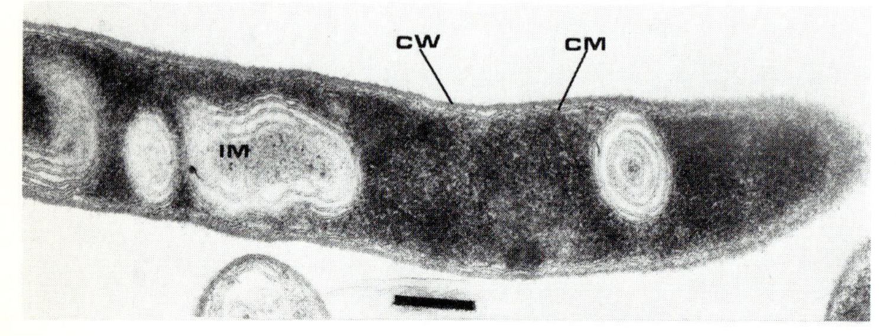


(c)

FIGURE 14.2

Methanogen cell walls containing pseudomurein. (a) The structure of the repeating unit of pseudomurein. (b) Electron micrograph of a thin section of *Methanobacterium*. CW, cell wall; CM, cell membrane; IM, intracytoplasmic membranes. The bar indicates 0.15 $\mu$m. (c) Electron micrograph of a thin section of *Methanobrevibacter*, showing the inner dense layer, intermediate layer, and outer amorphous layer. The arrow indicates apparent attachment of intracytoplasmic membranes to the site of a nascent septum. The bar indicates 0.32 $\mu$m. From J. G. Zeikus and V. G. Bowen, "Comparative Ultrastructure of Methanogenic Bacteria", *Can. J. Microbiol.* **21,** 121–129 (1975).

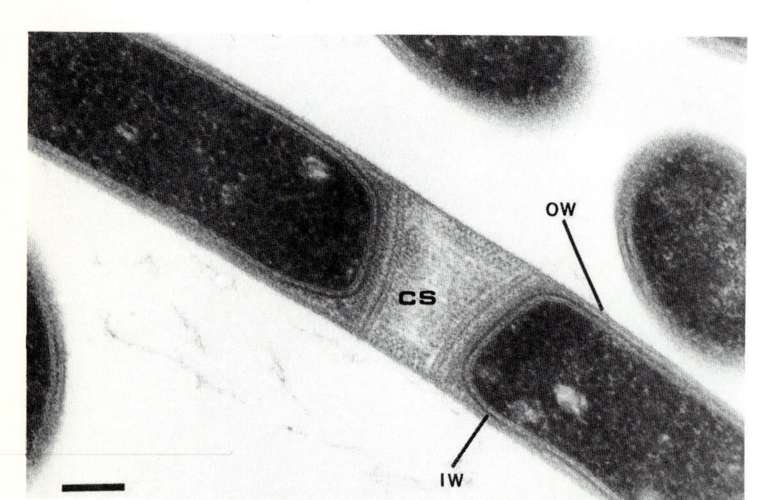


FIGURE 14.3 (left)

Electron micrograph of a thin section of *Methanospirillum*. OW, proteinaceous outer wall layer; IW, inner wall layer; CS, intercalary spacer. The bar indicates 0.13 $\mu$m. From J. G. Zeikus and V. G. Bowen, "Comparative Ultrastructure of Methanogenic Bacteria", *Can. J. Microbiol.* **21,** 121–129 (1975).

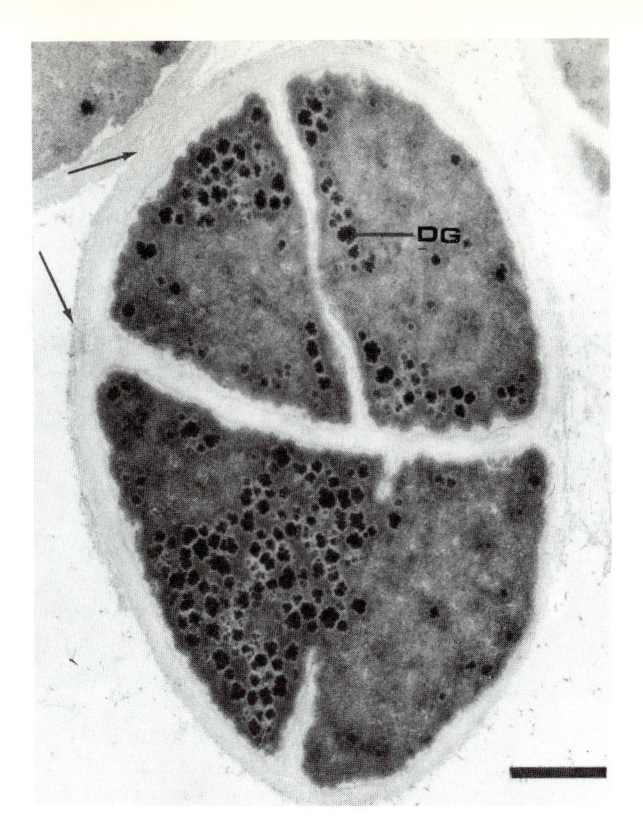


## Unique Cofactors

Methanogens contain several cofactors not found in other bacteria (Figure 14.5). Three of them (methanopterin, methanofuran, and CoM) are carriers of the $C_1$ unit during its reduction from $CO_2$ to $CH_4$ (Figure 14.6); Factor 420 probably functions as a hydrogen carrier in these reductions; and Factor 430, an unusual nickel tetrapyrrole, is the prosthetic group of methyl-CoM reductase, the last enzyme in the reduction pathway. Some methanogens require CoM as a growth factor.

FIGURE 14.4 (left)

Electron micrograph of a thin section of *Methanosarcina*. Arrows point to areas of the wall showing its lamellar structure. DG, electron-dense granules of unknown composition. The bar indicates 0.49 $\mu$m. From J. G. Zeikus and V. G. Bowen, "Comparative Ultrastructure of Methanogenic Bacteria" *Can. J. Microbiol.* **21,** 121–129 (1975).

FIGURE 14.5 (below)

Unique coenzymes of methanogens: (a) Factor 420; (b) coenzyme M; (c) Factor 430; (d) methanopterin; (e) methanofuran.

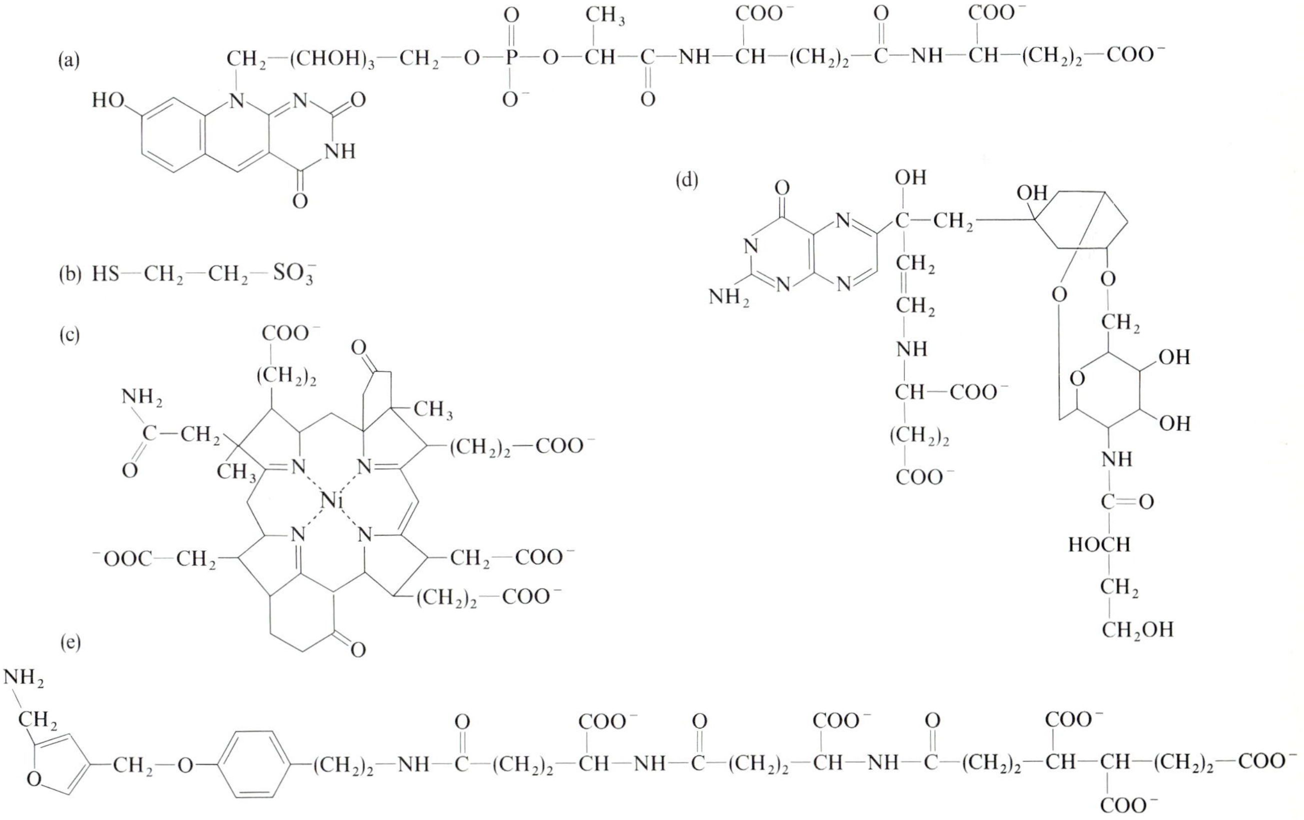

(a) $CH_2—(CHOH)_3—CH_2—O—P(=O)(O^-)—O—CH(CH_3)—C(=O)—NH—CH(COO^-)—(CH_2)_2—C(=O)—NH—CH(COO^-)—(CH_2)_2—COO^-$

HO, N, N, O, NH, O

(b) $HS—CH_2—CH_2—SO_3^-$

(c) $COO^-$, $(CH_2)_2$, O, $NH_2$, $C—CH_2$, O, $CH_3$, $(CH_2)_2—COO^-$, $CH_3$, N, N, Ni, N, N, $^-OOC—CH_2$, $CH_2—COO^-$, $(CH_2)_2—COO^-$, O

(d) O, N, N, OH, $C—CH_2$, OH, $NH_2$, N, N, $CH_2$, $CH_2$, NH, $CH—COO^-$, $(CH_2)_2$, $COO^-$, O, O, $CH_2$, O, OH, OH, NH, C=O, HOCH, $CH_2$, $CH_2OH$

(e) $NH_2$, $CH_2$, O, $—CH_2—O—C_6H_4—(CH_2)_2—NH—C(=O)—(CH_2)_2—CH(COO^-)—NH—C(=O)—(CH_2)_2—CH(COO^-)—NH—C(=O)—(CH_2)_2—CH(COO^-)—CH(COO^-)—(CH_2)_2—COO^-$

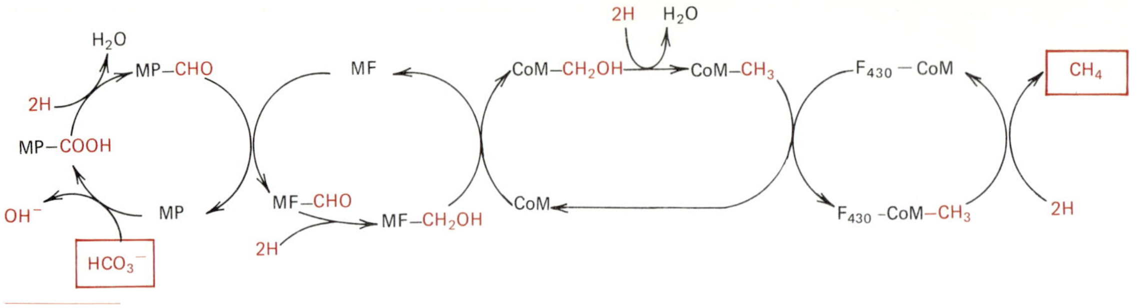


FIGURE 14.6

The pathway of methanogenesis: MP, methanopterin; MF, methanofuran; CoM, coenzyme M; $F_{430}$, Factor 430.

## Energy Metabolism

The range of compounds that serve as energy source for the methanogens is limited (Table 14.2). $H_2$ and $CO_2$ are substrates used by most methanogens; formate, acetate, methanol, and methylamines can also be utilized by some strains. The mechanism of coupling methanogenesis to synthesis of ATP is unknown, but the absence of cytochromes and quinones in most methanogens indicates that classical electron transport is not involved. Energy conservation is nevertheless probably mediated by a chemiosmotic membrane potential, because ATP concentration diminishes rapidly in the presence of agents that collapse the potential. One attractive model for the generation of the protonmotive force is shown in Figure 14.7. Hydrogen oxidation is presumed to occur on the outside of the cytoplasmic membrane, while $CO_2$ reduction occurs inside. Since the former produces protons and the latter consumes them, the net result is equivalent to pumping two protons out of the cell for every $H_2$ oxidized. Additionally, if bicarbonate ion rather than $CO_2$ is the molecule transported and condensed with methanoprotein (as shown in Figure 14.6), the hydroxyl ion production is equivalent to one additional proton pumped per methane produced.

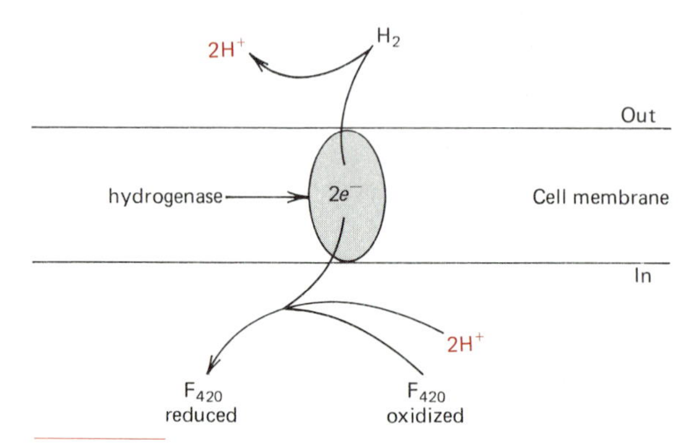


FIGURE 14.7

A mechanism for generating a protonmotive force during methanogenesis.

In addition to coupling $H_2$ oxidation to the reduction of $CO_2$, methanogens seem to be universally capable of using elemental sulfur as electron acceptor, a form of anaerobic respiration termed *sulfur reduction*, characteristic of one group of eubacteria (see Chapter 20), and of some of the thermoacidophilic archaebacteria discussed below. Methanogens of Groups I and II virtually cease methanogenesis when grown in the presence of sulfur; Group III methanogens continue methanogenesis simultaneously with sulfur reduction.

**TABLE 14.2**

**Energy-Yielding Reactions Performed by Methanogens**

| Reaction | $\Delta G^{\circ\prime}$ (kcal/mole $CH_4$) |
|---|---|
| $4H_2 + HCO_3^- + H^+ \rightarrow CH_4 + 3H_2O$ | −33 |
| $4HCOO^- + 4H^+ \rightarrow CH_4 + 3CO_2 + 2H_2O$ | −35 |
| $4CH_3OH \rightarrow 3CH_4 + CO_2 + H_2O$ | −76 |
| $4CH_3NH_3^+ + 3H_2O \rightarrow 3CH_4 + 4NH_4^+ + HCO_3^- + H^+$ | −75 |
| $CH_3COO^- + H^+ \rightarrow CH_4 + CO_2$ | −9 |

## Carbon Assimilation

Most methanogens grow well with $CO_2$ as sole carbon source; thus they are autotrophs. The pathway by which $CO_2$ is incorporated into cell material is uncertain; it is probably similar to the *carbon monoxide pathway* (see Figure 22.24) found in some sulfur-reducing bacteria and some clostridia. Enzyme assay and the patterns of labeling of cell constituents by radioactive $CO_2$ rule out the other known pathways of autotrophic $CO_2$ fixation: the ribulose diphosphate pathway (Calvin-Benson cycle, page 95) and the reductive tricarboxylic acid cycle (page 379). However, one of the two distinctive reactions of the reductive TCA cycle, namely the reductive carboxylation of acetyl-CoA, is apparently an important step in $CO_2$ fixation (Figure 14.8). The same data rule out the two pathways (the ribulose monophosphate pathway and the serine pathway; see Chapter 16) by which methophiles fix $C_1$ units more reduced than $CO_2$.

Acetate is an important precursor of cell carbon in all methanogens. However, the distribution of its label differs markedly among species, suggesting that the various groups of methanogens may have quite different patterns of intermediary metabolism (Figure 14.8).

## Ecology

Methanogens are ubiquitous in highly reducing habitats ($E_h < -0.33$ V). In some sediments methanogens are predominantly endosymbiotic within a variety of anaerobic protozoa, reaching cell densities of $10^{10}$ bacteria per milliliter of protozoal cytoplasm.

FIGURE 14.8

Pathways of acetate incorporation into amino acids in Group I methanogens (colored lines) and Groups III methanogens (black lines).

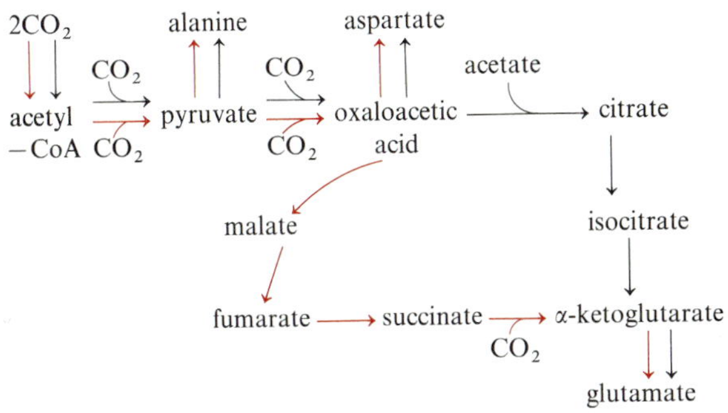

2$CO_2$ → acetyl-CoA; acetyl-CoA + $CO_2$ → pyruvate → alanine; pyruvate + $CO_2$ → oxaloacetic acid → aspartate; oxaloacetic acid + acetate → citrate → isocitrate → α-ketoglutarate → glutamate; oxaloacetic acid → malate → fumarate → succinate + $CO_2$ → α-ketoglutarate

**TABLE 14.3**

**Products of the Fermentation of Glucose by *Selenomonas rumanantium* in Pure Culture and In Co-culture with *Methanobacterium rumanantium***

| | Moles per 100 Moles Glucose | |
|---|---|---|
| PRODUCT | Selenomonas ALONE | Selenomonas + Methanobacterium |
| Lactate | 156 | 68 |
| Acetate | 46 | 99 |
| Propionate | 27 | 20 |
| Formate | 4 | 0 |
| Methane | 0 | 51 |
| $CO_2$ | 42 | 48 |

Source: From M. Chen and M. J. Wolin, *App. Env. Microbiol.* **34,** 756 (1977).

The impact of the methanogens on their environment is substantial. As mentioned earlier, they convert fermentation end products such as formate and acetate to a gaseous product that can diffuse into an aerobic zone, thereby preventing sequestration of substantial amounts of organic material that are only slowly metabolized anaerobically. Another important effect of methanogens on their environment stems from their efficient hydrogenase, the $K_m$* of which is low enough to maintain a vanishingly low partial pressure of $H_2$ where active methanogenesis is occurring. This low $P_{H_2}$ in turn allows a number of fermentative anaerobes to reoxidize NADH by means of an NADH-coupled hydrogenase:

$$NADH + H^+ \rightleftharpoons NAD^+ + H_2$$

Such hydrogenases are widely distributed among anaerobes, but they cannot reoxidize NADH when the organism is grown in pure culture because the equilibrium lies far to the left. However, when grown in mixed culture with methanogens, the low $P_{H_2}$ which results from the continual removal of hydrogen by methanogenesis shifts the equilibrium and the reaction proceeds. This in turn shifts the fermentation balance towards more oxidized end products (Table 14.3), with correspondingly higher yields of ATP and biomass. This phenomenon has been termed *interspecies hydrogen transfer.*

* The $K_m$ of an enzyme is a measure of its affinity for its substrate; the lower the value of the $K_m$, the higher the affinity. Numerically, the $K_m$ is equal to the substrate concentration at which the rate of the enzymatic reaction is half of its maximal rate at saturating substrate concentrations.

## THE HALOPHILES

Highly saline environments (salt lakes, brines) harbor large populations of a small and distinctive group of bacteria: immotile cocci (*Halococcus*) and polarly flagellated rods (*Halobacterium*). Despite the differences of cell form, these organisms share a number of properties, many of which are clearly adaptations to the high salinity and high light intensities of their natural habitat. The *minimum* NaCl concentration that permits growth is 2 to 2.5 M, the optimum 4 to 5 M; the $Mg^{2+}$ requirement is also very high (about 0.025 M). The intracellular salt concentration is at least as high as the extracellular one, although the intracellular ionic composition is different, being composed largely of $Na^+$, $K^+$, and $Cl^-$. The biochemical machinery of the cell (enzymes, ribosomes) is not only salt-tolerant, but salt-requiring, and functions effectively only at near-saturated salt concentrations.

A characteristic feature of the extreme halophiles is their content of red carotenoids, which are incorporated into the cell membrane. The carotenoids of *Halobacterium* have been shown to protect the cell from photochemical damage by the high light intensities characteristic of the natural environment (see Chapter 8).

Extreme halophiles are aerobic chemoheterotrophs with complex nutritional requirements, and are commonly cultured in peptone-containing media; amino acids are the preferred carbon and energy sources. These bacteria frequently develop as colored patches on salted dried fish and hides as a result of treatment with salt that contains them. They also give the ponds in which salt is recovered by evaporation from sea water their characteristic brick-red color.

### The Halophile Genome

The genome of both *Halobacterium* and *Halococcus* contain two components with different percent G + C values. The majority of the DNA has a G + C content of 66 to 68 percent while the minority component (associated principally with a large plasmid at least in *Halobacterium*) has a G + C content of 57 to 60 percent. There is some evidence that the plasmid encodes the ability to make gas vacuoles (a common property of the halophiles) and the pigment bacteriorhodopsin (see below).

The total genome size of *Halobacterium* is about $2.5 \times 10^9$. It contains a large number of different repeated sequences up to about 5,000 base pairs in length, each present in up to 20 or more copies. They are found on both the chromosome and the plasmid. At least some are similar in molecular structure to eubacterial transposable elements (p. 279), and have been shown to be highly mobile.

### Cell Walls of Halophiles

The principal constituent of the cell wall of *Halobacterium* is a large (MW = 200,000), acidic glycoprotein. Removal of this glycoprotein either by dilution of the suspending medium (monovalent cations are presumably needed to neutralize the high density of negative charges due to the large number of acidic amino acids) or by proteinase treatment results in a loss of cell shape and, in dilute media, osmotic lysis. Thus this protein is the principal structural component of the wall. Its glycan component consists of 22 to 24 disaccharides (glucosylgalactose) linked via O-glycosidic bonds to threonine residues; 12 to 14 trisaccharides (containing glucose, galactose, and a hexuronic acid), also O-linked to threonines; and a single hetero-oligosaccharide (containing a number of N-acetyl sugars) in N-glycosidic linkage to asparagine.

The assembly of these oligosaccharides and their linkage to the protein, like the assembly of the peptidoglycan in eubacteria, occurs while attached to a polyisoprenoid phosphate carrier. After assembly of the oligosaccharide, the complex moves through the membrane and transfers the oligosaccharide to the protein. Hydrolysis of the pyrophosphate regenerates the carrier, which simultaneously moves back through the membrane. The antibiotic bacitracin inhibits regeneration; hence in its presence walls are made that lack sugar residues. Growth of *Halobacterium* in the presence of bacitracin has the same result as growth in the presence of proteases: the rod-shaped cells become spherical.

In addition to the glycoprotein, the cell envelope of halobacteria contains nonglycosylated protein and glycolipid.

The wall of *Halococcus* is chemically and structurally quite different from that of *Halobacterium*. It resembles the wall of Gram-positive eubacteria (Figure 14.9) and, unlike the proteinaceous wall of *Halobacterium*, it retains its integrity when the salt concentration of the suspending medium is reduced. Chemically, however, the halococcal wall bears little resemblance to eubacterial walls. Its principal constituent is a complex sulfated heteropolysaccharide composed of several neutral sugars, uronic acids, and amino sugars (many of which are N-acetylated). The glycan strands are probably cross-linked by glycine residues that bridge the

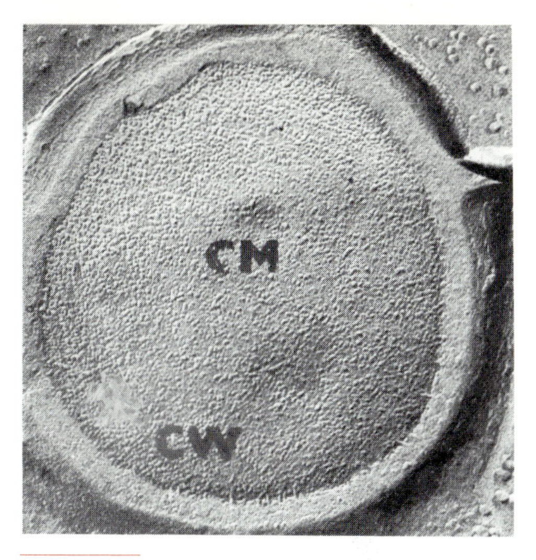


FIGURE 14.9

Electron micrograph of a freeze-fractured preparation of *Halococcus*, showing the interior of the cell membrane (CM) and the thick wall (CW) ($\times$ 69,700). From M. Kocur, B. Smid, and T. Martinec, "The fine structure of extreme halophilic cocci," *Microbios* 5, 101–107 (1972).

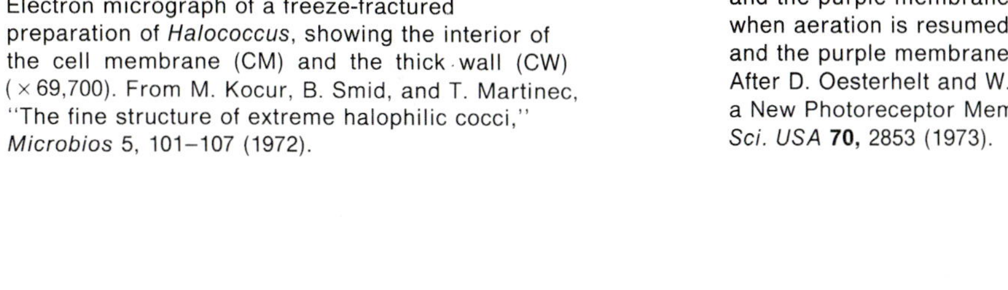

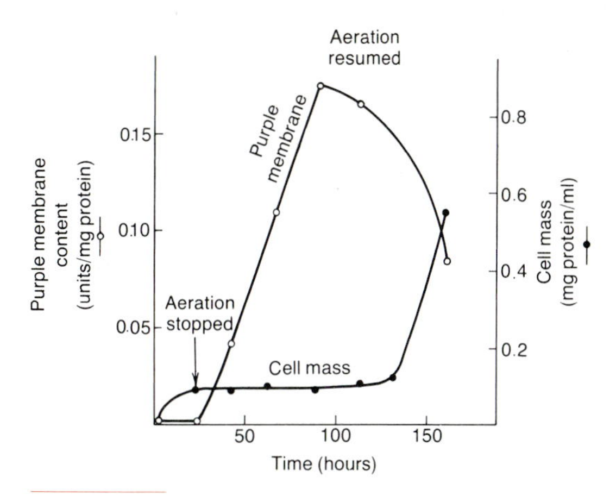


FIGURE 14.10

The effect of $O_2$ on the growth and synthesis of purple membrane in a liquid culture of *Halobacterium*. When the culture is no longer aerated, growth ceases, and the purple membrane content of the cells starts to rise; when aeration is resumed, growth once again starts and the purple membrane content of the cells declines. After D. Oesterhelt and W. Stoeckenius, "Functions of a New Photoreceptor Membrane," *Proc. Natl. Acad. Sci. USA* **70,** 2853 (1973).

amino groups of amino sugars and the carboxyl groups of uronic acids. If this postulated structure is in fact correct, the halococcal polymer represents a third type of peptidoglycan, along with eubacterial murein and the pseudomurein of the Group I methanogens.

FIGURE 14.11

Electron micrograph of a freeze-etched preparation of *Halobacterium* cells in 4.3 M NaCl, showing surface structure of the cell membrane ($\times$ 35,000). The areas composed of purple membrane (pm) are recognizable by their smooth surfaces. Courtesy of Dr. W. Stoeckenius.

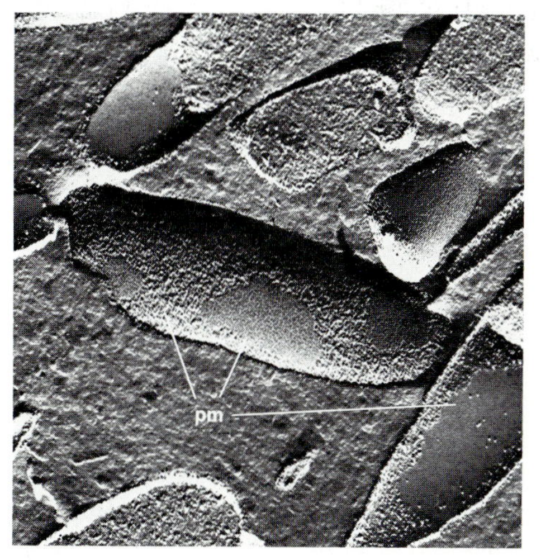


## Photophosphorylation in *Halobacterium*

Most halophiles have an obligately respiratory metabolism; however, some strains of *Halobacterium* can also generate ATP by a novel form of photophosphorylation. Realization of this capability developed out of a series of experiments, performed chiefly by W. Stoeckenius and his collaborators, on the *purple membrane* of *Halobacterium.* When competent strains of halobacteria are subjected to oxygen limitation, they synthesize a chemically modified cell membrane. In aerobically grown cells, the membrane is red, as a result of its high carotenoid content. A shift to anaerobiosis stops growth and induces the synthesis of a new component, the purple membrane (Figure 14.10). Purple membrane is laid down in discrete patches that can account for as much as half of the total membrane area. It is readily distinguishable in electron micrographs of freeze-fractured cells (Figure 14.11).

Purple membrane is composed of lipid (25 percent) and a single species of protein, termed *bacteriorhodopsin* because of its similarities to the vertebrate visual pigment rhodopsin (although the two pigments are quite similar structurally, their photochemistry differs in important ways). Bacteriorhodopsin has seven regions of alpha-helical structure oriented across the membrane; hence the protein is in contact with both the cytoplasm and the external medium.

The chromophore of bacteriorhodopsin is the carotenoid *retinal* (Figure 14.12), linked via a Schiff base to the epsilon amino group of a lysine residue.

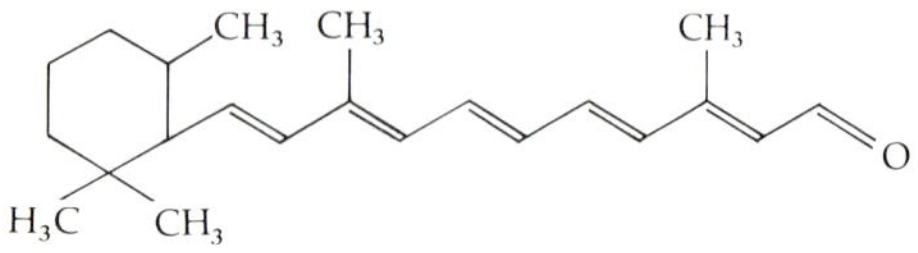

FIGURE 14.12

The structure of retinal, the $C_{20}$ carotenoid which is the light-sensitive chromophore of the vertebrate visual chromoprotein, rhodopsin, and the purple membrane chromoprotein, bacteriorhodopsin, of *Halobacterium*.

Illumination of the protein induces conformational changes in the retinal, and concommitant deprotonation of the Schiff base (Figure 14.13). Return of the retinal to its ground state is accompanied by reprotonation of the Schiff base. Protonation consumes protons from the inside of the membrane, and deprotonation releases them to the outside. Therefore, in the light *bacteriorhodopsin acts as a proton pump, converting light energy to a proton-motive force across the cell membrane*, which can be used to maintain cellular ATP levels.

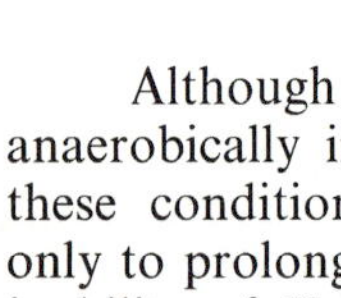

$$\text{lysine—N=R} \;\; \underset{H^+_{out}}{\overset{H^+_{in}}{\rightleftharpoons}} \;\; \text{lysine—NH}^+\text{=R}$$

FIGURE 14.13

The reversible protonation of the Schiff base by which retinal (R) is linked to apobacteriorhodopsin.

Although *Halobacterium* can generate ATP anaerobically in the light, it cannot grow under these conditions; photophosphorylation is used only to prolong viability of nongrowing cells. The inability of *Halobacterium* to grow anaerobically in the light reflects the requirement for molecular oxygen to form retinal from beta-carotene. Photoheterotrophic growth thus occurs only if the partial pressure of oxygen is low enough to induce the synthesis of bacteriorhodopsin but high enough to permit retinal biosynthesis. Since the solubility of $O_2$ in brine is very low ($O_2$ concentrations of less than 0.2 mg/liter have been measured in *surface* waters of the Great Salt Lake), $O_2$-limitation must be a common form of environmental stress encountered by halophiles.

## THE THERMOACIDOPHILES

The thermoacidophiles are a heterogeneous group (Chapter 13) defined by their ability to grow at high temperature and low pH. Several subgroups have been identified: *Sulfolobus*, *Thermoplasma*, and the *Thermoproteus* subgroup.

### *Sulfolobus*

Hot acid springs and soils all over the world contain substantial numbers of the facultative chemoautotroph *Sulfolobus*, an irregularly lobate spherical organism. Temperature optima for growth vary among isolates from 63° to 80° C; the pH optimum for growth is normally around 2.0. The range of conditions over which *Sulfolobus* will grow is fairly wide: temperature of 55° to 85° C; pH of 1 to 5.9. The G + C content is 60 to 68 percent.

The cell wall of *Sulfolobus* is composed principally of lipoprotein and carbohydrate (containing both neutral sugars and amino sugars), which form a distinct layer outside the cell membrane (Figure 14.14).

Although *Sulfolobus* can grow on organic compounds (probably fermentatively), in its natural habitats it probably grows as a respiratory chemoautotroph. Geothermal steam or hot water leaches substantial amounts of iron and of sulfide, which is rapidly oxidized to elemental sulfur by oxygen or ferric iron, either chemically or biologically. *Sulfolobus* rapidly oxidizes $H_2S$, but coupling of this oxidation to growth has not been shown. The principal growth substrate in hot springs (and presumably in hot soils as well) is elemental sulfur; *Sulfolobus* grows on the surface of droplets or crystals of sulfur (Figure 14.14), oxidizing it to sulfuric acid, which is largely responsible for the acidity of these habitats.

The identity of the natural acceptor of the electrons removed from sulfur is unclear; in the laboratory, molecular oxygen, molybdate, or ferric ion will serve. The solubility of $O_2$ in water near its boiling point is extremely low; in nature ferric ion is probably the major electron acceptor. Its reduction potential is close to that of $O_2$ (0.77 V versus 0.82 V), and hence its reduction by sulfur is thermodynamically capable of being coupled to growth.

## *Thermoplasma*

*Thermoplasma* is a facultative anaerobe that uses a small number of mono- and disaccharides as sources of carbon and energy. All strains have an absolute requirement for an unusual growth factor, which can be satisfied by an oligopeptide (MW about 1,000) from yeast extract. *Thermoplasma* is found only in the refuse piles from coal mines, which contain residual coal and substantial amounts of iron pyrite (FeS). Oxidation of the FeS by chemoautotrophic bacteria acidifies and heats the pile, creating a favorable growth environment for *Thermoplasma* (pH 0.5 to 4.5; temperature 37° to 65° C). The heating is probably also necessary to pyrrolyze the large organic molecules in coal spoils to smaller ones utilizable by *Thermoplasma.* It is not known if the growth factor supplied from pyrrolized coal is the same as the one found in yeast. Since the only known habitat of *Thermoplasma* is the result of recent human activity, its original habitat remains a mystery. Extensive attempts to isolate *Thermoplasma* from acidic hot springs or animal stomachs have uniformly failed. Perhaps its true natural habitat is coal veins exposed to the atmosphere by geological processes, or even subterranean coal deposits (bacteria are known to inhabit deep hydrocarbon bearing rock strata that are perfused with water).

*Thermoplasma* lacks a cell wall (Figure 14.15), but its cell membrane contains large amounts of lipopolysaccharide and glycoprotein, both of which contain mannose as the principal sugar monomer. The structure of the lipopolysaccharide is thought to be

$$\text{mannose}_{24}\text{-glucose-glycerol diether}$$

The glycan moiety of the glycoprotein, a highly branched structure containing over 60 mannose units, is probably linked via an N-acetylglucosamine moiety to an asparagine residue. Together, these two compounds probably form a layer of polysaccharide on the outer membrane surface. The glycan strands may interact with themselves or with membrane proteins to confer substantial rigidity of the membrane.

The presence of a rigid glycan pellicle also helps to explain another peculiarity of *Thermoplasma*; namely, its motility by means of flagella, a process that requires a rigid surface layer (Chapter 6).

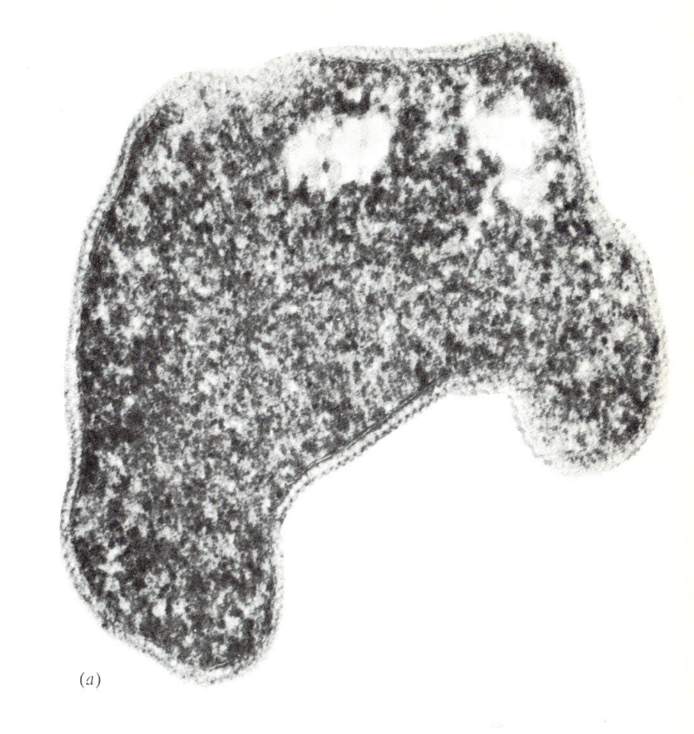


FIGURE 14.14
*Sulfolobus acidocaldarius.* (a) Electron micrograph of a thin section (×83,000). From T. D. Brock, K. M. Brock, R. T. Belly, and R. L. Weiss, *Arch. Mikrobiol.* **84**, 64 (1972).
(b) Fluorescent photomicrograph of cells stained with acridine orange attached to elemental sulfur crystal (×750).

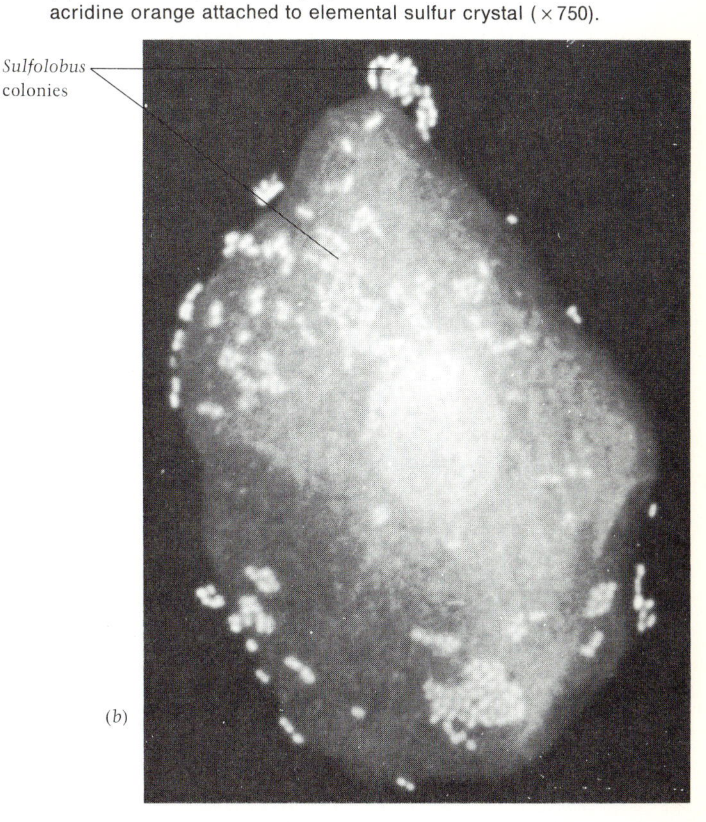

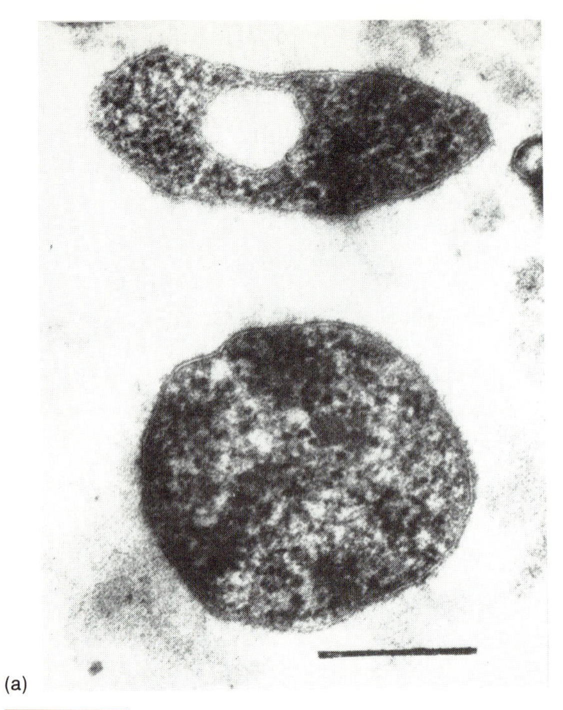

(a)

(b)

FIGURE 14.15

*Thermoplasma acidophilum.* (a) electron micrograph of a thin section, showing the absence of a defined cell wall, although a layer of amorphous material (presumably the glycan moieties of lipopolysaccharide and glycoproteins) is visible external to the cytoplasmic membrane. (b) Electron micrograph of a negatively stained cell, showing the flagellum. The bars indicate 0.25 $\mu$m. From F. T. Black, E. A. Freundt, O. Vinther, and C. Christiansen, "Flagellation and swimming motility of *Thermoplasma* acidophilum," J. Bacteriol, **137,** 456–460 (1979).

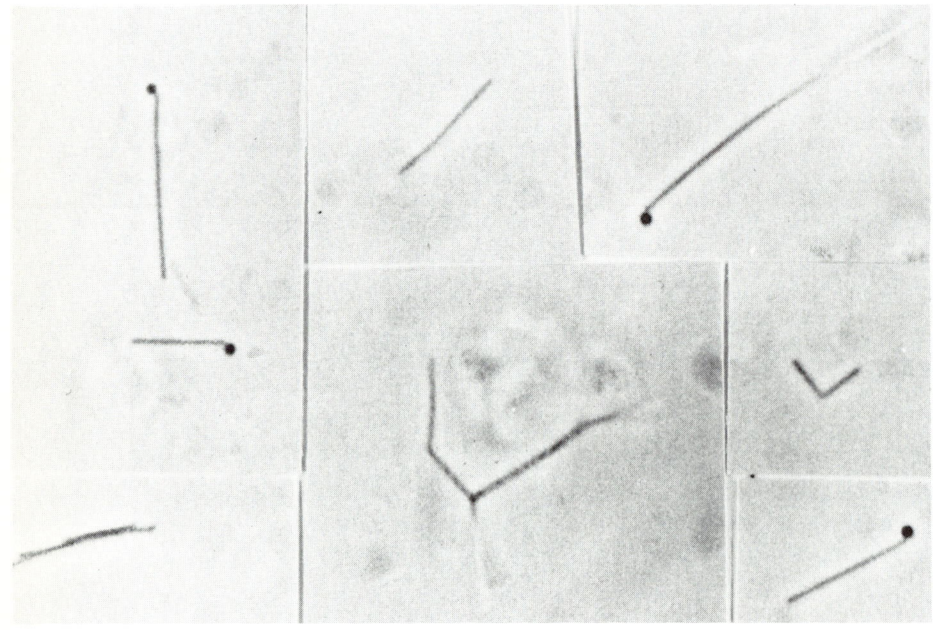

FIGURE 14.16

Phase-contrast light micrograph of cells of *Thermoproteus.* $\times 3000$. From W. Zillig, K. O. Stetter, W, Shäfer, D. Janekovic, S. Wunderl, J. Holz, and P. Palm, "Thermoproteales: A novel type of extremely thermoacidophilic anaerobic archaebacteria isolated from Icelandic solfataras," *Zbl. Bakt. Hyg.* I, Abt. Orig. C2, 205–227 (1981).

*Thermoplasma* DNA has a G + C content of about 46 percent, and a genome size estimated to be $8 \times 10^8$ to $10 \times 10^8$ daltons. If it truly does lie in the lower end of this range, it would be the smallest known genome of a free-living microorganism.

### The *Thermoproteus* Group

The *Thermoproteus* group consists of thermoacidophiles whose metabolism is principally or solely respiratory, with elemental sulfur serving as terminal electron acceptor. The product of sulfur reduction is $H_2S$, which is also a required nutrient for organisms in this group (perhaps as a source of cell sulfur). They are all strict anaerobes; however, although growth is prevented by traces of oxygen, viability is maintained for long periods of aerobic incubation. Their G + C content is 50 to 56 percent. Their glycerol lipids include both diphytanyl diethers and dibiphytanyl tetraethers. The cell wall appears to consist of glycoprotein. To date the only habitats that have yielded organisms of this group are geothermal areas of Iceland, where they are ubiquitous. Two genera within the group have been characterized: *Thermoproteus* and *Desulfurococcus.*

*Thermoproteus* strains are long thin rods, with a maximum diameter of 0.5 $\mu$m and a length up to 80 $\mu$m. They are often sharply bent in the middle to produce a V-form and are occasionally branched (Figure 14.16). Their metabolism is strictly respiratory, with a variety of sugars, alcohols, organic acids (including formate), and CO serving as respiratory substrates. Organic growth factors are not required. They grow over a temperature range of 78° to 95°C, and a pH range of 2.5 to 6.

*Desulfurococcus* strains are spherical cells, which may be flagellated. They are capable of respiratory or fermentative growth at the expense of substrates present in yeast extract or tryptone, but cannot use any of a variety of sugars, alcohols, or organic acids. They grow over a temperature range of 75° to 95° C and a pH range of 5 to 7.

## FURTHER READING

**Books**

BROCK, T. D., *Thermophilic Microorganisms and Life at High Temperature.* New York: Springer-Verlag, 1978.

KANDLER, O., *Archaebacteria.* New York: Gustav Fischer, 1982.

# Chapter 15
# The Photosynthetic Eubacteria

The Gram-negative procaryotes include three distinct and well-defined groups of photosynthetic eubacteria. The *cyanobacteria* or blue-green bacteria perform oxygenic photosynthesis and possess a pigment system similar in basic respects to that of photosynthetic eucaryotes. This group has long been treated by botanists as one of the major divisions of the algae. However, the typically procaryotic cell structure of these organisms, which was clearly established about 1960, identifies them unambiguously as bacteria. This large and structurally diverse group includes many different types of filamentous and unicellular organisms. Movement, when it occurs, is by gliding.

*Purple bacteria* and *green bacteria* perform anoxygenic photosynthesis and possess unique pigment systems that confer on them spectral properties unlike those of all other phototrophs. Green bacteria are small, immotile, rod-shaped organisms or filamentous gliding ones; the purple bacteria are unicellular rods, cocci, or spirilla, frequently motile by means of flagella. Because of the evident structural resemblance of these organisms to nonphotosynthetic bacteria, their taxonomic position has neven been questioned; they have been included among the bacteria since their discovery in the mid-nineteenth century. However, the nature of their metabolism remained controversial for many decades.

About 1885 W. Engelmann suggested that the purple bacteria might be photosynthetic organisms, as a result of his discovery that they are phototactic, and that their growth is favored by light. Repeated attempts to demonstrate oxygen production by these organisms in the light failed; the absence of this property appeared at the time to be strong evidence against Engelmann's hypothesis. Moreover, S. Winogradsky had shown that some purple bacteria can oxidize $H_2S$ to sulfate with transient intracellular accumulation of elemental sulfur, an unusual property also possessed by certain chemoautotrophic bacteria (see Chapter 16). About 1905 W. Molisch observed that other purple bacteria can grow, either in the light or in the dark, in complex organic media, and do not oxidize $H_2S$. The seemingly irreconcilable reports of Engelmann, Winogradsky, and Molisch remained without a coherent explanation until 1930, when C. B. van Niel first recognized and defined the various metabolic versions of anoxygenic photosynthesis and demonstrated that it is the characteristic mode of energy-yielding metabolism in both purple and green bacteria.

The discovery of a novel mode of photophosphorylation in *Halobacterium* (see Chapter 14) opened a new chapter in the study of phototrophic growth. It is now clear that at least two very different chemical mechanisms for trapping light energy in a biologically useful form have evolved in the context of the procaryotic cell: the chlorophyll-dependent electron transport systems of the eubacteria (and their descendants, the chloroplasts); and the bacteriorhodopsin-containing proton pump of the halobacteria.

## COMMON PROPERTIES OF PHOTOSYNTHETIC EUBACTERIA

All known photosynthetic eubacteria are capable, at least facultatively, of autotrophic growth; therefore they are able to fix $CO_2$. However, this common phenotypic trait conceals a fundamental biochemical dichotomy occurring in the pathway by which $CO_2$ is assimilated: the cyanobacteria and purple bacteria employ the ribulose bisphosphate pathway (the Calvin-Benson cycle; see Chapter 4), while the green bacteria utilize a novel pathway, the *reductive tricarboxylic acid cycle* (see below). There are also differences in the role that photoassimilation of organic compounds plays in the metabolism of procaryotic phototrophs. All of them possess this capability, but its quantitative contribution to cellular carbon varies substantially. Many purple bacteria can grow wholly as photoheterotrophs; organic compounds can serve as sole source of cell carbon. However many green bacteria and most cyanobacteria derive the bulk of their cell carbon from $CO_2$ even when they are provided with any of a variety of organic materials.

### Organization of the Photochemical Apparatus

The molecular organization of the photochemical apparatus is basically the same in all phototrophic eubacteria (Chapter 4). It consists of three major components: a primary light harvesting *pigment antenna*, a *reaction center* (at which the photochemical event occurs), and an *electron transport system*.

The primary photochemical event of photosynthesis is initiated by the absorption of the energy of a photon by a molecule in the pigment antenna. This energy is transferred among molecules within the antenna until it reaches the reaction center where it promotes one of the pi electrons of the chlorophyll tetrapyrrole into an outer orbital. As a consequence the chlorophyll molecule becomes for a brief period a powerful reductant, and is immediately oxidized by a closely associated electron acceptor. Thus, the energy originally present in the photon is conserved as a separation of electric charge which can be used to do chemical work (Figure 15.1).

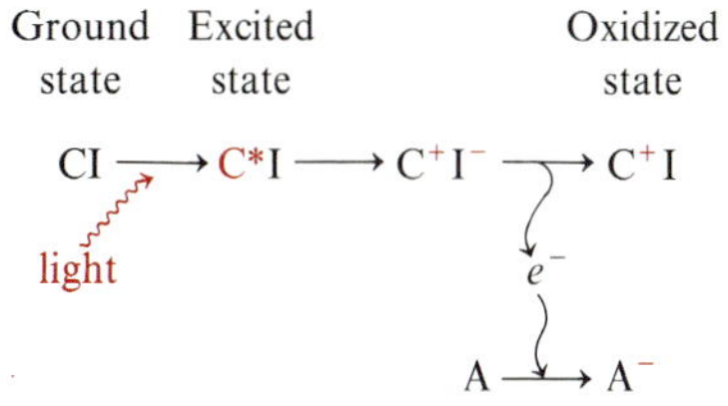

Ground state | Excited state | Oxidized state

$$CI \xrightarrow{\text{light}} C^*I \longrightarrow C^+I^- \longrightarrow C^+I$$

$$e^-$$

$$A \longrightarrow A^-$$

FIGURE 15.1
The primary photochemical events
C = reaction center chlorophyll
I = immediate electron acceptor (chlorophyll or pheophytin)
A = first stable reduced electron carrier (iron sulfur protein or quinone)

One of the ways that chemical work is done is through a process termed *cyclic photophosphorylation* in which electrons are passed through a closed loop of electron carrier molecules back to the chlorophyll molecule, now an oxidant, that lost the electron in the primary photochemical event (Chapter 4). The photosynthetic electron transport system consists of quinones and cytochromes located in the cell or thylakoid membrane. By passage of electrons through the system, protons are pumped across the membrane, thereby generating a proton-motive force that can be used to generate ATP. All phototrophic eubacteria are capable of performing cyclic photophosphorylation.

**TABLE 15.1**

**Constituents of the Photochemical Systems of Eubacteria**

| Component | Cyanobacteria | Green Bacteria | Purple Bacteria |
|---|---|---|---|
| Reaction center pigment | Chlorophyll *a* | Bacteriochlorophyll *a* | Bacteriochlorophyll *a* or *b* |
| Antenna pigment | Phycobiliproteins and chlorophyll *a* | Bacteriochlorophyll *c*, *d*, or *e* | Bacteriochlorophyll *a* or *b* |
| Principal carotenoids | Bicyclic | Aryl | Aliphatic (often methoxylated) |
| Distinctive lipids | mono- and digalactosyl diglycerides | monogalactosyl diglycerides and triglyco diglycerides | Ornithine lipids |
| Unsaturated fatty acids | + or − | − | − |

**TABLE 15.2**

**Chemical Differences among Chlorophylls**

| | $R_1$ | $R_2$ | $R_3$ | $R_4$ | $R_5$ | $R_6$ | $R_7$ |
|---|---|---|---|---|---|---|---|
| Chlorophyll *a* | | | | | | | |
| | $-CH{=}CH_2$ | $-CH_3$ | $-C_2H_5$ | $-CH_3$ | $-C(=O)-O-CH_3$ | Phytyl | $-H$ |
| Bacteriochlorophylls *a*, *b*, and *g* | | | | | | | |
| *a* | $-C(=O)-CH_3$ | $-CH_3$[a] | $-C_2H_5$[a] | $-CH_3$ | $-C(=O)-O-CH_3$ | Phytyl, geranyl-geranyl, or Farnesyl | $-H$ |
| *b* | $-C(=O)-CH_3$ | $-CH_3$[b] | $=CH-CH_3$[b] | $-CH_3$ | $-C(=O)-O-CH_3$ | Phytyl | $-H$ |
| *g* | $-CH{=}CH_2$ | $-CH_3$[b] | $=CH-CH_3$[b] | $-CH_3$ | $-C(=O)-O-CH_3$ | Geranyl-geranyl | $-H$ |
| Bacteriochlorophylls *c*, *d*, and *e* | | | | | | | |
| *c* | $-CH(OH)-CH_3$ | $-CH_3$ | $-C_2H_5$ | $-C_2H_5$ | $-H$ | Farnesyl | $-CH_3$ |
| *d* | $-CH(OH)-CH_3$ | $-CH_3$ | $-C_2H_5$ | $-C_2H_5$ | $-H$ | Farnesyl | $-H$ |
| *e* | $-CH(OH)-CH_3$ | $-C(H){=}O$ | $-C_2H_5$ | $-C_2H_5$ | $-H$ | Farnesyl | $-CH_3$ |

[a] No double bond between C—3 and C—4; additional —H atoms at C—3 and C—4.
[b] No double bond between C—3 and C—4; additional —H atom at C—3.

## DIFFERENCES AMONG THE MAJOR GROUPS OF PHOTOTROPHIC EUBACTERIA

The three principal groups of phototrophic eubacteria are distinguished by three properties: the molecular constituents of the photochemical apparatus, the structural arrangement of the photochemical apparatus, and the mechanism of photochemical generation of reductant.

### Chemistry of the Photochemical Apparatus

The photochemical apparatus always consists of a set of pigments and electron carriers embedded in or associated with a membrane system, but the chemical composition of its components vary as does the membrane that contains them (Table 15.1).

PHOTOSYNTHETIC PIGMENTS Each photosynthetic eubacterial group has its own distinctive combination of photosynthetic pigments. These various pigments are discussed in the following sections.

*Chlorophylls* are cyclic tetrapyrroles within which a magnesium ion is chelated (Figure 15.2); *in vivo* the chlorophylls are bound noncovalently to membrane proteins. The chlorophyll of cyanobacteria is *chlorophyll a*, the same one found in all eucaryotic phototrophs. The chlorophylls of purple and green bacteria are termed *bacteriochlorophylls*. Figure 15.2 and Table 15.2 show the chemical differences between the main bacteriochlorophylls and chlorophyll *a*. The bacteriochlorophylls can be divided into two classes on the basis of similarity of chemical structure: *a*, *b*, and *g* comprise one subclass and *c*, *d*, and *e* comprise the other.

Purple bacteria contain only one form of bacteriochlorophyll, either *a* or *b*, depending on the particular species. Some molecules of it participate in light-harvesting while others are in the photochemical reaction centers. All green bacteria contain two types of bacteriochlorophyll in unequal amounts. Depending on the species the predominant form is either bacteriochlorophyll *c*, *d*, or *e*; it has a light-harvesting role. The minor form in all green bacteria is bacteriochlorophyll *a*; it is in the photochemical reaction centers.

In spite of their chemical differences the three light-harvesting chlorophylls that occur in green bacteria (bacteriochlorophylls *c*, *d*, and *e*) all closely resemble chlorophyll *a* in spectral properties (Table 15.3). In contrast, the two forms of bacteriochlorophyll (*a* and *b*) that play the corresponding role in purple bacteria share the chemical feature of saturation of one double bond in the ring system (between C—3 and C—4) that profoundly affects their spectral properties, changing the positions of all the major absorption peaks relative to those of other chlorophylls. The long wavelength peak of

FIGURE 15.2

The chemical structure common to chlorophylls. The nature of variable substituents ($R_1$ through $R_7$) in chlorophyll a and in the various bacteriochlorophylls is shown in Table 15.2.

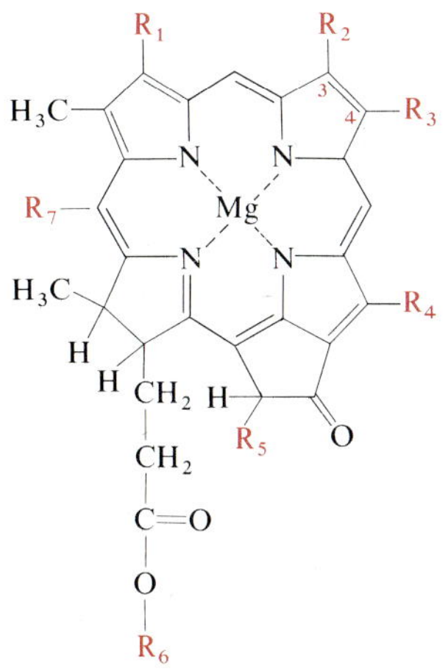

**TABLE 15.3**

**The Positions of the Long Wavelength Peaks of the Chlorophylls of Procaryotes in Organic Solvents (Ether or Acetone) and in the Intact Cells**

| Biological Group | Position of Peaks (nm) in: ORGANIC SOLVENTS | CELLS | Magnitude of *in vivo* shift (nm) |
|---|---|---|---|
| Cyanobacteria | | | |
| Chlorophyll *a* | 662 | 680–685 | 18–23 |
| Purple bacteria | | | |
| Bacteriochlorophyll *a* | 773 | 850–910 | 78–137 |
| Bacteriochlorophyll *b* | 795 | 1,020–1,035 | 225–240 |
| Green bacteria | | | |
| Bacteriochlorophyll *c* | 660 | 750–755 | 90–95 |
| Bacteriochlorophyll *d* | 654 | 725–735 | 71–79 |
| Bacteriochlorophyll *e* | 647 | 715–725 | 68–78 |
| Bacteriochlorophyll *a*[a] | 773 | 805–810 | 32–37 |

[a] This bacteriochlorophyll, the only one common to all green bacteria, is always a minor pigment in this group, and is represented in the cellular absorption spectrum by a peak that is very small relative to that of the major, group-specific bacteriochlorophyll (*c*, *d*, or *e*).

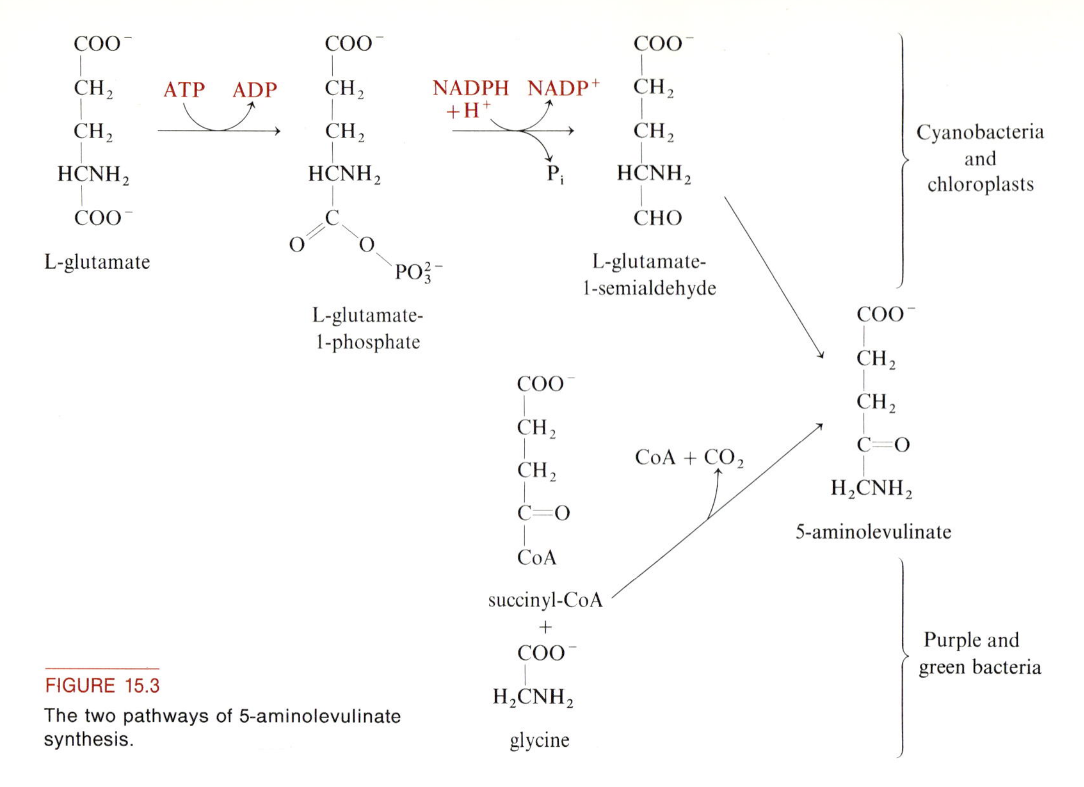


FIGURE 15.3

The two pathways of 5-aminolevulinate synthesis.

bacteriochlorophylls *a* and *b* lies over 100 nm farther toward the red than that of other bacterial and plant chlorophylls, being situated very close to the infrared region.

Both chlorophyll *a* and the bacteriochlorophylls are synthesized from 5-aminolevulinate (ALA) by substantially the same biochemical pathways. There is, however, a dichotomy among the phototrophic eubacteria with respect to the biosynthesis of ALA. In purple and green bacteria, ALA is a product of a condensation between glycine and succinyl-CoA, whereas in cyanobacteria and chloroplasts it is formed from L-glutamic acid (Figure 15.3).

*Pheophytins* are chlorophylls without a chelated magnesium atom. Pheophytin *a* is the intermediate electron carrier in the reaction center ("I" in Figure 15.1) of plant and cyanobacterial photosystem II. The photosystem of green sulfur bacteria and the cyanobacterial photosystem I contain bacteriochlorophyll *c* or additional chlorophyll *a* molecules respectively instead of pheophytin. In the photosystems of purple and of green non sulfur bacteria, the carrier is bacteriopheophytin *a*, or bacteriopheophytin *b* in those purple bacteria that contain bacteriochlorophyll *b*.

The *phycobiliproteins*, which are the major light-harvesting pigments of both cyanobacteria and red algae, are water-soluble proteins that contain covalently bound linear tetrapyrroles (bilins) as chromophores (Figure 15.4). They absorb light in a broad region near the middle of the visible spectrum, and belong to three principal spectral classes (Figure 15.5). The two blue pigments, allophycocyanin and phycocyanin, which have absorption maxima at relatively long wavelengths, occur universally in cyanobacteria and red algae; the red pigment, phycoerythrin, which absorbs at shorter

FIGURE 15.4

Structures of the chromophores of phycobiliproteins: (a) phycocyanobilin, the chromophore of phycocyanin and allophycocyanin; (b) phycoerythrobilin, the chromophore of phycoerythrin. Both are covalently linked to the proteins with which they are associated.

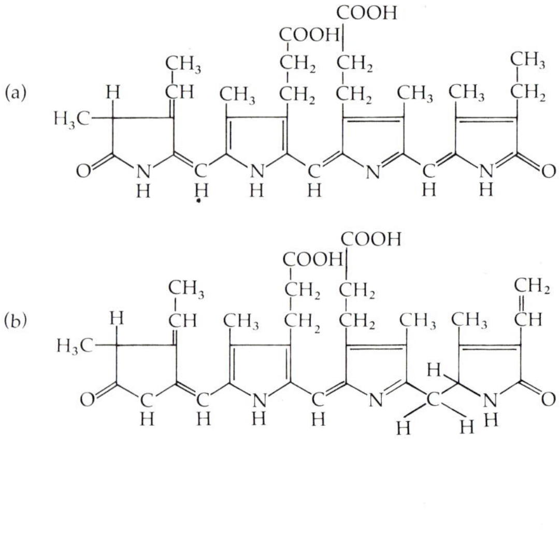

wavelengths, is formed by some, but not all, members of each group. Phycobiliproteins can constitute as much as 60 percent of the cell's soluble proteins. Under conditions of nitrogen limitation in the absence of stored cyanophycin (p. 178), phycobiliproteins are degraded by specific proteases; thus these proteins play a secondary role as nitrogenous reserve material.

*Carotenoids* (Figure 15.6) are always associated with the photosynthetic apparatus and have two functions in photosynthesis. They may serve as light-harvesting pigments, absorbing light in the blue-green region of the spectrum, between 400 and 550 nm; their relative contribution to this function is major in some photosynthetic organisms, minor

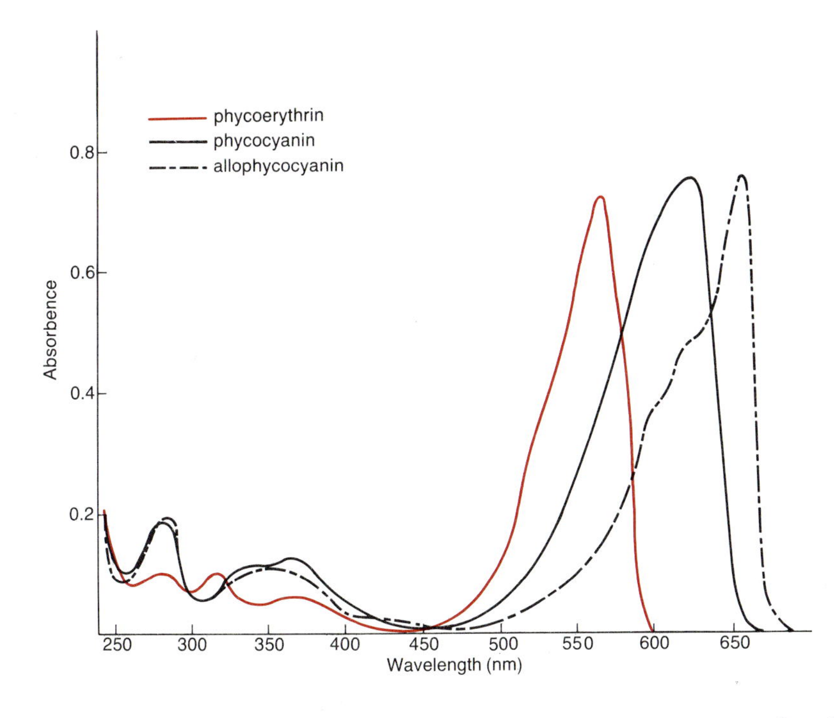


FIGURE 15.5

The absorption spectra of phycobiliproteins, isolated from a filamentous cyanobacterium, and adjusted to the same peak heights at the maxima. After A. Bennet and L. Bogorad, "Properties of Subunits and Aggregates of Blue-Green Algal Biliproteins," *Biochemistry* **10,** 3625 (1971).

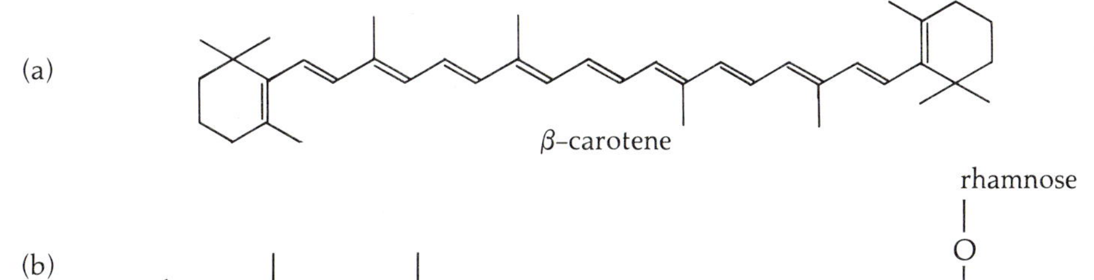


(b)

rhamnose

O

OH

HO

myxoxanthophyll

(c)

lycopene

(d)

isorenieratene

FIGURE 15.6

Representative carotenoids of photosynthetic eubacteria: (a) a bicyclic carotenoid; (b) a carotenoid glycoside; (c) an aliphatic carotenoid; (d) an aryl carotenoid.

in others. In addition, they play an indispensable role as quenchers of chlorophyll-catalyzed photooxidation, thus protecting the photosynthetic apparatus from photooxidative damage (Chapter 8).

Among phototrophs, carotenoid composition tends to be both complex and group specific. Eucaryotic phototrophs all contain as major carotenoids bicyclic carotenes and related oxygen-containing carotenoids. Such pigments are likewise characteristic of cyanobacteria; many of these organisms also contain carotenoid glycosides, which are group specific.

In purple bacteria, carotenoid composition is extraordinarily diverse; over 30 pigments of this type occur in the different members of the group, and none is common to all. However, the carotenoids of purple bacteria are group-specific compounds in the sense that they do not occur in other photosynthetic organisms. Most of them are aliphatic compounds, which often bear methoxyl groups; a few are aryl carotenoids, bearing an aromatic ring at one end of the chain. Aryl carotenoids are also characteristic of nearly all green bacteria.

FIGURE 15.7

Lipids of photosynthetic eubacteria. a) monogalactosyl-diglyceride; b) digalactosyl-diglyceride; c) an ornithine lipid from *Rhodospirillum rubrum*. $R_1$ and $R_2$ denote hydrocarbon chains.

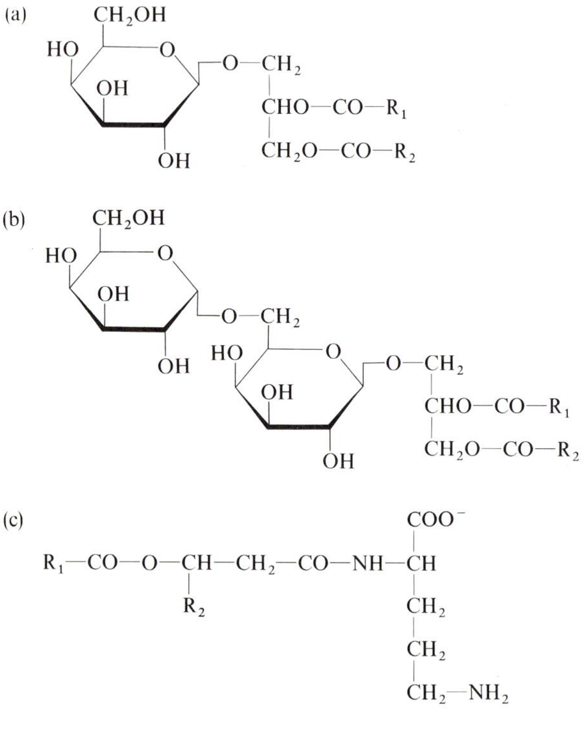

(a) $CH_2OH$, HO, O, OH, OH; $O-CH_2$; $CHO-CO-R_1$; $CH_2O-CO-R_2$

(b) $CH_2OH$, HO, O, OH, OH; $O-CH_2$; HO, O, OH, OH; $O-CH_2$; $CHO-CO-R_1$; $CH_2O-CO-R_2$

(c) $COO^-$; $R_1-CO-O-CH-CH_2-CO-NH-CH$; $R_2$; $CH_2$; $CH_2$; $CH_2-NH_2$

Lipids of the Photosynthetic Apparatus In photosynthetic eucaryotes, two classes of glycolipids, monogalactosyl and digalactosyl diglycerides (Figure 15.7) are specifically located in the chloroplast thylakoids and constitute about 80 percent of the total lipid content of these structures. They are largely esterified with α-linolenic acid, a polyunsaturated fatty acid that occurs only in the chloroplast.

Both of these galactolipids are similarly present in the thylakoids of all cyanobacteria, although they are not always esterified with α-linolenic acid. The cellular fatty acid composition of cyanobacteria is remarkably varied; some have a high content of α-linolenic acid; some contain other polyunsaturated fatty acids; and some have the fatty acids that are characteristic of bacteria in general, i.e., exclusively saturated and monounsaturated compounds.

The green bacteria contain only one of these galactolipids, monogalactosyl-diglyceride, which is the principal lipid constituent of the chlorosome membrane that surrounds the antenna bacteriochlorophyll (see below). In addition, they have in their cell membrane a novel glycolipid that contains rhamnose, galactose, and a third, unidentified, sugar.

The purple bacteria contain phospholipids as the principal lipids of both their cell and intracytoplasmic membranes, but in addition they contain novel ornithine lipids; the structure of one of these is shown in Figure 15.7.

Neither the purple nor green bacteria contain polyunsaturated fatty acids.

## Location of the Photochemical Apparatus in Phototrophic Eubacteria

The three major groups of phototrophic eubacteria produce distinctive structures in which the photochemical apparatus is located (Figure 15.8). In purple bacteria these are extensive invaginations of the cytoplasmic membrane, which may be tubular, lamellar, or vesicular in form (see Figure 6.8). All components of the photochemical apparatus (antenna chlorophyll, reaction centers, and electron transport systems) are located in these membranous structures.

The cytoplasmic membrane of green bacteria is characteristically not invaginated, and it houses only the reaction centers (which contain bacteriochlorophyll *a*) and the electron transport systems. The antenna chlorophylls (bacteriochlorophyll *c*, *d*, or *e*) are contained in vesicular sacs termed *chlorosomes* (Figure 15.9) that are tightly joined to the cell membrane through a *base plate* containing

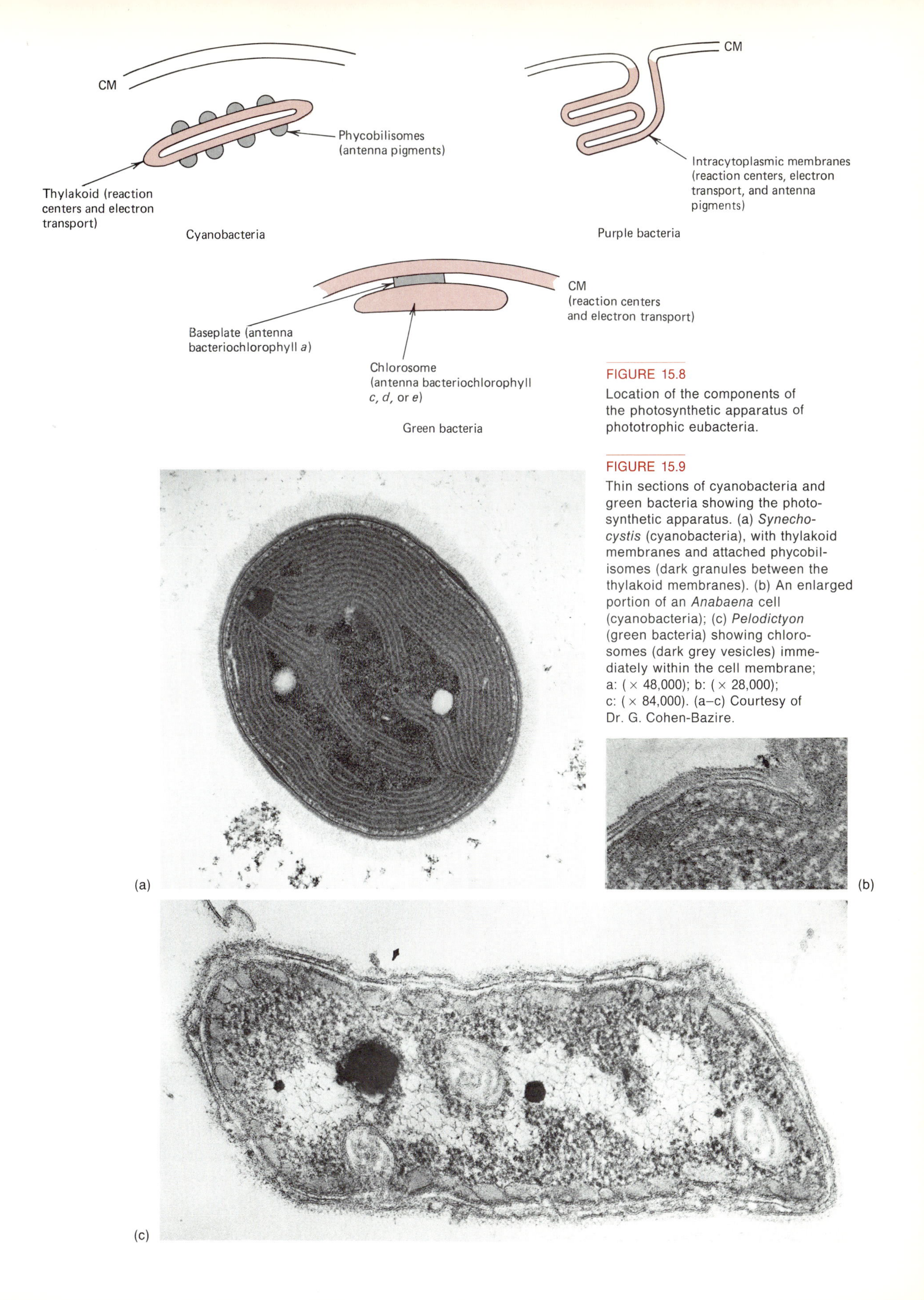


FIGURE 15.8

Location of the components of the photosynthetic apparatus of phototrophic eubacteria.

FIGURE 15.9

Thin sections of cyanobacteria and green bacteria showing the photosynthetic apparatus. (a) *Synechocystis* (cyanobacteria), with thylakoid membranes and attached phycobilisomes (dark granules between the thylakoid membranes). (b) An enlarged portion of an *Anabaena* cell (cyanobacteria); (c) *Pelodictyon* (green bacteria) showing chlorosomes (dark grey vesicles) immediately within the cell membrane; a: (× 48,000); b: (× 28,000); c: (× 84,000). (a–c) Courtesy of Dr. G. Cohen-Bazire.

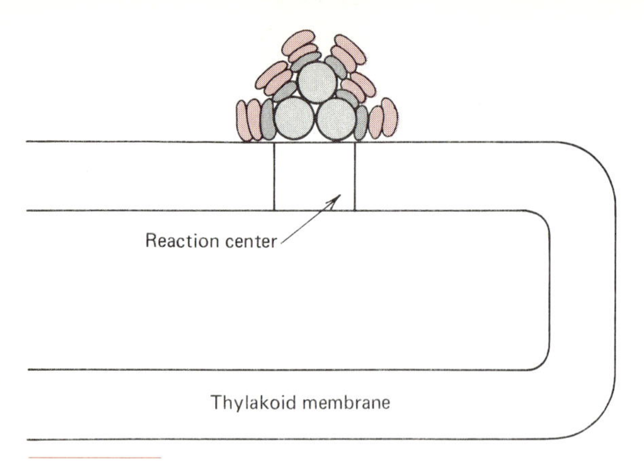


FIGURE 15.10
The molecular structure of a phycobilisome. Phycoerythrin (colored discs) and phycocyanin (grey discs) channel absorbed light energy to the reaction center via allophycocyanin (circles).

a special class of bacteriochlorophyll *a* that is not part of the reaction center but rather functions in energy transfer from the antenna to the reaction center. Chlorosomes are surrounded by a monolayer membrane about 4 nm thick composed of lipid and protein.

The cyanobacteria contain specialized unit membrane–bound sacs, termed *thylakoids*, that are structurally distinct from the cell membrane (Figure 15.9). They house the reaction centers and the electron transport systems as well as the antenna chlorophyll *a*. The major antenna pigments, *phycobiliproteins*, are aggregated into hemispherical structures termed *phycobilisomes* that are attached to the cytoplasmic surface of the thylakoid membranes. Within the phycobilisome, the various individual phycobiliproteins are arranged in a precise order with allophycocyanin closest to the reaction centers and phycoerythrin farthest from it (Figure 15.10). This arrangement ensures that energy transfer within the phycobilisome is towards the reaction centers, because the excited state of phycoerythrin (which absorbs short wavelength light) is more energetic than the excited state of phycocyanin (which absorbs long wavelength light).

## Photochemical Generation of Reductant

All organisms growing with a source of cell carbon more oxidized than that of the cell material (empirically about $CH_2O$) must reduce their carbon source in the process of its assimilation. The intermediate reductant, participating in the individual reactions in which carbon compounds undergo reduction, is nearly always reduced pyridine nucleotide (NADH or NADPH). However, the ultimate source of reducing power (i.e., the compound that reduces $NAD^+$ or $NADP^+$ to form the reduced pyridine nucleotides) is quite variable. In most heterotrophs a single compound serves as carbon source and electron donor, with the oxidation of some molecules of it serving to provide the reductant to drive the assimilation of others. In autotrophs the carbon source ($CO_2$) is different from the electron donor (any of a variety of inorganic compounds). In many cases the electron donors (e.g., the water and reduced sulfur compounds utilized by the phototrophic eubacteria) are not sufficiently powerful reductants to reduce directly the pyridine nucleotides required for $CO_2$ assimilation; additional energy must be supplied to generate a more powerful reductant from these weak reductants. The three groups of phototrophs utilize different biochemical mechanisms to couple the photochemically derived energy to generation of reductant (Figure 15.11).

### Reductant Generation in the Cyanobacteria

The first chemically stable product of the oxidation of photosystem I (denoted "$A^-$" in Figure 15.1) in the cyanobacteria is a reduced iron-sulfur protein with a midpoint reduction potential of about −530 mV. This powerful reductant can reduce $NAD(P)^+$ ($E_0' = -320$ mV) via the intermediate electron carrier, ferredoxin.

The function of water in cyanobacterial photosynthesis is to reduce the oxidized photosystem I; this reaction cannot occur directly because water is a weak reductant ($E_0' = +810$ mV) incapable of reducing oxidized photosystem I ($E_0' = +520$ mV). Rather, photosystem I is reduced by electrons from excited photosystem II, and oxidized photosystem II ($E_0' = +860$ mV) is reduced by water.

### Reductant Generation in the Green Bacteria

Like the cyanobacteria, the excited photosystem of green bacteria reduces a low-potential iron-sulfur protein, which in turn reduces ferredoxin. The reduced ferredoxin participates directly in $CO_2$-fixing redox reactions, and can also reduce pyridine nucleotides. Oxidized photosystem is reduced with electrons derived from reduced sulfur compounds. Because these compounds characteristically have low reduction potentials (about −200 mV), there is no need for a second photosystem.

### Reductant Generation in the Purple Bacteria

The first stable reduced intermediate in the photosystem of purple bacteria is a quinone with a reduction potential of approximately −100 mV, which

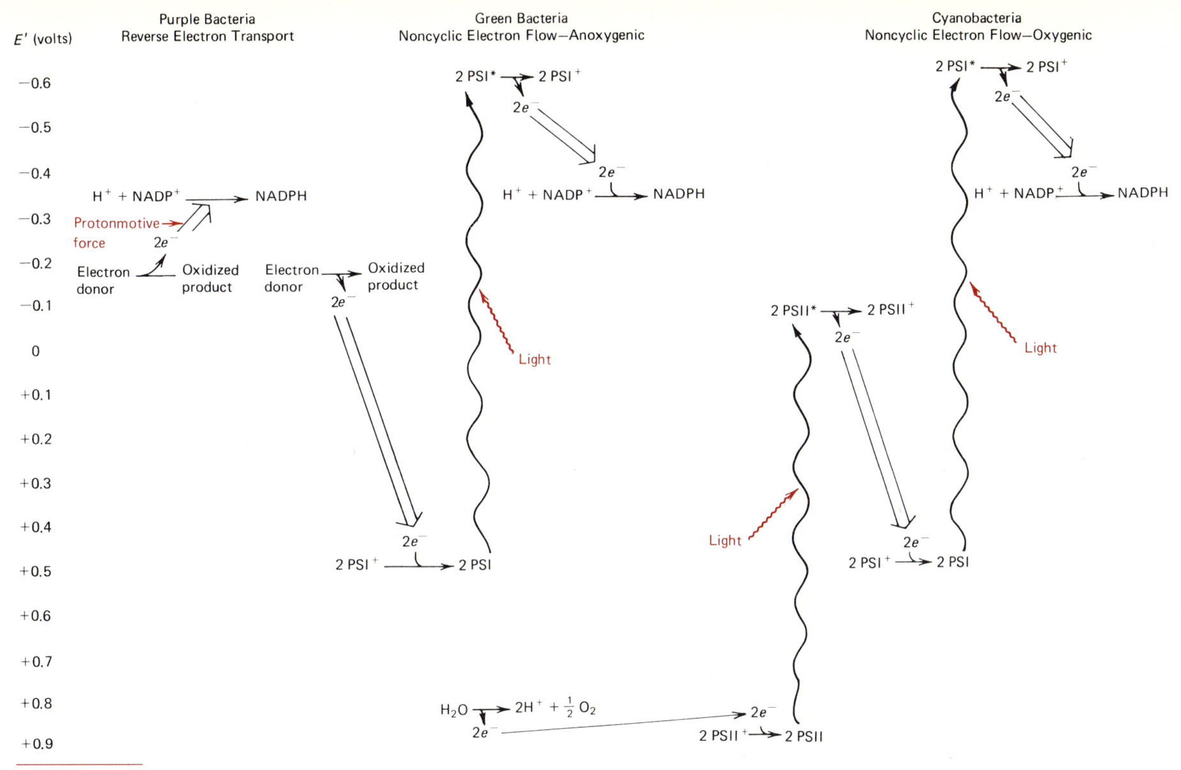


FIGURE 15.11

Photochemical generation of reductant among the photosynthetic eubacteria. Broad arrows denote membrane-bound electron transport systems.

thus is incapable of reducing pyridine nucleotides. Two mechanisms are utilized to accomplish this reduction. When molecular hydrogen is the electron donor, pyridine nucleotides may be directly reduced by it in a reaction catalyzed by the enzyme *hydrogenase.* However, when compounds such as reduced sulfur compounds with higher reduction potentials (and therefore incapable of directly reducing pyridine nucleotides) are the electron donors, the purple bacteria reduce $NAD(P)^+$ by reverse electron transport (Chapter 4). In this case, the photosystem does not provide reductant; rather it generates a protonmotive force that drives reverse electron flow.

## THE CELLULAR ABSORPTION SPECTRA OF PHOTOSYNTHETIC EUBACTERIA

Although not all pigments in the photosynthetic apparatus are equally effective in harvesting light (e.g., chlorophyll *a* is far less effective than the phycobiliproteins found in cyanobacteria), the cellular absorption spectra of photosynthetic organisms provide a rough indication of the spectral regions that are utilized for the performance of photosynthesis. Figure 15.12 compares the cellular absorption spectra of several different photosynthetic procaryotes; in each case, the specific contributions to light absorption made by chlorophylls, carotenoids, and phycobiliproteins are indicated. It is evident that the cellular absorption spectra characteristic of each group of photosynthetic procaryotes are distinctive, and to a considerable extent complementary.

In cyanobacteria, light is absorbed largely between 550 and 700 nm (by phycobiliproteins and by chlorophyll *a*). In green bacteria the major absorption band lies considerably farther toward the red region, between 700 and 800 nm; it is attributable to the light-harvesting bacteriochlorophylls (*c*, *d*, or *e*) characteristic of this group. In purple bacteria the major absorption bands lie largely in the near-infrared region, being represented by one or more peaks, attributable to either bacteriochlorophyll *a* or *b*. The position of this band in purple bacteria that contain bacteriochlorophyll *b* is situated beyond 1,000 nm, very close to the spectral limit beyond which light can no longer mediate photochemical reactions, as a result of the low energy content of the light quanta.* Broadly speaking, the cellular absorption spectra of photo-

* The energy content of light quanta is an inverse function of wavelength.

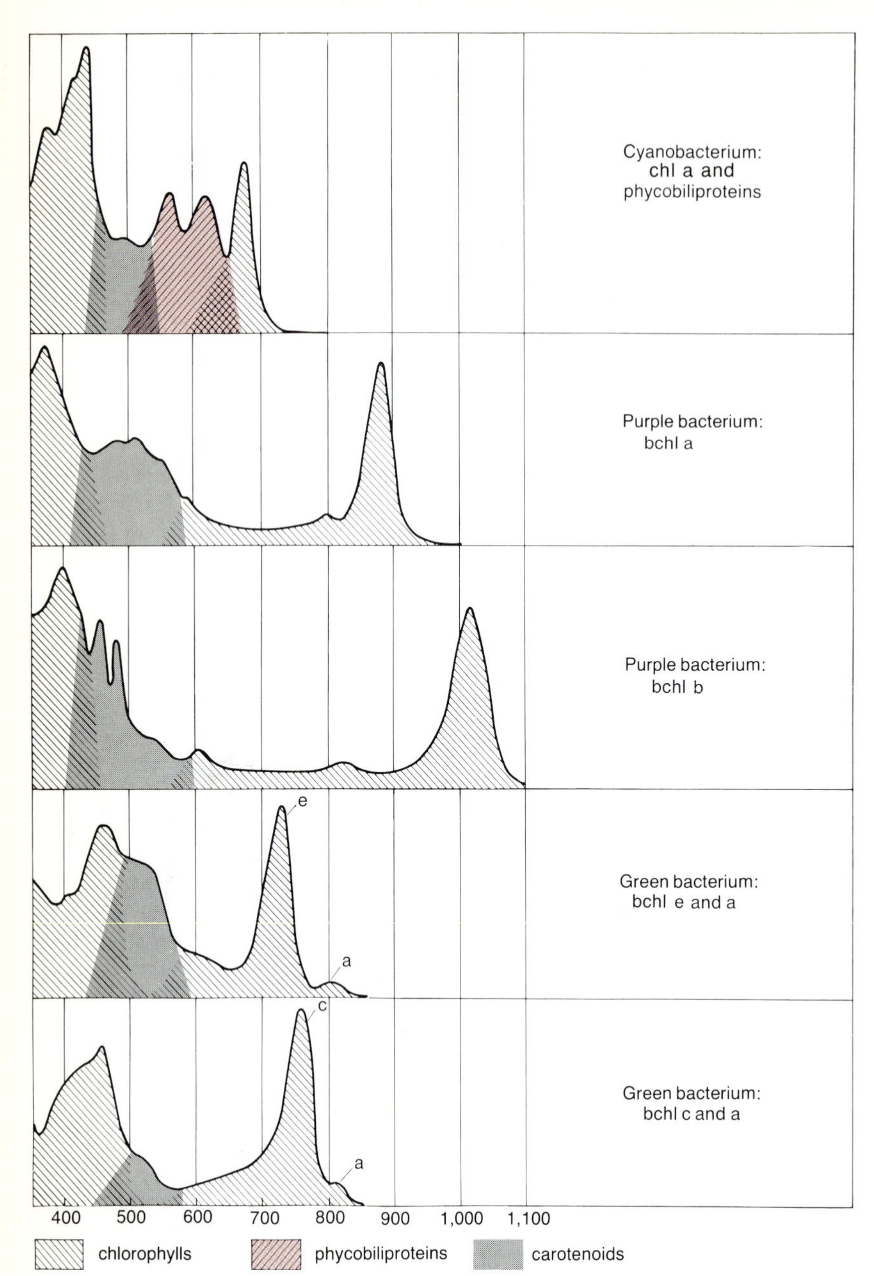


FIGURE 15.12

Cellular absorption spectra of five representative photosynthetic eubacteria, to show the characteristic differences in the positions of the major absorption bands. The approximate contributions to cellular light absorption by the major classes of photosynthetic pigments, and the types of chlorophyll present in each organism are indicated on the figure. The double peak of phycobiliprotein light absorption in the spectrum of the cyanobacterium illustrated reflects the presence of both phycoerythrin (maximum: 565 nm) and phycocyanin (maximum: 625 nm). Allophycocyanin absorption (maximum: 650 nm) is masked by the phycocyanin peak.

synthetic eucaryotes resemble those of cyanobacteria, though only red algae show major peaks in the region between 550 and 630 nm, where phycobiliproteins absorb. The differences in the light-absorbing properties of the various groups of photosynthetic organisms are of profound ecological significance, as will be discussed later (p. 380).

In all phototrophs the chlorophyll peaks *in vivo* occur at longer wavelengths than the peaks of the extracted pigments; however, the magnitude of this *in vivo* wavelength shift is not constant, and it is far greater in purple and green bacteria than in cyanobacteria and eucaryotes (Table 15.3). This shift is caused by a modification of the intrinsic spectral properties of chlorophylls *in vivo*, which results from the way they are associated with the proteins of the photosynthetic apparatus. Thus, the intrinsic spectral properties of the bacteriochlorophylls only in part account for the ability of purple and green bacteria to perform photosynthesis with light of very long wavelengths; this is also determined to a considerable extent by the nature of the chlorophyll-protein complexes present in the photosynthetic apparatus.

## The Colors of Photosynthetic Eubacteria

The common names of the three groups of photosynthetic procaryotes are not always well correlated with the color of their cells, as judged visually. Since the major chlorophyll absorption bands of purple bacteria lie in the infrared, to which the eye is blind, the visible color of these organisms is determined largely by their carotenoid complement. They appear brown, reddish, or purple. The green bacteria appear yellow-green, orange, or brown, again depending on their carotenoid composition.

The phycobiliproteins of cyanobacteria contribute largely to light absorption in the visible region, and the visible color of the cells is therefore much influenced by the phycobiliprotein complement. If phycoerythrin is absent, the cells appear blue-green; if it is present, they may appear red, violet, brown, or almost black.

# THE CYANOBACTERIA

The cyanobacteria are a structurally diverse assemblage of Gram-negative eubacteria characterized by their ability to perform oxygenic photosynthesis. Their structural diversity is paralleled by their genetic heterogeneity, reflected in G + C values (35 to 71 percent) and genome sizes (MW of $1.6 \times 10^9$ to $8.6 \times 10^9$ daltons) that span nearly the entire range found among all bacteria. Motility, when it occurs, is by gliding.

With respect to nutritional and metabolic properties, however, the group is relatively uniform. All are photoautotrophs; growth factors are rarely required, although some marine cyanobacteria require vitamin $B_{12}$. Assimilation of $CO_2$ occurs through the Calvin-Benson cycle, with the formation and deposition of glycogen as an intracellular reserve material.

Many cyanobacteria are *obligate photoautotrophs*, being wholly incapable of dark growth at the expense of organic sources of carbon and energy. The chemotrophic growth rate of those strains that can grow in the dark is very low relative to that in the light; it occurs only at the expense of glucose and a few other sugars, which are dissimilated by aerobic respiration. The limited range of utilizable sources of carbon and energy reflects the universal absence of a functional TCA cycle. Cyanobacteria lack a key enzyme of this cycle, α-ketoglutarate dehydrogenase, a metabolic peculiarity that they share with many obligately anaerobic chemoheterotrophs and aerobic chemoautotrophs. Respiratory metabolism occurs exclusively through the oxidative pentose phosphate cycle, the reactions of which are in large part common to those of the Calvin-Benson cycle (Figure 15.13). The obligate photoautotrophy characteristic of many cyanobacteria appears to be caused by the absence of the specific permeases necessary for the uptake of exogenous sugars by the cell, because the enzymatic machinery of the oxidative pentose phosphate cycle is present in all cyanobacteria. This pathway permits the generation of ATP in the dark, through endogenous respiration of stored glycogen.

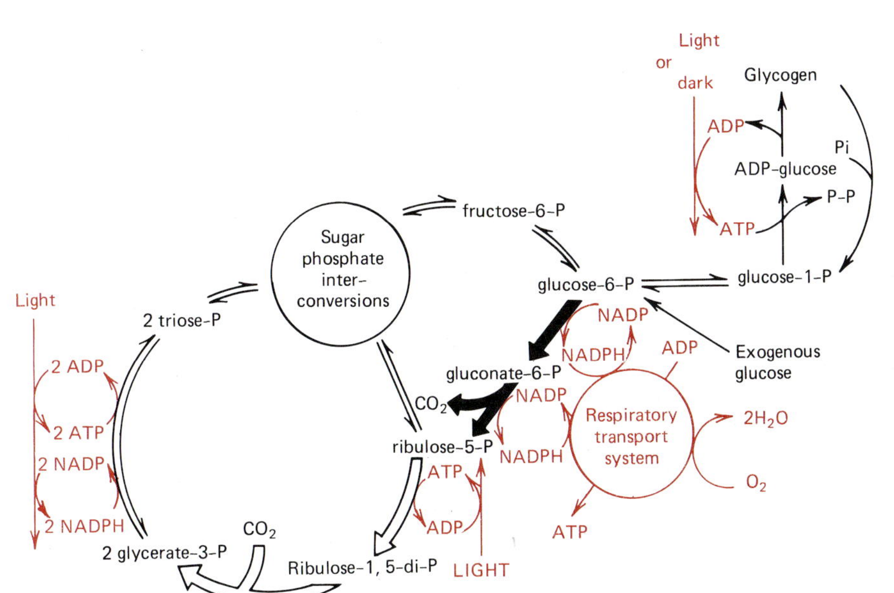


FIGURE 15.13

The pathways of light and dark metabolism in cyanobacteria, showing the central role of the pentose phosphate cycle. Heavy black arrows indicate reactions specifically operative in the dark; heavy white arrows are those specifically operative in the light. After R. Y. Stanier and G. Cohen-Bazire, "Phototrophic prokaryotes: the cyanobacteria." *Ann. Rev. Microbiol.* **31,** 225–274 (1977).

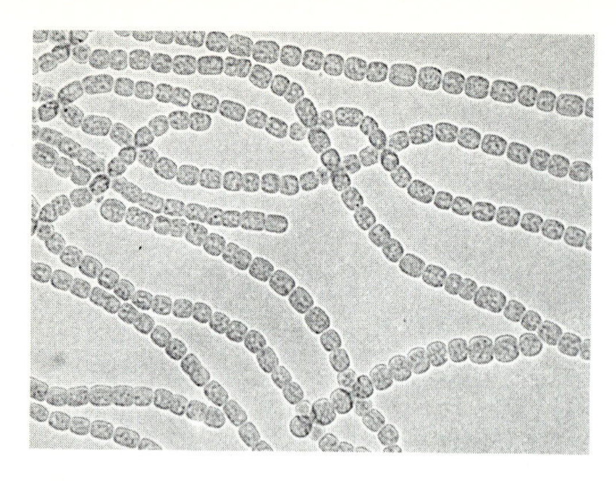

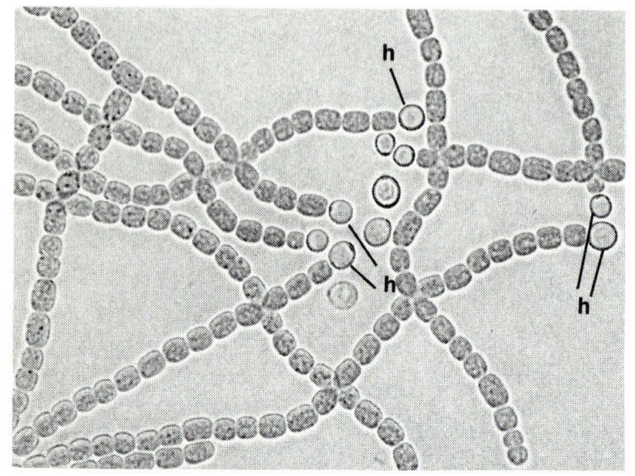


FIGURE 15.14

The effect of the nitrogen source on heterocyst formation in a cyanobacterium, *Anabaena* sp. (a) Filaments from a culture grown with ammonia as a nitrogen source; (b) filaments from a culture grown with $N_2$ as a nitrogen source; h, heterocyst (×616). Courtesy of R. Rippka.

## Nitrogen Fixation

Cyanobacteria are the only organisms able to perform oxygenic photosynthesis that can also fix nitrogen; many (though not all) are vigorous nitrogen fixers. Such organisms have the simplest known nutritional requirements; they can grow in the light in a mineral medium exposed to air using $CO_2$ as a carbon source and $N_2$ as a nitrogen source.

The coexistence in a single organism of the processes of oxygenic photosynthesis and nitrogen fixation presents an obvious paradox, since nitrogen fixation is an intrinsically anaerobic process; the key enzyme, nitrogenase, is rapidly and irreversibly inactivated *in vivo* by exposure even to low partial pressures of oxygen. With few exceptions, nitrogen-fixing cyanobacteria are filamentous organisms, which produce a specialized type of cell, the *heterocyst*, within which $N_2$ fixation takes place.

Heterocystous cyanobacteria do not form heterocysts when grown with a combined nitrogen source (nitrate or ammonia); nitrogenase synthesis is also repressed under these conditions. However, both nitrogenase synthesis and heterocyst formation occur when cultures are deprived of a combined nitrogen source. These processes do not require the presence of $N_2$ for their initiation, since they occur in an illuminated suspension of filaments placed in an atmosphere containing argon instead of nitrogen. However, in the absence of $N_2$ heterocyst differentiation is arrested at an intermediate, or *proheterocyst* stage. The formation of a mature heterocyst requires about 30 hours. Normally, about 5 to 10 percent of the cells in the filaments develop into heterocysts, following removal of combined nitrogen from the medium (Figure 15.14). Little is known of the regulation of heterocyst differentiation or nitrogenase synthesis. However, it has recently been discovered that at least two genetic rearrangements (deletion of segments of DNA) in the vicinity of the nitrogenase genes accompany heterocyst formation.

The differentiation of a heterocyst from a vegetative cell is accompanied by the synthesis of a new thick outer wall layer, extensive reorganization of the thylakoids which become concentrated near the two poles of the cell, and the formation of constricted and specialized polar connections at the points where the heterocyst is attached to adjacent vegetative cells (Figure 15.15). Mature heterocysts have an almost normal content of chlorophyll *a*, but are devoid of phycobiliproteins, the principle antenna pigments of photosystem II (Figure 15.16). Although heterocysts retain photosystem I activity, they lack photosystem II and ribulose-bisphosphate carboxylase (a key enzyme of the Calvin-Benson cycle); they therefore can neither fix $CO_2$ nor produce $O_2$ in the light.

Heterocysts have a significant respiration rate; respiratory substrates include $H_2$ generated in the course of nitrogen fixation (nitrogenase produces one mole of $H_2$ for every mole of $N_2$ reduced). The lack of photosynthetic $O_2$ generation, coupled with hydrogen-dependent respiration, helps to ensure that the partial pressure of $O_2$ within the heterocyst remains very low.

While the lack of photosystem II contributes to the ability of heterocyst to maintain reducing conditions necessary for $N_2$ fixation, it renders the heterocyst dependent upon adjacent vegetative cells for a source of reductant required for nitrogenase to function. Reductant is probably provided in the form of mono- or disaccharides although other compounds such as pyruvate or even reduced ferre-

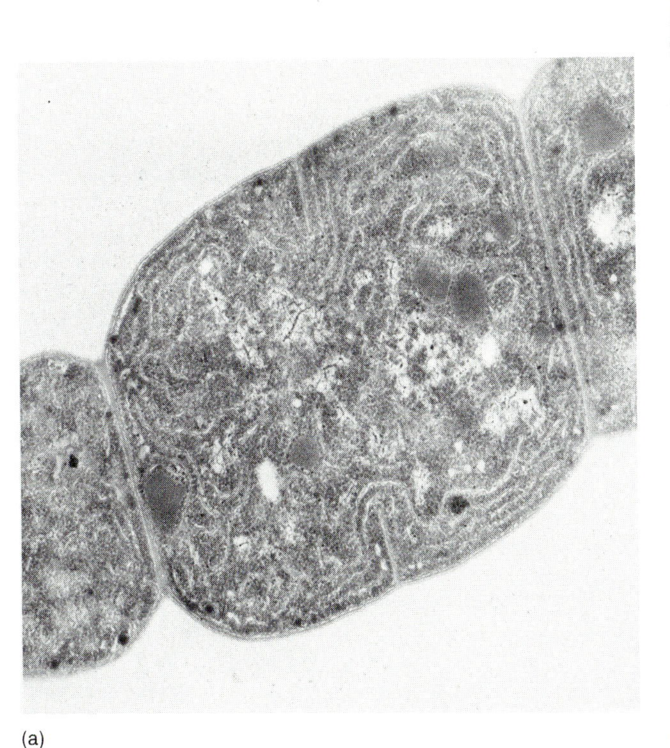

(a)

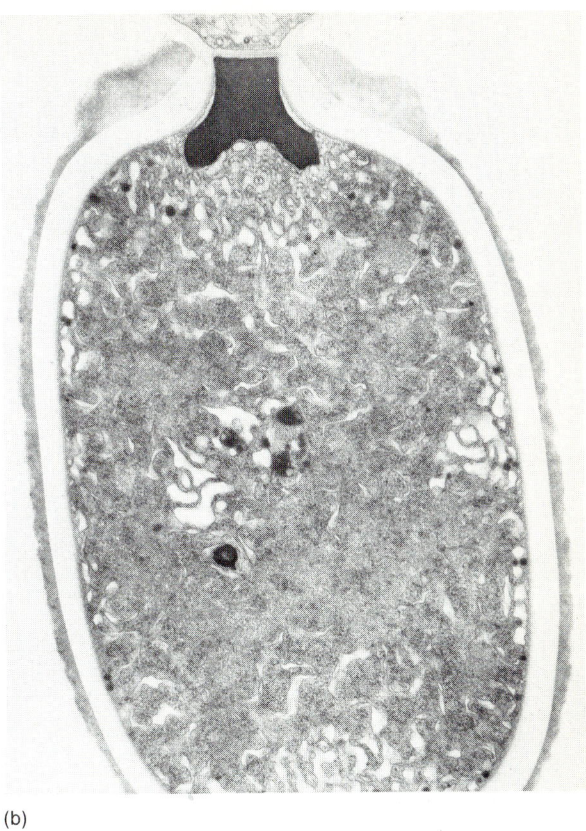

(b)

FIGURE 15.15

*Anabaena cylindrica*. (a) vegetative cells; (b) heterocyst, showing the sparse and disorganized thylakoids, thickened wall, and apical plug. a: ( × 41,000); b: ( × 22,500). Courtesy of Dr. M. Roussard-Jacquemin.

FIGURE 15.16

Photomicrographs of heterocyst-containing filaments of *Anabaena cylindrica*, to show the distribution of chlorophyll and phycocyanin. (a) Transmission image taken with blue light, preferentially absorbed by chlorophyll. The densities of vegetative cells and heterocysts (arrows) are similar, showing that they do not differ significantly in chlorophyll content. (b) Fluorescence image, taken under conditions which specifically reveal the fluorescence of phycocyanin. Vegetative cells are brilliantly fluorescent, whereas heterocysts (arrows) are barely visible, showing that they contain little phycocyanin. Courtesy of Dr. Marcel Donze.

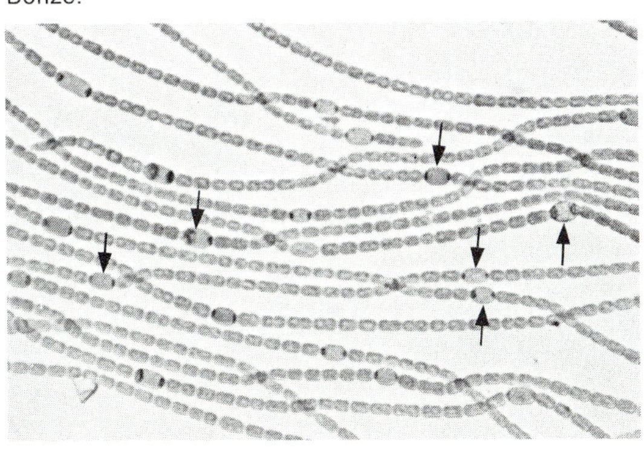

(a)

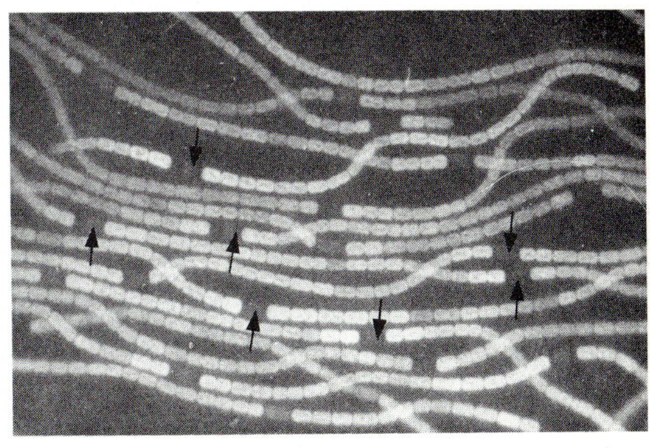

(b)

(a)

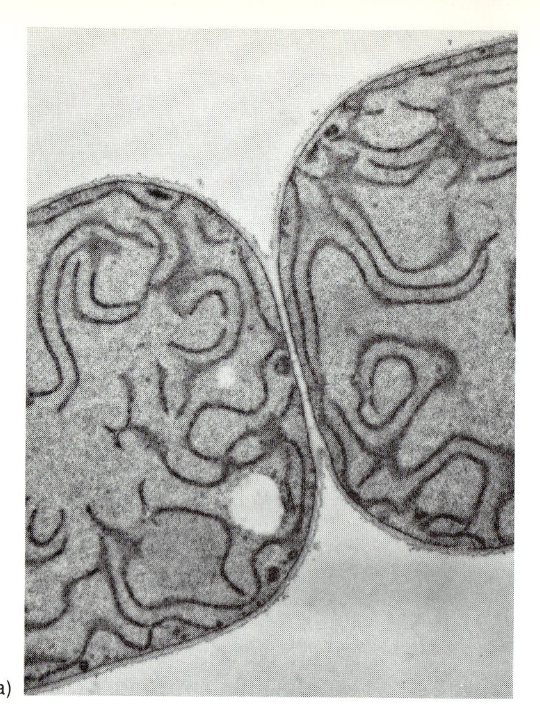

FIGURE 15.17

Microplasmadesmata. (a) Electron micrograph of a longitudinal thin section of *Anabaena cylindrica*. Microplasmadesmata are visible as faint connections traversing the septum between vegetative cells. (b) Electron micrograph of a transverse thin section of two adjacent trichomes of an oscillatorean cyanobacterium. The plane of the section grazes a septum (s) showing the pores (pd) through which microplasmadesmata penetrate. b: dense granules; t: thylakoids. (a) ×40,000. (b) ×81,000. (a) From T. H. Giddings, Jr. and L. A. Staehelin, "Plasma membrane architecture of *Anabaena cylindrica:* occurrence of microplasmadesmata and changes associated with heterocyst development and the cell cycle," *Cytobiol.* **16,** 235–249 (1978). (b) From H. C. Lamont, "Sacrificial cell death and trichome breakage in an oscillatoriacean blue-green alga: the role of murein," *Arch. Mikrobiol.* **69,** 237–259 (1969).

(b)

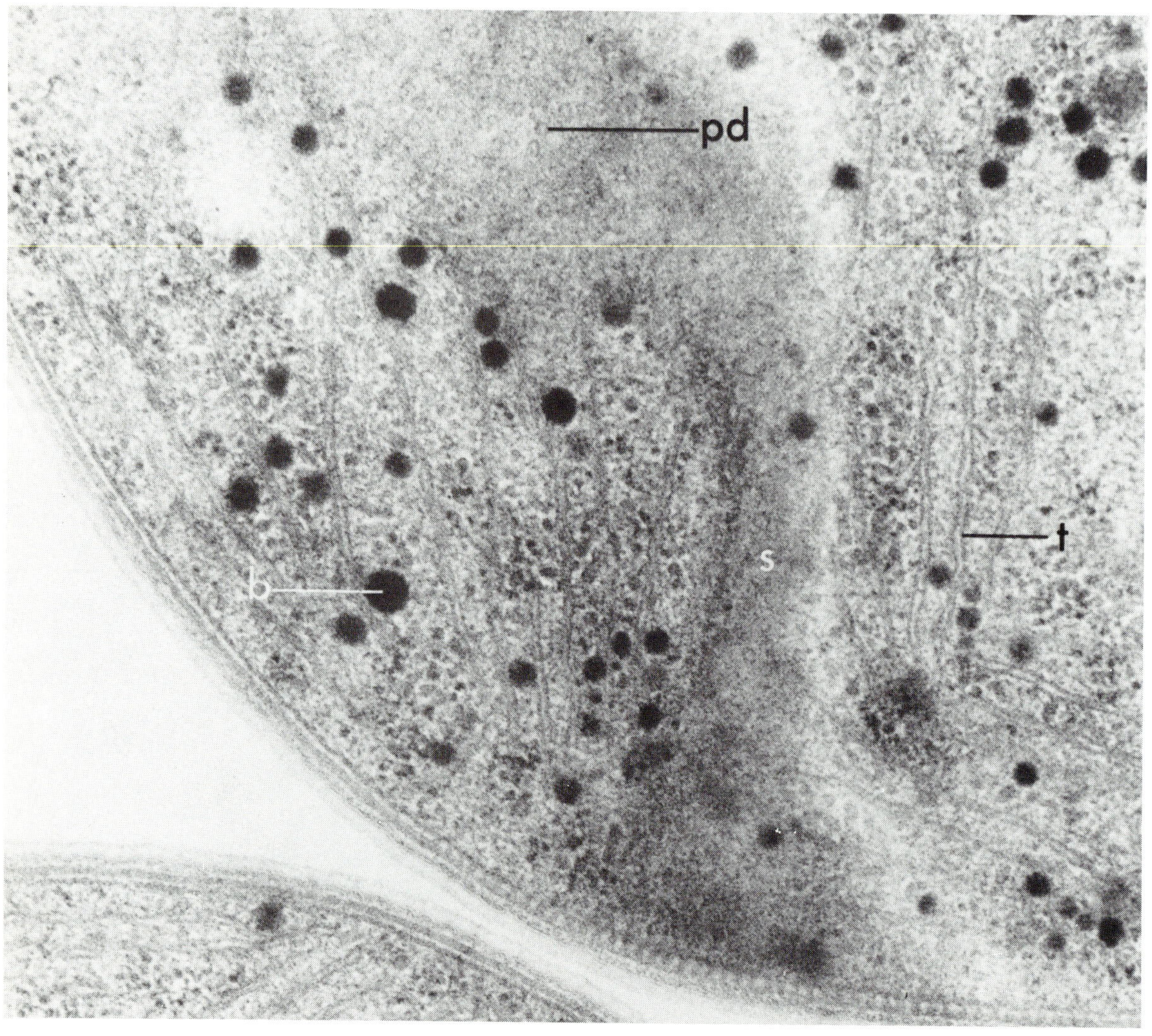

doxin itself may flow from vegetative cells to heterocysts. Cells within a cyanobacterial filament (including heterocysts) are connected by minute channels that traverse the septa between adjacent cells. These *microplasmadesmata* (Figure 15.17) provide cytoplasmic continuity among cells of the filament, allowing the exchange of metabolites required by the sequestering of $N_2$ fixation into heterocysts. (Microplasmadesmata, although in lesser numbers, are also found in nonheterocystous cyanobacteria; they thus must serve other functions in addition to their presumed role in $N_2$ fixation.) The metabolic interactions between heterocysts and vegetative cells are shown schematically in Figure 15.18.

In some cyanobacteria with relatively short filaments, there may be a single terminal heterocyst per filament. Other cyanobacteria produce intercalary as well as terminal heterocysts. As growth and cell division lengthen the filament, the heterocysts become more widely separated, and new heterocysts differentiate equidistant from the old ones. The signals that initiate development of heterocysts and the mechanism by which their proper spacing is achieved are unknown; possibly the signal is simply nitrogen starvation of the cells that are far from an existing heterocyst.

The ability of a filamentous cyanobacterium to fix nitrogen aerobically depends on its ability to form heterocysts; nonheterocystous organisms cannot simultaneously perform nitrogen fixation and oxygenic photosynthesis. However, many nonheterocystous filamentous cyanobacteria have the ability to make nitrogenase in the absence of combined nitrogen if conditions are initially reducing. The dilemma posed by this observation is explained by the recent demonstration that many nonheterocystous filamentous cyanobacteria are capable of *facultative anoxygenic photosynthesis* (see below). Thus nitrogenase synthesis, and consequently nitrogen fixation, functions in these organisms under anaerobic growth conditions.

Among the unicellular cyanobacteria, nitrogen fixation is virtually absent, the only known exception being found in strains belonging to the genus *Gleothece* that are capable of fixing nitrogen while growing photosynthetically in air. This organism produces a nitrogenase that is as $O_2$-sensitive as other nitrogenases; the mechanism by which it is protected *in vivo* is unknown.

## Anoxygenic Photosynthesis

During the course of an ecological study of a hypersaline lake on the shore of the Gulf of Elat in Israel, M. Shilo and his co-workers discovered that *Oscillatoria limnetica* is capable of anaerobic, sulfide-dependent photoassimilation of $CO_2$. Sulfide inhibits the functioning of photosystem II and induces an enzyme system that allows sulfide to donate electrons to photosystem I; elemental sulfur, the oxidized product, accumulates as extracellular granules. In the dark, small amounts of ATP can

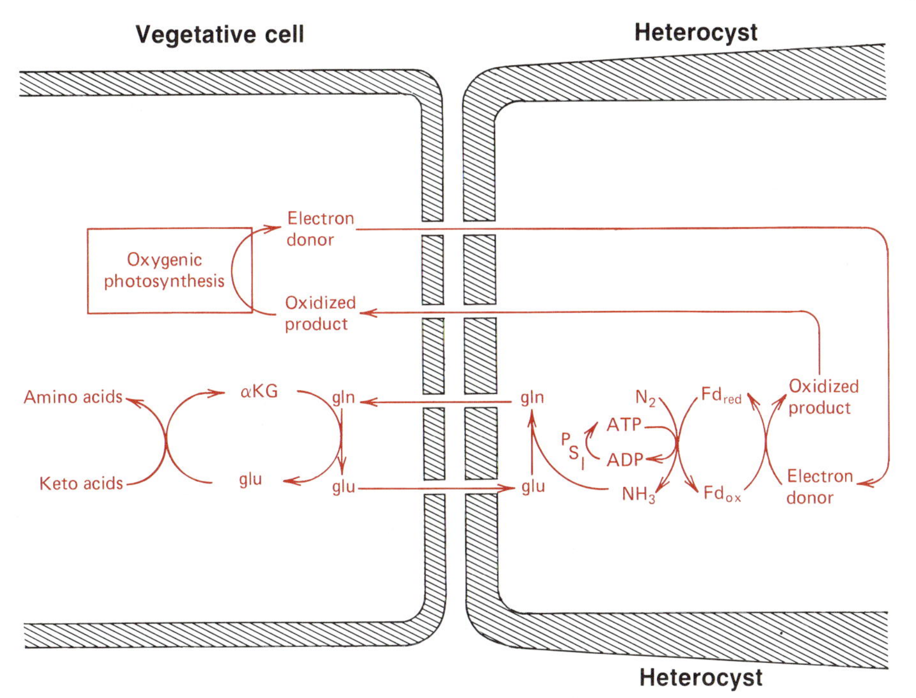


FIGURE 15.18
Metabolic exchanges among vegetative cells and heterocysts. Abbreviations: Fd, ferredoxin; PSI, photosystem I; glu, glutamate; gln, glutamine; αKG, α-ketoglutaric acid.

be generated from stored polyglucose reserves by either of two means: anaerobic respiration, using the accumulated elemental sulfur as terminal electron acceptor; or a homolactic fermentation. Thus under anaerobic conditions in the presence of high concentrations of sulfide, *O. limnetica* exhibits a pattern of photochemical metabolism that is functionally equivalent to normal metabolism of the green bacteria (Figure 15.11). This type of facultative anoxygenic photosynthesis is widely distributed among nonheterocystous filamentous cyanobacteria.

## Regulation of Pigment Synthesis

Synthesis of photosynthetic pigments and other components of the photosynthetic apparatus by cyanobacteria is constitutive. Even after many transfers on sugar-containing media in the dark, facultatively chemotrophic strains retain a normal pigment complement and can initiate immediate growth when returned to a mineral medium in the light. However, many phycoerythrin-producing strains exhibit an interesting response to chromatic illumination, known as *complementary chromatic adaptation.* When grown in green light, these strains contain a high ratio of phycoerythrin to phycocyanin whereas when grown in red light, they synthesize very little phycoerythrin (Figure 15.19). These specific light-induced changes of phycobiliprotein content enable the cells to absorb most effectively the wavelengths of light that are available. The mechanism of chromatic adaptation is not known, but there are indications that it is mediated by a regulatory light-sensitive pigment, similar to but not identical with *phytochrome,* a biliprotein that is an important photoregulator of plant growth and differentiation.

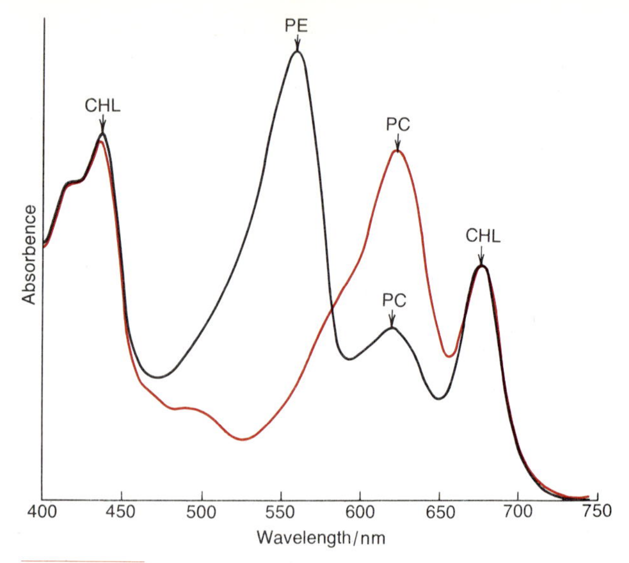


FIGURE 15.19

The effect of chromatic illumination on phycobiliprotein synthesis by a cyanobacterium, *Pseudanabaena* sp. Absorption spectra of cells grown in green light (black line) and in red light (colored line). The positions of the absorption maxima of chlorophyll *a* (CHL), phycocyanin (PC), and phycoerythrin (PE) are indicated by arrows. Cells grown in red light contain virtually no phycoerythrin; those grown in green light have a high content of phycoerythrin, and a much reduced phycocyanin content. Courtesy of Mrs. N. Tandeau de Marsac.

**TABLE 15.4**

**Major Groups of Cyanobacteria**

| Group | Growth Habit | Multiple Fission | Heterocysts |
|---|---|---|---|
| Unicellular cyanobacteria | Unicellular | − | − |
| *Pleurocapsa* group | Unicellular[a] | + | − |
| *Oscillatoria* group | Filamentous | − | − |
| Heterocystous cyanobacteria | Filamentous | − | + |

[a] Cells may be held together in aggregates by external wall layers; these aggregates may resemble filaments.

## Constituent Groups of Cyanobacteria

The broad outlines of a coherent taxonomy of cyanobacteria have just recently emerged. Part of the reason for our delayed understanding of relationships among members of this group is that until recently, classification schemes for this group were exclusively based on morphological observations made in the field. It is now clear that many structural traits are critically dependent upon culture conditions; field observations are consequently unreliable, and only observations on pure cultures can form the basis of a systematic understanding of this group.

An additional difficulty has faced cyanobacterial taxonomists: the lack of a significant range of physiological diversity among members of the group. This physiological uniformity has forced a near total reliance on morphological traits in defining biological groups, a practice that has proved inadequate, since the genetic variability within several morphological groups is quite large. Development of a sound taxonomy of the group will almost certainly require a detailed study of genetic relationships.

Based on studies of the still relatively limited number of cyanobacterial strains available in pure culture, four major groups can be defined on the basis of their pattern of development and the presence or absence of heterocysts (Table 15.4).

THE UNICELLULAR CYANOBACTERIA Three major subgroups are recognizable on the basis of the pattern of cell division (Table 15.5 and Figure 15.20). Cells may be rods or cocci, and may be ensheathed or not. If a sheath is present, it is loosely adherent to the cells, and clearly visible in the light microscope (Figure 15.20). Binary fission occurs within the sheath, and following division each daughter cell synthesizes a new sheath. Consequently, cells of this type are held together in aggregates by a multilaminate sheath. Because aggregates of cells produced by successive binary fissions in a single plane may appear to be the result of division in two or three planes (e.g. *Gloeothece* in Figure 15.20), determination of the pattern of binary fissions requires microscopic observation of successive divisions.

**TABLE 15.5**

**The Unicellular Cyanobacteria**

| Subgroups | Percent G + C | Aerobic $N_2$ Fixation | Sheath |
|---|---|---|---|
| Division by budding | | | |
| *Chamaesiphon* | 47 | − | − |
| Division by binary fission in one plane | | | |
| *Gloeothece* | 40–43 | + | + |
| *Synechococcus* | 39–43<br>47–56<br>66–71 | − | − |
| Division by binary fission in two or three planes | | | |
| *Synechocystis* | 35–37<br>42–48 | − | − |
| *Gloeocapsa* | 40–46 | − | + |

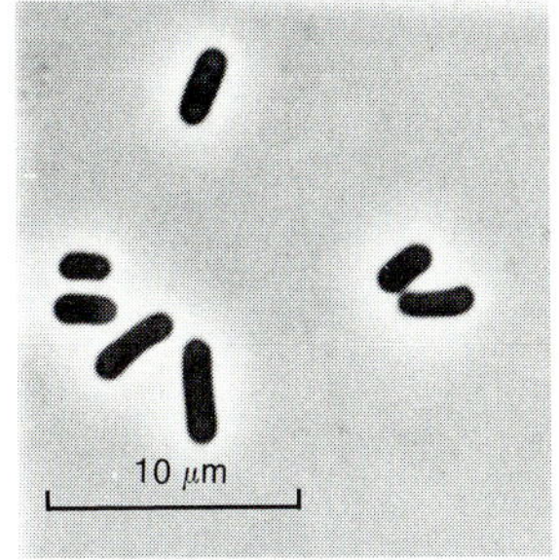


(a)

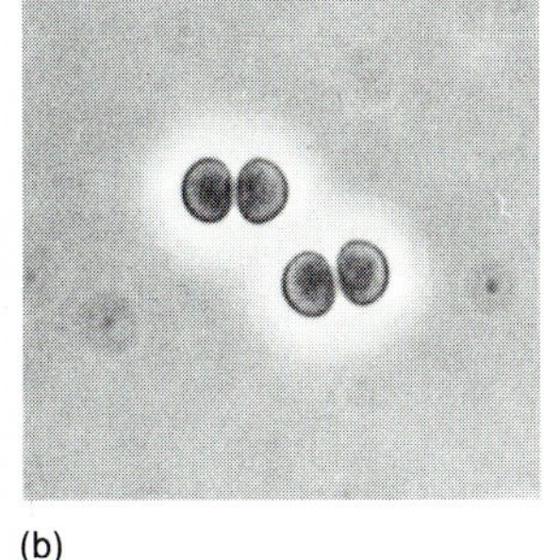

(b)

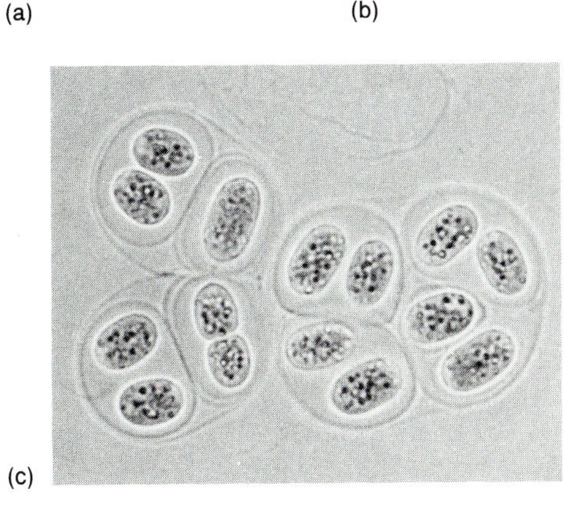

(c)

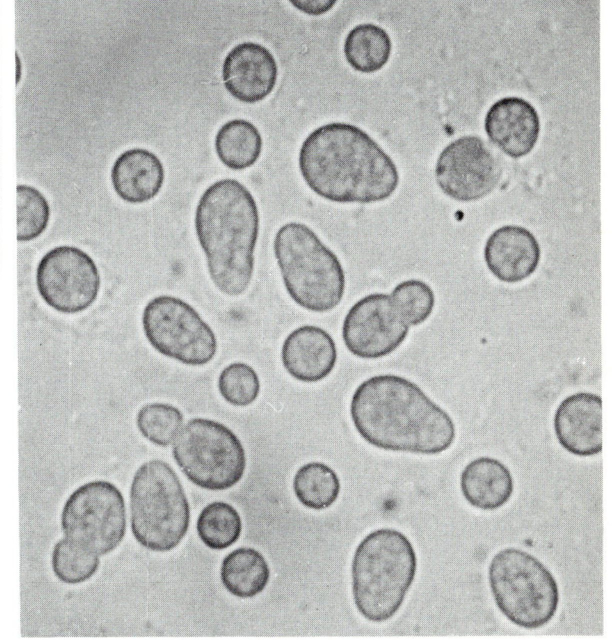

(d)

FIGURE 15.20

Light micrographs of some representative unicellular cyanobacteria. (a) *Synechococcus* (phase contrast); (b) *Synechocystis* (phase contrast); (c) *Gloeothece* (bright field); (d) *Chamaesiphon* (bright field). (a) and (b) × 1730; (c) × 935; (d) × 2000. (a)–(c) Courtesy of R. Rippka and R. Kunisawa. (d) From J. Waterbury and R. Y. Stanier, "Two unicellular cyanobacteria which reproduce by budding," *Arch. Microbiol.* **115,** 249–257 (1977).

(a)

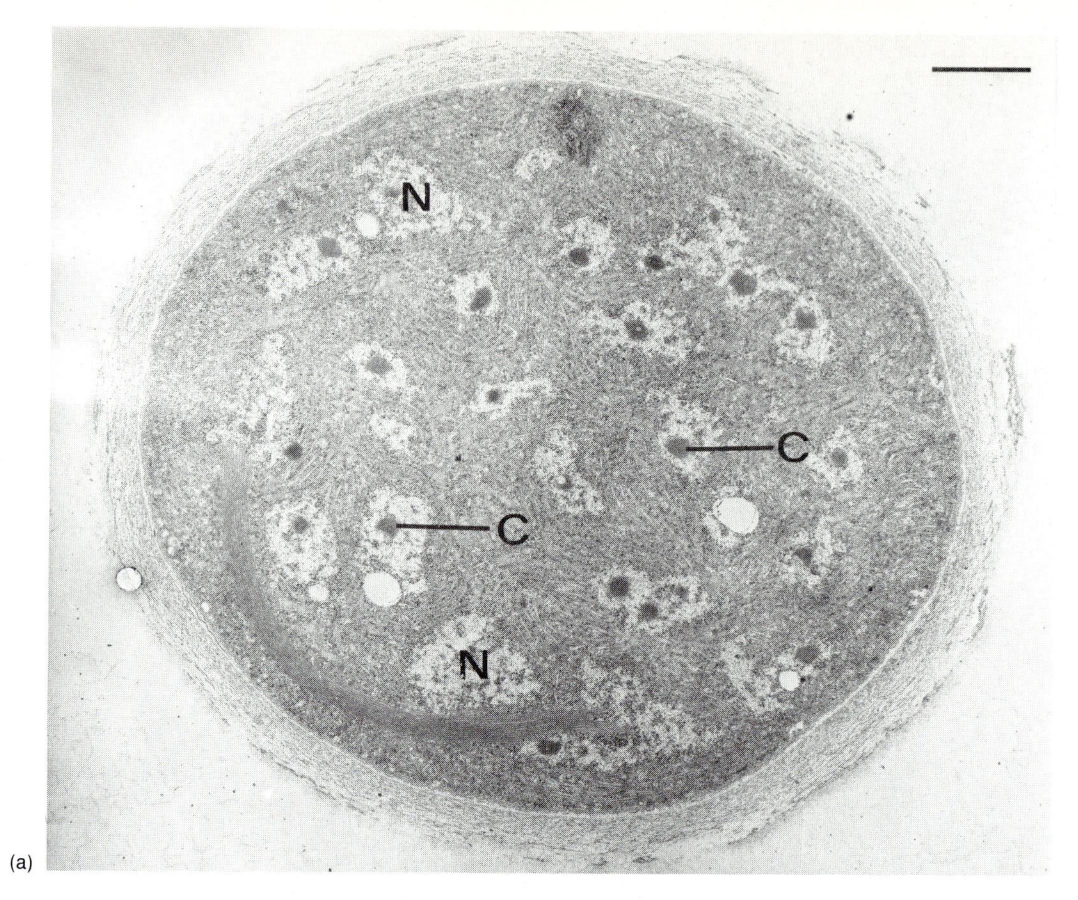


(b)

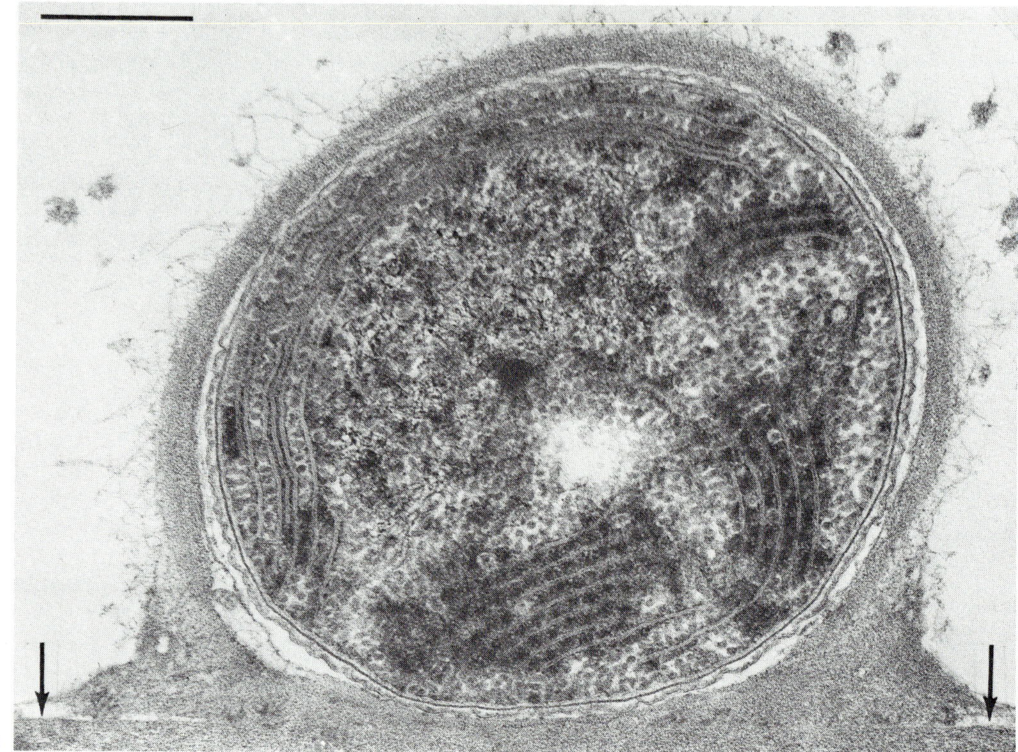

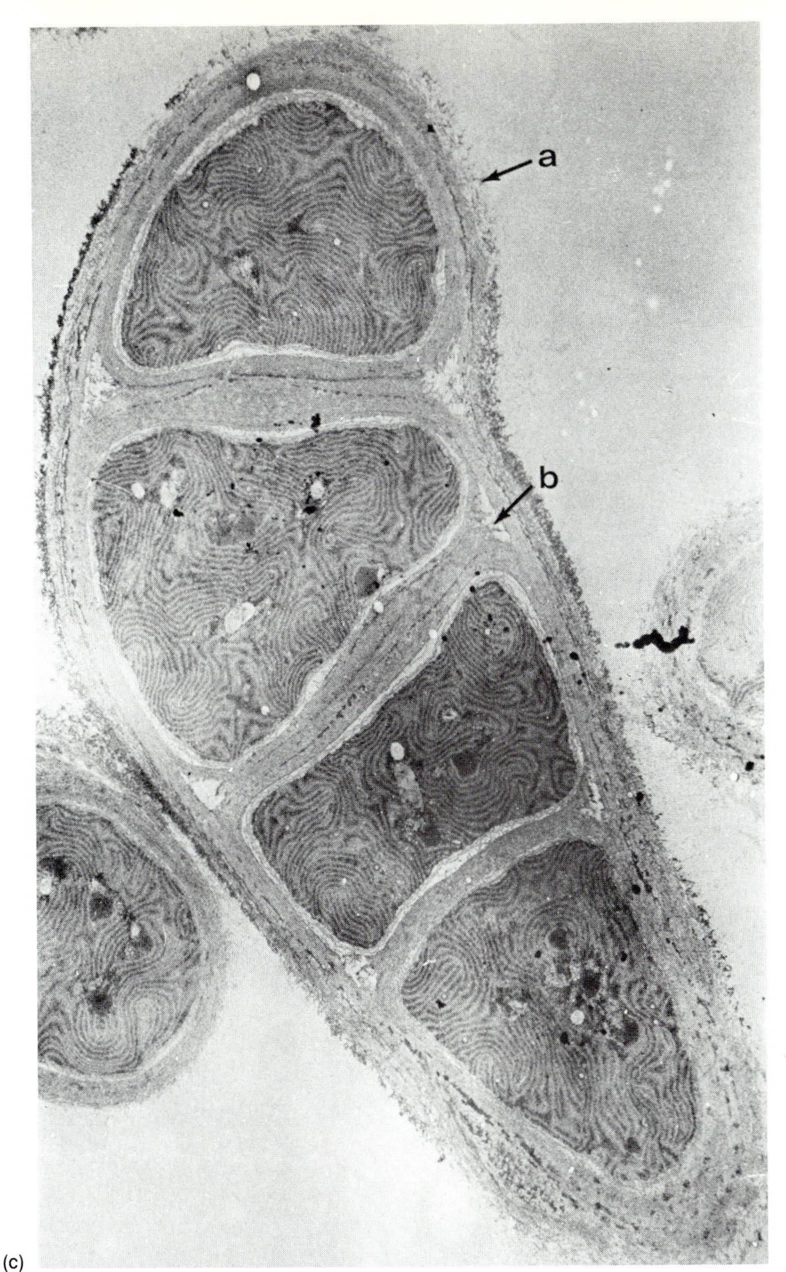


FIGURE 15.21

Electron micrographs of thin sections of pleurocapsalean cyanobacteria, showing the fibrous outer wall layer. (a) Multinucleate *Dermocarpa* cell just prior to multiple fission. N, nucleii; C, carboxysomes; (b) *Dermocarpella* attached by its fibrous layer to the substratum (arrows); (c) *Pleurocapsa*, showing the fibrous layers deposited at each cell generation. Old layers either stretch (a) or tear (b) as the cells grow. Bar equals 1 μm. From J. B. Waterbury and R. Y. Stanier, "Patterns of Growth and Development in Pleurocapsalean Cyanobacteria," *Microbiol Rev*. **42,** 2–44 (1978).

The wide range of percent G + C values found among various isolates of *Synechococcus*, and to a lesser extent among isolates of *Synechocystis*, indicates the need to subdivide these genera. In both genera, the percent G + C values are clustered within discrete smaller ranges; each of these subgroups probably deserves separate generic status.

The occurrence of a number of traits, including motility (rare among unicellular cyanobacteria), synthesis of phycoerythrin, and dark aerobic chemoheterotrophic growth is sporadic in distribution and does not correlate well with generic assignments.

THE PLEUROCAPSA GROUP The *Pleurocapsa* group is distinguished by a mode of reproduction that is rarely encountered among other bacteria: *multiple fission.* Multiple fission is a series of successive binary fissions without intervening cell growth; hence the cell undergoing this mode of division is cleaved into a number of daugther cells, termed *baeocytes*, that are much smaller than the mother cell. The number of baeocytes produced per reproductive cell varies from 4 to over 1,000. This number is partially under genetic control, and partially subject to environmental conditions. In general, favorable growth conditions lead to an early onset of multiple fission, and consequently to a smaller number of baeocytes per mother cell.

Baeocytes may be motile; if so, they lose their motility before commencing growth. Baeocytes are motile only if they lack the fibrous wall layer external to the outer membrane. This layer (Figure 15.21) is characteristic of the vegetative cells of all pleurocapsalian cyanobacteria, but its synthesis does not always accompany multiple fission. When it does, the baeocytes are immotile; when its synthesis is delayed, the baeocytes are motile, remaining so only until synthesis of the fibrous layer begins.

Among some of these organisms, multiple fission is the only mechanism of cell division; their developmental cycle is accordingly quite simple. Other members of the group divide by binary fission to produce cell aggregates held together by their common fibrous wall layer (Figure 15.21). In these organisms, the developmental cycle is more complex, depending on whether or not there is a regular alternation of division planes, and if specialized reproductive cells are formed. These variations in life cycle, coupled with baeocyte motility, form the basis for classification in this group (Table 15.6; Figures 15.22 and 15.23).

Most members of the *Pleurocapsa* group are facultative chemoheterotrophs, most synthesize phycoerythrin, and many produce nitrogenase under anaerobic conditions. Their range of G + C values is relatively narrow: 38 to 47 percent. They are widespread in soil or in fresh and salt water, particularly in the intertidal and splash zones of oceans, where they adhere by their fibrous wall layer to rocks, shells, or other algae (Figure 15.21).

FIGURE 15.22

Representative life cycle of the *Pleurocapsa* group. Heavy colored arrows indicate multiple fission; heavy black arrows indicate binary fission.

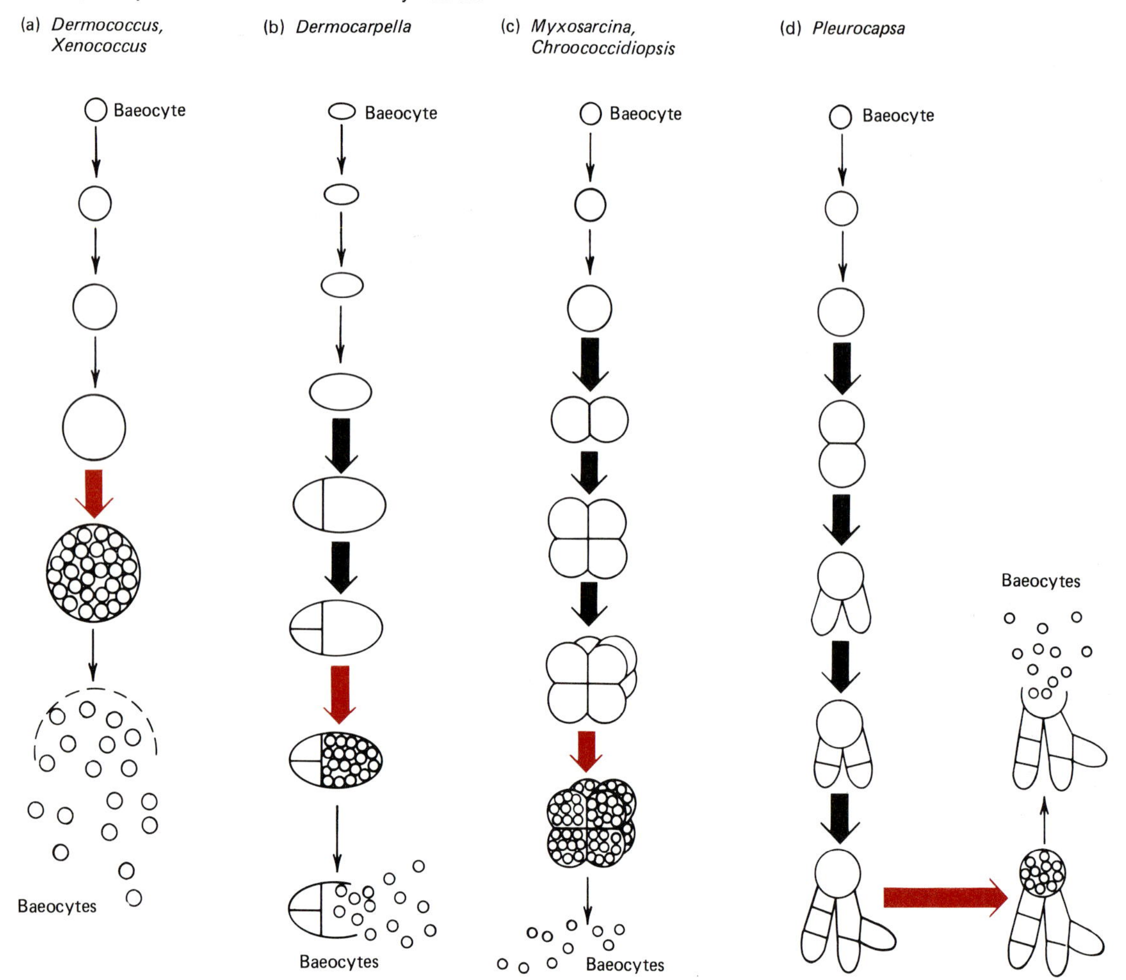

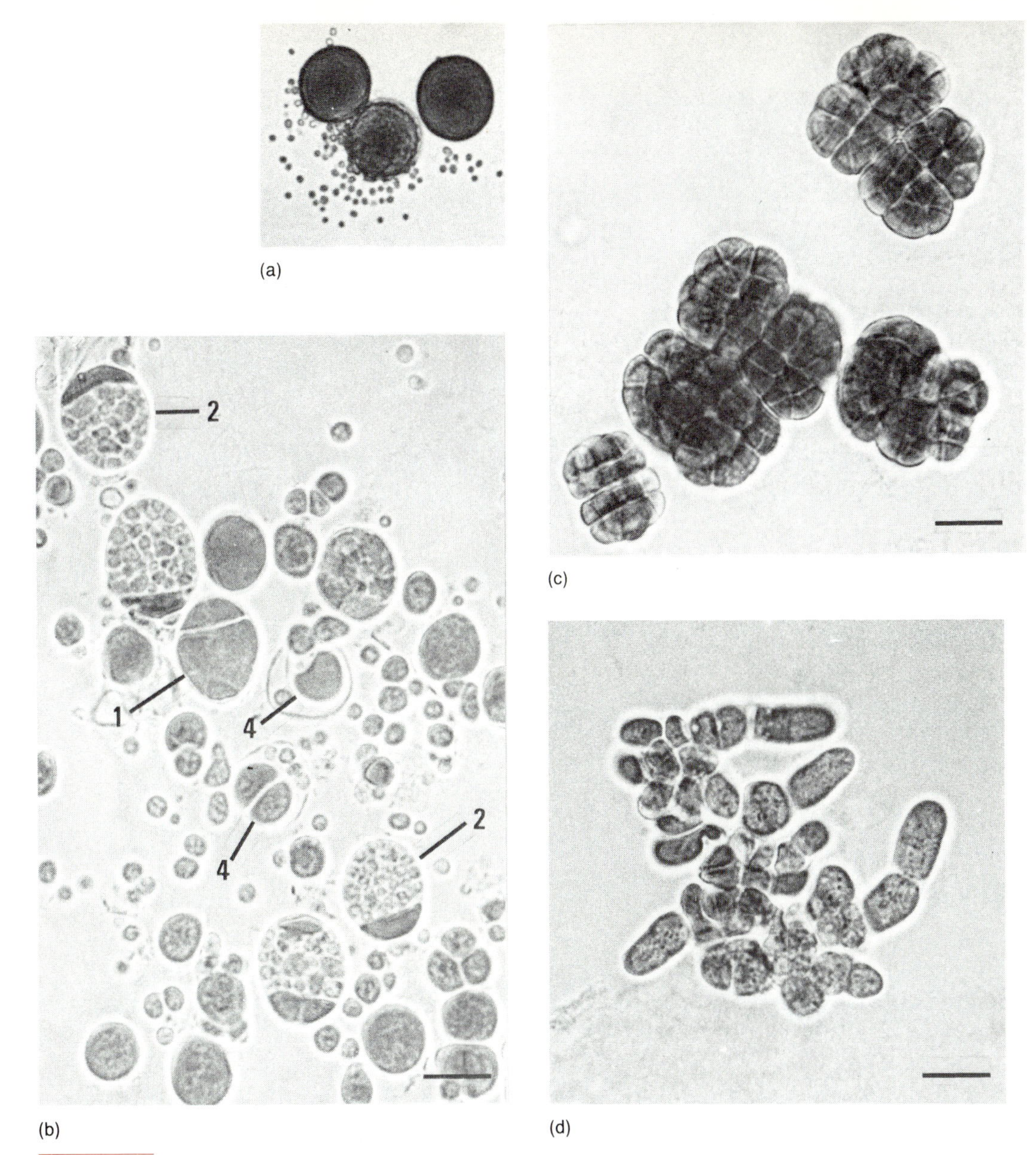


FIGURE 15.23

Light micrographs of representatives of the *Pleurocapsa* group. (a) *Dermocarpa*, showing large cells undergoing multiple fission to produce baeocytes; (b) *Dermocarpella*, showing various division stages: (1) A cell that has just undergone transverse fission to produce a basal non-reproductive cell and an apical reproductive cell; (2) an individual cell following multiple fission of the apical reproductive cell; and (4) the basal cell retained within the parental fibrous layer following release of the baeocytes; (c) *Myxosarcina;* (d) *Pleurocapsa*. (a) $\times 400$; (b)–(d) $\times 1000$. From J. B. Waterbury and R. Y. Stanier, "Patterns of Growth and Development in Pleurocapsalean Cyanobacteria," *Microbiol. Rev.* **42,** 2–44 (1978).

**TABLE 15.6**

**The *Pleurocapsa* Group**

| | Binary Fission | Motile Baeocytes | Special Reproductive Cells | Regular Alternation of Division Planes |
|---|---|---|---|---|
| *Dermocarpa* | − | + | NA[a] | NA |
| *Xenococcus* | − | − | NA | NA |
| *Dermocarpella* | + | + | + | +[b] |
| *Myxosarcina* | + | + | − | + |
| *Chroococcidiopsis* | + | − | − | + |
| *Pleurocapsa* | + | + | −[c] | − |

[a] NA = not applicable

[b] If any additional binary fissions occur after the one that produces a reproductive cell.

[c] However, not all cells in an aggregate undergo multiple fission.

FIGURE 15.24

Electron micrograph of a grazing thin section of trichome of an oscillatorean cyanobacterium, showing the rows of junctional pores (jp) on either side of every septum (s). From H. C. Lamont, "Sacrificial cell death and trichome breakage in an oscillatoriacean blue-green alga: the role of murein," *Arch. Mikrobiol.* **69,** 237–259 (1969).

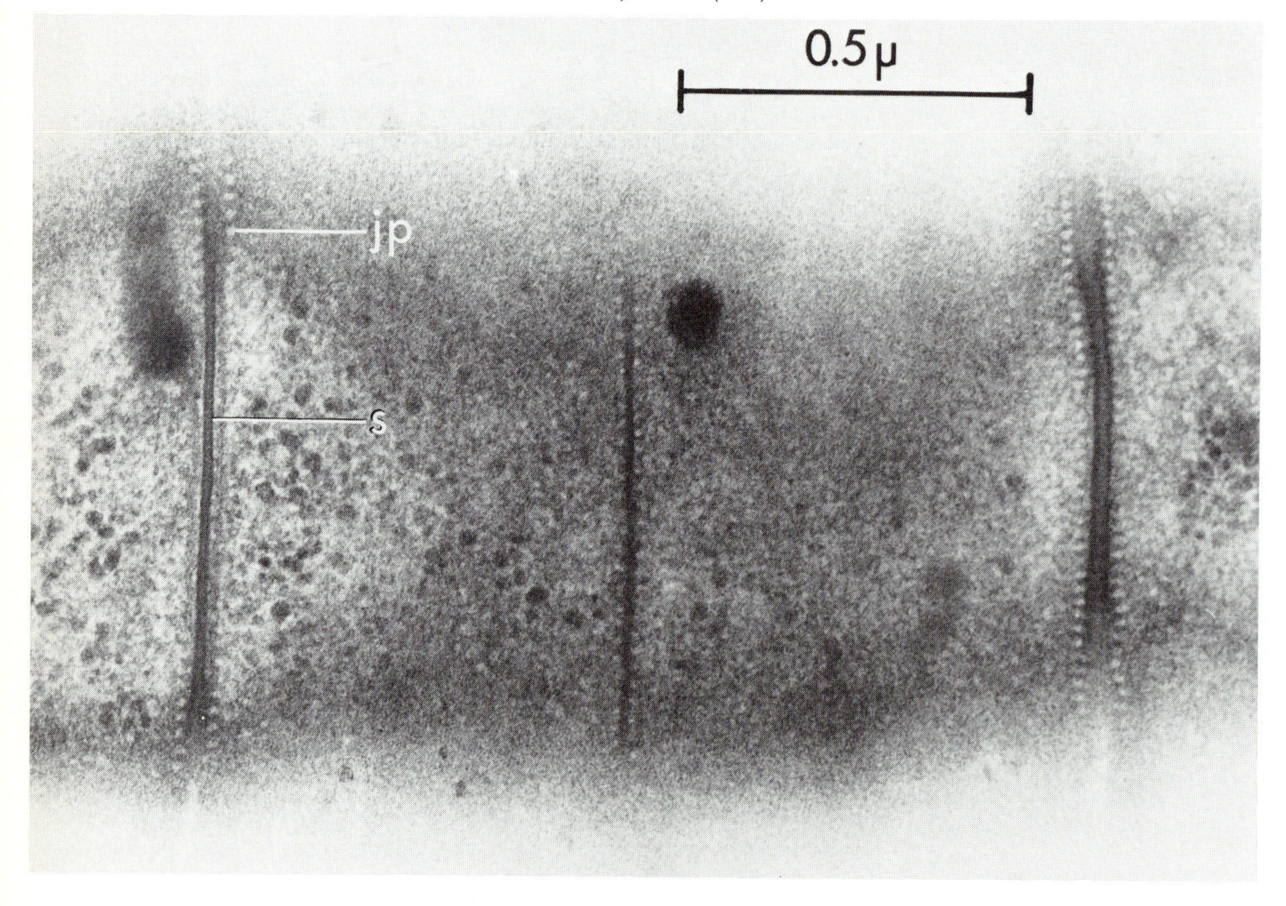

THE OSCILLATORIA GROUP The *Oscillatoria* group comprises those filamentous cyanobacteria that do not form heterocysts. Cell division is by binary fission; reproduction is by breakage of the filament. Two mechanisms of filament breakage are known. In *Pseudanabaena* a plane of weakness runs down the central plane of each septum; when the filament is subjected to sufficient stress, two daughter filaments are formed by tears through this plane. In *Oscillatoria* no such delamination of the septum occurs; rather, each septum is bordered on both sides by a row of *junctional pores*, perforations through the murein that resemble the perforations in a row of postage stamps (Figure 15.24). Filament breakage occurs by a tear along one of these rows of junctional pores, usually preceded by death of the cell (termed a *necridium*) at the torn junction.

Motility is the rule among oscillatorian cyanobacteria, unless the trichome is surrounded by an especially heavy sheath; such heavy sheaths are not found in the well-characterized members of this group, but are found in some strains of uncertain generic affiliation.

On the basis of trichome shape and presence of polar gas vacuoles, three genera are recognized (Table 15.7 and Figure 15.25). However, many strains of oscillatorian cyanobacteria do not fit into these genera; their taxonomic situation is confused, and it is not clear how many additional genera will be needed to accommodate them. All three presently recognized genera appear to be widely distributed in soil and water (both fresh and salt), although the gas vacuoles characteristic of *Pseudanabaena* suggest that it is predominantly aquatic. *Spirulina* is also predominantly aquatic and is characteristically found in highly alkaline saline waters. Indeed, in some Mexican and African lakes its abundance colors the water deep green. Both the preconquest Aztec of Mexico and the Kanenbou of Chad independently learned to harvest and dry *Spirulina* for food from these lakes, probably the only direct use of bacteria for human food.

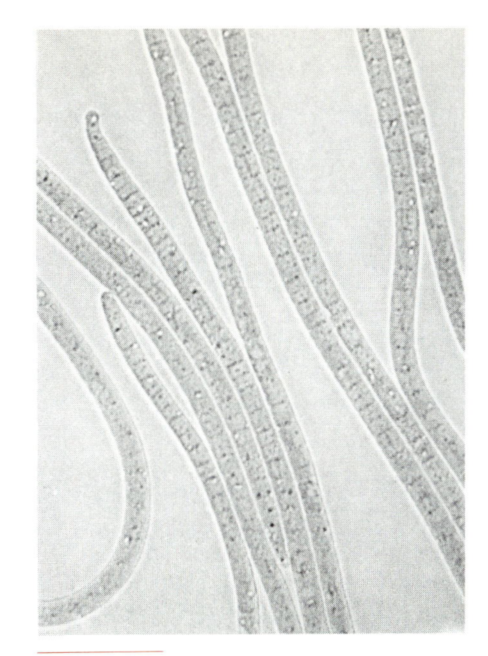

FIGURE 15.25
Brightfield light micrograph of *Oscillatoria* sp. ×748. Courtesy of R. Rippka.

THE HETEROCYSTOUS CYANOBACTERIA The filamentous heterocystous cyanobacteria form heterocysts when deprived of a source of combined nitrogen. They also may form specialized resting cells termed *akinetes* (Figure 15.26). Under favorable conditions the akinetes can germinate to produce a new filament.

Reproduction in this group occurs by two mechanisms: filament breakage may occur, producing daughter filaments; or morphologically distinct short trichomes, termed *hormogonia*, may be formed. Although the mechanisms of filament breakage have not been studied in this group they probably resemble those found in the oscillatorian cyanobacteria. Hormogonia are readily distinguished from broken filaments on the basis of their small size, motility, or gas vacuolation (Figure 15.27).

**TABLE 15.7**
**The Oscillatoria Group**

| | Trichome Shape | Constrictions between Cells | Polar Gas Vacuoles | Percent G + C |
|---|---|---|---|---|
| *Spirulina* | Helical[a] | − | − | 44–54 |
| *Oscillatoria* | Straight | − | − | 40–50 |
| *Pseudanabaena* | Straight | + | + | 44–52 |

[a] Flat spirals on solid media.

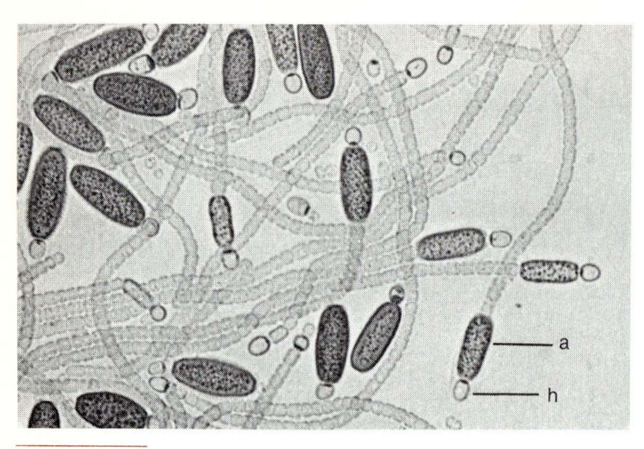


FIGURE 15.26

Brightfield light micrograph of *Cylindrospermum*, showing akinetes (a) adjacent to terminal heterocysts (h). ×339. Courtesy of R. Rippka.

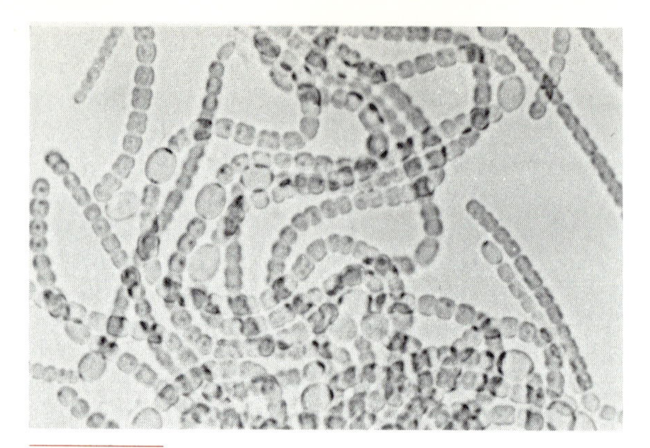

FIGURE 15.27

Phase-contrast micrograph of *Nostoc* sp., showing hormonogonia (thin filaments) among the thicker vegetative filaments (×252). Courtesy of Dr. J. C. Meeks.

FIGURE 15.28

Brightfield light micrograph of *Fischerella*, a heterocystous cyanobacterium that divides in two planes to form branched filaments. ×467. Courtesy of R. Rippka.

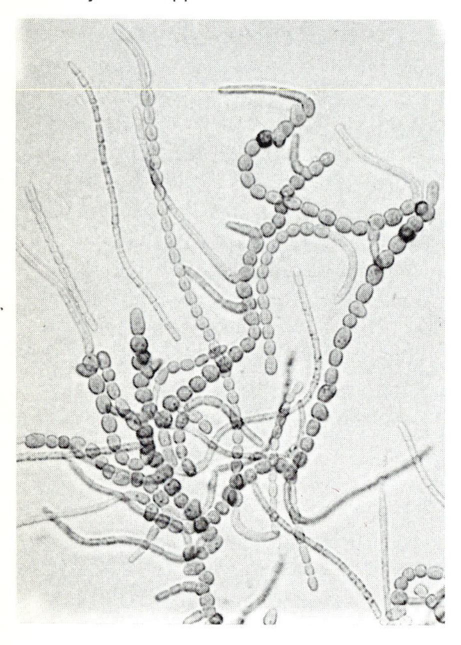

FIGURE 15.29

Electron micrograph of a thin section of *Gloeobacter*, showing the multilamellar sheath surrounding the cells, and the densely-packed row of columnar phycobilisomes immediately within the cytoplasmic membrane. The cell labelled (A) contains a cyanophycin granule; the one labelled (B) contains a polyphosphate granule. From R. Rippka, J. Waterbury, and G. Cohen-Bazire, "A cyanobacterium which lacks thylakoids," *Arch. Microbiol.* **100,** 419–436 (1974).

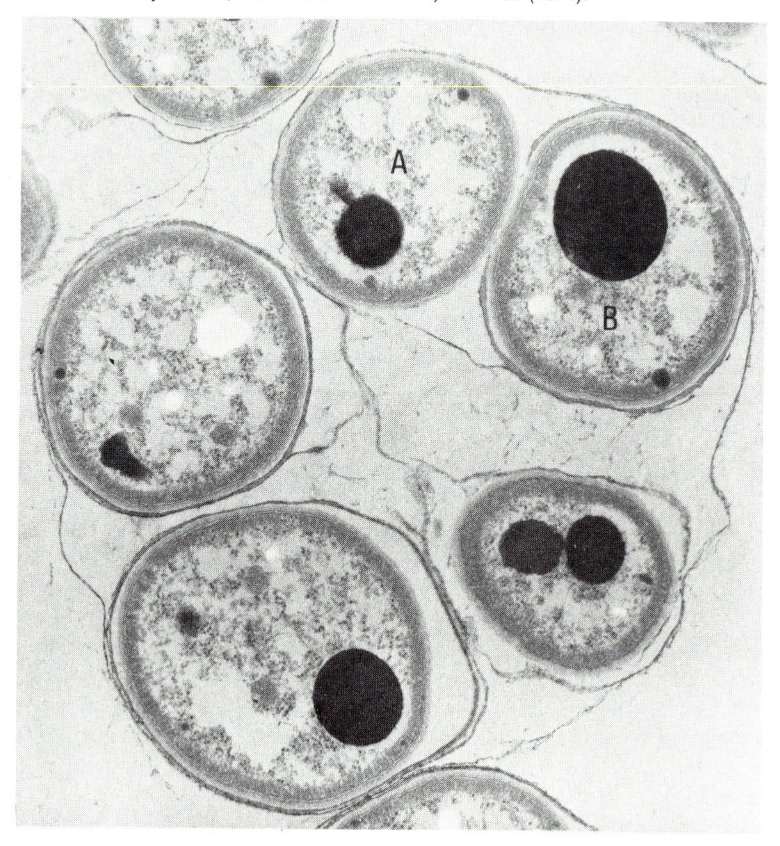

**TABLE 15.8**

**Heterocystous Cyanobacteria**

| Genus | Division in More than One Plane | Hormogonia Produced | Heterocyst Location: VEGETATIVE TRICHOME | Heterocyst Location: HORMOGONIA | Vegetative Filament Tapered | Special Features |
|---|---|---|---|---|---|---|
| *Anabaena* | – | – | T or I[a] | NA[b] | – | Cells not discoidal |
| *Nodularia* | – | – | T or I | NA | – | Discoidal cells |
| *Cylindrospermum* | – | – | T | NA | – | Akinetes always adjacent to heterocysts |
| *Nostoc* | – | + | T or I | T (both ends) | – | Often produces chains of akinetes |
| *Syctonema* | – | + | T or I | T (one end) | – | Discoidal cells |
| *Calothrix* | – | + | T or I | T (one end) | + | Terminal heterocyst always at basal (larger) end |
| *Chlorogloeopsis* | + | + | T or I | T or I | – | Division in multiple planes produces aggregates |
| *Fischerella* | + | + | T or I | I | + | Division in alternate planes produces branched filaments |

[a] T = terminal; I = intercalary.
[b] NA = not applicable.

The cells of heterocystous cyanobacteria divide by binary fission in the plane transverse to the filament, thus elongating it. However, in two genera, fission also occurs in a second plane, thus producing a branched filament or an irregular aggregate of cells (Figure 15.28).

Recognized genera of heterocystous cyanobacteria are described in Table 15.8; representative organisms were shown in Figures 15.14, 15.26 and 15.28.

**ORGANISMS RELATED TO CYANOBACTERIA** Two recently isolated organisms, *Gloeobacter* and *Prochloron*, are clearly related to the cyanobacteria, but do not share completely their distinctive pigment composition and structure. *Gloeobacter* is a typical unicellular cyanobacterium in all respects except fine structure: it lacks thylakoids and its phycobilisomes are atypical. The chlorophyll *a* and photosynthetic reaction centers are located in the cell membrane, immediately beneath which the cylindrical phycobilisomes form a cortical layer (Figure 15.29). It is immotile, incapable of chemoheterotrophic growth, and has a G + C content of 64 percent.

*Prochloron* (Figure 15.30) is a unicellular symbiont of certain colonial ascidians of tropical inter- and subtidal seas. The *Prochloron* cells adhere loosely to the surface of the host or grow as clusters in the cloacal cavity. All attempts to obtain sustained growth *in vitro* have failed. Nevertheless, the organisms can be easily removed from its host to produce suspensions of high purity with which biochemical analysis is feasible. The distinctive feature of *Prochloron* is the possession of chlorophyll *a* and *b* within a procaryotic cell structure; this pigment composition is characteristic of green algae and higher plants. No phycobilins are detectable, nor are carotenoid glycosides (although in other respects the carotenoid composition is similar to that of typical cyanobacteria). The photosynthetic pigments are located in thylakoid membranes that, like those of the cyanobacteria, are independent of the cell membrane and not enclosed within an outer membrane as in chloroplasts. Unlike typical cyanobacterial thylakoids, those of *Prochloron* are double, consisting of two closely appressed thylakoid sacs (Figure 15.30). *Prochloron* has a G + C content of about 41 percent and a genome size of $3.6 \times 10^9$. Carbon dioxide

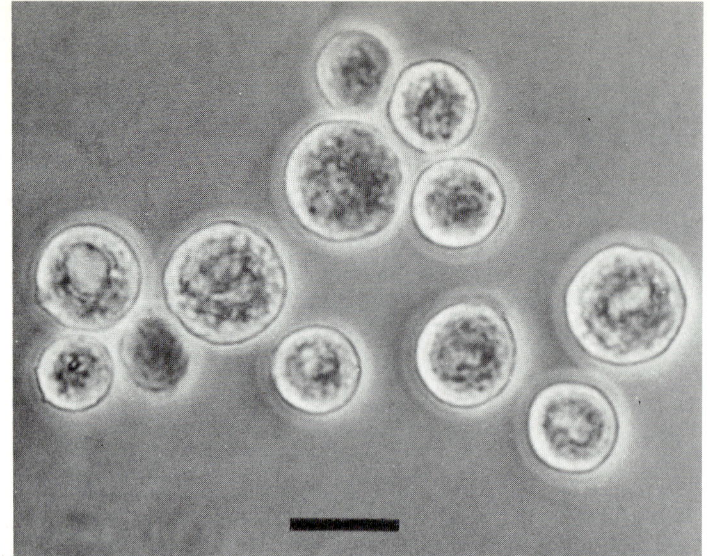
(a)

FIGURE 15.30

*Prochloron* (a) phase-contrast light micrograph. Bar represents 10 $\mu$m. (b) Electron micrograph of a thin section, showing the characteristic pairwise arrangement of thylakoids ($\times$ 19,000). (a) Courtesy of Dr. R. Lewin; (b) From J. M. Whatley, "The fine structure of *Prochloron*," *New Phytol*. **79,** 309–313 (1977).

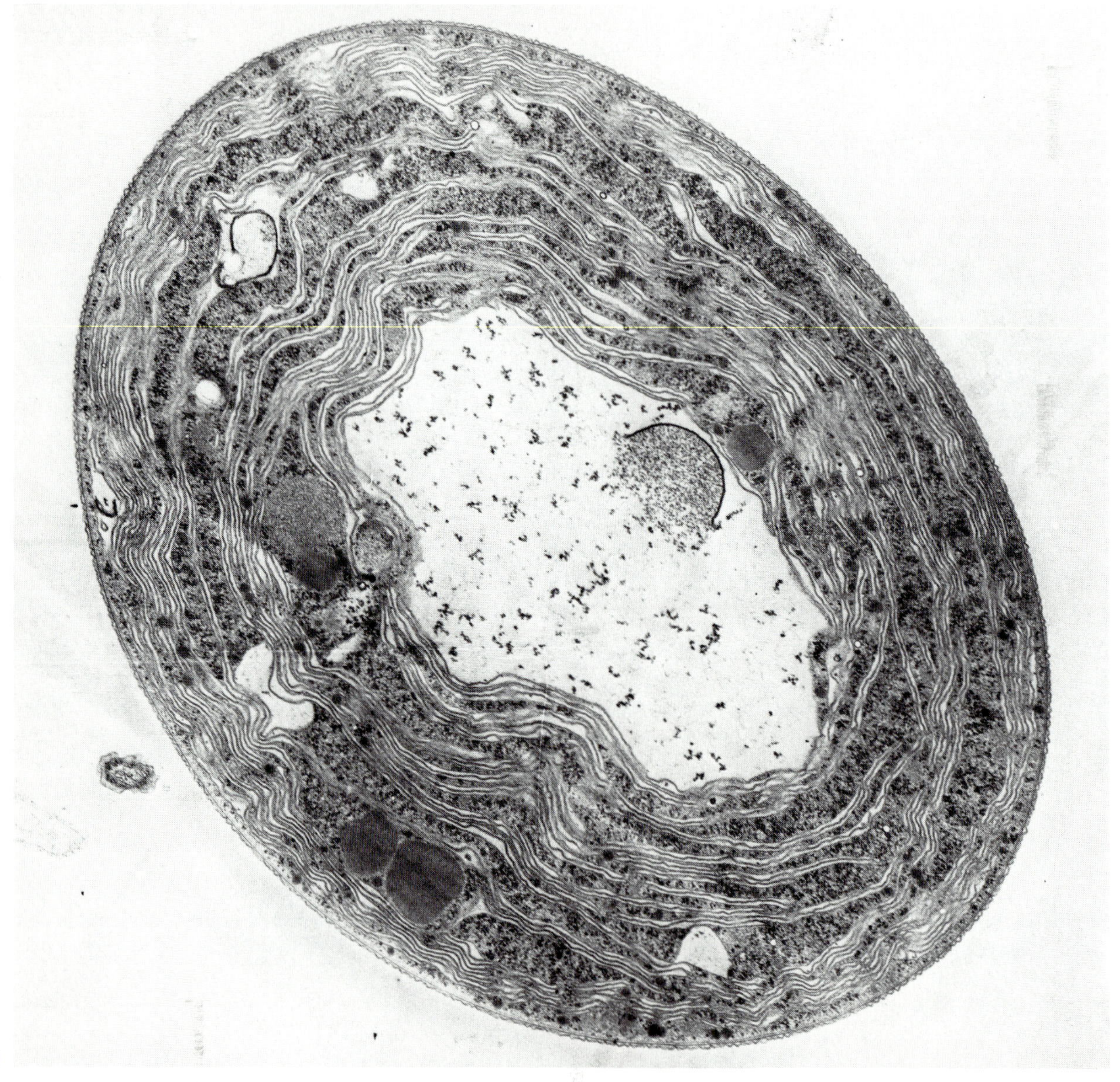
(b)

fixation in the light, accompanied by oxygen evolution, has been demonstrated. A relationship between *Prochloron* and the cyanobacteria, not obvious on the basis of structural and chemical properties, has been shown by 16S rRNA fingerprinting (Chapter 13), the closest relatives being heterocystous cyanobacteria.

## Ecology

The cyanobacteria occupy a far wider range of habitats than do other photosynthetic procaryotes. They occur in all environments that support the growth of algae: the sea, fresh water, and soil. They also develop in certain habitats from which photosynthetic eucaryotes are largely or completely excluded. Nitrogen-fixing representatives are conspicuous in environments where combined nitrogen is a limiting nutrient, notably in tropical soils, and in soils exposed by retreating glaciers or newly created by volcanic activity. Certain thermophilic cyanobacteria grow abundantly in neutral or alkaline hot springs, where they are the predominant members of the photosynthetic population. The temperature ranges of thermophilic cyanobacteria vary, but some unicellular forms can grow at temperatures up to 75° C. They are excluded by their relatively high pH range from acid hot springs, of which the characteristic photosynthetic inhabitant is a red alga, *Cyanidium caldarum*, which has a low pH optimum and is the only truly thermophilic photosynthetic eucaryote. However, its temperature maximum (approximately 56° C) is considerably below that of many thermophilic cyanobacteria.

Deserts are an extreme environment in which the microbial photosynthetic population consists almost entirely of unicellular cyanobacteria. These organisms grow in microfissures just below the surface of rocks, where small amounts of moisture are trapped and where sufficient light penetrates to permit photosynthesis. The ability to tolerate extreme fluctuations of temperature is important to their survival in the desert habitat.

In lakes that have undergone eutrophication (enrichment with mineral nutrients, notably phosphate and nitrate), a massive development of unicellular and filamentous cyanobacteria characteristically occurs during the warmer months of the year. These are largely gas vacuolate forms; in calm weather, the population floats to the surface, accumulating there to produce a so-called "bloom." Subsequent death and decomposition of the bloom promotes a massive development of chemoheterotrophic microorganisms, which may have catastrophic effects on the animal population of the lake because they deplete the dissolved oxygen supply.

Cyanobacteria are also common in symbiotic associations with eucaryotic organisms. Their role in these associations may be to provide fixed carbon to a heterotrophic partner (see Chapter 25), or fixed nitrogen to another photoautotrophic organism. An agriculturally important symbiosis between a heterocystous cyanobacterium and a phototrophic eucaryote is that between the small floating aquatic fern *Azolla* and its partner *Anabaena azollae*. The fern (Figure 15.31) grows on the surface of still water in the tropical and temperate zones. On its leaves are mucilage-containing cavities, initially open to the outside but later closed, that harbor filaments of *Anabaena* (Figure 15.32). It is possible to grow the host free of its symbiont; under such conditions, *Azolla* becomes dependent on an exogenous source of fixed nitrogen. This symbiosis provides a nitrogen source in the traditional practice of rice cultivation in Southeast Asia, and more recently it has been employed in Africa. The *Azolla* is grown on the surface of paddy water among the rice plants, where its heavy growth may result in an amount of $N_2$ fixed per square meter equivalent to that fixed in a terrestrial field by the legume-*Rhizobium* symbiosis (see p. 552). In Southeastern Asia a vigorous cottage industry has flourished for centuries to provide a stable source of inoculum for the paddies.

FIGURE 15.31

*Azolla* frond from a culture grown in the absence of combined nitrogen. From G. A. Peters and B. C. Mayne, "The Azolla, Anabaena Azollae Relationship I. Initial Characterization of the Association." *Plant Physiol.* **53**, 813 (1974).

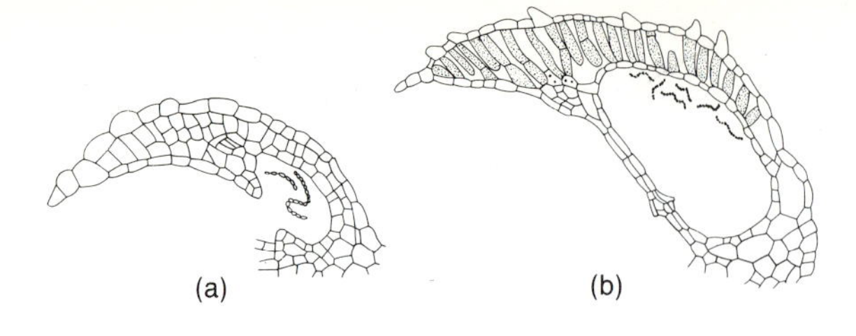


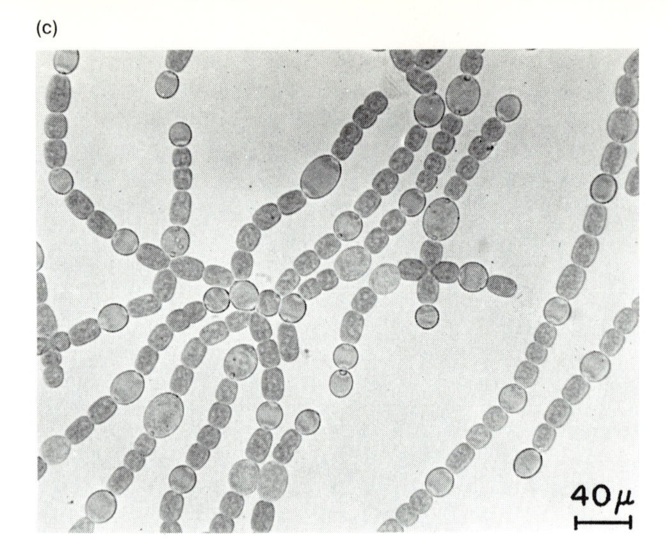


FIGURE 15.32

*Anabaena azollae* in *Azolla* leaves. (a) Section through an immature leaf (×215) showing *Anabaena* filaments in the developing leaf cavity; (b) section through a mature leaf (×80) showing *Anabaena* filaments in the leaf cavity; (c) *Anabaena* filaments isolated by micromanipulation from an *Azolla* leaf cavity. (a) and (b) from G. M. Smith, *Cryptogamic Botany*, 2nd ed. (1938). (c) From G. A. Peters and B. C. Mayne, "The Azolla, Anabaena Azollae Relationship I. Initial Characterization of the Association." *Plant Physiol.* **53,** 813 (1974).

## THE PURPLE BACTERIA

Taxonomically speaking, the purple bacteria are a small group of Gram-negative eubacteria, consisting of only about 30 species. They are unicellular and reproduce by binary fission or, in a few species, by budding. Most are motile by flagella; a few are immotile. Gas vacuoles are formed by some. Nitrogen fixation under anaerobic conditions is common. Despite the small size of this group, it is genetically diverse, because the mean DNA base composition ranges from 46 to 73 percent G + C; rRNA sequence analysis confirms this conclusion (Chapter 13).

All purple bacteria are, at least potentially, photoautotrophs, capable of growing anaerobically in the light with $CO_2$ as the carbon source and reduced inorganic compounds as the electron donor. Under these conditions, the Calvin-Benson cycle is the principal pathway of carbon assimilation. However, the purple bacteria can also develop photoheterotrophically under anaerobic conditions in the light at the expense of organic compounds, of which acetate is the most widely utilized. Under these circumstances, cell material is derived largely from the organic substrate, although $CO_2$ may also be assimilated.

Concomitant $CO_2$-uptake becomes important if the organic substrate is more reduced than cell material, because the reductive assimilation of $CO_2$ provides an acceptor for electrons derived from the organic substrate during its oxidation. This point may be illustrated by considering the paths of photometabolism by purple bacteria of two fatty acids, acetate and butyrate. Both compounds are rapidly assimilated by many purple bacteria under anaerobic conditions in the light, and they are initially converted by all purple bacteria, in large part to a reserve material, poly-β-hydroxybutyric acid, through the sequences of reactions shown in Figure 15.33. The photosynthetic assimilation of acetate occurs in the absence of $CO_2$ but butyrate assimilation requires its presence.

The conversion of acetate to poly-β-hydroxybutyrate is a reductive process:

$$2n\text{CH}_3\text{COOH} + 2n\text{H} \longrightarrow (\text{C}_4\text{H}_6\text{O}_2)_n + 2n\text{H}_2\text{O}$$

All purple bacteria possess the machinery of the TCA cycle and can thus generate reductant by the anaerobic oxidation of acetate through this cycle:

$$\text{acetate} + 2\text{H}_2\text{O} \longrightarrow 2\text{CO}_2 + 8\text{H}$$

This permits a reductive conversion of acetate to poly-β-hydroxybutyrate, the balanced equation for the overall reaction being

$$9n\text{CH}_3\text{COOH} \longrightarrow 4(\text{C}_4\text{H}_6\text{O}_2)_n + 2n\text{CO}_2 + 6n\text{H}_2\text{O}$$

As this equation shows, carbon assimilation is remarkably efficient, some 90 percent of the organic substrate being converted to cellular reserve material. This high efficiency is possible only because the photochemical reactions (cyclic photophosphorylation) can furnish potentially unlimited quantities of ATP, necessary for the initial activation of acetate (i.e., formation of acetyl-CoA).

The formation of poly-β-hydroxybutyrate from butyrate is an oxidative process:

$$n\text{CH}_3\text{CH}_2\text{CH}_2\text{COOH} \longrightarrow (\text{C}_4\text{H}_6\text{O}_2)_n + 2n\text{H}$$

It can therefore proceed anaerobically *only if a hydrogen acceptor is available.* The role of acceptor is played by $CO_2$, which is assimilated via the Calvin-Benson cycle and converted to glycogen, the other storage material formed by all purple bacteria. Symbolizing glycogen as $(CH_2O)_n$, the coupled photoassimilation can be represented as

$$2nC_4H_8O_2 + nCO_2 \longrightarrow 2(C_4H_6O_2)_n + (CH_2O)_n + nH_2O$$

butyrate → PHB, glycogen

The anaerobic photoassimilation of butyrate is thus obligatorily coupled with $CO_2$ assimilation, both processes being driven by ATP derived from cyclic photophosphorylation; operation of the TCA cycle, essential for the anaerobic assimilation of acetate, plays no role in the process.

The synthesis of poly-$\beta$-hydroxybutyrate does not, in itself, represent a *de novo* synthesis of cell material. In order for this reserve product to be used as a general source of cell constituents, the constituent acetyl units must be converted to pyruvate. Like many anaerobic chemoheterotrophic bacteria, the purple bacteria can synthesize pyruvate from acetyl units by the ferredoxin (Fd)-mediated reaction:

$$\text{acetyl-CoA} + CO_2 + FdH_2 \longrightarrow CH_3COCOOH + Fd + CoA$$

Sugar phosphates and dicarboxylic acids can be synthesized from pyruvate via phosphoenolpyruvate. The synthesis of dicarboxylic acids involves a second reductive carboxylation:

$$\text{P-enolpyruvate} + CO_2 + NADH + H^+ \longrightarrow \text{malate} + NAD^+ + P_i$$

Under many growth conditions, this alternative pathway of $CO_2$ fixation becomes of considerable quantitative importance, relative to fixation of $CO_2$ via the Calvin-Benson cycle, in purple bacteria. However, the acetyl-CoA-malate fixation pathway is a noncyclic one, and its operation therefore depends on the availability of acetyl-CoA, either from an endogenous or from an exogenous source. The pathways of carbon assimilation from organic sources and from $CO_2$ are thus varied and relatively complex in this group.

With few exceptions, purple bacteria do not appear to be able to synthesize ATP by fermentative means in the dark. In *Chromatium*, an interesting mechanism for the anaerobic generation of ATP in the dark, capable of providing the cell with maintenance energy, has been discovered; the conversion of the intracellular glycogen store to the other intracellular reserve material, poly-$\beta$-hydroxybutyrate. Glycogen is decomposed (probably by the Embden-Meyerhof pathway) to pyruvate, which is in turn converted to $CO_2$ and acetyl-CoA and thence to poly-$\beta$-hydroxybutyrate. Since the synthesis of poly-$\beta$-hydroxybutyrate from acetyl-CoA does not require an input of ATP, the overall reaction results in a net ATP gain, derived from substrate-level phosphorylation during the conversion of glycogen to pyruvate. The overall reaction can be represented as

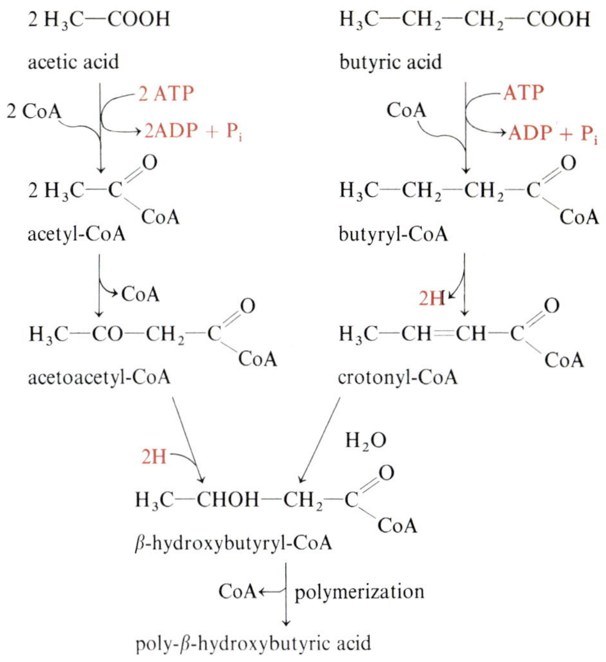

$$2nADP + 2nP_i + (C_6H_{10}O_5)_n + nH_2O \longrightarrow (C_4H_6O_2)_n + 2nCO_2 + 6nH + 2nATP$$

It depends on the availability of a suitable hydrogen acceptor; in *Chromatium* this role is assumed by the intracellular deposits of elemental sulfur, which are reduced to $H_2S$:

$$3nS + 6nH \longrightarrow 3nH_2S$$

2 H₃C—COOH (acetic acid) + 2 CoA, 2 ATP → 2ADP + Pᵢ → 2 H₃C—CO—CoA (acetyl-CoA) → (− CoA) H₃C—CO—CH₂—CO—CoA (acetoacetyl-CoA) → (+ 2H) H₃C—CHOH—CH₂—CO—CoA (β-hydroxybutyryl-CoA)

H₃C—CH₂—CH₂—COOH (butyric acid) + CoA, ATP → ADP + Pᵢ → H₃C—CH₂—CH₂—CO—CoA (butyryl-CoA) → (− 2H) H₃C—CH=CH—CO—CoA (crotonyl-CoA) → (+ H₂O) H₃C—CHOH—CH₂—CO—CoA (β-hydroxybutyryl-CoA)

β-hydroxybutyryl-CoA → (CoA released; polymerization) poly-β-hydroxybutyric acid

FIGURE 15.33
The conversions of acetic and butyric acids to poly-$\beta$-hydroxybutyric acid by purple nonsulfur bacteria.

## Constituent Groups of Purple Bacteria

It is customary to recognize two subgroups among the purple bacteria; the distinctions between them are both physiological and ecological (Table 15.9). Most *purple sulfur bacteria* are strict anaerobes with a predominantly photoautotrophic mode of metabolism, based on the use of $H_2S$ as an electron donor. *Purple nonsulfur bacteria* have a predominantly photoheterotrophic mode of metabolism.

**TABLE 15.9**

**Characteristics that Distinguish the Two Subgroups of Purple Bacteria**

| | Purple Sulfur Bacteria | Purple Nonsulfur Bacteria |
|---|---|---|
| Principal mode of photosynthesis | Photoautotrophic | Photoheterotrophic |
| Range of photoassimilable organic substrates | Narrow | Broad |
| Aerobic growth | –[a] | +[a] |
| Ability to oxidize $H_2S$ | + | +[ab] |
| Accumulation of $S^0$ as intermediate in $H_2S$ oxidation to $SO_4^{2-}$ | + | –[a] |
| $H_2S$ toxicity | Usually low | Usually high |
| Percent G + C | 45–70 | 61–72 |

[a] A few exceptions exist.
[b] Normally only at low concentrations.

They are sensitive to $H_2S$, their growth being inhibited by low concentrations of sulfide, even though many can oxidize sulfide anaerobically in the light if the concentration is kept very low. Most members of both groups of purple bacteria can grow photoautotrophically with $H_2$ as electron donor.

Whereas the purple sulfur bacteria are obligate phototrophs, many purple nonsulfur bacteria can grow well aerobically in the dark. Such strains possess an aerobic electron transport chain, and are thus endowed with respiratory capacity. A few of them can also grow (though very slowly) anaerobically in the dark, through the fermentation of pyruvate or sugars.

The purple nonsulfur bacteria typically occur in freshwater lakes or ponds, where organic matter is present but sulfide is either absent or present at low concentrations. The typical habitats of the purple sulfur bacteria are sulfide-rich waters, where sulfide is generated by the activity of sulfate-reducing bacteria.

## Purple Sulfur Bacteria

The characteristics that distinguish the genera of purple sulfur bacteria are shown in Table 15.10 and some typical representatives are illustrated in Figure 15.34. The characteristic photometabolism of these organisms involves assimilation of $CO_2$, largely through the Calvin-Benson cycle, ATP being provided by cyclic photophosphorylation, reducing power by reverse electron transport. The electron donor, $H_2S$, is oxidized via elemental sulfur to sulfate. The overall reaction can be presented schematically as

$$2CO_2 + H_2S + 2H_2O \longrightarrow 2(CH_2O) + H_2SO_4$$

Some (but not all) purple sulfur bacteria can use other reduced inorganic sulfur compounds ($S^0$, thiosulfate, sulfite) in place of $H_2S$ as exogenous reductants. The biochemistry of the oxidation of these reduced sulfur compounds by purple bacteria is complex and not well established. It probably is similar to the respiratory oxidation of these compounds by chemoautotrophic bacteria (Chapter 16).

The oxidation of $H_2S$ by the purple sulfur bacteria always leads to a massive but transient accumulation of elemental sulfur, since this first step is much more rapid than the ensuing oxidation of $S^0$ to $SO_4^{2-}$. In most of these organisms the elemental sulfur is deposited within the cell, as refractile globules. However, *Ectothiorhodospira* excretes sulfur into the medium, and subsequently reabsorbs it prior to further oxidation.

The photometabolism of purple sulfur bacteria is never obligatorily photoautotrophic, since all these organisms can photoassimilate some organic compounds, acetate being a universal substrate. Some of them require a small amount of $H_2S$ for photoheterotrophic growth, using it as a cellular sulfur source, since they are unable to perform an assimilatory sulfate reduction. The only organic growth factor required is vitamin $B_{12}$, an essential nutrient for a few species.

A number of purple sulfur bacteria, including species of *Thiocapsa*, *Chromatium*, *Thiocystis*, and *Amoebobacter*, have been shown to be capable of chemoautotrophic growth under low partial pressures of oxygen, with reduced sulfur compounds as electron donor. Only *Thiocapsa* is capable of chemotrophic growth under full atmospheric oxygen tension; it can grow aerobically either autotrophically or heterotrophically.

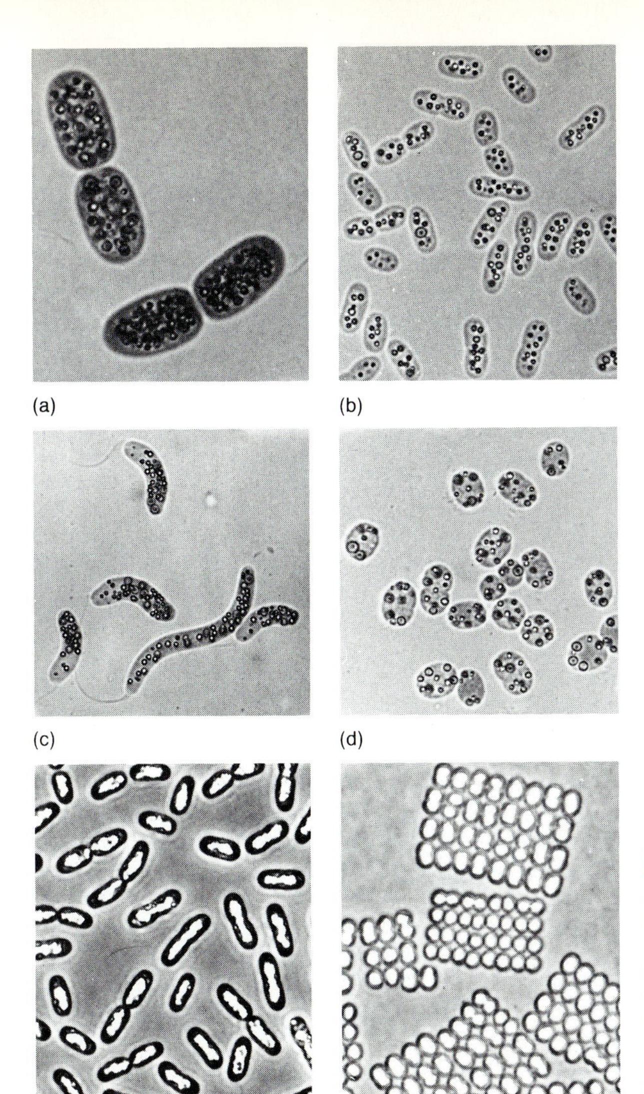


FIGURE 15.34

Photomicrographs of some representative purple sulfur bacteria. (a) *Chromatium okenii*, (×1400); (b) *Chromatium vinosum*, (×1400); (c) *Thiospirillum jenense*, (×1190); (d) *Thiocystis gelatinosa*, (×1400); (e) *Thiodictyon elegans*, (×1400); (f) *Thiopedia rosea*, (×1400). (a)–(d) Brightfield illumination; (e) and (f) phase contrast. Note the intracellular sulfur granules in (a)–(d). The phase-bright intracellular areas in (e) and (f) are gas vacuoles. Courtesy of Dr. N. Pfennig.

**TABLE 15.10**

**The Genera of Purple Sulfur Bacteria**

| Genus | Cell arrangement and shape | Motility | Gas Vacuoles | Site of Sulfur Deposition |
|---|---|---|---|---|
| *Thiospirillum* | Single; helical | + | − | Intracellular |
| *Ectothiorhodospira* | Single; vibroid | + | − | Extracellular |
| *Chromatium* | Single; cylindrical | + | − | Intracellular |
| *Thiocystis* | Single; spherical | + | − | Intracellular |
| *Thiocapsa* | Single or cubical packets; spherical | − | − | Intracellular |
| *Lamprocystis* | Single; spherical | + | + | Intracellular |
| *Thiodictyon* | Single or loose networks; cylindrical | − | + | Intracellular |
| *Thiopedia* | Flat rectangular plates; ovoid | − | + | Intracellular |
| *Amoebobacter* | Single; spherical | − | + | Intracellular |

## TABLE 15.11

**The Genera of Purple Nonsulfur Bacteria**

| Genus | Cell Shape | Flagellar Insertion | Intracytoplasmic Membranes | Prosthecae | Exospores | Mode of Cell Division |
|---|---|---|---|---|---|---|
| *Rhodospirillum* | Helical | Polar | Vesicular or Lamellar | – | – | Fission |
| *Rhodopseudomonas* | Rod | Polar | Lamellar | – | – | Budding directly from cell pole |
| *Rhodomicrobium* | Ovoid | Peritrichous | Lamellar | + | + | Budding from hyphal tip |
| *Rhodopila* | Coccoid or ovoid | Polar | Vesicular | – | – | Fission |
| *Rhodocyclus* | Curved rod | Polar or immotile | Tubular | – | – | Fission |
| *Rhodobacter* | Ovoid or rod | Polar or immotile | Vesicular | – | – | Fission |

### Purple Nonsulfur Bacteria

The distinguishing properties of the genera of purple nonsulfur bacteria are listed in Table 15.11, and photomicrographs of some typical representatives are shown in Figure 15.35. The only purple bacteria that reproduce by budding rather than by binary fission are members of this subgroup; they include *Rhodomicrobium* and *Rhodopseudomonas* species.

The range of organic compounds that can be photoassimilated by purple nonsulfur bacteria is quite wide: it includes fatty acids, other organic acids, primary and secondary alcohols, carbohydrates, and even aromatic compounds. Species capable of respiratory metabolism can grow aerobically in the dark, by the oxidation of the same range of organic substrates that they photoassimilate anaerobically in the light. This does not necessarily involve the operation of the same metabolic pathways, however. As previously discussed, the photoassimilation of butyrate is obligatorily coupled with $CO_2$-assimilation and does not involve the operation of the TCA cycle. When it serves as a substrate for dark respiratory metabolism, much of the substrate is oxidized via the TCA cycle to provide ATP by oxidative phosphorylation.

FIGURE 15.35

Photomicrographs of some representative purple non-sulfur bacteria ($\times$ 1400). (a) *Rhodospirillum rubrum;* (b) *Rhodocyclus purpureus;* (c) *Rhodobacter sphaeroides;* (d) *Rhodomicrobium vanielli.* (b) from N. Pfennig, "*Rhodocyclus purpureus* gen. nov. and sp. nov., a Ring-Shaped, Vitamin $B_{12}$-Requiring Member of the Family Rhodospirillaceae." *Int. J. Syst. Bact.* **28,** 283–288 (1978).

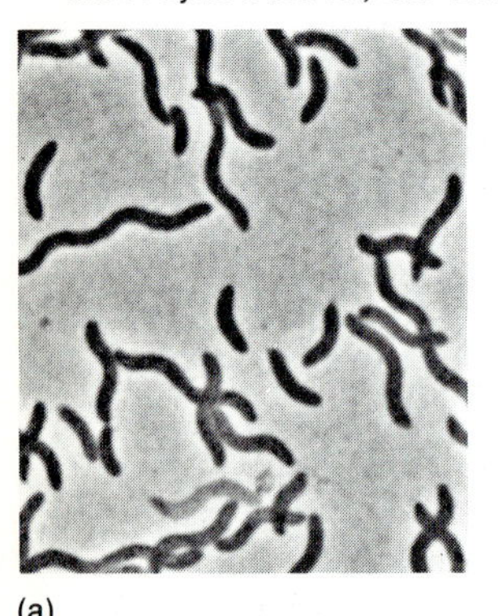

(a)

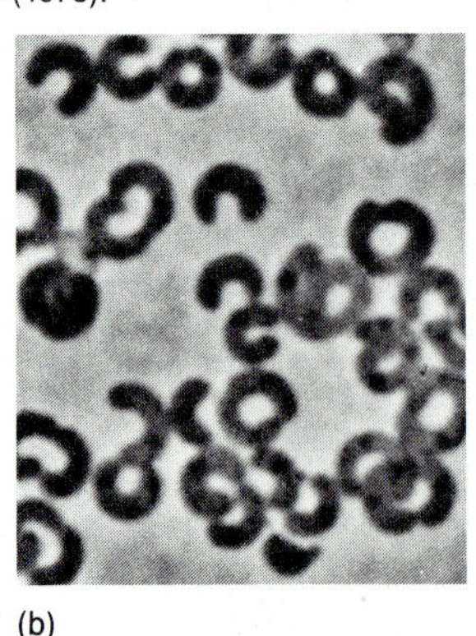

(b)

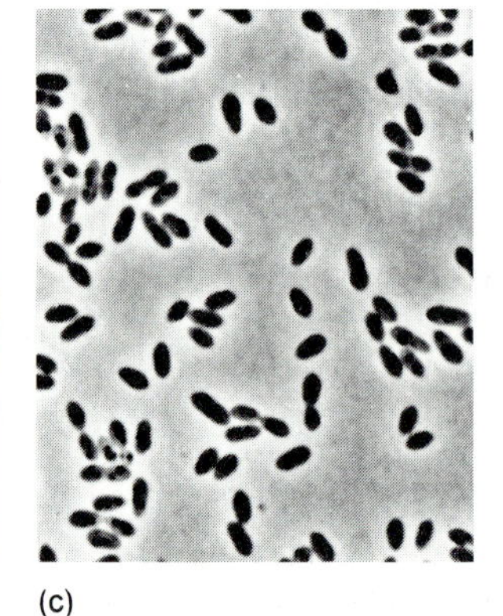

(c)

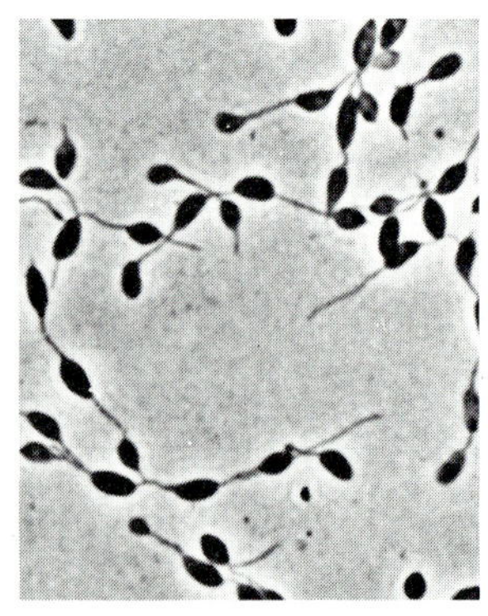

(d)

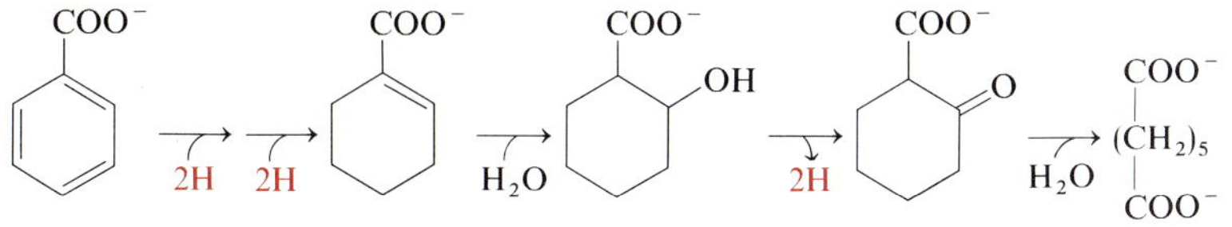

FIGURE 15.36

The anaerobic metabolism of benzoate by purple bacteria.

The rule that photoassimilable organic substrates can also be respired by purple bacteria has one interesting exception. Some of these organisms can photoassimilate benzoate anaerobically in the light, but are completely unable to use it as a respiratory substrate. The photometabolism of benzoate occurs through a unique *reductive* pathway, the initial steps of which convert benzoate to a saturated dicarboxylic acid, pimelate (Figure 15.36). The enzymes that catalyze these reactions are exceedingly oxygen-sensitive, the photoassimilation of benzoate being immediately arrested if cells are exposed even to traces of $O_2$. The absence from purple bacteria of the enzymes that catalyze an oxygenative pathway of benzoate dissimilation, characteristic of benzoate-utilizing aerobic chemoheterotrophs (p. 93) accounts for their inability to use this compound as a respiratory substrate.

As previously mentioned, purple nonsulfur bacteria are frequently capable of photoautotrophic growth with reduced inorganic sulfur compounds. A number of species utilize thiosulfate; and many others are capable of oxidizing $H_2S$, provided that its concentration is kept low. Some species oxidize $H_2S$ only to elemental sulfur, which is excreted into the medium; others oxidize it to sulfate, with or without the intermediate accumulation of $S^o$. The metabolism of reduced sulfur compounds in this microbial group is accordingly quite varied.

Several species of purple nonsulfur bacteria, most notably *Rhodobacter capsulatus*, have been shown to grow as aerobic chemoautotrophs with $H_2$ as electron donor, and there are suggestions that thiosulfate may also be used as an electron donor.

Other purple bacteria (e.g., *Rhodopseudomonas palustris* and *Rhodobacter sphaeroides*) are capable of denitrification, using a variety of organic compounds as their energy source. Indeed under some conditions, denitrification in conjunction with nitrogen fixation may supply reduced nitrogen for cell growth. *R. sphaeroides* cannot assimilate nitrate directly, but under photoheterotrophic growth conditions with nitrate as sole nitrogen source, tracer experiments have shown that the coupled processes of denitrification and nitrogen fixation can provide sufficient combined and reduced nitrogen to support growth.

Most of the purple nonsulfur bacteria require vitamins, and their growth rate is frequently improved by the provision of amino acids. Various combinations of biotin, thiamin, and niacin are the typical vitamin requirements; a requirement for vitamin $B_{12}$, characteristic of some purple sulfur bacteria, occurs only in *Rhodocyclus purpureus*.

**EFFECTS OF $O_2$ ON GROWTH AND PIGMENT SYNTHESIS IN PURPLE NONSULFUR BACTERIA** None of the purple nonsulfur bacteria are killed by exposure to air; however, some of these organisms cannot use $O_2$ as a terminal electron acceptor, and therefore they cannot grow aerobically in the dark. Others grow at least as rapidly under aerobic conditions in the dark as they do under anaerobic conditions in the light. However, aerobic growth leads rapidly to an almost complete loss of the photosynthetic pigment system. This is a consequence of the fact that, even at relatively low partial pressures, $O_2$ is a potent repressor of pigment synthesis by purple bacteria, exerting this effect even in the presence of light. Light itself is not required for pigment synthesis, as shown by the fact that species able to grow fermentatively maintain a high pigment content through many generations of heterotrophic growth in the dark.

The aerobic growth of purple bacteria consequently leads, either in the dark or in the light, to a progressive dilution of the cellular pigment content. This is a purely physiological phenomenon, immediately reversed when cells are returned to anaerobic growth conditions. Consequently, the photosynthetic development of all purple bacteria, anaerobes and facultative aerobes alike, is possible only in an $O_2$-free environment. Under anaerobic conditions in the light, both the growth rate and the differential rate of bacteriochlorophyll synthesis are governed by light intensity. As light intensity is increased, the growth rate increases and the cellular bacteriochlorophyll content declines (Figure 15.37).

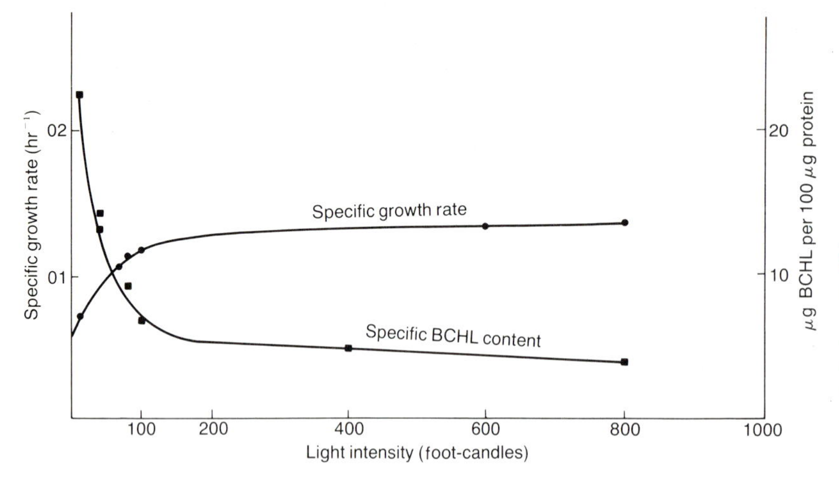


FIGURE 15.37
The effect of light intensity on the growth rate and specific cellular content of bacteriochlorophyll of *Rhodospirillum rubrum*, growing photoheterotrophically in the absence of $O_2$.

## THE GREEN BACTERIA

The green bacteria comprise an even smaller taxonomic group of Gram-negative eubacteria than the purple bacteria (Table 15.12). The span of DNA base composition is comparatively narrow: 48 to 58 percent G + C. With respect to its physiological and nutritional properties, the group shows interesting parallels to the purple bacteria. Most members—the *green sulfur bacteria*—are counterparts of the purple sulfur bacteria. However, the recently isolated thermophilic green bacterium, *Chloroflexus*, closely resembles the purple nonsulfur bacteria in its metabolic and nutritional properties.

### The Green Sulfur Bacteria

Green sulfur bacteria are small, usually immotile bacteria; five genera are recognized on the basis of structural characters; some of these are illustrated in Figure 15.38. The members of this subgroup are strictly anaerobic photoautotrophs that use $H_2S$, other reduced inorganic sulfur compounds, or $H_2$ as electron donors. Elemental sulfur arising from $H_2S$ oxidation is deposited extracellularly (as in *Ectothiorhodospira*), prior to oxidation to sulfate. Since green sulfur bacteria cannot use sulfate as a sulfur source, they require sulfide to meet biosynthetic needs when growing with $H_2$ as an electron donor; some also require vitamin $B_{12}$. Nitrogen

**TABLE 15.12**
**Genera of Green Bacteria**

| | Cell Form And Arrangement | Gliding Motility | Gas Vacuoles | Prosthecae |
|---|---|---|---|---|
| Green sulfur bacteria | (G + C = 48 to 58 percent) | | | |
| *Chlorobium* | Straight or curved rods, single or short chains | − | − | − |
| *Prosthecochloris* | Ovoid, single or short chains | − | − | + |
| *Pelodictyon* | Chains of rods, forming nets | − | + | − |
| *Ancalochloris* | Spherical | − | + | + |
| *Chloroherpeton* | Unicellular filaments | + | − | − |
| Green nonsulfur bacteria | (G + C = 53 to 55 percent) | | | |
| *Chloroflexus* | Long filaments composed of rod-shaped cells | + | − | − |
| *Chloronema* | Long filaments composed of rod-shaped cells | + | + | − |
| *Oscillochloris* | Trichomes of discoidal cells | + | + | − |

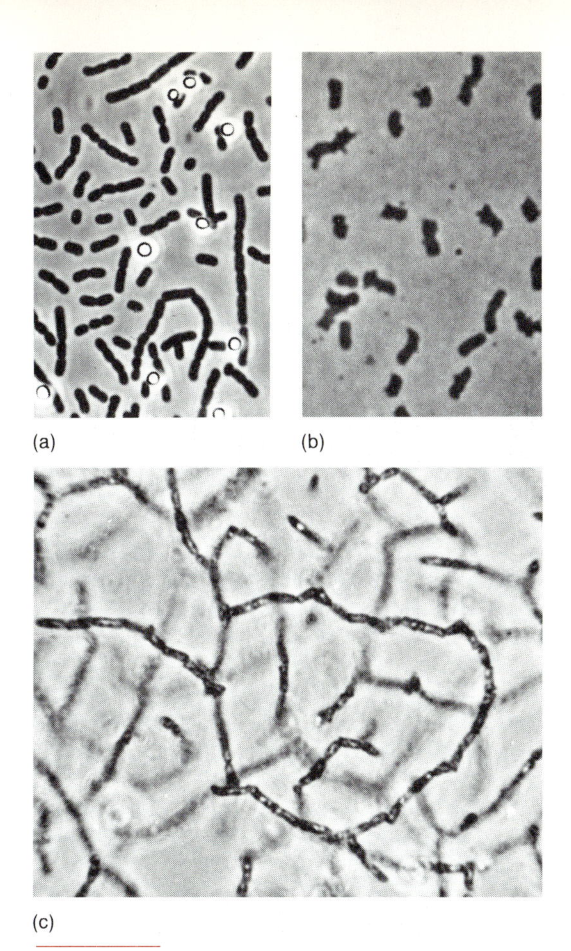


FIGURE 15.38

Photomicrographs (phase contrast) of green sulfur bacteria. (a) *Chlorobium limicola* ($\times$ 1,500); note extracellular sulfur granules. (b) *Prosthecochloris aestuarii*, ($\times$ 2,300); the prosthecae are just detectable by light microscopy, conferring an irregular outline on the profile of the cells. (c) *Pelodictyon clathratiforme* ($\times$ 1,500), showing the characteristic net formation; the phase-bright areas in some of the cells are gas vacuoles. Courtesy of Dr. N. Pfennig.

fixation is of common occurrence. In all these respects, the analogies to purple sulfur bacteria are evident. Indeed, purple and green sulfur bacteria commonly coexist in illuminated, sulfide-rich anaerobic aquatic environments and have essentially overlapping natural distributions.

There is a marked difference, however, with respect to carbon nutrition in the two groups. None of the green sulfur bacteria can grow photoheterotrophically, using organic compounds as their sole or principal carbon source in the absence of an inorganic reductant. They can photoassimilate acetate, but only if $H_2S$ and $CO_2$ are simultaneously provided. These organisms do not synthesize poly-$\beta$-hydroxybutyrate as a reserve material, and acetate is assimilated exclusively through the reductive synthesis of pyruvate, from acetyl-CoA and $CO_2$, thus serving directly as a precursor of cell material.

The pathway of assimilation of $CO_2$ in the green sulfur bacteria has long been a matter of controversy. The presence of ferredoxin-linked enzymes capable of catalyzing the reductive carboxylation of acetate and of succinyl-CoA (reactions that are exceedingly unfavorable with pyridine nucleotides as electron donor) suggested to B. Buchanan and his colleagues that $CO_2$ was assimilated by a reversal of the tricarboxylic acid cycle (the *reductive TCA cycle*; Figure 15.39). Recent labeling studies have confirmed this insight. Hence the green bacteria are unique among eubacterial autotrophs in their use of this pathway; most others use the Calvin-Benson cycle for $CO_2$ fixation. Recent indications that the archaebacterium *Sulfolobus* also uses the reductive TCA cycle suggests that it may be an ancient pathway, and that the Calvin-Benson cycle is a more recent acquisition.

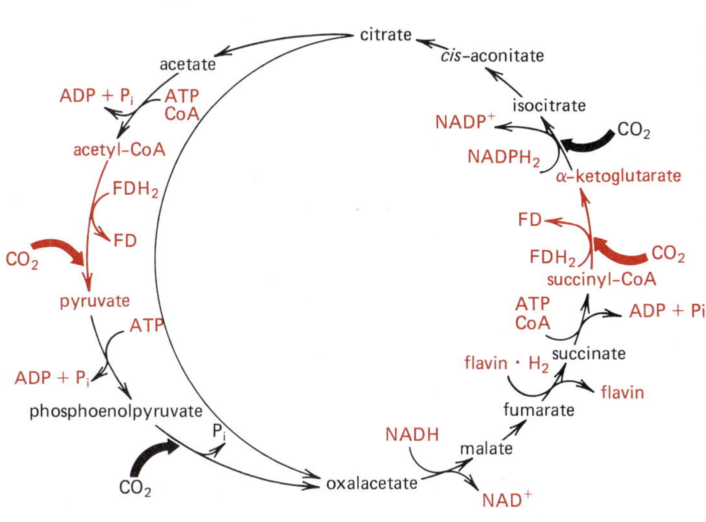


FIGURE 15.39

The reductive TCA cycle, showing the two reductive carboxylations driven by reduced ferredoxin ($FDH_2$). Redrawn from M. C. W. Evans, B. B. Buchanan, and D. I. Arnon, "A New Ferredoxin-Dependent Carbon Reduction Cycle in a Photosynthetic Bacterium," *Proc. Natl. Acad. Sci. USA* **55,** 928–934 (1966).

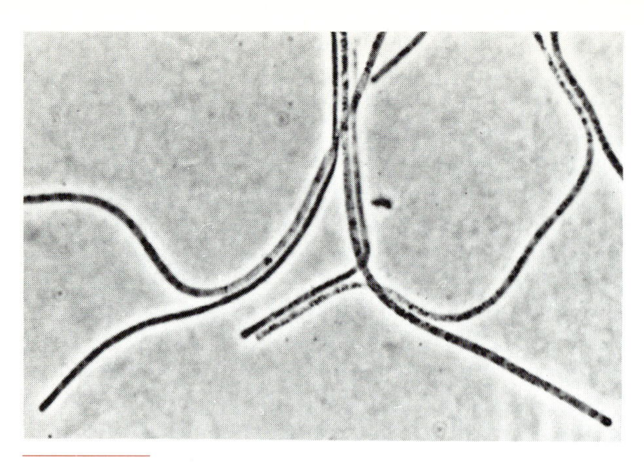

FIGURE 15.40

*Chloroflexus aurantiacus*, a filamentous, gliding green bacterium (phase contrast × 1,040). Courtesy of B. K. Pierson and R. W. Castenholz.

### Green Nonsulfur Bacteria: The *Chloroflexus* Group

Although the green nonsulfur bacteria differ from green sulfur bacteria in their structure, nutrition, metabolism, and ecology, they possess two properties that clearly identify them as green bacteria: the presence in the cells of chlorosomes and of bacteriochlorophyll *c* or *d* and *a* as the major and minor chlorophyllous pigments, respectively.

*Chloroflexus* is a filamentous gliding organism, whose filaments sometimes attain a length of 300 μm (Figure 15.40). They are thermophiles and develop abundantly in neutral or alkaline hot springs at temperatures in the range of 45° to 70° C. The masses of intertwined filaments form orange to dull green mats several millimeters thick, often closely associated with unicellular thermophilic cyanobacteria of the *Synechococcus* type. Their natural habitat is often partly aerobic; since the synthesis of bacteriochlorophylls by *Chloroflexus* is repressed by oxygen, the bacteriochlorophyll content of the mats is often low, and is largely masked by the orange carotenoids that are abundantly formed by *Chloroflexus* under all growth conditions. Hence, the filaments of which the mats are composed had long been interpreted as gliding, nonphotosynthetic bacteria. Growth is most rapid in complex media, incubated anaerobically in the light; under these conditions the filaments have a high content of bacteriochlorophylls *c* and *a*. No growth occurs anaerobically in the dark. Growth is good in complex media incubated aerobically either in the light or in the dark, though under these conditions the bacteriochlorophyll content of the cells is very low. The nutritional requirements of *Chloroflexus* are complex and not yet precisely determined.

*Chloroflexus* is evidently a typical photoheterotroph and facultative photoautotroph or chemoheterotroph. It grows in hot springs that have a low content of organic matter, where it appears to derive organic nutrients from the cyanobacteria with which it is naturally associated. The two organisms can be successfully maintained in the laboratory as two-membered cultures, grown in the light on a mineral medium.

Although *Chloroflexus* is the only green nonsulfur bacterium currently available in pure culture, several other organisms are assigned to the group (Table 15.12). Two of them, *Chloronema* and *Oscillochloris*, have been identified as members by spectral and electron microscopic studies on impure material. A third, *Heliothrix*, is a gliding filamentous organism, thought to be related to *Chloroflexus* on the basis of rRNA fingerprinting. *Heliothrix*, currently available only in two-membered culture, has only bacteriochlorophyll *a* as chlorophyllous pigment, and lacks chlorosomes; accordingly it appears to be a naturally occurring variant. It has a very narrow natural distribution, being known only from a few hot springs in eastern Oregon and Yellowstone National Park.

## ECOLOGICAL RESTRICTIONS IMPOSED BY ANOXYGENIC PHOTOSYNTHESIS

For the performance of photosynthesis, anoxygenic phototrophs require anaerobic conditions and either organic compounds or reduced inorganic compounds other than water. These limitations do not apply to cyanobacteria and photosynthetic eucaryotes. The purple and green sulfur bacteria are hence confined to a limited range of special habitats, and their quantitative contribution to photosynthetic productivity in the biosphere is negligible. They are exclusively aquatic and grow in bodies of water that provide the indispensable combination of anaerobiosis, light, and the nutrients specific for these organisms. These conditions occur principally in two types of aquatic environments, similar in chemical respects but differing markedly in the quality of light available. One consists of shallow ponds, relatively rich in organic matter, $CO_2$, $H_2$, and often $H_2S$ produced by anaerobic bacteria in the underlying sediment. Except near the air-water interface, occupied by cyanobacteria and algae, the water is essentially oxygen-free. Hence, purple and green bacteria can grow close to the water surface,

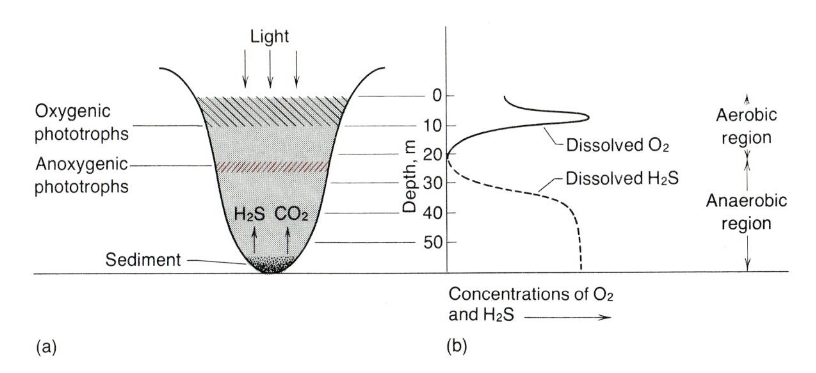


FIGURE 15.41

Diagram of the structure of a stratified lake, showing (a) the vertical distributions of oxygenic phototrophs and anoxygenic phototrophs and (b) the relative concentrations in the water profile of dissolved oxygen and $H_2S$.

where light intensity is high, but they are usually covered by a growth of oxygenic phototrophs. It is in this environment where the ability of these bacteria to absorb light of very long wavelength, transmitted by the overlying aerobic phototrophs, becomes of critical importance for their survival. The light used for photosynthesis is almost entirely absorbed by bacteriochlorophylls, in the far red and near-infrared regions.

The second environment in which purple and green bacteria abound occurs at a considerable depth in lakes, particularly so-called *meromictic* lakes, which are characterized by a permanent stratification of the water. The warmer, aerobic upper layer is underlain at depths of 10 to 30 meters by a stagnant layer that is cold and oxygen-free. The anoxygenic phototrophs occur in a narrow horizontal band, situated just within the anaerobic layer (Figure 15.41). Water samples from this depth are often brightly colored as a result of their content of purple and green bacteria, the population density being far greater than that of oxygenic phototrophs in the upper, aerobic layers. However, at the depth where the purple and green bacteria find the anaerobic conditions necessary for development, *the overlying water column itself becomes an effective light filter*, transmitting only green and blue-green light, of wavelengths between 450 and 550 nm. The role of light-harvesting pigments is largely assumed by carotenoids, not by bacteriochlorophylls; and anoxygenic phototrophs from this environment typically have a very high carotenoid content.

## BACTERIOCHLOROPHYLL IN AEROBIC EUBACTERIA

Recent work from several Japanese laboratories has demonstrated that a variety of aerobic bacteria synthesize significant amounts of bacteriochlorophyll *a*. Ecological studies of marine organisms indicated that, on the basis of spectral evidence, about 1 percent of the intertidal and pelagic aerobes, and up to 6 percent of those isolated from beach sand, make bacteriochlorophyll.

The best studied of these bacteria is a recent marine isolate, the strict aerobe *Erythrobacter*. Strains of this organism produce from about 0.1 to nearly 0.5 $\mu$g bacteriochlorophyll per mg dry weight; this value is nearly as high as that produced by anaerobic cultures of purple bacteria (typically in the range of 0.2 to 1.0 $\mu$g bacteriochlorophyll per mg dry weight). Pigment synthesis requires $O_2$; microaerophilic growth conditions lead to severe repression of bacteriochlorophyll synthesis, and some repression of carotenoid synthesis. Vesicular intracytoplasmic membrane systems are also seen in sections of aerobically grown cells, but in reduced number in $O_2$-limited cultures.

The bacteriochlorophyll appears to be functional in *Erythrobacter*; light-dependent reduction of cytochromes and inhibition of $O_2$ consumption indicate that photochemical ATP generation occurs. The magnitude of inhibition of oxygen uptake (to nearly 60 percent) suggests that light-driven ATP synthesis may meet the majority of the cell's energy needs. The resultant sparing of organic electron donors may allow more efficient incorporation of organic compounds into cell material, possibly important under the conditions of carbon limitation common in marine environments.

## HELIOBACTERIUM

An extremely unusual phototrophic eubacterium has recently been isolated from aerobic soil by H. Gest and his co-workers, and named *Heliobacterium chlorum*. The cells are rod-shaped and move by gliding. Growth is relatively rapid as an anaerobic photoheterotroph, biotin is required, and

$N_2$ is fixed. Chemotrophic growth has not been demonstrated.

The single photopigment of *Heliobacterium* is a new bacteriochlorophyll termed bacteriochlorophyll *g* (Figure 15.2, Table 15.2). Its cellular location has not been determined; however, the lack of extensive intracytoplasmic membranes and chlorosomes suggests that it is probably located largely within the cytoplasmic membrane.

Analysis of the 16S rRNA suggests that *Heliobacterium* has affinities with the clostridia, rather than the purple or green bacteria (Chapter 13).

## FURTHER READING


### Books

CARR, N. G., and B. A. WHITTON, eds., *The Biology of Cyanobacteria.* Boston: Blackwell, 1982.

CLAYTON, R. K., and W. R. SISTROM, eds., *The Photosynthetic Bacteria.* New York: Plenum Press, 1978.

ORMEROD, J. G., ed., *The Phototrophic Bacteria: Anaerobic Life in the Light.* Berkeley: University of California Press, 1983.

### Reviews

CIFERRI, O., "*Spirulina,* the Edible Microorganism," *Microbiol. Rev.* **47,** 551 (1983).

RIPPKA, R., J. DERUELLES, J. B. WATERBURY, M. HERDMAN, and R. Y. STANIER, "Generic Assignments, Strain Histories and Properties of Pure Cultures of Cyanobacteria," *J. Gen. Microbiol.* **111,** 1 (1979).

WATERBURY, J. B., and R. Y. STANIER, "Patterns of Growth and Development in Pleurocapsalean Cyanobacteria," *Microbiol. Rev.* **42,** 2 (1978).

# Chapter 16
# The Chemoautotrophic and Methophilic Eubacteria

Two highly specialized physiological groups of aerobic Gram-negative bacteria will be discussed in this chapter. They are the *chemoautotrophs*, which can derive the energy required for growth from the oxidation of inorganic compounds, and the *methophiles*, which can derive both the energy and the carbon required for growth from methane and other one-carbon organic compounds. Although readily distinguishable from one another by their different energy sources, the two groups show some interesting similarities.

Both the chemoautotrophs and the methophiles are in turn made up of several physiological subgroups, distinguished by their specific energy sources, and also to some extent by the degree of their specialization for the autotrophic (or methophilic) way of life. They are all widely distributed in soil and water, and play important roles in the cycles of elements in the biosphere (see Chapter 27). The total number of species represented is small, but both groups are diverse in structural as well as physiological respects.

## THE CHEMOAUTOTROPHS

By definition, a chemoautotroph can grow in a strictly mineral medium in the dark, deriving its carbon from $CO_2$ and its ATP and reducing power from the respiration of an inorganic substrate. This mode of life, which exists only among procaryotes, was discovered between 1880 and 1890 by

S. Winogradsky whose pioneering studies on several of the principal subgroups provided a solid foundation for all later work on chemoautotrophy. Winogradsky showed that two other remarkable properties are characteristic of the chemoautotrophs:

**1.** High specificity with respect to the inorganic energy source.
**2.** Frequent inability to use organic compounds as energy and carbon sources; indeed, their growth is sometimes adversely affected by organic compounds.

## Utilizable Substrates

The inorganic materials capable of supporting chemoautotrophic growth include $H_2S$ and other reduced forms of sulfur, ammonia, nitrite, molecular hydrogen, carbon monoxide, and ferrous iron ($Fe^{2+}$). In the biosphere these substrates are in part produced through the metabolic activities of other organisms and in part of geochemical origin. It should be noted that certain of them are chemically unstable under aerobic conditions: $H_2S$ is readily oxidized to elemental sulfur by contact with air; ferrous iron also undergoes autooxidation in neutral or alkaline solutions, although it is stable under acid conditions. The chemical instability of $H_2S$ and $Fe^{2+}$ has seriously impeded the isolation and study of some organisms that use these substrates.

The substrate specificities of the chemoautotrophs permit the recognition of five major subgroups (Table 16.1). *Nitrifying bacteria* use reduced inorganic nitrogen compounds as energy sources. The substrate specificity within this subgroup is very high; its members either oxidize ammonium to nitrite, or nitrite to nitrate; none can oxidize both these reduced nitrogen compounds. *Sulfur-oxidizing bacteria* use $H_2S$, elemental sulfur, or its partially reduced oxides, as energy sources; all these substances are converted to sulfate. One member of this group can in addition use ferrous iron as an energy source. *Iron bacteria* can oxidize reduced iron and manganese, but not reduced sulfur compounds; however, their status as true chemoautotrophs remains in some doubt. The *hydrogen bacteria* use molecular hydrogen as an energy source; and *carboxydobacteria* use carbon monoxide as an energy source.

## The Nitrifying Bacteria

In the middle of the nineteenth century circumstantial evidence indicated that the oxidation of ammonium to nitrate in natural environments is a microbial process. However, many attempts to isolate the causal agents using conventional culture media failed completely. This problem was solved in 1890 by S. Winogradsky, who succeeded in isolating pure cultures of nitrifying bacteria using strictly inorganic media. The causal agents proved to be small, Gram-negative, rod-shaped bacteria: *Nitrosomonas*, which oxidizes ammonium to nitrite ($NO_2^-$); and *Nitrobacter*, which oxidizes nitrite to nitrate ($NO_3^-$). The combined activities of these two organisms effected the complete conversion of ammonium to nitrate. The nitrifying bacteria develop best under neutral or alkaline conditions; since the oxidation of ammonium to nitrite results in considerable acid formation, the growth medium for *Nitrosomonas* must be well buffered (for example, by the addition of insoluble carbonates). Growth of both organisms is slow (minimal gen-

**TABLE 16.1**

**Physiological Groups of Aerobic Chemoautotrophs**

| Group | | Oxidizable Substrate | Oxidized Product | Terminal Electron Acceptor |
|---|---|---|---|---|
| Nitrifying bacteria | Ammonia oxidizers | $NH_4^+$ | $NO_2^-$ | $O_2$ |
| | Nitrite oxidizers | $NO_2^-$ | $NO_3^-$ | $O_2$ |
| Sulfur oxidizers[a] | | $H_2S$, $S^0$, $S_2O_3^{2-}$ | $SO_4^{2-}$ | $O_2$; sometimes $NO_3^-$ |
| Iron bacteria | | $Fe^{2+}$ | $Fe^{3+}$ | $O_2$ |
| Hydrogen bacteria | | $H_2$ | $H_2O$ | $O_2$; sometimes $NO_3^-$ |
| Carboxydobacteria | | CO | $CO_2$ | $O_2$ |

[a] One species can also use $Fe^{2+}$ as an energy source.

**TABLE 16.2**

**Some Genera of Nitrifying Bacteria**

| Energy-Yielding Reaction | Cell Form | Flagella | Intracytoplasmic Membranes | DNA Base Composition: Percent G + C | Obligate Autotrophy | Genus |
|---|---|---|---|---|---|---|
| $NH_4^+ \longrightarrow NO_2^-$ | Rod | Subpolar[a] | Lamellar | 47–50 | + | *Nitrosomonas* |
| | Tight spiral | Peritrichous[a] | None | 54 | + | *Nitrosospira* |
| | Sphere | Peritrichous[a] | Lamellar | 50–51 | + | *Nitrosococcus* |
| | Irregular, lobed | Peritrichous[a] | Vesicular | 54–55 | + | *Nitrosolobus* |
| $NO_2^- \longrightarrow NO_3^-$ | Rod, often pear-shaped[b] | Polar[b] | Lamellar | 60–62 | + or − | *Nitrobacter* |
| | Long, slender rod | — | None | 58 | + | *Nitrospina* |
| | Sphere | Polar | Tubular | 61 | + | *Nitrococcus* |

[a] Some strains are nonmotile.
[b] Reproduction by budding; all other nitrifiers reproduce by binary fission.

eration times approximating 24 hours), and the growth yields are low. Winogradsky showed that these bacteria are obligate autotrophs, incapable of development in the absence of their specific inorganic energy source.

In recent years a few other ammonium and nitrite oxidizers have been discovered. They resemble the two classical prototypes physiologically, but they are remarkably diverse in structural respects (Table 16.2). The group is a small one, consisting of several genera distinguished by structural properties and by the specific oxidizable substrate. Both the ammonium oxidizers and the nitrite oxidizers have narrow ranges of DNA base composition, the values for the latter being significantly higher than for the former.

The diversity of the nitrifying bacteria in gross cell structure is paralleled by a curious diversity in fine structure. In some genera the cell membrane is devoid of intrusions; in others there are extensive intrusions, which may be vesicular, lamellar, or tubular (Figure 16.1).

Obligate chemoautotrophy is the rule in the nitrifying bacteria, with the exception of some *Nitrobacter* strains that have been shown to use acetate as a carbon and energy source; however, these strains grow much more slowly with acetate than with nitrite.

The biochemistry of ammonium and nitrite oxidation has been elucidated only recently (Figure 16.2). Ammonium is oxidized to hydroxylamine by a mono-oxygenase, for which cytochrome $P_{450}$ is the electron donor. Oxidized $P_{450}$ is reduced during the oxidation of hydroxylamine. Thus two of the electrons removed during the oxidation of ammonium to nitrite are lost to ATP-generating electron transport, because they are required to reduce $P_{450}$. In contrast to the complexity of ammonium oxidation, the oxidation of nitrite to nitrate is an enzymatically simple and direct process. The immediate electron acceptor in these oxidations appears to be a cytochrome, possibly cytochrome *c* for hydroxylamine and cytochrome *a* for nitrite.

## Sulfur Oxidizers

In the course of his pioneering studies on chemoautotrophy, Winogradsky examined the properties of the filamentous gliding bacteria of the *Beggiatoa-Thiothrix* group, which occur characteristically in certain sulfide-rich environments and often contain massive inclusions of elemental sulfur. He showed that these organisms can oxidize $H_2S$, initially to elemental sulfur which accumulates in the cells; the stored sulfur is subsequently further oxidized to sulfate. A variety of other aerobic bacteria (Figure 16.3) have been subsequently shown to oxidize $H_2S$ in a similar manner, with transient intracellular sulfur deposition (Table 16.3). However, few of the organisms listed in Table 16.3 have been isolated in pure culture because of the technical difficulty of growing them aerobically at the expense of $H_2S$. Current knowledge of the group is based almost entirely on work with unicellular sulfur oxidizers which have small cells and do not accumulate sulfur within the cell (Table 16.4). These members of the group can be purified and grown without difficulty, as a result of their ability to oxidize chem-

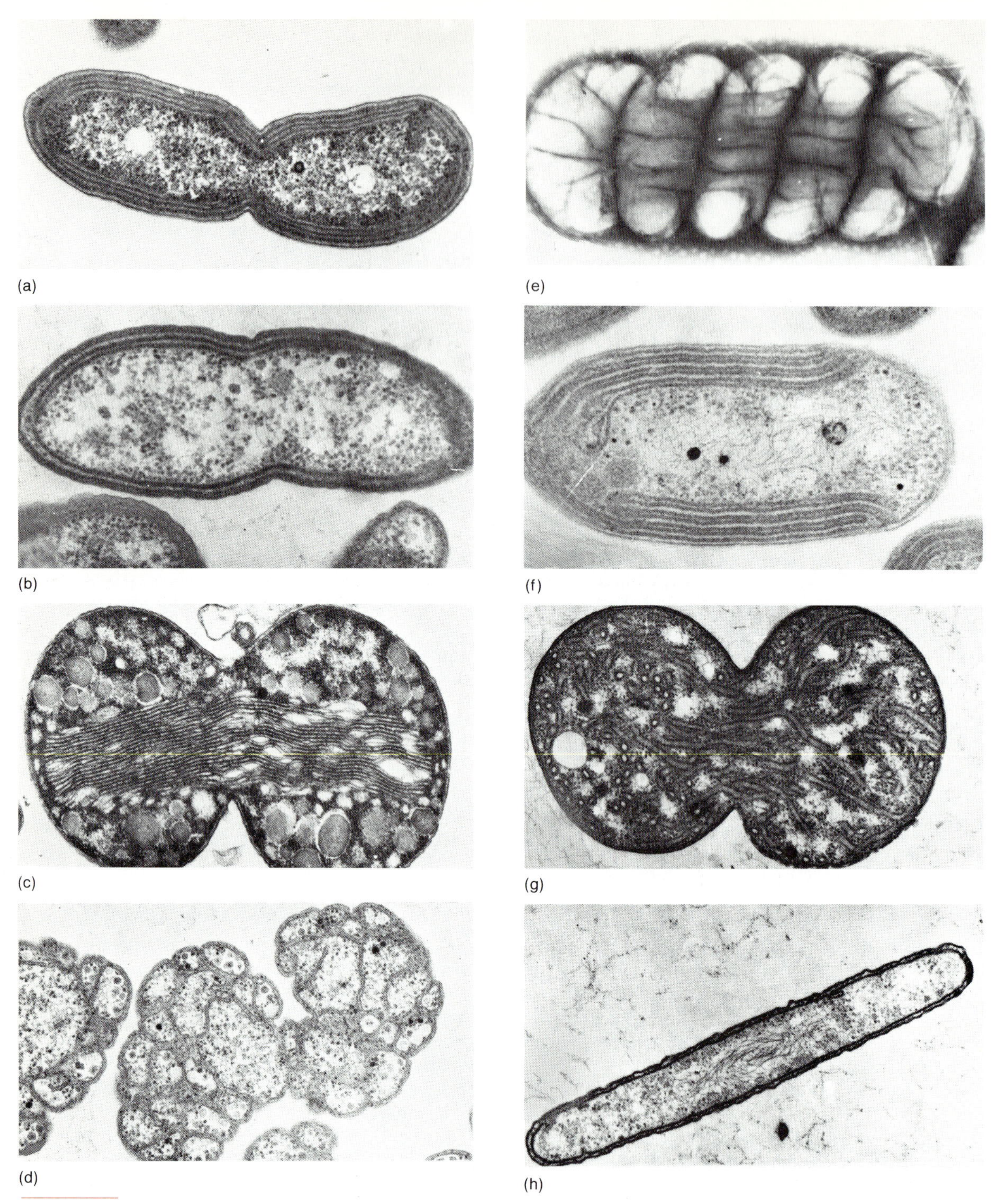


FIGURE 16.1

Electron micrographs of thin sections of nitrifying bacteria. (a) *Nitrosomonas europea* (× 32,500). (b) *Nitrosomonas* sp. (× 39,600). (c) *Nitrosocystis oceanus* (× 23,800). (d) *Nitrosolobus multiformis* (× 22,000). (e) *Nitrosospira briensis* (× 35,300). (f) *Nitrobacter winogradskyi* (× 63,200). (g) *Nitrococcus mobilis* (× 21,000). (h) *Nitrospina gracilis* (× 37,500). From S. W. Watson and M. Mandel. "Comparison of the Morphology and Deoxyribonucleic Acid Composition of 27 Strains of Nitrifying Bacteria," *J. Bacteriol.* **107,** 563 (1971).

(a) $NH_4^+ + O_2 + 2H^+ \longrightarrow NH_3OH^+ + H_2O$

$P_{450}$ reduced → $P_{450}$ oxidized

$$NH_3OH^+ \xrightarrow[2e^-]{3H^+} (NOH) \xrightarrow{H_2O} NO_2^- + 3H^+$$

$P_{450}$ oxidized → $P_{450}$ reduced

(b) $NO_2^- + H_2O \xrightarrow[2e^-]{} NO_3^- + 2H^+$

FIGURE 16.2

Oxidation of ammonia (a) and nitrite (b).

**TABLE 16.3**

**Bacteria That Oxidize $H_2S$ with Formation of Intracellular Sulfur Deposits**

1. Filamentous gliding organisms:
   *Beggiatoa, Thiothrix, Thioploca*
2. Very large unicellular gliding organisms:
   *Achromatium*
3. Large unicellular rod-shaped or spiral organisms, immotile or motile by flagella:
   Cells rod-shaped, immotile: *Thiobacterium*
   Cells rod-shaped, polar flagella: *Macromonas*
   Cells round or ovoid, peritrichous flagella: *Thiovulum*
   Cells spiral, polar flagella: *Thiospira*

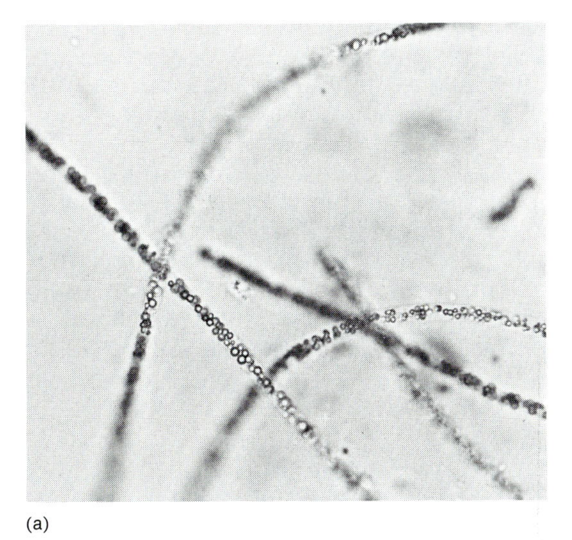

(a)

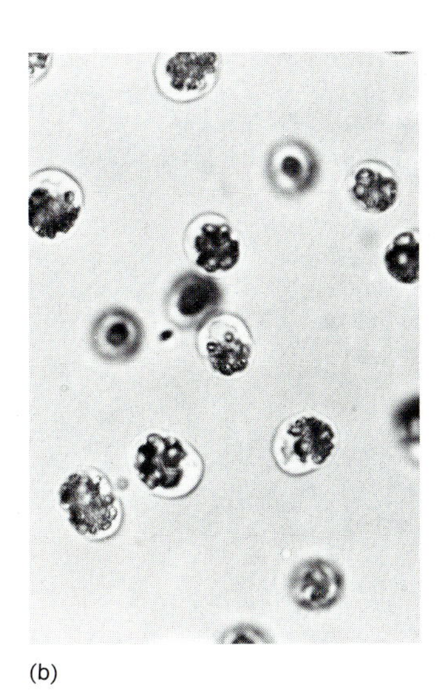

(b)

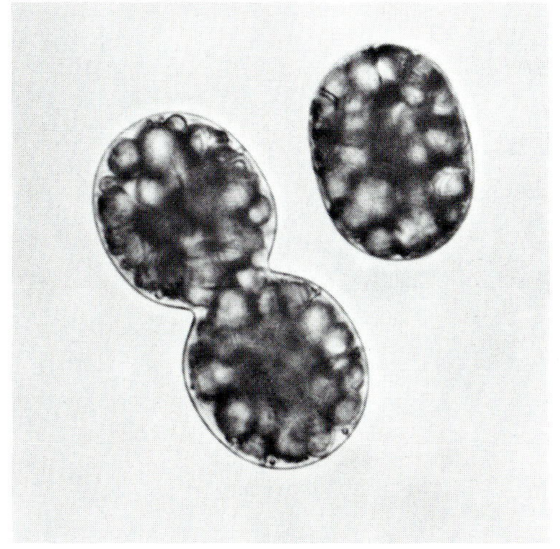

(c)

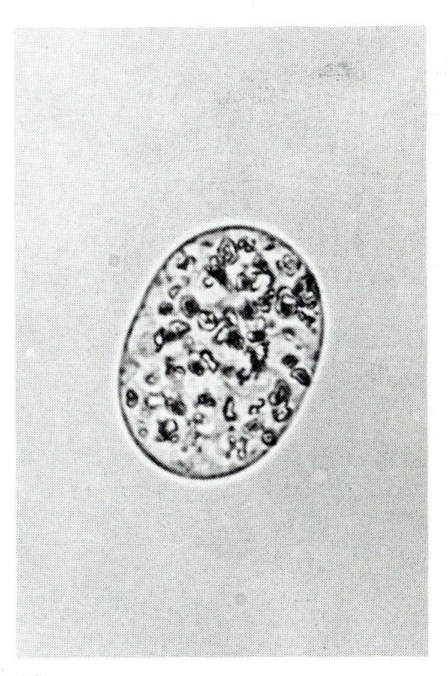

(d)

FIGURE 16.3

Some large, colorless, sulfur-oxidizing bacteria that accumulate sulfur internally. (a) Filaments of *Beggiatoa* (×900). (b) *Thiovulum* (×701). (c), (d) *Achromatium* (×700). The cells of this very large bacterium contain numerous calcium carbonate inclusions, shown in (c); the cell in (d) has been treated with dilute acetic acid, which has dissolved the inclusions of calcium carbonate, revealing the sulfur granules which are also present. Courtesy of Dr. J. W. M. La Rivière. (a) From J. W. M. La Rivière, "The Microbial Sulfur Cycle and Some of Its Implications for the Geochemistry of Sulfur Isotopes," *Geologischer Rundschau* **55,** 568 (1966). (c) and (d) from W. E. de Boer, J. W. M. La Rivière, and K. Schmidt, "Some Properties of *Achromatium oxaliferum*," *Antonie van Leeuwenhoek* **37,** 553 (1971).

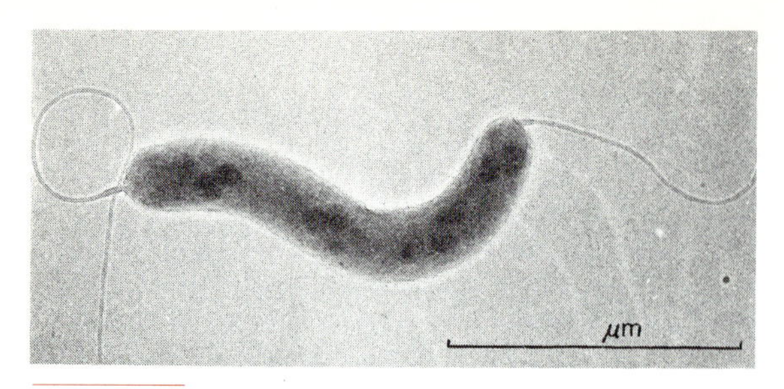


FIGURE 16.4

The colorless sulfur-oxidizing bacterium *Thiomicrospira*. Electron micrograph, showing the polar flagella. From J. G. Kuenen, and H. Veldkamp, "*Thiomicrospira pelophila*, gen. n., sp. n., a New Obligately Chemolithotrophic Colorless Sulfur Bacterium," *Antonie van Leeuwenhoek* **38,** 241 (1972).

**TABLE 16.4**

**Bacteria That Oxidize $H_2S$ with Formation of Extracellular Sulfur Deposits; Cell Size is Small**

| | Thiobacillus | Thiomicrospira |
|---|---|---|
| Cell form | Rods | Spirals |
| Flagella | Polar | Polar |
| DNA base composition (percent G + C) | 34–70 | 48 |
| pH range | V[a] | 5.0–8.5 |
| Autotrophy | V | Obligate |

[a] V: variable among strains

ically stable reduced forms of sulfur, notably thiosulfate and elemental sulfur. Most of them are small, polarly flagellated rods, placed in the genus *Thiobacillus*; they occur widely in both marine and terrestrial environments. The spiral organism, *Thiomicrospira* (Figure 16.4) occurs in marine mud. The range of *Thiobacillus* DNA base composition is extremely wide, indicating substantial genetic heterogeneity. None of the small-celled sulfur oxidizers contain the extensive membranous intrusions that are characteristic of many nitrifying bacteria.

The chemoautotrophic growth of these organisms is rapid, some having generation times as short as 2 hours when growing at the expense of thiosulfate. A common and striking feature of the group is their extreme acid tolerance; some species can grow at a pH as low as 1 to 2 and fail to grow at a pH above 6. These organisms are often found in special environments in which the pH is maintained at a low level by their metabolic activities, since the oxidation of reduced sulfur compounds to sulfate results in considerable acid formation.

A specialized, man-made environment in which thiobacilli are abundant is the acid drainage water discharge from mines that contain metal sulfide minerals, notably iron pyrite ($FeS_2$). The predominant species in this habitat are the strongly acidophilic species *Thiobacillus thiooxidans*, which rapidly oxidizes elemental sulfur, and *T. ferrooxidans*, which can derive energy from the oxidation both of reduced sulfur compounds and $Fe^{2+}$.

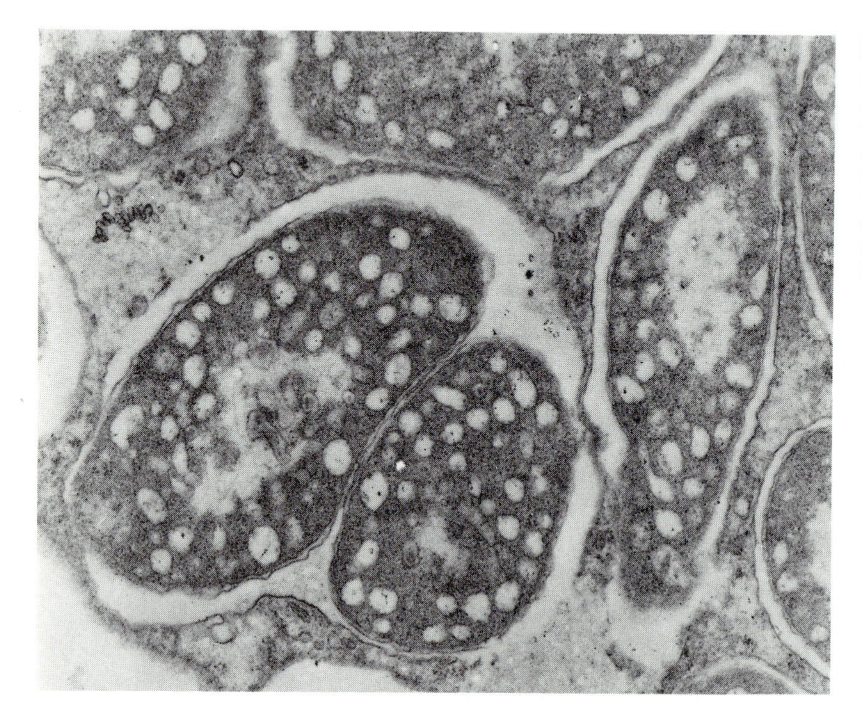

FIGURE 16.5

Electron micrograph of a thin section of a vestimentiferan tube worm from a hydrothermal vent in the east Pacific, showing endosymbiotic bacteria (×24,700).
From C. M. Cavanaugh, "Symbiotic chemoautotrophic bacteria in marine invertebrates from sulfide-rich habitats," *Nature*, **302,** 58–61 (1983).

An unusual habitat containing enormous numbers of small-celled sulfur oxidizers has been recently discovered—the immediate vicinity of geothermal vents in the abyssal depths of the ocean. At these depths no light penetrates, and most of the deep ocean floor is a lightly populated desert on which life depends on the steady, but sparse rain of organic detritus from the illuminated layers. However, in the areas where the sea floor is spreading, the conditions are quite different. Geothermal activity injects large volumes of sulfide-rich hot water into the ocean. The immediate vicinity of such vents contains an abundance of invertebrate life, which apparently depends on the primary productivity of free-living *Thiomicrospira* and unidentified thiobacilli that are symbiotic within the tissues of some of the invertebrates (Figure 16.5). Recent evidence suggests that such endosymbioses are not confined to deep hydrothermal vents, but are common among invertebrate inhabitants of shallow, sulfide-rich habitats as well.

Obligate chemoautotrophy is not the rule among sulfur oxidizers as it is among nitrifying bacteria. Several thiobacilli can grow with organic carbon and energy sources; however, the utilizable substrates appear to be confined to glucose and a few amino acids. One of the thiobacilli, *T. intermedius*, has an absolute requirement for a reduced source of sulfur, which can be met either by thiosulfate or by a sulfur-containing amino acid. Another of the thiobacilli, *T. perometabolis*, lacks enzymes of $CO_2$ assimilation; accordingly it is incapable of autotrophic growth and must be provided with an organic carbon source, the assimilation of which is driven by sulfur oxidation.

The biochemistry of sulfur oxidation has been extensively studied in several strains of thiobacilli. Two alternative pathways have been suggested. In the first (Figure 16.6) sulfide is oxidized to polysulfide, then to elemental sulfur by membrane-bound enzyme systems. The oxidation of elemental sulfur to sulfite requires the participation of reduced glutathione as a carrier. Sulfite oxidation may proceed by either of two routes: direct oxidation by sulfite oxidase, or activation with AMP, to form adenylphosphosulfate (APS) followed by phosphorolysis with phosphate or pyrophosphate to yield ADP or ATP respectively. Substrate-level phosphorylation in the latter pathway of sulfite oxidation thus increases the efficiency of energy conservation.

Alternatively, sulfide oxidation may occur by a reversal of the reactions known to be used by the sulfur reducers to reduce sulfate to sulfide (Figure 20.9).

Because few large-cell sulfur oxidizers have been isolated in pure culture, little is known of their physiology. A few strains of *Beggiatoa* have been isolated, and their capacity for autotrophic growth studied. Despite the difficulties attendant on the use of $H_2S$ as electron donor, and the measurement of growth of a filamentous organism, it is becoming clear that the genus is not physiologically uniform. Some strains are capable of autotrophic growth with $H_2S$, using the Calvin-Benson cycle to fix $CO_2$. Others appear to lack the ability to fix $CO_2$ and thus are obligate heterotrophs. Among these, some can probably gain energy from the oxidation of $H_2S$ and can couple this energy generation to the assimilation of the organic carbon source, a form of metabolism termed *mixotrophy*. Other heterotrophic strains of *Beggiatoa* oxidize $H_2S$ but appear to gain no energy from the oxidation; two possible advantages to sulfide oxidation have been proposed for these organisms: it may aid the establishment of symbiotic associations with rice plants; and the accumulation of elemental sulfur may provide an electron acceptor under anaerobic conditions.

A symbiotic association between *Beggiatoa* and rice has been described. Since rice is grown in flooded paddies, there is often substantial $H_2S$ generation in the anaerobic soil of the root zone. Rice seedlings are sensitive to the toxic effects of sulfide, and may depend on *Beggiatoa* to lower its concentration in the rhizosphere (the area immediately adjacent to the roots). The plants transport $O_2$ to the roots (which accounts for their ability to grow in flooded soil), some of which diffuses into the rhizosphere to create the microaerophilic conditions preferred by *Beggiatoa*. The plants also ex-

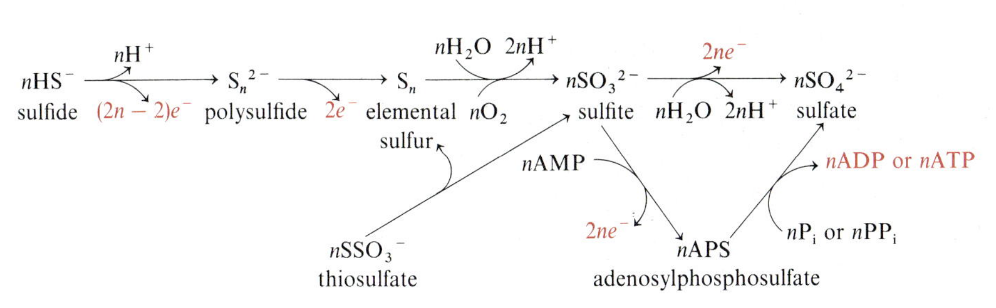


FIGURE 16.6
The oxidation of reduced compounds of sulfur.

crete catalase, which stimulates the growth of the catalase-negative *Beggiatoa.*

In habitats with high light intensities, *Beggiatoa* has been shown to glide into the anaerobic dark zone during the day, emerging only when the light intensity diminishes (presumably an adaptation to avoid high $O_2$ tensions due to high rates of cyanobacterial photosynthesis). Under anaerobic conditions, some strains can couple the oxidation of organic compounds or stored poly-$\beta$-hydroxybutyrate to the reduction of intracellular sulfur, producing $H_2S$. Thus the advantage to these strains of aerobic $H_2S$ oxidation may be to allow the accumulation and storage of an endogenous electron acceptor for later use in an anaerobic respiration.

## The Iron Bacteria

Certain freshwater ponds and springs have a high content of reduced iron salts. It has long been known that a distinctive bacterial flora is associated with such habitats. These *iron bacteria* form colonies that are heavily encrusted with ferric hydroxide. However, since most iron springs are neutral or alkaline, ferrous iron undergoes rapid spontaneous oxidation, so it has proven very difficult to ascertain the role, if any, that iron oxidation plays in the metabolism of these bacteria. The most conspicuous iron bacteria are filamentous, ensheathed bacteria of the *Sphaerotilus* group (Chapter 17) in many of which the sheaths are encrusted with iron oxide. They can be readily grown as chemoheterotrophs and so isolated in pure culture. Although it is possible to show that such pure cultures will accumulate iron oxide on their sheaths, there is no evidence that such deposition is a physiologically significant process or that these bacteria can develop as chemoautotrophs. Chemoautotrophic growth seems more probable in the case of another structurally distinctive iron bacterium, *Gallionella.* All attempts to obtain cultures of *Gallionella* in organic media have failed, but it has been grown in a mineral medium containing ferrous sulfide. The use of this virtually insoluble ferrous salt minimizes the rate of spontaneous oxidation under neutral conditions. In these cultures, *Gallionella* will form cottony colonies attached to the wall of the vessel.

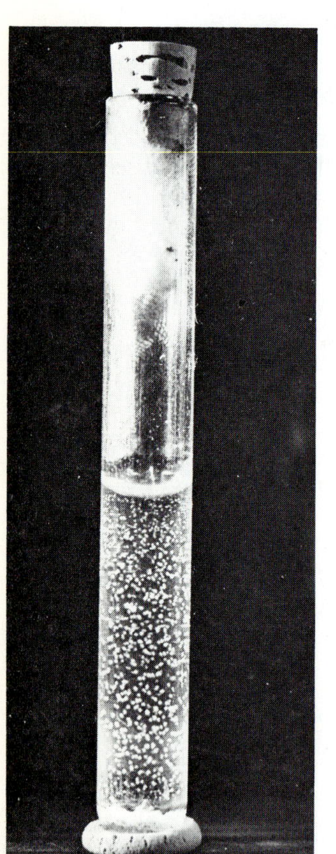

(a)

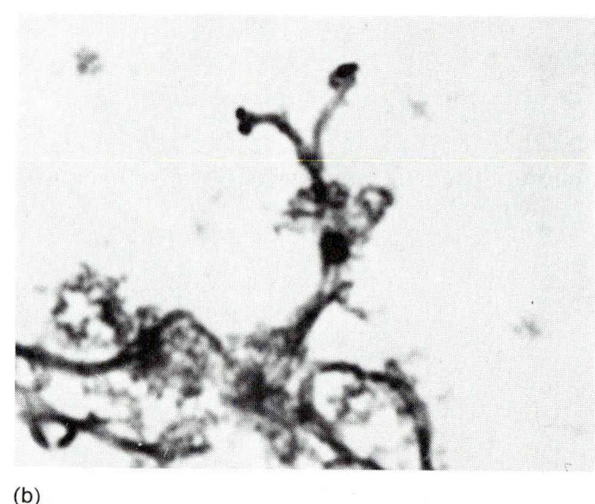

(b)

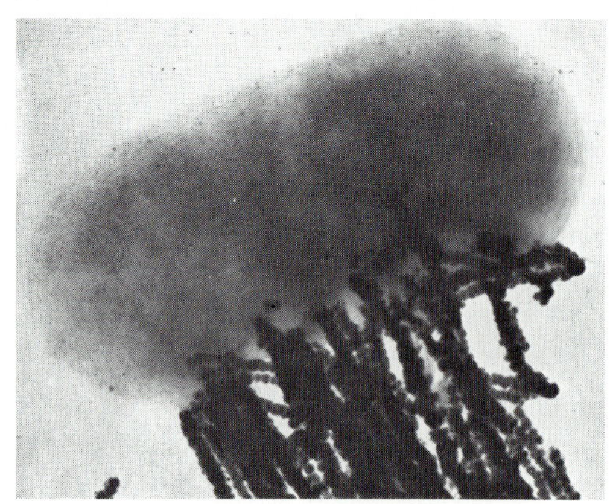

(c)

FIGURE 16.7

The iron bacterium, *Gallionella.* (a) Flocculent colonies (consisting largely of ferric hydroxide) growing attached to the glass in a liquid culture. (b) Light micrograph of the edge of a colony, showing cells attached to the tips of a branched stalk ($\times$ 2,430). (c) Electron micrograph of a single cell, attached to the tip of the stalk, which is impregnated with ferric hydroxide. (a) and (b) reproduced from S. Kucera and R. S. Wolfe, "A Selective Enrichment Method for *Gallionella Ferruginea,*" *J. Bacteriol.* **74,** 347 (1957). (c) from R. S. Wolfe, "Iron and Manganese Bacteria," in *Principles and Applications in Aquatic Microbiology,* H. Heukelekien and N. C. Dondero, eds., p. 82 (New York: John Wiley, 1964).

Much of the colony is inorganic: the small, bean-shaped bacterial cells are located at the branched tips of the excreted stalks, which are heavily impregnated with ferric hydroxide (Figure 16.7).

However, a rigorous demonstration of chemoautotrophic growth by *Thiobacillus ferrooxidans* at the expense of ferrous ion oxidation has been possible because this bacterium grows autotrophically at low pH values, where $Fe^{2+}$ is chemically stable. The cytoplasmic pH is maintained at a value of about 6.5, so this organism has an environmentally imposed pH gradient across its cytoplasmic membrane of over 4 pH units; a proton motive force of this magnitude is more than sufficient to allow ATP synthesis. Under such conditions, iron oxidation must provide electrons for pyridine nucleotide reduction and for reduction of the protons that enter in the process of ATP synthesis, reverse electron transport, and other cell activities that are driven by the membrane potential (such as flagellar motility, some permeases, etc.). There is, however, no necessity for the operation of a respiratory proton pump, normally an integral feature of respiratory electron transport systems.

## The Hydrogen Bacteria

Many species of aerobic bacteria possess the ability to grow chemoautotrophically with molecular hydrogen. In contrast to other groups of chemoautotrophs, the hydrogen bacteria are all nutritionally versatile organisms that can use a wide range of organic compounds as carbon and energy sources. Most of these bacteria were formerly placed in a special genus *Hydrogenomonas*. However, their facultative chemoautrotrophy does not appear to justify a generic separation from similar organisms that are obligate chemoheterotrophs, and they are now classified in a series of genera that contain phenotypically similar nonautotrophic bacteria (Table 16.5).

The observation that genes specifying hydrogen utilization (although not those encoding enzymes of carbon dioxide fixation) are located on transmissible plasmids in a number of species of hydrogen bacteria suggests that the ability to use molecular hydrogen may be a recent addition to the genome of these bacteria.

The enzymology of hydrogen oxidation has been extensively studied, principally by H. G. Schlegel and his collaborators. Most hydrogen bacteria have a single hydrogenase which catalyzes the reaction

$$H_2 \rightleftharpoons 2H^+ + 2e^-$$

**TABLE 16.5**

**Genera That Contain One or More Species Capable of Chemoautotrophic Growth with Hydrogen**

| Gram-Negative Bacteria | Gram-Positive Bacteria |
|---|---|
| *Alcaligenes* | *Bacillus* |
| *Aquaspirillum* | *Arthrobacter* |
| *Azospirillum* | *Mycobacterium* |
| *Derxia* | *Nocardia* |
| *Flavobacterium* | |
| *Microcyclus* | |
| *Paracoccus* | |
| *Pseudomonas* | |
| *Rhizobium* | |
| *Spirillum* | |

The electron acceptor in this reaction is an unidentified component of the electron transport chain, probably with reduction potential near zero millivolts. In addition to this membrane-bound hydrogenase coupled directly to an electron transport chain, some species of *Alcaligenes* and *Nocardia* have a second, soluble hydrogenase that catalyzes the reaction

$$NAD^+ + H_2 \rightleftharpoons NADH + H^+$$

The selective advantage of producing two hydrogenases is obscure because either alone would seem to be adequate to support growth at the expense of molecular hydrogen. A possible explanation might lie in the differential activities of these enzymes with changing concentrations of $H_2$. If the partial pressure of $H_2$ is very low, as it often is in nature, the soluble hydrogenase cannot function because its equilibrium shifts to the left. Thus at low tensions of $H_2$ growth depends on the particulate hydrogenase to provide electrons for respiratory ATP synthesis, and for pyridine nucleotide reduction via reverse electron transport. However, if the partial pressure of $H_2$ is high the $NAD^+$-coupled reaction would obviate the energy requirement imposed by NADH formation via reverse electron transport (see below). Thus the soluble hydrogenase appears to be a specific adaptation to growth in the presence of a high partial pressure of $H_2$.

## The Carboxydobacteria

A variety of bacteria are capable of oxidizing carbon monoxide to carbon dioxide. Although this is a strongly exergonic reaction, not all of these bacteria are able to grow at the expense of carbon monoxide; those that can are termed *carboxydobacteria*, and are, like the hydrogen bacteria, widely scattered among diverse bacterial taxa. They include some strains of *Alcaligenes*, *Azotobacter*, *Derxia*, and *Pseudomonas*. Indeed, most carboxydobacteria are also hydrogen bacteria (although the converse is not true), and the enzyme systems that catalyze $H_2$ oxidation and CO oxidation are often induced simultaneously under autotrophic growth conditions.

The biochemistry of carbon monoxide oxidation is still poorly understood. At least some carboxydobacteria, however, appear to utilize a soluble carbon monoxide oxidoreductase which catalyzes the reaction

$$CO + H_2O \rightleftharpoons CO_2 + 2H^+ + 2e^-$$

The electron acceptor in this reaction is an unidentified component of the electron transport chain with a reduction potential of about zero millivolts, perhaps ubiquinone or cytochrome *b*.

## The Metabolic Basis of Chemoautotrophy

There is good evidence that all aerobic chemoautotrophic eubacteria assimilate $CO_2$ through the reactions of the Calvin-Benson cycle (see Chapter 5). When grown chemoautotrophically, cells contain high levels of the two enzymes specific to this pathway, ribulose bisphosphate carboxylase and phosphoribulokinase. However, in facultatively autotrophic thiobacilli and in hydrogen bacteria, synthesis of these enzymes is often largely or partly repressed when cells grow with organic substrates. Many obligately autotrophic thiobacilli and nitrifying bacteria possess carboxysomes, the specialized procaryotic organelles that contain ribulose bisphosphate carboxylase.

In order to drive the reactions of primary carbon assimilation, the chemoautotrophs must obtain both ATP and reducing power (reduced pyridine nucleotide) through the oxidative dissimilation of the inorganic substrate.

## Energy Conservation and Pyridine Nucleotide Reduction

Chemoautotrophic oxidations have traditionally been regarded as respirations that differ from heterotrophic oxidations only with respect to the site at which electrons enter the respiratory chain. However, it is becoming clear that this mechanism is not universal. In some groups of chemoautotrophic bacteria, energy conservation is probably achieved by a mechanism substantially the same as in heterotrophic respiration: the operation of an electron transport chain results in the vectoral transport of protons across the cell membrane. The chemiosmotic potential that results can then be converted to chemical bond energy by the membrane-bound ATPase (Chapter 4). The chemoautotrophs in which this mechanism appears to operate are the sulfur oxidizers, the hydrogen bacteria, the carboxydobacteria, and the ammonia oxidizers. In all of these bacteria the electrons released by substrate oxidation enter the electron transport chain at a reduction potential substantially more positive than that of pyridine nucleotide, despite the fact that some of these electron donors (specifically $H_2$ and CO) have reduction potentials that would allow them to reduce $NAD^+$ directly (Table 16.6). As mentioned above even those hydrogen bacteria that possess an $NAD^+$-coupled hydrogenase contain also a particulate hydrogenase that feeds the electron transport chain at a potential near zero volts.

**TABLE 16.6**

**Reduction Potentials for Electron Donating Reactions of Chemoautotrophic Respiration**

| Half-reaction | $E_0'$ (V) |
|---|---|
| $CO + H_2O \longrightarrow CO_2 + 2H^+ + 2e^-$ | −0.54 |
| $HSO_3^- + H_2O \longrightarrow SO_4^{2-} + 3H^+ + 2e^-$ | −0.52 |
| $H_2 \longrightarrow 2H^+ + 2e^-$ | −0.41 |
| $NAD(P)H \longrightarrow NAD(P)^+ + H^+ + 2e^-$ | −0.32 |
| $HS^- \longrightarrow S^o + H^+ + 2e^-$ | −0.27 |
| $NH_2OH + H_2O \longrightarrow NO_2^- + 5H^+ + 4e^-$ | +0.07 |
| $NO_2^- + H_2O \longrightarrow NO_3^- + 2H^+ + 2e^-$ | +0.43 |
| $Fe^{2+} \longrightarrow Fe^{3+} + e^-$ | +0.77 |
| $H_2O \longrightarrow \frac{1}{2}O_2 + 2H + 2e^-$ | +0.82 |

Two species of chemoautotrophs appear to use a different mechanism: the acidophilic iron bacteria (*Thiobacillus ferrooxidans*) and the nitrite oxidizers. The mechanism of the former group is, in a broad sense, respiratory, because ATP is generated via a membrane-bound ATPase. The unique characteristic of ATP-generating metabolism in this group of bacteria is the origin of the proton motive force that drives the process: the intrinsic difference between the neutral cytoplasm and the acidic environment is the sole source of potential energy. A respiration-linked proton pump is not utilized; that is, electron transport in these cells does not pump protons out of the cell. Rather it serves only to reduce intracellular protons and, thereby, main-

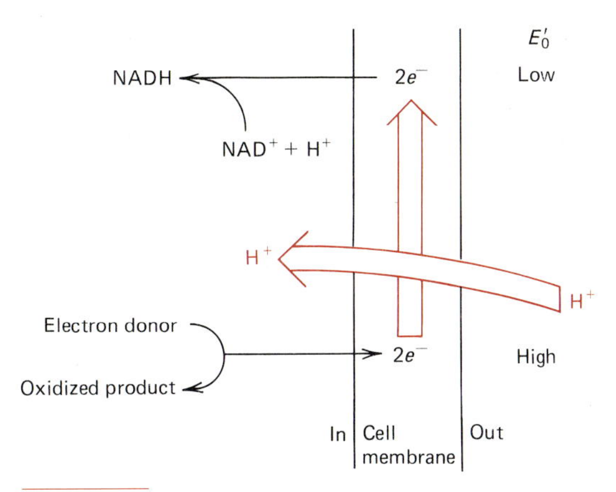


FIGURE 16.8

Reverse electron transport as the mechanism for pyridine nucleotide reduction in chemoautotrophic bacteria.

tain the cytoplasm near neutrality. Neutralization is catalyzed by the cytochrome oxidase reaction

$$4H^+ + 4e^- + O_2 \longrightarrow 2H_2O$$

In contrast, ATP generation by nitrite oxidizers may be achieved solely by substrate-level phosphorylation. Although the reaction by which the hypothetical phosphorylation occurs is unknown, attempts to detect proton pumping coupled to nitrite oxidation have failed. Alternatively, the electron transport chains and ATPase may be located in intracytoplasmic membrane systems with no connection to the extracellular milieu, analogous to the thylakoids of cyanobacteria. Proton pumping would consequently not be detected by measurements taken on cell suspensions.

A common biochemical problem unites all chemoautotrophs except those hydrogen bacteria which possess an $NAD^+$-linked hydrogenase; namely, the necessity to utilize reverse electron transport to reduce pyridine nucleotides. This mechanism (Figure 16.8) serves to allow a thermodynamically unfavorable reaction (electron flow from a donor of relatively high reduction potential to a low potential acceptor—see Table 16.6) to be coupled to a strongly exergonic reaction (proton entry). Since a substantial proportion of proton entry must be utilized to drive reverse electron transport rather than ATP synthesis, the growth yields of chemoautotrophic bacteria are typically quite low (Table 16.7).

**TABLE 16.7**

**Growth Yields of Some Chemoautotrophic Bacteria**

| Organism | Growth Yield[a] |
|---|---|
| *Pseudomonas facilis* | 12 g/mol. $H_2$ |
| *Thiobacillus neapolitanus* | 4 g/mol. $S_2O_3^{2-}$ |
| *Thiobacillus ferrooxidans* | 0.35 g/g-atom $Fe^{2+}$ |

[a] Expressed as grams (dry weight) of cell material synthesized per mole or gram-atom of substrate oxidized.

## The Phenomenon of Obligate Autotrophy

Primarily on the basis of his experience with nitrifying bacteria, Winogradsky proposed that an inability to grow at the expense of organic substrates is an intrinsic property of chemoautotrophs. Current knowledge shows that this is not correct: obligate autotrophy, although well-nigh universal among the nitrifiers, is a variable character among the sulfur oxidizers, and is completely absent from hydrogen bacteria and carboxydobacteria. Many attempts have been made to find a biochemical explanation for obligate autotrophy, and for its nonrandom distribution among the major physiological groups of chemoautotrophs. All such attempts have failed, and it seems probable that there is not a single, simple mechanism; indeed the term obligate autotroph is currently out of fashion, in large part due to the demonstration that under appropriate conditions organic compounds may contribute a significant portion of the cell carbon of "obligate autotrophs." For instance, a number of amino acids and acetate are assimilated at relatively high rates. However, little if any $CO_2$ is produced from either carbon atom of C-acetate; the metabolism of this compound is strictly assimilatory in obligate autotrophs. Furthermore, acetate provides only a small fraction (10 to 15 percent) of newly synthesized cell carbon. In comparable experiments with facultative autotrophs (e.g., *Thiobacillus intermedius*), acetate carbon makes a much larger fractional contribution to newly synthesized cell carbon. Biochemical studies have shown that the failure of many obligate autotrophs to utilize acetate more effectively is attributable to the absence of a functional TCA cycle: these organisms lack a key enzyme, α-ketoglutarate dehydrogenase, and have unusually low levels of both succinic and malic dehydrogenase. In the absence of an α-ketoglutarate dehydrogenase, the other enzymes associated with the cycle cannot mediate an oxidation of acetyl-CoA; they function as two separate pathways, which have purely biosynthetic roles (Figure 16.9). The dicarboxylic acid branch (fed by carboxylation of phosphoenolpyruvate) operates in the reverse of its customary direction, to provide precursors of the amino acids of the aspartate family and of

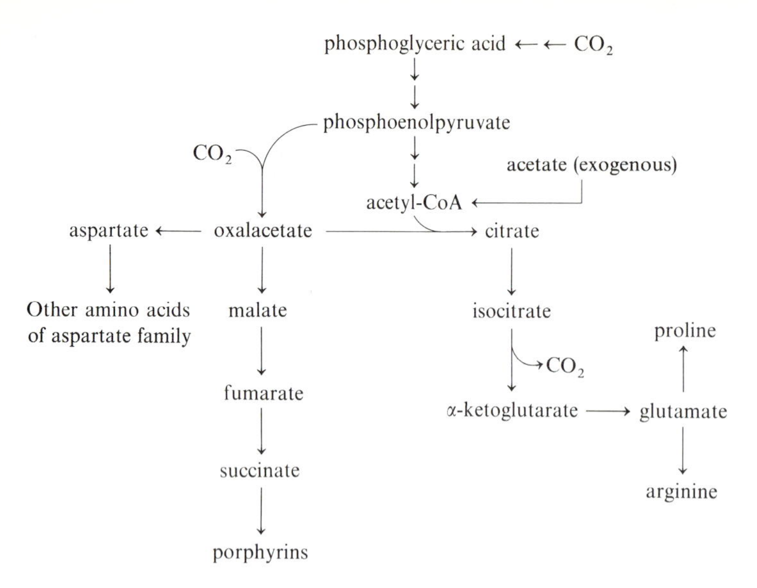


FIGURE 16.9
The biosynthetic roles of reactions normally associated with the operation of the tricarboxylic acid in organisms that cannot convert α-ketoglutarate to succinate. Note that carbon from exogenous acetate can enter the amino acids of the glutamate family via citrate and α-ketoglutarate, but cannot enter those of the aspartate family via succinate and oxalacetate as it does in organisms with a functional TCA cycle.

porphyrins; the citrate branch (fed by oxalacetate and acetyl-CoA) operates in its customary direction to produce α-ketoglutarate, the precursor of amino acids of the glutamate family. The relatively small contribution of exogenous acetate to newly synthesized cell carbon, accordingly, reflects the fact that it enters only those biosynthetic pathways for which acetyl-CoA is a specific precursor: in addition to the pathway shown in Figure 16.9, these include the pathways of lipid synthesis and of leucine synthesis.

For an organism with a strictly respiratory mode of metabolism, the absence of α-ketoglutarate dehydrogenase prevents dissimilation of most organic substrates, with the possible exception of glucose and a few other sugars, which can be oxidized to $CO_2$ through the pentose phosphate pathway; this mode of sugar dissimilation occurs in many cyanobacteria, a group which similarly lacks a functional TCA cycle.

## Carbon Reserve Materials in Chemoautotrophs

Most chemoautotrophs can accumulate intracellular stores of one or both of the organic reserve polymers (glycogen and poly-β-hydroxybutyrate) characteristic of procaryotes. These substances presumably provide (as in other bacteria) endogenous carbon and energy reserves. Hence, there are good reasons to believe that chemoautotrophs possess the biochemical machinery for the metabolism of these reserve materials. In one of the thiobacilli, *T. neapolitanus*, utilization of accumulated polyglucose is fermentative, via the heterolactic pathway (see Figure 23.5). Fermentative metabolism of endogenous reserves allows survival and possibly cell division under conditions of sulfide starvation, or of $O_2$ limitation. The presence of phosphoketolase (one of the key enzymes of the heterolactic pathway) in other thiobacilli suggests that this pathway of glycogen dissimilation may be widespread among members of the group.

## Growth Inhibition by Organic Compounds

Winogradsky concluded from his physiological studies on the nitrifying bacteria that organic compounds are not simply unutilizable, but are actually toxic to these organisms. However, later work has not confirmed this contention; organic substances are not in general growth inhibitory for chemoautotrophs when they are added to cultures at concentrations tolerated by chemoheterotrophs. However, some examples of specific inhibition, particularly by amino acids, have been described. Growth of some thiobacilli is arrested by low concentrations (1 to 10 mM) of single amino acids, the specific patterns of inhibition varying somewhat from strain to strain. In every such case that has been carefully analyzed, the growth inhibition is attributable to the end-product inhibition of an enzyme that mediates an early step in a branched biosynthetic pathway; growth can be restored by adding other end products of the affected pathway. Thus, phenylalanine inhibition is relieved by tyrosine and tryptophan. Entirely comparable cases of specific inhibition have been observed in chemoheterotrophic bacteria. The phenomenon is, accordingly, not specific to chemoautotrophs.

## THE METHOPHILES*

The distinguishing property of methophilic organisms is the ability to derive carbon and energy from the metabolism of one-carbon compounds or compounds containing two or more carbon atoms that are not directly linked to one another (e.g., dimethyl ether, dimethylamine, etc.). By far the most abundant compound of this class in nature is the gas methane ($CH_4$), which occurs in coal and oil deposits and is continuously produced on a large scale in anaerobic environments by the methanogenic bacteria (Chapter 14). Methanol is also abundant in nature; it is formed during the breakdown of pectins and other naturally occurring compounds that contain methyl esters or ethers. Methylamines and their oxides occur in plant and animal tissue.

Dissimilation of these *$C_1$ compounds*, as they are termed, is almost always respiratory, being mediated by strict aerobes, although denitrification at the expense of methanol is characteristic of some methophiles, and there is ecological evidence that sulfate reduction is sometimes coupled to methane oxidation. Other exceptions to this rule are the dissimilation of methanol by methanogens and the photoassimilation of methanol and formate by purple bacteria.

A variety of methophilic procaryotes are known; the only eucaryotes capable of methophilic growth are some yeasts. The procaryotic methophiles fall into two primary physiological subgroups: the *methanotrophs* and the *methylotrophs*. Methanotrophs are all able to grow at the expense of methane; many are also able to utilize methanol, formaldehyde, or dimethyl ether, but only a few are capable of utilizing a wider range of organic compounds. The methylotrophs are generally more versatile nutritionally, being frequently capable of growing with a variety of organic compounds; however, they cannot use methane as carbon and energy source.

### The Metabolism of Methyl Compounds

The oxidation of methane to formaldehyde is shown in Figure 16.10. The initial attack is via a monooxygenase, for which the electron donor is NADH. The electron-accepting prosthetic group of the methanol dehydrogenase reaction has recently been

* The term "methylotrophs" has traditionally been used to denote this group. However, the term is more usefully applied to a specific subgroup, to distinguish it from the other subgroup, the "methanotrophs." We will accordingly use the new word *methophile* as the more generic term.

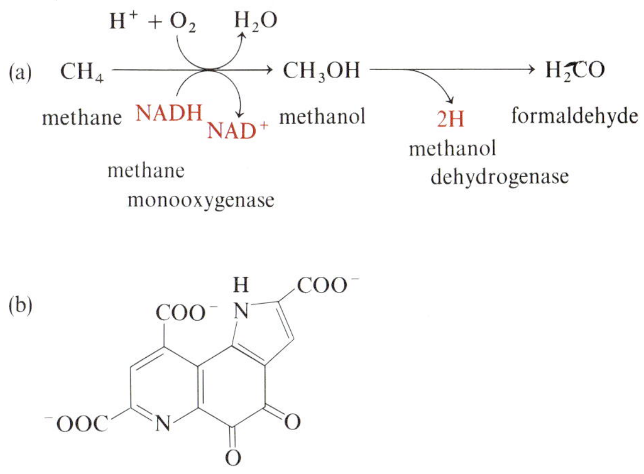

FIGURE 16.10

Methane and methanol oxidation (a), and the structure of pyrrolo-quinoline quinone (b).

identified as an unusual quinone: pyrrolo-quinoline quinone (PQQ; Figure 16.10) with a reduction potential of about +0.09 V. PQQ in turn reduces a component of the electron transport chain, probably cytochrome *c*.

There are two pathways by which formaldehyde is oxidized (Figure 16.11): one via formate, and the other via a hexulose. The latter, a cyclic pathway termed the *dissimilatory ribulose monophosphate pathway*, is a variant of the pathway through which many methylotrophs assimilate formaldehyde. Some methylotrophs utilize the formate and cyclic pathways simultaneously.

Two enzymatic patterns of direct oxidation of formaldehyde to formate are found among methophiles. In some cases, its oxidation is catalyzed by an $NAD^+$-linked formaldehyde dehydrogenase. However, in other cases, this oxidation is catalyzed by a nonspecific PQQ-linked dehydrogenase active with both methanol and formaldehyde.

When methanotrophs that use the PQQ-linked dehydrogenase oxidize methane, no net reduction of pyridine nucleotides occurs: methane monooxygenase oxidizes the NADH generated by formate dehydrogenase, and the other oxidations in the pathway are coupled to the reduction of PQQ. Reduced pyridine nucleotides needed for biosynthesis must hence be generated by reverse transport, as in chemoautotrophs.

Di-, tri-, and tetramethylamine are dissimilated by a series of successive oxidative demethylations (catalyzed in most cases by monooxygenases) yielding formaldehyde and, ultimately, methylamine. Methylamine may also be oxidatively

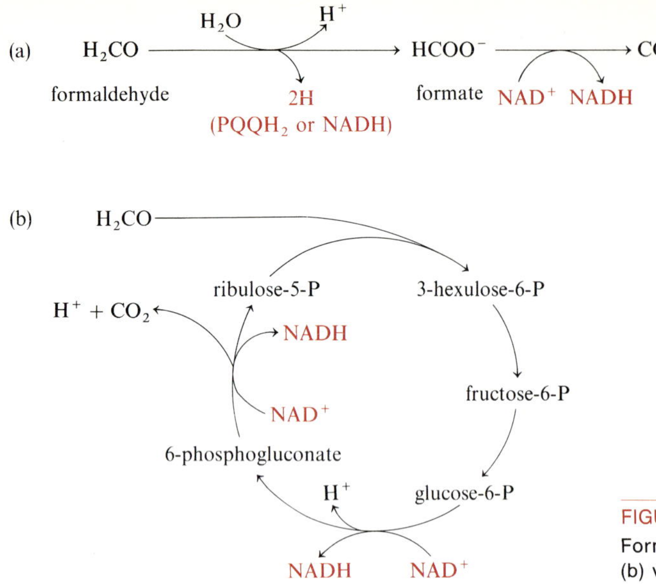

(a) $H_2CO$ → $HCOO^-$ → $CO_2$

formaldehyde; $H_2O$; $H^+$; 2H ($PQQH_2$ or NADH); formate; $NAD^+$ NADH

(b) $H_2CO$

ribulose-5-P; 3-hexulose-6-P; $H^+ + CO_2$; NADH; fructose-6-P; $NAD^+$; 6-phosphogluconate; $H^+$; glucose-6-P; NADH; $NAD^+$

FIGURE 16.11

Formaldehyde oxidation (a) via formate; (b) via the dissimilatory RMP pathway.

deaminated to ammonium and formaldehyde; alternatively it may be oxidized by the cyclic mechanism:

$$(CH_3)NH_3^+ + \text{glutamate} \longrightarrow \text{N-methylglutamate} + NH_4^+$$

$$\text{N-methylglutamate} \longrightarrow \text{glutamate} + H_2CO + 2H^+ + 2e^-$$

The electron acceptor for N-methylglutamate dehydrogenase is $NAD^+$ in some methylotrophs, and an unidentified component of the electron transport chain in others.

## Carbon Assimilation by Methophiles

All the methophiles are capable of growth with $C_1$ compounds serving as sole source of carbon. Pathways by which they assimilate $C_1$ compounds into cellular components differ dramatically. A few facultative methylotrophs are autotrophs that assimilate $CO_2$ via the Calvin-Benson cycle, but most methylotrophs and all methanotrophs assimilate $C_1$ substrates via one of two unique pathways for which formaldehyde is the immediate precursor. These are the *ribulose monophosphate pathway* and the *serine pathway*.

FIGURE 16.12

The ribulose monophosphate pathway.

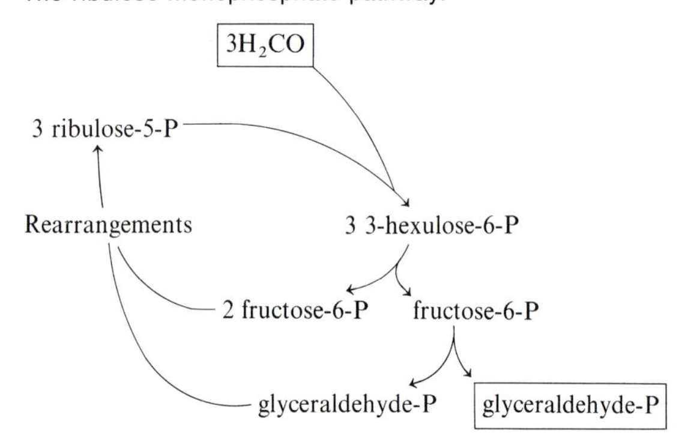


The ribulose monophosphate pathway is cyclic and characterized by two key enzymes: *hexulose phosphate synthase*, which catalyzes the condensation of ribulose-5-phosphate and formaldehyde to form an unusual 3-hexulose; and *hexulose phosphate isomerase*, which isomerizes the 3-hexulose to a 2-hexulose (fructose). Both enzymes are unknown outside the methophiles. Following these primary assimilatory reactions are a series of rearrangements, the details of which vary among species, which regenerate the $C_1$ acceptor, ribulose-5-phosphate. This pathway is shown schematically in Figure 16.12. In certain respects it is quite similar to the Calvin-Benson cycle (or *ribulose bisphosphate pathway*), but differs from it by lacking the reduction steps because the $C_1$ unit incorporated in the monophosphate pathway is already at the oxidation state of cell material.

The serine pathway (Figure 16.13) is also cyclic, but is composed of a completely different set of reactions from those in the ribulose bisphosphate and ribulose monophosphate cycles. The serine pathway accomplishes the assimilation of both formaldehyde and $CO_2$ in the approximate ratio of 2 to 1. $C_1$ units derived from formaldehyde are transferred, following their addition to tetrahydrofolic acid (THF) and reduction, to the amino acid glycine, with the formation of serine:

$$\text{methylene-THF} + \text{glycine} \longrightarrow \text{THF} + \text{serine}$$

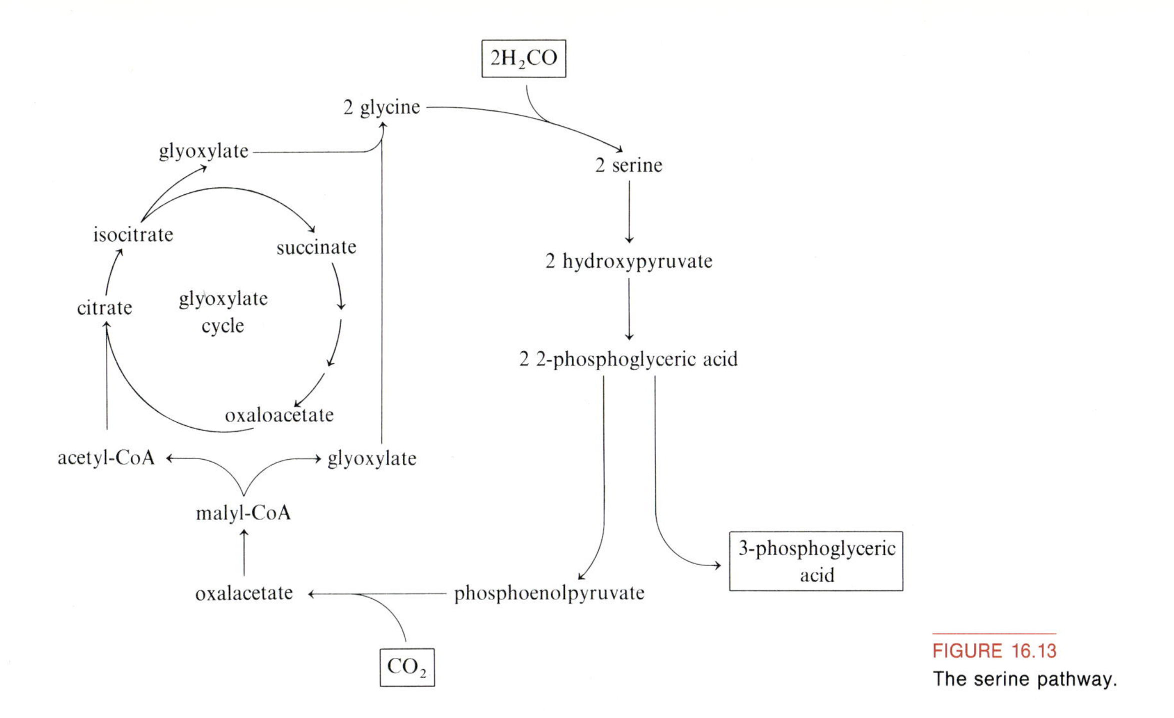


FIGURE 16.13
The serine pathway.

Serine is then converted to phosphoglyceric acid, some of which is assimilated, and some of which serves to regenerate glycine, the $C_1$ acceptor. The set of reactions by which glycine is regenerated are complex, requiring the assimilation of $CO_2$ in the course of synthesis of the intermediate, malate. Cleavage of malyl-CoA yields glyoxylate and acetyl-CoA. In species that possess the enzymes of the glyoxylate cycle, acetyl-CoA is converted to glyoxylate via isocitrate (as shown in Figure 16.13). However, some methophiles lack a key enzyme, isocitrate lyase, of the glyoxylate cycle; how acetyl-CoA is converted to glyoxylate in these strains is unknown.

## The Methanotrophs

Methanotrophic organisms are all Gram-negative strict aerobes, although they characteristically grow more rapidly under reduced oxygen tension. All are unicellular rods, vibrios, cocci, or pear-shaped cells (Figure 16.14); they are often motile by means of polar flagella. They all form some type of resting structure, and they are characterized by complex systems of internal membranes when grown with methane. These membrane systems are of two types: type I consists of stacks of flattened vesicles, and type II of peripheral lamellae (Figure 16.15). The type of internal membrane system correlates with

**TABLE 16.8**

**Characteristics of Type I and Type II Methanotrophs**

| Characteristic | Type I | Type II |
|---|---|---|
| Cytomembranes | Stacked, flattened vesicles | Peripheral lamellae |
| Carbon assimilation | RMP pathway | Serine pathway |
| Principal fatty acid[a] | 16:0 | 18:1 |
| Nitrogen fixation | + or − | + |
| TCA cycle | Incomplete (α-ketoglutarate dehydrogenase lacking) | Complete |
| Resting cells | Cysts | Exospores or cysts |

[a] Number of carbon atoms: numbers of double bonds

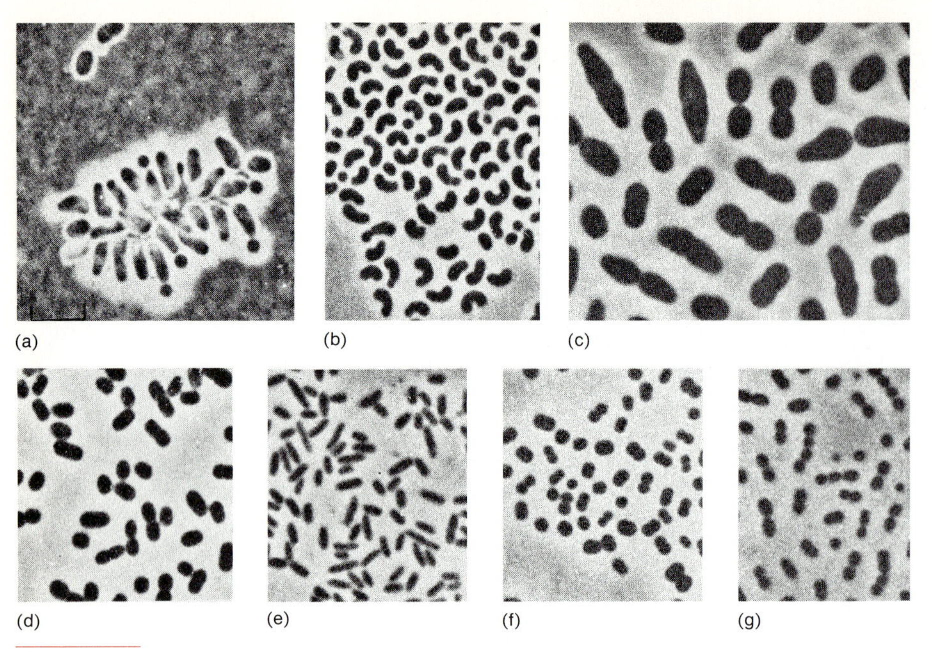

FIGURE 16.14

Phase-contrast photomicrographs (all × 1,600) of some methanotrophs
(a) *Methylosinus*; (b) *Methylocystis*; (c) *Methylobacter*; (d), (e) two species of *Methylomonas*; (f), (g), two species of *Methylococcus*. From R. Whittenbury, K. C. Phillips, and J. F. Wilkinson, "Enrichment, Isolation and Some Properties of Methane-Utilizing Bacteria," *J. Gen. Microbiol.* **61,** 205 (1970).

several other characteristics (Table 16.8), allowing the methanotrophs to be divided into two distinct subgroups.

Most methanotrophs are obligate methophiles; that is, they are incapable of growth with compounds that contain carbon-carbon bonds. Most are capable of growth with several $C_1$ compounds, including methanol and formaldehyde; in their natural habitats, however, methane is nearly always the most abundant growth substrate. One type II methanotroph (*Methylobacterium*) is facultative, and grows more rapidly with multicarbon substrates than with methane.

Despite the inability of most methanotrophs to utilize organic compounds with carbon-carbon bonds as growth substrates, they are capable of a wide variety of *co-oxidations* (co-oxidation is the gratuitous oxidation of a nongrowth substrate concomitantly with the oxidation of the growth substrate). In the case of the methanotrophs, co-oxidation is a consequence of the remarkable lack of substrate specificity of methane monooxygenase. In addition to its capacity to hydroxylate methane it can hydroxylate many alkanes and aromatic compounds, and can form epoxides from alkenes. The products of these reactions are not further metabolized by methanotrophs and thus they accumulate in the medium. There is considerable interest in the potential for commercial exploitation of some of these transformations.

## Resting Stages of Methanotrophs

Two types of desiccation-resistant resting cells are formed by methanotrophs: *exospores* and *cysts.* Exospores (Figure 16.16) are made by the type II genera *Methylosinus* and *Methylobacterium.* They are produced by budding from one pole of the cell and, when mature, have a complex wall and a fibrous capsule. They are refractile, and exhibit no metabolic activity. After budding the mother cell is incapable of growth and division, even when transferred to fresh medium.

Cysts are produced by the remaining type II genus of methanotroph (*Methylocystis*) and by all type I genera. During cyst formation, the entire vegetative cell enlarges, becomes spherical, and elaborates additional wall layers. Extensive accumulation of poly-$\beta$-hydroxybutyrate may occur. There is some variation in the details of the structure of the cysts of different genera; those formed

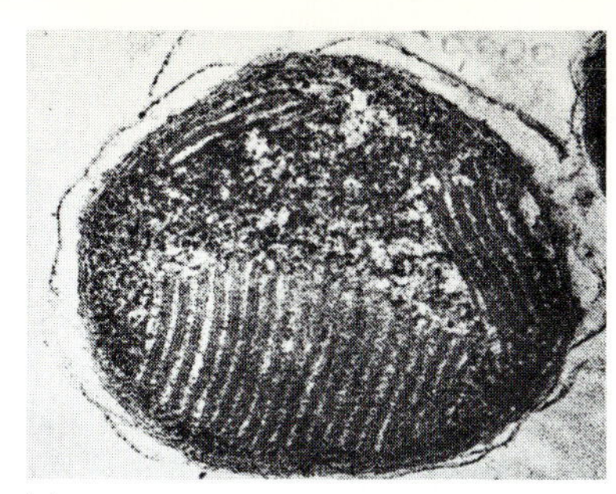

(a)

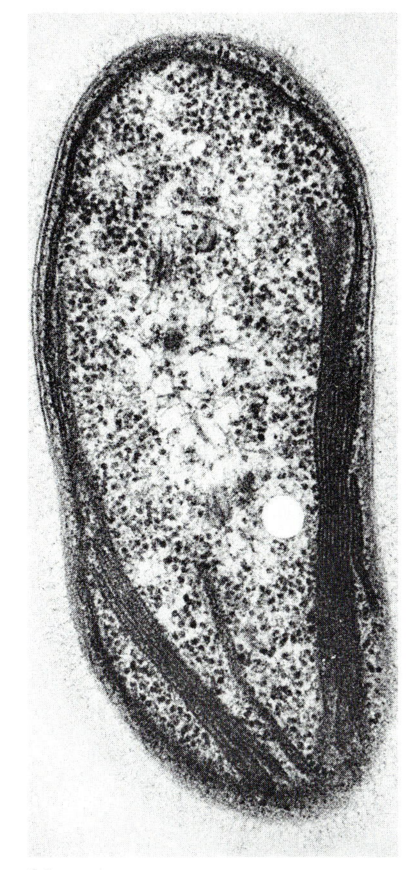

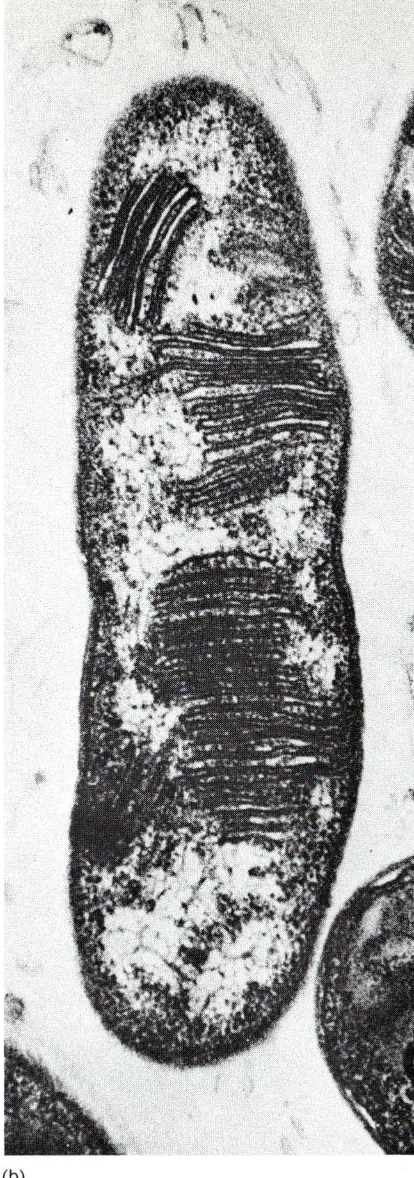

(c) (b)

FIGURE 16.15

Electron micrographs of thin sections of three methanotrophs, showing the two types of membrane systems characteristic of these organisms. (a) *Methylococcus* (type I membrane system) ($\times$45,900). (b) *Methylomonas* (type I membrane system) ($\times$22,900). (c) *Methylosinus* (type II membrane system) ($\times$45,900). From S. L. Davies, and R. Whittenbury, "Fine Structure of Methane and Other Hydrocarbon-Utilizing Bacteria," *J. Gen. Microbiol.* **61,** 227 (1970).

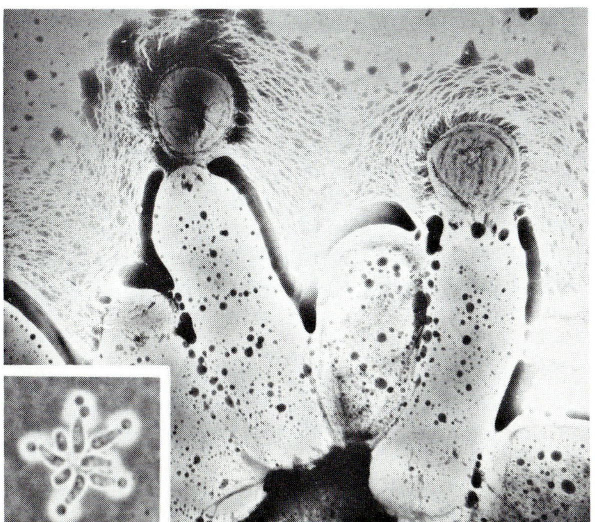

FIGURE 16.16

The budding of exospores from one cell pole in *Methylosinus*. Electron micrograph of a negatively stained preparation of whole cells, showing the wrinkled appearance of the surface of the exospore, and the fine fibers which extend from it ($\times$12,000). Insert: a similar group of cells forming exospores, negatively stained with India ink and observed by phase-contrast microscopy ($\times$1,400). From R. Whittenbury, S. L. Davies, and S. L. Davey, "Exospores and Cysts Formed by Methane-Utilizing Bacteria," *J. Gen. Microbiol.* **61,** 219 (1970).

**TABLE 16.9**

**Properties of Some Methanotrophs**

| Genus | Cell Structure | Resting Structure | Membrane Type | Obligate Methanotrophy | Percent G + C |
|---|---|---|---|---|---|
| *Methylococcus* | Nonmotile coccus | Cyst | I | + | 62—64 |
| *Methylomonas* | Polarly flagellated rod | Cyst | I | + | 50—54 |
| *Methylobacter* | Polarly flagellated rod | Cyst | I | + | 50—54 |
| *Methylocystis* | Nonmotile rod or vibrio | Cyst | II | + | 62—63 |
| *Methylosinus* | Polarly flagellated rod, often pear-shaped | Exospore | II | + | 62—63 |
| *Methylobacterium* | Polarly flagellated rod | Exospore | II | − | 58—66 |

by the type I genus *Methylobacter* are indistinguishable from the cysts made by *Azotobacter* (see Chapter 17), while those of most other methanotrophs have a somewhat less complex wall. The properties of some methanotrophic genera are summarized in Table 16.9.

### The Methylotrophs

Methylotrophic organisms are able to grow at the expense of one or more of the compounds typically used by methophiles, but cannot use methane. Methylotrophs are a diverse group, including both Gram-negative and Gram-positive genera. None makes cysts or exospores, and none has the complex intracellular membrane systems that characterize methanotrophs growing on methane.

A single obligate methylotroph (*Methylophilus*) is known. It is a Gram-negative, polarly flagellated rod capable of rapid growth with methanol. Some strains can also utilize formaldehyde or methylamines. Carbon is assimilated via the ribulose monophosphate pathway.

Facultative methylotrophy is a relatively widely distributed trait among heterotrophic bacteria (Table 16.10). It may also be common among chemoautotrophs; several thiobacilli and nitrifying bacteria can drive $CO_2$ assimilation via the Calvin-Benson cycle by formate oxidation.

**TABLE 16.10**

**Some Bacterial Genera Containing One or More Facultatively Methylotrophic Species (Capable of Growth with Methanol and/or Methylamines)**

| Gram-Negative Bacteria | Gram-Positive Bacteria |
|---|---|
| *Pseudomonas* | *Bacillus* |
| *Klebsiella* | *Arthrobacter* |
| *Alcaligenes* | *Mycobacterium* |
| *Acinetobacter* | *Streptomyces* |
| *Hyphomicrobium* | |
| *Rhodopseudomonas* | |
| *Microcyclus* | |
| *Paracoccus* | |

## ORIGINS OF CHEMOAUTOTROPHS AND METHOPHILES

In view of the simplicity of their nutritional requirements, chemoautotrophs and methophiles were once regarded as "primitive" organisms, possibly representative of the earliest forms of life on earth. This notion of their place in evolution is now untenable, for two reasons. First, their biochemical machinery (and even their cellular fine structure) is at least as complex as that of most chemoheterotrophic bacteria. Second, there is now good evidence to support the view that the earliest living organisms arose on an anaerobic earth, where the oceans contained an abundance of preformed organic matter. The shift to an oxygen-rich biosphere occurred much later (probably about 2 billion years ago); this major geochemical change can be plausibly explained only as a consequence of the evolution of oxygenic photosynthesis. In this evolutionary scenario the appearance of aerobic chemoautotrophs and methophiles on earth was dependent on the development of oxygenic photosynthesis. It is therefore conceivable that the chemoautotrophs and methophiles arose from procaryotic ancestors that performed either oxygenic or anoxygenic photosynthesis, by loss of the photosynthetic apparatus and adaptation to a new function of the photosynthetic electron transport chain. Unusual properties shared by some members of these two major procaryotic assemblages, one photosynthetic and other nonphotosynthetic, include several elaborate and distinctive types of internal membrane systems; absence of a functional tricarboxylic acid cycle; presence of the Calvin-Benson cycle or its analogue, the pentose phosphate cycle; and location of a key enzyme of the Calvin-Benson cycle (ribulose bisphosphate carboxylase) in carboxysomes.

## FURTHER READING

### Books

ANTHONY, C., *The Biochemistry of Methylotrophs.* New York: Academic Press, 1982.

CRAWFORD, R. L., and R. S. HANSON, eds., *Microbial Growth on $C_1$ Compounds.* Washington, D.C.: American Society for Microbiology, 1984.

STROHL, W. R., and O. H. TUOVINEN, eds., *Microbial Chemoautotrophy.* Columbus: Ohio State University Press, 1984.

### Reviews

BOWIEN, B. and H. G. SCHLEGEL, "Physiology and Biochemistry of Aerobic Hydrogen-Oxidizing Bacteria," *Ann. Rev. Microbiol.* **35,** 405 (1981).

DALTON, H., "Oxidation of Hydrocarbons by Methane Monooxygenases from a Variety of Microbes," *Adv. Appl. Microbiol.* **26,** 71 (1980).

HANSON, R. S., "Ecology and Diversity of Methylotrophic Organisms," *Adv. Appl. Microbiol.* **26,** 3 (1980).

HEGEMAN, G., "Oxidation of Carbon Monoxide by Bacteria," *Trends in Biochem. Sci.* **5,** 256 (1980).

KELLY, D. P., "Biochemistry of the Chemolithotrophic Oxidation of Inorganic Sulphur," *Phil. Trans. R. Soc. Lond. B* **298,** 499 (1982).

# Chapter 17
# Gram-Negative Aerobic Eubacteria

The groups of eubacteria that are photosynthetic were discussed in Chapter 15 and those that synthesize ATP by the oxidation of inorganic compounds or reduced $C_1$ compounds were discussed in Chapter 16. Chapters 17 through 25 describe the major groups of eubacteria that are chemoheterotrophs that use organic compounds containing more than one carbon atom as energy and carbon sources. These organic substrates may be dissimilated either by respiration or by fermentation. This chapter presents a survey of groups that possess an aerobic respiratory metabolism and that if motile, bear flagella. Most gliding bacteria are also organisms with an aerobic respiratory metabolism; they are described separately in Chapter 18.

The characteristics that distinguish the major groups to be reviewed in this chapter are shown in Table 17.1. Primary taxonomic differentiation is based on structural properties, particularly cell shape and the mode of flagellar insertion (polar or peritrichous). However, certain groups (and many constituent genera within groups) are distinguished by additional properties of a physiological or ecological nature.

The great majority of the bacteria belonging to the groups listed in Table 17.1 are dependent on $O_2$ as a terminal electron acceptor, and hence are *strict aerobes*, for which molecular oxygen is always an essential nutrient. However, some of these bacteria are an exception to this rule; because they can use nitrate in place of $O_2$ as a terminal electron acceptor, they can grow anaerobically. Anaerobic growth through nitrate respiration can be readily distinguished from fermentative growth by its strict dependence on the provision of nitrate in quantities sufficient to meet

**TABLE 17.1**

**Major Subgroups among Aerobic Gram-Negative Chemoheterotrophs**

| Cell Shape | Flagellar Insertion | Percent G + C | Other Distinctive Properties | Group | Genera Included |
|---|---|---|---|---|---|
| Rods | Polar | 58–70 | None | Aerobic pseudomonads | *Pseudomonas*<br>*Xanthomonas*<br>(*Zoogloea*) |
| Rods | Peritrichous or subpolar | 58–70 | Some form nodules or galls on plants | *Rhizobium* group | *Alcaligenes*<br>*Rhizobium*<br>*Bradyrhizobium*<br>*Agrobacterium* |
| Rods | Polar or subpolar | 59–65 | Prosthecate; some have a special division cycle | Prosthecate bacteria | *Caulobacter*<br>*Asticcacaulis*<br>*Hyphomicrobium*<br>*Hyphomonas*<br>*Stella*<br>*Prosthecomicrobium*<br>*Ancalomicrobium*<br>*Prosthecobacter* |
| Rods | Peritrichous, rarely polar | 57–70 | Free-living, aerobic nitrogen fixers | *Azotobacter* group | *Azotobacter*<br>*Azomonas*<br>*Beijerinckia*<br>*Derxia* |
| Rods | Polar or peritrichous | 55–64 | Oxidize organic substrates incompletely | Acetic acid bacteria | *Gluconobacter*<br>*Acetobacter* |
| Rods | Polar or subpolar | 69–70 | Form sheaths | Sheathed bacteria | *Sphaerotilus*<br>*Leptothrix*<br>*Haliscomenobacter* |
| Helical | Polar | 30–65 | None | *Spirillum* group | *Spirillum*<br>*Aquaspirillum*<br>*Oceanospirillum*<br>*Azospirillum*<br>*Campylobacter*<br>*Bdellovibrio* |
| Cocci or short rods | Nonflagellate | 40–52 | None | *Moraxella* group | *Neisseria*<br>*Branhamella*<br>*Moraxella*<br>*Acinetobacter* |
| Rods | Lateral or polar | 39–43 | Branched fatty acids in phospholipids | *Legionella* group | *Legionella* |
| Ovoid or cocci | Polar or subpolar | 50–58 | No murein in wall | *Planctomyces* group | *Planctomyces*<br>*Pasteuria* |

respiratory needs. Nitrate respiration may result in vigorous gas formation, because $N_2$ or $N_2O$ are often the major end product of nitrate reduction.

The respiratory electron transport chains of Gram-negative chemoheterotrophs differ widely in composition, particularly with respect to cytochrome components. The oxidase test, frequently employed in the identification of these organisms, reveals differences with respect to the nature of the cytochromes in the transport chain. Oxidase-positive aerobes immediately form colored products when their cells are mixed with a solution of *p*-phenylenediamine or a related oxidizable amine; the reaction is associated with the presence of a cytochrome of the *c* type in the transport chain. Oxidase-negative aerobes, which do not give this reaction, have transport chains without a cytochrome *c* component.

The respiratory dissimilation of most organic substrates, whether linked to the reduction of $O_2$ or of nitrate, leads to the formation of $CO_2$ as the principal or sole oxidized product, and involves the operation of the tricarboxylic acid cycle as the mechanism of terminal oxidation. However, terminal oxidation cannot occur until the primary substrate has been converted to acetyl-CoA or to other metabolites that are intermediates of the tricarboxylic acid cycle. The ability to use a wide and varied range of organic compounds as sole sources of carbon and energy, which is characteristic of many of the aerobic bacteria discussed in this chapter, therefore depends on their possession of numerous special metabolic pathways for substrate oxidation, all of which converge on the tricarboxylic acid cycle (Chapter 4). Some of these pathways occur in many of the groups listed in Table 17.1.

Most of the organic substrates that can be dissimilated aerobically by denitrifying bacteria can also be dissimilated by these organisms under anaerobic conditions, with nitrate as a terminal electron acceptor. However, this rule does not hold for aromatic substrates and for certain other substrates, such as aliphatic hydrocarbons, because one or more early steps in substrate breakdown are usually catalyzed by oxygenases, for which molecular oxygen is an obligatory cosubstrate. As a result, these pathways can function only under aerobic conditions.

One of the most striking collective properties of the aerobic chemoheterotrophs is their nutritional range; there are probably very few naturally occurring organic compounds that cannot be used by some of these bacteria as principal sources of carbon and energy. This can be very simply shown by enrichment culture experiments. When a synthetic medium containing almost any organic compound as sole source of carbon and energy is inoculated with soil or water and incubated aerobically, the predominant bacterial population that develops is largely composed of Gram-negative chemoheterotrophs. In many natural environments these bacteria appear to be the principal agents for the aerobic mineralization of organic material. Members of the *Azotobacter* group have an additional natural role as agents of nitrogen fixation: they are virtually the only chemoheterotrophs that can fix nitrogen under full atmospheric tensions of $O_2$.

The acetic acid bacteria provide a striking exception to the general rule that the respiratory dissimilation of organic susbtrates yields $CO_2$ as the principal end product. These organisms oxidize many organic compounds only partially, with the accumulation of large amounts of organic products; the formation of acetic acid from ethanol is the most characteristic of these partial substrate oxidations.

The Gram-negative aerobic chemoheterotrophs include a considerable number of bacteria that are either pathogenic for, or live in close association with, plants or animals. Some of these associations will be discussed in the following pages.

## THE AEROBIC PSEUDOMONADS

Among the many groups of Gram-negative flagellated rods that contain DNA with a base composition in the range of 58 to 70 percent G + C (Table 17.1), the organisms known as *aerobic pseudomonads* have received the most extensive study. It must be emphasized that the present limits of this group are somewhat arbitrary, its distinction from such organisms as *Alcaligenes*, *Agrobacterium*, *Rhizobium*, and the acetic acid bacteria being based on characters the taxonomic significance of which has not yet been carefully evaluated. The primary criteria for assigning a Gram-negative, aerobic chemoheterotroph to this particular group is the mode of flagellar insertion, which is polar (Figure 17.1). However, some rods with polar flagella are excluded (e.g., prosthecate bacteria, some acetic acid bacteria, and some members of the *Azotobacter* group) on the basis of special characters. About 30 species are now recognized among the aerobic pseudomonads; with the exception of yellow-pigmented plant pathogens, assigned to the genus *Xanthomonas*, and the floc-forming *Zoogloea*, they are all placed in the genus *Pseudomonas*. Nucleic acid hybridization has revealed that the aerobic pseudomonads are a group of considerable internal heterogeneity, the constituent species being assignable to a total of five major

FIGURE 17.1

Phase contrast photomicrograph of cells of *Pseudomonas aeruginosa* (× 1,100). Inset (upper right): flagella stain of *Pseudomonas stutzeri*, showing the polar monotrichous flagellation characteristic of many *Pseudomonas* species (× 1,290). Courtesy of N. J. Palleroni.

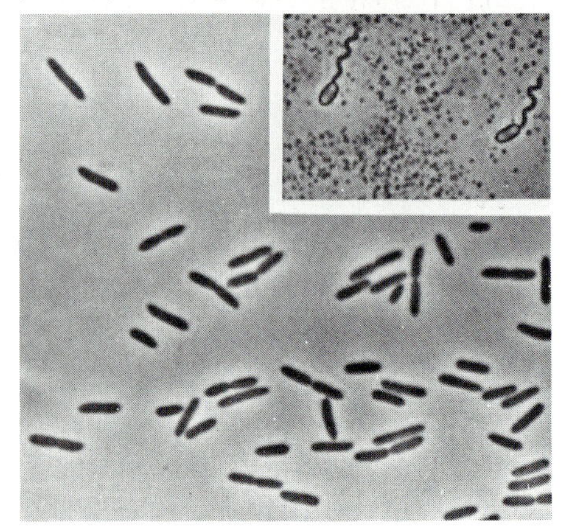

**TABLE 17.2**

**Differential Properties of the Major Subgroups of Aerobic Pseudomonads**

| Subgroup | Percent G + C | Accumulation of Cellular Reserve Material | Pigmentation | | Growth Factors Required |
|---|---|---|---|---|---|
| | | | SOLUBLE, FLUORESCENT | CELLULAR, YELLOW | |
| Fluorescent group | 60–67 | – (PHB)[a] | + | – | – |
| Pseudomallei group | 67–69 | + (PHB) | – | – | – |
| Acidovorans group | 61–69 | + (PHB) | – | – | – |
| Diminuta group | 65–67 | + | – | + or – | + |
| Xanthomonas group | 64–69 | – | – | + | + |

[a] Poly-$\beta$-hydroxybutyrate

and isolated genetic homology groups (see p. 320) that probably deserve separate generic status. We shall describe here some of the species representative of these homology groups (Table 17.2).

## The Fluorescent Pseudomonads

A somewhat variable but distinctive property of fluorescent pseudomonads is the production of a yellow-green, water-soluble pigment that diffuses into the medium and is fluorescent under ultraviolet light. Its synthesis is specifically stimulated by iron deprivation, reflecting its function as a *siderophore*, or iron chelator that solubilizes ferric hydroxide making soluble iron available. The structure of the pigment has been determined in *P. fluorescens* (where it is termed *pyoverdin*) and an unidentified plant-associated pseudomonad (where it is termed *pseudobactin*). The two pigments contain the same chromophore (Figure 17.2), linked to a short peptide of six (plant pseudomonad) or seven (*P. fluorescens*) amino acids via an amide linkage to the epsilon amino group of lysine.

In addition to the fluorescent pigment, the blue phenazine pigment *pyocyanin* (Figure 17.2) is characteristic of the species *P. aeruginosa*. The function of pyocyanin is unclear; there are, however, indications that it has antimicrobial activity, and it may hence confer some selective advantage to natural populations of *P. aeruginosa*.

The fluorescent pseudomonads do not require growth factors, and they do not synthesize poly-$\beta$-hydroxybutyrate as a cellular reserve material. They are common members of the microflora of soil and water, and are all nutritionally highly versatile, each being able to use 60 to 80 different organic compounds as sole sources of carbon and energy. For this reason, they have been much studied by microbial biochemists as biological material for the elucidation of the special metabolic pathways involved in the dissimilation of different classes of organic compounds. They have also recently become accessible to genetic study, following the discovery of conjugational and transductional systems of genetic transfer in *P. putida* and *P. aeruginosa*, and natural transformation systems in other members of the *P. aeruginosa* DNA-DNA homology group (see p. 259) such as *P. stuzeri*. One outcome of this work has been the discovery that the genetic determinants governing certain of the special pathways of substrate dissimilation (e.g., the dissimilatory pathways for terpenes such as camphor, aromatics such as naphthalene and toluene, and for hydrocarbons) are carried on plasmids, transmissible from strain to strain (Chapter 11).

*P. aeruginosa*, which has a considerably higher temperature maximum than *P. fluorescens* and *P. putida*, is sometimes pathogenic for humans. It belongs to the category of opportunistic pathogens, which do not normally inhabit animal hosts, but

FIGURE 17.2

*Pseudomonas* pigments: (a) the chromophore of pyoverdin and pseudobactin; (b) pyocyanin.

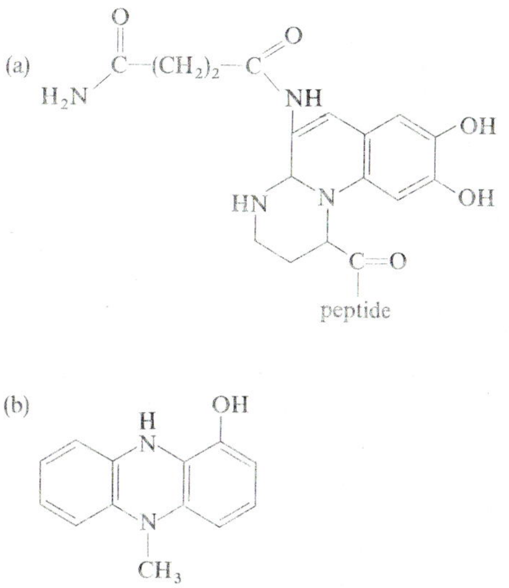

which can establish infections in individuals whose natural resistance has been reduced. Thus, *P. aeruginosa* typically causes infections, frequently fatal, in victims of severe burns, in cancer patients who have been treated with immunosuppressive drugs, and in victims of cystic fibrosis.

The fluorescent pseudomonads also include organisms that are pathogenic for plants; the many varieties, which differ in host range, are assigned to one species, *P. syringae*. These plant pathogens are true parasites, readily distinguishable from the free-living soil and water species by their physiological and biochemical properties. They are less versatile nutritionally, and their growth rates, both in synthetic and in complex media, are much lower. They are the only oxidase-negative members of the fluorescent group.

### The Pseudomallei Group

Like the fluorescent group, members of the pseudomallei group are nutritionally versatile organisms that do not require growth factors. Although usually pigmented, they never produce a yellow-green

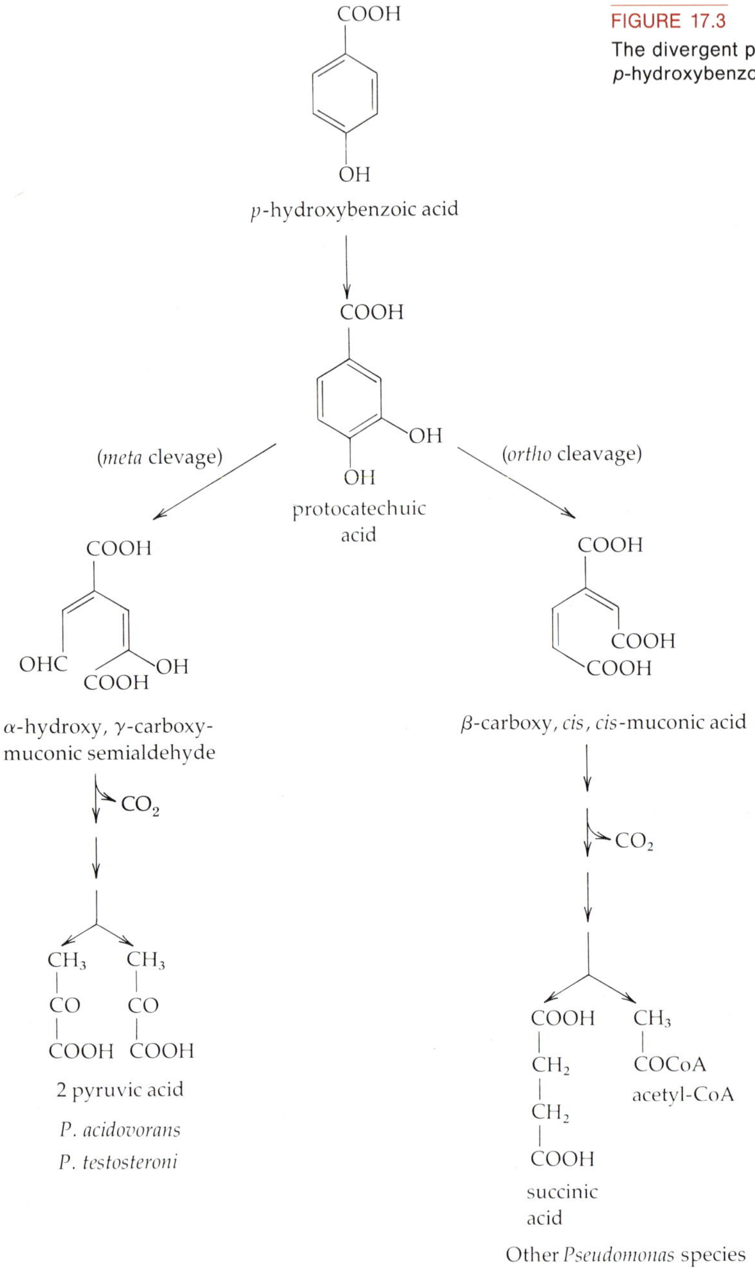

COOH
OH
*p*-hydroxybenzoic acid

COOH
OH
OH
protocatechuic acid

(*meta* clevage)

(*ortho* cleavage)

COOH
OHC
OH
COOH
α-hydroxy, γ-carboxy-muconic semialdehyde

$CO_2$

$CH_3$ $CH_3$
CO CO
COOH COOH
2 pyruvic acid
*P. acidovorans*
*P. testosteroni*

COOH
COOH
COOH
β-carboxy, *cis*, *cis*-muconic acid

$CO_2$

COOH
$CH_2$
$CH_2$
COOH
succinic acid

$CH_3$
COCoA
acetyl-CoA

Other *Pseudomonas* species

FIGURE 17.3
The divergent pathways for the oxidation of *p*-hydroxybenzoate among aerobic pseudomonads.

diffusible fluorescent pigment; and they all synthesize poly-$\beta$-hydroxybutyrate as a reserve material. The prototype of this group, *P. pseudomallei*, was originally discovered as the agent of *melioidosis*, a frequently fatal tropical disease of humans and other mammals. Even in the tropical areas where melioidosis is endemic, it is a relatively rare disease, typically contracted through the contamination of wounds with soil or mud. In fact, *P. pseudomallei* appears to be, like *P. aeruginosa*, an opportunistic pathogen that is a normal member of the microflora of soil and water in the tropics. However, the closely related species *P. mallei* is a true parasite, causing a disease of horses known as glanders. *P. mallei* is unable to survive in nature in the absence of its specific animal host. It is the only aerobic pseudomonad that is permanently immotile: its inclusion in the group is based on its close genetic relationship and phenotypic similarity to *P. pseudomallei.*

The pseudomallei group also contains several species that occur in soil and are occasionally pathogenic for plants; they are exemplified by *P. cepacia*, notable for its extreme nutritional versatility; it can use any of over 100 different organic compounds as a carbon and energy source.

### The Acidovorans Group

The soil bacteria of the acidovorans group, comprising the species *P. acidovorans* and *P. testosteroni*, are nonpigmented; they accumulate poly-$\beta$-hydroxybutyrate and do not require growth factors. The nutritional spectrum of these bacteria is distinctive. Many of the organic substrates commonly utilized by the fluorescent and pseudomallei groups cannot support their growth; these include glucose and other aldose sugars, polyamines such as putrescine and spermine, and several amino acids, notably arginine and lysine. Nevertheless, they can grow at the expense of several organic acids and amino acids rarely, if ever, utilized by members of other groups: these include glycollate, muconate, and norleucine.

Another interesting property of the acidovorans group is the dissimilation of widely utilized substrates via metabolic pathways that are unlike those found in other aerobic pseudomonads. For example, *p*-hydroxybenzoate, dissimilated through the $\beta$-ketoadipate (*ortho* cleavage) pathway by all other aerobic pseudomonads capable of utilizing this substrate, is dissimilated through a special (*meta* cleavage) pathway by the acidovorans group (Figure 17.3).

### The Diminuta Group

Pseudomonads of the diminuta group were first recognized as distinct from other pseudomonads by virtue of their unusually tightly coiled flagella (wavelength of about 0.6 $\mu$m, less than a third the value typical of most polar monotrichously flagellated bacteria). In addition to this unusual property, members of this group are quite fastidious nutritionally, being unable to utilize many of the compounds used by other pseudomonads, and requiring pantothenate, biotin, and $B_{12}$. They are unusual among bacteria in being incapable of assimilatory nitrate reduction; nitrogen must hence be provided in a reduced form.

*P. vesiculare* differs from *P. diminuta* in its ability to use a variety of sugars as sole source of carbon and energy, its independence from cystine as a growth factor, and its production of an orange-yellow pigment (probably a carotenoid).

### The *Xanthomonas* Group

Certain plant pathogenic pseudomonads have long been placed in a special genus, *Xanthomonas*, distinguished by the production of yellow cellular pigments. These pigments are unusual brominated aryl alkanes termed *xanthomonadins* (Figure 17.4); pigments of this type are not known to occur in any other bacteria. A closely related organism, *Pseudomonas maltophilia*, does not produce xanthomonadins, but, like *Xanthomonas*, requires organic growth factors.

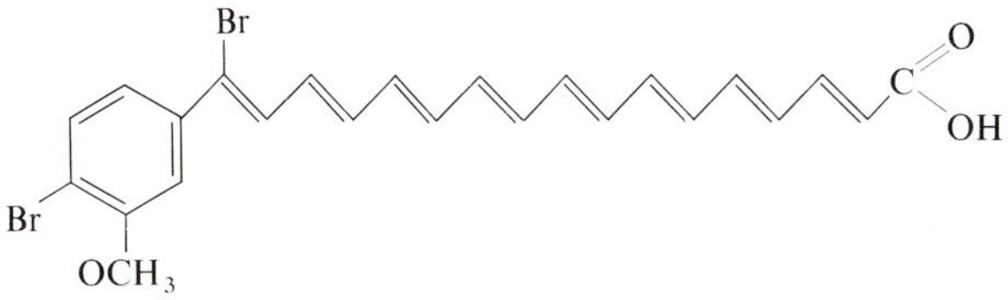

FIGURE 17.4
The structure of a xanthomonadin.

### The *Zoogloea* Group

The pseudomonads also contain a group of strains that characteristically inhabit polluted waters and aerobic sewage digestors, forming large clumps of cells held together by copious amounts of a fibrillar extracellular polyglucose. These organisms, termed *Zoogloea*, are distinguished by their characteristic floc formation under some conditions of growth (Figure 17.5). They have a G + C content of 63–65 percent, and many isolates require growth factors. Their genetic relationship to the other pseudomonads has not been determined.

(a)

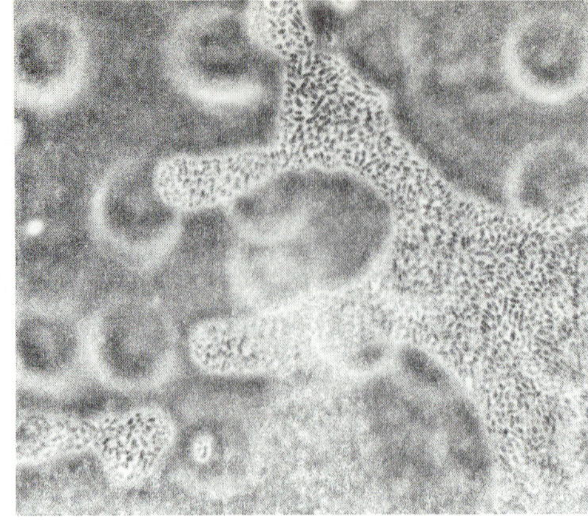

(b)

FIGURE 17.5

Flocs of *Zoogloea ramigera* in broth (× 23). (b) Phase contrast photomicrograph of projections at the periphery of a floc of *Zoogloea ramigera*, showing cells contained in slime layer (× 1,100). From K. Crabtree and E. McCoy, "*Zoogloea ramigera* Itzigsohn, Identification and Description," *Int. J. Syst. Bacteriol.* **17**, 1 (1967).

## THE *RHIZOBIUM* GROUP

The Gram-negative aerobic chemoheterotrophs with rod-shaped cells include many representatives in which flagellar insertion is not polar, and which are hence excluded by definition from the aerobic pseudomonads. These bacteria normally bear few (1 to 4) flagella and the flagellar insertion, which is not easy to determine unambiguously, is often described as "subpolar," "lateral," or as "degenerately peritrichous." Apart from the practical difficulty of determining this character, there are considerable grounds for doubting whether or not it possesses the taxonomic importance that has traditionally been ascribed to it. An extension of the study of genetic interrelationships by nucleic acid hybridization and sequencing, which has been so successful in revealing the internal relationships of the aerobic pseudomonads, to these nonpolarly flagellated rods will probably aid considerably in determining their taxonomic status, which is now most unclear.

At the present time, the members of the genera *Rhizobium*, *Bradyrhizobium*, and *Agrobacterium* are distinguished by their special relationships (described below) to plants. The genus *Alcaligenes* (57–70 percent G + C) is used as a repository for other Gram-negative, rod-shaped aerobes in which flagellar insertion is nonpolar.

### The rhizobia

It has long been known that the fertility of agricultural land can be maintained by a "rotation of crops." If a given plot of soil is sown year after year with a grass, such as wheat or barley, its productivity begins to decline but can be restored by interrupting this annual cycle with a crop of some leguminous plant such as clover or alfalfa. Roman writers on agriculture recognized that leguminous plants possess this ability to restore or maintain soil fertility, which is not shown by other types of plants. It was also known that the leguminous plants have peculiar nodular structures on their roots (Figure 17.6). The plant anatomists of the seventeenth and eighteenth centuries who examined these nodules in some detail interpreted them as pathological structures analogous to the galls formed on the shoots of some plants as a result of infestation by insects.

About the middle of the nineteenth century, a new interpretation of the nature of root nodules was offered. At this time, the development of chemical methods enabled scientists to start analyzing the problems of soil fertility and plant growth in chemical terms, and one of the early results of these studies was the elucidation of the role that leguminous plants play in the maintenance of soil fertility. It was found that most plants are limited in their growth by the amount of combined nitrogen in the soil but that leguminous plants are not. Furthermore, by total nitrogen analyses it could be shown that when leguminous plants are grown on nitrogen-poor soil, there is a net increase in the amount of fixed nitrogen in the soil. Since the only possible source of this extra nitrogen is the atmosphere, such experiments demonstrated that leguminous plants, unlike other higher plants, can fix atmospheric nitrogen. Hence, the growth of a crop of legumes on a nitrogen-poor soil results in an increase in the total fixed nitrogen content of the soil, particularly if the crop is plowed under. This is the chemical basis for the long-established practice of crop rotation.

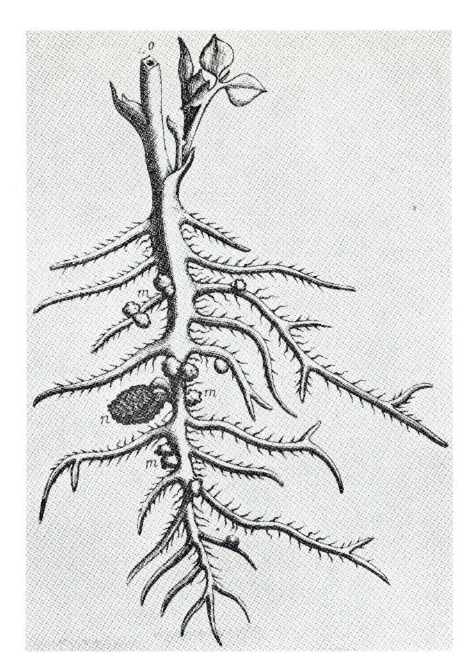


FIGURE 17.6

A seventeenth-century drawing by Malpighi of the root of a leguminous plant, showing the root nodules, m. The large dark object, n, is the coat of the seed from which the plant has developed.

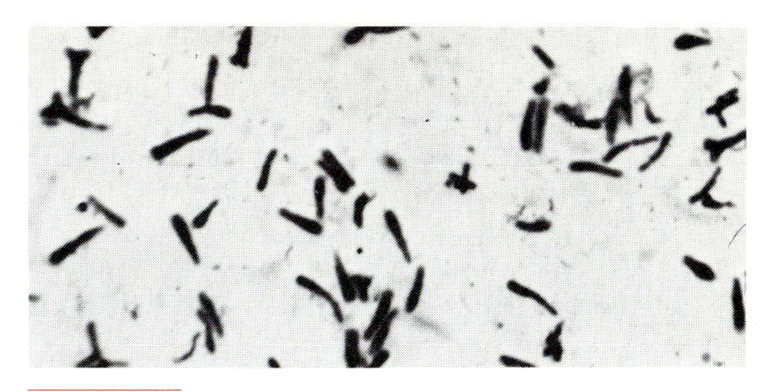

FIGURE 17.7

A stained smear of the contents of a root nodule, showing bacteroids ( × 1,050). Courtesy of H. G. Thornton and the Rothamsted Experimental Station, United Kingdom.

Once these facts had been established, the question naturally arose as to whether the peculiar nodulations on the roots of leguminous plants had any connection with their ability to fix nitrogen. Occasionally, leguminous plants fail to form nodules, and analyses showed that such plants do not fix nitrogen. When the contents of nodules were examined microscopically, they were found to contain large numbers of "bacteroids": small, rod-shaped, or branched bodies similar in size and shape to bacteria (Figure 17.7). These facts suggested that the nitrogen-fixing ability of leguminous plants is not a property of the plants as such but results from infection of their roots by bacteria in the soil, such infection leading to the formation of nodules. About 1885 the correctness of this hypothesis was established by showing that if seeds are treated with chemical disinfectants so as to sterilize their surface without impairing their capacity to germinate, and then grown in pots of sterile soil, they never form nodules. The growth of such plants is strictly dependent on a supply of combined nitrogen in the soil. Nodulation can be induced by adding to the soil crushed nodules from plants of the same species. Once nodulation has occurred, the growth of the plants becomes independent of a supply of combined nitrogen (Figure 17.8). The final proof came in 1888, when M. W. Beijerinck succeeded in isolating and cultivating the bacteria present in the nodules and demonstrated that sterile seeds produced the characteristic nodules when treated with pure cultures of the isolated bacteria.

The agricultural importance of nitrogen fixation led to extensive work on the nodule bacteria. These organisms are Gram-negative motile rods that are classified in the genera *Rhizobium* and *Bradyrhizobium*. It was soon found that the nodule bacteria isolated from the roots of the various kinds of leguminous plants resemble one another closely in their morphological and cultural properties. However, when inoculated back into plants they show a considerable degree of host specificity. The nodule bacteria isolated from the roots of lupines cannot evoke nodule formation on peas, and vice

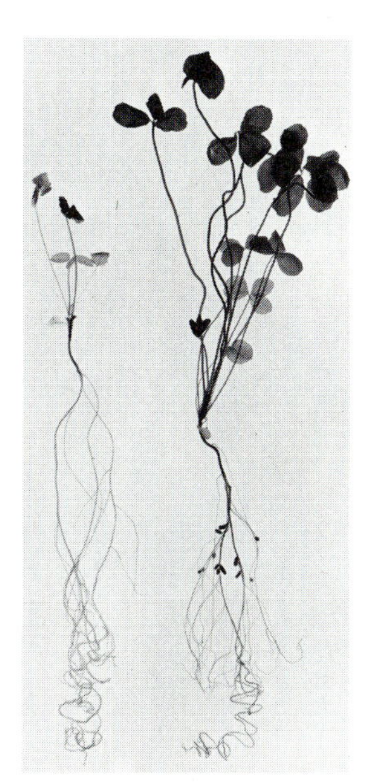

FIGURE 17.8

The effect of nodulation on plant growth. Two red clover plants grown in a medium deficient in combined nitrogen. The one at left, without nodules, shows very poor growth as a result of nitrogen deficiency. The plant at right, with nodules, shows normal growth. Courtesy of H. G. Thornton and the Rothamsted Experimental Station, United Kingdom.

versa. In contrast, the nodule bacteria from peas, lentils, and broad beans can evoke nodulation in every member of this group of legumes. There are thus differences between the nodule bacteria of peas and those of lupines, which can be detected in terms of their host specificity. The nodule bacteria can be classified into a series of cross-inoculation groups. Strains of any one group have the same host range, which differs from those of the other groups.

The nodule bacteria are normally present in soil. Their numbers vary with the nature of the soil and on its previous agricultural use. Hence, frequently a leguminous crop will develop poorly in a given plot because the nodule bacteria specific for it are either absent or present in such small numbers that effective nodulation does not occur. Nodulation can be ensured by inoculating the seed with a pure strain of nodule bacteria belonging to the correct cross-inoculation group. Bacterial cultures of proved effectiveness were first made commercially available at the beginning of the twentieth century, and seed inoculation is now a routine agricultural operation. This is by far the most important contribution that the science of soil bacteriology has made to agricultural practice.

Whereas the soil under a nonleguminous crop, such as wheat, may have fewer than 10 rhizobial cells per gram, the same soil will contain between $10^5$ and $10^7$ per gram following the development of a flourishing legume crop. The ability of legume plants to stimulate the growth of rhizobia in the soil extends as far as 10 to 20 mm from the roots. The effect is highly specific: bacteria other than *Rhizobium* and *Bradyrhizobium* show little or no stimulation, and growth of the rhizobial species able to infect that particular leguminous plant is stimulated more than the growth of other species. The substances responsible for this stimulation have not been identified. It has been experimentally established that the high number of rhizobial cells in the rhizosphere (the region of soil closely surrounding the root) of legumes is a result of stimulation of the growth of free-living cells rather than liberation of bacteroids from nodules, by showing that the increase occurs in the absence of active nodulation.

The number of nodules formed on the roots of the legumes is directly proportional to the density of rhizobia in the soil, up to about $10^4$ cells per gram. Above this number, no further increase takes place, and nodule formation may even decline. When the number of nodules is limited their size is proportionately larger. As a result the total *volume* of nitrogen-fixing tissue per acre of legu-

FIGURE 17.9

(a) A newly infected root hair. The bacterial infection thread can be seen passing up a root hair, which has curled at the tip as a result of infection. Courtesy of H. G. Thornton and the Rothamsted Experimental Station, United Kingdom. (b) Infection thread crossing a central tissue cell of a nodule aged one to two days. From D. J. Goodchild and F. J. Bergersen, "Electron Microscopy of the Infection and Subsequent Development of Soybean Nodule Cells," *J. Bacteriol.* **92,** 204 (1966).

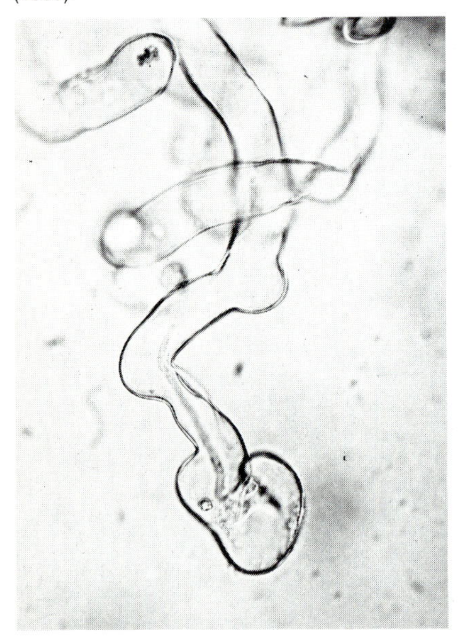

(a)

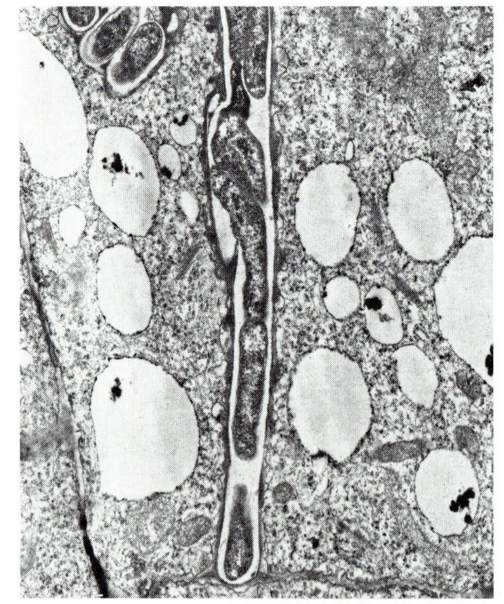

(b)

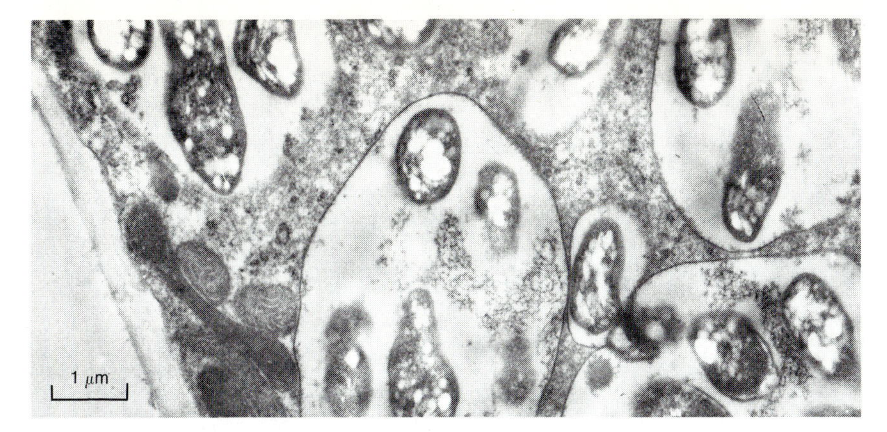


FIGURE 17.10

Mature nodule cell with large membrane envelopes containing four to six bacteroids. No further bacterial growth occurs. From D. J. Goodchild and F. J. Bergersen, "Electron Microscopy of the Infection and Subsequent Development of Soybean Nodule Cells," *J. Bacteriol.* **92,** 204 (1966).

minous plants remains fairly constant.

The infection of plants by rhizobia begins with the penetration of a root hair by a group of rhizobial cells, and involves the invagination of the root hair membrane. A tube is thus formed containing bacteria and lined with cellulose produced by the host cell. This tube is called the *infection thread* (Figure 17.9). The infection thread penetrates the cortex of the root, passing through the cortical cells rather than between them.

As the thread passes through a cell, it may branch to produce vesicles that contain bacteria; the walls of the thread and vesicles are continuous with the host cell membrane. The bacteria are finally liberated into the cytoplasm of the host cell enclosed, either singly or in small groups, in a membranous envelope (Figure 17.10).

Development of the nodule itself is initiated when the infection thread reaches a tetraploid cell of the cortex. This cell, along with neighboring diploid cells, is stimulated to divide repeatedly, forming the young nodule. The rhizobial cells invade only tetraploid plant cells, the uninfected diploid tissue becoming the cortex of the nodules (Figure 17.11). In young nodules the bacteria occur mostly as rods but subsequently acquire irregular shapes, becoming branched, club-shaped, or spherical (the typical bacteroids). At this stage they become incapable of cell division, but they may persist as metabolically active bacteroids for long periods. Ultimately, most of them lyse and their contents are absorbed by the plant.

Throughout the period during which the bacteroids persist, they actively fix atmospheric dinitrogen. The reductant and ATP necessary for $N_2$ reduction are derived from photosynthate provided

FIGURE 17.11

(a) Section of a root nodule. The dark cells are filled with bacteria. (b) Section of a nodule at high magnification, showing the individual bacteria in the infected cells. Courtesy of H. G. Thornton and the Rothamsted Experimental Station, United Kingdom.

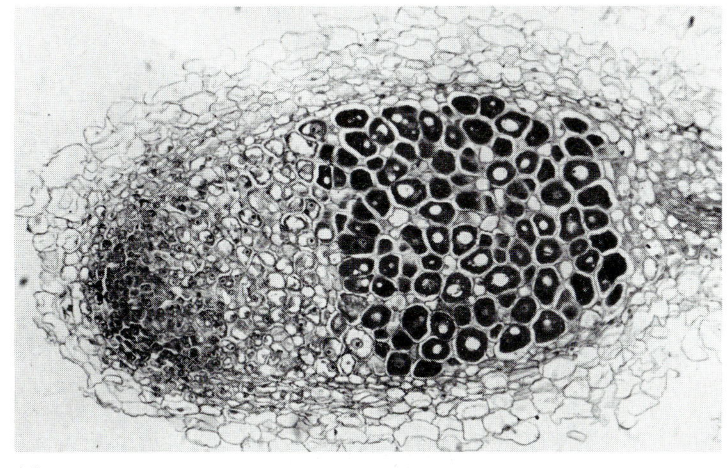

(a)

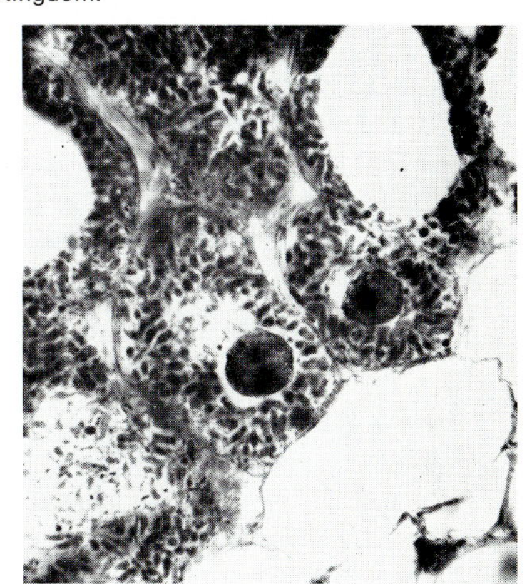

(b)

**TABLE 17.3**

**The Two Genera of Rhizobia**

| Genus | Percent G + C | Flagella | | Growth on Complex Media | Preferred Carbon Sources |
|---|---|---|---|---|---|
| | | NUMBER | INSERTION | | |
| *Rhizobium* | 59–63 | 2–6 | Peritrichous | Rapid | Glucose fructose |
| *Bradyrhizobium* | 62–66 | 1 | Subpolar | Slow | Pentoses |

by the plant, largely as dicarboxylic acids such as succinate, malate, and fumarate; the fixed nitrogen is excreted from the nodules to the plant's vascular tissue as ammonia. Because the rhizobial metabolism of dicarboxylic acids, which provides the substantial amounts of ATP needed by nitrogenase (approximately 15 to 20 moles ATP hydrolyzed per mole of $N_2$ fixed), is provided by aerobic respiration within the bacteroids, a continuous supply of $O_2$ must be assured. However, nitrogenase from bacteroids is extremely sensitive to inactivation by oxygen (a property shared by all known nitrogenases). These apparently incompatible requirements, for a rapid flux of $O_2$ combined with continuously low partial pressures, are met by the synthesis within the nodule of *leghemoglobin*, which becomes concentrated in the plant cell cytoplasm immediately surrounding the vacuoles that enclose the bacteroids. Leghemoglobin, like its analogue in vertebrate animals (hemoglobin), rapidly and reversibly binds molecular oxygen with high affinity. Hence, it maintains $O_2$ at a sufficiently low partial pressure that nitrogenase can function, yet ensures that a continuous supply is available for respiration. Neither the plant nor bacterium is individually capable of leghemoglobin synthesis; apparently, the apoprotein is encoded by a plant gene, and the synthesis of the heme moiety is under the control of bacterial genes.

Two genera of rhizobia are recognized (Table 17.3): *Rhizobium* and *Bradyrhizobium.* Nucleic acid reannealing studies indicate that both genera are internally heterogeneous, but closely related to each other and to *Agrobacterium.*

### The genus *Agrobacterium*

Organisms of the genus *Agrobacterium* cause galls or tumors on the stems or roots of many families of dicotyledonous plant. These bacteria are common in the soil. They are motile by means of one to four lateral flagella, and their G + C content is 59 to 63 percent. The principal species are *A. tumifaciens*, which causes *crown gall*, a tumor of the plant "crown" (the junction of stem and root); and *A. rhizogenes*, which causes an abnormal proliferation of root tissue termed *hairy root disease.*

The process of tumor formation by *A. tumifaciens* has been extensively studied. Tumorigenic strains all harbor a large plasmid termed the *Ti* (for tumor-inducing) plasmid. When cells of *A. tumifaciens* enter plant tissue through a wound, part of the Ti plasmid (termed *T-DNA*) is transferred to plant cells, and integrated into their genome. As a consequence of this gene transfer, the plant cells are said to be *transformed* because they act somewhat like cancerous cells of animals. Transformed cells acquire along with the new genes two novel characteristics: they synthesize elevated levels of plant growth hormones, rendering their growth independent of an extracellular supply of phytohormones; and they synthesize one of several related unusual guanido amino acids termed *opines* (Figure

FIGURE 17.12

Opines whose chemical structure is known.

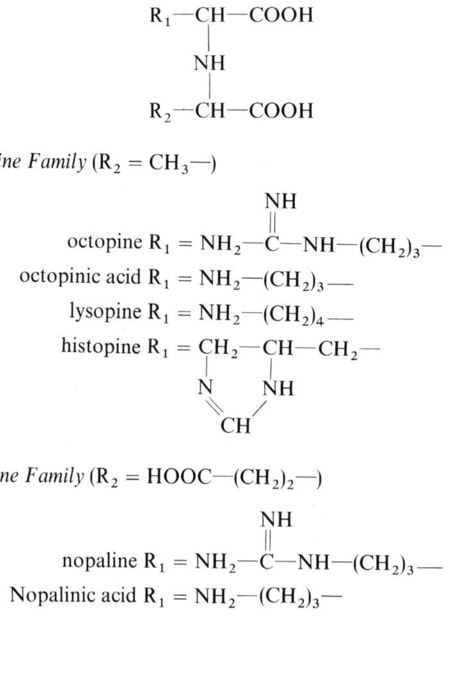

$$R_1\text{—CH—COOH} \quad | \quad \text{NH} \quad | \quad R_2\text{—CH—COOH}$$

*Octopine Family* ($R_2 = CH_3$—)

octopine $R_1 = NH_2\text{—C(=NH)—NH—}(CH_2)_3\text{—}$

octopinic acid $R_1 = NH_2\text{—}(CH_2)_3\text{—}$

lysopine $R_1 = NH_2\text{—}(CH_2)_4\text{—}$

histopine $R_1 = CH_2\text{—CH—}CH_2\text{—}$ (imidazole ring: N, NH, CH)

*Nopaline Family* ($R_2 = HOOC\text{—}(CH_2)_2\text{—}$)

nopaline $R_1 = NH_2\text{—C(=NH)—NH—}(CH_2)_3\text{—}$

Nopalinic acid $R_1 = NH_2\text{—}(CH_2)_3\text{—}$

| Vitamin Requirements | Representative Species | Plant Host Range |
|---|---|---|
| Biotin (some) | *R. leguminosarum* | Pea, vetch, lentil |
| None | *B. japonicum* | Soybeans |

17.12). Once a cell has been transformed by integration of T-DNA into its genome, its progeny continue to exhibit these characteristics even in the absence of agrobacterial cells.

The function of the opines synthesized in plant tumors is unknown. However, it appears that Ti plasmids encode the enzymes that catabolize opines, in addition to those that synthesize them. Thus a strain of *A. tumifaciens* that harbors a particular Ti plasmid has the selectively advantageous biochemical ability to utilize as sources of carbon and energy the opine produced by the tumor it caused.

Although not as well studied, *A. rhizogenes* probably causes plant disease by a similar mechanism. It too transfers part of a large plasmid (the *Ri* plasmid) to susceptible plant cells, inducing their proliferation and synthesis of opines.

## PROSTHECATE BACTERIA

The prosthecate bacteria are characterized by the formation of *prosthecae* (singular: *prostheca*), filiform or conical extensions of the cell that are contained within the peptidoglycan layer and outer membrane; hence the interior of these cellular extensions is continuous with the cytoplasm of the major portion of the cell (Figure 17.13). The principal constituent genera of this group are listed in Table 17.4.

The function of prosthecae has been a matter of some debate. In certain genera e.g., *Hyphomicrobium* and *Hyphomonas*, the prosthecae clearly serve a reproductive function: these organisms reproduce by budding at the tips of the prosthecae (Figure 17.14). However, in other prosthecate bacteria the prosthecae play no role in reproduction. The function of prosthecae in these organisms is not known, but they do *increase the surface area of the cell, thereby increasing the efficiency of nutrient uptake*, and they do *retard sedimentation* of cells in an aqueous environment, a property that probably increases the survival of these immotile, obligately aerobic aquatic organisms. The possible role of prosthecae in nutrient absorption is supported by the effect of changes in the concentration of nutrients on the length of prosthecae of several of these organisms (e.g., *Caulobacter* and *Hyphomicrobium*): their length increases dramatically when nutrients (particularly phosphorus) are present in growth-limiting concentrations. While these functions may provide the selective advantage that accounts for the occurrence of the prosthecae on some bacteria, it is questionable if they can account for the short,

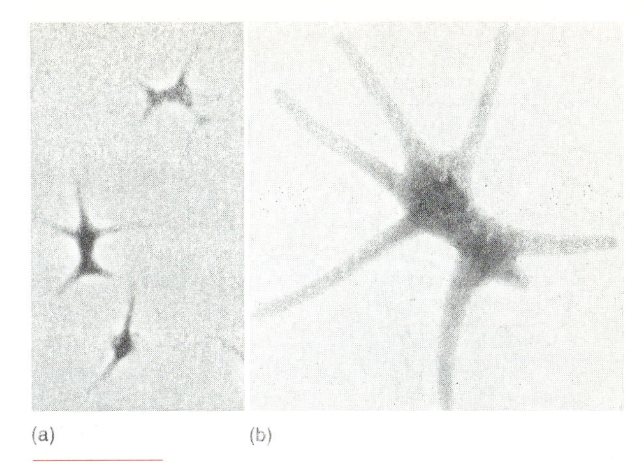


FIGURE 17.13

A prosthecate, freshwater bacterium, *Ancalomicrobium adetum.* (a) Phase contrast micrograph of a group of cells (×2,180). (b) Electron micrograph of a single cell, showing continuity of prosthecae with the body of the cell (×8,530). From J. T. Staley, "*Prosthecomicrobium* and *Ancalomicrobium:* New Prosthecate Freshwater Bacteria," *J. Bacteriol.* **95,** 1921 (1968).

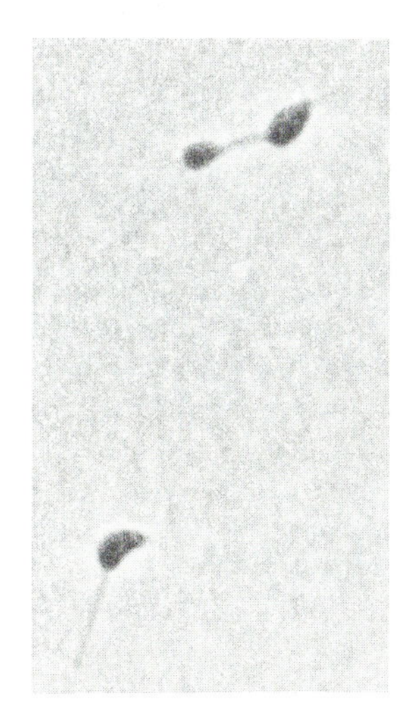

FIGURE 17.14

Phase contrast micrograph of *Hyphomicrobium vulgare* (×2,450). Micrograph courtesy of Peter Hirsch.

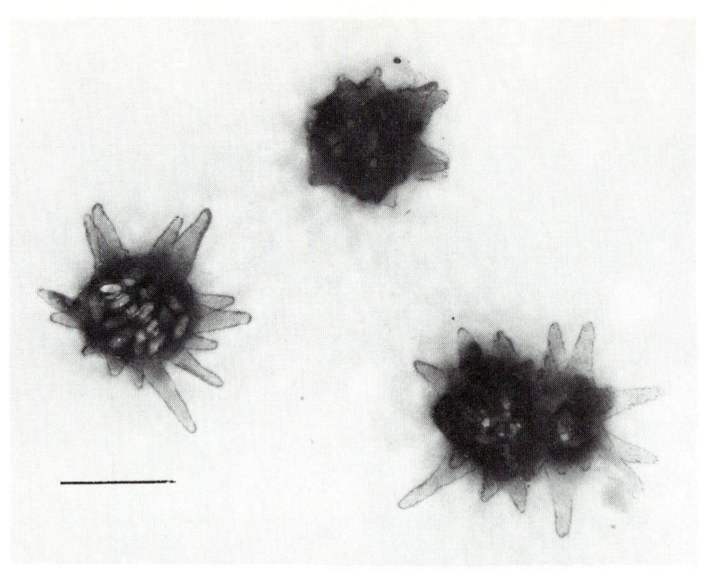

(a)

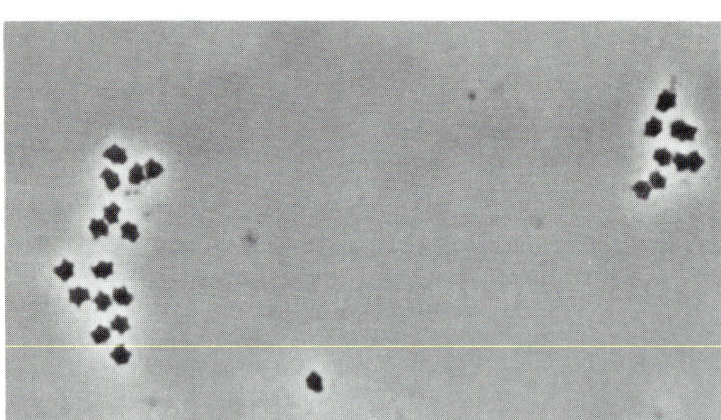

(b)

FIGURE 17.15

Some prosthecate bacteria with numerous short prosthecae. (a) Electron micrograph of a negatively-stained preparation of *Prosthecomicrobium pneumaticum*. The light areas within the cells are gas vacuoles. Bar equals 1.0 $\mu$m. (b) light micrograph of *Stella* ( × 1,280). (a) from J. T. Staley, "The Genera *Prostheco-microbium* and *Ancalomicrobium*", pp. 456–460; (b) from P. Hirsch and H. Schlesner, "The genus *Stella*", pp. 461–465. Both in M. P. Starr, H. Stolp, H. G. Trüper, A. Balows, and H. G. Schlegel (eds), *The Prokaryotes*. New York: Springer-Verlag (1981).

**TABLE 17.4**
**The Principal Genera of Prosthecate Bacteria**

| Genus | Shape | Cell Division | Location of Prosthecae | Number of Prosthecae |
|---|---|---|---|---|
| *Hyphomicrobium* | Rod | Budding from tip of hypha | Polar or bipolar | 1–2 |
| *Hyphomonas* | Rod | Budding from tip of hypha | Polar or bipolar | 1–2 |
| *Stella* | Flat | Fission (symmetrical) | Around cell periphery | 6–8 |
| *Prosthecomicrobium* | Cocci | Budding from cell body | Peritrichous | Many |
| *Ancalomicrobium* | Irregular | Budding from cell body | Peritrichous | 2–8 |
| *Prosthecobacter* | Fusiform | Fission (symmetrical) | Polar | 1 |
| *Caulobacter* | Rod or vibrio | Fission (assymetrical) | Polar | 1 |
| *Asticcacaulis* | Rod | Fission (assymetrical) | Polar or lateral | 1–2 |

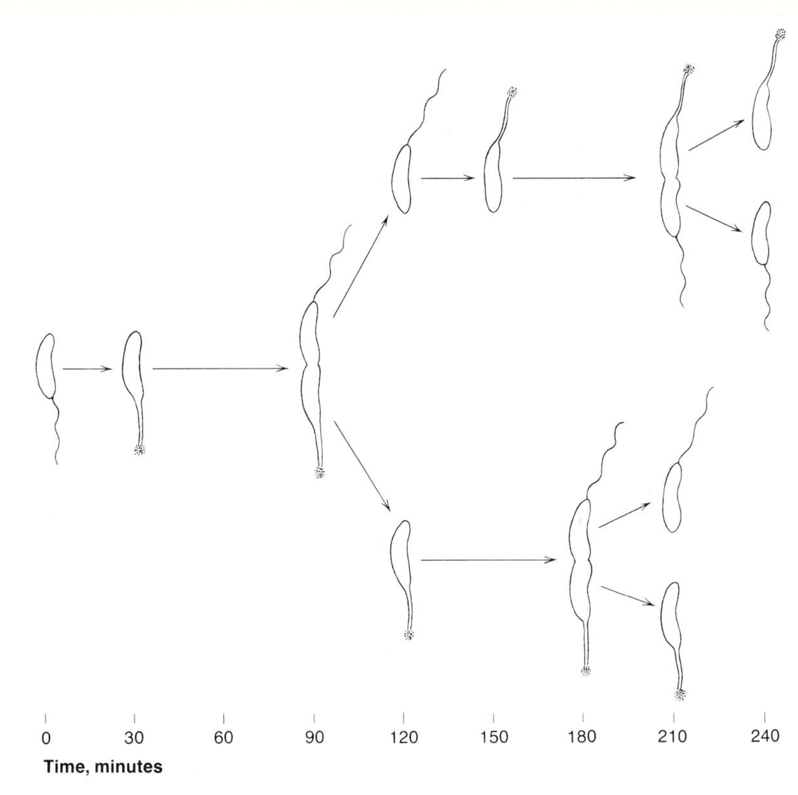


FIGURE 17.16

A diagrammatic representation of clonal growth in *Caulobacter*, based on continuous microscopic observations. Note that the time required for the division of a swarmer cell is considerably longer than that required for the division of its stalked sibling. After J. L. S. Poindexter. "Biological Properties and Classification of the Caulobacter group," *Bacteriol. Rev.* **28,** 231 (1962).

conical appendages of *Stella* or *Prosthecomicrobium* (Figure 17.15).

The prosthecate bacteria are primarily aquatic organisms that occur in both fresh and marine waters. They are adapted to growth in habitats in which the concentration of nutrients is extremely low (*oligotrophic* habitats); indeed microscopic examination of such waters often shows prosthecate bacteria to be the predominant microflora. One member, *Hyphomicrobium*, of this group is a facultative methylotroph; it can grow by metabolizing compounds containing as many as four carbons but it grows best when provided with the single carbon substrate, methanol. Because it is a vigorous denitrifier, *Hyphomicrobium* can be readily enriched by incubating samples anaerobically in a medium containing methanol and nitrate. The otherwise similar organism *Hyphomonas* is incapable of methylotrophic growth.

A characteristic feature of two groups of prosthecate bacteria, the caulobacters (*Caulobacter* and *Asticcacaulis*) and the hyphomicrobia (*Hyphomicrobium* and *Hyphomonas*), is a life cycle in which cell division produces two unequal products: one immotile prosthecate cell, and one motile nonprosthecate *swarmer* cell. The prosthecate cell can immediately initiate another cycle of cell division, but the swarmer cell cannot divide until it has grown a prostheca and lost its flagellum, an event that usually occurs after a period of active motility. The life cycle of *Caulobacter* is shown in Figure 17.16.

Another characteristic feature of many prosthecate bacteria is the synthesis of *holdfasts*, local-

| Holdfast | Percent G + C | Motility |
|---|---|---|
| + or − (at cell pole) | 59–67 | + (swarmer only) |
| + or − (at cell pole) | 60–62 | + (swarmer only) |
| − | 59–60 | − |
| − | 65–70 | + or − |
| − | 70 | − |
| + (at tip of (prostheca) | 54–60 | − |
| + (at tip of prostheca) | 63–67 | + (swarmer only) |
| + (at cell pole) | 55–61 | + (swarmer only) |

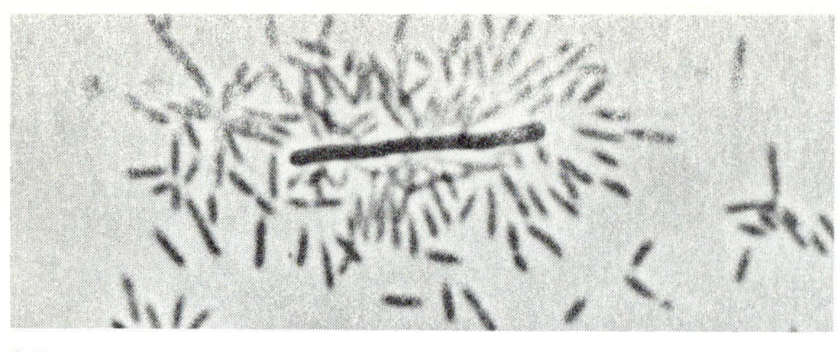

(a)

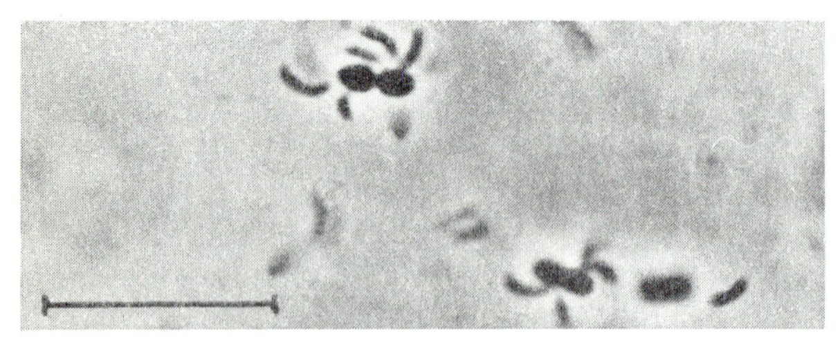

(b)

FIGURE 17.17

The attachment of *Caulobacter* to (a) *Bacillus and* (b) *Azotobacter* (scale marker, 1 μm). From J. L. S. Poindexter, "Biological Properties and Classification of the *Caulobacter* group," *Bacteriol. Rev.* **28**, 231 (1962).

ized patches of adhesive material secreted at one point on the cell surface (typically either the tip of the prostheca or the pole of the cell body). Via holdfasts these bacteria can attach nonspecifically to a variety of solid substrates, including the cell walls of other microorganisms (Figure 17.17). Under natural conditions, it is probable that they mostly grow attached to larger microorganisms (algae, protozoa, other bacteria), utilizing organic materials secreted by the organisms to which they adhere. Relative to other aerobic chemoheterotrophs from the same aquatic habitats (e.g., aerobic pseudomonads), their growth rates (even in complex media) are low. The minimal generation time is never less than two hours, and for many species as long as four to six hours. Most caulobacters require organic growth factors, and their ranges of utilizable substrates are less broad than those of aerobic pseudomonads. Hence, their capacity for attachment to other microorganisms appears to be an important factor for successful competition in nature with other aerobic chemoheterotrophs.

## THE *AZOTOBACTER* GROUP

The rod-shaped organisms of the *Azotobacter* group possess a property that does not occur in any other group of Gram-negative chemoheterotrophs: the ability to fix nitrogen under aerobic growth conditions. In view of the extreme sensitivity of nitrogenase to oxygen inactivation, the existence of this ability in bacteria that are strict aerobes appears paradoxical. It is evident that the *Azotobacter* group must possess special mechanisms for the protection of nitrogenase, since facultatively anaerobic nitrogen-fixing chemoheterotrophs (e.g., *Bacillus polymyxa*) can maintain nitrogenase activity only when growing in the absence of oxygen. The azotobacters have extraordinarily high respiratory rates, far in excess of those of all other aerobic bacteria, and this may prevent $O_2$ penetration to the intracellular sites of nitrogenase activity. It has also been suggested that nitrogenase exists in the *Azotobacter* cell in a special conformational state which renders it oxygen-resistant.

The members of the *Azotobacter* group have oval to rod-shaped cells, which are large (as much as 2 μm wide) in most species. They form poly-β-hydroxybutyrate as a reserve material. Cultures on solid media have a characteristic mucoid appearance (Figure 17.18), since these organisms produce large amounts of extracellular polysaccharide. The members of the genera *Azotobacter*, *Azomonas*, and *Derxia* (Table 17.5) are common in temperate regions in neutral or alkaline soils and waters. In tropical regions the far more acid-tolerant members of the genus *Beijerinckia* are the prevalent members of the aerobic, nitrogen-fixing soil microflora; they can grow at pH values as low as 3, and thus are well adapted to the relatively acid soils

FIGURE 17.18

A streaked plate of *Azotobacter vinelandii*, showing the smooth, glistening colonies typical of *Azotobacter*. Courtesy of O. Wyss; reproduced from his *Elementary Microbiology* (New York: John Wiley, 1963).

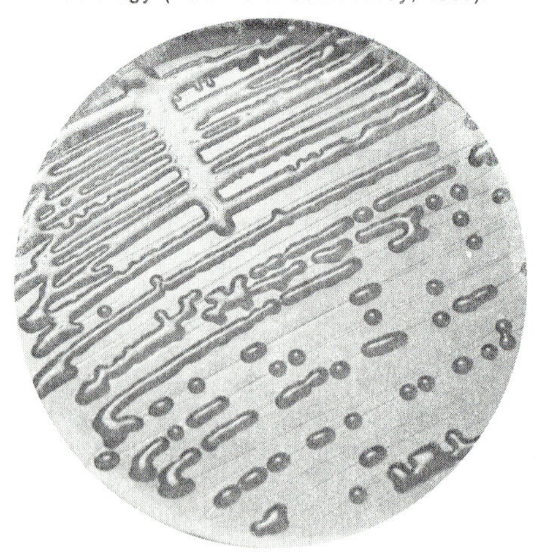

**TABLE 17.5**

**Properties of the Principal Genera of the *Azotobacter* Group**

| Genus | Mode of Flagellar Insertion | Percent G + C | Formation of Cysts | Minimal pH for Growth |
|---|---|---|---|---|
| *Azotobacter* | Peritrichous | 63–66 | + | 5.5 |
| *Azomonas* | Peritrichous or polar | 53–59 | – | 4.5 |
| *Beijerinckia* | Peritrichous or absent | 55–60 | – | 3.0 |
| *Derxia* | Polar | 70 | – | 5.5 |

characteristic of this climatic zone.

The members of the genus *Azotobacter* (but not of the other three genera) produce distinctive resting cells known as *cysts* (Figure 17.19). These structures, which arise by the deposition of additional outer layers around the vegetative cell wall, are resistant to desiccation but not to heat.

## THE ACETIC ACID BACTERIA

Bacteria of the acetic acid group are rod-shaped organisms that are distinguishable from other aerobic Gram-negative chemoheterotrophs by a series of physiological and metabolic characteristics, which seem to be at least in part a reflection of their ecology. Acetic acid bacteria occur on the surface of plants, particularly flowers and fruits. They develop abundantly as a secondary microflora in decomposing plant material under aerobic conditions, following an initial alcoholic fermentation of sugars by yeasts. Under these circumstances, they use ethanol as an oxidizable substrate, converting it to acetic acid. Many carbohydrates and primary and secondary alcohols can also serve as energy sources, their oxidation characteristically resulting in the transient or permanent accumulation of partly oxidized organic products. Since most members of the group have relatively complex growth factor requirements, they are usually grown in media with a complex base (e.g., yeast extract), supplemented with an oxidizable substrate. These bacteria are markedly acidophilic, growing at pH values as low as 4, with an optimum between pH 5 and 6.

By virtue of their capacity to convert many organic compounds almost stoichiometrically to partly oxidized organic end products, the acetic acid bacteria are a group of considerable industrial importance. Their major industrial use is in the manufacture of vinegar, by the acetification of ethanol-containing materials (e.g., wine, cider).

Two genera (Table 17.6) are distinguished by differences in flagellation and in oxidative capacities. *Gluconobacter* spp. do not have a functional

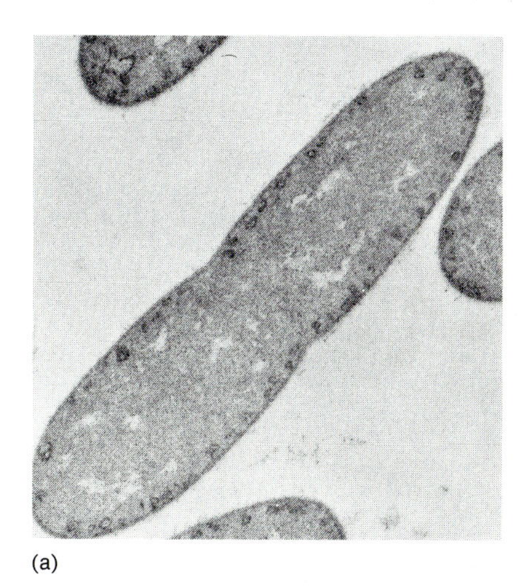

(a)

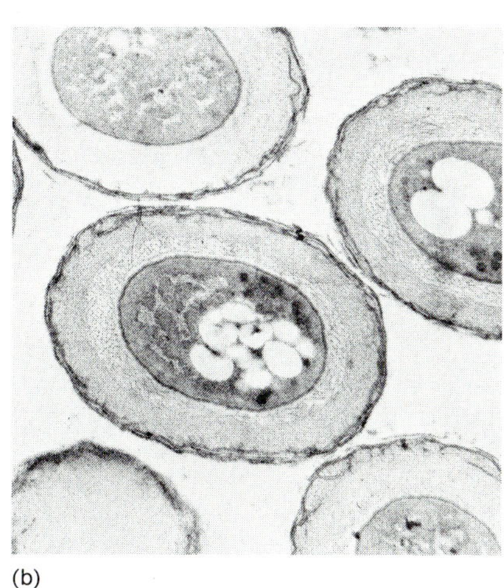

(b)

FIGURE 17.19

Electron micrographs of thin sections of vegetative cells and cysts of *Azotobacter vinelandii*. (a) Vegetative cells (× 17,800). (b) Cysts (× 10,700). Both from O. Wyss, M. G. Neumann, and M. D. Socolofsky, "Development and Germination of the *Azotobacter* Cyst.," *J. Biophys. Biochem. Cytol.* **10,** 555 (1961).

**TABLE 17.6**

**The Distinguishing Characteristics of the Genera *Gluconobacter* and *Acetobacter***

| Genus | Flagellar Insertion | Percent G + C | Presence of Tricarboxylic Acid Cycle |
|---|---|---|---|
| *Gluconobacter* | Polar or absent | 60–64 | – |
| *Acetobacter* | Peritrichous or absent | 55–65 | + |

tricarboxylic acid cycle, and cannot oxidize acetate; hence, they oxidize ethanol (and other substrates convertible to ethanol) with the stoichiometric accumulation of acetic acid; they are known in the vinegar industry as *underoxidizers. Acetobacter* spp. possess a functional tricarboxylic acid cycle, and hence they can oxidize acetate to $CO_2$. When growing at the expense of ethanol, they convert this substrate rapidly to acetic acid, which is then oxidized more slowly to completion, a property to which they owe the name of *overoxidizers.* Apart from this difference, the oxidative metabolism of all acetic acid bacteria is similar, and it has a number of unusual features.

Sugars are oxidized to $CO_2$ exclusively through the pentose phosphate pathway; the Entner-Doudoroff pathway, otherwise common in aerobic chemoheterotrophs, does not operate in this group. The metabolism of pyruvate is also unusual. Whereas in most aerobes, pyruvate is oxidized to acetyl-CoA and $CO_2$, the acetic acid bacteria decarboxylate it nonoxidatively to acetaldehyde. Since acetaldehyde is also the first intermediate in ethanol dissimilation, it lies at the point of metabolic convergence between the oxidation of substrates metabolized through pyruvate and the oxidation of ethanol (Figure 17.20).

In addition to the oxidative pentose phosphate pathway, the acetic acid bacteria oxidize glucose through a second, nonphosphorylative pathway that results in the accumulation of partly oxidized products—gluconate and ketoacids derived from it (Figure 17.20). Growth at the expense of glucose is therefore always accompanied by the conversion of a large part of the substrate to these acidic derivatives.

In addition to the oxidation of primary alcohols with accumulation of the corresponding car-

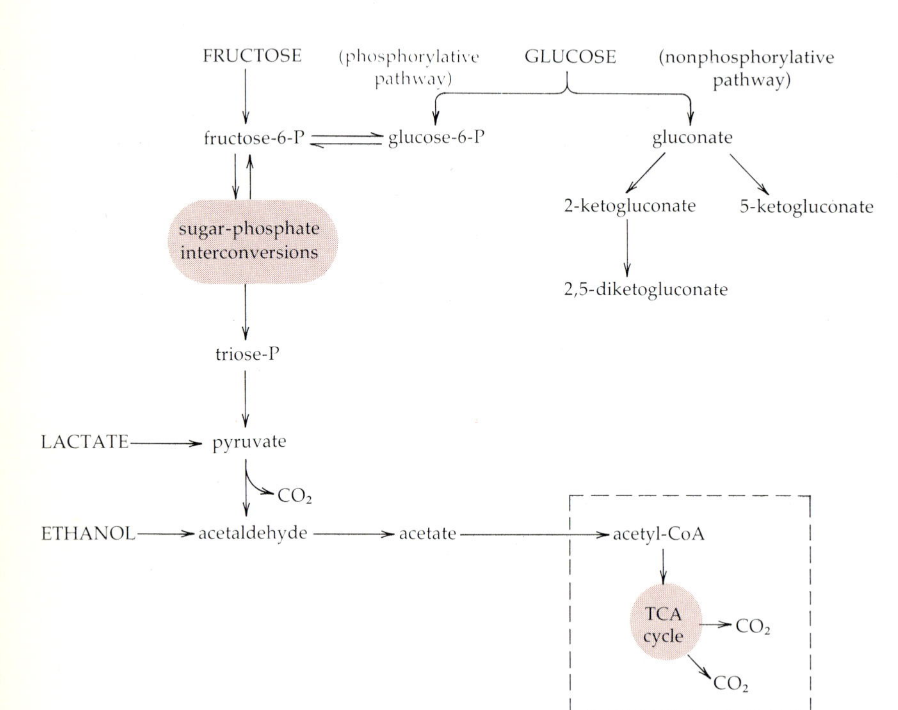


FIGURE 17.20

Major pathways of oxidative metabolism in acetic acid bacteria.
The terminal reactions of acetate metabolism (square box) do not occur in *Gluconobacter*. Primary subtrates are capitalized.

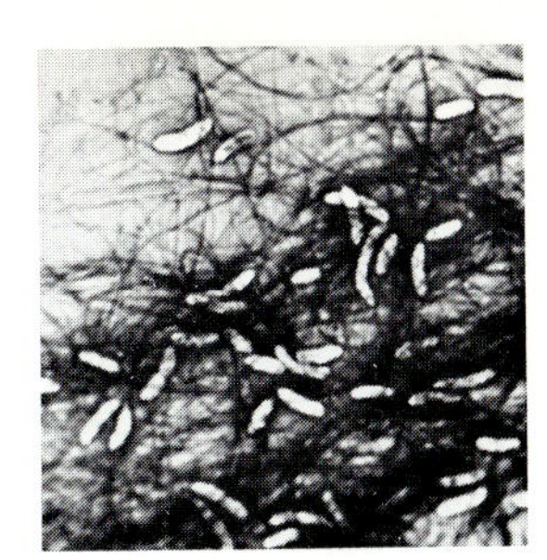

FIGURE 17.21
Extracellular formation of cellulose by *Acetobacter xylinum* (×860). The bacterial cells are entangled in a mesh of cellulose fibrils. From J. Frateur, "Essai sur la systématique des acétobacters," *La Cellule* **53**, 3 (1950).

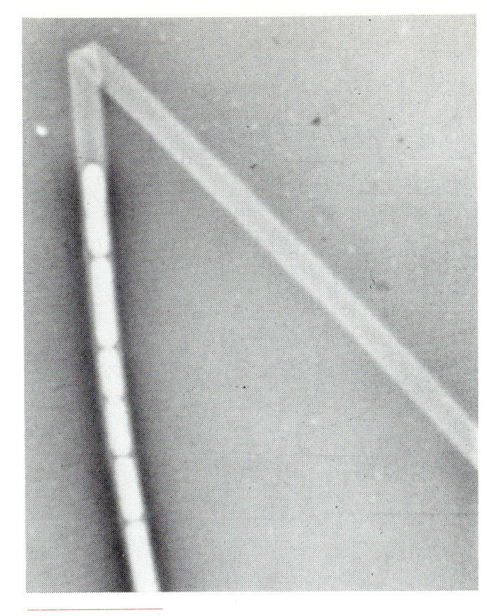

FIGURE 17.22
*Sphaerotilus*, showing a chain of cells enclosed within the sheath. From J. L. Stokes, "Studies on the Filamentous Sheathed Iron Bacterium *Sphaerotilus natans*," *J. Bacteriol.* **67**, 281 (1954).

boxylic acids, exemplified by the oxidation of ethanol to acetic acid, the acetic acid bacteria can oxidize many secondary alcohols to ketones. Only polyalcohols can be attacked in this manner, since the secondary alcoholic group that undergoes oxidation must be adjacent to another alcoholic group, primary or secondary.

The species *Acetobacter xylinum* is one of the few procaryotes that synthesizes large amounts of the polysaccharide, cellulose. This substance is formed as a result of growth at the expense of glucose and certain other sugars, and is deposited outside the cells in the form of a loose, fibrillar mesh (Figure 17.21). In liquid cultures this organism forms a tough, cellulosic pellicle which encloses the cells and can attain a thickness of several centimeters.

## THE SHEATHED BACTERIA

Sheathed bacteria are rod-shaped aquatic organisms that grow as chains of cells enclosed in tubular sheaths (Figure 17.22), often attached to solid substrates by basal holdfasts. Reproduction occurs by the liberation of cells from the open apex of the sheath. Three principal genera are distinguishable (Table 17.7). *Sphaerotilus* forms thin sheaths, normally without encrustations of metal oxides. It is found in slowly running streams contaminated with sewage or other organic matter, where it grows as long, slimy, attached tassels. It also develops in aerobic sewage digestors. *Leptothrix* is common in uncontaminated fresh water containing metal salts, where its sheaths are heavily encrusted with hydrated ferric or manganic oxides (Figure 17.23). *Haliscomenobacter* is abundant in aerobic sewage treatment systems. Its long, thin cells are enclosed in a barely visible sheath. Although it is clear that the oxidation of ferrous and manganous salts is mediated by *Leptothrix* spp., the process appears to be the result of a nonspecific reaction, catalyzed by sheath proteins, and consequently leading to a massive deposit of the metal oxides on or in the sheath. Such a process evidently cannot provide the enclosed cells with ATP; and all current evidence suggests that *Leptothrix* spp., like *Sphaerotilus*, are obligate chemoheterotrophs.

TABLE 17.7
**The Sheathed Bacteria**

| Genus | Cell Diameter (μm) | Motility | Holdfasts | Oxidation of | | PHB Accumulated | Percent G + C |
|---|---|---|---|---|---|---|---|
| | | | | $Fe^{2+}$ | $Mn^{2+}$ | | |
| *Sphaerotilus* | 1.2–2.5 | + | + | + | − | + | 69–71 |
| *Leptothrix* | 0.6–1.4 | + | + or − | + | + | + | 69–71 |
| *Haliscomenobacter* | 0.35–0.45 | − | − | − | + | − | 48–50 |

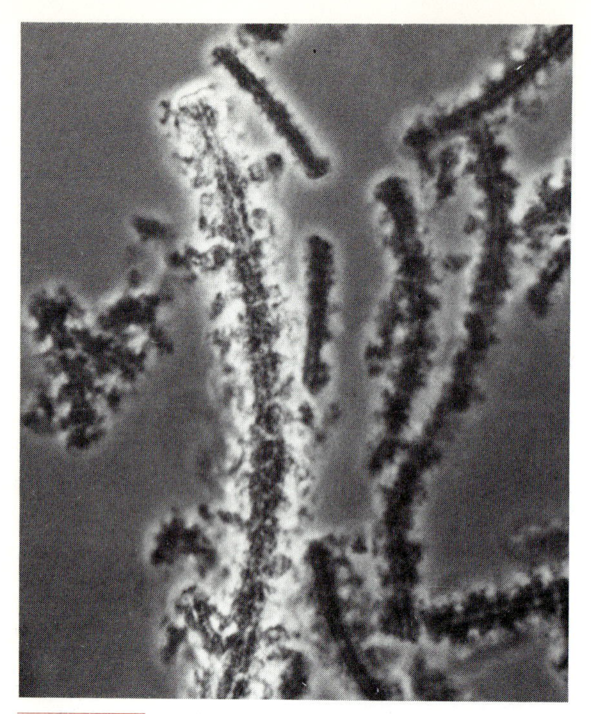

FIGURE 17.23

Phase-contrast light micrograph of *Leptothrix discophora* grown in the presence of $Fe^{2+}$ and $Mn^{2+}$, showing the deposition of ferric hydroxide and manganic dioxide (× 1,560). From E. G. Mulder and M. H. Deinema, "The Sheathed Bacteria," pp. 425–440 in M. P. Starr, H. Stolp, H. G. Trüper, A. Ballows, and H. G. Schlegel (eds), *The Prokaryotes*. Berlin: Springer-Verlag (1981).

## THE *SPIRILLUM* GROUP

The spirilla are aerobic chemoheterotrophs characterized by the possession of vibrioid or helical cells, bearing bipolar flagellar tufts (Figure 17.24). They are all aquatic bacteria, common in both freshwater and marine environments. The group is composed of four genera (Table 17.8).

The substrates most commonly used as carbon and energy sources include a limited range of amino acids and organic acids; sugars are rarely utilized. A physiological property, shared to some degree by all spirilla is a *preference for low oxygen tensions*. The microaerophilic tendencies of these bacteria are shown by their behavior in wet mounts; the highly motile cells accumulate in a dense, narrow band, located at some distance from the edge of the cover slip (Figure 17.25). Despite this, most spirilla can form colonies on the surface of agar media. However, *Spirillum volutans*, a species with very large cells, cannot be isolated on streaked plates exposed to air. It is an *obligate microaerophile*, which can initiate growth only in an oxygen-depleted environment containing 3 to 9 percent $O_2$, instead of the normal atmospheric concentration of 20 percent.

A vibrioid organism of uncertain affinity for which the name *Azospirillum* has been proposed is often placed in this group. It has a single polar flagellum when grown in liquid media, but also produces peritrichous flagella of significantly different

FIGURE 17.24

A single cell of the large spirillum, *S. volutans* (phase contrast). Note the flagellar tufts at one pole of the cell. From N. R. Krieg, "Cultivation of *Spirillum volutans* in a Bacteria-Free Environment," *J. Bacteriol.* **90,** 817 (1965).

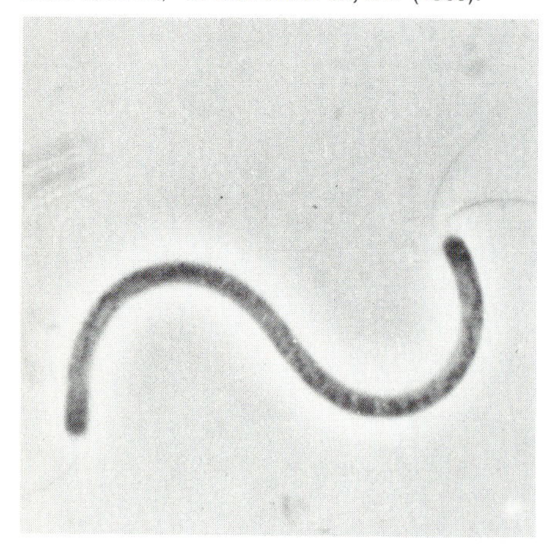

FIGURE 17.25

Typical aerotactic pattern of *Spirillum* in a wet mount, showing the accumation of the cells in a very narrow band some distance from the air-water interface. From N. R. Krieg, "Cultivation of *Spirillum volutans* in a Bacteria-free Environment," *J. Bacteriol.* **90,** 817 (1965).

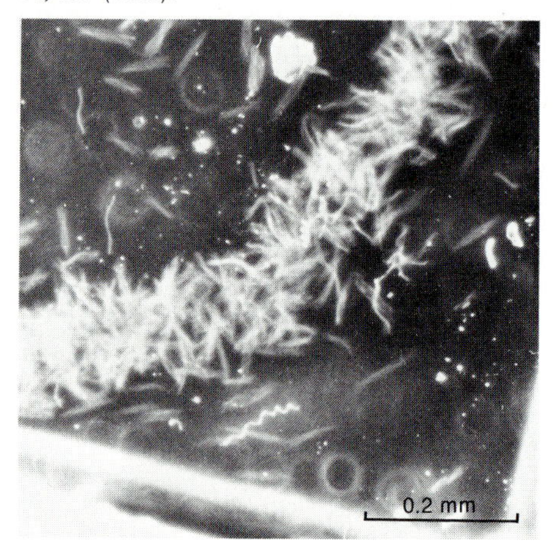

**TABLE 17.8**

**Differential Properties of the *Spirillum* Group**

| Genus | Cell Diameter (μm) | Percent G + C | Obligately Microaerophilic | Fixes $N_2$[a] | Requires NaCl | Denitrification | Habitat |
|---|---|---|---|---|---|---|---|
| *Spirillum* | 1.4–1.7 | 38 | + | – | – | – | Fresh water |
| *Aquaspirillum* | 0.2–1.4 | 50–65 | – | + or – | – | + or – | Fresh water |
| *Oceanospirillum* | 0.3–1.2 | 42–48 | – | – | + | – | Marine |
| *Azospirillum* | 1.0–1.7 | 69–71 | – | + | – | + | Soil |

[a] Can fix $N_2$ only under microaerophilic conditions.

wavelength when grown on solid media. It is a common soil organism, particularly numerous in the rhizosphere of a variety of plants, including many forage grasses and agronomically significant grains including corn and sorghum. Since *Azospirillum* fixes nitrogen if the oxygen tension is low, its economic importance to agriculture may be considerable.

This group also contains two genera of parasitic bacteria with very small curved or helical cells, bearing single polar flagella (Table 17.9). The members of the genus *Campylobacter* are animal parasites or pathogens with complex nutritional requirements. They are pronouncedly microaerophilic and, like most spirilla, utilize amino acids and organic acids as energy sources, being unable to grow at the expense of carbohydrates. They can also be isolated from sediments with low oxygen tensions, and from anaerobic sediments, where they use a variety of oxidized nitrogen and sulfur compounds (including elemental sulfur but not sulfate) as terminal electron acceptor. Hydrogen and formate are the preferred electron donors under anaerobic conditions; acetate is a good carbon source.

The bdellovibrios, of which *Bdellovibrio bacteriovorus* is the type species (see Figure 17.26), are very small, Gram-negative bacteria bearing a single, polar flagellum. They attack and kill other Gram-negative bacteria, multiplying inside the periplasm. As isolated from nature they are obligate parasites; rare, host-independent variants can be selected, however, from cultures of host-grown cells. These host-independent strains can grow *in vitro* on peptone media supplemented with B vitamins.

The life cycle of *Bdellovibrio* is unique. It begins in a violent collision with a host cell: the speed of the *Bdellovibrio* cell is so great (as high as 100 cell lengths per second) that a host cell many times its size is carried a considerable distance by the momentum of impact. The parasite immediately attaches to the host cell wall by its nonflagellated end, and rotates about its long axis at speeds exceeding 100 revolutions per second. Shortly thereafter, the host cell rounds up, a pore appears in the cell wall at the site of parasite attachment, and the *Bdellovibrio* cell enters the space between the host's cell wall and cell membrane.

The entry step appears to require enzymatic action of the parasite, which liberates proteases, lipases, and a lysozyme-like muramidase. In addition, the rapid rotation of the *Bdellovibrio* cell may contribute a mechanical drilling effect to the overall penetration process.

**TABLE 17.9**

**The *Campylobacter-Bdellovibrio* Group**

| Genus | Cell Structure | Percent G + C | Ecological Properties |
|---|---|---|---|
| *Campylobacter* | Curved or helical rods, 0.3–0.8 by 0.5–5 μm, single polar flagellum | 29–35 | Parasites of mammals; anaerobic or microaerobic sediments |
| *Bdellovibrio* | Curved or helical rods, 0.3–0.4 by 0.8–1.2 μm, single sheathed polar flagellum | 42–50 | Parasites of other Gram-negative bacteria |

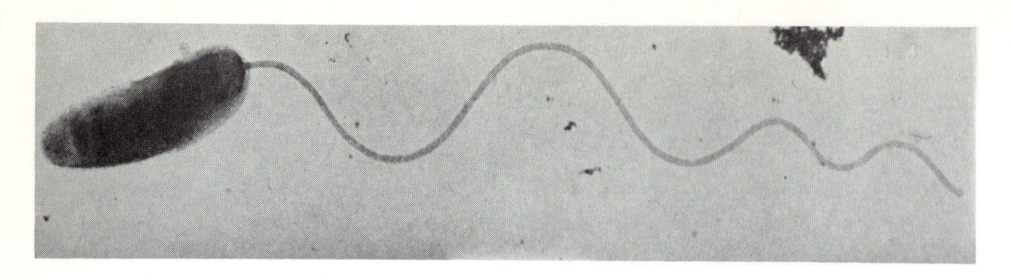

FIGURE 17.26

Electron micrograph of a cell of *Bdellovibrio*, showing the unusually thick single polar flagellum (uranyl acetate stain; ×29,000). From R. J. Seidler and M. P. Starr, "Structure of the Flagellum of *Bdellovibrio bacteriovorus*," *J. Bacteriol.* **95,** 1952 (1968).

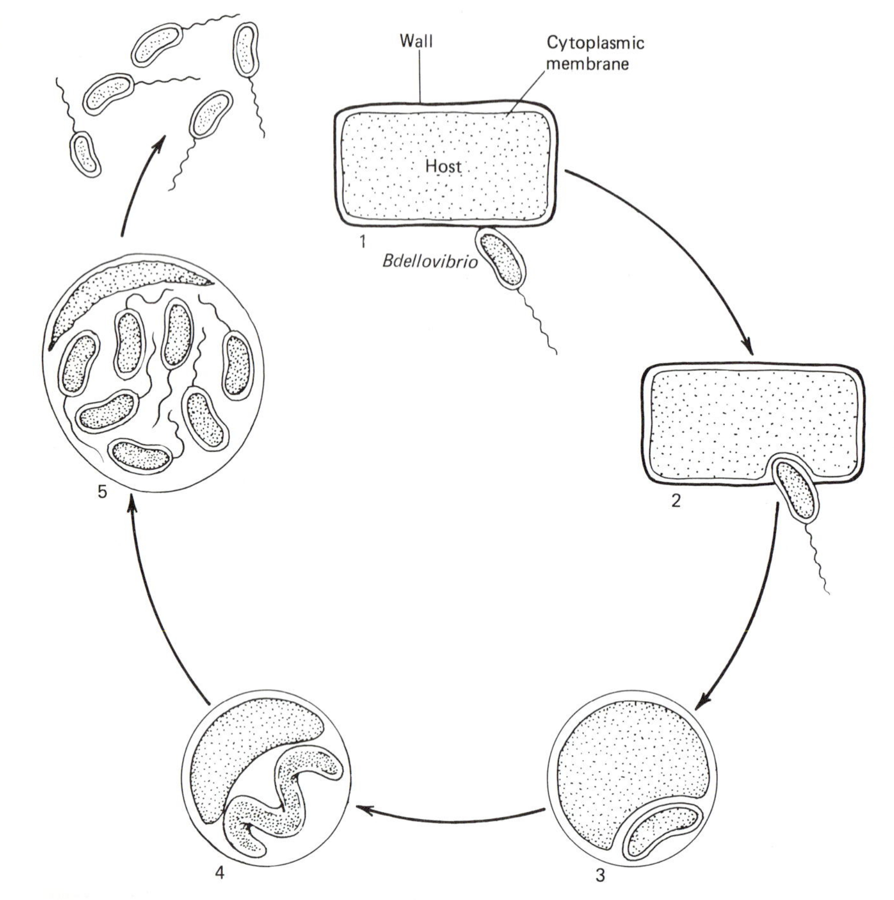


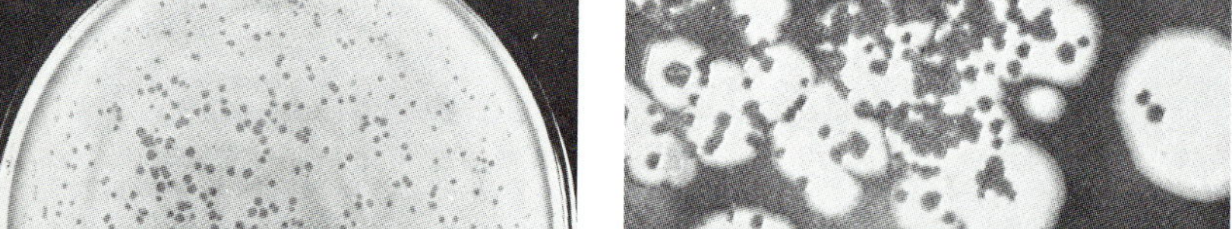


FIGURE 17.27

Life cycle of *Bdellovibrio*. Attachment to the host cell (1) is quickly followed by penetration (2). Growth of the parasite occurs without concommitant cell division, leading to a helical *Bdellovibrio* (3 and 4). Ultimately the helical organism divides into a number of short rods or vibrios (curved rods), and flagella are synthesized (5). The weakened host cell wall (evident from the swelling and loss of shape which occur early in infection) then ruptures and the *Bdellovibrio* Progeny cells are released.

(a)

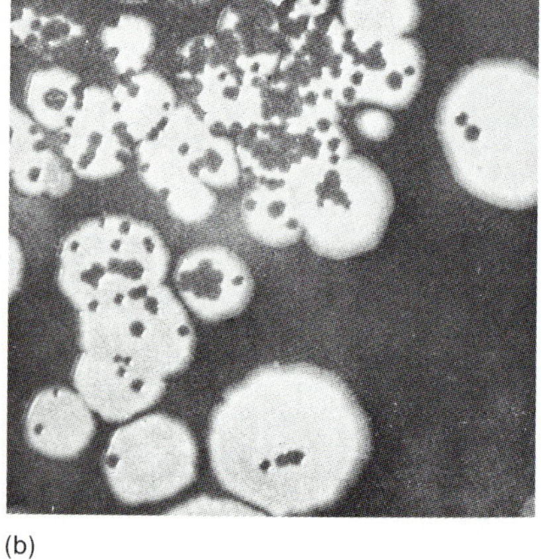

(b)

FIGURE 17.28

Macroscopically visible lysis of susceptible bacteria by *Bdellovibrio*. (a) Plaque formation in a lawn of *Pseudomonas putida* (poured plate). (b) Partial lysis of surface colonies of *Escherichia coli* on a nutrient agar plate streaked with a mixture of host and parasite. From H. Stolp and M. P. Starr, "*Bdellovibrio bacteriovorus* gen. et. sp. n., a Predatory, Ectoparasitic, and Bacteriolytic Microorganism," *Antonie van Leeuwenhoek* **29,** 217 (1963).

The penetration process is extremely rapid, being completed in a few seconds. There is a lapse of 5 to 10 minutes, however, between attachment and the initiation of penetration, during which period changes occur in the host's cell wall and perhaps in the parasite as well. The rounding up of the host cell, which may occur before penetration is apparent, suggests that the muramidase of the parasite has converted it into a spheroplast; indeed, a host cell can be lysed without penetration, if attacked by many bdellovibrios simultaneously. The rounded host cells are not osmotically sensitive, however, so the term spheroplast is not entirely accurate; they have been termed bdelloplasts.

The *Bdellovibrio* cell, which loses its flagellum during the penetration process, commences growth in the space between the cell wall and cell membrane. The membrane, although never penetrated by the parasite, becomes porous and leaks cell constituents which serve as nutrients for the parasite. The *Bdellovibrio* cell elongates into a helical filament several times its original length, depending on the size of the host cell, and then divides by multiple fission into a number of flagellated progeny cells; the entire process of multiplication takes about 4 hours. By this time, the host cell wall has undergone further decomposition, and the *Bdellovibrio* progeny are readily liberated (Figure 17.27).

On the surface of a plate covered with a growth of host cells, *Bdellovibrio* will form lytic plaques, superficially similar to those produced by a phage infection (Figure 17.28). In fact, this remarkable bacterium, discovered by H. Stolp in 1962, was isolated during a search for phages active against bacterial plant pathogens. Stolp observed that plaques developed on his plates after several days of incubation, long after any phage plaques should have appeared. Material isolated from the late-appearing plaques was found to contain a large number of rapidly moving vibrioid cells, which constituted the first isolation of *Bdellovibrio*.

By screening for organisms that form late-appearing plaques on lawns of Gram-negative bacteria, it has been possible to detect bdellovibrios in a wide variety of natural materials. They have been found in soil samples from many parts of the world, in numbers ranging from $10^2$ to $10^5$ per gram, as well as in sewage and—in much lower numbers—in pond water and seawater.

The nutrition and physiology of the bdellovibrios have been studied in axenic cultures of host-independent variants. They have been found to lack the ability to catabolize carbohydrates but to be strongly proteolytic, apparently using peptides and amino acids as carbon and energy sources. They are obligate aerobes, respiring substrates via the tricarboxylic acid cycle. The host-dependent strains derive all of their nutrients from the host rather than from the medium.

In Chapter 7 it was stated that most bacteria give a value for $Y_{ATP}$ (the yield of cell mass in grams dry weight per mole of ATP) of about 10. In contrast, bdellovibrios give $Y_{ATP}$ values of 20 to 30; this remarkable efficiency has been shown by S. Rittenberg and co-workers to be due, in part, to the ability of *Bdellovibrio* to assimilate nucleotides directly from the host, conserving the energy-rich phosphate bonds. They are also able to assimilate the host's fatty acids, incorporating some directly into lipids and converting others to their own specific types.

## THE *MORAXELLA* GROUP

The members of the *Moraxella* group are nonflagellated short rods or cocci (Figure 17.29). The DNA base compositions range from approximately 40 to 50 percent G + C.

Generic distinctions are shown in Table 17.10. An unusual property shared by the parasitic members of the group is their marked penicillin sensi-

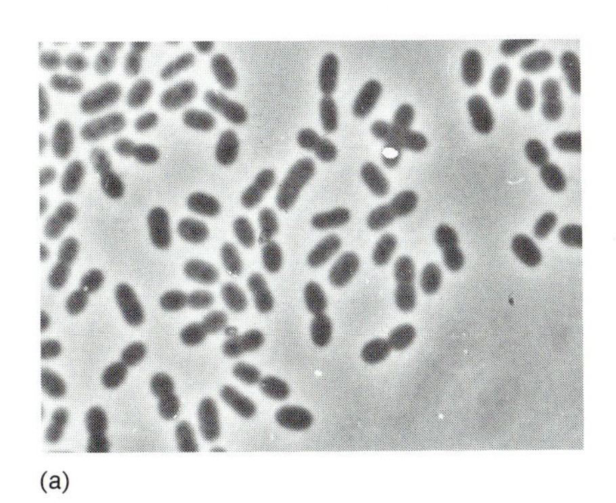

(a)

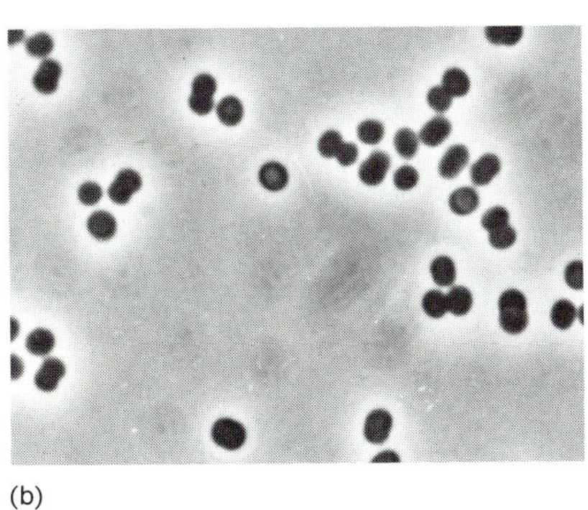

(b)

FIGURE 17.29

The *Moraxella-Neisseria* group (phase contrast, ×2,200). (a) *Moraxella osloensis*. (b) *Neisseria catarrhalis*.

**TABLE 17.10**

**The Distinguishing Properties of the Genera in the *Moraxella* Group**

| Genus | Cell Shape | Planes of Division | Percent G + C | Resistance to Penicillin, 10 Units/ml | Growth Factors Required | Oxidase Reaction |
|---|---|---|---|---|---|---|
| *Neisseria* | Cocci | 2 | 47–52 | – | + | + |
| *Branhamella* | Cocci | 2 | 40–45 | – | + | + |
| *Moraxella* | Rods | 1 | 40–46 | – | V[a] | + |
| *Acinetobacter* | Rods[b] | 1 | 39–47 | + | – | – |

[a] V: variable within group.
[b] Rod in exponential phase; becomes spherical in stationary phase.

tivity; except in strains that have acquired plasmid encoded penicillin resistance, growth is inhibited by penicillin at levels of 1 to 10 units/ml, whereas the usual inhibitory concentration for Gram-negative bacteria lies between 100 and 1,000 units/ml. The coccoid organisms of the genera *Neisseria* and *Branhamella* are parasites found on the mucous membranes of mammals; the genus *Neisseria* includes two human pathogens, responsible for gonorrhea and some cases of meningitis. As shown in Table 17.10, the two genera are very similar in phenotypic respects, but they differ slightly in their ranges of DNA base composition. Their separation is based primarily on genetic considerations: studies by DNA transformation have shown no genetic affinities between the members of the two genera, whereas *Branhamella* is genetically related to the rod-shaped organisms of the genus *Moraxella*. The rod-shaped organisms of the genera *Moraxella* and *Acinetobacter*, although similar in many phenotypic respects and in DNA base composition, give no evidence of genetic relationships, either by transformation or by nucleic acid hybridization in vitro. *Moraxella*, like the *Neisseria-Branhamella* group, consists of parasites of the mucous membranes of vertebrates, whereas the *Acinetobacter* group are free-living bacteria with an exceptionally wide natural distribution. The two genera are most easily distinguished by the oxidase test and penicillin sensitivity (Table 17.10).

The members of the *Acinetobacter* group are nutritionally versatile chemoheterotrophs. The range of substrates used as sole carbon and energy sources parallels that of the aerobic pseudomonads; and like these organisms, the *Acinetobacter* group are common in soil and water, from which they can be selected by similar enrichment procedures. The outcome of such enrichment cultures appears to be largely determined by a secondary environmental factor, aeration. In unshaken liquid enrichment cultures the aerobic pseudomonads often predominate, since as a result of their motility and aerotactic behavior, these organisms can occupy the air-water interface and develop there, excluding the acinetobacters. The same enrichment medium, receiving the same inoculum, will often yield predominantly acinetobacters if mechanically agitated to ensure full aeration; under these conditions, the selective advantage possessed by aerobic pseudomonads through their ability to occupy the well-aerated surface layer of an unshaken medium disappears.

The acinetobacters cannot use hexoses as carbon and energy sources, a characteristic which they share with the *Branhamella-Moraxella* group. Nevertheless, many *Acinetobacter* strains can produce acid from glucose and other sugars, when grown in a complex medium that contains these carbohydrates. Acid production results from the oxidation of aldose sugars to the corresponding sugars acids, as exemplified by the reaction

$$\text{glucose} + O_2 \longrightarrow \text{gluconic acid}$$

It is mediated by a single, highly nonspecific, aldose dehydrogenase, which can oxidize at least 10 different sugars, including pentoses, hexoses, and disaccharides.

## THE *LEGIONELLA* GROUP

The *Legionella* group contains aquatic rod-shaped organisms that are obligate aerobes, motile by means of one to several lateral or polar flagella, with complex growth factor requirements (typically iron salts, cysteine, and several other amino acids). They are incapable of utilizing sugars as energy source; the preferred substrates are amino acids. Their cellular fatty acids are largely *branched*, usually bearing a single methyl group on the second or third carbon from the hydrophobic end; nearly 80 percent of the fatty acids of the cell membrane may be so methylated. The G + C content is 39 to 43 percent. DNA-DNA hybridization studies indicate that the group is internally heterogeneous. They are common inhabitants of freshwater systems, particularly warm ones (their temperature optimum is about 36° C), and are now recognized to be common inhabitants of artificial warm water habitats such as evaporative cooling and hot water systems.

This group was discovered only very recently, as a consequence of an outbreak of respiratory disease among a group of American Legion members. Since its identification as a clinical entity, and the isolation of the causative agent *Legionella*, Legionnaires' disease has been recognized as a fairly common form of pneumonia in immunocompromised patients, and a frequently fatal one if untreated. It is fortunately almost completely incapable of person-to-person transfer; the primary route of infection appears to be inhalation of aerosols produced by cooling towers, hot water taps, etc.

## THE *PLANCTOMYCES* GROUP

The *Planctomyces* group contains several poorly characterized bacteria that divide by budding. The cells are typically ovoid, with a holdfast at the nonreproductive end of the cell. In *Planctomyces*

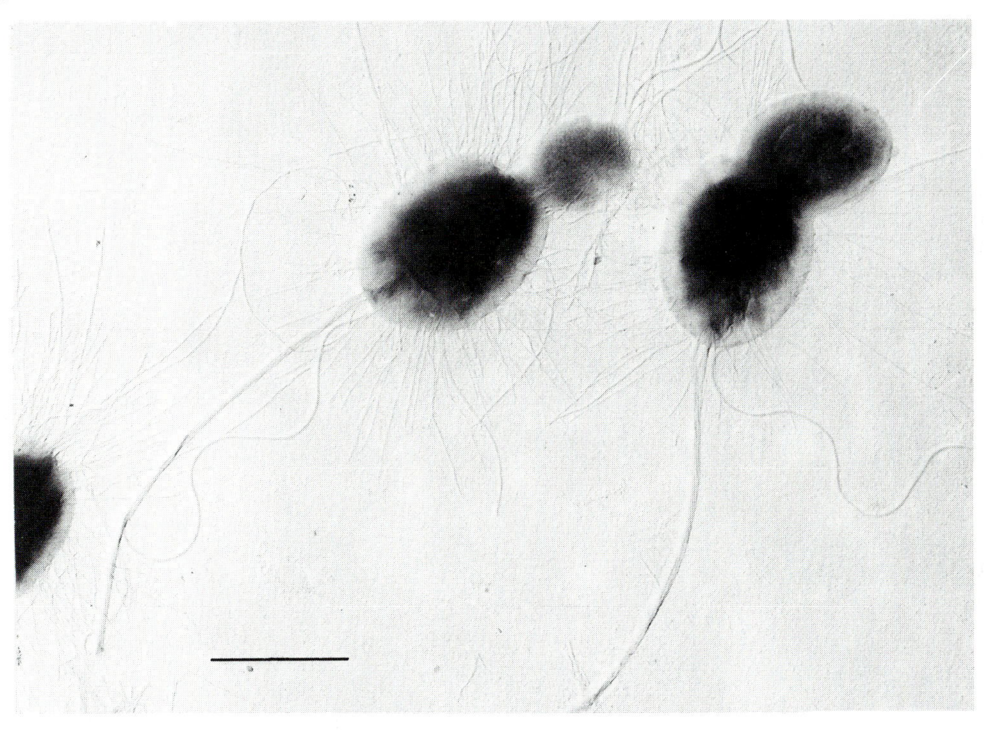

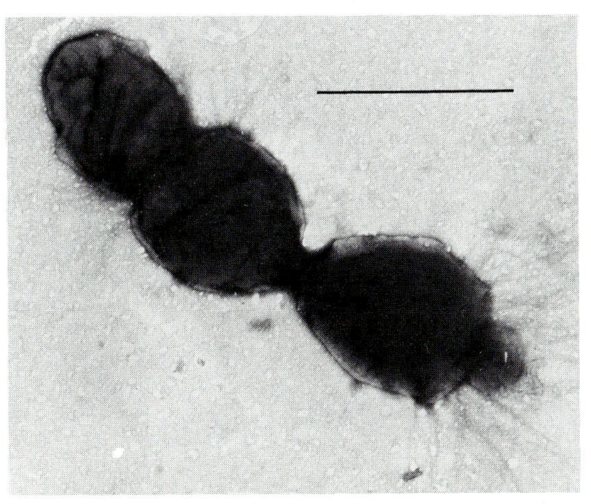



FIGURE 17.30

Electron micrographs of negatively-stained representatives of the *Planctomyces* group. (a) *Planctomyces maris*, showing two fimbriated cells each bearing a non-cellular stalk and flagellum at one pole, and a bud at the other. (b) *Pasteuria ramosa*, showing two cells attached to each other by their polar holdfasts. Both cells are in the process of budding from the pole opposite the holdfast. The bars represent 1.0 $\mu$m. Courtesy of Dr. J. T. Staley.

strains, the holdfast is at the end of a noncellular appendage termed a stalk (not to be confused with the stalks of prosthecate bacteria that are extensions of the cell), whereas in *Pasteuria* strains there is no stalk (Figure 17.30). They are obligately aerobic aquatic bacteria, and have been observed in field samples of a variety of fresh, brackish, and marine waters.

The most striking characteristic of this group is the lack of murein in the cell wall; it is the only known walled eubacterial group to lack this polymer. Instead, these organisms have a wall composed principally of a glycoprotein rich in glutamate. Analysis of their 16S rRNA suggests a distant relationship to other eubacterial groups (Chapter 13).

## FURTHER READING


### Books

THORNSBERRY, C., A. BALLOWS, J. C. FEELEY, and W. JAKUBOWSKI, (eds.) *Legionella; Proceedings of the Second Symposium.* Washington, D.C.: American Society for Microbiology, 1984.

### Reviews

KRIEG, N. R., and P. B. HYLEMON, "The Taxonomy of the Chemoheterotrophic Spirilla," *Ann. Rev. Microbiol.* **30,** 303 (1976).

MOORE, R. L., "The Biology of *Hyphomicrobium* and Other Prosthecate, Budding Bacteria," *Ann. Rev. Microbiol.* **35,** 567–594 (1981).

POINDEXTER, J. S., "The Caulobacters: Ubiquitous Unusual Bacteria," *Microbiol. Rev.* **45,** 123 (1981).

SHAPIRO, L., "Differentiation in the *Caulobacter* Cell Cycle," *Ann. Rev. Microbiol.* **30,** 377 (1976).

SMIBERT, R. M., "The Genus *Campylobacter*," *Ann. Rev. Microbiol.* **32,** 673 (1978).

VAN VEEN, W. L., E. G. MULDER, and M. H. DEINMA, "The *Sphaerotilus–Leptothrix* Group of Bacteria," *Microbiol. Rev.* **42,** 329 (1978).

WHITTENBURY, R., and C. S. DOW, "Morphogenesis and Differentiation in *Rhodomicrobium vannielii* and Other Budding and Prosthecate Bacteria," *Ann. Rev. Microbiol.* **41,** 754 (1977).

# Chapter 18
# The Gliding Eubacteria

Gliding motility is widespread among the Gram-negative eubacteria, and it also occurs in at least two Gram-positive bacteria, the phototroph *Heliobacterium* and the filamentous sulfur-reducer *Desulfonema.* A number of other gliding organisms have already been discussed in Chapter 15 (the cyanobacteria and green nonsulfur bacteria), and a few were discussed in Chapter 16 (gliding sulfur bacteria); the remaining ones, which are discussed in this chapter, are largely aerobic chemoheterotrophs. They can be subdivided into three major groups: the myxobacteria, unicellular organisms with a high G + C content in their DNA, most of which form fruiting bodies; the cytophagas, unicellular organisms with a low G + C content that do not form fruiting bodies; and the filamentous gliding bacteria.

Gliding motility requires contact with a solid substratum, which can include other microbial cells as well as inanimate objects. It is often associated with a rotary motion around the long axis of the cell. Unicellular organisms typically glide at rates of 1 to 20 $\mu$m/minute, although some cytophagas can glide as fast as 150 $\mu$m/minute. The filamentous gliders often are much faster, some gliding in excess of 600 $\mu$m/minute; in general their rate of movement is proportional to the length of their filaments. For comparison, unicellular bacteria, motile by means of flagella, typically swim at speeds in the range of 1,000 to 5,000 $\mu$m/minute.

Although a variety of mechanisms have been proposed to account for gliding motility, none explains the phenomenon completely. It now appears that different mechanisms account for gliding motility in various microbial groups. Only one of these has been established with any certainty.

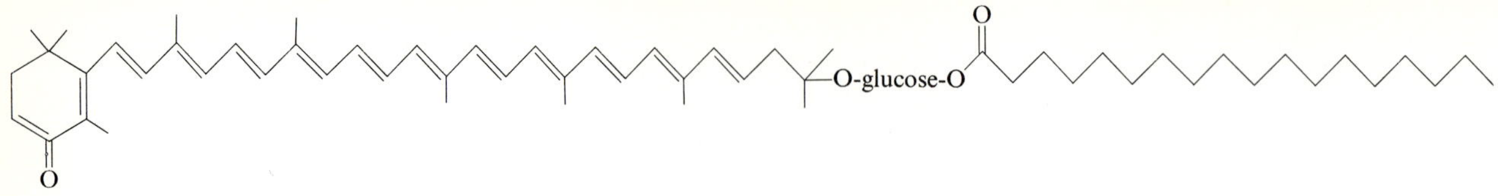

O-glucose-O

(a)

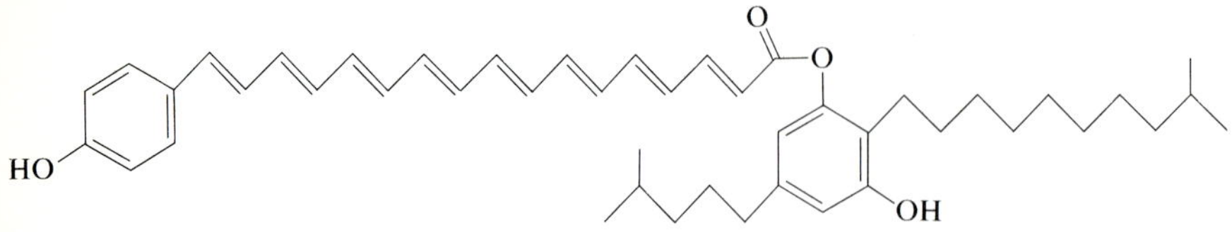

HO OH

(b)

R
NH
$^{-}O_3S$
OH

(c)

FIGURE 18.1

Unusual constituents of gliding bacteria: (a) a mycobactin; (b) a flexirubin; (c) a sulfonolipid (R may be a hydrogen or an acyl group).

In *Myxococcus*, gliding occurs in a thin film of water that covers the solid substrata. As a result of assymetric surface tension forces at the ends of individual cells, caused by localized excretion of a surfactant at the posterior (trailing end) of a cell, the cell glides forward.

The gliding bacteria contain a number of unusual chemical constituents (Figure 18.1). The myxobacteria contain high concentrations of unusual fatty acid esters of carotenoid glycosides, termed *myxobactins*, that give these organisms their typical bright color. Members of the cytophaga group also normally contain novel pigments; these are phenyl esters of phenylpolyeneoic acids termed *flexirubins*. Most members of the cytophaga group contain a unique class of sulfonolipids in their cell envelope (probably in their cell membrane or outer membrane), and some contain the unique sugar 3-*O*-methyl-D-xylose in their lipopolysaccharides.

## THE MYXOBACTERIA

The myxobacteria are differentiated from other gliding chemoheterotrophs by their high G + C content (67 to 71 percent) and unicellular morphology. The majority of them also have a special developmental cycle in which a fruiting body is formed. The vegetative cells are rods, which are usually quite flexible. In some cases this flexibility may be a simple consequence of the rather high length-to-width ratio of individual cells. However, in the case of *Myxococcus* flexibility appears to depend on the structure of the murein layer. Evidence suggests that the murein layer in these organisms is discontinuous, being organized as discrete patches of peptidoglycan, apparently held together by an uncharacterized lipoprotein material.

On solid substrates these organisms form flat, spreading colonies with irregular borders, made up of small groups of advancing cells (Figure 18.2). The migrating cells produce a tough slime layer that underlies and gives coherence to the colony. Under favorable conditions, the vegetative cells aggregate at a number of points in the inner area of the colony, and fruiting bodies then differentiate from these

FIGURE 18.2

Edge of the growth of a *Sorangium* species on agar (× 19). Courtesy of M. Dworkin and H. Reichenbach.

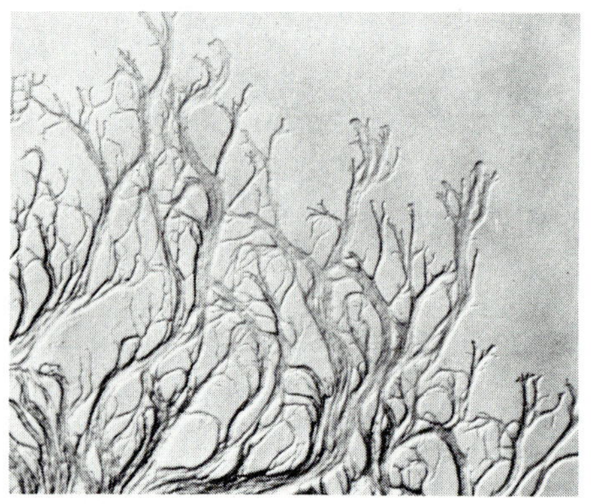

cellular aggregates. Each fruiting body is made up of slime and bacterial cells; when mature, it acquires a definite size, form, and color. As the fruiting bodies mature, the cells within them become converted to resting cells, known as *myxospores*. Upon subsequent germination, each myxospore gives rise to a vegetative rod.

The myxobacteria are soil organisms and are usually detected in nature through the development of their fruiting bodies on solid substrates: the bark of trees, decomposing plant material, and the dung of animals. They can be isolated from soil by a number of special enrichment methods; for example, placing sterilized dung particles on the surface of a large sample (50 to 200 g) of moistened soil. The myxobacteria develop on the particles, forming fruiting bodies on the dung surface after one to three weeks of incubation.

Myxobacteria are strict aerobes that fall into two nutritional subgroups: bacteriolytic and cellulolytic organisms. The majority of species are bacteriolytic, growing at the expense of bacteria and other microorganisms, living or dead. Living host cells are killed by antibiotics secreted by the myxobacteria; these substances have not yet been characterized chemically. The host cells are then lysed through the action of extracellular enzymes, which include proteases, nucleases, and lipases. Myxobacterial growth occurs at the expense of the soluble hydrolytic products, principally oligopeptides. The favorable nature of dung as a substrate is largely attributable to its high bacterial content. The bacteriolytic myxobacteria are most readily cultivated on the surface of agar, over which a heavy suspension of host cells has been spread. Growth in liquid media is often poor. The minimal nutritional requirements of these organisms are not well known, since many species will not grow readily except at the expense of microbial cells.

The most easily cultivable representatives (*Myxococcus* spp.) can grow in a defined medium containing a mixture of amino acids as source of carbon, nitrogen, and energy; carbohydrates are not utilized.

A few species, all of the genus *Polyangium*, have quite different nutrient requirements. They are active cellulose decomposers and grow in a medium with a mineral base, supplemented either with cellulose or with its hydrolytic products (the soluble sugars cellobiose and glucose).

Most myxobacteria (exceptions: some *Myxococcus* spp., the cellulose-decomposing *Polyangium* spp.) do not readily form fruiting bodies on media that support good vegetative growth. Fructification is generally favored by cultivation on media that are poor in nutrients, but the specific factors that control the process are not understood. Since the classification of these bacteria is very largely based on the structure of their fruiting bodies, the difficulty of obtaining reproducible fructification can be a serious obstacle to their identification.

The distinguishing properties of some of the principal genera are shown in Table 18.1. A primary division can be made on the basis of the structure of the vegetative cells and myxospores. In the genera listed under group I the vegetative cells have tapered ends [Figures 18.3 (a) and (b)], and the myxospores are considerably more refractile than the vegetative cells. In the genera listed under group II the vegetative cells are of uniform diameter, with rounded ends [Figure 18.3 (c)]; the myxospores usually do not differ from vegetative cells, either in form or in refractility. When growing on agar plates, the representatives of group II form colonies that erode and penetrate the substrate. Among the representatives of group I, erosion of agar by the colonies does not usually occur.

The formation of myxospores in the genus *Myxococcus* involves a striking morphogenetic change: the long, slender vegetative cells shorten and round up to produce spherical myxospores, which are much more refractile than the vegetative cells. The fruiting bodies of most species are glistening, colored droplets, uniformly filled with myxospores.

In *Cystobacter*, *Melittangium*, and *Stigmatella* the myxospores are shortened refractile rods that develop within *sporangioles*, sacks that enclose a large number of individual myxospores. The sporangioles of *Cystobacter* lie on the surface of the substrate, embedded in a mass of slime; those of *Melittangium* and *Stigmatella* develop at the apices of a colorless stalk, which consists largely of hardened slime. The stalk is unbranched and bears a single sporangiole in *Melittangium*; it is branched and bears multiple sporangioles in *Stigmatella*.

Fructification of *Polyangium* and *Chondromyces* leads to the formation of fruiting bodies analogous in form to those of *Cystobacter* and *Stigmatella*, respectively. However, the myxospores within the sporangioles are nonrefractile and cannot be distinguished microscopically from vegetative cells.

*Nannocystis* is unusual in several ways. It lacks the carotenoid glycosides characteristic of myxobacteria. Unlike other members of group II, it produces spherical, refractile myxospores, but like them, is able to erode agar; indeed, when grown

## TABLE 18.1

**Properties of Some Genera of Fruiting Myxobacteria**

| Genus | Form of Myxospore | Myxospores Refractile | Structure of Fruiting Body | |
|---|---|---|---|---|
| | | | MYXOSPORES IN SPORANGIOLES | SPORANGIOLES BORNE ON STALKS |
| Group I: Vegetative cells tapered | | | | |
| *Myxococcus* | Sphere | + | − | |
| *Cystobacter* | Rod | + | + | − |
| *Melittangium* | Rod | + | + | +, simple |
| *Stigmatella* | Rod | + | + | +, branched |
| Group II: Vegetative cells of uniform width | | | | |
| *Nannocystis* | Sphere | + | + | − |
| *Polyangium* | Rod | − | + | − |
| *Chondromyces* | Rod | − | + | +, branched |

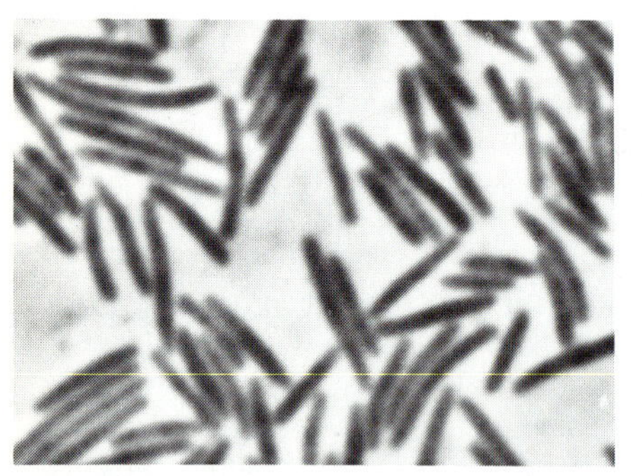
(a)

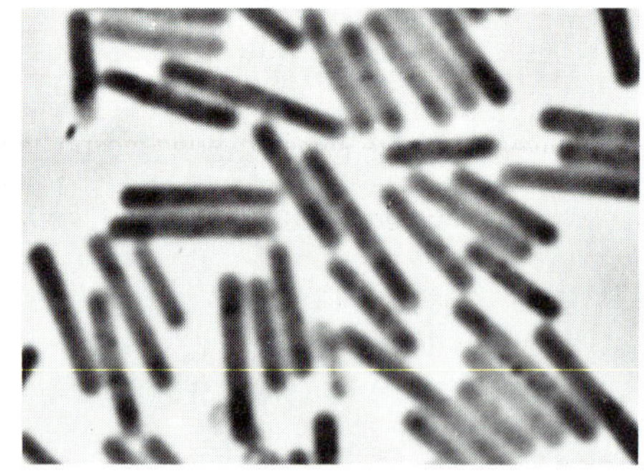
(c)

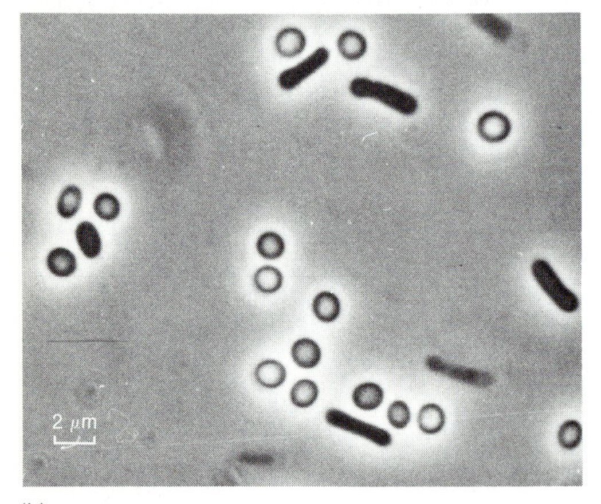

(b)

FIGURE 18.3

Vegetative cells of myxobacteria: (a) *Myxococcus xanthus* (phase contrast, ×2,050); (b) microcysts and shortened rods from a fruiting body of *Myxococcus fulvus* (phase contrast, ×2,210); (c) *Chondromyces crocatus* (phase contrast, ×2,050). (a) and (c) Courtesy of Dr. H. McCurdy; (b) courtesy of M. Dworkin and H. Reichenbach.

under conditions of low nutrient concentrations, agar erosion becomes extensive, the agar layer becoming honeycombed with tunnels. *Nannocystis* is one of the most common myxobacteria in soil. Like many actinomycetes, it produces large amounts of the aromatic compound *geosmin* (see Figure 24.16), which has a pronounced earthy odor, and is partially responsible for the characteristic smell of freshly turned soil.

In *Stigmatella* and *Chondromyces* the mature fruiting body has a treelike form, numerous sporangioles being formed at the tips of the much-branched stalk. Successive stages in the differentiation of such a complex fruiting body are shown in Figures 18.4 and 18.5. The range of fruiting body structure in the genera described is illustrated in Figure 18.6.

The properties of myxospores have been examined primarily in the genus *Myxococcus*. Although they are more heat-resistant than vegetative cells, the difference is not great. However, they are

| Colonies Erode Agar Surface | Nutritional Character | |
|---|---|---|
| | BACTERIOLYTIC | CELLULOLYTIC |
| Variable | + | − |
| − | + | − |
| − | + | − |
| − | + | − |
| | | |
| + | + | − |
| + | + (some) | + (some) |
| + | + | − |

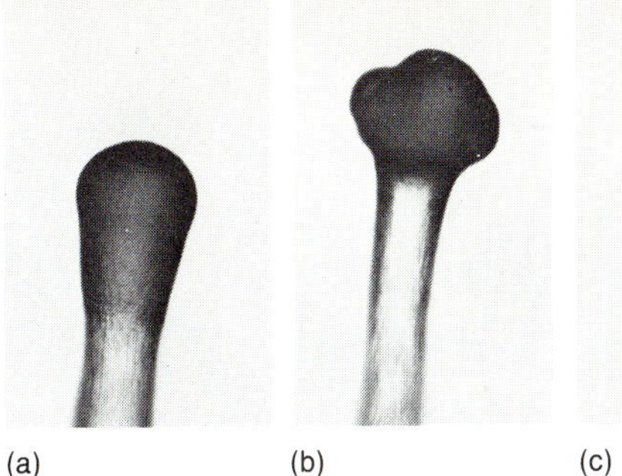

(a) (b)

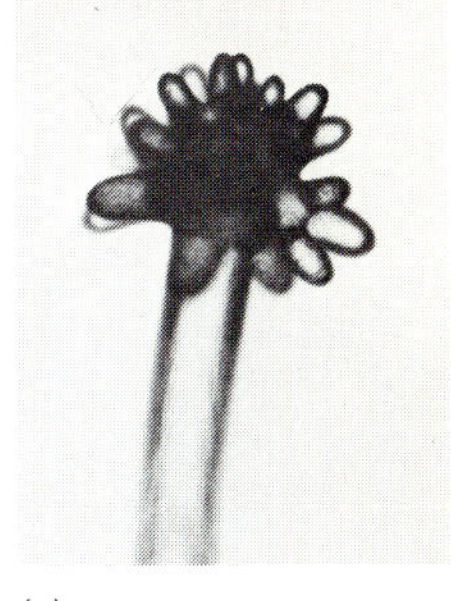

(c)

(d)

FIGURE 18.4

Successive stages (a)–(d) in the development of the apical region of a fruiting body of *Chondromyces apiculatus*, taken over a period of four hours at 27° C. Frames from film E 779, *Publ. wiss. Filmen, Sekt. biologie*, **7,** 245 (1974), prepared by the Institut für den wissenschaftlichen Film, Göttingen. Courtesy of Dr. Hans Reichenbach.

FIGURE 18.5 (below)

Scanning electron micrographs of the process of fruiting body formation in *Chondromyces.* (a) Aggregation of vegetative cells. Note slime covering most of the colony. (b and c) Early stages of fructification. (d) Composite picture showing progressively later stages, with a mature fruiting body to the right (approximately 10–15 ×). Courtesy of P. Grilione and J. Pangborn.

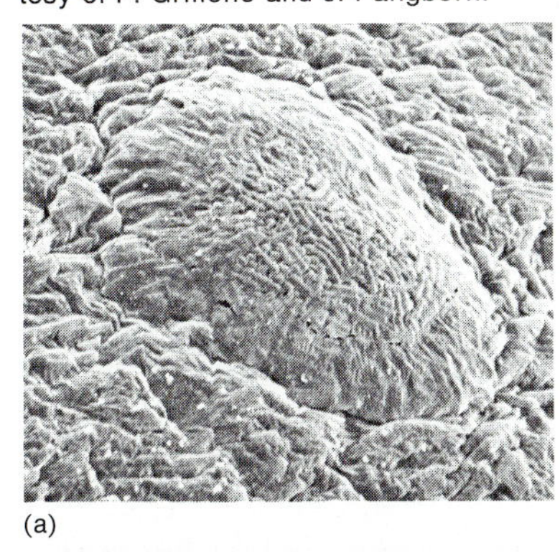

(a)

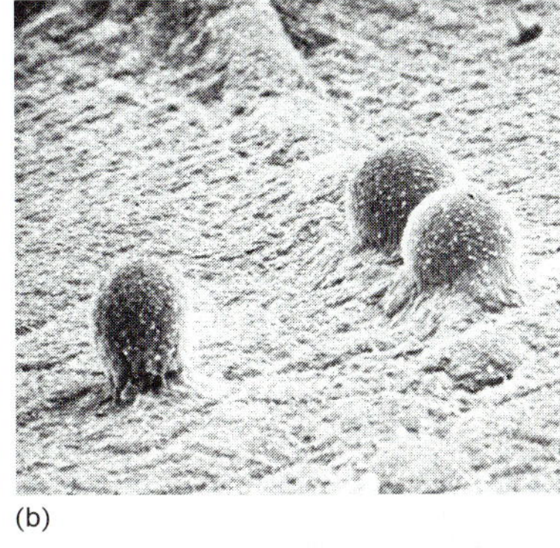

(b)

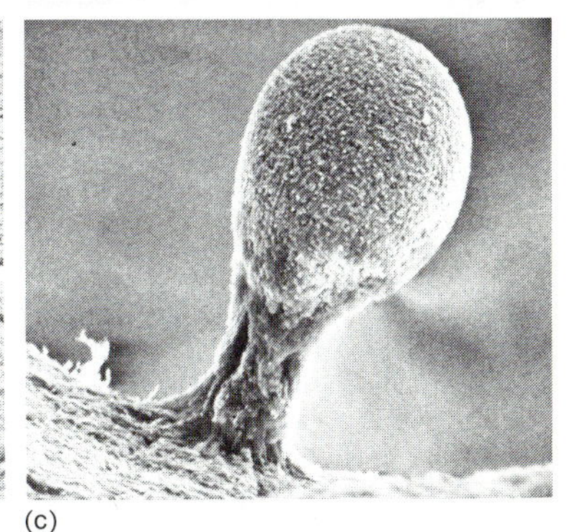

(c)

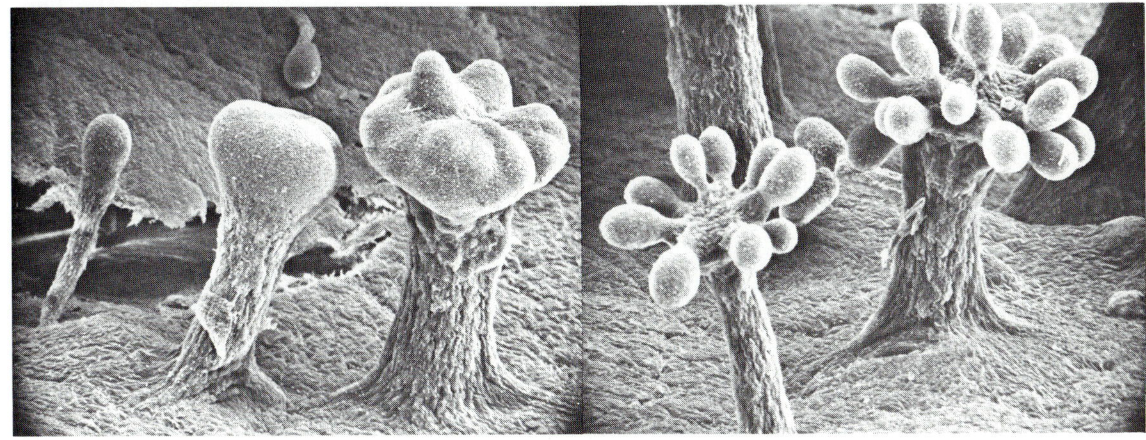

(d)

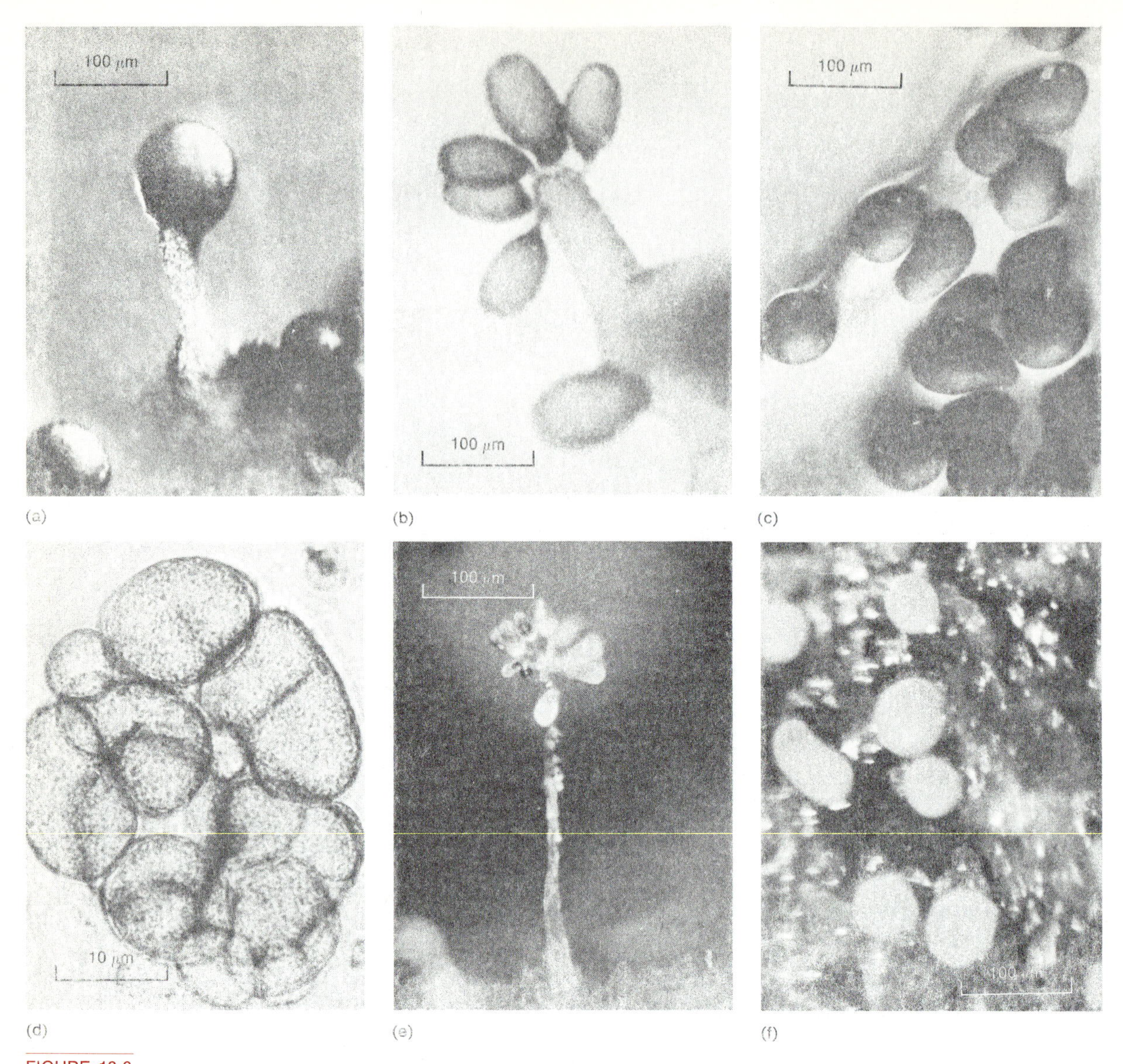


FIGURE 18.6

Fruiting bodies of myxobacteria: (a) *Melittangium lichenicolum* (×232); (b) *Stigmatella aurantiaca* (×318); (c) *Cystobacter fuscus* (×106); (d) *Polyangium* sp. (×573); (e) *Chondromyces pediculatus* (×94); (f) *Myxococcus xanthus* (×32). (a)–(d) Courtesy of M. Dworkin and H. Reichenbach; (e) and (f) courtesy of H. McCurdy.

much more resistant to desiccation and ultraviolet light, and can survive for months or years in the dry state. The formation of these resting structures in fruiting bodies that are raised off the surface of the substrate no doubt facilitates their physical dispersion. In the genus *Myxococcus* the myxospore is both the resting structure and the unit of dispersion. However, in sporangiole-forming genera the unit of dispersion is not the individual myxospore, but the sporangiole, which contains many myxospores. This fact is most evident if one compares the respective modes of germination. The fruiting bodies of most *Myxococcus* spp. are deliquescent, and each of the myxospores liberated by their breakdown can germinate under favorable conditions, giving rise through subsequent growth to a vegetative colony. In sporangiole-forming genera, germination of myxospores is accompanied by a rupture of the wall of the enclosing sporangiole, with the release of hundreds of vegetative cells (Figure 18.7). The cells remain in association and give rise to a single vegetative colony. Sporangiole germination thus permits the rapid buildup of a vegetative population, prior to the initiation of cell division. Such large initial populations probably facilitate production of concentrations of the extracellular hydrolytic enzymes sufficient to initiate rapid digestion of their macromolecular growth substrates (cellulose or microbial cells).

Another adaptation that serves to maintain high population densities is a regulatory system by which motility appears to depend on cell-to-cell contact. Cells glide when they are in direct contact with other cells, as large aggregates or "rafts." They can also move singly. However, this cannot be sustained for long; periodic contact with other cells seems to be necessary for continued movement. Thus individual cells can move out from the periphery of a colony, but if they go too far they become immotile until the advancing edge of the growing colony reaches them.

In *Myxococcus*, myxospore formation can be induced experimentally in vegetative cells, without the usual preliminary events of aggregation and fructification. The addition of high concentrations of certain primary or secondary alcohols such as butanol or glycerol to a suspension of vegetative cells causes a massive conversion to myxospores within two hours; starvation for specific amino acids (methionine, or phenylalanine and tyrosine) has the same effect.

## Nonfruiting Myxobacteria

The death of natural populations of green algae and cyanobacteria is often caused by myxobacteria. Many of these algicidal myxobacteria have not been observed to form fruiting bodies, so that their taxonomic position is uncertain; however, the DNA base composition of these organisms is similar to that of fruiting myxobacteria, lying in the range of 69 to 71 percent G + C. They are assigned to the genus *Lysobacter*. In contrast to bacteriolytic fruiting myxobacteria, the lysobacters have simple nutrient requirements; they grow well in liquid media of defined composition, and can use glucose or other carbohydrates as sole sources of carbon and nitrogen.

Some lysobacters produce extracellular enzymes that destroy the peptidoglycan layer of their prey. The lytic enzymes include a protease of very low molecular weight (8,000 daltons) and broad substrate specificity, which hydrolyzes peptide bonds of peptidoglycans. Other lysobacters kill only by cell-to-cell contact (Figure 18.8); the enzymatic mechanism of the attack is not known.

FIGURE 18.7

Sporangiole germination in *Chondromyces apiculatus*. (a) Mature fruiting body, bearing sporangioles (× 104). (b) Germination of a detached sporangiole on the surface of an agar plate, showing the emergence of a large population of vegetative cells, derived from the enclosed myxospores (phase contrast, × 120). (c) The empty wall of a germinated sporangiole (phase contrast, × 485). Courtesy of Dr. Hans Reichenbach.

(a)

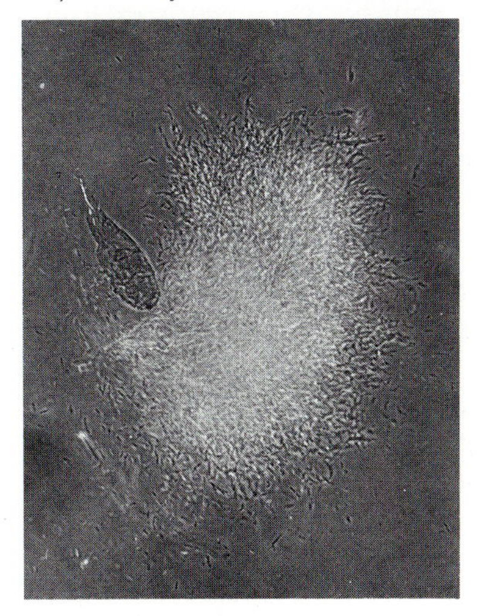

(b)

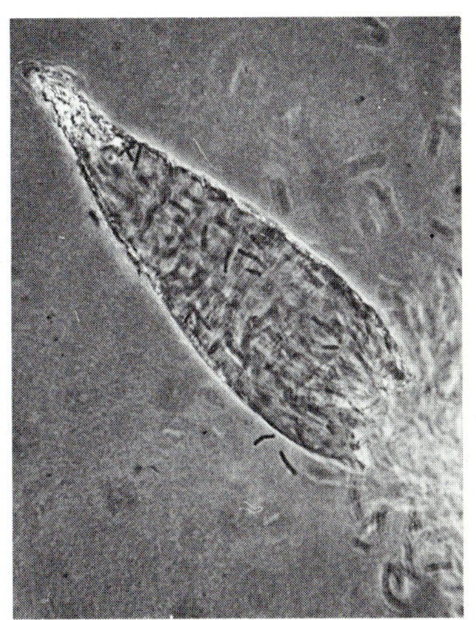

(c)

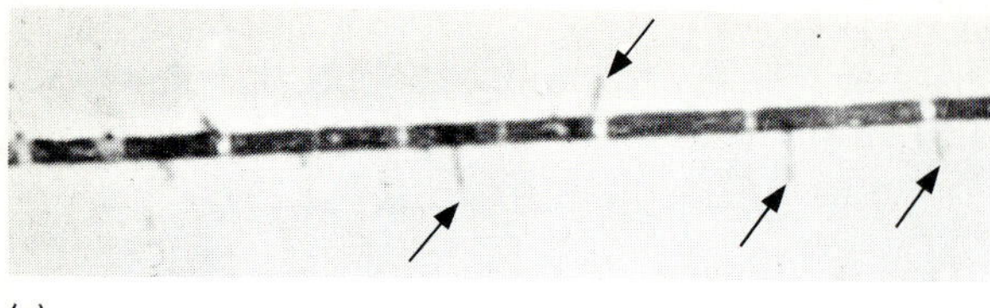

(a)

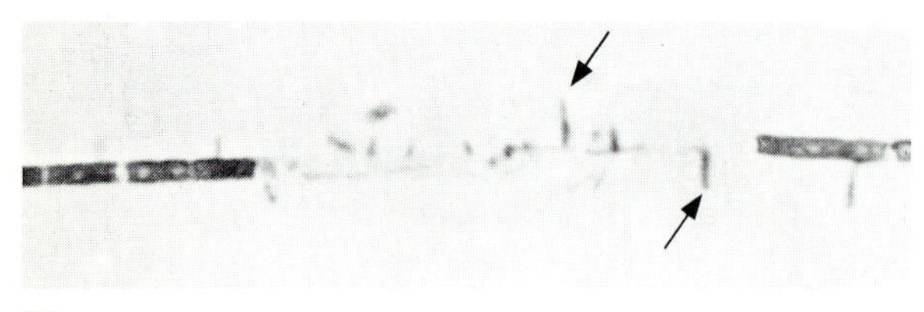

(b)

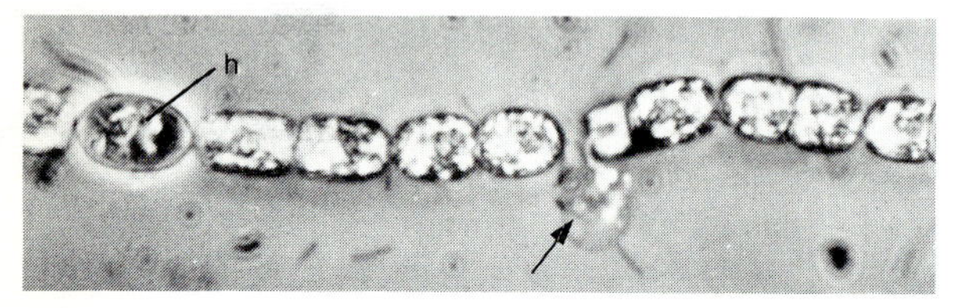


(c)

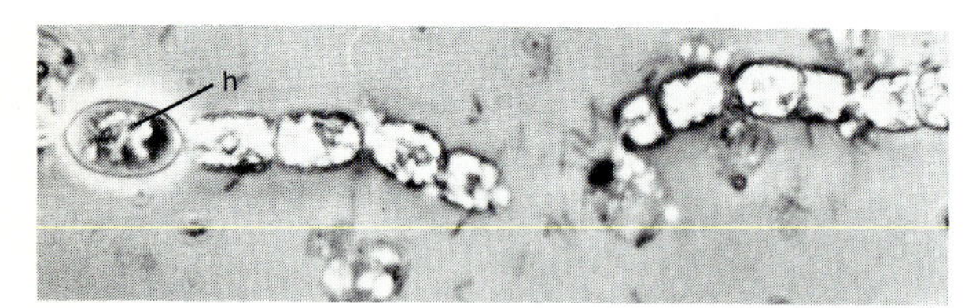


(d)

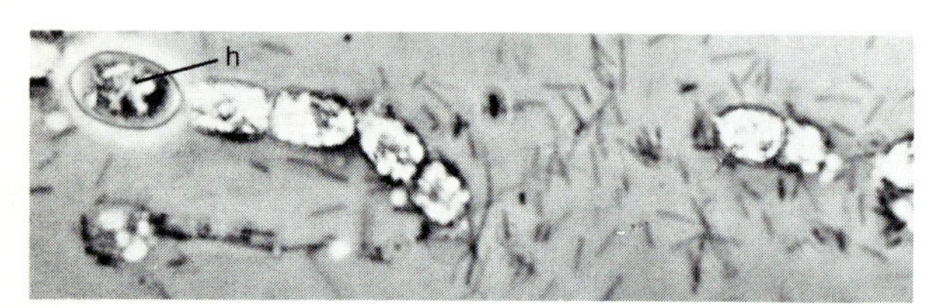


(e)

FIGURE 18.8

The attack by a myxobacterium on filaments of two cyanobacteria. (a) Filament of *Oscillatoria redekei*, showing polar attachment of myxobacterial cells (arrows) at various points along the filament. (b) Filament of *O. redekei* after lysis of some of the component cells; the myxobacterium is still attached to the lysed cells (arrows). (c)–(e) Time lapse sequence, showing lysis by the myxobacterium of a filament of *Anabaena flosaquae*. This cyanobacterium contains gas vacuoles, which accounts for the phase-bright appearance of the vegetative cells. The filament contains a heterocyst (h), which is not susceptible to attack. In (c) the arrow points to a lysed vegetative cell, the destruction of which has caused a break in the filament. All × 1,260. From M. J. Daft, and W. D. P. Stewart, "Light and Electron Microscope Observations on Algal Lysis by Bacterium CP-1," *New Phytol.* **72,** 799 (1973).

## THE *CYTOPHAGA* GROUP

The gliding bacteria of the *Cytophaga* group do not form fruiting bodies, and they differ markedly from the myxobacteria in their DNA base composition, which lies in the range of 30 to 50 percent G + C. Some of these bacteria produce chains of cells, 100 $\mu$m or more in length, a character that does not occur in myxobacteria. However, most cytophagas grow as single, slender rods, and cannot be distinguished by vegetative cell structure from myxobacteria. Resting cells are formed only in the genus *Sporocytophaga*, which produces spherical, refractile microcysts, similar in structure and development to the myxospores of *Myxococcus*. The base composition of the DNA is therefore a character of primary importance to distinguish the members of the *Cytophaga* group from the myxobacteria, and in particular from *Lysobacter*.

The principal genera of the *Cytophaga* group (Table 18.2) are distinguished primarily by their nutritional properties. *Cytophaga* and *Sporocytophaga* spp. can hydrolyze and grow at the expense of complex polysaccharides, whereas *Flexibacter* spp. are not polysaccharide decomposers. Most of these bacteria are strict aerobes; a few *Cytophaga* and *Flexibacter* spp. are facultative anaerobes and can ferment carbohydrates. The major fermentation products are fatty acids and succinic acid; carbon dioxide is required in substrate amounts for fermentative growth. The only strictly fermentative member of this group, *Sphaerocytophaga*, is an $O_2$-tolerant organism that also requires high $CO_2$ tensions for growth.

The most active aerobic cellulose-decomposing bacteria in soil are certain species of *Cytophaga* and *Sporocytophaga*. They can be readily enriched from soil in a medium with a mineral base containing filter paper as the sole source of carbon and energy. Upon initial isolation, these bacteria are unable to grow at the expense of any other organic substrate than cellulose; by subsequent selection, mutants able to grow with the soluble sugars, glucose and cellobiose, can be obtained. The cellulolytic ability of these organisms is remarkable; when streaked on a sheet of filter paper placed on the surface of a mineral agar plate, they completely destroy its fiber structure, the attacked areas being converted into slimy colored patches filled with bacterial cells (Figure 18.9). Another distinctive property of the cellulose-decomposing soil cytophagas is the necessity for direct contact with the cellulosic substrate; this behavior suggests that the primary attack on cellulose is mediated by an exoenzyme which is nondiffusible, remaining bound to

**TABLE 18.2**

**Principal Genera of the *Cytophaga* Group**

| | Vegetative Structure | Formation of Microcysts | Percent G + C | Growth at Expense of Cellulose, Chitin, or Agar | Relations to Oxygen |
|---|---|---|---|---|---|
| *Cytophaga* | Single rods | − | 33–42 | + | Aerobic |
| *Sporocytophaga* | Single rods | + | 36 | + | Aerobic |
| *Flexibacter* | Rods, single or in chains | − | 31–43 | − | Aerobic |
| *Sphaerocytophaga* | Single rods | − | 33–41 | − | Anaerobic |

the cell surface. As a consequence, the cells in a culture containing cellulose adhere closely to the cellulose fibers, often in a very regular alignment, the rod-shaped cells being oriented parallel to the polysaccharide fibrils (Figure 18.10).

Other species of *Cytophaga* grow at the expense of the polysaccharides chitin and agar (Figure 18.11). Chitin is a polymer of N-acetylglucosamine, and is common in arthropod exoskeletons and in the cell walls of some fungi; agar is the commercial name of a complex mixture of polygalactans that serve as structural polymers in certain marine algae and that do not occur in terrestrial environments. The agarolytic cytophagas are, in consequence, mostly organisms of marine origin, whereas chitinolytic members of the group occur both in soil and in sea water.

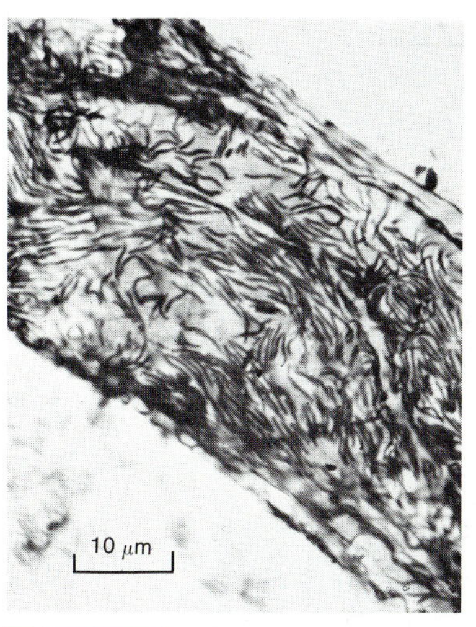


FIGURE 18.10

Cellulose fiber heavily attacked by *Cytophaga* (stained preparation). Note the characteristic regular arrangement of the cells. From S. Winogradsky, *Microbiologie du Sol*. Paris: Masson, 1949. Reprinted with permission of M. Manigault and the publisher.

FIGURE 18.9

Agar plate covered with a layer of filter paper and streaked with a culture of *Cytophaga*. Note that the filter paper has been completely dissolved where growth has occurred.

The hydrolysis of chitin and agar is usually mediated by inducible, extracellular enzymes, and hence does not require direct contact with the substrate. The chitinolytic and agarolytic cytophagas are also much less specialized nutritionally than the cellulose-decomposing species. They can all use a wide range of soluble sugars as carbon and energy sources, and most can grow in complex nitrogenous media (e.g., peptone) in the absence of carbohydrate.

The soil and freshwater cytophagas make flexirubins as their principal pigment. Marine strains, although unable to make this pigment, are nevertheless colored as a consequence of their synthesis of carotenoids.

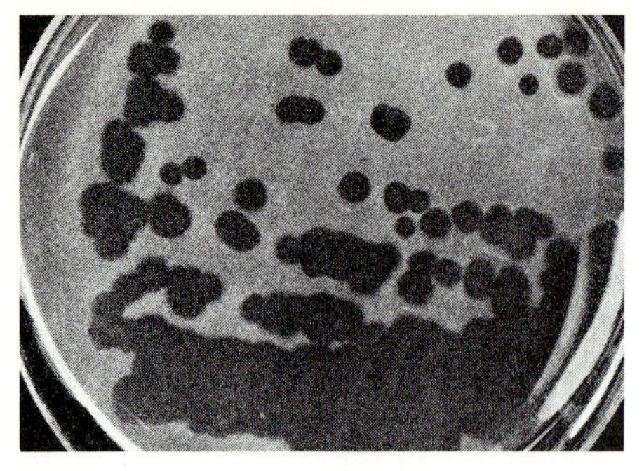

(a)

(b)

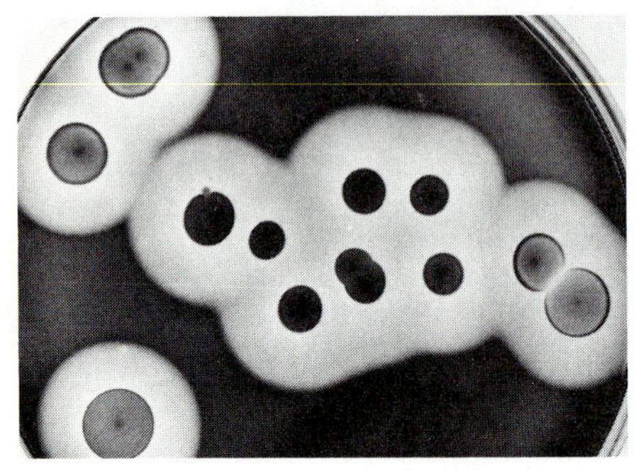

(c)

FIGURE 18.11

The decomposition of chitin and agar by members of the cytophaga group. (a) Plate culture of *Cytophaga johnsonae*, growing on an agar medium containing a suspension of the insoluble polysaccharide, chitin. The chitin has been decomposed by extracellular enzymes in the clear areas beneath the spreading, translucent colonies. (b) and (c) Two photographs of a plate culture of *Cytophaga fermentans*, growing on a complex medium containing 1 percent agar. (b) is photographed by reflected light, to reveal the depressions resulting from agar decomposition around the colonies. In (c) the plate has been flooded with an $I_2$-KI solution, which is decolorized in the areas of agar decomposition, and thus reveals the extent of the diffusion zones of agarase around the colonies. (a) Courtesy of Dr. H. Veldkamp; (b) and (c) from H. Veldkamp, "A Study of Two Marine, Agar-Decomposing, Facultatively Anaerobic Myxobacteria," *J. Gen. Microbiol.* **26,** 331 (1961).

Most cytophagas are nonpathogenic. However, *Cytophaga columnaris* is an important pathogen of fish. Infections caused by this organism can reach epidemic proportions in fish hatcheries, causing massive mortalities in warmer months of the year. Certain obligately anaerobic cytophagas, named *Sphaerocytophaga*, occur in enormous numbers in the human gingival crevices, where they possibly contribute to periodontal disease.

## FILAMENTOUS, GLIDING CHEMOHETEROTROPHS

In addition to the *Cytophaga* group, the gliding bacteria with DNA of low G + C content include a number of filamentous chemoheterotrophs. The principal genera are listed in Table 18.3.

*Saprospira* and *Vitreoscilla* are organisms that develop as flexible, gliding filaments as much as 500 $\mu$m long, made up of cells 2 to 5 $\mu$m in length. The filaments of *Saprospira* are helical (Figure 18.12), those of *Vitreoscilla* straight (Figure 18.13). Both groups occur largely in aquatic environments. The genus *Simonsiella* is distinguished by the formation of flattened, ribbon-shaped filaments (Figure 18.14), which are motile only when the broad surface of the filament is in contact with the substrate. These bacteria are aerobic members of the microflora of the oral cavity of humans and other animals. Their nutritional requirements are not precisely known; they develop best in complex media, supplemented with serum or blood.

FIGURE 18.12

Phase contrast photomicrographs of *Saprospira* (×451): (a) *S. albida;* (b) *S. grandis*. Courtesy of Ralph Lewin.

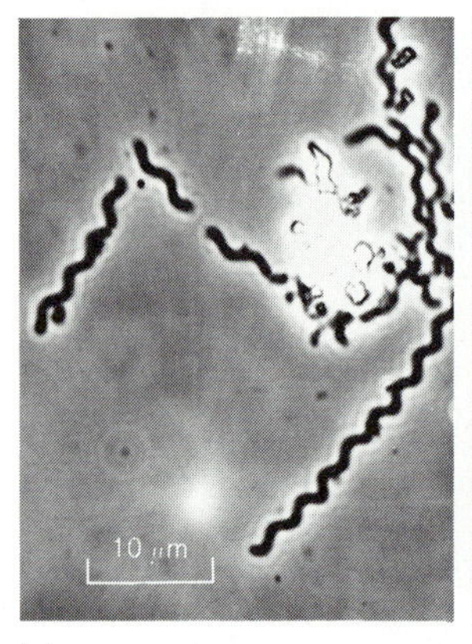


(a)

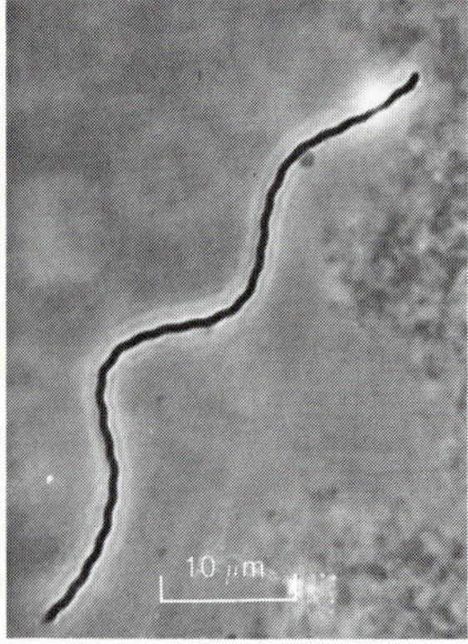


(b)

**TABLE 18.3**

**Major Genera of Filamentous, Gliding Chemoheterotrophs**

| Genus | Shape of Filament | Motility of Filament | Presence of Holdfast | Mode of Reproduction | Percent G + C |
|---|---|---|---|---|---|
| *Saprospira* | Helical cylinder | + | − | Fragmentation of filament | 44–46 |
| *Vitreoscilla* | Straight cylinder | + | − | Fragmentation of filament | 44–45 |
| *Simonsiella* | Flattened ribbon of cells | + | − | Fragmentation of filament | 41–55 |
| *Leucothrix* | Straight cylinder | − | + | Release of single gliding cells from filament apex | 46–51 |

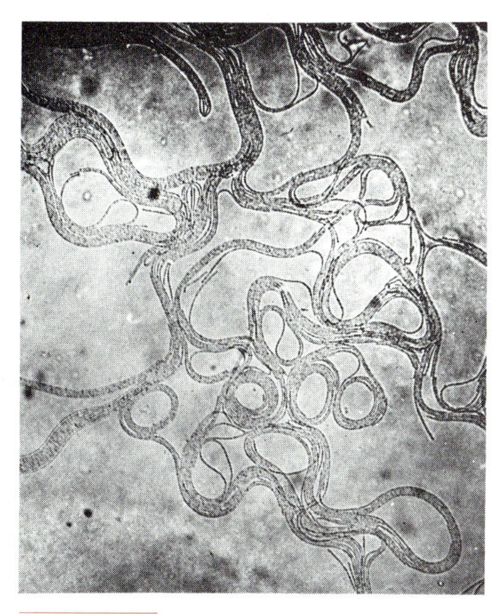

FIGURE 18.13
*Vitreoscilla* filaments growing on the surface of an agar plate. Courtesy of E. G. Pringsheim.

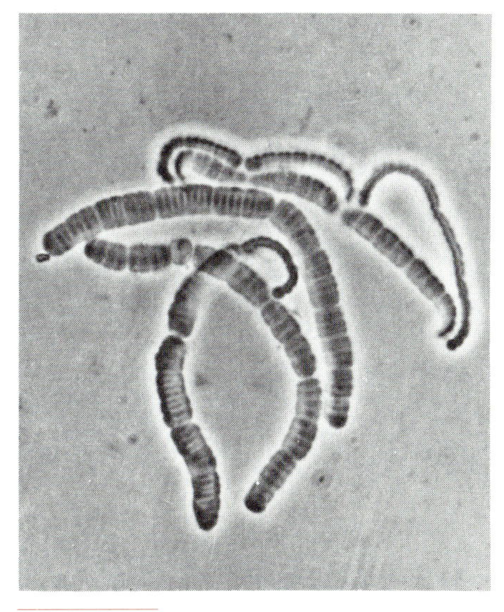

FIGURE 18.14
*Simonsiella*, a gliding bacterium which forms ribbon-shaped filaments of flattened cells (phase contrast, ×440). Some of the filaments are viewed on edge, and they appear much thinner than the filaments that lie flat. Courtesy of Mrs. P. D. M. Glaister.

The most highly differentiated of the gliding chemoheterotrophs is *Leucothrix*, a marine organism that grows as an epiphyte on seaweeds, and is also found in decomposing algal material. The very long filaments are immotile and are attached to substrates by an inconspicuous basal holdfast. Reproduction occurs not by random fragmentation of the filament, as in *Saprospira*, *Vitreoscilla*, and *Simonsiella*, but by the breaking off of ovoid cells, singly or in short chains, from the apical end of the filament (Figure 18.15). These reproductive cells (sometimes termed *gonidia*) are capable of gliding movement. When released in large numbers, they aggregate to form rosettes, held together by the holdfasts that are located at one pole of the cell. Subsequent outgrowth of the cells in such rosettes gives rise to colonies made up of numerous radiating filaments, all attached to a central mass of holdfast material. *Leucothrix* can grow in a seawater medium containing a variety of simple organic compounds (principally organic acids) as sole sources of carbon and energy.

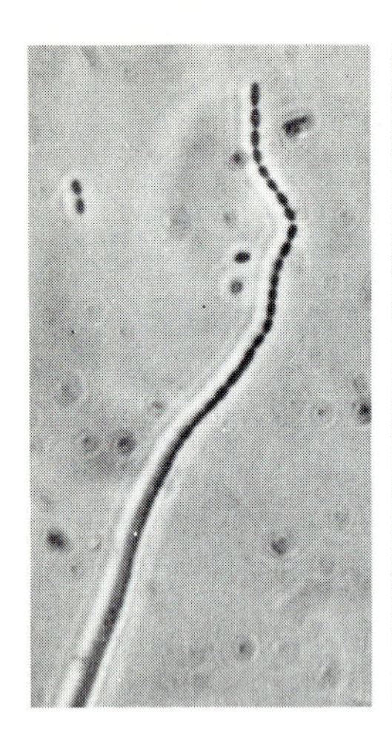

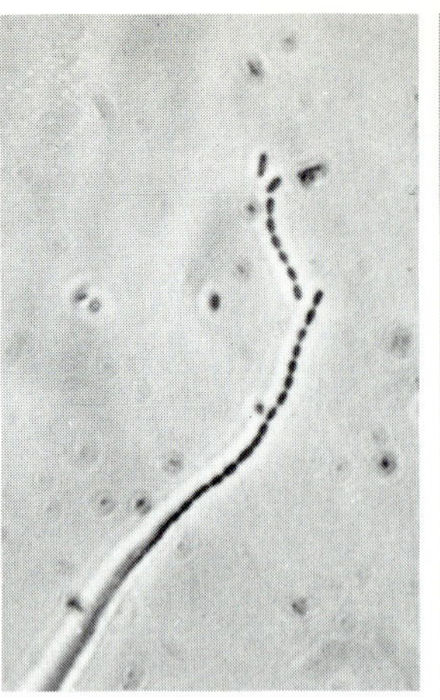

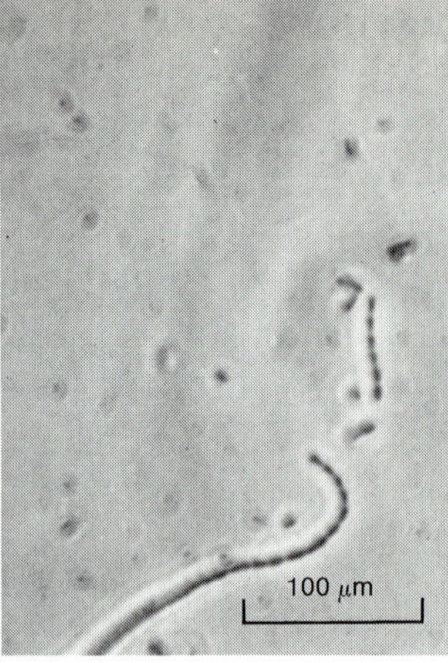


FIGURE 18.15

Successive pictures of a *Leucothrix* filament showing liberation of gonidia (phase contrast, ×309). From Ruth Harold and R. Y. Stanier, "The Genera *Leucothrix* and *Thiothrix*," *Bacteriol. Rev.* **19,** 49 (1955).

## FURTHER READING


### Books

ROSENBERG, E., ed., *Myxobacteria: Development and Cell Interactions.* New York: Springer-Verlag (1984).

### Reviews

KAISER, D., C. MANIOL, and M. DWORKIN, "Myxobacteria: Cell Interactions, Genetics, and Development," *Ann. Rev. Microbiol.* **33,** 595–639 (1979).

REICHENBACH, H., "Taxonomy of the Gliding Bacteria," *Ann. Rev. Microbiol.* **35,** 339–364 (1981).

Chapter 19

# The Enteric Group and Related Eubacteria

The organisms treated in this chapter constitute one of the largest well-defined groups among the Gram-negative, nonphotosynthetic eubacteria. They have small, rod-shaped cells, either straight or curved, not exceeding 1.5 $\mu$m in width. Some are permanently immotile; motile representatives include organisms with peritrichous flagella, with polar flagella, and with "mixed" flagellation (see p. 167). They can be distinguished from all other Gram-negative eubacteria of similar structure by the property of *facultative anaerobiosis*. Under anaerobic conditions, energy is provided by fermentation of carbohydrates; under aerobic conditions, a wide range of organic compounds can serve as substrates for respiration.

The classical representative is *Escherichia coli*, one of the characteristic members of the normal intestinal flora of mammals, and an important pathogen, causing intestinal and urinary tract infections. Closely related to this organism are the other so-called "coliform" bacteria (genera *Salmonella* and *Shigella*). They are pathogens, responsible for such intestinal infections as bacterial dysentery, typhoid fever, and some bacterial food poisonings.

Clearly related to these coliform bacteria, but of different ecology, are the genera *Enterobacter*, *Serratia*, and *Proteus*, which occur primarily in soil and water, and the plant pathogens of the genus *Erwinia*. Together with the coliform bacteria, these genera constitute the enteric group as classically defined.

In recent years it has been recognized that certain bacteria pathogenic for animals which were formerly placed in the ill-defined genus *Pasteurella* are in fact members of the enteric group. These organisms are now classified in the genus *Yersinia*. They include *Yersinia pestis*, the agent of bubonic plague, an infection markedly different in both mode of transmission and symptomatology from the enteric infections.

The bacteria so far discussed are either immotile or peritrichously flagellated. The primary importance that was for so long accorded to the mode of flagellar insertion as a taxonomic character impeded the recognition that some polarly flagellated bacteria are also allied to the enteric group, and most appropriately treated as members of it. These are all aquatic bacteria, which occur either in freshwater (*Vibrio*, *Aeromonas*) or marine environments (*Vibrio*, *Photobacterium*). Many of the marine forms show mixed flagellation. Some are animal pathogens; these include two species that cause intestinal diseases (*Vibrio cholerae*, *V. parahemolyticus*).

As yet, there is no generally accepted collective name for this entire assemblage. It will be termed here the *enteric group*, but it must be emphasized that this designation includes genera (the polarly flagellated group) that fall outside the traditional confines of the enterobacteria.

## COMMON PROPERTIES OF THE ENTERIC GROUP

Most members of the enteric group can use a considerable number of simple organic compounds as substrates for respiratory metabolism: organic acids, amino acids, and carbohydrates are universally utilized. Under aerobic conditions, all these bacteria grow well in conventional complex bacteriological media, the nitrogenous constituents of which (amino acids and peptides) provide oxidizable substrates. Under anaerobic conditions, however, growth is usually dependent on the provision of a fermentable carbohydrate, although some species are capable of nitrate or fumarate respiration (Chapter 20). Some monosaccharides, disaccharides, and polyalcohols are fermented by all members of the enteric group. The utilization of polysaccharides is less common; however, pectin is attacked by many of the plant pathogens (*Erwinia*), and chitin and alginic acid by many of the marine species.

Although it is customary to grow enteric bacteria on complex media, the minimal nutritional requirements of these organisms are usually simple. In many genera no growth factors are required (e.g., *Escherichia*, *Enterobacter*, *Serratia*, most *Salmonella* species). Auxotrophic representatives usually have very simple growth factor requirements. A requirement for nicotinic acid is particularly frequent, occurring in many species of the genera *Proteus*, *Erwinia*, and *Shigella*. A specific requirement for tryptophan exists in *Salmonella typhi*, and for methionine in some *Photobacterium* species.

The regulation of amino acid biosynthesis has been studied in many members of the enteric group, and this work has revealed distinctive regulatory patterns that appear to set these organisms apart from other bacteria. For example, the initial step in the biosynthesis of amino acids of the aspartate family, conversion of aspartic acid to aspartylphosphate, is always mediated in the enteric group by three isofunctional aspartokinases, whose activity and synthesis are each independently regulated by different end-products of this branched pathway (see p. 304). This particular mode of regulation of the aspartate pathway has not been demonstrated in any bacteria outside the enteric group.

The mean DNA base composition for members of the enteric group is rather wide, extending from 37 to 63 percent G + C (Table 19.1). The ranges within each genus are narrow. The values for the closely related organisms of the three genera *Escherichia*, *Salmonella*, and *Shigella* are not significantly different. The total span of base composition for the "classical" enteric bacteria (37 to 59 percent G + C) is close to that for the polarly flagellated members (39 to 63 percent).

### Fermentative Metabolism

Sugar fermentation by enteric bacteria occurs through the Embden-Meyerhof pathway. The products vary, both qualitatively and quantitatively. These fermentations have one characteristic biochemical feature, however, which is not encountered in any other bacterial fermentations. This is a special mode of cleavage of the intermediate, pyruvic acid, to yield formic acid:

$$CH_3COCOO^- + CoA \longrightarrow CH_3COCoA + HCOO^-$$

Formic acid is, therefore, frequently a major fermentative end product. It does not always accumulate, however, since some of these bacteria possess the enzyme system *formic hydrogenlyase*,* which splits formic acid to $CO_2$ and $H_2$:

$$HCOO^- \longrightarrow CO_2 + H_2$$

* The formic hydrogenlyase system is composed of two distinct enzymes: formic dehydrogenase and hydrogenase. The carriers that mediate electron flow between the two enzymes have not been identified.

**TABLE 19.1**

**The Ranges of Mean DNA Base Composition Characteristic of the Members of the Enteric Group**

| | | Percent Guanine + Cytosine in DNA (40, 45, 50, 55, 60) |
|---|---|---|
| Peritrichous flagellation or immotile | *Escherichia, Salmonella, Shigella* | ←→ |
| | *Yersinia* | ←→ |
| | *Proteus* | ←→ |
| | *Providencia* | ←→ |
| | *Erwinia* | ←→ |
| | *Enterobacter* | ←→ |
| | *Serratia* | ←→ |
| Polar or "mixed" flagellation | *Vibrio* | ←→ |
| | *Aeromonas* | ←→ |
| | *Photobacterium* | ←→ |

In such organisms, formic acid is largely replaced as a fermentative end product by equimolar quantities of $H_2$ and $CO_2$.*

The most frequent mode of fermentative sugar breakdown in the enteric group is the *mixed-acid fermentation*, which yields principally lactic, acetic, and succinic acids; formic acid (or $CO_2$ and $H_2$); and ethanol. This fermentation is characteristic of the genera *Escherichia, Salmonella, Shigella, Proteus, Yersinia, Photobacterium* and *Vibrio*, and it occurs in some *Aeromonas* species. The ratios of the end products may vary considerably, both from strain to strain and within a single strain grown under different environmental conditions (e.g., at different pH values). This variability reflects the fact that the end products arise from pyruvic acid through three independent pathways (Figure 19.1; Table 19.2).

In some enteric bacteria, sugar fermentation gives rise to an additional major end product, 2,3-butanediol, which is formed from pyruvic acid by a fourth independent pathway (Figure 19.2). This *butanediol fermentation* is characteristic of *Enterobacter* and *Serratia*, most species of *Erwinia*, and some *Aeromonas* and *Photobacterium* species. The formation of butanediol is accompanied by increased formation of the reduced end product ethanol (Table 19.2), since the formation of butanediol from glucose results in a net generation of reducing power:

$$C_6H_{12}O_6 \longrightarrow CH_3CHOHCHOHCH_3 + 2CO_2 + 2H$$

Many bacteria of this group produce a mixture of end products including substantial quantities of neutral compounds (butanediol and ethanol) as well as organic acids (e.g., *Photobacterium phosphoreum*, Table 19.2). By convention, a fermentation is designated as the mixed acid type if sufficient acid is produced to decrease the pH of a specified medium to a value of about 4.5 or less, regardless of whether butanediol is produced; a fermentation is designated as the butanediol type if acid production is insufficient to reduce the pH to this extent, and if large amounts of butanediol are produced.

The formation of gas as a result of sugar fermentation is a character of considerable differential value in the enteric group, since it distinguishes the gas formers of the genus *Escherichia* from the pathogens of the *Shigella* group and *Salmonella typhi* which ferment sugar without gas production. In a simple mixed-acid fermentation, gas can be formed only by the cleavage of formate; gas production therefore reflects the possession of formic hydrogenlyase. This system is not, of course, essential

* Formation of molecular hydrogen as an end product of sugar fermentation is also characteristic of many sporeformers of the genera *Clostridium* and *Bacillus* (see Chapter 22). The biochemical mechanism responsible for its production is, however, different. In sporeformers, hydrogen is formed as a *direct* product of pyruvic acid cleavage:

$$CH_3COCOOH + CoA \longrightarrow CH_3COCoA + H_2 + CO_2$$

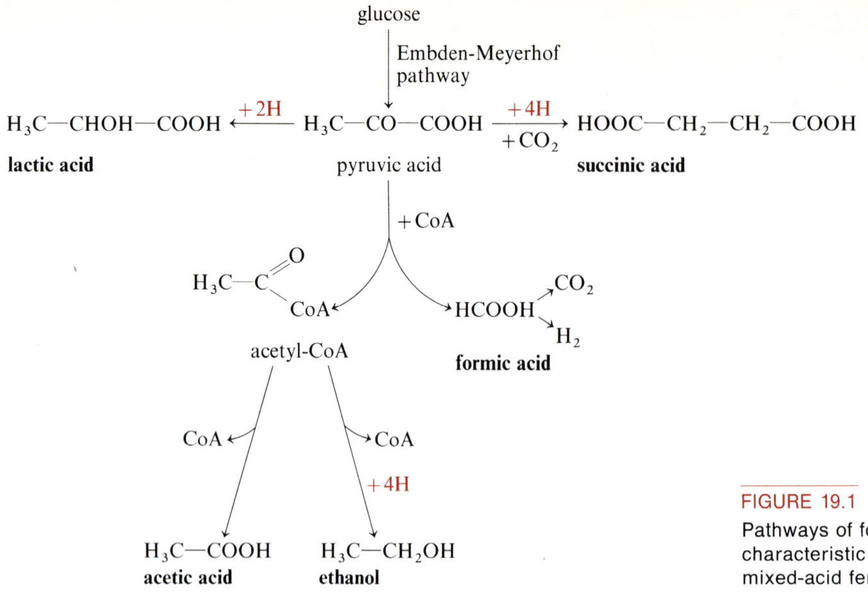

glucose

Embden-Meyerhof pathway

$H_3C{-}CHOH{-}COOH \xleftarrow{+2H} H_3C{-}CO{-}COOH \xrightarrow[+CO_2]{+4H} HOOC{-}CH_2{-}CH_2{-}COOH$

**lactic acid** pyruvic acid **succinic acid**

+CoA

$H_3C{-}C(=O){-}CoA$ acetyl-CoA

$HCOOH \rightarrow CO_2 + H_2$ **formic acid**

CoA CoA +4H

$H_3C{-}COOH$ **acetic acid**

$H_3C{-}CH_2OH$ **ethanol**

FIGURE 19.1

Pathways of formation from pyruvic acid of the characteristic end products (boldface) of a mixed-acid fermentation.

**TABLE 19.2**

**Products of Glucose Fermentation by Representative Enteric Bacteria**

| | Products, moles per 100 Moles of Glucose Fermented | | | | |
|---|---|---|---|---|---|
| | MIXED-ACID FERMENTATIONS | | | BUTANEDIOL FERMENTATIONS | |
| | *Escherichia coli* | *Aeromonas punctata* | *Photobacterium phosphoreum* | *Enterobacter aerogenes* | *Seratia marcescens* |
| Ethanol | 50 | 64 | 8 | 70 | 46 |
| 2,3-Butanediol | — | — | 0.6 | 66 | 64 |
| Acetic acid | 36 | 62 | 44 | 0.5 | 4 |
| Lactic acid | 79 | 43 | 42 | 3 | 10 |
| Succinic acid | 11 | 22 | 6 | — | 8 |
| Formic acid | 2.5 | 105 | 94 | 17 | 48 |
| $H_2$ | 75 | — | 55 | 35 | — |
| $CO_2$ | 88 | — | 48 | 172 | 116 |
| Total acid formed | 129 | 232 | 186 | 20 | 70 |

for fermentative metabolism and can be lost by mutation without effect on fermentative capacity. In fact, experience has shown that such strains of *Escherichia coli* exist in nature. Hence, although gas production is a useful differential character in the enteric group, it is by no means an infallible one.

The bacteria that perform a butanediol fermentation also differ with respect to the possession of formic hydrogenlyase. Members of the genus *Enterobacter* almost always contain formic hydrogenlyase and are vigorous gas producers; members of the genus *Serratia* do not contain the system and produce little or no visible gas, as judged by the customary criterion (formation of a bubble in an inverted vial placed in the fermentation tube). This may appear paradoxical, since the formation of butanediol from sugars is accompanied by a considerable net production of $CO_2$. However, this gas is very soluble in water, so that most (or all) of the $CO_2$ produced tends to remain dissolved in the medium. When $CO_2$ is the sole gaseous product of a bacterial fermentation, special cultural methods may be required to demonstrate its formation, a point discussed in connection with the lactic acid bacteria (Chapter 23).

Another character of considerable diagnostic importance in the enteric group is the ability to ferment the disaccharide, lactose, which depends on possession of a galactoside permease and the enzyme $\beta$-galactosidase. Strains lacking permease but

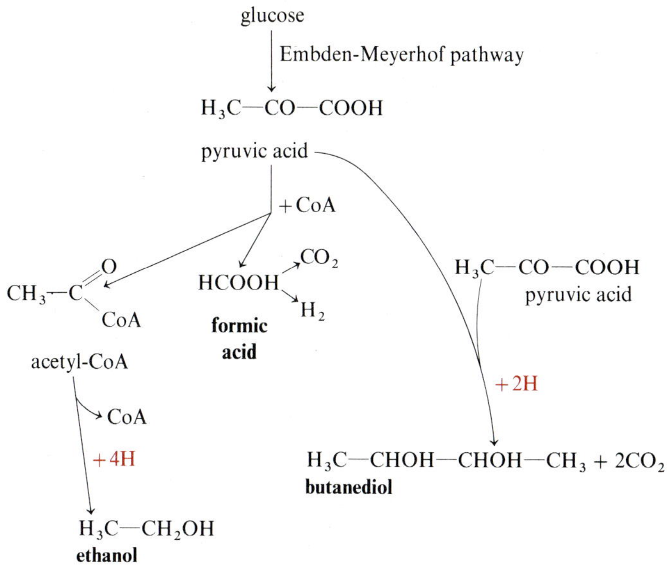

glucose

↓ Embden-Meyerhof pathway

$H_3C—CO—COOH$

pyruvic acid

↓ +CoA

$CH_3—C(=O)—CoA$ **acetyl-CoA**

$HCOOH \rightarrow CO_2 + H_2$ **formic acid**

$H_3C—CO—COOH$ pyruvic acid

+2H

$H_3C—CHOH—CHOH—CH_3 + 2CO_2$ **butanediol**

↓ CoA, +4H

$H_3C—CH_2OH$ **ethanol**

FIGURE 19.2

Pathways of formation from pyruvic acid of the characteristic end products (boldface) of a butanediol fermentation.

containing β-galactosidase cannot take up lactose at a sufficient rate to produce a prompt and vigorous fermentation and will normally be classified as nonfermenters of this sugar. Lactose fermentation is characteristic of *Escherichia* and *Enterobacter* and is absent from *Shigella*, *Salmonella*, and *Proteus*. It should be noted that some *Shigella* strains produce β-galactosidase but cannot ferment lactose because they lack galactoside permease.

The fermentative characteristics of the various genera are summarized in Table 19.3.

## Some Physiological Characters of Differential Value

A few physiological and biochemical characters are of considerable use in distinguishing major subgroups within the enteric group. One is the oxidase reaction, the mechanism of which was discussed in Chapter 17 (see p. 403). The "classical" enteric group and the genus *Yersinia* are all oxidase negative. Most polar flagellates, however, are oxidase positive.

Nearly all groups of enteric bacteria synthesize glycogen as the sole organic cellular reserve material. The formation of poly-β-hydroxybutyrate as a cellular reserve material is a property confined to the marine bacteria of the genera *Vibrio* and *Photobacterium*, although it does not occur in all species.

The members of the genera *Vibrio* and *Photobacterium* can be readily distinguished from all other members of the enteric group by their *salt*

**TABLE 19.3**

**Summary of Fermentative Patterns in the Enteric Group and Related Organisms**

I. MIXED-ACID FERMENTATION
- A. Produce $CO_2 + H_2$ (contain formic hydrogenlyase)
  - *Escherichia*
  - *Proteus*
  - *Salmonella* (most spp.)
  - *Photobacterium* (some spp.)[a]
  - *Vibrio* (some spp.)[a]
  - *Aeromonas* (some spp.)
  - *Providencia* (some spp.)
- B. No gas produced (formic hydrogenlyase absent)
  - *Shigella*
  - *Salmonella typhi*
  - *Yersinia*[a]
  - *Vibrio* (most spp.)[a]
  - *Aeromonas* (some spp.)
  - *Photobacterium* (some spp.)[a]
  - *Providencia* (some spp.)

II. BUTANEDIOL FERMENTATION
- A. Produce $CO_2 + H_2$ (contain formic hydrogenlyase)
  - *Enterobacter*
  - *Aeromonas hydrophila*
  - *Photobacterium phosphoreum*
- B. Produce only $CO_2$ (formic hydrogenlyase absent); visible gas formation slight or undetectable
  - *Serratia*
  - *Erwinia herbicola* and *E. carotovora*
  - *Vibrio alginolyticus*

[a] Some butanediol may also be produced.

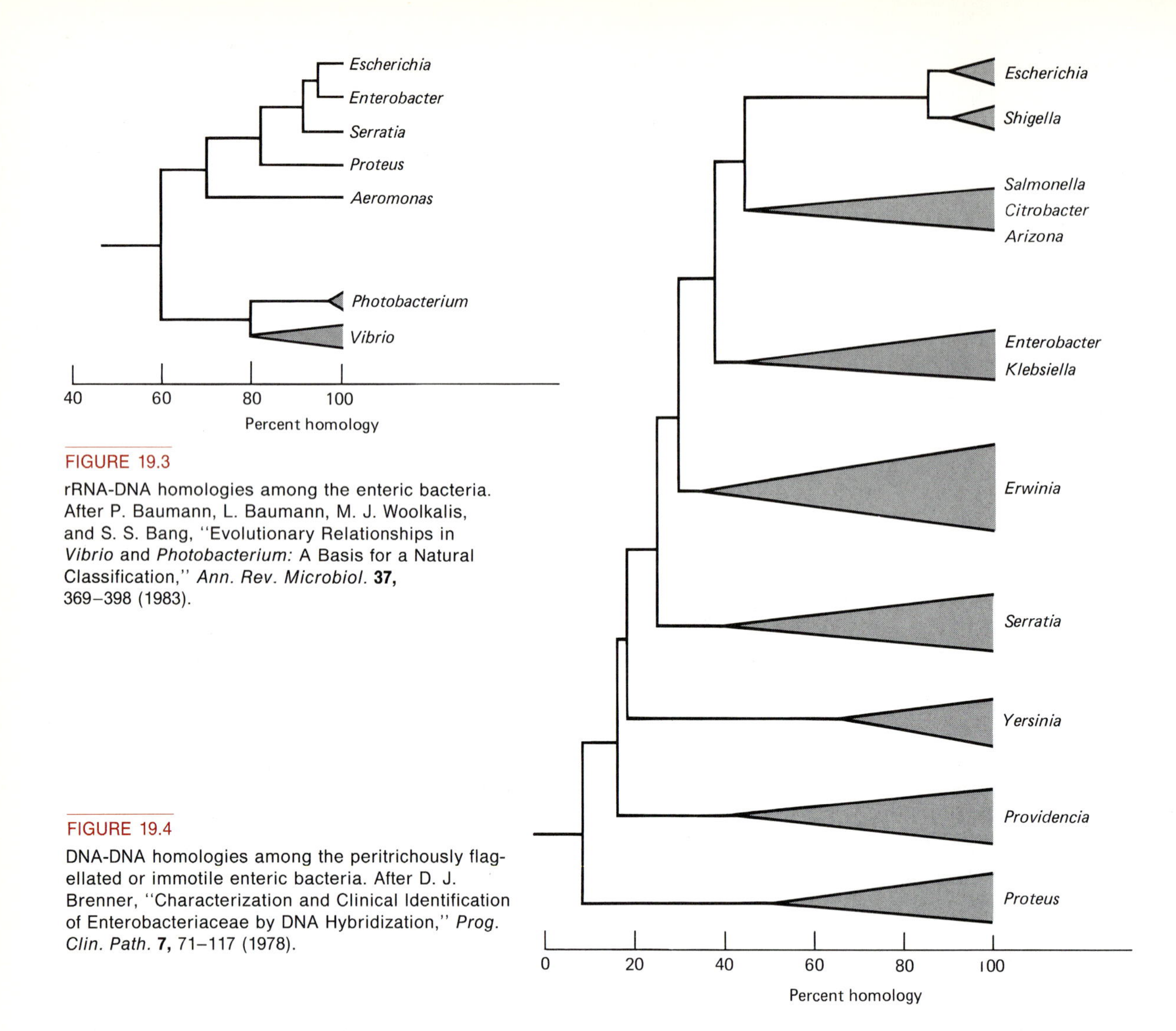


FIGURE 19.3

rRNA-DNA homologies among the enteric bacteria. After P. Baumann, L. Baumann, M. J. Woolkalis, and S. S. Bang, "Evolutionary Relationships in *Vibrio* and *Photobacterium:* A Basis for a Natural Classification," *Ann. Rev. Microbiol.* **37,** 369–398 (1983).

FIGURE 19.4

DNA-DNA homologies among the peritrichously flagellated or immotile enteric bacteria. After D. J. Brenner, "Characterization and Clinical Identification of Enterobacteriaceae by DNA Hybridization," *Prog. Clin. Path.* **7,** 71–117 (1978).

*requirements.* Indigenous marine bacteria have an absolute requirement for sodium ions, no growth occurring if sodium salts are omitted from the medium. Furthermore, the magnitude of this requirement is considerable, concentrations of $Na^+$ ranging from about 100 to 300 mM being necessary to assure growth at the maximal rate. A specific $Na^+$ requirement cannot be demonstrated for most nonmarine bacteria (apart from extreme halophiles). Furthermore, both $Mg^{2+}$ and $Ca^{2+}$ are required for marine bacteria at much higher concentrations than those which satisfy these essential mineral requirements for other bacteria. In all these respects, the members of *Vibrio* and *Photobacterium* are typical marine bacteria; they cannot grow in media that contain low concentrations of $Na^+$, $Mg^{2+}$, and $Ca^{2+}$, which are adequate to support growth at maximal rates of other enteric bacteria. A single species of *Vibrio, V. cholerae,* can grow in the absence of sodium; however, its growth rate and cell yield are both substantially reduced.

## GENETIC RELATIONSHIPS AMONG THE ENTERIC BACTERIA

Extensive studies of the nucleic acid homology among the enteric bacteria have confirmed that the group is a natural one. Figures 19.3 and 19.4 show dendrograms based on rRNA-DNA and DNA-DNA reannealing studies done on the organisms described in this chapter. Comparisons of ribosomal RNA sequence comparisons clearly distinguish two groups; (I) *Vibrio* and *Photobacterium,* and (II) *Aeromonas* and the "classical" enteric bacteria. It is also clear that *Vibrio* itself is a heterogeneous group; the extent of divergence of rRNA sequences within it is as great as that of the entire group of peritrichously flagellated or immotile enterobacteria.

By DNA-DNA hybridization studies, the genera *Escherichia* and *Shigella* have been shown to be quite closely related; possibly they constitute a single species. Other genera within group II are considerably more heterogeneous.

## TAXONOMIC SUBDIVISION OF THE ENTERIC GROUP

A simplified scheme for the subdivision of the enteric group is shown in Tables 19.4 and 19.5. A reasonably satisfactory primary separation can be made on the basis of the mode of flagellar insertion and the oxidase reaction. The oxidase negative groups which, if motile, possess only peritrichous flagella, include the classical representatives of the group, together with *Yersinia*. As shown in Table 19.4, DNA base composition, together with a few biochemical and physiological characters, permits the recognition of four major subgroups. The second major group of enteric bacteria consists of organisms that bear polar flagella or display mixed flagellation and, with some exceptions, are oxidase positive (Table 19.5). A primary separation within this group can be made on the basis of ionic requirements, which distinguish *Vibrio* and *Photobacterium* from *Aeromonas*. One character of considerable utility in this group is flagellar structure: the polar flagella of *Vibrio* are relatively thick, being enclosed by a sheath that is made up of an extension of the cell membrane. The polar flagella of *Aeromonas* and *Photobacterium* are not sheathed.

### Group I: *Escherichia-Salmonella-Shigella*

The members of group I are all inhabitants of the intestinal tract of humans and other vertebrates. The principal generic distinctions are shown in Table 19.6. It should be noted that certain strains of intestinal origin—the so-called *paracolon group*—have characters intermediate between those shown for the genera in Table 19.6, so that distinctions are not always as clear-cut as the table suggests. The additional genera, *Arizona* and *Citrobacter*, were created for the intermediate forms of the paracolon group, but this generic hypertrophy really does nothing to help the problem of differentiation.

Highly detailed, intraspecific subdivisions among the species of group I have been made on the basis of the immunological analysis of the surface structures of the cell. The extreme specificity of antigen-antibody reactions makes it possible to recognize differences in these respects between strains of a bacterial species that are indistinguishable on the basis of other phenotypic criteria. Three classes of surface antigens have been extensively explored among the enteric bacteria of group I: the *O antigens*, which are the polysaccharide components of the lipopolysaccharides in the outer membrane, the *K antigens*, which are capsular polysaccharides; and the *H antigens*, which are the flagellar proteins. Many of these organisms possess two sets of genetic determinants for different flagellar antigens; these are subject to alternate phenotypic expression, a phenomenon known as *phase variation*. On any given cell the flagella are of one antigenic type, but as the organism multiplies, variants of the alternate type arise with a certain probability. Cultures of such biphasic strains thus contain two specific sets of H antigens. The regulation of phase variation is discussed on p. 300.

In the genus *Salmonella* the detailed analysis of O and H antigenic structure (including phase variation of the H antigens) has made it possible to distinguish several thousand different *serotypes;* comparable analyses of *Escherichia* and *Shigella* are less extensive. The principal utility of these systems of antigenic classification is not taxonomic, but *epidemiological*. The serotype of a pathogenic *Salmonella* strain is a marker that permits its recognition (and hence allows one to follow its transmission) where other phenotypic characters do not.

*Escherichia coli* and some members of the paracolon group are components of the normal adult human intestinal flora. They are also important pathogens; diarrhea caused by toxin-producing *E. coli* (Chapter 31) is the major cause of infant mortality in the third world. The genera *Salmonella* and *Shigella* comprise pathogens that cause a wide variety of enteric diseases in humans and other animals. Entry occurs through the mouth, and although the small intestine is the primary locus of infection, some of these pathogens may subsequently invade other body tissues and cause more generalized damage in the infected host. The members of the genus *Shigella* are the agents of a specifically human enteric disease, bacterial dysentery. In the genus *Salmonella* both the host range and the variety of diseases produced are much broader. *Salmonella typhi* and *S. paratyphi*, the agents of typhoid and paratyphoid fevers, are specific pathogens of humans, whereas certain other species are

**TABLE 19.4**

**Taxonomic Subdivision of the Peritrichously Flagellated Enteric Bacteria and Related Immotile Forms**

| Major Subgroup | Percent G + C | Motility | Production of: BUTANEDIOL | Production of: $H_2 + CO_2$ | Production of: UREASE | Production of: TRYPTOPHAN DEAMINASE | Constituent Genera | Other Generic Names Frequently Applied to Some Members of Group |
|---|---|---|---|---|---|---|---|---|
| I | 50–53 | V[a] | – | V | – | – | *Escherichia, Salmonella, Shigella* | *Arizona* and *Citrobacter* for "intermediate" types |
| II | 50–59 | V | + | V | – | – | *Enterobacter, Serratia, Erwinia* | *Klebsiella* and *Hafnia* |
| III | 37–42 | + | – | V | V | + | *Proteus, Providencia* | |
| IV | 46–47 | V | – | – | + | – | *Yersinia* | |

Note: Straight rods, immotile or motile by peritrichous flagella; oxidase negative.
[a] V denotes variable within group.

**TABLE 19.5**

**Taxonomic Subdivision of Polarly Flagellated Enteric Bacteria**

| Genus | Percent G + C | Sheathed Polar Flagellum | Poly-β-Hydroxy Butyrate Accumulated Intracellularly; Exogenous β-Hydroxybutyrate Not Used | Bioluminescence | $Na^+$ Requirement | Fermentative Characters: BUTANEDIOL PRODUCED | Fermentative Characters: PRODUCTION OF $H_2 + CO_2$ |
|---|---|---|---|---|---|---|---|
| *Aeromonas* | 57–63 | – | – | – | – | V[a] | V |
| *Vibrio* | 38–54 | +[b] | –[c] | V | +[d] | – | V |
| *Photobacterium* | 40–44 | – | + | V | + | V | V |

Note: Straight or curved rods; motile by polar flagella, some showing "mixed" polar-peritrichous flagellation; mainly oxidase positive.
[a] V denotes variable within group.
[b] May also have mixed flagellation when grown on solid media.
[c] A number of *Vibrio* strains accumulate PHB; but they are distinguished from *Photobacterium* by their ability to use exogenous β-hydroxybutyrate as carbon and energy source.
[d] Except for *V. cholerae*, whose growth is stimulated by $Na^+$.

specific pathogens of other mammals or of birds. The great majority of the *Salmonella* group, however, have a low host specificity. They exist, often without causing disease symptoms, in the intestine and in certain tissues of animals or birds. If these forms gain access to and develop in foods, their subsequent ingestion by humans can give rise to *food poisoning* (Chapter 31). Outbreaks of food poisoning often have an epidemic character, because food preparation on a large scale provides many favorable opportunities for the growth of these organisms.

### Group II: *Enterobacter-Serratia-Erwinia*

*Enterobacter aerogenes*, the prototype of group II, is common in soil and water, and sometimes also occurs in the intestinal tract. Similar bacteria, distinguished from *E. aerogenes* by permanent immotility and the presence of capsules, occur in the respiratory tract; they are often classified in a separate genus, *Klebsiella*. A biochemical property that distinguishes some (though not all) *Enterobacter* strains from other enteric bacteria is the ability to fix nitrogen. This property can be expressed only under anaerobic growth conditions, since the nitrogenase of these bacteria is rapidly denatured in the cells in the presence of oxygen. Other enteric bacteria do not possess nitrogenase.

*Serratia*, also a common soil and water organism, differs from *Enterobacter* principally by its failure to produce formic hydrogenlyase (little or no visible gas formed during sugar fermentation) and by its inability or weak ability to ferment lactose (Table 19.7). Many (but by no means all) strains of *Serratia* produce a characteristic red cellular pigment, *prodigiosin*, a tripyrrole derivative (Figure 19.5).

Relative to the enteric bacteria so far discussed, the representatives of the genus *Erwinia* constitute a very heterogeneous group. Three principal subgroups are now recognized, exemplified by the species *Erwinia amylovora*, *E. carotovora*, and *E. herbicola;* many intermediate forms are known, however.

*E. amylovora* is the agent of fire blight, a necrotic disease of pears and related plants. This species is notable for its limited range of utilizable sugars and its requirement for organic growth factors, characters absent from other erwinias. *E. carotovora* causes soft rots of the storage tissues of many plants, an action attributable in part to its ability to produce pectolytic enzymes, which destroy the pectic substances that serve as intracellular cementing materials in plant tissues. *E. herbicola*,

FIGURE 19.5

The structure of prodigiosin, the red pigment formed by *Serratia*.

H H N N N $CH_3$ $OCH_3$ $C_5H_{11}$

**TABLE 19.6**

**Internal Differentiation of the Major Genera of Group I**

| Characteristics | *Escherichia* | *Salmonella* | *Shigella* |
|---|---|---|---|
| Pathogenicity for man or animals | V[a] | V | + |
| Motility | V | + | − |
| Gas ($CO_2 + H_2$) from fermentation of glucose | + | + | − |
| Fermentation of lactose | + | − | − |
| β-galactosidase | + | − | V |
| Utilization of citrate as carbon source | − | + | − |
| Production of indole from tryptophan | + | − | V |

Note: Phenotypes are idealiized; exceptions exist for most traits in all three genera.
[a] V denotes a variable reaction within the group.

**TABLE 19.7**

**Internal Differentiation of the Major Genera of Group II**

| Characteristics | *Enterobacter* | *Serratia* | *Erwinia* |
|---|---|---|---|
| Motility | V[a] | + | + |
| $CO_2 + H_2$ formed by glucose fermentation | + | − | − |
| Lactose fermentation | + | − | V[a] |
| β-galactosidase | + | + | + |
| Gelatin liquefied | − | + | + |
| Pectinolytic enzymes produced | − | − | V[a] |
| Yellow cellular pigments | − | − | V[a] |
| Red cellular pigments | − | + | − |
| Plant pathogens or parasites | − | − | + |

Note: Phenotypes are idealized; exceptions exist for most traits in all three genera.
[a] V denotes a variable reaction within the group.

which produces yellow cellular pigments, occurs commonly on the leaf surfaces of healthy plants; some strains are plant pathogens. Similar pigmented strains have occasionally been isolated from human sources, although their pathogenicity for humans remains uncertain.

## Group III: *Proteus-Providencia*

The members of the genus *Proteus* are probably soil inhabitants, although they are found in particular abundance in decomposing animal materials. The relatively low G + C content of their DNA distinguishes most species from the groups so far discussed, from which they are also distinguishable by certain physiological properties. These include strong proteolytic activity (gelatin is rapidly liquefied) and ability to hydrolyze urea. Most *Proteus* spp. are very actively motile and can spread rapidly over the surface of a moist agar plate, a phenomenon known as *swarming.* A curious feature of the swarming phenomenon is its periodicity: it occurs in successive waves, separated by periods of growth and cell division. This produces a characteristic zonate pattern of development on an agar plate (Figure 19.6).

*Providencia* strains are biochemically quite similar to *Proteus*, but they differ in habitat. They are most frequently isolated as human pathogens, commonly from infected urinary tracts, occasionally from wounds or burns. They are also often isolated from the intestinal tract of people with diarrhea, although their possible etiological role in this disease has not been demonstrated.

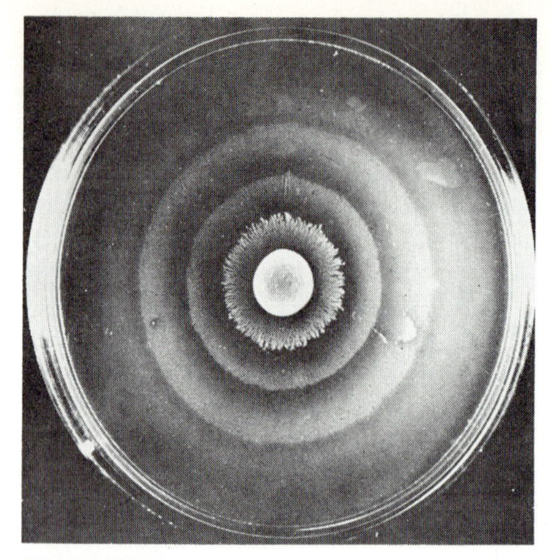

FIGURE 19.6

Swarming of *Proteus* on the surface of nutrient agar plate. The plate was inoculated in the center with a drop of a bacterial suspension and was photographed after incubation at 37° C for 20 hours. From H. E. Jones and R. W. A. Park, "The Influence of Medium Composition on the Growth and Swarming of *Proteus*," *J. Gen. Microbiol.* **47,** 369 (1967).

## Group IV: *Yersinia*

The genus *Yersinia* contains several species that are agents of disease in rodents. *Yersinia pestis* can be transmitted by fleas from its normal rodent hosts to humans; it is the cause of human *bubonic plague*, a disease that has been responsible throughout history for massive epidemics with a very high mortality. In humans the disease can also be transmitted through the respiratory route (*pneumonic plague*). Both in their mode of transmission and in their symptoms, the diseases caused by *Yersinia* species are entirely different from the major enteric diseases.

The members of the genus *Yersinia* carry out a mixed-acid fermentation without production of $H_2$ and $CO_2$; in this respect, they resemble the *Shigella* group. They produce $\beta$-galactosidase and have a powerful urease. Motility is a variable character; *Y. pestis* is permanently immotile. The G + C content of the DNA is significantly lower than that of the *Escherichia-Salmonella-Shigella* group. A cultural character that distinguishes them from other enterobacteria is their relatively slow growth on complex media.

When grown at 25° C most yersinias exhibit no requirement for growth factors; only *Y. pestis* requires them (methionine and phenylalanine); however, growth at 37° C generates vitamin requirements (biotin and thiamin) in all yersinias. The production of butanediol, a characteristic of some species, is also heat-sensitive; it occurs at 25° C but not at 37° C.

## The Polar Flagellates: *Aeromonas-Vibrio-Photobacterium*

Among the polarly flagellated facultative anaerobes three groups may be recognized (Table 19.5).

The genus *Aeromonas* contains organisms that vary with respect to the nature of sugar fermentation. *Aeromonas hydrophila* performs a butanediol fermentation, accompanied by $H_2$ and $CO_2$ production, similar to that of *Enterobacter. Aeromonas punctata* and *A. shigelloides* perform a mixed-acid fermentation without gas production, similar to that of *Shigella. Aeromonas shigelloides* also closely resembles the *Shigella* group in DNA base composition and has been shown to share certain somatic antigens with this enteric group. Some authors place it in a separate genus, *Plesiomonas.*

*Aeromonas hydrophila* and *A. punctata* are widespread in fresh water; the former is capable of causing disease in both frogs and fish. *A. shigelloides* appears to be an inhabitant of the intestinal tract, and it has been implicated in human gastroenteritis.

The genus *Vibrio* contains marine bacteria that are straight or curved rods. They are distinguished from *Photobacterium* by their sheathed polar flagellum and their inability to accumulate poly-$\beta$-hydroxybutyrate. Two species cause disease in humans: *V. cholerae* causes the water-transmitted disease *cholera*, a major cause of mortality in some parts of the developing world; and *V. parahemolytica* is a frequent cause of gastroenteritis in Japan, associated with the consumption of raw fish.

Bacteria of the genera *Vibrio* and *Photobacterium* are among the most abundant in marine environments. They occur in seawater, in the intestinal tract, and on the body surfaces of marine animals. Many can decompose chitin and alginic acid.

The property of *bioluminescence* is common to several species of both *Vibrio* and *Photobacterium.* The light emitted by these organisms is blue-green in color, being confined to a rather narrow spectral band, with a maximum near 490 nm. Cells emit light continuously, provided that oxygen is present. As shown in Figure 19.7, an unaerated cell suspension rapidly becomes dark, as a result of the depletion of the dissolved oxygen supply by bacterial respiration. Indeed, the production of light by a suspension of luminous bacteria is one of the most sensitive methods known for the detection of traces of oxygen.

Light emission can be obtained in cell-free extracts of luminous bacteria, and biochemical studies have revealed the nature of the reaction involved. It is mediated by the enzyme *luciferase*, which is a mixed function oxidase. The substrates for the light-emitting reaction are reduced flavin mononucleotide ($FMNH_2$), molecular oxygen, and a long-chain saturated aldehyde, containing more than eight carbon atoms (R—CHO). The overall reaction catalyzed by luciferase can be formulated as follows:

$$FMNH_2 + O_2 + R{-}CHO \longrightarrow FMN + H_2O + R{-}COOH + h\nu$$

The immediate product of $FMNH_2$ oxidation is electronically excited flavin mononucleotide, FMN*. Light emission results from the return of FMN* to the ground state:

$$FMN^* \longrightarrow FMN + h\nu$$

Bioluminescence is widespread in the animal kingdom, occurring in such diverse groups as jellyfish, earthworms, fireflies, squid, and fish. The emission of light by these animals very often appears to be a recognition device, promoting schooling, mating, or the attraction of prey. In most cases the luminescence is produced by the tissues of the animal itself, but in some species of squid, and in

FIGURE 19.7

Luminous bacteria photographed by their own light: left, a streaked plate of *Photobacterium phosphoreum;* right, two flasks containing a suspension of the same organism in a sugar medium. A stream of air was passed continuously through the flask on the right during the photographic exposure. The bacteria in the unaerated flask on the left had exhausted the dissolved oxygen and had ceased to luminesce except at the surface, where organisms were exposed to the air.

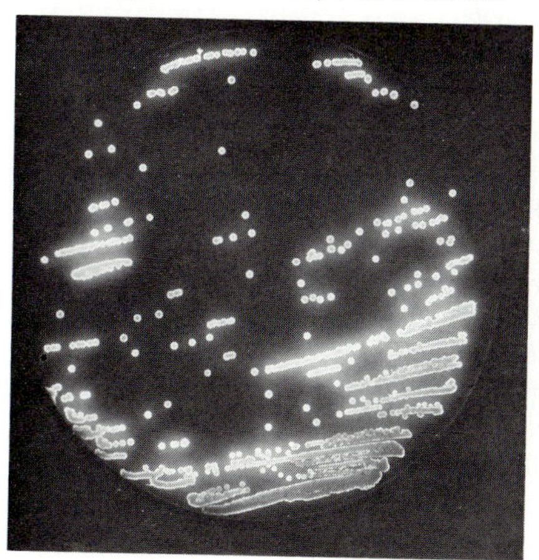

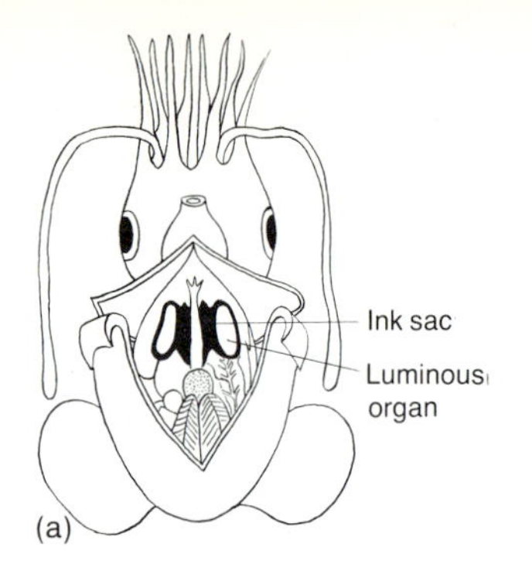


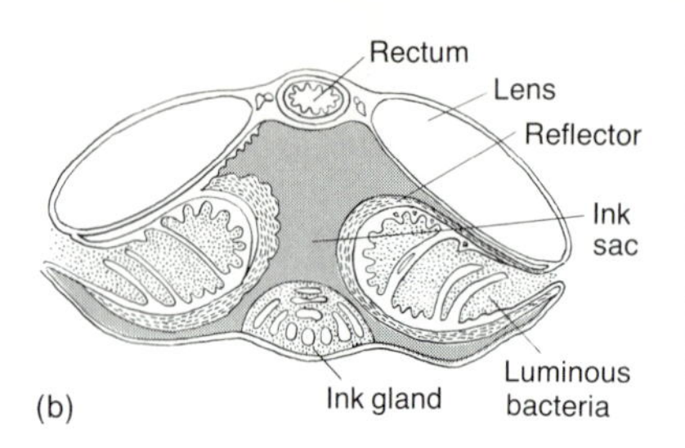


FIGURE 19.8

The cuttlefish, *Euprymna morsei*. (a) Male with opened mantle, showing the luminous organs embedded in the ink sac. (b) Cross section through the luminous organs, showing the reflectors, lens, and the open chambers containing the luminous bacteria. After T. Kishitani, "Studien über Leuchtsymbiose von Japanischen Sepien," *Folia Anat. Japan.* **10,** 315 (1932).

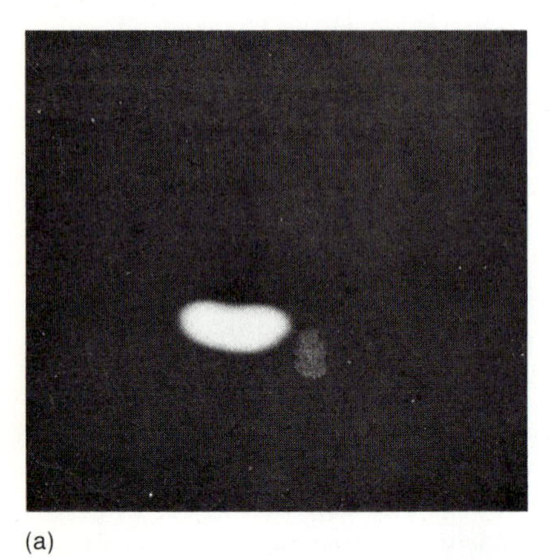

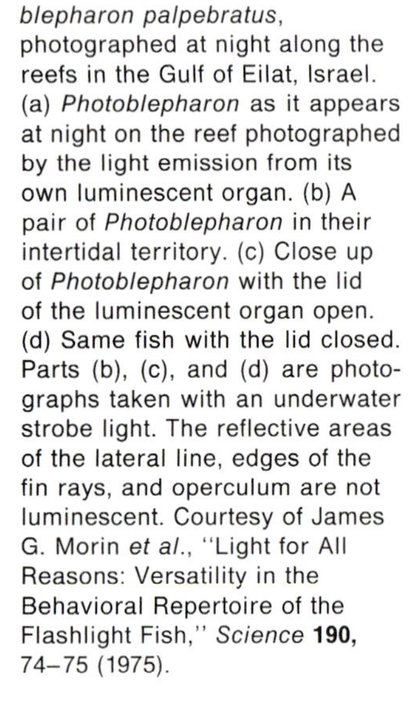

FIGURE 19.9

The flashlight fish, *Photoblepharon palpebratus*, photographed at night along the reefs in the Gulf of Eilat, Israel. (a) *Photoblepharon* as it appears at night on the reef photographed by the light emission from its own luminescent organ. (b) A pair of *Photoblepharon* in their intertidal territory. (c) Close up of *Photoblepharon* with the lid of the luminescent organ open. (d) Same fish with the lid closed. Parts (b), (c), and (d) are photographs taken with an underwater strobe light. The reflective areas of the lateral line, edges of the fin rays, and operculum are not luminescent. Courtesy of James G. Morin *et al.*, "Light for All Reasons: Versatility in the Behavioral Repertoire of the Flashlight Fish," *Science* **190,** 74–75 (1975).

certain fishes, it is produced by luminous bacteria living ectosymbiotically in special glands of the host.

Among the squids (mollusks belonging to the class Cephalopoda), symbiotic luminous bacteria have been identified in a number of species of one suborder, Myopsida. The myopsid squids, also called cuttlefishes, are characterized by their strongly calcified shell. Figure 19.8 shows a male of the genus *Euprymna*, the light organs of which are quite typical of the myopsid squids. The luminous glands are embedded in the ink sac and are partially enclosed in a layer of reflective tissue; just above the glands are lenses made up of hyaline cells which transmit light. In some squids the animal can control the emission of light by a muscular contraction that squeezes the ink sac, pushing it between the light source and the lens. The luminous organs are open to the exterior; all evidence suggests that young animals are infected externally and that transmission via the egg does not occur.

A large number of unrelated species of fish also possess light organs that consist of open glands containing luminous bacteria. In a few species the organ is provided with a reflecting layer of tissue. The most complex organs are found in the two closely related genera, *Photoblepharon* and *Anomalops*, both of which harbor luminous bacteria in special pouches under the eyes (Figure 19.9). What makes these forms particularly spectacular is their ability to control the emission of light; in *Photoblepharon*, this is accomplished by drawing up a fold of black tissue over the pouch like an eyelid, while in *Anomalops* the light organ itself can be rotated downward against a pocket of black tissue.

The complexity of the organs that have been evolved to control the symbiotic light emission implies that luminescence has great adaptive value for the host; in most cases, it is believed to serve as a recognition device. The functions performed by the host on behalf of the luminous bacteria are undoubtedly those of providing nutrients and protection.

### *Zymomonas*

The bacteria of the genus *Zymomonas* are polarly flagellated, Gram-negative rods that occur in fermenting plant materials. Like enteric bacteria, they are facultative anaerobes that have both respiratory and fermentative capacity. However, the sugar fermentation characteristic of *Zymomonas* is a unique one that sharply distinguishes these bacteria from the enteric group. Only glucose, fructose, and sucrose can be fermented; they are converted to equimolar quantities of ethanol and $CO_2$ (Figure 19.10). As in yeast, pyruvate is decarboxylated nonoxidatively, with the formation of acetaldehyde, subsequently reduced to ethanol. However, the conversion of glucose to pyruvate occurs through the Entner-Doudoroff pathway, not the Embden-Meyerhof pathway.

FIGURE 19.10
The pathway of glucose fermentation by *Zymomonas*.

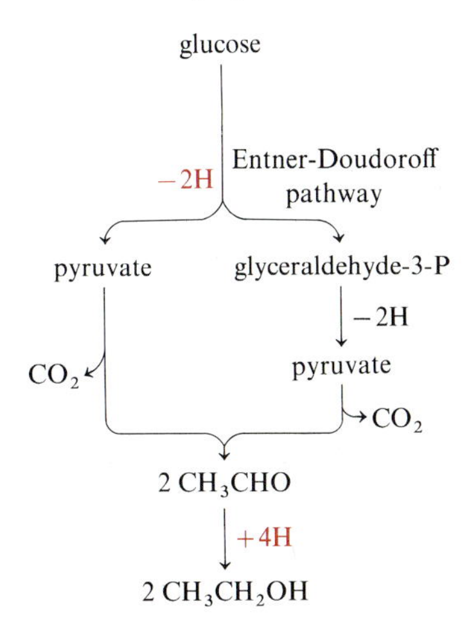


## COLIFORM BACTERIA IN SANITARY ANALYSIS

The enteric diseases caused by the coliform bacteria are transmitted almost exclusively by the fecal contamination of water and food materials. Transmission through contaminated water supplies is by far the most serious source of infection and was responsible for the massive epidemics of the more serious enteric diseases (particularly typhoid fever and cholera) that periodically scourged all countries until the beginning of the present century. Today these diseases are almost unknown in most parts of the Western world, although cholera occurs sporadically in the countries bordering the Mediterranean and Carribean. Their eradication was achieved primarily by appropriate sanitary controls. An essential part of this operation was *the development of bacteriological methods for ascertaining the occurrence of fecal contamination in water and foodstuffs.*

It is seldom possible to isolate enteric pathogens directly from contaminated water because they are usually present in small numbers, unless contamination from an infected individual has been recent and massive. To demonstrate the fact of fecal contamination, it is sufficient to show that the sample under examination contains bacteria known to be specific inhabitants of the intestinal tract, even though they may themselves not be agents of disease. The bacteria that have principally served as indices of such contamination are the fecal streptococci (discussed in Chapter 23) and *E. coli*. The methods of sanitary analysis developed by bacteriologists differ somewhat from country to country.

One method for detecting *E. coli* is to inoculate dilutions of the sample under test into tubes of lactose broth, which are then incubated at 37°C, and examined after one and two days for acid and gas production. Cultures showing acid and gas formation are then streaked on a special medium, with a composition that facilitates recognition of *E. coli* colonies. One of the media most commonly used is a lactose-peptone agar containing two dyes, eosin and methylene blue (EMB agar). On this medium, *E. coli* produces blue-black colonies with a greenish metallic sheen, whereas the other principal member of the group capable of fermenting lactose with acid and gas production, *Enterobacter aerogenes* (not necessarily indicative of fecal contamination) produces pale pink mucoid colonies without a sheen (Figure 19.11). For a final distinction between these two organisms, a series of physiological tests, known as the *IMViC tests*, can be performed on material from an isolated colony. The typical results obtained with the two species are shown in Table 19.8. Of these four tests, the Methyl Red ("M" of *IMViC*) and Voges-Proskauer ("Vi") tests are the most significant, since they indirectly reveal the mode of fermentative sugar breakdown. Both are performed on cultures grown in a glucose-peptone medium. The Methyl Red test affords a measure of the final pH: this indicator is yellow at a pH of 4.5 or higher and red at lower pH values. A positive test (red color) is therefore indicative of substantial acid production, characteristic of a mixed-acid fermentation. The Voges-Proskauer test is a color test for acetoin, an intermediate in the formation of butanediol from pyruvic acid; a positive reaction is therefore indicative of a butanediol fermentation. The test for indole production from tryptophan ("T"), performed on a culture grown in a peptone medium rich in tryptophan, is a test for the presence of the enzyme tryptophanase, which splits tryptophan to indole, pyruvate, and ammonia. This enzyme is present in many bacteria of the enteric group (including *E. coli*) but is not found in *Enterobacter aerogenes.* The citrate utilization test ("C") determines ability to grow in a synthetic medium containing citrate as the sole carbon source. This ability is lacking in most strains of *E. coli*, as a result of the absence of a citrate permease.

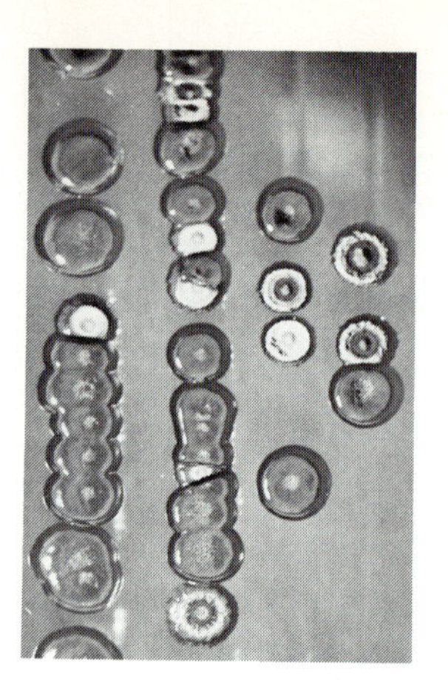

FIGURE 19.11

A plate of EMB agar streaked with a mixture of *Escherichia coli* and *Enterobacter aerogenes*. The colonies of *E. coli* are relatively small and appear light as a result of their metallic sheen. Courtesy of N. J. Palleroni.

The first step of the analytical procedure described above is relatively nonspecific, since many bacteria, not even necessarily members of the enteric group, can grow at 37° C in lactose broth with acid and gas production. A much more specific primary enrichment of *E. coli* can be achieved by the Eijkman method: use of a lactose broth, incubated at 46°C. This slight elevation of incubation temperature elimates most *Enterobacter aerogenes* and other organisms that ferment lactose with gas production but permits growth of most *E. coli.*

**TABLE 19.8**

**IMViC Tests for the Differentiation between *Escherichia coli* and *Enterobacter aerogenes***

| | Typical Reactions | | | |
|---|---|---|---|---|
| | INDOLE | METHYL RED | VOGES-PROSKAUER | CITRATE |
| *Escherichia coli* | + | + | − | − |
| *Enterobacter aerogenes* | − | − | + | + |

## FURTHER READING

### Book


VON GRAEVENITZ, A. and S. J. RUBIN, *The Genus* Serratia. Boca Raton, Fla.: CRC Press, 1980.

# Chapter 20
# Gram-Negative Anaerobic Eubacteria

The Gram-negative eubacteria include a number of species that are obligately anaerobic. Most are strict anaerobes, being rapidly killed by traces of oxygen and incapable of growing except in media with low values of $E_h$. Two distinctive physiological assemblages can be recognized within this group: the *Gram-negative fermentative bacteria*, and the *sulfur-reducing bacteria*. Collectively the Gram-negative fermentative bacteria are capable of fermenting a wide range of sugars, amino acids, and other organic acids; some are capable of fumarate- or nitrate-linked respiration. They are characteristic symbionts within the alimentary tract of homeothermic animals.

The sulfur-reducing bacteria are respiratory organisms that use sulfate or elemental sulfur rather than $O_2$ as electron acceptor. Some are capable of autotrophic growth, and some can ferment a limited range of organic acids. They are found in large numbers in anaerobic sediments, and in low numbers in intestinal tracts of animals.

## THE GRAM-NEGATIVE FERMENTATIVE EUBACTERIA

The intestinal tract, oral mucosa, and rumen of warm-blooded animals have a characteristic microbial flora that varies somewhat among host animals, but is always dominated by certain anaerobic eubacteria. Some of these that are Gram-positive are discussed in Chapter 23; the gram-negative ones are discussed here. Since many representatives of this group are strict anaerobes that do not form oxygen-resistant resting stages, they require special care in their isolation and study; consequently, it is only recently that they have become well known. Probably the group is more diverse than we currently appreciate.

### Fermentation Patterns of Gram-Negative Eubacteria

A variety of different fermentations are performed by the Gram-negative eubacteria; they can be broadly divided into six types: clostridial-type amino acid fermentations, discussed in Chapter 22; the homolactic fermentation, which is discussed in Chapter 23; two types of propionate fermentations; the butyrate fermentation; and the succinate fermentation.

**Propionate Fermentations** Several different fermentative pathways yield propionate as a major end product. Some of these include as intermediates two compounds (fumarate and succinate) that have an axis of symmetry; i.e., carbon atoms C1 and C4 are indistinguishable as are carbon atoms C2 and C3. Hence a substrate that is radioactively labeled in a specific carbon atom is converted to fermentative products that are labeled in two carbons, a process termed *randomization.* These pathways are thus termed *randomizing* (Figure 20.1). The other pathway of propionate fermentation does not pass through a symmetrical intermediate, and hence is termed the *nonrandomizing pathway* (Figure 20.2). Propionate fermentations also produce acetate as an end product. When the substrate is pyruvate or lactate, acetate formation provides the only

FIGURE 20.1

Propionate formation via randomizing (succinate) pathways among the gram-negative eubacteria. (a) The *Selenomonas* pathway. (b) The *Veillonella* pathway.

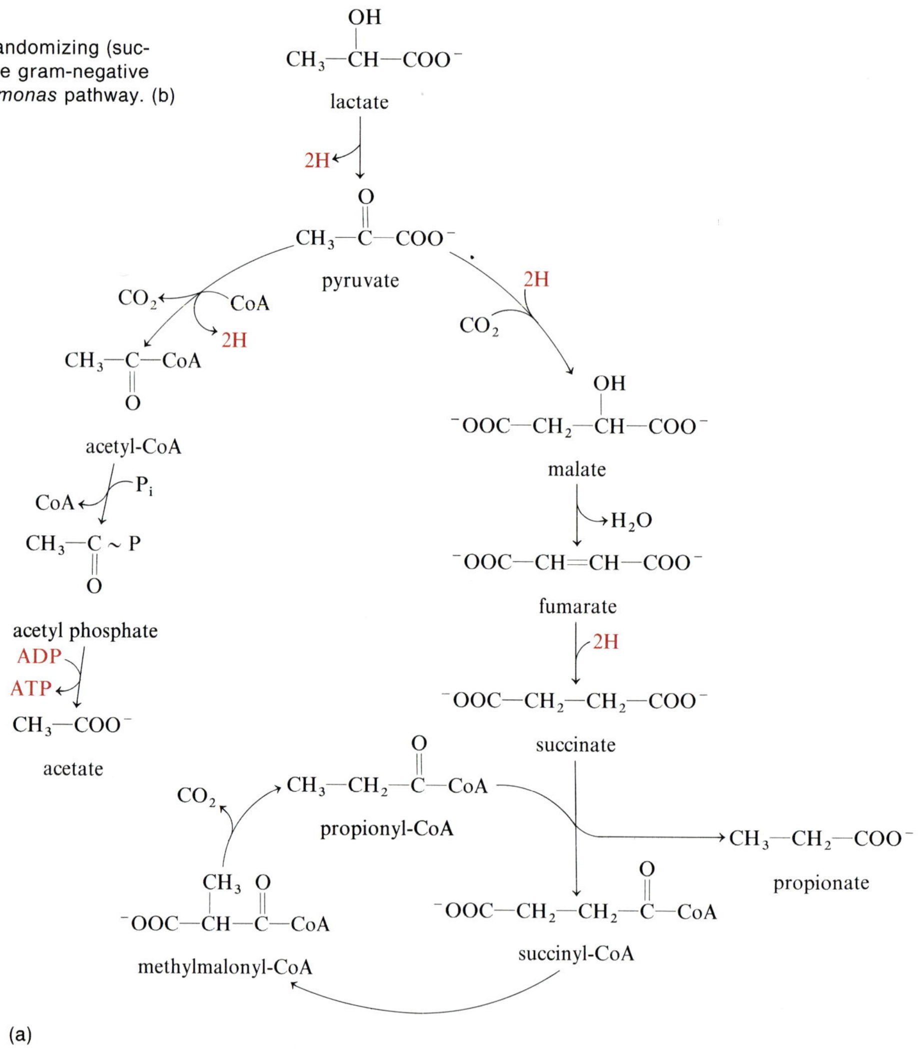

lactate
2H
pyruvate
CoA
$CO_2$
2H
acetyl-CoA
$P_i$
CoA
acetyl phosphate
ADP
ATP
$CH_3—COO^-$
acetate
2H
$CO_2$
$^-OOC—CH_2—CH(OH)—COO^-$
malate
$H_2O$
$^-OOC—CH=CH—COO^-$
fumarate
2H
$^-OOC—CH_2—CH_2—COO^-$
succinate
$CH_3—CH_2—C(=O)—CoA$
propionyl-CoA
$CO_2$
$^-OOC—CH(CH_3)—C(=O)—CoA$
methylmalonyl-CoA
$^-OOC—CH_2—CH_2—C(=O)—CoA$
succinyl-CoA
$CH_3—CH_2—COO^-$
propionate

(a)

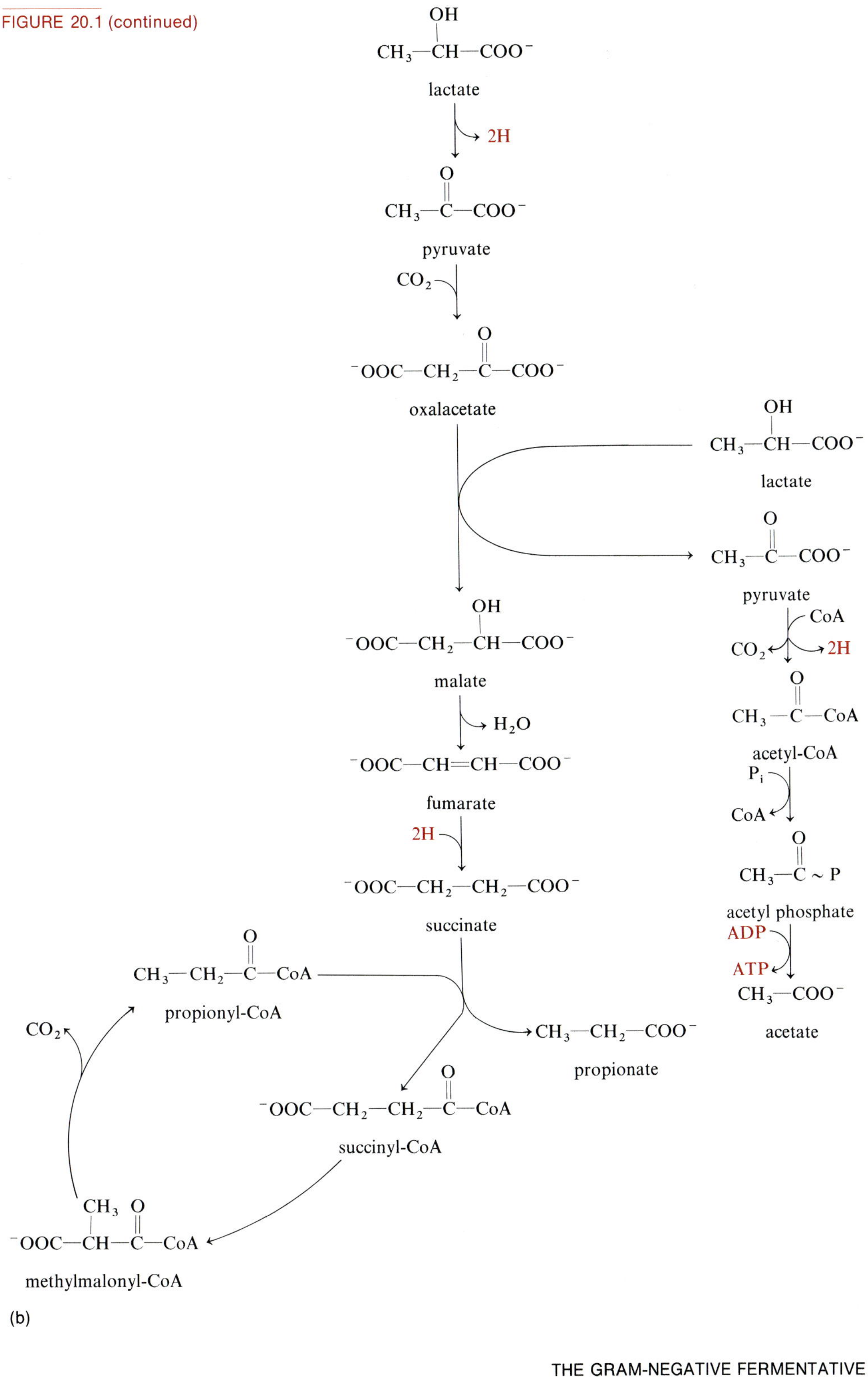

FIGURE 20.1 (continued)

$CH_3-CH(OH)-COO^-$

lactate

→ 2H

$CH_3-C(=O)-COO^-$

pyruvate

$CO_2$ →

$^-OOC-CH_2-C(=O)-COO^-$

oxalacetate

$CH_3-CH(OH)-COO^-$

lactate

$CH_3-C(=O)-COO^-$

pyruvate

CoA

$CO_2$ ← → 2H

$CH_3-C(=O)-CoA$

acetyl-CoA

$P_i$

CoA ←

$CH_3-C(=O)\sim P$

acetyl phosphate

ADP

ATP ←

$CH_3-COO^-$

acetate

$^-OOC-CH_2-CH(OH)-COO^-$

malate

→ $H_2O$

$^-OOC-CH{=}CH-COO^-$

fumarate

2H →

$^-OOC-CH_2-CH_2-COO^-$

succinate

$CH_3-CH_2-C(=O)-CoA$

propionyl-CoA

→ $CH_3-CH_2-COO^-$

propionate

$CO_2$ ←

$^-OOC-CH_2-CH_2-C(=O)-CoA$

succinyl-CoA

$^-OOC-CH(CH_3)-C(=O)-CoA$

methylmalonyl-CoA

(b)

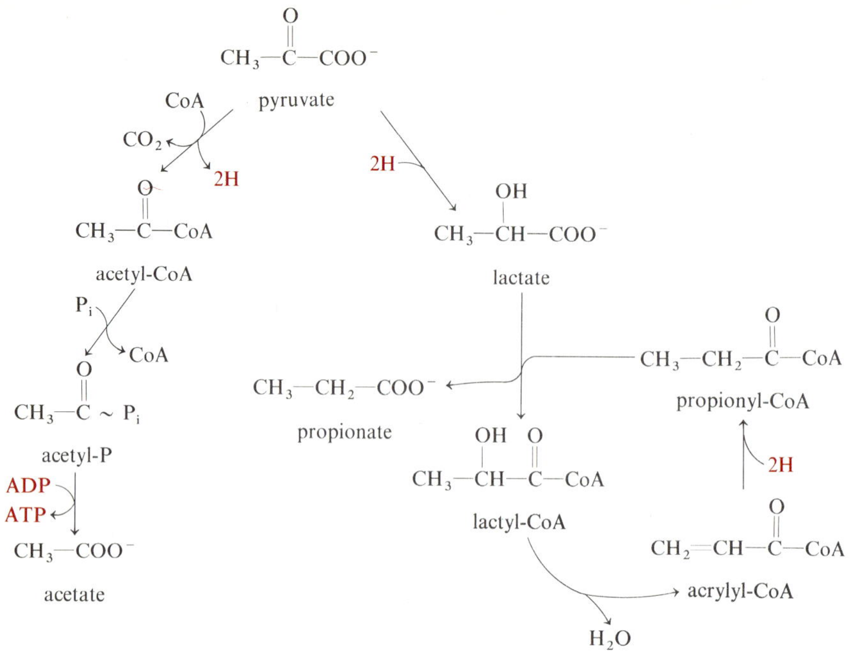

FIGURE 20.2
Propionate formation via the nonrandomizing (acrylate) pathway.

reactions by which substrate-level phosphorylation generates ATP. When sugars are the substrate, additional substrate-level phosphorylations occur, and cell yields are enhanced.

THE BUTYRATE FERMENTATION A number of bacteria produce butyrate as the sole or principal end product of fermentation (Figure 20.3). In all cases the penultimate product is butyryl-CoA, the high-energy bond of which can generate ATP by either of two pathways: the CoA moiety may be exchanged for a phosphate group in anhydride linkage, which can then be transferred to ADP [Figure 20.3(a)]; or the CoA moiety may be first transferred to acetate, generating acetyl-CoA which can generate ATP by a comparable process [Figure 20.3(b)].

THE SUCCINATE FERMENTATION Another quite common fermentation mediated by Gram-negative anaerobic bacteria is the succinate fermentation. In this fermentation a variety of substrates are converted to pyruvate which is then metabolized to a mixture of succinate and acetate (Figure 20.4).

## Fumarate Respiration

In those fermentations in which fumarate is reduced to succinate (namely, the succinate fermentation and the randomizing pathway of propionate formation), the possibility exists for coupling the reduction to electron transport. The fumarate/succinate half cell (with a reduction potential of +33 mV) produces sufficient free energy from the oxidation of NADH to generate ATP. A number of anaerobes (including facultative anaerobes) produce a short electron transport chain consisting of a menaquinone ($E'_0 = -75$ mV) and a *b*-type cytochrome ($E'_0 = -20$ mV) that couples NADH oxidation to reduction of fumarate. Obligate anaerobes that do so include representatives of Gram-negative (*Bacterioides*, *Selenomonas*, and *Veillonella*), and Gram-positive (*Propionibacterium* and *Ruminococcus*; see Chapter 23) genera.

## Nitrate Respiration

Another respiration found among anaerobic bacteria is nitrate respiration. Unlike denitrification, the characteristic reduced end product is ammonia rather than $N_2$ (although one species of *Propionibacterium* is known to produce $N_2$). Nitrate reduction is coupled to NADH oxidation via the same carriers as fumarate reduction. In some cases growth can occur with $H_2$ and $NO_3^-$ as the sole source of energy. Obligate anaerobes capable of nitrate respiration include species of the Gram-negative genera *Selenomonas* and *Veillonella*, and the Gram-positive genus, *Propionibacterium*.

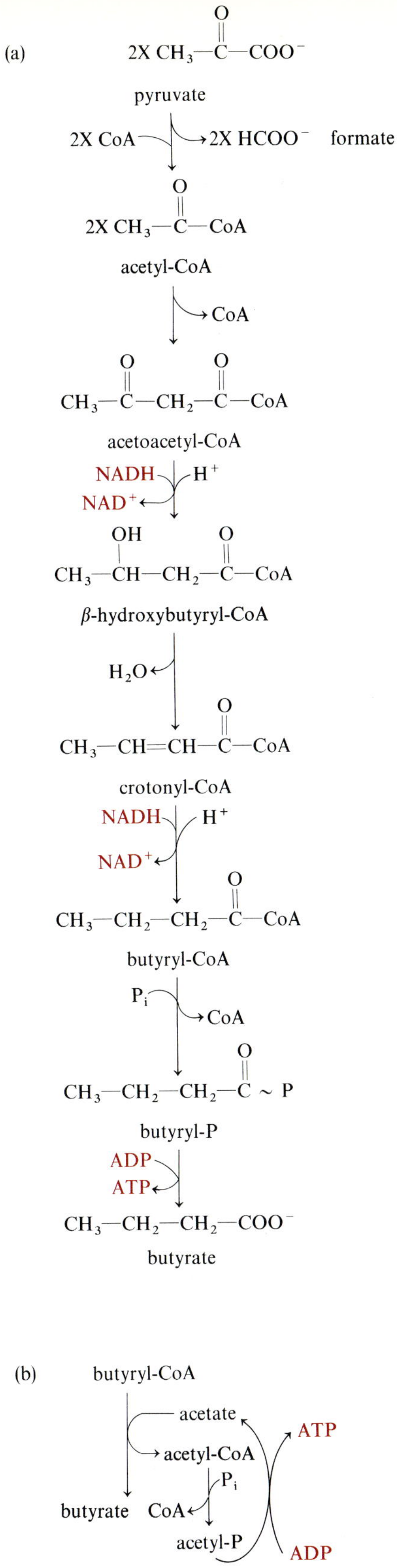

FIGURE 20.3

The butyrate fermentation.

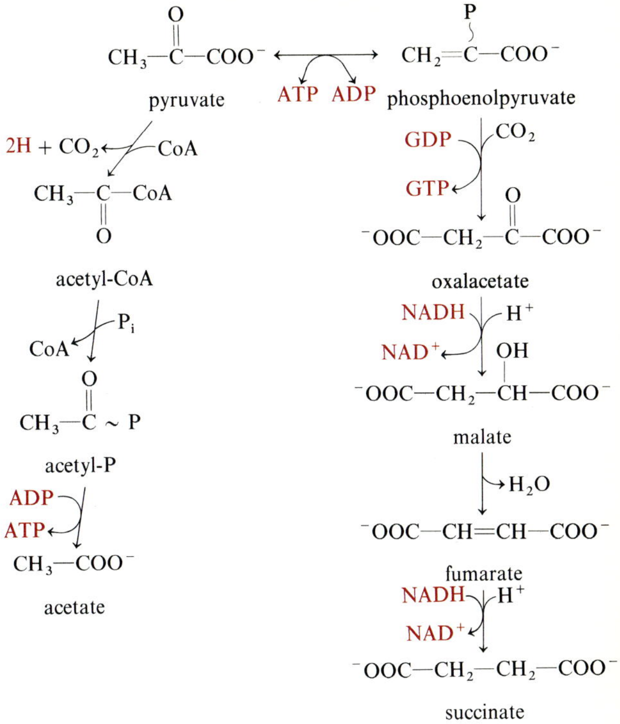

(a)

$2X\ CH_3{-}CO{-}COO^-$ (pyruvate)

2X CoA → 2X $HCOO^-$ (formate)

$2X\ CH_3{-}CO{-}CoA$ (acetyl-CoA)

→ CoA

$CH_3{-}CO{-}CH_2{-}CO{-}CoA$ (acetoacetyl-CoA)

NADH, $H^+$ → $NAD^+$

$CH_3{-}CH(OH){-}CH_2{-}CO{-}CoA$ (β-hydroxybutyryl-CoA)

→ $H_2O$

$CH_3{-}CH{=}CH{-}CO{-}CoA$ (crotonyl-CoA)

NADH, $H^+$ → $NAD^+$

$CH_3{-}CH_2{-}CH_2{-}CO{-}CoA$ (butyryl-CoA)

$P_i$ → CoA

$CH_3{-}CH_2{-}CH_2{-}CO \sim P$ (butyryl-P)

ADP → ATP

$CH_3{-}CH_2{-}CH_2{-}COO^-$ (butyrate)

(b)

butyryl-CoA → butyrate; acetate → acetyl-CoA

acetyl-CoA + $P_i$ → CoA + acetyl-P

acetyl-P + ADP → acetate + ATP

$CH_3{-}CO{-}COO^-$ (pyruvate) ⇄ $CH_2{=}C(P){-}COO^-$ (phosphoenolpyruvate) [ATP, ADP]

pyruvate → $2H + CO_2$, + CoA → $CH_3{-}CO{-}CoA$ (acetyl-CoA)

acetyl-CoA + $P_i$ → CoA + $CH_3{-}CO \sim P$ (acetyl-P)

acetyl-P + ADP → ATP + $CH_3{-}COO^-$ (acetate)

phosphoenolpyruvate + GDP + $CO_2$ → GTP + $^-OOC{-}CH_2{-}CO{-}COO^-$ (oxalacetate)

NADH, $H^+$ → $NAD^+$

$^-OOC{-}CH_2{-}CH(OH){-}COO^-$ (malate)

→ $H_2O$

$^-OOC{-}CH{=}CH{-}COO^-$ (fumarate)

NADH, $H^+$ → $NAD^+$

$^-OOC{-}CH_2{-}CH_2{-}COO^-$ (succinate)

FIGURE 20.4 (above)

The succinate fermentation.

## Constituent Groups of Gram-Negative Fermentative Eubacteria

Subdivision of this group is based primarily on morphology, and secondarily on fermentation patterns (Tables 20.1 and 20.2). The spherical members of the group are distributed among three genera (Table 20.1), the largest of which is *Veillonella.* DNA-DNA homology studies on many isolates have revealed that this genus is composed of seven quite distinct species. All are characterized by low G + C content of their DNA and their ability to ferment organic acids including pyruvate, lactate, malate, fumarate, and oxalacetate to propionate via a randomizing pathway [Figure 20.1(b)]. In addition, all veillonellas share the property of fluorescing red or pink when irradiated with long wavelength ultraviolet light; the compound responsible for the fluorescence is not known.

*Megasphaera* also ferments organic acids (lactate and pyruvate) to propionate, but by the nonrandomizing pathway (Figure 20.2). It also can ferment glucose, producing caproate as the predominant end product.

**TABLE 20.1**

**The Coccoidal Gram-Negative Fermentative Bacteria**

| | *Veillonella* | *Acidaminoccus* | *Megasphaera* |
|---|---|---|---|
| Percent G + C | 36–44 | 56–57 | 53–54 |
| Fermentable substrates | | | |
| Sugars | – | – | + |
| Amino acids | – | + | – |
| Organic acids | + | – | + |
| Fermentation | Propionate (randomizing) | Butyrate | Propionate (nonrandomizing) |
| Habitat | Mouth, intestinal tract | Intestinal tract | Rumen, intestinal tract |

**TABLE 20.2**

**The Rod-Shaped Gram-Negative Fermentative Bacteria**

| | *Bacteroides* | *Fusobacterium* | *Leptotrichia* | *Succinivibrio* | *Selenomonas* |
|---|---|---|---|---|---|
| Shape | Rod | Fusiform | Rod | Spiral | Curved rod or spiral |
| Flagellation | – | – | – | Polar | Tuft on concave side |
| Percent G + C | 28–61 | 26–34 | 25 | ND[a] | 54–61 |
| Fermentation pathway | Variable: succinate, either propionate pathway | Butyrate | Lactate | Succinate | Propionate (randomizing) |
| Fermentable substrates | | | | | |
| Sugars | + | + | + | + | + |
| Amino acids | + | + | – | – | + |
| Organic acids | + | + | + | – | + |
| Habitat | Intestinal tract, oral cavity, rumen, sediments | Intestinal tract, oral cavity | Oral cavity | Rumen | Rumen, oral cavity, intestinal tract |

[a] ND = not determined.

*Acidaminococcus* ferments amino acids, particularly glutamate, to acetate and butyrate, probably by a pathway similar to the one that occurs in certain clostridia (page 490). They require a number of specific amino acids and some vitamins as growth factors.

The rod-shaped Gram-negative anaerobes have been divided into numerous genera on the basis of morphological and physiological features; some of these genera are described in Table 20.2. One of these, *Bacteroides* is a heterogeneous assemblage; it includes strains that vary in G + C content by more than 30 percent, and by nutritional type. Collectively, *Bacteroides* are the predominant organisms in the intestinal tract of nonruminant mammals; they are also found in a variety of other habitats, including anaerobic sewage digesters and sediments.

*Fusobacterium* and *Leptotrichia* are rods with pointed and rounded ends, respectively, that ferment sugars by the butyrate (Figure 20.3) or homolactic (Chapter 23) pathways. A characteristic feature of *Leptotrichia* visible in the electron microscope is the presence of regularly spaced projections on the cell surface that originate from the outer membrane (Figure 20.5). *Fusobacterium* contains the unique compound *lanthionine* (Figure 20.6) as the murein diaminoacid.

The motile members of this group include two helical or vibrioid organisms that differ physiologically and morphologically (Table 20.2). *Selenomonas* has a distinctive cell structure (Figure 20.7): the cells are usually crescent-shaped and bear a tuft of flagella arranged in a row on the concave side of the cell. *Selenomonas* performs a randomizing propionate fermentation [Figure 20.1 (a)].

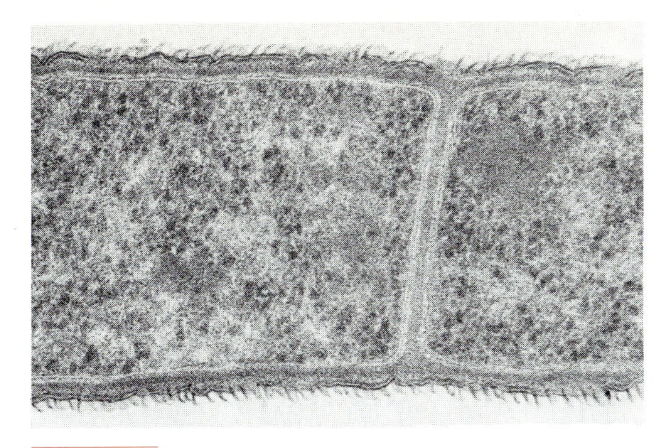

FIGURE 20.5

Electron micrograph of a thin section of *Leptotrichia buccalis* showing the distinctive scale-like projections from the outer leaf of the outer membrane (×66,700). From M. A. Listgarten and C.-H. Lai, "Unusual Cell Wall Ultrastructure of *Leptotrichia buccalis*;" *J. Bacteriol.* **123,** 747–749 (1975).

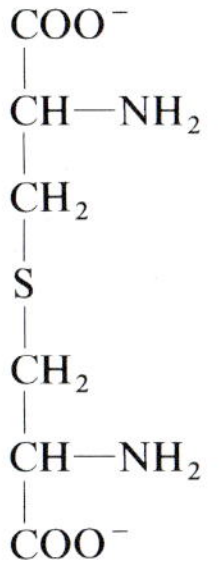

$$\begin{array}{l} COO^- \\ | \\ CH—NH_2 \\ | \\ CH_2 \\ | \\ S \\ | \\ CH_2 \\ | \\ CH—NH_2 \\ | \\ COO^- \end{array}$$

FIGURE 20.6

Lanthione, the diaminoacid of fusobacterial peptidoglycan.

FIGURE 20.7

Electron micrograph of a negative stained cell of *Selenomonas ruminantium*, showing the tuft of flagella inserted on the concave side of the cell (×10,144). From V. V. Kingsley and J. F. M. Hoeniger, "Growth, Structure, and Classification of *Selenomonas*," *Bact. Rev.* **37,** 479–521 (1973).

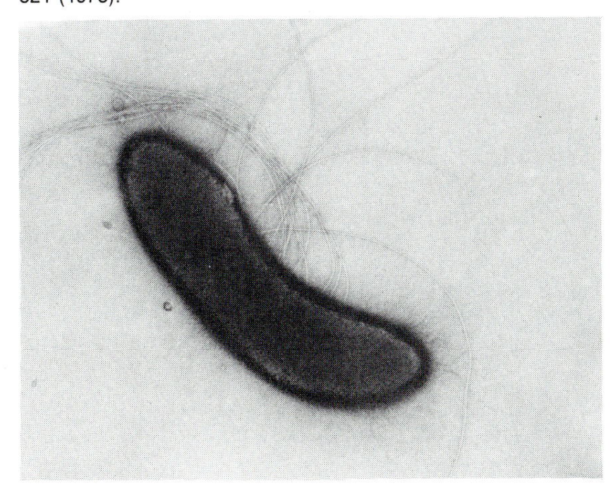

## THE SULFUR-REDUCING BACTERIA

The sulfur-reducing bacteria are strict anaerobes that are capable of anaerobic respiration utilizing a variety of oxidized compounds of sulfur as electron acceptor.* This is an uncommon ability among the eubacteria; in addition to the organisms considered here it includes only the saprophytic campylobacters (see page 421), which are capable of facultative anaerobic respiration with elemental sulfur as electron acceptor, and the Gram-positive filamentous gliding organism *Desulfonema*, which is capable of sulfate-linked respiration. The endospore former *Desulfotomaculum* (see page 493) is capable of nonrespiratory sulfate reduction accompanying the fermentation of organic compounds.

### The Pathway of Sulfate Reduction

The nature of the first few steps of dissimilatory sulfate reduction has been clearly established (Figure 20.8). Sulfate is activated by esterification to an adenyl group by the enzyme ATP sulfurylase to form adenosylphosphosulfate and pyrophosphate. The equilibrium of this reaction lies strongly towards ATP and sulfate, so the isolated reaction does not proceed to any significant extent. However, *in vivo* one of the products, pyrophosphate, is rapidly hydrolyzed by the action of pyrophosphatase, thereby allowing the other product, adenosylphosphosulfate, to accumulate in significant quantities. Adenosylphosphosulfate is then reductively cleaved to yield AMP and bisulfite.

The subsequent steps of the pathway, by which bisulfite is reduced to sulfide, are still only incompletely understood; Figure 20.8 presents the most widely accepted scheme, in which bisulfite is reduced sequentially to trithionite, thiosulfate, and sulfide. Alternatively, the reduction of bisulfite may occur in three successive two-electron transfers to enzyme-bound intermediates, analogous to the mechanism by which sulfate is reduced during its assimilation by most bacteria (see page 107). If this is the case, trithionite and thiosulfate are byproducts rather than intermediates of the pathway.

The enzyme *bisulfite reductase* is present in all sulfur-reducing bacteria. It catalyzes the reduction of bisulfite to trithionite, also producing

* This group was formerly termed the *sulfate-reducing bacteria*. However, the recent discovery that there are strains capable of reducing a variety of oxidized forms of sulfur, but not sulfate, makes a new terminology necessary. We are using the designation *sulfur-reducing bacteria* in a broad sense, to denote any organism that can couple the oxidation of an energy source to the reduction of sulfur in any of its oxidation states.

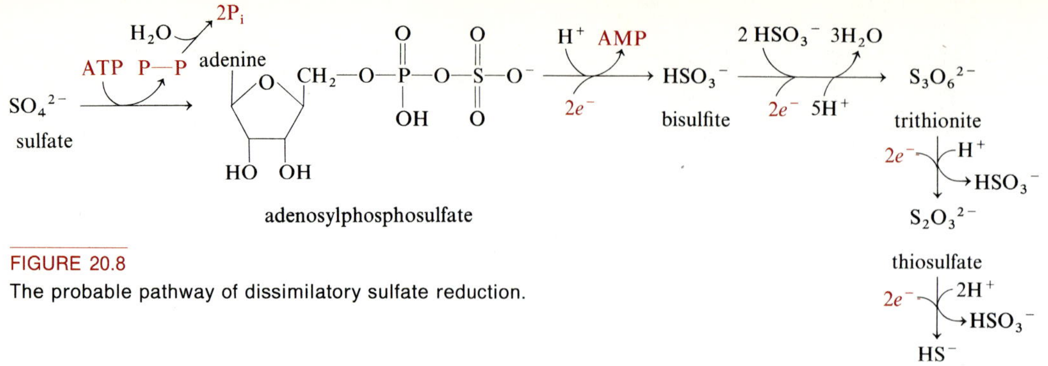

$SO_4^{2-}$ sulfate → (ATP, P—P; $H_2O$, $2P_i$) adenosylphosphosulfate → ($2e^-$; $H^+$, AMP) $HSO_3^-$ bisulfite → ($2\ HSO_3^-$, $2e^-$, $5H^+$; $3H_2O$) $S_3O_6^{2-}$ trithionite → ($2e^-$, $H^+$; $HSO_3^-$) $S_2O_3^{2-}$ thiosulfate → ($2e^-$, $2H^+$; $HSO_3^-$) $HS^-$

FIGURE 20.8
The probable pathway of dissimilatory sulfate reduction.

variable amounts of thiosulfate and sulfide. The controversy about the actual route of bisulfite reduction centers on the question of the amount of sulfide that bisulfite reductase produces *in vivo.*

Two different classes of bisulfite reductases occur in sulfur reducers; the difference has taxonomic significance. One type is a green protein termed *desulfoviridin*; the other is a reddish-brown protein termed *desulforubidin.* Both enzymes contain nonheme iron and an unusual class of prosthetic group termed *siroheme*, an iron-chelating tetrapyrrole in which two of the pyrrole rings are reduced.

## Diversity of Sulfur-Reducing Bacteria

A wide variety of sulfur reducers are now known. They are present in large numbers in sulfate-containing sediments, especially marine and estuarine ones because these sediments are constantly perfused with sulfate-containing seawater. In addition, sulfate reducers occur in smaller numbers in the rumen and the intestinal tract of nonruminant animals. Like the methanogens, they are important ecologically because they are terminal members of the anaerobic food chain. The principal genera of sulfur-reducing bacteria are described in Table 20.3; some representatives species are shown in Figure 20.9.

All sulfur-reducing bacteria, with the exception of *Desulfuromonas*, can use sulfate, sulfite, or thiosulfate as electron acceptors. Many can also use elemental sulfur or fumarate, and *Desulfobulbus* can use nitrate (producing ammonia). In contrast, *Desulfuromonas* cannot use sulfate, sulfite, thiosulfate, or nitrate, but can use elemental sulfur, polysulfide, cystine, oxidized glutathione, and fumarate. The pathway by which this organism

**TABLE 20.3**
**The Sulfur-Reducing Bacteria**

| | *Desulfovibrio* | *Desulfomonas* | *Desulfobacter* | *Desulfococcus* | *Desulfobulbus* | *Desulfosarcina* | *Desulfuromonas* |
|---|---|---|---|---|---|---|---|
| Shape | Curved rod or spiral | Rod | Coccus | Ellipsoidal | Ellipsoidal | Packets of cocci | Rod |
| $SO_4^{2-}$ used as electron acceptor | + | + | + | + | + | + | – |
| Acetate oxidized | V[a] | – | + | + | – | + | + |
| Higher fatty acids and aromatics oxidized | – | – | + | – | – | + | – |
| Desulfoviridin | +[b] | + | + | – | – | – | – |
| Motility | +[b] | – | – | V[a] | V[a] | V[a] | + |
| Percent G + C | 46–61 | 66–67 | 57 | 46 | 60 | 51 | 50–63 |

[a] V = variable.
[b] Occasional exceptions exist.

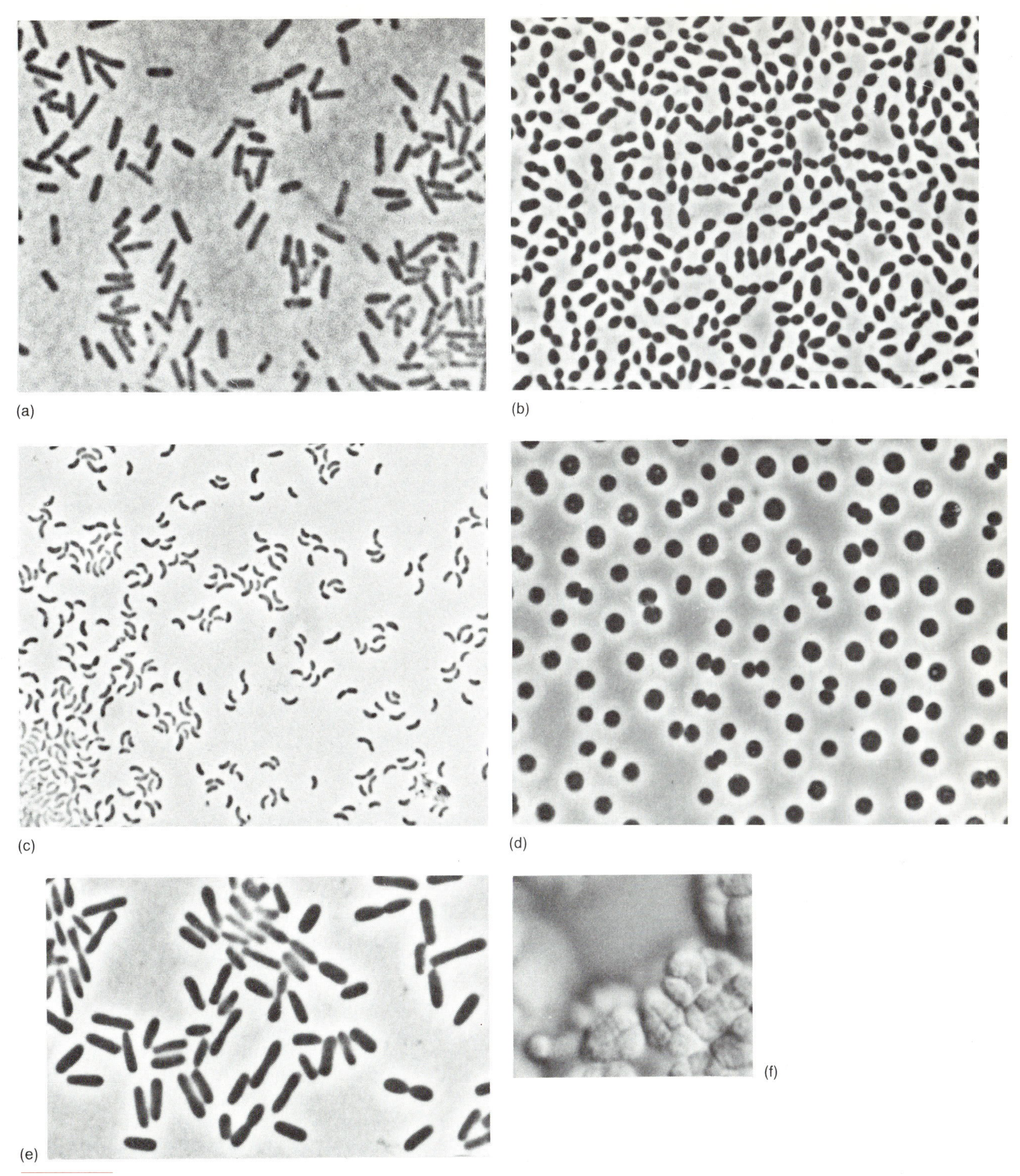


FIGURE 20.9

Representative sulfur-reducing bacteria. (a) *Desulfuromonas acetoxidans* (b) *Desulfobulbus propionicus* (c) *Desulfovibrio vulgaris* (d) *Desulfococcus multivorans* (e) *Desulfobacter* sp. (f) *Desulfosarcina variabilis*. (a)–(e) phase contrast; (f) interference contrast. (a) 2,850 × (b) 1,600 × (c) 1,086 × (d) 1,630 × (e,f) 2,000 × . (a) Courtesy of Dr. Norbert Pfennig; (b)–(f) Courtesy of Dr. Friedrich Widdel.

reduces sulfur is unknown. The rumen organism *Wolinella* and the campylobacters appear to be physiologically similar to *Desulfuromonas*; probably the group of bacteria that uses elemental sulfur but not sulfate as terminal oxidant is large and diverse.

Electron donors for all the sulfur-reducing bacteria are the fermentative end products produced by other anaerobes: lactate, ethanol, short-chain fatty acids (e.g., propionate, butyrate), formate, and $H_2$. There is, however, a marked dichotomy within the sulfate reducers with regard to ability to use acetate as electron donor. Many have only an incomplete TCA cycle, and therefore cannot oxidize acetate; rather they accumulate it as an end product of the oxidation of lactate, malate, or other organic compounds. Other sulfate-reducing bacteria, and *Desulfuromonas*, have a complete TCA cycle and oxidize acetate completely to $CO_2$.

$H_2$ plays a central role in the metabolism of those sulfur-reducing bacteria that cannot oxidize acetate. It is generated by a soluble hydrogenase from hydrogen atoms derived from the oxidation of compounds such as lactate, and then diffuses across the cell membrane; its subsequent fate depends on the external conditions. If sulfate is available, it is oxidized by a periplasmic hydrogenase that transfers the electrons inwards reducing sulfate intracellularly. The result, termed *hydrogen cycling* (Figure 20.10), is equivalent to the respiratory pumping of protons, and results in a protonmotive force (this mechanism is very similar to that proposed to explain ATP generation by methanogens; see Figure 14.7). However, if sulfate is not available, and the habitat contains methanogens, interspecies hydrogen transfer can allow growth of the sulfur reducer by substrate-level phosphorylation.

FIGURE 20.10
Hydrogen cycling in the sulfur-reducing bacteria.

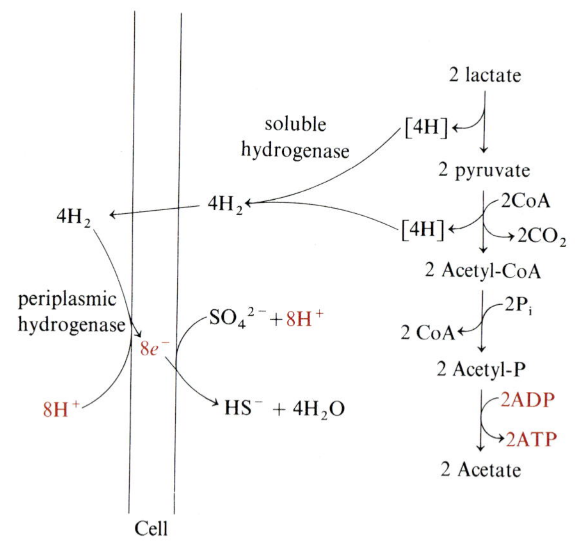


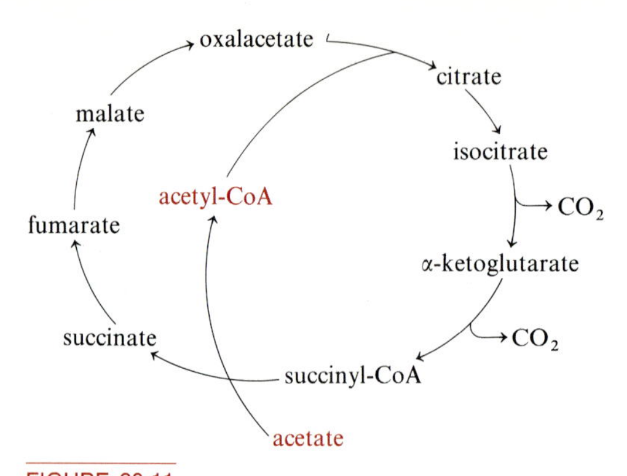


FIGURE 20.11
Oxidation of acetate via the TCA cycle in sulfur-reducing bacteria.

In those groups that oxidize acetate, the first step of the process is formation of acetyl-CoA. In aerobic heterotrophs this is an energetically expensive process, catalyzed by the enzyme *acetyl-CoA synthetase:*

$$\text{acetate} + \text{ATP} + \text{CoA} \longrightarrow \text{acetyl-CoA} + \text{AMP} + \text{PP}$$

However, in at least some acetate-oxidizing sulfur reducers, acetyl-CoA is synthesized by the action of a CoA transferase that catalyzes the transfer of CoA moiety from succinyl-CoA produced in the TCA cycle (Figure 20.11), thus conserving the equivalent of two molecules of ATP.

In addition to being able to oxidize short-chain fatty acids and other organic acids, *Desulfococcus* and *Desulfosarcina* can utilize a variety of long-chain fatty acids (up to 18 carbons) and several aromatic compounds.

Although the ability to use $H_2$ or formate as electron donor is widespread among the sulfur reducing bacteria, most strains with this ability require an organic carbon source. The only exceptions are *Desulfosarcina*, which can grow chemoautotrophically with $CO_2$ and $H_2$ or formate, and *Desulfovibrio baarsii*, which can grow with $CO_2$ and formate. $CO_2$ is assimilated by the *carbon monoxide pathway* (see Figure 22.24), which occurs in some clostridia, *Acetobacterium*, and probably also in methanogens and some other autotrophic archaebacteria.

In addition to respiratory metabolism, some sulfur-reducing bacteria are capable of fermenting pyruvate or lactate to propionate and acetate. Organisms with this capability include *Desulfococcus*, *Desulfobulbus*, and *Desulfosarcina*.

Some sulfur-reducing bacteria have growth factor requirements. Many accumulate poly-β-hydroxybutyrate, glycogen, or polyphosphate. Nitrogenase is made by many strains.

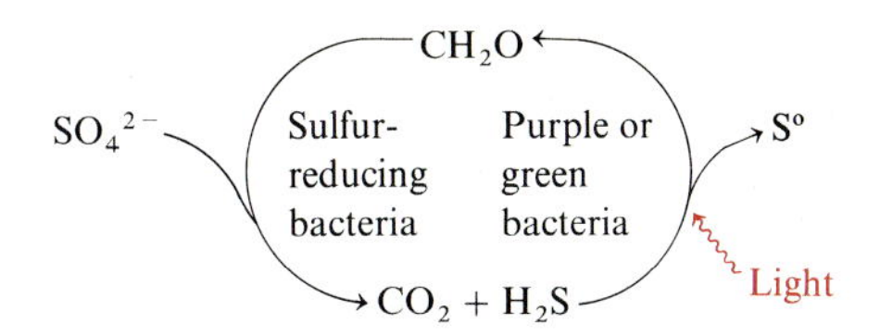


FIGURE 20.12
The formation of elemental sulfur from sulfate in a sulfuretum.

## Ecological Activities

The sulfur-reducing bacteria are widely distributed, and produce large quantities of sulfide under appropriate conditions. Sulfide is quite toxic, and consequently these bacteria are occasionally responsible for massive mortality of fish and estuarine waterfowl. In waterlogged (and hence largely anaerobic) soils, sufficient sulfide may be produced to damage plants. This is a particular problem in the cultivation of rice, which is normally grown in flooded paddies to control weeds; the role of *Beggiatoa* in preventing the accumulation of $H_2S$ and thereby protecting rice plants was discussed on page 389. Sulfur-reducing bacteria also contribute to the corrosion of metal pipes in anaerobic soils or waters.

Sulfide production by the sulfur-reducing bacteria can provide a source of reductant to support the growth of either aerobic chemoautotrophs, or anaerobic green or purple bacteria in illuminated habitats. In some shallow stagnant lakes fed with sulfate-rich spring water, a quantitative conversion of sulfate to elemental sulfur is catalyzed by the combined activities of sulfur-reducing bacteria and phototrophic bacteria (Figure 20.12). Such habitats are termed *sulfureta* (singular *sulfuretum*), and can become so rich in suspended elemental sulfur that the water becomes densely turbid. Most known geological deposits of elemental sulfur were formed in this way, as is revealed by their isotopic composition.*

* Sulfur, like most other elements, has several stable isotopes. Chemical processes, for instance the oxidation of sulfide by $O_2$, do not discriminate among isotopes. Hence geochemically produced elemental sulfur would be expected to have the same isotope ratios as the average for all sulfur atoms. Enzymes, however, characteristically catalyze reactions with light isotopes more rapidly than reactions with heavy isotopes. Consequently, biologically produced products tend to be slightly enriched in lighter isotopes.

# FURTHER READING


**Book**

POSTGATE, J. R. *The Sulphate-Reducing Bacteria.* Cambridge: Cambridge University Press, 1984.

**Reviews**

MACY, J. M., and I. PROBST, "The Biology of Gastrointestinal Bacteroides," *Ann. Rev. Microbiol.* **33,** 561–594 (1979).

ODOM, J. M., and H. D. PECK, JR., "Hydrogenase, Electron-Transfer Proteins, and Energy Coupling in the Sulfate-Reducing Bacteria *Desulfovibrio*," *Ann. Rev. Microbiol.* **38,** 551–593 (1984).

## Chapter 21

# Gram-Negative Eubacteria: Spirochetes, Rickettsias and Chlamydias

In the preceding six chapters the Gram-negative eubacteria have been divided into groups principally on the basis of their modes of energy metabolism. In this chapter the remainder of the Gram-negative eubacteria are described: the *spirochetes*, defined by their unique cellular structure; and the *rickettsias* and *chlamydias*, two groups of parasitic bacteria distinguished by their life cycles.

## THE SPIROCHETES

The spirochetes are a heterogenous group of bacteria that share a distinctive cell structure: their cell body is helical, and is intertwined with an organelle termed the *axial filament*, a bundle of fibrils that winds around the cell body between the murein layer and the outer membrane (Figures 21.1 and 21.2). The individual fibrils that compose the axial filament originate near the ends of the cell, and are normally more than half the length of the cell; they therefore overlap in the middle. *Leptospira* is an exception; its axial filament is discontinuous, because the fibrils originating at each end are less than half the length of the cell.

The fibrils of the axial filament have a fine structure essentially the same as that of flagella: a basal body containing a series of disks; a hook; and the filament itself. The number of disks varies among the spirochetes: *Leptospira* typically has two pairs of disks, similar to other Gram-negative bacteria, whereas most other spirochetes have a single pair of disks; their basal bodies accordingly resemble those of Gram-positive bacteria. The

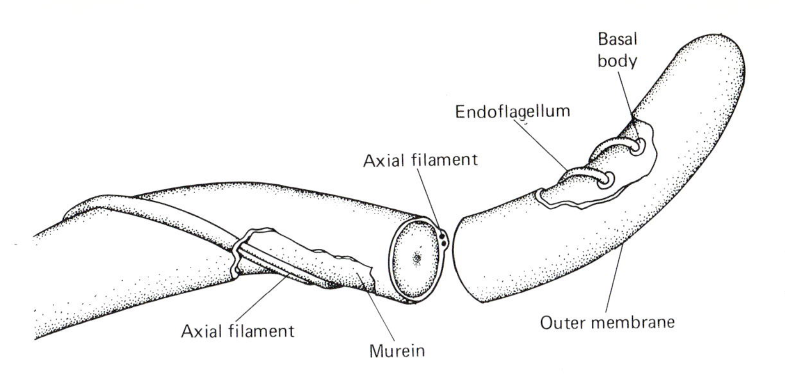


FIGURE 21.1

The cell structure of a spirochete. After S. C. Holt, "Anatomy and Chemistry of Spirochetes," *Microbiol. Rev.* **42,** 114–160 (1978).

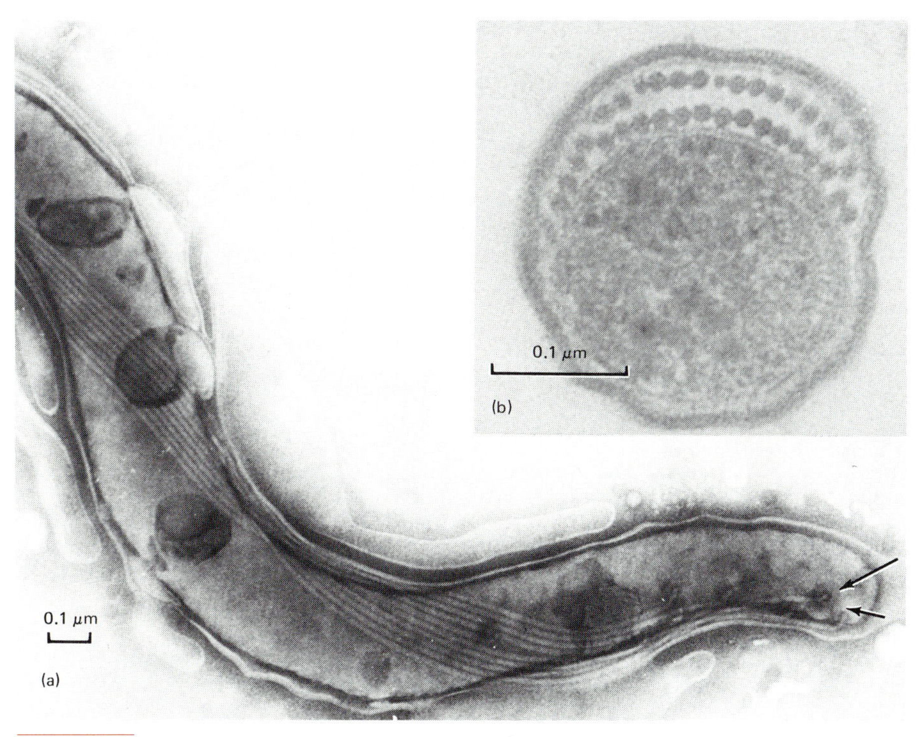


FIGURE 21.2

The structure of the spirochetal cell as shown by electron micrographs. (a) End of a cell, negatively stained with phosphotungstic acid, showing the relationship of the axial filament to the protoplast (× 51,000). The insertion points of two endoflagella are just visible at the pole of the cell (arrows). (b) Cross section of a large spirochete, showing the location of the endoflagella underlying the outer membrane (× 183,000). From M. A. Listgarten and S. S. Socransky, "Electron Microscopy of Axial Fibrils, Outer Envelope and Cell Division of Certain Oral Spirochetes," *J. Bacteriol.* **88,** 1087 (1964).

filaments are often enveloped in a proteinaceous sheath, unlike the eubacterial flagellum which, if sheathed, is surrounded by an extension of the outer membrane (see Chapter 6). The close structural and chemical resemblance between the axial fibril and the flagellum, as well as their similar role in motility (see below), suggests that they are homologous organelles. The terms *endoflagella* and *periplasmic flagella* have accordingly been proposed to describe axial fibrils.

The number of endoflagella per cell varies over a large range. Typically, *Leptospira* and *Spirochaeta* have 2, *Treponema* from 2 to 16, and *Borrelia* 30 to 40; some of the large spirochetes not yet available in pure culture have over 200. The numbers inserted at each end of the cell are approximately equal.

## Motility of Spirochetes

Spirochetes are all very actively motile, exhibiting a variety of translational and nontranslational movements. Among the latter are twisting, lashing, and writhing movements that are characteristic of the group. Since these movements are very rapid, and since the cells are often so thin as to be at the edge of detection with the light microscope, they look as if they were instantaneously moving from one position to another. Translational motility occurs both on solid substrata and in suspension. On surfaces, spirochetes can glide, a motility often termed *creeping* rather than gliding. They also exhibit an "inchworm" kind of motility, in which one end of the cell attaches to the surface, and the other end attaches nearby; the first end then detaches and reattaches at a distance. Repetition of these steps results in relatively rapid movement.

Translational motility through a liquid is quite rapid, and, unlike flagellår motility, is most effective at high viscosities. Flagellated bacteria show a rapid decrease in rate of movement above viscosities of about 5 cP, and are completely immotile above 60 cP; spirochetes, in contrast, swim faster as the viscosity is increased past 300 to 500 cP (approximately the viscosity of engine oil), and are not immobilized until viscosities reach about 1,000 cP. Indeed, *Leptospira* has been shown to be positively tactic to increasing viscosity. Apparently, spirochetal motility is especially effective in habitats such as thick muds and the mucous membranes of animals, both of which may support large numbers of spirochetes.

The axial filament has long been suspected to play a role in motility, a suspicion that was strengthened when the structural resemblance between their constituent fibers and bacterial flagella was recognized. The isolation of immotile mutants with altered axial fibrils (straight rather than coiled) confirms the role of endoflagella in motility.

Nevertheless, it is still unclear how motion is actually effected. Because of the structural homology between flagella and endoflagella, it is tempting to propose that they share a similar mechanism, namely rotation from the base of the filament. How endoflagellar rotation would result in cell movement is controversial. However it is clearly not due

**TABLE 21.1**

**Distinguishing Characteristics of Spirochetes in Pure Culture**

| | *Spirochaeta* | *Treponema* | *Borrelia* | *Leptospira* |
|---|---|---|---|---|
| Cell diameter | 0.2–0.8 μm | 0.1–0.4 μm | 0.2–0.5 μm | 0.1 μm |
| Relationship to $O_2$ | Obligate or facultative anaerobe | Anaerobe | Microaerophile | Aerobe |
| Energy metabolism | Fermentative or facultatively respiratory | Fermentative | Fermentative | Respiratory |
| Percent G + C | 51–65 | 25–54 | Unknown | 35–53 |
| Carbon and energy source | Sugars | Sugars or amino acids | Sugars | Fatty acids or fatty alcohols |
| Requires fatty acids for lipid synthesis | − | + | + | + |
| Monogalactosyl diglyceride | + | + | + | − |
| 3-*O*-Methyl-mannose in cell envelope | − | − | − | + |
| Murein diamino acid | Ornithine | Ornithine | Ornithine | Diaminopimelate |
| Habitat | Sediments | Oral cavity, intestinal and genital tracts, rumen | Systemic pathogen of humans and arthropods | Soil, mammalian kidney |

to rotation of the cell as a whole; rather a *helical wave* is propagated down the cell. Two fundamental mechanisms are envisaged: rotation of flexible endoflagella may cause the rigid helical murein-bound protoplast to rotate within the outer membrane (whose counterrotation would be resisted by shear forces with the medium); or rotation of the rigid helical endoflagella may cause the pliable cell body to flex in the form of a helical wave.

## Cell Division in the Spirochetes

Most spirochetes are unicellular, and divide by binary fission. One of the first detectable steps in division is the appearance of a new set of endoflagella originating at the middle of the cell. A septum then is laid down between the basal bodies, and the two daughter cells separate. However, such coupling between cell division and cell separation is not universal; there are strains of *Spirochaeta* that are multicellular filaments up to 250 μm long.

## Diversity of Spirochetes

Relatively few spirochetes are available in pure culture; they are described in Table 21.1 and Figure 21.3. Even these few representatives constitute a heterogeneous group in terms of their metabolism, habitats, percent G + C, and rRNA sequences.

Lipids are a major part of the spirochete cell mass (about 20 percent of the dry weight). All but leptospires contain monogalactosyl-diglyceride as one of their polar lipids; *Borrelia* also contains cholesterol.

Only *Spirochaeta* is capable of *de novo* fatty acid synthesis; all other spirochetes are consequently fatty acid auxotrophs. Most of them require long chain (>15 carbons) fatty acids, and

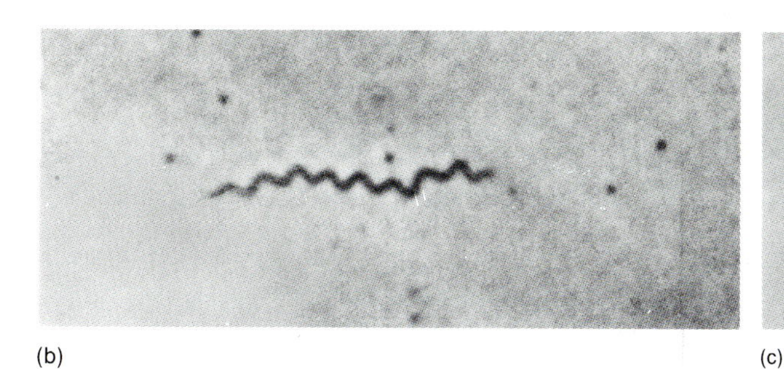

(a)

FIGURE 21.3

Phase-contrast light micrographs of representative spirochetes. (a) *Spirochaeta plicatilis*. (b) *Treponema pallidum*. (c) *Borrelia anserina*. (d) *Leptospira interrogans*. (All ×2200). Courtesy of Dr. D. A. Kuhn, from *Bergey's Manual of Determinative Bacteriology*, 8th ed., Baltimore: Williams and Wilkens (1974).

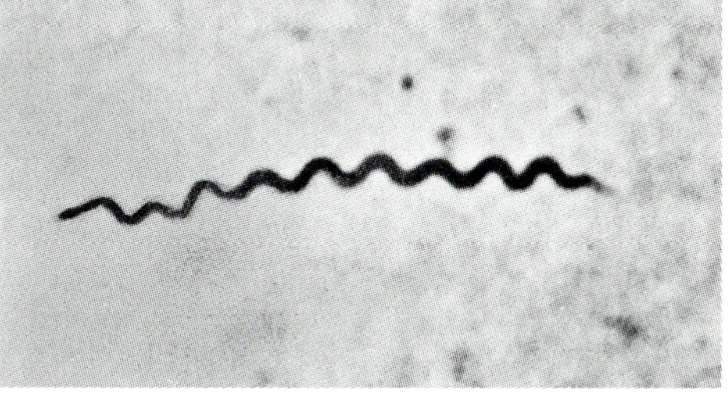

(b) (c)

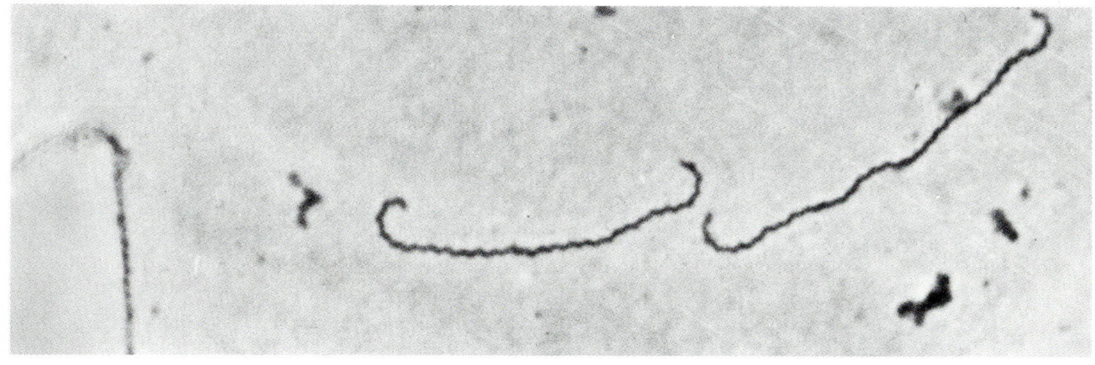

(d)

some require a mixture of saturated and unsaturated ones. Some of the treponemes, however, appear to be able to elongate short fatty acids, and can consequently grow with isobutyrate and *n*-valerate as lipid precursors.

The envelope of many spirochetes contains substantial amounts of polysaccharide of varying composition. The presence of 3-O-methylmannose in this carbohydrate fraction is characteristic of *Leptospira*.

*Spirochaeta* are free-living, and characteristically inhabit anaerobic or microaerobic sediments. They have a high sulfide tolerance, important in sediments that support active sulfur reduction. They ferment carbohydrates (including cellobiose but not cellulose) via the Embden-Meyerhof pathway to pyruvate, which is cleaved by a clostridial-type clastic reaction (page 488) to acetyl-CoA. Fermentation products are ethanol, acetate, $CO_2$, and $H_2$ (Figure 21.4). One species of *Spirochaeta*, *S. zuelzerae*, produces lactate and some succinate rather than ethanol.

Although *Spirochaeta* is unable to grow fermentatively with substrates other than carbohydrates, they can apparently gain maintenance energy from fermenting branched amino acids by the scheme shown in Figure 21.5. After transamination has produced an α-ketoacid, the pathway is formally equivalent to acetate production from pyruvate; however, the enzymatic mechanism for the oxidative decarboxylation of the α-ketoacid is different, and does not produce $H_2$. The absence of substrate amounts of an appropriate electron acceptor probably prevents growth at the expense of this fermentation.

Two species of *Spirochaeta*, *S. aurantia* and the halophile *S. halophila*, are facultative aerobes, and can synthesize components of a respiratory electron transport chain. However, they lack a complete TCA cycle and consequently accumulate acetate and $CO_2$ from carbohydrates.

*Treponema* includes a number of parasitic but not pathogenic anaerobes. They include both amino acid–fermenting and carbohydrate-fermenting strains. The carbohydrate-fermenting strains produce acetate and butyrate as major products, sometimes accompanied by varying amounts of succinate or lactate, and ethanol and butanol. The amino acid–fermenting strains produce principally acetate, with varying amounts of propionate, butyrate, or lactate.

Pathogenic *Treponema* strains and *Borrelia* are both microaerophilic, parasitic spirochetes. *Treponema* includes the aetiologic agents of four chronic human diseases spread by direct contact (venereal syphilis, endemic syphilis, yaws, and pinta) collectively known as *treponematoses*. *Borrelia* causes relapsing fever (tick- or louse-borne) in humans. Both organisms have complex and poorly understood nutritional requirements. The borrelias have been cultivated on complex media including N-acetylglucosamine, a required growth factor; the pathogenic treponemes have not

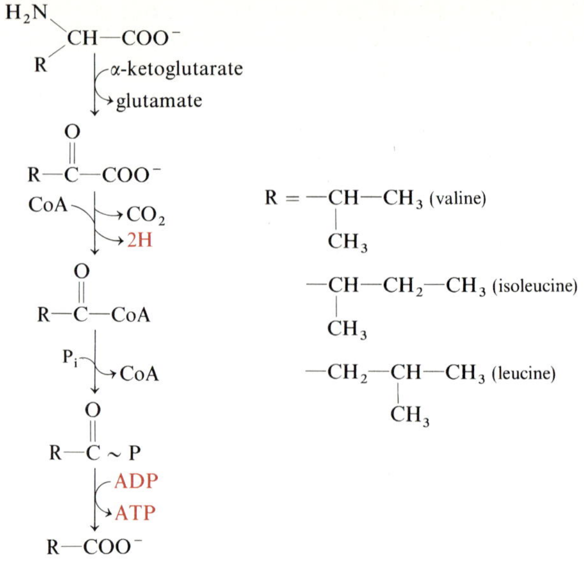

$$H_2N(R)CH{-}COO^- \xrightarrow[\alpha\text{-ketoglutarate} \rightarrow \text{glutamate}]{} R{-}CO{-}COO^- \xrightarrow[\text{CoA};\ \rightarrow CO_2,\ 2H]{} R{-}CO{-}CoA \xrightarrow[P_i;\ \rightarrow \text{CoA}]{} R{-}CO{\sim}P \xrightarrow[\text{ADP} \rightarrow \text{ATP}]{} R{-}COO^-$$

R = —CH(CH₃)—CH₃ (valine); —CH(CH₃)—CH₂—CH₃ (isoleucine); —CH₂—CH(CH₃)—CH₃ (leucine)

FIGURE 21.5

Branched amino acid fermentation by *Spirochaeta*.

FIGURE 21.4

Hexose fermentation by *Spirochaeta*. Reducing equivalents are denoted [H], since the actual carriers are not known.

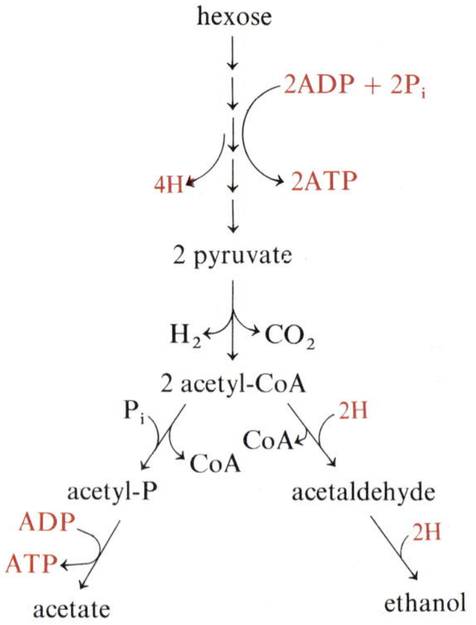

hexose → (2ADP + 2P$_i$ → 2ATP; → 4H) → 2 pyruvate → ($H_2$, $CO_2$) → 2 acetyl-CoA

2 acetyl-CoA → ($P_i$ in, CoA out) → acetyl-P → (ADP → ATP) → acetate

2 acetyl-CoA → (2H in, CoA out) → acetaldehyde → (2H) → ethanol

been cultured axenically. They are normally cultured by inoculation into host animals, or in tissue culture with animal cells.

*Borrelia* is fermentative, performing a homolactic fermentation of glucose. The pathogenic treponemes are possibly respiratory; *T. pallidum* (the causative agent of venereal syphilis) has been reported to contain cytochromes, and to convert glucose to acetate and $CO_2$.

The leptospires are the only obligately aerobic spirochetes. They require fatty acids not only as precursors of membrane lipids, but also as their respiratory substrate; they cannot respire carbohydrates or other compounds. They have a characteristic morphology: the posterior end of swimming cells is always hooked, while the anterior end is straight or helical; immotile cells may be hooked at both ends (Figure 21.3).

There are two distinct groups of leptospires, which probably each deserve generic status. One consists of the free-living strains; they are characteristically soil organisms, but may frequently be isolated from water. The other group contains parasitic strains, which characteristically inhabit the mammalian kidney, where they colonize the proximal convoluted tubules. They are principally parasites of rodents and domestic animals, and may be present in large numbers without apparent harm to the host. They are opportunistic pathogens of humans, who acquire them by drinking water contaminated with urine from infected animals.

### Spirochetes Symbiotic with Invertebrate Animals

Two very different symbiotic habitats are characterized by microbial floras with large numbers of morphologically distinctive spirochetes: the *crystalline style* of bivalve and gastropod molluscs; and the hindgut of termites and wood-eating roaches. These spirochetes are characteristically large (0.4 to 3.0 $\mu$m diameter) and have many endoflagella. None of them has been obtained in pure culture.

Many molluscs have a digestive system that contains an organ termed the crystalline style. The style is a gelatinous rod consisting of mucoprotein, with a variety of digestive enzymes embedded in it. It is housed in the *style sac*, and one end protrudes into the stomach. Cilia in the style sac rotate the style, abrading it against the chitinous *gastric shield* to release the contained enzymes, and simultaneously reeling in the food-bearing mucous strand from the gills (Figure 21.6). Most molluscs with a crystalline style harbor large numbers of a distinctive spirochete, *Cristispira* (Figure 21.7). *Cristispira* is quite large; healthy cells may be mistaken for spirilla, and unhealthy cells, with their outer membrane distended by the axial filament, have been mistaken for trypanosome protists. The principal habitat of *Cristispira* is the style itself, but variable numbers may be found free in the style sac or elsewhere in the intestinal tract.

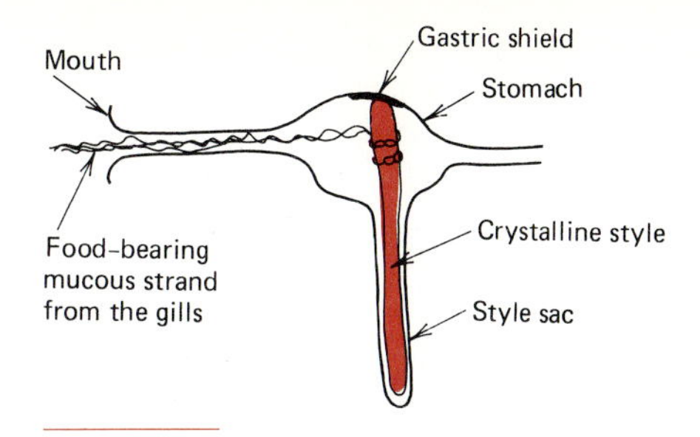


FIGURE 21.6
Schematic diagram of the digestive tract of style-bearing gastropods.

The hindgut of termites and wood-eating roaches is a fermentation chamber in which ingested cellulose is fermented by the microbial inhabitants. The microbial flora of many of these insects is characterized by flagellate protozoa and by large spirochetes, many of which are attached to the surface of the protozoa (Figure 21.8). Several genera have been proposed for these spirochetes; their collective name is *pillotinas* after the generic name of one of them. In some cases it is clear that the attached pillotinas provide motility for the host protozoan; the flagella steer, while the propulsive force is due to coordinated swimming motions by the spirochetes. The pillotinas are a morphologically diverse group; they characteristically have a large number of endoflagella, and their outer membrane is often crenulated or grooved (Figure 21.9).

## THE RICKETTSIAS

It is known from electron microscopic studies that virtually all classes of metazoan organisms support chronic infection by bacteria, frequently endosymbiotic. Most of these bacteria have not been further characterized; even those that cause human disease are still relatively poorly understood. Those that cause arthropod disease, or mammalian disease transmissible by arthropods, are placed in the *rickettsias*.

Most of the rickettsias are obligate intracellular parasites, and have not been cultured axenically. They are generally cultured either by inoculation into the yolk sac of chicken eggs, or by infecting

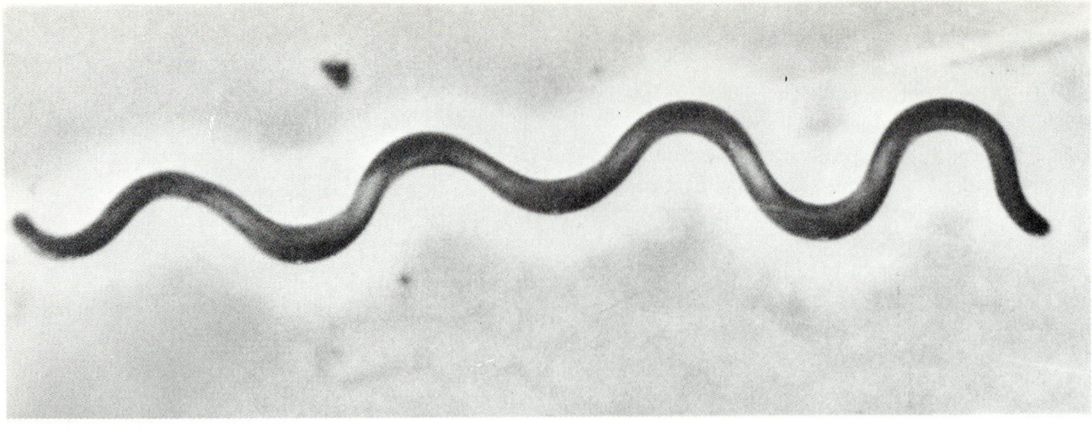

(a)

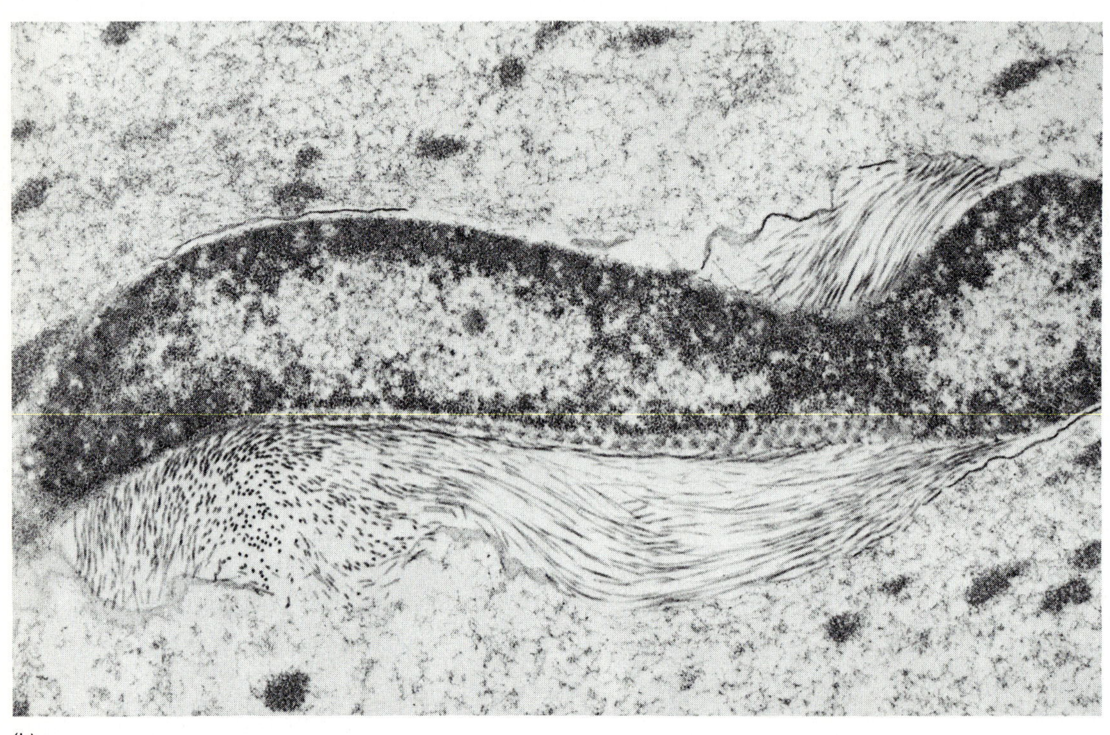

(b)

FIGURE 21.7

*Cristispira.* (a) Phase-contrast light micrograph, ×2200. (b) Electron micrograph of a thin section of the terminal portion of a cell showing the numerous endoflagella and a row of basal bodies (×26,180). (a) Courtesy of Dr. D. A. Kuhn, from *Bergey's Manual of Determinative Bacteriology*, 8th ed., Baltimore: Williams and Wilkens (1974). (b) Courtesy of P. W. Johnson and J. M. Sieburth, University of Rhode Island and Biological Photo Service.

FIGURE 21.8

Interference contrast photomicrograph of the anterior end of the protozoan *Mixotricha paradoxa*, showing the wavy contour of the cell, a result of numerous adherent spirochetes. Courtesy of Dr. Sidney L. Tamm.

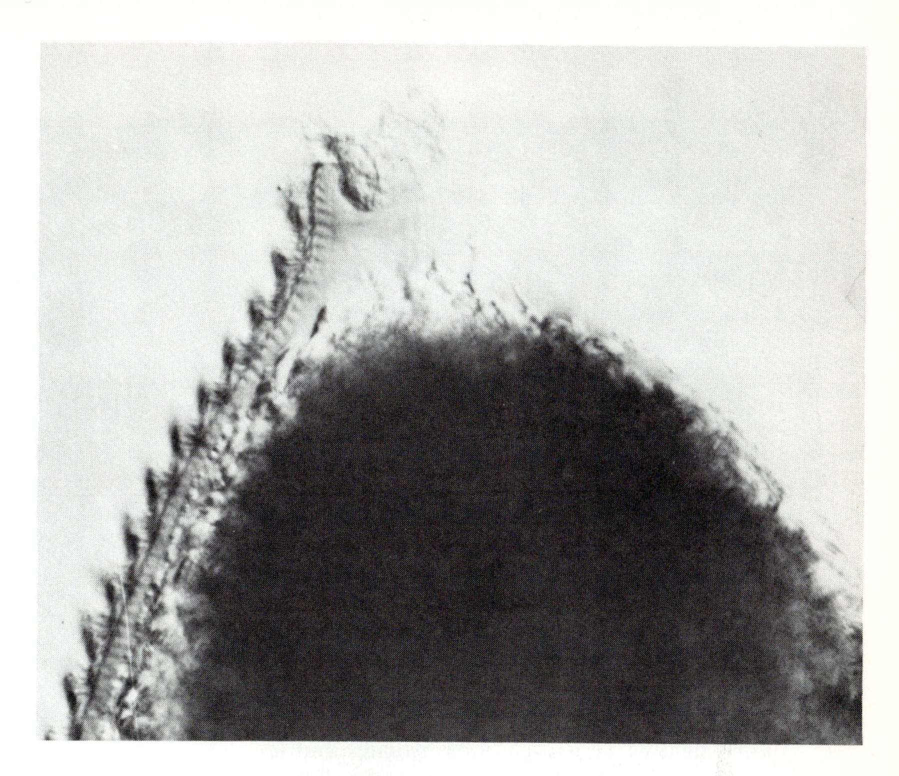

FIGURE 21.9

Electron micrographs of transverse thin sections of pillotinas from the hindgut of the termite *Reticulitermes flavipes*. (a) a cell showing a groove where the outer membrane is in local contact with the peptidoglycan layer. (b) a cell whose crenulated outer membrane is also in local contact with the murein. Both cells have numerous endoflagella in the periplasm. (a) 81,000 ×; (b) 39,000 ×. From J. A. Breznak, "Hindgut Spirochetes of Termites and *Cryptocercus punctulatus*," in N. R. Krieg and J. G. Holt, eds., *Bergey's Manual of Systematic Bacteriology*, Vol. 1, 67–70, Baltimore, Md.: Williams and Wilkins (1984).

(a)

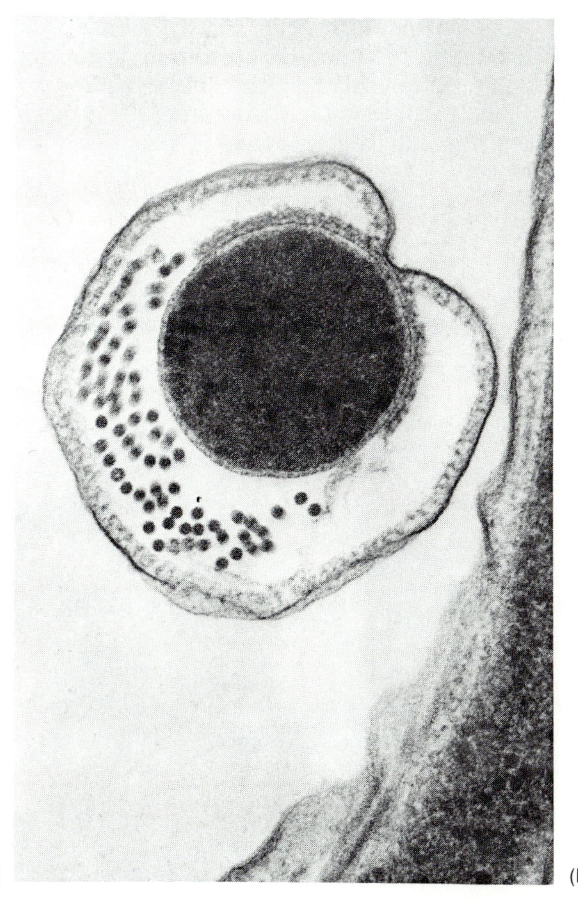

(b)

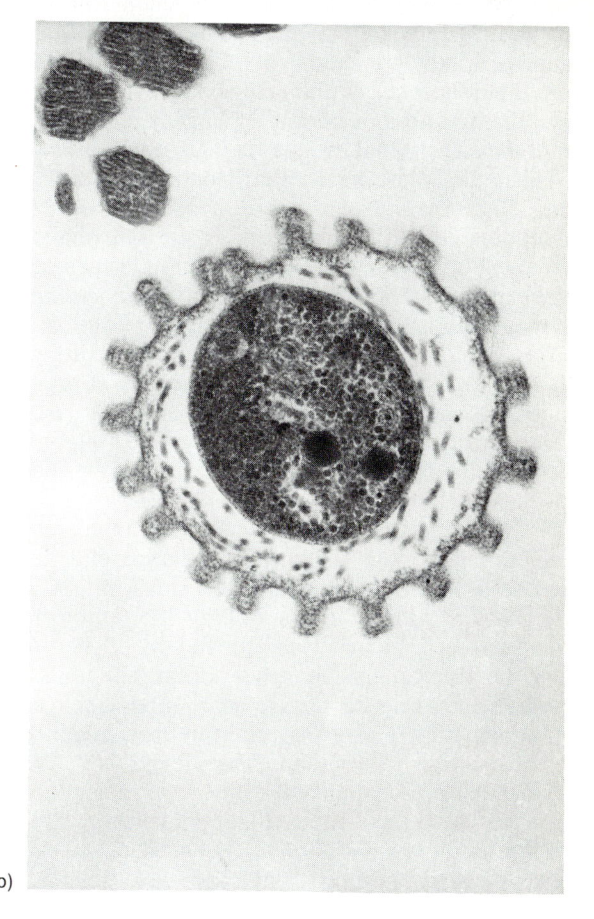

**TABLE 21.2**

**The Rickettsias**

| | *Rickettsia* | *Coxiella* | *Rochalimaea* |
|---|---|---|---|
| Site of multiplication | Cytoplasm | Phagolysosome | Exterior surface of host cell |
| Obligate intracellular parasite | + | + | − |
| Can be cultivated axenically | − | − | + |
| Spore formation | − | + | − |
| Human diseases | Typhus, scrub typhus, spotted fever | Q fever | Trench fever |

host cells in tissue culture; biochemical studies may then be performed on bacteria separated by physical techniques from their host cells.

The three genera that have been well studied are described in Table 21.2. They are small rods with the fine structure typical of Gram-negative eubacteria. They are respiratory, preferring compounds such as TCA cycle intermediates and (especially) glutamate as substrate. The genera *Rickettsia* and *Coxiella* are obligate intracellular parasites that differ in their location within the host cell. Both enter their host cells by inducing phagocytosis, even by cells that are not normally phagocytic (e.g., the endothelial cells that line the vascular system). *Rickettsia* is able to escape the phagosome by degrading the phagosome membrane with lipases; this occurs simultaneously with phagocytosis, so that phagosome-enclosed bacteria are not an intermediate stage in the establishment of infection. *Coxiella* remains within the phagosome, which then fuses with a lysosome. Apparently lysosomal fusion is required to activate *Coxiella*, and metabolism is most rapid at pH 4.5, approximately that of the phagolysosome. Presumably some alterations of the phagolysosomal membrane occur and permit the transfer of nutrients into the vesicle.

*Rickettsia* has been shown to have an adenylate exchange system that will exchange endogeneous ADP with exogeneous ATP. Thus while they are growing within the host cell much or all of the energy needed for ricketsial growth may be met by host metabolism and phosphorylation. The ability of rickettsial cells to respire compounds such as glutamate may thus be important to provide maintenance energy rather than energy for growth.

*Rochalimaea*, unlike the other two rickettsias, is neither an obligate intracellular parasite nor difficult to culture axenically. It grows attached to the outer surface of host cells, and apparently lacks the ability of *Rickettsia* and *Coxiella* to induce phagocytosis. *Rochalimaea* shows significant (about 30 percent) DNA-DNA homology to one species of *Rickettsia.*

*Coxiella* forms endospores (Figure 21.10) that are resistant to drying and other environmental stresses. They are substantially smaller than the vegetative cells, appear to have a reduced metabolic rate, and they lack dipicolinic acid, a compound characteristic of the endospores of Gram-positive bacteria (see Chapter 22). Dust from hides or pelts of infected animals may contain large numbers of highly infectious spores from dried fecal material of arthropod parasites such as ticks. This is the

FIGURE 21.10

Electron micrograph of a thin section of *Coxiella burnetii*, showing the endospore formed within the envelope of the mother cell. (×30,100). From T. F. McCaul and J. C. Williams, "Developmental cycle of *Coxiella burnetii:* structure and morphogenesis of vegetative and sporogenic differentiations," *J. Bacteriol.*, 147, 1063–1076 (1981).

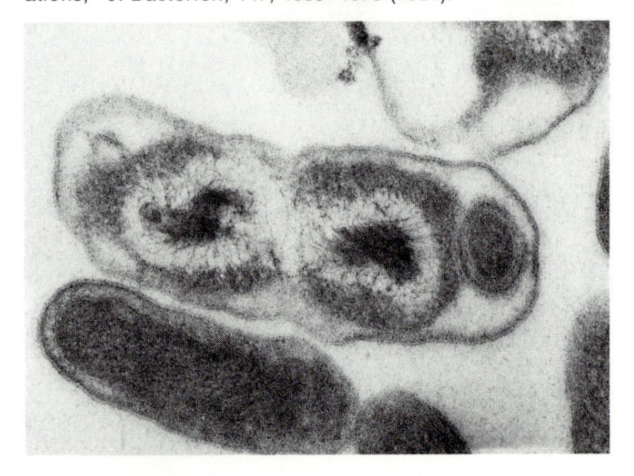

primary route of human infection, and may be an important route of transmission among animals.

Since neither *Rochalimaea* nor *Rickettsia* form spores, and lose viability rapidly outside their host cells, more elaborate modes of transmission are required. Two major types occur. Many *Rickettsia* strains (the scrub typhus and spotted fever rickettsias) are principally arthropod parasites, and are transmitted *transovarially*; that is, the host germ tissue becomes infected and the infection is transmitted to the next generation via the egg. These organisms cause principally arthropod diseases, and infection of humans, or domestic or wild animals is incidental to the survival of the parasite. In contrast, *Rochalimaea* and the typhus group of *Rickettsia* are apparently not transovarially transmitted. They thus depend on their animal host to infect new arthropods.

A variety of other endosymbiotic bacteria have been incompletely described. Some are tick symbionts and cause diseases of domestic animals (genera *Cowdria* and *Ehrlichia*). They grow within vacuoles like *Coxiella*, but it is not known if the vacuoles are phagolysosomes. They infect endothelial cells or leucocytes. The genera *Bartonella* and *Granhamella* infect erythrocytes. The former causes human disease transmitted by sand flies and the latter causes wild animal disease transmitted by fleas.

## THE CHLAMYDIAS

The chlamydias are Gram-negative eubacteria that are obligate intracellular parasites, and that, like *Coxiella*, have a life cycle that includes a resistant stage that mediates transmission. The growing form is termed a *reticulate body* or *initial body*, and the infectious form an *elementary body*. This confusing terminology was devised prior to the full realization of the bacterial nature of the chlamydias (they were for a long time considered to be intermediate between viruses and bacteria). To emphasize the homology of cellular processes in the chlamydias and other bacteria, we shall call the actively growing cell (the reticulate body) the *vegetative cell*, and the infectious form or elementary body a *chlamydiospore*.

The chlamydial life cycle is shown in Figures 21.11 and 21.12. The chlamydiospore is small (about 0.2 to 0.4 $\mu$m diameter), with a rigid cell wall; they lack detectable metabolic activity. When it contacts a host cell, it induces phagocytosis by the host cell; the rest of the life cycle occurs within the phagosome. Some component of the chlamydiospores inhibits the fusion of phagosome with lysosomes; presumably there are other changes in the phagosome membrane that allow permeation by host metabolites needed by the parasite. The chlamydio-

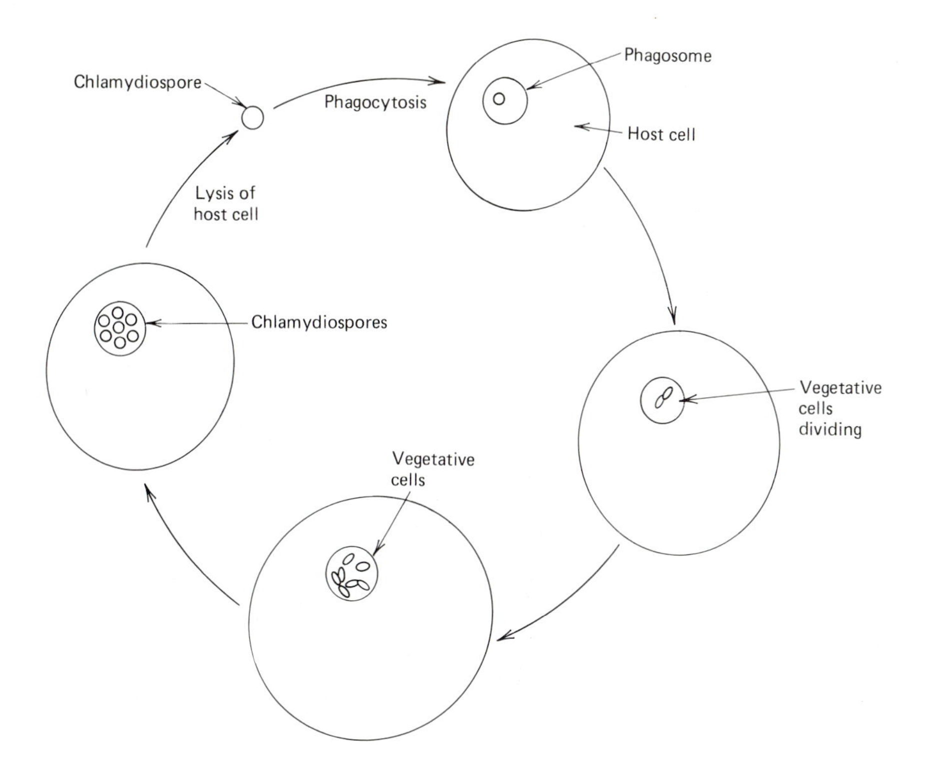


FIGURE 21.11
The chlamydial life cycle.

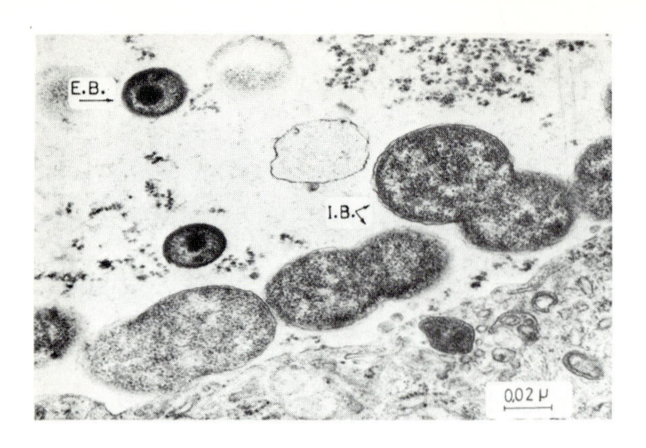


FIGURE 21.12

Electron micrograph of a thin section of a cell infected by *Chlamydia trachomatis*, showing the vegetative cells (I.B.) dividing within a phagosome. A few chlamydiopores (E.B.) are visible. From B. Gutter, Y. Asher, Y. Cohen, and Y. Becker, "Studies on the developmental cycle of *Chlamydia trachomatis:* Isolation and characterization of the initial bodies". *J. Bacteriol.* **115,** 691–702 (1973).

spore enlarges, loses its rigidity, and begins macromolecular synthesis. Since the chlamydiospore is very low in RNA, particularly rRNA, initial protein and RNA synthesis is probably devoted to increasing the number of ribosomes. Following a period of growth and division by binary fission, the vegetative cells convert to chlamydiospores.

One of the distinctive characteristics of the chlamydias is their apparent complete lack of ATP-generating pathways; they are accordingly *energy parasites*. They have an ADP-ATP exchange system like that of *Rickettsia*, and probably similar systems for the exchange of the other nucleotide triphosphates.

Although *Chlamydia* is sensitive to antibiotics that inhibit murein synthesis, chemical tests for murein have been negative. There is apparently no muramic acid (or other amino sugar) in the wall, so that the rigidity of the chlamydiospore, and the sensitivity of vegetative cells to antibiotics that inhibit peptidoglycan cross-linking, have an unknown basis. In other respects the cell envelope is similar to that of other Gram-negative bacteria, including the presence of lipopolysaccharide in the outer membrane.

The G + C content of the DNA is 41 to 45 percent; the genome is quite small (MW about $7 \times 10^8$).

Currently a single genus is recognized, with two species. *Chlamydia psittaci* is an avian parasite, causing mainly gastrointestinal infections. Inhalation of dried fecal material containing chlamydiospores can cause respiratory disease in humans. *C. trachomatis* is the agent of two common human diseases: the venereal disease *lymphogranuloma venereum* and the conjunctival infection *trachoma.* As with the rickettsias, however, field observations indicate that a variety of other organisms may be hosts to chlamydias, and it seems probable that additional genera will need to be recognized.

## FURTHER READING

### Books


MARCHETTE, N. J., *Ecological Relationships and Evolution of the Rickettsiae.* Boca Raton, Fla.: CRC Press, 1982.

ROSEBURY, T., *Microbes and Morals.* New York: Viking, 1971. A popular account of the natural history of syphilis and the treponematoses.


### Reviews


BACA, O. G., and D. PARETSKY, "Q Fever and *Coxiella burnetii:* A Model for Host-Parasite Interactions," *Microbiol. Rev.* **47,** 127–149 (1983).

BECKER, Y., "The Chlamydia: Molecular Biology of Procaryotic Parasites of Eucaryocytes," *Microbiol. Rev.* **42,** 274–306 (1978).

HARWOOD, C. S., and E. CANALE-PAROLA, "Ecology of Spirochetes," *Ann. Rev. Microbiol.* **38,** 161–192 (1984).

HOLT, S. C., "Anatomy and Chemistry of Spirochetes," *Microbiol. Rev.* **42,** 114–160 (1978).

JOHNSON, R. C., "The Spirochetes," *Ann. Rev. Microbiol.* **31,** 89–106 (1977).

MOULDER, J. W., "Looking at Chlamydiae without Looking at Their Hosts," *Am. Soc. Microbiol. News* **50,** 353–362 (1984).

SCHACHTER, J., and H. D. CALDWELL, "Chlamydiae," *Ann. Rev. Microbiol.* **34,** 285–309 (1980).

WINKLER, H., "Rickettsiae: Intracytoplasmic Life," *Am. Soc. Microbiol. News* **48,** 184–187 (1982).

## Chapter 22

# Gram-Positive Eubacteria: Unicellular Endosporeformers

Many Gram-positive bacteria share the ability to form a distinctive type of dormant cell known as an *endospore*. Endospores (Figure 22.1) can be readily recognized microscopically by their intracellular site of formation, their extreme refractility, and their resistance to staining by basic aniline dyes that readily stain vegetative cells. They are not normally formed during active growth and division; their differentiation begins when a population of vegetative cells passes out of the exponential growth phase as a consequence of nutrient limitation. Typically, one endospore is formed in each vegetative cell. The mature spore is liberated by lysis of the vegetative cell in which it has developed. Free endospores have no detectable metabolism, but for many years (often decades) they retain the potential capacity to germinate and develop into vegetative cells. This state of total dormancy is known as *cryptobiosis*. Endospores are highly resistant to heat, ultraviolet and ionizing radiation, and many toxic chemicals. Their heat resistance is frequently exploited in the isolation of spore-forming bacteria; these organisms can be selected by subjecting inocula to a thermal pretreatment sufficient to kill most vegetative cells.

The unicellular endosporeformers all reproduce by binary transverse fission, and with few exceptions they are rod-shaped. They all have the cell envelope typical of gram-positive eubacteria; however, many give a negative or variable reaction to the Gram stain, particularly in stationary phase. The Gram stain is accordingly of limited value in the taxonomy of this group. Motility is widespread, but not universal, and is effected by means of peritrichous flagella.

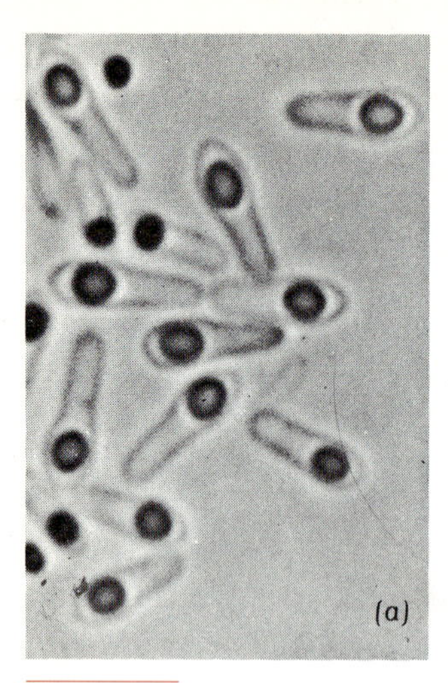


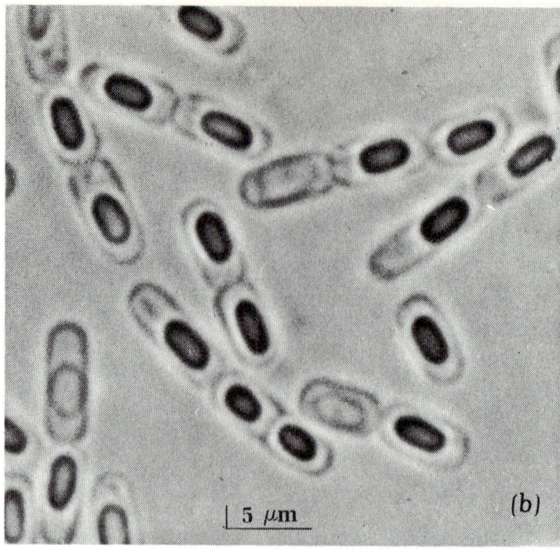


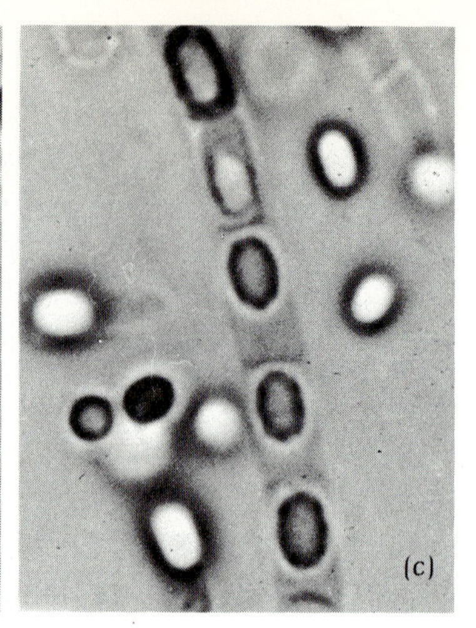


FIGURE 22.1

Sporulating cells of *Bacillus* species: (a) unidentified bacillus from soil; (b) *B. cereus;* (c) *B. megaterium*. From C. F. Robinow, in *The Bacteria*, Vol. 1, ed. I. C. Gunsalus and R. Y. Stanier (New York: Academic Press, 1960), p. 208.

The sporeformers are chemoheterotrophs; dissimilation of organic substrates occurs by aerobic respiration, nitrate respiration, or fermentation. Growth factors are required by some, but not all.

The typical habitat of spore-forming bacteria is soil. A few species are pathogenic for either insects or vertebrates. Most pathogenic sporeformers cause disease by toxin production; few of them are able to invade animal tissues.

## THE ENDOSPORE

The ability to produce endospores, normally not expressed during the vegetative growth of a spore-forming bacterium, constitutes a complex process of differentiation that is initiated as the population passes out of exponential growth and approaches the stationary phase. The process leads to the synthesis, *within* most vegetative cells, of a new type of cell, quite different from the mother cell in fine structure, chemical composition, and physiological properties. After release from the mother cell, the endospore normally enters a long period of dormancy; however, if subjected to appropriate stimuli, it can germinate and grow out into a typical vegetative cell. The fascinating problems of endospore formation and germination have been intensively and extensively studied by many microbiologists, and only a relatively brief summary of the vast literature on this subject can be presented here. For reasons of convenience, most of the experimental work on endospores has been conducted with members of the genus *Bacillus;* however, both the formation and the germination of endospores in other sporeformers appear on present information to involve basically similar events.

### Endospore Formation

The structural events associated with spore formation have been elucidated by a combination of light and electron microscopic observations. A synthesis based on both types of observations will be presented here (Figure 22.2 and 22.3).

At the end of exponential growth, each cell contains two nuclear bodies. These coalesce to form an *axial chromatin thread*, the first definite sign of the onset of sporulation. A transverse septum is then formed near one cell pole, which separates the cytoplasm and the DNA of the smaller cell (destined to become the spore) from the rest of the cell contents. Septum formation is not accompanied, as in normal cell division, by the development of a transverse wall; instead, the membrane of the larger cell rapidly grows around the smaller cell, which thus becomes completely engulfed within the cytoplasm of the larger cell, to produce a so-called *forespore*. In effect, the forespore is a protoplast, enclosed by two concentric sets of unit membranes: its own bounding membrane, and the membrane of the mother cell which has grown around it. At this stage, the development process becomes irreversible: the cell is said to be "committed" to undergo sporulation. By phase contract microscopy, the forespore appears as a dark, nonrefractile area that is free of granular inclusions.

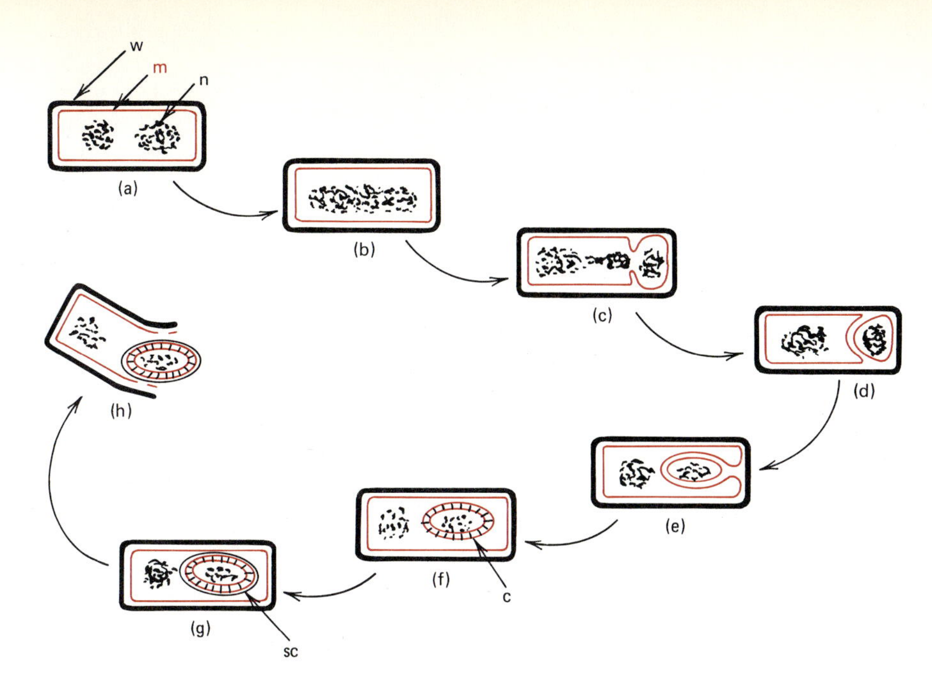


FIGURE 22.2

A diagrammatic representation of the cytological changes accompanying endospore formation in *Bacillus cereus*. (a) Vegetative cell, containing two nuclear bodies (stippled areas). (b) Condensation of the nuclear material. (c) Beginning of transverse wall formation. (d) Completion of the transverse wall; the forespore with its nuclear material is now cut off from the vegetative cell. (e) Engulfment of the forespore. (f) Synthesis of the spore cortex. (g) Synthesis of the spore coat. (h) Liberated spore. w: cell wall; m: cell membrane; n: nuclear area; c: cortex; sc: spore coat. After I. E. Young and P. FitzJames, "Chemical and Morphological Studies of Bacteria Spore Formation, I," *J. Biophys. Biochem. Cytol.* **6,** 467 (1959).

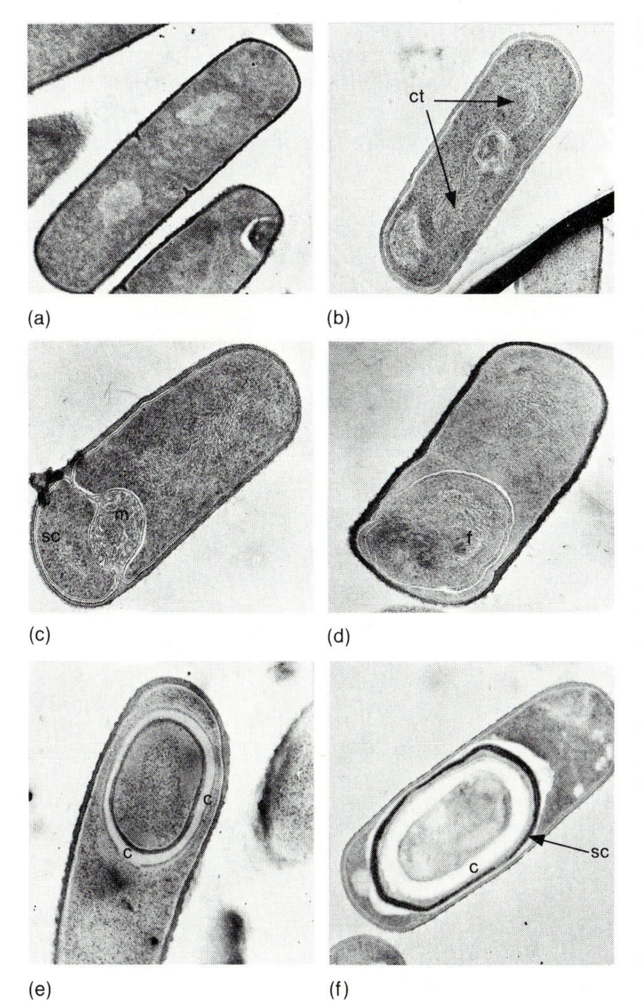


FIGURE 22.3

Electron micrographs of thin sections of *Bacillus subtilis*, showing the sequence of structural changes associated with endospore formation. (a) Vegetative cell in course of exponential growth. (b) Condensation of the nuclei, to form an axial chromatin thread, ct. (c) Formation of transverse septum (containing a mesosome, m) near one cell pole, delimiting the future spore cell, sc, from the rest of the cell. (d) Formation of a forespore f, completely enclosed by the cytoplasm of the mother cell. (e) The developing spore is surrounded by the cortex, c. (f) The terminal stage of spore development: the mature spore, still enclosed by the mother cell, is now surrounded by both cortex, c, and spore coat, sc. From A. Ryter, "Étude Morphologique de la Sporulation de *Bacillus subtilis*," *Ann. Inst. Pasteur* **108,** 40 (1965).

Once the forespore has been engulfed by the mother cell, there is a rapid synthesis and deposition of new structures that enclose it. The first to appear is the *cortex*, which develops between the two membranes surrounding the forespore. Shortly afterward, a more electron-dense layer, the *spore coat*, begins to form exterior to both membranes surrounding the cortex. In the *B. cereus* group an additional, looser and thinner layer, the *exosporium*, forms outside the spore coat. Once the spore coat is synthesized, the maturing spore begins to become refractile, although it is not yet heat-resistant. The development of heat resistance closely follows two major chemical changes: a massive uptake of $Ca^{2+}$ ions by the sporulating cell, and the synthesis in large amounts of dipicolinic acid, a compound absent from vegetative cells. The time sequence of the development of refractility and heat resistance and

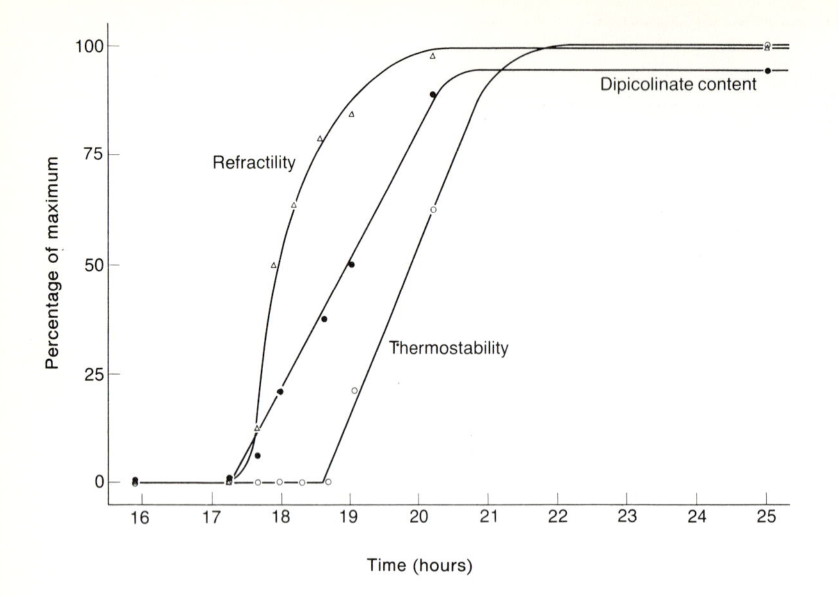


FIGURE 22.4

The increases in the refractility, thermostability of the cells, and dipicolinate content of the population that occur during sporulation in a culture of *Bacillus cereus*. All values are plotted against the age of the culture in hours. After T. Hasimoto, S. H. Black, and P. Gerhardt, "Development of fine structure, thermostability, and dipicolinate during sporogenesis in a bacillus." *Can. J. Microbiol.* **6,** 203 (1960).

the synthesis of dipicolinic acid are shown in Figure 22.4. In mature spores the molar ratio of dipicolinate to $Ca^{2+}$ is close to unity, which suggests that dipicolinate occurs as a Cachelate. This complex represents 10 to 15 percent of the spore dry weight, and it is located within the spore protoplast. Dipicolinic acid is formed by a single oxidative reaction (Figure 22.5) from an intermediate (dihydrodipicolinic acid) in the biosynthetic pathway leading to the amino acid lysine (Figure 5.16).

The cortex is largely composed of a unique peptidoglycan, containing three repeating N-acetylglucosamine-muramic acid dimers differing with respect to substitutions on the lactic acid moiety of muramic acid: a muramic lactam subunit, without any attached amino acids; an alanine subunit, bearing only an L-alanyl residue; and a tetrapeptide subunit, bearing the sequence L-ala-D-glu-meso-DAP-D-ala (Figure 22.6). These subunits represent, respectively, approximately 55, 15, and 30 percent of the total. There is very little cross-linking between tetrapeptide chains. The distinctiveness of the cortex peptidoglycan is further shown by the fact that *B. subtilis* and *B. sphaericus*, which synthesize chemically different vegetative cell wall peptidoglycans (see Table 22.1), contain essentially similar cortex peptidoglycans.

The outer spore coat, which represents 30 to 60 percent of the dry weight of the spore, is largely composed of protein and accounts for about 80 percent of the total spore protein. The spore coat proteins have an unusually high content of cysteine and of hydrophobic amino acids, and are highly resistant to treatments that solubilize most proteins.

After the completion of spore development, the spore protoplast, accordingly, contains a high content of Ca dipicolinate and is enclosed by newly synthesized outer layers of unique chemical structure (the cortex and the spore coat, sometimes also an exosporium), which account for a large fraction of the spore dry weight. When liberated by autolysis of the mother cell, the mature endospore is highly dehydrated, shows no detectable metabolic activity,

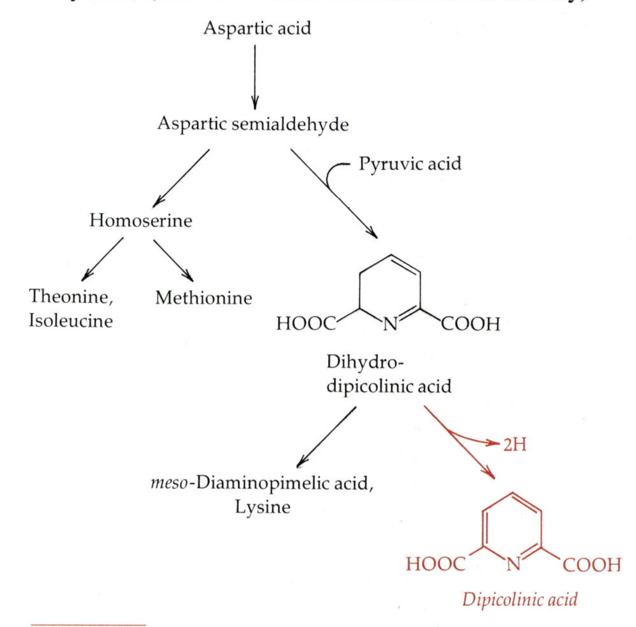


FIGURE 22.5

Pathway of biosynthesis of dipicolinic acid showing its relationship to the pathways of biosynthesis of the amino acids: threonine, isoleucine, methionine, and lysine.

**TABLE 22.1**

**Distribution of Peptidoglycan Types in Spore-Forming Bacteria**

| Diamino Acid in Position 3 | Interpeptide Bridge | |
|---|---|---|
| *Meso*-diaminopimelic acid | Absent | *Bacillus* (most spp.)<br>*Clostridium* (most spp.) |
| L-lysine | Present | *Bacillus* (group III) |
| L-lysine | Absent | *Clostridium* (some spp.) |
| L, L-diaminopimelic acid | Present | *Clostridium* (some spp., including *C. perfringens*) |

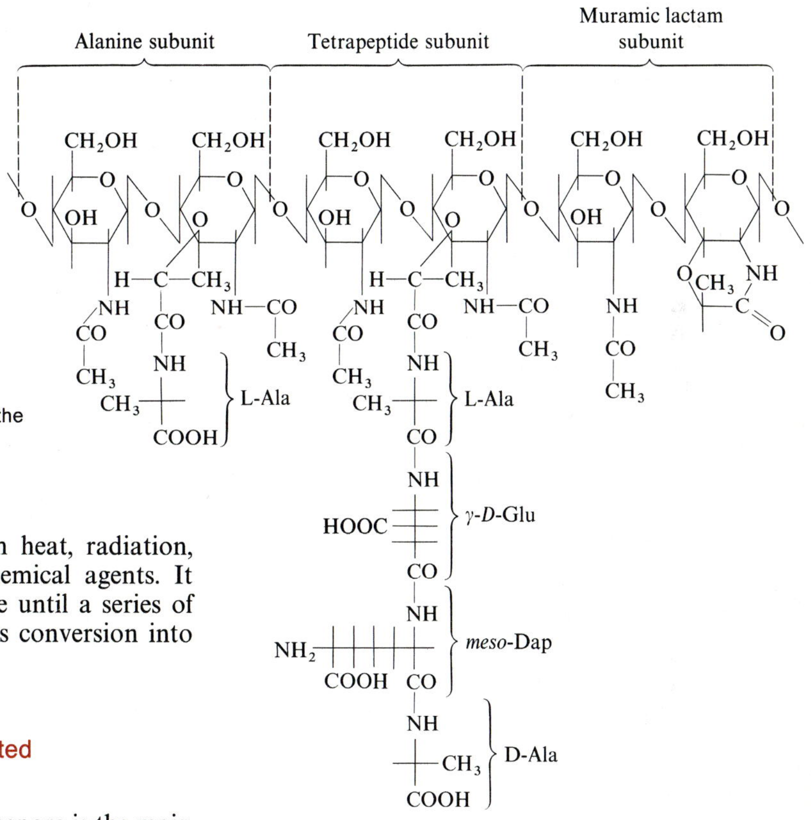

FIGURE 22.6
Structures of the repeating subunits of the spore cortex peptidoglycan.

and is effectively protected from heat, radiation, and attack by enzymatic or chemical agents. It remains in this cryptobiotic state until a series of environmental triggers initiate its conversion into a vegetative cell.

## Other Biochemical Events Related to Sporulation

Although the synthesis of an endospore is the main enterprise of a sporulating cell, it is by no means the only one. One striking concomitant event, characteristic of *Bacillus thuringensis*, is the formation of a bipyramidal parasporal protein crystal adjacent to each endospore (see below). Various noncrystalline parasporal structures of defined form have been described in other species of *Bacillus* and *Clostridium.*

In many sporeformers, both aerobic and anaerobic, the onset of sporulation is accompanied by the synthesis of a distinctive class of antimicrobial substances: peptides with molecular weights of approximately 1,400 daltons. Many of these peptide antibiotics have been characterized chemically and functionally. They can be assigned to three classes: *edeines,* linear basic peptides that inhibit DNA synthesis; *bacitracins,* cyclic peptides that inhibit cell wall synthesis; and the *gramicidin-polymyxin-tyrocidin*-type peptides, which are linear or cyclic and modify membrane structure or function. As shown in Figure 22.7, many of them contain amino acids of D-configuration, amino acids that do not occur in proteins (e.g., diaminobutyric acid, ornithine), and even constituents which are not amino acids (e.g., the polyamine spermidine, present in edeines).

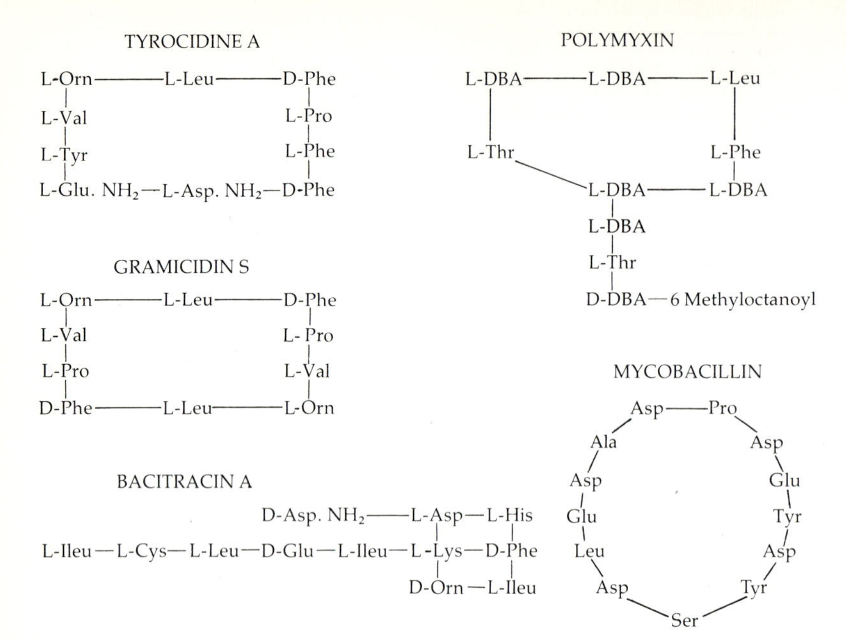


FIGURE 22.7

Structures of certain antibiotics synthesized during the sporulation process by sporeforming bacteria. Abbreviations are: ornithine, orn; valine, val; tyrosine, tyr; glutamine, glu · $NH_2$; asparagine, asp · $NH_2$; phenylalanine, phe; proline, pro; leucine, leu; isoleucine, ileu; glutamic acid, glu; lysine, lys; histidine, his; threonine, thr; α-diaminobutyric acid, DBA; aspartic acid, asp; and serine, ser.

The production of peptide antibiotics occurs rather early in the sporulation process (Figure 22.8). The biosynthesis of these compounds involves a novel assembly mechanism, in which the amino acid sequence is determined and the peptide bond formed by specific enzymes; neither tRNAs nor ribosomes participate. The role of these compounds in sporulation is unknown, but it has been suggested that they may be effectors that control various stages of the differentiation process.

## Activation, Germination, and Outgrowth of Endospores

Freshly formed endospores will remain largely dormant even if placed in optimal conditions for germination. The state of dormancy can be broken by a variety of treatments, collectively termed *activation.* Perhaps the most general mechanism for activating spores is *heat shock:* exposure for several hours to an elevated, but sublethal temperature (e.g., 65° C). Heat activation is not accompanied by any detectable change in the appearance of the spores: it simply enables them to germinate when subsequently placed in an environment favorable for this process. Furthermore, heat activation is reversible: if spores are subsequently placed at a lower temperature for some days, their induced germinability declines. A much slower activation takes place upon the storage of spores, even at relatively low temperatures (5°C) or under dry conditions; this activation is irreversible, an increasing fraction of the spore population becoming capable of germination as storage is prolonged.

When activated spores are placed under favorable conditions, germination can take place. This process is very rapid and is expressed by a loss of refractility, a loss of resistance to heat and other deleterious agents, and the unmasking of metabolic activity, as evidenced by a sudden onset of metabolism. These processes are accompanied by the liberation, as soluble materials, of about 30 percent of the spore dry weight. This spore exudate consists largely of Ca-dipicolinate (derived from the spore protoplast) and peptidoglycan fragments (derived from the cortex). The cortex is rapidly destroyed, only the outer spore coat remaining.

Germination of activated spores requires a chemical trigger; the specific substances that are active are numerous and varied, including L-alanine, ribosides (adenosine, inosine), glucose, Ca-dipicolinate, and various inorganic anions and cations. The

FIGURE 22.8

The time course of growth, bacitracin synthesis, and formation of sporangia and free spores by a culture of *Bacillus licheniformis.*

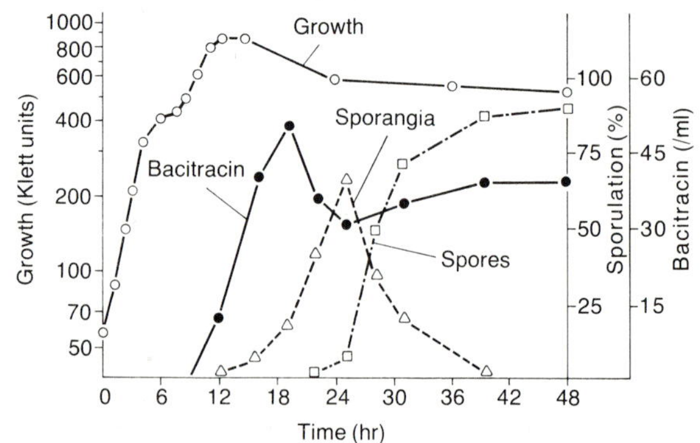

particular germination requirements often differ from species to species, and maximal germination may require a combination of several germinants. Furthermore, germination can be accomplished mechanically by subjecting spores to physical treatments that erode or crack the spore coat.

The process of germination does not appear to be accompanied by significant macromolecular synthesis; the appearance of metabolic activity is a consequence of the unmasking of preexisting but inactive enzymes in the spore protoplast. If germination occurs in a medium that does not contain the nutrients required for vegetative growth, no further change will occur. However, if the nutrients required for macromolecular synthesis are present upon germination, the germinated spore will proceed to grow out into a vegetative cell. Outgrowth involves an initial swelling of the spore within its spore coat, accompanied by a rapid synthesis of a vegetative cell wall, the foundation layer of which is present around the spore protoplast at the time of germination. The newly formed vegetative cell then emerges from the spore coat (Figure 22.9), elongates, and proceeds to undergo the first vegetative division.

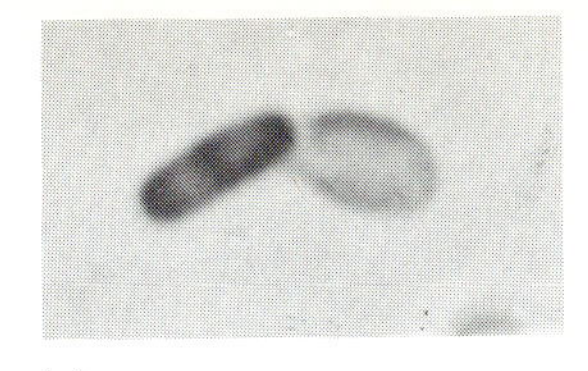

(a)

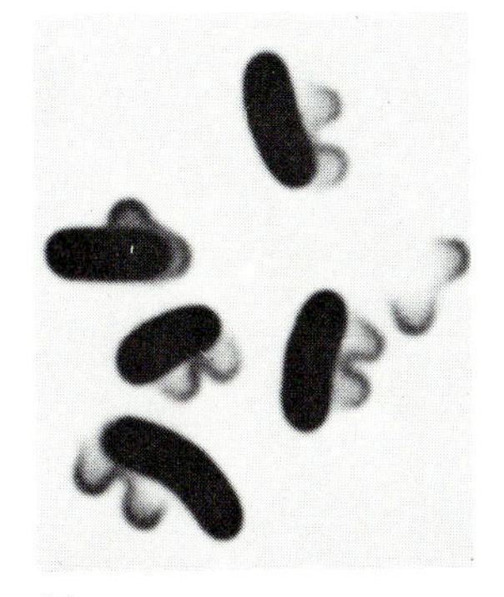

(b)

FIGURE 22.9
Spore germination in (a) *Bacillus polymyxa* and (b) *B. circulans* (stained preparations, ×5,260). Courtesy of C. F. Robinow and C. L. Hannay.

## Classification of the Endosporeformers

The primary taxonomic subdivision of the sporeformers (Table 22.2) is based on their morphology and oxygen relations. The genera *Clostridium* and *Desulfotomaculum* are strict anaerobes, although *Clostridium* spp. show some variation with respect to oxygen sensitivity. As a rule, vegetative cells are rapidly killed by exposure to air, although spores are not. The distinction between *Clostridium* and *Desulfotomaculum* is metabolic. *Clostridium* spp. synthesize ATP by fermentative means, the modes of fermentation being remarkably diverse with respect both to substrates and products. *Desulfotomaculum* spp. synthesize ATP solely by substrate-level phosphorylation; their metabolism is accordingly also fermentative. However, they are capable of using sulfate as a terminal electron acceptor, the redox balance of their fermentation being maintained by sulfate reduction.

The aerobic unicellular endosporeformers are placed in the genus *Bacillus*. They are normally rod-shaped; one organism with spherical cells was formerly placed in a separate genus, *Sporosarcina*. However, it is closely related to *Bacillus*, and separate generic status does not seem warranted.

The endosporeformers that are mycelial (i.e., they grow as long branched filaments enclosed in a common cell wall) are placed in the genus *Thermoactinomyces*. These organisms were long thought

**TABLE 22.2**
**Properties of the Genera of Endosporeformers**

| Genus | Vegetative Cell Shape | Relations to Oxygen | Dissimilatory Metabolism | Percent G + C |
|---|---|---|---|---|
| *Bacillus* | Rods | Aerobes; some facultative anaerobes | Aerobic respiration, also fermentation or denitrification in some | 33–51 |
| *Clostridium* | Rods | Strict anaerobes | Fermentation | 22–28 |
| *Desulfotomaculum* | Rods | Strict anaerobes | Fermentation (with $SO_4^{2-}$ as terminal electron acceptor) | 42–46 |
| *Sporolactobacillus* | Rods | Aerotolerant | Fermentation | 39–47 |
| *Thermoactinomyces* | Mycelial | Aerobes | Respiration | 53 |

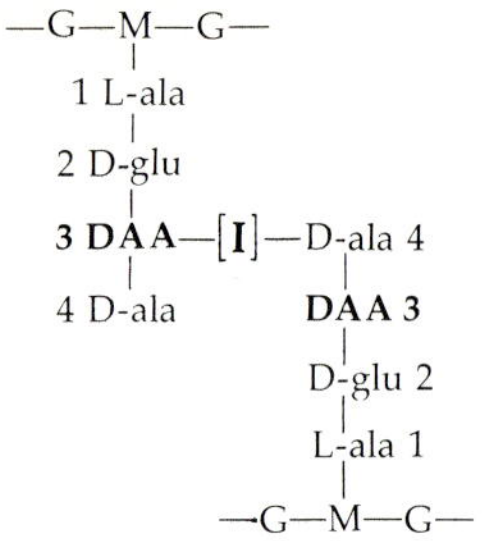

```
—G—M—G—
   |
 1 L-ala
   |
 2 D-glu
   |
 3 DAA—[I]—D-ala 4
   |          |
 4 D-ala    DAA 3
              |
            D-glu 2
              |
            L-ala 1
              |
          —G—M—G—
```

FIGURE 22.10

Peptidoglycan structure in spore-forming bacteria. The variable structural features (boldface) are the diamino acid (DAA) in position 3 of the peptide chain and the presence or absence of additional amino acids, comprising an interpeptide bridge, I, between cross-linked peptide chains. M, G, ala, and glu represent respectively, N-acetylglucosamine, N-acetylmuramic acid, alanine, and glutamic acid.

to be members of the actinomycete group on the basis of their mycelial growth; however their relatively low G + C content and their synthesis of endospores clearly places them in the *Bacillus* group.

### Peptidoglycan Structure

Variability of the cell wall peptidoglycan structure, which is extremely marked in Gram-positive bacteria of the actinomycete line (Chapter 23), occurs to a limited extent in sporeformers (Figure 22.10 and Table 22.1). Most bacilli and clostridia synthesize a peptidoglycan in the vegetative cell wall of the type that is well-nigh universal in Gram-negative procaryotes, containing *meso*-diaminopimelic acid as a diamino acid, directly cross-linked to D-alanine; the distribution of other forms of wall peptidoglycans is shown in Table 22.1

## THE AEROBIC SPOREFORMERS

### The Genus *Bacillus*

Most bacilli are versatile chemoheterotrophs capable of utilizing a considerable range of simple organic compounds (sugars, amino acids, organic acids) as respiratory substrates, and in some cases also of fermenting carbohydrates. A few species require no organic growth factors; others may require amino acids, B vitamins, or both. The majority are mesophiles, with temperature optima in the range of 30° to 45°C; however, the genus also contains a number of thermophilic representatives that grow at temperatures as high as 65°C.

The mesophilic species of the genus *Bacillus* can be divided into three principal subgroups, distinguished by the structure and intracellular location of the endospore (Table 22.3).

Within group I, a further distinction can be made between the *B. subtilis* and *B. cereus* groups on the basis of cell size and on the presence or absence of poly-$\beta$-hydroxybutyrate as a cellular reserve material (Table 22.4).

Most species of group I can grow anaerobically at the expense of sugars. They carry out a distinctive fermentation, in which 2,3-butanediol, glycerol, and $CO_2$ are the major end products, accompanied by small amounts of lactate and ethanol. The fermentation can be approximately represented as:

$$3 \text{ glucose} \longrightarrow 2 \text{ butanediol} + 2 \text{ glycerol} + 4\, CO_2$$

Glucose is initially dissimilated through the Embden-Meyerhof pathway, to the level of triose phosphate, at which point a metabolic divergence occurs. Part of the triose phosphate is converted

**TABLE 22.3**

**Subgroups of Mesophilic *Bacillus* Species**

| | Structure of Sporulating Cell | | | |
|---|---|---|---|---|
| Group | ENDOSPORE SHAPE | ENDOSPORE POSITION IN VEGETATIVE CELL | VEGETATIVE CELL DISTENDED BY SPORE | Major Representatives |
| I | Oval | Central | – | *B. subtilis* group<br>*B. cereus* group<br>*B. fastidiosus* |
| II | Oval; spore coat thick and ridged | Central | + | *B. polymyxa* group |
| III | Spherical | Terminal | + | *B. pasteurii* group |

TABLE 22.4

**Distinguishing Properties of the *B. subtilis* and *B. cereus* Groups**

*B. subtilis* Group: Vegetative Cells < 0.8 $\mu$m Wide; do not form Poly-$\beta$-Hydroxybutyrate as Reserve Material

| | Anaerobic Growth by | |
|---|---|---|
| | FERMENTATION OF SUGARS | DENITRIFICATION |
| *B. subtilis* | − | − |
| *B. licheniformis* | + | + |

*B. cereus* Group: Vegetative Cells > 1.0 $\mu$m Wide; form Poly-$\beta$-Hydroxybutyrate as Reserve Material

| | Motility | Fermentation of Sugars | Require Growth Factors | Pathogenicity |
|---|---|---|---|---|
| *B. cereus* | + | + | + | None |
| *B. anthracis* | − | + | + | Pathogenic for mammals |
| *B. thuringiensis* | + | + | + | Pathogenic for insects |
| *B. megaterium* | + | − | − | None |

to pyruvate, from which butanediol and $CO_2$ are produced; the rest is reduced to glycerol (Figure 22.11).

*B. subtilis*, unlike most other species of group I, cannot grow anaerobically at the expense of glucose, probably because it cannot reduce triose phosphate to glycerol; in the presence of air this species metabolizes glucose with the formation of large amounts of 2,3-butanediol.

*B. licheniformis* can also grow anaerobically at the expense of nonfermentable organic substrates when furnished with nitrate, since it is a vigorous denitrifier, the only one in the genus.

A distinctive species-cluster in group I consists of *B. cereus*, one of the most abundant aerobic sporeformers in soil, and two related pathogens, *B. anthracis* and *B. thuringensis*. In all three species the endospores are enclosed by a loose outer coat known as the *exosporium* (Figure 22.12), which is not formed by other bacilli. Certain strains of *B.*

FIGURE 22.11

The glycerol-butanediol fermentation.

3 glucose
6ATP
6ADP
6 triose-P
2 NADH
2 NAD$^+$
2 glycerol-P
$2P_i$
2 glycerol
8 ADP
4NAD$^+$
8 ATP
4NADH
4 pyruvate
$4CO_2$
2NADH
2NAD$^+$
2 butanediol

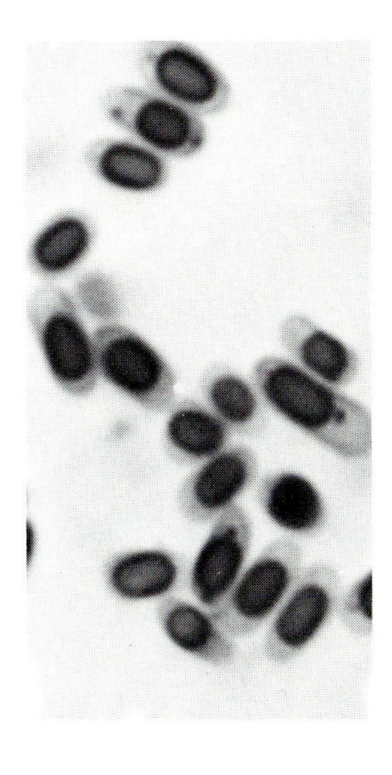

FIGURE 22.12

Mature endospores of *Bacillus cereus* (×3,600). Each stained spore is surrounded by a less deeply stained exosporium. Courtesy of C. F. Robinow.

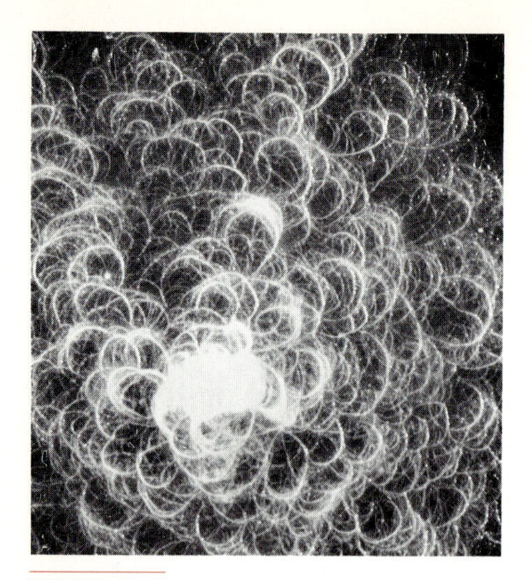

FIGURE 22.13

A colony of *Bacillus cereus var. mycoides* (×1.29). Courtesy of David Cornelius and C. F. Robinow.

The protein of which the crystal is composed is toxic for insects; after ingestion, it dissolves in the alkaline gut contents of the caterpillar and causes a loosening of the epithelial gut wall, with a consequent diffusion of liquid from the gut into the blood. This leads to rapid paralysis. Since the parasporal protein of *B. thuringiensis* is toxic for a wide range of lepidopterous larvae, but nontoxic for vertebrates, preparations of sporulating cells of this organism have found an extensive application in agriculture as a biological insecticide (Chapter 33).

The large-celled sporeformers of group I include one species, *B. fastidiosus*, that is remarkable for its extreme nutritional specialization. This organism, a strict aerobe, can be specifically enriched by virtue of its ability to perform an oxidative dissimilation of the purine, uric acid, which serves as its sole source of carbon, energy, and nitrogen. Of the many other organic compounds that have been

*cereus* produce distinctive loosely spreading colonies, superficially resembling those of fungi (Figure 22.13). *B. anthracis* is the agent of anthrax, a disease of cattle and sheep that is also transmissible to humans. It is one of the few spore-forming bacteria that is a true parasite, in the sense that it is able to develop massively within the body of the animal host. Apart from its pathogenic properties, the biochemical mechanisms of which have been extensively studied (Chapter 31), it differs from *B. cereus* by its permanent immotility.

*B. thuringiensis* is the causal agent of paralytic disease of the caterpillars of many lepidopterous insects, paralysis resulting from the ingestion of plant materials that carry on their surface spores or sporulating cells of the bacterium. Each sporulating cell of *B. thuringiensis* produces, adjacent to the spore, a regular bipyramidal protein crystal (Figure 22.14), which is liberated, along with the spore, by autolysis of the parent cell (Figure 22.15).

FIGURE 22.15

Electron micrograph of free crystals and a free spore surrounded by an exosporium of *Bacillus thuringiensis* (metal-shadowed preparation, ×7,500). From C. L. Hannay and P. FitzJames, "The Protein Crystals of *B. thuringiensis*." *Can. J. Microbiol.* **1,** 694 (1955).

FIGURE 22.14

A chain of sporulating cells of *Bacillus thuringiensis* (phase contrast, ×3,900). Each cell contains, in addition to the bright, refractile spore, a less refractile bipyramidal crystalline inclusion. Courtesy of P. FitzJames.

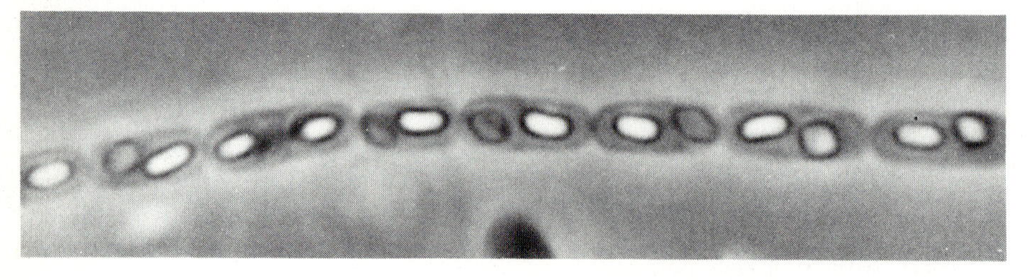

tested, only intermediates in the pathway of uric acid dissimilation (Figure 22.16) can be used as sole source of carbon and energy.

The two principal *Bacillus* species of group II, *B. polymyxa* and *B. macerans*, form spores with a distinctive, thick outer coat which bears a series of raised ridges, giving a crenulated profile to the spore (Figure 22.17). Both are facultative anaerobes, dissimilating starch and pectins as well as monosaccharides; good growth occurs only in the presence of a utilizable carbohydrate. *B. polymyxa* carries out a butanediol fermentation, chemically distinct from that of the group I *Bacillus* species; in addition to 2,3-butanediol, the main products are ethanol, $CO_2$, and $H_2$. The products of sugar fermentation of *B. macerans* are ethanol, acetone, acetate, formate, $CO_2$, and $H_2$. The production of $H_2$ as a major end product and the failure to produce glycerol distinguish the sugar fermentations of these species from those of group I. Another distinctive property of both *B. polymyxa* and *B. macerans* is the ability to fix $N_2$ when growth under anaerobic conditions; they are the only *Bacillus* species known to possess this property.

The *Bacillus* species of group III (*B. sphaericus* and *B. pasteurii*), which are distinguished by the formation of spherical spores located terminally in the sporulating cell, form a distinct subclass of bacilli in several physiological and metabolic respects. As shown in Table 22.1, their wall peptidoglycan is chemically distinct from that of other aerobic sporeformers. They lack fermentative capacity and are unable to utilize carbohydrates effectively as respiratory energy sources. The principal substrates of respiratory metabolism are amino acids and organic acids. Many (though not all) produce large amounts of the enzyme urease, which catalyzes a hydrolysis of urea:

$$CO(NH_2)_2 + H_2O \longrightarrow CO_2 + 2NH_3$$

Although this reaction does not result in ATP formation, it plays an important physiological role for the ureolytic bacilli. The principal ureolytic species, *B. pasteurii*, cannot grow on conventional complex media (e.g., nutrient broth) at neutral pH, unless such media are supplied with urea. Although urea is a neutral substance, its hydrolysis is accompanied by a considerable production of alkali, because two moles of ammonia are formed per mole of urea decomposed. Hence, a urea-containing medium inoculated with *B. pasteurii* rapidly becomes highly alkaline. The specific urea requirement of this bacterium can be replaced by ammonum, in conjunction with a high initial pH (8.5). No other monovalent cation will replace ammonum, and it is required for respiration as well as

FIGURE 22.16
The metabolism of uric acid by *Bacillus fastidiosus*.

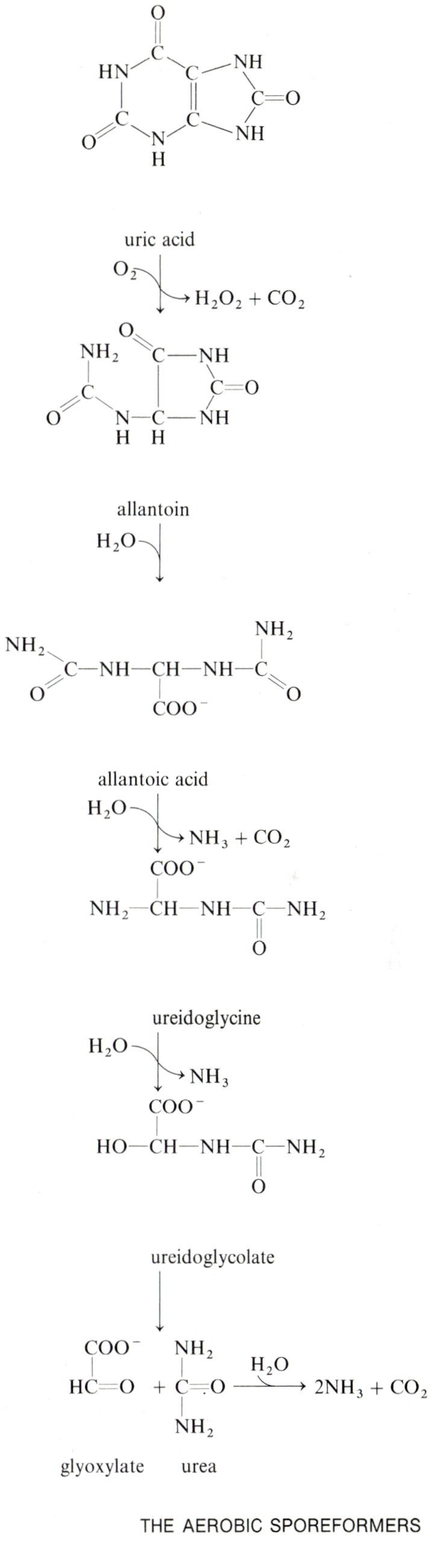

uric acid

$O_2 \rightarrow H_2O_2 + CO_2$

allantoin

$H_2O$

allantoic acid

$H_2O \rightarrow NH_3 + CO_2$

$NH_2{-}CH(COO^-){-}NH{-}C(=O){-}NH_2$

ureidoglycine

$H_2O \rightarrow NH_3$

$HO{-}CH(COO^-){-}NH{-}C(=O){-}NH_2$

ureidoglycolate

$HC(=O){-}COO^- + (NH_2)_2C{=}O \xrightarrow{H_2O} 2NH_3 + CO_2$

glyoxylate urea

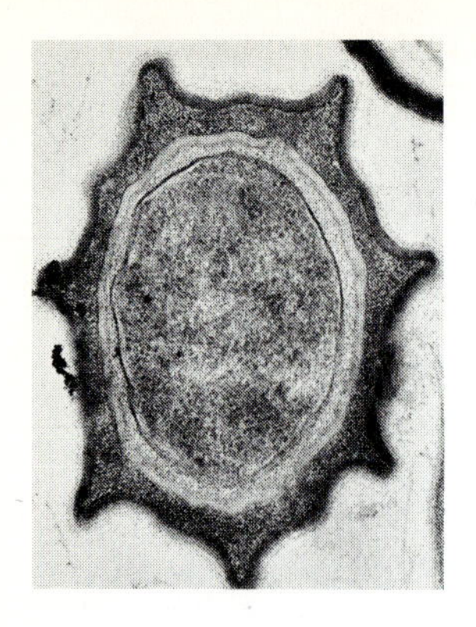

FIGURE 22.17

Electron micrograph of a transverse section of a spore of *Bacillus polymyxa*, showing the elaborate contour of the spore coat (×32,400). Courtesy of R. G. E. Murray.

growth. Therefore, in reality, the urea requirement of *B. pasteurii* reflects a specific requirement for ammonium and for a high pH. The spherical aerobic sporeformers (formerly *Sporosarcina*) closely resemble *B. pasteurii* in metabolic and physiological respects as well as in cell wall composition. They are genetically related, as shown by nuclei acid hybridization experiments.

The ecological advantage conferred on these sporeformers by their strong ureolytic ability can be readily demonstrated by a simple enrichment experiment. If a peptone medium containing 2 to 5 percent of urea is inoculated with unheated soil, the population that develops consists exclusively of ureolytic sporeformers, despite the fact that a wide variety of chemoheterotrophic aerobic bacteria can grow in pure culture on this medium. In an enrichment culture all competing bacteria are killed by the high concentration of free ammonia produced from urea by the ureolytic sporeformers.

## Thermophilic Bacilli

A special physiological group among the aerobic sporeformers consists of the extreme thermophiles, which are capable of growing at temperatures as high as 65° C and often fail to grow at temperatures below 45° C. There are probably a number of species with this attribute, but their taxonomy has not been studied in detail. The characteristic environment for these organisms is decomposing plant material, in which the heat generated by microbial metabolic activity cannot be readily dissipated.

## Lipid Composition of the Bacilli

The bacilli have quite complex patterns of membrane lipids. Their fatty acids are saturated or mono-unsaturated (rarely di-unsaturated), and many of them are branched, bearing a methyl group on the penultimate (*iso*-branching) or antepenultimate (*anteiso*-branching) carbon. Only *Bacillus acidocaldarius* lacks these fatty acids; instead, it contains cyclohexane fatty acids.

In addition to unusual fatty acids, many bacilli contain a variety of unusual lipids, including several glycosyl diglycerides (some containing sugar phosphates or sulfonates), phospholipids containing an amino acid, and cyclic peptidolipids. The latter resemble the cyclic peptide antibiotics; however, there is no evidence that these lipids are functionally or biosynthetically related to known antibiotics. Some of these novel lipids are shown in Figure 22.18.

## The Genus *Thermoactinomyces*

Decaying vegetable matter that is heated by the metabolism of resident microorganisms or by the sun contains a distinctive flora of endosporeformers, including thermophilic bacilli and the mycelial *Thermoactinomyces*. Indeed, damp, well-aerated stacks of vegetation may support such rapid growth of these microorganisms that the metabolic heat released raises the temperature to a point at which spontaneous chemical oxidations occur, and the

FIGURE 22.18

Some unusual lipids found in *Bacillus* spp.: (a) a cyclohexyl fatty acid; (b) a cyclic peptidolipid; (c) a sulfonoglycosyl lipid. R denotes hydrocarbon side chains.

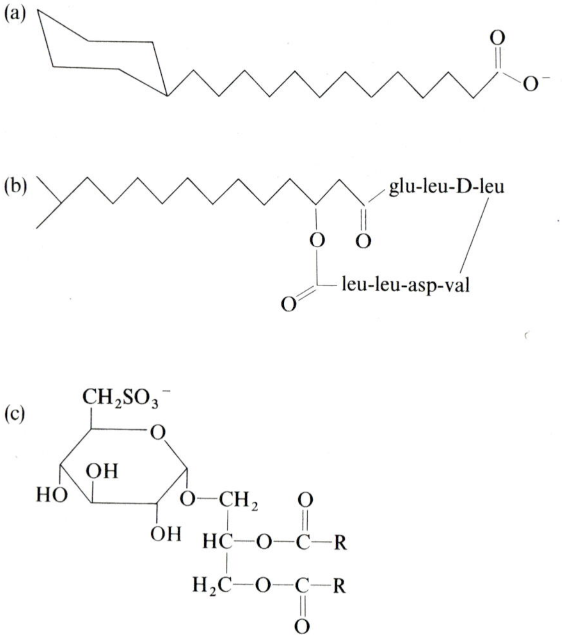

(a)

(b) glu-leu-D-leu / leu-leu-asp-val

(c) $CH_2SO_3^-$

pile ignites. Such *spontaneous combustion* has caused the loss of many barns storing damp hay or straw. The large number of endospores formed in damp hay may also pose a health hazard; although the organisms themselves are not pathogenic, repeated inhalation of their spores may cause a severe allergic reaction.

*Thermoactinomyces* grows very much like many of the actinomycetes (Chapter 24); vegetative growth results in the mycelium ramifying throughout the substratum. As the mycelium expands, the center of the colony becomes starved and sporulation is initiated. In common with many of the actinomycetes, *Thermoactinomyces* produces spores on specialized hyphae that extend away from the substratum; these are collectively termed the *aerial mycelium* to distinguish them from the *substrate mycelium.* Production of an aerial mycelium is presumably an adaptation to increase the efficiency of dissemination of the spores, which are produced in short side branches of the aerial hyphae (Figure 22.19). Endospores are also produced on the substrate mycelium.

## THE ANAEROBIC SPOREFORMERS

Anaerobic sporeformers with a fermentative mode of metabolism (genus *Clostridium*) were discovered by Pasteur in the middle of the nineteenth century when he demonstrated that some of these organisms carry out a fermentation of sugars accompanied by the formation of butyric acid. Shortly afterward it was recognized that clostridia are also the principal agents of the anaerobic decomposition of proteins ("putrefaction"). Toward the end of the nineteenth century it became evident that some clostridia are agents of human or animal disease. Like other members of the group, the pathogenic clostridia are normal soil inhabitants, with little or no invasive power; the diseases they produce result from the production of a variety of highly toxic proteins (exotoxins). Indeed, botulism (caused by *C. botulinum*) is a pure intoxication, resulting from the ingestion of foods in which these organisms have previously developed and formed exotoxins. The other principal clostridial diseases, tetanus

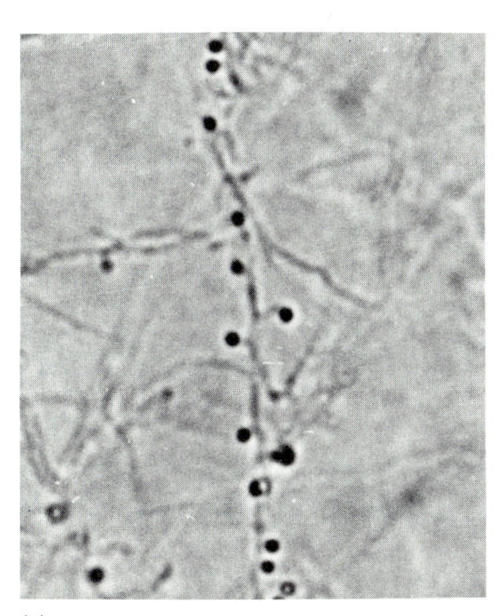

(a)

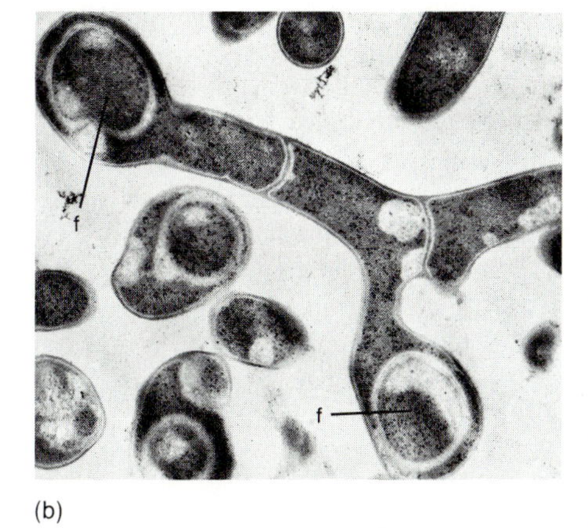


(b)

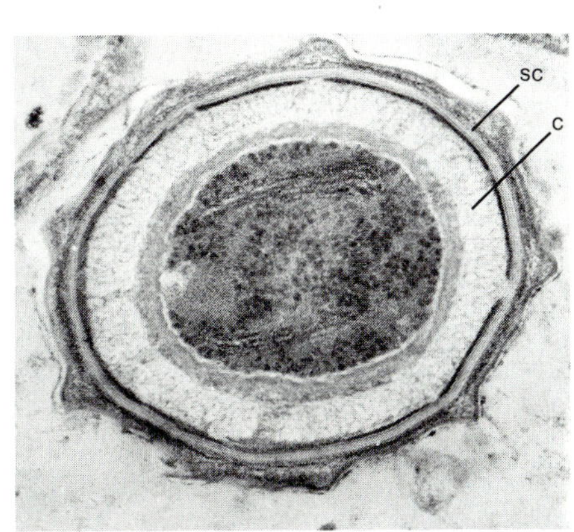


(c)

FIGURE 22.19

*Thermoactinomyces.* (a) Phase contrast light micrograph of a portion of the mycelium, showing the refractile endospores, borne on short side branches ($\times$ 1,360). (b) Electron micrograph of a thin section of a sporulating hypha, showing two forespores, f, enclosed by the cytoplasm of enlarged hyphal tips ($\times$ 20,000). (c) Electron micrograph of a thin section of a mature endospore, showing the cortex, c, and elaborate spore coat, sc, which surround the spore cell ($\times$ 50,600). From J. Lacey, "*Thermoactinomyces sacchari* sp. nov., a Thermophilic Actinomycete Causing Bagassosis," *J. Gen. Microbiol.* **66,** 327 (1971).

(cause by *C. tetani*) and gas gangrene (caused by several species including *C. perfringens*) are the results of wound infections; tissue necrosis leads to the development of an anaerobic environment which permits localized growth and toxin formation by these organisms. Some clostridial toxins (those responsible for botulism and tetanus) are potent inhibitors for nerve function. Others (those responsible for gas gangrene) are enzymes that cause tissue destruction; they include lecithinases, hemolysins, and a variety of proteases (Chapter 31).

Although over 60 *Clostridium* species have been described, the conventional taxonomic treatment of this group is not satisfactory, and it is now clear that the group is polyphyletic (see Chapter 13). The highly diverse mechanisms of dissimilatory metabolism that occur in this genus provide a sound basis for its taxonomic subdivision, and the group will be described here primarily in terms of these properties.

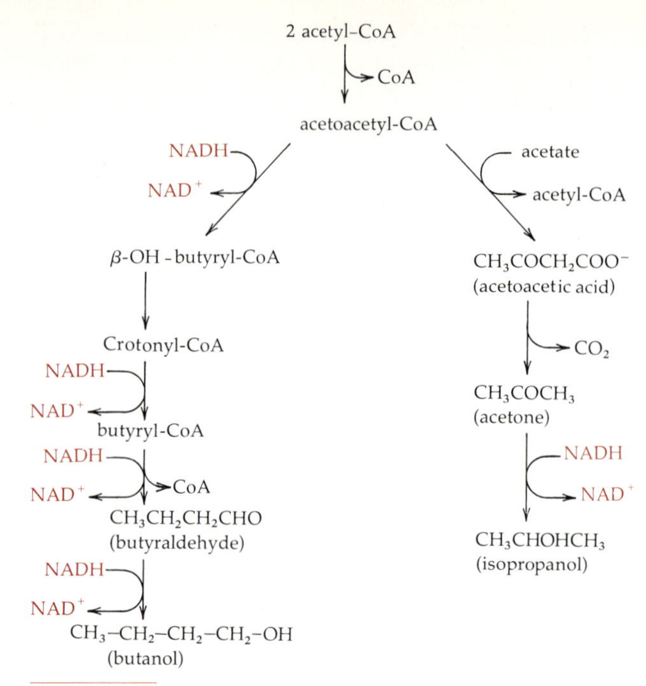


FIGURE 22.20

Pathways for the formation of butanol, acetone, and isopropanol from acetyl-CoA.

## The Butyric Acid Clostridia

Many clostridia perform a fermentation of soluble carbohydrates, starch or pectin, with the formation of acetic and butyric acids, $CO_2$ and $H_2$. These *butyric acid clostridia* grow poorly (if at all) in complex media devoid of a fermentable carbohydrate. Two other characters are distinctive of this subgroup: they synthesize as cellular reserve material a starchlike polysaccharide, detectable microscopically by its deep purple color in cells treated with an iodine solution; and many fix nitrogen very actively, a property otherwise absent from the clostridia.

The butyric fermentation is initiated by a conversion of sugars to pyruvate through the Embden-Meyerhof pathway. The pathways for the formation of end products from pyruvate are shown in Figure 20.3. However the mechanism for acetyl-CoA production by these clostridia differs from that shown in Figure 20.3; pyruvate is oxidized to acetyl-CoA and $CO_2$, with ferredoxin as electron acceptor. Hydrogenase then reoxidizes ferredoxin, producing $H_2$. This system is termed the *clastic system.*

Some of the butyric acid clostridia form additional neutral compounds (butanol, acetone, isopropanol, small amounts of ethanol) from sugars. With the exception of ethanol (formed by the reduction of acetyl-CoA), these neutral products arise by divergences from the normal pathway of butyrate formation, as shown in Figure 22.20; and their formation is accompanied by a reduction in the amounts of butyrate and $H_2$ formed, as shown by comparative fermentation balances (Table 22.5). In these modified butyric acid fermentations the neutral products typically arise in the later stages of growth. Their production is accompanied by reutilization of some of the $H_2$ initially produced, which serves as a reductant of $NAD^+$. The accumulation of neutral end products is, accordingly, favored by the maintenance of a high partial pressure of $H_2$ in the cultures, and it can be largely prevented if the $H_2$ is removed as it is formed. The *acetone-butanol fermentation*, effected by *C. acetobutylicum* (Table 22.5), has been operated on an industrial scale (Chapter 33).

## The Anaerobic Dissimilation of Amino Acids by Clostridia

A large number of *Clostridium* spp. can grow well in complex media containing peptones or yeast extract, in the absence of a fermentable carbohydrate. These organisms are collectively responsible for the putrefactive decomposition of nitrogenous compounds in nature; they also include the principal pathogenic clostridia (*C. botulinum, C. tetani, C. perfringens*). Growth in complex media is accompanied by the formation of ammonia, $CO_2$, $H_2S$, fatty acids, and a variety of other volatile substances, often having unpleasant odors. The nature

**TABLE 22.5**

**The Products of Fermentation of Glucose by Three Species of Butyric Acid Bacteria**

| | *C. lactoacetophilum* | *C. acetobutylicum* | *C. butylicum* |
|---|---|---|---|
| Butyric acid | 73 | 4 | 17 |
| Acetic acid | 28 | 14 | 17 |
| $CO_2$ | 190 | 221 | 204 |
| $H_2$ | 182 | 135 | 78 |
| Ethanol | — | 7 | — |
| Butanol | — | 56 | 59 |
| Acetone | — | 22 | — |
| Isopropanol | — | — | 12 |

Note: Amounts are moles/100 moles of glucose fermented.

of the specific fermentable substrates and the pathways involved in their dissimilation long remained unknown; they have been elucidated during the past 40 years, largely through the work of L. H. Stickland and of H. A. Barker and his associates.

Probably the most widespread mechanism of amino acid dissimilation among these organisms is the fermentation of pairs of amino acids, one of which acts as an electron donor, undergoing oxidation, while the other acts as an electron acceptor, undergoing reduction. This general mode of amino acid dissimilation is known as the Stickland reaction. As shown in Table 22.6, amino acids can be assigned to two series in terms of their roles in the Stickland reaction. Some serve uniquely as electron donors, others as electron acceptors; only tryptophan and tyrosine can play both roles.

The Stickland reaction can be illustrated by a simple example: the coupled fermentation of alanine and glycine, neither of which can be fermented singly by most clostridia. The overall reaction can be represented as:

$$\underset{\text{(alanine)}}{CH_3CHNH_2COOH} + \underset{\text{(glycine)}}{2CH_2NH_2COOH} \longrightarrow 3NH_3 + 3CH_3COOH + CO_2$$

**TABLE 22.6**

**Classification of Amino Acids in Terms of Their Roles as Electron Donors and Electron Acceptors in the Stickland Reaction**

| Electron Donors | | Electron Acceptors | |
|---|---|---|---|
| AMINO ACID | RELATIVE RATE OF OXIDATION | AMINO ACID | RELATIVE RATE OF REDUCTION |
| Alanine | 100 | Glycine | 100 |
| Leucine | 100 | Proline | 100 |
| Isoleucine | 100 | Hydroxyproline | 100 |
| Norleucine | 100 | Ornithine | 100 |
| Valine | 76 | Arginine | 80 |
| Histidine | 37 | Tryptophan | 67 |
| Phenylalanine | 28 | Tyrosine | 25 |
| Tryptophan | 17 | Cysteine | 22 |
| Tyrosine | 16 | Methionine | 15 |
| Serine | 16 | | |
| Asparagine | 13 | | |

Note: The relative rates of oxidation (or reduction) shown were determined for the species *C. sporogenes*.

The various reactions which comprise this conversion are shown in Figure 22.21. Alanine, the electron donor, is oxidatively deaminated to pyruvate, which then undergoes thiolytic cleavage to acetyl-CoA and $CO_2$; ATP is then formed through the conversion of acetyl-CoA to acetate. NADH is reoxidized by a reductive deamination of glycine to acetate. The ATP yield is thus one mole for three moles of amino acid dissimilated. The Stickland reaction allows virtually all the constituent amino acids of proteins to be utilized as energy sources.

Many clostridia are able to ferment specific, single amino acids, a mode of dissimilatory metabolism that may or not be accompanied by the ability to perform the Stickland reaction (Table 22.7). The pathway for the dissimilation of glutamate, fermented by many species, is outlined in Figure 22.22. The initial attack on glutamate involves a rearrangement of the carbon skeleton and leads to the formation of a branched dicarboxylic acid, citramalate, which is cleaved to pyruvate and acetate. The subsequent pathway of pyruvate dissimilation is biochemically similar to that of the butyric acid fermentation (Figure 20.3), the synthesis of ATP occurring during the conversion of acetyl-CoA to acetate.

A considerable number of clostridia that ferment amino acids can also ferment carbohydrates, these substrates undergoing a typical butyric acid fermentation. It is evident, for example, that a glutamate-fermenting *Clostridium* that can convert glucose to pyruvate through the Emden-Meyerhof pathway possesses the remaining enzymatic machinery necessary for the performance of a butyric acid fermentation. However, many amino acid fermenters are wholly unable to ferment carbohydrates; such organisms are exemplified by *C. tetani* and *C. histolyticum.* There is, accordingly, a broad spectrum with respect to fermentable substrates extending from butyric acid bacteria that have little or no ability to ferment amino acids to organisms such as *C. tetani* and *C. histolyticum* that are unable to ferment carbohydrates.

Many of the clostridia that ferment amino acids are proteolytic organisms and produce a wide diversity of proteases; several hydrolytic enzymes of this type with different substrate specificities are often produced by a single organism. Proteolysis is of course a necessary preliminary step in the production of fermentable substrates from proteins by members of this group which obtain energy through amino acid fermentations. It should be noted, however, that by no means all amino acid-fermenting clostridia are proteolytic organisms. Nonproteolytic species are consequently dependent on the availability of free amino acids as growth substrates.

FIGURE 22.21

Mechanism of the Stickland reaction, with alanine as electron donor and glycine as electron acceptor.

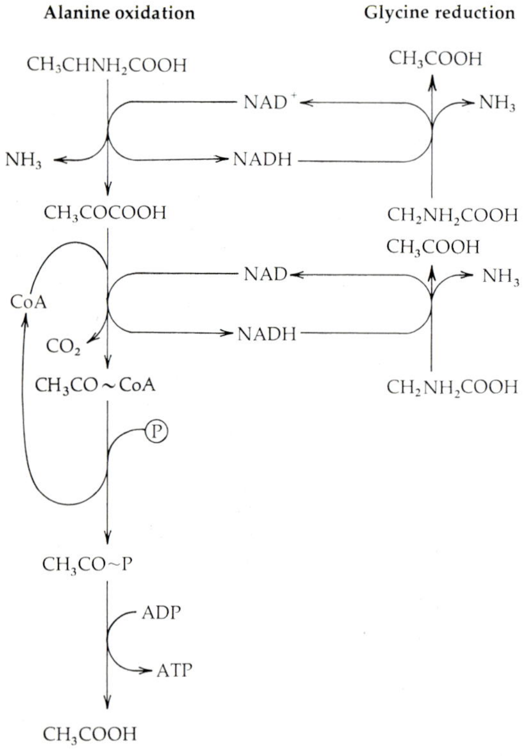

Alanine oxidation

Glycine reduction

$CH_3CHNH_2COOH$ → $NH_3$ ; NAD⁺ → NADH → $CH_3COCOOH$

$CH_3COCOOH$ + CoA → $CO_2$ ; NAD → NADH → $CH_3CO{\sim}CoA$

$CH_3CO{\sim}CoA$ + Ⓟ → CoA → $CH_3CO{\sim}P$

$CH_3CO{\sim}P$ + ADP → ATP → $CH_3COOH$

$CH_2NH_2COOH$ + NADH → NAD⁺ + $NH_3$ → $CH_3COOH$

$CH_2NH_2COOH$ + NADH → NAD + $NH_3$ → $CH_3COOH$

glutamate → β-methyl-aspartate → mesaconate ($-NH_3$) → citramalate ($+H_2O$) → pyruvate + acetate

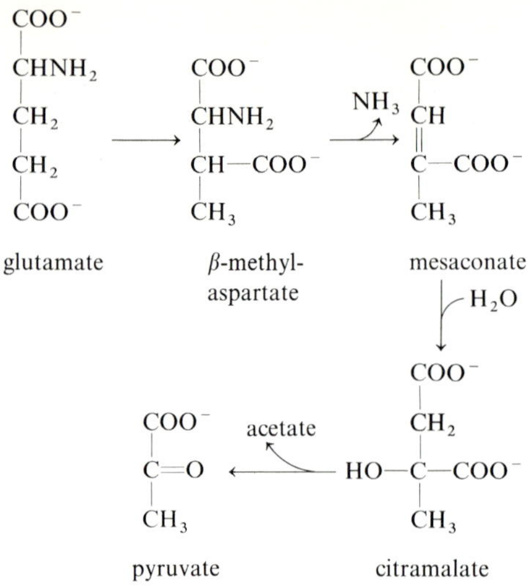

FIGURE 22.22

Conversion of glutamate to pyruvate by clostridia. The pyruvate is fermented to acetate and butyrate.

**TABLE 22.7**

**The Amino Acids Fermented by Some Species of *Clostridium***

| Species | Single Amino Acids That Can Be Fermented | | | | | | | | | | | | | |
|---|---|---|---|---|---|---|---|---|---|---|---|---|---|---|
| | Alanine | Arginine | Aspartate | Cysteine | Glutamate | Histidine | Leucine | Lysine | Methionine | Phenylalanine | Serine | Threonine | Tyrosine | Ability to Perform Stickland Reaction |
| *C. botulinum* | | | | | | | | | | | + | | | + |
| *C. cochlearium* | | | | | + | | | | | | | | | − |
| *C. perfringens* | | | | | + | | | | | | + | + | | |
| *C. propionicum* | + | | | + | | | | | | | + | + | | − |
| *C. sporogenes* | | + | | + | | | + | | + | + | + | + | | + |
| *C. sticklandii* | | + | | | | | | + | | + | + | | | + |
| *C. tetani* | | | + | | + | + | | | | + | | | | − |
| *C. tetanomorphum* | | | + | + | + | + | | | | + | | | + | − |

## The Fermentation of Nitrogen-Containing Ring Compounds

Some clostridia can obtain energy by the fermentation of heterocyclic compounds including purines, pyrimidines, and nicotinic acid. The fermentation of purines (guanine, uric acid, hypoxanthine, xanthine) is carried out by *C. acidiurici* and *C. cylindrosporum*, nutritionally highly specialized species, which are unable to ferment other substrates. The fermentation products consist of acetate, glycine, formate, $CO_2$, or other products. Only one mole of acetate per mole of purine fermented can be derived directly from a $C_2$ fragment because purines contain only two contiguous carbon atoms. However, the yield of acetate is often greater than one mole, which shows that it must be formed in part from $C_1$ precursors. Acetate synthesis from $CO_2$ is a characteristic of certain other clostridial fermentations, discussed below.

## Carbohydrate Fermentations by Clostridia That Do Not Yield Butyric Acid as a Product

A number of clostridia utilizing carbohydrates as energy sources dissimilate them by pathways other than the butyric acid pathway. These organisms include cellulose-fermenting clostridia, most of which are highly specialized with respect to substrates; some species can ferment only cellulose. The products include ethanol, formate, acetate, lactate, and succinate.

The species *C. thermoaceticum* ferments glucose and other soluble sugars with the formation of acetate as the sole end product; the formation of this product is virtually quantitative, almost three moles of acetate being produced per mole of glucose decomposed. No known pathway of glucose dissimilation permits a direct formation of acetate from all six carbon atoms of the substrate. In fact, only two-thirds of the acetate produced is directly derived from glucose carbon through the reactions of glycolysis (Figure 22.23); one-third of the acetate is produced by a complex process of synthesis from $CO_2$, involving the participation of tetrahydrofolate and a corrinoid coenzyme (a vitamin $B_{12}$ derivative) as carriers of the $C_1$ and $C_2$ intermediates (Figure 22.24). One $CO_2$ is reduced to the methyl level, with tetrahydrofolate (THF) as the $C_1$ carrier for all but the first step. The methyl group is transferred to a corrinoid coenzyme ("$B_{12}$"). The second $CO_2$ is reduced to an enzyme-bound intermediate exchangeable with carbon monoxide ([CO]). This carboxyl precursor is then transferred to the methyl-"$B_{12}$" to form a bound acetyl group that is finally transferred to CoA.

Although labeling studies show that $CO_2$ can be incorporated into both positions of acetate, it is

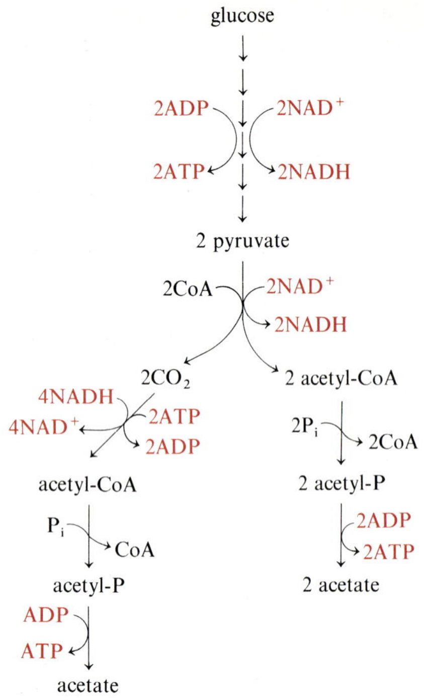

FIGURE 22.23

The homoacetate fermentation of glucose by *Clostridium thermoaceticum*.

FIGURE 22.24

Synthesis of acetyl-CoA from $CO_2$ by acetogenic clostridia. THF, tetrahydrofolate; "$B_{12}$," a corrinoid coenzyme derived from vitamin $B_{12}$; [CO], an enzyme-bound $C_1$ fragment with the valence of carbon monoxide.

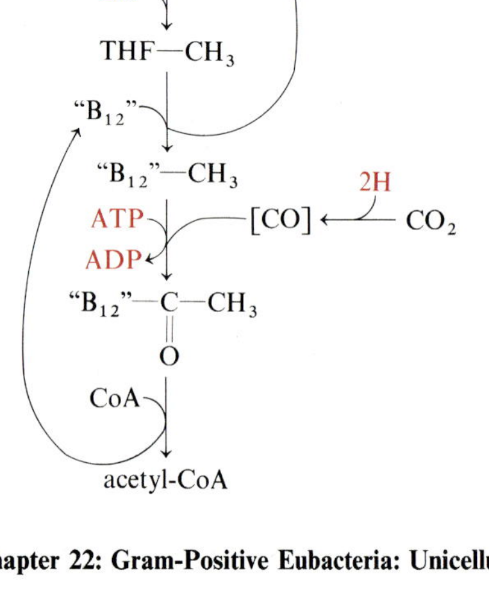

likely that during the fermentation of sugars the carboxyl group of acetate comes from the carboxyl of pyruvate without passing through $CO_2$; pyruvate presumably acts directly as the source of enzyme-bound carbon monoxide. The labeling of both carbon atoms of acetate by radioactive $CO_2$ is largely due to an exchange of the carboxyl of pyruvate with $CO_2$. However, some clostridia are capable of acetate formation from $CO_2$ and $H_2$; in this case, the carboxyl group almost certainly comes directly from $CO_2$ as shown in Figure 22.24. As mentioned in Chapters 14 and 20, both the methanogens and the autotrophic sulfur-reducing bacteria probably assimilate $CO_2$ via an enzyme-bound carbon monoxide; however the methyl carbon atom is derived from methyl-CoM in the methanogens, and probably from a methyl-pterin in the sulfur reducers.

## The Ethanol-Acetate Fermentation by *Clostridium kluyveri*

A most remarkable clostridial fermentation is that performed by *C. kluyveri*. This organism grows only at the expense of a mixture of ethanol and acetate as its energy sources. The main organic products of the fermentation are two higher fatty acids, butyrate and caproate; in addition, some $H_2$ is produced. If $H_2$ production is neglected, the fatty acid synthesis can be represented by the two equations:

$$CH_3CH_2OH + CH_3COOH \longrightarrow CH_3(CH_2)_2COOH + H_2O$$

$$2CH_3CH_2OH + CH_3COOH \longrightarrow CH_3(CH_2)_4COOH$$

The mechanism of net ATP synthesis associated with this fermentation has been discovered only recently and will be outlined for the case of butyrate synthesis (Figure 22.25).

Ethanol is dehydrogenated in two steps to acetyl-CoA, ferredoxin being the initial acceptor for the second dehydrogenation. Reduced ferredoxin can transfer electrons either to $NAD^+$ with formation of NADH, or to protons with formation of $H_2$. As in other clostridia, the synthesis of butyrate then occurs through the cyclic reactions shown in Figure 22.25.

In the absence of $H_2$ formation, the oxidation of ethanol thus produces the exact quantities of acetyl-CoA and NADH required for butyrate synthesis; hence, under these conditions, a net synthesis of ATP cannot occur. However, the formation of $H_2$ diverts part of the electrons from acetaldehyde oxidation that would otherwise serve for

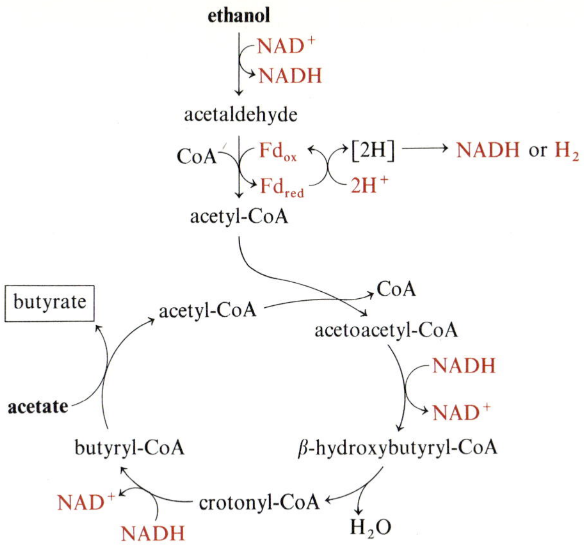

FIGURE 22.25

The butyrate fermentation of ethanol and acetate. Fd denotes ferredoxin.

$NAD^+$ reduction, and the molar ratio of NADH to acetyl-CoA produced during ethanol oxidation becomes less than 2:1. As a result, some acetyl-CoA is available for ATP synthesis by conversion to acetate via acetylphosphate. It can be calculated that, for every mole of $H_2$ produced, 0.5 mole of acetyl-CoA becomes available for ATP synthesis. An experimental determination of the balance for this fermentation showed that approximately 0.25 mole of $H_2$ was produced per mole of ethanol used. The ATP yield under these circumstances was therefore roughly 0.12 mole of ATP produced per mole of ethanol oxidized.

## The Genus *Desulfotomaculum*

The anaerobic endosporeformers of the genus *Desulfotomaculum* can be clearly distinguished from the clostridia by their higher percent G + C and their ability to use sulfate as a terminal electron acceptor. In most metabolic respects *Desulfotomaculum* is indistinguishable from the Gram-negative sulfate-reducing bacteria (see Chapter 20). However, *Desulfotomaculum* appears not to be able to generate a protonmotive force as a consequence of electron transport to sulfate; it is thus able to synthesize ATP only via substrate-level phosphorylation, and sulfur reduction in this organism is accordingly not respiratory.

Most *Desulfotomaculum* species have a nutritional spectrum similar to that of *Desulfovibrio:* they utilize fermentation end products such as lactate, and convert them quantitatively to acetate and $CO_2$ (because they lack a complete TCA cycle, required for the oxidation of acetate). Since only two moles of ATP can be formed in the course of oxidation of one mole of lactate to acetate, the activation of sulfate in these organisms can only require the equivalent of a single ATP. In the respiratory sulfate-reducing bacteria, sulfate is activated in a process that consumes the equivalent of two ATP molecules (see Figure 20.9). In *Desulfotomaculum* the initial step is the same: sulfate is activated with ATP to yield APS and pyrophosphate; however, the pyrophosphate is cleaved by acetate rather than water, thus conserving its bond energy (Figure 22.26). The overall energy yield is one mole of ATP for every two moles of lactate oxidized to acetate.

One species of *Desulfotomaculum, D. acetoxydans*, has a complete TCA cycle and is capable of growth at the expense of acetate, which is oxidized to $CO_2$. This species activates acetate by CoA transfer from succinyl-CoA in the same fashion as the respiratory sulfate-reducing bacteria (Figure 20.12).

FIGURE 22.26

Lactate fermentation in *Desulfotomaculum*. APS: adenylphosphosulfate.

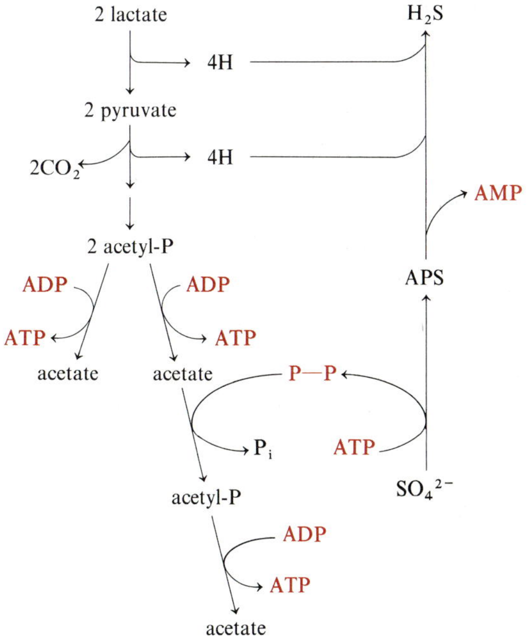

### The Genus *Sporolactobacillus*

The sporolactobacilli perform a homolactic fermentation of hexoses identical to that of many lactobacilli (Figure 23.4), and, also like the lactobacilli, are somewhat aerotolerant despite their lack of catalase. They are motile by means of peritrichous flagella. Their natural habitat appears to be the rhizosphere of plants. Ellipsoidal spores are produced terminally or subterminally within swollen sporangia.

## FURTHER READING


**Books**

BERKELEY, R. C. W., and M. GOODFELLOW, *The Aerobic Endospore-Forming Bacteria: Classification and Identification.* New York: Academic Press, 1981.

GOTTSCHALK, G., *Bacterial Metabolism*, 2nd ed. New York, Heidelberg, and Berlin: Springer-Verlag, 1985.

HALVORSON, H. O., R. HANSON, and L. L. CAMPBELL, *Spores V.* Washington, D.C.: American Society for Microbiology, 1972.

HURST, A., and G. W. GOULD, *The Bacterial Spore*, Vols. 1 and 2. New York: Academic Press, 1969 and 1983.

**Reviews**

ARONSON, A. I., and P. FITZ-JAMES, "Structure and Morphogenesis of the Bacterial Spore Coat," *Bacteriol. Rev.* **40,** 360–402 (1976).

BARKER, H. A., "Amino Acid Degradation by Anaerobic Bacteria," *Ann. Rev. Biochem.* **50,** 23–40 (1981).

VOGELS, G. S., and C. VAN DER DRIFT, "Degradation of Purines and Pyrimidines by Microorganisms, *Bacteriol. Rev.* **40,** 403–468 (1976).

# Chapter 23
# Gram-Positive Fermentative Eubacteria

The asporogenous Gram-positive bacteria contain a number of anaerobic or facultatively anaerobic organisms; some of these are members of the actinomycete group, discussed in the next chapter, but many of them are unicellular bacteria unrelated to the mycelial procaryotes. Most are related to the endospore-forming bacteria (Chapter 13), and like them include a variety of physiological types. The only true facultative anaerobe is *Staphylococcus*, capable of fermentation or respiration. The remainder of these organisms are fermentative, although they include some (notably the lactic acid bacteria) that are capable of growth in the presence of air, but whose metabolism is fermentative even aerobically; such organisms are termed *aeroduric anaerobes*. The fermentative members of this group closely resemble the Gram-negative anaerobes (Chapter 20) in their physiology, nutrition, and natural distribution, and like them are classified principally on the basis of their fermentation patterns and morphology. Within this rather heterogeneous group it is customary to recognize the *lactic acid bacteria* as a distinct subgroup characterized by the production of large amounts of lactic acid from the fermentation of sugars.

## THE GENUS *STAPHYLOCOCCUS*

The members of the genus *Staphylococcus* are facultative anaerobes and ferment sugars with the formation of lactic acid as one of the major products. Their DNA base composition (30 to 40 percent G + C) is in the same range as that of many spherical lactic acid bacteria. However, they can be readily distinguished from these organisms by several criteria: possession of catalase and other heme pigments; capacity for respiratory metabolism; and much less restricted requirements for carbon and energy (growth will occur on complex media in the absence of carbohydrates). Many also produce carotenoid pigments, absent from all lactic acid bacteria. These organisms are typical members of the normal microflora of the skin, and some are potential pathogens causing either infections or food poisoning (Chapter 31).

## THE LACTIC ACID BACTERIA

The lactic acid bacteria are immotile, rod-shaped or spherical organisms (Figure 23.1), united by an unusual constellation of metabolic and nutritional properties. The name derives from the fact that ATP is synthesized through fermentations of carbohydrates, which yield lactic acid as a major (and sometimes as virtually the sole) end product.

The lactic acid bacteria are all aerotolerant anaerobes that grow readily on the surface of solid media exposed to air. However, they are unable to synthesize ATP by respiratory means, a reflection of their failure to synthesize cytochromes and other heme-containing enzymes. Although they can perform limited oxidations of a few organic compounds, mediated by flavoprotein enzymes, either oxidases or peroxidases, these oxidations are not accompanied by ATP formation. The growth yields of lactic acid bacteria are, accordingly, largely unaffected by the presence or absence of air, the fermentative dissimilation of sugars being the source of ATP under both conditions.

One consequence of the failure to synthesize heme proteins is that the lactic acid bacteria are catalase negative, and hence cannot mediate the decomposition of $H_2O_2$ according to the reaction

$$2H_2O_2 \longrightarrow 2H_2O + O_2$$

The absence of catalase activity, readily demonstrated by the absence of $O_2$ formation when cells are mixed with a drop of dilute $H_2O_2$, is one of the most useful diagnostic tests for the recognition of these organisms, since they are virtually the only bacteria devoid of catalase that can grow in the presence of air.

The inability of lactic acid bacteria to synthesize heme proteins is correlated with an inability to synthesize heme, the porphyrin which is the prosthetic group of these enzymes. However, certain lactic acid bacteria acquire catalase activity when grown in the presence of a source of heme (e.g., on media containing red blood cells). Such

FIGURE 23.1

The form and arrangement of cells in three genera of lactic acid bacteria: (a) *Lactobacillus;* (b) *Streptococcus;* (c) *Pediococcus* (phase contrast, ×2,180).

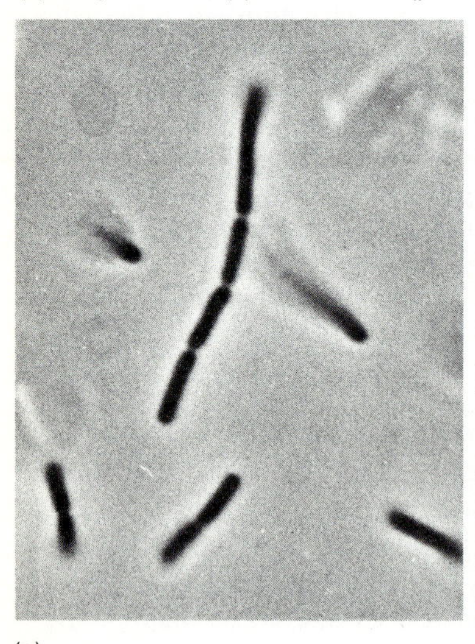

(a)

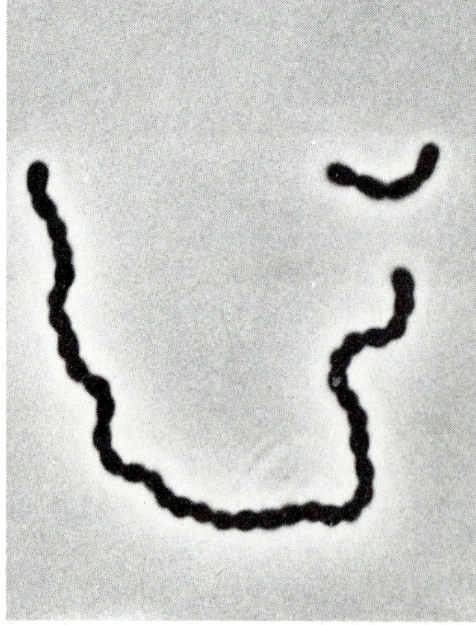

(b)

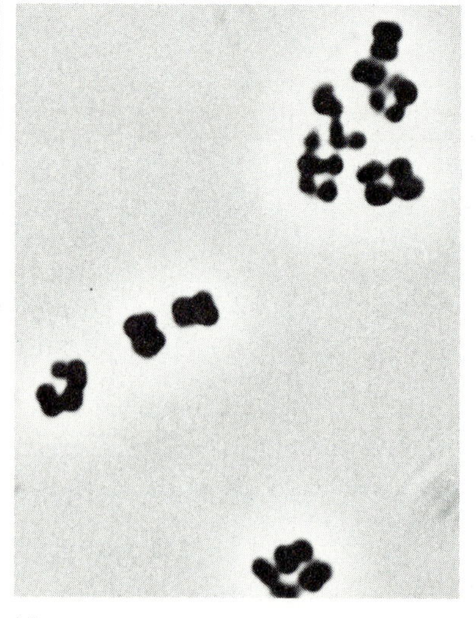

(c)

species synthesize a protein termed *pseudocatalase*, that can combine with exogenously supplied heme to produce an enzyme with the properties of catalase. Pseudocatalase is a Mn-containing enzyme that has weak catalase activity even in the absence of heme; it has been shown to prolong the viability of stationary-phase cells incubated under aerobic conditions, but appears to be unimportant to actively growing cells.

In addition to its function in pseudocatalase, manganese plays another important role in protecting many of this group against the toxic effects of oxygen. High levels (in the range of 10 to 25 mM) of $Mn^{2+}$ are accumulated intracellularly. At these high concentrations, $Mn^{2+}$ is functionally equivalent to the enzyme superoxide dismutase (Chapter 8). The chemical details of this activity have not yet been elucidated; a possible series of reactions is

$$2Mn^{2+} + 2O_2{\cdot}^- \longrightarrow 2MnO_2{}^+$$
$$2MnO_2{}^+ \longrightarrow MnO_2 + O_2 + Mn^{2+}$$

The lack of superoxide dismutase, and a requirement for large amounts of manganese in the growth medium, is a common pattern among the lactic acid bacteria, and indicates that $Mn^{2+}$ may be widely employed as a replacement for superoxide dismutase in the group. Indeed, the only member that has been shown to have detectable superoxide dismutase activity is *Streptococcus*; most of the rest accumulate large amounts of intracellular $Mn^{2+}$. The few that have neither are markedly oxygen-sensitive.

The inability to synthesize heme is only one manifestation of the *extremely limited synthetic abilities* characteristic of the lactic acid bacteria. With one exception (*Streptococcus bovis*), these organisms have complex growth-factor requirements; they require B vitamins, a considerable number of amino acids, and purine and pyrimidine bases. As a result of their complex nutritional requirements, lactic acid bacteria are usually cultivated on media containing peptone, yeast extract, or other digests of plant or animal material. These must be supplemented with a fermentable carbohydrate to provide an energy source.

Even when growing on very rich media, the colonies of latic acid bacteria (Figure 23.2) always remain relatively small (at most, a few millimeters in diameter). They are rarely pigmented; as a result of the absence of cytochromes, the growth has a characteristic chalky white appearance. The small colony size of these bacteria is attributable primarily to low growth yields, a consequence of their exclusively fermentative metabolism. Some species can produce quite large colonies when grown on sucrose-containing media, as a result of the massive synthesis of extracellular polysaccharides (either dextran or levan) at the expense of this disaccharide; in this special case, much of the volume of the colony consists of polysaccharide. Since dextrans and levans are synthesized only from sucrose, the species in question form typical small colonies on media containing any other utilizable sugar. In the isolation of the spherical lactic acid bacteria, which can grow in media that have an initial pH of 7 or above, the incorporation of finely divided $CaCO_3$ in the plating medium is useful as a buffer, and also because the colonies can be readily recognized by the surrounding zones of clearing, caused by acid production (Figure 2.5).

Another distinctive physiological feature of lactic acid bacteria is *their high tolerance of acid.* Although the spherical lactic acid bacteria can ini-

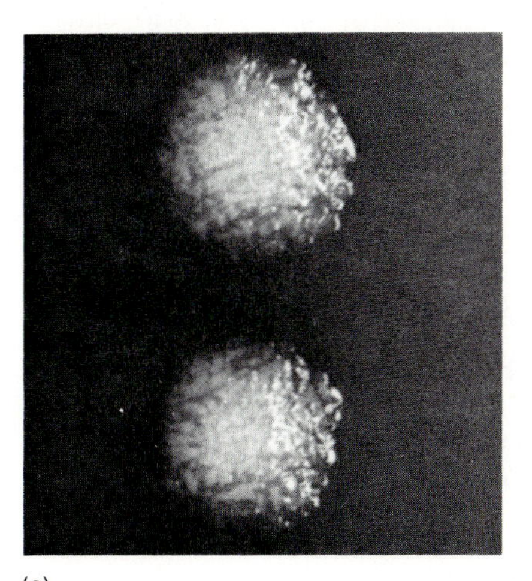

(a)

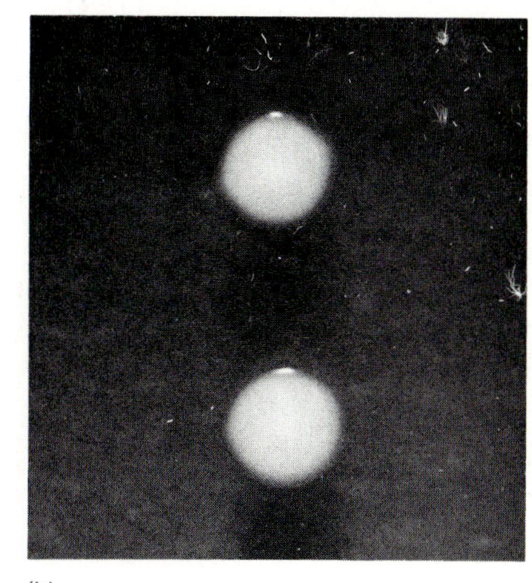

(b)

FIGURE 23.2

Colonies of lactic acid bacteria: (a) *Lactobacillus plantarum* and (b) *Streptococcus lactis* ($\times$ 9.6).

tiate growth in neutral or alkaline media, most of the rod-shaped forms cannot grow in media with an initial pH greater than 6. Growth of all lactic acid bacteria continues until the pH has fallen, through fermentation, to a value of 5 or less.

The capacity of lactic acid bacteria to produce and tolerate a relatively high concentration of lactic acid is of great selective value, since it enables them to eliminate competition from most other bacteria in environments that are rich in nutrients. This is shown by the fact that lactic acid bacteria can be readily enriched from natural sources through the use of complex media with a high sugar content. Such media can, of course, support the growth of many chemoheterotrophic bacteria, but competing organisms are largely eliminated as growth proceeds by the accumulation of lactic acid, formed through the metabolic activity of the lactic acid bacteria.

As a result of their extreme physiological specialization, the lactic acid bacteria are confined to a few characteristic natural environments. Some live in association with plants and grow at the expense of the nutrients liberated through the death and decomposition of plant tissues. They occur in foods and beverages prepared from plant materials: pickles, sauerkraut, ensilaged fodder, wine, and beer. A lactic fermentation of the sugar initially present occurs during the preparation of pickles, sauerkraut, and ensilage. In beer and wine the lactic acid bacteria are potential spoilage agents, which sometimes grow and produce an undesirable acidity.

Other lactic acid bacteria constitute part of the normal flora of the animal body and occur in considerable numbers in the nasopharynx, the intestinal tract, and the vagina. These forms include a number of important pathogens of humans and other mammals, all belonging to the genus *Streptococcus.*

A third characteristic habitat of the lactic acid bacteria is milk, to which they gain access either from the body of the cow or from plant materials. The normal souring of milk is caused by certain streptococci, and both rod-shaped and spherical lactic acid bacteria play important roles in the preparation of fermented milk products (butter, cheeses, buttermilk, yogurt).

Because of their activities in the preparation of foods and as agents of human and animal disease, the lactic acid bacteria are a group of major economic importance (Chapter 33).

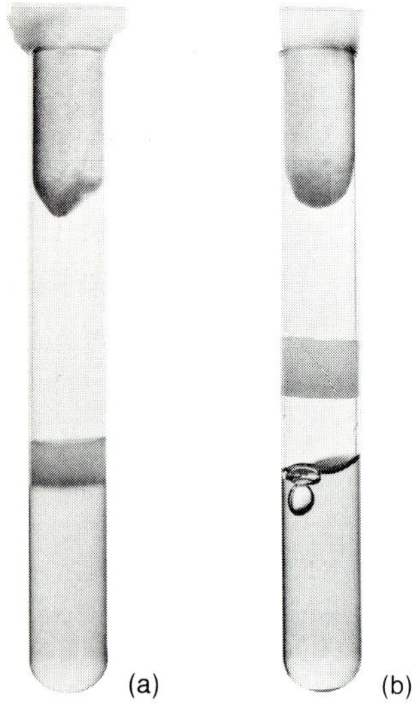

(a) (b)

FIGURE 23.3
The demonstration of $CO_2$ production by lactic acid bacteria in tubes of a sugar-rich medium with agar seals: (a) *Streptococcus lactic;* (b) *Leuconostoc mesenteroides.*

## Patterns of Carbohydrate Fermentation in Lactic Acid Bacteria

It was shown by S. Orla Jensen in about 1920 that lactic acid bacteria can be divided into two biochemical subgroups, distinguishable by the products formed from glucose. *Homofermenters* convert glucose almost quantitatively to lactic acid; *heterofermenters,* to an equimolar mixture of lactic acid, ethanol, and $CO_2$. The mode of glucose fermentation can be most simply determined by the detection of $CO_2$ production. However, since the amount produced by heterofermenters is small (one mole per mole of glucose fermented), its detection requires a special procedure: growth in a well-buffered medium of high sugar content, sealed with an agar plug to trap the $CO_2$ formed (Figure 23.3).

The metabolic explanation of the dichotomy in fermentative patterns among lactic acid bacteria was discovered many years later. Homofermenters dissimilate glucose through the Embden-Meyerhof pathway (Figure 23.4). However, heterofermenters cannot utilize this pathway, since they lack a key enzyme, fructose-bisphosphate aldolase, which mediates sugar-phosphate cleavage. These organisms dissimilate glucose through the oxidative pentose phosphate pathway (Figure 23.5); this fermentation results in a strictly equimolar ratio of the three end products: lactic acid, ethanol, and $CO_2$. A major difference between the two pathways is their net ATP yields: two moles per mole of glucose fermented by the homofermentative pathway and only one mole by the heterofermentative pathway.

Two heterofermentative lactobacilli, *L. brevis* and *L. buchneri*, cannot grow anaerobically with glucose, because they are not able to effect the reduction of acetyl-phosphate to ethanol, essential to maintain overall redox balance. They can, however,

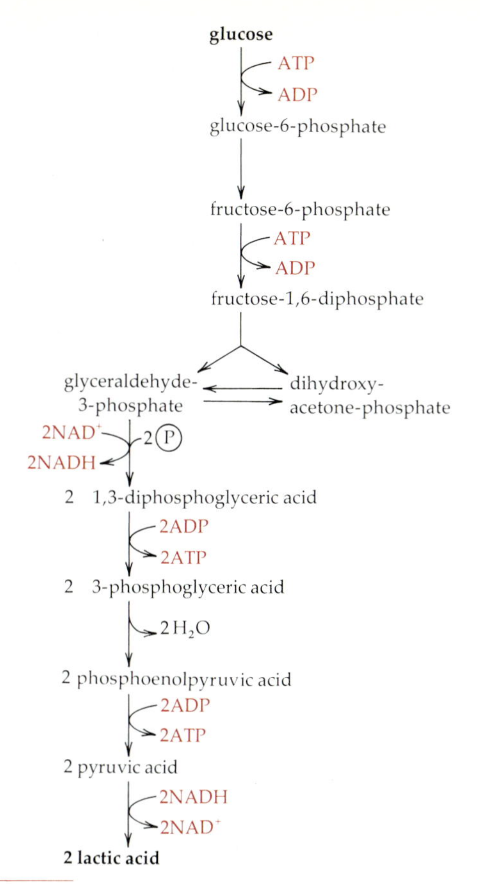


FIGURE 23.4
The EMP pathway by which homofermentative lactic acid bacteria convert glucose to lactic acid.

ferment glucose aerobically, reoxidizing NADH at the expense of $O_2$ by means of a flavoprotein enzyme. The overall reaction for glucose fermentation under these conditions becomes

$$\text{glucose} + O_2 \longrightarrow \text{lactate} + \text{acetate} + CO_2 + 2H_2O$$

Both these species can grow anaerobically at the expense of another hexose, fructose, because they possess a mannitol dehydrogenase, which mediates reduction of this ketosugar to the polyalcohol, mannitol:

$$\text{fructose} + \text{NADH} + H^+ \rightleftharpoons \text{mannitol} + \text{NAD}^+$$

This reduction maintains redox balance under anaerobic conditions, the overall equation for fructose fermentation being

$$3\ \text{fructose} \longrightarrow \text{lactate} + \text{acetate} + CO_2 + 2\ \text{mannitol}$$

Many other heterofermenters of the genera *Lactobacillus* and *Leuconostoc* also contain mannitol dehydrogenase and produce mannitol as a product of fructose fermentation, with a concomitant formation of some acetate rather than ethanol.

Several homofermentative lactic acid bacteria (*Streptococcus* spp. and *Lactobacillus* spp.) will produce fermentation products other than lactic acid (normally formate, ethanol, and acetate) under some conditions. This shift is not a switch to the pentose-phosphate pathway of sugar metabolism; rather it reflects a change in the way pyruvate is metabolized: less lactate is produced, and the rest of the pyruvate is converted to acetyl-CoA. The metabolic basis of this shift is now known: lactate dehydrogenase is activated by fructose bisphosphate, and pyruvate formate-lyase is inhibited by triose phosphate (Figure 23.6). Thus under conditions of nutrient excess, the concentrations of intermediates of sugar catabolism are high, and pyruvate is converted quantitatively to lactate. However, under starvation conditions, some pyruvate is metabolized to ethanol and acetate, presumably an adaptation that allows more efficient use of the limiting amount of sugar (because ATP is generated during the conversion of pyruvate to acetate).

Another adaptation that increases the efficiency of energy conservation by some lactic acid bacteria (*Streptococcus* spp.) is the symport of lactic

FIGURE 23.5
The pathway of glucose dissimilation by heterofermentative lactic acid bacteria.

glucose
ATP
ADP
glucose-6-phosphate
NAD+
NADH
6-phosphogluconic acid
NAD+
NADH
$CO_2$
ribulose-5-phosphate
(P)
glyceraldehyde-3-phosphate
acetyl phosphate
NADH
NAD+
(P)
NADH
NAD+
(P)
1,3-diphosphoglyceric acid
acetaldehyde
ADP
ATP
NADH
NAD+
3-phosphoglyceric acid
ethanol
$2H_2O$
phosphoenolpyruvic acid
ADP
ATP
pyruvic acid
NADH
NAD+
lactic acid

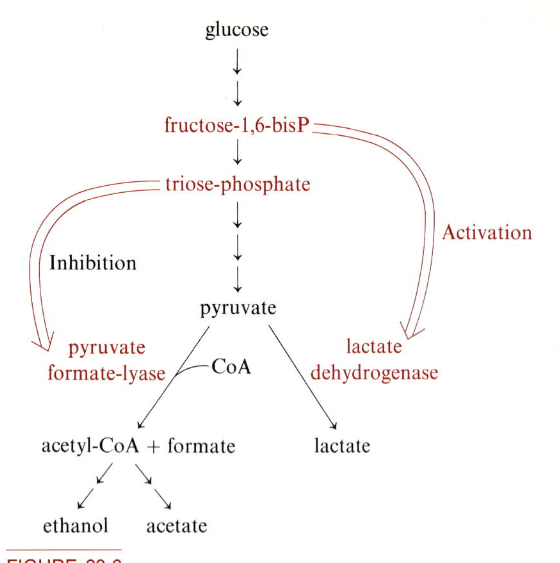


FIGURE 23.6
Regulation of pyruvate metabolism by glycolytic intermediates in the lactic acid bacteria.

acid with a proton during its excretion from the cell. This is only feasible at neutral pH values and relatively low external concentrations of lactate; however, under favorable conditions it may increase the ATP yield by as much as 50 percent (two protons per hexose is probably equivalent to one ATP). It is likely that this is of considerable significance in natural populations; an increased growth yield of *Streptococcus* has been obtained by coculturing it with a lactate-utilizing pseudomonad.

Lactic acid bacteria differ with respect to isomers of lactic acid that they produce. This is determined by the stereospecificity of the lactic dehydrogenases which mediate pyruvate reduction:

$$CH_3COCOOH + NADH + H^+ \rightleftharpoons CH_3CHOHCOOH + NAD^+$$

Some species contain only D-lactic dehydrogenase, and hence form the D-isomer; others contain only L-lactic dehydrogenase, and hence form the L-isomer. Certain species contain two lactic dehydrogenases of differing stereospecificity and form racemic lactic acid.

## Subdivision of the Lactic Acid Bacteria

Lactic acid bacteria with spherical cells are placed in three different genera, distinguished in part on structural and in part on biochemical grounds. *Pediococcus* divides in two perpendicular planes, to produce tetrads of cells, and consists of homofermenters. *Streptococcus* and *Leuconostoc* divide in one plane, to produce chains of cells; the former are homofermenters, the latter heterofermenters. As shown in Table 23.1, these three genera also differ with respect to the isomers of lactic acid that they produce.

All rod-shaped lactic acid bacteria are placed in one genus, *Lactobacillus*. However, this genus is divided into three subgenera, distinguished by the properties shown in Table 23.2. The heterofermenters (subgenus *Betabacterium*) ferment sugars exclusively by the pentose phosphate pathway, and always form racemic lactic acid. The homofermenters are placed in two subgenera, *Thermobacterium* and *Streptobacterium*. In the former, sugars are dissimilated exclusively through the Embden-Meyerhof pathway, and neither pentoses nor gluconic acid can be fermented. The thermobacteria have high temperature maxima and minima. In *Streptobacterium*, hexoses are similarly dissimilated exclusively through the Embden-Meyerhof pathway; however, these organisms also contain the enzymes of the oxidative pentose-phosphate pathway, and they dissimilate gluconic acid and pentoses through this metabolic route. They are, accordingly, *facultative* homofermenters, in contrast to the *obligate* homofermenters of the subgenus *Thermobacterium*. Both temperature minima and maxima are lower in *Streptobacterium* than in *Thermobacterium*. Although constant for each species, the isomer of lactic acid formed is not a characteristic property of each subgenus.

**TABLE 23.1**

**Taxonomic Subdivision of the Lactic Acid Bacteria**

| Genus | Cell Shape and Arrangement | Mode of Glucose Fermentation | Configuration of Lactic Acid | Percent G + C |
|---|---|---|---|---|
| *Streptococcus* | Spheres in chains | Homofermentative | L- | 33–44 |
| *Leuconostoc* | Spheres in chains | Heterofermentative | D- | 38–44 |
| *Pediococcus* | Spheres in tetrads | Homofermentative | DL- | 34–44 |
| *Lactobacillus* | Rods | V[a] | V[a] | 35–51 |

[a] V: variable among strains.

TABLE 23.2

**Distinguishing Properties of the Three Subgenera of *Lactobacillus***

| Subgenus | Fermentation of Sugars and Pathway Employed: HEXOSES | Fermentation of Sugars and Pathway Employed: GLUCONATE, PENTOSES | Isomer of Lactic Acid Produced | Temperature Range: 15° C | Temperature Range: 45° C |
|---|---|---|---|---|---|
| *Thermobacterium* | Embden-Meyerhof | Not attacked | L-, D-, or DL- | − | + |
| *Streptobacterium* | Embden-Meyerhof | Pentose-phosphate | L-, or DL- | + | − |
| *Betabacterium* | Pentose-phosphate | Pentose-phosphate | DL- | V[a] | V |

[a] Variable from one species to another.

## OTHER GRAM-POSITIVE ANAEROBES

The remaining Gram-positive, asporogenous anaerobes are a heterogeneous group of organisms; some of the principal genera are described in Table 23.3.

Among the Gram-positive anaerobic cocci, *Sarcina* is unusual in several respects. The cells are quite large, and the plane of division alternates regularly among three perpendicular planes to produce cubical packets of cells (Figure 23.7). One species, *S. ventriculi*, typically has a thick fibrous layer external to the murein; it is composed principally of cellulose (Figure 23.8). *S. ventriculi* is also very acid-tolerant, with growth occurring at pH values as low as 1. It has been isolated from human and animal stomachs, where it may be one of very few bacteria able to grow (the human stomach typically has a pH of about 2). The two species of *Sarcina*, *S. ventriculi* and *S. maxima*, both ferment sugars via the Embden-Meyerhof pathway, but with the formation of different end products (Table 23.4); *S. maxima* carries out a typical butyric acid fermentation, whereas *S. ventriculi* produces a mixture of lactate, acetate, ethanol, and $H_2$.

Two of the rod-shaped members of this group have traditionally been classified with corynebacteria or proactinomycetes (Chapter 24): *Propionibacterium* because its pleomorphic morphology is similar to that of the corynebacteria, and *Bifidobacterium* because its branched rods resemble the rudimentary mycelium of the proactinomycetes. Sequence analysis of their rRNAs confirms that they are related to the members of the actinomycete group, although distantly (Chapter 13).

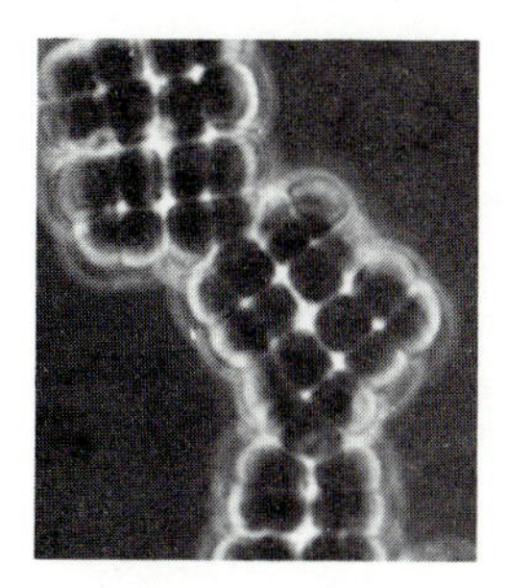

FIGURE 23.7

*Sarcina maxima* (phase contrast, × 1,630). From S. Holt and E. Canale-Parola, "Fine Structure of *Sarcina maxima* and *Sarcina ventriculi*," *J. Bacteriol.* **93,** 399 (1967).

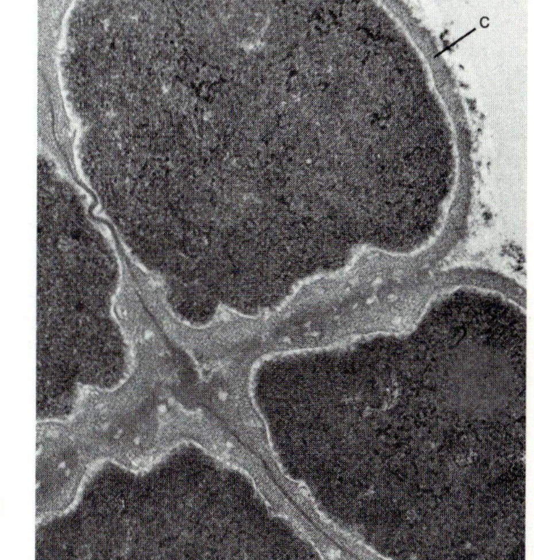


FIGURE 23.8

Electron micrograph of a thin section of a group of cells of *Sarcina ventriculi*, showing the heavy cellulose layer, c, that encloses each cell ( × 27,900). Courtesy of S. Holt and E. Canale-Parola.

**TABLE 23.3**

**Gram-Positive Asporogenous Anaerobic Eubacteria That Do Not Perform Homo- or Heterolactic Fermentations**

| | *Peptococcus* | *Ruminococcus* | *Sarcina* | *Propionibacterium* | *Bifidobacterium* |
|---|---|---|---|---|---|
| Motility | – | – | – | – | – |
| Cell shape | Spherical | Spherical | Spherical (in cubical packets) | Rod (often pleomorphic) | Rod (often branched) |
| Percent G + C | 36–37 | 40–45 | 28–31 | 57–67 | 59–66 |
| Substrates fermented | | | | | |
| Sugars | – | + | + | + | + |
| Organic acids | – | – | – | + | – |
| Amino acids | + | – | – | – | – |
| Purines | + | | | | |
| Characteristic fermentation[a] | V[b] | Succinate; or ethanol and acetate | Ethanol and acetate; or butyrate | Propionate (randomizing pathway) | Lactate and acetate |
| Habitat | Genital and intestinal tracts | Rumen | Soil, intestinal tract | Skin, dairy products, genital and intestinal tracts | Skin, intestinal tract |

[a] See Chapter 20, Figures 20.3 and 20.4 for the butyrate and succinate fermentations; see Chapter 22, Figures 22.23 and 22.24 for the homoacetate fermentation; see Figure 23.9 for the propionate fermentation.
[b] V = variable.

However, the propionibacteria are clearly distinguishable from the corynebacteria by a variety of characteristics (e.g., physiology, murein type, lack of the mycolic acids that characterize the corynebacteria), and the branching of *Bifidobacterium* is a consequence of starvation for amino sugars, required growth factors for these organisms, and is hence unrelated to the branching growth habit of the proactinomycetes.

**TABLE 23.4**

**Products of glucose fermentation by *Sarcina***

| | *Sarcina ventriculi* | *Sarcina maxima* |
|---|---|---|
| Ethanol | 100 | 0 |
| Lactic acid | 10 | 0 |
| Acetic acid | 60 | 40 |
| Butyric acid | 0 | 77 |
| $CO_2$ | 190 | 197 |
| $H_2$ | 140 | 223 |

Note: Amounts are moles/100 moles of glucose fermented.

*Acetobacterium* has a metabolism similar to that of the acetogenic clostridia (page 492); it ferments hexoses quantitatively to acetate, as a consequence of its ability to synthesize acetate from $CO_2$. It is capable of anaerobic chemoautotrophic growth with $H_2$ and $CO_2$; the mechanism of ATP synthesis under these conditions is not known, but is probably respiratory.

Bacteria of the genus *Propionibacterium* were first isolated from Swiss cheese (they play an important role in the ripening). They develop as a secondary microflora, fermenting the lactate initially produced in the curd by lactic acid bacteria, with formation of propionate, acetate, and $CO_2$. The two fatty acids give this cheese its distinctive flavor, and the $CO_2$ produces the characteristic holes. Subsequent work has shown that the primary natural habitat of propionic acid bacteria is the rumen of herbivores, where they ferment the lactate produced by other members of the rumen population, and the skin of humans and other animals. In addition to fermenting lactate, these organisms can ferment a variety of sugars. Although they cannot grow exposed to air, requiring anaerobic conditions or

| *Eubacterium* | *Butyrivibro* | *Acetobacterium* |
|---|---|---|
| + or − | + | + |
| Rod | Curved rod | Rod |
|  | 36–37 | 39 |
| + | + | + |
| + | − | + |
| − | − | − |
| Butyrate | Butyrate | Homoacetate |
| Rumen, intestinal tract, soil, spoiled food | Rumen, intestinal tract | Sediments |

low tensions of $O_2$, they contain heme pigments, both cytochromes and catalase. Their metabolism is fermentative; sugars are dissimilated through the Embden-Meyerhof pathway, with formation of propionate, acetate, and $CO_2$, accompanied by some succinate. The formation of succinate is strongly influenced by the content of $CO_2$ in the growth medium; its formation occurs through a carboxylation of the glycolytic intermediate phosphoenol pyruvate, to yield oxalacetate, subsequently reduced to succinate:

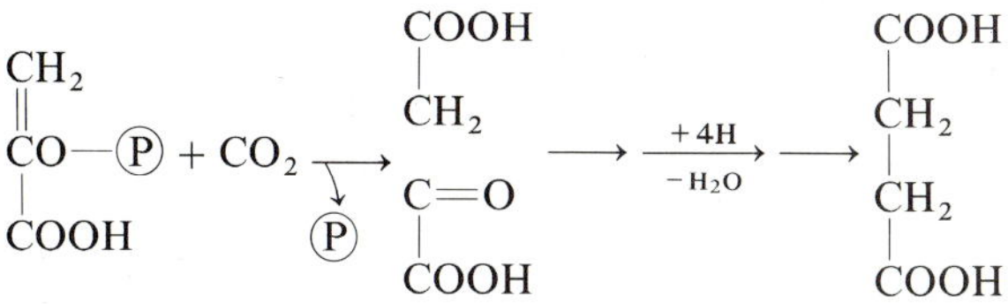

$$\begin{array}{c} CH_2 \\ \| \\ CO-Ⓟ \\ | \\ COOH \end{array} + CO_2 \xrightarrow[\text{Ⓟ}]{} \begin{array}{c} COOH \\ | \\ CH_2 \\ | \\ C{=}O \\ | \\ COOH \end{array} \longrightarrow \xrightarrow[-H_2O]{+4H} \longrightarrow \begin{array}{c} COOH \\ | \\ CH_2 \\ | \\ CH_2 \\ | \\ COOH \end{array}$$

This is a separate mechanism from the transcarboxylation that produces oxalacetate from pyruvate during the course of propionate formation (Figure 23.9).

When lactate is the fermentable substrate, it is initially oxidized to pyruvate. Part of the pyruvate is further oxidized to acetyl-CoA and $CO_2$, ATP being produced in the conversion of acetyl-CoA to acetate. The formation of the oxidized products of the fermentation, acetate and $CO_2$, is balanced by a concomitant reductive formation of propionate, via a randomizing pathway (Figure 23.9).

The aeroduric anaerobes of the genus *Bifidobacterium*, which constitute a major fraction of the intestinal microflora of breast-fed infants, resemble the lactic acid bacteria in several respects. They are catalase negative and have complex nutritional requirements; and they ferment sugars, with the formation of lactic acid as a major end product. However, the overall equation is a distinctive one, corresponding neither to a homolactic nor to a typical heterolactic fermentation:

$$2 \text{ glucose} \longrightarrow 3 \text{ acetate} + 2 \text{ L-lactate}$$

The biochemical route of this fermentation is unique (Figure 23.10). As in the Embden-Meyerhof pathway, glucose is converted to fructose-6-phos-

FIGURE 23.9

The randomizing pathway for propionate formation by *Propionibacterium*.

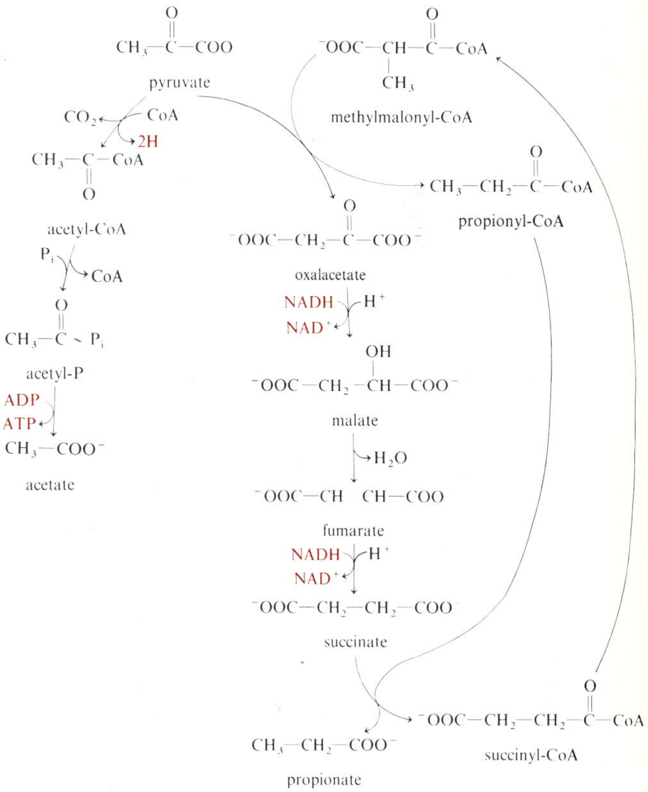

phate. Fructose-6-phosphate undergoes a $C_2$-$C_4$ split, accompanied by an uptake of inorganic phosphate. A complex series of sugar phosphate interconversions is then initiated by a reaction between erythrose-4-phosphate and fructose-6-phosphate, from which two moles of acetyl phosphate and two moles of glyceraldehyde-3-phosphate are ultimately produced. Acetyl-phosphate is converted to acetate; triose-phosphate is converted via pyruvate to L-lactate. Energetically, this fermentation is slightly more favorable than the homolactic fermentation, since it yields five moles of ATP for every two moles of glucose fermented.

The cells of bifidobacteria are typically swollen, irregular, and branched. The complex nutritional requirements of these organisms include a requirement for N-acetylglucosamine or a $\beta$-substituted disaccharide containing this amino sugar (e.g., N-acetyllactosamine). These compounds are present in milk, which accounts for the fact that this is the most favorable medium for bifidobacteria and probably explains their predominance in the intestinal flora of breast-fed babies. When cultivated in a medium containing an excess of N-acetylglucosamine, the cells of bifidobacteria assume a much more regular rod form. Hence the branched, swollen cells characteristic of these organisms probably reflect the fact that they are usually grown with a limiting supply of N-acetylglucosamine, an essential peptidoglycan precursor.

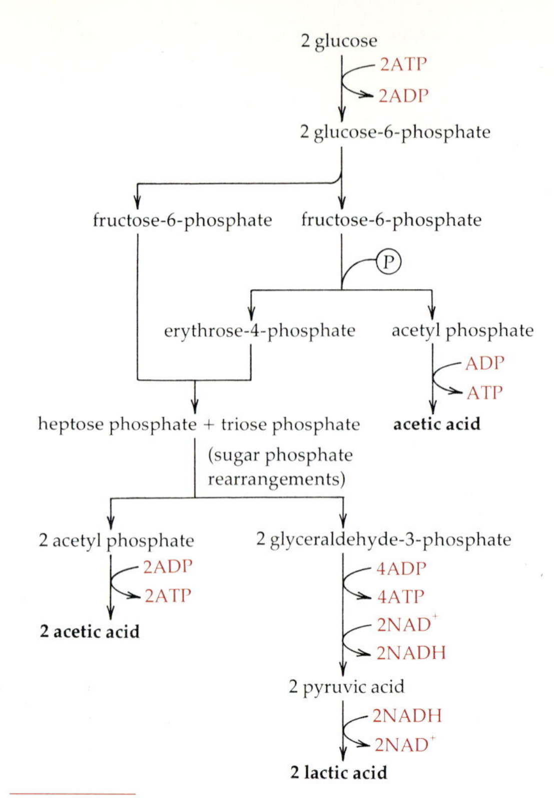


FIGURE 23.10

The pathway of glucose fermentation by *Bifidobacterium;* end products are shown in boldface type.

## FURTHER READING

### Reviews

LONDON, J., "The Ecology and Taxonomic Status of the Lactobacilli," *Ann. Rev. Microbiol.* **30,** 279–301 (1976).

POUPARD, J. A., I. HUSAIN, and R. F. NORRIS, "Biology of the Bifidobacteria," *Bacteriol. Rev.* **37,** 136–165 (1973).

## Chapter 24

# Gram-Positive Eubacteria: The Actinomycetes

Among the bacteria, a mycelial growth habit is common only among the *actinomycetes*, a diverse group of Gram-positive eubacteria. Some of these organisms, termed *sporoactinomycetes* or *euactinomycetes*, develop only in the mycelial state, and they reproduce through the formation of unicellular spores, differentiated either singly or in chains at the tips of hyphae. This group is a large and complex one, containing many genera distinguished by a combination of structural and chemical properties. The broad outlines of a coherent taxonomy of this group are only beginning to emerge, and there are many organisms whose relationships are unclear. In this chapter, only the reasonably well characterized groups will be discussed.

In other actinomycetes, mycelial development is less complete: it may be transitory, occurring only during active growth and fragmenting immediately when the growth rate slows; it may be limited, in the extreme case appearing as branched single cells; or it may occur only under special growth conditions. These organisms are often termed *proactinomycetes*. However it is now clear that this morphological trait is distributed sporadically throughout several recognizable biological groups that also contain unicellular organisms. The term proactinomycete is hence useful only to describe a particular growth habit, and has no taxonomic meaning.

The actinomycetes also include a number of representatives that are unicellular, although they are often irregular in shape.

## CHARACTERISTICS OF ACTINOMYCETES

### Motility

If motile, actinomycetes bear flagella. However, motility is rather rare, although it is the rule in a few groups. The mycelium of the sporoactinomycetes is always permanently immotile; motility when it occurs in this group is confined to the spores, sometimes called *zoospores* by analogy to the motile zoospores of algae and fungi.

### Cell Walls

The cell walls of actinomycetes are extraordinarily diverse; however, within groups the wall composition is often constant, and it has been accorded a major role in the taxonomy of these bacteria. In all, nearly 60 varieties of murein have been described in the actinomycetes.

Variations occur at many places in the murein structure: the acetyl group on the amide nitrogen of muramic acid may be replaced by a glycolyl group; the L-alanine attached to the lactyl moiety of muramic acid may be replaced with L-glycine; a variety of diaminoacids may be found in the third position of the tetrapeptide side chain; cross-bridges with variable composition may occur; and a variety of polysaccharides or lipids may be covalently bound to the murein.

### Developmental Patterns in Mycelial Actinomycetes

The life cycle of sporoactinomycetes includes two distinct phases: a stage of vegetative mycelial growth, and a stage of spore formation. Sporulation is initiated by starvation for essential nutrients. Because the portion of mycelium in the center of a colony may be deprived of nutrients even while the edge is still actively growing, these two stages often occur concurrently. A mature colony may contain within it a complete temporal developmental sequence: the mycelium at the periphery is actively growing, while closer to the center sporulation has begun and in the center sporulation is complete and the vegetative mycelium has lysed (Figure 24.1).

The actinomycete mycelium may be entirely within the substratum; such a mycelium is termed *substrate mycelium.* This is the most common situation among those actinomycetes that do not form spores. However many sporoactinomycetes, and a few proactinomycetes, form in addition a structurally distinct *aerial mycelium,* which typically extends away from the substratum (Figure 24.1). In such organisms spores, if they occur, are usually formed at the tips of the aerial hyphae.

The substrate mycelium is typically septate, although the multinucleate cells are normally quite long (e.g., 20 $\mu$m or more). The hyphae are of constant diameter because, as in most other bacteria, cell growth is accomplished by elongation with

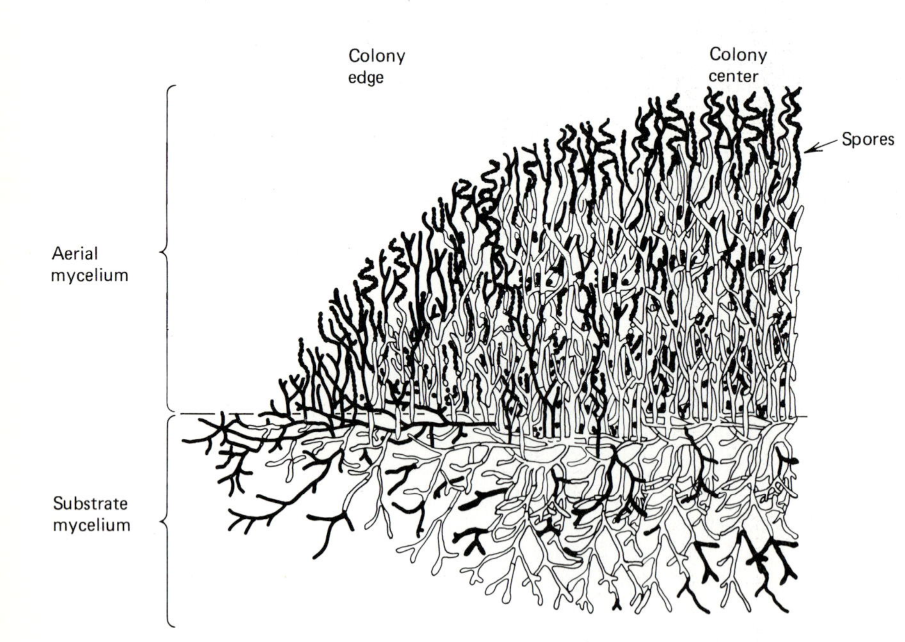


FIGURE 24.1
Cross-section of an actinomycete colony, showing the substrate mycelium, aerial mycelium, chains of spores, and lysis of hyphae in the center of the colony. Hyphae are approximately 1 $\mu$m in diameter. Shaded hyphae are living; white hyphae are lysed. After H. Wildermuth, "Development and Organization of the Aerial Mycelium in *Streptomyces coelicolor*," *J. Gen. Microbiol.* **60,** 43–50 (1970).

TABLE 24.1

**Principle Groups of Actinomycetes**

| Group | Percent G + C | Substrate Mycelium Formed | Aerial Mycelium Formed | Sclerotia Formed | Spores Formed | Spores Motile | Murein Diamino Acid | Other Wall Components |
|---|---|---|---|---|---|---|---|---|
| Actinobacteria | 60–77 | – | – | – | – | – | V | V |
| Nocardioform bacteria | 51–70 | V[a] | V | – | – | – | *meso*-DAP | Arabinogalactan and mycolic acids |
| *Dermatophilus* group | 57–72 | + | – | V | + | V | *meso*-DAP | V |
| Streptomycetes | 66–78 | + | + | Rare | + | – | LL-DAP | None |
| Actinoplanetes | 71–73 | + | – | – | + | V | *meso*-DAP or 3-hydroxy DAP | Xylose and arabinose |

[a] Variable among strains.

no increase in cell diameter. However, a few actinomycetes grow in width as well as length, and deposit cross walls longitudinally as well as across the mycelium. The result is a mass of cells forming a tissuelike *thallus* or sclerotium (plural *thalli, sclerotia*).

## MAJOR GROUPS OF ACTINOMYCETES

Based on their life cycles and chemical composition, five groups of actinomycetes can be recognized (Table 24.1).

### The Actinobacteria

The actinobacteria are a heterogeneous group of predominantly unicellular organisms; mycelial development is either absent or rudimentary. The group contains cocci (*Micrococcus*) and irregular rods that may show some branching. At least eight peptidoglycan types (Chapter 6) are found in the group, and many genera contain wall constituents in addition to the murein. Most actinobacteria are respiratory strict aerobes; however, a few are fermentative. They may be isolated from a variety of habitats. Despite their diversity, however, rRNA sequencing suggests that they constitute a coherent phylogenetic group. Some actinobacteria are described in Table 24.2.

MICROCOCCUS The micrococci are aerobic, nonmotile cocci that occur as tetrads, formed by a regular alternation of the plane of cell division as in *Sarcina*, as diplococci (pairs of cells), or as irregular aggregates. They are usually brightly pigmented as a consequence of the synthesis of large amounts of carotenoids; red, pink, and yellow are the most common colors. Teichoic acids (page 158) are often not detectable; these are one of the few Gram-positive eubacteria that lack these wall constituents.

The natural habitat of *Micrococcus* was for a long time unclear; they can be isolated in small numbers from almost any habitat. They are partic-

TABLE 24.2

**Some Genera of Actinobacteria**

| Genus | Percent G + C | Metabolism | Motility | Murein Diamino Acid | Cell Wall Carbohydrate | Habitat |
|---|---|---|---|---|---|---|
| *Micrococcus* | 66–75 | Respiratory | – | Lysine | None | Skin |
| *Arthrobacter* | 59–66 | Respiratory | V[a] | Lysine | Galactose | Soil |
| *Cellulomonas* | 71–77 | Facultative | V | Ornithine | Rhamnose | Soil |
| *Actinomyces* | 60–73 | Fermentative | – | Lysine or ornithine | V | Oral cavity |

[a] Variable among strains.

ularly prominent among the microorganisms suspended in air or associated with dust particles, and are hence often encountered as contaminants deposited on petri plates from the air. Recently they have been recovered in fairly large numbers from human and animal skin, where they are a principal component of the aerobic flora. Their ubiquity is now thought to be primarily due to contamination by organisms from skin.

ARTHROBACTER Organisms of the genus *Arthrobacter* constitute a large fraction of the aerobic chemoheterotrophic bacterial population of soil, and are important agents for the mineralization of organic matter. The most distinctive property of *Arthrobacter* spp. is the succession of changes in cell form that accompany growth (Figure 24.2). In cultures that have entered the stationary phase the cells are spherical and of uniform size, resembling

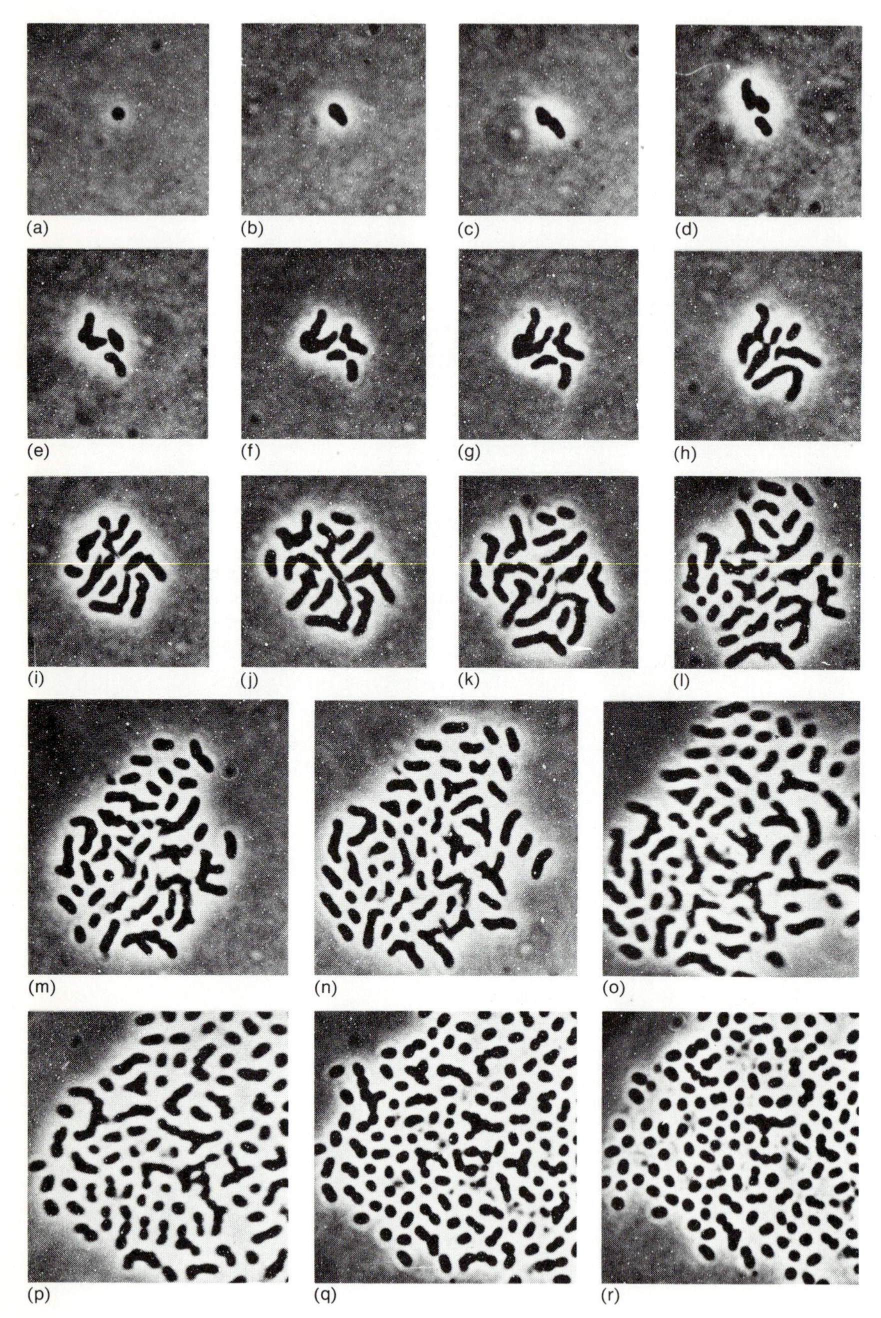


FIGURE 24.2
The cellular life cycle of *Arthrobacter*, shown by successive photomicrographs of the growth of a microcolony from a single coccoid cell on agar over a period of 40 hours (phase contrast, ×1,020). From H. Veldkamp, G. Van den Berg, and L. P. T. M. Zevenhuizen, "Glutamic Acid Production by *Arthrobacter globiformis*," *Antonie van Leeuwenhoek* **29,** 35 (1963).

micrococci. When growth is reinitiated, these cells elongate into rods, which undergo binary fission. From these rods, thinner outgrowths may develop near one or both poles of the cell, producing branched forms that resemble early developmental stages of proactinomycetes. Return to the coccoid state may occur either by multiple fragmentation (as in proactinomycetes) or by a progressive shortening of the rods through successive binary fissions.

During vegetative growth many arthrobacters display characteristic *V-forms*, a distinctive angular relationship of daughter cells ( Figure 24.3). This is the result of an unusual postfission movement of daughter cells relative to each other, often termed *snapping fission* (Figure 24.4). The mechanism of this distinctive behavior is now known: during division only the inner layer of the two-layered wall participates in septum formation; thickening of the completed septum results in considerable pressure that causes a localized rupture of the outer layer, which tears around the circumference of the septum forming a hinge opposite to the point of original rupture (Figures 24.5 and 24.6).

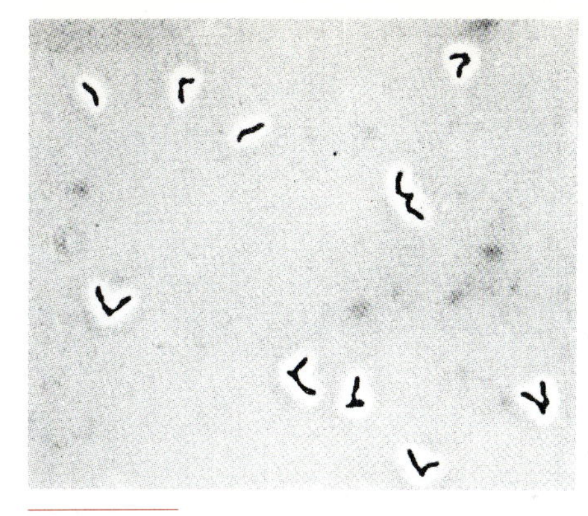

FIGURE 24.3

The typical angular arrangement of dividing cells of *Arthrobacter*, brought about by "snapping" postfission movement (phase contrast, × 1,400). From T. A. Krulwich and J. L. Pate, "Ultrastructural Explanation for Snapping Post-Fission Movements in *Arthrobacter crystallopoietes*," *J. Bacteriol.* **105,** 408 (1971).

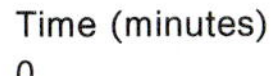


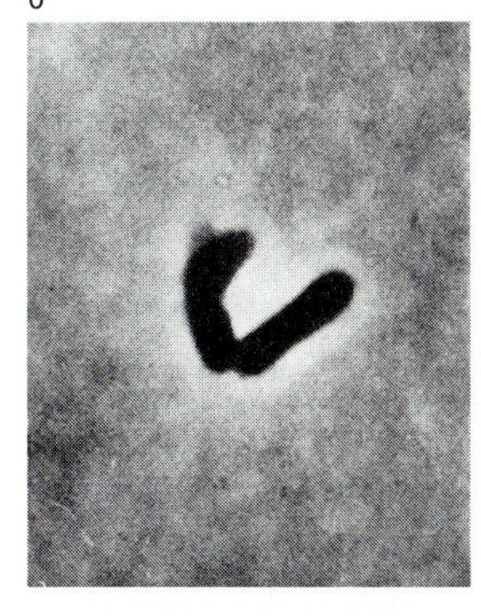

15

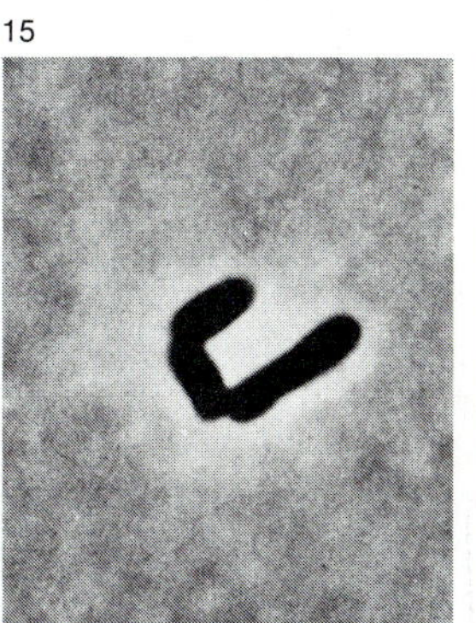

45

FIGURE 24.4

Postfission movement in *Arthrobacter* illustrated by time-lapse pictures of a group of cells growing on agar (phase contrast, × 1,720). From M. P. Starr and D. A. Kuhn, "On the Origin of V-forms in *Arthrobacter atrocyaneus*," *Arch. Mikrobiol.* **42,** 289 (1962).

FIGURE 24.5 (below)

A diagrammatic representation of the mechanism of transverse wall growth and snapping post-fission movement in *Arthrobacter;* pm, cell membrane; il, inner wall layer; ol, outer wall layer. Only the inner wall layer (cross-hatched) participates in transverse wall formation. (a) Initiation of transverse wall formation. (b) Completion of transverse wall; (c) Extension of inner wall layer, placing tension on outer layer; (d) Unilateral rupture of outer wall layer, causing snapping movement. From T. A. Krulwich and J. L. Pate, "Ultrastructural Explanation for Snapping Post-Fission Movements in *Arthrobacter crystallopoietes*," *J. Bacteriol.* **105,** 408 (1971).

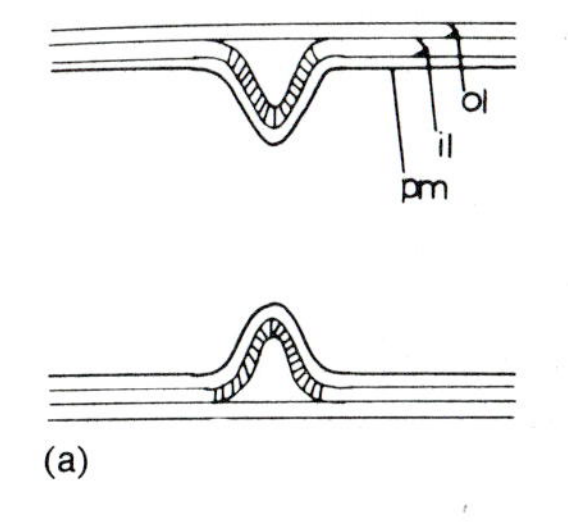


(a)

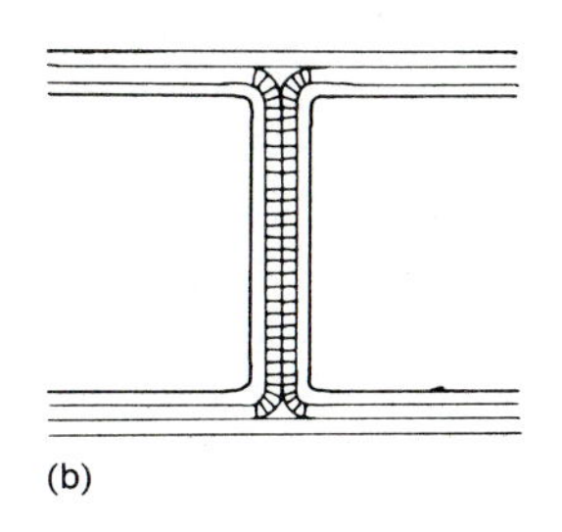

(b)

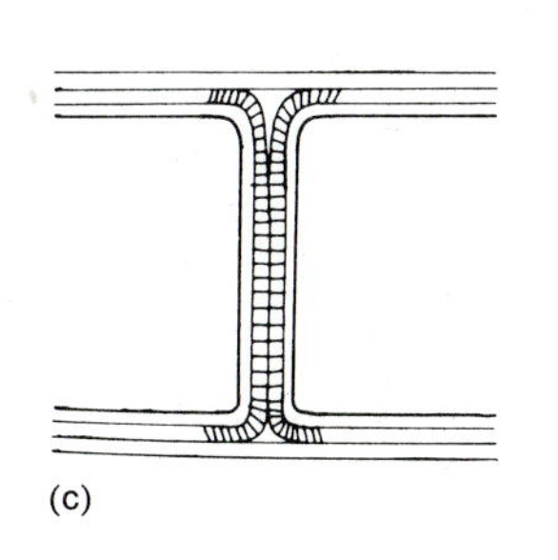

(c)

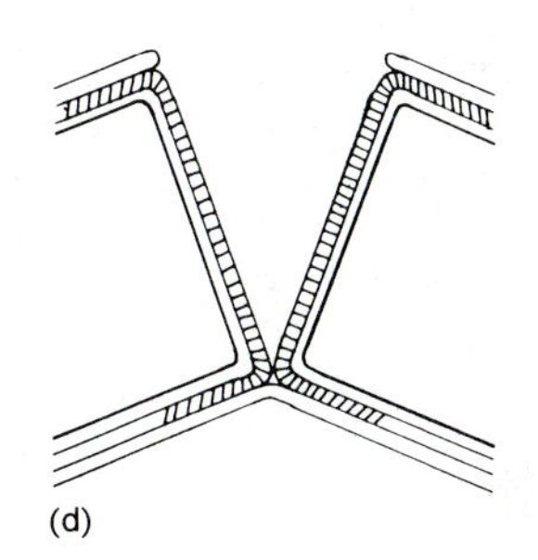

(d)

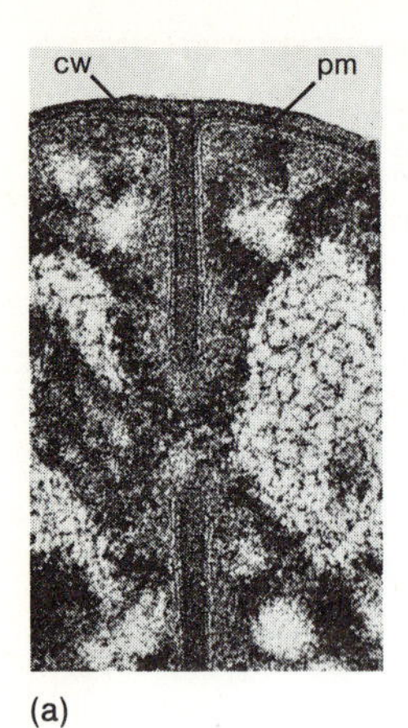


(a)

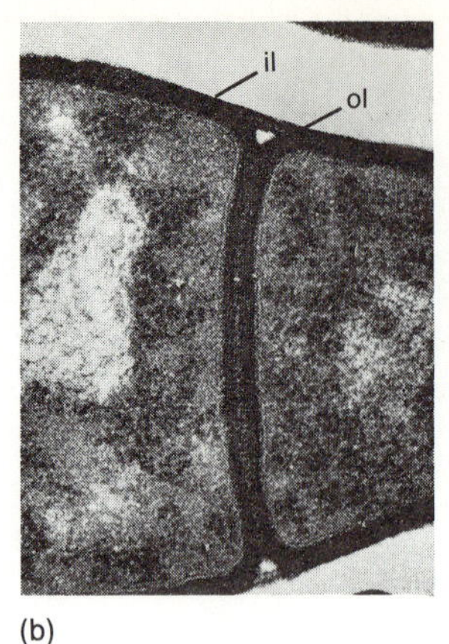


(b)

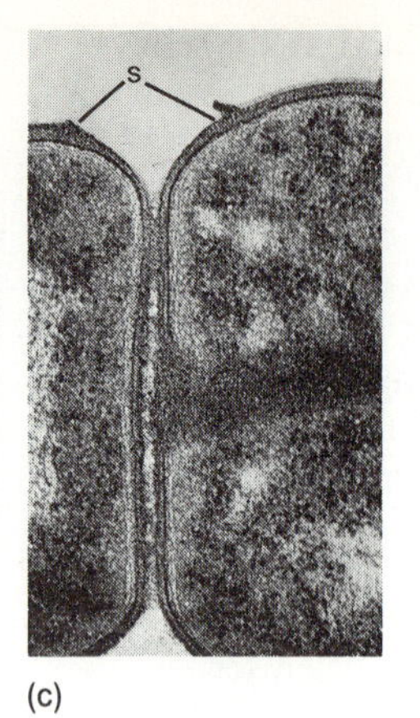


(c)

FIGURE 24.6

Electron micrographs of thin sections of dividing cells of *Arthrobacter*, showing the two-layered structure of the cell wall. (a) Cell in which the transverse wall has been almost completed: cw, cell wall; pm, cell membrane. (b) Cell in which transverse wall formation is complete, showing the separation between the inner (il) and outer (ol) wall layers. (c) Cell which has just undergone snapping postfission movement, showing the separation, s, between the two parts of the previously continuous outer wall layer. From T. A. Krulwich and J. L. Pate, "Ultrastructural Explanation for Snapping Post-Fission Movements in *Arthrobacter crystallopoietes*," *J. Bacteriol.* **105,** 408 (1971).

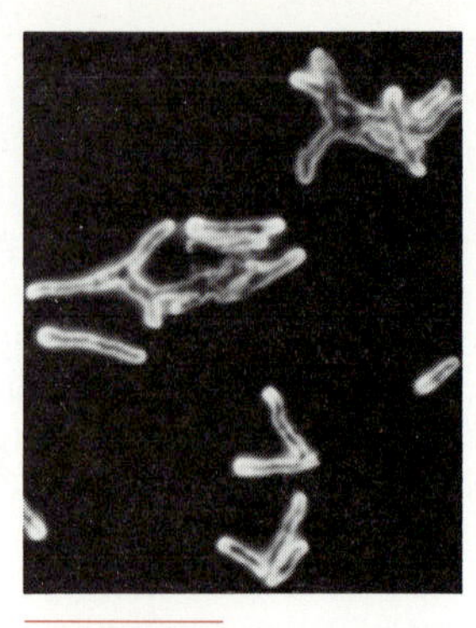

FIGURE 24.7

*Actinomyces israelii* from a broth culture, showing branched cells and short mycelial fragments (dark field illumination, × 1,120). From J. M. Slack, S. Landfried, and M. A. Gerencer, "Morphological, Biochemical and Serological Studies on 64 Strains of *Actinomyces israelii*," *J. Bacteriol.* **97,** 873 (1969).

In their nutritional properties *Arthrobacter* show interesting analogies to aerobic pseudomonads; most species can utilize a wide and varied range of simple organic compounds as principal carbon and energy sources. However, the majority require growth factors.

CELLULOMONAS The distinctive feature of actinobacteria of the genus *Cellulomonas* is the ability to utilize cellulose as source of carbon and energy. They are morphologically very similar to *Arthrobacter*, although they do not undergo the distinctive rod-coccus morphogenesis; however cells in stationary phase are markedly shorter than growing cells, and a small proportion may be truly coccoidal. They are facultative anaerobes, although fermentative growth is very poor. Most strains are motile by one or a few subpolar or lateral flagella. They require biotin and thiamin.

ACTINOMYCES Organisms of the genus *Actinomyces* inhabit the mouth and throat of humans and other animals; some are pathogenic. They are aeroduric anaerobes that have complex nutritional requirements, often including elevated partial pressures of $CO_2$. The products of sugar fermentation include formate, succinate, lactate, and acetate. They often show a pronounced tendency to mycelial growth in young cultures; however the mycelium is typically transient, breaking up into rods or branched fragments (Figure 24.7).

## The Nocardioform Bacteria

The nocardioform group includes four closely related genera (Table 24.3) with a distinctive and complex cell wall. The principal cell wall carbohydrate is a copolymer of arabinose and galactose termed *arabinogalactan*, covalently bound by phosphodiester bonds to the acetyl or glycolyl moieties of the murein. Esterified to the arabinogalactan are the distinctive lipids of this group, the 2-branched 3-hydroxy fatty acids termed *mycolic acids* (Figure 24.8 and Table 24.3). Mycolic acids may also be esterified to the disaccharide trehalose, typically as dimycolates. Many other complex lipids are found in the wall of these organisms.

The large amounts of lipid may substantially reduce the permeability of the walls of nocardioform bacteria. These bacteria may consequently be difficult to stain; when the lipid content is especially high, staining may require heating the cells to nearly 100° C to allow the stain to permeate the cell. Once stained, decolorization may be equally difficult. This is the basis of the *acid-fast stain*, a stain of substantial clinical importance because pathogenic nocardioform bacteria are typically acid-fast, whereas few other organisms are. In general, acid-fastness cor-

$$
\begin{array}{lll}
\text{OH} & & \\
| & & \\
\text{CH}- & \text{CH}- & \text{COO}^- \\
| & | & \\
\text{R}_2 & \text{R}_1 &
\end{array}
$$

FIGURE 24.8

The general structure of mycolic acids. $R_1$ and $R_2$ are alkyl groups that may be saturated or unsaturated.

## TABLE 24.3

**Some Nocardioform Actinomycetes**

| Genus | Percent G + C | N-substituted Muramic Acid | Growth Habit | Mycolic Acids | |
|---|---|---|---|---|---|
| | | | | TOTAL NUMBER OF CARBON ATOMS | NUMBER OF DOUBLE BONDS |
| *Corynebacterium* | 51–59 | Acetyl | Unicellular | 22–38 | 0–2 |
| *Mycobacterium* | 62–70 | Glycolyl | Unicellular, occasional rudimentary mycelium | 60–90 | 1–2 |
| *Rhodococcus* | 59–69 | Glycolyl | Variable—unicellular to mycelial | 34–66 | 0–4 |
| *Nocardia* | 64–69 | Glycolyl | Usually mycelial | 46–60 | 0–3 |

relates with the size of the mycolic acids. The cell surface of acid-fast nocardioforms may be so hydrophobic that colonies have a waxy appearance, and growth in liquid media may be poorly dispersed unless detergent is added (Figure 24.9).

The morphology of nocardioform bacteria is quite variable. *Corynebacterium* strains are pleomorphic rods, typically club-shaped. They divide by binary fission, often accompanied by postfission movements. Most *Mycobacterium* strains are also unicellular rods, although they are more regular in shape than the corynebacteria. Some mycobacteria develop as mycelial organisms, but the mycelium typically fragments early in the growth phase to form rods or branched rods. *Rhodococcus* is quite variable morphologically; some strains have a growth habit very like *Arthrobacter*, with a pronounced rod-to-coccus morphogenesis as they enter stationary phase. Others develop an extensive mycelium that fragments into cocci or short rods at the cessation of growth. Most *Nocardia* develop almost exclusively as mycelia (Figure 24.10), often producing aerial as well as substrate mycelia. However, spores are not produced; reproduction is by mycelial fragmentation.

Most members of this group are strictly respiratory aerobes; however *Corynebacterium* is a facultative anaerobe, capable of fermenting carbohydrates.

Habitats of this group are varied; typically the corynebacteria and some mycobacteria are animal parasites, and the nocardias, rhodococci, and other mycobacteria are soil organisms. Several of the animal parasites cause human disease: *Cory-*

FIGURE 24.9

The characteristic appearance of cultures of *Mycobacterium tuberculosis*. (a) Colony growing on the surface of an agar plate (×7). (b) Cordlike aggregations of cells from a liquid culture (×345). Courtesy of Professor N. Rist, Institut Pasteur, Paris.

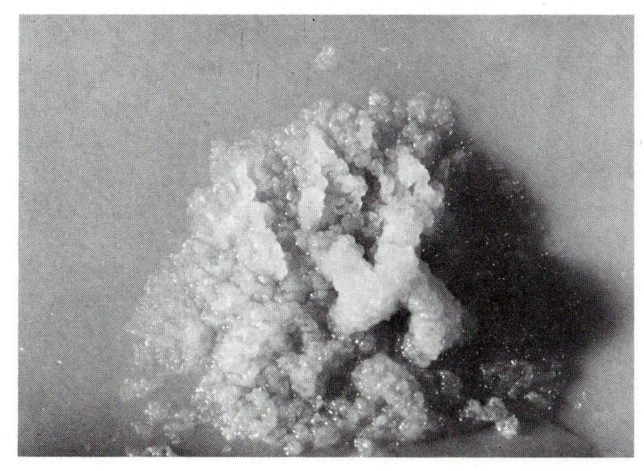

(a)

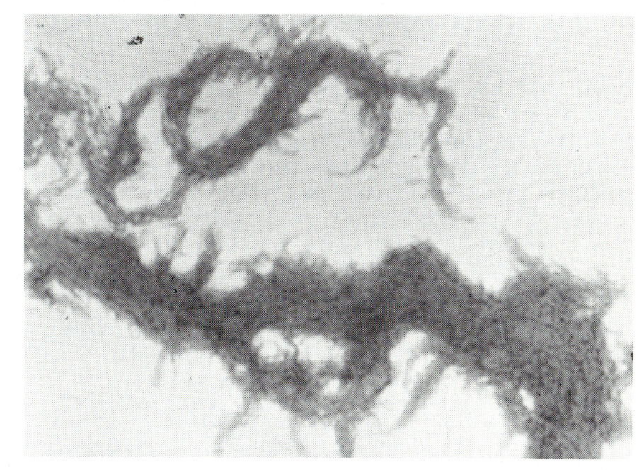

(b)

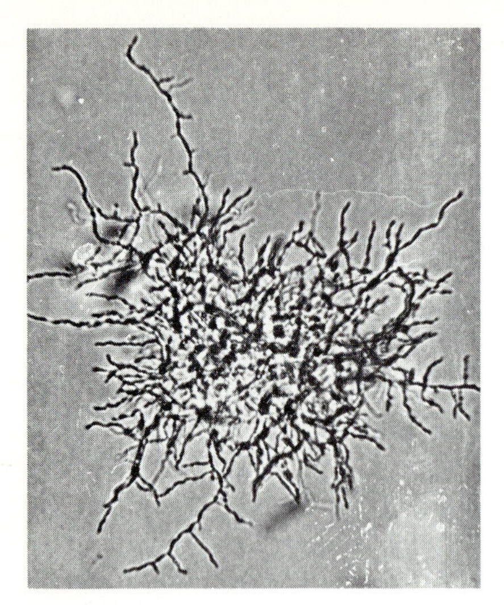

FIGURE 24.10

Young surface colonies on agar plates of *Nocardia asteroides* (×648). Courtesy of Ruth Gordon and H. Lechevalier.

*nebacterium diphtheriae* causes diphtheria, and *Mycobacterium tuberculosis* and *M. leprae* cause tuberculosis and leprosy (Hanson's disease), respectively. *Rhodococcus* and *Nocardia* also contains some pathogens of humans and domestic animals.

## The *Dermatophilus* Group

The actinomycetes that form spores from a tissue-like mass of cells derived by a series of cell divisions in different planes are placed in the *Dermatophilus* group (Table 24.4). Two of these organisms (*Dermatophilus* and *Geodermatophilus*) develop as substrate mycelia that divide in random planes to form sclerotia (Figure 24.11); the third (*Frankia*) develops as a uniseriate mycelium that forms terminal or occasional intercalary swellings that divide to form the sporangial mass (Figure 24.12). The spores of *Dermatophilus* and *Geodermatophilus* are motile by means of tufts of polar flagella, and those of *Geo-*

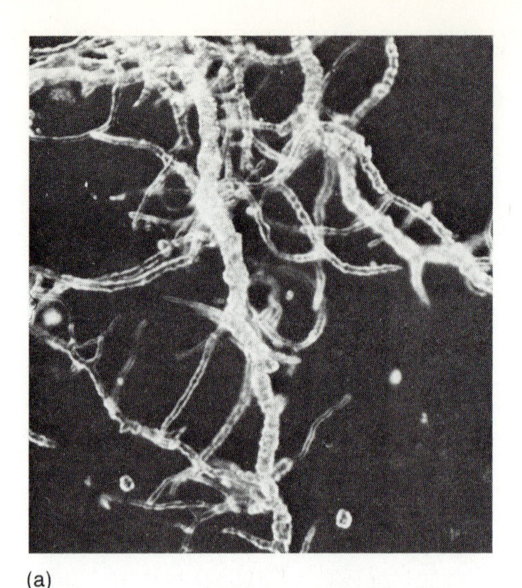

(a)

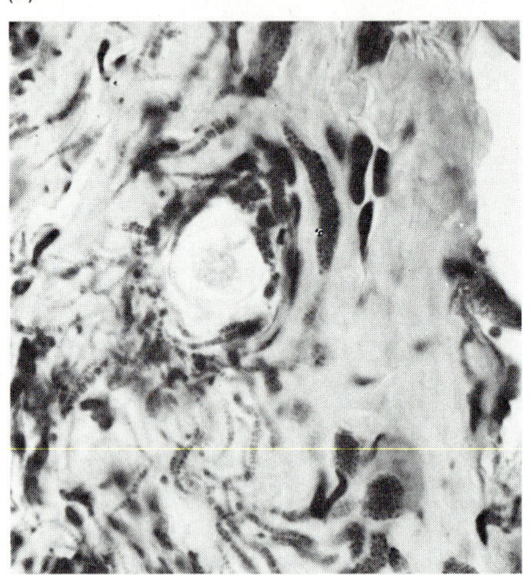

(b)

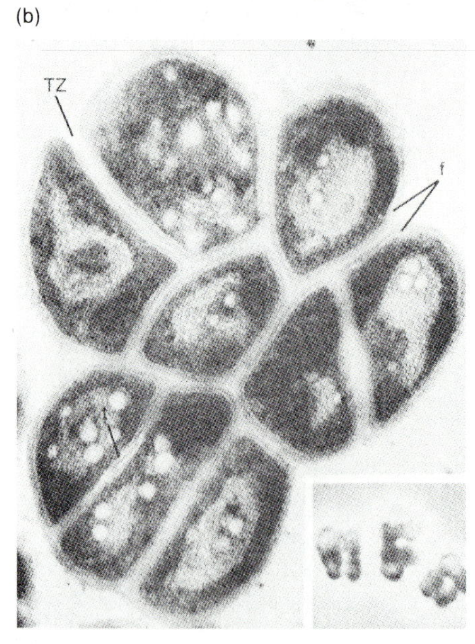


(c)

FIGURE 24.11

*Dermatophilus* and *Geodermatophilus*. (a) Darkfield light micrograph of *Dermatophilus*, showing the mycelium and developing sclerotia. (b) Brightfield light micrograph of a section of skin infected with *Dermatophilus*, showing stained sclerotia and sporangial masses. (c) Electron micrograph of a thin section through a sclerotium of *Geodermatophilus* showing its complex wall with an electron-lucent inner layer (t) and fibrous outer layer (f). Adjacent cells are separated by a transparent zone (TZ). Inset: phase-contrast light micrograph of *Geodermatophilus*. (a) ×497; (b) ×1,800; (c) ×21,600. (a) and (b) From D. J. Dean, M. A. Gordon, C. W. Severinghaus, E. T. Kroll, and J. R. Reilly, "Streptothricosis: A new zoonotic disease," *N. Y. State J. Med.* 61, 1283–1287 (1961); (c) From E. E. Ishiguro and R. S. Wolfe, "Control of morphogenesis in *Geodermatophilus*: ultrastructural studies," J. Bacteriol. **104**, 566 (1970).

**TABLE 24.4**

**The Members of the *Dermatophilus* Group**

| Genus | Percent G + C | Cell Wall Sugars | Sclerotia Formed | Spores Motile | Spores Capable of Division | Relations with Oxygen | Habitat |
|---|---|---|---|---|---|---|---|
| *Dermatophilus* | 57–59 | Madurose | + | + | – | Facultative anaerobe | Skin |
| *Geodermatophilus* | 73–75 | None | + | + | + | Aerobe | Soil |
| *Frankia* | 68–72 | Madurose, fucose, and xylose | – | – | – | Microaerophile | Plant roots |

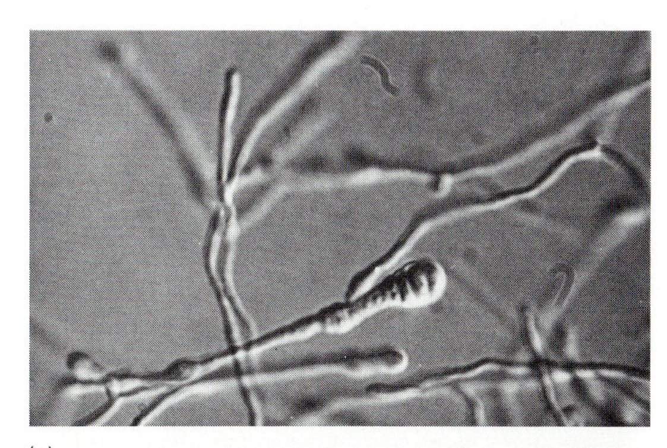

(a)

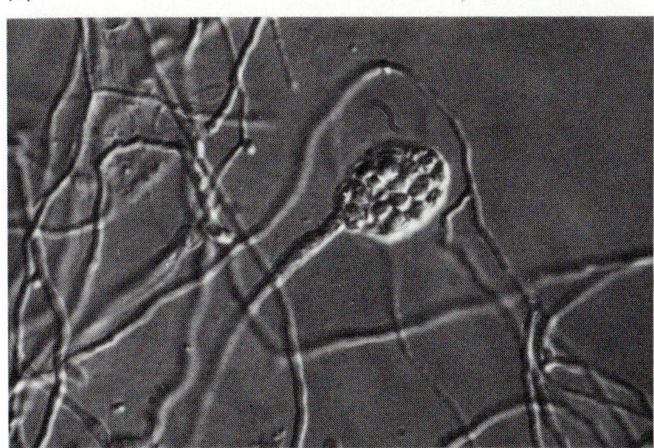

(b)

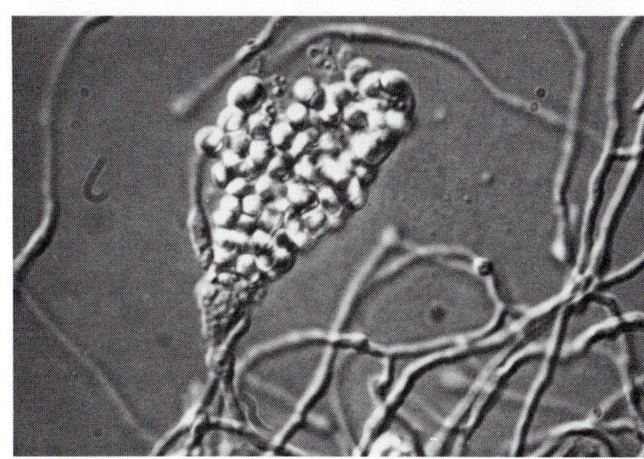

(c)

(d)

FIGURE 24.12

*Frankia*. (a) and (b) interference contrast light micrographs showing substrate mycelium and developing sporangia. (c) interference contrast light micrograph of a mature sporangium releasing spores. (d) electron micrograph of a thin section through a sporangium, showing mature spores at the apex and developing spores at the base of the sporangium. (a) ×1100; (b) ×900; (c) ×1100, (d) ×11,534. (a)–(c) from W. Newcomb, D. Callahan, J. G. Torrey, and R. L. Peterson, "Morphogenesis and fine structure of the actinomycetous endophyte of nitrogen-fixing root nodules of *Comptonia* peregrina," *Bot. Gaz.* 140 (Suppl.) S22–S34 (1979). (d) Courtesy of Dr. W. Newcomb.

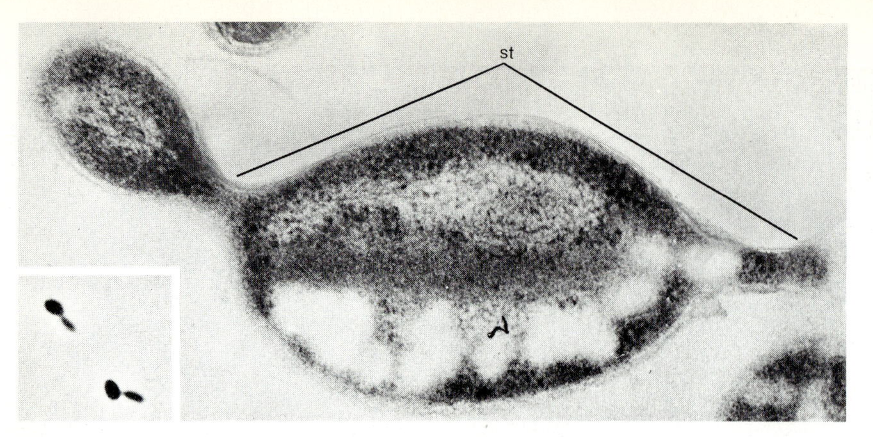


FIGURE 24.13

Dividing spores of *Geodermatophilus*. Electron micrograph of a thin section of a budding cell with two polar stalks, st, from one of which a daughter bud is being formed (× 47,600). Insert: phase contrast photomicrograph of budding spores. From E. E. Ishiguro and R. S. Wolfe, "Control of Morphogenesis in *Geodermatophilus:* Ultrastructural Studies," *J. Bacteriol.* **104,** 566 (1970).

*dermatophilus* have the unique ability to multiply as unicellular, budding organisms (Figure 24.13). The growth habit is nutritionally controlled; in the absence of a nutrient of unknown chemical composition the zoospores multiply by budding; when the unknown factor is present, they cease budding and develop as a rudimentary mycelium composed of a thallus of cuboidal cells.

Although *Dermatophilus* and *Geodermatophilus* are similar morphologically and developmentally, they are phylogenetically rather distant. Analysis of 16S rRNA sequences show that *Geodermatophilus* represents a distinct line of descent among the actinomycetes, while *Dermatophilus* is closely related to the actinobacteria. A comparable analysis of *Frankia* rRNA has not yet been performed.

The habitats of the three members of this group are distinctive: *Geodermatophilus* is a soil organism like many other actinomycetes; *Dermatophilus* is a parasite of animal and occasionally human skin, on which it is a pathogen that develops in the uncornified epidermis; and *Frankia* infects the roots of a variety of nonleguminous plants.

*Frankia* has a very broad host range; some species in at least 17 genera in seven orders of plants are known to be susceptible (Table 24.5). When infected, the roots produce $N_2$-fixing nodules, of which two types are known. Nodules of plants in the Casuarinales and Myricales are composed of negatively geotropic roots; nodules of other plants are composed of repeatedly branching roots (Figure 24.14). Within the nodules, cells may be filled with actinomycete hyphae whose tips are swollen, vesicular masses (Figure 24.15). These vesicles are believed to be the site of $N_2$ fixation; their relationship (if any) to sporogenous tissue is unclear.

As early as 1896 it was shown that alder plants (genus *Alnus*) can grow well in a medium free of combined nitrogen only if nodulated; by the early 1960s similar experiments had been done with members of a number of other genera listed in Table 24.5. By analogy with the legumes, it was supposed that the nodules were fixing atmospheric nitrogen, and this was eventually confirmed by the use of $^{15}N$. Other parts of the plant do not fix nitrogen.

**TABLE 24.5**

**Plants Nodulated by *Frankia***

| Order | Family | Genus |
|---|---|---|
| Casuarinales | Casuarinaceae | *Casuarina* |
| Myricales | Myricaceae | *Myrica, Comptonia* |
| Fagales | Betulaceae | *Alnus* |
| Rhamnales | Rhamnaceae | *Ceanothus, Discaria, Colletia, Trevoa* |
| | Elaeagnaceae | *Elgaeagnus, Hippophae, Shepherdia* |
| Coriariales | Coriariaceae | *Coriaria* |
| Rosales | Rosaceae | *Rubus, Dryas, Purshia, Cercocarpus* |
| Cucurbitales | Datiscaceae | *Datisca* |

The fixation process resembles that of legumes in many respects. In both legumes and nonlegumes fixation is inhibited by carbon monoxide and by high levels of oxygen or hydrogen, and is poor when the plants are deficient in cobalt or molybdenum. Hemoglobin has been detected spectroscopically in the nodules of three groups of nonlegumes, *Alnus, Myrica,* and *Casuarina.*

Nodulation can be readily induced in most genera by applying suspensions of crushed nodules to the roots; untreated roots show little or no nodulation in control experiments. Cross-inoculation experiments using crushed nodules show that they possess group specificity. Cross inoculation is often possible between species of a given genus of host plant, but usually not between species of different

(a)

(b)

(c)

FIGURE 24.14

Root nodules produced by *Frankia*-infected plants. (a) and (b) negatively geotropic infected roots from *Casuarina* and *Myrica* respectively. (c) corraloid nodules from *Ceanothus*. (a) and (b) ×0.75; (c) ×1.43. (a) and (b) from J. H. Becking, "The genus *Frankia*," pp. 1991–2003 in M. P. Starr, H. Stolp, H. G. Trüper, A. Ballows, and H. G. Schlegel (eds), *The Prokaryotes*. Berlin: Springer-Verlag (1981).

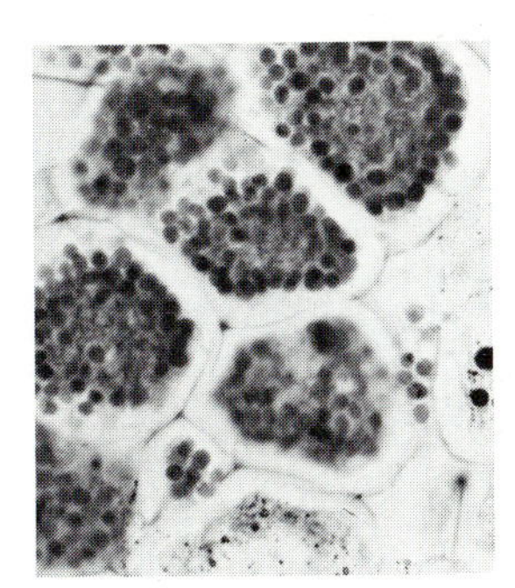

FIGURE 24.15

Cortical region of a transverse section of a root nodule of *Alnus glutinosa*, showing the dark stained vesicular tissue of the endosymbiotic *Frankia*. From G. Bond, "The Root Nodules of Non-leguminous Angiosperms," in *Symbiotic Associations: Thirteenth Symposium of the Society for General Microbiology* (New York: Cambridge University Press, 1963). Section prepared by E. Boyd; micrograph by W. Anderson.

genera or families. Such specificities are reminiscent of the cross-inoculation groups of rhizobia.

Recently the endophyte has been isolated and grown in pure culture. Growth is exceptionally slow (visible colony formation may take more than a month). Microaerophilic conditions are preferred, and the nutritional requirements are complex. Reinoculation of axenic host seedlings has allowed the unequivocal demonstration that the symbiont is responsible for the nitrogen-fixing ability of the nodulated plant.

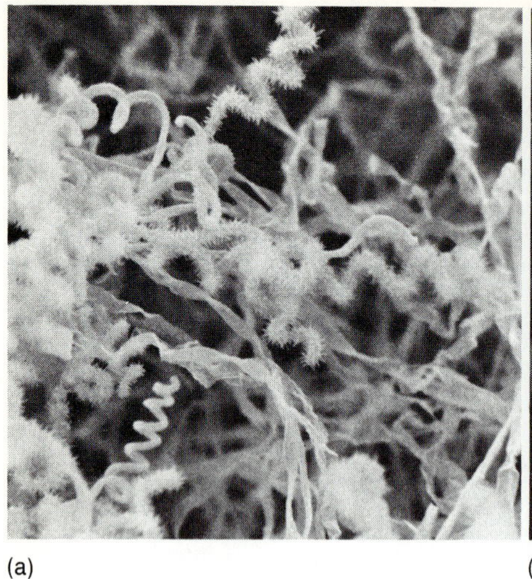

(a)

(b)

FIGURE 24.16

The surface of a *Streptomyces* colony, as observed with the scanning electron microscope. (a) General view of the aerial mycelium (× 1,740). (b) A helically wound chain of conidia. The individual conidia (not distinguishable in this figure) bear spiny appendages (× 5,800). Reproduced with permission from S. Kimoto and J. C. Russ, "The Characteristics and Applications of the Scanning Electron Microscope," *Am. Scientist* **57,** 112 (1969).

FIGURE 24.17

Two of the volatile compounds produced by streptomycetes that contribute to the odor of soil.

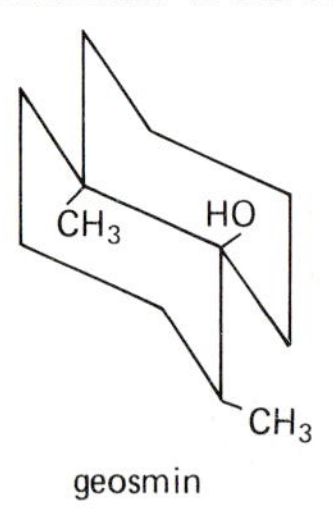


geosmin

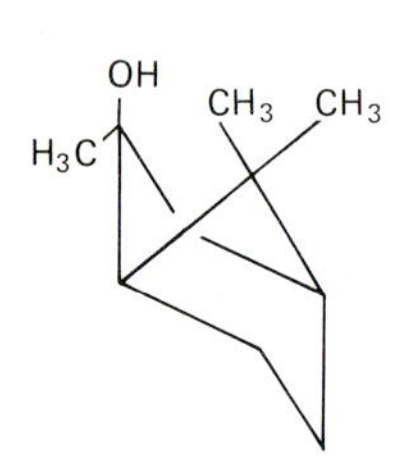


2-methyl-isoborneol

FIGURE 24.18

Portions of the aerial mycelium of two streptomycetes, illustrating two different kinds of arrangement of the sporulating hyphae. Photograph (a) × 2,940, courtesy of H. Lechevalier; (b) × 2,550, courtesy of Peter Hirsch.

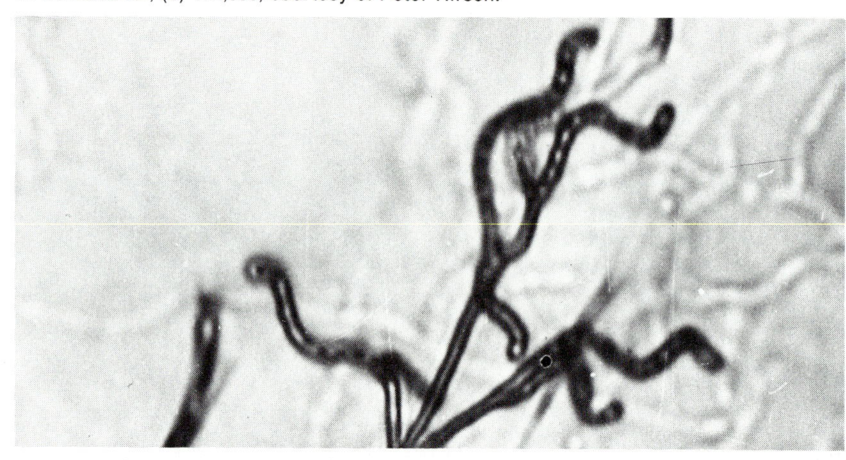

(a)

(b)

## The Streptomycetes

The streptomycetes develop as fully mycelial organisms, and reproduce by the formation of immotile spores at the tips of aerial hyphae (Figure 24.16). They are abundant in soil, and are largely responsible for the odor of damp earth (along with gliding bacteria of the genus *Nannocystis*), which is due to the production of a number of volatile substances (Figure 24.17). Streptomycete colonies on laboratory media often smell very strongly of earth.

Streptomycete colonies are typically firm and compact in the early stages of development, and may be difficult to pick for subculturing. Later, an aerial mycelium is formed that covers the substrate mycelium with a loose, cottony growth. The aerial mycelium, and the spores formed from it by repeated fission events, are covered with a hydrophobic sheath that probably assists their dispersal.

The streptomycetes are strict aerobes and are quite versatile nutritionally. In addition to being able to use a wide range of soluble organic compounds as carbon and energy source, many are capable of using complex polymeric substances such as latex, chitin, lignocellulose, and peptidoglycan. They are consequently important organisms in the mineralization of these compounds. Growth factors are not required.

The range of minor morphological variation in the mode of spore formation (Figure 24.18), surface ornamentation of spores (Figure 24.19),

FIGURE 24.19

Electron micrographs of the spores of six different *Streptomyces* species, which illustrate various types of surface structure and ornamentation: (a) *S. cacaoi*, showing smooth spores; (b) *S. hirsutus*, showing spiny spores with obtuse spines; (c) *S. griseoplanus*, which has warty spores; (d) *S. aureofaciens*, of a smooth but special "phalangiform" type; (e) *S. fasciculatus*, showing spiny spores with long acute spines; (f) S. *flavoviridis*, showing hairy spores. Courtesy of H. D. Tresner, Lederle Laboratories; reproduced in part from E. B. Shirling and D. Gottlieb, "Cooperative Description of Type Cultures of *Streptomyces. II.* Species Descriptions from First Study," *Int. J. Syst. Bacteriol.* **18,** 69–189 (1968).

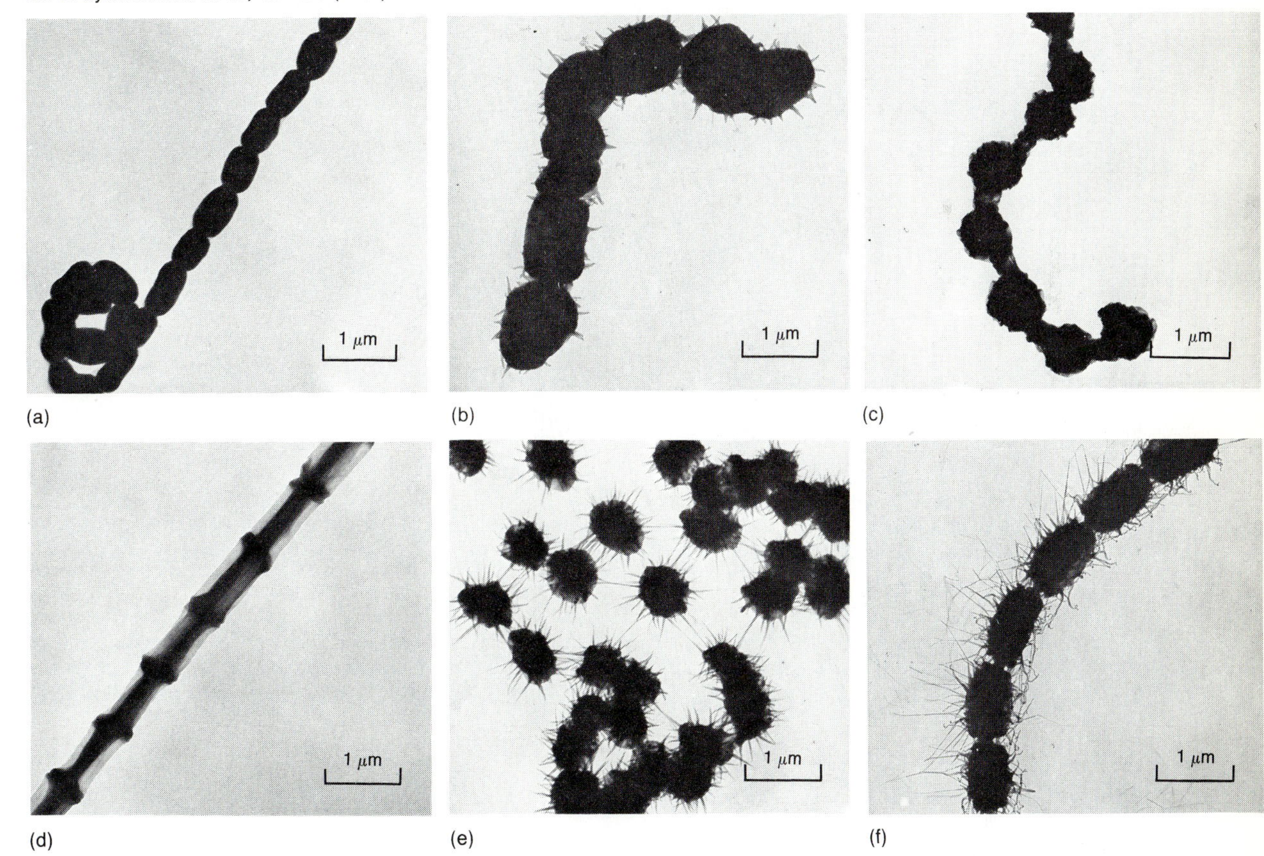

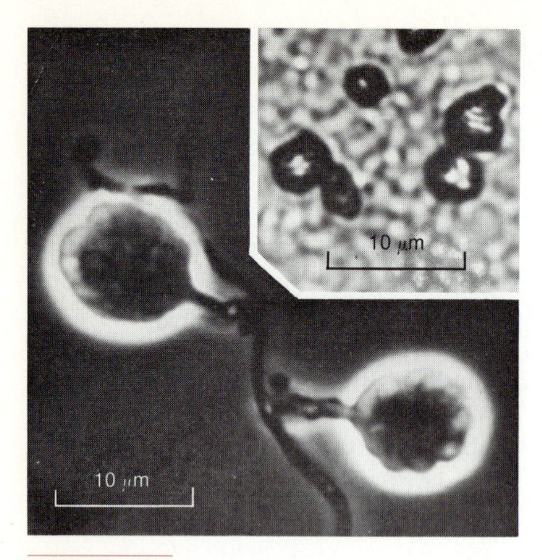


FIGURE 24.20

Spore vesicles of *Actinoplanes:* (a) (inset) A group of mature spore vesicles viewed on the surface of a colony (bright field illumination). (b) Two mature spore vesicles attached to a hypha, mounted in water (phase contrast illumination). From H. Lechevalier and P. E. Holbert, "Electron Microscopic Observation of the Sporangial Structure of a Strain of *Actinoplanes*," *J. Bacteriol.* **89,** 217 (1965).

and physiological characteristics is very large. A number of strains are valuable economically, as a consequence of their production of therapeutically useful antibiotics (Chapter 33). Consequently many hundreds of species in scores of genera have been described, often on very slender grounds.

A few streptomycetes form sclerotia, within which the cells are cemented together by a material that contains L-2,3 diaminopropionic acid. These organisms have been named *Chainia*; however, they are otherwise typical *Streptomyces* species, and the single characteristic of sclerotium formation is questionable as the basis for generic distinction. Furthermore, the ability to form sclerotia is rapidly lost in laboratory culture.

## The Actinoplanetes

The actinoplanetes form spores within a sporangium that arises directly from the substrate mycelium (Figure 24.20). The sporangium arises by the growth and coiling of the sporangial mycelium within the sporangium wall (Figure 24.21); the process is hence very similar to that of the streptomycetes, the sporangial wall being analogous to the fibrous sheath of the streptomycete

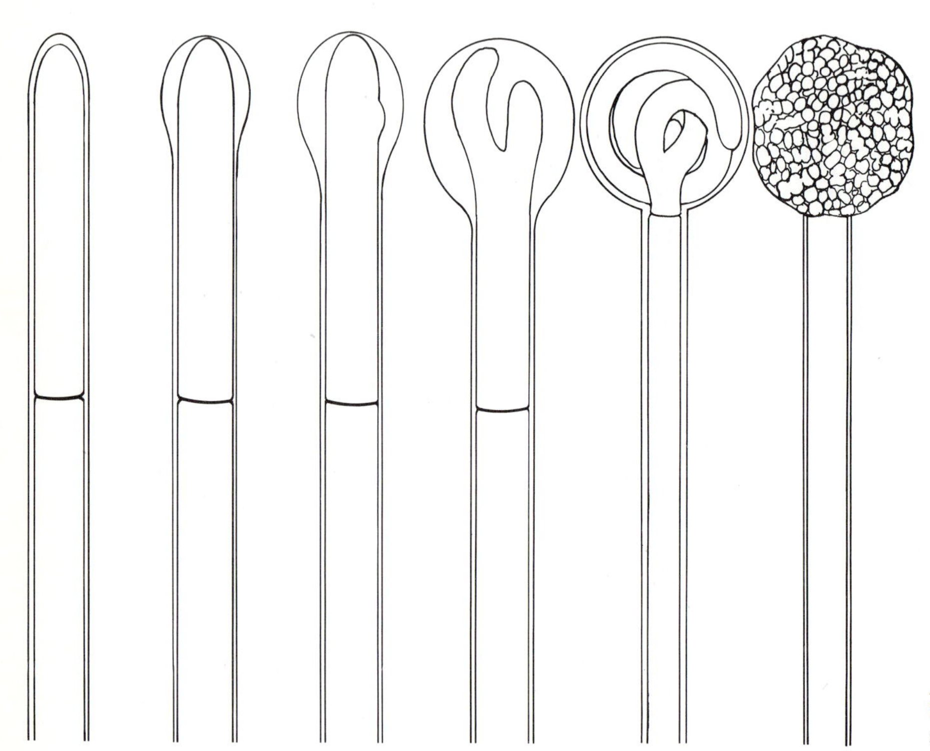

FIGURE 24.21

Sporulation in *Actinoplanes*.

aerial mycelium. The number of spores formed within the sporangium is quite variable, ranging from 1 to more than 1,000. Most actinoplanetes produce spores that are motile by means of a tuft of polar flagella.

The organism *Micromonospora* (Figure 24.22) has a very different developmental pattern; it forms single spores not enclosed in a sporangial wall. However, rRNA sequencing indicates that it is closely related to the actinoplanetes.

The actinoplanetes are common water and soil organisms. With the exception of a few *Micromonospora* strains they are strict aerobes. *Micromonospora* includes some strictly anaerobic cellulose-fermenting representatives isolated from the termite hindgut.

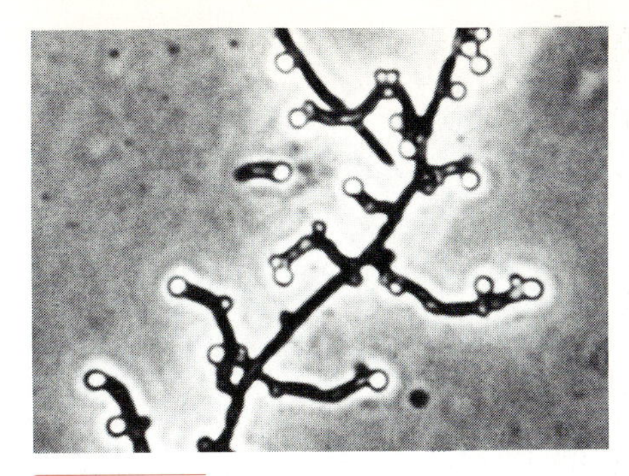

FIGURE 24.22
*Micromonospora chalcea*, showing spherical spores borne singly at the tips of hyphae (phase contrast, ×2,300). Courtesy of G. M. Leudemann and the Schering Corporation.

## FURTHER READING


### Books

GOODFELLOW, M., M. MORDARSKI, and S. T. WILLIAMS, *The Biology of the Actinomycetes*. New York: Academic Press, 1983.

### Review

KALALOUTSKII, L. V., and N. S. AGRE, "Comparative Aspects of Development and Differentiation in Actinomycetes," *Bacteriol. Rev.* **40,** 469–524 (1976).

Chapter 25

# The Mollicutes

The distinguishing property of the mollicutes* group of eubacteria is their lack of a defined cell wall. Consequently, they share a unique constellation of characteristics: sensitivity to osmotic lysis, resistance to penicillin and other antibiotics that inhibit cell wall synthesis, pleomorphic shape, and easily deformable cells that allow them to be squeezed through membrane filters with a pore size small enough to retain most walled bacteria. All mollicutes share the additional properties of being parasites of eucaryotic organisms, and having complex growth factor requirements that typically include fatty acids, amino acids, purines and pyrimidines, vitamins, and sterols in all but *Acholeplasma* spp. The fatty acids and sterols are usually provided by supplementing media with serum, which contains these compounds as soluble, nontoxic lipoproteins.

Colonies that develop on solid media are often small, in some cases microscopic; they typically have a nippled or "fried-egg" appearance (Figure 25.1). The raised center is a nearly spherical mass of cells partly embedded in the agar; it is surrounded by a thin film of surface growth.

Although the mollicutes lack a cell wall, they all have substantial amounts of polysaccharide associated with the cell membrane (Figure 25.2). The detailed structure of this material is not known for any member of the group; however it appears in all cases to be composed principally of hexoses, often including hexosamines (particularly glucosamine and galactosamine) and N-acetylglucosamine. This layer has been termed a

* The common name for this group has traditionally been the *mycoplasmas*. However, this usage invites confusion, since it is often not clear whether "mycoplasma" is being used generically or to refer to a member of the genus *Mycoplasma*. We will thus adopt the more recently coined term *mollicutes* as the designation for members of the larger group.

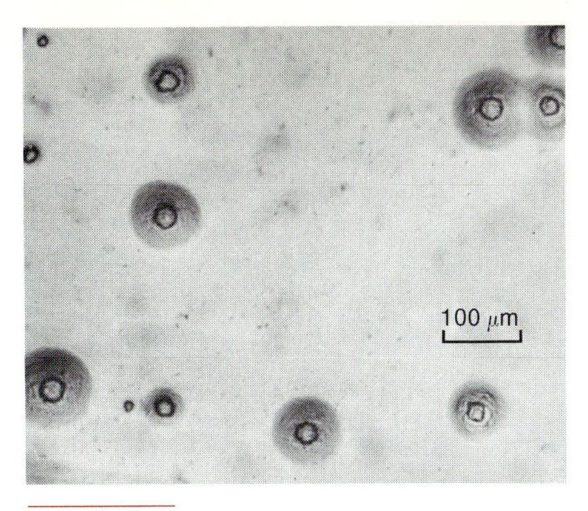


FIGURE 25.1
Characteristic colony structure of a mollicute (×79). Courtesy of M. Shifrine.

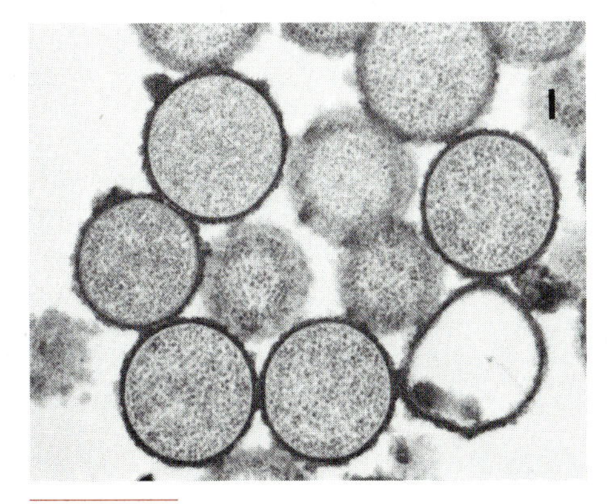

FIGURE 25.2
Electron micrograph of a thin section of *Ureaplasma urealytium* stained with ruthenium red to visualize extracellular carbohydrate. The bar indicates 0.1 μm. From J. Robertson and E. Smook, "Cytochemical evidence of extramembranous carbohydrates on *Ureaplasma urealyticum*," *J. Bacteriol.* **128,** 658–660 (1976).

capsule, but it is difficult to remove from the membrane, suggesting that it may be covalently bound to hydrophobic constituents. Probably it plays a structural role, partially compensating for the lack of a peptidoglycan wall. A similar role has been proposed for the lipopolysaccharide of the wall-less archaebacterium *Thermoplasma* (Chapter 14).

The evolutionary origin of this group was obscure until recently; it was presumed to be polyphyletic; i.e., the group was presumed to contain organisms derived as stable L-forms (Chapter 6) from a variety of different bacteria. However, the many similarities among mollicutes, including unusually low G + C values (23 to 36 percent) and small genome size (MW of $0.5 \times 10^9$ to $1.0 \times 10^9$), suggested that they might instead have a common origin. This suspicion has been confirmed by 16S rRNA sequencing (Chapter 13): the mollicutes are a coherent phylogenetic group closely related to the clostridia. Five genera are recognized in the mollicutes (Table 25.1).

## METABOLISM OF THE MOLLICUTES

With a single exception (*Ureaplasma*), the mollicutes are fermentative. Growth substrates include carbohydrates and amino acids, particularly arginine, the metabolism of which proceeds via the *arginine dihydrolase* or *arginine deiminase* pathway (Figure 25.3), a pathway that also occurs in a variety of other bacteria.

The failure of mollicutes to respire reflects their lack of cytochromes and quinones; however many mollicutes do have a rudimentary electron transport chain that consists of a flavoprotein-coupled NADH oxidase that reduces $O_2$ directly. Only in *Acholeplasma* is this a membrane-bound system; in other members of the group it is cytoplasmic. There is no evidence that a protonmotive force is generated by electron transport, even by *Acholeplasma*; presumably the function of the oxidase is to reoxidize the NADH formed during carbohydrate fermentation.

**TABLE 25.1**

**The Mollicutes**

| Genus | Percent G+C | Sterol Requirement | Growth Substrates | Metabolism | Motility | Morphology | Habitat |
|---|---|---|---|---|---|---|---|
| *Mycoplasma* | 23–41 | + | Carbohydrates, arginine | Fermentative | + (gliding) or – | Pleomorphic | Animals |
| *Ureaplasma* | 27–30 | + | Urea | Respiratory | – | Cocci | Animals |
| *Acholeplasma* | 27–36 | – | Carbohydrates | Fermentative | – | Pleomorphic | Animals |
| *Anaeroplasma* | 29–34 | + | Carbohydrates | Fermentative | – | Cocci | Rumen |
| *Spiroplasma* | 25–31 | + | Carbohydrates, arginine | Fermentative | + (swimming) | Helical | Arthropods, plants |

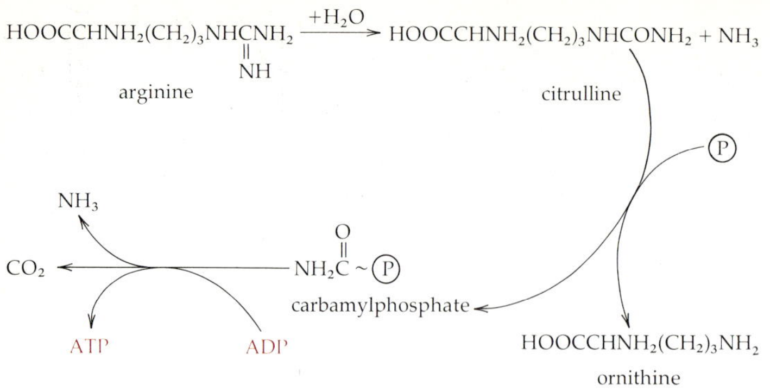

$$\underset{\text{arginine}}{HOOCCHNH_2(CH_2)_3NHC(=NH)NH_2} \xrightarrow{+H_2O} \underset{\text{citrulline}}{HOOCCHNH_2(CH_2)_3NHCONH_2} + NH_3$$

citrulline + (P) → $\underset{\text{ornithine}}{HOOCCHNH_2(CH_2)_3NH_2}$ + $\underset{\text{carbamylphosphate}}{NH_2C(=O) \sim (P)}$

carbamylphosphate + ADP → $CO_2$ + $NH_3$ + ATP

FIGURE 25.3 (left)

Mechanism of the so-called "arginine dihydrolase" reaction, which permits the generation of ATP by substrate-level phosphorylation.

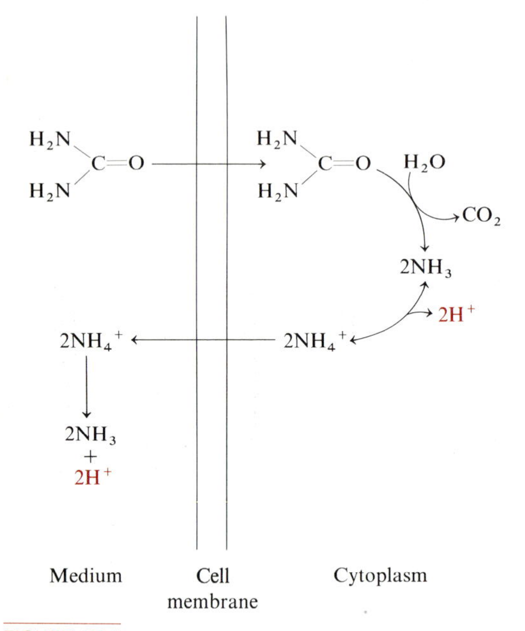


FIGURE 25.4

Proposed mechanism for chemiosmotic coupling during urea catabolism by *Ureaplasma*.

In *Ureaplasma* energy is conserved via a chemiosmotic mechanism. This organism depends on the hydrolysis of urea for its growth; the mechanism of proton pumping is not fully established, but a plausible model has been proposed (Figure 25.4): hydrolysis of urea generates ammonia, which leaves the cell as ammonium, thereby carrying a proton with it.

## CELL SHAPE AND REPRODUCTION

All mollicutes are pleomorphic to at least some extent (Figure 25.5). However it is becoming clear that much of the extensive pleomorphism observed by early workers was a result of suboptimal growth conditions; several members of this group exhibit a fairly constant morphology, and reproduce by binary fission. Indeed, one of them (*Spiroplasma*) grows as quite regular spirals (Figure 25.6). Others (*Ureaplasma, Anaeroplasma*) are typically unicellular cocci or "coccobacilli" (very short rods). Only in *Mycoplasma* and *Acholeplasma* do healthy growing cells exhibit extensive variation in form. In these organisms, reproduction takes two forms: unicellular cocci may divide by fission; or they elongate into branching filaments that then fragment into many cocci (Figure 25.7). Cultures of mycoplasmas and acholeplasmas thus may contain a mixture of cocci, short filaments, and longer branched filaments.

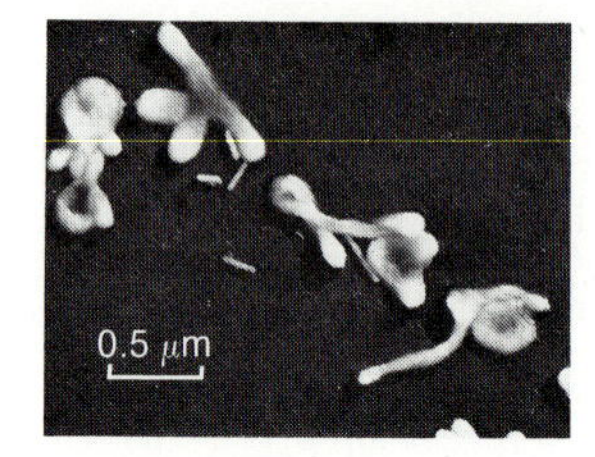


FIGURE 25.5

Electron micrograph of cells of *Mycoplasma* spp., showing characteristic pleomorphism. From E. Klieneberger-Nobel and F. W. Cuckow, "A Study of Organisms of the Pleuropneumonia Group by Electron Microscopy," *J. Gen. Microbiol.* **12**, 99 (1955).

FIGURE 25.6

Electron micrograph of negatively-stained dividing cells of *Spiroplasma citri* (× 10,064). Courtesy of Dr. I.-M. Lee.

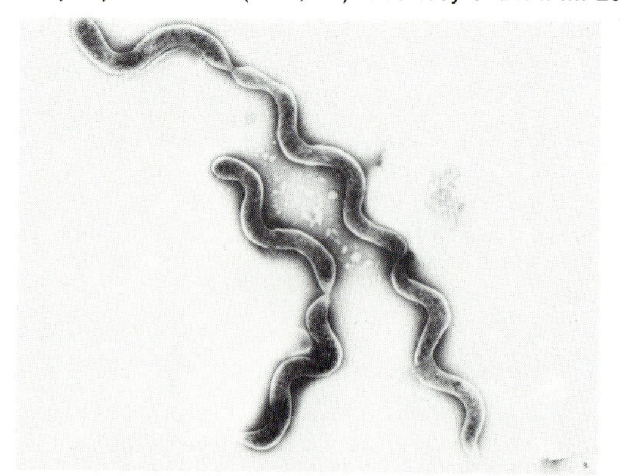

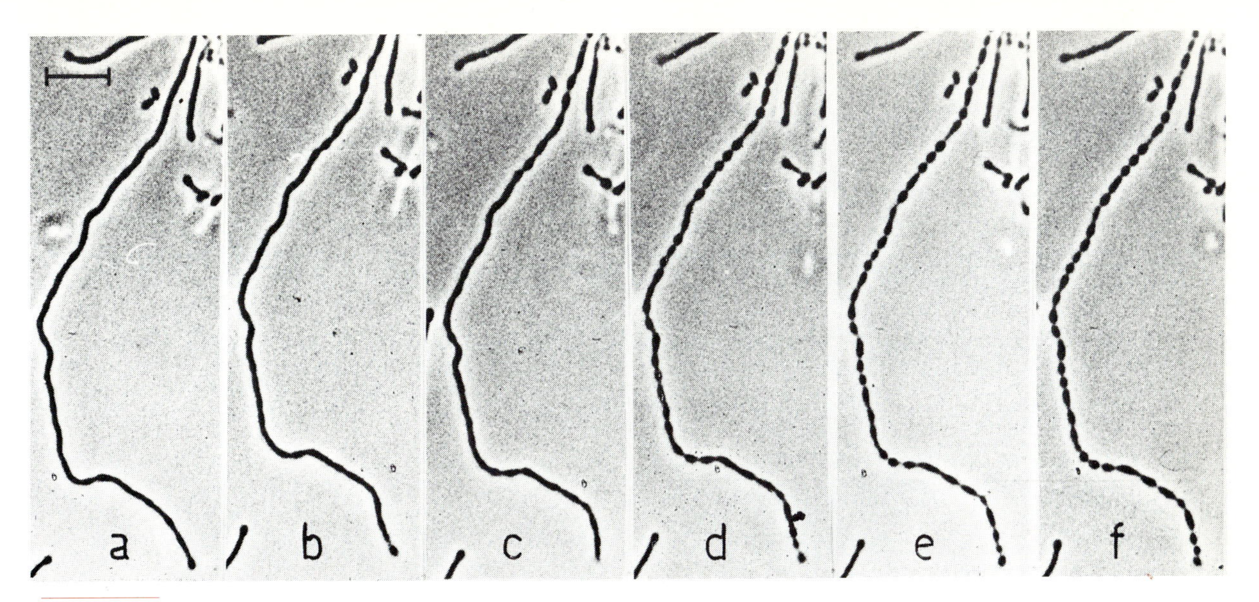


FIGURE 25.7

Reproduction by filament fragmentation in *Mycoplasma hominis*. The bar represents 10 μm. (a) 0 minutes; (b) 2.8 minutes; (c) 3.3 minutes (d) 4.8 minutes; (e) 5.3 minutes; (f) 5.5 minutes. From W. Bredt, H. H. Heunert, K. H. Höfling, and B. Milthaler, "Microcinematographic Studies of *Mycoplasma hominis* Cells," *J. Bacteriol.* **113,** 1223–1227 (1973).

## *Mycoplasma*

The mycoplasmas are parasites of animal mucous membranes, principally those of the respiratory and genital tracts, and the synovial membranes of joint capsules. Infection may cause disease, e.g., pneumonia or arthritis. They are facultative anaerobes, and some prefer reduced $O_2$ tensions. Most ferment either arginine or carbohydrates; some are capable of fermenting both. The products of carbohydrate fermentation always include lactate, with variable but usually small amounts of acetate, pyruvate, and butanediol. The genus is quite large; over 60 species are currently recognized.

A few mycoplasmas are capable of unidirectional gliding motility; they cannot reverse direction. They have a polarized cell organization that includes at the anterior (leading) end of the cell a protrusion of the cell membrane surrounding a rodlike structure (Figure 25.8). The rod consists of a bundle of parallel proteinaceous fibers; there is some evidence suggesting that the constituent proteins are actin-like.

FIGURE 25.8

Electron micrograph of a thin section of *Mycoplasma* sp. showing the terminal structure (TS) at the anterior end of the cell. The bar indicates 0.1 μm. From J. E. Peterson, A. W. Rodwell, and E. S. Rodwell, "Occurrence and Ultrastructure of a Variant (*rho*) form of *Mycoplasma*," *J. Bacteriol.* **115,** 411–425 (1973).

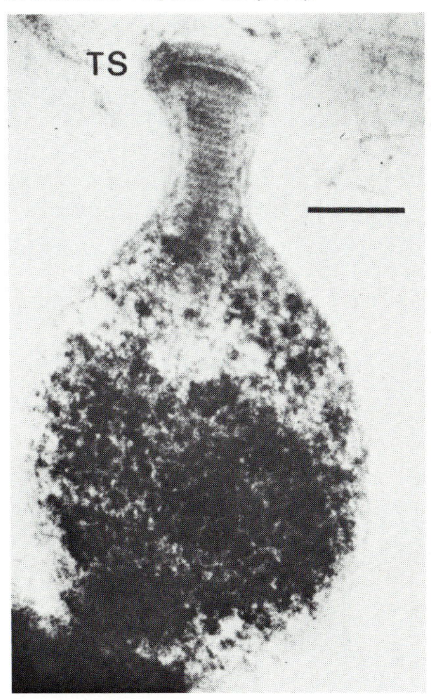


## *Acholeplasma*

The acholeplasmas are also animal parasites, and may be widely distributed in the tissues of many different vertebrates. They have consequently been a nuisance in the *in vitro* culture of animal cells, because the prevention of bacterial contamination of tissue cultures has usually relied on the inclusion of penicillin in the medium, a measure ineffective against mollicutes. Such contamination can damage the cultured animal cells because the production of $H_2O_2$ by the NADH oxidase of contaminating acholeplasmas may be cytotoxic.

*Acholeplasma* is sharply distinguished from all other mollicutes by its ability to grow in the absence of sterols. However, if sterols are available *Acholeplasma* will incorporate them into the cell membrane. The independence from exogenous sterols does not reflect an ability to synthesize them; rather it reflects the ability of these mollicutes to organize their membranes as functioning, stable structures without the strengthening effect of sterols. Some acholeplasmas synthesize and incorporate into their membranes large amounts of carotenoids which serve a sterol-like function; others apparently do not have a sterol replacement in their neutral lipid fraction.

### *Spiroplasma*

The genus *Spiroplasma* contains mollicutes that are parasitic on arthropods and plants. They are particularly common in the hemolymph, gut, and salivary glands of insects, like leafhoppers, that feed on plants. Presumably plants become infected during feeding of these insects. A wide variety of plant diseases are caused by spiroplasma infection of plant vascular tissue. These organisms are also encountered in arthropods, e.g., rabbit ticks, that feed on animals, and in some cases may cause disease in experimentally infected laboratory animals. Consequently there is a growing suspicion that spiroplasmas may be important etiologic agents of human and animal disease.

Spiroplasmas are helical, normally less than 0.2 $\mu$m in diameter, 3 to 5 $\mu$m long. They are motile, with the same range of movements that are characteristic of spirochetes: swimming, flexing, and creeping. In wet mounts they are easily mistaken for spirochetes; only by observing them in an electron microscope can they be distinguished by their lack of cell walls and axial filaments. The basis of the motility of these organisms remains a mystery; however the finding that they contain 3.6 nm fibrils, apparently contractile ones, suggests that motility may be associated with these cytoplasmic contractile microfilaments.

### *Anaeroplasma*

The anaeroplasmas are strictly anaerobic mollicutes that inhabit the bovine and ovine rumen. They ferment carbohydrates to a mixture of acids (acetate, formate, succinate, lactate, and propionate), ethanol, and $CO_2$. Some are bacteriolytic; they lyse walled bacteria by excreting lytic enzymes. Their lipid requirements are complex and poorly understood; sterols are required, as are esterified fatty acids in the form of phospholipids or lipopolysaccharide.

### *Ureaplasma*

The ureaplasmas are the only nonfermentative mollicutes. They are microaerophiles, and depend on the hydrolysis of urea for their energy. They never grow to high cell densities; $10^7$ cells per milliliter is characteristic of growth in liquid media, and tiny colonies less than 60 $\mu$m diameter are formed on solid media. They grow best if the gas phase is enriched in $CO_2$, possibly because carbonic acid partially counteracts the alkalinization that accompanies ammonia production. Ureaplasmas normally inhabit the mouth and respiratory and genital tracts of humans and animals.

## FURTHER READING


**Book**

BARILE, M., and S. RAZIN, eds., *The Mycoplasmas.* New York: Academic Press, 1979.

**Reviews**

MANILOFF, J., "Evolution of Wall-less Prokaryotes," *Ann. Rev. Microbiol.* **37,** 477–499 (1983).

RAZIN, S., "The Mycoplasmas," *Microbiol. Rev.* **42,** 414–470 (1978).

WHITCOMB, R. F., "The Genus *Spiroplasma,*" *Ann. Rev. Microbiol.* **34,** 677–709 (1980).

# Chapter 26
# The Protists

Microorganisms with a eucaryotic cell structure are termed *protists*, of which three major groups can be recognized: *algae*, *protozoa*, and *fungi*. Each of these groups is very large and internally diverse. The more highly specialized representatives—e.g., a seaweed, a ciliate, and a mushroom—can be readily assigned to the algae, the protozoa, and the fungi, respectively. However, there are many protists for which the assignment is arbitrary: numerous transitions exist between algae and protozoa and between protozoa and fungi. For this reason, the three major groups of protists cannot be sharply distinguished in terms of simple sets of clear-cut differences. Broadly speaking, the algae may be defined as organisms that perform oxygenic photosynthesis and possess chloroplasts. Some of them are unicellular microorganisms; some are filamentous, colonial, or coenocytic; and some have a plantlike structure that is formed through extensive multicellular development, with little or no differentiation of cells and tissues. In organismal terms, accordingly, the algae are highly diverse, and by no means all fall into the category of microorganisms. The brown algae known as kelps may attain a total length of as much as 50 m. The protozoa and fungi are nonphotosynthetic organisms, and the difference between them is essentially one of organismal structure; protozoa are predominantly unicellular, whereas fungi are predominantly coenocytic and grow in the form of a filamentous, branched structure known as a *mycelium*.

For historical reasons that have been discussed in Chapter 3, the algae and fungi were traditionally regarded as "plants" and have been largely studied by botanists, while the protozoa were traditionally regarded

**TABLE 26.1**

**Major Groups of Algae**

| Group Name | Pigment System: CHLOROPHYLLS | Pigment System: OTHER SPECIAL PIGMENTS | Composition of Cell Wall | Nature of Reserve Materials |
|---|---|---|---|---|
| Green algae: division Chlorophyta | a + b | — | Cellulose | Starch |
| Euglenids: division Euglenophyta | a + b | — | No wall | Paramylum and fats |
| Dinoflagellates and related forms: division Pyrrophyta | a + c | Special carotenoids | Cellulose | Starch and oils |
| Chrysophytes and diatoms: division Chrysophyta | a ± c | Special carotenoids | Wall composed of two overlapping halves, often containing silica (some have no walls) | Leucosin and oils |
| Brown algae: division Phaeophyta | a + c | Special carotenoids | Cellulose and algin | Laminarin and fats |
| Red algae: division Rhodophyta | a | Phycobilins | Cellulose | Starch |

as "animals" and have been largely studied by zoologists. As a result of this specialization, the many interconnections between the three groups have tended to be overlooked. We shall attempt in this chapter to provide a unified account of the properties of protists that emphasizes possible evolutionary interrelationships.

## THE ALGAE

The primary classification of algae is based on cellular, not organismal, properties: the chemical nature of the wall, if present; the organic reserve materials produced by the cell; the nature of the photosynthetic pigments; and the nature and arrangement of the flagella borne by motile cells. In terms of these characters, the algae are arranged in a series of divisions, summarized in Table 26.1.

The divisions are not equivalent to one another in terms of the range of organismal structure of their members. For example, the Euglenophyta (euglenid algae) consist entirely of unicellular or simple colonial organisms, while the Phaeophyta (brown algae) consist only of plantlike, multicellular organisms. The largest and most varied group, the Chlorophyta (green algae), from which the higher plants probably originated, spans the full range of organismal diversity, from unicellular organisms to multicellular representatives with a plantlike structure.

The common cellular properties of each algal division suggest that its members, however varied their organismal structure may be, are representatives of a single major evolutionary line. Evolution among the algae thus in general appears to have involved *a progressive increase in organismal complexity in the framework of a particular variety of eucaryotic cellular organization.* Although it is possible to perceive these evolutionary progressions *within* each algal division, the relationships *between* divisions are completely obscure. The primary origin of the algae as a whole is accordingly an unsolved problem.

### The Photosynthetic Flagellates

In many algal divisions, the simplest representatives are motile, unicellular organisms, known collectively as *flagellates.* The cell of a typical flagellate,

| Numbers and Type of Flagella | Range of Structure |
|---|---|
| Generally two identical flagella per cell | Unicellular, coenocytic, filamentous; plantlike multicellular forms |
| One, two , or three flagella per cell | All unicellular |
| Two flagella, dissimilar in form and position on cell | Mostly unicellular, a few filamentous forms |
| Two flagella, arrangement variable | Unicellular, coenocytic, filamentous |
| Two flagella, of unequal length | Plantlike multicellular forms |
| No flagella | Unicellular; plantlike multicellular forms |

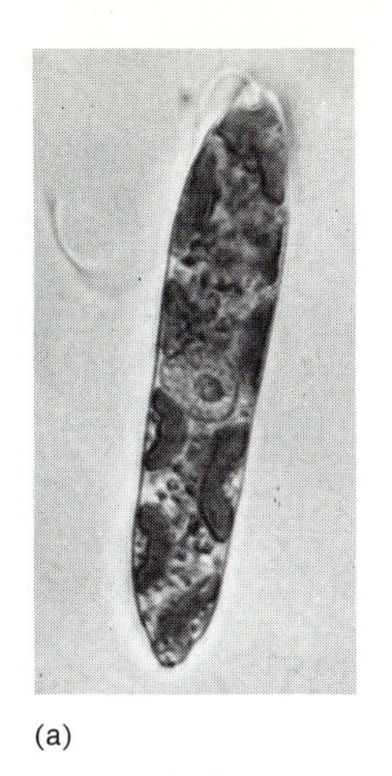
(a)

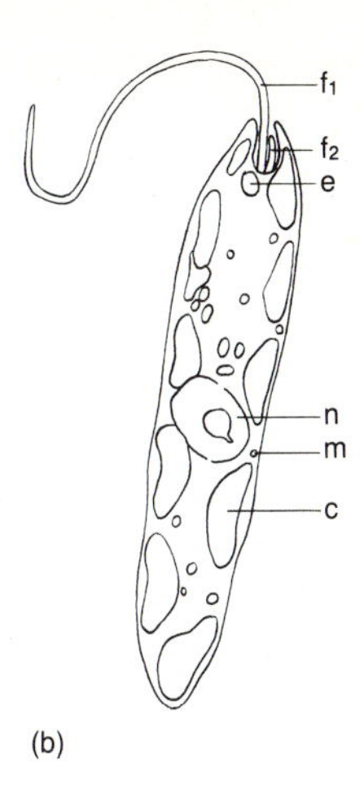

(b)

FIGURE 26.1

*Euglena gracilis*. (a) Photomicrograph of fixed cell (× 1,000). Courtesy of Gordon F. Leedale. (b) Schematic drawing of the same cell, to show principal structural features: n, nucleus; c, chloroplast; m, mitochondrion; e, eyespot; $f_1$, $f_2$, the two flagella of unequal length, originating within a small cavity of the anterior end of the cell.

illustrated by *Euglena* (Figure 26.1), has a very marked polarity: it is elongated and leaf-shaped, the flagella usually being inserted at the anterior end. In the Euglenophyta, to which *Euglena* belongs, there are two flagella of unequal length, which originate from a small cavity at the anterior end of the cell. Many chloroplasts and mitochondria are dispersed throughout the cytoplasm. Near the base of the flagellar apparatus is a specialized organelle, the *eyespot*, which is red, owing to its content of special carotenoid pigments; the eyespot serves as a photoreceptor to govern the active movement of the cell in response to the direction and intensity of illumination. The cell of *Euglena*, unlike that of many other flagellates, is not enclosed within a rigid wall; its outer layer is an elastic *pellicle*, which permits considerable changes of shape. Cell division occurs by *longitudinal fission* [Figure 26.2(a)]. About the time of the onset of mitosis, there is a duplication of the organelles of the cell, including the flagella and their basal apparatus; cleavage subsequently occurs through the long axis, so that the duplicated organelles are equally partitioned between the two daughter cells. This mode of cell division is characteristic of all flagellates except those belonging to the Chlorophyta, such as *Chlamydomonas*, where each cell undergoes *two or more multiple fissions* to produce four smaller daughter cells, liberated by rupture of the parental cell wall [Figure 26.2(b)]. Even in such cases, however, the internal divisions take place in the longitudinal plane. As we shall see in a subsequent section, longitudinal division also occurs in the nonphotosynthetic flagellate protozoa and is one of the primary characters that distinguish these organisms from the other major group of protozoa that possess flagellalike locomotor organelles, the ciliates.

Most multicellular algae are immotile in the mature state. However, their reproduction frequently involves the formation and liberation of motile cells, either asexual reproductive cells (*zoospores*) or gametes. Figure 26.3 shows the liberation of zoospores from a cell of a filamentous member of the Chlorophyta, *Ulothrix*; it can be seen that these zoospores have a structure very similar to that of the *Chlorogonium* cell, illustrated in Figure 26.2(c). The structure of the motile reproductive cells of multicellular algae thus often reveals their relationship to a particular group of unicellular flagellates.

## The Nonflagellate Unicellular Algae

By no means are all unicellular algae flagellates; several algal divisions also contain unicellular members that are either immotile or possess other means of movement. Many of these unicellular

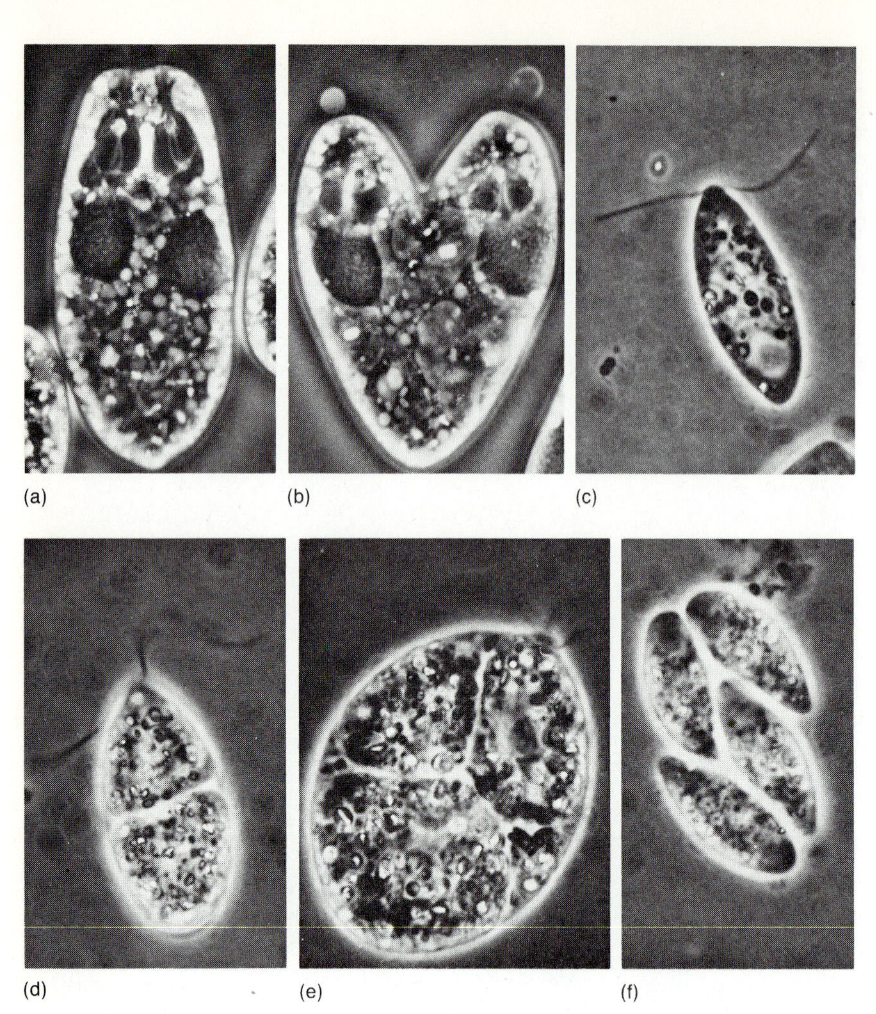


FIGURE 26.2
Longitudinal and multiple fission in flagellate algae. (a), (b) Two cells of *Euglena gracilis* in the course of longitudinal fission (phase contrast, × 1,240). Reproduced from G. F. Leedale, in *The Biology of Euglena,* ed. D. E. Buetow, (New York: Academic Press, 1968). In (a), division of the nucleus and of the locomotor apparatus at the anterior end of the cell is complete. In (b), cell cleavage has begun. (c), (d), (e), (f) Four steps in the cellular life cycle of *Chlorogonium tetragamum*, a green alga that reproduces by multiple fission (phase contrast, × 1,430). (c) Newly liberated daughter cell. (d) Two-celled stage. (e) Four-celled stage. (f) Four daughter cells just after liberation from the mother cell. Courtesy of Paul Kugrens, Department of Botany, University of California, Berkeley.

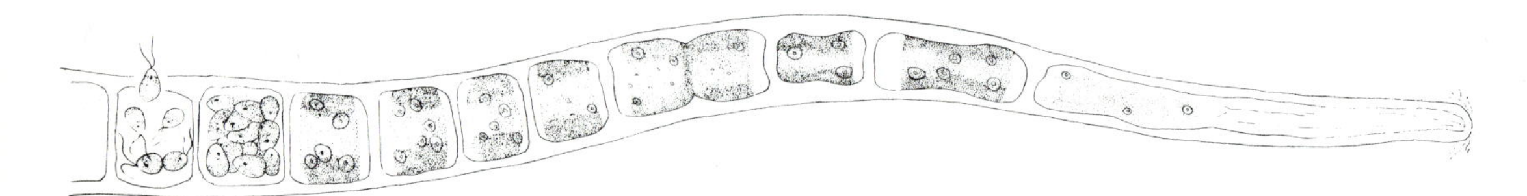

FIGURE 26.3
The filamentous green alga, *Ulothrix* ( × 1,250). At left, the formation and liberation of biflagellate zoospores.

nonflagellate algae possess strikingly specialized and elaborate cells, which may be illustrated by considering two groups, the *desmids* and the *diatoms.*

The desmids, members of the Chlorophyta, have flattened, relatively large cells, with a characteristic bilateral symmetry (Figure 26.4). Asexual reproduction involves the synthesis of two new half-cells in the equatorial plane, followed by cleavage between the new half-cells to produce two bilaterally symmetrical daughters, each of which has a cell consisting of an "old" and a "new" half.

The diatoms (Figure 26.5), members of the Chrysophyta, have organic walls impregnated with silica. The architecture of the diatom wall is exceedingly complex; it always consists of two overlapping halves, like the halves of a petri dish. Division is longitudinal, each daughter cell retaining half of the old wall and synthesizing a new half.

Although devoid of flagella, some desmids and diatoms can move slowly over solid substrates. The mechanism of desmid locomotion is not known.

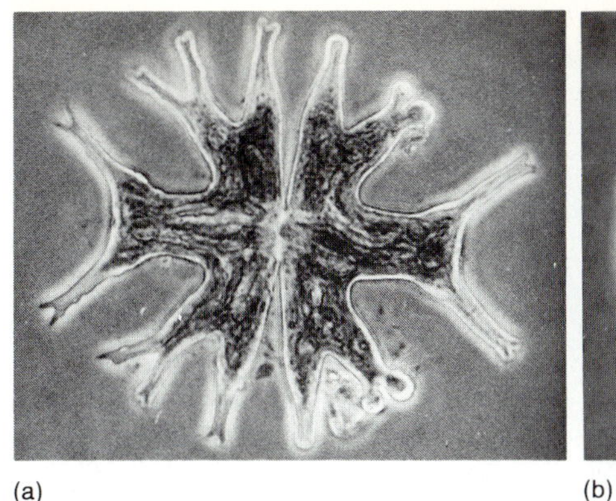
(a)

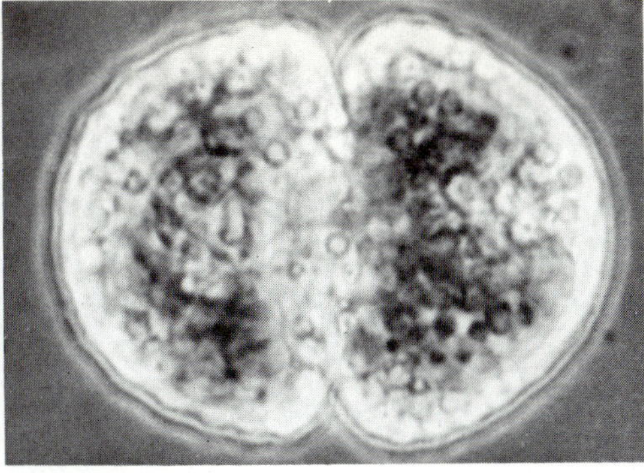
(b)

FIGURE 26.4
Two phase contrast photomicrographs of living desmid cells: (a) *Micrasterias* (×360); (b) *Cosmarium* (× 1,560). Material from the collection of algae of the Department of Botany, University of California, Berkeley, provided by R. Berman.

The locomotion of diatoms is accomplished by a special modification of ameboid movement. In motile diatoms, there is a narrow longitudinal slot in the wall, known as a *raphe*, through which the protoplast can make direct contact with the substrate. Movement is brought about by directed cytoplasmic streaming in the canal of the raphe, which pushes the cell over the substrate.

Many fossil diatoms are known, because the siliceous skeleton of the wall (Figure 26.6) is practically indestructible, and as diatoms are one of the major groups of algae in the oceans, large fossil deposits of diatom walls have accumulated in many areas. These deposits, known as *diatomaceous earth*, have industrial uses as abrasives and filtering agents.

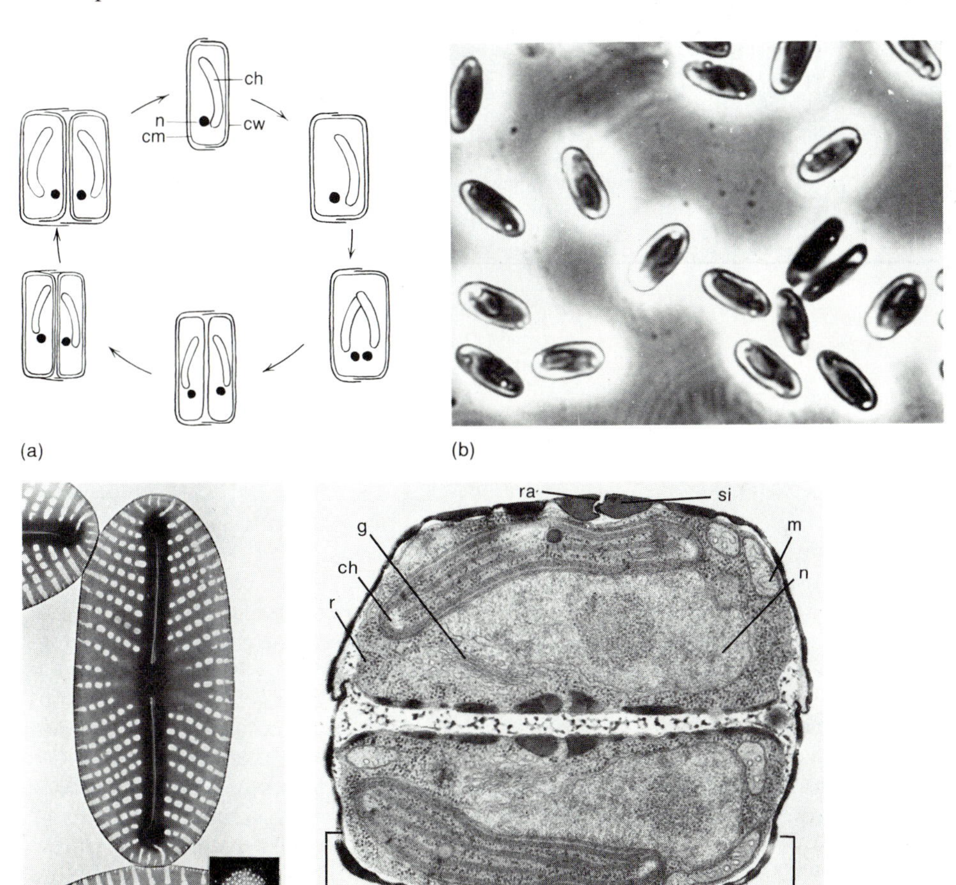


FIGURE 26.5
The diatom *Navicula pelliculosa*. (a) Diagrammatic representation of the division cycle. (b) Living cells, phase contrast illumination (× 1,320). (c) Electron micrograph of the wall (×9,800). Insert depicts fine structure of one of the wall pores (×56,000). (d) Transverse section of a dividing cell (×23,800): ch, chloroplast; g, Golgi apparatus; n, nucleus; m, mitochondrion; r, ribosomes; ra, raphe; si, silica in wall; cw, cell wall; cm, cell membrane. Courtesy of M. L. Chiappino and B. E. Volcani, University of California, San Diego.

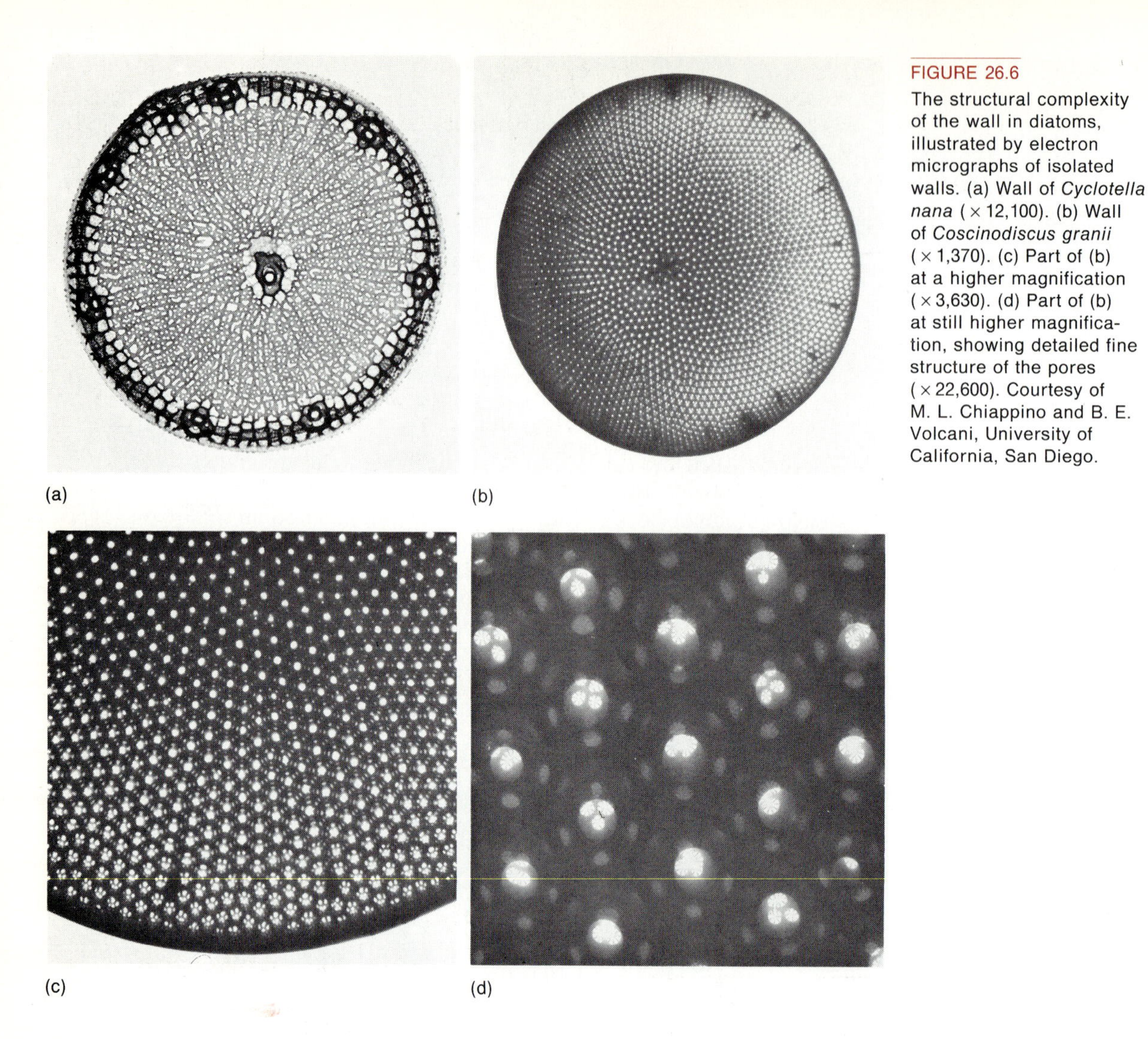


FIGURE 26.6
The structural complexity of the wall in diatoms, illustrated by electron micrographs of isolated walls. (a) Wall of *Cyclotella nana* (× 12,100). (b) Wall of *Coscinodiscus granii* (× 1,370). (c) Part of (b) at a higher magnification (× 3,630). (d) Part of (b) at still higher magnification, showing detailed fine structure of the pores (× 22,600). Courtesy of M. L. Chiappino and B. E. Volcani, University of California, San Diego.

## The Natural Distribution of Algae

Most algae are aquatic organisms that inhabit either fresh water or the oceans. These aquatic forms are principally free-living, but certain unicellular marine algae have established durable symbiotic relationships with specific marine invertebrate animals (e.g., sponges, corals, various groups of marine worms) and grow within the cells of the host animal. Some terrestrial algae grow in soil or on the bark of trees. Others have established symbiotic relationships with fungi, to produce the curious, two-membered natural associations termed *lichens*, which form slowly growing colonies in many arid and inhospitable environments, notably on the surface of rocks. Some of the symbiotic relationships into which algae have entered will be described in Chapter 28.

The marine algae play a very important role in the cycles of matter on earth, since their total mass (and consequently their gross photosynthetic activity) is at least equal to that of all land plants combined and is probably much greater. This role is by no means evident, because the most conspicuous of marine algae, the seaweeds, occupy a very limited area of the oceans, being attached to rocks in the intertidal zone and the shallow coastal waters of the continental shelves. The great bulk of marine algae are unicellular floating (*planktonic*) organisms, predominantly diatoms and dinoflagellates, distributed through the surface waters of the oceans. Although they sometimes become abundant enough to impart a definite brown or red color to local areas of the sea, their density is usually so low that there is no gross sign of their presence. It is the enormous total volume of the earth's oceans which they occupy that makes them the most abundant of all photosynthetic organisms.

## Nutritional Versatility of Algae

The ability to perform photosynthesis confers on many algae very simple nutrient requirements; in the light they can grow in a completely inorganic medium. However, many algae have specific vitamin requirements, a requirement for vitamin $B_{12}$ being particularly common. In nature the source of these vitamins is probably bacteria that inhabit the same environment. The ability to perform photosynthesis does not necessarily preclude the utilization of organic compounds as the principal source of carbon and energy, and many algae have a mixotrophic metabolism.

Even when growing in the light, certain algae (e.g., the green alga *Chlamydobotrys*) cannot use $CO_2$ as their principal carbon source and are therefore dependent on the presence of acetate or some other suitable organic compound to fulfill their carbon requirements. This is caused by a defective photosynthetic machinery: although these algae can obtain energy from their photosynthetic activity, they cannot obtain the reducing power to convert $CO_2$ to organic cell materials.

Many algae that perform normal photosynthesis in the light, using $CO_2$ as the carbon source, can grow well in the dark at the expense of a variety of organic compounds; such forms can thus shift from photosynthetic to respiratory metabolism, the shift being determined primarily by the presence or absence of light. Algae completely enclosed by cell walls are osmotrophic and dependent on dissolved organic substrates as energy sources for dark growth. However, a considerable number of unicellular algae that lack a cell wall, or are not completely enclosed by it, can phagocytize bacteria or other smaller microorganisms and thus employ a phagotrophic mode of nutrition as well. It is not correct, accordingly, to regard the algae as an *exclusively* photosynthetic group; on the contrary, many of their unicellular members possess and can use the nutritional capacities characteristic of the two major subgroups of nonphotosynthetic eucaryotic protists, the protozoa and fungi.

## The Leucophytic Algae

Loss of the chloroplast from a eucaryotic cell is an irreversible event, which results in a permanent loss of photosynthetic ability. Such a change appears to have taken place many times among unicellular algal groups with a mixotrophic nutrition, to yield nonpigmented counterparts, which can be clearly recognized on the basis of other cellular characters as *nonphotosynthetic derivatives of algae*. Such organisms, known collectively as *leucophytes*, exist in many flagellate groups, in diatoms, and in nonmotile groups among the green algae. The recognition of leucophytes is often easy, since they may have preserved a virtually complete structural identity with a particular photosynthetic counterpart. In some cases, this structural near-identity may include the preservation of vestigial, nonpigmented chloroplasts, as well as a pigmented eyespot. There can be little doubt accordingly that these nonphotosynthetic organisms are close relatives of their structural counterparts among the algae and have arisen from them by a loss of photosynthetic ability in the recent evolutionary past. Indeed, the transition can be demonstrated experimentally in certain strains of *Euglena*, which yield stable, colorless races when treated with the antibiotic streptomycin or when exposed to small doses of ultraviolet irradiation or to high temperatures (Figure 26.7). These colorless races cannot be distinguished from the naturally occurring nonphotosynthetic euglenid flagellates of the genus *Astasia*.

The classification of the leucophytes raises a difficult problem. In terms of cell structure, they can be easily assigned to a particular division of algae, as nonphotosynthetic representatives, and this classification is no doubt the most satisfactory one. However, since they are nonphotosynthetic unicellular eucaryotic protists, they can alternatively be regarded as protozoa, and they are, in fact, included among the protozoa by zoologists. The leucophytes accordingly provide the first and by far the most striking case of a group, or rather a whole series of groups, which are clearly transitional between two major assemblages among the protists.

FIGURE 26.7

The loss of chloroplasts in *Euglena gracilis* as a result of ultraviolet irradiation. (a) A light-grown plate culture of *E. gracilis*. (b) A light-grown culture of the same organism, after exposure to brief ultraviolet irradiation. Most of the cells have given rise to clones devoid of chloroplasts (pale colonies). Courtesy of Jerome A. Schiff.

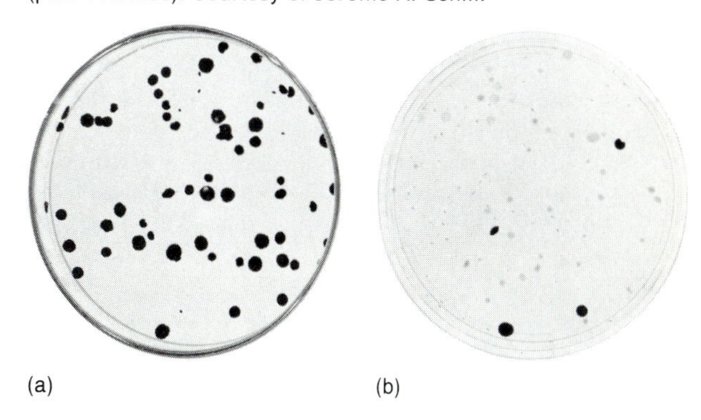

## THE PROTOZOA

### The Origins of the Protozoa

The protozoa are a highly diverse group of unicellular, nonphotosynthetic protists, most of which show no obvious resemblances to the various divisions of algae. Nevertheless, the various kinds of leucophytes, which are recognizably of algal origin, provide a plausible clue concerning the evolutionary origin of many groups among the protozoa. The loss of photosynthetic function abruptly reduces the nutritional potentialities of an organism; leucophytes are therefore immediately confined to a more restricted range of environments than their photosynthetic ancestors. Specific features of cellular construction that possessed adaptive value in the context of photosynthetic metabolism become superfluous; the eyespot is the most obvious example. Hence, one could expect that loss of photosynthetic ability would be followed by a series of evolutionary changes in the structure of the cell which better fit the organism for an osmotrophic or phagotrophic mode of life. Beyond a certain point, these changes would make the algal origin of the organism unrecognizable, and it would then be classified without question as a protozoan.

One group of protists, the dinoflagellates, has several features of cell structure that permit the biologist to recognize a dinoflagellate origin even in organisms that have evolved very far from the typical unicellular photosynthetic flagellate members of this algal group (Figure 26.8). The motile cell of a dinoflagellate has two flagella, which differ in structure and arrangement. One lies in a groove or girdle around the equator of the cell; the other extends away from the cell in a posterior direction. The dinoflagellate nucleus is also unusual; its division is highly specialized (Chapter 3), and the chromosomes remain visible in interphase.

Most photosynthetic dinoflagellates are unicellular planktonic organisms, widely distributed in the oceans, and characteristically brown or yellow in color as a result of the possession of a distinctive set of photosynthetic pigments. Many (the so-called "armored" dinoflagellates) possess very elaborate cell walls, composed of a series of plates, which do not completely enclose the protoplast. There is a very pronounced tendency to phagotrophic nutrition among these photosynthetic members of the group, because the wall structure permits pseudopodial extension and the engulfment of small prey. A few filamentous algae, completely enclosed by walls, can be recognized as of dinoflagellate origin, since they form zoospores with the characteristic flagellar arrangement.

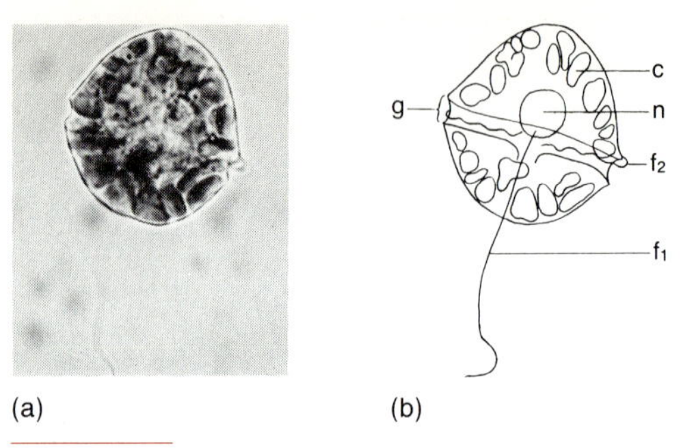


FIGURE 26.8

A photosynthetic dinoflagellate, *Glenodinium foliaceum*. (a) A living cell ($\times 1{,}000$). (b) A diagrammatic drawing of the cell: c, chloroplast; n, nucleus; g, girdle; $f_1$ and $f_2$, flagella.

A much more extensive series of specialized forms can be traced among the nonphotosynthetic members of this flagellate group. Many of the free-living unicellular dinoflagellates are nonphotosynthetic phagotrophic organisms. Some preserve close structural similarities to photosynthetic members of the group; others, such as the large marine organism, *Noctiluca*, have a highly specialized cellular organization not found in any photosynthetic member of the group. However, the most far-reaching modifications of cell structure within the dinoflagellates are to be found among its parasitic members, most of which occur in marine invertebrates. *Hematodinium*, which occurs in the blood of certain crabs, is completely devoid of flagella. *Chytriodinium*, which parasitizes the eggs of copepods, develops as a large, saclike structure within the egg, subsequently giving rise by multiple internal cleavage to numerous motile spores with a typical dinoflagellate structure. Were it not for the retention of the distinctive nuclear organization (and, in the case of *Chytriodinium*, the flagellar structure of the spores), neither of these parasitic protists could be recognized as belonging to the same group as the photosynthetic dinoflagellates. *Hematodinium* could be classified with the sporozoan protozoa and *Chytriodinium* with the primitive group of fungi known as chytrids.

Accordingly, within this one small flagellate group, it is possible to reconstruct some major patterns of evolutionary radiation that are probably characteristic of protists as a whole (Figure 26.9).

In the light of the preceding discussion, the protozoa can best be regarded as comprising a number of groups of nonphotosynthetic, typically motile, unicellular protists, some of which have probably derived at various times in the evolutionary past from one or another group among the unicellular algae (see Table 26.2).

**TABLE 26.2**

**Primary Subdivisions of the Protozoa**

| | |
|---|---|
| I. Class Mastigophora: | The flagellate protozoa. Motile by means of one or more flagella. Cell division always longitudinal. Included in this class are the "phytoflagellates" (i.e., unicellular motile representatives of the various algal divisions) as well as the "zooflagellates," nonphotosynthetic organisms not recognizable as leucophytes. These forms are largely osmophilic. |
| II. Class Rhizopoda: | The ameboid protozoa. Motile by means of pseudopodia. It should be noted that the distinction from class I on the basis of locomotion is not absolute, since many of the *Rhizopoda* can also form flagella. Reproduction by transverse fission. Phagotrophic. |
| III. Class Sporozoa: | A very diverse group of parasitic protozoa. Immotile or showing gliding movement. Reproduction by multiple fission. Osmophilic. Some examples are discussed in Chapter 32. |
| IV. Class Ciliata: | The ciliates. Motile by means of numerous cilia, organized into a coordinated locomotor system. The cell has two nuclei, differing in structure and function. Division always transverse. Phagotrophic. |

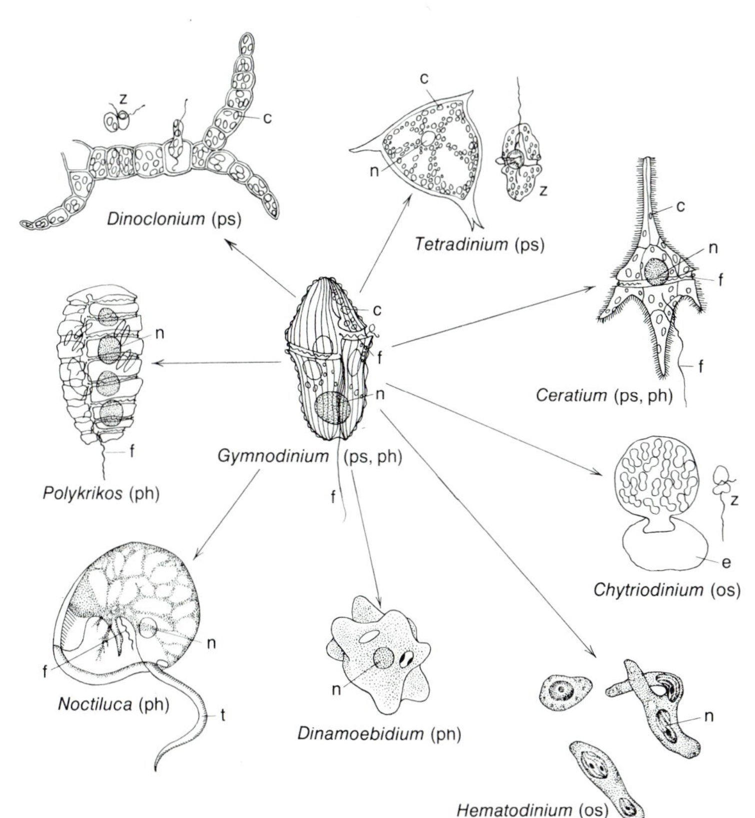


FIGURE 26.9

The different evolutionary trends that are represented among dinoflagellates. *Gymnodinium* is a relatively unspecialized photosynthetic dinoflagellate, which is both photosynthetic, ps, and phagotrophic, ph. *Ceratium* is a more specialized photosynthetic dinoflagellate, characterized by a very complex wall with spiny extensions, comprised of many plates. *Tetradinium* and *Dinoclonium* are nonmotile, strictly photosynthetic organisms, which reproduce by multiple cleavage to form typical dinoflagellate zoospores. *Polykrikos*, *Noctiluca*, and *Dinamoebidium* are three free-living phagotrophic dinoflagellates. *Polykrikos* is a coenocytic, multinucleate organism, the cell of which bears a series of pairs of flagella. *Noctiluca* has one small flagellum, and bears a large and conspicuous tentacle. *Dinamoebidium* is an ameboid organism. *Chytriodinium* and *Hematodinium* are parasitic dinoflagellates whose nutrition is osmotrophic, os. *Chytriodinium* parasitizes invertebrate eggs and reproduces by cleavage of a large saclike structure into dinoflagellate zoospores. *Hematodinium* is a blood parasite in crabs: n, nucleus; f, flagellum; c, chloroplast; z, zoospore; e, parasitized invertebrate egg; t, tentacle.

### The Flagellate Protozoa: The Mastigophora

The Mastigophora are protozoa that always bear flagella as the locomotor organelles. In contrast to the ciliates, in which cell division is transverse, flagellate protozoa undergo longitudinal division. This mode of division has already been described for a photosynthetic flagellate, *Euglena*. In addition to leucophytes, this protozoan group includes many representatives that show no resemblance to photosynthetic flagellates and are for the most part parasites of animals.

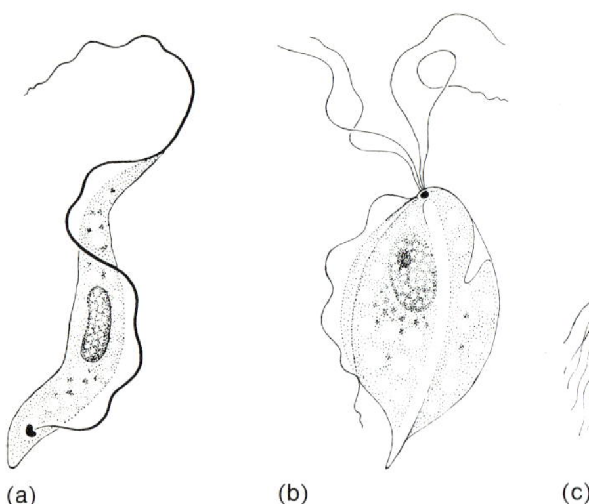


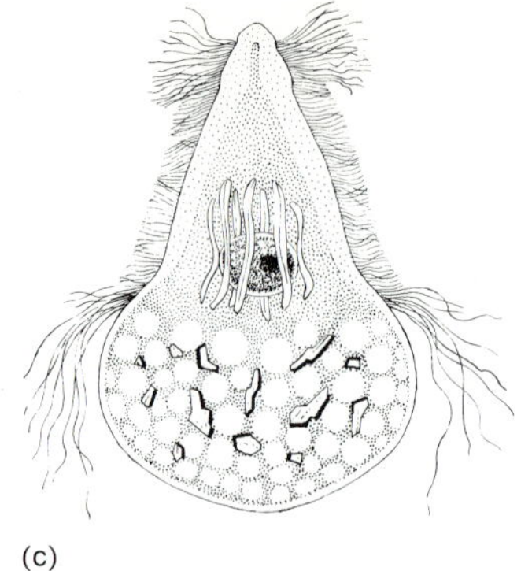


FIGURE 26.10

Some nonphotosynthetic flagellate protozoa (Mastigophora). (a) A trypanosome. The leaf-shaped cell and the long undulating membrane, to which the flagellum is attached, are characteristic of this organism. (b) *Trichomonas*. (c) *Trichonympha*.

The trypanosomes are frequently parasitic in vertebrates, where they develop in the bloodstream, being transmitted from host to host by the bite of insects. They include important agents of human disease, such as the agent of African sleeping sickness, transmitted by the tsetse fly. The cell is slender and leaf-shaped, its single flagellum being directed posteriorly and attached through part of its length to the body of the cell, to form an *undulating membrane* [Figure 26.10(a)]. The trypanosomes are osmotrophic protozoa, which absorb their nutrients from the blood of the host.

Other parasitic flagellates inhabit the gut of vertebrates or invertebrates. The trichomonads, which have four to six flagella [Figure 26.10(b)] are harmless inhabitants of the gut of vertebrates. Several very highly specialized groups of flagellate protozoa inhabit the gut of termites; one of the most striking of these organisms, *Trichonympha*, is illustrated in Figure 26.10(c).

### The Ameboid Protozoa: The Rhizopoda

The Rhizopoda are protozoa in which ameboid locomotion is the predominant mode of cell movement, although some of them are able to produce flagella as well. The simplest members of this group are amebae, which have characteristically amorphous cells as a result of the continuous changes of shape brought about by the extension of pseudopodia. Most amebae are free-living soil or water organisms that phagocytize smaller prey. A few inhabit the animal gut, including forms that cause disease (amebic dysentery). Other members of the Rhizopoda have a well-defined cell form, as the result of the formation of an exoskeleton or shell (typical of the foraminifera) or an endoskeleton (typical of the heliozoa and radiolaria). Several members of the Rhizopoda are illustrated in Figure 26.11.

### The Ciliate Protozoa: The Ciliophora

The ciliate protozoa are a very large and varied group of aquatic, phagotrophic organisms that are particularly widely distributed in fresh water. The ciliates share a number of fundamental cellular characters that distinguish them sharply from all other protists. This suggests that despite the very great internal diversity of this group, it is one class of protozoa that may have had a single common evolutionary origin.

The common characters of ciliates can be summarized as follows:

**1.** At some time in the life history, the cell is motile by means of numerous short, hairlike projections, structurally homologous with flagella, which are termed *cilia*.

**2.** Each cilium arises from a basal structure, the kinetosome, which is homologous with the kinetosome of a flagellum; however, in ciliates the kinetosomes are interconnected by rows of fibrils called *kinetodesmata* to form very elaborate compound locomotor structures termed *kineties*. This internal system persists, even if the cell is devoid of cilia.

**3.** Cell division is transverse, not longitudinal as in flagellates. Ciliates show a marked polarity, with posterior and anterior differentiation of the cell, so the transverse mode of cell division necessarily entails an elaborate process of morphogenesis each time division occurs, during which the anterior daughter cell resynthesizes posterior structures, while the posterior daughter cell resynthesizes anterior structures. The morphogenetic transformations are generally almost complete when the two daughter cells separate.

**4.** Each individual contains two dissimilar nuclei, a large *macronucleus* and a much smaller *micronucleus*, which differ in function as well as in structure.

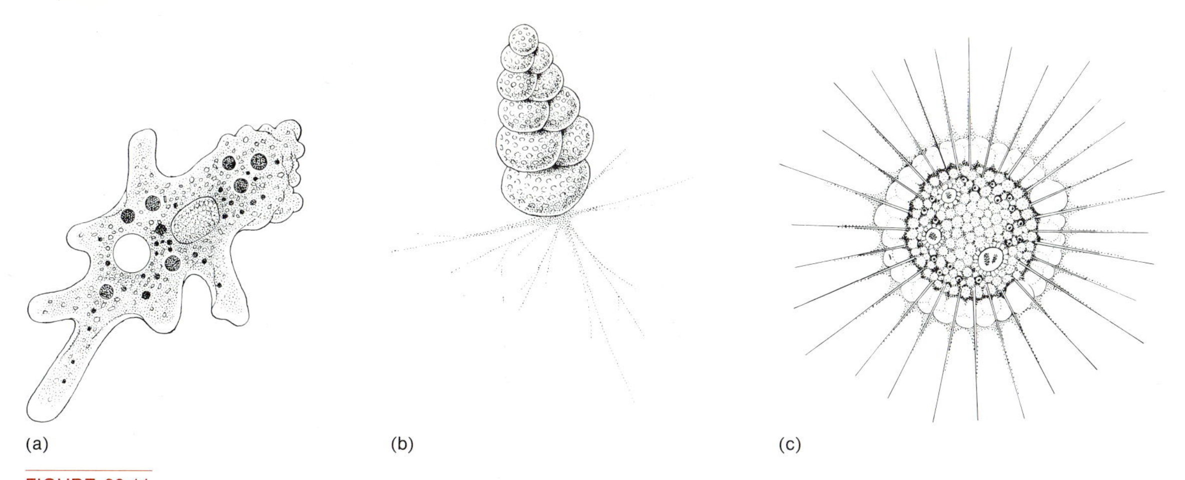
(a) (b) (c)

FIGURE 26.11
Some ameboid protozoa (*Sarcodina*). (a) An ameba. (b) A foraminiferan. Note the many-chambered shell, from which the pseudopodia extend. (c) A heliozoan.

We may illustrate the distinctive character of the ciliates by considering the properties of a simple member of the group, *Tetrahymena pyriformis* (Figure 26.12). It has a pear-shaped body about 50 $\mu$m long, enclosed by a semirigid pellicle. The surface is covered with hundreds of cilia, arranged in longitudinal rows of kineties. The beating of the cilia, which propels the organism, is rhythmic and coordinated.

Near the narrow anterior end of the cell is the mouth or *cytostome*. It consists of an oral aperture, a mouth cavity that extends some distance into the cell, an undulating membrane, and three membranelles. The undulating membrane and membranelles are composed of specialized, adherent cilia, the movements of which sweep food particles into the mouth cavity. Captured food enters food vacuoles which are formed by successive phagocytic events at the base of the mouth cavity. These food vacuoles then circulate within the cell as a result of cytoplasmic streaming until the food material has been digested and the soluble products absorbed; undigested material is ejected from the cell by exocytosis at by a posteriorly located pore known as the *cyto-*

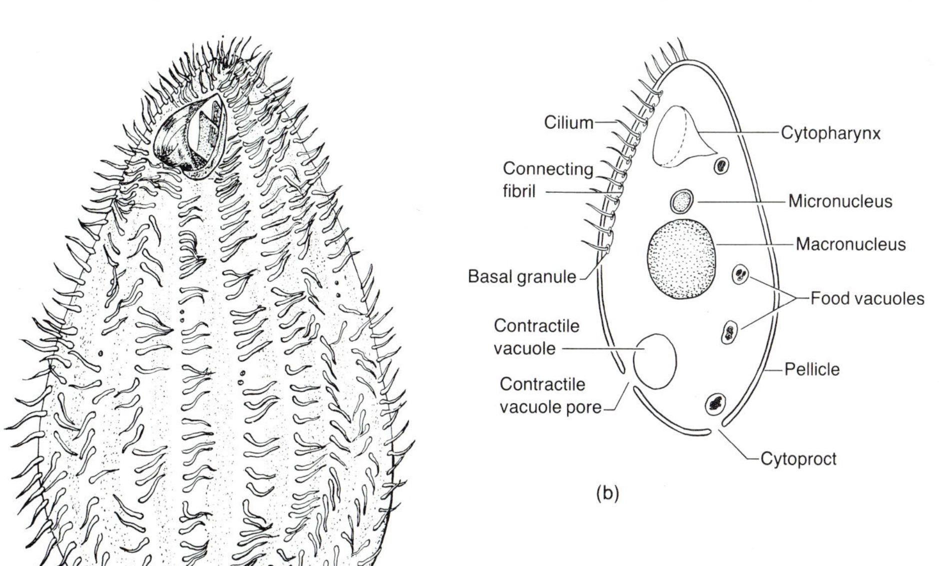

(a) (b)

FIGURE 26.12
The ciliate protozoan, *Tetrahymena*. (a) A general view, showing external appearance. (b) Diagrammatic cross section, showing main structural features of the cell.

*proct.* In nature *Tetrahymena* is normally a predator and feeds on smaller microorganisms. However, in the laboratory it can be grown in pure culture on a medium that contains only soluble nutrients. Under such conditions, the liquid nutrients are still taken in through the mouth, in the form of vacuoles.

Although its natural environment is a dilute one, with an osmotic pressure far below that of the contents of the cell, *Tetrahymena* is able to maintain water balance by the operation of a *contractile vacuole.* This structure, located near the posterior end of the cell, is formed by the coalescence of smaller vacuoles in the cytoplasm; when it reaches a certain critical size, it discharges its liquid contents into the environment through a pore in the pellicle and then starts to grow in volume again.

As mentioned above, typical ciliates have two dissimilar nuclei in the cell. The larger *macronucleus,* which is polyploid, is necessary for normal cell division and growth and is therefore sometimes referred to as the "vegetative nucleus." Some strains of *Tetrahymena* have only this kind of nucleus; they can reproduce indefinitely by binary fission but cannot undergo sexual reproduction. Other strains possess also a small, diploid *micronucleus,* which plays an essential role in sexual reproduction. In *Tetrahymena* the first step in cell division is an elongation of the macronucleus parallel to the long axis of the cell. At the same time, a structural reorganization of the cytoplasm begins. Its principal feature is the formation of a *second cytostome* just posterior to the future plane of cell division. A furrow then develops across the center of the cell, which becomes dumbbell-shaped. If a micronucleus is present, it divides mitotically, and the two daughter nuclei migrate respectively to the anterior and posterior portions of the cell. Finally, the elongated macronucleus divides, and the two daughter cells separate.

*Tetrahymena* is among the simplest of ciliates. The foregoing account suffices to show what an extraordinarily elaborate and complex biological organization has been evolved in this protozoan group within the framework of unicellularity. The ciliates represent the apex of biological differentiation on the unicellular level, but they appear to be a terminal evolutionary group. The development of more complex biological systems took place through the establishment of multicellularity and involved the differentation of specialized cell types during the growth of the individual organism, characteristic of all plants and animals.

FIGURE 26.13

Successive stages in the development of a fungal mycelium from a reproductive cell or conidium ( × 85). After C. T. Ingold, *The Biology of Fungi* (London: Hutchinson, 1961).

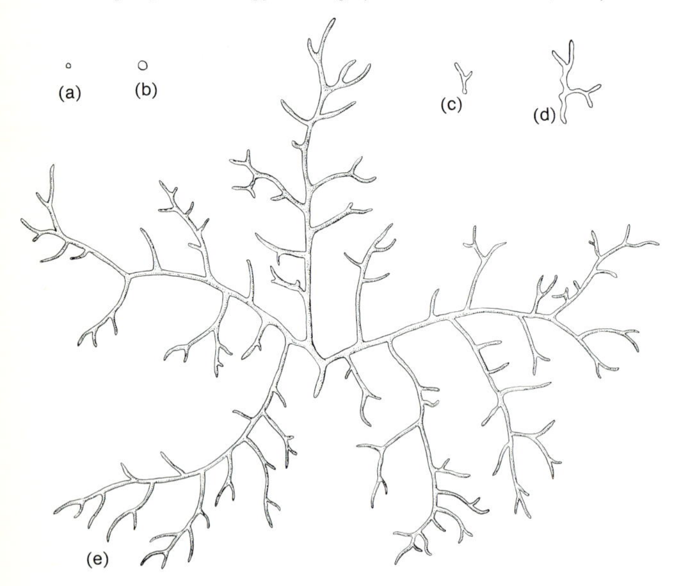


## THE FUNGI

Like the protozoa, the fungi are nonphotosynthetic. Although some of the more primitive aquatic fungi show resemblances to flagellate protozoa, the fungi as a whole have developed a highly distinctive biological organization that can be regarded as an adaptation to life in their most common habitat, the soil. We shall start by considering the main features of this type of biological organization.

Most fungi are coenocytic organisms and have a vegetative structure known as a *mycelium* (Figure 26.13). The mycelium consists of a multinucleate mass of cytoplasm enclosed within a rigid, much-branched system of tubes, which are fairly uniform in diameter. The enclosing tubes represent a protective structure that is homologous with the cell wall of a unicellular organism. A mycelium normally arises by the germination and outgrowth of a single reproductive cell, or spore. Upon germination, the fungal spore puts out a long thread, or *hypha,* which branches repeatedly as it elongates to form a ramifying system of hyphae which constitutes the mycelium. Fungal growth is characteristically confined to the tips of the hyphae; as the mycelium extends, the cytoplasmic contents may disappear from the older, central regions. The size of a single mycelium is not fixed; as long as nutrients are available, outward growth by hyphal extension can continue, and in some of the Basidiomycetes a single mycelium may be as much as 50 ft in diameter. Usually, asexual reproduction occurs by the formation of uninucleate or multinucleate spores which are pinched off at the tips of the hyphae. Neither the spores nor the mycelium of higher fungi are capable of movement. However, the internal contents of a mycelium show streaming movements, which cannot be translated into progression over the substrate, because the cytoplasm is completely enclosed within its wall.

Since a mycelium is capable of almost indefinite growth, it frequently attains macroscopic dimensions. In nature, however, the vegetative mycelium of fungi is rarely seen, because it is normally embedded in soil or other opaque substrates. Many fungi (the mushrooms) form specialized, spore-bearing fruiting structures, however, which project above soil level and are readily visible as macroscopic objects. Such structures were known long before the beginning of scientific biology, although their nature and mode of formation were not clearly understood until the nineteenth century. The superficial resemblance of these fruiting structures to plants was undoubtedly a very important factor in the decision of the early biologists to assign fungi to the plant kingdom, despite their nonphotosynthetic nature.

Since fungi are always enclosed by a rigid wall, they are unable to engulf smaller microorganisms. Most fungi are free-living in soil or water and obtain their energy by the respiration or fermentation of soluble organic materials present in these environments. Some are parasitic on plants or animals. A number of soil forms are predators and have developed ingenious traps and snares, composed of specialized hyphae, which permit them to capture and kill protozoa and small invertebrate animals such as the soil-inhabiting nematode worms. After the death of their prey, such fungi invade the body of the animal by hyphal growth and absorb the nutrients contained in it.

The fungi comprise three major groups: the Phycomycetes, the Ascomycetes, and the Basidiomycetes. A fourth group, the Fungi Imperfecti, has been set aside to include those species for which the sexual stage, and hence the correct classification, is not yet known.

## The Aquatic Phycomycetes

Although soil is by far the most common habitat of the fungi as a whole, many are aquatic. These fungi are known collectively as *water molds* or *aquatic Phycomycetes*. They occur on the surface of decaying plant or animal materials in ponds and streams; some are parasitic and attack algae or protozoa. It is these fungi which show the closest resemblances to protozoa; they produce motile spores or gametes, furnished with flagella, and in the simpler forms the vegetative structure is not mycelial. This description applies, for example, to many of the fungi known as *chytrids*.

The developmental cycle of a typical simple chytrid, which occurs in ponds on decaying leaves, is shown in Figure 26.14. The mature vegetative structure consists of a sac about 100 $\mu$m in diameter which is anchored to the solid substrate by a number of fine, branched hyphae known as *rhizoids*. The sac is a *sporangium*, within which reproductive cells, or spores, are produced. The enclosed cytoplasm contains many nuclei, formed by repeated nucleur division during vegetative growth. Each nucleus eventually becomes surrounded with a distinct volume of cytoplasm, bounded by a membrane. The sporangium then ruptures to release uninucleate flagellated zoospores, each of which can settle down and grow into a new organism. The

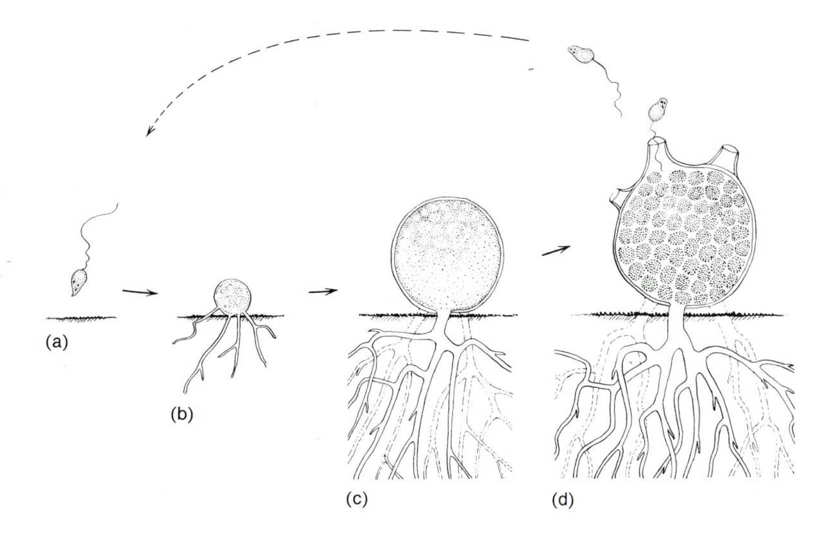


FIGURE 26.14

The life cycle of a chytrid. The flagellated zoospore (a) settles down on a solid surface. As development begins (b), a branching system of rhizoids is formed, anchoring the fungus to the surface. Growth results in the formation of a spherical zoosporangium, which cleaves internally to produce many zoospores (c). The zoosporangium ruptures to liberate a fresh crop of zoospores (d).

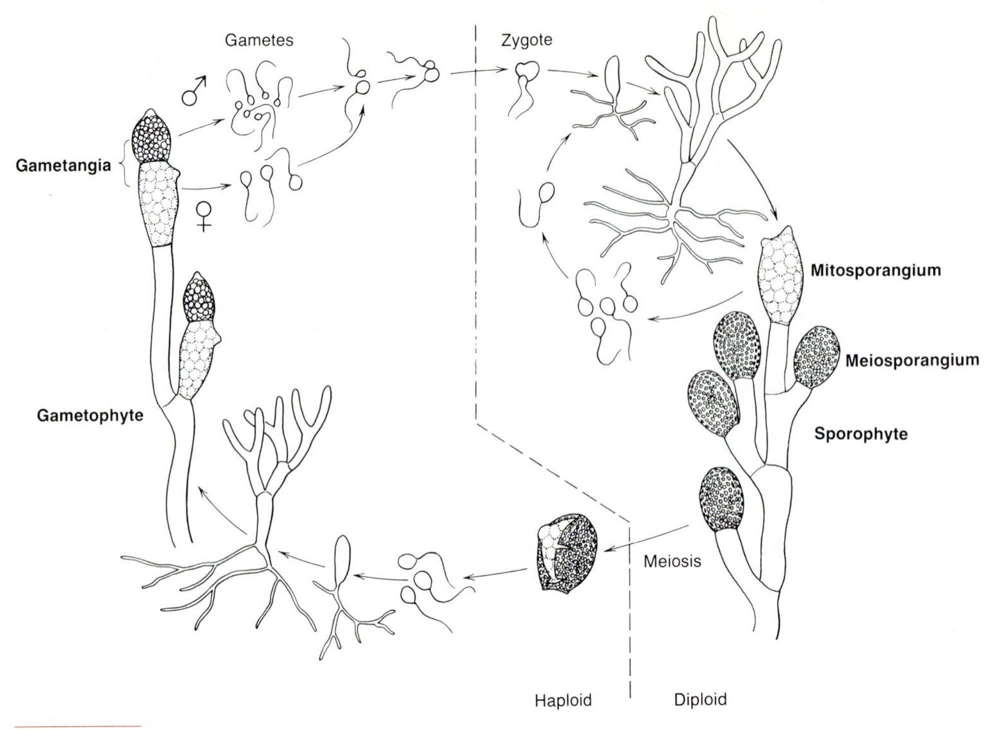


FIGURE 26.15
The life cycle of *Allomyces*, an aquatic phycomycete with a well-marked alternation of haploid and diploid generations. From a drawing made by Raphael Rodriguez and reprinted by permission of Arthur T. Brice.

rhizoids serve to anchor the developing sporangium to the substrate and to absorb the nutrients required for its growth.

The aquatic Phycomycetes are a varied group with respect to their mechanisms of reproduction and life cycles. The range of this variation can be well illustrated by comparing a chytrid with another aquatic phycomycete, *Allomyces. Allomyces* shows a well-marked alternation of haploid and diploid generations (Figure 26.15). We shall describe first the diploid *sporophyte*. When mature it looks like a microscopic tree, with a basal system of anchoring rhizoids from which springs a much-branched mycelium bearing two different kinds of sporangia. The *mitosporangia* have thin, smooth, colorless walls, whereas the *meiosporangia* have thick, dark-pitted walls. Upon maturation, both kinds of sporangia liberate flagellated spores, but the subsequent development of these spores is very different. The mitospores derived from mitosporangia are diploid and germinate into more sporophytic individuals. The *meiospores* derived from meiosporangia are haploid, because meiosis takes place during the maturation of the meiosporangium; they give rise to haploid or *gametophytic* individuals.

The gametophyte is grossly similar in structure to the sporophyte, but instead of bearing meiosporangia and mitosporangia, it produces male and female *gametangia*, which are generally borne in pairs. The female gametangium looks very much like a mitosporangium, whereas the male gametangium is distinguished by its brilliant orange color. The gametangia rupture to liberate male and female gametes, a considerable number arising from each gametangium. Both male and female gametes are motile, moving by means of flagella, but they can be readily distinguished from one another by size and color. The female gamete is larger than the male and is colorless, and the male has an orange oil droplet at the anterior end. The gametes fuse in pairs to form biflagellate zygotes which eventually settle down and develop once more into sporophytes.

## The Terrestrial Phycomycetes

The Phycomycetes also include a group known as the *terrestrial Phycomycetes*, which are inhabitants of soil. These organisms differ from all aquatic

Phycomycetes in not possessing motile flagellated reproductive cells. *They are thus permanently immotile.* This is a property they share with all the higher groups of fungi. The absence of motility characteristic of the higher fungi is understandable in terms of their ecology: motile reproductive cells are of value only when dispersion occurs through water. The reproductive cells of soil-inhabiting fungi are dispersed primarily through the air.

As a typical example of a terrestrial Phycomycete, we may take *Rhizopus*. The mycelium is differentiated into branched rhizoids that penetrate the substrate, horizontal hyphae known as *stolons* that spread over the surface of the substrate, bending down at intervals to form tufts of rhizoids, and erect sporangiophores that emerge from the stolons in tufts [Figure 26.16 (a)]. The unbranched sporangiophore enlarges at the tip to form a rounded sporangium which becomes separated from the rest of the sporangiophore by a cross wall. Within this sporangium, large numbers of spherical spores are formed. These asexual *sporangiospores* are eventually released by rupture of the surrounding wall and are dispersed by air currents. They give rise on germination to a new vegetative mycelium.

*Rhizopus* also reproduces sexually, but sexual reproduction can occur only when two mycelia of opposite sex come into contact with one another. Fungi that show this phenomenon are known as *heterothallic fungi* in contrast to *homothallic fungi* (such as *Allomyces*) that can produce both kinds of sex cells on a single mycelium. In *Rhizopus* the two kinds of mycelia between which sexual reproduction can take place are known as + and − strains, because there are no morphological indications of maleness and femaleness. As the hyphae from a + and a − mycelium meet, each produces a short side branch at the point of contact. This side branch then divides to form a special cell, the *gametangium*. The two gametangia, which are in direct contact with one another, fuse to form a large zygospore, surrounded by a thick, dark wall. The whole sequence of events is shown in Figure 26.16 (b). It can be seen that the behavior of both partners in the sexual act is identical; hence, there is no basis for a designation as "male" or "female." Upon germination of the zygospore, meiosis occurs, and a hypha emerges and produces a sporangium. The haploid spores from this sporangium in turn develop into the typical vegetative mycelium.

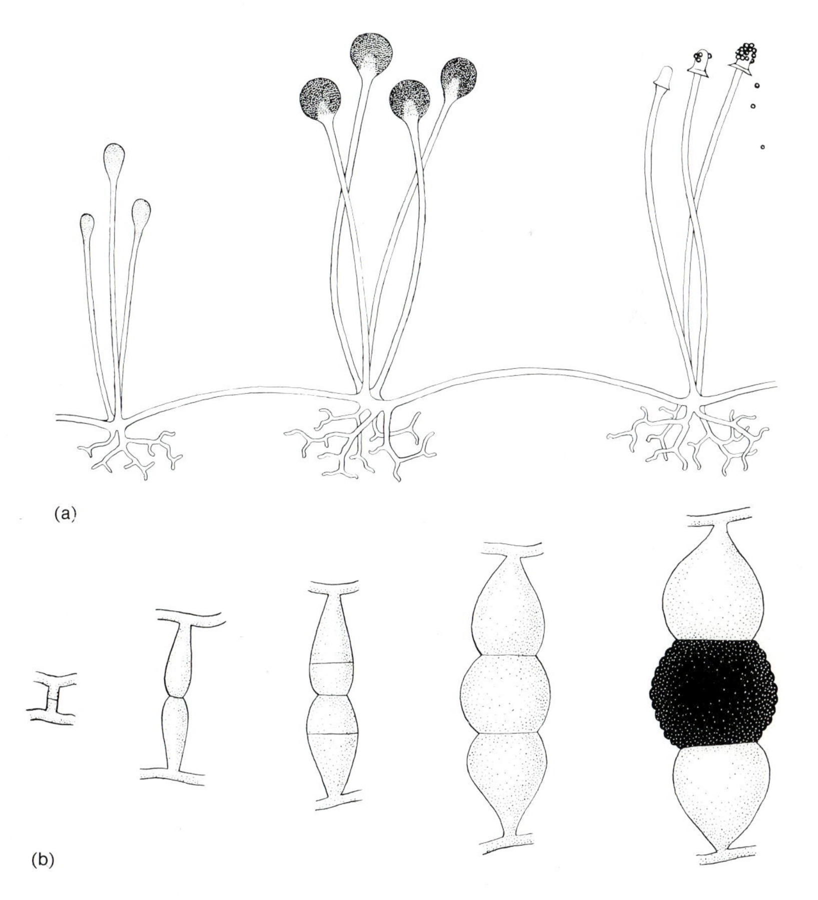


FIGURE 26.16
(a) The vegetative stage of *Rhizopus*, a terrestrial phycomycete. (b) Sexuality in *Rhizopus*. Successive stages of sexual fusion and the formation of a zygospore.

## Distinctions between Phycomycetes and Higher Fungi

Despite the considerable differences among them, all Phycomycetes share two properties that readily distinguish them from the remaining classes of fungi (Ascomycetes, Basidiomycetes, and Fungi Imperfecti). First, *their asexual spores are always endogenous*, formed inside a saclike structure, the zoosporangium of the aquatic types or the sporangium containing immotile sporangiospores of the terrestrial types. In the other groups of fungi, the asexual spores are always exogenous, being formed free at the tips of hyphae (Figure 26.17). Second, *the mycelium in Phycomycetes shows no cross walls* except in regions where a specialized cell, such as a sporangium or gametangium, is formed from a hyphal tip. Such a mycelium is known as a *nonseptate mycelium*. In the remaining groups of fungi, distinct (although incomplete) cross walls occur at regular intervals along the hyphae. Thus, on the basis of these two simple criteria, one can readily distinguish a Phycomycete from any other type of fungus.

Since the mycelium of Phycomycetes is nonseptate, it is clear that these organisms are coenocytic. The regular occurrence of cross walls in the mycelium of other groups of fungi suggests, in contrast, that they are cellular organisms. This is not true, however. The cross walls do not divide the cytoplasm into a number of separate cells: each cross wall has a central pore, through which both cytoplasm and nuclei can move freely. There is thus just as much cytoplasmic continuity in the septate fungi as in the Phycomycetes, and both groups are, in fact, coenocytic.

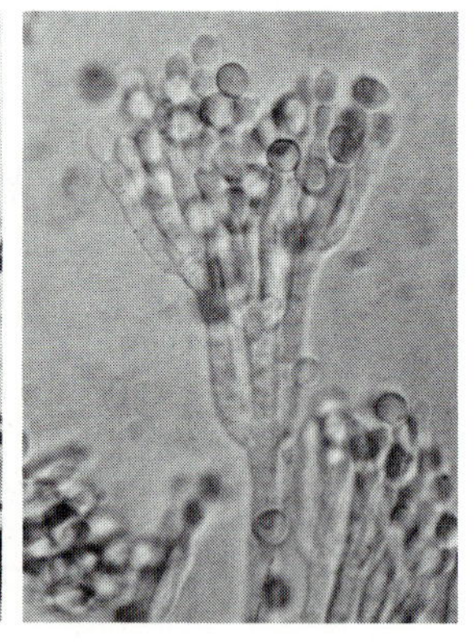

FIGURE 26.17

*Penicillium*. Left, edge of a colony at relatively low magnification, showing spore heads. Right, conidiophore at high magnification, showing branched structure and terminal chains of spherical conidia. Courtesy of Dr. K. B. Raper.

## The Ascomycetes and Basidomycetes

The fungi with septate mycelia and exogeneous asexual spores are broadly classified into two groups, Ascomycetes and Basdiomycetes, on the basis of their sexual development. Following zygote formation in these fungi, there is an immediate meiotic division to form four or eight haploid sexual spores, which are borne in or on structures known as *asci* and *basidia*. The formation of an ascus is characteristic of fungi of the class Ascomycetes, and the formation of a basidium is characteristic of fungi of the class Basidomycetes. In Ascomycetes, the diploid zygote develops into a saclike structure, the *ascus*, while the nucleus undergoes meiosis, often followed by one or more mitotic divisions. A wall is formed around each daughter nucleus and the neighboring cytoplasm to produce four, eight, or more ascospores within the ascus (Figure 26.18). Eventually the ascus ruptures, and the enclosed spores are liberated.

In Basidiomycetes, the zygote enlarges to form a club-shaped cell, the *basidium*; at the same time, the diploid nucleus undergoes meiosis. The subsequent course of events is strikingly different from that which occurs in an ascus. No spores are formed within the basidium; instead, a slender projection known as a *sterigma* develops at its upper end, and a nucleus migrates into this sterigma as the latter enlarges. Eventually, a cross wall is formed near the base of the sterigma, the cell thus cut off being a basidiospore. The same process is repeated for the remaining three nuclei in the basidium, so that a mature basidium bears on its surface four basidiospores (Figure 26.19). Basidiospore discharge is a remarkable phenomenon. After the basidiospore has matured, a minute droplet of liquid appears at the point of its attachment to the basidium. This droplet grows rapidly until it is about one-fifth the size of the spore, and then, quite suddenly, both spore and droplet are shot away from the basidium.

## The Fungi Imperfecti

The classification of the septate fungi into Ascomycetes and Basidiomycetes has one practical disadvantage. Obviously, the assignment of a fungus to its correct class is possible only if one has observed the sexual stage of its life cycle. If one happens to deal with a fungus that is incapable of sexual reproduction, or in which the sexual stage. is unknown, it cannot be assigned either to the Ascomycetes or to the Basidiomycetes. Since heterothallism is very common in the higher fungi, it often happens that a single isolate of an ascomy-

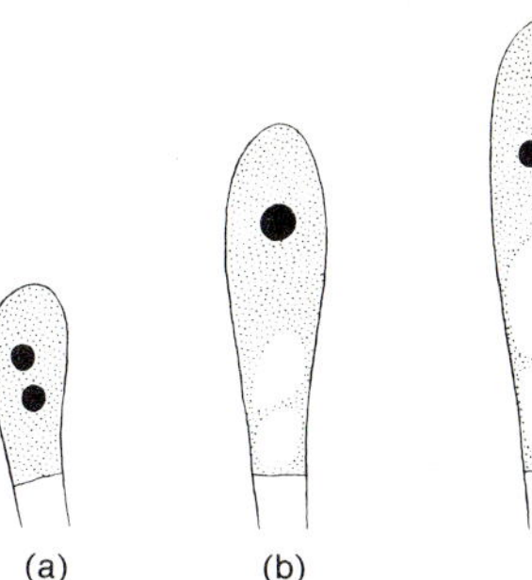
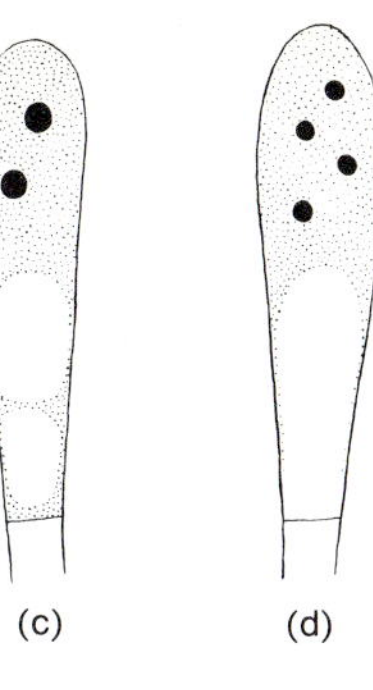
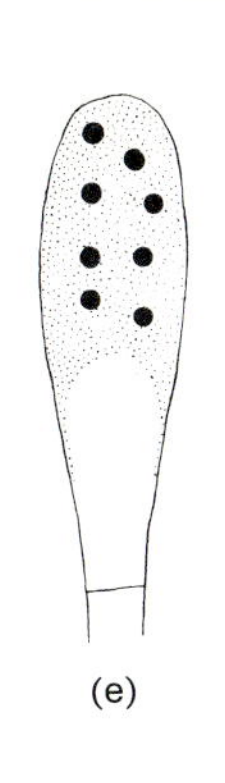
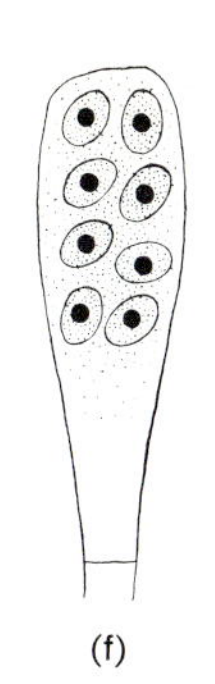


FIGURE 26.18

Successive stages in the formation of an ascus: (a) binucleate fusion cell; (b) nuclear fusion; (c), (d), (e), nuclear divisions; (f) ascospore formation.

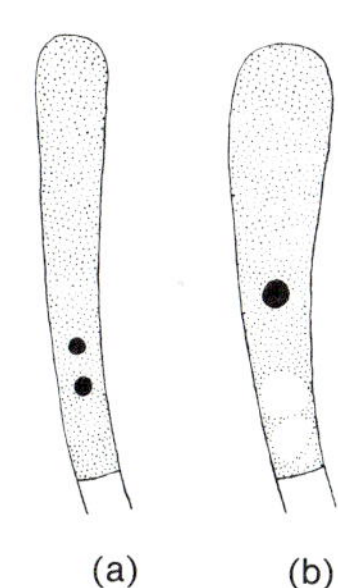

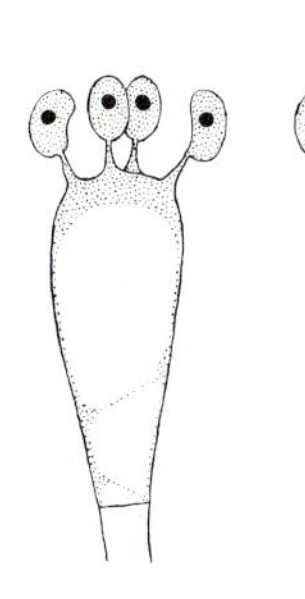
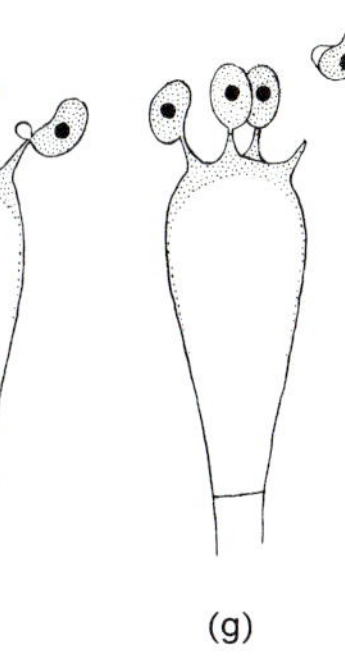


FIGURE 26.19

Successive stages in basidium formation and basidiospore discharge: (a) binucleate cell; (b) nuclear fusion; (c), (d) nuclear division; (e) formation of basidiospores; (f), (g) basidiospore discharge.

cete or basidiomycete will never undergo sexual reproduction, which requires the presence of another strain of opposite mating type. Accordingly, it has been necessary to create a third class, the *Fungi Imperfecti*, for those kinds in which a sexual stage has not so far been observed. It should be realized that the Fungi Imperfecti is essentially a provisional taxonomic group; from time to time the sexual stage is discovered in a fungus originally assigned to this group, and the organism in question is then transferred to either the Ascomycetes or the Basidiomycetes.

## The Yeasts

Among the Ascomycetes, Basidiomycetes, and Fungi Imperfecti, the characteristic vegetative structure is the coenocytic mycelium. Nonetheless, there are a few groups in these classes that have largely lost the mycelial habit of growth and have become unicellular. Such organisms are known collectively as *yeasts:* A typical yeast consists of small, oval cells that multiply by forming buds. The buds enlarge until they are almost equal in size to the mother cell, nuclear division occurs, and then a cross wall is formed between the two cells (Figure 26.20). Although the yeasts constitute a minor branch of the higher fungi in terms of number of species, they are very important microbiologically.

FIGURE 26.20

A sequence of photomicrographs of a budding cell of the ascomycetous yeast, *Wickerhamia,* showing nuclear division and transverse wall formation (phase contrast, × 1,770): n, nucleus; v, vacuole; tw, transverse wall. From P. Matile, H. Moore, and C. F. Robinow, in *The Yeasts*, Vol. 1, ed. A. N. Rose and J. S. Harrison (New York: Academic Press, 1969), p. 219.

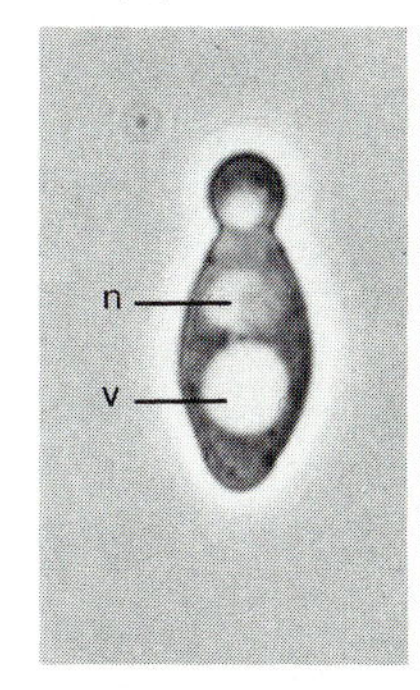

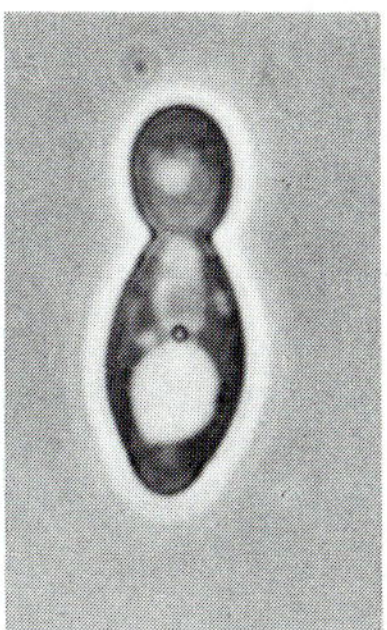
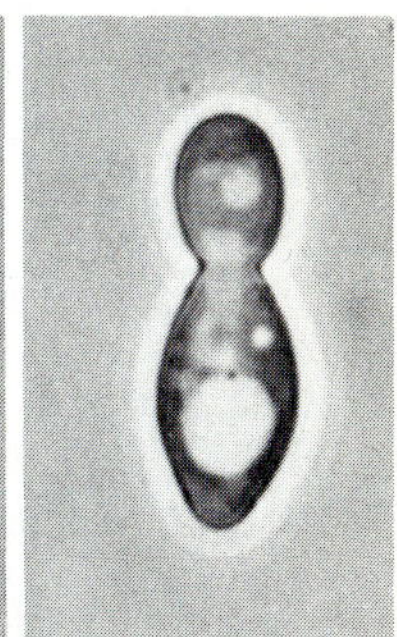
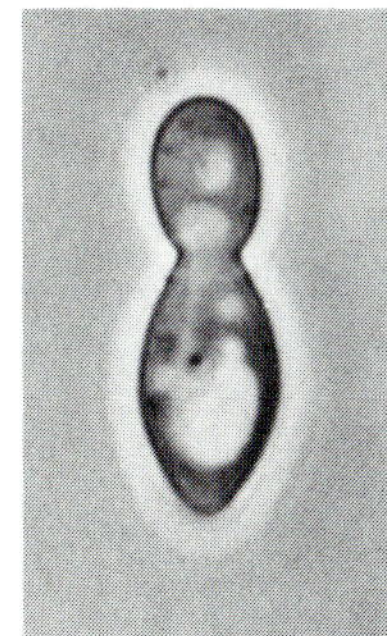
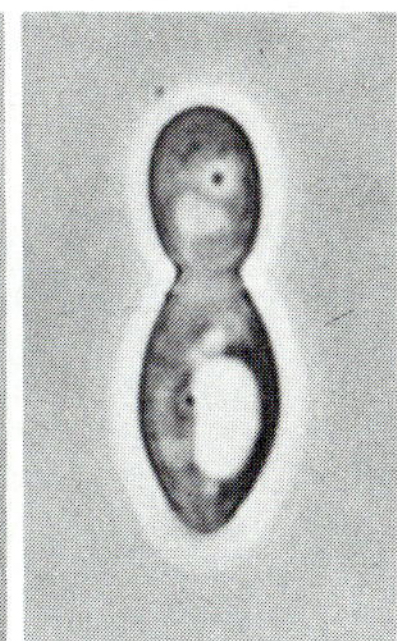
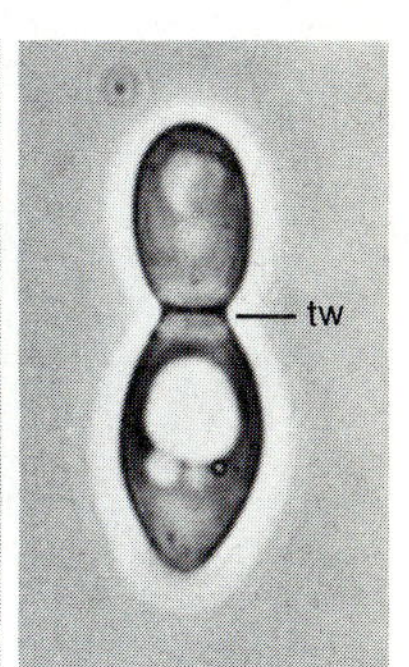

Most yeasts do not live in soil but have instead become adapted to environments with a high sugar content, such as the nectar of flowers and the surface of fruits. Many yeasts (the fermentative yeasts) perform an alcoholic fermentation of sugars and have been long exploited by man (see Chapter 33).

Yeasts are classified in all three classes of higher fungi: Ascomycetes, Basidiomycetes, and Fungi Imperfecti. The principal agent of alcoholic fermentation, *Saccharomyces cerevisiae*, is an ascomycetous yeast. Budding ceases at a certain stage of its growth, and the vegetative cells become transformed into asci, each containing four ascospores. For a long time it was believed that ascospore formation in *S. cerevisiae* was not preceded by zygote formation because pairing of vegetative cells prior to the formation of ascospores could never be observed. Eventually, however, it was discovered that zygote formation takes place at an unexpected stage of life cycle—immediately after the germination of the haploid ascospores. Pairs of germinating ascospores, or the first vegetative cells produced from them, fuse to form diploid vegetative cells. Diploidy is then maintained throughout the entire subsequent period of vegetative development, and meiosis occurs immediately prior to the formation of ascospores. Thus, *S. cerevisiae* exists predominantly in the diplophase. Other ascomycetous yeasts do not share this pattern of behavior but form zygotes by fusion between vegetative cells immediately before ascospore formation. The germinating ascospores then gives rise to haploid vegetative progeny.

Although budding is the predominant mode of multiplication in yeasts, there are a few that multiply by binary fission, much like bacteria; these are placed in a special genus, *Schizosaccharomyces*.

In ascomycetous yeasts, the vegetative cell or zygote becomes entirely transformed into an ascus at the time of ascospore formation. Yeasts of the genus *Sporobolomyces* form basidiospores, and in this case the entire vegetative cell becomes transformed into a basidium. Just as in the mushrooms, basidiospore discharge in *Sporobolomyces* is a violent process, and the colonies of this yeast are readily detectable on plates that have been incubated in an inverted position because the portion of the glass cover underlying a *Sporobolomyces* colony becomes covered with a desposit of discharged spores that form a mirror image of the colony (Figure 26.21).

## THE SLIME MOLDS

We conclude this survey of the protists by discussing the *slime molds*, which are not classified as true fungi, although they possess certain characteristics that resemble those of the fungi. The best-known representatives of the slime molds are the Myxomycetes, organisms that are found most commonly growing on decaying logs and stumps in damp woods. The vegetative structure, known as *plasmodium*, is a multinucleate mass of cytoplasm unbounded by rigid walls, which flows in ameboid fashion over the surface of the substrate, ingesting smaller microorganisms and fragments of decaying plant material. An actively moving plasmodium is characteristically fan-shaped, with thickened ridges of cytoplasm running back from the edge of the fan; it resembles a spreading layer of thin, colored slime (Figure 26.22). As long as conditions are favorable for vegetative development, the plasmodium continues to increase in bulk with accompanying repeated nuclear divisions. Eventually, the organism may become a mass of cytoplasm containing thousands of nuclei and weighing several hundred grams. Fruiting occurs when a plasmodium migrates to a relatively dry region of the substrate. Out of the undifferentiated plasmodium there is then produced a fruiting structure that is often of remarkable complexity and beauty [as illustrated by the case of *Ceratiomyxa* (Figure 26.23)]. As this fruiting body develops, small, uninucleate sections of the plasmodium become surrounded by walls to form large numbers of uninucleate spores, borne on the fruiting structure. After liberation, the spores germinate to produce uniflagellate *ameboid gametes* which fuse in pairs to form *biflagellate zygotes*. After some time, a zygote loses its flagella and develops into a new plasmodium. The vegetative nuclei in a growing plasmodium are diploid, meiosis taking place just prior to the formation of spores in the fruiting body.

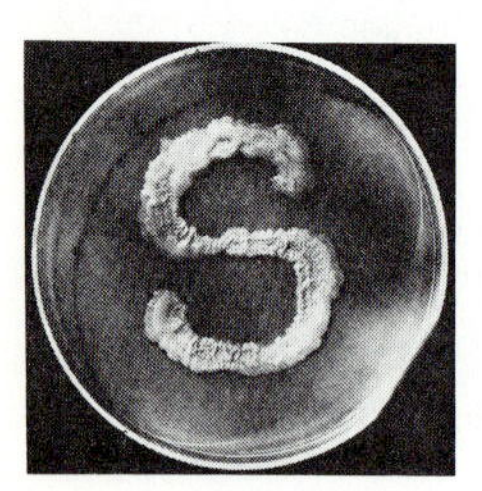

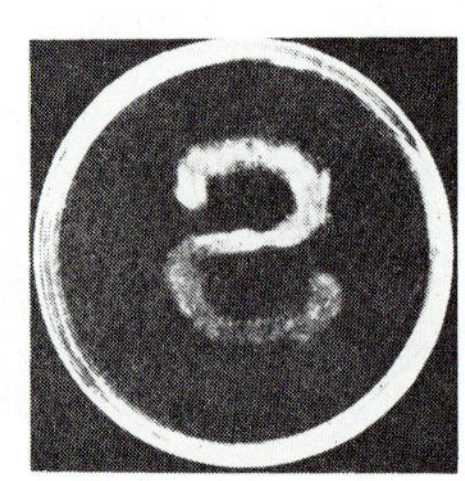

FIGURE 26.21

The formation of a mirror image of a colony of *Sporobolomyces* by basidiospore discharge in a petri dish incubated in the inverted position: (left) the colony on the agar surface, streaked in the form of an S; (right) the deposit of basidiospores formed on the lid of the petri dish as a result of spore discharge from the colony. From A. H. R. Buller, *Researches on Fungi*, Vol. 5 (New York: Longmans Green, 1933), p. 175.

FIGURE 26.22

The plasmodium of a myxomycete, *Didymium*, growing at the expense of bacteria on the surface of an agar plate. Courtesy of Dr. K. B. Raper.

FIGURE 26.23

Fruiting bodies of a myxomycete, *Ceratiomyxa*, on a piece of wood. From C. M. Wilson and I. K. Ross, "Meiosis in the Myxomycetes," *Am. J. Botany* **42,** 743 (1955).

It is, of course, the fruiting stage of a myxomycete that at once reminds one of a true fungus; at first sight, the amorphous, plasmodial vegetative stage appears to resemble little, if at all, the branched, mycelial vegetative stage of the fungi but suggests, rather, a relationship to the ameboid protozoa. In fact, the plasmodium and the mycelium are basically similar structures. Both are coenocytic, and in both the cytoplasm can flow, although in the mycelium cytoplasmic streaming is confined within the walls of branched tubes. The superficial difference between a plasmodium and a mycelium is essentially caused by the fact that in a plasmodium the cytoplasm is not bounded by rigid walls and is thus free to flow in any direction.

The slime molds also include a small group, the Acrasieae or *cellular slime molds* (Figure 26.24), which show far greater resemblances to the unicellular ameboid protozoa than do the true Myxomycetes. The vegetative stage of an acrasian consists of small, uninucleate amebae, which multiply by binary fission and can in no way be distinguished, at this stage of their life history, from other small ameboid protozoa. Nevertheless, when conditions are favorable, thousands of these isolated amebae are capable of aggregating and cooperating, without ever losing their cellular distinctness, in the construction of an elaborate fruiting body. The first sign of approaching fructification is the aggregation of the vegetative cells to form a macroscopically visible heap. This heap of cells gradually differentiates into a tall stalked structure that bears a rounded head of asexual spores. At all stages in the formation of this fruiting body, the cells remain separate; some individuals form the stalk, which is surrounded and given rigidity by a cellulose sheath, while others migrate up the outside of the rising stalk to form the spore head. As this matures, each ameba in it rounds up and becomes surrounded by

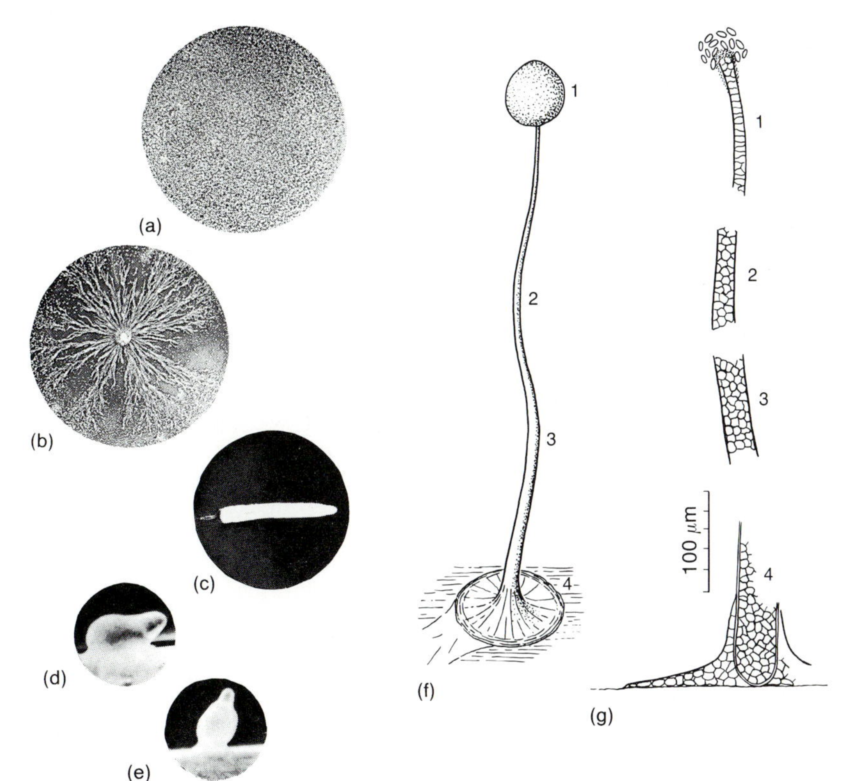


FIGURE 26.24

The life cycle of *Dictyostelium*, a representative of the Acrasieae: (a) a uniform mass of vegetative amebae; (b) aggregation of the amebae to a fruiting center; (c) motile mass of aggregated cells; (d), (e) early stages in formation of the fruiting body; (f) a mature fruiting body; (g) magnified sections through various regions of the fruiting body. From K. B. Raper, "Isolation, Cultivation, and Conservation of Simple Slime Molds," *Quart. Rev. Biol.* **26,** 169 (1951).

a wall. These spores, following their release, germinate and give rise to individual ameboid vegetative cells once more. This remarkable kind of life cycle, in which a communal process of fructification is imposed on a unicellular phase of vegetative development, occurs in one procaryotic group, the myxobacteria (Chapter 18).

## THE PROTISTS: SUMMING UP

It is not possible in one chapter to do justice to the extraordinary profusion and biological variety of the protists; only a few representatives of each major subgroup have been somewhat summarily described. For more detailed information about these organisms, the reader should consult specialized books dealing with the algae, the protozoa, or the fungi (see the bibliography at the end of the chapter). There is, unfortunately, no single book that provides a more extended survey of the entire biological group. Comprehension of the comparative biology of protists is further impeded by major terminological difficulties, because botanists and zoologists have applied entirely different names to structures common to all three subgroups. Moreover, the taxonomic treatments adopted for flagellate algae and leucophytes by zoologists and botanists are widely different. Here we have tried to bridge these differences and to provide a broader account than is customary of the protists, in terms of their possible evolutionary interrelationships.

## FURTHER READING


**Books**

*Algae:*

BOLD, H. C., and M. J. WYNNE, *Introduction to the Algae*, 2nd ed. Englewood Cliffs N.J.: Prentice-Hall, Inc., 1985.

STUART, W. D. P., ed., *Algal Physiology and Biochemistry*. Berkeley: University of California Press, 1974.

*Protozoa:*

GRELL, K. G., *Protozoology*. Heidelberg: Springer-Verlag, 1973.

LEVANDOWSKY, M. and S. H. HUTNER, eds., *Biochemistry and Physiology of Protozoa*. New York: Academic Press, 1979.

SLEIGH, M., *The Biology of Protozoa*. Amsterdam: Elsevier, 1973.

*Fungi:*

ALEXOPOULOS, C. J., and C. W. MIMS, *Introductory Mycology*, 3rd ed. New York: John Wiley, 1979.

BURNETT, J. H. *Fundamentals of Mycology*. New York: St. Martin's Press, 1968.

LOOMIS, W. F., *The Development of Dictyostelium discoideum*. New York: Academic Press, 1982.

# Chapter 27
# Microorganisms as Geochemical Agents

The current chemical state of the elements on the outer surface of the earth is, to a considerable extent, a consequence of the chemical activities of living organisms. This fact is dramatically illustrated by the changes that occurred in the earth's atmosphere as life evolved. Oxygen was probably completely missing from the prebiotic atmosphere, and only began to appear after oxygenic photosynthesis evolved. Then for some time it was present only in low concentrations set by the competing rates of its formation by oxygenic photosynthesis and its utilization by oxidation of reduced minerals and by microbial respiration. Only about 2.5 billion years ago was the present concentration of oxygen ($\sim$20 percent) in the atmosphere established. Oxygen, an essential nutrient for most species of eucaryotes is clearly a product of life rather than a prerequisite to its appearance.

Similarly, $N_2$, the major component of the earth's atmosphere, is a product of biological activity, in this case exclusively procaryotic. Essentially all atmospheric $N_2$ is the product of the anaerobic respiratory process termed denitrification (Chapter 4).

The quantity of the various gases in the atmosphere, and indeed, of many other compounds on the earth's surface represents the net balance between their rates of formation and utilization in biological and geological processes. Such transformation occurs in all regions of the earth that contain living organisms, collectively known as the *biosphere*. The oceans, freshwaters, land surface of the continents, and lower portion of the atmosphere comprise the biosphere. This thin film of life on the earth's surface exists in a more or less steady state, maintained by a cyclic turnover of the elements necessary for life, powered by a continuous input of energy from the sun. An exception is the hydrothermal vent communities (Chapter 16).

The various steps in the cyclic turnover of elements are brought about by different types of organisms. Thus, the continued existence of any particular group of organisms depends on the chemical transformation carried out by others. A break in the cycle at any point would dramatically affect all life. All the major elements necessary for life (carbon, oxygen, nitrogen, sulfur, and phosphorus) are transformed cyclically.

The cyclic nature of transformations in the biosphere can be summarized as follows. Through solar-energy conversion by photosynthesis, $CO_2$ and other inorganic compounds are withdrawn from the environment and are accumulated in the organic constituents of living organisms. The major producers of organic matter through photosynthesis are unicellular algae (principally diatoms and dinoflagellates) in the ocean and seed plants on land. Organic material thus accumulated provides, either directly or indirectly, the energy sources for all other forms of life.

Insofar as photosynthetic organisms serve as food sources for animals or microorganisms, the elements of major biological importance remain, at least in part, in the organic state, during the transformations which lead to their incorporation into the cells and tissues of the primary consumers of the products of photosynthesis. The primary consumers may themselves provide food sources for other organisms so that these elements may persist in organic food chains, made up of many types of nonphotosynthetic organisms. Before they can be again utilized by photosynthetic organisms they usually must be converted once more to inorganic form. This conversion, known as *mineralization*, is brought about largely by the decomposition of plant and animal remains and excretory products by microorganisms, principally fungi and bacteria. It is estimated that 90 percent of the mineralization of organic carbon atoms (i.e., their conversion to $CO_2$) is the result of the metabolic activities of these two groups of microorganisms. The remaining 10 percent results from the metabolism of all other organisms, as well as the combustion of fuels and other materials. The overwhelming contribution of microorganisms to this process reflects the ubiquity of microorganisms, their significant contribution to the total bulk of living material (their biomass), their high rates of growth and metabolism, and their collective ability to degrade a vast variety of naturally occurring organic materials.

## THE FITNESS OF MICROORGANISMS AS AGENTS OF GEOCHEMICAL CHANGE

### The Distribution of Microorganisms in Space and Time

The omnipresence of microorganisms throughout the biosphere is a consequence of their ready dissemination by wind and water. Surface waters, the floors of oceans over the continental shelves, and the top few inches of soil are teeming with microorganisms that are ready to decompose organic matter that may become available to them. It has been estimated that the top 6 in. of fertile soil may contain more than 2 tons of fungi and bacteria per acre. Any handful of soil contains many different kinds of microbes. Even on a single soil particle, the conditions may change from hour to hour and from facet to facet, presenting at different times microscopic ecological niches for different types to develop.

Upon the addition of plant or animal tissues to the soil, the organic compounds within are rapidly attacked by microorganisms that are capable of digesting and oxidizing these compounds. As oxygen is consumed, conditions may become anaerobic in the immediate proximity of the dead tissue and fermentative organisms develop. The products of fermentation then diffuse to regions in which oxygen is still present or they may be oxidized anaerobically by organisms capable of reducing nitrates, sulfates, or carbonates. Ultimately, the organic compounds will be completely converted to $CO_2$ or assimilated, the condition will again become fully aerobic, and autotrophs will develop at the expense of such reduced inorganic products as ammonia, sulfide, and hydrogen. Thus, the inorganic products of the decomposition of the plants or animals are eventually completely oxidized. This sequence of events, which occurs on a microscopic scale on a particle of soil, can be observed on a macroscopic scale in nature. When a tree falls into a swamp or a whale decomposes on a beach, the eventual chemical results are essentially the same. Seasonal and climatic conditions may retard or accelerate the cyclic turnover of matter. In cold climates, decomposition is most rapid in the early spring; in semiarid areas, it is largely restricted to the rainy season.

In nature, only those microorganisms that are favored by the local and temporary environment reproduce, and their growth ceases when they have altered their environment. Most of them are eventually consumed by such ever-present predators such as the protozoa, but a few cells of each type persist to initiate a new burst of growth when conditions again become favorable for their development.

### The Metabolic Potential of Microorganisms

The relatively enormous catalytic power of microorganisms contributes to the major role they play in the chemical transformations occurring on the earth's surface. Because of their small size, bacteria and fungi possess a large surface-volume ratio compared with higher animals and plants. This permits a rapid exchange of substrates and waste products between the cells and their environment.

Per gram of body weight, the respiratory rates of some aerobic bacteria are hundreds of times greater than that of humans. On the basis of the known metabolic rates of microorganisms, one can estimate that the metabolic potential of the microorganisms in the top 6 in. of an acre of well-fertilized soil at any given instant is equivalent to the metabolic potential of some tens of thousands of human beings.

An even more important factor influencing the chemical role that microorganisms play in nature is their high rate of reproduction in favorable environments.

### The Metabolic Versatility of Microorganisms

The remarkable ability of microorganisms to degrade a vast variety of organic compounds has led to a widely held conviction that has been termed the principle of microbial infallibility, a principle that was clearly stated by E. F. Gale in 1952: "It is probably not unscientific to suggest that somewhere or other some organism exists which can, under suitable conditions, oxidise any substance which is theoretically capable of being oxidised."* With the increasing production of plastics as well as synthetic insecticides, herbicides, and detergents, it has become clear that some substances are remarkably resistant to microbial attack; they persist and accumulate in nature. Even certain naturally occurring organic compounds are somewhat resistant; they accumulate and constitute the organic fraction of soil known as *humus* which confers the deep brown or black color to fertile soils. Because of the importance of humus to agriculture, this complex mixture of persistent organic compounds has been studied extensively. In large degree, it appears to consist of degradation products of a particularly stable component of woody plants known as *lignin*. The remarkable stability of humus has been demonstrated by radiocarbon dating; humus from certain soils is thousands of years old.

These exceptions aside, most organic compounds that are no longer a part of a living organism are rapidly mineralized by microorganisms in the biosphere.

Although some nonphotosynthetic microorganisms (e.g., the *Pseudomonas* group) can attack many different organic compounds, the metabolic versatility of the microbial world *as a whole* is not primarily a reflection of the metabolic versatility of its individual members. Any single bacterial species is only a limited agent of mineralization. Highly specialized physiological groups of microorganisms play important roles in the mineralization of specific classes of organic compounds. For example, the decomposition of cellulose, which is one the most abundant constituents of plant tissues, is mainly brought about by organisms that are highly specialized nutritionally. Among the aerobic bacteria capable of decomposing cellulose, the gliding bacteria that belong to the *Cytophaga* group are perhaps the most important. The cytophagas can rapidly dissolve and oxidize this insoluble compound, but cellulose is the only substance they can use as carbon source.

It will be recalled that the autotrophic bacteria, responsible for the oxidation of reduced inorganic compounds in nature, are also highly specific. Each type of autotroph is capable of oxidizing only one class of inorganic compounds and, in some cases (the nitrifying bacteria), only one compound.

## THE CYCLES OF MATTER

The turnover of the elements that compose living organisms constitute the *cycles of matter*. All organisms participate in various steps of these cyclic conversions, but the contribution of microorganisms is particularly important, both quantitatively (as discussed previously) and qualitatively. For example, certain steps in the nitrogen cycle are exclusively brought about by procaryotes.

## THE PHOSPHORUS CYCLE

Considered from a chemical point of view, the phosphorus cycle is simple, because phosphorus occurs in living organisms only in the +5 valence state,

* Gale, E. F., *The Chemical Activities of Bacteria*. New York: Academic Press, 1952.

either as free phosphate ions ($PO_4^{3-}$) or as organic phosphate constituents of the cells. Most organic phosphate compounds cannot be taken into the living cell; instead, phosphorus requirements are met by the uptake of phosphate ion. Organic phosphate compounds are then synthesized within the cell, and upon the death of the organism, phosphate ion is rapidly released by hydrolysis.

In spite of the rapid functioning of the phosphorus cycle and the relative abundance of phosphates in soils and rocks, phosphate is a limiting factor for the growth of many organisms because much of the earth's supply of phosphates occurs as insoluble calcium, iron, or aluminum salts. Freshwaters often contain phosphate ions in mere trace amounts, being available to animals only after it has been concentrated by the phytoplankton.

Soluble phosphates are constantly being transferred from terrestrial environments to the sea as a consequence of leaching, a transfer which is largely unidirectional. Only small quantities are returned to the land, principally by the deposits of guano by marine birds. Thus, the availability of phosphate for terrestrial forms of life depends on the continued solubilization of insoluble phosphate deposits, a process in which microorganisms play an important role. Their acidic metabolic products (organic, nitric and sulfuric acids) solubilize the phosphate of calcium phosphate, and their production of $H_2S$ dissolves ferric phosphates.

The steady state functioning of the phosphorus cycle has a geological component: phosphates that are leached into the oceans in soluble form are slowly returned to the land masses in insoluble form by geological transformation of ocean floors into continents.

## THE OXYGEN CYCLE

The cycling of oxygen between its two principal reservoirs, gaseous $O_2$ and water, is relatively uncomplicated. Gaseous oxygen is generated from water almost exclusively by oxygenic photosynthesis carried out by higher plants, algae, and cyanobacteria. The reverse molecular conversion is mediated by all organisms that carry out aerobic respiration, of which, as stated earlier, microorganisms are quantitatively the most important. Combustion of fossil fuels is, of course, an increasingly important route by which atmospheric oxygen is utilized.

The other principle product of respiration, $CO_2$, is a minor reservoir of oxygen but a critically important intermediate in cyclic interconversions of carbon-containing compounds.

## THE CARBON CYCLE

The concentration of $CO_2$ in the atmosphere, like the concentration of $O_2$, is largely set by the competing processes of photosynthesis and respiration, although other processes contribute (Figure 27.1). However, in striking contrast to the large amount of oxygen present in the atmosphere, the concentration of $CO_2$ is quite low, only about 0.03 percent by volume. Indeed, under favorable environmental conditions of light intensity and temperature, the rate of photosynthesis and therefore the rate of plant growth, is limited by the concentration of $CO_2$ available to the plant.

When $CO_2$ is dissolved in slightly alkaline water, bicarbonate ($HCO_3^-$) and carbonate ($CO_3^{2-}$) ions are formed:

$$CO_2 + OH^- \rightleftharpoons HCO_3^-$$
$$HCO_3^- + OH^- \rightleftharpoons H_2O + CO_3^{2-}$$

Therefore, bicarbonate serves as the reservoir of carbon for photosynthesis in aquatic environments. The bicarbonate concentration of ocean waters ($\sim$0.002 M) acts as a reservoir for $CO_2$ for the atmosphere; the oceans trap a large fraction of the $CO_2$ produced on land, keeping its atmospheric concentration at a relatively low and constant level.

The importance of the carbon cycle can best be emphasized by the estimate that the total $CO_2$ contained in the atmosphere. If it were not replenished, it would be completely exhausted in less than 20 years at the present rate of photosynthesis. This estimate does not appear too radical when it is realized that the carbon contained in a single giant redwood tree is equivalent to that present in the atmosphere over an area of approximately 40 acres. On land, seed plants are the principal agents of photosynthetic activity. A minor contribution is made by the algae. In the oceans, however, it is the unicellular photosynthetic organisms that play the most important role. The large plantlike algae (seaweeds) are confined in their development to a relatively narrow coastal strip. Since light of photosynthetically effective wavelengths is largely filtered out at a depth of about 50 ft, these sessile algae cannot grow in deeper waters. Because they are free-floating, the microscopic algae of the ocean (known as the *phytoplankton*) are capable of developing in the surface layers wherever the environment is favorable. Their growth is largely limited by the relative scarcity of two elements: phosphorus and nitrogen. Where these elements are made available as phosphates and nitrates by the runoff of rain water from continents and subsequent distribution by ocean currents, profuse development of

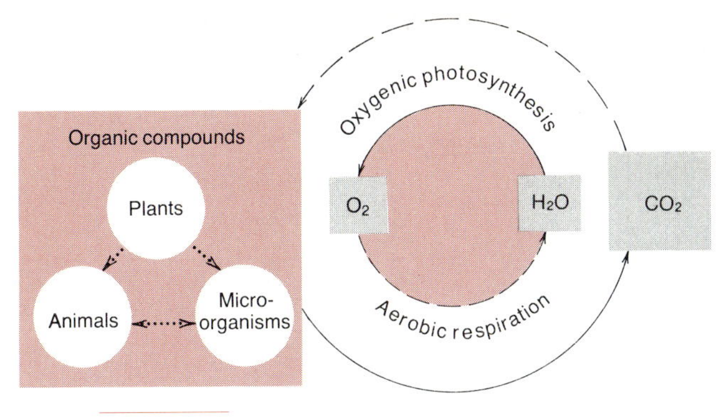


FIGURE 27.1
The carbon cycle. Oxidations are shown as solid arrows, reductions as broken arrows, and reactions with no valence change as dotted arrows.

phytoplankton occurs. According to one estimate the total annual fixation of carbon in the oceans amounts to approximately $1.2 \times 10^{10}$ tons, whereas that on the land is about $1.6 \times 10^{10}$ tons.

Although oxygenic photosynthesis is by far the most important means of reducing $CO_2$ to organic matter, other processes contribute to a small extent. These include photosynthesis by the purple and green bacteria, $CO_2$ reduction by the chemoautotrophs, and traces of $CO_2$ fixed in the metabolism of most organisms.

## The Mineralization Process: Carbon Dioxide Formation and the Reduction of Oxygen

The biological conversion of organic carbon to $CO_2$ with the concomitant reduction of molecular oxygen involves the combined metabolic activity of many different kinds of microorganisms. The complex constituents of dead cells must be digested, and the products of digestion must be oxidized by specialized organisms that can use them as nutrients. Many aerobic bacteria (pseudomonads, bacilli, actinomycetes), as well as fungi, carry out complete oxidations of organic substances derived from dead cells. However, it should be remembered that even those organisms that produce $CO_2$ as the only waste product of the respiratory decomposition of organic compounds usually use a large fraction of the substrate for the synthesis of their own cell material. In anaerobic environments, organic compounds are decomposed initially by fermentation, and the organic end products of fermentation are then further oxidized by anaerobic respiration, provided that suitable inorganic hydrogen acceptors (nitrate, sulfate, or $CO_2$) are present.

## The Sequestration of Carbon: Inorganic Deposits

The carbonate ions in the oceans combine with dissolved calcium ions and become precipitated as calcium carbonate. Calcium carbonate is also deposited biologically in the shells of protozoa, corals, and mollusks. This is the geological origin of the calcareous rock (limestone) that is an important constituent of the surface of continents. Calcareous rock is not directly available as a source of carbon for photosynthetic organisms, and hence its formation causes a depletion of the total carbon supply available for life. Nevertheless, much of this carbon eventually reenters the cycle through weathering. The formation and solubilization of calcium carbonate are brought about primarily by changes in hydrogen ion concentration, and microorganisms contribute indirectly to both processes as a consequence of pH changes that they produce in natural environments. For example, such microbial processes as sulfate reduction and denitrification cause an increase in the alkalinity of the environment, which favors the deposition of calcium carbonate in the ocean and other bodies of water. Microorganisms also play an important role in solubilizing calcareous deposits on land, similar to the role they play in solubilizing phosphates, by production of acid during nitrification, sulfur oxidation, and fermentation.

As a general principle, anaerobic environments tend to serve as sinks in which organic materials accumulate because fewer organic materials can be metabolized anaerobically than aerobically. But methanogenesis, the anaerobic production of a gaseous product, provides a major route by which organic material can escape from an anaerobic environment to an aerobic one where it can be metabolized further to $CO_2$ and $H_2O$. Sulfate-reducing bacteria also play an important role in oxidizing products of fermentation. The sorts of conversions that occur anaerobically in marine sediments are summarized in Figure 27.2.

## The Sequestration of Carbon: Organic Deposits

A high moisture content, causing oxygen depletion and the accumulation of acidic substances, is sometimes particularly favorable for the accumulation of humus. This phenomenon is most pronounced in *peat bogs*, where, in the course of time, deposits of undecomposed organic matter known as *peat* accumulate. These deposits may extend for hundreds of feet below the surface of the bog. In the course of geological time the compression of peat

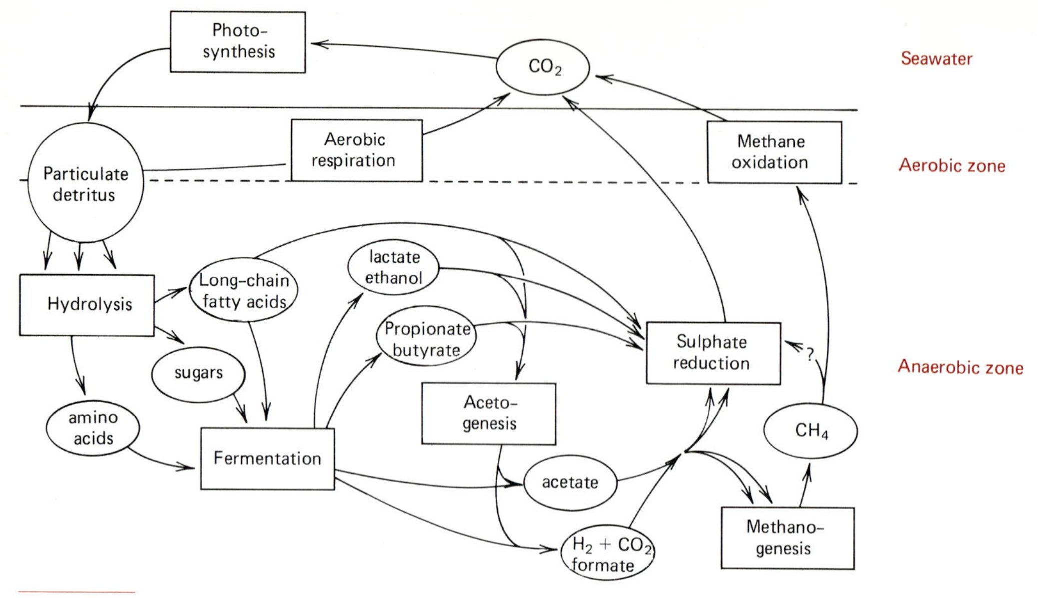


FIGURE 27.2

Degradation and cycling of organic matter in sediments in relation to bacterial sulphate reduction and methanogenesis. After T. H. Blackburn, "The Microbial Nitrogen Cycle," in Krumbein, W. E., ed., *Microbial Geochemistry*, Boston: Blackwell Publications (1983).

deposits, probably aided by other physical and chemical factors, has resulted in the formation of coal. Much carbon has thus been sequestered from the biosphere in the form of peat and coal deposits. A second kind of sequestration of carbon in organic form has occurred in deposits of natural oil and gas (methane).

Since the Industrial Revolution, human exploitation of the stored deposits of organic carbon in the earth's crust has resulted in their very rapid mineralization. Although substantial deposits still remain to be exploited, it is estimated that at current rates of consumption most of the petroleum and natural gas will be used up within a few decades.

A consequence of this very rapid burning of fossil fuels has been an increase in the rate of production of $CO_2$ over the rate at which it is utilized in biological fixations.* Over the past 100 years the net increase of $CO_2$ in the atmosphere has been about 15 percent. Although this increase is relatively small, if it were to continue, its impact could be profound because atmospheric $CO_2$ tends to prevent the loss of radiant energy from the earth, thus causing its average temperature to increase, perhaps to dangerous levels ("the greenhouse effect"). The consequences of continued rapid burning of fossil fuels are potentially serious but they are by no means certain. The functioning of the components of the carbon cycle might be, to a certain extent, self-adjusting: only a 1 percent increase in photosynthesis is needed to balance fossil fuel burning, wood cutting, and forest fires.

* Fixation is a process in which a gaseous compound is converted to a nongaseous one.

## THE NITROGEN CYCLE

Although molecular nitrogen ($N_2$) is abundant, constituting about 80 percent of the earth's atmosphere, it is chemically inert and therefore not a suitable source of the element for most living forms. All plants and animals, as well as most microorganisms, depend on a source of combined, or fixed, nitrogen in their nutrition. Combined nitrogen in the form of ammonia, nitrate, and organic compounds is relatively scarce in soil and water, often constituting the limiting factor for the development of living organisms. For this reason, the cyclic transformation of nitrogenous compounds is of paramount importance in supplying required forms of nitrogen to the various nutritional classes of organisms in the biosphere. The main features of the nitrogen cycle are illustrated schematically in Figure 27.3.

## Nitrogen Fixation

The turnover of nitrogen through its cycle is estimated to be between $10^8$ and $10^9$ tons per year. The vast supply of nitrogen gas ($N_2$) in the atmosphere and the relative scarcity of combined nitrogen on the earth's surface suggest that the process of nitrogen fixation is the rate limiting step. This process is largely a biological one, and bacteria are the only organisms capable of causing it (see Chapter 5). Some nitrogen is fixed by lightning, ultraviolet light, electrical equipment, and the internal combustion engine, but these nonbiological processes are quantitatively insignificant, together accounting for only about 0.5 percent of nitrogen fixation.

The most important agents of nonsymbiotic nitrogen fixation are heterocyst-forming blue-green bacteria such as *Anabaena* and *Nostoc*. A wide variety of other bacteria are also capable of fixing nitrogen; these include both aerobic bacteria (the *Azotobacter* group, *Azospirillum*, and *Bacillus polymyxa*) and anaerobic bacteria (photosynthetic bacteria and *Clostridium* spp.).

Even industrial manufacture of fertilizer by the Haber process contributes less than biological fixation. Thus, the majority of all fixed nitrogen on the earth is the product of the metabolic activities of certain bacteria (Table 27.1).

**TABLE 27.1**

**Types of Nitrogen-Fixing Bacteria**

| | Phototrophic | Chemotrophic |
|---|---|---|
| Free-living, aerobic | Cyanobacteria | *Azotobacter* Group<br>*Mycobacterium*<br>Methane oxidizers<br>*Thiobacillus* |
| Free-living, anaerobic | Cyanobacteria<br>Purple bacteria<br>Green bacteria | *Clostridium*<br>*Klebsiella*<br>*Bacillus*<br>*Desulfovibrio*<br>*Desulfotomaculum*<br>Methanogenic bacteria |
| Symbiotic, aerobic | Cyanobacteria (+fungi, ferns) | *Rhizobium* (+legumes, grass)<br>*Azospirillum* (+grass)<br>*Frankia* (+alder, hawthorn, etc.) |
| Symbiotic, anaerobic | None known | *Citrobacter* (+termites) |

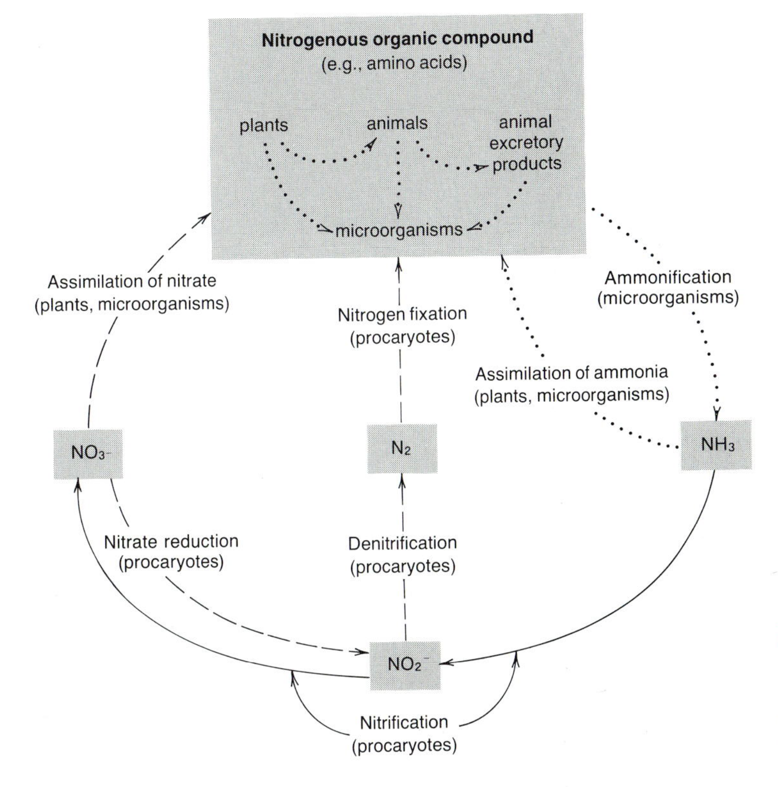


FIGURE 27.3

The nitrogen cycle. The oxidations of nitrogen are shown as solid arrows, the reductions as broken arrows, and reactions with no valence change as dotted arrows.

**TABLE 27.2**

**Efficiencies of Some Nitrogen-Fixing Systems**

| Organism or System | Pounds of N Fixed/Acre/year[a] |
|---|---|
| SYMBIOTIC | |
| Alfalfa: *Rhizobium* | >264 |
| Clover: *Rhizobium* | 220 |
| Lupin: *Rhizobium* | 132 |
| NONSYMBIOTIC | |
| Cyanobacteria | 22 |
| *Azotobacter* spp. | 0.26 |
| *Clostridium pasteurianum* | 0.22 |

[a] Recalculated from E. N. Mishustin and V. K. Shil'nikova, *Biological Fixation of Atmospheric Nitrogen* (London: Macmillan, 1971).

Biological nitrogen fixation is mediated in part by free-living bacteria (*nonsymbiotic nitrogen fixation*), but the symbiotic fixers are quantitatively more important. The most thoroughly studied of the symbiotic fixers are representations of the genus *Rhizobium* because they form associations with agronomically important leguminous crops. Evaluated in terms of nitrogen fixed per acre of soil per year the contribution of legume-*Rhizobium* associations is far greater than that made by nonsymbiotic nitrogen fixers (Table 27.2).

Because of the critical agronomic importance of fixed nitrogen, the current world food crisis, and the fact that manufacture of nitrogen fertilizers by the Haber process requires large expenditures of energy, biological nitrogen fixation has become an intensive subject of investigation. Among the important goals of these investigations are the development of new plants capable of harboring nitrogen-fixing symbionts. Although the current range of such plants is wide, it does not include the world's major food crops, wheat and rice, nor the major forage crops.

## The Utilization of Fixed Nitrogen

The immediate product of nitrogen fixation, ammonia, is usually utilized directly by the fixing organism or its symbiont but in some cases, e.g., the *Azolla-Nostoc* symbiont (Chapter 17), greater amounts of ammonia are formed than are utilized, the excess being excreted into the environment. Regardless of whether fixed nitrogen is released directly or only after it has passed through plant and animal tissues and been reconverted to ammonia (*ammonification*), it enters the environment in the form of ammonium. In the soil ammonium is rapidly immobilized because it is tightly bound to clay particles. Only after it has been oxidized to nitrate ion (*nitrification*) does nitrogen become freely diffusable in the soil, and it is in this form that plants and many soil bacteria assimilate nitrogen.

Within the cell, nitrate again is reduced to ammonia. This process of nitrate reduction proceeds only to the extent to which nitrogen is required for growth; ammonia is not excreted. It is this feature in particular that distinguishes *nitrate assimilation* by plants (and also by microorganisms) from *nitrate reduction*, a process of anaerobic respiration which is limited to procaryotes (Figure 27.3).

## The Transformations of Organic Nitrogen by Which Ammonia Is Formed

The organic nitrogenous compounds synthesized by algae and plants serve as the nitrogen source for the animal kingdom. During their assimilation by animals, the complex nitrogenous compounds of plants are hydrolyzed to a greater or lesser extent, but the nitrogen remains largely in reduced organic form. Unlike plants, however, animals do excrete a significant quantity of nitrogenous compounds in the course of their metabolism. The form in which this nitrogen is excreted varies from one group of animals to another. Invertebrates predominantly excrete ammonia; but vertebrates principally excrete organic nitrogenous compounds. In reptiles and birds, uric acid is the major form in which nitrogen is excreted; in mammals, urea is the principal form. The urea and uric acid excreted by animals are rapidly mineralized by special groups of microorganisms, with the formation of $CO_2$ and ammonia.

Only part of the nitrogen stored in organic compounds through plant growth is converted to ammonia by animal metabolism and the microbial decomposition of urea and uric acid. Much of it remains in plant and animal tissues and is liberated only on the death of these organisms. Whenever a plant or animal dies, its body constituents are immediately attacked by microorganisms, and the nitrogenous compounds are decomposed with the liberation of ammonia. Part of the nitrogen is assimilated by the microorganisms themselves and thus converted into microbial cell constituents. Ultimately, these constituents are converted to ammonia following the death of the microbes.

The first step in this process of ammonification is the hydrolysis of the proteins and nucleic acids with the liberation of amino acids and organic nitrogenous bases, respectively. These simpler compounds are then attacked by respiration or fermentation.

Protein decomposition under anaerobic conditions (*putrefaction*) usually does not lead to an immediate liberation of all the amino nitrogen as ammonia. Instead, some of the amino acids are converted to amines. The putrefactive decomposition is characteristically brought about by anaerobic spore-forming bacteria (genus *Clostridium*). In the presence of air the amines are oxidized by other bacteria with the liberation of ammonia.

### Nitrification

Through all the transformations that nitrogen undergoes from the time of its reductive assimilation by plants until its liberation as ammonia, the nitrogen atom remains in the reduced form. The conversion of ammonia to nitrate (*nitrification*) is brought about in nature by two highly specialized groups of obligately aerobic chemoautotrophic bacteria in which, respectively, nitrification occurs in two steps. In the first step, ammonia is oxidized to nitrite; in the second, nitrite is oxidized to nitrate. As a result of the combined activities of these bacteria, the ammonia liberated during the mineralization of organic matter is rapidly oxidized to nitrate. Thus nitrate is the principal nitrogenous material available in soil for the growth of plants. The practice of soil fertilization with manure depends on the microbial mineralization of organic matter and results in the conversion of organic nitrogen to nitrate through ammonification and nitrification. Injection into the soil of liquified anhydrous ammonia, which is one of the modern methods used for fertilization, is an even more direct means by which the nitrate content of soil is increased. Ammonia, which can be synthesized chemically from molecular nitrogen, is the most concentrated form of combined nitrogen available because it contains about 82 percent nitrogen by weight. Nitrates are very soluble compounds and are therefore easily leached from the soil and transported by water; hence, a certain amount of combined nitrogen is constantly removed from the continents and carried down to the oceans. In some special localities, notably in the semiarid regions of Chile, deposits of nitrate have accumulated in the soil as a result of the runoff and evaporation of surface water. Such deposits are a valuable source of fertilizer, although their importance has diminished greatly in the course of the last 50 years as a result of the development of chemical methods for making nitrogen compounds from atmospheric nitrogen.

Nitrates have played an important role not only in the development of agriculture but also in the destructive activities of humans. Gunpowder, which was the only explosive used for war before the invention of nitroglycerine (dynamite), is a mixture of sulfur, carbon, and saltpeter ($KNO_3$). During the Napoleonic wars, largely as a result of the British blockade, a shortage of nitrate for gunpowder production occurred in France. This led to the development of "nitrate gardens," in which nitrate was obtained by the mineralization of organic matter. A mixture of manure and soil was spread on the surface of the ground and frequently turned to permit aeration. After the manure had decomposed, nitrate was extracted from the residue.

### Denitrification

Many aerobic bacteria can use nitrate in place of oxygen as a final electron acceptor if conditions are anaerobic.

Thus, whenever organic matter is decomposed in soil or water and oxygen is exhausted as a result of aerobic microbial respiration, certain of these aerobes will continue to respire the organic matter if nitrate is present, i.e., by anaerobic respiration. As a consequence, nitrate is reduced. Some bacteria (e.g., *Escherichia coli*) are only able to reduce nitrate to the level of nitrite but a variety of other bacteria (Table 27.3) are able to mediate a cascade of two subsequent anaerobic respirations by which nitrite ion is reduced to nitrous oxide gas ($N_2O$) and subsequently to dinitrogen gas ($N_2$). By this process, termed *denitrification*, combined nitrogen is removed from soil and water, releasing $N_2$ gas to the atmosphere.

Denitrification is a process of major ecological importance. It depletes the soil of an essential nutrient for plants, thereby decreasing agricultural productivity. Such losses are particularly important from fertilized soils. Although precise values are not available, under certain conditions a large amount of fixed nitrogen fertilizer may be lost through denitrification; in rare cases this may approach 80 percent.

Nevertheless, not all the consequences of denitrification are detrimental. Denitrification is vital to the continued availability of combined nitrogen on the land masses of the earth. The highly soluble nitrate ion is constantly leached from the soil, and it is eventually carried to the oceans. Without denitrification, the earth's supply of nitrogen, including $N_2$ of the atmosphere, would eventually accumulate in the oceans, precluding life on the

**TABLE 27.3**

**Genera of Bacteria Containing Representatives Capable of Mediating Denitrification**

A. Photosynthetic Eubacteria
*Rhodobacter*

B. Chemoautotrophic Eubacteria
*Thiobacillus, Thiomicrospira, Thermothrix*

C. Methophilic Eubacteria
*Hyphomicrobium*

D. Gram-Negative Respiratory Eubacteria
*Agrobacterium, Alcaligenes, Aquaspirillum, Branhamella, Campylobacter, Chromobacterium, Flavobacterium, Gluconobacter, Kingella, Neisseria, Paracoccus, Pseudomonas, Rhizobium, Spirillum*

E. Gliding Eubacteria
*Cytophaga*

F. Gram-Positive Eubacteria: Endospore Formers
*Bacillus*

G. Gram-Positive Eubacteria: The Actinomycete Group
*Corynebacterium*

H. Gram-Positive Anaerobic Eubacteria
*Propionibacterium*

After Jeter, R. M. and J. L. Ingraham, "The Denitrifying Prokaryotes" in Starr, M. P., H. Stolp, H. G. Truper, A. Ballows and H. G. Schlegel, *The Prokaryotes*. Berlin, Heidelberg, New York: Springer-Verlag, 1981.

land masses except for a fringe near the oceans. Denitrification also maintains the potability of freshwaters, because high concentrations of nitrate ions may be toxic.

Some bacteria, notably certain clostridia, reduce nitrite to ammonium ion by a process that is not linked to an electron transport chain (and hence is not an anaerobic respiration). Moreover this process yields quantities of ammonia far in excess of the amount required for growth (and hence is not an assimilatory reduction of nitrite). The value of this reduction to the anaerobic organisms that mediate it derives from the diversion of electrons from NADH to nitrite, rather than to an organic compound. As a consequence, the organic products of fermentation are more oxidized and the yield of ATP via substrate-level phosphorylation is increased. The ecological impact of the reduction of nitrite ion (derived from nitrate ion) to ammonia is considerable: the process is competitive with denitrification, but it does not deplete soil and water of their complement of nitrogen. The process is quantitatively important: the majority of nitrate added to most soils is reduced to ammonia by fermentative bacteria rather than to $N_2$ by denitrifiers.

## THE SULFUR CYCLE

Sulfur, an essential constituent of living matter, is the tenth most abundant element in the earth's crust. It is available to living organisms principally in the form of soluble sulfates or reduced organic sulfur compounds. Reduced sulfur in the form of $H_2S$ also occurs in the biosphere as a result of microbial metabolism and, to a limited extent, of volcanic activity. Except under anaerobic conditions, however, its concentration is low because it is oxidized rapidly in the presence of oxygen, either spontaneously or by bacteria. However, even from anaerobic environments the amount of $H_2S$ released into the atmosphere is relatively low, because $S^{2-}$ is rapidly immobilized in the form of an insoluble FeS precipitate as is evident from the black coloration of anaerobic sediments.

The turnover of sulfur compounds is referred to as the *sulfur cycle*. The biological aspects of this cycle are shown in Figure 27.4. In certain respects, it resembles the nitrogen cycle already described.

In addition to the biological sulfur cycle, important nonbiological transformations of gaseous forms of sulfur occur in the earth's atmosphere. It is estimated that some 90 million tons of sulfur in the form of biologically generated $H_2S$ are released to the atmosphere annually; an additional 50 million tons are contributed in the form of $SO_2$ by the burning of fossil fuels; and about 0.7 million tons in the form of $H_2S$ and $SO_2$ come from the earth's volcanic activity. In the atmosphere, $H_2S$ is rapidly oxidized by atomic oxygen (O), molecular oxygen ($O_2$), or ozone ($O_3$) to $SO_2$ which may dissolve in water to form sulfurous acid ($H_2SO_3$), or be oxidized by a second and slower series of reactions (requiring hours or days) to $SO_3$. When dissolved in water, $SO_3$ becomes sulfuric acid ($H_2SO_4$). Some sulfuric acid is neutralized by the small quantities of ammonia in the atmosphere, but much of it returns along with unoxidized $H_2SO_3$ to the earth's surface in acid form where it causes considerable damage to stone structures and sculptures and to unbuffered aquatic ecosystems. The rate of generation of acidic sulfur compounds increases as more fossil fuels are being burned. The problem is particularly acute in areas of high population density and even now it is causing the rapid destruction of much stone sculpture, and even the killing of the flora and fauna of lakes owing to a dramatic increase in their acidity.

### The Assimilation of Sulfate

Sulfate is almost universally used as a nutrient by plants and microorganisms. The assimilation of sulfate resembles the assimilation of nitrate. Like the nitrogen atom of nitrate, the sulfur atom of sulfate must become reduced in order to be incorporated into organic compounds, because in living organisms, sulfur occurs primarily in reduced form as —SH or —S—S— groups, but other linkage groups such as the ester sulfates ($C—O—SO_3^-$) and sulfonates ($C—SO_3^-$) also occur. In adenylphosphosulfate, an intermediate in the pathway by which sulfate is reduced to $S^{2-}$ (Chapter 5), sulfur occurs as a phosphoester ($P—O—SO_3^-$).

### The Transformation of Organic Sulfur Compounds and Formation of $H_2S$

When sulfur-containing organic compounds are mineralized, sulfur is liberated in the reduced inorganic form as $H_2S$. The latter process resembles ammonification, in which nitrogen is liberated from organic matter in its reduced inorganic form as ammonia.

### The Direct Formation of $H_2S$ from Sulfate

The utilization of sulfate for the synthesis of sulfur-containing cell constituents and the subsequent decomposition of these compounds results in an overall reduction of sulfate to $H_2S$. $H_2S$ is also formed more directly from sulfate through the activity of the sulfate-reducing bacteria. These obligately anaerobic bacteria oxidize organic compounds and molecular hydrogen by using sulfate as an oxidizing agent. Their role in the sulfur cycle may therefore be compared to the role of the nitrate-reducing bacteria in the nitrogen cycle. The activity of the sulfate-reducing bacteria is particularly apparent in the mud at the bottom of ponds and streams, in bogs, and along the seashore. Since seawater contains a relatively high concentration of sulfate, sulfate reduction is an important factor in the mineralization of organic matter on the shallow ocean floors. Signs of the process are the odor of $H_2S$ and the pitch-black color of the mud in which it occurs. The color of black mud is caused by the accumulation of ferrous sulfide. Some

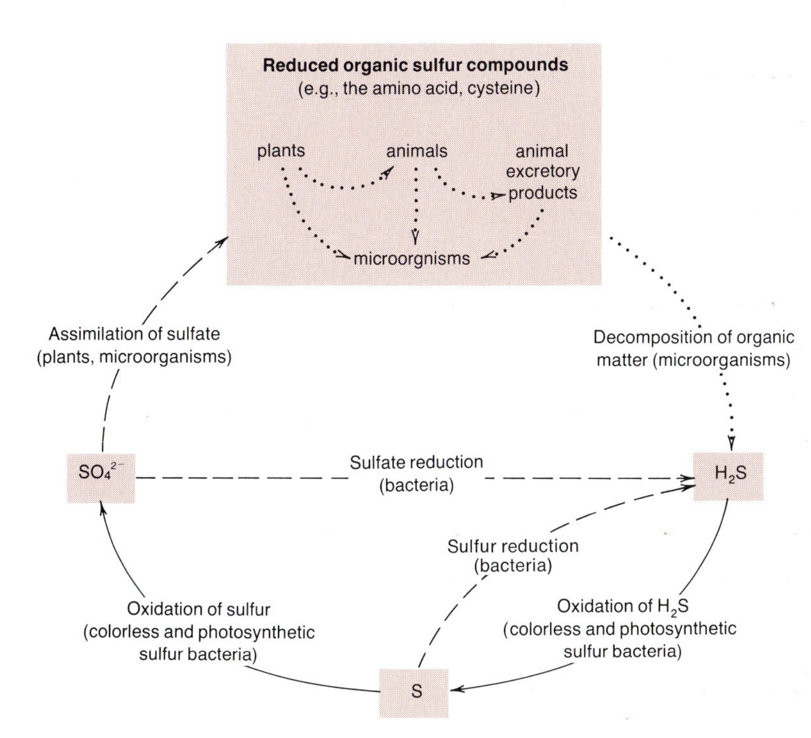


FIGURE 27.4
The sulfur cycle: the oxidations of the sulfur atom are shown as solid arrows, the reductions as broken arrows, and reactions with no valence change as dotted arrows.

coastal areas, where an accumulation of organic matter leads to a particularly massive reduction of sulfate, are practically uninhabitable because of the odor and the toxic effects of $H_2S$.

### The Oxidation of $H_2S$ and Sulfur

The $H_2S$ that is produced in the biosphere as a result of the decomposition of sulfur-containing compounds, of sulfate reduction, and of volcanic activity is largely converted to sulfate. Only a small part of it becomes sequestered in the form of insoluble sulfides or, after spontaneous oxidation with oxygen, as elemental sulfur.

The biological oxidation of $H_2S$ and of elemental sulfur is brought about by photosynthetic and chemoautotrophic bacteria. It can be effected either aerobically by the colorless sulfur bacteria or anaerobically by the photosynthetic purple and green sulfur bacteria. Since these oxidations result in the production of hydrogen ions, they result in the local acidification of soils. Elemental sulfur is commonly added to alkaline soil to increase its acidity.

## THE CYCLE OF MATTER THROUGH GEOLOGICAL TIME

The integration of the various reactions that constitute the cycle of matter results in a balanced production and consumption of the biologically important elements in the biosphere. Probably the cycles of matter as we know them today have operated without significant change for at least 2.5 billion years. However, there are good reasons to believe that the cycle of matter was considerably different at an early period in the history of the earth, when biological systems first developed on the planet.

As discussed at the beginning of this chapter, when the earth was formed about 4.8 billion years ago, the elements which are the principal constituents of living organisms were probably present on the primitive earth in their reduced forms. Molecular oxygen was absent from the atmosphere. It is now generally believed that the emergence of living systems was preceded by a period of chemical organic synthesis involving reactions between the reduced components of the atmosphere catalyzed by ultraviolet light and lightning. The products of these reactions accumulated and underwent further reactions in the primitive oceans. These reactions have been reproduced in the laboratory. If the presumed components of the primitive earth's atmosphere (methane, ammonia, and water) are exposed to irradiation by ultraviolet light or to electrical discharge, a vast array of organic compounds are formed. Among the organic compounds are the constituents of living matter, including sugars, amino acids, purines, and pyrimidines. If phosphate is present in the mixture, nucleotides (including ATP) are formed. If $H_2S$ is present, sulfur-containing amino acids are formed. Under the reducing conditions that prevailed and in the absence of living organisms, these compounds would have accumulated on the primitive earth.

The formation of ever more complex organic compounds and molecular aggregates as a result of chemical interactions eventually resulted in the development of self-duplicating systems, and biological evolution began its course. The first living systems probably had very limited synthetic abilities and depended on fermentative reactions for the generation of energy. With their growth and expansion the preexisting store of organic raw materials gradually became depleted, favoring the emergence of organisms with an increasing degree of synthetic ability. At a relatively early stage in this primary biochemical evolution, the supply of energy-rich organic compounds must have become limiting, and the further course of biological evolution therefore depended on the acquisition by some organisms of the ability to use light as an energy source. The development of a mechanism for the performance of photosynthesis was therefore one of the earliest and most important steps in biochemical evolution. The first photosynthetic organisms must have been anaerobes, with modes of photosynthetic metabolism analogous to those of the contemporary purple and green bacteria.

The early evolution of photosynthetic organisms culminated in the emergence of forms such as the cyanobacteria that were able to use water as reductant for the photosynthetic assimilation of $CO_2$. Once this point had been attained, the oxidized product of water, molecular oxygen, began to accumulate in the atmosphere, creating the conditions necessary for the evolution of organisms that obtain energy by aerobic respiration. As a consequence of the presence of molecular oxygen, the oxidized forms of nitrogen and sulfur (nitrate and sulfate) became predominant in the biosphere, and the stage was at last set for the establishment of the cycle of matter as we know it today.

## THE INFLUENCE OF HUMANS ON THE CYCLES OF MATTER

The emergence of humans as members of the biological community did not at first significantly affect the cycle of matter on earth. However, the rapid increases in the total size and local density of human populations that have occurred since the Industrial Revolution, coupled with the ever-increasing power of the human species to modify its environment, have begun to change the picture. Within the past century these factors have led to local environmental changes comparable in scale to those produced by major geological upheavals in the past history of the earth. The spread of agriculture, the denudation of forests, the mining and burning of fossil fuels, and the pollution of the environment with human and industrial wastes have profoundly affected the distribution and growth of other forms of life.

### Sewage Treatment

As a result of the concentration of the human population in large cities, which has proceeded at an ever-increasing pace during the past 150 years, the disposal of organic wastes, both domestic and industrial, has become a major ecological problem. The discharge of untreated urban wastes into adjoining rivers and lakes presents two hazards: the contamination of drinking water by microbial agents of enteric diseases and the depletion of the dissolved oxygen supply as a result of the microbial decomposition of organic matter, leading to the destruction of animal life. For reasons of public health and of conservation, society has been forced to develop methods of *sewage treatment*, which result in the destruction of pathogens and the mineralization of the organic components of sewage prior to its discharge.

In a typical sewage treatment plant, the sewage is first allowed to settle. The precipitate, or *sludge*, undergoes a slow anaerobic decomposition, in which methanogenic bacteria play an important part. The soluble organic compounds in the supernatant liquid are mineralized under aerobic conditions. This is sometimes achieved by spraying the liquid on a bed of loosely packed rocks, over which it trickles by gravity. In the *activated sludge process*, air is forced through the sewage, and a floc or precipitate is formed, the particles of which contain actively oxidizing microorganisms. After the organic matter has been largely oxidized, the sludge is allowed to settle. The sludges produced both by anaerobic decomposition and in the activated sludge process consist largely of bacteria, which have grown at the expense of nutrients in the sewage. These residues are eventually dried and used as fertilizer, either directly or after being ashed.

In all processes of sewage treatment the goal is to produce a final liquid effluent in which the biologically important elements have been restored to the inorganic state. Sewage treatment accordingly involves intensive operation of a substantial segment of the cycles of matter under more or less controlled conditions. Even an effluent in which mineralization is complete may, however, produce undesirable ecological effects. If it is discharged into a lake, the consequent enrichment of lake water with nitrates and phosphates can cause an enormous increase in algal productivity, so that the water becomes colored and turbid at certain times of the year. If algal growth is sufficiently massive, the subsequent decomposition of algal organic matter may deplete the dissolved oxygen supply in the lake, with catastrophic effects on its animal life. Such progressive biological degradation of freshwater environments, first encountered in relatively small lakes near urban communities, then became serious in some of the Great Lakes, particularly Lake Erie. Recently this trend has been reversed.

The threat to inland waters presented by the discharge of mineralized sewage, along with the realization that water containing nitrate ions is dangerous to human health, has prompted the U.S. government to issue new water quality standards. Drinking water must contain not more than 10 mg of nitrate per liter.

### The Dissemination of Synthetic Organic Chemicals

In recent decades the chemical industry has produced an ever-increasing variety of synthetic organic chemicals that are being used on an ever-increasing scale as textiles, plastics, detergents, insecticides, herbicides, and fungicides. Some of these, such as textiles and plastics, are virtually completely resistant to microbial decomposition; in nature they tend to remain as permanent unsightly litter. The function of insecticides, herbicides, and fungicides requires their distribution in nature. Some of them are also remarkably resistant to microbial decomposition. Representative examples of their persistence in soil are shown in Table 27.4. In other

**TABLE 27.4**

**Persistence of Pesticides in Soil**

| Common Name | Chemical Name | Period of Persistence |
|---|---|---|
| INSECTICIDES | | |
| Aldrin | 1,2,3,4,10,10-hexachloro-1,4,4,5,8,8-hexahydro-endo-1,4-exo-5,8-dimethano-napthalene | >9 years |
| Chlordane | 1,2,4,5,6,7,8,8-octachloro 2,3,3a,4,7,7-hexahydro-4,7-methanoindene | >12 years |
| DDT | 2,2-*bis* (*p*-chlorophenyl)-1,1,1-trichlorethane | 10 years |
| HCH | 1,2,3,4,5,6-hexachloro-cyclohexane | >11 years |
| HERBICIDES | | |
| Monuron | 3-(*p*-chlorophenyl)-1,1-dimethylurea | 3 years |
| Simazene | 2-chloro-4,6-*bis*(ethylamino)-s-triazine | 2 years |
| FUNGICIDES | | |
| PCP | Pentachlorophenol | >5 years |
| Zineb | Zinc ethylene-1,2-*bis*-dithiocarbamate | >75 days |

environments, such as more anaerobic ones (for example, the bottoms of lakes) they persist for even longer periods.

Many of these compounds are toxic for forms of life other than those they are designed to control, and the long-term ecological effects of their dissemination are difficult (if not impossible) to predict, but it is already clear that their accumulation in nature presents very real hazards to many species.

It is now recognized as desirable that any synthetic organic compound widely disseminated in the natural environment should be susceptible to microbial decomposition. That this is the case comes from painful experience derived from the dissemination of a variety of synthetic organic compounds including detergents. During the 1950s alkylbenzene sulfonates (Figure 27.5) became major ingredients of household detergents. The side chains (R) of the compounds were branched aliphatic residues rendering the entire molecule remarkably refractory to microbial decomposition. Accordingly, they passed through sewage treatment plants largely unaltered, and subsequently contaminated supplies of potable water causing it to foam. During the early 1960s the manufacturing process was altered in order to synthesize benzene sulfates with linear aliphatic side chains (linear alkylbenzene sulfonates); since these are highly susceptible to microbial decomposition, the problem was largely solved.

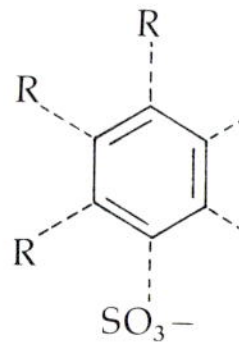

R R R R R $SO_3-$

FIGURE 27.5
Generalized structure of alkylbenzene sulfonates. The alkyl R group (either branched or linear) can be at any of the positions indicated on the dotted bonds.

## FURTHER READING

### Books


Atlas, R. M., and R. Bartha, *Microbial Ecology. Fundamentals and Applications.* Reading, Massachusetts: Addison-Wesley, 1981.

Delwiche, C. C. ed., *Denitrification, Nitrification and Atmospheric Nitrous Oxide.* New York: John Wiley, 1981.

Garrels, R. M., F. T. Mackenzie and C. Hunt, *Chemical Cycles and the Global Environment.* Los Altos, California: Wm. Kaufman, 1975.

Krumbein, W. E., ed., *Microbial Geochemistry.* Boston: Blackwell Publications, 1983.

Payne, W. J. *Denitrification.* New York: John Wiley, 1981.


### Reviews


Atlas, R. M., "Microbial Degradation of Petroleum Hydrocarbons: An Environmental Perspective," *Microbiol. Rev.* **45,** 180 (1981).

Revelle, R. "Carbon Dioxide and World Climate," *Sci.* Am. **247,** 35 (1982).

# Chapter 28
# Symbiosis

Each group of organisms has had to adapt itself during its evolution not only to the nonliving environment, but also to the other organisms that surrounded it. Adaptation to the environment sometimes involves the acquisition of special metabolic capacities that endow their possessor with the unique ability to occupy a particular physicochemical niche. The nitrifying bacteria, for example, can grow in a strictly inorganic environment with ammonia or nitrite as the oxidizable energy source; in the absence of light, no other living organisms are capable of developing in this particular environment, and the nitrifying bacteria are thus largely freed from biological competition. Withdrawal into a unique physiochemical niche is one means, and a highly effective one, of meeting the challenge of biological competition. A second method, however, which has been adopted by large numbers of microorganisms, has been to meet the challenge *by adapting to existence in continued close association with some other form of life*. This is the biological phenomenon known as *symbiosis*.

## TYPES OF SYMBIOSES

The symbiotic associations that microorganisms form with plants and animals, as well as with other microorganisms, vary widely in their degree of intimacy. In terms of the closeness of the associations, symbioses may be roughly divided into two categories: *ectosymbioses* and *endosymbioses*. In ectosymbioses the microorganism remains external to the cells of its host,* in endosymbioses the microorganism grows within the cells of its host. The distinction, however, is not always clear-cut; in lichens, for example, the fungal partner forms a projection that penetrates the cell wall, but not the cell membrane, of its algal partner.

Symbioses also differ with respect to the relative advantage accruing to each partner. In *mutualistic symbioses* both partners benefit from the association; in *parasitic symbioses* one partner benefits, but the second gains nothing and often suffers more or less severe damage. It is sometimes difficult to determine whether a given symbiosis is mutualistic or parasitic. The degree to which each partner is benefited or harmed can only be evaluated by comparing the fitness of the two members when living independently with their fitness when living in association. Furthermore, the nature of a particular symbiosis can shift under changing environmental conditions, so that a relationship that starts out as mutualistic may become parasitic, or vice versa.

The fact that two organisms have evolved a symbiosis implies that at least one partner derives some advantage from the relationship. The extent to which this partner depends on symbiosis for its existence, however, varies considerably. At one extreme are the microorganisms that populate the *rhizosphere*—the region that includes the surface of the roots, together with the soil immediately surrounding the root hairs, of higher plants. These microorganisms live successfully in other regions of the soil, but they attain higher cell densities in the rhizosphere, where they derive advantages from their proximity to the root hairs. At the other extreme are the obligate parasites, which have never been successfully cultivated outside their hosts.

Thus, symbioses vary with respect to the degree of intimacy (ectosymbiosis vs. endosymbiosis), the balance of advantage (mutualism vs. parasitism), and the extent of dependence (facultative vs. obligate symbiosis).

## Mutualistic Symbioses

In the following pages we shall describe a variety of symbioses in which microorganisms have established mutually advantageous associations with other microorganisms, with plants, and with animals. We shall confine our discussion to a few examples, to illustrate the varying types of relationships that may properly be regarded as mutualistic.

In mutualistic endosymbioses the microbial symbionts live within the cells of their hosts. In many such associations the microorganism leads a permanently intracellular existence and is passed from one generation of the host to the next in the cytoplasm of the egg. In other associations the microorganism remains intracellular through only a part of the life cycle of the host; at some stage it is liberated into the extracellular environment, from which the next generation of host becomes infected.

Among the mutualistic ectosymbioses we can discern two broad types of associations. First, there are the types in which the microbial symbiont lives on the external surface of its host. Some photosynthetic bacteria, for example, attach themselves to the surface of other, nonphotosynthetic bacteria (Figure 28.1) and certain flagellated protozoa, which inhabit the termite gut, bear on their surface a mantle of spirochetes (Figure 28.2).

Second, many microbial symbionts inhabit the body cavities of their hosts. These associations are still considered to be ectosymbioses, because

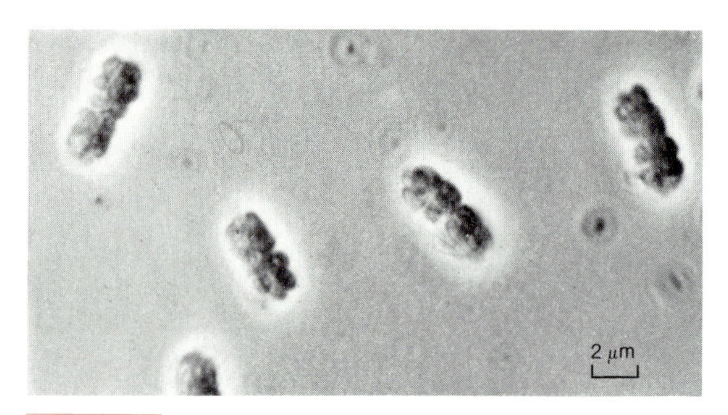


FIGURE 28.1
The ectosymbiosis of green bacteria with a larger, colorless bacterium (the so-called "chlorochromatium" association). Each object is a rod-shaped sulfur-reducing bacterium coated with the regularly arranged smaller cells of a *Chlorobium* (phase contrast, ×2,190). Courtesy of Norbert Pfennig.

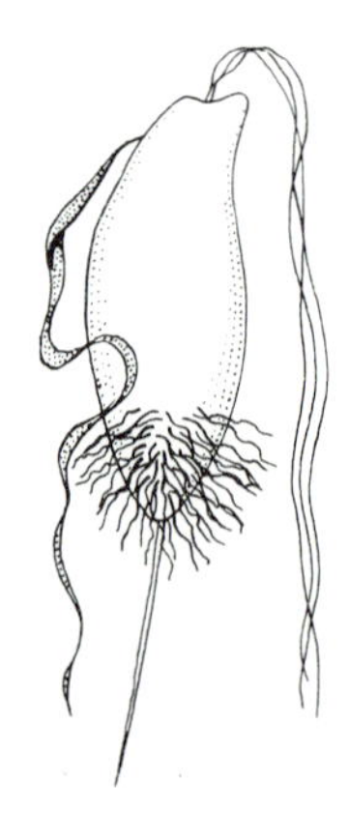

FIGURE 28.2
Spirochetes attached to posterior end of termite flagellate, *Glyptotermes* sp. From H. Kirby, Jr., *Univ. Calif. Publ. Zool.* **45,** 247 (1945).

* The term *host* refers to the larger of two symbionts.

the body cavities, although internal to the whole organism, are external to the tissues and are continuous with the external surfaces of the host. The most familiar examples are the microorganisms that inhabit the digestive tract of mammals; to the same category belong the luminous bacteria that populate the light-emitting organs of some fishes and mollusks (see page 449).

Of all the ectosymbioses, none is more remarkable than the cultivation of fungi by certain insects, notably the higher termite, *Termes*, and the wood-boring "ambrosia beetles." The termites have special chambers in their nests that contain "fungus gardens," on which the young nymphs and larvae browse. The ambrosia beetles have evolved elaborate methods of infecting their tunnels with fungal spores (Figure 28.3), so that the tunnels become lined with mycelium, on which the beetles feed. These practices closely parallel human cultivation of food plants, and in fact all these activities are symbiotic.

The picture of one symbiont literally devouring its partner may at first glance appear to contradict our concept of symbiosis as "existence in continued close association." When one of the partners is a microorganism, however, we are dealing with a *population*, not with an individual. Thus, the cultivated fungus garden of the termite benefits as a population, although a fraction of the population at any given moment is being devoured. In such ectosymbioses, as well as in all endosymbioses, the relationship is one of *reciprocal exploitation*, involving a dynamic balance between offensive and defensive activities of the two members.

## Parasitic Symbioses

We have defined the parasitic symbioses as those in which one of the partners does not profit from the association and may suffer more or less severe damage. In the case of microbial symbioses with plants and animals, it is usually a simple matter to demonstrate the advantage accruing to the microorganism, but it is often difficult or impossible to evaluate the effect on the host. *Infectious disease*, in which the host is progressively weakened and may eventually die, is obviously a parasitic symbiosis. When, however, there is no overt damage to the host, it is difficult to tell whether the relationship is a parasitic or a mutualistic one. For example, the "normal flora" of the mammalian intestinal tract was long assumed to be parasitic, only the microorganisms benefiting from the association. Work with *germ-free animals*, however, has revealed a number of subtle but important benefits which the intestinal flora confers on the host (see page 586).

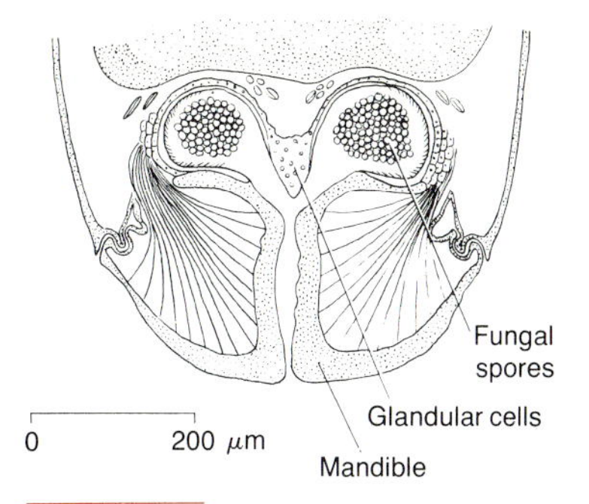


FIGURE 28.3
Section through the head of the ambrosia beetle, *Xyleborus monographus*, showing the special pockets for the storage of fungal spores. After Schedl.

## Parasitism as an Aspect of Ecology

The central problem of ecology is to discover the factors that determine the survival of a species. Other than the reproductive potential of a species, (i.e., the number of viable offspring produced per parent per unit of time), the factors that affect survival all fall into one of two categories: those that affect the available food supply and those that affect the rate of destruction of individuals. For any species other than humans, who have invented extraordinary ways of destroying themselves, the possible fates of an individual are restricted. Animals, other than a few domestic ones, rarely die of old age or from accidental mishaps. Most are either eaten by their natural predators or destroyed by their pathogenic parasites. The distinction between a predator and a pathogenic parasite is a fine one, however, for both satisfy their nutritional needs at the expense of their victim.

Since the normal fate of most individuals is to be killed either by predator or by parasite, each species constitutes a link in a biological *food chain.* Microbial cells, for example, are food for many species of plankton, the minute plants and animals that drift in the oceans in great abundance. The plankton serve as the major food source for many marine invertebrates and fishes, which are fed upon in turn by larger fishes and some marine mammals. The largest animals, which seemingly stand at the ends of the food chains, must eventually die and be devoured by microorganisms, so that the food chains are actually *cyclic* in nature.

The fact that every organism is part of a food chain means that—for any heterotroph—the source of nutrients is, directly or indirectly, another organism. The factors that affect the available food supply of a species are numerous and complex, whether the organism feeds as a predator or as a parasite. For example, an excessively large population may exhaust the food supply to such an extent that the next generation will find an extreme shortage and hence will survive in much smaller numbers. The food supply may also be affected by a change in climate, by long-term geochemical evolution, or by changing competition with other predators or parasites.

The ability of a species to resist being eaten by others depends both on its own defense mechanisms (e.g., protective coloration, armor, ability to fly or burrow, immunity to disease) and on the properties of its predators or parasites. A new predator or parasite may appear on the scene, for example, or an existing one may evolve more efficient feeding habits.

The living world is thus organized into a large number of intersecting food chains. Each chain consists of a number of species, the populations of which have reached equilibrium in terms of rate of reproduction and destruction. The equilibrium size of a population may shift abruptly if any one of its complex determinants changes. In modern times, the most significant factors affecting this ecological equilibrium have been the activities of humans. By building dams, destroying forests, polluting streams, slaughtering game, spreading poisons, and transporting parasites, human kind has exterminated many species and changed the ecology of many others.

## THE FUNCTIONS OF SYMBIOSIS

A symbiont substitutes for part or all of the nonliving environment that free-living organisms occupy; among the myriads of symbioses that have evolved we can find examples in which almost every known environmental function is furnished by one or another symbiont for its partner. For convenience, we shall discuss these functions under four headings: protection, provision of a favorable position, provision of recognition devices, and nutrition.

To determine the functions fulfilled by the partners in a symbiosis it is necessary—in all but the most obvious cases—to separate the partners and study their requirements in isolation. In many cases, this has not yet been achieved, either because the symbionts cannot be separated without damaging them or because the isolated partners cannot be cultivated. Symbionts that have defied attempts to cultivate them in isolation are said to be *obligate symbionts*. The classification of a symbiont as "obligate" is provisional, since it is always possible that identification of the function performed by its partner will permit its eventual cultivation.

### Protection

Endosymbionts, as well as those ectosymbionts that live in the body cavities of animals, are protected from many adverse environmental conditions. These habitats protect the symbionts from dessication and—in the case of warm-blooded hosts—from extremes of temperature.

The microbial symbionts of plants and animals also perform functions that protect their hosts. Most notable is the protection that the normal flora of vertebrates offers against invasion by pathogenic (disease-producing) microorganisms; germ-free animals are much more susceptible to infection than their normal counterparts, as we shall describe on page 587. The removal of toxic substances is another function that many microbes perform for their symbiotic partners; in some insects, for example, bacteria harbored in the excretory organs break down uric acid and urea to ammonia, which the bacteria themselves assimilate.

### Provision of a Favorable Position

A symbiotic association may provide one partner with a position that is favorable with respect to the supply of nutrients. Many of the marine ciliated protozoa, for example, are found only on the body surfaces of crustacea, where the host's respiratory and feeding currents assure the microbe of a constant supply of food (Figure 28.4). No less spectacular is the favorable position provided by many marine invertebrates for their photosynthetic algal symbionts. Some of these hosts are phototactic, carrying their photosynthetic partners toward the light. Others, such as the tridacnid clams, house their algal symbionts in special organs that act as lenses to gather light.

These clams (family Tridacnidae) have several unique anatomical features, the most prominent of which is the location and thickening of the mantle, the epithelial tissue that lines the shell. Unlike the mantle of all other clams, the mantle of the Tridacnidae is greatly extended along the dorsal, or nonhinged part of the shell. The mantle, olive-green in color because of its dense population of algal symbionts, is so thick that it prevents the shell from closing, and its surface is covered with conical protuberances (Figure 28.5).

Sections through the mantle tissue reveal the nature and function of its conical protuberances. Each protuberance contains one or more lenslike structures, the *hyaline organs*, made up of transparent cells. Each hyaline organ is surrounded by a dense mass of algae (Figure 28.6); the function of the lenslike hyaline organ is to permit light to penetrate deeply into the mass of algae.

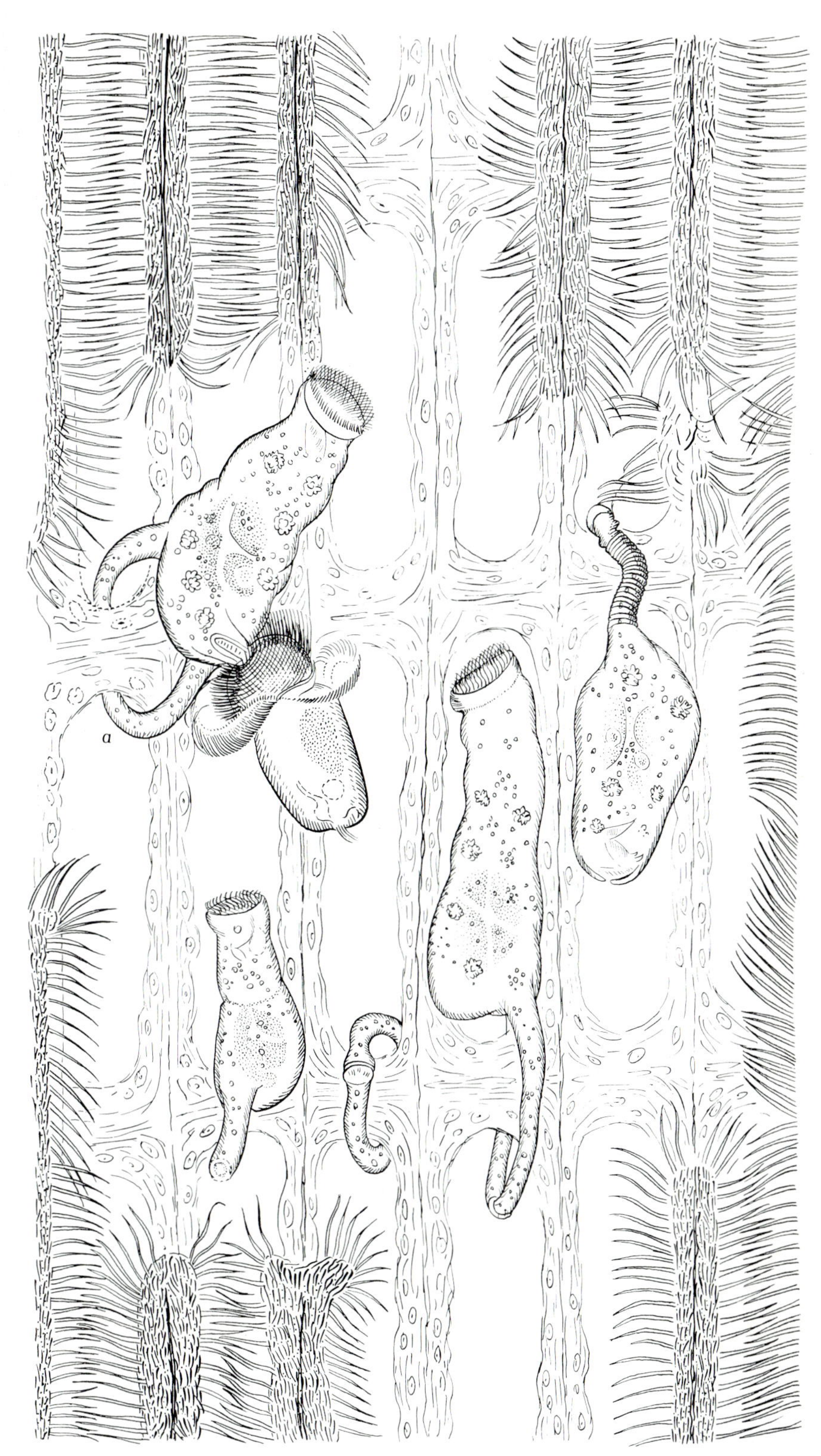


FIGURE 28.4

The ciliate protozoon, *Ellobiophyra donacis*, "padlocked" to the gills of the bivalve, *Donax vittatus*. In (a) *Ellobiophyra* is seen in the process of reproduction by budding. From E. Chatton and A. Lwoff, "*Ellobiophyra donacis* Ch. et Lw., peritriche vivant sur les branchies de l'acephale *Donax vittatus* da Costa," *Bull. Biol. Belg*. **63,** 321 (1929).

FIGURE 28.5

Underwater photograph of *Tridacna maxima,* showing characteristic exposure of mantle to sunlight. From P. V. Fankboner, "Intracellular Digestion of Symbiotic Zooxanthellae by Host Amoebocytes in Giant Clams (*Bivalvia: Tridacnidae*), with a note on the Nutritional Role of the Hypertrophied Siphonal Epidermis," *Biol. Bull.* **141,** 222 (1971).

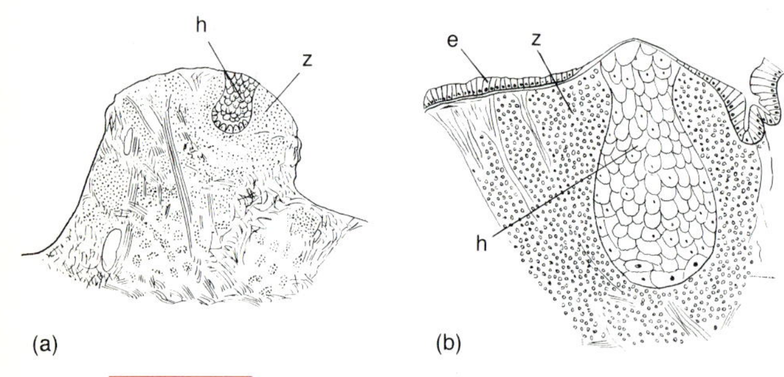


FIGURE 28.6

Endosymbiotic algae of the clam, *Tridacna crocca:* (a) section through a protuberance on the inner fold of the dorsal mantle edge; (b) enlarged view of a hyaline organ. e, epithelium; h, hyaline organ; z, zooxanthellae (endosymbiotic algae). After M. J. Yonge.

The tridacnid clams have thus evolved a highly specialized system for cultivating algae within their own tissues. The algae grow extracellularly in haemal channels that lie perpendicular to the exposed surface of the mantle. Senescent and dead algal cells are eventually digested by phagocytic cells of the host.

There is much anatomical evidence to suggest that the tridacnid clams rely heavily on their algae as a source of food. The digestive system is reduced, for example, and the feeding organs are so altered that they screen out all but the most minute particles. Finally, the kidneys are vastly increased in size, presumably to handle the excretion of products formed in the phagocytes by digestion of the algae. The tridacnids thus represent an extreme example of an evolutionary response to symbiosis.

## Provision of Recognition Devices

Bioluminescence is widespread in the animal kingdom, occurring in such diverse groups as jellyfish, earthworms, fireflies, squid, and fish. The emission of light by these animals very often appears to be a recognition device, promoting schooling, mating, or the attraction of prey. In most cases the luminescence is produced by the tissues of the animal itself, but in some species of squid, and in certain fishes, it is produced by luminous bacteria living ectosymbiotically in special glands of the host. Some of these symbioses were discussed earlier, page 449.

## Nutrition

By far the most common function of symbionts is to provide nutrients for their partners. The provision of nutrients may be indirect, as in the case of fungi that infect plant roots and thereby increase the water-absorbing capacity of the root system. Usually, however, nutritional support is direct, the symbiont furnishing one or more essential nutrients to its partner.

Possibly the most dramatic and extensively studied example is *nitrogen fixation.* As previously discussed (Chapter 27), the capacity to fix nitrogen is exclusively procaryotic, but many groups of eucaryotes—both plants and animals—have entered into symbiotic associations with nitrogen-fixing bacteria. Nitrogen fixation by root-nodule bacteria was discussed in some detail in Chapters 17 (*Rhizobium*) and 24 (*Frankia*).

Cellulose, as a major plant constituent, provides the principal carbon and energy source for grazing animals as well as for wood-boring insects. Some of these animals are incapable of digesting cellulose; in the ruminants and in at least one group of insects, the termites and wood-eating roaches, cellulose digestion is performed on behalf of the host by symbiotic bacteria and protozoa. The digestion of other complex carbohydrates is often carried out by microbial symbionts living in the digestive tracts of animals.

One of the most intriguing detective stories in biology has been the discovery and elucidation of the endosymbioses between microorganisms and their insect hosts. In 1888 F. Blochmann recognized that certain special cells of cockroaches contain symbiotic bacteria, and soon entomologists discovered bacterial, fungal, and protozoan endosymbionts in a variety of other insects.

The significance of these symbioses became clear when techniques were developed for ridding the insects of their symbionts; such animals require one or more B vitamins to develop normally. The importance of the symbionts to the well-being of their hosts is emphasized by the elaborate mechanisms that insects have evolved for transmitting the symbionts to their young. These mechanisms are described below.

Many symbioses occur between photosynthetic and respiratory partners. In such cases, the metabolic functions of the two partners are often complementary with respect to the metabolism of carbon and oxygen; in effect, therefore, these symbiotic associations carry out a complete cycle of the two elements, according to the scheme described in Chapter 27.

## THE ESTABLISHMENT OF SYMBIOSES

As we shall see later on, the evolution of a symbiosis is usually characterized by a progressively greater interdependence of the two partners. This in turn places a premium on the development of mechanisms to ensure the continuity of the symbiosis from generation to generation. Such mechanisms are of two kinds: those in which the host transmits its symbionts directly to its progeny at each generation and those in which each new generation is reinfected.

### Direct Transmission

The simplest type of direct transmission is found in the endosymbioses of protozoa with algae. The protozoan and its intracellular algal symbiont divide at more or less the same rate, so that each daughter cell of the host receives a proportionate share of algal cells. In some instances cell division is precisely regulated: the host cell, containing two algal symbionts, divides to yield two daughter cells, each containing one symbiont. The symbiont then divides, restoring the number per cell to two.

In sexually reproducing animals, direct transmission may be accomplished by infection of the egg cytoplasm. This may require an extremely elaborate sequence of host cell movements, morphological changes, and interactions. In certain insects, for example, the microbial symbionts are contained within specialized cells, the *mycetocytes*, which make up organs called *mycetomes*. The mechanism for transferring the symbionts from the mycetome to the egg varies from family to family and may be very complex. A common sequence of events is the following. The symbionts are liberated from the mycetocytes and—being nonmotile in all cases—are passively transported to the ovary by way of the lymph. At the ovary, the symbionts are taken up by special epithelial cells and from these are ultimately transferred to the egg cells.

In other cases, the germ cells become infected early in the embryonic development of the insect. If the insect matures as a male, the germ cells become testes and the symbionts disintegrate. If the insect matures as a female, however, the germ cells become ovaries and the symbionts multiply, so that each egg contains large numbers of them.

The cycle is completed during embryogenesis of the progeny, when events occur that lead to the formation of the symbiont-filled mycetome. The way in which mycetomes are formed varies from family to family, being particularly complicated in those insects that carry several different symbionts, each of which must eventually be housed in its own special type of mycetocyte or mycetome. In every case, the development of the mycetome involves a process of differentiation comparable to that which leads to the formation of any other animal organ. Differentiation is initiated by a series of regulated nuclear divisions in the region of the egg that contains the symbiont mass, and culminates in the formation of the mycetome of the adult insect.

### Reinfection

Again, it is among the insects that we find the most elegant examples of mechanisms designed to ensure infection of the progeny with symbionts from the mother. Each group of insects has evolved its own set of specialized devices: in some insects a direct anatomical connection between the intestine and vagina guarantees that intestinal symbionts will be copiously smeared on the surface of the eggs as they pass through the ovipositor. When the young larvae hatch, they infect themselves immediately by eating part or all of the eggshell.

Two groups of flies are viviparous: *Glossina* (the tsetse flies) and a large group called *Pupipara*, which are themselves ectosymbionts of mammals and birds. The larvae of these flies are retained in the uterus, where they are nourished by the secretions of greatly developed accessory glands, the "milk glands." The symbiotic microorganisms are localized in these glands and are delivered to the larvae during feeding.

Two mechanisms of transmission are particularly intriguing because they involve a stereotyped, genetically determined *behavior pattern* of the newly hatched larvae. In one, the hatchlings suck up drops of bacterial suspension that exude from the mother's anus during the period of brood care. In the other, discovered in the plant-juice sucking insect *Coptosoma*, the female deposits a bacterium-filled "cocoon" or capsule between each pair of eggs. When the eggs hatch, each larva sinks its proboscis into a cocoon and sucks up a supply of symbionts (Figure 28.7).

Equally complex are the mechanisms that have evolved in the plant kingdom for the initiation of root nodules when the bacterium, *Rhizobium*, infects its leguminous host. The plant root excretes a number of substances, among which is tryptophan. The bacterial cells in the soil convert tryptophan to the plant growth hormone, indoleacetic acid, and also produce an extracellular polysaccharide capsular material that induces the plant root to excrete the enzyme polygalacturonase. The bacteria then commence penetration of those root hairs that have been induced to grow abnormally by the indoleacetic acid; it is possible that polygalacturonase plays some role in mediating the penetration. The process of nodulation was discussed in greater detail in Chapter 17.

In general, the more interdependent the symbiotic partners, the more we can expect to find that evolution has produced means for ensuring their continued association. In contrast, the formation of loose associations often seems to depend entirely on chance, and both partners may also be free-living.

FIGURE 28.7

*Coptosoma scutellatum*. (a) Eggs deposited on a vetch leaf. (b) The eggs seen from below, showing the symbiont-filled cocoons lying between each pair of eggs. (c) A newly hatched larva sucking the symbiont suspension from a cocoon. (d) Enlarged view of egg and cocoon. After H. J. Müller.

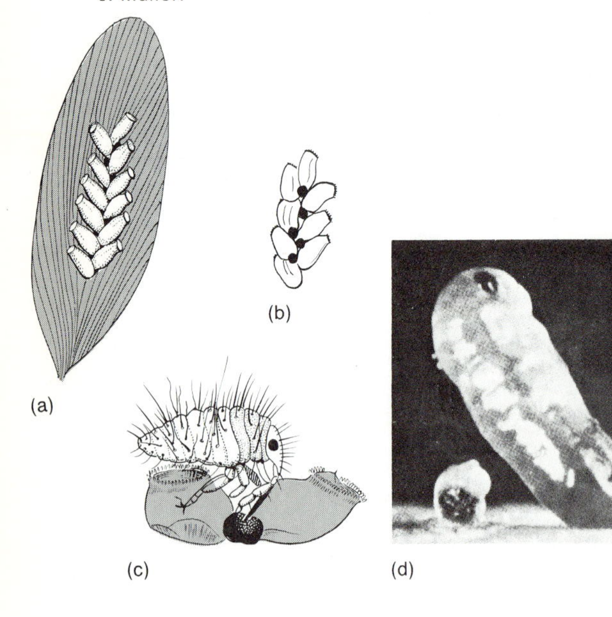


## THE EVOLUTION OF SYMBIOSES

Natural selection acts on symbiotic associations as well as on individual organisms; symbioses thus have their own phylogenies. In the absence of fossil evidence, such phylogenies are necessarily speculative, but certain trends can nevertheless be deduced from the nature of contemporary symbioses.

It seems inescapable that symbioses evolve in the direction of increasing intimacy. Starting with a loose association, in which one or both organisms finds an optimal environment in the vicinity of the other, an ectosymbiosis may gradually develop. At a later state in its evolution the relationship may become endosymbiotic, the small organism penetrating the host tissues and ultimately the host cells.

Once a symbiosis has been established, selection operates to increase its efficiency. An increased degree of adaptation to one highly specialized environment necessarily implies, however, a decreased degree of adaptation to other environments. The result is a high degree of specialization; the symbiont not only loses its ability to live freely, but it also becomes increasingly *specific* with respect to its choice of partner. Today we find many extreme cases, particularly in the endosymbioses, where neither partner can grow without the other.

Indeed, as discussed in Chapter 3, it is entirely conceivable that the eucaryotic cell originally arose as an endosymbiosis between two (or more) primitive cell types. The similarities between chloroplasts and endosymbiotic cyanobacteria, for example, are numerous (Figure 28.8), as are the similarities between mitochondria and endosymbiotic bacteria.

## SYMBIOTIC ASSOCIATIONS BETWEEN PHOTOSYNTHETIC AND NONPHOTOSYNTHETIC PARTNERS

In a great number of mutualistic symbioses, one of the partners is a photosynthetic organism. The function served by the nonphotosynthetic partner varies widely: in some cases, such as the nitrogen-fixing root-nodule bacteria, it is the provision of nutrients; in other cases, such as the fungal partners of lichens, it appears to be protection; and in still other cases, such as the tridacnid clams which house algae, it appears to be the provision of a favorable position.

The contribution of the photosynthetic partner, however, is always the provision of nutrients, i.e., the carbohydrates formed by the fixation of carbon dioxide. The movement of carbohydrate

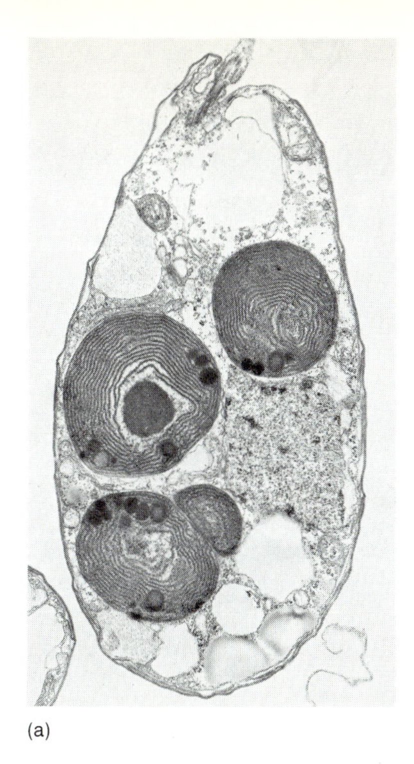

(a)

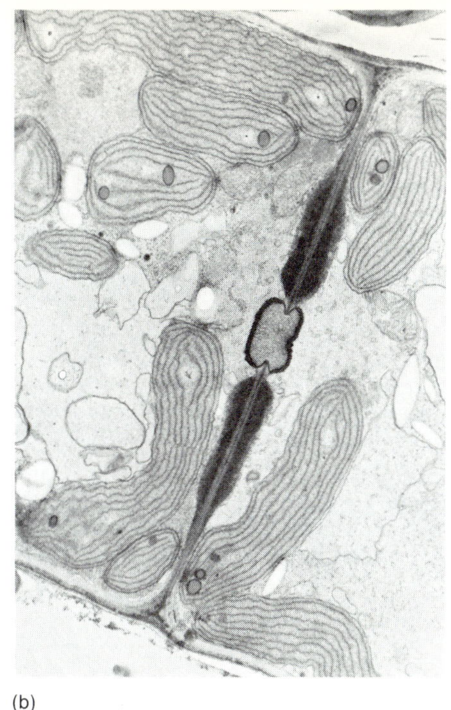

(b)

FIGURE 28.8

(a) Electron micrograph of a thin section of the flagellated protozoon, *Cyanophora paradoxa*, containing several endosymbiotic cyanobacteria (×4,370). Courtesy of William T. Hall. (b) Electron micrograph of a thin section of *Pseudogloiophloea confusa*, a red alga, containing several chloroplasts. (The dark structure in the center is a cross-septum of a multicellular filament, ×7,200.) Courtesy of J. Ramus.

from one symbiont to the other has been studied by allowing $^{14}C$-labeled $CO_2$ to be assimilated in the light, following with time the incorporation of label into metabolites of each partner. In symbiotic associations of algae with invertebrate animals, as well as in associations of algae with fungi (lichens), isotope studies have revealed a number of important adaptations that facilitate the *unidirectional transport of carbohydrate* from the photosynthetic to the nonphotosynthetic partner. For example, the symbiotic algae excrete a much greater proportion of their fixed carbon than do related free-living algae; in many cases this excretion ceases soon after the symbiotic alga is isolated, indicating a specific stimulatory effect by the nonphotosynthetic partner.

The excreted carbohydrate is usually different from the major intracellular carbohydrates of the alga: in most cases it is carbohydrate that the nonphotosynthetic partner, but not the alga itself, can utilize. For example, the green algae of lichens excrete polyols, such as ribitol, which are not metabolizable by the algae themselves but are rapidly utilized by the fungal components of the lichens (Table 28.1). This phenomenon explains the uni-

**TABLE 28.1**

**Carbohydrate Movement from Photosynthetic to Nonphotosynthetic Symbiont**

| Photosynthetic Donor | | | Nonphotosynthetic Recipient | |
|---|---|---|---|---|
| ORGANISM | CARBOHYDRATE RELEASED | | IMMEDIATE FATE OF CARBOHYDRATE | ORGANISM |
| Zoochlorellae | Maltose, glucose | ⟶ | Glycogen, pentoses | Marine invertebrates |
| Zooxanthellae | Glycerol | ⟶ | Lipids, proteins | Marine invertebrates |
| Lichen algae | | | | |
| Chlorophyceae | Polyols | ⟶ | Polyols | Lichen fungi |
| Cyanobacteria | Glucose | ⟶ | Mannitol | Lichen fungi |
| Higher plants | Sucrose | ⟶ | Trehalose, glycogen, polyols | Mycorrhizal fungi |

Source: Modified, with permission, from Table 8 in D. Smith, L. Muscatine, and D. Lewis, "Carbohydrate Movement from Autotrophs to Heterotrophs in Parasitic and Mutualistic Symbioses," *Biol. Rev.* **44,** 17 (1969).

directional flow of excreted carbohydrate; the utilization of the excreted material by the nonphotosynthetic partner creates a *concentration gradient*, such that carbohydrate must flow steadily from the alga to its partner. In some cases, the excreted carbohydrate is one that may also be metabolized by the alga; in such cases, the unidirectional flow is maintained by the rapid conversion of the carbohydrate in the fungus to a form that only the latter can utilize.

In *mycorrhizas* (associations between fungi and the roots of higher plants), carbohydrate is again found to move from the photosynthetic to the nonphotosynthetic partner. Here the transported carbohydrate appears to be sucrose, which is the form in which carbohydrate is also translocated within the plant. Movement to the fungus thus represents a *diversion of the translocation stream*; in part, this can be accounted for by the rapid conversion of sucrose to fungal carbohydrates such as trehalose and polyols, which the plant cannot utilize. It is possible, however, that the diversion is brought about through the release of plant hormones, many of which are known to be produced by fungi.

## SYMBIOSES IN WHICH THE PHOTOSYNTHETIC PARTNER IS A HIGHER PLANT

Microorganisms are found in a number of different symbiotic associations with higher plants. As ectosymbionts, they inhabit the surfaces of leaves (the *phyllosphere*), as well as the soil immediately surrounding the roots (the *rhizosphere*). As endosymbionts, fungi invade the roots to form the associations known as *mycorrhizas*; and certain bacteria invade the roots to form *nitrogen-fixing nodules*. (see pages 408 and 514).

### The Rhizosphere

The regions of the soil immediately surrounding the roots of a plant, together with the root surfaces, constitute that plant's *rhizosphere*. Operationally, it can be defined as the region, extending a few millimeters from the surface of each root, in which the microbial population of the soil is influenced by the chemical activities of the plant. The major effect observed is a quantitative one: the numbers of bacteria in the rhizosphere usually exceed the numbers in the neighboring soil by a factor of 10 and often by a factor of several hundred.

There is also a qualitative effect. Short Gram-negative rods predominate in the rhizosphere, while Gram-positive rods and coccoid forms are less numerous in the rhizosphere than elsewhere in the soil. No specific association of a particular bacterial species with a particular plant has, however, been established.

The reason for the relative abundance of bacteria in the rhizosphere must certainly be the excretion by plant roots of organic nutrients, which selectively favor certain nutritional types of bacteria. However, no clear-cut nutritional relationships have been discovered, although many organic products excreted by plant roots have been identified. Our state of knowledge concerning the effects of the microbial population of the rhizosphere on the plant is even less satisfactory; despite numerous claims, it remains to be established that the plant benefits from the association. Many free-living soil bacteria, however, perform functions essential for plants, such as nitrogen fixation and the mineralization of organic compounds, so it seems reasonable to assume that some plants do profit from the proximity of some microorganisms.

### Mycorrhizas

The roots of most higher plants are infected by fungi. As in so many symbioses, a dynamic condition of mutual exploitation results, both partners benefiting as long as a balance between invasive and defensive forces is maintained. As a result of the infection, the plant root is structurally modified in a characteristic way. The composite root-fungus structure is called a *mycorrhiza*.

The formation of a mycorrhiza begins with the invasion of the plant root by a soil fungus; growth of the fungus toward the root is stimulated by the excretion into the soil of certain organic compounds by the plant. The fungal mycelium penetrates the root cells by means of projections called *haustoria*, and develops intracellularly. In some mycorrhizas the fungus forms intracellular branching structures called *arbuscules* (Figure 28.9); in others it forms characteristic coils.

Depending on the host, the fungus either maintains its intracellular state or undergoes digestion. In the latter case the fungal mycelium persists mainly in the form of *intercellular* hyphae. In all mycorrhizas, however, a large fraction of the mycelium remains in the soil, the intercellular forms tending to produce a compact sheath around the root.

With few exceptions, mycorrhizas are not species specific. A given fungus may be associated with any of several plant hosts, and in most cases a

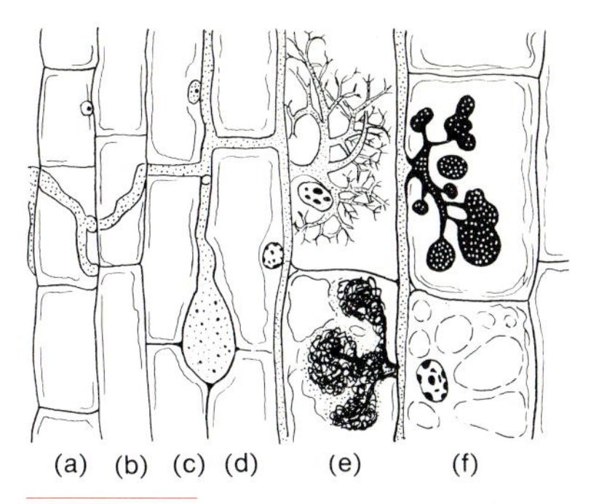


FIGURE 28.9

Drawing showing the penetration of the root of *Allium* by a mycorrhizal fungus. In the first two cell layers, (a) and (b), the fungal mycelium is intracellular. In the third and fourth layers, (c) and (d), it is intercellular; a vesicular storage organ is shown between these layers. In the fifth and sixth layers, (e) and (f), the fungus has formed intracellular branching structures (arbuscules). In (f) the arbuscules are undergoing digestion by the host cells. From F. H. Meyer, "Mycorrhiza and Other Plant Symbioses," in *Symbiosis*, Vol. 1, ed. S. M. Henry, (New York: Academic Press, 1966).

given plant may form mycorrhizas with any of a number of soil fungi. One species of pine tree, for example, has been found to associate with any of 40 different fungi. A great many free-living soil fungi are capable of forming mycorrhizas. In an experiment performed with pure cultures of free-living fungi and sterile plant roots, over 70 fungal species were found to form mycorrhizas, and many times that number are undoubtedly capable of doing so in nature.

A typical mycorrhiza is shown in Figure 28.10. The stocky, club-shaped appearance results from several effects of the fungus on the root: cell volumes increase but root elongation is inhibited, and lateral root formation is stimulated by *auxins* (plant growth hormones) produced by the fungi.

The mutualistic nature of the mycorrhiza symbiosis can be readily demonstrated in many cases. The fungi that participate are characteristically those that are unable to use the complex polysaccharides that are the principal carbon sources for microorganisms in forest soils and humus. By invasion of plant roots, these fungi avail themselves of simple carbohydrates such as glucose. In fact, the auxins excreted by the fungi induce a dramatic flow of carbohydrate from the leaves to the roots of the host plant.

The plant also benefits from the association. Many forest trees become stunted and die when deprived of their mycorrhiza. Stunted trees can be restored to health by the introduction of suitable mycorrhizal fungi into the soil. The fungus seems to facilitate the absorption of water and minerals from the soil; the absorbing surface of the plant's root system is increased manyfold by the fungal hyphae. The function of a mycorrhiza as an absorbing organ has been confirmed by comparing the uptake of minerals from the soil by plants with and without mycorrhiza. Pines, for example, absorb two to three times more phosphorus, nitrogen, and potassium when mycorrhiza are present than when they are absent.

## SYMBIOSES IN WHICH THE PHOTOSYNTHETIC PARTNER IS A MICROORGANISM

### Endosymbionts of Protozoa

Many protozoa of the Ciliophora and of the Rhizopoda are hosts to endosymbiotic algae. In freshwater forms the algae are generally green types belonging to the Chlorophyta; in the marine forms the algae are generally yellow or brown types belonging to the dinoflagellates. The two groups of symbionts are called *zoochlorellae* and *zooxanthellae*, respectively.

FIGURE 28.10

Mycorrhiza of *Fagus sylvatica*, showing the club-shaped apices of roots and hyphae radiating from the surface. From F. N. Meyer, "Mycorrhiza and Other Plant Symbioses," in *Symbiosis*, Vol. 1, ed. S. M. Henry (New York: Academic Press, 1966).

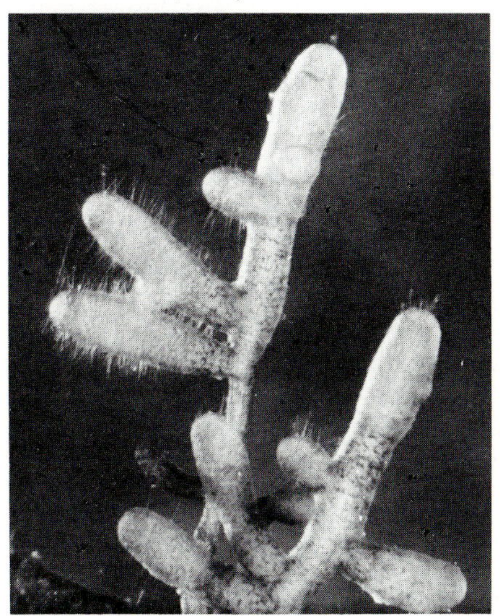

Figure 28.11 shows zooxanthellae liberated by crushing a foraminiferan protozoon. As found in their hosts, both zoochlorellae and zooxanthellae are invariably coccoid. When cultured free of their hosts, however, zooxanthellae are sometimes observed to form swarming zoospores, which are typical dinoflagellates. Each protozoan cell harbors from 50 to several hundred algae; maintenance of the symbiosis is ensured by similar growth rates of the two partners. Endosymbionts resist digestion by the host. This resistance is undoubtedly related to their location in the host cytoplasm. It should be recalled that microorganisms are taken into the cells of phagotropic protozoa by phagocytosis and localized inside food vacuoles formed by invaginations of the cell membrane. Their digestion is effected by hydrolytic enzymes liberated into these food vacuoles from lysosomes. The endosymbionts in the cytoplasm are not contained in food vacuoles, and thus are isolated from the digestive enzymes of the lysosomes.

As we discussed at the beginning of this chapter, symbioses between photosynthetic and respiratory partners are particularly successful because together the two organisms can carry out a full carbon cycle and a full oxygen cycle. The photosynthetic partner uses light energy to convert carbon dioxide to organic products, while liberating $O_2$ from water; the nonphotosynthetic partner uses the $O_2$ to respire the organic products, producing carbon dioxide as a byproduct. This is presumably the basis for the extremely common occurrence of algal-protozoan endosymbioses.

Many protozoan hosts exhibit *phototaxis* when they harbor a photosynthetic endosymbiont. In paramecia it has been shown that the alga is the photoreceptor; the movements of the protozoon seem to be controlled by the intracellular concentration of oxygen produced by algal photosynthesis, since phototaxis is exhibited only when the external supply of oxygen is limiting.

Cyanobacterial endosymbionts (termed *cyanellae*) are found in a few genera of freshwater protozoa.

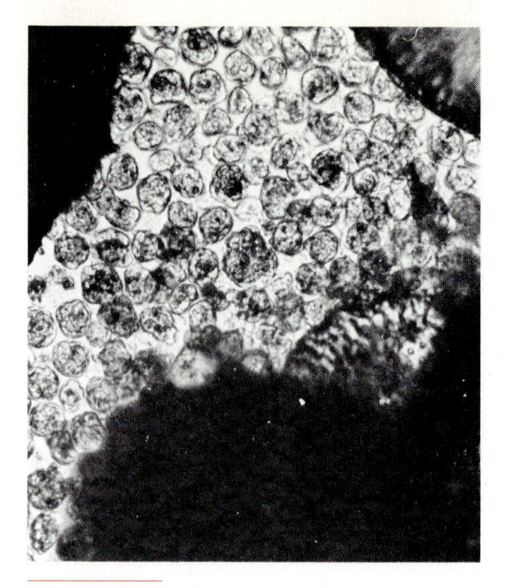

FIGURE 28.11
Zooxanthellae escaping from a crushed foraminiferan. From J. McLaughlin and P. Zahl, "Endozoic Algae," in *Symbiosis*, Vol. 1, ed. S. M. Henry (New York: Academic Press, 1966). Photograph made by J. J. Lee and H. D. Freudenthal.

## Symbioses with Fungi: The Lichens

A lichen is a composite organism, consisting of a specific fungus, usually an ascomycete, living in association with one—or sometimes two—species of algae or cyanobacteria.* The symbionts form a vegetative body, or *thallus*, of which both the gross structure and the fine structure are characteristic for each lichen "species."

In terms of gross structure, the lichen thalli are divided into three types. The *crustose* lichens adhere closely to their substrate (either rocks or the bark of trees). The *foliose* lichens are leaflike and are more loosely attached to the substrate. The *fructicose* lichens form pendulous strands or upright stalks. Figure 28.12 shows a representative of each type, together with a cross section showing the internal organization of the thallus. The bulk of the thallus is made up of fungal hyphae. In most species these are differentiated into distinct tissues: a closely packed *cortex*, a loosely packed *medulla*, and (in the foliose lichens) attachment regions or *rhizinae*. The algal cells are usually found in a thin layer just below the cortex; in a few species of lichens, however, the fungal hyphae and algal cells are distributed throughout the thallus.

Electron micrographs of thin sections show that in most lichens each algal cell is penetrated by one or more fungal haustoria. In some lichens the haustoria penetrate deeply into the algal cells, the membrane of the algal cell invaginating to form a sheath around the haustorium (Figure 28.13). In all cases, the haustoria penetrate only the algal cell wall. The fungi in a few lichens do not have haustoria; instead, there is an intimate contact between the algal and fungal cell walls, which in these species are very thin.

* In discussing the lichen symbioses, the general term "alga" will be used to refer to the photosynthetic partner.

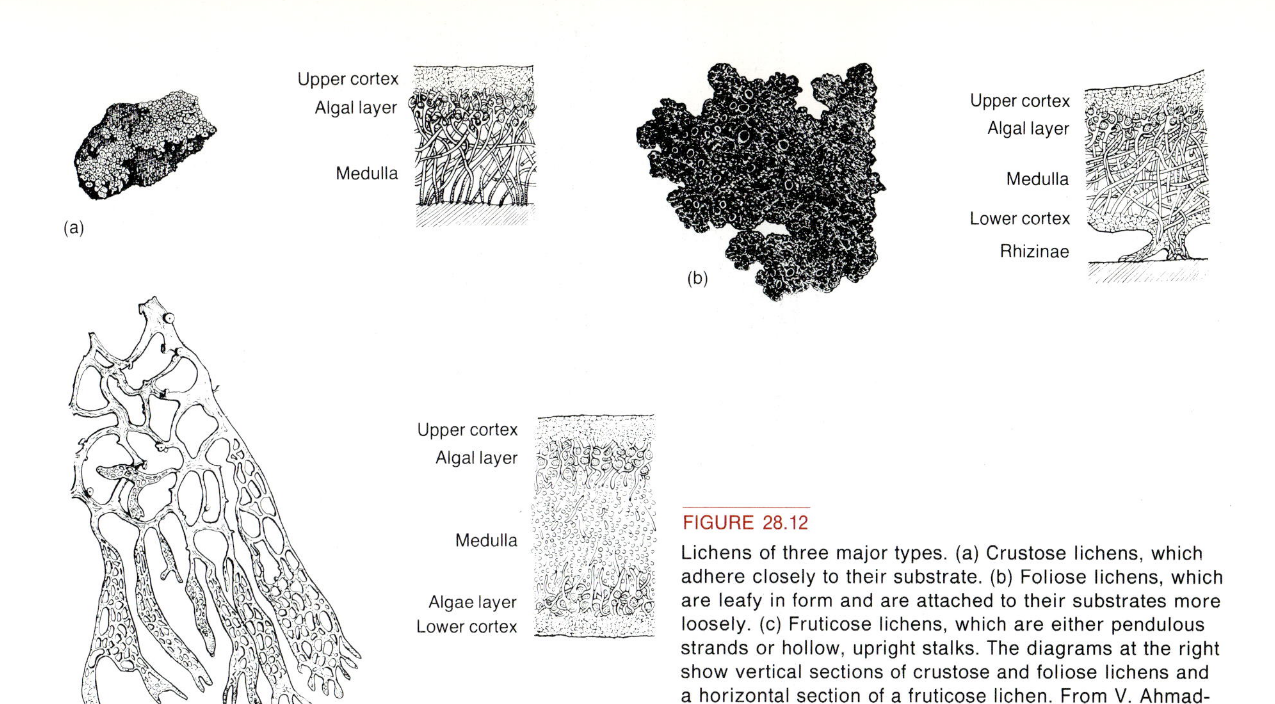


FIGURE 28.12

Lichens of three major types. (a) Crustose lichens, which adhere closely to their substrate. (b) Foliose lichens, which are leafy in form and are attached to their substrates more loosely. (c) Fruticose lichens, which are either pendulous strands or hollow, upright stalks. The diagrams at the right show vertical sections of crustose and foliose lichens and a horizontal section of a fruticose lichen. From V. Ahmadjian, *The Lichen Symbiosis*. Waltham, Mass.: Blaisdell, 1967.

FIGURE 28.13

Electron micrograph of a section through the lichen, *Lecanora rubina*, showing the penetration of an algal cell (*Trebouxia*) by a fungal haustorium. The haustorium has penetrated the outer layer of the cell wall but not the inner layer or the membrane of the algal cell. From J. B. Jacobs and V. Ahmadjian, "The Ultrastructure of Lichens. I. A General Survey," *J. Phycol.* **5,** 227 (1969).

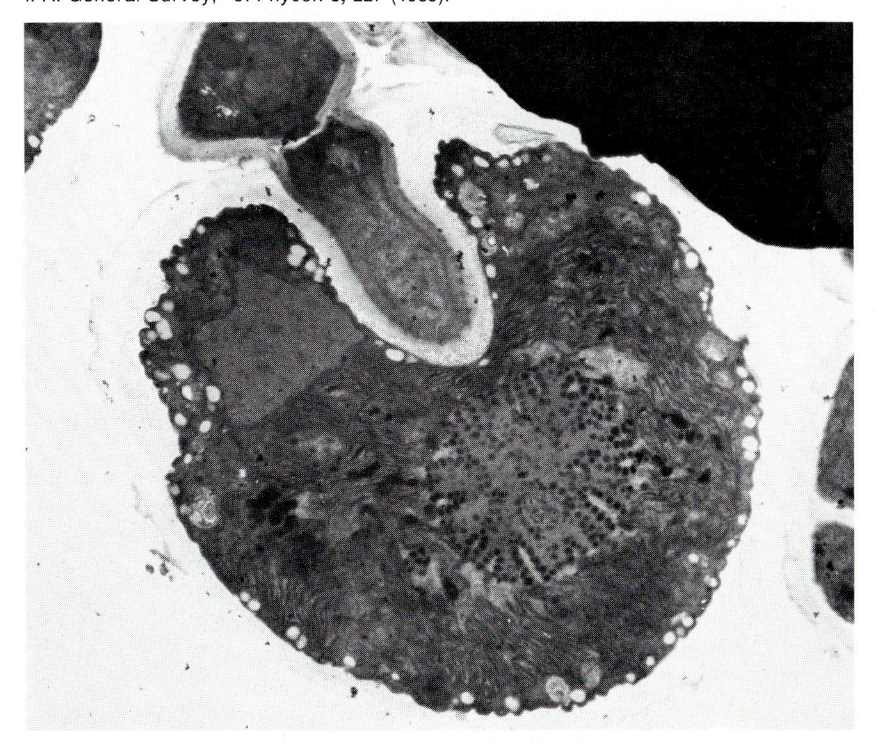

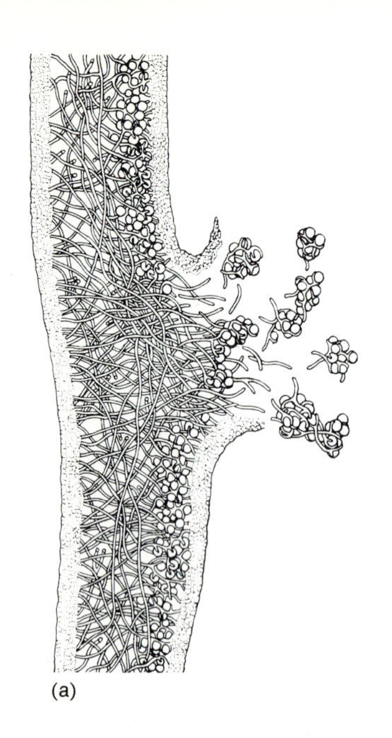


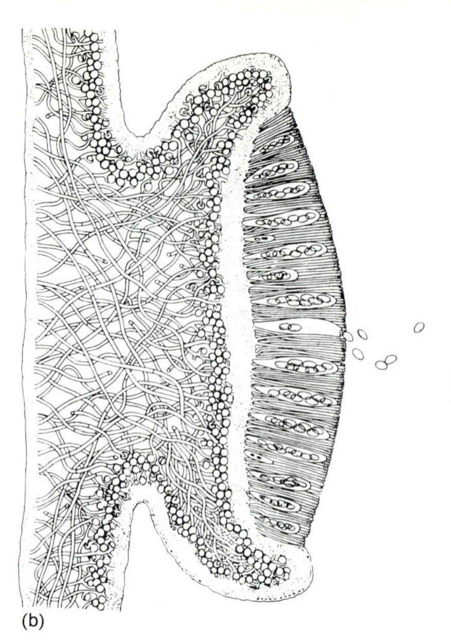


FIGURE 28.14

Lichen reproduction: (a) by the liberation of soredia, made up of fungal threads and hyphae; (b) by the liberation of fungal spores (in this case, ascospores). On germination, a fungal mycelium will develop, which may form a lichen if it comes into contact with an algal cell. From V. Ahmadjian, "The Fungi of Lichens," *Scientific American*, **208,** 122 (1963).

Most lichens propagate by the liberation of *soredia:* small fragments, composed of algal cells and fungal hyphae [Figure 28.14(a)]. In addition, lichens liberate fungal spores [Figure 28.14(b)]. There is considerable evidence, as will be discussed later, that the hyphae produced when these spores germinate make contacts with free-living algal cells and initiate the formation of new lichen thalli.

THE MORPHOLOGICAL CONSEQUENCES OF SYMBIOTIC EXISTENCE It is relatively easy to separate the symbiotic partners of many lichens and to grow them in pure culture, making possible a comparison of the morphology of each partner as a free-living organism and as a symbiont.

The photosynthetic partner may be severely modified by "lichenization." Certain filamentous cyanobacteria, for example, fail to form normal filaments when in the thallus; instead, each cell is separated and surrounded by fungal tissue. When isolated from the thallus, the bacterium regains its filamentous growth habit. The green algae found in lichens are so modified that they never produce their characteristic zoospores while part of the lichen thallus.

The fungal partner forms fruiting structures (ascospores and asexual spores) when it is lichenized, but with rare exceptions does not do so when it is isolated and cultivated in the free-living state. In the free-living state it is also incapable of forming a thallus with cortex, medulla, or other tissues. Thus, each partner affects the morphology of the other in a highly specific way.

SPECIES SPECIFICITY OF THE SYMBIOTIC PARTNERS The association of fungus with alga in a lichen is not specific. Thus, a given algal species may be found associated with any one of a variety of lichen fungi, and—conversely—a given fungus may be found associated with any one of a variety of algae. Altogether, photosynthetic symbionts belonging to approximately 30 genera have been found in lichens; most of them are members of the green algae or cyanobacteria. One green alga, *Trebouxia*, is found in more than half of the described lichens; among the cyanobacteria, *Nostoc* is the most common representative.

The lichen fungi have been classified into several hundred genera; most of the fungi in lichens are Ascomycetes, but a few imperfect fungi and a few Basidiomycetes have also been found.

It is difficult to make any statement about the taxonomic relationships of lichen fungi to free-living Ascomycetes. Historically, the lichens have been studied and classified by specialists who have given them a unique set of names based on the morphology of the composite plant. When a lichen is experimentally separated into its two components, the fungus retains the name of the lichen. Thus, the lichen *Cladonia cristatella* is said to be composed of the fungus, *Cladonia cristatella*, and its algal symbiont, *Trebouxia erici.*

No attempt has been made to integrate the taxonomies of lichen fungi and free-living fungi. Although the morphology of the lichen is dominated by the fungus, and the description of the lichen includes many fungal characters (e.g., shape and number of ascospores), some descriptive features of the lichen may well be expected to vary according to which alga is present. It thus seems highly possible that several different lichen "species" contain the same fungus; if so, the practice of assigning the name of the lichen to the isolated fungal component is bound to prove misleading.

**The Formation and Maintenance of the Symbiosis** Many lichen fungi have been isolated and grown in pure culture. Under conditions of low nutrient supply, the hyphae encircle almost any rounded object of appropriate size which they encounter. If the object is an algal cell, it is penetrated by haustoria. This type of experimentally observable association is thought to mimic the first stages of true lichenization in nature. Further stages in lichenization have been achieved experimentally by allowing mixed cultures of a fungus and an alga, isolated from the same lichen, to grow together under conditions of progressive dessication. After several months, structures typical of the parental lichen thallus are formed, including fungal fruiting bodies and asexual spores.

When a lichen is cultivated, it will continue its normal development only as long as the growth conditions are unfavorable for the independent growth of the individual components. A low supply of nutrients and alternate periods of wetting and drying are favorable for the maintenance of the symbiosis. If the lichen is placed on a rich nutrient medium with adequate moisture over a prolonged period, the union breaks down, and the algae grow out in their characteristic free-living form.

**The Physiology of the Composite Organism** The lichen association seems to have evolved by selection for the ability to withstand extreme drought as well as for the ability to scavenge essential minerals. These deductions follow from the ecology of lichens, as well as from the experimental observations discussed above. Lichens are found in nature colonizing exposed rock surfaces and tree trunks where other forms of life are unable to gain a foothold.

A lichen can remain viable in the dry state for months; when submerged, its water content can change from 2 percent of its dry weight to 300 percent of its dry weight in 30 seconds. Its ability to scavenge minerals is probably related to its production and excretion of *lichen acids*, organic compounds that have the ability both to dissolve minerals and to chelate them. Chelation, the process of binding metal atoms to organic ligands, undoubtedly plays an important role in the solubilization and uptake of minerals by lichens.

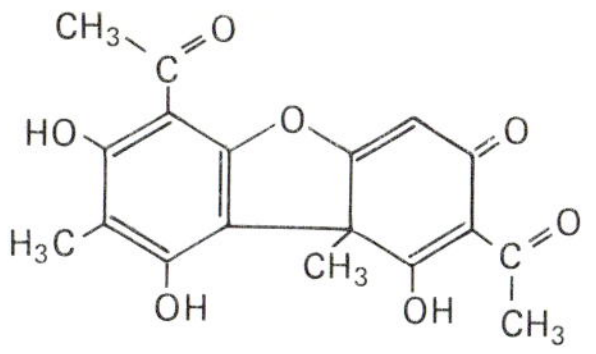

FIGURE 28.15
Usnic acid.

More than 100 different lichen acids have been described. Most of them contain two or more phenylcarboxylic acid substituents, with aliphatic side chains. They are not produced by isolated lichen algae, nor—with a very few exceptions—by isolated lichen fungi. A few nonlichen fungi produce similar compounds, however, and it seems likely that the biosynthesis of these compounds in the composite organism is mediated by the fungus.

The production of the lichen acid is thus a creative manifestation of symbiosis. The acids are often excreted in large quantities, crystallizing on the surface of the lichen. In addition to their role as chelating agents, described above, it has been suggested that they inhibit the growth of other microorganisms. Many of the lichen acids do possess strong antibiotic activity, and one of them—usnic acid (Figure 28.15)—is widely used in some European countries as a chemotherapeutic drug for external application.

Lichens grow extremely slowly; an annual increment in radius of 1 mm or less is typical. The range of growth rates is wide, however, and in some species may average 2 to 3 cm per year. Despite their slow growth, lichens form a significant part of the vegetation in some areas; in fact, they are a primary source of fodder for reindeer and caribou in artic regions.

The ability of lichens to scavenge nutrients at very low concentrations, normally advantageous, becomes injurious to these organisms in regions of industrial air pollution. In such regions the lichen population is greatly reduced or even totally eliminated.

**The Significance of the Lichen Symbiosis** Both partners of the lichen symbiosis are capable of free-living existence, as shown by the fact that under conditions favorable for the growth of the free-living forms, the symbiosis breaks down. The association is thus of mutual benefit *only in very special ecological situations* (i.e., in environments where nutrients are extremely scarce and where desiccation is frequent).

The benefit to the fungus of the symbiosis under such conditions is clear: it depends on the alga for its source of organic nutrients. Tracer experiments have confirmed that carbon dioxide fixed by the alga passes rapidly into the fungal mycelium.

In those lichens that contain cyanobacteria as symbionts, the fungus also benefits directly from the atmospheric nitrogen fixed by the bacteria.

The contribution of the fungus to the association is less clear, but there is good reason to believe that it facilitates the uptake of both water and minerals and may also protect the alga from desiccation as well as from excess light intensities. Free-living algae, however, are able to grow to a limited extent in the ecological niche inhabited by lichens, and the algal partner may thus benefit less than does the fungus.

### Endosymbioses of Algae with Aquatic Invertebrates

Endosymbiotic algae have been recorded in over 100 genera of aquatic invertebrates, particularly in the coelenterates (jellyfish, corals, sea anemones, hydra), the platyhelminths (flatworms, principally the planarians), the Porifera (sponges), and the molluscs (clams, squid). They are often found in the cytoplasm of cells concerned with digestion (e.g., in the amebocytes of sponges or the phagocytic blood cells of certain clams).

Some of these animal hosts acquire their symbionts with their food, either directly, as in the case of sponges that feed on algae, or indirectly, as in the case of carnivorous animals that find the algae in the tissues of their prey. Once infection has taken place, however, a permanent symbiosis is usually established in which intracellular growth of the algae is restricted.

In most coelenterates, and in certain other invertebrates, the algae are transmitted to the next generation through the cytoplasm of the egg. In such cases it has not been possible to obtain symbiont-free animals, so the significance of the symbiosis is not known. Nevertheless, experiments with isotopically labeled $CO_2$ show that organic compounds photosynthesized by the algae are utilized by the tissues of the host, and it can be shown that the molecular oxygen generated by algal photosynthesis is several times more than that which would be necessary to provide for the respiratory needs of the host-alga complex. Although the environment of the animal provides dissolved oxygen as well as organic material (principally as plankton), the mechanisms that have evolved for ensuring symbiosis suggest that it is of great ecological significance.

In certain molluscs the photosynthetic symbionts are not algal cells, but *intact, surviving chloroplasts* which are liberated when the algal cells undergo digestion by the host. The animal cell thus becomes photosynthetic, by acquiring a plant organelle. In a sense, a symbiotic relationship can be said to exist between the animal cell and the chloroplast; although the latter does not grow and divide in its new "host," it may persist and function for several months.

The algae found in aquatic invertebrates are of very few types. They are either green algae or dinoflagellates: *zoochlorellae* and *zooxanthellae*, respectively. The algae are presumed to benefit by their intracellular habitat, which supplies a rich supply of essential nutrients. An example of this type of symbiosis (the tridacnid clams) was described previously (see page 562).

## SYMBIOTIC ASSOCIATIONS BETWEEN TWO NONPHOTOSYNTHETIC PARTNERS

In many symbioses involving microorganisms neither partner is photosynthetic. The examples to be discussed here are mutualistic. Parasitic examples include *Bdellovibrio* parasitism of other bacteria (discussed in Chapter 17) and the organisms that cause infectious disease (discussed in Chapters 31 and 32). The predominantly mutualistic symbioses between humans and their resident microbial populations (the *normal flora*) are discussed in Chapter 29.

In the mutualistic nonphotosynthetic symbioses a microorganism may be associated with another microorganism (e.g., the bacterial endosymbionts of protozoa) or with a metazoan partner. In some cases, the microbial symbiont provides its host with a metabolic function, such as the synthesis of a growth factor or the digestion of a complex carbohydrate; in other cases, the microbial symbiont protects its host from invasion by pathogenic parasites. The host in turn furnishes its symbiont with protection, a favorable position, nutrients, or a combination of these functions.

## SYMBIOSES IN WHICH BOTH PARTNERS ARE MICROORGANISMS

### Bacterial Endosymbionts of Protozoa

Bacterial endosymbionts are extremely widespread in protozoa; they have been described in amebae, flagellates, ciliates, and sporozoa. None of them has been cultivated outside its host, but their bacterial nature has been clearly established on the basis of their morphology, staining properties, and mode of cell division. Some multiply in the nucleus of the host and others in the cytoplasm.

In most cases, the contribution that the bacterium makes to the symbiosis is unknown. In one case, however, its contribution is clear: the bacterial endosymbiont provides its host with amino acids and other growth factors that most protozoa require as exogenous nutrients. The infected host, a trypanosomatid flagellate named *Crithidia oncopelti*, can grow in a simple synthetic medium containing glucose as carbon source together with adenine, methionine, and several vitamins as growth factors. In contrast, another species of *Crithidia* requires not only the above nutrients but also 10 other amino acids (including lysine), heme, and several additional vitamins. Radioisotope studies showed that in C. *oncopelti* lysine is synthesized via the diaminopimelic acid pathway, characteristic of bacteria. Final proof of the role of the endosymbiotic bacterium found in this protozoan was obtained by fractionating the *Crithidia* cells and showing that diaminopimelic acid decarboxylase, the last enzyme of the biosynthetic pathway leading to lysine, is located in the fraction consisting of the cells of the endosymbiont.

Perhaps the most fascinating, and certainly the most extensively studied, protozoan symbiosis is that of *Paramecium aurelia* and its endosymbiont, *kappa*. In the first of a series of investigations extending over 20 years, T. M. Sonneborn and collaborators showed that most strains of *P. aurelia* fall into two general classes: killers and sensitives. The former liberate toxic particles to which killers are immune but which are lethal for sensitive strains. The ability to liberate toxic particles is genetically controlled by the cytoplasm of the host, rather than by its nucleus; at conjugation, when cytoplasm is exchanged, a sensitive cell mated with a killer is itself converted to a killer.

In attempts to identify the genetic material in the cytoplasm, J. Preer used X rays to inactivate it. Surprisingly, the data yielded a calculated target size for the genetic element so large that it should be visible with the light microscope. Staining experiments were then performed, and the feulgen stain—which is specific for DNA—revealed that the genetic element responsible for liberation of toxic particles is a bacterium-like endosymbiont that divides by binary fission in the cytoplasm of the paramecium.

Kappa, as the endosymbiont was designated, has the morphological and chemical properties of a small bacterium, and can be eliminated from its host by a variety of physical and chemical agents, including many antibiotics. Its loss is irreversible, and the host continues to propagate normally without it. Kappa can be transmitted to sensitive paramecia through extracts prepared from killers, but so far it has not been cultivated outside its host.

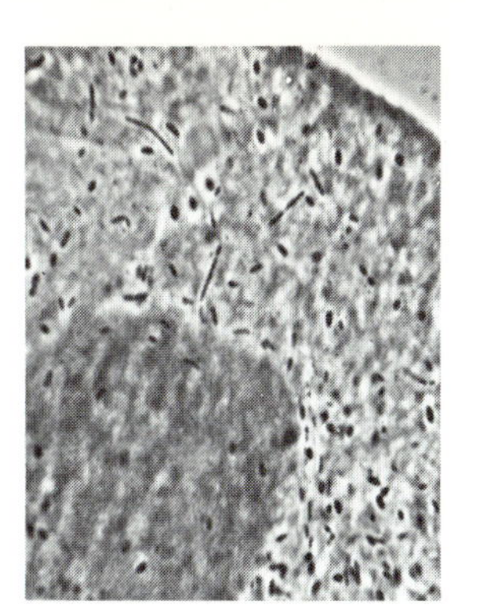

FIGURE 28.16

Stained preparation of unsectioned *Paramecium* containing kappa symbionts (dark rod-shaped bodies) in its cytoplasm. The dark area is the host nucleus (dark phase contrast, ×540). From G. H. Beale, A. Jurand, and J. B. Preer, Jr., "The Classes of Endosymbionts of *Paramecium aurelia*," *J. Cell. Sci.* **5**, 65 (1969):

Kappa contains DNA and can undergo mutations, including mutation to antibiotic resistance. Its reproduction is dependent on the presence in the host nucleus of a particular gene called the *K gene*. *Paramecium aurelia* is a diploid organism; the K gene can mutate to the recessive allele, k, and hence a cell may have the genotype KK, Kk, or kk. When a cross between two Kk killers produces a kk segregant, kappa can no longer reproduce and is diluted out during ensuing divisions of the kk host cell. Ultimately, the kk cell gives rise to a clone of sensitive paramecia.

Cells that are infected with kappa harbor several hundred to a thousand of these endosymbionts in their cytoplasm (Figure 28.16). When preparations of purified kappa cells are observed by phase contrast microscopy, some of the cells are found to contain refractile (R) bodies (Figure 28.17). Kappa cells containing R bodies are called *brights*, and those lacking them are called *nonbrights*.

Preer has shown that the toxic particles liberated into the medium are whole, bright kappa cells, which have lost the ability to reproduce further. When brights are fractionated, the toxic activity is found associated with the R bodies, which are seen in the electron microscope to be tightly rolled ribbons of protein (Figure 28.18). It is not clear whether the toxin is the R body itself or a second protein associated with it. The latter seems more likely, in view of the fact that the R body is very stable, while the toxin is quite unstable.

Both the toxin and the R body of kappa have been found to be produced as a consequence of the induction of a defective prophage that is present in the genome of all kappa cells. Cells in which the prophage has been induced are found to contain phage heads and tails as well as circular DNA mole-

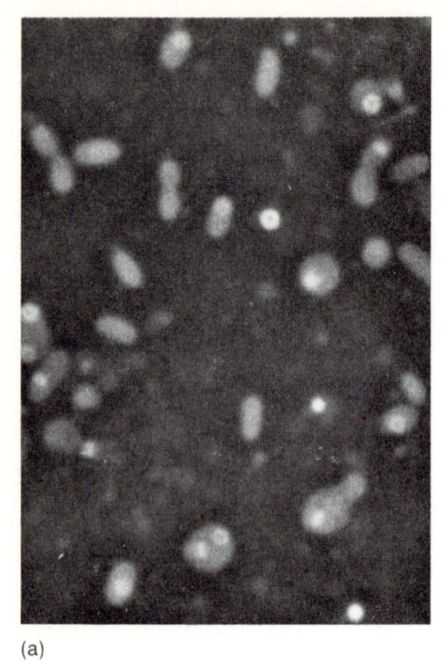

(a)

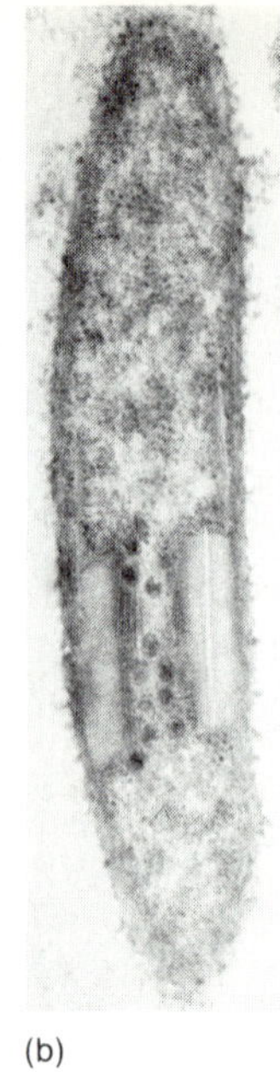

(b)

FIGURE 28.17

(a) Unfixed, purified preparation of kappa. The rods with uniform color are nonbright kappas; those containing a light spherical refractile body are bright kappas (bright phase contrast, ×4,700). (b) Electron micrograph showing longitudinal section through a bright kappa. Note dark-staining spherical phagelike structures inside the coiled refractile body. Surrounding the refractile body and extending beyond it on either side is a fine membrane, the sheath (×32,100). (a) Courtesy of J. Preer; (b) from J. R. Preer, Jr. and A. Jurand, "The Relation between Virus-like Particles and R Bodies of *Paramecium aurelia*," *Genet. Res.* **12,** 331 (1968).

cules; they are not lysed, but cease to reproduce further. The phage heads are always found in close contact with R bodies, and it is very possible that R bodies and toxin are coded by phage genes.

In addition to kappa, a number of other bacterial endosymbionts have been found in killer stocks of *Paramecium aurelia* isolated from nature. These have also been designated by Greek letters. One of them, called *alpha,* has been shown to be a long, spiral gliding organism with strong affinities to *Cytophaga*; it reproduces mainly in the nucleus of its host. The others are eubacteria for which three new genera have been proposed: *Caedobacter*, containing nonflagellated cells, includes the endosymbionts kappa, mu, gamma, and nu; *Lyticum*, containing large, heavily peritrichously flagellated cells, includes lambda and sigma; and *Tectobacter*, containing sparsely peritrichously flagellated cells, includes only delta. Some representatives of these groups are shown in Figure 28.19.

One basis of the mutualistic relationship between *Paramecium* and its bacterial endosymbionts has been clarified by the discovery that one such endosymbiont synthesizes the folic acid required by its host. The equilibrium between the host and endosymbiont is a precarious one, however, and

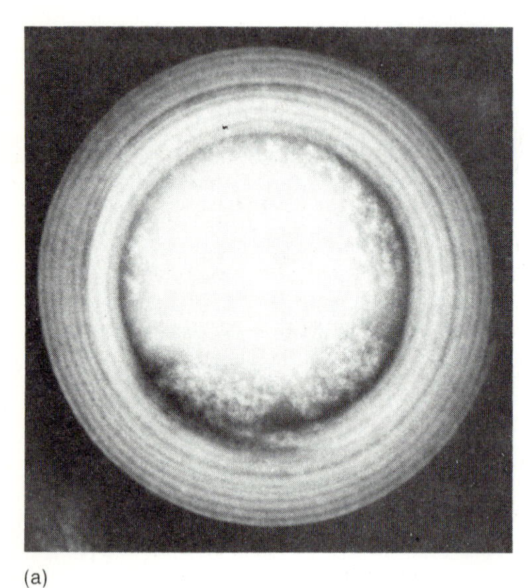

(a)

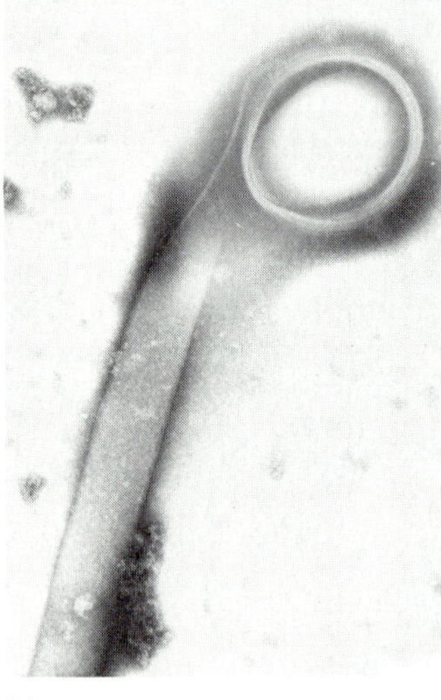

(b)

FIGURE 28.18

Electron micrographs of R bodies of kappa, negatively stained with phosphotungstic acid. (a) Intact, coiled R body (×119,000). (b) Unrolling R body (×33,800). (a) From J. R. Preer, Jr., L. B. Preer, and A. Jurand, "Kappa and Other Endosymbionts in *Paramecium aurelia*," *Bacteriol. Rev.* **38,** 113 (1974); (b) from J. R. Preer, Jr., et al. "The Classes of Kappa in *Paramecium aurelia*," *J. Cell Sci.* **11,** 581 (1972).

may shift in favor of one or the other. Thus, when bearers of endosymbionts are first cultivated in axenic medium, there is often an unbalanced increase in the reproduction of the endosymbiont leading to the death of the host. Conversely, the rapid growth of the protozoan, once it is established in culture, may lead to the loss of the slower growing endosymbiont by dilution.

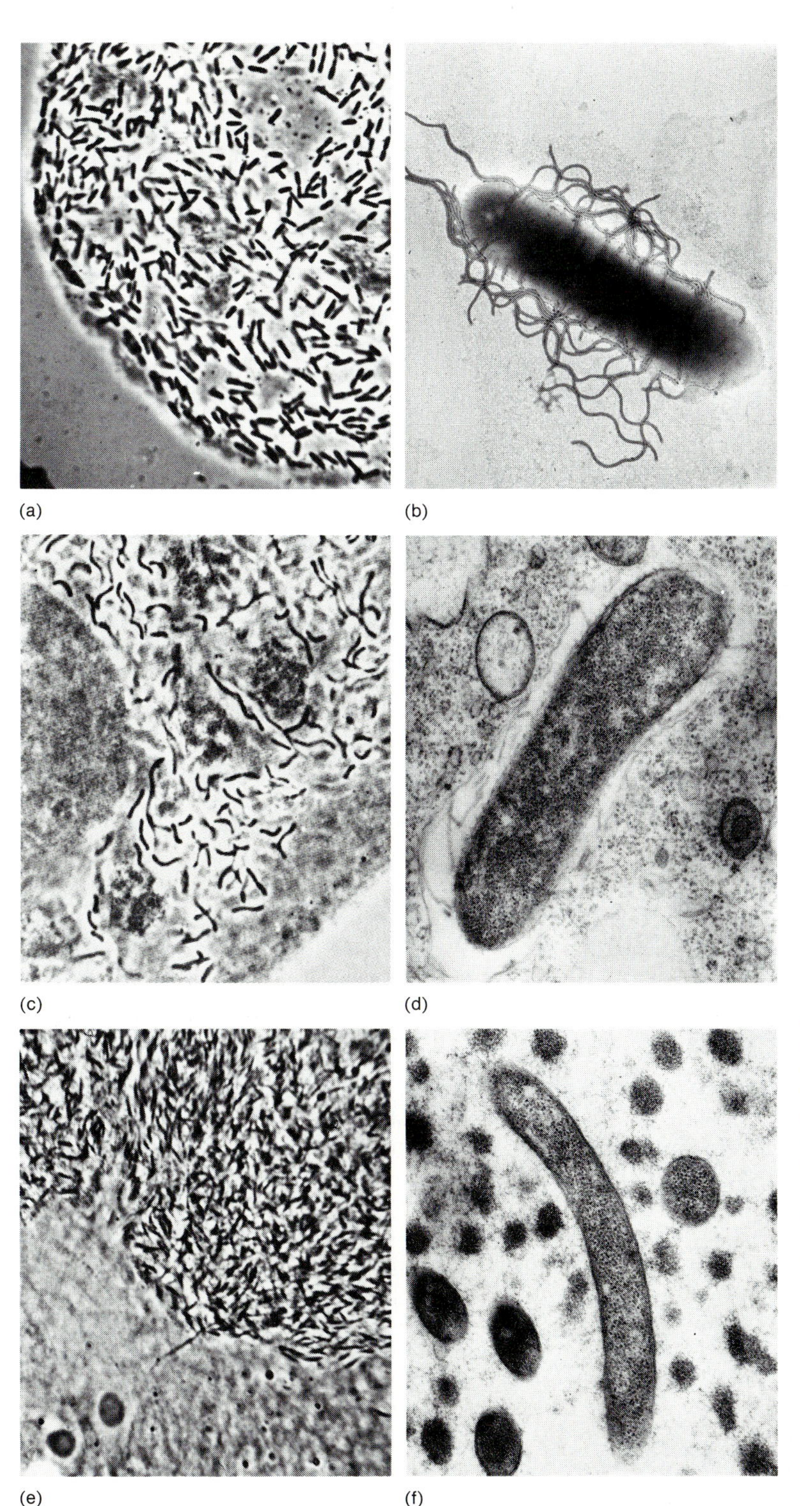


FIGURE 28.19

Some representative bacterial endosymbionts of *Paramecium aurelia*. (a) Lambda in host cytoplasm, stained unsectioned preparation (dark phase contrast, ×750). (b) Isolated lambda, negatively stained electron micrograph showing flagella, (×11,100). (c) Sigma in host cytoplasm, stained unsectioned preparation (dark phase contrast, ×729). (d) Sigma, electron micrograph of section through host cytoplasm showing symbiont and flagella (×22,000). (e) Alpha in host macronucleus, stained unsectioned preparation (dark phase contrast, ×870). (f) Alpha, electron micrograph of thin section of host macronucleus (×27,000). From G. H. Beale, A. Jurand, and J. B. Preer, Jr. "The Classes of Endosymbionts of *Paramecium aurelia*," *J. Cell Sci.* **5,** 65 (1969).

## SYMBIOSES BETWEEN MICROORGANISMS AND METAZOAN HOSTS

### Ectosymbioses of Protozoa with Insects: The Intestinal Flagellates of Wood-Eating Termites and Roaches

The woody tissue of trees, consisting mainly of cellulose and lignin, is unavailable as a source of food for most animals; in general, animals do not possess the enzymes necessary to degrade these polymers. Nevertheless, many species of insects obtain the bulk of their food from wood by virtue of an ectosymbiotic relationship with cellulose- and lignin-digesting microorganisms.

Both the termites and cockroaches, which have evolved from a common ancestral group, include some species that eat wood. All the wood-eating species of both groups harbor in their gut immense numbers of flagellated protozoa belonging to the polymastigotes and hypermastigotes. The flagellates are packed in a solid mass within a saclike dilation of the hindgut; it has been reported that they constitute over one-third of the body weight of the insect in some cases. The flagellates are responsible for cellulose digestion, of which the insects themselves are incapable. The flagellates, in turn, are themselves hosts to extracellular spirochetes (Figure 28.2) and to intracellular bacteria, and it is possible that some—if not all—of the cellulases produced by the flagellates derive from their intracellular symbionts. Nitrogen fixation also occurs in the termite gut, and is similarly assumed to reflect the activity of nitrogen-fixing bacteria. Whether these occur free in the gut, or as intracellular symbionts of the flagellates, is unknown.

The mode of transmission of the flagellated symbionts from one insect generation to the next differs in the two groups. The newly hatched nymphs of termites feed on fecal droplets that exude from the adults; the droplets are laden with symbionts, which infect the young insects. The newly hatched nymphs of cockroaches eat dry fecal pellets that are excreted by the adults; the pellets are laden with flagellate *cysts*, which are able to withstand desiccation. The cysts germinate in the gut of the nymphs, reestablishing the symbiosis.

One remarkable feature of the transmission cycle in cockroaches is that *the encystment of the flagellates is regulated by hormones of the insect.* The hatching of eggs in this insect coincides with the peak of the molting season, and protozoan cyst formation is induced by the molting hormone, *ecdysone.* This mechanism ensures that the flagellates will survive dessication in the fecal pellets and be available for infection of the hatching nymphs.

The flagellates enter a sexual cycle following encystment, nuclear and cytoplasmic divisions giving rise to one male and one female gamete from each cyst. Ultimately, these fuse to form a zygote. In an extensive series of studies, L. Cleveland established that sexuality in flagellates is induced by ecdysone, at concentrations of the hormone well below those required to induce molting of the insect. The adaptive significance of this regulation is not clear. It may reflect an obligatory coupling of gametogenesis with encystment.

### Endosymbioses of Fungi and Bacteria with Insects

Microbial endosymbioses are extremely widespread among insects. P. Buchner, the German biologist whose pioneering work on symbiosis has spanned more than half a century, discovered a striking correlation between the diet of insects and the presence of symbionts: symbionts are never found in insects that have a nutritionally complete diet, but are present in all insects that have a nutritionally deficient diet during their developmental stages. Thus, no carnivorous insect has symbionts, whereas insects that live on blood or on plant sap all contain symbionts. The main function of the symbiont is thus to provide the host with one or more growth factors that are lacking in the insect's diet.

Certain apparent exceptions prove this rule. Mosquitoes, for example, contain no symbionts, although they suck blood. It is only the adult female, however, that takes a blood meal; the larvae and pupae have a nutritionally complete diet consisting of microorganisms and organic debris. Conversely, the granary weevil, *Sitophilus granarius,* contains symbionts although it feeds on nutritionally rich grains. This genus, however, inherits its symbionts from its wood-eating ancestors; it is able to survive and reproduce if freed of its symbionts, *provided that it is fed a nutritionally rich diet.* Without symbionts its choice of food is severely restricted.

THE MICROBIAL ENDOSYMBIONTS The microbial endosymbionts of insects include both bacteria and yeasts. Most of these have been identified as such solely on the basis of their appearance and mode of reproduction in the host. A few, however, have been successfully isolated and grown in pure culture. For example, one of the symbionts of *Rhodnius,* a kissing bug, has been isolated and identified as an actinomycete of the genus *Nocardia.* Other isolated insect symbionts have proved to be coryne-

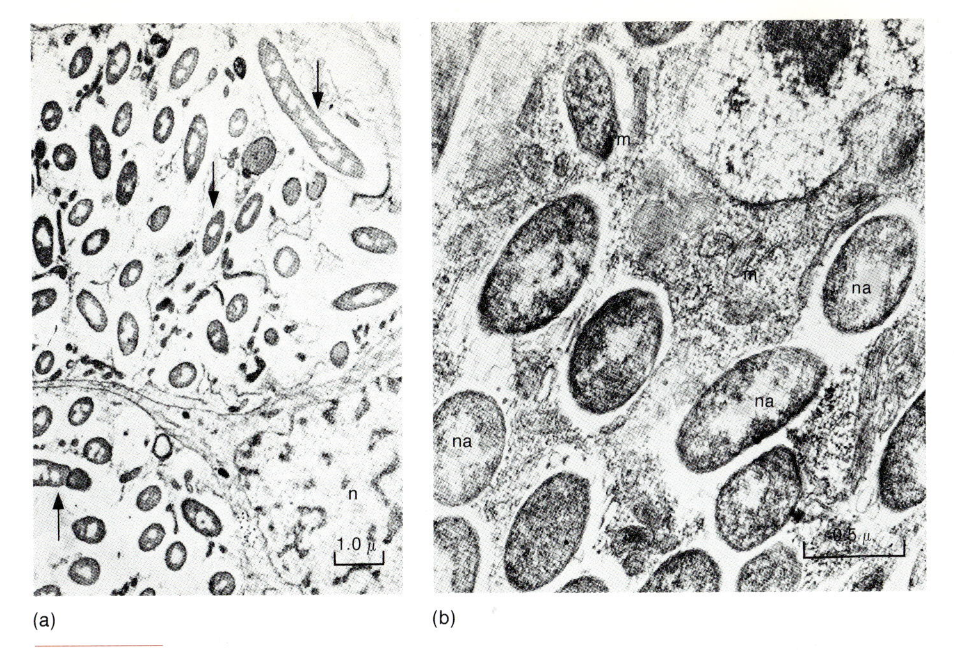


FIGURE 28.20

Mycetocytes of the insect, *Sitophilus granarius;* electron micrographs of thin sections. (a) Low magnification, × 3,330; arrows indicate bacterial endosymbionts. (b) High magnification, × 19,500. m, mitochondria; n, mycetocyte nucleus; na, nuclear area of endosymbiotic bacterium. From I. Grinyer and A. J. Musgrave, "Ultra-structure and Peripheral Membranes of the Mycetomal Microorganisms of *Sitophilus granarius* (L.) (Coleoptera)," *J. Cell Sci.* **1,** 181 (1966).

form bacteria or Gram-negative rods. Some yeasts have also been successfully isolated, notably from the long-horned beetles (*Cerambycidae*) and the deathwatch beetles (*Anobiidae*). Although most insects are monosymbiotic, it is not uncommon for a particular species to harbor two or more different microorganisms. The relationship between insects and their endosymbionts appears to be highly specific; an insect species can often be identified reliably by observing the nature of its symbionts.

THE LOCALIZATION OF THE ENDOSYMBIONTS The microbial endosymbionts are housed within specialized cells of the insect (Figure 28.20). These are called *mycetocytes* when they harbor yeasts and *bacteriocytes* when they harbor bacteria. Some authors refer to both as mycetocytes, and we will use this terminology here.

In some insects the mycetocytes are scattered randomly throughout a normal tissue, such as the wall of the midgut or the *fat body*, a loose, discontinuous tissue lining the body cavity. In many insects, however, the mycetocytes are restricted to special organs called *mycetomes*, the only function of which is to house the endosymbionts. It is possible to trace an evolutionary series of steps between ectosymbiosis, in which the symbionts develop in the lumen of the insect gut, and endosymbiosis in mycetomes. Figure 28.21 shows schematically the principal parts of the insect digestive tract. Figure 28.22 illustrates the localization of endosymbionts in out-pocketings or *blind sacs* of the insect midgut. Figure 28.23 shows how, in a series of species of anobiid beetles, the blind sacs have evolved to become progressively more independent of the midgut. In the most primitive endosymbioses the symbionts are found both extracellularly in the gut lumen and intracellularly in the blind sacs. In the

FIGURE 28.21

Schematic diagram of the digestive tract of the insect.

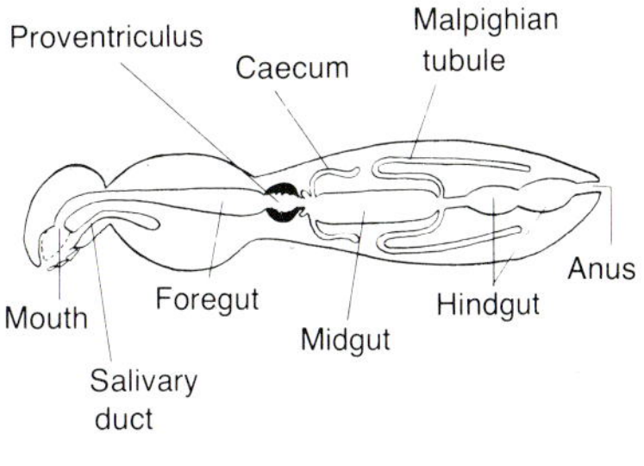

Proventriculus
Caecum
Malpighian tubule
Anus
Mouth
Foregut
Midgut
Hindgut
Salivary duct

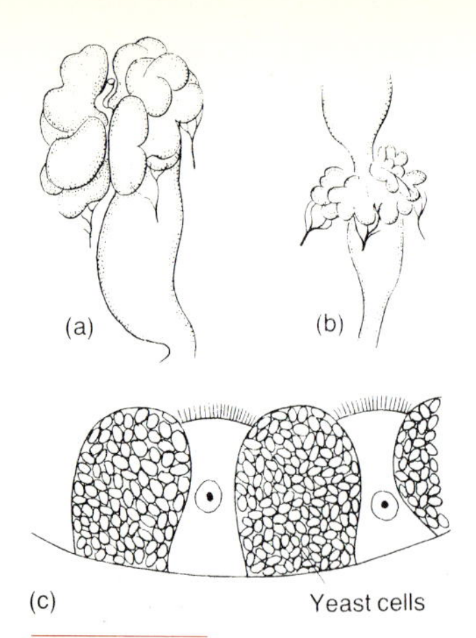


FIGURE 28.22

Blind sacs of the midgut of *Sitodrepa panicea*, an anobiid beetle: (a) larva; (b) adult; (c) epithelium of the blind sac of the larval midgut, showing yeast-filled mycetocytes separated by sterile cells with brush borders. After A. Koch.

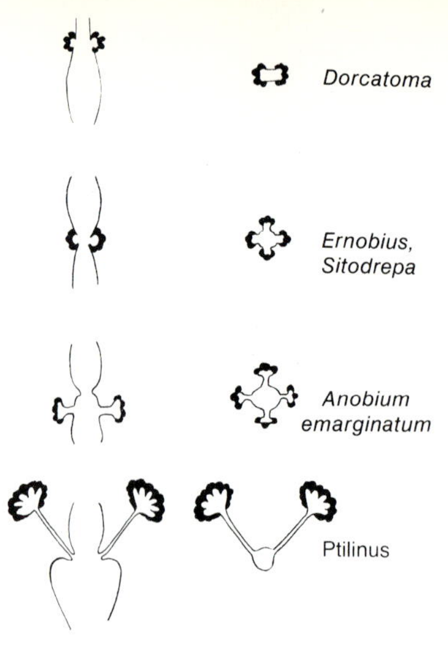


FIGURE 28.23

Blind sacs of the midgut of a series of anobiid beetles, showing evolutionary development of the blind sacs as independent organs. Left column: longitudinal sections. Right column: cross sections. After A. Koch.

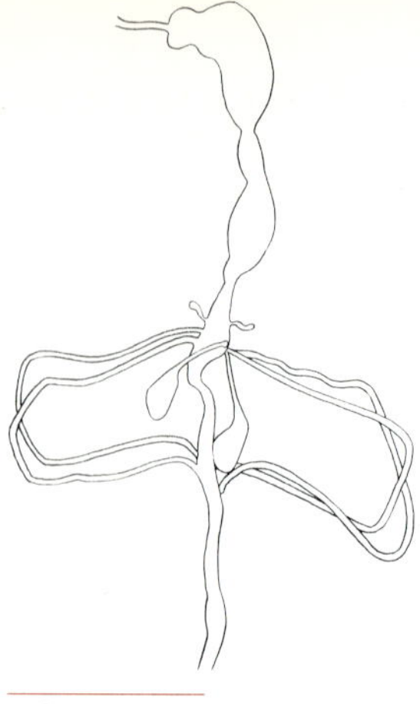

FIGURE 28.24

The adult gut of *Apion pisi*, showing the transformation of two of the six Malpighian tubules into mycetomes. After A. Koch.

most advanced forms the symbionts are completely isolated, and the blind sacs have evolved into independent organs, or mycetomes.

In some insects the mycetocytes are localized in the Malpighian vessels, the excretory organs of the insect. In certain genera of the *Curculionidae* (the family that includes weevils, snout beetles, and curculios), two of the six Malpighian vessels have become anatomically specialized for this purpose and have evolved into club-shaped mycetomes (Figure 28.24). In a number of other insects the mycetomes are detached from the gut, forming essentially independent structures in the body cavity.

### THE SIGNIFICANCE OF THE INSECT ENDOSYMBIOSES

The essential role played by the endosymbionts in the nutrition of the host can be demonstrated by artificial elimination of the symbionts and study of the behavior of the symbiont-free insects. Elimination has been accomplished by a variety of ingenious methods. In insects that smear their eggs with symbionts, the egg surface can be sterilized. In insects with well-defined and isolated mycetomes, such as the stomach disc of *Pediculus*, the louse, the mycetome can be surgically removed. Some insects can be freed of their symbionts by the use of high temperatures or of antibiotics. In some cases, growth of symbiont-free insects is severely retarded, and the adult stage may not be reached (Figure 28.25). In other cases, the principal effect is to disturb the reproductive system: the female organs may be damaged, or their formation may be completely blocked.

In many such experiments the loss of the symbionts can be totally compensated for by the provision of vitamins, particularly the B vitamins. In cockroaches (family Blattidae) it has also been shown that symbionts provide the host with some essential amino acids. Feeding the young insects $^{14}$C-labeled glucose led to the appearance of labeled tyrosine, phenylalanine, isoleucine, valine, and arginine in symbiotic, but not in symbiont-free, individuals. The injection of $^{35}$S-labeled sulfate similarly showed that the methionine and cysteine of the cockroach are synthesized by the symbiotic bacteria.

FIGURE 28.25

The effect of symbiont loss on the growth of larvae of *Sitodrepa panicea*: (a) symbiont-free larva on normal diet; (b) symbiont-free larva on normal diet plus 25 percent dried yeast; (c) normally infected larva on normal diet without supplementation. After A. Koch.

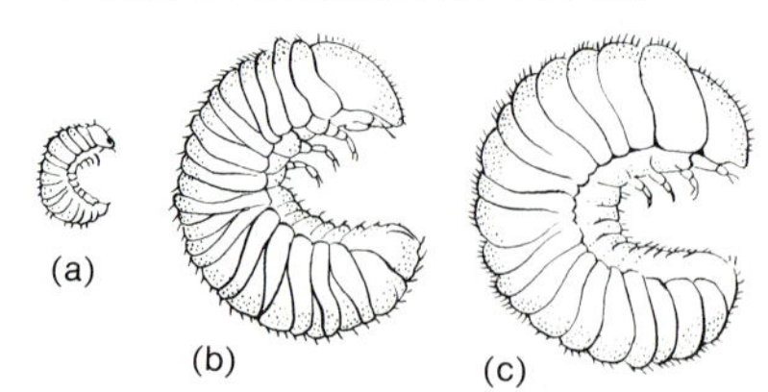

The bacterial symbionts of some insects also appear to aid the host in the breakdown of nitrogenous waste products (uric acid, urea, and xanthine).

### THE EVOLUTION OF THE INSECT-MICROBE SYMBIOSES

The evolutionary relationship between a specific symbiosis and the diet of the host can be clearly traced in the termites. The fossil record shows that the termites and the cockroaches split off from a common ancestor about 300 million years ago. The most primitive group of termites, *Mastotermes*, harbors endosymbiotic bacteria that are identical in type and location with those harbored by all genera of cockroaches. *Mastotermes* also harbors the intestinal flagellated protozoa, which, by digesting cellulose, allow their hosts to feed on wood. All higher termites, however, have lost the endosymbiotic bacteria and rely exclusively on intestinal flagellate protozoa for all symbiotic functions.

The enzymatic activities of the symbiotic microorganisms (synthesis of growth factors and digestion of cellulose) have allowed the insects to enter new ecological niches. The ability of insects to live by sucking blood, sucking plant sap, or boring in wood is entirely dependent on their development of organs to house symbionts and to their development of mechanisms for transmitting these symbionts from generation to generation.

The microbial symbionts have also undergone adaptive evolutionary changes. Such adaptation has frequently been accompanied by the loss of ability to grow in the free-living state. The evolutionary changes that have occurred are also indicated by the specificity exhibited when symbiont-free insects are infected with "foreign" symbionts. The anobiid beetle *Sitodrepa*, for example, can be reinfected either with its normal symbiotic yeast or with a foreign yeast. The former infects only the normal mycetocytes, whereas the latter infects all the epithelial cells of the blind sacs and midgut. Furthermore, the foreign yeast is not transmitted to the adult stage during morphogenesis. It is of particular interest that a beetle that harbors its normal symbiont *cannot* be infected with the foreign yeast that easily infects symbiont-free individuals; the normal symbiont appears to confer on its host an immunity to infection by related microorganisms, a phenomenon frequently observed in symbiotic relationships of vertebrates as well as of invertebrates.

## The Ruminant Symbiosis

The ruminants are a group of herbivorous mammals, that includes cattle, sheep, goats, camels and giraffes. Ruminants, like other mammals, cannot make cellulases. They have evolved an ectosymbiosis with microorganisms, however, which enables them to live on a diet in which the major source of carbon is cellulose.

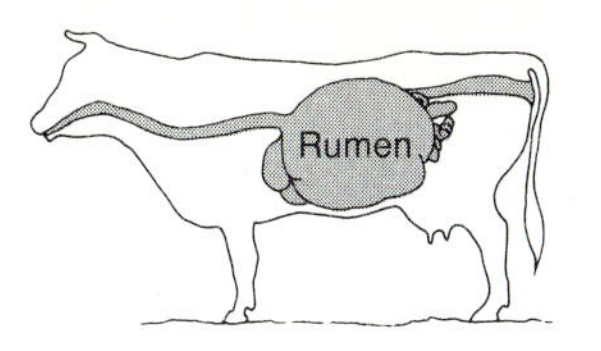


FIGURE 28.26
Diagram of the intestinal tract of the cow, to show the rumen.

The digestive tract of a ruminant contains four successive stomachs. The first two, collectively known as the *rumen* (Figure 28.26), are essentially a vast incubation chamber teeming with bacteria and protozoa. In the cow the rumen has a capacity of about 100 liters. The plant materials ingested by the cow are mixed with a copious amount of saliva and then passed into the rumen, where they are rapidly attacked by bacteria and protozoa. The total microbial population is enormous, and the population density is of the same order as that of a heavy laboratory culture of bacteria ($10^{10}$ cells per milliliter). Many different microorganisms are present (Figure 28.27), and the full details of their biochemical activities are not yet understood. However, the net effect is clear: the cellulose and other complex carbohydrates present in the ingested fodder are broken down with the eventual formation of simple fatty acids (acetic, propionic, and butyric) and gases (carbon dioxide and methane). The fatty acids are absorbed through the wall of the rumen into the bloodstream, circulating in the blood to the various tissues of the body where they are respired. The cow gets rid of the gases formed in the rumen by belching at frequent intervals. The microbial population of the rumen grows rapidly, and the microbial cells pass out of the rumen with undigested plant material into the lower regions of the cow's digestive tract. The rumen itself produces no digestive enzymes, but the lower stomachs secrete proteases, and as the microbial cells from the rumen reach this region they are destroyed and digested. The resulting nitrogenous compounds and vitamins are absorbed by the cow. For this reason, the nitrogen requirements of the cow and other ruminants are much simpler than those of other groups of mammals. Whereas humans require many amino acids (the so-called essential amino acids) preformed in the diet, the ruminant can grow on ammonia or urea, which are excretion products in most mammals. These simple nitrogenous com-

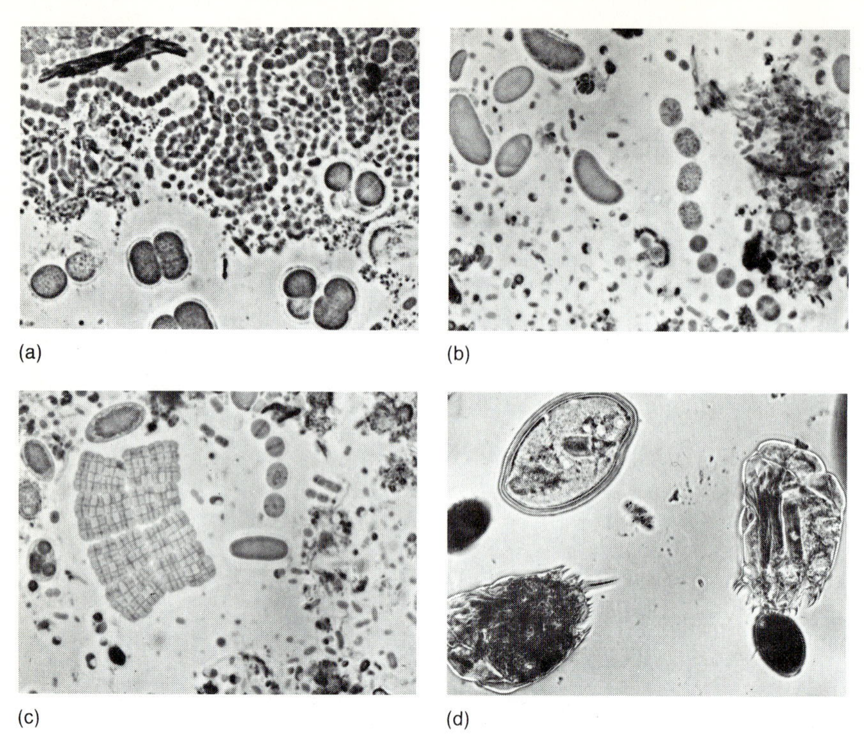

FIGURE 28.27

Some microorganisms from the rumen of the sheep. (a), (b), (c) Bacteria (ultraviolet photomicrographs, ×732). From J. Smiles and M. J. Dobson, "Direct Ultraviolet and Ultraviolet Negative Phase-contrast Micrography of Bacteria from the Stomachs of the Sheep," *J. Roy. Micro. Soc.* **75,** 244 (1956). (d) Ciliate protozoa (×9). Courtesy of J. M. Eadie and A. E. Oxford.

pounds are built up into microbial proteins by the rumen population.

The evolution of the rumen has involved both structural and functional modifications of the gastrointestinal tract. The principal structural modification is the development of a complex stomach, of which the largest compartments are essentially fermentation vats. The functional modifications that ruminants have undergone are even more profound. In the first place, the salivary glands do not secrete enzymes, the saliva being essentially a dilute salt solution (principally sodium bicarbonate and sodium phosphate) that provides a suitable nutrient base for the microbes of the rumen. In the second place, the lower fatty acids have very largely replaced sugar as the primary energy-yielding substrate. This, in turn, has led to changes in the enzymatic makeup of nearly all the tissues in the body, which respire fatty acids far more rapidly than do the tissues of nonruminants. Finally, the source of amino acids and vitamins has become very largely internalized (microorganisms instead of ingested food materials).

For the microorganisms that have taken up residence in the rumen, the situation also offers advantages; they are provided with an environment always rich in fermentable carbohydrates, well buffered by the saliva, and maintained at a constant favorable temperature, the body temperature of the cow. Individually, their ultimate fate is to fall prey to the proteolytic enzymes in the lower regions of the digestive tract; for the species, however, the rumen provides a safe and relatively constant ecological niche.

The rumen association is in a delicately balanced equilibrium, easily disturbed by slight changes of the environment. The principal failure to which this symbiosis is liable is a mechanical one. The gas production in the rumen of a cow is some 60 to 80 liters per day and, since the total volume of the rumen is only 100 liters, steady belching is necessary to get rid of the accumulating gases. For reasons that are not fully understood, certain diets lead to foaming of the rumen contents, and when this happens the belching mechanism of the cow fails to function properly. This causes a painful and, if untreated, eventually fatal affliction known as "bloat," (i.e., distention of the rumen by the trapped gases).

METABOLIC ACTIVITIES OF RUMEN BACTERIA Since the redox potential ($E_h$) of the rumen contents is steadily maintained at $-0.35$ V, all the microbial processes that occur in the rumen are anaerobic ones. As the ruminant grazes, the rumen receives a steady flow of finely ground plant materials mixed with saliva. The plant materials consist chiefly of cellulose, pectin, and starch, together with some protein and lipid. The first stage in the process is the digestion of these polymeric macromolecules. A great deal of attention has been given to the identification of the microorganisms responsible for the digestion of cellulose, since this is the major digestive process in the rumen. Between 1 and 5 percent of the bacterial cells in the rumen have been found to be cellulolytic; they produce extracellular cellulases that hydrolyze cellulose, glucose appearing as the final product of digestion.

The cellulose-digesting bacteria of the rumen, like the other principal ruminant microorganisms, are all strict anaerobes. Several different species of cellulose-digesting bacteria have been isolated and described: *Bacteroides succinogenes*, *Ruminococcus flavofaciens*, *R. albus*, and *Butyrovibrio fibrisolvens*. *B. succinogenes* requires ammonium $CO_2$, and straight and branched fatty acids. It forms succinic acid and lesser amounts of acetic acid. *Ruminococcus* produces principally succinic acid, and *Butyrovibrio* produces principally butyric acid.

The great bulk of the bacterial population, however, is noncellulolytic. These organisms rapidly utilize the glucose and cellobiose produced by the cellulolytic species; their great efficiency in scavenging these molecules presumably accounts for their predominance over the cellulolytic forms. Furthermore, many of the rumen bacteria (including some of the cellulolytic species) are capable of digesting starch, pectin, proteins, and lipids. Indeed, only the lignin of the ingested plant material escapes digestion by the rumen flora.

The products of digestion of polysaccharides, proteins, and lipids are fermented by the rumen bacteria. In the rumen these metabolic activities lead to the accumulation of the gases $CO_2$ and methane and the fatty acids acetic, propionic, and butyric acid. When the predominant rumen bacteria are isolated and studied as pure cultures, however, not only are the above listed products formed but also hydrogen gas, together with large amounts of formic, lactic, and succinic acids. By adding radioactive isotopes of these compounds to rumen contents, the reasons for their failure to accumulate in the rumen have been discovered. Thus, hydrogen gas is quantitatively combined with carbon dioxide to form methane, by the archaebacterium *Methanobacterium ruminantium*; the lactate is fermented to acetate plus smaller amounts of propionate and butyrate, by such organisms as *Peptostreptococcus elsdenii;* and the formate is converted first to carbon dioxide and hydrogen, by a variety of bacteria, and then to methane by *M. ruminantium*. Finally, the succinate is rapidly decarboxylated to propionate by *Veillonella alcalescens* and other bacteria.

The net result is the formation of carbon dioxide, methane, and acetic, propionic, and butyric acids, in remarkably constant proportions. Carbon dioxide accounts for 60 to 70 percent of the gases, methane accounting for the remainder; acetic acid represents 47 to 60 percent, propionic acid represents 18 to 23 percent, and butyric acid represents 19 to 29 percent of the fatty acids, respectively.

**The Rumen Protozoa** Protozoa were seen microscopically in rumen contents as early as 1843, but almost 100 years elapsed before they were successfully isolated and cultivated *in vitro*. Representatives of several genera have now been cultivated, notably the oligotrichous ciliates *Diplodinium*, *Entodinium*, *Epidinium*, *Metadinium*, and *Ophryoscolex*, and the holotrichous ciliates *Isotricha* and *Dasytricha*. Unfortunately, attemps to grow these protozoa in axenic (bacteria-free) culture have not yet been successful.

All such cultures to date have been contaminated with bacteria, intracellular as well as extracellular, so a final conclusion cannot be drawn from the enzymological and metabolic experiments that have been reported. Nevertheless, some suggestions about the role of these organisms come from experiments in which the protozoa were maintained alive for extended periods of time in the presence of high levels of bactericidal antibiotics. These experiments have implicated species of *Diplodinium* and *Metadinium* in the digestion of cellulose and species of *Entodinium* and *Epidinium* in the digestion of starch. Many of these protozoa are active predators on the rumen bacteria.

### Ectosymbioses of Microorganisms with Birds: The Honey Guides

The honey guides, a group of birds belonging to the genus *Indicator*, are found in Africa and India. Their name accurately describes their behavior: they literally guide honey badgers, as well as humans, to the nests of wild bees, where they wait for their follower to break open the hive. When the badger (or human) has departed, the honey guide proceeds to feed on the remnants of honeycomb that have been left exposed.

This behavior became all the more remarkable when it was discovered that these birds do not possess enzymes for digesting beeswax. Instead, they harbor in their intestines two microorganisms that carry out the digestion for them: a bacterium, *Micrococcus cerolyticus*, and a yeast, *Candida albicans*. The micrococcus is a highly specialized symbiont, depending on a growth factor that is produced in the small intestine of the honey guide.

## FURTHER READING


### Books

AHMADJIAN, V., and M. E. HALE, eds., *The Lichens.* New York: Academic Press, 1973.

BROWN, D. H., D. L. HAWKSWORTH, and R. H. BAILEY, eds., *Lichenology: Progress and Problems.* New York: Academic Press, 1976.

HALE, M. E., *The Lichens.* Dubuque, Iowa: Wm. C. Brown Co., 1969.

HENRY, S. M., ed., *Symbiosis*, Vols. 1 and 2. New York: Academic Press, 1966, 1967.

MARGULIS, L., *Symbiosis in Cell Evolution: Life and Its Environment on the Early Earth.* San Francisco: W. H. Freeman and Co., 1981.

*Symbiosis: 29th Symposium of the Society for Experimental Biology.* New York: Cambridge University Press, 1975.

### Reviews

ALEXANDER, M., "Why Microbial Predators and Parasites Do Not Eliminate Their Prey and Hosts," *Ann. Rev. Microbiol.* **35,** 113 (1981)

BARNETT, H. L., and F. L. BINDER, "The Fungal Host-Parasite Relationship," *Ann. Rev. Phytopathol.* **11,** 273 (1973).

BREZNAK, J. A., "Intestinal Microbiota of Termites and Other Xylophagous Insects," *Ann, Rev. Microbiol.* **36,** 323 (1982).

HUNGATE, R. E., "The Rumen Microbial Ecosystem," *Ann. Rev. Microbiol.* **29,** 39 (1975).

PISTOLE, T. G., "Interaction of Bacteria and Fungi with Lectins and Lectin-like Substances," *Ann. Rev. Microbiol.* **35,** 85 (1981).

PREER, J. R., L. B. PREER, and A. JURAND, "Kappa and Other Endosymbionts in *Paramecium aurelia*," *Bacteriol. Rev.* **38,** 113 (1974)

TAYLOR, D. L., "Algal Symbionts of Invertebrates," *Ann. Rev. Microbiol.* **27,** 171 (1973).

WOLIN, M. J., "The Rumen Fermentation: A Model for Microbial Interactions in Anaerobic Ecosystems," in *Advances in Microbial Ecology*, Vol. 3, ed. M. Alexander, pp. 49–77. New York: Plenum Press, 1979.

# Chapter 29
# Nonspecific Host Defense

Diseases caused by microorganisms and metazoan parasites are termed *infectious diseases*. Today they are a secondary, yet important cause of death in highly developed countries and the leading cause of death in many lesser developed countries. Those species that cause disease are called *pathogens* or *pathogenic species*. With the exception of the viruses, only a minority of microbial species are pathogenic, but these exert a great impact on natural ecosystems and human affairs.

## PHYSICAL AND CHEMICAL BARRIERS TO INFECTION

With the exception of the few patients who live in specially constructed germ-free environments, humans come into daily contact with pathogens, yet few of these contacts result in disease. Maintenance of the relatively good state of health enjoyed by most individuals living in highly developed countries depends on a complex set of defenses against infection that includes physical barriers to invasion by microorganisms, antimicrobial chemicals, and specialized host cells that eliminate most microorganisms that invade body tissues.

### Body Surfaces

Skin and mucous membranes, which are impervious to most microorganisms, constitute the primary barrier against *infection*, i.e., invasion and growth of microorganisms in body tissues. A burn or wound that destroys the integrity of these barriers almost always results in infection, at least a local one, but most wounds are quickly sealed by blood clots, which themselves constitute an important secondary barrier to infection. Although mucous membranes of the eyes, lungs, intestines, and urinary tract are intrinsically more susceptible than skin to penetration by microorganisms, these tissues have the additional protection afforded by *lavaging*, i.e., being washed by the fluids that recurrently move across these body surfaces.

### The Role of pH

The low pH of certain body surfaces constitutes an added barrier to microorganisms. Most dramatic of these is the acidity (pH $\sim$ 2) within the stomach of humans and certain animals which kills the majority of microorganisms that are ingested. However, a few pathogens, including *Shigella* spp., are remarkably acid-resistant; they survive passage through the stomach and are thereby able to infect the intestinal wall. The mildly acid pH of skin and the vagina cannot kill microorganisms, but by inhibiting their growth it too serves as a barrier to infection.

### Antimicrobial Compounds

Blood and other body fluids contain a variety of antibacterial proteins. The most important of these is the class of proteins termed *antibodies* (Chapter 30), individual members of which confer resistance to specific pathogenic microorganisms. The second most important class of protective proteins, collectively termed *complement* (Chapter 30), is also found in blood: these trigger inflammation at a site of infection and kill Gram-negative bacteria. Blood, tears, and saliva also contain *lysozyme*, an enzyme that can kill Gram-positive bacteria (Chapter 6) by dissolving their peptidoglycan layer, causing them to lyse. Yet another protective protein, the basic polypeptide *beta-lysin*, is released from *platelets* (cell fragments that initiate the clotting process); it can kill some Gram-positive bacteria by a mechanism that is still unknown.

### Sequestration of Iron

The bodies of vertebrates contain large amounts of iron, but very little is available as a nutrient for microbial growth because most iron is contained within cells, tightly bound to hemoglobin, myoglobin, and cytochromes. Only a small amount is found extracellularly in the blood, but this too is unavailable to microorganisms because it is bound to *transferrin*, the protein that transports iron from the small intestine, where it is absorbed, to tissues, in which it is used. Excess iron is stored intracellularly, tightly bound to a protein termed *ferritin*. Through the action of these proteins, the concentration of free iron in blood or other tissues is normally less than $10^{-18}$ M, far lower than the concentration necessary for growth of microorganisms. Thus, most pathogens have evolved specific mechanisms by which they release iron from these various host proteins in order to use it themselves (see Chapter 31).

## THE PROTECTIVE ROLE OF HOST MICROFLORA

Prior to birth, a mammal is normally completely free of microorganisms. However, as it passes through the birth canal its skin becomes covered with them, and during the first few days after birth, microorganisms enter both the upper respiratory tract and the gastrointestinal tract. Some of these microorganisms are able to grow and survive in their new environment, i.e., they are able to *colonize* their host, thereby constituting its *normal microflora* (Table 29.1). Although some of the bacteria that make up the normal flora are pathogens, they rarely cause disease unless they are introduced into another region of the body, usually one that is not protected by colonization. For example, *Escherichia coli* normally grows harmlessly in the colon, but can cause serious kidney infections when it enters the urinary tract. Even normally harmless members of the normal flora can cause serious disease in an *immunocompromised host*, i.e., an individual with a defective immune system (Chapter 30).

Microorganisms that comprise the normal flora compete effectively for the limited space and nutrients available in their environment, thereby limiting the growth of individual members of the flora and rendering it difficult for new microorganisms to colonize the host. A reduction in the populations of normal flora can promote the growth of pathogens by reducing competition. For example, excessive cleansing with soaps can diminish the normal flora of the vagina, allowing the pathogenic yeast *Candida albicans* to colonize and grow abundantly. Similarly, treatment with a broad-spectrum antibiotic (Chapter 33) reduces the normal flora of mucous membranes, sometimes allowing anti-

**TABLE 29.1**

**Examples of Normal Human Microflora of the Skin, Nose, Mouth, Throat, and Vagina**

| Habitat | Species |
|---|---|
| Skin | *Corynebacterium xerosis, Micrococcus luteus, Physosporium orbicularis, P. ovale, Propionibacterium acnes, Staphylococcus aureus, S. epidermidis, Streptococcus anginosus.* |
| Mouth, nose, and throat | *Actinomyces israelii, A. naeslundii, A. odontolyticus, A. viscosus, Bacteroides coagulans, B. corrodens, B. melaninogenicus, B. ochraceus, B. oralis, B. pneumosintes, Corynebacterium diphtheriae, C. pseudodiphtheriticum, C. xerosis, Fusobacterium mortiferum, F. necrophorum, F. nucleatum, F. plauti, Haemophilus haemolyticus, H. influenzae, H. parahaemolyticus, H. parainfluenzae, H. paraphrohaemolyticus, H. paraphrophilus, Lactobacillus acidophilus, L. brevis, L. cellobiosis, L. fermentum, L. plantarum, L. salivarius, Mycobacterium gastri, M. gordonae, M. peregrinum, M. scrofulaceum, M. terrae, M. triviale, Neisseria meningitides, N. mucosa, N. sicca, N. subflava, Peptococcus aerogenes, P. asaccharolyticus, Staphylococcus aureus, S. epidermidis, Streptococcus anginosis, S. equisimilis, S. mitis, S. pneumoniae, S. pyogenes, S. salivarius, S. sanguis, Treponema denticola, T. macrodentium, T. orale, T. vincentii, Veillonella alcalescens, V. parvula.* |
| Vagina | *Bacteriodes corrodens, Clostridium ghoni, Haemophilus paraphrohaemolyticus, H. paraphrophilus, Lactobacillus acidophilus, L. jensenii, Peptococcus aerogenes, P. anaerobicus, P. asaccharolyticus, Staphylococcus saprophyticus, Streptococcus anginosus, S. refringens.* |

biotic-resistant pathogens to colonize and grow abundantly. Under such conditions, *C. albicans* sometimes colonizes the mouth or vagina causing the diseases, *oral* or *vaginal candidiasis*, which are serious but rarely life-threatening. However, treatment with certain antibiotics, including clindamycin, can so disrupt the intestinal microflora that *pseudomembranous colitis*, a very serious disease of the colon, can result from the proliferation of certain drug-resistant bacteria (often *Clostridium difficile*).

### Germ-Free Animals

Direct evidence of the critical importance of a normal microflora in host defense comes from the study of *germ-free animals* which are now readily obtainable owing to the availability of specialized techniques and equipment designed to deliver mammals aseptically by caesarean section and to rear them in sterile environments. Upon attaining sexual maturity, they mate and produce germ-free litters, thus allowing the maintenance of colonies of animals that lack a microbial flora. Such animals are superficially normal, but they possess poorly developed lymphoid tissue (Chapter 30), an abnormally thin intestinal wall, and a greatly enlarged cecum. They have an abnormally low concentration of antibodies in their blood, and they must be supplied with vitamin K in their diet, a vitamin that is normally synthesized by the intestinal microflora.

Germ-free animals also exhibit an abnormal susceptibility to microbial disease. When exposed to a pathogen, germ-free animals are far less resistant in most cases than animals with a normal microflora: the number of cells necessary to infect a germ-free animal is typically much smaller than the number necessary to cause disease in a normal host. However, infections caused by the protozoan agent of amebic dysentery, *Entamoeba histolytica*, is a notable exception; germ-free animals are almost completely resistant to this microorganism, probably because the intestinal tract of germ-free animals lacks the bacteria that normally serve as food for these amebae.

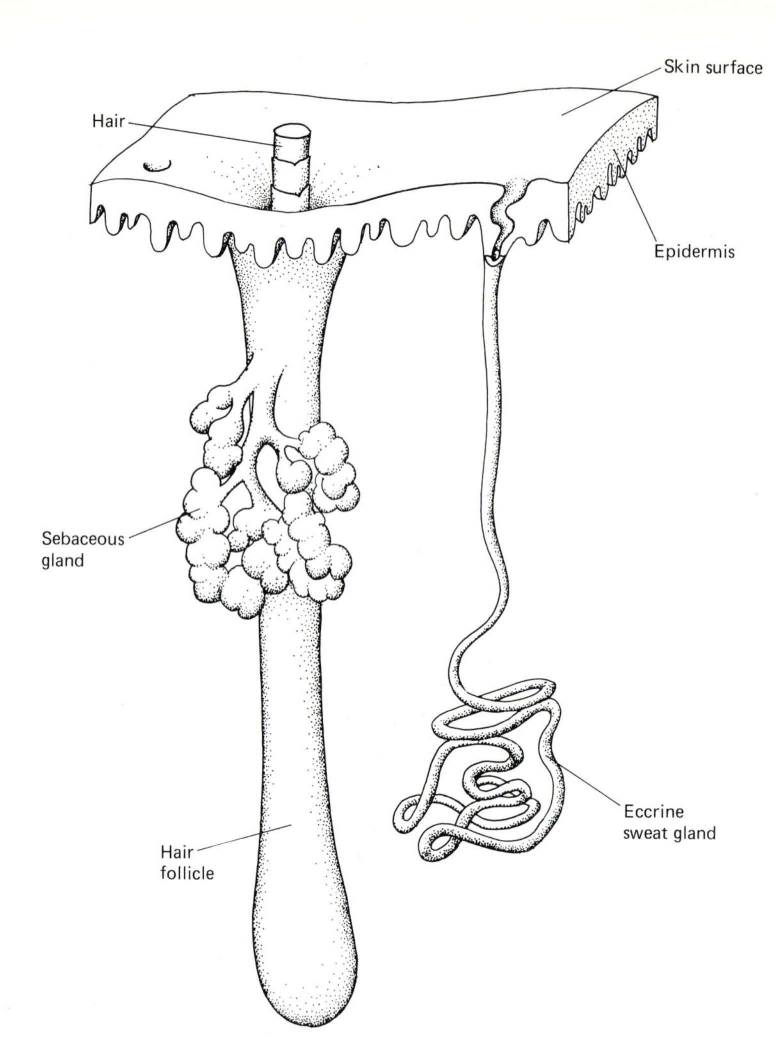


FIGURE 29.1
Diagram of human skin. A hair follicle and its associated sebaceous gland and an eccrine sweat gland are shown embedded in the dermis.

## Normal Skin Flora

Human skin possesses two contrasting environments for microbial colonization (Figure 29.1). The more accessible of these, the surface, is hostile because it is relatively dry and salty, but even here, certain components of the normal flora, some Gram-positive cocci including *Micrococcus* spp. and *Staphylococcus epidermidis* proliferate, as do some pathogens. Notably, the important pathogen *S. aureus* is commonly found on the skin of healthy individuals. Other organisms commonly isolated from human skin are *Acinetobacter calcoaceticus* and enteric bacteria including *Escherichia coli.*

The second environment of the skin, a smaller but more hospitable one, is provided by skin glands, composed principally of the *eccrine sweat glands* and *sebaceous glands* (Figure 29.1). Eccrine sweat glands occur in most areas of skin and are present from birth; sebaceous glands occur largely on the face, back and chest and develop during adolescence. Unlike the skin surface, the lumen of these glands has a high water content and a low oxygen tension. The most numerous microorganism found there is an anaerobic rod, the Gram-positive bacterium *Propionibacterium acnes.* Nonpathogenic members of the genera *Corynebacterium* and *Streptococcus* are also present. Skin glands are also inhabited by large numbers of yeasts belonging to the genus *Physosporium,* the major species of which are *P. ovale* and *P. orbiculare*. They are small (3 to 4 $\mu$m in diameter) and utilize, for metabolism, fatty acids that are present in these glands.

Although the normal skin flora usually plays a protective role, it plays a pathogenic one in the disease *acne vulgaris* that commonly occurs during adolescence when sebaceous glands become plugged. Normally these glands produce a substance termed *sebum,* which is formed when cells in the lining of the gland die, become released into the gland's internal cavity, and lyse; there bacteria metabolize some of the cellular components, producing sebum. In some individuals, sebum triggers an inflammatory response, causing redness and swelling that blocks the gland's duct (Figure 29.1). Such a plugged sebaceous gland is called a blackhead or pimple. *Propionobacterium acne* has been tentatively identified as the organism responsible for producing the substances in sebum that cause inflammation. This organism is extremely sensitive to the antibiotic tetracycline, a drug that in low doses often exerts a beneficial effect on the disease.

### Normal Flora of the Mouth and Upper Respiratory Tract

Microorganisms are normally removed quickly from the lungs by a film of mucous that moves up the bronchi and trachea until it flows into the esophagus where it is swallowed. As a consequence, the lungs of healthy people are almost completely free of bacteria. In contrast, both the mouth and the respiratory tract above the trachea are inhabited by large numbers of microorganisms (Table 29.1): nonpathogenic *Streptococcus* spp. colonize surfaces of the mouth; *S. salivaris* and *S. sanguinis* colonize the tongue; *S. mitis* colonizes the cheek. *S. mutans* colonizes teeth by producing a brittle capsule of dextran, termed *dental plaque*, in which this species grows in association with other bacteria including *Peptococcus* spp., *Actinomyces* spp. and *Veillonella* spp. Where the teeth meet the gingiva (gums), anaerobic bacteria in the genera *Bacteroides*, *Fusobacterium*, and *Treponema* grow. Inhabitants of the upper respiratory tract include nonpathogenic organisms in the genera *Streptococcus*, *Corynebacterium*, *Neisseria* and *Haemophilus*. The pathogens *Staphylococcus aureus*, *Haemophilus influenzae*, and *Neisseria meningitidis* are commonly found there as well, even in healthy individuals.

### Normal Intestinal Flora

The stomach and upper small intestine are sparsely inhabited by bacteria; typically there are only about $10^4$ per ml of contents, but the numbers of bacteria increase markedly in the lower small intestine and they become quite large in the terminal colon where there are more than $10^{11}$ bacteria per ml of contents. The composition of the microflora of the colon is determined to a large degree by the food that the host consumes, as is illustrated dramatically by the succession of microfloras that occurs during infancy. The colon of breast-fed infants is inhabited almost exclusively by *Bifidobacterium* spp., but soon after weaning, these largely disappear and the predominant microorganisms become obligate anaerobes belonging to the genus *Bacteroides*. Other obligate anaerobes here include species of *Eubacterium*, *Peptostreptococcus*, *Fusobacterium*, *Coprococcus*, *Ruminococcus*, *Clostridium*, and *Peptococcus* (Table 29.2). Facultative anaerobes are present in lesser numbers. Even the most abundant of these, *Escherichia coli*, which is sometimes considered to be a typical member of the colonic microflora, constitutes only about 0.1 percent of the total population. Other facultative organisms normally present are *Klebsiella pneumoniae*, *Enterobacter aerogenes*, *Streptococcus faecalis*, and the yeast *Candida albicans*. In addition to the known microbial components of the colonic microflora, many others are probably present because micrographs of specimens from the colon reveal microorganisms that do not resemble any known species.

## THE ROLE OF PHAGOCYTIC CELLS IN THE ANIMAL HOST

The critically important role of phagocytic cells in defending against invading microorganisms is brought into focus by the consequences of radiation. Following a large dose of ionizing radiation, e.g., X rays or gamma rays, the body's physical and chemical barriers against invasion may remain unaffected, but soon thereafter the microorganisms constituting the normal flora of mucous membranes invade underlying tissues, spread throughout the body, and cause death. This dramatic loss of resistance to microbial invasion results from the killing of cells in bone marrow that are unusually sensitive to radiation. These cells are essential because they generate all types of cells in the blood, including the *leukocytes* (white blood cells), which constitute the host's major defense against infection. Without these cells, the host can remain healthy only in a germ-free environment.

### Leukocytes

Five types of leukocytes are found in normal blood (Figure 29.2): *polymorphonuclear neutrophils* (PMNs), *eosinophils*, *basophils*, *lymphocytes*, and *monocytes*. Their principal functions and concentrations in blood are shown in Table 29.3. Polymorphonuclear neutrophils or *neutrophils*, as they are commonly called, are recognized by the distinctive shape of their nucleus: it is divided into segments connected by thin bridges of nuclear material. Ba-

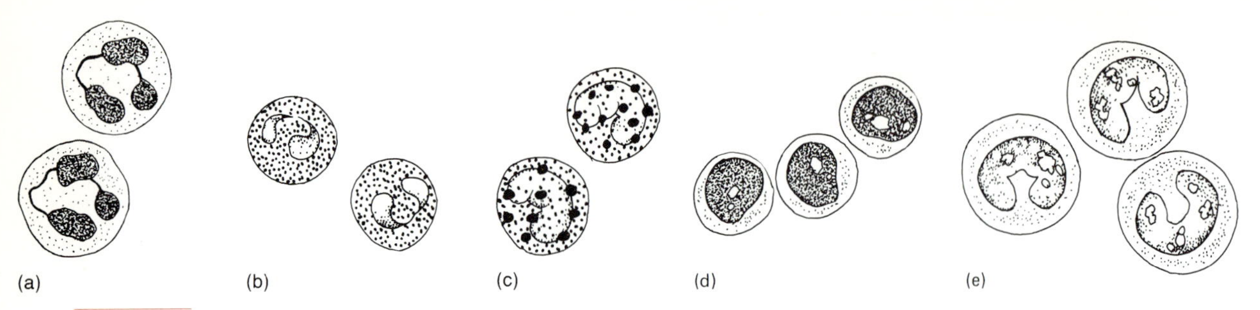


FIGURE 29.2
Human leukocytes: (a) neutrophils, (b) eosinophils, (c) basophils, (d) lymphocytes, (e) monocytes.

sophils, eosinophils, and neutrophils all possess high concentration of cytoplasmic granules. Collectively, these granule-containing leukocytes are called *granulocytes*, the subtypes of which are distinguished by their staining properties. The granules in basophils stain intensely with basic dyes such as hematoxylin, while those in eosinophils stain intensely with eosin. Lymphocytes are recognized by the morphology of their nucleus; it is quite round and is stained intensely. Monocytes are recognized as being the largest cells normally found in blood; they also are distinguished by the distinctive appearance of their nuclei which may be indented or even horseshoe-shaped.

Monocytes have the property of migrating from blood into most tissues of the body including the lung, liver, and lymphoid tissue; there they are called *macrophages*, which can be seen in normal lung, liver, and lymphoid tissue. Monocytes accompanied by PMNs also infiltrate areas of infection and inflammation where they actively phagocytize and thereby kill the invading microorganisms. Eosinophils infiltrate areas of infection to a lesser extent. They phagocytize microorganisms but are

**TABLE 29.2**

**Examples of Adult Human Intestinal Microflora**

| Approximate Concentration per ml of Feces | Species |
|---|---|
| $6 \times 10^{10}$ | *Bacteroides fragilis* |
| $2 \times 10^{10}$ | *Eubacterium aerofaciens* |
| | *Peptostreptococcus productus* |
| $1 \times 10^{10}$ | *Fusobacterium prausnitzii* |
| $5 \times 10^{9}$ to $1 \times 10^{10}$ | *Coprococcus eutactus, Eubacterium rectale, Ruminococcus bromii, Bifidobacterium adolescentis, B. longum, Gemmiger formicilis, E. siraeum* |
| $1 \times 10^{9}$ to $5 \times 10^{9}$ | *R. torques, E. eligens, Bacteroides eggerthii, Clostridium leptum, E. biforme, Bifidobacterium infantis, Coprococcus comes, Bacteroides capillosus, R. albus, E. formicigenerans, E. ballii, E. ventriosum, F. russii, R. obeum, Clostridium ramosum, Lactobacillus leichmannii* |
| $5 \times 10^{8}$ to $1 \times 10^{9}$ | *R. callidus, Butyrivibrio crossotus, Acidaminococcus fermentans* |
| $1 \times 10^{8}$ to $5 \times 10^{8}$ | *Coprococcus catus, E. hadrum, E. cylindroides, E. ruminantium, E. limosum, Bacteroides praeacutus, F. mortiferum, F. naviforme, F. nucleatus, R. flavefaciens, Clostridium innocuum, Escherichia coli, Streptococcus morbillorum* |
| Less than $10^{8}$ | *Enterobacter aerogenes, Klebsiella pneumoniae, Streptococcus faecalis* |

**TABLE 29.3**

**Normal Leukocyte Concentrations and Functions**

| Type | Percent of Total Leukocytes[a] | Function |
|---|---|---|
| Granulocytes | | |
| Neutrophils | 55–70 | Migration to site of inflammation, phagocytosis of bacteria |
| Eosinophils | 1–3 | Migration to site of inflammation, killing of helminthic larvae |
| Basophils | 0–1 | Histamine release |
| Lymphocytes | 25–35 | Antibody synthesis, regulation of immune response, killing of foreign eukaryotic cells and cancer cells |
| Monocytes | 3–7 | Migration into tissues to become phagocytes termed macrophages |

[a] The normal range varies slightly from laboratory to laboratory. The total leukocyte count is usually 5,000 to 10,000 per ml. of blood.

less active than neutrophils or macrophages. However, they are particularly effective in eliminating the larvae of helminthic parasites.

Lymphocytes and basophils exert their antimicrobial activities only indirectly. Lymphocytes are the cells that synthesize antibodies; they also mediate a type of defense termed *cellular immunity*, which is discussed in Chapter 30. Basophils, along with a type of cell found in connective tissue, termed *mast cells*, contain *histamine* which, when released, causes *inflammation*, a fundamental process that protects against infection (see below).

## Phagocytosis

*Phagocytosis* is the endocytosis (Chapter 3) of a particle such as a bacterium or virus. This process does not occur if the surface of the object is negatively charged as the bacterial envelope normally is. Phagocytosis becomes possible after positively charged proteins, either antibodies or a component (C3b) of complement (see Chapter 30), bind to the cell surface. Certain antibodies and C3b play an additional role in phagocytosis: they bind to specific receptors present on the surfaces of both neutrophils and macrophages and also bind to components on the bacterial cell. Consequently, these proteins can attach a bacterial cell to a host phagocyte. This attachment is facilitated when the bacterial cell becomes trapped between the phagocyte and a surface such as a mucous membrane. The cytoplasmic membrane of the phagocyte then invaginates and eventually engulfs the attached bacterium, trapping it in a pocket of membrane that pinches off within the cytoplasm (Figure 29.3). The membrane-bound vacuole in which the bacterium is contained is termed a *phagosome*. Later, the phagosome fuses with a lysosome, producing a new vacuole termed a *phagolysosome*. Lysosomes contribute to the fused vacuole a variety of hydrolytic enzymes including lysozyme, phospholipase, ribonuclease, deoxyribonuclease, and several proteases. Together these function rapidly to kill the entrapped microorganism.

Some granules of phagocytes also contain enzymes of a pathway (Figure 29.4) leading to the synthesis of the highly lethal radical, superoxide ($O_2^-$), and the less lethal but toxic compound, hydrogen peroxide ($H_2O_2$). The first enzyme in this pathway, *NADPH oxidase*, catalyzes a reaction between NADPH and $O_2$ to form superoxide; the second enzyme, *superoxide dismutase*, converts superoxide to hydrogen peroxide. Neutrophils contain an enzyme, *myeloperoxidase*, that lengthens the pathway by one additional step, producing yet another toxic compound, hypochlorous acid, as a product of a reaction between hydrogen peroxide and chloride ion. The various bacteriocidal activities of phagocytosis are remarkably effective: most bacteria phagocytized by macrophages or PMNs are killed within 30 minutes by products of this pathway.

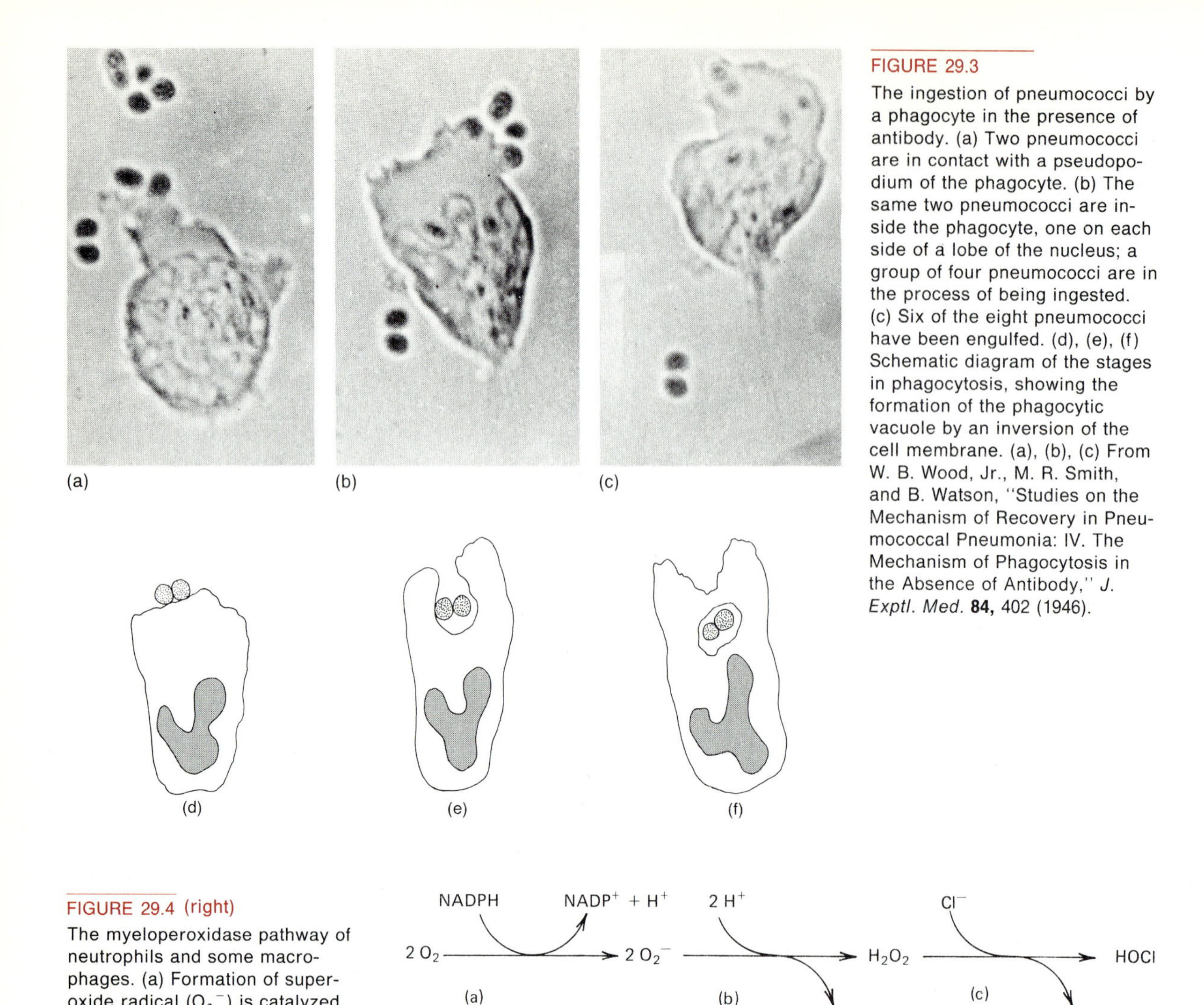


FIGURE 29.3

The ingestion of pneumococci by a phagocyte in the presence of antibody. (a) Two pneumococci are in contact with a pseudopodium of the phagocyte. (b) The same two pneumococci are inside the phagocyte, one on each side of a lobe of the nucleus; a group of four pneumococci are in the process of being ingested. (c) Six of the eight pneumococci have been engulfed. (d), (e), (f) Schematic diagram of the stages in phagocytosis, showing the formation of the phagocytic vacuole by an inversion of the cell membrane. (a), (b), (c) From W. B. Wood, Jr., M. R. Smith, and B. Watson, "Studies on the Mechanism of Recovery in Pneumococcal Pneumonia: IV. The Mechanism of Phagocytosis in the Absence of Antibody," *J. Exptl. Med.* **84,** 402 (1946).

FIGURE 29.4 (right)

The myeloperoxidase pathway of neutrophils and some macrophages. (a) Formation of superoxide radical ($O_2^-$) is catalyzed by NADPH oxidase. (b) Superoxide dismutase catalyzes formation of hydrogen peroxide. (c) Myeloperoxidase catalyzes formation of hypochlorous acid.

## INFLAMMATION

Inflammation is a general response to tissue injury or to the presence of foreign material. It is characterized by the symptoms of pain, redness, heat, and swelling. Microscopic examination of inflamed tissue reveals *edema*, the accumulation of fluid, and the presence of an abnormally large number of leukocytes.

Inflammation is classified as *acute* or *chronic* depending on which types of leukocytes predominate. One of the first signs of acute inflammation, which usually occurs during early stages of infection, is adhesion of neutrophils to the *endothelial cells* that line small blood vessels. Neutrophils then insert pseudopods between endothelial cells and migrate to the space between them and the basement membrane (Figure 29.5). Finally, they pass through the basement membrane toward the site of inflammation where they often accumulate in large numbers. Eosinophils also migrate into a region of acute inflammation but are usually outnumbered by PMNs.

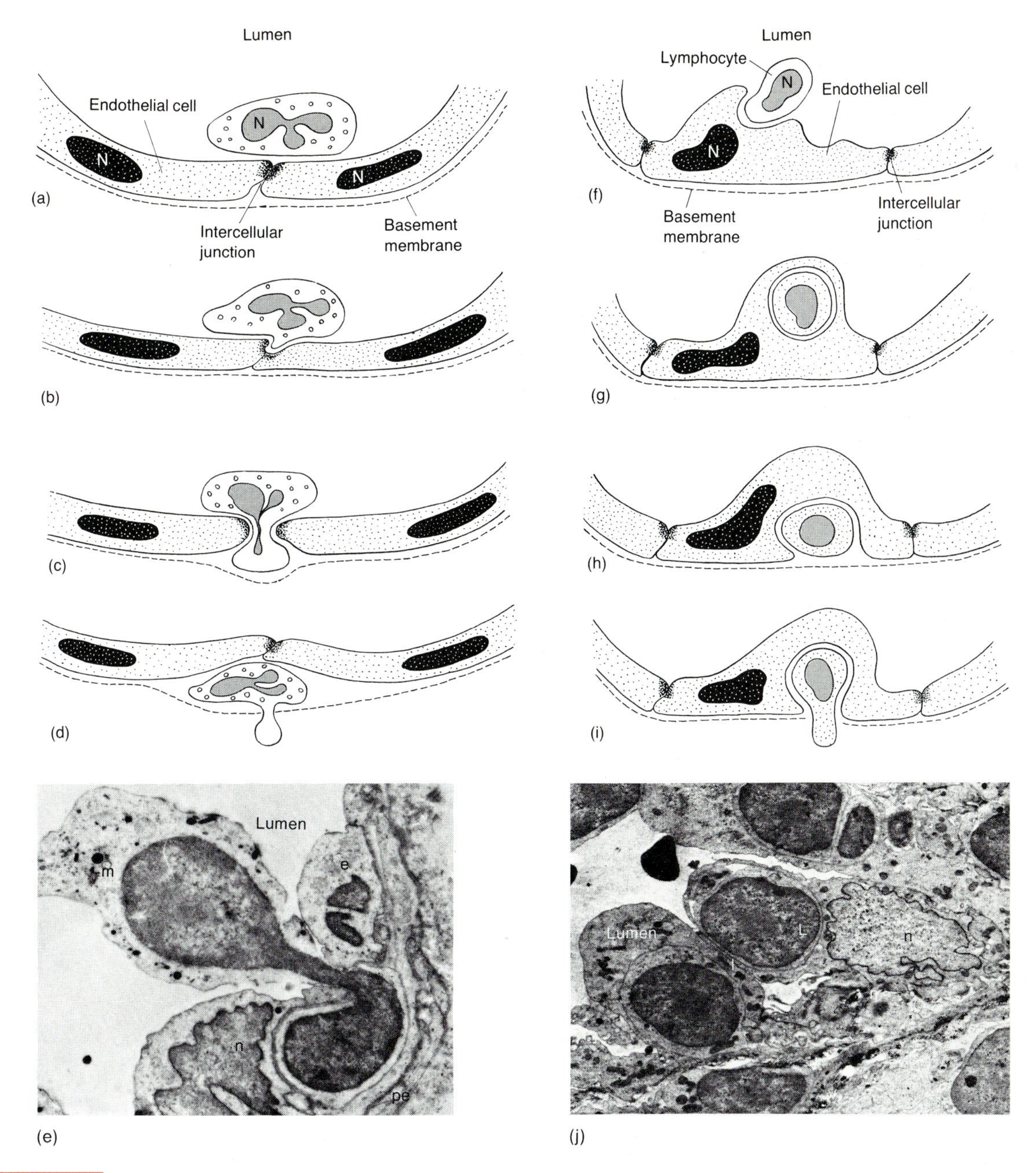


FIGURE 29.5
(a), (b), (c), (d) Stages in the migration of a granulocyte through the venule wall. The cell penetrates an intercellular junction and remains extracellular at all times. (e) Part of an inflamed venule. Cell m is a monocyte, which is penetrating an intercellular junction by the same mechanism; e, endothelium; n, nucleus of an endothelial cell; pe, periendothelial sheath. (f), (g), (h), (i) Stages in the migration of a small lymphocyte through an endothelial cell of a venule. The lymphocyte is totally intracellular at one stage in the passage. (j) Part of a venule from a normal lymph node; a lymphocyte, L, is completely enclosed by the cytoplasm of an endothelial cell; N, nucleus of the endothelial cell. From V. T. Marchesi and J. L. Gowans, "The Migration of Lymphocytes through the Endothelium of Venules in Lymph Nodes: An Electron Microscope Study," *Proc. Royal Soc. B*, **159,** 283 (1964).

**TABLE 29.4**

**Examples of Chemical Mediators of Inflammation**

| Substance | Source | Effects |
|---|---|---|
| Histamine | Mast cells, basophils | Increased blood flow, edema, itching |
| C3a | Complement fixation | Histamine release |
| C5a | Complement fixation | Histamine release, chemoattraction |
| Prostaglandins PGE and PGF | Platelets? | Edema, pain |
| Leukotrienes LTC4 and LTD4 | Monocytes? | Edema |
| Bradykinin | Blood clotting | Edema, pain |
| Lymphokines | Lymphocytes | Chemoattraction |

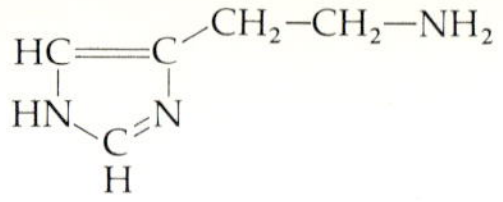

histamine

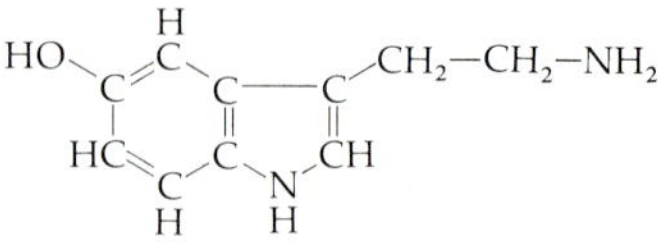

serotonin

FIGURE 29.6
The structures of histamine and serotonin. Histamine is the decarboxylation product of histidine; serotonin is the decarboxylation product of 5-hydroxytryptophan.

Chronic inflammation, which occurs during many persistent infections, is characterized by the presence of an abnormally large number of lymphocytes within a tissue. Macrophages also are usually present in large numbers as are basophils in some types of chronic inflammation. The terms acute inflammation and chronic inflammation are not mutually exclusive. For example, when an inflamed appendix is removed from a patient with appendicitis, the signs of both types of inflammation can often be seen.

## Chemical Mediators of Inflammation

In humans, a number of compounds mediate inflammation (Table 29.4), but the primary mediator of acute inflammation is *histamine* (Figure 29.6), which is always present in cytoplasmic granules of *mast cells*, which underlie skin and surround blood vessels. Histamine is also present in the granules of circulating basophils. Within these cells, histamine exerts no physiological effect, but when it is released by exocytosis, it causes blood vessels to dilate and become more permeable. It also increases the sensitivity of sensory nerves. Together, these physiological effects of histamine account for the symptoms associated with inflammation: dilation of blood vessels increases localized blood flow, causing redness and the sensation of heat; increased vascular permeability permits protein and water to flow out of blood vessels, causing edema. Serotonin (5-hydroxytryptamine), which is present in platelets, plays a major role in inflammation in some rodent species, but not in humans.

Inflammation is also mediated by *prostaglandins* (PGs) and *leukotrienes* (LTs), both of which are synthesized from the same precursor, the polyunsaturated fatty acid, *arachidonic acid* (Figure 29.7). Although it is not yet known how their synthesis is regulated, it is clear that they play an important role in triggering the sensation of pain as well as other aspects of inflammation. Indeed, many analgesic drugs, including aspirin, act by blocking the synthesis of prostaglandins; they have little effect on the synthesis of leukotrienes.

Other important mediators of inflammation are peptide compounds including *bradykinin*, which is produced by proteolytic cleavage of a serum protein in a reaction catalyzed by enzymes that become activated when blood clots or tissues are injured. Bradykinin both causes pain and increases the permeability of blood vessels, thereby producing edema. The polypeptide fragments, C3a and C5a, produced during complement fixation (Chapter 30), can cause inflammation indirectly: they bind to mast cells, triggering the release of histamine.

## Chemotaxis during Inflammation

As stated, neutrophils and eosinophils migrate through the walls of blood vessels into tissues during the acute phase of inflammation. One or more days later, lymphocytes and macrophages also migrate to the site, thereby initiating chronic inflammation. These cells are attracted to the site of inflammation by certain substances termed *chemotactic factors* that are released there. During the acute phase, complement fragment C5a is probably the most important chemotactic factor, but others—including leukotriene $B_4$, factors produced during blood clotting, and substances released by bacteria—also have chemotactic activity. During chronic inflammation, lymphocytes and macrophages are attracted by substances termed *lymphokines* (Chapter 30) that are released from lymphocytes.

FIGURE 29.7

Pathways leading to synthesis of the inflammatory prostaglandins, $PGE_2$ and $PGF_{2\alpha}$, and the inflammatory leukotrienes, $LTC_4$ and $LTD_4$. Arachidonic acid, the precursor for both pathways, is produced from membrane lipids by the action of phospholipase. 5HPETE; 5-hydroperoxy-6,8,11,14-tetraenoic acid.

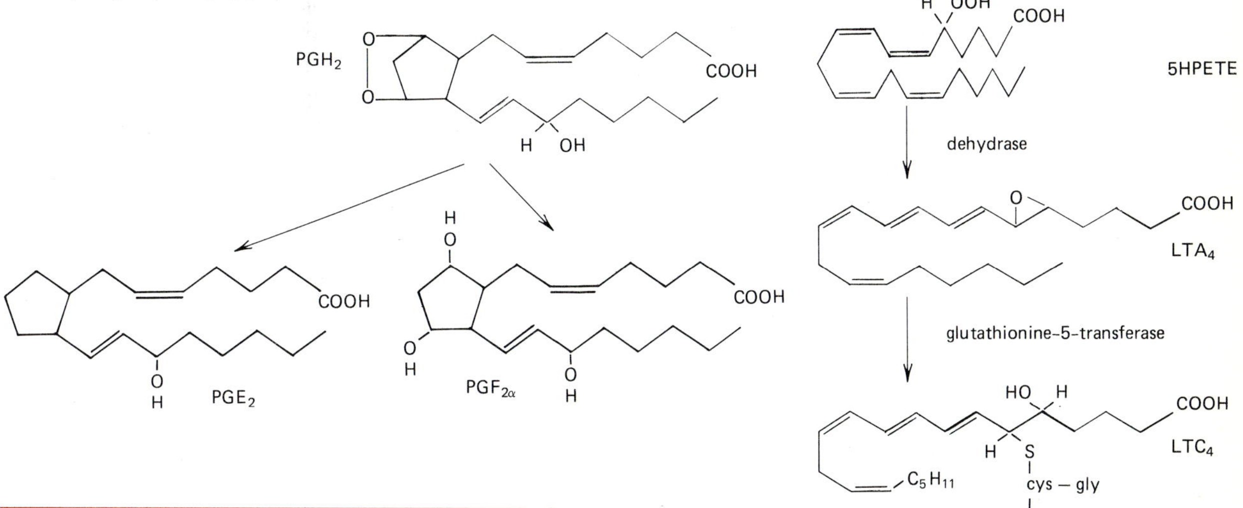

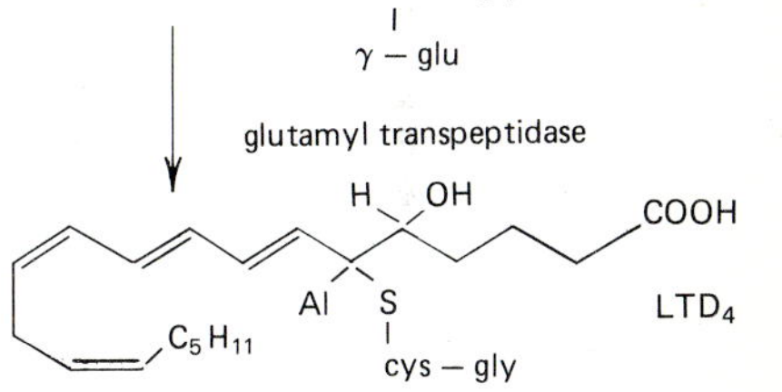

TABLE 29.5

**Types of Human Interferons**

| Class | Major Source | Number of Species Known | Stimulus for Production |
|---|---|---|---|
| IFN-α | Leukocytes | 8 | Virus, double-stranded RNA, lipopolysaccharide |
| IFN-β | Fibroblasts | 2 | Virus, double-stranded RNA, polysaccharide |
| IFN-γ | Lymphocytes | 1 | Antigens |

## NONSPECIFIC DEFENSE AGAINST VIRUSES

Soon after mammalian cells are infected by viruses, they begin to synthesize and secrete antiviral proteins, termed *interferons* (IFNs), which initiate a series of events finally resulting in inactivation of the virus. Interferons bind to specific receptors on the surface of uninfected cells, stimulating them to synthesize at least two enzymes, a *2′,5′-oligoadenylate synthetase* and a *protein kinase* (Figure 29.8). The former enzyme catalyzes the synthesis of an unusual polymer, 2′,5′-oligoadenylate, that activates an intracellular endoribonuclease, which in turn cleaves and thereby inactivates viral RNA. The protein kinase becomes active only in the presence of double-stranded RNA, which, with the exception of the retroviruses (Chapter 32), is always formed as an intermediate in the replication of RNA viruses. Once activated, the enzyme catalyzes the phosphorylation of one of the factors (eIF2) required for the initiation of synthesis of proteins. In its phosphorylated form, eIF2 is inactive, so that synthesis of all proteins, including viral proteins, stops.

Humans and other mammals produce more than ten distinct interferons that are classified as alpha-interferons (IFN-α), beta-interferons (IFN-β), or gamma-interferons (IFN-γ), depending on their chemical properties and the types of cells that synthesize them (Table 29.5). Synthesis of interferons appears to be the body's immediate defense against viral infection. Owing to the rapid rate at which these compounds are synthesized during the first few days following a viral infection, they afford an immediate, if temporary, protection during the period before specific antiviral antibodies appear. Interferons may also provide some protection against cancer: when exposed to chemical carcinogens, experimental animals develop fewer tumors if they are given interferon throughout the period of exposure than if they are not.

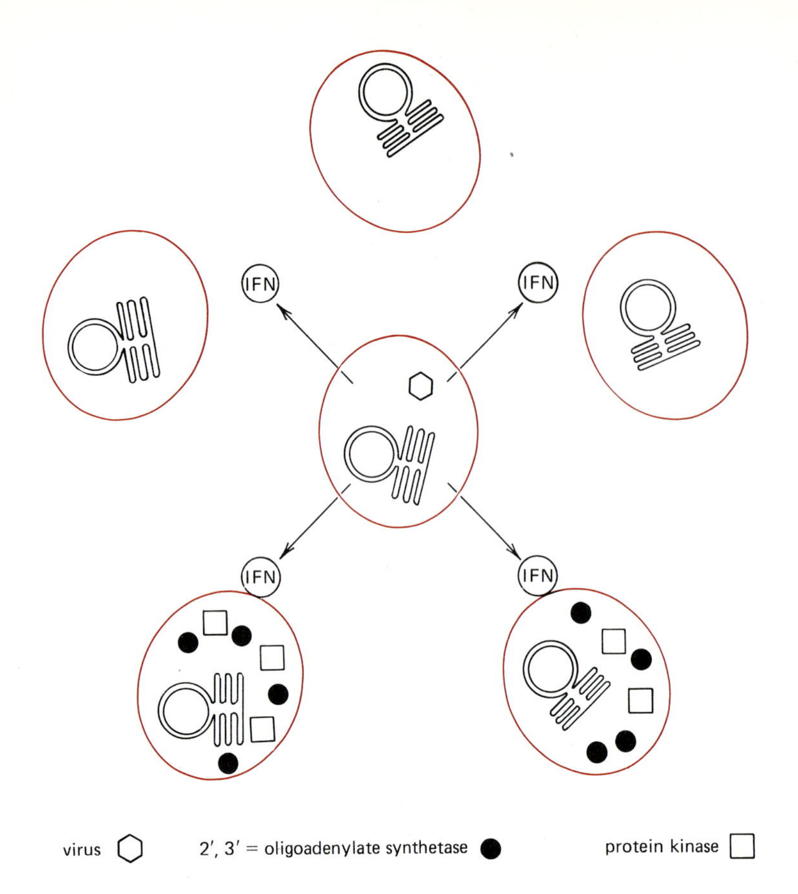


FIGURE 29.8

Induction of an antiviral state by interferon. A cell infected by virus secretes interferon (IFN), which binds to the surface of an uninfected cell. This stimulates the target cell to synthesize two enzymes, 2′,5′-oligoadenylate synthetase and a protein kinase, that induce an antiviral state in the target cell (see text). The intracellular structure, appearing here and in following chapters, represents the nucleus and endoplasmic reticulum.

## FURTHER READING

### Books

LEE, J. B., ed., *Prostaglandins.* New York: Elsevier, 1982.

SITES, D. P., J. D. STOBO, H. H. FUDENBERG, and J. V. WELLS, *Basic and Clinical Immunology,* 5th ed. Los Altos, Calif.: Lange Medical Publications, 1984.

TIZARD, I. R., *Immunology: An Introduction.* Philadelphia: Saunders, 1984.

### Reviews

BEAMAN, L., and B. L. BEAMAN, "The Role of Oxygen and Its Derivatives in Microbial Pathogenesis and Host Defense," *Ann. Rev. Microbiol.* **38,** 27 (1984).

CROSA, J. H., "The Relationship of Plasmid-Mediated Iron Transport and Bacterial Virulence," *Ann. Rev. Microbiol.* **38,** 69 (1984).

GORDON, J., and M. A, MINKS, "The Interferon Renaissance: Molecular Aspects of Induction and Action," *Microbiol. Rev.* **45,** 244 (1981).

HAMMARSTROM, S., "Leukotrienes," *Ann. Rev. Biochem.* **52,** 355 (1983).

SAMUELSSON, B., "Leukotrienes," *Science* **220,** 568 (1983).

# Chapter 30
# The Immune System

The Greek historian Thucydides described, in *The History of the Peloponesian War*, a plague that decimated the population of Athens in 430 B.C. The disease began with a cough and a headache; then it spread to other organs, causing high fever and a severe blistery rash, which was accompanied by an intense burning sensation. Although Thucydides' remarkably detailed account does not precisely fit any modern disease, the plague he described was clearly viral, perhaps related to smallpox or adult chickenpox. Thucydides also remarked that the sick were cared for by those who had recovered from the disease, because "no one was afflicted twice." Clearly the Athenians, and probably other ancient cultures as well, understood that immunity to certain diseases is acquired while recovering from them. During the following centuries, it became well established that recovery from some diseases (e.g., smallpox) always confers lifelong immunity to that disease. However, the basis for this immunity remained a mystery until 1890, when S. Kitasato and E. von Behring published their studies on the tetanus toxin (Chapter 31). They reported that animals became immune to the toxin after they had been injected with a small amount of a denatured form of it. Furthermore, they reported that blood from these animals contained a substance capable of inactivating native tetanus toxin, and that their serum (the fluid remaining after blood clots), when transferred to other animals, conferred immunity to the disease tetanus. These discoveries led to practical methods both for vaccinating humans against tetanus and for preventing the onset of disease in an individual who had been exposed to the bacterium that causes tetanus but had not been previously immunized.

These discoveries also led to the theory of *humoral immunity* that attributed immunity to soluble factors in blood, among which are proteins termed *antibodies* that are formed as a consequence of exposure to pathogens or foreign substances. This theory dominated immunology for many years, and little attention was paid initially to the theory of *cellular immunity*, proposed by E. Metchnikoff in 1883, that attributed immunity to specific cellular processes including phagocytosis. Although antibodies were shown to be responsible for immunity to tetanus and to many other diseases, it became apparent that in some diseases, most notably tuberculosis, antibodies cannot confer immunity: rather, recovery from this disease and the immunity that follows depends on the antibacterial activities of macrophages and lymphocytes (Chapter 29).

Gradually, it became clear that lymphocytes are responsible for the development of both humoral and cellular immunity: some are responsible for synthesizing antibodies; others are responsible for regulating the immune response and killing foreign cells. Thus modern immunology developed from a synthesis of the two opposing theories.

Lymphocytes, like all blood cells, develop from cells termed *stem cells* in bone marrow. Before development is complete, immature lymphocytes migrate from bone marrow into the bloodstream. Some mature into lymphocytes that are termed *T-cells* (T-lymphocytes) because the final stage of maturation occurs in the thymus gland. Others are transported to lymphoid tissue where they mature into *B-cells* (B-lymphocytes). These are named for the bursa of Fabricius, the pocket of lymphoid tissue where maturation occurs in birds. In mammals, the site where immature lymphocytes become B-cells is not known with certainty, but it seems likely that at least some maturation occurs in pockets of lymphoid tissue, termed *Peyer's patches*, that are located in the intestinal wall.

Although the morphology of T-cells appears to be identical to that of B-cells when viewed in the light microscope, there are major differences in the properties of these cells: B-cells synthesize antibodies, some of which remain bound to the B-cell surface; T-cells neither synthesize antibodies nor possess them on their surfaces. Instead, they synthesize proteins termed *T-cell receptors* (see below), all of which remain bound to their surfaces. T-cells also regulate the production of antibodies by B-cells and participate in cellular immunity by detecting and killing both virus-infected host cells and some foreign cells.

Lymphocytes are widely distributed throughout the body. Most are located in the spleen, in lymph nodes, in tonsils, in the appendix, and in the wall of the intestines. At these sites, B-cells and T-cells are intimately associated with macrophages. Both types of lymphocytes are transported throughout the body via the bloodstream. They migrate through the walls of blood vessels into tissues, then enter small lymphatic vessels that drain to lymph nodes. By migrating out of lymph nodes, lymphocytes can return to the bloodstream by way of large lymph vessels like the thoracic duct that empties into the vena cava.

## ANTIBODIES AND ANTIGENS

The blood of normal individuals contains an enormous number, probably greater than $10^6$, of chemically distinct molecules termed *immunoglobulins* (antibodies), making it possible to acquire humoral immunity to a large number of diseases. The functions of antibodies that mediate humoral immunity can be divided into two types: (1) specific binding to pathogens or to toxins, and (2) interactions with cellular or molecular components of the host's immune system. Antigen binding occurs within a relatively small region of the antibody; the other interactions involve a much larger region. The former region is highly variable from one antibody to another, but the latter region is largely responsible for both the structural similarity of all antibody molecules and for the structural differences between the five classes of antibodies discussed below. Each antibody class participates in its own set of interactions that contribute to humoral immunity.

Immunoglobulins are large glycoproteins with molecular weights ranging from 150,000 to 900,000. All share the *unit antibody structure* diagrammed in Figure 30.1. This symmetrical Y-shaped molecule is composed of two identical polypeptides termed *heavy chains* (MW approximately 50,000) and two identical polypeptides termed *light chains* (MW approximately 23,000). The stem of the Y-shaped structure is formed by approximately one-half of each heavy chain, and the two chains are covalently joined by disulfide bonds between cystine residues. The stem region of heavy chains also has polysaccharide side chains. Each arm of the Y is composed of one light chain and approximately one-half of a heavy chain, again joined by a disulfide bond. The $NH_2$-termini of both chains are at the tip. Light and heavy chains also possess intrachain disulfide bonds that create loops about 60 amino acid residues in circumference. Each loop together with approximately 25 amino acids on each side of the loop is termed a *domain*. Light

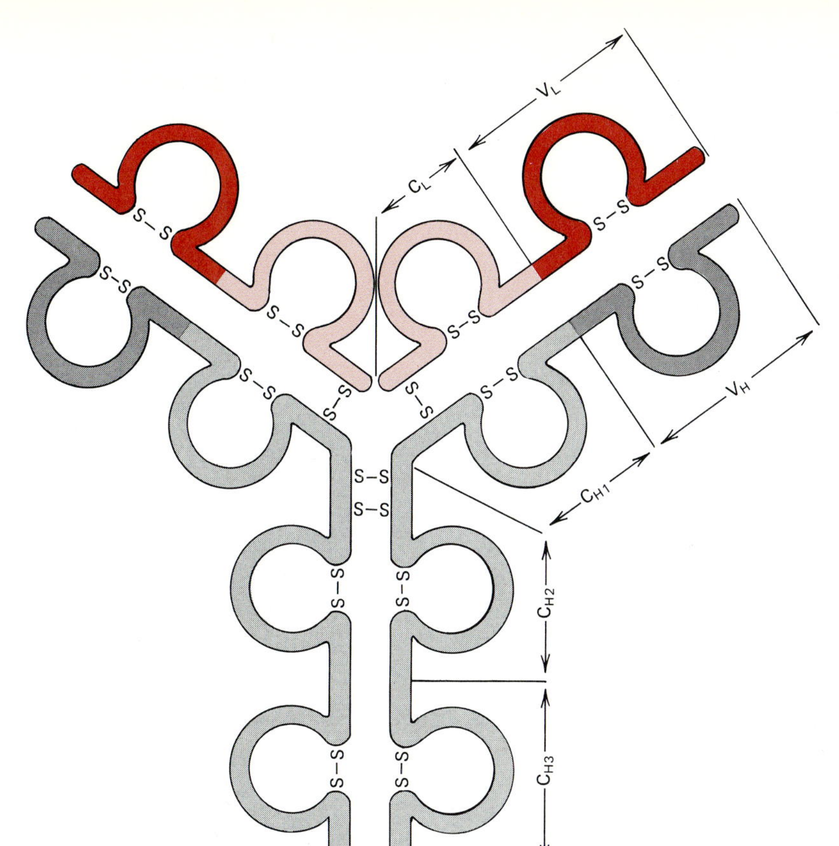


FIGURE 30.1

Immunoglobulin unit structure. (a) Diagram of primary structure: two light polypeptide chains and two heavy polypeptide chains are cross-linked by disulfide bonds, forming a symmetrical Y-shaped molecule. An intrachain disulfide bond creates a loop within each domain of these chains. All light chains contain a single variable domain ($V_L$) and a single constant domain ($C_L$). Heavy chains contain a variable domain ($V_H$) and either three or four constant domains—$C_{H1}$, $C_{H2}$, $C_{H3}$, and in some classes $C_{H4}$. (b) Diagram of three-dimensional relationships between immunoglobulin chains. From E. W. Silverton, M. A. Navia and D. R. Davies, "Three-dimensional Structure of an Intact Human Immunoglobulin," *Proc. Nat. Acad. Sci. USA*, **74,** 5140 (1977).

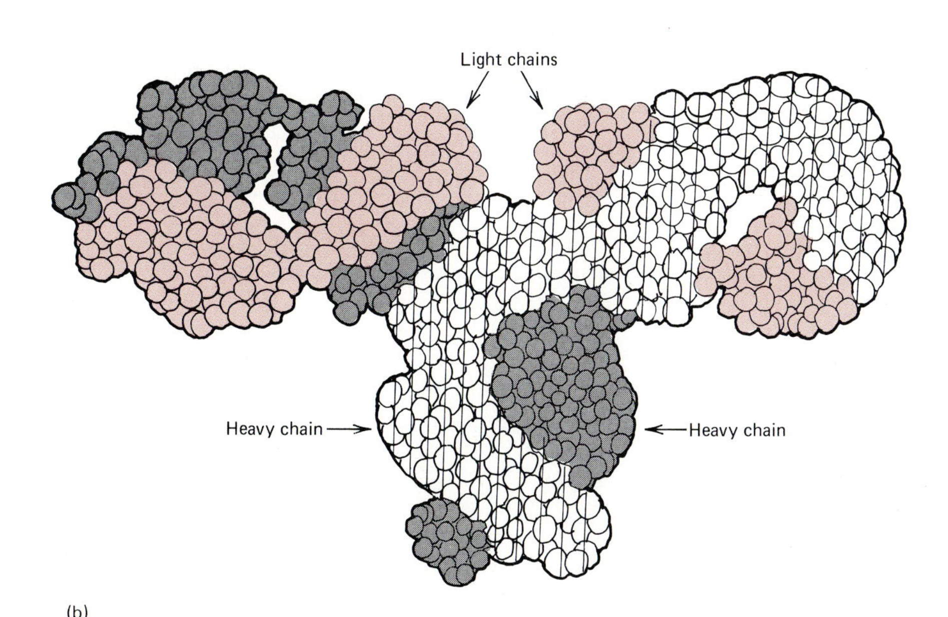


chains are composed of two domains, and heavy chains are composed of either four or five domains. Comparison of the amino acid sequences of two domains from a heavy or light chain reveals a similarity in their sequences: the amino acid at a particular place in one domain is identical to the amino acid at the same place in another domain approximately 30 percent of the time.

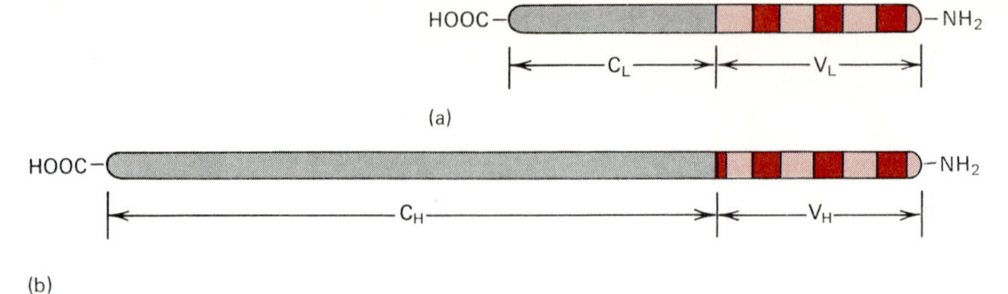


FIGURE 30.2
Location of hypervariable regions (color) within immunoglobulin chains: (a) light chains; (b) heavy chains.

## Constant and Variable Domains

Two *types* of light chains, *kappa* ($\kappa$) and *lambda* ($\lambda$), can be distinguished by the amino acid sequence of the COOH-terminal domain. In human antibodies, the COOH-terminal domains of all $\kappa$ chains are identical, and this region is termed the $\kappa$ constant ($C_\kappa$) region. There are four very similar sequences in COOH-terminal domains of human $\lambda$ chains, and these define the subtypes—$\lambda_1$, $\lambda_2$, $\lambda_3$, and $\lambda_4$—with their corresponding constant regions—$C_{\lambda 1}$, $C_{\lambda 2}$, $C_{\lambda 3}$, and $C_{\lambda 4}$. Comparison of the amino acid sequences from two $\kappa$ chains or two $\lambda$ chains reveals that many parts of these sequences are similar or identical. However, there are three regions termed *hypervariable regions* where the sequences usually differ: amino acids 24—34, 50—55, and 89—97 (Figure 30.2). These are in the $NH_2$-terminal domain ($V_L$ domain) which is denoted the $V_\kappa$ domain in $\kappa$ chains and the $V_\lambda$ domain in $\lambda$ chains.

In heavy chains, the $NH_2$-terminal domain has a pattern of variability similar to that of the $V_\kappa$ and $V_\lambda$ domains and is denoted the $V_H$ domain: it contains four hypervariable regions—amino acids 31–36, 51–67, 86–90, and 101–114 (Figure 30.2). Other domains of heavy chains are termed constant domains and are numbered starting with the domain adjacent to the variable domain. The constant domains of a heavy chain form the constant ($C_H$) region, and the amino acid sequence of this region determines the *classes* of heavy chains. The heavy chain class has great physiological significance as it determines the *immunoglobulin class* of the antibody molecule. In humans, there are five classes of heavy chains—$\gamma$, $\alpha$, $\mu$, $\delta$, and $\epsilon$—that determine, respectively, the five immunoglobulin classes—IgG, IgA, IgM, IgD, and IgE. The immunoglobulin class of an antibody determines many of its properties including its half-life, distribution in the body, and interaction with other components of the host defense system (Table 30.1).

## IgG

In humans, IgG is the most abundant antibody in blood and in the tissue fluid surrounding cells. It is also the only antibody that is normally transported across the placenta from the mother's blood into the fetal circulation, where it persists for several months. This IgG, termed *maternal antibody*, gives the infant some resistance to pathogens to which the mother has acquired immunity. Other properties of IgG are listed in Table 30.1.

**TABLE 30.1**

**Properties of Human Immunoglobulins**

| | IgG | IgA | IgM | IgD | IgE |
|---|---|---|---|---|---|
| Heavy chain class | $\gamma$ | $\alpha$ | $\mu$ | $\delta$ | $\epsilon$ |
| Molecular formula | $\gamma_2\kappa_2$ or $\gamma_2\lambda_2$ | $\alpha_2\kappa_2$, $\alpha_2\lambda_2$, $\alpha_4\kappa_4$, or $\alpha_4\lambda_4$ | $\mu_{10}\kappa_{10}$ or $\mu_{10}\lambda_{10}$ | $\delta_2\kappa_2$ or $\delta_2\lambda_2$ | $\epsilon_2\kappa_2$ or $\epsilon_2\lambda_2$ |
| Approximate molecular weight | 150,000 | 160,000 or 360,000[a] | 900,000 | 170,000 | 180,000 |
| $C_H$ domains | 3 | 3 | 4 | 3 | 4 |
| Approximate half-life in blood (days) | 22 | 6 | 5 | 3 | 2 |
| Concentration in serum (mg/100 ml) | 800–1,500 | 150–200 | 40–120 | 1.5–40 | 0.002–0.005 |
| Complement pathway activated | Classic or alternate | Alternate | Classic | Neither | Neither |

[a] Molecular weight of dimeric secretory IgA.

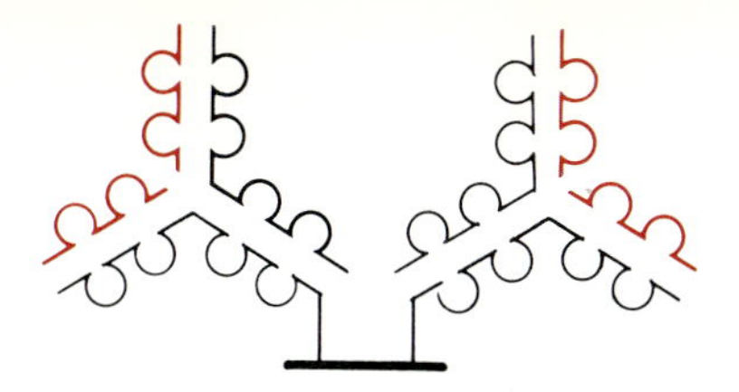

(a)

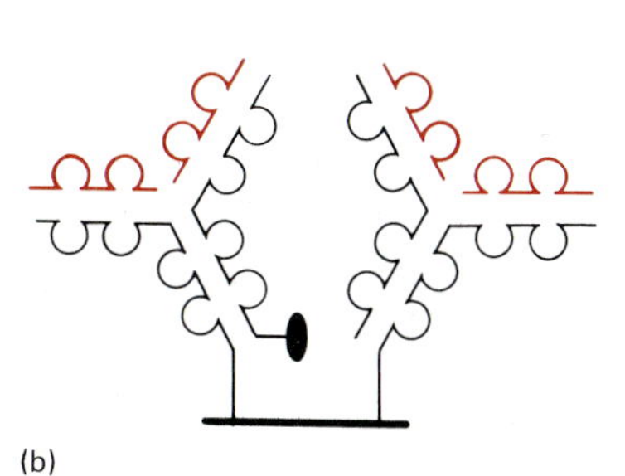

(b)

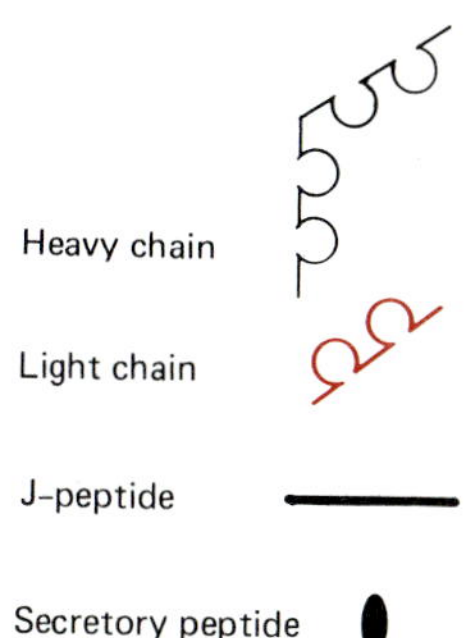


FIGURE 30.3

Structures of the common forms of IgA: (a) forms of IgA in blood; (b) secretory IgA.

The two light chains ($\kappa$ or $\lambda$) and two heavy chains of IgG are covalently joined as diagrammed in Figure 30.1 (a), and are folded into the globular protein depicted in Figure 30.1 (b). The $\gamma$ heavy chains contain three constant domains, and minor variations in these domains define the four subclasses—$\gamma_1$, $\gamma_2$, $\gamma_3$, and $\gamma_4$—which are found, respectively, in the IgG subclasses—IgG1, IgG2, IgG3, and IgG4.

## IgA

IgA is the second most abundant type of human antibody in blood, but is the most abundant one in mucous and other secretions. There are two subclasses, IgA1 and IgA2, determined by the subclasses of $\alpha$ heavy chains, $\alpha_1$ and $\alpha_2$. Some IgA exists as the unit structure of Figure 30.1, but most IgA is a dimer of unit antibodies joined by a polypeptide termed the *J chain* (Figure 30.3). A second polypeptide termed *secretory protein* is attached to those IgA molecules that are transported by epithelial cells into tears, saliva, and mucous of the respiratory and gastrointestinal tracts. Such IgA, termed *secretory antibody*, is also present in *colostrum*, the fluid secreted by the breast at the start of lactation.

## IgM

IgM is the third most abundant human antibody. It appears to have been the first type to arise during evolution, being the most abundant antibody in primitive vertebrates. It is the largest antibody, with a molecular weight of 900,000, and for this reason is confined mostly within blood vessels. IgM exists as a pentamer of unit structures linked by disulfide bonds. In addition, two of the unit structures are linked by the J chain (Figure 30.4). In any one IgM molecule, the 10 light chains are all identical and the 10 heavy chains are all identical. All $\mu$ heavy chains possess a fourth constant domain, $C_{H4}$.

FIGURE 30.4

Structure of IgM.

## IgD

IgD is composed of two $\kappa$ or $\lambda$ chains and two $\delta$ chains. Most IgD is bound to the cytoplasmic membrane of mature B-cells, which also have membrane-bound IgM that exists in the unit antibody structure rather than its usual pentameric structure. This binding is facilitated by a sequence of hydrophobic amino acids at the COOH-terminus of most $\delta$ and some $\mu$ chains. These bound antibodies probably play a role in triggering the final step of B-cell maturation. The concentration of free IgD in blood is very low, and IgD is unusually sensitive to proteolytic enzymes and to heat.

## IgE

Most IgE is bound to the cytoplasmic membranes of basophils in blood and mast cells in connective tissue (Chapter 29) where it mediates the release of histamine that causes some immediate-type allergic reactions. IgE contains two $\kappa$ or two $\lambda$ chains and two $\epsilon$ chains, which have four $C_H$ domains. The terminal constant domain serves to bind IgE tightly to cell surfaces, and the concentration of free IgE is lower than that of any other antibody.

## Antigens and Haptens

The primary role of antibodies in host defense is to bind to toxins or to molecules on the surface of a pathogen in a highly specific way that is discussed below. Most macromolecules to which antibodies bind also stimulate production of the antibodies to which they bind, i.e., they are *immunogenic*. Such molecules are termed *antigens*; they need not be produced by a pathogen. Indeed most, if not all, proteins and complex polysaccharides are antigens, as are some nucleic acids. Antibodies also bind to some small molecules, termed *haptens*: these differ fundamentally from antigens because they do not stimulate antibody synthesis.

Antigens bind to antibodies at the *antigen binding site*, within the *Fab regions* (Figure 30.5) of the unit antibody structure: a pocket is formed by the folding of both the $V_H$ region and the $V_L$ region. Here, amino acids in the hypervariable regions of both light and heavy chains can form noncovalent bonds (hydrophobic, ionic, and hydrogen bonds) with a part of the antigen molecule termed an *antigenic determinant* or *haptenic group*. Antigenic determinants are usually regions that project from the antigen's surface. Furthermore, when a hapten is covalently attached to the surface of an antigen, the hapten becomes a new antigenic determinant.

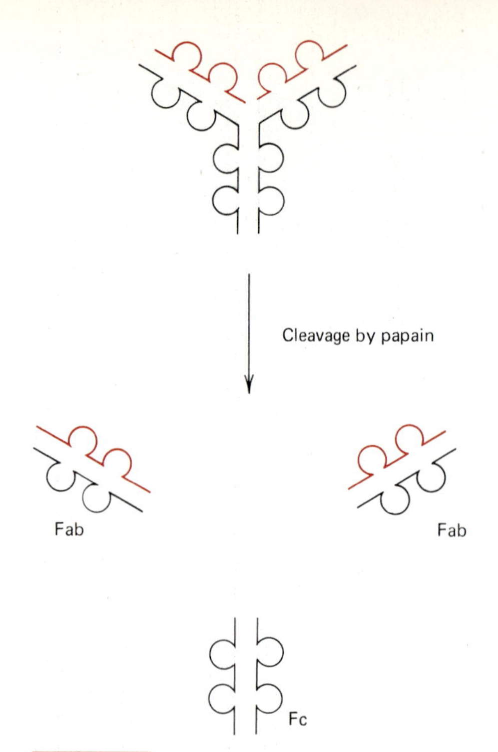


FIGURE 30.5
Proteolytic cleavage of IgG by papain. Hydrolysis of a peptide bond near the center of each heavy chain produces two antigen-binding fragments (Fab) and a single crystallizable fragment (Fc).

# ANTIBODY SOURCES

As a consequence of the remarkable specificity with which they bind antigens, antibodies can be used as reagents for detecting specific antigens or haptens. They can also be used as relatively safe therapeutic agents in the prevention and treatment of some infectious diseases, and they have great potential for the detection and treatment of cancer because antigens that are not found on normal cells are present on the surface of some cancer cells.

## Immunization

Antibodies that bind to a specific antigen are often produced by immunization of domestic animals or human volunteers. The antigen is usually deposited by injection in a tissue from which some of it is carried through lymph vessels to lymph nodes where antibody synthesis takes place. In order to enhance their efficiency in stimulating antibody synthesis, antigens that are injected into domestic animals are often mixed with substances, termed *adjuvants*, that enhance the immune response. Following repeated injections at intervals of several weeks to further increase the concentration of specific antibody, blood is removed and allowed to clot. The fluid fraction that remains after removal of the clot contains the antibodies and is termed *serum*: when it is obtained from an immunized animal, it is termed *antiserum*. For example, serum from an animal immunized against partially denatured tetanus toxin is called tetanus antiserum.

Although serum is a major and convenient source of antibodies, three properties limit its usefulness: (1) injection of serum from one species into another can cause serious allergic reactions; (2) serum always contains a large and unknown mixture of antibodies in addition to the one of interest; and (3) even those antibodies that bind the antigen of interest are a population of distinct molecules.

## Hybridomas

During the past several years, the limitations of serum as a source of antibodies have been overcome with the development of techniques to manipulate and culture mammalian cells that synthesize antibodies *in vitro*. Normally each cell and its progeny produce a single antibody species termed *monoclonal antibody*. A convenient source of such antibodies is the cancer *multiple myeloma* that occasionally occurs in rodents or humans. In many cases, the disease originates with a single cancerous B-cell that makes a single antibody species, and the cell's progeny spread to produce multiple tumors in the victim's bones.

Antibodies produced by myeloma cells are of limited usefulness because their binding specificity is usually not known. However, monoclonal antibody with a useful binding specificity can be produced by fusing, in the presence of polyethylene glycol, myeloma cells to normal B-cells that synthesize the antibody of interest. The primary fusion cell contains a set of chromosomes from each diploid parent cell. During subsequent divisions, chromosomal assortment gives rise to a variety of genetically stable combinations of chromosomes in progeny cells. Some of these new cell lines, termed *hybridomas*, possess the ability, inherited from the myeloma parent, to grow indefinitely in laboratory cultures and also possess the genetic information to produce the antibody of the B-cell parent.

For example, to isolate a mouse hybridoma cell line that produces monoclonal antibody against *Mycobacterium tuberculosis*, killed bacteria are injected into a mouse. Following repeated injections, the mouse's spleen, the richest source of lymphocytes, is removed, and from it a cell suspension is prepared and mixed with cultured myeloma cells under conditions that promote cell fusion. Cells from this mixture are diluted and plated on semisolid medium (that promotes growth of myeloma cells), and colonies that develop are screened to identify those that produce antimycobacterial antibody (Figure 30.6).

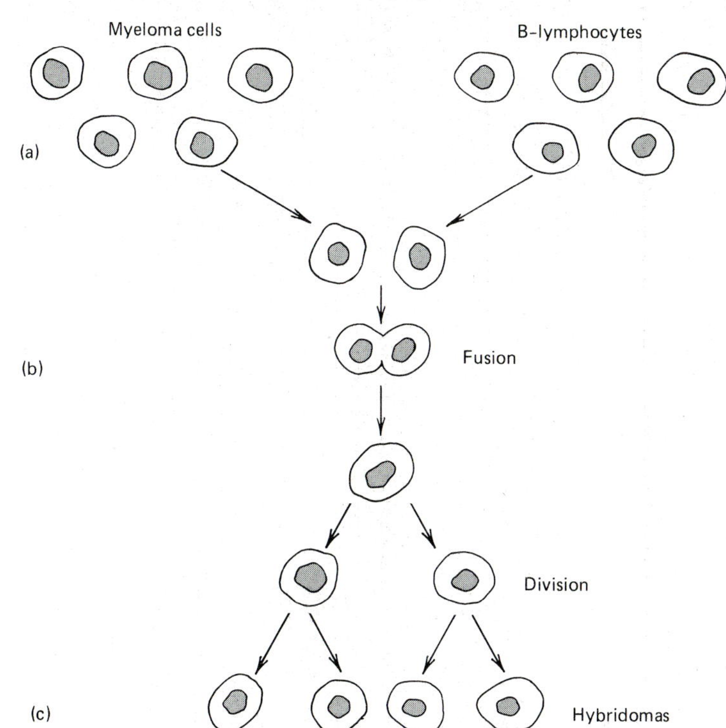


FIGURE 30.6

Formation of hybridomas. (a) Myeloma cells are mixed with lymphocytes from an immunized individual. (b) A myeloma cell and a lymphocyte fuse, producing a tetraploid cell. (c) This cell divides, and some of the progeny are hybridomas; i.e., like myeloma cells they grow indefinitely, but produce antibody from a gene inherited from a lymphocyte.

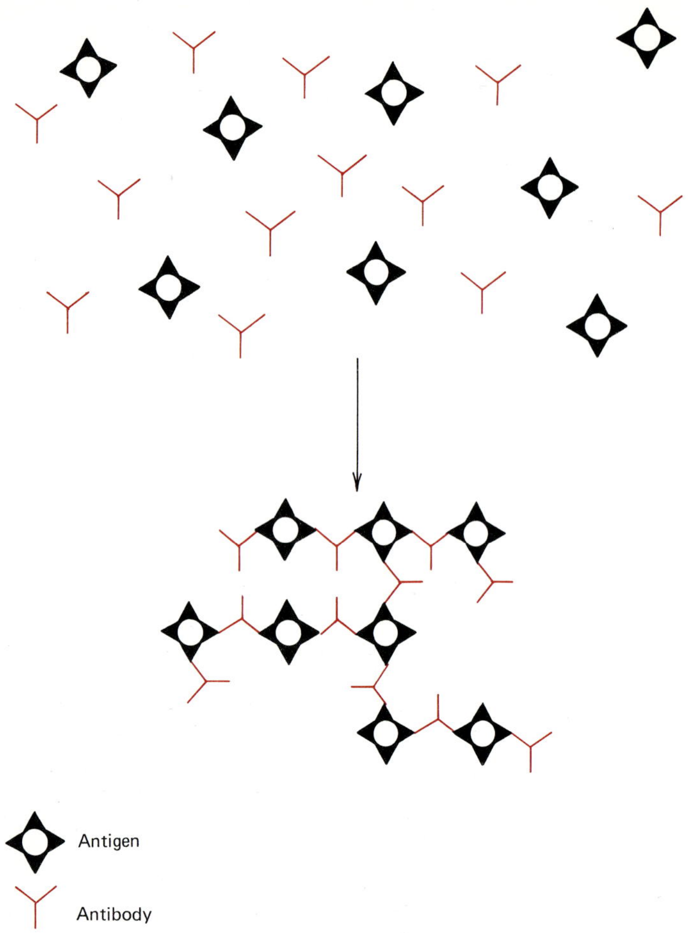

FIGURE 30.7
Formation of immune complexes. Antibodies cross-link antigens forming aggregates of antibody and antigen termed *immune complexes*.

Antigen

Antibody

## CONSEQUENCES OF ANTIGEN-ANTIBODY BINDING IN THE HOST

In the absence of other components of the immune system, pathogens are usually not killed by the binding of an antibody, although such binding can interfere with adhesion of bacteria (Chapter 31) or with adsorption of a virus to its host cells (Chapter 9). The principal function of the binding of antibody to the surface of a pathogen is to identify it as a foreign cell. Then other components of the immune system can eliminate it.

### Toxin and Virus Neutralization

Immunity to diseases like tetanus and diphtheria, in which pathogenesis is largely the consequence of the action of a toxin, depends on antibodies in blood that bind to the toxin and inactivate it, a process termed *toxin neutralization.* The resulting antibody-toxin complexes cannot adsorb to toxin receptors on host cells (Chapter 31). Similarly, antibodies bound to the surface of a virus prevent adsorption, and such antibody-mediated inactivation is termed *virus neutralization.*

### Immune Complex Formation and Agglutination

All antibodies have at least two antigen-binding sites, and most antigens have at least two antigenic determinants. Thus, antibodies can cross-link antigens producing aggregates termed *immune complexes* (Figure 30.7). Immune complexes are sometimes designated according to their physical properties. For example, when an immune complex containing a soluble antigen becomes large enough to precipitate, it is called a *precipitin*; when a com-

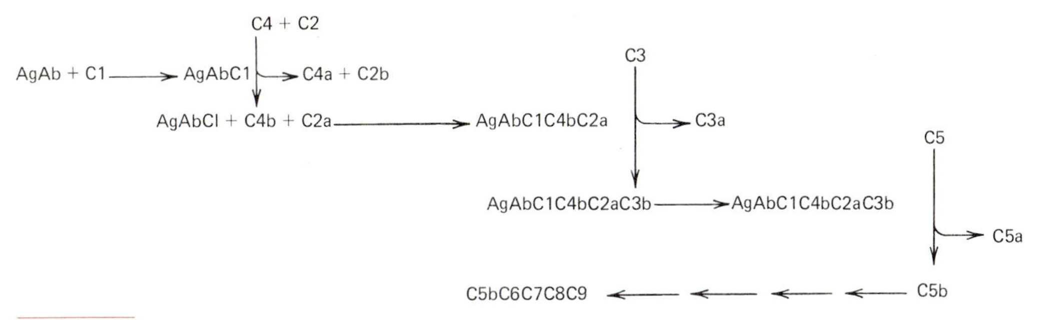


FIGURE 30.8

The classic complement pathway. Immune complexes containing IgM, IgG1, or IgG3 bind and activate C1, initiating the pathway.

plex is formed by cross-linking cells or other particles, it is called an *agglutinin*. Immune complexes are more readily phagocytized than are free antigens, but their formation can also be damaging to the host. For example, immune complexes that form in blood can be deposited in small blood vessels, thereby obstructing them.

### The Classic Complement Fixation Pathway

Blood contains a group of nine proteins (designated C1 through C9), collectively termed *complement*, that function in a complex series of reactions (termed *complement fixation*) to kill some kinds of foreign cells (Figure 30.8). The first reaction sequence to be demonstrated, termed the *classic pathway*, is triggered by the binding of a single IgM molecule (or of two molecules of IgG1, IgG2, or IgG3) to an antigen molecule, resulting in a conformational shift in antibody structure that exposes a receptor for C1 on the $C_{H2}$ domain. C1 is composed of three subunits: C1q, C1r and C1s. Binding of the C1q subunit to adjacent complement receptors results in a conformational change of the C1 complex. As a consequence, C1s becomes proteolytically active and cleaves C2 into fragments C2a and C2b and also cleaves C4 into fragments C4a and C4b. C4b binds to antigen-antibody-C1 complexes. The resulting AgAbC1C4b complex is stable, but binding activity of free C4b is quickly lost. C2a adsorbs to bound C4b to form a proteolytically active complex that cleaves C3 into fragments C3a and C3b. C3b then adsorbs to bound C4aC2b, forming the complex C4bC2aC3b that cleaves C5 into fragments C5a and C5b. C6 and C7 rapidly bind to C5b, forming a C5bC6C7 complex that possesses an unstable membrane-binding site, but once bound to a membrane, this complex is stable. Then C8 and C9 bind, forming the complex C5bC6C7C8C9 that creates a pore in the membrane. If the cell is eucaryotic, its cytoplasmic contents rapidly leak through the pore, causing cell death. If the cell is a Gram-negative bacterium, lysozyme in blood enters pores made in the outer membrane and digests the peptidoglycan of the bacterium, causing it to lyse. In contrast, Gram-positive bacteria are resistant to the cytolytic action of complement because they lack an external membrane.

### The Alternate Complement Pathway

In the absence of antibodies that bind to Gram-negative bacteria, complement is still bactericidal. This is the consequence of the *alternate complement pathway* (Figure 30.9) that begins with cleavage of C3 into fragments C3a and C3b by an enzyme normally present in blood. These fragments are produced at a slow rate and usually do not trigger the next step, the cleavage of C5, because free C3b is rapidly broken down into inactive fragments. However, C3b is stabilized by binding to lipopolysaccharides of bacterial outer membranes or to IgA and IgG antibodies in immune complexes. A protein in blood called *factor B* adsorbs to bound C3b, leading to formation of an active enzyme denoted C3bBb that is further stabilized by a second blood protein, *properidin*, resulting in efficient formation of C5b from the cleavage of C5. The steps that follow in the alternate pathway are identical to the terminal steps of the classic pathway.

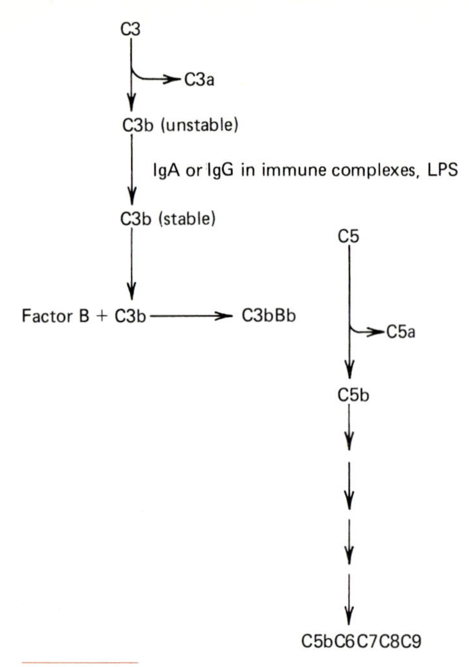


FIGURE 30.9

The alternate complement pathway. C3b produced at a low rate by enzymes present in blood is stabilized when bound to lipopolysaccharide or to IgA or IgG in immune complexes. Factor B binds to and activates C3b.

### Opsonization

*Opsonization* is the stimulation of phagocytosis, e.g., that which occurs when antibodies bind to antigens. This occurs if the bound antibodies belong to the subclasses IgG1 and IgG3 because these have a site (in the $C_{H3}$ domain) that binds to a receptor on the surface of macrophages, thereby forming a bridge between the phagocyte and the antigen.

Opsonization is also mediated by the complement fixation pathway. Both neutrophils and macrophages possess, on their surfaces, a receptor for C3b. Therefore, C3b that is bound to the surface of a pathogen can form a bridge that facilitates phagocytosis. Indeed, opsonization appears to be the most important function of complement: people who lack C3 (as a consequence of a genetic disorder) are usually much more susceptible to bacterial infections than are people who lack some other complement protein. Surprisingly, people who lack C6, C7, C8, or C9 are often healthy.

### Inflammation

The binding of an antibody to an antigen triggers inflammation (Chapter 29) by two separate routes. The first is mediated by IgE on the surface of mast cells and basophils. When two adjacent molecules of IgE become linked together by binding to a single antigen molecule, a sequence of intracellular events is initiated: cytoplasmic granules fuse with the cell membrane and release, into the extracellular environment, their complement of histamine, which causes the characteristic symptoms of inflammation.

The complement fixation pathway also mediates inflammation: fragments C3a and C5a bind to mast cells, triggering release of histamine, and to platelets, triggering release of histamine. Fragment C5a contributes to inflammation in a second way: being a very potent chemotactic factor, it causes macrophages, neutrophils, eosinophils, and basophils, to concentrate at the site of complement fixation.

## CONSEQUENCES OF ANTIBODY-ANTIGEN BINDING IN VITRO

Antibodies are useful reagents for a variety of laboratory procedures including the detection or quantitation of antigens. Conversely, antigens can be used to detect and quantitate antibodies. *In vitro* reactions involving antibodies and antigens, termed *serological reactions,* are widely used in clinical diagnosis, epidemiology, and basic research. Some of these uses are discussed in the following sections.

### Agglutination Reactions

Agglutination reactions can often be detected with the unaided eye when microscopic particles aggregate into large clumps. In the first agglutination reactions studied, blood cells or bacteria were employed as the particles: these are naturally coated with antigens that can be cross-linked. Recently, techniques have been developed to coat microscopic latex spheres with antigens, providing synthetic particles useful in agglutination reactions.

The best known use of agglutination reactions is the typing of human blood, a technique discovered in 1900 by K. Landsteiner: he demonstrated that four types of human blood—A, B, AB, and O—can be distinguished by simple agglutination tests (Table 30.2). For example, blood from a type B individual is clumped by serum from a type A individual but not by serum from a type B individuals. Before blood typing made it possible to match the blood of a donor with that of a pa-

**TABLE 30.2**

**Agglutination Reactions in the ABO System of Human Blood Types**

| Source of Serum | Agglutination (+) or No Agglutination (−) With Blood from Individual of Type: A | B | AB | O |
|---|---|---|---|---|
| Type A individual | − | + | + | − |
| Type B individual | + | − | + | − |
| Type AB individual | − | − | − | − |
| Type O individual | + | + | + | − |

tient, transfusions were rarely performed because of the fatal reactions that often occurred when untyped blood was used.

An excellent example of a diagnostic test based on the agglutination of microscopic latex spheres is one of the modern pregnancy tests; it detects the greatly elevated concentrations of the hormone, human chorionic gonadotropin (HCG), that occur in urine and blood early in pregnancy. First a sample (usually urine) from the patient is mixed with a solution of antibody specific for HCG. In the second step, latex spheres coated with HCG are added. If present at high enough concentration in the first step, HCG will bind to most HCG-specific antibodies, thereby preventing them from agglutinating the latex spheres (Figure 30.10).

FIGURE 30.10

The latex bead agglutination inhibition test. (a) In a negative test, latex beads coated with antigen are clumped by specific antibodies. (b) In a positive test, antigen present in a sample binds to and thereby inactivates these specific antibodies. No clumping occurs when the latex beads are added.

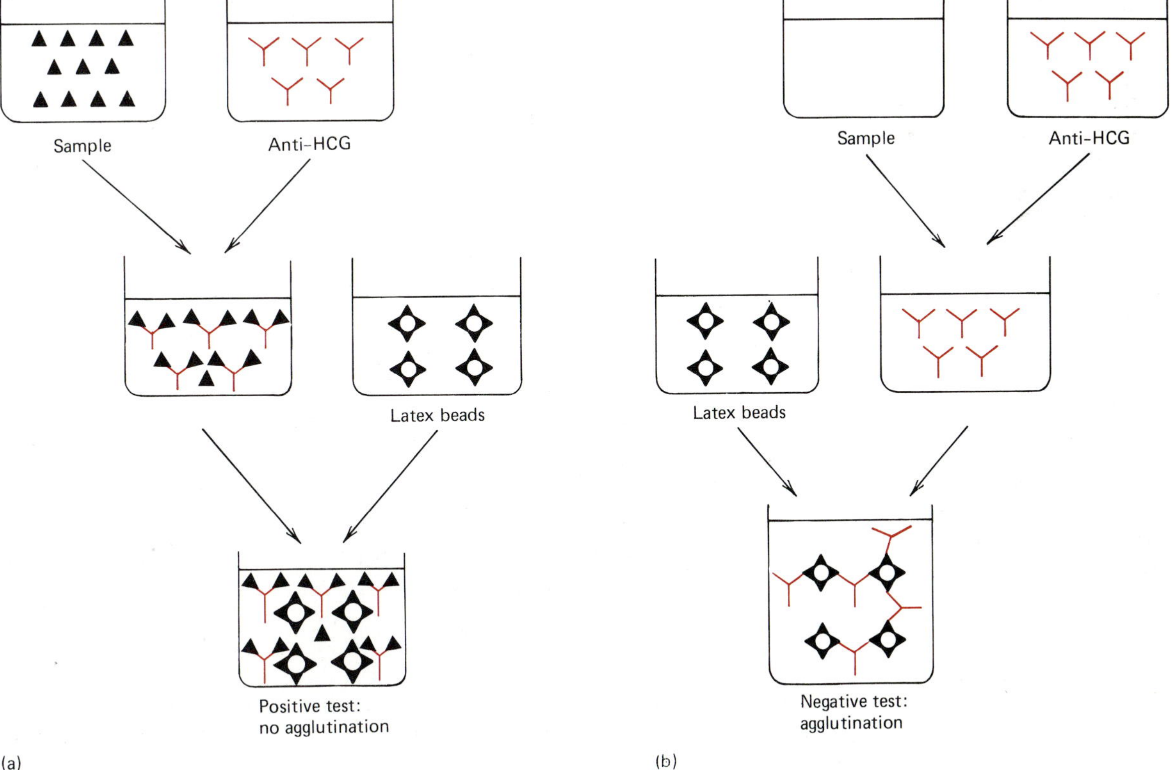

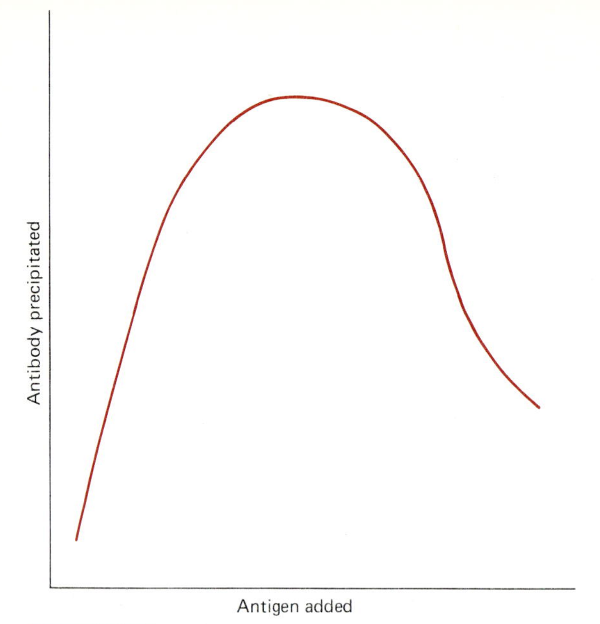


FIGURE 30.11
Precipitin formation. When increasing amounts of antigen are added to a series of tubes containing a fixed amount of antibody, the amount of insoluble immune complexes (precipitin) at first increases, then declines.

## Immunoprecipitation

Immunoprecipitation, the serological reaction by which precipitins are formed, occurs in two stages: (1) first, antibodies bind to antigens in a reaction that takes a few seconds or minutes, then (2) the constant regions of antibodies in these immune complexes bind to each other in a reaction that takes several hours and results in the formation of a visible precipitate. However, the second reaction usually does not occur among immune complexes containing a single antibody molecule saturated with antigen. Consequently, in the procedure in which increasing concentrations of antigen are added to a series of tubes containing a fixed concentration of antibody, the maximum amount of precipitin forms in tubes that receive approximately equal amounts of antigen and antibody (Figure 30.11). Those concentrations of antigen that result in precipitation of nearly all antigen and specific antibody define the *zone of equivalence.*

## Immunodiffusion

When a solution of antibodies and a solution of antigens are placed in nearby depressions (termed *wells*) of a gel, remarkably thin bands of precipitin can be visualized where antibodies and antigens meet in the zone of equivalence. Such techniques, termed *immunodiffusion*, are often used to determine whether two antigens share common antigenic determinants. For example, if serum containing antibody against a certain antigen is placed in one of three wells that form a triangular array in a gel, and the other two wells are filled with a solution of the antigen, a continuous precipitin forms in a v-shaped pattern [Figure 30.12(a)]. This v-shaped precipitin is termed the *reaction of identity*, and it demonstrates that the antibodies bind to the same antigenic determinants in each of the antigen samples. Alternatively, if one of the wells is filled with a different antigen that shares some (but not all) determinants with the first antigen, a y-shaped band of precipitin [Figure 30.12(b)] termed a *reaction of partial identity*, forms. The stem of the y (called a *spur*) is formed by cross-linking those antigenic determinants present in the first antigen but absent in the second. If two completely unrelated antigens are added to the wells, either a single straight band forms between two wells or two separate bands form, creating an x-shaped pattern, termed a *reaction of nonidentity* [Figure 30.12 (c)].

## Immunoelectrophoresis

Many individual components in a complex mixture of antigens can be visualized when the techniques of electrophoresis and immunoprecipitation are combined in a procedure termed *immunoelectrophoresis.* First, the mixture of antigens is separated in a gel by applying an electrical field (i.e., by electrophoresis). Then a solution of antibodies is placed in a narrow groove that is parallel to the direction of electrophoresis. The consequence is a set of crescent-shaped immunodifussion precipitins. A clinically important example of this technique is the resolution of major blood proteins in serum (Figure 30.13).

(a)

(b)

(c)

FIGURE 30.12
Characteristic patterns of precipitin formation during immunodiffusion. In each case, antibody is placed in the top well, and samples of antigen are placed in the lower wells. (a) Reaction of identity. (b) Reaction of partial identity. (c) Reaction of nonidentity. Courtesy of P. Baumann.

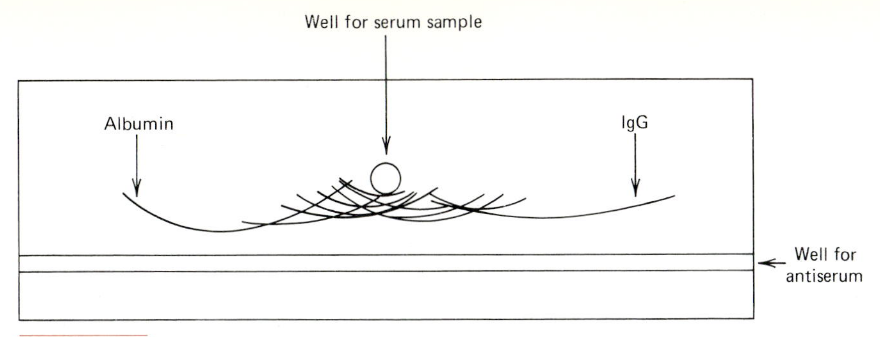


FIGURE 30.13

Immunoelectrophoresis. Serum proteins placed in the wells are separated along the horizontal axis by electrophoresis through a thin slab of agar. Antiserum to blood proteins (e.g. albumin and IgG) is then placed in the horizontal trough. It diffuses into the agar, forming bands of precipitin with specific serum proteins.

## Complement Fixation

Two properties of complement make it a useful reagent for the detection of immune complex formation: (1) complement fixation is triggered by such complexes and (2) complement fixation irreversibly inactivates essential complement proteins. An assay that measures the residual activity of complement after it is incubated in the presence of antigens and antibodies is termed a *complement fixation test.* This technique has been widely used to detect a specific antigen and is usually performed by mixing the sample with antiserum (lacking complement) from an animal immunized with the antigen. When immune complexes have had time to form, complement (usually from a guinea pig) is added. After some complement fixation has had time to occur, *sensitized erythrocytes* (sheep erythrocytes previously coated with complement-fixing antibodies) are added. Extensive lysis of erythrocytes results if immune complexes did not form in the first stage of the test because the antigen was not present; less extensive lysis indicates that the antigen was present in the sample.

## Radioimmunoassays

In some of the most sensitive serological reactions that have been developed, one of the reactants—antibody, antigen, or hapten—is made radioactive. In such techniques, termed *radioimmunoassays,* the radioactivity associated with precipitins is detected. Such assays are frequently used to measure concentrations of a hormone: a small amount of radioactive hormone is added to the sample and to each of a series of standards with known concentrations of hormone. Then the same amount of hormone-binding antibodies is added to each mixture. After immune complexes containing hormone have formed, they are separated from unbound hormone by adding a second antibody that cross-links constant domains of the original antibodies, thereby forming precipitins (Figure 30.14). The amount of radioactivity that remains in solution is an increasing function of the initial hormone concentration; the absolute concentration in the sample can be estimated by comparing its residual radioactivity with that of the standards.

FIGURE 30.14

A radioimmunoassay for a hormone. (a) A sample is mixed with a small amount of radioactive hormone and specific antibody is added. (b) A second antibody is added to precipitate the first, and the bound radioactivity is quantitated. The concentration of hormone can then be determined from a standard curve constructed by performing the test with samples of known hormone concentration.

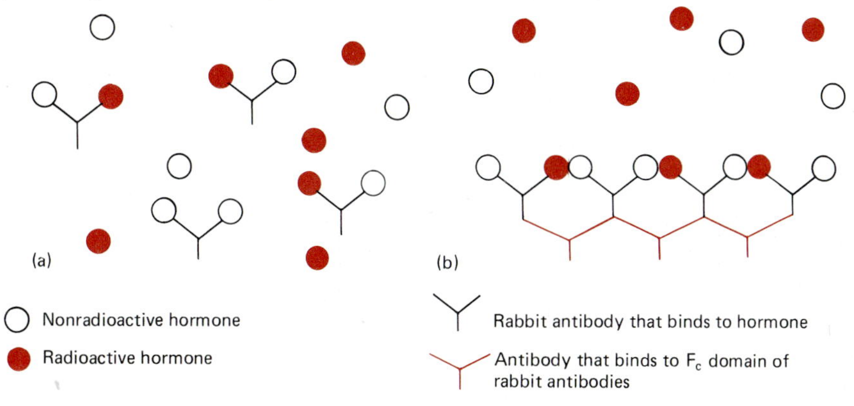

## Techniques Employing Conjugated Antibodies

A number of radioimmunoassays have been replaced by equally sensitive techniques that employ antibody molecules covalently linked to a nonradioactive molecule that is easily detectable by its chemical or physical properties. For example, fluorescein isothiocyanate can be covalently linked to the $\epsilon$-amino group of lysine, producing fluorescent antibodies, many of which still bind normally to antigens. The complex of these antibodies with antigens can be detected by their fluorescence that is easily seen in the ultraviolet microscope, rather than by their radioactivity. An alternative way to visualize antibodies is to link them chemically to electron-dense molecules such as ferritin that are easily seen in electron micrographs.

Highly sensitive quantitative assays have been developed by covalently binding an enzyme to an antibody. Such hybrid molecules can be used to measure the concentration of antigens, haptens, or antibodies in techniques termed *enzyme-linked immunosorbent assays* (ELISAs). The enzymes most commonly used in such techniques are horseradish peroxidase, alkaline phosphatase, and $\beta$ galactosidase: these are unusually stable and can be assayed with great sensitivity.

The ELISA technique can be used to measure the concentration of antibodies that bind to a particular antigen by first incubating the sample with an excess of antigen attached to the surface of a shallow plastic reaction vessel. After it is rinsed with buffer to remove unbound antibody, the amount of antibody that is bound is quantitated by adding a solution of enzyme-conjugated antibody that binds to constant domains of antibodies in the sample. Excess conjugated antibody is rinsed away, and the activity of bound enzyme is determined (Figure 30.15). This activity is approximately proportional to the amount of antigen-binding antibody in the sample; the original concentration of such antibodies can be estimated from a series of control assays employing known concentrations of specific antibody.

Alternatively, the amount of a particular antigen in a sample can be determined by incubating it with an excess of specific antibody fixed to the surface of a reaction vessel. The amount bound can be determined by adding an excess of specific antibody conjugated with an enzyme, rinsing away unbound antibody, and measuring the enzyme activity that remains.

# THE BASIS OF ANTIBODY DIVERSITY

The most remarkable property of antibodies is their diversity. Current estimates indicate that each individual synthesizes more than $10^6$ different kinds of immunoglobulins. This diversity is largely the result of variations in the sequence of amino acids in variable domains of both light and heavy chains.

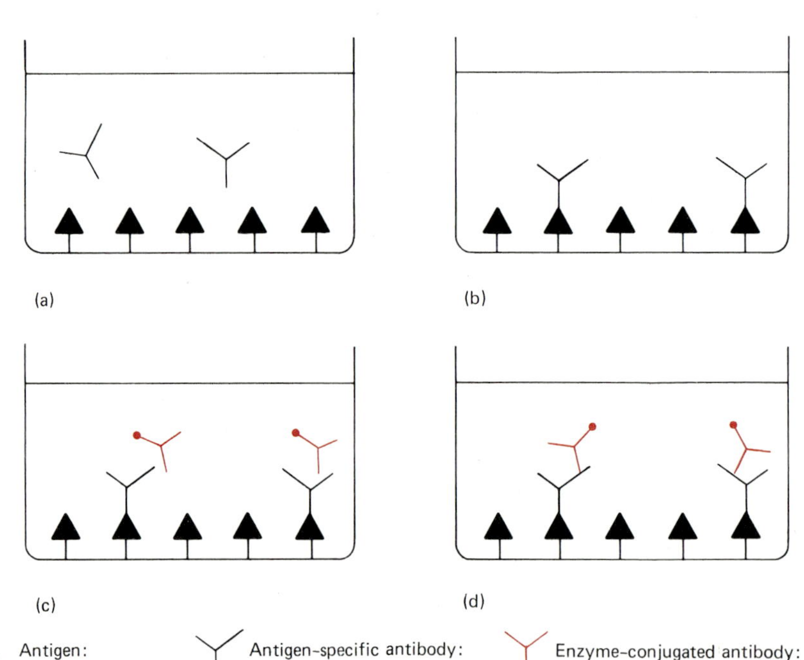


FIGURE 30.15

Diagram of an enzyme-linked immunosorbent assay (ELISA) for quantitating the amount of antibody that binds to a particular antigen. (a) The sample is added to a vessel with antigen attached to the surface. (b) Antibody in the sample binds specifically. (c) A solution containing an enzyme conjugated with an antibody that binds to and Fc domain of antibodies in the sample is added. (d) After rinsing away all unbound antibodies, enzyme activity is assayed. In this example, the intensity of color, measured photometrically, is approximately proportional to the amount of antigen-specific antibody in the sample.

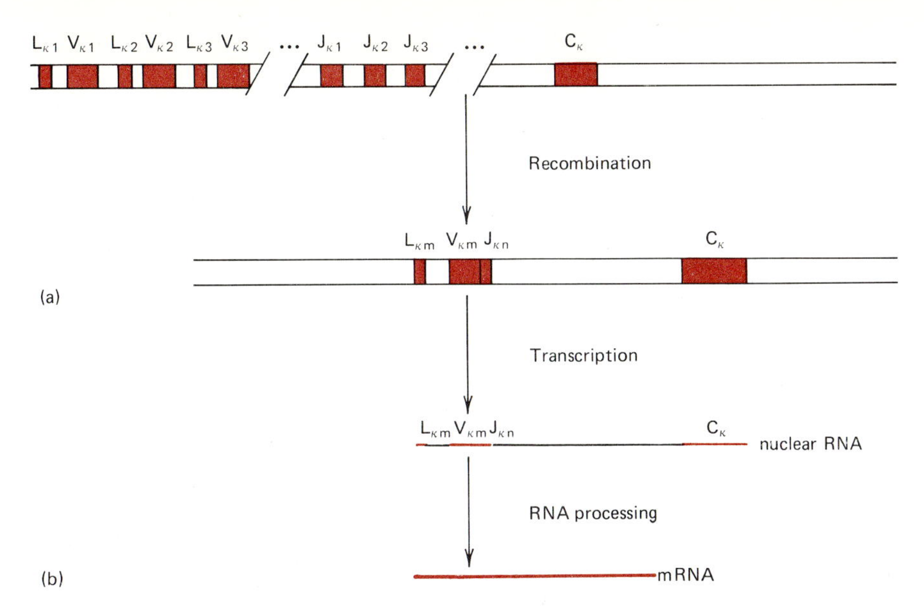


FIGURE 30.16

Organization and expression of $\kappa$ chain genes on human chromosome 2. (a) During B-cell development, recombination within the chromosome joins a V segment to a J segment. (b) Following transcription, introns are removed from RNA, which can then be translated into a $\kappa$ chain.

## The "Germ Line" and "Somatic Mutation" Theories

For many years, two quite different theories have attempted to explain the enormous diversity of antibodies. One, the *germ line theory*, holds that each human cell contains a separate gene encoding each antibody chain that the individual is capable of synthesizing. The other, the *somatic mutation theory*, holds that each cell contains a small number of genes encoding antibody chains and that each of these is especially susceptible to mutation, so that the multiple mutations accumulating in mature B-cells confer on the organism the ability to produce a wide variety of different antibodies. At present a different theory, one published in 1965 by W. Dreyer and J. Bennett, is generally accepted as the correct explanation for most antibody diversity: they proposed that variable regions and constant regions of antibodies are encoded by separate groups of genes.

Current knowledge of the genetic basis for antibody diversity comes largely from applications of recombinant DNA technology (Chapter 12): fragments of DNA that encode parts of antibody genes have been cloned in bacteria. The sequence of bases in the DNA of some of these fragments has been determined, and they have been mapped by DNA hybridization techniques to a particular chromosomal location. These studies have demonstrated that the information encoding antibodies is located on three chromosomes: one chromosome contains all information for $\kappa$ chains, a second contains that for $\lambda$ chains, and a third contains that for all heavy chains. In addition, the information encoding antibody chains on each of these chromosomes is partitioned into at least three sets of genes, termed *segments*, that encode a portion of an antibody chain, and a functional gene for antibody synthesis is generated by recombination that must occur during B-cell development in order to join segments encoding different parts of immunoglobulin variable regions. How this generates much of the known antibody diversity is discussed more fully below.

## The Generation of $\kappa$ Chain Diversity

All segments encoding $\kappa$ chains have been shown to be on chromosome 2 in humans and on chromosome 6 in mice. In stem cells of bone marrow, the information required to synthesize functional $\kappa$ chain is separated into four sets of segments: $L_{\kappa}$ segments encoding most of the leader peptide of the antibody chain, $V_{\kappa}$ segments encoding most of the variable domain, $J_{\kappa}$ segments encoding a short region of the variable domain near the V-C junction, and a $C_{\kappa}$ segment encoding the constant domain. The leader peptide, which is cleaved from the light chain as it is transported into endoplasmic reticulum, is required for antibody secretion. Each $L_{\kappa}$ segment is separated from a $V_{\kappa}$ segment by a short intervening sequence (*intron*). The number of distinct $V_{\kappa}$ sequences is approximately 20 on the human chromosome and approximately 200 on the mouse chromosome. In both cases, the L-V pairs are clustered on the chromosome. The five $J_{\kappa}$ sequences, each of which codes for approximately 13 amino acids of the $\kappa$ chain, are in a separate cluster, and the single $C_{\kappa}$ segment is nearby (Figure 30.16).

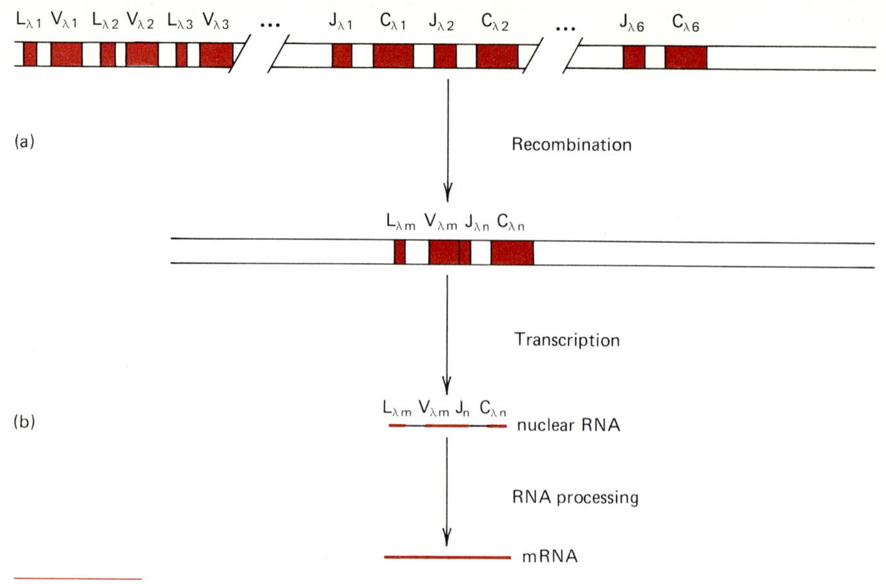


FIGURE 30.17

Organization and expression of $\lambda$ chain genes on human chromosome 22. (a) Recombination within the chromosome joins a L-V pair with a J-C pair. (b) Following transcription, introns are removed from RNA, which can then be translated into a $\lambda$ chain.

During the development of B-cells, recombination randomly joins the right end of a $V_\kappa$ segment with the left end of a $J_\kappa$ segment, a process termed *V-J joining*, thereby generating many different V-J combinations. Transcription starts to the left of the L segment that is paired with the now joined V segment and proceeds through the $C_\kappa$ segment, producing a nuclear RNA (nRNA) that is converted into mRNA by a poorly understood process in which RNA transcribed from the intron between the L and V segments is spliced out as is that from the region between the J segment and the $C_\kappa$ segment (Figure 30.16).

## The Generation of $\lambda$ Chain Diversity

The genetic information for $\lambda$ chains is encoded by four sets of segments located on chromosome 22 in humans and on chromosome 16 in mice. Again, each $L_\lambda$ segment is separated from its paired $V_\lambda$ segment by an intron, and these pairs form a cluster on the chromosome. The major difference between the organization of $\lambda$ chain genes and $\kappa$ chain genes is that each $J_\lambda$ segment is paired with its own $C_\lambda$ segment. In undifferentiated cells, these pairs of segments form a second cluster on the chromosome. There are four such pairs on the mouse chromosome and at least six on the human chromosome. As in the case of $\kappa$ genes, a functional $\lambda$ gene is produced during B-cell development by random recombination between the right end of a $V_\lambda$ segment and the left end of a $J_\lambda$ segment, thereby generating a large number of different genes (Figure 30.17).

## Generation of Heavy Chain Diversity

All genes for heavy chains are on chromosome 14 in humans and chromosome 12 in mice. As in the light chain genes, heavy chain $L_H$ segments are separated from their $V_H$ segments by an intron, and these $L_H$-$V_H$ pairs are clustered on the chromosome. Nearby is a separate cluster of at least nine segments (in humans) termed D segments that encode 6–17 amino acids near the $V_H$–$C_H$ junction. Further along the chromosome is a cluster of at least four $J_H$ segments in mice and at least six in humans. The heavy chain J segments code for 16–21 amino acids at the COOH-terminal end of the variable domain. Beyond the $J_H$ segments is a cluster of at least nine segments in humans and eight in mice that encode the constant regions of heavy chains. In mice the order of these segments is $C_\mu$, $C_\delta$, $C_{\gamma 3}$, $C_{\gamma 1}$, $C_{\gamma 2b}$, $C_{\gamma 2a}$, $C_\epsilon$, and $C_\alpha$ (Figure 30.18). In humans, the order is not completely known, but it is known that $C_\mu$ is first, followed by $C_\delta$.

To produce a functional gene for heavy chain synthesis, recombination during B-cell development must join the right end of a V segment with the left end of a D segment and must also join the

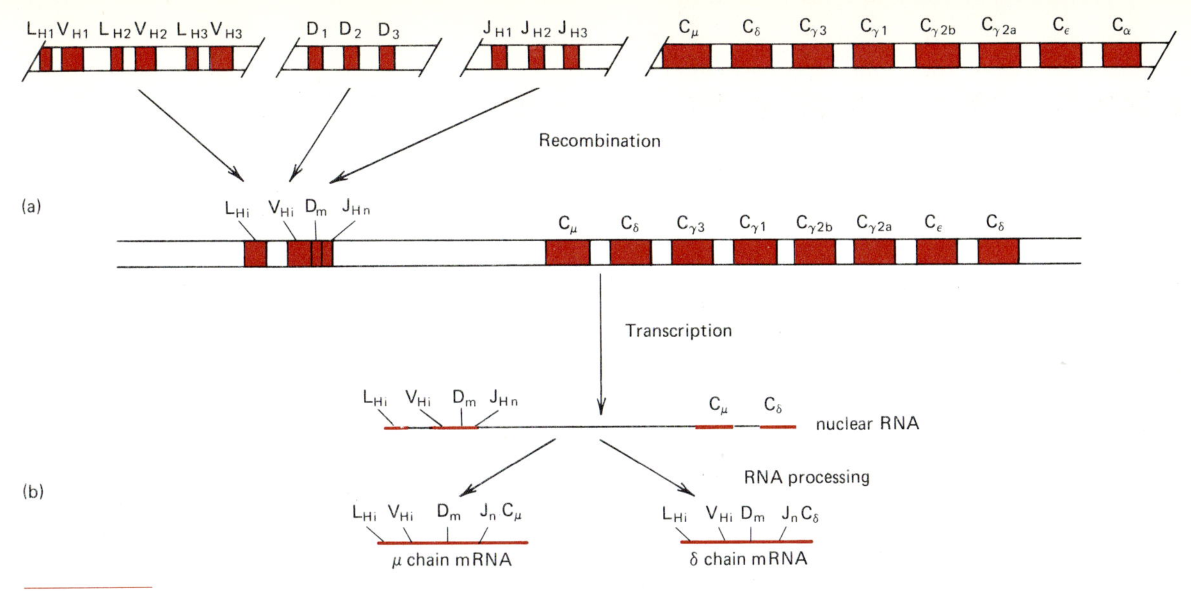


FIGURE 30.18

Organization and expression of heavy chain genes on human chromosome 14. (a) Recombination within the chromosome joins a V segment to a D segment and joins this D segment to a J segment. (b) Following transcription, introns are removed from RNA to produce two species of mRNA: one that codes for IgM and one that codes for IgD.

right end of this D segment with the left end of a J segment (Figure 30.18). Following transcription in immature B-cells, RNA copied from the region between the J segment and the $C_\mu$ segment is usually spliced out, leading to IgM synthesis. However, sometimes the splice extends through RNA transcribed from the $C_\mu$ segment (Figure 30.18), resulting in IgD synthesis. Both of these immunoglobulins are bound to the surface of B-cells. Following stimulation by specific antigen and T-cell factors (see below), a further recombination event occurs within the chromosome encoding heavy chains: the right end of the J segment is joined to the left end of a constant segment (Figure 30.19). This constant segment determines the antibody class synthesized by the mature B-cell.

## How Many Different Antibodies?

The number of distinct antibodies that can be produced as a result of the recombinational events of B-cell development is not known with certainty, but it is possible to estimate a lower bound: if the number of $V_H$ segments is greater than 200 on the human chromosome (and this seems likely), then the number of V-D-J combinations for human heavy chains is greater than $200 \times 11 \times 6 = 13{,}200$. These can pair randomly with light chains, of which well over 100 possibilities can be generated by V-J joining. Therefore, it appears that humans are able to produce more than $10^6$ antibodies, differing in their variable regions, from a relatively small amount of genetic information.

FIGURE 30.19

The final step in formation of heavy chain genes. Recombination fuses the J segment with one of the C segments, producing in this example a functional gene for IgG synthesis.

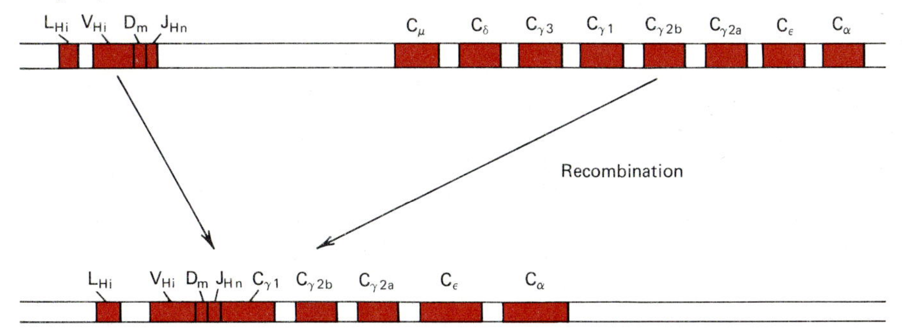

Antibody diversity is augmented by two processes: somatic mutation and imprecise joining of either the V and J segments encoding parts of light chains or of the V, D, and J segments encoding parts of heavy chains. Somatic mutations occur during B-cell development and are restricted largely to V, D, and J segments by unknown mechanisms. Imprecise joining during the recombination that generates light or heavy chain genes can result in addition or loss of a small number of amino acids in one of the hypervariable regions. Both of these processes appear to be significant sources of antibody diversity.

## FUNCTIONS OF T-CELLS

Although T-cells do not produce antibodies, they do produce antigen-binding proteins, termed *T-cell receptors*, that play an important role in the immune response. Unlike antibodies, the T-cell receptor always remains firmly bound to the cytoplasmic membrane of the cell that produces it. Each T-cell receptor is a dimer composed of two nonidentical polypeptides. These, like the light chains of antibody molecules, possess a variable domain and a constant domain. In addition they possess a region of hydrophobic amino acids at the COOH-terminal end of the polypeptides that anchors the receptor in the membrane (Figure 30.20).

FIGURE 30.20

Diagram of the T-cell receptor. Many copies of a specific antigen-binding protein, the T-cell receptor, are anchored in the cytoplasmic membrane by hydrophobic amino acids. The receptor contains two nonidentical polypeptides termed the $\alpha$ chain and the $\beta$ chain. Reprinted by permission from H. Saito, D. M. Kranz, Y. Takagaki, A. C. Hayday, H. N. Eisen, and S. Tonegawa, "Complete Primary Structure of a Heterodimeric T-cell Receptor Deduced from cDNA Sequence", *Nature*, **309**, 757 (1984), copyright 1984 Macmillan Journals Limited.

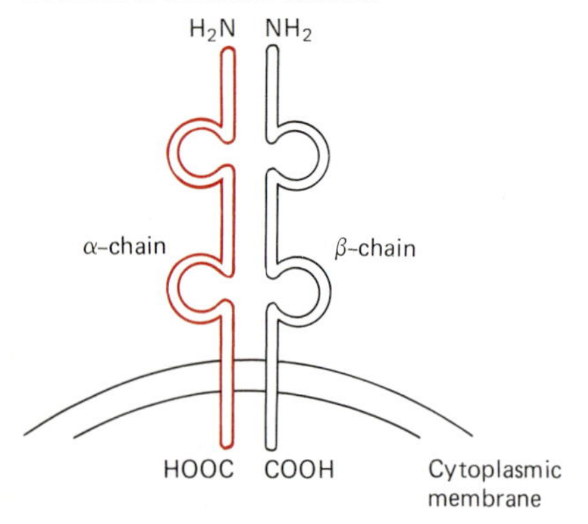


T-cells are classified as *regulator* or *effector* T-cells on the basis of their functional roles: regulator T-cells control the response of B-cells to antigens, and effector T-cells mediate other immune functions such as the attraction of leukocytes to sites of inflammation and the rejection of transplanted organs.

### Effector T-Cells

Some T-cells, termed T-killer ($T_K$) cells, are able to kill eucaryotic cells that possess surface antigens recognized by the T-cell receptor. Cell death does not involve antibodies, but does require contact between a T-cell and its target.

Other effector cells, *delayed-type hypersensitivity* ($T_{DTH}$) cells, release substances termed *lymphokines* in response to specific antigens. Lymphokines attract macrophages and other leukocytes to a site of inflammation. They also activate macrophages and inhibit their migration away from a site of inflammation (Table 30.3). $T_{DTH}$ cells (and probably $T_K$-cells) are essential in host defense against some viral, protozoal, and fungal diseases, as well as against tuberculosis. They may also play a role in the detection and killing of cancer cells, a process termed *immune surveillance*.

### Regulator T-Cells

With the exception of the immune response to some polysaccharide antigens, antibody production requires the assistance of both macrophages and regulator T-cells belonging to the class termed *T-helper* ($T_H$) *cells*. One role of macrophages is to bind antigen on their surfaces, so that $T_H$-cells with receptors for the antigen bind to it. This initiates what is perhaps the most important and least understood chain of events in the immune response: these macrophages release a protein termed *interleukin*-1, which stimulates immunological functions of T-cells, B-cells, and macrophages. The antigen-stimulated $T_H$-cells release a lymphokine (Chapter 29), termed *interleukin*-2, that stimulates antigen-stimulated $T_H$-cells to divide and to release a second lymphokine, *B-cell growth factor* (BCGF).

Antibody production also requires the binding of antigen to antibodies attached to the surface of B-cells, which, in the presence of interleukin-1 and BCGF, then divide. However, B-cells do not secrete antibody unless they are stimulated to differentiate into *plasma cells* by additional lymphokines released from T-cells. Plasma cells are large B-lymphocytes whose major activity is synthesis and secretion of antibodies.

**TABLE 30.3**

**Examples of Lymphokines, Proteins That Are Secreted by Lymphocytes and That Regulate Cellular Activities**

| Lymphokine | Regulatory Effect(s) |
|---|---|
| $\gamma$-Interferon | Activates macrophages and T-cells |
| Interleukin-2 | Stimulates activated T-cells |
| Macrophage-activating factor (MAF) | Stimulates phagocytic activities of macrophages |
| Macrophage chemotactic factor (MCF) | Attracts macrophages to a site of inflammation |
| Migration-inhibition factor (MIF) | Inhibits migration of macrophages away from a site of inflammation |

The other class of regulator T-cells, *T-suppressor* ($T_S$) *cells,* plays an essential role in preventing the immune system from attacking host tissues: $T_S$-cells with receptors for an antigen present in host tissue interfere in an unknown way with the activation of $T_H$-cells that also possess receptors for the self antigen (Figure 30.21).

## Histocompatibility Antigens

Certain antigens, termed *histocompatibility antigens,* that are normally present on the surface of many mammalian cells, strongly influence regulatory T-cell responses. These antigens are divided into 2 groups on the basis of the type of immune response that they provoke: class I histocompati-

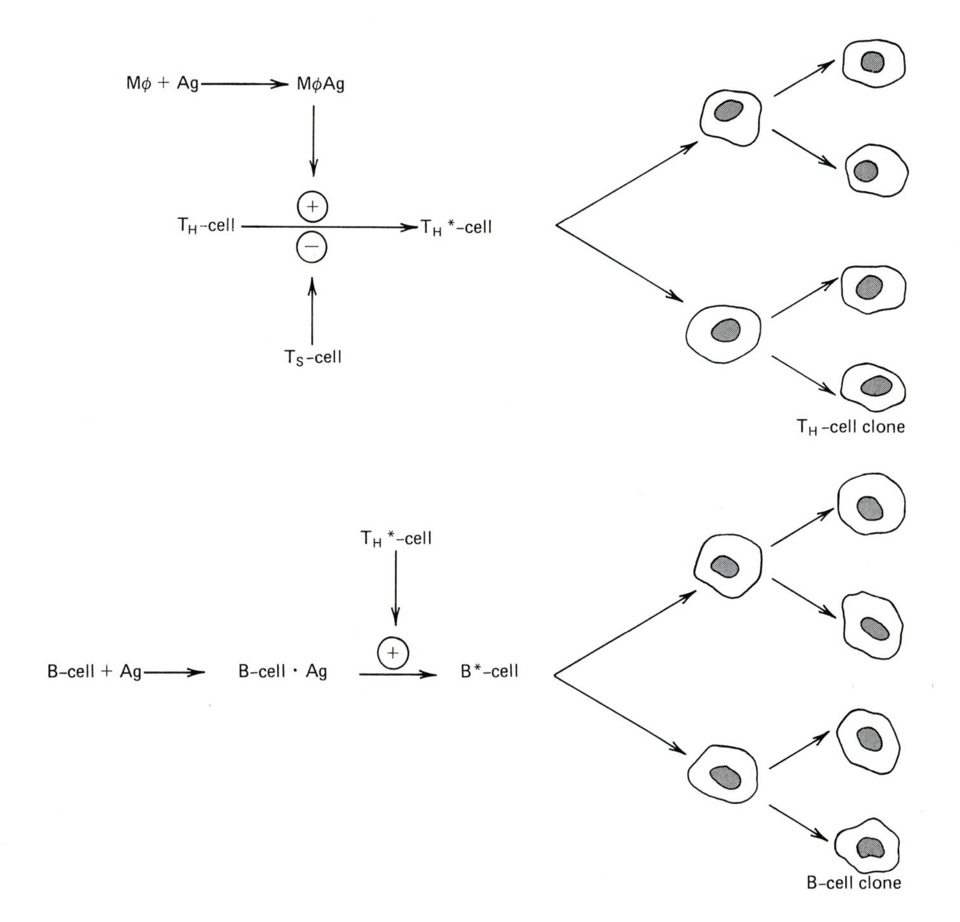


FIGURE 30.21
Regulation of the immune system by $T_H$ and $T_S$ cells. Activation of $T_H$-cells requires antigen bounds to the surface of a macrophage. Activated $T_H$-cells ($T_H^*$-cells) divide and also stimulate B-cells bound to the same antigen to divide. Some of these B-cells differentiate into antibody-producing cells. If the individual has $T_S$-cells specific for the antigen, they can block the initial activation of $T_H$-cells.

bility antigens are found on all nucleated host cells and stimulate antibody production when injected into a host with different class I antigens. Class II histocompatibility antigens occur on lymphocytes and macrophages, but they do not usually stimulate antibody production. Rather, they are required by T-cells in order to recognize both macrophages and B-cells. It seems likely that part of the T-cell receptor must recognize a class II antigen on an adjacent cell before it can secrete lymphokines necessary for an immune response.

## IMMUNIZATION

Immunization is the artificial induction of immunity to disease. The first successful method of immunization, termed *variolation*, was discovered centuries ago in Asia and was used to produce immunity to smallpox. The procedure was quite simple: material from a pustule on an infected person was scratched into the skin of the person to be immunized. In most cases, this produced a mild case of smallpox without the scarring that was common in naturally acquired cases, but approximately 3 percent of those variolated died (as contrasted with a 30 percent fatality rate in naturally acquired smallpox, which first infects the respiratory tract). Variolation was introduced into England in 1721 by Lady M. Montagu, who became aware of the procedure while living in Turkey. However, it was never widely used in England despite her vigorous attempts to promote it.

During the eighteenth century, in parts of rural England, the belief arose that people who acquired from cattle a mild disease called cowpox, became immune to smallpox. In 1774 an English farmer, B. Jesty, successfully immunized his children against smallpox by inoculating them with pustular material from an infected cow; later they survived a smallpox epidemic. In 1778, the physician E. Jenner inoculated his son with material from a cowpox pustule, beginning a 20-year study that led to publication in 1798 that the procedure (termed vaccination, after the Latin name for cowpox, vaccinia) was safe and effective. However, Jenner's procedure was only slowly adopted, and epidemics of smallpox continued throughout the world until 1977, when the last case of smallpox was reported. This dreaded disease now appears to be extinct.

The remarkably successful vaccination program designed by the World Health Organization to eradicate smallpox depended on several special features of the disease: (1) all infections in susceptible individuals produce a disease that is easily recognized; (2) no asymptomatic carriers of the disease exist; (3) humans are the only host of this pathogen; and (4) the pathogen cannot survive outside of its host. Unfortunately, few diseases possess all of these properties.

### Passive Immunization

Resistance to many diseases can be produced by injecting into an individual a preparation containing antibodies that bind specifically to antigens produced by a pathogen. Such resistance is termed *passive immunity*. It has the advantage of becoming effective within several hours after the injection, whereas vaccines require at least several days to produce resistance. However, the duration of passive immunity following a single treatment is relatively brief: resistance rarely lasts longer than several months.

During the first half of this century, antiserum, usually produced in a domestic animal, was widely used both for prevention and treatment of certain diseases. These injections were sometimes successful, but many allergic reactions were produced when antiserum from domestic animals was used. Currently, most preparations used for passive immunization contain antibodies termed *human immune globulins* that are purified either from the serum of a person who has a high degree of immunity to a particular disease or, in the case of diseases to which immunity is common, from a collection of serum samples taken from many persons. Preparations of human immune globulin are used for treatment or prevention of several diseases including diphtheria, hepatitis A, hepatitis B, rabies, and tetanus. Furthermore, individuals with a genetic defect in their ability to produce antibodies are currently given repeated injections of human immune globulin, which greatly increases their resistance to common infectious diseases.

### Active Immunization

Antigenic material that produces specific immunity by stimulating an immune response is termed a *vaccine*. Three different kinds of vaccines are widely used today: (1) killed pathogens, (2) avirulent (*attenuated*) strains of pathogenic species, and (3) chemically modified toxins (*toxoids*). The most widely used vaccine of the first type is *pertussin*, the vaccine against the childhood disease whooping cough (caused by *Bordetella pertussis*). The first successful vaccine against polio, developed by J. Salk, was also a killed pathogen, i.e., poliovirus, but it has been replaced by the Sabin oral polio vaccine, which is of the second type. Other vaccines composed of killed pathogens are listed in Table 30.4.

**TABLE 30.4**

**Examples of Vaccines That Are Effective in Producing Immunity to Specific Diseases**

| Disease | Type of Vaccine |
|---|---|
| VIRAL DISEASES | |
| Hepatitis B | Viral antigen purified from the blood of individuals with chronic hepatitis |
| Influenza | Formalin-killed type A influenza virus strains |
| Measles | Attenuated strain |
| Mumps | Attenuated strain |
| Polio | Attenuated strains (Sabin oral vaccine); formalin-killed virus (Salk vaccine) |
| Rabies | Killed virus |
| Rubella | Attenuated strain |
| Yellow fever | Attenuated strain |
| BACTERIAL DISEASES | |
| Cholera | Killed pathogen |
| Diphtheria | Toxoid |
| *Haemophilis influenzae* infections | Purified surface antigen |
| Tetanus | Toxoid |
| Typhoid fever | Killed pathogen |
| Whooping cough | Killed pathogen |

### Attenuated Strains

Strains of a pathogen that have lost virulence are termed *attenuated* strains. With the exception of the cowpox virus, which is probably a naturally attenuated strain of smallpox virus, attenuated strains of pathogens used for vaccination have been developed in the laboratory. L. Pasteur was the first to develop one and demonstrate that it could be safely and effectively used as a vaccine: in 1879 he discovered that a culture of *Pasteurella septica*, the cause of chicken cholera, had lost virulence during several weeks' storage. He then showed that chickens exposed to the attenuated strain acquired immunity to virulent strains.

In 1881, Pasteur deliberately set out to produce an attenuated strain of *Bacillus anthracis*, the cause of anthrax (Chapter 1). He stored a culture for weeks, but this was ineffective. He then tried a variety of other procedures and found that growing cultures at 42° C caused attenuation. We now know the reason for Pasteur's success: virulence of *B. anthracis* depends on the presence of a plasmid that cannot replicate at 42° C.

Most vaccines that are used against viral diseases are strains that have been attenuated during extended periods of cultivation in the laboratory. Apparently, mutations that occur during this period are responsible for the loss of virulence. Vaccines against measles, mumps, polio (oral polio vaccine), and rubella are examples of live, attenuated vaccines (Table 30.4).

### Toxoids

Immunity to tetanus or diphtheria (Chapter 31) can only be conferred by antibodies that neutralize the toxins produced by the bacteria causing these diseases. Active toxins cannot be used to immunize against these diseases because the amount required to stimulate an immune response is lethal. However, these toxins can be chemically modified to nontoxic forms termed *toxoids* that retain the ability to stimulate production of toxin-neutralizing antibodies.

### Kinetics of Immunization

Usually, an individual who has never been exposed to an antigen lacks detectable amounts of the corresponding antibody. Then, following exposure by injection or some other method, a *primary response* occurs: the level of antibody usually remains unde-

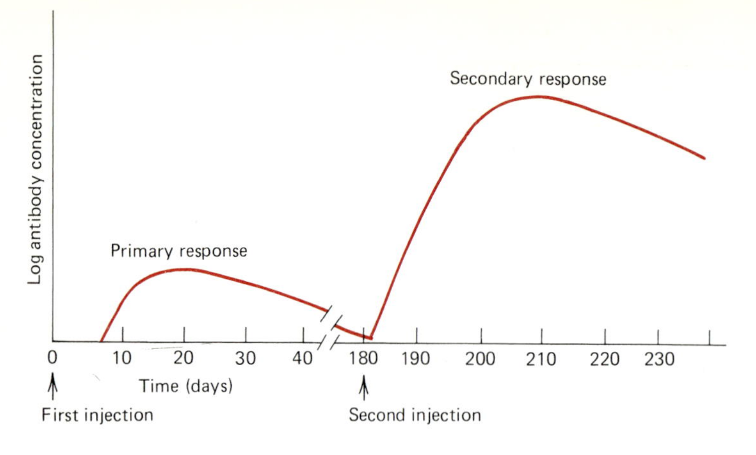


FIGURE 30.22

Kinetics of vaccination. At day 0, an individual is vaccinated for the first time with specific antigen. A rise in the amount of specific antibody in his blood is typically first detected after several days (the primary response). When the individual is vaccinated again with the same antigen, the rise in specific antibody is almost immediate (secondary response).

tectable for several days, and then it rises over a period of one or more weeks, reaching a peak, and finally, gradually decreasing (Figure 30.22).

When the individual is exposed again to the same antigen, a response termed a *secondary* or *anamnestic response* follows in most cases: the level of antibody begins to rise much sooner than it did before, and it reaches a much higher value (Figure 30.22).

The physiological basis of the secondary response is the proliferation of specific $T_H$-cells and B-cells that occurs during the primary response. These cells have a long lifespan and are therefore able to serve as *memory cells* in the secondary response. It is these cells that are responsible in many cases for acquired immunity to an infectious disease.

## HYPERSENSITIVITY AND AUTOIMMUNITY

In disorders termed *hypersensitivity reactions* or *allergies* (Table 30.5), an exaggerated response of the immune system to foreign antigens produces symptoms of disease. These may be mild (e.g., sneezing), or they may be alarming (e.g., severe wheezing) and can be triggered by a wide variety of substances including dust, pollen, certain foods, and insect saliva. In a group of related but much rarer disorders, termed *autoimmune diseases*, the immune system of an individual responds to one of the body's own components, termed a *self-antigen*, and thereby causes disease.

### Anaphylaxis

The most common type of allergy in humans is termed *anaphylaxis*: its symptoms vary widely, depending on the part of the body involved and the extent of the reaction. Examples of anaphylaxis include the usually mild reactions of hay fever, mosquito bites, and hives as well as fatal reactions to bee stings. This type of allergy occurs only in an individual who has previously become sensitized to an antigen by synthesizing relatively large amounts of antigen-specific IgE during a prior exposure. Symptoms occur during subsequent exposures and are initiated by the antigen's binding to IgE on mast cells and basophils, triggering release of histamine which causes the acute inflammation (Chapter 29) characteristic of anaphylaxis occurring in the skin or mucous membranes. When released throughout the body, histamine causes a large loss of fluid from the bloodstream, resulting in shock. In severe anaphylaxis, leukotrienes are also released (by an unknown mechanism): they cause the smooth muscle in bronchial walls to contract, resulting in wheezing and in some cases asphyxiation. Before their chemical identity was known, these substances that cause bronchial constriction were termed *slow-reacting substance of anaphylaxis* (SRS-A).

### Antibody-Dependent Cytotoxicity

In *antibody-dependent cytotoxicity*, cells of the body are killed as a consequence of antibodies binding to antigens on or near the cell surface. One example is the hemolytic anemia caused by certain drugs that bind to the surface of red blood cells, initiating their lysis through the complement pathway.

### Immune Complex Disorders

Any condition in which formation of immune complexes in the body causes damage is termed an *immune complex disorder*. For example, a person who receives a second injection of antiserum prepared in a horse or other animal sometimes develops a disease termed *serum sickness* that is caused by immune complexes formed between the individual's

**TABLE 30.5**

**Types of Allergy (Hypersensitivity)**

| Type | Mechanism | Examples |
|---|---|---|
| Anaphylaxis | Binding of antigen to IgE on surface of mast cells triggers release of histamine. | Allergy to bee sting; hay fever; hives |
| Antibody-dependent cytotoxicity | Antibodies bind to antigens on the surface of cells, initiating complement fixation. | Transfusion reaction |
| Immune complex disease | Immune complexes form in tissues and initiate inflammation. | Serum sickness |
| Delayed-type hypersensitivity | Specific T-cells release lymphokines that attract other leukocytes | Allergy to poison oak and poison ivy |

antibodies and foreign proteins of the antiserum. These complexes become lodged in the walls of blood vessels where they trigger inflammation mediated by the complement pathway. The neutrophils that are attracted to the site release lysosomal contents (Chapter 29) that damage blood vessels.

### Delayed Hypersensitivity

Hypersensitivity initiated by T-cells that release lymphokines which attract other leukocytes to a site of inflammation is termed *delayed hypersensitivity* (as contrasted with the above antibody-mediated types that are collectively termed *immediate hypersensitivity* reactions). The symptoms of delayed hypersensitivity, beginning one to three days after exposure, are redness, itching, and formation of small blisters. These contain large numbers of lymphocytes, macrophages, and sometimes basophils, all attracted to the site by lymphokines. Histamine is not involved in this type of allergic response.

### Autoimmune Diseases

An abnormal response to self-antigens has been demonstrated to be the cause of a few human disorders termed *autoimmune diseases* (Table 30.6).

**TABLE 30.6**

**Examples of Diseases in Which the Immune System Damages the Host**

| Disease | Role of the Immune System |
|---|---|
| DISEASES IN WHICH AN AUTOIMMUNE ETIOLOGY IS ACCEPTED | |
| Graves's disease | An abnormal antibody binds to receptors on thyroid cells, thereby stimulating them. |
| Hashimoto's disease | T-cells invade thyroid tissue and initiate inflammation. |
| Goodpasture's syndrome | Antibodies bind to membranes of kidney and lung, initiating inflammation. |
| Myasthenia gravis | Antibodies bind to acetylcholine receptor on muscle cells. |
| Pernicious anemia | Antibodies bind to parietal cells in the stomach wall, thereby triggering cell destruction. |
| DISEASES IN WHICH AN AUTOIMMUNE ETIOLOGY HAS BEEN SUGGESTED BUT NOT PROVED | |
| Juvenile-onset diabetes mellitus | Antibodies bind to beta cells of the islet of Langerhans in the pancreas. |
| Rheumatoid arthritis | Antibodies bind to constant domains of human IgG. |
| Systemic lupus erythematosis | Antibodies bind to RNA and DNA. |

Most of these are caused by the abnormal production of an antibody that binds to a host component, resulting in tissue damage. For example, in Goodpasture's syndrome, antibodies bind to membranes in the lung and kidney, resulting in severe damage to both. However, autoimmune disease can be produced in the absence of tissue destruction: Graves' disease is caused by antibodies that bind to receptors on cells of the thyroid gland, stimulating the release of abnormally large amounts of thyroid hormone. In contrast with these antibody-mediated disorders, the disease, *Hashimoto's thyroiditis*, is caused by a cellular immune response initiated by sensitized T-cells that invade and destroy thyroid tissue.

In a second group of human diseases, an abnormal response of the immune system is implicated, but not proven, to be the cause (Table 30.6). These include *rheumatoid arthritis* and *juvenile-onset diabetes*. Although antibodies that bind to human antigens have been demonstrated, their role in causing these diseases is controversial: in none of these has a viral etiology been ruled out.

## FURTHER READING


### Books

EISEN, H. N., *Immunology*, 3rd ed. New York: Harper & Row, 1984.

FUDENBERG, H. H., D. P. SITES, J. L. CALDWELL, and J. V. WELLS, *Basic* and *Clinical Immunology*, 5th ed. Los Altos, Calif.: Lange Medical Publications, 1984.

TIZARD, I. R., *Immunology: An Introduction*. Philadelphia: Saunders, 1984.

### Reviews

Behbehani, A. M., "The Smallpox Story: Life and Death of an Old Disease," *Microbiol. Rev.* **47,** 455 (1983).

HONJO, T., "Immunoglobulin Genes," *Ann. Rev. Immunol.* **1,** 499 (1983).

JOINER, K. A., E. J. BROWN and M. M. FRANK, "Complement and Bacteria: Chemistry and Biology in Host Defense," *Ann. Rev. Immunol.* **2,** 461 (1984).

LEDER, P., "The Genetics of Antibody Diversity," *Scientific American* **246,** 102 (1982).

REID, K. B. M., and R. R. PORTER, "The Proteolytic Activation System of Complement," *Ann. Rev. Biochem.* **50,** 433 (1981).

SMITH, H. R., and A. D. STEIMBERG, "Autoimmunity—A Perspective," *Ann. Rev. Immunol.* **1,** 175 (1983).

TONEGAWA, S., "Somatic Generation of Antibody Diversity," *Nature* **302,** 575 (1983).

YELTON, D. E., and M. D. SCHARFF, "Monoclonal Antibodies: A Powerful New Tool in Biology and Medicine," *Ann. Rev. Biochem.* **50,** 657 (1981).

### Original Article

SAITO, H., D. M. KRANZ, Y. TAKAGAKI, A. C. HAYDAY, H. N. EISEN, and S. TONEGAWA, "Complete Primary Structure of a Heterodimeric T-Cell Receptor Deduced from cDNA Sequences," *Nature* **309,** 757 (1984).

# Chapter 31
# Microbial Pathogenesis

Individual strains within a pathogenic species often vary widely with respect to their ability to cause disease, i.e., with respect to their *pathogenicity*. Strains that can cause disease are described as *virulent*, and those that cannot are described as *avirulent*. All virulent organisms possess one or more special properties that contribute to their ability to cause disease. In a few species, pathogenicity is largely or entirely due to *toxigenicity*, the ability to produce compounds termed *toxins*; these are either proteins or lipopolysaccharides that produce specific harmful effects on the host. For example, the disease *botulism* can be caused merely by ingesting a purified toxin produced by *Clostridium botulinum.*

Completely unlike the toxin-dependent virulence of *C. botulinum*, virulence of most pathogens depends at least in part on *invasiveness*, i.e., the ability to proliferate in host tissue, or on the ability to *colonize* a surface such as skin or membranes of the respiratory, gastrointestinal, or genitourinary tracts.

Growth of microorganisms within host tissues is termed *infection.* It sometimes occurs in the absence of disease, in which case it is termed *silent infection*, but often it produces inflammation and other symptoms of disease. Many pathogens cause disease by a combination of toxigenicity and invasiveness or of toxigenicity and colonizing ability. However, some invasive pathogens cause disease that results from an abnormal reaction of the host's defense system, termed a *hypersensitivity reaction.* In such an instance, damage is caused by the host's own immune system (Chapter 30) rather than by microbial toxins.

## BACTERIAL TOXINS

Bacterial toxins are divided by chemical properties into two groups: *exotoxins*, which are soluble proteins found in cell extracts or in the growth medium, and *endotoxins*, which are the lipopolysaccharide components of the outer membranes of Gram-negative bacteria. Like most proteins, nearly all exotoxins are quite heat-labile, being inactivated by boiling for only a few minutes, but there are important exceptions. Exotoxins are conventionally classified into three types (Table 31.1): (1) *enterotoxins*, which stimulate cells of the gastrointestinal tract in an abnormal way; (2) *cytotoxins*, which kill host cells by enzymatic attack; and (3) *neurotoxins*, which interfere with normal transmission of nerve impulses.

Unlike exotoxins, all endotoxins are heat-stable. They are remarkably similar in chemical structure, and in their effects on the host. The active moiety of endotoxins, lipid A (Chapter 4), can cause the same reactions in the host as can the intact endotoxin, namely fever, shock, diarrhea, and sometimes internal hemorrhage or abortion. Remarkably, endotoxin exerts quite a different response when administered at sublethal levels: it confers enhanced resistance to bacterial infections by triggering the release of *interleukin*-1 from host cells (Chapter 30). This lymphokine (Chapter 29) has been tentatively identified as *endogenous pyrogen*, which causes fever by its action on the hypothalamus.

**TABLE 31.1**

**Examples of Bacterial Toxins Involved in Human Diseases**

| Toxin | Pathogen | Mode of Action |
|---|---|---|
| NEUROTOXINS | | |
| Botulinum toxin | *Clostridium botulinum* | Binds to motor neurons and prevents release of the neurotransmitter, acetylcholine, at myoneural junctions. |
| Tetanus toxin | *Clostridium tetani* | Binds to nerve cells and blocks inhibitory impulse transmission. |
| CYTOTOXINS | | |
| Diphtheria toxin | *Corynebacterium diphtheriae* | Penetrates host cells and inactivates protein elongation factor to stop protein synthesis. |
| Streptolysin O, Streptolysin S | *Streptococcus pyogenes* | Destroys integrity of lysosomal membrane in phagocytes, causing intracellular release of hydrolytic enzymes. |
| ENTEROTOXINS | | |
| Cholera toxin | *Vibrio cholerae* | Stimulates abnormal pumping of electrolytes and water into the colon by elevating intracellular cyclic-AMP concentration. |
| Heat-labile enterotocin (LT) | *Escherichia coli* | Identical in function to the cholera toxin. |
| Heat-stable enterotoxin (ST) | *Escherichia coli* | Stimulates abnormal pumping of electrolytes and water into the colon by increasing intracellular cyclic-GMP concentration. |
| Shiga toxin | *Shigella dysenteriae* | Stops protein synthesis in cells of the intestinal mucosa by interfering with the 60S ribosomal subunit. |

### Identification of Bacterial Toxins

A few years after the discoveries that the important human diseases diphtheria and tetanus are caused by bacteria, a simple procedure demonstrated that death, when it results from these diseases, is caused by microbial toxins: a cell-free preparation made from a culture of the pathogenic bacterium was injected into experimental animals: they then developed a fatal disease nearly identical to the naturally acquired one. During the past century, many other toxins have been identified and the mechanisms by which most of them damage host tissue are at least partially understood. Properties of toxins that are largely responsible for the prominent symptoms of some infectious diseases are summarized in Table 31.1; some of these are discussed in the following sections.

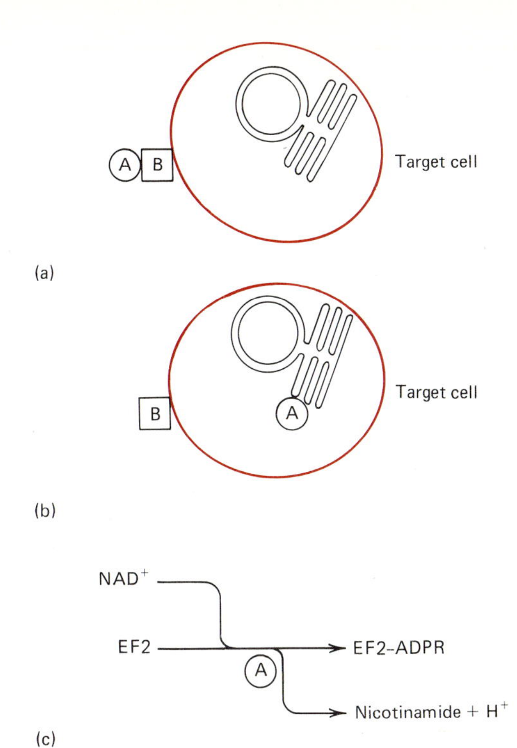


FIGURE 31.1

Cytotoxicity of diphtheria toxin. (a) Component B of the dimeric toxin (AB) binds to a specific receptor on the membrane of a target cell, and (b) component A enters the cell. (c) Within the cytoplasm, component A catalyzes ADP-ribosylation of elongation factor, EF2 (required for protein synthesis), thereby inactivating it.

## EXAMPLES OF TOXIN-CAUSED PATHOGENESIS

In addition to botulism and tetanus, several other diseases owe their prominent symptoms to the action of bacterial toxins. These are discussed below; the infectious diseases in which toxins play a lesser role are discussed in Chapter 32.

### Diphtheria

Before a vaccine for diphtheria was available, this disease was a leading cause of death in children. For unknown reasons, the disease was most common in temperate climates, and most cases occurred during the fall or winter. The causative agent, *Corynebacterium diphtheriae*, is a Gram-positive rod that frequently inhabits the upper respiratory tract. Virulent strains secrete the *diphtheria toxin*, but avirulent strains, which are common, do not; they lack the gene that encodes diphtheria toxin. Curiously, this gene is carried within the genome of a temperate bacteriophage, *phage beta*, that infects *C. diphtheriae*. Only those strains that have been infected and thereby carry the toxin gene within the integrated prophage produce the toxin. This acquisition of virulence by *C. diphtheriae* is an example of a phenomenon termed *lysogenic conversion* (Chapter 9), in which expression of genes within a prophage alters the phenotype of the bacterial host. Virulent strains produce the toxin only when the iron concentration is restricted, a condition prevalent in human tissues and secretions. Consequently, virulent strains produce the toxin when they grow on mucous membranes of the upper respiratory tract.

From one to seven days after infection by a virulent strain, a mild sore throat and a slightly elevated temperature develop. Soon a grey-to-white film forms on the back of the throat. The pathogen grows within this film, composed largely of dead host cells, and secretes the toxin, some of which diffuses into the bloodstream and becomes disseminated throughout the host. The action of diphtheria toxin on host cells is remarkably specific (Figure 31.1). One of the subunits, *component B* (MW 41,000), of the dimeric protein mediates the binding of the complete toxin molecules to specific receptor sites on the outside surface of the cytoplasmic membrane and the subsequent transport of the other subunit, *component A* (MW 21,000) into the cell. Within the cytoplasm, component A catalyzes the transfer of the ADP-ribosyl moiety of NAD (Chapter 4) onto *elongation factor 2* (EF2), an essential component of the cell's machinery for protein synthesis (Chapter 5). The ADP-ribosylated form of EF2 is inactive. Since the action of component A is catalytic, one molecule of component A within a cell can inactivate enough EF2 molecules to stop protein synthesis, eventually causing death of the affected cell. Destruction of cells in the heart, kidney, and other vital organs far from the site of infection leads to death of the host.

### Tetanus

Tetanus is another disease that was once relatively common but is now quite rare as a consequence of widespread vaccination. It is caused by *Clostridium tetani*, a Gram-positive, spore-forming, strict anaerobe commonly found in soil and in the colon of many mammals including humans. This organism has very low invasiveness and is completely innocuous when growing in the colon. However, spores of this pathogen sometimes enter a wound when an object such as a splinter or nail contaminated by soil punctures the skin and carries spores into damaged tissue. If the oxygen tension within the wound is low enough, the spores germinate and the organism grows, secreting toxin composed of a single polypeptide (MW 160,000), which is transported from the site of infection to nerve cells where it binds. By an unknown mechanism, the toxin interferes with the normal function of inhibitory neurons which are required for normal muscle tone. As a result, motor neurons become hyperactive, producing prolonged muscle spasms termed *tetany*. Interestingly, the gene encoding tetanus toxin is not carried on the bacterial chromosome. Rather, it is on a plasmid normally found in *C. tetani*.

Symptoms of tetanus develop from three days to several weeks after the pathogen is inoculated into a wound. The disease is characterized by rigidity (spasms) of muscles, which can be restricted to those in a single limb or can involve all skeletal muscles in the body. In severe cases, muscle spasms can constrict the throat or stop respiration and therefore cause death.

### Cholera

Throughout history, cholera has been a major endemic disease in the Indian subcontinent and parts of Southeast Asia. Occasionally, it has spread to other regions, causing epidemics such as the one that claimed the life of the Russian composer, P. Tchaikovsky, in Leningrad (St. Petersberg) in 1893. The disease is caused by *Vibrio cholerae*, a motile Gram-negative rod, and is acquired by ingesting food or water contaminated by fecal material. After an incubation period of two to five days, diarrhea and abdominal pain begin suddenly. Vomiting sometimes occurs, but fever is rare. The diarrhea can be profuse, exceeding 15 liters during 24 hours and causing a loss of fluid and ions (sodium, bicarbonate, and potassium) that is often fatal in the absence of prompt treatment. When vomiting does not occur, the lost water and electrolytes can usually be replaced by the oral administration of water containing (per liter) 20 g D-glucose, 3.5 g sodium chloride, 2.5 g sodium bicarbonate, and 1.5 g potassium chloride. Glucose is required for efficient uptake of electrolytes in the small intestine. In a seriously dehydrated patient or one who is vomiting, lost fluid and electrolyte must be replaced intravenously.

*V. cholerae* colonizes the intestines by adhering to the intestinal mucosa without invading the mucosal tissue. There it secretes *cholera toxin*, a potent enterotoxin composed of five molecules of *subunit B* (MW approximately 11,000) and a single molecule of *subunit A* (MW approximately 29,000). Subunit B mediates binding of the toxin to a glycosylated lipid (ganglioside $G_{M1}$) on the surface of mucosal cells. Then subunit A is cleaved by a proteolytic enzyme, producing two fragments, A1 and A2. Fragment A1 enters the cell and catalyzes the transfer of the ADP-ribosyl moiety of NAD onto the regulatory subunit (RS) of adenylate cyclase, the enzyme that produces 3′,5′-cyclic AMP (cAMP). Adenylate cyclase is activated by GTP bound to RS, but this activation is normally brief because RS hydrolyzes GTP. However, ADP-ribosylated RS cannot hydrolyze GTP. Thus the net effect of the toxin is to cause cAMP to be produced at an abnormally high rate (Figure 31.2), which stimulates mucosal cells to pump large amounts of chloride into the intestinal contents. Water , sodium, and other electrolytes then follow, owing to the formation of the osmotic and electrical gradients caused by the loss of chloride; the lost water and electrolytes in the mucosal cells are replaced from the blood. Thus, the toxin-damaged mucosal cells become pumps for water and electrolytes, causing the diarrhea and loss of electrolytes typical of cholera.

### Staphylococcal Food Poisoning

Staphylococcal food poisoning, one of the three most common types of bacterial food poisoning (Table 31.2), is caused by strains of *Staphylococcus aureus* that produce an exotoxin called an enterotoxin although it is not known whether it acts directly on the stomach wall or acts on the central nervous system following absorption in the upper gastrointestinal tract.

Staphylococcal food poisoning is caused by ingestion of improperly stored foods in which *S.*

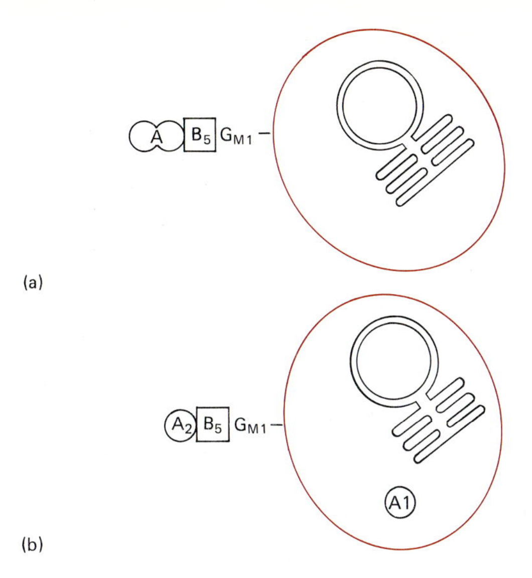


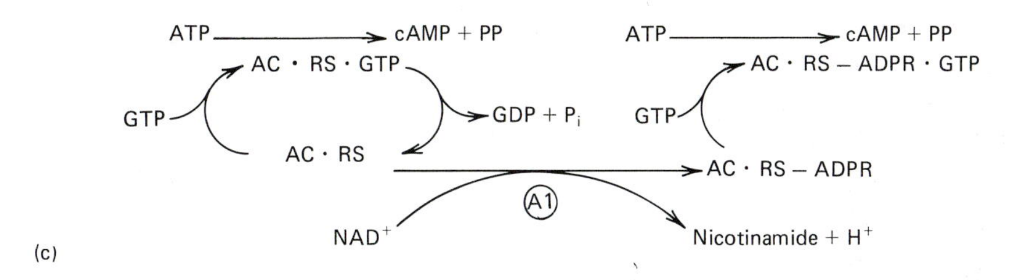


FIGURE 31.2

Effects of cholera toxin on cells in the intestinal mucosa. (a) Subunit B of the hexameric toxin ($AB_5$) binds to ganglioside ($G_{M1}$) on the surface of the target cell. (b) Subunit A is cleaved into fragments A1 and A2, one of which (A1) enters the cell, where (c) in the cytoplasm, it catalyzes ADP-ribosylation of the regulatory subunit (RS) of the host's adenylate cyclase (AC). ADP-ribosylated RS binds to GTP but does not hydrolyze it. Therefore the enzyme is constantly active, causing the intracellular concentration of cyclic AMP (cAMP) to become abnormally high, which, in turn, cause mucosal cells to pump electrolytes and water into the intestinal lumen.

**TABLE 31.2**

**Important Bacterial Food Poisonings**

| Type | Cause | Onset of Symptoms | Duration of Symptoms | Symptoms |
|---|---|---|---|---|
| Botulism | *Clostridium botulinum* | 12 hr to several days | Many days | Flaccid paralysis sometimes preceded by vomiting. |
| Clostridial food poisoning | *Clostridium perfringens* | 8 to 24 hr | <24 hr | Severe abdominal pain and diarrhea; sometimes nausea and vomiting. |
| Staphylococcal food poisoning | *Staphylococcus aureus* | 1 to 7 hr | <12 hr | Severe nausea and vomiting; sometimes diarrhea. |
| Enteric bacterial food poisoning | Usually *Escherichia coli* or *Salmonella typhimurium* | 7 hr to several days | Several days | Abdominal pain and diarrhea. Nausea, vomiting and fever are common. |

*aureus* has grown; such foods contain enterotoxin along with living or dead staphylococci. Soon after ingestion of contaminated foods, usually within 1 to 7 hours, abdominal pain, nausea, and vomiting which can be violent begin. Diarrhea is sometimes present, but fever rarely is. Recovery is usually complete by 12 hours after the onset of symptoms, except in those cases in which severe dehydration occurs as a result of prolonged vomiting. This toxin is remarkably heat stable, remaining active after being heated to 100°C.

Approximately 50 percent of the people in the United States carry *S. aureus* as a part of their normal skin flora so it is virtually impossible to prevent some contamination of foods during routine handling. However, enterotoxin does not form if foods are stored below 4° C. Most poisonings are caused by consumption of contaminated meat or confectionary products containing cream, but about 10 percent of them are caused by consumption of improperly stored milk, in which case the source of contamination is usually dairy cattle rather than humans.

### Clostridial Food Poisoning

Another common form of food poisoning is caused by *Clostridium perfringens*, a Gram-positive, spore-forming, strict anaerobe which is ubiquitous, even occurring commonly in the colon of many mammals, including humans. It is not practicable to prevent contamination of all food with these spores. However, food poisoning occurs only after ingestion of a large number of live vegetative cells of *C. perfringens.* Therefore, contaminated food usually becomes dangerous only after storage at a favorable temperature and low oxygen tension, conditions that allow spores to germinate and vegetative cells to proliferate. These conditions are likely to be met when a large amount of food is heated near boiling, stimulating spores to germinate, and then cooled slowly, allowing vegetative cells to proliferate. Therefore, it is not surprising that, on average, about 35 individuals are affected per episode of clostridial food poisoning. Between 8 and 24 hours after ingestion of contaminated food, abdominal pain and diarrhea begin, sometimes accompanied by vomiting but rarely by fever. Usually recovery is complete within 24 hours.

### Food Poisonings Caused by Enteric Bacteria

Enteric bacteria are a major cause of food or water poisonings in all countries. The most common of these are *Escherichia coli* and *Salmonella typhimurium*, but *Shigella* spp. and other *Salmonella* spp. also cause a significant number of cases. In all these infections, live bacteria must be ingested to produce disease and following this, symptoms begin after variable periods of time: usually 7 to 48 hours in the case of *S. typhimurium* and one to four days in the case of *E. coli.* Abdominal pain and diarrhea are the most prominent symptoms but vomiting and fever are also common, persisting for several days to several weeks. Most adults recover, but the attendant loss of fluid can lead to the death of children and elderly people. If vomiting does not occur, lost body fluid can be replaced by the ingestion of an electrolyte-glucose solution like the one used in the treatment of cholera.

These enteric bacteria produce disease by first colonizing the intestinal tract. There they cause diarrhea by producing toxins; endotoxin may contribute, but the major agents are enterotoxins. Many strains of *E. coli* that cause infantile or traveler's diarrhea, termed *enterotoxigenic E. coli* (ETEC) strains, produce either a heat-labile toxin (LT) or a heat-stable toxin (ST). LT is structurally and functionally similar to the cholera toxin, but ST, a much smaller protein (1,000 to 6,000 daltons), is completely unrelated. Rather than affecting adenylate cyclase, it activates guanylate cyclase which catalyzes the synthesis of 3′,5′-cyclic guanosine monophosphate (cGMP), a compound that acts similarly to cAMP. The genes encoding LT and ST are both encoded on plasmids.

Some strains of *E. coli* that do not produce LT or ST are able to cause diarrheal disease by secreting a *shiga toxin* (Table 31.1), so-called because it is similar to the toxin produced by many strains of *Shigella dysenteriae.* This toxin kills cells in the intestinal epithelium by inactivating the 60S subunits of their ribosomes. In *E. coli* strains that cause hemorrhagic diarrhea, this toxin is encoded in the genome of certain temperate phages (Chapter 9), but in *S. dysenteriae*, it is chromosomally encoded.

Many strains of *Salmonella*, a few of *Shigella*, and *E. coli* can invade the superficial layers of the intestinal wall, where they penetrate and grow within epithelial cells (see below). In the case of *E. coli*, the ability to do so depends on plasmid-encoded functions. Invasive strains produce fever by releasing endotoxin into the tissues of the host.

### Botulism

Botulism is a rare, but often fatal disease caused by *Clostridium botulinum*, a Gram-positive spore-forming strict anaerobe that grows in soil and anoxic waters. Its spores are widely disseminated by winds and occasionally they contaminate food. However, with the exception of infants, *C. botulinum* does not grow in the gastrointestinal tract of humans. Botulism in adults is almost always a pure intoxication. Spores can germinate and vegetative growth can ensure in improperly preserved food if it is nonacidic, protected from oxygen, and lacks nitrite. The most common source of botulism food poisoning is canned food that has not been heated sufficiently to kill spores. Toxin is produced during vegetative growth, but is only released by the cell lysis that accompanies sporulation. Free toxin, the cause of this food poisoning, is not inactivated by the acid pH of the stomach and is actually activated by the proteolytic enzyme, trypsin, which is present in the stomach. Remarkably, this toxin is one of the few large molecules that are absorbed intact from the gastrointestinal tract. It is transported via the bloodstream to nerve cells where it binds, preventing release of acetylcholine, the neurotransmitter that triggers contraction of skeletal muscle. Symptoms of botulism—blurred vision, difficulty in swallowing and speaking, and increasing weakness—usually begin between 12 and 36 hours after ingestion of the toxin. Nausea and vomiting are also present in about 50 percent of cases. Death often results from paralysis of the muscles required for respiration.

*C. botulinum* occasionally colonizes the intestine of infants, causing *infantile botulism*, which is characterized by muscular weakness that can progress to respiratory paralysis. Infantile botulism is responsible for only a minority of cases of "crib death," but in the United States, the number of deaths due to infantile botulism is probably greater than the number due to botulism food poisoning. Honey, which often contains *C. botulinum* spores, is a common source of the pathogen in the case of infantile botulism. Consequently it is recommended that honey not be given to children less than one year old.

Virulent *C. botulinum* strains are divided into six groups: types A through F. Each group is characterized by an antigenically distinct toxin. The toxins differ somewhat in potency but all act by the same mechanism of pathogenesis. Most cases of botulism in the United States are caused by types A, B, and E. In types C and D, the toxin gene is on the chromosome of a temperate virus rather than on the bacterial chromosome.

## Toxic Shock Syndrome

*Toxic shock syndrome* is characterized by low blood pressure, fever, and an extensive skin rash; diarrhea and vomiting are also common. More than 80 percent of cases occur in women under 30 years old. The onset is usually during menstruation and is associated with the use of tampons that are not changed frequently. The disease is apparently caused by strains of *Staphylococcus aureus* that sometimes grow in the vagina and secrete a toxin. Little is known about the toxin, but it clearly can penetrate the vaginal wall to enter the bloodstream.

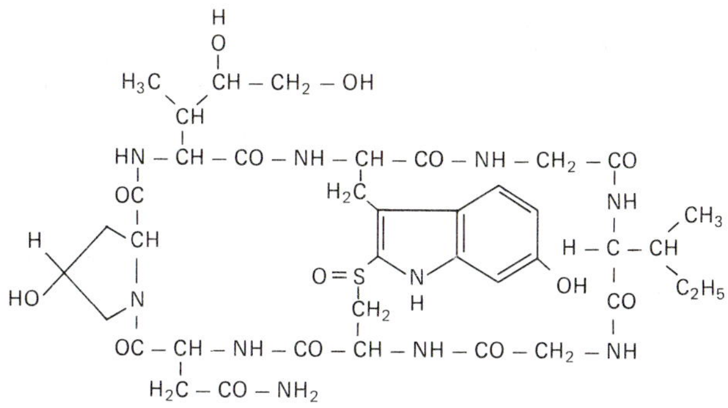

FIGURE 31.3
The structure of α-amanitin, a mycotoxin that inhibits mammalian RNA polymerase.

## Diseases Caused by Mycotoxins

Many species of fungi produce poisonous compounds termed *mycotoxins*. Most of these are heat-stable compounds of relatively low molecular weight. For example, some mushrooms of the genus *Amanita* produce *α-amanitin*, a heat-stable cyclic peptide (Figure 31.3) that is deadly because it inhibits mammalian RNA polymerase.

The *aflatoxins* (Figure 31.4) are formed by *Aspergillus flavus*, a mold that grows on a variety of plant materials. If stored while damp, peanuts

FIGURE 31.4
The structure of some aflatoxins.

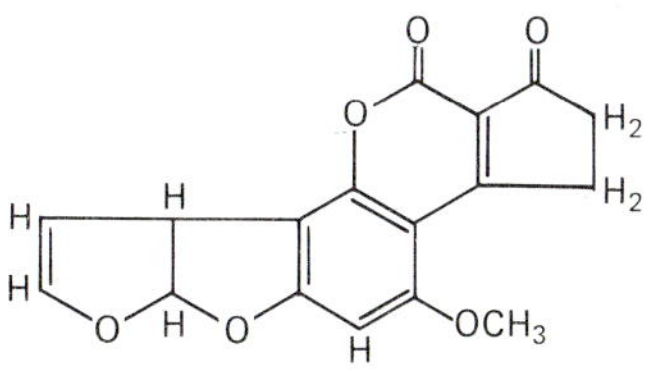

Aflatoxin B1

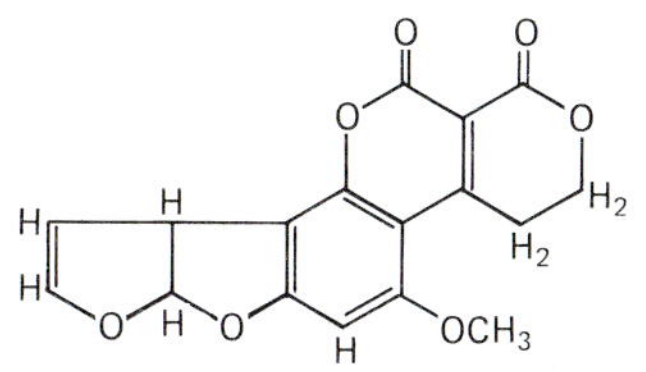

Aflatoxin G1

or grain may become contaminated with enough aflatoxin to cause severe liver damage when ingested. In many parts of Asia and Africa, aflatoxins are a serious public health problem, but in the United States, rigid standards of food storage combined with the imposition of maximum permissible levels of toxins in foodstuffs have effectively prevented aflatoxin poisoning.

Aflatoxins affect the host by binding to DNA and inhibiting RNA synthesis. They are also potent mutagens and carcinogens, causing hepatomas (cancers of the liver) in experimental animals. Indeed, a high correlation exists between the incidence of liver cancer and dietary intake of aflatoxins in Africa and Asia. However, in the United States, most liver cancer occurs in persons previously infected with hepatitis B virus.

## BACTERIAL COLONIZATION AND INVASION

To establish a host-parasite relationship with humans or other mammal, a bacterial pathogen must be able to colonize a surface or invade a tissue of the host. Colonization depends on specialized molecular structures on the bacterial cell surface, termed *adhesion factors* or *adhesins*, that bind to specific receptor sites on host cells. Invasion depends on the pathogen's being able to grow in host tissue and to survive or evade the action of phagocytic cells in the host. Both colonization and invasion depend on the microorganisms' successful competition with the host's normal microflora for essential nutrients.

### Iron Uptake

Most bacteria grow poorly, if at all, in media that contain less than $10^{-8}$ M free iron, and the concentration of free iron in most human tissue is less than $10^{-18}$ M (Chapter 29). Thus, most pathogens possess special mechanisms for acquiring iron. Many do so by secreting low molecular weight compounds, termed *siderophores* (Figure 31.5), which bind free iron tightly, thereby removing bound iron from transferrin or other iron-binding compounds that are present in the host. The pathogen then takes up iron while it is complexed to a siderophore. In the case of enteric bacteria that synthesize a siderophore termed *enterobactin*, the siderophore-iron complex is dissociated by a protein in the outer membrane. Then the iron is transported into the cell.

Some pathogens sequester iron without producing siderophores. For example, some *Neisseria* spp. synthesize an outer membrane protein that removes iron directly from transferrin.

### Adhesion

In order to colonize mucous membranes of the gastrointestinal, genitourinary, or respiratory tracts, bacteria must adhere to mucosal cells because the surfaces of these membranes are recurrently washed with fluids that sweep away unattached organisms. Several components of the bacterial envelope have been tentatively identified as adhesins; these include the capsule, the lipopolysaccharide layer, certain outer membrane proteins, flagella, and pili.

However, in only a few cases has a specific structure clearly been shown to play a role in adhesion. One such case is the role of pili in the adhesion of *Neisseria gonorrhoeae* to the genitourinary tract. Strains that genetically lose the ability to produce pili also lose the ability to bind tightly to the urethral membrane, and they lose the ability to colonize the urethra. In the case of *E. coli*, the majority of strains that produce diarrheal disease have specific pili, termed *type* I *pili*; these strains adhere more strongly to the intestinal mucous membrane than do strains that have lost the ability to produce these pili. However, type I pili are not always associated with pathogenicity; they are also present on many avirulent strains that colonize the large intestine. Other strains of *E. coli* that cause a significant amount of diarrheal disease in humans and domestic animals possess other kinds of pili termed *colonization factor antigens type* I and II (CFA/I and CFA/II); these also function as adhesins.

Genes encoding CFA pili are probably always plasmid-borne, and strains that lose this plasmid also lose both virulence and ability to adhere to the intestinal mucous membrane. Further evidence that these pili play an essential role in pathogenesis comes from the observation that animals immunized against purified CFA/I and CFA/II are protected from strains that bear these pili.

### Intracellular Growth

Pathogens that have the ability to penetrate host cells are shielded from the host's immune system during the intracellular phase of growth. Examples include the obligate intracellular parasites, the chlamydias and the rickettsias (Chapter 21), the

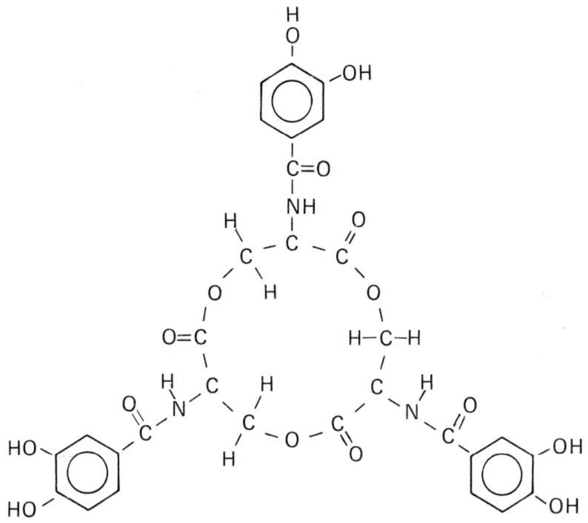

(a)

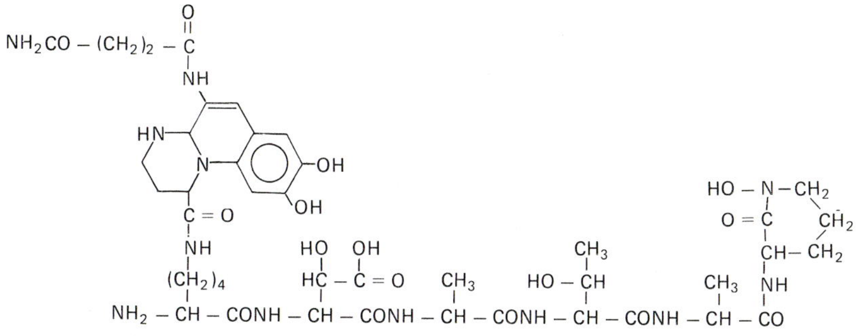

(b)

FIGURE 31.5

Examples of siderophores, low molecular weight iron-binding compounds secreted by microorganisms. (a) Enterobactin, a siderophore synthesized by many virulent enteric bacteria including strains of *Escherichia coli.* (b) Pseudobactin, a siderophore produced by *Pseudomonas* sp.

strains of *Salmonella* spp., *Shigella* spp., and *E. coli* that penetrate intestinal epithelial cells, and parasites of the genus *Plasmodium* (Chapter 32) that grow in both liver cells and red blood cells to produce the disease malaria (Chapter 32).

## Resistance to Phagocytosis

Nonpathogenic bacteria that are deposited in a wound are usually rapidly eliminated by phagocytic cells of the host. However, invasive pathogens possess special properties that protect them from elimination by host defenses. Many pathogens, produce a capsule that confers resistance to phagocytosis as was first demonstrated in studies on *Streptococcus pneumoniae:* virulent strains possess capsules and are resistant to phagocytosis, whereas avirulent strains lack capsules and are easily phagocytized. In many other species as well (including *Streptococcus pyogenes* and *Neisseria meningitidis*), virulence is also associated with capsule formation. Specific proteins on the surface of pathogens have also been shown to confer resistance to phagocytosis: these include the *M-protein* of *S. pyogenes* and pili of *N. gonorrhoeae.*

Some pathogens are relatively easily phagocytized but are not killed within phagocytic cells. These pathogens kill or grow inside phagocytes, and thus they produce diseases of long duration, termed *chronic diseases.* Pathogens in this group include *Mycobacterium tuberculosis* and *M. leprae,* which cause tuberculosis and leprosy, respectively. Other important examples are listed in Table 31.3.

TABLE 31.3

**Some Pathogens That Survive within Host Phagocytes**

| Pathogen | Disease |
|---|---|
| *Brucella abortus* | Brucellosis (undulant fever) |
| *Chlamydia trachomatis* | Lymphogranuloma venerium |
| *Franciscella tularensis* | Tularemia |
| *Listeria monocytogenes* | Listeriosis |
| *Mycobacterium leprae* | Leprosy |
| *Mycobacterium bovis and M. tuberculosis* | Tuberculosis |
| *Nocardia asteroides* | Nocardiosis |
| *Salmonella typhi* | Typhoid fever |
| *Yersinia pestis* | Bubonic plague |

Some pathogens resist phagocytosis by secreting proteins that interfere with host cell functions. For example, *Staphylococcus aureus* and *Streptococcus pyogenes* produce exotoxins, termed *leukocidins*, that kill phagocytes, thereby causing pyogenic (pus-forming) infections. Other pathogens evade phagocytosis by secreting chemical substances that initiate blood clotting, thereby producing a barrier between pathogens and phagocytes; e.g., *Staphylococcus aureus* secretes the protein *coagulase*. Still other pathogens secrete chemical substances that interfere with the normal chemotactic response of phagocytes.

### Antigenic Variation and Antigenic Mimicry

Some pathogenic protozoa including trypanosomes (Chapter 32), evade the host's immune defense by periodically changing their surface antigens during the course of infection, a process termed *antigenic variation*. When a new surface antigen appears, the host is defenseless until the immune system responds by producing new antibodies. Other pathogens evade the host's immune system by producing surface molecules that are antigenically similar or even identical to one of the host's macromolecules, a strategy termed *antigenic mimicry*. For example, some invasive strains of *E. coli* produce a capsule composed of a polysaccharide (*designated* type K5) that is identical to a portion of the heparin molecule normally present in host tissue. The mechanism (*tolerance*) that prevents the host's immune system from producing antibodies active against components of its own tissue (Chapter 30) may act to protect these pathogens.

## VIRUSES AND CANCER

The characteristic tissues of animals are formed by the *regulated limited growth* of their component cells. As a rare event, a cell may escape normal regulatory constraints and divide in an uncontrolled manner, forming an abnormal mass of tissue. Such masses are termed *neoplasms* or *tumors*.

Tumors are classified by their pattern of growth into two groups: those that do not invade surrounding tissue are termed *benign*. They grow by displacing adjacent cells but rarely kill the organism unless they occur in the brain. On the other hand, *malignant* tumors, termed *cancers*, invade and destroy surrounding tissue as they grow. They also release cells into the bloodstream or into the lymphatic circulation that can establish new neoplastic foci, termed *metastases*.

Tumors, both benign and malignant, are usually named by appending *-oma* to a term describing their appearance when examined with the light microscope (Table 31.4). Cancers formed by layers of cells are called *carcinomas*; those that arise in connective tissue or blood vessels are termed *sarcomas*. Several cancers are named after the specific cell types in which they arise. For example, *hepatomas* arises in liver hepatocytes, *melanomas* arise in skin melanocytes, and *lymphomas* arise in lymphocytes. An exception to this system of nomenclature occurs in the case of the cancers, termed *leukemias*, that arise in the bone marrow cells which produce leukocytes (Chapter 29).

The first evidence of a causal relationship between viruses and cancer was obtained in 1908 when V. Ellerman and O. Bang demonstrated that a type of leukemia that affects chickens could be transmitted to healthy birds by injecting them with a cell-free filtrate of the blood of a leukemic bird. Three years later, P. Rous demonstrated that a chicken sarcoma could be similarly transmitted, and he established that the active agent in the filtrates was a virus, now called *Rous sarcoma virus* (RSV), and known to be a member of the retrovirus family of RNA viruses (Chapter 32 and Table 9.1).

At first, the discovery of avian *oncogenic* (tumor-causing) *viruses* received little attention, but in 1932, when R. Shope showed that rabbit papilloma (a malignant tumor related to benign human warts) was also caused by a virus, interest in the phenomenon increased sharply because the possibility that at least some human cancers might have a similar cause became widely considered. This interest increased in 1936 when J. Bittner demonstrated that a virus of mice termed *mammary tumor virus* (MTV) which is transmitted in milk from a fe-

**TABLE 31.4**

**Examples of Human Neoplasms**

| Neoplasm | Description |
|---|---|
| BENIGN NEOPLASMS | |
| Papillomas | Warts (caused by a virus) |
| Adenomas | Benign tumors formed by cells arranged into glandular structures |
| Fibromas | Benign tumors formed by connective tissue cells |
| MALIGNANT NEOPLASMS | |
| Carcinomas | Malignant tumors formed by cells organized into sheets or layers |
| Sarcomas | Malignant tumors formed by poorly differentiated cells of connective tissue, muscle, bone, or blood vessels |
| Leukemias | Malignant neoplasms of bone marrow cells that normally produce leukocytes |
| Lymphomas | Neoplasms formed by lymphoid tissue |
| Hepatomas | Malignant tumors arising in liver cells |
| Melanomas | Malignant tumors arising in melanocytes, the pigment cells of skin |

male mouse to her offspring, can cause mammary cancer. Bittner's work led to the understanding of several important aspects of virus-caused cancer: first, an animal that is infected with a tumor virus during infancy may not develop a tumor until adulthood; second, an oncogenic virus does not always cause a tumor to form; other factors such as the environment or physiology of the host are important. For example, female mice exposed as infants to MTV develop tumors at a high frequency during pregnancy; mice that lack the high levels of estrogens characteristic of pregnancy are not likely to develop tumors. However, even male mice exposed to MTV will develop mammary tumors if given injections of estradiol over a long period of time.

The intensive search for oncogenic viruses has yielded only about 30 that cause cancer in experimental animals or in cell culture (see below). Most belong to one family of RNA viruses, the retroviruses (Table 31.5), but oncogenic viruses also occur in each of the families of mammalian double-stranded DNA viruses: poxviruses, adenoviruses, herpesviruses, and papovaviruses.

## The Role of DNA Viruses in Human Cancer

Although many viruses have been found in specimens from human cancer tissue, most of these appear to have infected the tissue after the cancer began. An important exception is the herpesvirus, *Epstein-Barr* (EB) *virus*, which was first isolated from a patient in Africa with a type of malignant lymphoma termed *Burkitt's lymphoma*; almost all subsequent cases studied have been associated with EB virus. Surprisingly, EB virus is also strongly implicated as the cause of another cancer, nasopharyngeal carcinoma, which is common in Hong Kong and surrounding regions but rare elsewhere.

Most types of cells in culture cannot be infected with EB virus, but the lymphocytes of primates can be. These cells are not killed; rather they continue to grow and produce virus indefi-

**TABLE 31.5**

**Selected Retroviral Oncogenes**

| Retrovirus | Oncogene | Function |
|---|---|---|
| Rous sarcoma virus | v-src | Tyrosine kinase |
| Abelson murine leukosis virus | v-abl | Tyrosine kinase |
| Feline sarcoma virus (GR strain) | v-fgr | Tyrosine kinase |
| Maloney murine sarcoma virus | v-mos | Serine kinase? |
| Simian sarcoma virus | v-sis | Cellular growth factor |
| Avian erythroblastosis virus | v-erb-B | Membrane receptor for growth hormone |
| MC29 avian leukosis virus | v-myc | Nuclear protein |
| Harvey murine sarcoma virus | v-Ha-ras | GTP-binding protein |
| Kirsten murine sarcoma virus | v-Ki-ras | GTP-binding protein |

nitely. Furthermore, when transplanted back into monkeys, these "virus-transformed" cells produce malignant tumors.

In the United States, EB virus causes the common disease of young people, *infectious mononucleosis* (Chapter 32). The virus can often be isolated from the throat weeks after the symptomatic phase of the illness and sometimes can be isolated from lymphocytes years after the illness. Yet Burkitt's lymphoma and nasopharyngeal carcinoma are rare diseases in this country. Therefore, it appears that either there is a subtle difference in those strains of EB virus that cause cancer or that some environmental factors, absent in this country, are required along with EB virus for carcinogenesis.

Another DNA virus implicated as a cause of cancer is *hepatitis B virus,* which is associated with hepatoma (cancer of the liver). Although this cancer accounts for only about 2 percent of cancer deaths in the United States, it causes more than 20 percent of cancer deaths in parts of Africa and Asia. This geographical bias was originally attributed to the presence of dietary aflatoxins, but hepatitis B virus has since been identified as the likely cause: nearly all individuals with this cancer have antibodies against hepatitis B virus, indicating past infection, and the incidence of hepatomas in a country is highly correlated with the incidence of hepatitis B infections. Furthermore, both viral DNA and viral protein have been detected in hepatoma cells.

## The Role of RNA Viruses in Human Cancer

The discovery of retroviruses as a cause of a variety of cancers in birds, rodents, cats, and monkeys stimulated the search for a link between these viruses and human cancer. S. Spiegelman and others have examined extracts of large numbers of human cancers for the presence of retroviruses, but in most cases no regular association between a particular type of cancer and retroviruses has emerged, with one notable exception: the association in Japan and certain other countries between *adult T-cell leukemia* and a retrovirus called *human T-cell leukemia virus* (HTLV). In some cases, all malignant cells produce the virus. However, many individuals who become infected with HTLV do not develop cancer. Why some infections result in cancer and others do not is not known.

## The Animal Cell Culture Model of Cancer

Methods for growing animal cells in culture have contributed greatly to our understanding of cancer. In a suitable medium, some types of animal cells grow for a limited period like a population of microorganisms, but when they come into contact with one another, growth and cell movement stop. This phenomenon is termed *contact inhibition:* it is a fundamental property of normal animal cells. However, cancer cells do not exhibit contact inhibition, rather they continue to grow in culture, forming disorganized masses of cells.

Attempts to propagate cells from an animal for long periods of time usually fail: Even when transferred repeatedly to fresh medium, growth rarely continues for more than 50 generations. However, occasionally a cell in the culture acquires by mutation the ability to grow indefinitely. The descendants of such an "immortal cell" are termed a *cell line.* Some cell lines behave in culture like cancer cells, but others continue to exhibit contact inhibition. Cultured cells that have lost contact inhibition are said to be *transformed.* A test for transformation provides a useful means of detecting carcinogenic chemicals as well as oncogenic viruses: cells are mixed with a chemical or a suspension of virions, incubated, and the number of masses of transformed cells is determined.

## Transformation by SV40

The papovavirus, *simian virus 40* (SV40), is the most thoroughly studied oncogenic DNA virus. Its frequency of transformation is typically between $10^{-3}$ and $10^{-5}$ transformed cells per virion. Transformation requires integration of the SV40 chromosome into one of the host cell chromosomes, but there are many sites on all of the chromosomes where this integration can occur. Therefore, transformation by SV40 superficially resembles lysogeny by phage mu (Chapter 9).

Isolation of temperature sensitive mutants (Chapter 10) has demonstrated that, of the five genes in the viral chromosome, two are required for transformation. One encodes the *small tumor antigen* and the other encodes the *large tumor antigen.* The small tumor antigen is required only briefly at an early stage of transformation of non-growing cells; it may stimulate cell division which is essential in the transformation process. On the other hand, functional large tumor antigen is required to maintain the transformed state.

## Transformation by Retroviruses

During normal replication of a retrovirus, a DNA copy of the viral chromosome becomes inserted into a host chromosome (Chapter 32). Hence, it is not surprising that cells transformed by retroviruses always contain a copy of the viral chromosome. In some retroviruses the frequency of transformation is 100 percent, but the site at which insertion occurs appears to be random. Thus, viral transformation is not the consequence of inactivating a normal cellular gene but rather is the consequence of the addition of new genetic information. In most cases studied to date, transformation results from the presence of a single viral gene termed an *oncogene.*

The oncogene of Rous Sarcoma Virus (termed *v-src*) encodes a protein of MW 60,000 (denoted *pp*-60-*v-src*) that is largely associated with the cytoskeleton, a network of protein microfilaments underlying the cytoplasmic membrane, and that phosphorylates tyrosine residues in certain proteins. One of these, *vinculin*, is a membrane protein associated with zones, termed *adhesion plaques*, where the membrane establishes contact with a surface. It is hypothesized that phosphorylated vinculin cannot function in establishing these contacts. The functions of most other proteins phosphorylated by *pp*-60-*v-src* are unknown.

An intensive search for the function of other oncogene products has yielded some clues as to how they act: the oncogene of *Maloney murine sarcoma virus*, denoted *v-mos*, also encodes a kinase which phosphorylates serine residues in certain cellular proteins and is largely found in the cytoplasm unbound to the cytoskeleton. The oncogene of *Simian sarcoma virus*, denoted *v-sis*, encodes a protein (MW approximately 28,000) that closely resembles *platelet derived growth factor* (PDGF), a protein released from platelets (Chapter 29) that stimulates cells to divide during the normal process of wound healing. The oncogene of *Avian erythroblastosis virus*, denoted *v-erb-B*, encodes a membrane protein (MW approximately 67,000) that functions as a receptor for *hemopoietic growth hormone*, a protein involved in the normal regulation of erythrocyte production. These and other examples of oncogene products are listed in Table 31.5.

## Cellular Oncogenes

Tests to determine if human cancer cells possess DNA sequences that are homologous to those of viral oncogenes indicate that all human cancers do have such sequences. However, normal cells also possess similar sequences, termed *proto-oncogenes*, that are homologous with parts of oncogenes. This startling discovery provides evidence for the theory, proposed by R. Huebner and G. Todero, that viral oncogenes were originally acquired from normal cellular genes and have subsequently evolved to become viral oncogenes.

Following the discovery of proto-oncogenes, fragments of DNA from human cancers were tested for their ability to transform cultures of animal cells. In this way, cancer cell genes that are able to transform cell lines were identified. These are termed *cellular oncogenes*. In only one case (T-cell leukemia) was an oncogene from a human cancer shown to be associated with retroviral genes. Hence, the majority of human cancers appear to arise, at least in part, from the activation of normal cellular genes (proto-oncogenes).

Several processes appear to be involved in activation of proto-oncogenes. The simplest example known is a single base-pair substitution mutation (GC → TA transition, Chapter 10) that converts a normal gene into the oncogene that causes, at least in part, some human bladder carcinomas. This oncogene is partially homologous to those of certain strains of murine sarcoma virus. This group of similar genes comprises the *ras* family of viral oncogenes and their proto-oncogene homologues of normal cells. It is these proto-oncogenes that have been converted into the oncogenes found in the majority of human cancers: those of the lung, colon, prostate, and breast. In most of these, the mechanism of cellular *ras* gene activation is more complex than the transition mutation discussed above, but little is presently known of the alternative pathways that produce oncogenes.

In one group of malignancies, the leukemias and lymphomas, a consistent pattern of proto-oncogene activation is emerging. Most of these cancers are characterized by chromosomal abnormalities, termed *translocations*, where an arm of one chromosome has been broken and rejoined to the arm of another chromosome. Such an abnormality was first discovered in *chronic myelocytic leukemia*, where an arm of chromosome 9 is translocated next to the genes for antibody light chain synthesis (Chapter 30) on chromosome 22. Furthermore, the cellular proto-oncogene *c-abl*, which shares homology with *v-abl*, the oncogene of the Abelson murine sarcoma virus, is translocated in this process to a new location adjacent to the antibody genes.

Recent studies on Burkitt's lymphoma offer insight into the mechanism by which EB virus, which does not appear to have an oncogene, might cause cancer. In this cancer, cells have a translocation involving the proto-oncogene *c-myc*, which shares homology with *v-myc*, the oncogene of the MC29 avian leukosis virus. The proto-oncogene, which is normally found on chromosome 8, is translocated next to genes for the synthesis of antibody light or heavy chains on chromosome 2, 14, or 22 (Chapter 30). Why this should activate *c-myc* is not known, but a plausible role of EB virus in causing Burkitt's lymphoma is promotion of translocations: EB virus is known to cause breaks to occur in host cell chromosomes, and these presumably result in an abnormally high frequency of translocations.

## FURTHER READING


### Books

DAVIS, B. D., R. DULBECCO, H. N. EISEN, and H. S. GINSBERG, *Microbiology*, 3rd ed. New York: Harper & Row, 1980.

JOKLIK, W., H. P. WILLETT, and D. B. AMOS, *Zinsser Microbiology*, 18th ed. East Norwalk, Conn.: Appleton-Century-Crofts, 1984.

WILSON, G., and H. M. DICK, *Topley and Wilson's Principles of Bacteriology, Virology and Immunity*, 7th ed. Baltimore: Williams and Wilkins, 1984.

*Viruses and Cancer:*

WEISS, R., N. TEICH, H. VARMUS, and J. COFFIN, *RNA Tumor Viruses*, 2nd ed. Cold Spring Harbor, N.Y.: Cold Spring Harbor Laboratory, 1984.

### Reviews

BORST, P., and G. A. M. CROSS, "Molecular Basis for Trypanosome Antigenic Variation," *Cell* **29,** 291 (1982).

EIDELS, L., R. L., PROIA, and D. A. HART, "Membrane Receptors for Bacterial Toxins," *Microbiol. Rev.* **47,** 596 (1983).

GILL, D. M., "Bacterial Toxins: A Table of Lethal Amounts," *Microbiol. Rev.* **46,** 86 (1982).

HOLMGREN, J., "Actions of Cholera Toxin and the Prevention and Treatment of Cholera," *Nature* **292,** 413 (1981).

LEVINE, M. M., J. B. KAPER, R. E. BLACK, and M. L. CLEMENTS, "New Knowledge of Pathogenesis of Bacterial Enteric Infections as Applied to Vaccine Development," *Microbiol. Rev.* **47,** 510 (1983).

MIDDLEBROOK, J. L., and R. B. DORLAND, "Bacterial Toxins: Cellular Mechanisms of Action," *Microbiol. Rev.* **48,** 199 (1984).

PAASO, B, and D. C. HARRISON, "A New Look at an Old Problem: Mushroom Poisoning," *Am. J. Med.* **58,** 505 (1975).

*Viruses and Cancer:*

BISHOP, J. M., "Cellular Oncogenes and Retroviruses," *Ann. Rev. Biochem.* **52,** 301 (1983).

COOPER, G. M., "Cellular Transforming Genes," *Science* **218,** 801 (1982).

### Original Articles

FINN, C. W., R. P. SILVER, W. H. HABIG, M. C. HARDEGREE, G. ZON, and C. F. GARON, "The Structural Gene for Tetanus Neurotoxin Is on a Plasmid," *Science* **224,** 881 (1984).

MCBRIDE, J. S., D. WALKER, and G. MORGAN, "Antigenic Diversity in the Human Malaria Parasite *Plasmodium falciparum*," *Science* **217,** 254 (1982).

O'BRIEN, A. D., J. W. NEWLAND, S. F. MILLER, R. K. HOLMES, H. W. SMITH, and S. B. FORMAL, "Shiga-like Toxin-Converting Phages from *Escherichia coli* Strains That Cause Hemorrhagic Colitis or Infantile Diarrhea," *Science* **226,** 694 (1984).

*Viruses and Cancer:*

BEASLEY, R. P., "Hepatocellular Carcinoma and Hepatitis B Virus," *Lancet* **2,** 1129 (1981).

DE KLEIN, A., A. G. VAN KESSEL, G. GROSVELD, C. R. BARTRAM, A., HAGEMEIJER, D. BOOTSMA, N. K. SPURR, N. HEISTERKAMP, J. GROFFEN, and J. R. STEPHENSON, "A Cellular Oncogene Is Translocated to the Philadelphia Chromosome in Chronic Myelocytic Leukemia," *Nature* **300,** 765 (1982).

SHAERITZ, D. A., "Integration of Hepatitis B DNA into the Genome of Liver Cells in Chronic Liver Disease and Hepatocellular Carcinoma," *New Engl. J. Med.* **305,** 1067 (1981).

SLAMON, D. J., K. SHIMOTOHNO, M. J. CLINE, D. W. GOLDE, and I. S. Y. CHEN, "Identification of the Putative Transforming Protein of the Human T-Cell Leukemia Viruses HTLV-I and HTLV-II", *Science* **226,** 61 (1984).

TSUJIMOTO, Y., G. YUNIS, L. ONORATO-SHOWE, I. ERIKSON, P. C. NOWELL, and C. M. CROCE, "Molecular Cloning of the Chromosomal Breakpoint of B-Cell Lymphomas and Leukemias with the t(11;14) Chromosome Translocation," *Science* **224,** 1403 (1984).

# Chapter 32
# Human Pathogens

Microbial diseases are quite heterogenous: they can affect any organ and thus can produce many different symptoms. Hence, they are related only by their common cause, pathogenic microorganisms (Chapter 31). Indeed, this is the only large group of disease of which the primary cause can in most cases be identified.

## EPIDEMIOLOGY OF INFECTIOUS DISEASES

The study of factors that determine the distribution and frequency of diseases is termed *epidemiology*. In the case of infectious diseases, such studies often provide both a basis for disease control and clues for diagnosis.

### Reservoirs of Infection

As part of their infectious cycle, all pathogens exist, at least temporarily, in one or more natural environments, termed *reservoirs of infection*, from which they are transmitted to humans. The major reservoir for most common infectious diseases is the human population. The principal reservoir for another large group of diseases, termed *zoonoses*, is a population of domestic or wild animals, and the reservoir for a third group of diseases is water or soil. Reservoirs for a number of infectious diseases are listed in Table 32.1.

**TABLE 32.1**

**Reservoirs of Infection**

| Significant Reservoir | Diseases |
|---|---|
| Human population | Acquired immunodeficiency syndrome (AIDS), amebic dysentery, campylobacter diarrhea, cholera, diphtheria, epidemic relapsing fever, epidemic typhus, giardiasis, gonorrhea, hepatitis A, hepatitis B, herpes simplex infections, leprosy, lymphogranuloma venereum, malaria, measles, mononucleosis, mumps, poliomyelitis, smallpox, streptococcal infections, staphylococcal infections, syphilis, trichomoniasis, trachoma, typhoid, tuberculosis, whooping cough |
| Animal populations | |
| Rodents | Bubonic plague, endemic typhus, endemic relapsing fever, leptospirosis, Rocky Mountain spotted fever, scrub typhus, tularemia |
| Livestock | Anthrax, brucellosis, leptospirosis, orf, Q fever, toxoplasmosis |
| Dogs or cats | Campylobacter diarrhea, ringworm, toxoplasmosis |
| Soil and water | Coccidiomycosis, legionnaires' disease, pseudomonas infections, sporotrichosis, tetanus |

### Modes of Transmission

Every infectious disease is transmitted to humans from its reservoir by a characteristic *mode of transmission* (Table 32.2). The most common modes are (1) ingestion of food or water contaminated by feces (*the oral-fecal route*), (2) contamination of the respiratory tract by droplets or other material containing respiratory secretions (*the respiratory route*), and (3) direct contact with another person, animal, or contaminated object. Another important mode of transmission is inoculation through the skin when a wound is produced by an inanimate object or by the bite of an arthropod or mammal.

## BACTERIAL PATHOGENS

One of the great achievements of medical science, occurring in this century, was the discovery of therapeutically effective antibacterial drugs (Chapter 33). Now, nearly all bacterial diseases can be cured if an accurate diagnosis is made early in the course of the infection. In the following discussions of the principal human bacterial diseases (Tables 32.3 and 32.4), emphasis is placed on features important in their recognition and control.

### Staphylococcal Diseases

Nearly all human staphylococcal diseases are caused by *Staphylococcus aureus*, a facultatively anaerobic coccus that produces clumps of cells as it grows. It occurs on the skin and nasal passages of healthy humans and domestic animals. Approximately 50 percent of these strains produce a heat-stable enterotoxin (Chapter 31) that causes food poisoning when ingested. *S. aureus* can also cause a wide variety of infections that are described as *pyogenic* (pus-forming). Examples include impetigo, boils, wound abscesses, and pneumonia. Impetigo is a superficial skin infection that is common in children and occasionally occurs in adults; boils (furuncles) are abscesses that form in hair follicles.

Pathogenic staphylococci produce a number of extracellular proteins that are important in pathogenesis, including *coagulase*, *leukocidin*, and *hemolysins*. Coagulase initiates formation of blood clots that can protect bacteria from phagocytosis. Leukocidins are cytotoxins that kill leukocytes; hemolysins are cytotoxins that lyse red blood cells *in vitro* and are also toxic to leukocytes.

### Streptococcal Diseases

Two species of streptococci, *S. pyogenes* and *S. pneumoniae*, cause most human streptococcal disease. The human population is the reservoir of these

**TABLE 32.2**

**Typical Modes of Transmission of Certain Diseases**

| Mode | Diseases |
|---|---|
| Bite | |
| Arthropod | African sleeping sickness, bubonic plague, Chagas' disease, leishmaniasis, malaria, relapsing fever, Rocky Mountain spotted fever, typhus (all forms), yellow fever |
| Mammal | Rabies |
| Direct contact | Gonorrhea, herpes simplex infections, impetigo, inclusion conjunctivitis, leprosy, lymphogranuloma venereum, ringworm, syphilis, trachoma |
| Inoculation into superficial wounds | Leptospirosis, tularemia |
| Inoculation into puncture wounds | Sporotrichosis, tetanus |
| Oral-fecal route | Amebic liver abscesses; bacterial, protozoal, and viral diarrheal diseases; hepatitis A; poliomyelitis; typhoid |
| Respiratory route | Chickenpox, diphtheria, influenza, measles, mumps, mycoplasma pneumonia, pneumonic plague, rhinovirus infections, rubella, tuberculosis, whooping cough |

**TABLE 32.3**

**Examples of Pathogenic Gram-Positive Bacteria**

| Group: Pathogen | Principal Disease(s) |
|---|---|
| Nocardioform bacteria | |
| *Corynebacterium diphtheriae* | Diphtheria |
| *Mycobacterium bovis* | Tuberculosis |
| *M. leprae* | Leprosy |
| *M. tuberculosis* | Tuberculosis |
| *Nocardia asteroides* | Infections in immunocompromised individuals |
| Endospore-forming bacteria: | |
| *Bacillus anthracis* | Anthrax |
| *Clostridium botulinum* | Botulism |
| *C. difficile* | Pseudomembraneous colitis |
| *C. perfringens* | Food poisoning, gangrene |
| *C. tetani* | Tetanus |
| Staphylococci | |
| *Staphylococcus aureus* | Impetigo, boils, wound infections, pneumonia, toxic shock syndrome |
| Lactic acid bacteria: | |
| *Streptococcus pneumoniae* | Middle ear infections, pneumonia, meningitis |
| *S. pyogenes* | Pharyngitis, rheumatic fever, glomerulonephritis, impetigo |

**TABLE 32.4**

**Examples of Pathogenic Gram-Negative Bacteria**

| Group: Pathogen | Principal Disease(s) |
|---|---|
| Aerobic motile rod-shaped bacteria: | |
| *Bordetella pertussis* | Whooping cough |
| *Pseudomonas aeruginosa* | Urinary tract infections, "swimmer's ear," burn infections, pneumonia |
| *Campylobacter jejuni* | Diarrhea |
| Aerobic nonmotile rod-shaped bacteria: | |
| *Brucella* spp. | Brucellosis |
| *Francisella tularensis* | Tularemia |
| *Legionella pneumophila* | Legionnaires' disease |
| Chlamydias: | |
| *Chlamydia psittaci* | Pharyngitis, pneumonia |
| *C. trachomatis* | Trachoma, inclusion conjunctivitis, infections of the genitourinary tract |
| Enteric bacteria: | |
| *Escherichia coli* | Urinary tract infections, diarrhea |
| *Klebsiella pneumoniae* | Urinary tract infections, pneumonia |
| *Salmonella typhi* | Typhoid fever |
| *S. typhimurium* | Diarrhea |
| *Yersinia enterocolitica* | Enterocolitis |
| *Y. pestis* | Bubonic plague |
| Facultatively anaerobic rod-shaped bacteria with polar flagella: | |
| *Vibrio cholerae* | Cholera |
| Facultatively anaerobic nonmotile rod-shaped bacteria: | |
| *Haemophilus influenza* | Pharyngitis, middle ear infections, meningitis |
| Mycoplasmas: | |
| *Mycoplasma pneumoniae* | Pneumonia |
| Neisserias: | |
| *Neisseria gonorrhoeae* | Gonorrhea, pelvic inflammatory disease, conjunctivitis, infections of joints |
| *N. meningitidis* | Pharyngitis, pneumonia, meningitis |
| Rickettsias: | |
| *Coxiella burneti* | Q fever |
| *Rickettsia mooseri* | Endemic typhus |
| *R. prowazekii* | Epidemic typhus |
| *R. rickettsii* | Rocky Mountain spotted fever |
| *R. tsutsugamushi* | Scrub typhus |
| Spirochetes: | |
| *Borrelia* spp. | Relapsing fever |
| *Leptospira interrogans* | Leptospirosis |
| *Treponema pallidum* | Syphilis |

two species of facultatively anaerobic cocci that grow in chains (Chapter 23). *S. pneumoniae*, commonly called pneumococcus, is present in the upper respiratory tract of most healthy individuals, but *S. pyogenes* is rarely present in healthy people.

The most important pneumococcal diseases are pneumonia, otitis media (infection of the middle ear), and meningitis (infection of the membranes surrounding the brain). Immunity depends on production of an antibody (Chapter 30) that binds specifically to pneumococcal capsular polysaccharide. However, there are more than 85 different antigenic types of pneumococci, and immunity to one does not protect the host from another.

The most common diseases caused by *S. pyogenes* are impetigo (which is similar in appearance to the impetigo caused by *Staphylococcus aureus*), and *streptococcal pharyngitis* (streptococcal sore throat). Many strains of *S. pyogenes* produce both extracellular enzymes that break down host macromolecules and *streptokinases*, enzymes that activate a host factor that dissolves blood clots. It has been suggested, but not proven, that these enzymes facilitate the spread of streptococci. Most strains also produce the cytotoxins, *streptolysin O* and *streptolysin S* (Table 31.1) that kill host leukocytes and red blood cells, thereby contributing to the pus formation characteristic of streptococcal disease. Strains that secrete these streptolysins can be readily identified because they produce clear zones of lysis, termed *β-hemolysis* [Figure 32.1(a)], when grown on blood agar. Some other streptococci produce by an unknown mechanism small zones of partial hemolysis, termed *α-hemolysis* [Figure 32.1(b)]. Some strains also secrete a toxin termed *erythrogenic toxin* that produces the characteristic rash of scarlet fever. There are three distinct types of erythrogenic toxin, and the gene encoding each is carried on the chromosome of a temperate bacteriophage (Chapter 9).

Immunity to *S. pyogenes* depends on antibodies that bind to a protein, termed *M protein*, that is located on the cell surface and that inhibits phagocytosis. There are more than 50 antigenically distinct types of M protein, and immunity to one type does not protect the host from infection by another.

Streptococcal pharyngitis is occasionally followed by *poststreptococcal glomerulonephritis*, a disease characterized by temporary kidney failure apparently caused by immune complexes (produced from fragments of streptococcal walls cross-linked by antibodies) that become lodged in the glomeruli of kidneys.

Streptococcal pharyngitis can also be followed by *rheumatic fever*, a disease characterized in part by an enlargement of the heart and temporary arthritis. In turn, rheumatic fever in some cases is followed by a more serious disease, *rheumatic heart disease*. Although it is clear that streptococcal infection plays a role in causing these two diseases, the mechanisms involved are not understood.

## Diseases Caused by Endospore-Forming Bacteria

The diseases caused by various endospore-forming bacteria are summarized in Table 32.3. One of these, *anthrax*, which was studied by R. Koch (Chapter 1) and by L. Pasteur (Chapter 30), is pri-

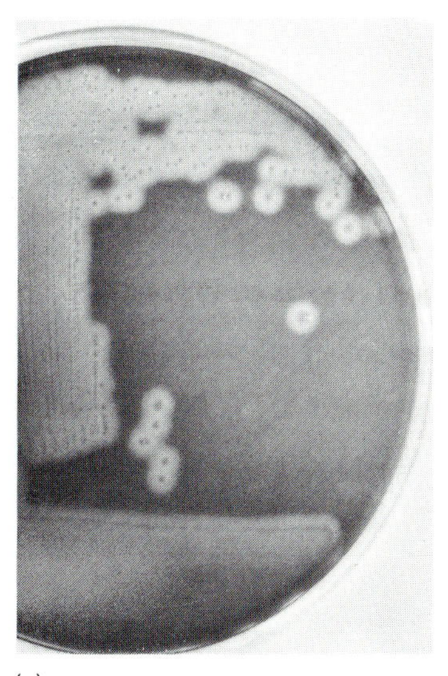

(a)

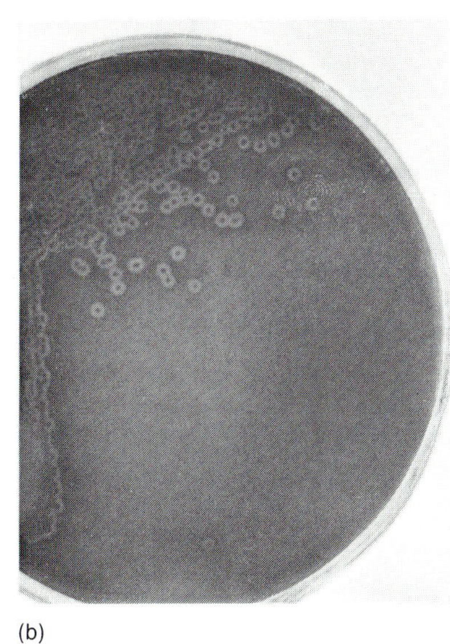

(b)

FIGURE 32.1

Colonies of hemolytic streptococci growing on blood agar plates: (a) β-hemolysis surrounding colonies of *Streptococcus pyogenes*; (b) α-hemolysis surrounding colonies of *S. salivarius*.

marily a disease of sheep. Rarely, it affects humans when spores of the causative agent, *Bacillus anthracis*, enter a wound or are inhaled. In the latter case, they cause a severe pneumonia termed *woolsorter's disease*.

Clostridial gangrene develops in necrotic (dead) tissue that has lost its blood supply. As a consequence, spores of certain obligate anaerobes, often *Clostridium perfringens*, can germinate and vegetative cells can proliferate there. As they do, they secrete hydrolytic enzymes and cytotoxins that kill and digest surrounding host cells, expanding the necrotic area in which the clostridial cells grow. Antibiotics are of little use in the treatment of gangrene because, without a blood supply in the affected tissue, there is no effective way to get them into the site of infection.

## Diseases Caused by Mycobacteria

Mycobacteria cause two of the most important diseases in history, *tuberculosis* and *leprosy*. These aerobic bacteria are termed *acid-fast* because they retain certain stains (e.g., carbol fuchsin) when treated with a mixture of ethanol and hydrochloric acid. This remarkable property results from a high lipid content of the cell wall: more than 50 percent of its dry weight is composed of unusual lipids of high molecular weight (e.g., mycolic acid, Chapter 24). Mycobacteria are also resistant to many antimicrobial drugs. Both of these diseases are chronic and often progressive, but they differ widely in their contagiousness and in the tissues infected.

When R. Koch discovered in 1881 that *tuberculosis* is caused by a bacterium, this disease, along with bacterial pneumonia, was the leading cause of death among adults. In developed countries, the frequency of tuberculosis declined dramatically in the twentieth century, partly as a result of improved working and living conditions, but largely as a result of public health measures designed to limit the spread of this disease. Unfortunately, today it is still a major cause of death in many less developed areas.

Most cases of human tuberculosis are caused by *Mycobacterium tuberculosis* acquired from other humans, but a significant number of cases are caused by *M. bovis*, which can be acquired from either humans or cattle. Tuberculosis is nearly always acquired by the respiratory route, resulting in pulmonary tuberculosis. During the primary stage of the disease, mycobacteria grow at the site of infection, becoming surrounded first by lymphocytes and macrophages and eventually by connective tissue that forms a firm structure, termed a *tubercle*. The disease is usually arrested at this stage, but mycobacteria remain alive within the tubercle and are not killed following phagocytosis (Chapter 31).

Sometimes mycobacterial cells escape from a tubercle and establish infections at new sites. Usually, these are in the lungs, but mycobacteria can on occasion spread to any organ of the body. This stage of the disease, termed *reactivation tuberculosis*, is often fatal if untreated.

Immunity to tuberculosis occurs by development of delayed-type hypersensitivity (Chapter 30) to mycobacterial surface antigens. Individuals who produce antibodies against *M. tuberculosis* but do not develop delayed-type hypersensitivity usually die of the disease.

In many countries, an attenuated strain (Chapter 30), called the *bacillus of Calmette-Guerin* (BCG), is used to induce hypersensitivity. However, it is not used in the United States because, like a natural infection, it causes the tuberculin skin test to become positive, rendering the test of little value for detection of actual exposures to the disease.

*Leprosy*, a severely disfiguring skin disease caused by the acid-fast bacterium *M. leprae*, is acquired by direct contact with infected persons or objects contaminated by them. The disease is found mostly in tropical countries, but is also fairly common in parts of China, Korea, and Mexico. Spread of leprosy to other areas is limited by its long incubation period (often three to five years) and by its low infectivity: repeated contact with an infected individual is usually required for transmission of this disease.

Infected individuals develop one of two distinct forms of the disease: those who develop delayed-type hypersensitivity to antigens on the surface of *M. leprae* have a mild, nonprogressive form termed *tuberculoid leprosy*, but those who fail to develop delayed-type hypersensitivity have a relentlessly progressive form termed *lepromatous leprosy*, in which large numbers of mycobacteria develop in the skin, killing underlying tissues and causing a progressive loss of fingers, facial features, and other structures.

## Listeriosis

Infection by the Gram-positive rod-shaped bacterium, *Listeria monocytogenes*, is termed *listeriosis*, a moderately rare disease that occurs mostly in persons over 55 years or under one year of age. The causative bacterium can be isolated from soils throughout the world and from the feces of humans and other animals. Listeriosis has been detected in wild and domestic animals, and transmission to humans by contact with contaminated tissues has been documented. In listeria infections during pregnancy, transmission across the placenta can occur, sometimes resulting in stillbirth. Listeriosis can also be acquired by ingestion of contaminated food: the source of an epidemic occurring in the United States in 1985 was identified as cheese from one factory. However, it is unusual for listeriosis to occur in epidemics, and the reservoir of infection in most cases is not known.

The symptoms of listeriosis are highly variable and can mimic those of several other diseases. Sometimes the infection is localized in the throat and surrounding tissues, producing illness that can resemble influenza or mononucleosis. The infection can also be localized in the membranes surrounding the brain, producing a life-threatening meningitis. Spread of *L. monocytogenes* through the bloodstream can produce a disease that resembles typhoid fever.

## Diseases Caused by Enteric Bacteria

Enteric bacteria (Chapter 19) cause a wide spectrum of diseases. The most common of these are the diarrheal diseases discussed in Chapter 31. They also cause urinary tract infections: *Escherichia coli* is by far the most common cause of bladder and kidney infections, but species of *Enterobacter*, *Klebsiella*, and *Proteus* are relatively common causes. The principal reservoir of all these infections is the normal human intestinal flora.

*Typhoid fever*, which is caused by *Salmonella typhi*, occurred in great epidemics during past centuries. The disease is acquired by ingesting food or water contaminated by feces of infected humans. The pathogen colonizes the small intestine, penetrates the intestinal wall, and proliferates in lymph nodes, spleen, and other lymphoid tissue. Within 6 to 14 days after exposure, headache and fever develop. The latter can continue for several weeks and rise above 40° C. In most cases, *S. typhi* is shed in the feces for several weeks but ceases to be shed before three months have elapsed. However, approximately 3 percent of those who recover continue to shed *S. typhi* for extended periods but show no symptoms of disease. In such individuals, termed *carriers*, the bacterium grows in the gall bladder and finds its way to the intestine through the bile duct.

*Plague*, perhaps the most notorious of all bacterial disease, has produced vast epidemics that decimated human populations. The first pandemic of plague was in the 6th century. The second, termed the *black death*, is estimated to have killed more than one-fourth of the people in Europe during the 14th century. The causative bacterium, *Yersinia pestis*, infects a variety of rodents which serve as the reservoir; it is transmitted to humans by fleas and other arthropods that feed on both humans and rodents. In most cases a high fever develops one to six days after the bite, accompanied by formation of greatly enlarged lymph nodes, termed *buboes*. Hence, this type of plague is termed *bubonic plague*. In more than 60 percent of untreated cases, death follows within three to five days. Plague can also be acquired by inhaling bacteria within droplets produced by the cough of an infected person. If untreated, this form of the disease, termed *pneumonic plague*, causes death in less than three days.

Plague is a rare disease in the United States: only a few cases occur each year and these are usually acquired in the southwestern and western states from infected ground squirrels.

## Diarrhea Caused by *Campylobacter*

Only in the past decade has it been recognized that *Campylobacter jejuni* is a major human pathogen. This microaerophile grows in environments where the partial pressure of $O_2$ is 22–114 mm Hg (3–15 percent). It is acquired from an infected person or a domestic animal (usually a cat) and colonizes the intestines where it produces symptoms that are almost identical to the diarrheas caused by enteric bacteria (Chapter 31). Remarkably, *C. jejuni* secretes an exotoxin that is antigenically similar to the cholera toxin (Chapter 31).

## Legionnaires' Disease

In 1976 *Legionella pneumophila* was identified as the causative agent of an outbreak of fatal pneumonia termed *Legionnaires' disease* that occurred earlier in that year at a convention of the American Legion in Philadelphia. *L. pneumophila* normally grows in soil or fresh water and is acquired from these reservoirs by inhaling bacteria in droplets or dust particles. Although exposure is common, it rarely produces pneumonia in healthy individuals: nearly all cases of Legionnaires' disease occur in patients with a deficient immune system or with some other predisposing disease.

## Tularemia

*Tularemia*, the disease caused by *Franciscella tularensis*, can be acquired from a wide variety of animals, but rabbits are the major reservoir in North

America. In most years, the number of cases reported in the United States is fewer than 200. Tularemia is transmitted to humans by several routes, including arthropod bites and contact between wounded human skin and infected animals. Following an incubation period of 1 to 10 days, a high fever suddenly develops. In most cases, an ulcer forms at the site of inoculation, and nearby lymph nodes become enlarged and tender.

### Brucellosis

*Brucellosis* is caused by various *Brucella* spp. that infect many domestic animals: Swine (infected by *B. suis*) and cattle (infected by *B. abortus*) are the major reservoirs from which the disease is transmitted to humans by inhalation, by direct contact, or by eating or drinking contaminated meat or dairy products. After an incubation period of several days to several weeks, fever, headache, and pains in joints begin and then gradually increase. If untreated, the symptoms usually disappear, only to recur several weeks later.

### Diseases Caused by *Pseudomonas*

*Pseudomonas aeruginosa*, an ubiquitous inhabitant of soil and fresh water, causes several diseases that can be difficult to treat because this bacterium is remarkably resistant to most antibiotics. It causes approximately 4 percent of urinary tract infections, most cases of "swimmer's ear" (an infection of the outer ear canal), infections following extensive burns, and pneumonias, particularly in patients with the genetic disease, *cystic fibrosis*.

### Diseases Caused by *Bordetella* and *Haemophilus* Species

*Whooping cough*, a childhood disease that was common before the development of pertussin vaccine, is caused by *Bordetella pertussis*, an aerobe that is acquired from other humans and grows in the throat. Following an incubation period of 7 to 16 days, symptoms that resemble a cold begin. Then during the next 7 to 14 days, a rapid, intense cough interrupted by a strident gasp of air (the "whoop") develops. With modern drug therapy, the fatality rate of whooping cough has fallen from 4 percent to approximately 0.6 percent.

*Haemophilus influenzae*, a common inhabitant of the throat in healthy persons, is a facultative anaerobe that requires both a source of heme-iron (factor X) and NAD (factor Y) for growth. This bacterium causes disease primarily in children where it sometimes grows in the throat and larynx, producing a severe epiglotitis (inflammation of the epiglottis) that can result in suffocation. In children from one to four years of age, *H. influenzae* is the major cause of meningitis, which is manifested first by irritability usually accompanied by vomiting and then by lethargy that progresses to a stupor. Untreated cases of this disease are invariably fatal.

### Neisserial Diseases

*Neisseria meningitidis* and *N. gonorrhoeae* are nonmotile Gram-negative cocci that characteristically occur in pairs. Infections caused by *N. meningitidis* can be acquired by inhalation of droplets from diseased individuals or healthy carriers and most commonly occur in the throat. Occasionally the infection spreads to other organs causing pneumonia, arthritis, meningitis, or septicemia (infection of blood).

*N. gonorrhoeae* causes *gonorrhea*, a disease that is transmitted by sexual contact except in rare instances where prepubescent females acquire it by contact with contaminated material. In males, the incubation period of two to eight days is followed by frequent urination accompanied by a burning sensation and a urethral discharge. In untreated cases, the discharge, which contains live bacteria and white blood cells, can persist for months.

In females, *N. gonorrhoeae* first colonizes the cervix, often without causing symptoms. However, during menstruation the pathogen can ascend into the uterus and fallopian tubes. Spread of this nonmotile bacterium may be assisted by the presence of sperm cells to which it is able to attach. The resulting infection of the fallopian tubes and surrounding tissues, termed *pelvic inflammatory disease*, can be fatal if untreated. The scarring of the fallopian tubes that results from pelvic inflammatory disease often causes sterility. *N. gonorrhoeae* can also infect joints, the throat, and the cornea; prior to the routine use of silver nitrate or erythromycin drops in the eyes of babies, gonorrheal eye infections were often acquired at birth from a mother with cervical gonorrhea.

### Mycoplasmal Diseases

*Mycoplasma pneumoniae*, like all mollicutes (Chapter 25), lacks a cell wall. It can grow in the human respiratory tract, causing pharyngitis or pneumonia. Initial symptoms of mycoplasma pneumonia are headache, weakness, and a low fever. However, a cough gradually becomes the predominant symptom. The disease can persist for weeks in untreated cases.

## Diseases Caused by Spirochetes

*Treponema pallidum*, the spirochete (Chapter 21) that causes *syphilis*, quickly loses viability outside the human body, and no reservoir other than the human population is known. Except in the case of congenital syphilis, which is acquired *in utero* from the mother, the disease is transmitted by contact, usually during sexual activity. *T. pallidum* multiplies in the skin at the site of contact for 2 to 10 weeks before the characteristic sign of *primary syphilis*, the *chancre*, a painless reddened ulcer with a hard rim, forms at the site. The chancre heals spontaneously, and in approximately one-third of cases the disease does not progress further. In the remaining cases, the infected person is asymptomatic for 2 to 10 weeks before the appearance of the characteristic rash of *secondary syphilis*. Both the chancre and the raised reddened patches forming the rash are infectious. The rash also disappears spontaneously, but in about one-third of untreated cases, *tertiary syphilis* develops, in which degenerative lesions form in skin, bone, liver, and the central nervous system.

*Relapsing fever*, another spirochetal disease, is caused by certain members of the genus *Borrelia* which parasitize rodents and humans. It can be transmitted from animals to humans by ticks and from person to person by body lice. Following an incubation period of 3 to 15 days, there is a rapid rise in temperature accompanied by chills, muscle and joint pain, and nausea. The temperature remains high for 3 to 5 days and then decreases to normal. However, the patient feels well for only 4 to 10 days before the initial symptoms return. After 2 to 10 cycles of illness and remission, recovery is usually complete. The periods of high temperature are associated with the presence of spirochetes in the blood; they disappear from blood during periods of remission.

*Leptospirosis* is caused by *Leptospira interrogans*, a spirochete that infects many wild and domestic animals. Humans acquire the disease usually when wounded skin or mucous membrane comes into contact with a diseased animal or with water contaminated by *L. interrogans*. Leptospirosis is largely an occupational disease, affecting primarily young males who work with livestock. The incubation period of 2 to 26 days ends with the sudden onset of chills, fever, headache, and severe muscle aches, often in the thighs. During this phase, large numbers of spirochetes are present in the blood. Both the temperature and the number of spirochetes decrease within 6 days as the concentration of specific antibody in blood rises. However, in some cases large numbers of immune complexes (Chapter 30) containing leptospiral components form in the blood, the liver, the kidneys, and other organs. The damage caused by these immune complexes can be severe enough to cause death.

## Rickettsial Diseases

Several important diseases of humans are caused by rickettsias (Chapter 21), small intracellular parasites that infect both arthropods and mammals. In most cases, an arthropod host (a flea, louse, tick, or mite) transmits the disease from one mammal to another, and with the single exception of epidemic typhus, humans are only an accidental host, playing no significant role in the propagation of the pathogen.

In previous centuries *epidemic typhus*, caused by *Rickettsia prowazekii*, decimated populations as it spread. It is transmitted only by the human body louse, which becomes infected by feeding on a diseased person. In the louse, *R. prowazekii* multiplies in the gut, is shed in feces, and eventually kills the host. The pathogen cannot be transmitted from louse to louse, and humans are usually infected by scratching areas of their skin contaminated with feces from infected lice. About 7 to 14 days later, a severe headache accompanied by chills and fever begins, followed 4 days later by a rash that spreads from the trunk to the extremities. In untreated cases, the temperature may remain high for two weeks or more, and the fatality rate is greater than 50 percent in the elderly.

*Endemic typhus* (caused by *R. mooseri*) and *scrub typhus* (caused by *R. tsutsugamushi*) differ from epidemic typhus in that they are not usually transmitted from human to human. Infected populations of rats are the principal reservoir of endemic typhus, a disease that occurs sporadically throughout the world and that is transmitted from rats to humans mostly by fleas. Scrub typhus is endemic in Southeast Asia and surrounding areas. It is transmitted to humans by mites that feed on infected rats and mice. In these mites, the pathogen can be transmitted transovarily from parent to offspring so that populations of infected mites are a secondary reservoir. The symptoms of both of these diseases closely resemble those of epidemic typhus but tend to be milder.

*Rocky Mountain spotted fever* is a rickettsial disease that occurs only in North and South America. It is caused by *R. rickettsii*, which usually infects rodents but can be transmitted to humans by ticks. In these arthropods, it produces a mild infection transmitted from parent to offspring through the egg. In humans, a severe, persistent headache accompanied by a high fever begins 2 to 4 days after being bitten by an infected tick and is followed several days later by a rash that begins on the extremities and spreads to the trunk. In untreated cases the temperature may remain very high for more than one week.

*Q fever* is caused by *Coxiella burnetii*, a rickettsia that infects a variety of wild and domestic animals and is transmitted among these primarily by arthropods. However, humans nearly always acquire this disease by contact with material from an infected domestic animal (usually a sheep, cow, or goat). After an incubation period of 9 to 20 days, a severe headache begins, accompanied by a fever, which may remain very high for more than a month if the case is not treated. Unlike other rickettsial diseases, a rash is absent during Q fever.

### Chlamydial diseases

Two groups of chlamydias (Chapter 22) cause diseases in humans. Members of the first, the *TRIC group*, cause *trachoma*, *inclusion conjunctivitis*, *lymphogranuloma venereum*, and some urinary tract infections called "nonspecific urethritis." Members of the second group cause *psittacosis*.

Trachoma, an infection of the cornea and conjunctiva, is endemic in the hot, dry regions of northern Africa and southwestern Asia. The disease is spread by direct contact, and the corneal scarring that results when the disease is untreated is the cause of more than 20 million cases of blindness throughout the world.

Chlamydial strains that cause both "nonspecific urethritis" in men and infection of the cervix (cervicitis) in women can also cause inclusion conjunctivitis, a disease usually acquired during birth from an infected mother; symptoms develop 5 to 14 days later, but fortunately, corneal scarring is rare. Chlamydial urethritis and cervicitis are among the most prevalent sexually transmitted diseases. In females, the infection sometimes spreads to the fallopian tubes, causing pelvic inflammatory disease.

Chlamydial strains that cause lymphogranuloma venereum are also transmitted sexually; after first infecting the urethra or cervix, they invade nearby lymph nodes. These can become greatly enlarged and can drain infectious fluid to the skin.

Psittacosis is a respiratory disease acquired from birds infected with *Chlamydia psittaci*. If limited to the upper respiratory tract, the disease is usually mild, but a severe pneumonia may result from infection of the lungs.

## FUNGAL DISEASES

Fungi cause three types of diseases, collectively termed *mycoses* (see Table 32.5). Most common are the *superficial mycoses* exemplified by *Candida albicans* infections of mucous membranes of the mouth or vagina and skin infections termed *dermatomycoses*. Much rarer are the *subcutaneous mycoses*, which result occasionally from puncture wounds, and the *deep mycoses*, which usually begin as lung infections.

### Dermatomycoses

Dermatomycoses are chronic diseases characterized by small raised patches of skin, which may become scaly or progress to form blisters. These diseases can be acquired both from humans and domestic animals and are among the most prevalent of all microbial diseases. They are classified by the location of the skin that is infected: *tinea capitis* involves the scalp, *tinea pedis* involves the feet, *tinea cruris* involves the groin, and *tinea corporis* involves other areas. Tineas can be caused by several species of *Microsporum*, *Trichophyton*, and *Epidermophyton* (Table 32.5). Some of these produce annular lesions termed *ringworm*. In tinea capitis, the fungus often invades hair shafts which then break, producing a bare patch of scalp.

### Subcutaneous Mycoses

The major subcutaneous mycosis of humans is caused by *Sporothrix schenckii*, a fungus that grows on living or decaying plant tissue. To cause disease, it must be introduced into a wound, usually by a thorn or splinter, where it grows and spreads along lymphatic channels, producing subcutaneous nodules that sometimes drain to the skin.

### (Deep) Systemic Mycoses

Most systemic mycoses are acquired by inhaling the spores of free-living fungi. For example, *Coccidiomycosis* is caused by *Coccidiodes immitis*, a fungus that thrives in dry regions of the southwestern states, California, and Mexico. In the soil and on most culture media, it grows as a mold, forming barrel-shaped arthrospores at the tips of hyphae (Figure 32.2). However, in humans it grows as a yeast, forming thick-walled spherules filled with endospores. The disease is acquired by inhaling arthrospores, usually contained in dust, and in

**TABLE 32.5**

**Examples of Pathogenic Fungi**

| Disease | Pathogen(s) |
|---|---|
| DERMATOMYCOSES | |
| Tinea capitis | *Microsporium audouinii, M. canis* |
| Tinea corporis | *Trichophyton violaceum, T. tonsurans, T. schoenleinii* |
| Tinea cruris | *Trichophyton rubrum, T. mentagrophytes, Epidermophyton floccosum* |
| Tinea pedis | *Trichophyton rubrum, T. mentagrophytes, Epidermophyton floccosum* |
| SUBCUTANEOUS MYCOSIS | |
| Sporotrichosis | *Sporothrix schenckii* |
| SYSTEMIC MYCOSES | |
| Blastomycosis | *Blastomyces dermatitidis* |
| Coccidiomycosis | *Coccidiodes immitis* |
| Cryptococcosis | *Cryptococcus neoformans* |
| Histoplasmosis | *Histoplasma capsulatum* |

parts of the San Joaquin Valley of California, over 50 percent of the inhabitants have been infected by *C. immitis.* Following an incubation period of 7 to 28 days, a cough and fever begin. In most cases the disease remains localized in the lung, and recovery occurs after a mild illness lasting one to two weeks. However, in approximately 1 percent of cases the disease spreads to other areas of the lungs or to the central nervous system, often causing death.

*Blastomycosis* is caused by *Blastomyces dermatitidis*, a fungus that grows as a budding yeast in humans but as a mycelial fungus on culture media (Figure 32.3). The disease is most common in the states bordering the Mississippi and Ohio rivers, and is rare in the western United States and other countries. Symptoms begin with a cough which becomes chronic and is sometimes accompanied by fever and loss of weight. Occasionally, the fungus spreads from the lungs to the skin.

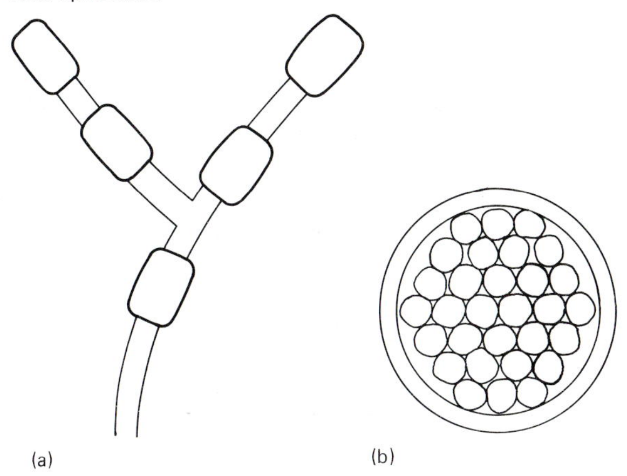


FIGURE 32.2

Morphology of *Coccidiodes immitis*. (a) Morphology when growing in soil or on culture medium. The barrel-shaped structures are arthrospores. (b) Appearance of a thick-walled spherule in human tissue. The spherule ruptures, liberating its endospores, which grow to produce new spherules.

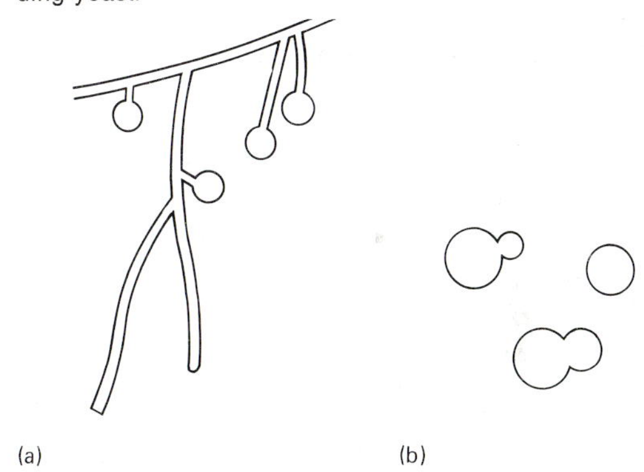


FIGURE 32.3

Morphology of *Blastomyces dermatitidis*. (a) Mycelial morphology when grown on culture medium. Large conidia are present at hyphal tips. (b) Morphology while growing in human tissue where the fungus reproduces as a budding yeast.

*Histoplasmosis* is caused by *Histoplasma capsulatum*, a fungus that grows as a small budding yeast in humans and on culture medium at 37° C. At 20° C it grows as a mold, producing small conidia (microconidia), that are connected to hyphae by short stalks and large conidia (macroconidia) at the hyphal tips (Figure 32.4). Most individuals infected by *H. capsulatum* do not develop signifi-

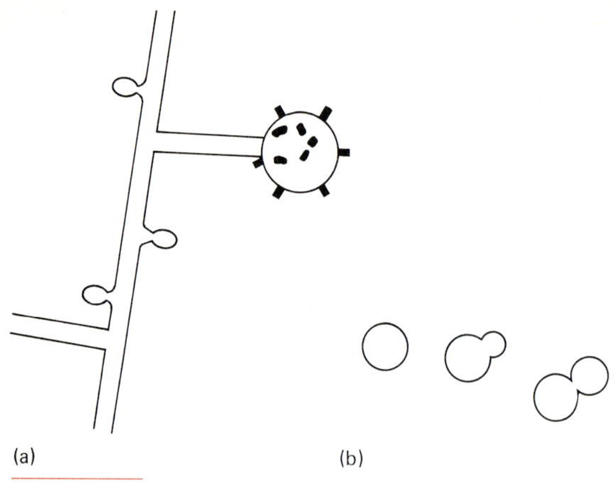


FIGURE 32.4

Morphology of *Histoplasma capsulatum*. (a) Mycelial morphology when grown at 20° C. Microconidia are attached to hyphae by short stalks and macroconidia appear at hyphal tips. (b) Morphology at 37° C, where it grows as a budding yeast.

cant disease, but a few develop a chronic lung disease characterized by cough, fever, shortness of breath, and chest pain.

*Cryptococcosis* is caused by *Cryptococcus neoformans*, a fungus that always grows as a large budding yeast. It can be isolated from decomposed pigeon droppings and other environments throughout the world: humans are infected by inhaling viable cells, but these respiratory infections are almost always asymptomatic. Occasionally, *C. neoformans* is carried in blood to the brain, causing a serious meningitis in which large numbers of encapsulated yeast cells occur in the cerebrospinal fluid.

Several mycoses occur almost entirely in persons with a defective immune system. These opportunistic mycoses include systemic candidiasis (caused by *Candida* spp.), aspergillosis (caused by *Aspergillus* spp.), and mucormycosis (caused by species of *Mucor* and *Rizopus*).

## PROTOZOAL DISEASES

Although fewer than 20 species of protozoa are human pathogens (Table 32.6), these are a major cause of severe illness and death. At any time, more than 100 million people have malaria, and each year more than one million die of this disease. Millions of others are infected by trypanosomes, leishmanias and amebas.

### Malaria

Human malaria is caused by any of four species of sporozoites: *Plasmodium falciparum*, *P. malariae*, *P. vivax*, and *P. ovale*. Each is transmitted from human to human by a female *Anopheles* mosquito, which as it bites injects saliva containing plasmodial sporozoites. These are carried in blood to the liver where they multiply intracellularly. After one to six weeks, they produce large numbers of cells, termed *merozoites*, that are released into the bloodstream where they attach to receptors on erythrocytes and penetrate them. Each species attaches to a specific receptor. For example, *P. vivax* attaches to the *Duffy blood group antigen*. Many natives of West Africa lack this antigen and are therefore resistent to *P. vivax*.

Inside an erythrocyte, the plasmodium enlarges as a uninucleate cell termed a *trophozoite*. Then its nucleus divides repeatedly, producing a *schizont* that has 6 to 24 nuclei (Figure 32.5). The schizont divides, producing mononucleated merozoites. In the case of *P. malariae*, the erythrocyte ruptures 72 hours after penetration; in other plasmodia, rupture occurs after 48 hours. The rapid rise in temperature and severe chills that are characteristic of malaria occur at the time of erythrocyte lysis. The liberated merozoites rapidly infect a new population of erythrocytes, initiating the next cycle of fever and chills. In some erythrocytes, me-

FIGURE 32.5

Growth of *Plasmodium vivax* within an erythrocyte. (a) A young trophozoite appears as a ring. (b) A trophozoite at an intermediate growth stage showing basophilic stippling. (c) A mature schizont. (d) A mature gametocyte.

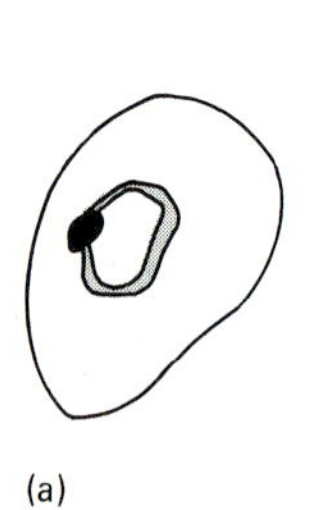


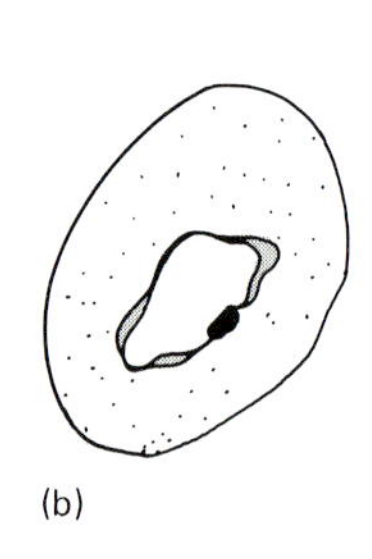


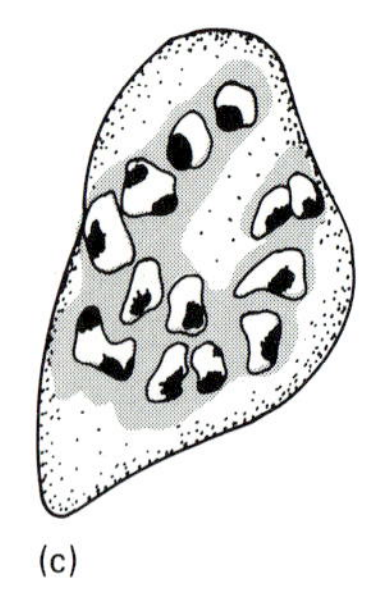


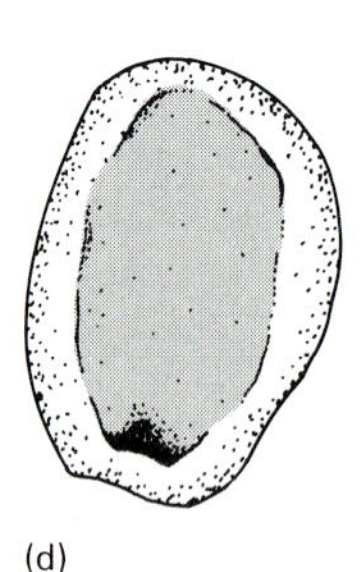

**TABLE 32.6**

**Examples of Pathogenic Protozoa**

| Group: Pathogen | Principal Disease(s) |
|---|---|
| Amebas: | |
| *Entamoeba histolytica* | Amebic dysentery, liver abscesses |
| *Naegleria fowleri* (an ameba-flagellate) | A rare, fatal encephalitis usually acquired while swimming in warm ponds |
| Flagellates: | |
| *Giardia lamblia* | Giardiasis (a diarrheal disease) |
| *Leishmania donovani* | Kala azar (visceral leishmaniasis) |
| *L. braziliensis* | Mucocutaneous leishmaniasis |
| *L. tropica* and *L. mexicana* | Cutaneous leishmaniasis |
| *Trichomonas vaginalis* | Trichomoniasis (an infection of the genitourinary tract) |
| *Trypanosoma cruzi* | Chagas' disease |
| *T. brucei* | African sleeping sickness |
| Sporozoa: | |
| *Plasmodium falciparum, P. malariae, P. ovale,* and *P. vivax.* | Malaria |
| *Pneumocystis carinii* | Pneumocystis pneumonia |
| *Toxoplasma gondii* | Toxoplasmosis |
| Ciliates: | |
| *Balantidium coli* | A rare diarrheal disease usually acquired from pigs |

rozoites differentiate into *gametocysts*, which do not rupture the host cell. These are ingested by the mosquito during its blood meal. In the insect gut, erythrocytes lyse, and gametocysts fuse to form *ookinetes* that mature into sporozoites. Finally, these migrate to the insect's salivary gland.

## Diseases Caused by Leishmanias

Several human diseases are caused by flagellates of the genus *Leishmania* (Figure 32.6). The primary reservoirs of these parasites are infected populations of rodents and canines, and leishmaniasis is transmitted from these animals to humans by sandflies (*Phlebotomus*): these acquire the parasite while feeding on an infected animal. Leishmanias grow in cells of the insect proboscis, emerging as flagellated spindle-shaped cells (*promastigotes*), which are inoculated into whatever mammals the sandfly bites. There, they lose their flagella, penetrate mononuclear phagocytes, and multiply intracellularly as small cells termed *amastigotes*.

*Cutaneous leishmaniasis* is caused by either *L. tropica* or *L. mexicana*, while *mucocutaneous leishmaniasis* (affecting both mucous membranes and skin) is caused by *L. braziliensis*. Following an incubation period that is usually two to six months but can be as short as two weeks, a small red papule, which may itch intensely, appears at the site of inoculation and grows, eventually forming an ulcer. Such sores are frequently found on the face or ears in cutaneous leishmaniasis and on oral or nasal mucosa in mucocutaneous leishmaniasis.

Cutaneous leishmaniasis occurs in almost all countries that border the Mediterranean, as well as in Asia Minor, the Sudan, Ethiopia, the Congo Basin, the west coast of Africa, central and northern India, Turkestan, and China. It also occurs in tropical regions of the Americas from Mexico to northern Argentina.

*Kala azar* (*visceral leishmaniasis*), caused by *L. donovani* (Figure 32.6), is a generalized, often fatal disease accompanied by intermittent fever and enlargement of the spleen and liver. Following a highly variable incubation period of two weeks to

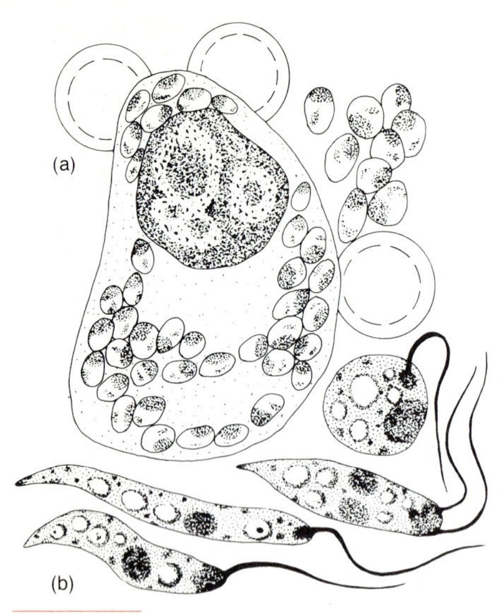


FIGURE 32.6

*Leishmania donovani.* (a) Large reticuloendothelial cell of spleen with amastigotes. (b) Promastigotes as seen in sandfly gut or in culture. Circles represent red blood cells for size comparison. From Jawetz, E., J. L. Melnick and E. A. Adelberg, *Review of Medical Microbiology*, 12th ed. (Los Altos, Calif.: Lange Medical Publications, 1976.)

18 months, the initial symptoms can mimic those of malaria, typhoid, or dysentery. In addition, joint pains, anemia, progressive emaciation, a reduced number of leukocytes, and edematous skin may develop.

Kala azar is found in India, north China, Turkestan, tropical Africa, countries that border the Mediterranean, western and middle Asia, and South America from Venezuela to nothern Argentina. Over 12 million people throughout the world are infected with either cutaneous or visceral leishmaniasis.

## Diseases Caused by Trypanosomes

*African sleeping sickness*, caused by the flagellated protozoan *Trypanosoma brucei*, is common both in western and central Africa where it is called Gambian sleeping sickness and in southeastern Africa where it is called Rhodesian sleeping sickness. The human population is the major reservoir of Gambian sleeping sickness, but it also can be acquired from a variety of wild and domestic animals. The major reservoir of Rhodesian sleeping sickness is the bushbuck, a species of antelope. Both forms of the diseases are transmitted by the tsetse fly as it bites humans and other animals. In addition to causing sleeping sickness, *T. brucei* contributes to malnutrition by causing disease in livestock in parts of Africa where cattle are an essential component of the food supply.

When a mammal is bitten by an infected tsetse fly, trypanosomes in the salivary gland are inoculated into the skin where they multiply, eventually reaching the bloodstream and aggregating in small blood vessels of the brain and heart. Following the incubation period, which is about two weeks in Rhodesian sleeping sickness but can be several years in Gambian sleeping sickness, the first symptom, a severe headache, begins.

Trypanosomes are able to grow in blood largely because they can evade the host immune response: the surface of a trypanosome is covered by a large number of identical glycoprotein antigen molecules. The host immune system can respond, after several days, to kill foreign cells covered with this antigen, but a single strain of trypanosome can produce scores of antigenic variants of the surface glycoprotein. Therefore, as antibodies to the parasite's glycoprotein surface antigen are produced, the concentration of parasites in blood falls, but three to eight days later the concentration increases again as a new antigenic variant arises. Consequently, the concentration of parasites in the host's blood oscillates as does the temperature, which is elevated when the concentration is high.

*Chagas' disease*, caused by *T. cruzi*, is endemic in large areas of Central and South America. It is transmitted by the blood-sucking bugs—*Panstrongylus megistis*, *Triatoma infestans*, and *Rhodnius prolixus*—that become infected while feeding on a diseased human, domestic animal, or wild animal. Initial symptoms of this disease—fever, fatigue, loss of appetite, and sometimes swelling of a single eyelid—begin gradually and then disappear within three months in most cases. However, viable trypanosomes persist in the tissues as a latent infection that can either remain quiescent or can become active again, causing a severe disease involving the heart or other organs.

### Amebic dysentery

*Entamoeba histolytica*, the causative agent of *amebic dysentery*, is acquired by ingesting food or water contaminated with feces. Usually no symptoms result from its presence in the colon, but about 10 percent of those colonized by *E. histolytica* experience abdominal pain accompanied by diarrhea sometimes containing bloody mucous. These symptoms of dysentery result from superficial invasion and ulceration of the intestinal mucosa. Occasion-

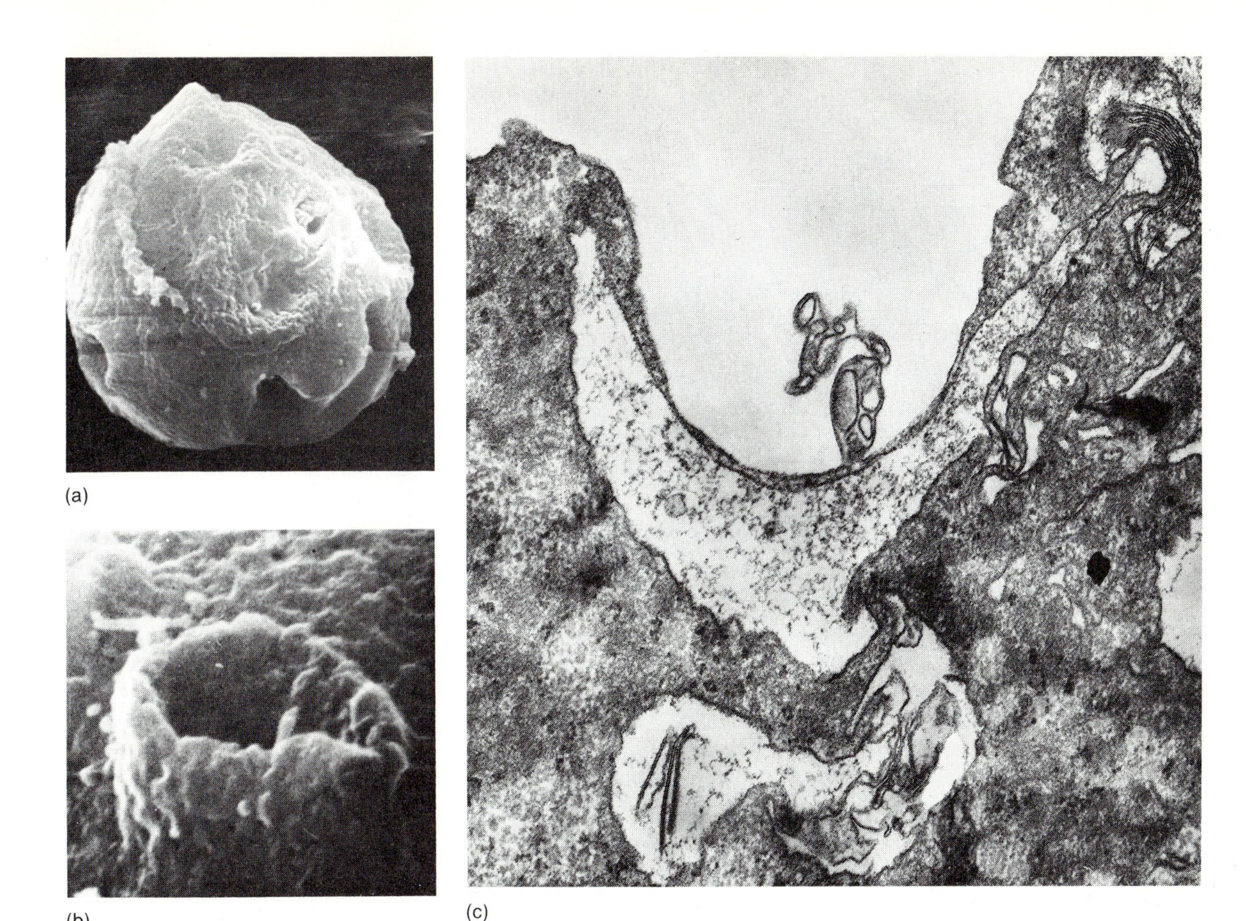


FIGURE 32.7

Surface lysosomes of *Entamoeba histolytica*. (a) Scanning electron micrograph of *E. histolytica* trophozoite, showing seven lysosomes in surface view, three of which have the trigger device in view ($\times$ 2,130). (b) Close-up view of a surface lysosome with a protruding trigger ($\times$ 6,800). (c) Electron micrograph of thin section of a surface lysosome, ($\times$ 36,600). From R. D. P. Eaton, E. Meerovitch, and J. W. Costerton, "The Functional Morphology of Pathogenicity in *Entamoeba histolytica*," *Ann. Trop. Med. and Parasitol.* **64,** 299 (1970).

ally, *E. histolytica* is carried by blood to the liver where it produces a life-threatening abscess.

Virulence of *E. histolytica* is correlated with a strain's ability to lyse mammalian cells. Such lysis requires direct contact and probably is caused by an unusual mechanism for exocytosis: enzymes present in lysosomes which lie beneath depressions in the cell membrane (Figure 32.7) are released when a vermiform (worm-shaped) appendage at the center of each depression is touched.

## Giardiasis

The relatively common intestinal disease *giardiasis* is caused by the flagellate, *Giardia lamblia*, which has a very distinctive appearance (Figure 32.8). It is acquired by ingestion of food or water contami-

FIGURE 32.8

*Giardia lamblia*. (a) "Face" and (b) "profile" of vegetative forms; (c) and (d) cysts (binucleate and quadrinucleate stages). Circle represents red blood cell for size comparison. From Jawetz, E., J. L. Melnick and E. A. Adelberg, *Review of Medical Microbiology*, 12th ed. (Los Altos, Calif.: Lange Medical Publications, 1976.)

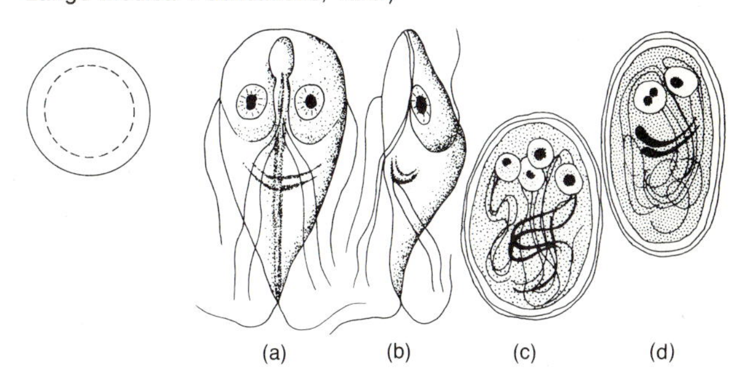

nated with feces. Although humans are the major reservoir, beavers and other wild animals have recently been implicated as carriers. *G. lamblia* colonizes the small intestine by attaching its unusual sucking disc to the intestinal mucosa. As unattached *G. lamblia* cells pass through the colon at a normal rate, they differentiate into thick-walled cysts that can survive for months in cold waters. However, the rapid expulsion of intestinal contents during periods of diarrhea can prevent the formation of cysts. Diarrheal disease develops only in a minority of people colonized by *G. lamblia* and is manifested by diarrhea, abdominal pain, and voluminous flatulence.

### Trichomoniasis

*Trichomoniasis* is a sexually transmitted disease caused by the flagellate *Trichomonas vaginalis* (Figure 26.10). Although some colonizations are asymptomatic, most are not, producing in males a burning sensation during urination and in females an itching or burning sensation of the vagina accompanied by a discharge.

### Toxoplasmosis

*Toxoplasmosis*, an infection caused by the sporozoan *Toxoplasma gondii*, can be acquired from a great variety of infected mammals and birds, either by ingestion of uncooked meat or material contaminated with their feces. Cysts of *T. gondii* release cells that penetrate intestinal mucosal cells and reproduce intracellularly, producing trophozoites that are released into the bloodstream. The trophozoites penetrate cells throughout the body in which they reproduce and produce cysts that persist throughout life. Most human cases of toxoplasmosis are either asymptomatic or manifested by a mild enlargement of lymph nodes throughout the body. Fever and fatigue occur in a minority of cases. Individuals with normal immunity usually recover quickly but continue to harbor intracellular *T. gondii* cysts for the remainder of their lives.

### Pneumocystis Pneumonia

*Pneumocystis carinii*, the etiologic agent of *Pneumocystis pneumonia* is a very common cause of asymptomatic infections: by age four, most children have antibodies that bind to *P. carinii* cells. Although symptomatic infections are rare in persons with normal immunity, they are often serious in people with a defect in cellular immunity (Chapter 30). For this reason, the disease is life-threatening when it occurs in patients with acquired immunodeficiency syndrome (AIDS), as it frequently does.

## VIRAL DISEASES

With the exception of the baculoviruses, all of the major groups of animal viruses (Table 9.1) contain pathogens that have been isolated from humans. Some groups (e.g., poxviruses) contain only a few viruses that are pathogenic to humans, but other groups (e.g., picornaviruses) contain a large number of human pathogens. In the following sections, certain important viral diseases of humans are discussed (Table 32.7).

**TABLE 32.7**

**Examples of Viruses that Cause Human Diseases**

| Group: Pathogen | Principal Disease(s) |
|---|---|
| Herpesviruses: | |
| Cytomegalovirus | Respiratory infections |
| Epstein-Barr virus | Mononucleosis |
| Herpes simplex viruses | Oral cold sores (fever blisters) and genital sores |
| Varicella virus | Chickenpox, shingles |
| Poxviruses: | |
| Orf virus | Contagious pustular dermatitis |
| Variola virus | Smallpox |
| Picornaviruses: | |
| Coxsackie viruses | Herpangina |
| Hepatitis A virus | Infectious hepatitis |
| Poliomyelitis virus | Poliomyelitis |
| Rhinoviruses | Most colds |
| Influenza viruses | Viral influenza and pneumonia |
| Parainfluenza viruses | Measles, mumps, rubella |
| Rhabdoviruses | Rabies |
| Reoviruses | Diarrheal diseases |
| Retroviruses: | |
| Human T-cell leukemia virus (HTLV) strains I and II | T-cell leukemia |
| HTLV-III | Acquired immunodeficiency syndrome (AIDS) |

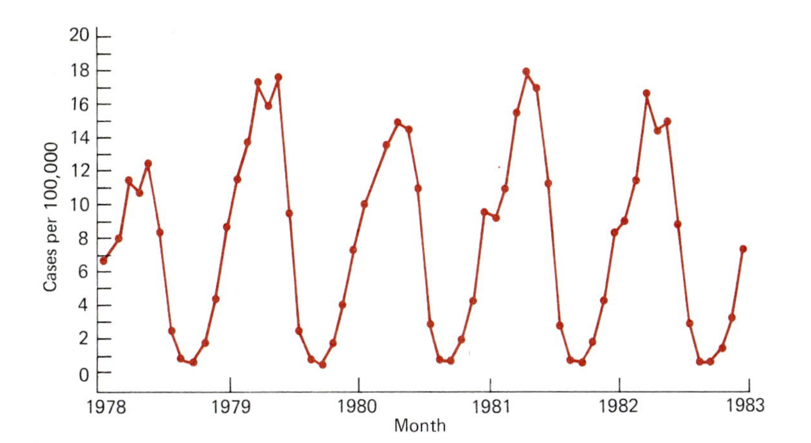


FIGURE 32.9
Reported cases of chickenpox per month in the United States from 1978 to 1983 (cases per 100,000 people). Courtesy of Center for Disease Control. Annual summary, 1982: *Morbidity Mortality Weekly Rept.* **31,** 21 (1983).

## Diseases Caused by Herpesviruses

*Herpesviruses* are large enveloped polyhedral viruses that contain linear double-stranded DNA. The chromosome (MW approximately $10^8$) is enclosed in an icosahedral capsid formed by 162 capsomers, which is in turn surrounded by an envelope composed of unit membrane derived from the nuclear membrane of a host cell. Both the replication of DNA and assembly of nucleocapsids of herpesviruses occur in the nucleus. Escape of virions from infected cells is discussed in Chapter 9.

Five herpesviruses cause diseases of humans: *varicella-zoster virus* causes *chickenpox*, which occurs primarily in children, and *zoster* (*shingles*), which occurs primarily in the elderly; *herpes simplex* type I and type II viruses cause cold sores (fever blisters) and genital herpes; *Epstein-Barr* (EB) *virus* causes *mononucleosis*; and *cytomegalovirus* causes a viral pneumonia. All these diseases are spread by direct contact or by droplets; the human population is their major reservoir of infection.

Chickenpox is acquired by the respiratory route or by contact with an infected person. The disease occurs in annual cycles, being common in the winter and spring but much less so during late summer and early autumn (Figure 32.9). Following an incubation period of 10 to 21 days, small blisters (vesicles) surrounded by a reddened area of skin appear, then burst and become covered by a scab. Unless they become infected, healing occurs, without scarring, within about 10 days.

However, people who recover from chickenpox are not completely free of the virus: some virions reside without replication in the nuclei of sensory neurons producing a *latent infection*, i.e., the pathogen remains but is dormant. Occasionally, these viruses migrate down the axons of sensory nerves and reinitiate viral replication, producing painful vesicles similar to those of chickenpox. This reactivated form of chickenpox, termed *shingles*, is usually restricted to the sensory distribution of a spinal nerve.

Herpes simplex viruses also persist in neurons after a primary infection. Occasionally they too migrate down axons to cause local skin infections that produce painful blisters. These are highly infectious, containing large amounts of virus.

Mononucleosis is most prevalent in the 15 to 25 year age group but is also common in children. It is acquired by contact or inhaling droplets from an infected person. The incubation period of one to two months is followed by the gradual onset of fatigue and fever, usually accompanied by a sore throat and enlargement of lymph nodes in the neck. These symptoms often persist for weeks; the host may continue to shed virus for months.

## Diseases Caused by Poxviruses

*Poxviruses* are the largest and most complex viruses known. Their virions have two exterior membranes surrounding a nucleoprotein core, which is indented by two lateral bodies (Figure 32.10) and contains double-stranded DNA. Poxviruses are the only DNA viruses that replicate in the cytoplasm, and their virions contain a virus-encoded *transcriptase* which catalyzes the synthesis of viral mRNA in the cytoplasm.

Poxviruses cause two skin diseases, *molluscum contageosum* and *contagious pustular dermatitis* (also termed *orf*). The former occurs mostly in children, and the latter is primarily a disease of sheep, from which it is occasionally acquired by humans. Another poxvirus, *variola*, is the cause of *smallpox*, the most important (as judged by lethality) viral disease in history, but a disease that now appears to be extinct (Chapter 30) as a consequence of widespread immunization using the cowpox (vaccinia) virus.

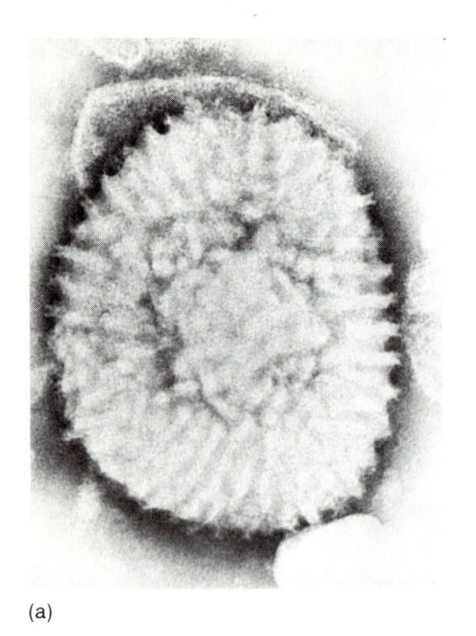

(a)

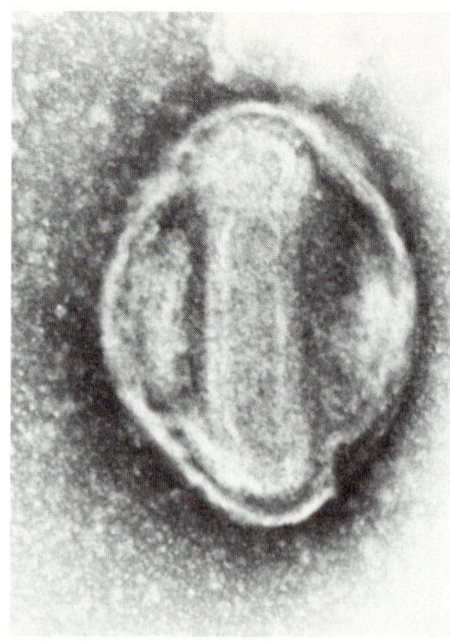

(b)

FIGURE 32.10

(a) A virion of the poxvirus vaccinia, negatively stained with phosphotungstic acid. The ridges on the surface may be long rodlets or tubules.
(b) A negatively stained vaccinia particle that has been centrifuged in a sucrose gradient. The particle has been partially disrupted, and it has lost its outer membrane. The remaining structure includes a biconcave inner core, containing the nucleic acid, two elliptical bodies, and a surrounding membrane.
From S. Dales, "The Uptake and Development of Vaccinia Virus in Strain L Cells Followed with Labeled Viral Deoxyribonucleic Acid," *J. Cell Biol.* **18,** 51 (1963).

Smallpox is highly contagious, being spread by the respiratory route. Following an incubation period of about 12 days, intense fatigue and high fever begin suddenly and a spotty red rash develops several days later. The spots become blisters that rupture, often resulting in the formation of scars. No effective treatment was discovered, and the fatality rate was approximately 30 percent.

## Serum Hepatitis

*Serum hepatitis* is caused by *hepatitis B* virus, an enveloped icosahedral DNA virus that does not belong to any of the major viral families. This virus is shed through the skin and into the urine of both symptomatic individuals and asymptomatic carriers, and it can be acquired by contact with an infected person, by ingestion, or through wounds (usually those produced by a hypodermic needle). This virus can also pass from the blood of an infected mother through the placenta to infect the fetus. It is estimated that there are more than 200 million carriers of hepatitis B virus, mostly in Africa and southern Asia. A high frequency of carriers occurs in the United States among male homosexuals and drug addicts.

Following an incubation period of 30 to 180 days, fatigue and fever gradually begin, followed by *jaundice* which is caused by the accumulation of *bilirubin*, a degradation product of hemoglobin, in the skin and other tissues. The destruction of liver cells can be extensive, and death follows from liver failure in 1 to 10 percent of cases. Many of those who recover continue to shed virus for years, and some develop a progressive degenerative liver disease termed *chronic active hepatitis*.

## Diseases Caused by Picornaviruses

*Picornaviruses* are small icosahedral viruses that contain a plus-strand RNA chromosome (Chapter 9). They replicate in the cytoplasm where the viral chromosome serves as mRNA. A unique feature of this group of viruses is their novel formation of viral proteins: the viral chromosome is translated into a single polypeptide which is proteolytically cut into six essential viral proteins. Four of these aggregate into a capsomer, of which 60 capsomers form the capsid.

There are two major groups of picornaviruses, the *rhinoviruses*, which cause most colds, and the *enteroviruses*, which usually cause mild gastrointestinal disease. More serious disease is caused by the enteroviruses that infect tissues of the central nervous system. The most important example is poliovirus (poliomyelitis virus), which is acquired by the oral-fecal route, replicating first in mucosal cells of the small intestine and spreading to nearby lymph nodes. Occasionally, poliovirus spreads to the spine or brain where it infects and kills motor neurons, causing permanent paralysis. Widespread vaccination programs have dramatically reduced the frequency of paralytic poliomyelitis (Figure 32.11).

The liver disease, *infectious hepatitis*, is caused by *hepatitis A* virus, a member of the enterovirus group. Infection by this virus is very common but

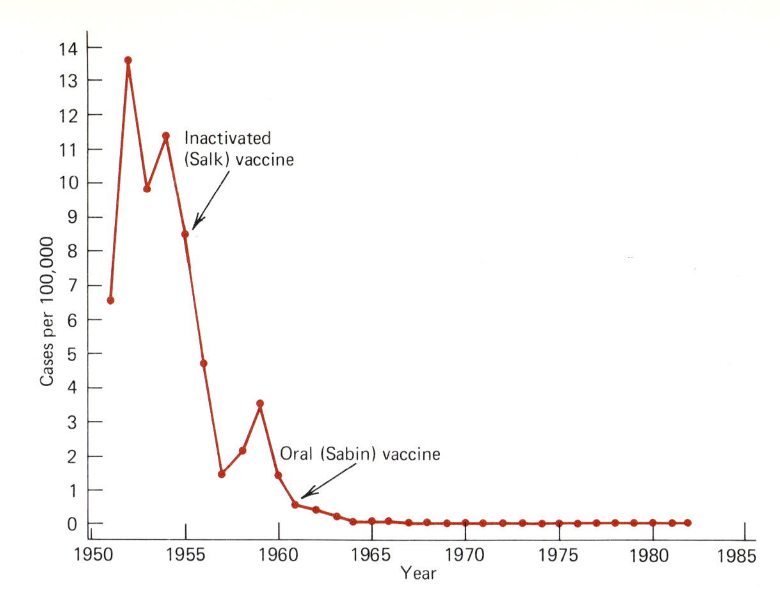


FIGURE 32.11

Reported cases of poliomyelitis per year in the United States since 1950 (cases per 100,000 people). Courtesy of Center for Disease Control. Annual summary 1982: *Morbidity Mortality Weekly Rept.* **31,** 52 (1983).

in most cases only mild intestinal symptoms result. Occasionally the virus spreads to the liver, and when it does the fatality rate is less than 1 percent. Those who recover do not become carriers of the virus, nor do they develop chronic active hepatitis.

## Influenza

The upper respiratory tract disease, *influenza,* is caused by *orthomyxoviruses* (also termed influenza viruses). These unusual viruses contain several minus-strand RNA chromosomes, each wound with protein into a helical nucleocapsid; the virion envelope has numerous projections on its outside surface (Figure 32.12). Most of these are the glycoprotein, *hemagglutinin,* which binds to glycoproteins that have a terminal neuraminic acid residue (Figure 32.13). Hemagglutinin is required for viral adsorption to host cells.

The three types of influenza virus, A, B, and C, are distinguished by the type of protein contained in the nucleocapsid. Type A viruses, which are both more common and cause more serious disease than the others, contain eight chromosomes, six of which each encodes a single viral protein, while each of the others encodes two proteins. Thus influenza virus is said to have a *segmented genome* because its genes are distributed on more than one chromosome.

In a host that is infected by two strains of influenza virus, a new combination of their chromosomes can be incorporated into a virion, giving rise to a new viral strain. The strains that cause worldwise epidemics (pandemics) have acquired a new hemagglutinin gene either from a rare human strain or a strain that normally infects some other species. Evolution of new strains is facilitated by the unusually broad host range of influenza viruses, including rodents and domestic animals as well as humans. A new strain producing a hemagglutinin that is different from those of common human strains may cause an epidemic because immunity to a strain of influenza virus requires the presence of antibody that binds to hemagglutinin. For example, the influenza virus strain that produced a

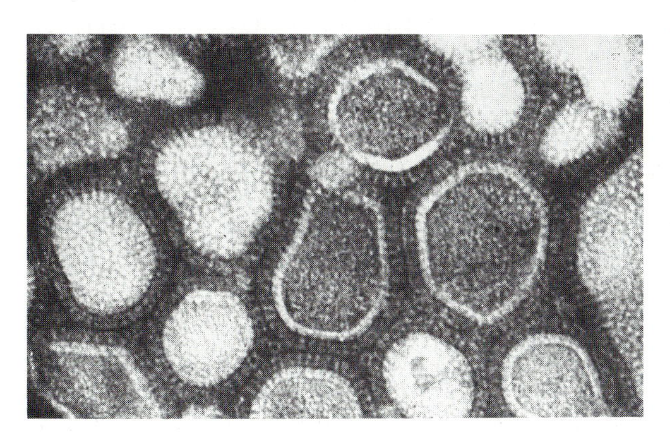

FIGURE 32.12

Particles of influenza virus, showing surface projections or "spikes." From R. W. Horne et al., "The Structure and Composition of the Myxoviruses. I. Electron Microscope Studies of the Structure of Myxovirus Particles by Negative Staining Techniques," *Virology* **11,** 79 (1960).

COOH
|
C=O
|
$CH_2$
|
HCOH
|
$H_2NCH$
|
HOCH
|
HCOH
|
$H_2COH$

FIGURE 32.13

Structure of neuraminic acid.

pandemic in 1918 acquired a hemagglutinin that is similar to the one that is common in swine influenza virus. It is estimated that more than a million people in the United States died of influenza during this epidemic.

Influenza is acquired from an infected person by the respiratory route. After an incubation period of one to three days, headache, fever, and muscle aches commence, followed in most cases by a cough and sore throat.

### Measles, Mumps, and Rubella

*Measles*, *mumps*, and *rubella* (German measles) are caused by *parainfluenza viruses* (paramyxoviruses), which possess a single minus-strand chromosome (Chapter 9) but are otherwise very similar to the orthomyxoviruses. These highly contagious diseases are spread by the respiratory route but are now rare diseases in many countries as a result of widespread vaccination of children (Figure 32.14).

### Rabies

*Rabies* is the only major disease of humans that is caused by a *rhabdovirus*. Virions of this family of viruses contain a single molecule of minus-strand RNA (Chapter 9) that is coiled inside a rod- or bullet-shaped capsid that is tightly wrapped in an envelope. Rabies virus can infect all warm-blooded animals, although bats are the most frequent host. Humans acquire rabies most often from skunks (sylvan rabies) or from dogs or cats (urban rabies). In almost all cases, the disease is transmitted by an animal bite.

Rabies virus replicates within nerve cells, ultimately killing them. Viruses migrate from neuron to neuron, starting near the bite and moving toward the brain. Symptoms of disease usually begin 2 to 16 weeks after exposure with fatigue, loss of appetite, and fever. Often there is a sensation of tingling or burning at the site of the wound. The disease quickly progresses to the stage of progressive paralysis. In approximately 50 percent of cases, intense and painful spasms of muscles of the throat and chest are produced by swallowing. For this reason, rabies has been called "hydrophobia." Death follows in nearly 100 percent of cases from destruction of regions in the brain that regulate breathing.

A safe and effective vaccine against rabies is now available. It is a killed suspension of virus produced in human diploid cells grown in the laboratory and can be used either before or after exposure to produce active immunity. However, to be effective it must be given soon after the person has been bitten.

### Diseases Caused by Rotaviruses

*Rotaviruses* are an important cause of human diarrheal disease. They have a very unusual genetic organization: each polyhedral capsid contains genetic information segmented into 10 or 11 double-stranded chromosomes. Therefore, rotaviruses are similar to reoviruses (Table 9.1).

### Diseases Caused by Togaviruses

*Togaviruses* are named for their membrane coat (toga) that envelopes the polyhedral nucleocapsid core containing plus-strand RNA. They are transmitted by arthropod bites and produce several diseases including *encephalitis* and *yellow fever*. The strains that cause encephalitis are transmitted by mosquitos from wild or domestic animals.

Yellow fever is transmitted by mosquitos of the genus *Aedes*. Monkeys are the primary reservoir of the disease, but during epidemics the disease is spread from person to person by mosquitos. After an incubation period of three to six days, symptoms begin with headache and chills. Fever lasts several days and can be high. Only in severe cases does the characteristic yellowing of the skin (jaundice) occur.

### Diseases Caused by Retroviruses

*Retroviruses* are enveloped RNA viruses with a highly unusual replication cycle: upon entering the cytoplasm of a host cell, the enzyme *reverse transcriptase*, which is brought into the cell inside the virion, first synthesizes a strand of DNA that is complementary to the viral (plus-strand) RNA. Next, reverse transcriptase synthesizes a second strand of DNA, but this one is complementary to the first DNA strand synthesized (Figure 32.15). The two DNA strands form a circular, double-stranded chromosome that migrates to the nucleus and becomes integrated into a host chromosome. Then the RNA polymerase of the host transcribes the integrated viral genes, producing viral RNA for incorporation into virions.

Only three retroviruses have been firmly established to cause diseases in humans: *human T-cell leukemia virus* strains 1 and 2 (HTLV-I and HTLV-II) cause some cases of leukemia (Chapter 31) and HTLV-III is at least in part the cause of acquired immunodeficiency syndrome (AIDS). Antibodies against this virus are found in the great majority of AIDS patients, but are rarely found in healthy individuals who have not had contact with AIDS patients. In some cases of AIDS, HTLV-III has been isolated.

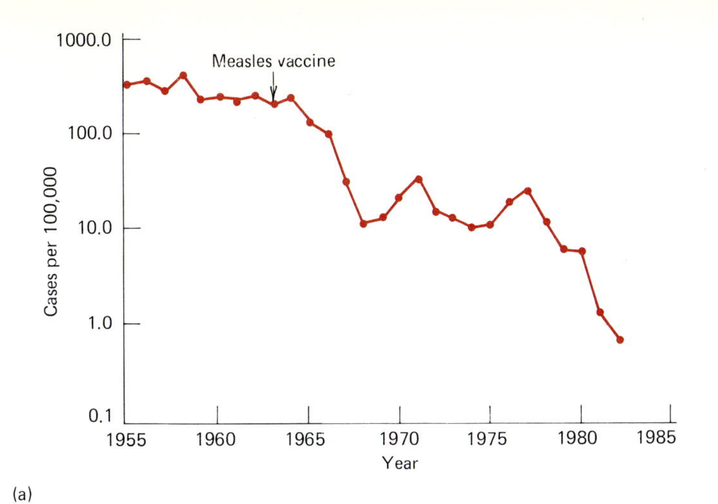


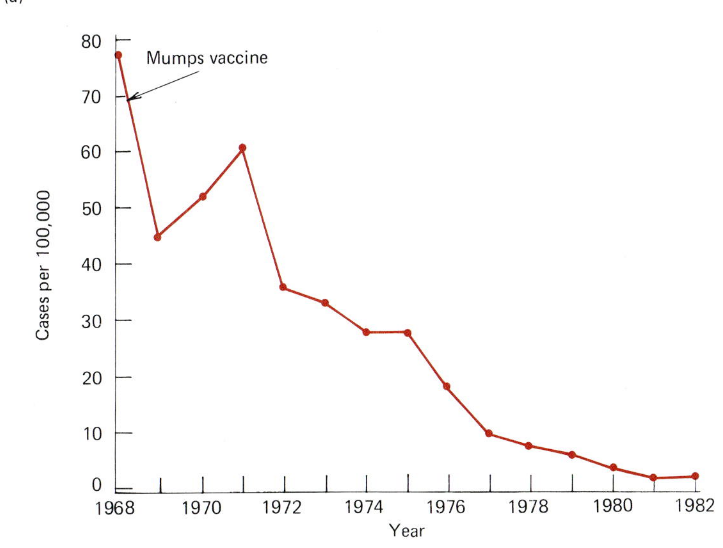


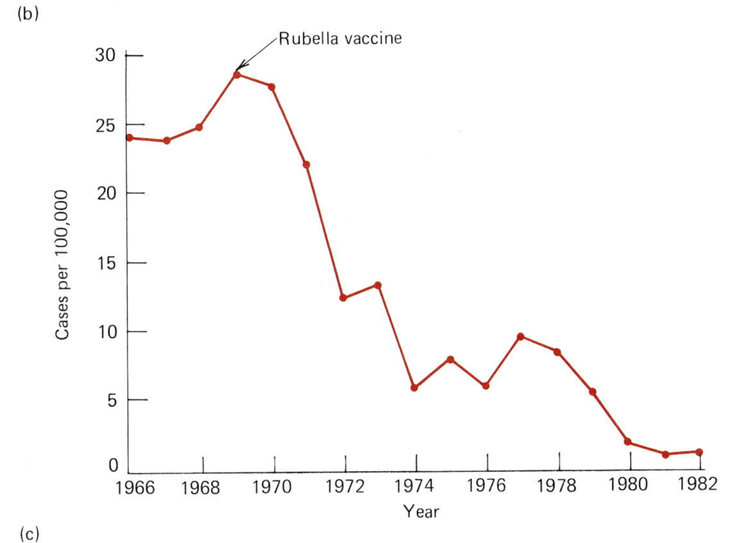


FIGURE 32.14

Cases of measles, mumps, and rubella occurring in the United States (cases per 100,000 people per year). (a) Cases of measles since 1954. From Annual summary 1982: *Morbidity Mortality Weekly Rept.* **31,** 48 (1983). (b) Cases of mumps since 1976. From Annual summary 1982: *Morbidity Mortality Weekly Rept.* **31,** 56 (1983). (c) Cases of rubella (German measles) since 1965. From Annual summary 1982: *Morbidity Mortality Weekly Rept.* **31,** 68 (1983).

AIDS patients often succumb to opportunistic infections such as pneumocystis pneumonia. This results from a severe deficiency of T-helper cells (Chapter 30). Apparently HTLV-III infects primarily T-helper cells, which are eventually killed by the virus.

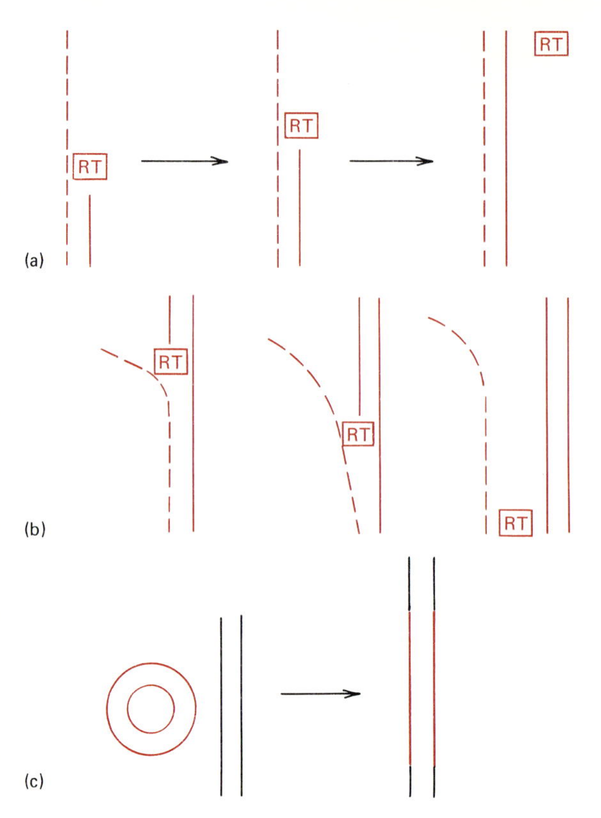


FIGURE 32.15

Replication of the retroviral chromosome. (a) In the cytoplasm, the RNA chromosome is copied by the enzyme, reverse transcriptase, that is contained within the virion. (b) Reverse transcriptase then synthesizes a strand of DNA that is complementary to the first DNA strand. In the process, the RNA chromosome is displaced, and the double-stranded DNA chromosome becomes circular. (c) The viral DNA chromosome migrates into the nucleus where it becomes linearly integrated in a host chromosome by a mechanism that may be similar to integration of the temperate bacteriophage mu (Chapter 9).

The mode of transmission for HTLV-I is not known, but AIDS is clearly transmitted primarily by direct contact between male homosexuals. It can also be transmitted by blood or blood products such as clotting factors. The risk of acquiring AIDS from a blood transfusion is approximately one per 30,000 transfusions, but this should decrease when methods of screening blood for antibodies against HTLV-III become widely used. Currently, fewer than 6 percent of AIDS cases in the United States appear to have been acquired during sexual contact between heterosexual partners, but this mode of transmission may become more common as the disease spreads.

## FURTHER READINGS

### Books


BROWN, H., and F. A. NEVAA, *Basic Clinical Parasitology.* East Norwalk, Conn.: Appleton-Century-Crofts, 1983.

DAVIS, B. D., R. DULBECCO, H. N. EISEN, and H. S. GINSBERG, *Microbiology,* 3rd ed. New York: Harper & Row, 1980.

JAWETZ, E., J. L. MELNICK, and E. A. ADELBERG, *Review of Medical Microbiology,* 16th ed. Los Altos, Calif: Lange Medical Publications, 1984.

JOKLIK, W., H. P. WILLETT, and D. B. AMOS, *Zinsser Microbiology,* 18th ed. East Norwalk, Conn.: Appleton-Century-Crofts, 1984.

PETERSDORF, R. G., R. D. ADAMS, E. BRAUNWALD, K. J. ISELBACHER, J. B. MARTIN, and J. D. WILSON, eds., *Harrison's Principles of Internal Medicine,* 10th ed. New York: McGraw-Hill, 1983.

WILSON, G., and H. M. DICK, *Topley and Wilson's Principles of Bacteriology, Virology and Immunity,* 7th ed. Baltimore: Williams and Wilkins, 1984.


### Reviews


BACA, O. G., and D. PARETSKY, "Q Fever and *Coxiella burneti:* A Model for Host-Parasite Interactions," *Microbiol. Rev.* **47,** 127 (1983).

CUKOR, G., and N. R. BLACKLOW, "Human Viral Gastroenteritis," *Microbiol. Rev.* **48,** 157 (1984).

DONELSON, J. E., and M. J. TURNER, "How the Trypanosome Changes its Coat," *Scientific American* **252** (2), 44 (1985).

DRUTZ, D. J., and A. CATANZARO, "Coccidiomycosis," *Ann. Rev. Resp. Dis.* **117,** 559 (1978).

JOKLIK, W. K., "Structure and Function of the Reovirus Genome," *Microbiol. Rev.* **45,** 483 (1983).

PUTNAK, J. R., and B. A. PHILLIPS, "Picornaviral Structure and Assembly," *Microbiol. Rev.* **45,** 287 (1981).

UNNY, S. K., and B. L. MIDDLEBROOKS, "Streptococcal Rheumatic Carditis," *Microbiol. Rev.* **47,** 97 (1983).

WIESNER, P. J., and S. E. THOMPSON, "Gonococcal Disease," *Dis. Mo.* **26,** 11 (1980).


### Original articles


LUZZATTO, L., "Genetics of Red Cells and Susceptibility to Malaria," *Blood,* **54,** 961 (1979).

POPOVIC, M., M. G. SARNGADHARN, E. READ, and R. C. GALLO, "Detection, Isolation, and Continuous Production of Cytopathic Retroviruses (HTLV-III) from Patients with AIDS and Pre-AIDS," *Science* **224,** 497 (1984).

SARNGADHARN, M. C., M. POPOVIC, L. BRUCH, J. SCHUPBACH, and R. C. GALLO, "Antibodies Reactive with Human T-lymphotropic Retrovirus (HTLV-III) in the Serum of Patients with AIDS," *Science* **224,** 506 (1984).

# Chapter 33
# The Exploitation of Microorganisms by Humans

The role of microorganisms in the transformations of organic matter was not recognized until the middle of the nineteenth century. Nevertheless, microbial processes have been used by humans since prehistoric times in the preparation of food, drink, and textiles; in many cases, these processes became controlled and perfected to an astonishing degree by purely empirical methods. The outstanding examples of *traditional microbial processes* are those used in the production of beer and wine; the pickling of certain plant materials; the leavening of bread; the making of vinegar, cheese, and butter; and the retting of flax. The rise of microbiology, which revealed the nature of these traditional processes, led not only to great improvements in many of them, but also to the development of entirely new industries based on the use of microorganisms that had previously not been exploited by humans.

## TRADITIONAL MICROBIAL PROCESSES UTILIZING YEASTS

Yeasts traditionally and still play an important role from a technical and industrial standpoint. Although many genera and species of yeast exist in nature and many of them are used industrially, the yeasts of greatest technical importance are strains of *Saccharomyces cerevisiae*. They are used in the manufacture of wine and beer and in the leavening of bread.

The manufacture of alcoholic beverages was already well established in early civilizations, most of which had myths about the origin of wine making that attributed the discovery to divine intervention. This suggests that even in very ancient times, the beginnings of the art were already shrouded in prehistoric darkness. The use of yeast as a leavening agent for bread originated in Egypt about 6,000 years ago, and spread slowly from there to the rest of the western world.

The discovery that alcohol can be distilled, and so concentrated, originated either in China or the Arab world. Distilleries began to appear in Europe in the middle of the seventeenth century. At first the alcohol manufactured was used only for human consumption, but with the industrial revolution, the demand for alcohol as a solvent and chemical raw material developed and the distilling industry grew very rapidly.

## The Making of Wine

The making of wine involves the fermentation of the soluble sugars (glucose and fructose) of the juices of grapes into $CO_2$ and ethyl alcohol. After the grapes are harvested, they are crushed to produce a raw juice or *must*, a highly acidic liquid containing 10 to 25 percent sugar by weight. In many parts of the world the mixed yeast flora on the grapes serves as the inoculum for the fermentation that converts the must into wine. In such a natural fermentation a complex succession of changes in the yeast population occurs; in the later stages the so-called true wine yeast, *Saccharomyces cerevisiae* var. *ellipsoides*, predominates. In other areas, California, for example, the must is first treated with sulfur dioxide, which virtually eliminates the natural yeast flora; it is then inoculated with the desired strain of wine yeast. The fermentation proceeds vigorously, usually being completed in a few days. Often it is necessary to control the rate of the fermentation or to cool the fermenting mixture, in order to prevent a rise of temperature which would affect the quality of the wine or even kill the yeast. Must from both red and white wine grapes (*Vitis vinifera*) is white and results in a white wine. Since the color of red grapes is in the skin, red wines are made by fermentation in the presence of the skins. The alcohol developed during fermentation extracts the color into the wine. Following fermentation the new wines must be clarified, stabilized, and aged to produce a satisfactory final product. These processes require months, and for high quality red wines, even years. During the first year, many wines (particularly red) undergo a second spontaneous fermentation, the *malo-lactic fermentation*, which can be caused by a variety of lactic acid bacteria (*Pediococcus*, *Leuconostoc*, or *Lactobacillus*). This fermentation converts malic acid, one of the two major organic acids of grapes, to lactic acid and $CO_2$, thus converting a dicarboxylic acid to a monocarboxylic acid, thereby reducing the acidity of the wine. Although the malo-lactic fermentation proceeds spontaneously, slowly, and undramatically (sometimes even without the winemaker's knowledge), it is absolutely vital to produce red wines of good quality from grapes grown in cool districts, which otherwise yield wines with too high an initial acidity to be palatable.

Certain special types of wine undergo additional microbial transformations. Sparkling wines (champagne types) undergo a second alcoholic fermentation under pressure at the expense of added sugar, either in the bottle or in bulk; the $CO_2$ thus produced carbonates the wine. The secondary fermentation is conducted with varieties of wine yeast which readily clump following fermentation and are consequently easily removed. Sherries (wines of the type produced in the Jerez district of Spain) are fortified with alcohol to about 15 percent, exposed to air, and allowed to develop a heavy surface growth of certain yeasts which impart the unique sherry flavor to the wine.

Some European sweet wines, notably those from the Sauternes district of France, undergo even more complex microbial transformations. Prior to picking, the grapes become spontaneously infected with a fungus, *Botrytis cinerea*. This infection causes water loss (thus increasing sugar content) and destruction of malic acid (thus decreasing the acidity of the grapes). Certain favorable changes of flavor and color occur. The resulting very sweet must from these infected grapes is fermented by so-called *glucophilic yeasts*, i.e., yeasts that rapidly ferment the glucose leaving residual fructose (the sweeter of the two sugars). The product is a sweet dessert-type wine.

Although the high alcohol content and low pH ($\sim 3.0$) of wines make them unfavorable substrates for growth of most organisms, they are subject to microbial spoilage. The problem of the "diseases" of wines was first scientifically explored by Pasteur, whose descriptions of the responsible organisms and recommendations for preventing their development are still valid today. The most serious spoilage problems are those that occur if wines are exposed to air. Film-forming yeasts and acetic acid bacteria grow at the expense of the alcohol, converting it to acetic acid, thus souring the wine. Serious diseases can also be caused by fer-

**TABLE 33.1**

**Partial List of Microorganisms Involved in the Production or Spoilage of Wines**

| Organism | Role(s) in Wine Making or Spoilage | Chemical or Physical Change Effected |
|---|---|---|
| *Saccharomyces cerevisiae* var. *ellipsoides* | 1. Primary alcoholic fermentation<br>2. Carbonation of sparkling wines by a secondary fermentation<br>3. Clouding of sweet wines | Glucose and/or fructose $\longrightarrow$ ethanol + $CO_2$ |
| *Pediococcus*, *Leuconostoc*, and *Lactobacillus* | Malo-lactic fermentation | 1. Malic acid $\longrightarrow$ lactic acid + $CO_2$<br>2. Flavor enrichment |
| Flor sherry yeasts (*Saccharomyces beticus*, *S. cheresiensis*, and *S. fermenti*) | Grow as heavy surface layer (Flor) to produce sherry flavor | 1. Oxidize ethanol to acetaldehyde<br>2. Produce flavor components |
| *Botrytis cinerea* | Grows in certain regions (e.g., Sauternes) on the surface of grapes used to produce sweet wines | 1. Desiccates grapes<br>2. Oxidizes malic acid to $CO_2$ and $H_2O$<br>3. Adds flavor and color |
| Acetic acid bacteria and film-forming yeasts | Spoil wine exposed to air | Oxidize ethanol to acetic acid |
| Lactic and bacteria, notably *Lactobacillus trichodes* | Spoil wine anaerobically | Produce "mousy" flavor |

mentative organisms in the absence of air. Rod-shaped lactic acid bacteria can grow anaerobically at the expense of residual sugar and impart a "mousy" taste to the wine. Wine yeasts can grow in sweet wines even after bottling; although such growth does not alter the flavor, the wine becomes cloudy and hence less attractive. Wine spoilage can be prevented by pasteurization, but this diminishes quality. Wines of low alcohol content that also contain sugar are particularly subject to spoilage, now commonly prevented either by chemical additives such as sulfur dioxide or by sterilization through filtration. The roles of certain microorganisms in the manufacture and spoilage of wines are summarized in Table 33.1.

## The Making of Beer

Beers are manufactured from grains which, unlike grape or other fruit musts, contain no fermentable sugars. The starch of the grains must be *saccharified* (hydrolyzed to the fermentable sugars, maltose, and glucose) prior to fermentation by yeasts. Three principal grains were traditionally used for the production of beer: barley in Europe, rice in the Orient, and corn in the Americas. In each case, a different solution to the saccharification of starch was found. In the case of barley, starch-hydrolyzing enzymes (amylases) of the grain itself were used. Barley seeds contain little or no amylase, but upon germination, large amounts of amylase are formed. Hence, barley is dampened, allowed to germinate, and is then dried and stored for subsequent use. Such dried, germinated barley, called *malt*, is dark in color as a result of the exposure to increased temperatures during drying and has more flavor than untreated barley seeds. The starch of barley remains largely unaffected by the malting process. Hence, the first step in beer making is the grinding of malt and its suspension in water to allow hydrolysis of the starch. Malt itself is sometimes used as the total source of starch, or, if a lighter beer is desired, unmalted barley or some other cereal grain is added to the saccharifying mixture. In the United States large quantities of rice are used in the manufacture of beer. Concomitantly with the hydrolysis of starch, other enzymatic processes occur, including the hydrolysis of proteins. After saccharification has reached the desired stage, the mixture is boiled to stop further enzymatic changes and it is then filtered. Hops (the pistillate inflorescence bracts of the vine, *Humulus lupus*) are added to the filtrate (*wort*) and contribute a soluble resin, which imparts the characteristic bitter flavor of beer and which also acts as a preservative against the growth of bacteria. The use of hops is a relatively recent modification of the art of beer making, having been introduced about the middle of the sixteenth century; even today, unhopped beer is made in some countries. After filtration, the hopped wort is ready for fermentation.

In contrast to wine fermentations, beer fermentations are always heavily inoculated with special strains of yeast derived from a previous fermentation. The fermentation proceeds at low temperatures for a period of 5 to 10 days. All yeasts used in making beer are *Saccharomyces cerevisiae*, but not all strains of *S. cerevisiae* can be used to make good beer. During the course of time, special strains with desirable properties have been selected, known as *brewer's yeast*. Before Pasteur, the selection and maintenance of good yeast strains was an empirical art. The success of a brewer depended largely on his ability to obtain a suitable strain and propagate it from batch to batch without its becoming too heavily contaminated by undesirable microorganisms. Good brewer's yeasts were developed over a period of centuries; they cannot be found in nature. Like cultivated higher plants, they are a product of human art, and in recognition of this fact, the brewer refers to other yeasts (including other strains of *S. cerevisiae*) as "wild yeasts."

Since Pasteur's time the recognition, testing, and maintenance of good strains of brewer's yeast have been placed on a scientific basis. The pioneer of this work was E. C. Hansen, who worked in the Carlsberg Brewery in Copenhagen. Strains of brewer's yeast fall into two principal groups known as *top* and *bottom* yeasts. Top yeasts are vigorous fermenters, acting best at relatively high temperatures (20° C), and are used for making heavy beers of high alcoholic content, such as English ales. Their name derives from the fact that during fermentation they are swept to the top of the vat by the rapid evolution of $CO_2$. In contrast, bottom yeasts are slow fermenters, act best at lower temperatures (12° to 15° C), and produce lighter beers of low alcoholic content of the type commonly made in the United States. Their name derives from the fact that the slower rate of $CO_2$ evolution allows them to settle to the bottom of the vat during fermentation.

The diseases of beer, like those of wine, were first scientifically studied by Pasteur. They occur most commonly following fermentation, either during maturation or following bottling. One agent is a wild yeast, *Saccharomyces pasteurianus*, which imparts a disagreeable bitterness to beer. Lactic acid bacteria sometimes make beer acidic and cloudy. They develop principally when the temperature becomes too high during maturation and storage. Acetic acid bacteria may at times also cause souring, particularly in barreled beer exposed to air. The principal methods of avoiding spoilage are the use of pure yeast strains as starters and pasteurization of the final product. Some beers are now sterilized by filtration prior to bottling, a process which avoids the slight damage to flavor that results from pasteurization.

Although wines from grapes and beer from barley are the characteristic fermented beverages of the western world, rice serves as the source of most fermented beverages (e.g., sake) in Asia. The problem of hydrolyzing rice starch as a preliminary to fermentation has been solved by the use of amylases from molds, principally *Aspergillus oryzae*. In the manufacture of sake the first step is the preparation of a culture of the mold. Mold spores, saved from a previous batch, are sown on steamed rice and are allowed to grow until the mass of rice is thoroughly permeated with mycelium. This material (*koji*), which serves both as a source of amylase and as an inoculum, is added to a larger batch of steamed rice mixed with water. Hydrolysis of the starch proceeds, and when sufficient sugar has accumulated, a spontaneous alcoholic fermentation begins. Lactic acid bacteria as well as yeasts are present in koji, so lactic acid is produced in addition to alcohol and $CO_2$. The production of alcoholic beverages from grain in Asia thus differs from the western process in two respects: saccharification is effected by microorganisms, and saccharification proceeds simultaneously with fermentation.

In the Americas as well as in certain regions of the Middle East, yet a third agent of saccharification is used—human saliva, which contains amylases. Indians in Central and South America prepare a corn beer by chewing the grains and spitting the mixture into a vessel, where it is allowed to undergo a spontaneous alcoholic fermentation.

## The Making of Bread

An alcoholic fermentation by yeasts is an essential step in the production of raised breads; this process is known as the leavening of bread (after the old word for yeast, "leaven"). The moistened flour is mixed with yeast and allowed to stand in a warm place for several hours. Flour itself contains almost no free sugar to serve as a substrate for fermentation, but there are some starch-splitting enzymes present that produce sufficient sugar to support leavening. In the highly refined flours commonly used in the United States, these enzymes have been destroyed and sugar must be added to the dough. The sugar is rapidly fermented by yeast. The carbon dioxide produced becomes entrapped in the dough, causing it to rise, while the alcohol produced is driven off during the baking process.

Yeast produces other more subtle changes in the physical and chemical properties of the dough, which affect the texture and flavor of the bread. This fact became evident when J. von Liebig, a German chemist of the last century, invented baking powder, a mixture of chemicals that produces carbon dioxide when moistened. Liebig anticipated that baking powder would replace yeast. Although Liebig's invention is widely used in other forms of baking, it did not supplant yeast as a leavening agent for bread.

The yeasts used in bread making all belong to the species *Saccharomyces cerevisiae* and have been derived historically from strains of top yeasts used in brewing. Until the nineteenth century, yeasts for bread making were obtained directly from the nearest brewery. The commercial production of compressed yeast by industry was greatly stimulated by the application of mass production techniques to bread making. A large modern bakery may use many hundreds of pounds of yeast daily, for about 5 pounds of yeast are required to leaven 300 pounds of flour. Much of the baker's yeast manufactured today is dried under controlled conditions that maintain viability of the yeast cells, a treatment which facilitates shipment and storage.

## TRADITIONAL MICROBIAL PROCESSES UTILIZING ACETIC ACID BACTERIA

When wine and beer are exposed to the air, they frequently turn sour. Souring is caused by the oxidation of alcohol to acetic acid, mediated by the strictly aerobic acetic acid bacteria. The spontaneous souring of wine is the traditional method of manufacturing vinegar. (The word vinegar is derived from the French "vinaigre" which literally means "sour wine.")

The manufacture of vinegar still remains largely empirical. The principal modifications introduced during the past century concern the mechanical rather than the microbiological aspects of the process. In the traditional *Orleans process*, which is still used in France, wooden vats are partially filled with wine, and the acetic acid bacteria develop as a gelatinous pellicle on the surface of the liquid. The conversion of ethanol to acetic acid takes several weeks—the rate of the process being limited by the slow diffusion of air into the liquid. The survival of this slow and inefficient method is attributable to the high quality of the product.

When the taste of the product is not of primary importance, vinegar is made by more rapid methods from cheaper raw materials (e.g., diluted distilled alcohol and cider). These methods are designed to accelerate oxidation by improved aeration and regulation of temperature, but they remain microbiologically uncontrolled. The oldest such method, developed early in the nineteenth century, utilizes a tank that is loosely filled with wooden shavings through which the liquid is circulated. The liquid is trickled into the tank and air is blown countercurrent to the liquid flow. The acetic acid bacteria develop as a thin film on the wooden shavings, thus providing a large surface of cells which are simultaneously exposed to the medium and to air. Once a bacterial population has become established on the shavings, successive batches of vinegar can be produced quickly; solutions initially containing 10 percent alcohol can be converted to acetic acid in four or five days. Much vinegar is still made by this method, but modern, stirred, deep tank fermentors, similar to those used to produce antibiotics, are now being introduced.

The oxidation of ethanol to acetic acid is an example of the *incomplete oxidations* carried out by acetic acid bacteria. Certain other incomplete oxidative conversions by acetic acid bacteria are industrially important. Gluconic acid, which is used by the pharmaceutical industry, is made by oxidation of glucose by acetic acid bacteria. Many sugar alcohols are converted to sugars by acetic acid bacteria. One such reaction in commercial use is the production of sorbose from sorbitol. Sorbose is used as a suspending agent for certain pharmaceuticals, and it is an intermediate in the manufacture of L-ascorbic acid (vitamin C).

## THE USES OF LACTIC ACID BACTERIA

Lactic acid bacteria produce large amounts of lactic acid from sugar. The resulting decrease in pH renders the medium in which they have grown unsuitable for the growth of most other microorganisms.

Growth of lactic acid bacteria, therefore, is a means of preserving food; in addition, they produce flavor components.

## Milk Products

The manufacture of such milk products as butter, cheese, and yogurt involves the use of microorganisms, among which the lactic acid bacteria are particularly important. The discovery of the roles played by microorganisms in the preparation of these foods has led to the development of a special branch of bacteriology known as *dairy bacteriology*.

Many lactic acid bacteria occur normally in milk and are responsible for its spontaneous souring. Milk souring provides a means of preserving this otherwise highly unstable foodstuff, and the manufacture of cheese and other fermented milk products undoubtedly began largely as a means of preservation.

The manufacture of cheese involves two main steps: *curdling* the milk proteins to form a solid material from which the liquid is drained away; and the *ripening* of the solid curd by the action of various bacteria and fungi, although certain fresh cheeses are essentially unripened.

The curdling process may be exclusively microbiological, since acid production of lactic acid bacteria is sufficient to coagulate milk proteins. However, an enzyme known as *rennin* (extracted from the stomachs of calves) which curdles milk is also often used for this purpose.

The subsequent ripening of the curd is a very complex process, and is highly variable, depending on the kind of cheese being made. The ripening process is chemically variable. In the young cheese, all nitrogen is present in the form of insoluble protein, but as ripening proceeds, the protein is progressively cleaved to soluble peptides and ultimately to free amino acids. The amino acids can be further decomposed to ammonia, fatty acids, and amines. In certain cheeses, protein breakdown is restricted. For example, in Cheddar and Swiss cheese, only 25 to 35 percent of the protein is converted to soluble products. In soft cheeses, such as Camembert and Limburger, essentially all the protein is converted to soluble products. In addition to changes in the protein components, ripening involves considerable hydrolysis of the fats present in the young cheese. The enzymes present in the rennin preparation contribute somewhat to the ripening process, but microbial enzymes in the cheese play the major role. The types of microorganism involved are varied. Hard cheeses are ripened largely by lactic acid bacteria, which grow throughout the cheese, die, autolyze, and release hydrolytic enzymes. Soft cheeses are ripened by the enzymes from yeasts and other fungi that grow on the surface.

Some microorganisms play highly specific roles in the ripening of certain cheeses. The blue color and unique flavor of Roquefort cheese is a consequence of the growth of a blue-colored mold, *Penicillium roqueforti*, throughout the cheese.* The characteristic holes in Swiss cheese are formed by carbon dioxide, a product of the propionic acid fermentation of lactic acid by species of *Propionibacterium*.

Butter manufacture is also in part a microbiological process, since an initial souring of cream, caused by milk streptococci, is necessary for subsequent separation of butterfat in the churning process. These organisms produce small amounts of acetoin which is spontaneously oxidized to *diacetyl*, the compound responsible for the flavor and aroma of butter. Since streptococci differ markedly in their ability to produce acetoin, it has become common practice to inoculate pasteurized cream with pure cultures of selected strains.

In many parts of the world, milk is allowed to undergo a mixed fermentation by lactic acid bacteria and yeasts which produces a sour, mildly alcoholic beverage (e.g., kefir and kumiss).

The roles of microorganisms in the manufacture of milk products are summarized in Table 33.2.

## The Lactic Fermentation of Plant Materials

Certain lactic acid bacteria are found characteristically on plant materials. These organisms are responsible for the souring that occurs in the preparation of pickles, sauerkraut, and Spanish-style olives. In these lactic acid fermentations, sugars initially present in the plant materials serve as the fermentable substrates. The lactic acid produced imparts flavor to the product as well as protecting it from further microbial attack.

The preservative value of a lactic acid fermentation is also exploited in the ensilaging of green cattle fodder. After plant materials have undergone fermentation in a silo, they may be kept indefinitely without risk of decomposition.

## Dextran Production

Some lactic acid bacteria belonging to the genus *Leuconostoc* produce large amounts of an extracellular polysaccharide known as dextran when

* In the United States a white mutant of *Penicillium roqueforti* is sometimes used to produce a mold-ripened cheese for those who find the flavor desirable but the color objectionable.

**TABLE 33.2**

**Microbiology of Milk Products**

| Product | Process | Microorganisms[a] |
|---|---|---|
| Buttermilk | Lactic acid fermentation | *Lactobacillus bulgaricus* |
| Yogurt | Lactic acid fermentation | *L. bulgaricus* + *Streptococcus thermophilus* |
| Kefir | Alcoholic and lactic acid fermentations | *Streptococcus lactis* + *L. bulgaricus* + lactose-fermenting yeasts |
| Cheeses (in general) | Initial lactic acid | |
| | fermentation temperature, 35° C | *S. lactis* or *S. cremoris* |
| | fermentation temperature, 42°C | Various thermophilic lactic acid bacteria, principally lactobacilli |
| Hard cheeses (e.g., Cheddar and Swiss) | Proteolysis and lipolysis | Various lactic acid bacteria within the cheese |
| Soft cheeses (e.g., Camembert, Brie, and Limburger) | Proteolysis and lipolysis | Surface growth, initially of fungi (*Geotrichum candidum* and *Penicillium* spp.), sometimes followed by *Bacterium linens* and *B. erythrogenes* |
| Swiss cheese | Propionic acid fermentation | *Propionibacterium* spp. |
| Roquefort | Lipolysis and production of blue mold pigment | *Penicillium roqueforti* |

[a] Microorganisms generally associated with the process.

grown with sucrose. Dextran is a polyglucose of high but variable molecular weight (15,000 to 20,000,000); the average molecular weight varies with the strain employed. These lactic acid bacteria first came to the attention of industrial microbiologists for their nuisance value; they occasionally develop in sugar refineries, and the large amounts of gummy polysaccharide produced may literally clog the works.

Dextran is now produced industrially, following the discovery that dextran derivatives that have been chemically cross-linked to make them insoluble in water can act as molecular sieves. Columns of such modified dextrans (marketed largely under the trade name of *Sephadex*) retard the passage of small molecules, and thus permit the physical fractionation of solutes that differ in molecular weight. Sephadex columns can be used for molecular weight determinations in the range of 700 to 800,000 daltons, after calibration with compounds of known molecular weight.

Another class of microbial polysaccharides now being produced industrially are the chemically complex extracellular polysaccharides synthesized by aerobic pseudomonads of the *Xanthomonas* group. These substances have the physical property of forming thixotropic gels, and in addition, are stable at relatively high temperatures. As a result, they have a wide variety of uses, among many others, as gelling agents for prepared foods such as salad dressings, ice cream, and frostings; as lubricants in the drilling of oil wells; and as gelling agents in paints with a water base.

## THE USES OF BUTYRIC ACID BACTERIA

### The Retting Process

Retting is a controlled microbial decomposition of plant materials designed to liberate certain components of the plant tissue. The oldest retting process, which has been used for several thousand years, is the retting of flax and hemp to free the bast fibers used in the making of linen, jute, and rope. These fibers, made up of cellulose, are held together in the plant stem by a cementing substance, pectin; their physical separation is difficult. The goal of retting is to bring about decomposition of the pectin, thus freeing the fibers without simultaneous decomposition of the fibers themselves. The plant stems are immersed in water; they become water logged and microbial decomposition begins. Initially, aerobic microorganisms develop and use up the dissolved oxygen, making the environment suitable for the subsequent development of the anaerobic butyric acid bacteria. These organisms rapidly attack the plant pectin, freeing the fibers. If retting is unduly prolonged, cellulose-fermenting bacteria will also develop and destroy the fibers. An analogous retting process is used in the preparation of potato starch. Its purpose is to free the starch-containing cells in the potato tuber from the pectin in which they are embedded.

### The Acetone-Butanol Fermentation

Certain *Clostridium* spp. have been used on a very large scale for the production of the industrial solvents, acetone and butanol. Many clostridia carry out a fermentation of sugars with the formation of carbon dioxide, hydrogen, and butyric acid. Some carry out further reactions, converting the butyric acid to butanol and the acetic acid to ethanol and acetone. The commercial development of the so-called *acetone-butanol fermentation* mediated by *Clostridium acetobutylicum* began in England just before World War I and expanded rapidly during the war because acetone was needed as a solvent in the manufacture of explosives. After World War I, the demand for acetone diminished, but the process survived because another major product of the fermentation, butanol, found a use as a solvent for the rapid drying of nitrocellulose paints in the growing automobile industry. A byproduct of the fermentation, the vitamin, riboflavin, also helped to maintain its commercial feasibility.

Today, this industry has virtually disappeared as a result of the development of competing methods, only in part microbiological, for the synthesis of the major products. Both acetone and butanol are produced in large amounts from petroleum; a microbiological process based on the use of yeasts is the principal source of riboflavin.

Recently with the rising cost of petroleum and with genetic improvements in the strains of clostridia, it again appears that the acetone-butanol fermentation might become commercially feasible.

The acetone-butanol fermentation made important technological contributions to industrial microbiology. It was the first large-scale process in which the exclusion of other kinds of microorganisms from the culture vessel was of major importance to the success of the operation. The medium used for the cultivation of *Clostridium acetobutylicum* is also favorable for the development of lactic acid bacteria; if these organisms begin to grow, they rapidly inhibit the further growth of the clostridia through lactic acid formation. An even more serious problem is infection with bacterial viruses, to which clostridia are highly susceptible. Thus, the acetone-butanol fermentation can be operated successfully only under conditions of careful microbiological control. The establishment of this industry led to the first successful use of pure culture methods on a mass scale, which were later improved and refined in connection with the industrial production of antibiotics.

## MICROBES AS SOURCES OF PROTEIN

Because of their rapid growth, high protein content, and ability to utilize organic substrates of low cost, microorganisms are potentially valuable sources of animal food. The growth of the science of animal nutrition has led to the development of a new industry, based on the cultivation of microorganisms for use as a supplement in animal feeds.

Yeasts and methylotrophic bacteria are the principal organisms that have been used and the generic name for the microbial product is *single-cell protein.* The proposed use of single-cell protein is as a supplement for animal feed to replace the other major supplements, soybeans and fishmeal. Since Europeans consume a high-meat diet, but live in an area that is unsuitable for growing soybeans, and must import fishmeal, they quite naturally have played a major role in the development of processes to manufacture single-cell protein.

### Production of Yeasts from Petroleum

The cost of raw material is a factor of paramount importance in the production of microorganisms for use as food, and cheap sources of carbohydrate (e.g., whey, molasses, paperplant waste) were initially used for growth of food yeasts. However, since aerobic growth conditions are used, all compounds that can support respiratory metabolism may serve. This led to the development of processes that utilize petroleum as a substrate. Petroleum is still very cheap, compared with other possible substrates, and since hydrocarbons are the most highly reduced of organic compounds, growth yields at their expense are extremely high.

The British Petroleum Corporation built an industrial unit in France for the cultivation of *Candida lipolytica* in an aqueous emulsion of crude petroleum. This yeast can oxidize aliphatic, unbranched hydrocarbons of chain length $C_{12}$ to $C_{18}$, compounds that comprise part of the complex mixture of alkanes present in crude petroleum. Their selective removal by the growth of *Candida lipolytica* produces a dewaxed petroleum that is much more easily refined. The economic feasibility of the

British Petroleum process depends on its twofold function: simplification of refining and protein production.

Other plants that produce yeast for a source of protein from petroleum have been built in Japan and Italy. In both countries concern that the product might contain residual petroleum hydrocarbons, including the carcinogen, benzypyrene, have caused the projects to be abandoned.

### Production of Bacteria from Petroleum

The use of methophilic bacteria to produce single-cell protein offers the advantage of using methane or methanol as substrate; these are inexpensive, abundant, and free of possible carcinogenic contaminants. Disadvantages of using these bacteria include their relatively slow growth rate (about five hours doubling time) and the fact that they must be cultivated at neutral pH rather than the relatively low pH tolerated by yeast; cultivation at neutral pH requires that rigorous and expensive procedures be employed to prevent contamination by other bacteria.

In 1976 Imperial Chemical Industries (ICI) began construction of a plant in Billingham, England, to make single-cell protein by growing the methylotroph *Methylophilus methylotrophyus* on methanol in a single huge fermentor ($1.5 \times 10^6$). The fermentor incorporates a variety of innovative features. Air is introduced in the base, and owing to the considerable hydrostatic pressure within the fermentor an increased concentration of dissolved oxygen and thereby an increased growth rate can be maintained. Methanol is injected through 3,000 ports inside the fermentor. The huge fermentor is operated with rigorous asepsis as a continuous growth chamber: a portion of the contents is removed constantly; cells are removed; and the re-sterilized liquid is returned to the fermentor. In spite of the technical success of the process, it is not yet clear if it will be able to compete economically with natural sources of protein.

### Production of Specific Amino Acids

The great potential value of microorganisms as foods or feed supplements lies in their high protein content. This makes them the best agents for the rapid and efficient conversion of other more readily available organic compounds into protein, of which the world is becoming critically short. This point becomes evident when protein production by cattle and by yeast is compared. A bullock weighing 500 kg produces about 0.4 kg of protein in 24 hours. Under favorable growth conditions, 500 kg of yeast produce over 50,000 kg of protein in the same period.

Many plant foods contain sufficient protein to supply the quantitative needs of mammals, but they cannot serve as sole sources of dietary protein because their proteins are deficient in certain specific amino acids required by mammals. Wheat protein is low in lysine, rice protein in lysine and threonine, corn protein in tryptophan and lysine, bean and pea protein in methionine. The addition of the deficient amino acid(s) to diets that contain a single source of vegetable protein will render them adequate. The practicality of fortifying diets of vegetable protein with individual amino acids has been amply demonstrated in numerous experiments with both animals and humans. Thus, the world shortage of certain specific amino acids—notably, lysine, threonine, and methionine—is more critical than the shortage of total protein. The microbial production of *specific amino acids* has therefore been intensively studied.

Since the metabolism of microorganisms is precisely regulated (Chapter 12), microorganisms normally synthesize quantities of amino acids just sufficient to meet their growth requirements. However, naturally occurring and mutant strains of some microorganisms have defective mechanisms for the regulation of specific biosynthetic pathways and, as a consequence, excrete large amounts of certain amino acids into the medium. Methods for the microbial production of nutritionally important amino acids are now available and are constantly being improved.

## THE MICROBIAL PRODUCTION OF CHEMOTHERAPEUTIC AGENTS

The period since World War II has seen the establishment and extremely rapid growth of a major new industry, the use of microorganisms for the synthesis of chemotherapeutic agents, particularly antibiotics and hormones. The development of this industry has had a dramatic and far-reaching social impact. Nearly all bacterial infectious diseases that were, prior to the antibiotic era, major causes of human death have been brought under control by the use of these drugs. In the United States, bacterial infection is now a less frequent cause of death than suicide or traffic accidents.

## The Rise of Chemotherapy

The importance of acquired immunity as a means of protection against specific bacteriological diseases was recognized shortly after the discovery of the role of microorganisms as the etiological agents of infectious diseases (see Chapter 1). For several decades thereafter control of infectious disease was based exclusively on the use of antisera and vaccines, and was largely preventative; usually, little could be done to cure infections after they had appeared.

A different kind of approach to the control of infectious disease was developed by the German physician-chemist Paul Ehrlich, who initiated an empirical search for synthetic chemicals that possess *selective toxicity* for pathogenic microorganisms. He coined the word *chemotherapy* to describe this approach to the control of infectious disease. A few years earlier, in 1905, H. Thompson had discovered that an arsenic-containing organic compound, termed *atoxyl*, was effective in treating trypanosomiasis, a protozoal disease common in parts of Africa and South America. Erlich set out systematically to modify the structure of *atoxyl* with the hope of finding a compound that would be selectively toxic to bacteria. His efforts produced one limited success; the 606th compound he tested, *salvarsan*, proved to be effective in the treatment of syphilis and other spirochetal infections, but it was far from an ideal drug: it produced severe side effects; its low solubility necessitated the intravenous injection of volumes as great as 800 ml; treatment had to be extended over periods of months and even years.

In spite of these difficulties, treatment of spirochetal infections with salvarsan remained the only effective chemotherapy of infectious disease until the 1930s.

The next significant advance in chemotherapy was also made empirically. Large numbers of aniline dyes were screened for antibacterial chemotherapeutic activity and one substance of this class, *prontosil*, was found to be effective. However, prontosil possessed no antibacterial action in vitro. Its antibacterial activity in infected animals was then shown to be attributable to a colorless breakdown product, *sulfanilamide*, formed in the animal body. Sulfanilamide possesses antibacterial activity both *in vitro* and *in vivo*. D. D. Woods observed that the inhibition of bacterial growth by sulfanilamide can be reversed by a structural analogue *p*-aminobenzoic acid (Figure 33.1).

Woods then made a brilliant series of deductions: that *p*-aminobenzoic acid is a normal constituent of the bacterial cell; that it has a coenzymatic function; and that this function is blocked by sulfanilamide as a result of its steric resemblance to *p*-aminobenzoic acid. In fact, *p*-aminobenzoic acid proved to be not a coenzyme, but a biosynthetic precursor of the coenzyme folic acid; sulfanilamide blocks its conversion to this end product. Sulfanilamide is selectively toxic because most bacteria must synthesize folic acid *de novo*, whereas mammals obtain it from dietary sources.

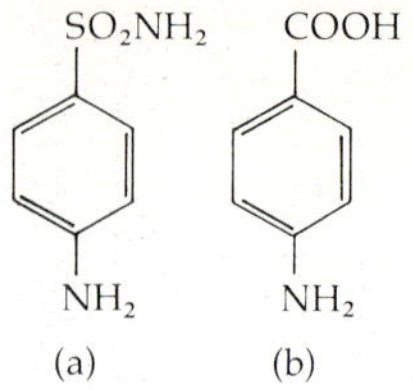

$SO_2NH_2$ $COOH$

$NH_2$ $NH_2$

(a) (b)

FIGURE 33.1
The structures of (a) sulfanilamide and (b) *p*-aminobenzoic acid.

Woods' work appeared to offer a *rational approach to chemotherapy* through the synthesis of analogues of known essential metabolites. In succeeding years, thousands of structural analogues of amino acids, purines, pyrimidines, and vitamins were synthesized and tested; but very few useful chemotherapeutic agents were discovered.

## The Discovery of Antibiotics

The great modern advances in chemotherapy have come from the chance discovery that many microorganisms synthesize and excrete compounds which are selectively toxic to other microorganisms. These compounds, called *antibiotics*, have revolutionized modern medicine.

The first chemotherapeutically effective antibiotic was discovered in 1929 by Alexander Fleming, a British bacteriologist who had long been interested in the treatment of wound infections. On returning from a vacation in the country, he noticed among a pile of petri dishes on his bench one that had been streaked with a culture of *Staphylococcus aureus* was also contaminated by a single colony of mold. As Fleming observed this plate he reportedly said, "That's funny" because the colonies immediately surrounding the mold were transparent and appeared to be undergoing lysis. He reasoned that the mold was excreting into the medium a chemical that caused the surrounding colonies to lyse. Sensing the possible chemotherapeutic significance of his observation, he isolated the mold, which proved to be a species of *Penicillium*, and established that culture filtrates contained an antibacterial substance which he called *penicillin*.

Although it has often been suggested that many bacteriologists must have observed petri dishes that were similarly contaminated and therefore similar in appearance to Fleming's now very famous one, such speculation is undoubtedly false. As subsequent experiments have shown, a highly unusual series of events must have occurred in order to produce the results seen on Fleming's plate: contamination by the mold must have occurred at the time the plate was streaked with bacteria (prior growth of either would have prevented growth of the other in the immediate vicinity); the inoculated petri dish must not have been incubated (if it had been, the bacterium would have outgrown the mold); the room temperature of the laboratory must have been below 68° F (a temperature that probably did occur during a brief cold storm in London in the summer of 1928).

Penicillin proved to be chemically unstable, and Fleming was unable to purify it. Working with impure preparations, he demonstrated its remarkable effectiveness in inhibiting the growth of many Gram-positive bacteria, and he even used it with success for the local treatment of human eye infections. In the meantime, the chemotherapeutic effectiveness of sulfanilamides had been discovered, and Fleming, discouraged by the difficulties of penicillin purification, abandoned further work on the problem.

Ten years later, a group of British scientists headed by H. W. Florey and E. Chain resumed the study of penicillin. Clinical trials with partly purified material were dramatically successful. By this time, however, Britain was at war; and the industrial development of penicillin was undertaken in the United States, where an intensive program of research and development was begun in many laboratories. Within three years penicillin was being produced on an industrial scale, an astonishing achievement in view of the many difficulties which had to be overcome. Penicillin remains one of the most effective chemotherapeutic agents for treatment of many bacterial infections.

Rather than being a single substance, penicillin proved to be a class of compounds, the production of the particular representative of which varied with the medium in which the mold was grown. The various penicillins differ with respect to composition of the side chain (Figure 33.2, R-group). The penicillin that was first isolated in Peoria, Illinois, designated *penicillin G*, carried a benzyl side chain (probably because corn steep liquor, which is rich in phenylalanine, and its degradation products were used as a component of the medium); the penicillin isolated soon thereafter in England and designated *penicillin F* carried an isopentanyl side chain. By varying the composition of the medium, a variety of penicillins collectively termed *biosynthetic penicillins* have been synthesized, but none proved superior to penicillin G. Later it became possible to remove the side chain of penicillin G and replace it by chemical means with a large variety of substituents, thereby producing a set of compounds termed *semisynthetic penicillins* (Figure 33.3). Some of these have proven quite effective. *Penicillin V* is resistant to acid and therefore can be administered orally without its being inactivated in the stomach; *ampicillin* is acid-resistant and effective against enteric bacteria; *oxacillin* is resistant to the action of $\beta$-lactamase, the enzyme produced by certain "penicillin-resistant" strains of bacteria.

The remarkable chemotherapeutic efficacy of penicillin for certain bacterial infections, primarily those caused by Gram-positive bacteria, prompted intensive searches both at universities and in industry for new antibiotics. Soon a second clinically important antibiotic, *streptomycin*, which is effective against both Gram-negative bacteria and *Mycobacterium tuberculosis*, was discovered by A. Schatz and S. Waksman.

Streptomycin was the first example of an antibiotic possessing a *broad spectrum of activity*, effective against many Gram-positive and Gram-negative bacteria. Other antibiotics with even broader spectra of activity (for example, the tetracyclines) have been subsequently discovered. Antibiotics have proved to be less useful in the treatment of fungal infections: antifungal antibiotics such as nystatin and amphotericin B are considerably less effective therapeutically than their antibacterial counterparts, at least in part because their toxicity is far less selective. Good antiviral antibiotics are yet to be found.

Since 1945, thousands of different antibiotics produced by fungi, actinomycetes, or unicellular bacteria have been isolated and characterized. A small fraction of these are of therapeutic value; about 50 are currently produced on a large scale for medical and veterinary use. Their nomenclature is complicated, one antibiotic often being sold under several different names. There are two reasons for this proliferation of names. First, many antibiotics are members of a group of compounds, all of which possess similar structures; a name is required for the *class of compounds*, as well as for each *individual representative*. Second, each manufacturer of an antibiotic assigns to it for marketing purposes a *trade name* which, by law, only he can use. To protect a trade name for exclusive use, the law requires that another name, available for general use, be also assigned to the antibiotic in question; this is called the *generic name*. The generation of multiple

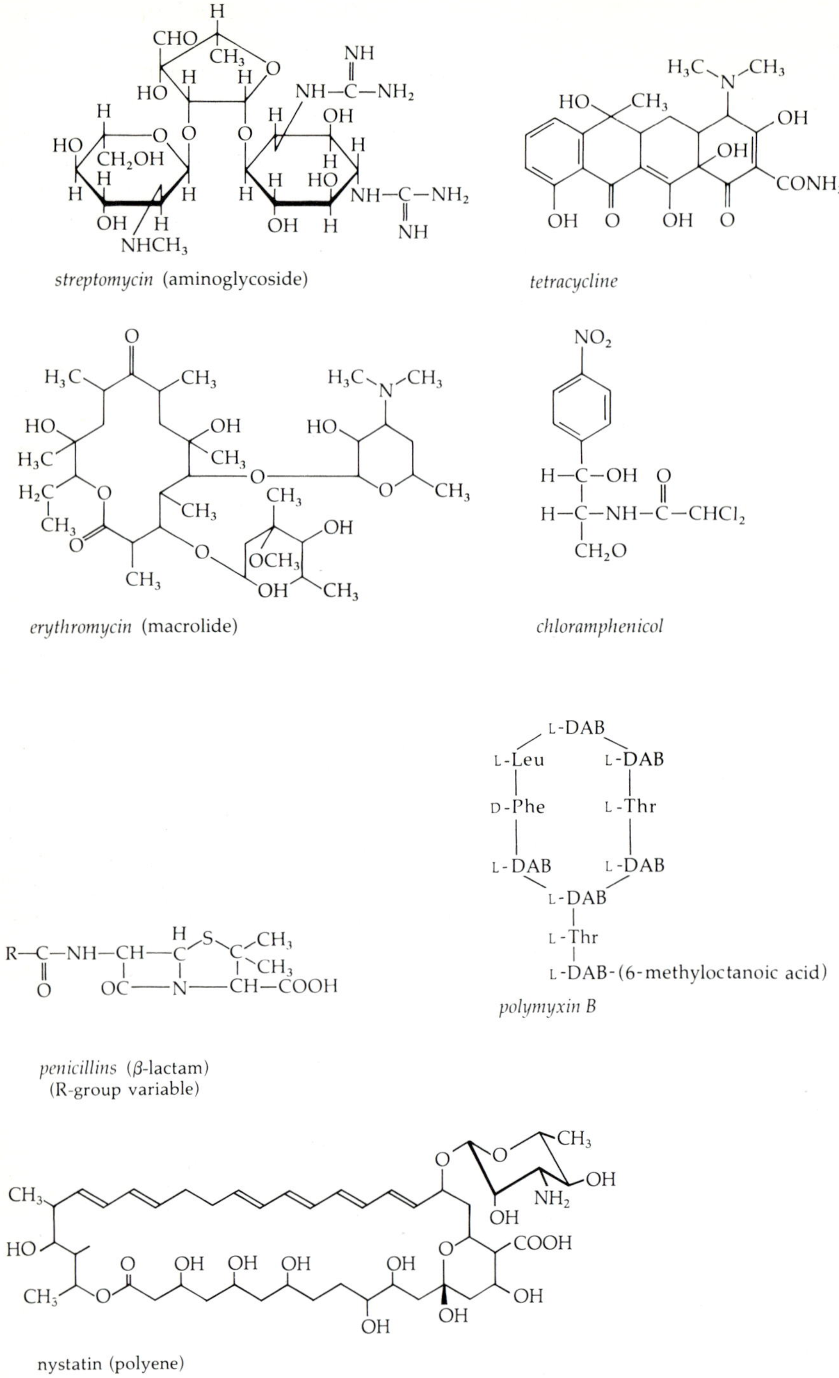

*streptomycin* (aminoglycoside)

*tetracycline*

*erythromycin* (macrolide)

*chloramphenicol*

*polymyxin B*

*penicillins* (β-lactam)
(R-group variable)

nystatin (polyene)

FIGURE 33.2
Structures of some antibiotics illustrating the wide diversity of the chemical classes. Polymyxin B is a cyclic polypeptide made up of the amino acid residues: leucine (leu), phenylalanine (phe), threonine (thr), and α,γ-diaminobutyric acid (DAB).

names can be illustrated by the example of an antibiotic which in the United States is given the generic name, *rifampin*. The generic name of the same compound in Europe is *rifampicin*. Its class name is rifamycin. It is sold under the trade names *Rifactin* and *Rifadin*, among others.

Antibiotics are exceedingly varied in chemical structure. Examples of some of the various chemical classes are shown in Figure 33.2. The generic names, sources, and mode of action of some antibiotics are shown in Table 33.3.

## Mode of Action of Antibiotics

The search for new antibiotics remains an empirical enterprise, and their physiological significance for the microorganisms that produce them is obscure.

| Name | Nature of acyl group |
| --- | --- |
| *Penicillin G* | 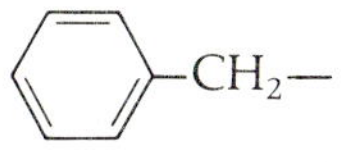 $C_6H_5-CH_2-$ |
| *Penicillin V* | 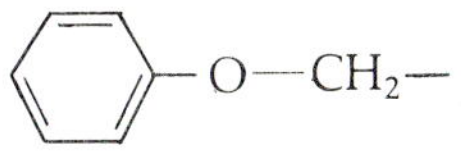 $C_6H_5-O-CH_2-$ |
| *Ampicillin* | 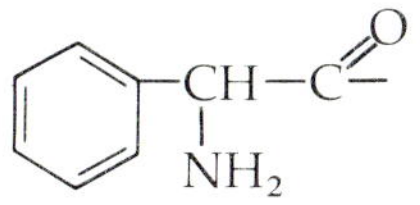 $C_6H_5-CH(NH_2)-C(=O)-$ |
| *Oxocillin* | 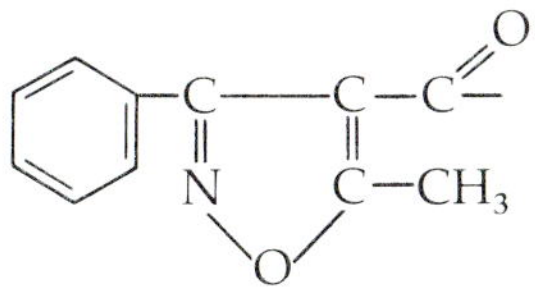 $C_6H_5-C{=}N-O-C(CH_3){=}C-C(=O)-$ (isoxazole ring) |

FIGURE 33.3

Some semisynthetic penicillins now in chemical use, showing the chemically introduced acyl substituents. (See Figure 33.2 for the general structure of penicillins.)

However, the reasons for their selective toxicity are in many cases now known. In general, antibiotics owe their selective toxicity to the fundamental biochemical differences between procaryotic and eucaryotic cells, their toxic effect being the consequence of their ability to inhibit one essential biochemical reaction specific either to the procaryotic or to the eucaryotic cell (Table 33.3).

## The Production of Antibiotics

The antibiotics were the first industrially produced microbial metabolites which were not *major* metabolic end products. The yields, calculated in terms of conversion of the major carbon source into antibiotic, are low and are greatly influenced by the composition of the medium and by the other cultural conditions. These facts have encouraged intense research directed toward improving yields. For this purpose, genetic selection has proved remarkably successful. The wild type strain of *Penicillium chrysogenum* first used for penicillin production yielded approximately 0.1 gram of penicillin per liter. From this strain a mutant was selected which produced 8 grams per liter under the same growth condition, a 60-fold improvement in yield. Subsequent strain selection following chemical mutagenesis has led to the development of new strains with even greater capacity for antibiotic production. By such *sequential genetic selection*, improvements of antibiotic yield as great as a 1,000-fold have often been obtained. Most genetic improvement has been empirical; large numbers of mutagenized clones are evaluated for their abilities to produce larger quantities of the antibiotic. However, with increased knowledge of the pathways of biosynthesis of antibiotics, more rational approaches are being exploited. It is now possible to select strains in which control of the synthesis of known precursors of an antibiotic has been altered by mutation. Such strains produce larger amounts of the precursor, and sometimes also larger amounts of the antibiotic end product.

The synthesis of antibiotics begins only after growth of the organisms that produce them has virtually ceased (Figure 33.4). They belong to a class of microbial products called *secondary metabolites*, because their synthesis is not associated with growth. The control mechanisms that trigger the synthesis of secondary metabolites as growth ceases are a fascinating but almost completely unexplored aspect of biochemical regulation.

Although the microorganisms used to produce antibiotics are all aerobes and are grown under conditions of vigorous aeration, the production process is generally referred to in the technical literature as a "fermentation." Antibiotics are produced by so-called *submerged cultivation methods*, using deep stainless steel tanks which must be subjected to continuous forced aeration and rapid

FIGURE 33.4

Temporal relationship between growth of *Penicillium chrysogenum* and its production of penicillin. After W. E. Brown and W. H. Peterson, "Factors Affecting Production of Penicillin in Semi-Pilot Plant Equipment," *Ind. Eng. Chem.* **42,** 1769 (1950).

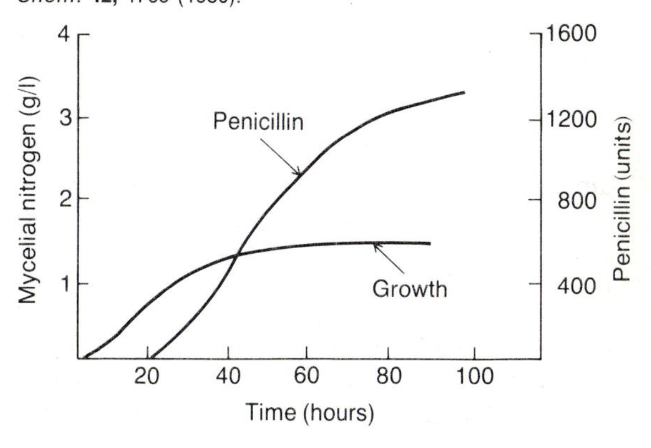

**TABLE 33.3**

**Properties and Uses of Certain Antibiotics**

| Chemical Class | Generic Name | Biological Source | Mode of Action |
|---|---|---|---|
| β-lactams | Penicillins<br>Cephalosporins | *Penicillium* spp.<br>*Cephalosporium* spp. | Inhibit synthesis of bacterial cell wall (peptidoglycan) |
| Macrolides | Erythromycin | *Streptomyces erythreus* | Inhibit 50S ribosome function |
| | Carbomycin | *S. halstidii* | |
| Aminoglycosides | Streptomycin<br>Neomycin | *S. griseus*<br>*S. fradiae* | Inhibit 30S ribosome function |
| Tetracyclines | Tetracycline[a] | *Streptomyces aureofaciens* | Inhibits binding of aminoacyl-t RNAs to ribosomes |
| Polypeptides | Polymyxin g | *Bacillus polymyxa* | Destroys cytoplasmic membrane |
| | Bacitracin | *B. subtilis* | Inhibits synthesis of bacterial cell wall (peptidoglycan) |
| Polyenes | Amphotericin B<br>Nystatin | *S. nodosus*<br>*S. nouresii* | Inactivate membranes containing sterols |
| — | Chloramphenicol[b] | *S. venezuelae* | Inhibits translation step of ribosome function |

[a] Made microbiologically and by chemical dehydrochlorination of chlorotetracycline.
[b] Now made by chemical synthesis.

mechanical agitation. The provision of adequate aeration is of great importance to yield, and the energy expended for aeration contributes appreciably to the cost of production.

When a microorganism is grown aerobically in tanks with capacities of tens of thousands of liters containing a rich, nonselective medium, the maintenance of a pure culture poses numerous special engineering problems. For the successful production of antibiotics, pure cultures are indispensable. This fact was first revealed during penicillin production. Many bacteria produce the enzyme, penicillinase, which catalyzes the hydrolytic cleavage of the four membered β-lactam ring of penicillin, with resulting loss of antibiotic activity. Consequently, contamination of a fermentor by penicillinase-producing bacteria can result in complete destruction of the accumulated penicillin.

In the manufacture of antibiotics the microbial product is sometimes subjected to subsequent chemical modification. One example is the chemical substitution of the acyl group of natural penicillins to produce the large variety of semisynthetic penicillins. Another example is the catalytic dehydrochlorination of chlorotetracycline to produce the more effective substance, tetracycline.

## Microbial Resistance to Antibiotics

The antibiotic era of medicine began abruptly some 30 years ago. How long it will last has become an open question. Although the search for new antibiotics continues undiminished, the rate of their discovery has declined sharply; most of the really effective antibiotics have probably already been discovered. Furthermore, strains of pathogens resistant to antibiotics have begun to develop at an alarming rate. Although most strains of staphylococci were sensitive to penicillin G when it was first introduced into medical practice, essentially all hospital acquired staphylococcal infections are now resistant to this antibiotic. A problem of even greater concern is the appearance of bacterial strains that are simultaneously resistant to several antibiotics, the so-called *multiply-resistant strains.* Between 1954 and 1964 the frequency of multiply-resistant strains of *Shigella* in Japanese hospitals rose from 0.2 percent to 52 percent.

Bacterial resistance to an antibiotic is sometimes acquired by the mutation of a chromosomal gene, which modifies the structure of the cellular target. A good example is mutationally acquired streptomycin resistance. This antibiotic deranges bacterial protein synthesis by attachment to one of

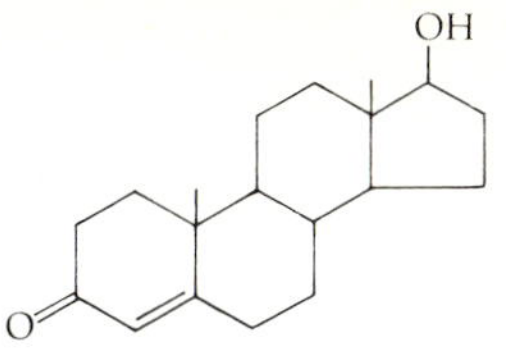

testosterone

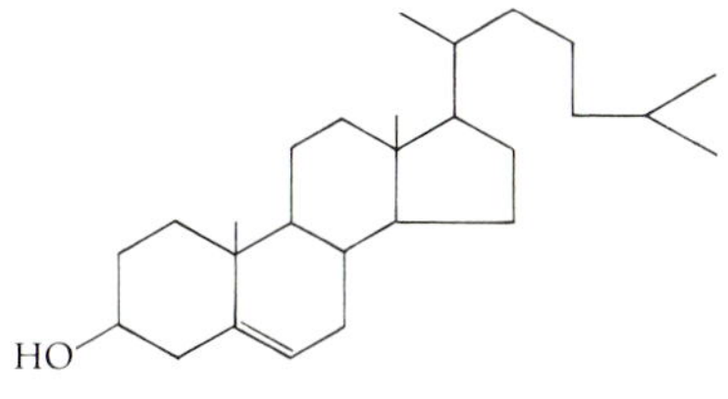

cholesterol

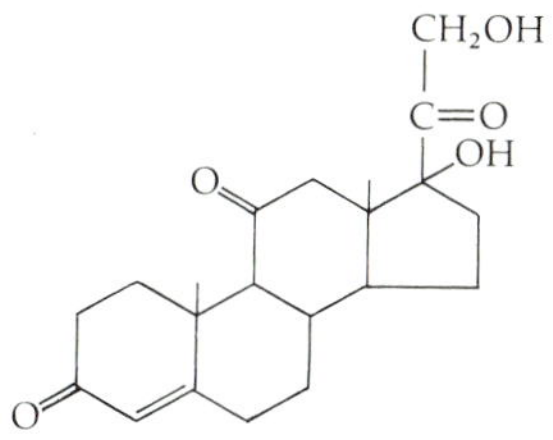

cortisone

FIGURE 33.5

The structure of cholesterol, a $C_{27}$ steroid, and of two mammalian steroid hormones for which cholesterol is a biosynthetic precursor—cortisone ($C_{21}$) and testosterone ($C_{19}$).

the proteins in the 30S subunit of the ribosome. Some mutations of the gene that encodes this ribosomal protein destroy the ability of the protein to bind streptomycin, but they do not substantially affect ribosomal function; the cell in which such a mutation occurs consequently becomes streptomycin-resistant.

Resistance can also be acquired as a result of infection of the bacterial cell by a plasmid belonging to the class of *resistance factors* (R factors, see Chapter 11). These plasmids often confer simultaneous resistance to several antibiotics. They carry genes that encode enzymes that catalyze chemical modifications of antibiotics, converting them to derivatives without antibiotic action. For example, streptomycin resistance conferred by an R factor may be caused either by adenylation, phosphorylation, or acetylation of the antibiotic; all these chemical modifications result in the loss of antibiotic activity. Multiply-resistant bacterial strains almost always owe their resistance to the presence of an R factor. Since these plasmids have wide host ranges, often being readily transferable between different bacterial genera, their increasing dissemination in natural bacterial populations is by far the most serious aspect of the problem of microbial resistance to antibiotics. Unless some solution to this problem can be found, the future therapeutic effectiveness of antibiotics is in jeopardy.

## Microbial Transformations of Steroids

Cholesterol (Figure 33.5) and chemically related steroids are structural components of most eucaryotic cellular membranes, of universal occurrence in metazoan animals. During the evolution of vertebrates, special pathways were evolved for the conversion of these universal cell constituents to new and functionally specialized steroids: the *steroid hormones*, which are potent regulators of animal development and metabolism. Steroid hormones are formed in specialized organs through the secondary metabolism of cholesterol, a $C_{27}$ steroid. The adrenocortical hormones are synthesized in the adrenal gland and are all $C_{21}$ compounds, such as *cortisone* (Figure 33.5); the sex hormones are synthesized in the ovary or the testis and are $C_{18}$ or $C_{19}$ compounds (Figure 33.5). Accordingly, by relatively slight chemical modifications of the basic steroid structure, vertebrates have evolved two new subclasses of steroid molecules with highly specific physiological functions and of great potency.

The elucidation of the structures and general functions of mammalian steroid hormones was completed about 30 years ago, but it was only in 1950 that possible chemotherapeutic uses for them became apparent, with the discovery that cortisone treatment can relieve dramatically the symptoms of rheumatoid arthritis. Today, cortisone and its derivatives are very widely used to treat a variety of inflammatory conditions, and additional uses for steroid hormones have emerged in the treatment of certain types of cancer and as oral contraceptives. The production of these compounds has now become a major industry.

Since the steroid hormones are produced by mammals in very small quantities, it was evident that their isolation from animal sources could not supply the clinical needs. Accordingly, the chemists turned their attention to the synthesis of those substances from plant sterols, which are abundant and can be cheaply prepared. One major chemical obstacle soon became apparent. All adrenocortical hormones are characterized by the insertion of an oxygen atom at position 11 on the ring system (Figure 33.6), by an organ-specific enzymatic hydroxylation of the biosynthetic precursor in the adrenal gland. Although it is easy to hydroxylate the steroid nucleus chemically, it is extremely difficult to insert a hydroxyl group at a specific position, and the specific 11-hydroxylation essential to the successful synthesis of cortisone from cheaper steroids became a major stumbling block to the development of a successful industrial process.

The discovery was then made that many microorganisms—fungi, actinomycetes, and bacteria—are capable of performing limited oxidations of steroids, which cause small and highly specific structural changes. The positions and nature of these changes are often characteristic for a microbial species, so that by the selection of an appropriate microorganism as an agent, it is possible to bring about any one of a large number of different modifications of the steroid molecule. Of particular practical importance is, of course, hydroxylation at the 11 position, which can be mediated by *Rhizopus* and other fungi. The introduction of a double bond by dehydrogenation between positions 1 and 2, mediated by a *Corynebacterium*, is another transformation of industrial importance, essential in the synthesis of a cortisone derivative, prednisolone.

The substrates for these microbial oxidative transformations are essentially insoluble in water. Furthermore, the limited transformation that the microorganism can effect does not provide it with either carbon or energy. The steroid substrates are, accordingly, added near the end of microbial growth, in the form of a finely dispersed suspension. The transformed products are released into the medium. In spite of the virtual insolubility of the substrates in water, many of these transformations proceed rapidly and with high yields.

## MICROBIOLOGICAL METHODS FOR THE CONTROL OF INSECTS

In Chapter 22 the formation of crystalline inclusions in the sporulating cells of certain *Bacillus* species was described. These bacilli (*Bacillus thuringensis* and related forms) are all pathogenic for the larvae (caterpillars) of certain insects, specifically, a very wide range of insects belonging to the Lepidoptera (butterflies and related forms). Following the isolation of the crystalline inclusions from sporulating bacterial cells, it was shown that all the primary symptoms characteristic of the natural disease of insects could be reproduced by feeding lar-

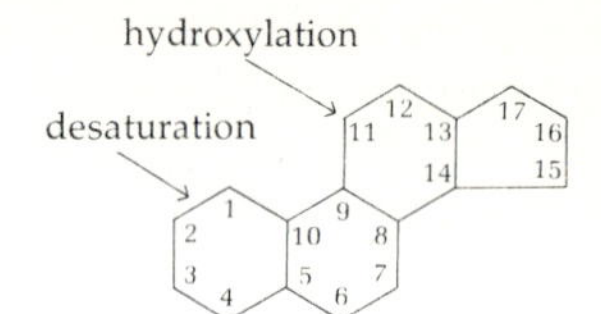


FIGURE 33.6
The ring system of steroids, showing the numbering of the carbon atoms and the specific sites of two commercially important chemical modifications that are mediated by microorganisms.

vae on leaves coated with the purified crystals. The crystals consist of a protein which is insoluble in water under neutral or mildly acid conditions but which can be dissolved in dilute alkali. The gut contents of larvae are, in general, alkaline, and when the ingested crystals reach the gut, they are dissolved and partially hydrolized. This modified protein attacks the cementing substances that keep the cells of the gut wall adherent, and as a consequence, the liquid in the gut can diffuse freely into the blood of the insect. The blood of the insect becomes highly alkaline, and this change in pH induces a general paralysis of the larva. Death, which ensues much later, appears to result from bacterial invasion of the body tissues.

The protein crystals possess a highly specific toxicity for the larvae of many Lepidoptera but are wholly nontoxic for other animals (including all the vertebrates) and for plants. They thus provide an ideal agent for the control of many serious insect pests that damage plant crops. Recognition of this fact has led recently to the development of a new microbiological industry: the large-scale production of the toxic protein for incorporation in dusting agents that can be used to protect commercial crops from the ravages of caterpillars. In industrial practice, the protein itself is not chemically isolated. Instead, the crystal-producing bacilli are grown on a large scale, harvested after the onset of sporulation with its accompanying crystal production, dried, and incorporated in a dusting powder.

With the increasing public concerns that chemical pesticides may pose significant hazards to the environment as well as to human health, research on the development of new and more effective microbiological methods for the control of insects has intensified. Viruses and fungi as well as a variety of bacteria are being evaluated as insecticidal agents.

## THE PRODUCTION OF OTHER CHEMICALS BY MICROORGANISMS

The widespread use of microorganisms in the chemical and pharmaceutical industries has come about because of the recognition that it is often cheaper to use a microorganism for the synthesis of a complex organic compound (for example, an antibiotic) than to synthesize it chemically. Microbial syntheses also have distinct advantages in the preparation of optically active compounds, since chemical synthesis leads to racemic mixtures which must subsequently be resolved.

As previously discussed, the microbial production of acetone and butanol, once the major source of these chemicals, has now been largely superseded by chemical synthesis. Nevertheless, the microbial production of many relatively simple and cheap organic compounds remains competitive with chemical methods of synthesis. These compounds include gluconic acid, produced by *Aspergillis niger* and acetic acid bacteria, and citric acid, produced by *A. niger*.

The production of two vitamins, vitamin $B_{12}$ and riboflavin, provides an instructive lesson in the economics of industrial microbiology. Both are now produced commercially by microbial means. Vitamin $B_{12}$ is produced by certain *Pseudomonas* spp. Although the yields are very low, this process remains competitive because of the very high price of the product: the structural complexity of vitamin $B_{12}$ virtually precludes a commercially feasible chemical synthesis. Riboflavin (vitamin $B_2$) is a much simpler compound, which can be readily prepared by chemical synthesis. It is still produced microbiologically, as a result of the discovery that certain plant pathogenic fungi (*Ashbya gossypii* and *Eremothecium ashbyi*) overproduce this vitamin and excrete the excess into the medium. By further genetic selection and improvement of culture methods, strains have been developed that produce so much riboflavin that the vitamin crystallizes in the culture medium.

## THE PRODUCTION OF ENZYMES BY MICROORGANISMS

The production of microbial enzymes, either pure or partly purified, is an important aspect of industrial microbiology. The uses of microbial enzymes in medicine and industry (Table 33.4) are remarkably diverse.

## THE IMPACT OF RECOMBINANT DNA TECHNOLOGY ON THE PRODUCTION OF USEFUL PRODUCTS BY MICROORGANISMS

The advent of recombinant DNA technology (Chapter 11) in the mid-1970s opened completely new possibilities for the production of useful products by microorganisms. Up until that time, a microorganism could only produce materials the synthesis of which was encoded in the cell's own

**TABLE 33.4**

**Partial List of Microbial Enzymes Produced Industrially**

| Name of Enzyme | Microbial Source | Uses | Reaction Catalyzed |
|---|---|---|---|
| Diastase | *Aspergillus oryzae* | Manufacture of glucose syrups; digestive aid | Hydrolysis of starch |
| Acid-resistant amylase | *A. niger* | Digestive aid | Hydrolysis of starch |
| Invertase | *Saccharomyces cereviseae* | Candy manufacture (prevents crystallization of sugar) | Hydrolysis of sucrose |
| Pectinase | *Sclerotina libertina* | Clarification of fruit juices | Hydrolysis of pectin |
| Protease | *A. niger* | Digestive aid | Hydrolysis of protein |
| Protease | *Bacillus subtilis* | Detergents; removal of gelatin from photographic film to recover silver | Hydrolysis of protein |
| Streptokinase | *Streptococcus* spp. | Promotes healing of wounds and burns | Hydrolysis of proteins |
| Collagenase | *Clostridium histolyticum* | Promotes healing of wounds and burns | Hydrolysis of protein (collagen) |
| Lipase | *Rhizopus* spp. | Digestive aid | Hydrolysis of lipids |
| Cellulase | *Trichoderma konigi* | Digestive aid | Hydrolysis of cellulose |

genome. Genetic selection might improve the level of production or slightly modify the chemical nature of the product, but production of totally new products, e.g., a new protein, was totally out of the question. Using techniques of recombinant DNA technology, it is at least theoretically possible to introduce any gene or set of genes into a microorganism and thereby cause it to produce the immediate gene products or products the synthesis of which is catalyzed by their action.

The initial emphasis of those involved in the development of these new capabilities has been to produce proteins that are medically useful. Already human insulin produced by strains of *E. coli* that carry human genes on a plasmid is commercially available. Soon other compounds, including human growth hormone to treat deficient children, factor VIII to treat hemophiliacs, and other medically important proteins, will almost certainly become available.

The capability of this new technology is clearly quite great.

## FURTHER READING


**Books**

BURGES, H. D. ed., *Microbial control of pests and plant diseases, 1970–1980.* New York: Academic Press, 1981.

FREITAS, Y. M. and F. FERNANDES, *Global Impacts of Applied Microbiology.* Bombay: The Examiner Press, 1971.

HARE, RONALD, *The Birth of Penicillin.* London: George Allen and Unwin, Ltd., 1970.

KELLY, D. P., N. G. CARR, *The Microbe.* Symp. 36 Part II, Soc. Gen. Microbiol. Cambridge: Cambridge Press, 1984.

REED, G., *Prescott and Dunn's Industrial Microbiology, 4th ed.* Westport: AVI Pub. Co., 1982.

**Reviews**

Amerine, M. S., and R. C. KUNKEE, "Microbiology of Wine Making," *Ann. Rev. Microbiol.* **22,** 323 (1968).

GADEN, E. L., "Production Methods in Industrial Microbiology", *Sci. Am.*, 245, 180 (1981).

ZEIKUS, J. G., "Chemical and Fuel Production by Anaerobic Bacteria," *Ann. Rev. Microbiol.* **34,** 423 (1980).

# Index

## A

## B

## C

# D

# E

## F



## G

## H



## I

## J



## K



## L



## M

## N

## O



## P

## Q



## R



## S

## T



## U



## V

## W



## X



## Y



## Z